※ Property of Robert K. George

350W

Handbook of Steel Construction

Seventh Edition

SECOND REVISED PRINTING, JULY 2000

Canadian Institute of Steel Construction
Institut canadien de la construction en acier
201 Consumers Road, Suite 300
Willowdale, Ontario M2J 4G8

Copyright © 2000

by

Canadian Institute of Steel Construction

All rights reserved. This book or any part thereof must not be reproduced in any form without the written permission of the publisher.

Seventh Edition

Second Revised Printing, July 2000

ISBN 0-88811-095-2

PRINTED IN CANADA

CONTENTS

Foreword v
Preface vi
Designations vii
General Nomenclature viii

PART ONE
CAN/CSA-S16.1-94 — Limit States Design of Steel Structures

PART TWO
CISC Commentary on CAN/CSA-S16.1-94

PART THREE
Connections and Tension Members

PART FOUR
Compression Members

PART FIVE
Flexural Members

PART SIX
Properties and Dimensions

PART SEVEN
CISC Code of Standard Practice and Miscellaneous Data

PART EIGHT
Selected Tables Based on CSA-G40.21 300W

PART NINE
General Index

FOREWORD

The Canadian Institute of Steel Construction is a national industry organization representing the structural steel, open-web steel joist and steel plate fabricating industries in Canada. Formed in 1930 and granted a Federal charter in 1942, the CISC functions as a non-profit organization promoting the efficient and economic use of fabricated steel in construction.

As a member of the Canadian Steel Construction Council, the Institute has a general interest in all uses of steel in construction. CISC works in close co-operation with the Steel Structures Education Foundation (SSEF) to develop educational courses and programmes related to the design and construction of steel structures. The CISC supports and actively participates in the work of the Standards Council of Canada, the Canadian Standards Association, the Canadian Commission on Building and Fire Codes and numerous other organizations, in Canada and other countries, involved in research work and the preparation of codes and standards.

Preparation of engineering plans is not a function of the CISC. The Institute does provide technical information through its professional engineering staff, through the preparation and dissemination of publications, through the medium of seminars, courses, meetings, video tapes, and computer programs. Architects, engineers and others interested in steel construction are encouraged to make use of CISC information services.

CISC is located at

201 Consumers Road, Suite 300
Willowdale, Ontario, M2J 4G8

and may be also contacted via one or more of the following:

Telephone: *(416) 491-4552*
Fax: *(416) 491-6461*
E–mail: *info@cisc-icca.ca*
Web site: *www.cisc-icca.ca*

PREFACE

This handbook has been prepared and published by the Canadian Institute of Steel Construction. It is an important part of a continuing effort to provide current, practical information to assist educators, designers, fabricators, and others interested in the use of steel in construction.

The First Edition of the CISC Handbook of Steel Construction was published in 1967 with Second through Sixth editions following each new edition of the CSA structural steel design standard, now called CAN/CSA-S16.1-94. All previous editions were based on what is now known as CSA-G40.21 grade 300W steel. The first printing of this Seventh Edition, while still based primarily on CSA Standard CAN/CSA-S16.1-94, had been made necessary by the introduction on May 1, 1997 of CSA-G40.21 350W as the basic steel grade for W and H shapes produced by Algoma. This second revised printing, however, reflects the decision made in 1999 by Algoma to stop producing these shapes, and is based in large part on CSA-G40.21 grade 350W, ASTM specifications A992 and A572 grade 50.

Part 1 is a reprint of CAN/CSA-S16.1-94 in its entirety. To assist with understanding the requirements of this CSA Standard, Part 2 is a Commentary on CAN/CSA-S16.1-94 prepared by CISC. Part 3 contains information on bolts and welds with tables for design and evaluation of various structural framing connections. As both 300W and 350W steels have the same specified ultimate tensile strength, resistances based on F_u are identical for both steel grades. Some values in Part 3 were left unchanged if the previous method or the values based on 300W steel were conservative or the resulting difference would be minor and more of an aggravation to the user if changed. Part 4 contains information on compression members. Part 5 contains information on flexural members. The Beam Selection and Uniform Load tables have been expanded to include all beams listed in Part 6. In Part 6, properties and dimensions data are provided for currently produced rolled and welded shapes.

The CISC Code of Standard Practice—Sixth Edition, November 19, 1999, leads the information found in Part 7.

Tables based on CSA-G40.21 300W steel which were previously provided in the Sixth Edition have been moved to a new Part 8 to assist in evaluating existing structures or designs based on 300W steel.

Permission to reprint portions of their publications, granted by the Canadian Standards Association and the American Institute of Steel Construction, Inc., is gratefully acknowledged. The contribution of Charles Albert and Ted Henderson, who helped make publication of this book possible, is sincerely appreciated.

Although no effort has been spared in an attempt to ensure that all data in this book is factual and that the numerical values are accurate to a degree consistent with current structural design practice, the Canadian Institute of Steel Construction does not assume responsibility for errors or oversights resulting from the use of the information contained herein. Anyone making use of the contents of this book assumes all liability arising from such use. All suggestions for improvement of this publication will receive full consideration for future printings.

M. I. Gilmor
Editor
June 2000

DESIGNATIONS

Standard designations should always be used to identify structural steel products on drawings and other documents. In Canada, the official designation is the metric (SI) designation, and examples of correct designations for most of the commonly used steel products are provided below. These designations should be used on all design drawings, for detailing purposes and for ordering material.

Shape	Example
Welded Wide Flange Shapes	WWF 900 x 169
W Shapes	W 610 x 113
Miscellaneous M Shapes	M 200 x 9.7
Standard Beams (S Shapes)	S 380 x 64
Standard Channels (C Shapes)	C 230 x 30
Miscellanous Channels (MC Shapes)	MC 250 x 12.5
Structural Tees — cut from WWF Shapes	WWT 250 x 138
— cut from W Shapes	WT 155 x 43
— cut from M shapes	MT 100 x 4.9
Bearing Piles (HP Shapes)	HP 250 x 62
Equal Leg angles — Imperial Series	L 102 x 102 x 9.5
Unequal Leg angles — Imperial Series	L 127 x 89 x 9.5
Plates (thickness x width)	PL 8 x 500
Square bars (side, mm)	Bar 25 □
Round bars (diameter, mm)	Bar 25 φ
Flat Bars (thickness x width)	Bar 5 x 60
Round Pipe (outside diameter x thickness)	DN 300 x 9.52†
Hollow Structural sections — Rectangular	HSS 152 x 102 x 9.5 Class C#
— Square	HSS 152 x 152 x 9.5 Class C#
— Round	HSS 141 x 9.5 Class C#
Cold Formed Channels	CFC 310 x 89 x 4.93
Super Light Beams (SLB Shapes)	SLB 100 x 5.4

\# *Class C or H*
† *This designation has been suggested by the U.S. National Institute of Building Sciences.*

GENERAL NOMENCLATURE

Explanations of the nomenclature used in many sections of this book appear in those specific sections. In addition, the following symbols are included here for convenience. See also pages 1–3 to 1–8, inclusive.

A	Area
A_b	Cross-sectional area of one bolt based on nominal diameter
A_f	Flange area
A_w	Web area; shear area; effective throat area of weld
a	Centre-to-centre distance between transverse web stiffebers; depth of conctere compression zone
a/h	Aspect ratio; ratio of distance between stiffeners to web depth
B_f	Bearing force in a member or component under factored load
B_r	Factored bearing resistance of a member or component
B_x	Bending factor with respect to axis x-x
B_y	Bending factor with respect to axis y-y
b	Width of stiffened or unstiffened compression elements; design effecetive width of concrete slab; overall flange width
C_e	Euler buckling load
C_f	Compressive force in a member or component under factored load; factored axial load
C_r	Factored compressive resistance of a member or component
C'_r	Compressive resistance of concrete acting at the centriod of the concrete area in compression
C_w	Warping torsional constant
C_y	Axial compressive load at yield stress
c	Distance from neutral axis to outer fiber of structural shape
D	Outside diameter of circular sections; diameter of rocker or roller; stiffener factor; fillet weld size (millimetres)
d	Depth; overall depth of a section; diameter of bolt or stud
E	Elastic modulus of steel (200 000 MPa assumed)
E_c	Elastic modulus of concrete
e	End distance; lever arm between the compressive resistance, C_r, and tensile resistance, T_r
e'	Lever arm between the compressive resistance, C'_r, of concrete and tensile resistance, T_r, of steel
F_{cr}	Critical plate buckling stress
F_s	Ultimate shear strength
F_u	Specified minimum tensile strength (Megapascals)
F_y	Specified minimum yield stress, yield point or yield strength
f'_c	Specified compressive strength of concrete at 28 days (Megapascals)
g	Transverse spacing between fastener gauge lines (gauge distance)
h	Clear depth of web between flanges; height of stud
I	Moment of inertia
I_x	Moment of inertia about axis x-x
I_y	Moment of inertia about axis y-y
J	St. Venant's torsion constant
K	Effective length factor
K_x	Effective length factor with respect to axis x-x
K_y	Effective length factor with respect to axis y-y
KL	Effective length
k	Distance from outer face of flange to web toe of fillet of rolled shapes
L	Length
L_{cr}	Maximum unbraced length adjacent to a plastic hinge

L_u	Maximum unsupported length of compression flange for which no reduction in factored moment resistance, M_r, is required
L_x	Unsupported length with respect to axis x-x
L_y	Unsupported length with respect to axis y-y
M_f	Bending moment in a member or component under factored load
M_{f1}	Smaller factored end moment of a beam-column; factored bending moment at a point of concentrated load
M_{f2}	Larger factored end moment of a beam-column
M_p	Plastic moment = ZF_y
M_r	Factored moment resistance of a member or component
M'_r	Factored moment resistance of a member of a given unbraced length greater that L_u
M_{rc}	Factored moment resistance of a composite beam
M_y	Yield moment = SF_y
m	Number of faying surfaces or shear planes in a bolted joint, equal to 1 for bolts in single shear and 2 for bolts in double shear
N	Length of bearing of an applied load
P	Concentrated load
Q_r	Sum of the factored resistances of all shear connectors between points of maximum and zero moment
q_r	Factored resistance of a shear connector
R	End reaction or concentrated transverse load applied to a flexural member
r	Radius of gyration
r_u	Radius of gyration with respect to axis u-u
r_v	Radius of gyration with respect to axis v-v
r_x	Radius of gyration with respect to axis x-x
r_y	Radius of gyration with respect to axis y-y
r_z	Radius of gyration with respect to axis z-z
S	Elastic section modulus
S_x	Elastic section modulus with respect to axis x-x
S_y	Elastic section modulus with respect to axis y-y
s	Centre-to-centre spacing (pitch) between successive fastener holes in line of stress
T_f	Tensile force in a member or component under factored load
T_r	Factored tensile resistance of a member or component; factored tensile resistance of the steel acting at the centroid of that part of the steel area in tension
t	Thickness
U	Amplification factor for stability analysis of beam-columns
V_f	Shear force in a member or component under factored load
V_r	Factored shear resistance of a member or component
V_s	Slip resistance of a bolted joint
W	Total uniformly distributed load; concentrated load
w	Web thickness; load per unit of length
Z	Plastic section modulus of a steel section
a	Load factor
g	Importance factor
l	Non-dimensional slenderness ratio in column formula
m	Coefficient related to the slip resistance of a bolted joint
f	Resistance factor
y	Load combination factor
w	Coefficient used to determine equivalent uniform bending effect in beam-columns
D	Deflection of a point of a structure
ASTM	American Society for Testing and Materials
CISC	Canadian Institute of Steel Construction

CPMA Canadian Paint Manufacturers' Association (*currently known as the Canadian Paint and Coatings Association*)
CSCC Canadian Steel Construction Council
CSA   Canadian Standards Association
SSEF Steel Structures Education Foundation
SSRC Structural Stability Research Council

PART ONE
CAN/CSA-S16.1-94
LIMIT STATES DESIGN OF STEEL STRUCTURES

General

This Standard is reprinted with the permission of the Canadian Standards Association and contains all erratum and revisions approved at time of printing.

CSA Standards are subject to periodic review and amendments will be published by CSA from time to time as warranted.

For information on requesting interpretations, see Note (5) to the Preface to CAN/CSA-S16.1-94.

National Standard of Canada   CAN/CSA-S16.1-94

Limit States Design of Steel Structures

Prepared by
Canadian Standards Association

Approved by
Standards Council of Canada

Technical Editor: Bill Glover
Managing Editor: Bernard Kelly

© Canadian Standards Association — 1994

All rights reserved. No part of this publication may be reproduced in any form, in an electronic retrieval system or otherwise, without the prior permission of the publisher.

ISSN 0317-5669
Published in December 1994 by Canadian Standards Association
178 Rexdale Boulevard, Rexdale (Toronto), Ontario, Canada M9W 1R3

Contents

Technical Committee on Steel Structures 1–xi
Preface 1–xiii

1. Scope and Application 1–1

2. Definitions and Symbols 1–2
2.1 Definitions 1–2
2.2 Symbols 1–3
2.3 Units 1–8

3. Reference Publications 1–9

4. Drawings 1–11
4.1 Design Drawings 1–11
4.2 Fabrication and Erection Documents 1–12
4.2.1 Connection Design Details 1–12
4.2.2 Shop Details 1–12
4.2.3 Erection Diagrams 1–12
4.2.4 Erection Procedures 1–12
4.2.5 Field Work Details 1–12

5. Material: Standards and Identification 1–12
5.1 Standards 1–12
5.1.1 General 1–12
5.1.2 Strength Levels 1–13
5.1.3 Structural Steel 1–13
5.1.4 Sheet Steel 1–13
5.1.5 Cast Steel 1–13
5.1.6 Forged Steel 1–13
5.1.7 Bolts 1–13
5.1.8 Welding Electrodes 1–14
5.1.9 Studs 1–14
5.2 Identification 1–14
5.2.1 Methods 1–14
5.2.2 Unidentified Structural Steel 1–14
5.2.3 Tests to Establish Identification 1–14
5.2.4 Affidavit 1–14

6. Design Requirements 1–15
6.1 General 1–15
6.1.1 Limit States 1–15
6.1.2 Structural Integrity 1–15
6.2 Requirements Under Specified Loads 1–15
6.2.1 Deflection 1–15
6.2.2 Camber 1–15
6.2.3 Dynamic Effects 1–16
6.2.4 Resistance to Fatigue 1–16
6.2.5 Prevention of Permanent Deformation 1–16

6.3 Requirements Under Factored Loads	1–16
6.3.1 Strength	1–16
6.3.2 Overturning	1–17
6.4 Expansion and Contraction	1–17
6.5 Corrosion Protection	1–17
7. Loads and Safety Criterion	**1–18**
7.1 Specified Loads	1–18
7.2 Safety Criterion and Effect of Factored Loads	1–18
8. Analysis of Structure	**1–19**
8.1 General	1–19
8.2 Continuous Construction	1–19
8.3 Simple Construction	1–20
8.4 Elastic Analysis	1–20
8.5 Plastic Analysis	1–20
8.6 Stability Effects	1–20
9. Design Lengths of Members	**1–21**
9.1 Simple Span Flexural Members	1–21
9.2 Continuous Span Flexural Members	1–21
9.3 Members in Compression	1–21
9.3.1 General	1–21
9.3.2 Failure Mode Involving Bending In-plane	1–22
9.3.3 Failure Mode Involving Buckling	1–22
9.3.4 Compression Members in Trusses	1–22
10. Slenderness Ratios	**1–22**
10.1 General	1–22
10.2 Maximum Slenderness Ratio	1–22
11. Width-Thickness Ratios: Elements in Compression	**1–22**
11.1 Classification of Sections	1–22
11.2 Maximum Width-Thickness Ratios of Elements Subject to Compression	1–23
11.3 Width and Thickness	1–23
12. Gross and Net Areas	**1–25**
12.1 Application	1–25
12.2 Gross Area	1–25
12.3 Effective Net Area	1–25
12.4 Pin-Connected Members in Tension	1–27
13. Member and Connection Resistance	**1–27**
13.1 General	1–27
13.2 Axial Tension	1–27
13.3 Axial Compression	1–27
13.4 Shear	1–28
13.4.1 Webs of Flexural Members with Two Flanges	1–28
13.4.2 Webs of Flexural Members not Having Two Flanges	1–29
13.4.3 Gusset Plates	1–29
13.4.4 Connecting Elements	1–29
13.4.5 Pins	1–30
13.5 Bending — Laterally Supported Members	1–30

13.6 Bending — Laterally Unsupported Members 1–30
13.7 Lateral Bracing for Members in Structures Analysed Plastically 1–31
13.8 Axial Compression and Bending 1–31
13.8.1 Member Strength and Stability — All Classes of Sections Except Class 1 Sections of I-Shaped Members 1–31
13.8.2 Member Strength and Stability — Class 1 Sections of I-Shaped Members 1–32
13.8.3 Value of U_1 1–32
13.8.4 Values of ω_1 1–33
13.9 Axial Tension and Bending 1–33
13.10 Load Bearing 1–33
13.11 Bolts in Bearing-Type Connections 1–34
13.11.1 General 1–34
13.11.2 Bolts in Shear 1–34
13.11.3 Bolts in Tension 1–34
13.11.4 Bolts in Combined Shear and Tension 1–34
13.12 Bolts in Slip-Critical Connections 1–34
13.12.1 General 1–34
13.12.2 Shear Connections 1–35
13.12.3 Connections in Combined Shear and Tension 1–35
13.13 Welds 1–35
13.13.1 General 1–35
13.13.2 Shear 1–36
13.13.3 Tension Normal to Axis of Weld 1–36
13.13.4 Compression Normal to Axis of Weld 1–37
13.14 Steel Plate Shear Walls 1–37

14. Fatigue 1–37
14.1 General 1–37
14.2 Fatigue Limit State 1–38
14.3 Live Load-Induced Fatigue 1–38
14.3.1 Calculation of Stress Range 1–38
14.3.2 Design Criteria 1–38
14.3.3 Fatigue Resistance 1–39
14.3.4 Detail Categories 1–39
14.4 Distortion-Induced Fatigue 1–39
14.4.1 General 1–39
14.4.2 Connection of Diaphragms, Cross-Frames, Lateral Bracing, or Beams 1–39

15. Beams and Girders 1–46
15.1 Proportioning 1–46
15.2 Rotational Restraint at Points of Support 1–46
15.3 Copes 1–46
15.4 Reduced Moment Resistance of Girders with Thin Webs 1–46
15.5 Flanges 1–46
15.6 Bearing Stiffeners 1–47
15.7 Intermediate Transverse Stiffeners 1–48
15.8 Lateral Forces 1–49
15.9 Web Crippling and Yielding 1–49
15.10 Openings 1–49
15.11 Torsion 1–50

16. Open-Web Steel Joists 1–51
16.1 Scope 1–51
16.2 General 1–51
16.3 Definitions 1–51
16.4 Materials 1–51
16.5 Drawings 1–51
16.5.1 Building Design Drawings 1–51
16.5.2 Joist Design Drawings 1–52
16.6 Design 1–52
16.6.1 Loading for Standard Open-Web Steel Joists 1–52
16.6.2 Loading for Special Open-Web Steel Joists 1–52
16.6.3 Design Assumptions 1–52
16.6.4 Verification of Joist Manufacturer's Design 1–53
16.6.5 Member and Connection Resistance 1–53
16.6.6 Width-Thickness Ratios 1–53
16.6.7 Tension Chord 1–53
16.6.8 Compression Chord 1–53
16.6.9 Webs 1–54
16.6.10 Spacers and Battens 1–55
16.6.11 Connections and Splices 1–55
16.6.12 Bearings 1–55
16.6.13 Anchorage 1–56
16.6.14 Deflection 1–56
16.6.15 Camber 1–56
16.6.16 Vibration 1–56
16.6.17 Welding 1–57
16.7 Stability During Construction 1–57
16.8 Bridging 1–57
16.8.1 General 1–57
16.8.2 Installation 1–57
16.8.3 Types 1–57
16.8.4 Diagonal Bridging 1–57
16.8.5 Horizontal Bridging 1–58
16.8.6 Attachment of Bridging 1–58
16.8.7 Anchorage of Bridging 1–58
16.8.8 Bridging Systems 1–58
16.8.9 Spacing of Bridging 1–58
16.9 Decking 1–58
16.9.1 Decking to Provide Lateral Support 1–58
16.9.2 Attachments 1–59
16.9.3 Diaphragm Action 1–59
16.9.4 Cast-In-Place Slabs 1–59
16.9.5 Installation of Steel Deck 1–59
16.10 Shop Painting 1–59
16.11 Manufacturing Tolerances 1–59
16.12 Inspection and Quality Control 1–60
16.12.1 Inspection 1–60
16.12.2 Identification and Control of Steel 1–60
16.12.3 Quality Control 1–60

16.13 Handling and Erection	1–61
16.13.1 General	1–61
16.13.2 Erection Tolerances	1–61

17. Composite Beams 1–61
17.1 Application	1–61
17.2 Definitions	1–61
17.3 General	1–62
17.3.1 Deflections	1–62
17.3.2 Vertical Shear	1–63
17.3.3 End Connections	1–63
17.4 Design Effective Width of Concrete	1–63
17.5 Slab Reinforcement	1–63
17.5.1 General	1–63
17.5.2 Parallel Reinforcement	1–63
17.5.3 Transverse Reinforcement, Solid Slabs	1–64
17.5.4 Transverse Reinforcement, Ribbed Slabs	1–64
17.6 Interconnection	1–64
17.7 Shear Connectors	1–65
17.7.1 General	1–65
17.7.2 End-Welded Studs	1–65
17.7.3 Channel Connectors	1–65
17.8 Ties	1–66
17.9 Design of Composite Beams with Shear Connectors	1–66
17.10 Design of Composite Beams without Shear Connectors	1–68
17.11 Unshored Beams	1–68
17.12 Beams During Construction	1–68

18. Concrete-Filled Hollow Structural Sections 1–68
18.1 Scope	1–68
18.2 Application	1–68
18.3 Axial Load on Concrete	1–69
18.4 Compressive Resistance	1–69
18.5 Bending	1–69
18.6 Axial Compression and Bending	1–70
18.6.1 Method 1: Bending Resisted by Composite Section	1–70
18.6.2 Method 2: Bending Assumed to be Resisted by the Steel Section Alone	1–70

19. General Requirements for Built-Up Members 1–70
19.1 Members in Compression	1–70
19.2 Members in Tension	1–73
19.3 Open Box-Type Beams and Grillages	1–73

20. Stability of Structures and Members 1–74
20.1 Structures	1–74
20.2 Members	1–74

21. Connections 1–75
21.1 Alignment of Members	1–75
21.2 Unrestrained Members	1–76
21.3 Restrained Members	1–76
21.4 Connections of Tension or Compression Members	1–77

21.5 Bearing Joints in Compression Members	1–77
21.6 Lamellar Tearing	1–77
21.7 Placement of Fasteners and Welds	1–77
21.8 Fillers	1–77
21.9 Welds in Combination	1–78
21.10 Fasteners and Welds in Combination	1–78
21.11 High-Strength Bolts (in Slip-Critical Joints) and Rivets in Combination	1–78
21.12 Connections Using Bolts	1–78
21.12.1 Connections Using Snug-Tightened High-Strength Bolts	1–78
21.12.2 Connections Using Pretensioned High-Strength Bolts	1–78
21.13 Welds	1–78
21.14 Special Fasteners	1–78
22. Bolting Details	**1–78**
22.1 High-Strength Bolts	1–78
22.2 A307 Bolts	1–79
22.3 Effective Bearing Area	1–79
22.4 Long Grips	1–79
22.5 Minimum Pitch	1–79
22.6 Minimum Edge Distance	1–79
22.7 Maximum Edge Distance	1–80
22.8 Minimum End Distance	1–80
22.9 Slotted or Oversize Holes	1–80
23. Structural Joints Using ASTM A325M, A490M, A325, or A490 Bolts	**1–80**
23.1 General	1–80
23.2 Bolts, Nuts, and Washers	1–81
23.3 Bolted Parts	1–81
23.4 Installation	1–83
23.4.1 Bolt Tension	1–83
23.4.2 Hardened Washers	1–83
23.4.3 Bevelled Washers	1–84
23.5 Turn-of-Nut Tightening	1–84
23.6 Tightening by Use of a Direct Tension Indicator	1–84
23.7 Inspection	1–84
24. Welding	**1–86**
24.1 Arc Welding	1–86
24.2 Resistance Welding	1–86
24.3 Fabricator and Erector Qualification	1–86
25. Column Bases	**1–87**
25.1 Loads	1–87
25.2 Resistance	1–87
25.2.1 Compressive Resistance of Concrete	1–87
25.2.2 Resistance to Pull-Out	1–87
25.2.3 Resistance to Transverse Loads	1–87
25.2.4 Moment Resistance	1–87
25.3 Finishing	1–88
26. Anchor Bolts	**1–88**
26.1 General	1–88

26.2 Bolt Resistance	1–88
26.2.1 Tension	1–88
26.2.2 Shear	1–88
26.2.3 Shear and Tension	1–88
26.2.4 Tension and Bending	1–89

27. Seismic Design Requirements 1–89

27.1 General	1–89
27.2 Ductile Moment-Resisting Frames	1–89
27.2.1 General	1–89
27.2.2 Beams	1–90
27.2.3 Columns (Including Beam-Columns)	1–90
27.2.4 Column Joint Panel Zone	1–91
27.2.5 Beam-to-Column Connections	1–91
27.2.6 Bracing	1–92
27.2.7 Fasteners	1–92
27.3 Moment-Resisting Frames with Nominal Ductility	1–92
27.4 Ductile Concentrically Braced Frames	1–92
27.4.1 General	1–92
27.4.2 Bracing Systems	1–93
27.4.3 Diagonal Bracing Members	1–93
27.4.4 Bracing Connections	1–93
27.4.5 Other Connections	1–94
27.4.6 Columns and Beams	1–94
27.5 Concentrically Braced Frames with Nominal Ductility	1–94
27.6 Ductile Eccentrically Braced Frames	1–95
27.6.1 Link Beam	1–95
27.6.2 Link Resistance	1–95
27.6.3 Length of Link	**1–95**
27.6.4 Link Rotation	1–96
27.6.5 Link Stiffeners	1–96
27.6.6 Lateral Support for Link	1–96
27.6.7 Link Beam-to-Column Connection	1–96
27.6.8 Brace-to-Link Beam Connections	1–97
27.6.9 Link Beam Resistance	1–97
27.6.10 Diagonal Braces	1–97
27.6.11 Columns	1–97
27.6.12 Roof Link Beam	1–98
27.6.13 Concentric Brace in Combination	1–98
27.7 Special Framing Systems	1–98
27.7.1 Steel Plate Shear Walls	1–98
27.7.2 Other Framing Systems	1–98

28. Fabrication 1–98

28.1 General	1–98
28.2 Straightness of Material	1–98
28.3 Gas Cutting	1–98
28.4 Sheared or Gas-Cut Edge Finish	1–98
28.5 Holes for Bolts or Other Mechanical Fasteners	1–99
28.6 Bolted Construction	1–99

28.7 Welded Construction 1–99
28.8 Finishing of Bearing Surfaces 1–99
28.9 Tolerances 1–99

29. Cleaning, Surface Preparation, and Priming ... 1–100
29.1 General Requirements 1–100
29.2 Requirements for Special Surfaces 1–101
29.3 Surface Preparation 1–101
29.4 Primer 1–101
29.5 One-Coat Paint 1–102

30. Erection ... 1–102
30.1 General 1–102
30.2 Temporary Loads 1–102
30.3 Adequacy of Temporary Connections 1–102
30.4 Alignment 1–102
30.5 Surface Preparation for Field Welding 1–102
30.6 Field Painting 1–102
30.7 Erection Tolerances 1–102
30.7.1 Elevation of Base Plates 1–102
30.7.2 Plumbness of Columns 1–103
30.7.3 Horizontal Alignment of Members 1–103
30.7.4 Elevations of Members 1–103
30.7.5 Members with Adjustable Connections 1–103
30.7.6 Column Splices 1–103
30.7.7 Joint Fit-Up 1–103
30.7.8 Special Tolerances 1–104

31. Inspection ... 1–104
31.1 General 1–104
31.2 Cooperation 1–104
31.3 Rejection 1–104
31.4 Inspection of High-Strength Bolted Joints 1–104
31.5 Third-Party Welding Inspection 1–104
31.6 Identification of Steel by Marking 1–104

Appendices
A — Standard Practice for Structural Steel 1–105
B — Effective Lengths of Columns 1–106
C — Criteria for Estimating Effective Column Lengths in Continuous Frames ... 1–107
D — Torsional-Flexural Buckling of Compression Members 1–109
E — Margins of Safety 1–111
F — Columns Subject to Biaxial Bending 1–112
G — Guide for Floor Vibrations 1–116
H — Wind Sway Vibrations 1–124
I — Recommended Maximum Values for Deflections for Specified Design Live and Wind Loads ... 1–125
J — Guide to Calculation of Stability Effects 1–127
K — Fatigue 1–129
L — Deflections of Composite Beams Due to Shrinkage of Concrete ... 1–135
M — Design Requirements for Steel Plate Shear Walls 1–138

… *Limit States Design of Steel Structures*

Technical Committee on Steel Structures

D.J.L. Kennedy	University of Alberta, Edmonton, Alberta	*Chairman*
G.L. Kulak	University of Alberta, Edmonton, Alberta	*Vice-Chairman*
M.I. Gilmor	Canadian Institute of Steel Construction, Willowdale, Ontario	*Secretary*
M. Archer-Shee	Canadian Welding Bureau, Dartmouth, Nova Scotia	
M. Aregawi	City of Toronto, Toronto, Ontario	
W.W. Baigent	Canron Construction — Fabrication East, Rexdale, Ontario	
P.C. Birkemoe	University of Toronto, Toronto, Ontario	
W. Blackwell	Blackwell Engineering Limited, Toronto, Ontario	
D.G. Calder	Calder Engineering Incorporated, North Vancouver, British Columbia	
M.P. Comeau	Campbell Comeau Engineering Ltd., Halifax, Nova Scotia	
R.W. Dryden	Butler Manufacturing Company, Burlington, Ontario	
S. Fox	Canadian Sheet Steel Building Institute, Cambridge, Ontario	
J.E. Henderson	Henderson Engineering Services, Milton, Ontario	
W.R. Hibbard	BP-TEC Engineering Group, Edmonton, Alberta	*Associate*
M.U. Hosain	University of Saskatchewan, Saskatoon, Saskatchewan	
W.S. Kendrick	Westmar Consultants Inc., North Vancouver, British Columbia	
D.G. Marshall	LeBlanc & Royle Telcom Incorporated, Oakville, Ontario	*Associate*
M.L. Mittleman	Algoma Steel Inc., Mississauga, Ontario	

December 1994

1-xi

C.J. Montgomery	The Cohos Evamy Partners, Edmonton, Alberta	
P.K. Ostrowski	Ontario Hydro, Toronto, Ontario	
A. Picard	Université Laval, Québec, Québec	
R.G. Redwood	McGill University, Montréal, Québec	
E.J. Rohacek	Morrison Hershfield Limited, North York, Ontario	
R.M. Schuster	University of Waterloo, Waterloo, Ontario	*Associate*
J. Springfield	Carruthers & Wallace Limited, Toronto, Ontario	
R.B. Vincent	The CANAM/MANAC Group Incorporated, Boucherville, Québec	
W.L. Glover	Canadian Standards Association, Rexdale, Ontario	*Administrator*

Preface

This is the fifth edition in Canada of CSA Standard CAN/CSA-S16.1, *Limit States Design of Steel Structures,* a general limit states design Standard for steel structures. The Standard is prepared in SI units and supersedes the SI edition published in 1989. The four limit states SI editions of 1994, 1989, 1984, and 1978 were based on the first limit states edition, written in imperial units, introduced in 1974. The five limit states design editions were preceded by seven working stress design editions published in 1969, 1965, 1961, 1954, 1940, 1930, and 1924. The last of these editions, S16-1969, was withdrawn in 1984 and therefore design must be carried out using limit states design principles.

 The sequence of twelve Standards marks a progression in technology, a change in the system of units, and more importantly, a change in the philosophy of design from working stress to limit states.

 This edition is simply entitled *Limit States Design of Steel Structures* because it is appropriate for the design of a broad range of structures. The scope recognizes that the requirements for the design of specific structures, such as bridges, are given in other Standards and that supplementary requirements may be necessary for some particular structures. The Standard sets out minimum requirements for the design of steel structures and, it is expected, will only be used by engineers competent in this field. The use of other Standards for design is proscribed.

 Although the basic limit states format, as set out in 1974, has proven itself in use and remains unaltered, a considerable number of technical changes reflecting the latest research developments have been incorporated in this edition. These are based on an increased understanding of the behaviour of structural materials, elements, and members and thus of the overall behaviour of structures. This increased understanding has itself been enhanced by the limit states approach as the designer explicitly recognizes the different possible modes of failure and designs against them. While the changes are not as extensive as those introduced in the 1989 edition, they are of significant import.

 Determination of effective lengths of columns has been simplified in that only members that buckle have effective lengths different from one. When members fail by in-plane bending, it is conservative, having taken PΔ effects into account, to use the actual length. Equations for determining effective net area for angles and when welds are used have been expanded. The column curves have been simplified and WWF shapes are now placed in the upper curve. New mandatory requirements for compression members that are likely to fail by torsional or torsional-flexural buckling are presented in Appendix D. The clauses on weld resistances have been reformatted and expressions introduced to give the variation of weld strength with the angle of loading. Fatigue requirements are now consistent with the latest bridge codes. The clauses on bracing have been considerably modified to include requirements for both strength and stiffness of the bracing system. The seismic requirements of Clause 27 have been somewhat relaxed for structures in low seismic zones and the requirements for eccentrically braced frames have been moved into the body of the code. A new appendix covering both the static and seismic design of steel plate shear walls has been introduced.

 The clauses in the Standard relating to fabrication and erection serve to show that design cannot be considered in isolation but that it is part of the design and construction sequence.

CAN/CSA-S16.1-94

This Standard has been adopted by the Associate Committee on the National Building Code as the reference Standard for steel structures in Section 4.6 of the *National Building Code* for 1995.

This Standard was prepared by the Technical Committee on Steel Structures for Buildings under the jurisdiction of the Standards Steering Committee on Structures, and was approved by these Committees. This Standard has been approved as a National Standard of Canada by the Standards Council of Canada.

December 1994

Notes:
(1) *Use of the singular does not exclude the plural (and vice versa) when the sense allows.*
(2) *Although the intended primary application of this Standard is stated in its Scope, it is important to note that it remains the responsibility of the users of the Standard to judge its suitability for their particular purpose.*
(3) *This publication was developed by consensus, which is defined by the CSA Regulations Governing Standardization as "substantial agreement reached by concerned interests. Consensus includes an attempt to remove all objections and implies much more than the concept of a simple majority, but not necessarily unanimity." It is consistent with this definition that a member may be included in the Technical Committee list and yet not be in full agreement with all clauses of the publication.*
(4) *CSA Standards are subject to periodic review, and suggestions for their improvement will be referred to the appropriate committee.*
(5) *All enquiries regarding this Standard, including requests for interpretation, should be addressed to Canadian Standards Association, Standards Development, 178 Rexdale Boulevard, Rexdale, Ontario M9W 1R3.*
  Requests for interpretation should
(a) define the problem, making reference to the specific clause, and, where appropriate, include an illustrative sketch;
(b) provide an explanation of circumstances surrounding the actual field condition; and
(c) be phrased where possible to permit a specific "yes" or "no" answer.
  Interpretations are published in CSA's periodical Info Update. *For subscription details, write to CSA Sales Promotion,* Info Update, *at the address given above.*

CAN/CSA-S16.1-94
Limit States Design of Steel Structures

1. Scope and Application

1.1
This Standard provides rules and requirements for the design, fabrication, and erection of steel structures. The design is based on limit states. The term "steel structures" refers to structural members and frames that consist primarily of structural steel components, including the detail parts, welds, bolts, or other fasteners required in fabrication and erection. This Standard also applies to structural steel components in structures framed in other materials.

1.2
Requirements for steel structures such as bridges, antenna towers, offshore structures, and cold-formed steel structural members are given in other CSA Standards.

1.3
This Standard applies unconditionally to steel structures, except that supplementary rules or requirements may be necessary for
(a) unusual types of construction;
(b) mixed systems of construction;
(c) steel structures that
   (i) have great height or spans;
   (ii) are required to be moveable or be readily dismantled;
   (iii) are exposed to severe environmental conditions or possible severe loads such as those resulting from vehicle impact or chemical explosion;
   (iv) are required to satisfy aesthetic, architectural, or other requirements of a nonstructural nature;
   (v) employ materials or products not listed in Clause 5; or
   (vi) have other special features that could affect design, fabrication, or erection; and
(d) tanks, stacks, other platework structures, poles, and piling.

1.4
Use of other standards for the design of members or parts of steel structures is neither warranted nor acceptable, except where specifically directed in this Standard.
  A rational design based on theory, analysis, and engineering practice acceptable to the regulatory authority may be used in lieu of the formulae provided in this Standard. In such cases, the design shall provide nominal margins (or factors) of safety at least equal to those intended in the provisions of this Standard (see Appendix E).

2. Definitions and Symbols

2.1 Definitions
The following definitions apply to this Standard:

Approved — approved by the regulatory authority.

Camber — the deviation from straightness of a member or any portion of a member with respect to its major axis. Frequently, camber is specified and produced in a member to compensate for deflections that will occur in the member when loaded. (See Clause 6.2.2.) Unspecified camber is sometimes referred to as bow.

Concrete — Portland cement concrete in accordance with CSA Standard A23.1.

Designer — the professional engineer responsible for the design.

Erection tolerances — tolerances related to the plumbness, alignment, and level of the piece as a whole. The deviations are determined by considering the locations of the ends of the piece. (See Clause 30.)

Fabrication tolerances — tolerances allowed from the nominal dimensions and geometry, such as the cutting to length, finishing of ends, cutting of bevel angles, and for fabricated members, out-of-straightness such as sweep and camber. (See Clause 28.)

Limit states — those conditions of a structure in which the structure ceases to fulfil the function for which it was designed. Those states concerning safety are called the ultimate limit states and include exceeding of load-carrying capacity, overturning, sliding, fracture, and fatigue. Those states that restrict the intended use and occupancy of the structure are called serviceability limit states and include deflection, vibration, and permanent deformation.

Factors

    Load factor, α — a factor, given in Clause 7.2, applied to a specified load for the limit states under consideration that takes into account the variability of the loads and load patterns and the analysis of their effects.

    Load combination factor, ψ — a factor, given in Clause 7.2, applied to factored loads other than dead load that takes into account the reduced probability of a number of loads from different sources acting simultaneously.

    Importance factor, γ — a factor, given in Clause 7.2, applied to factored loads that takes into account the consequences of collapse as related to the use and occupancy of the structure.

    Resistance factor, ϕ — a factor, given in the appropriate clauses in this Standard, applied to a specified material property or the resistance of a member, connection, or structure that, for the limit state under consideration, takes into account the variability of material properties, dimensions, workmanship, type of failure, and uncertainty in prediction of member resistance. To maintain simplicity of the design formulae in this Standard, the type of failure and the uncertainty in prediction of member resistance have been incorporated in the expressions of member resistance. (See Appendix E for a more detailed discussion.)

Loads

Gravity load — (newtons) is equal to the mass of the object (kilograms) being supported multiplied by the acceleration due to gravity, g (9.81 m/s^2).

Specified loads (D, E, L, T, and W) — those loads prescribed by the regulatory authority (see Clause 7.1).

Factored load — the product of a specified load and its load factor.

Mill tolerances — variations allowed from the nominal dimensions and geometry with respect to cross-sectional area, nonparallelism of flanges, and out-of-straightness such as sweep or camber in the product as manufactured and given in CSA Standard G40.20.

Regulatory authority — a federal/provincial/municipal ministry, department, board, agency, or commission that has responsibility for regulating, by statute, the use of products, materials, or services.

Resistance

Nominal resistance, R — the nominal resistance of a member, connection, or structure as calculated in accordance with this Standard based on the specified material properties and nominal dimensions.

Factored resistance, ϕR — the product of the nominal resistance and the appropriate resistance factor.

Sweep — the deviation from straightness of a member or any portion of a member with respect to its minor axis.

2.2 Symbols

The following symbols are used throughout this Standard. Deviations from them and additional nomenclature are noted where they appear.

A = area

A_b = cross-sectional area of a bolt based on its nominal diameter

A_c = transverse area of concrete between longitudinal shear planes; cross-sectional area of concrete in composite columns

A_{cv} = the critical area of two longitudinal shear planes, one on each side of the area A_c, extending from the point of zero moment to the point of maximum moment

A_f = flange area

A_g = gross area

A_m = area of fusion face

A_n = critical net area; tensile stress area of bolt

A_{ne} = effective net area

A'_{ne} = effective net area reduced for shear lag

A_p = concrete pull-out area

A_r = area of reinforcing steel

A_s = area of steel section including cover plates; area of bottom (tension) chord of steel joist; area of stiffener or pair of stiffeners

A_{sc} = area of steel shear connector

A_{st} = area of steel section in tension

A_w = web area; shear area; effective throat area of weld

a = centre-to-centre distance between transverse web stiffeners; depth of concrete compression zone

a' = length of cover plate termination

a/h = aspect ratio; ratio of distance between stiffeners to web depth

B = bearing force in a member or component under specified load

B_f = bearing force in a member or component under factored load

B_r = factored bearing resistance of a member or component

b = width of stiffened or unstiffened compression elements; design effective width of concrete or cover slab

C = compressive force in a member or component under specified load; axial load

C_e = Euler buckling strength = $\dfrac{\pi^2 EI}{L^2}$

C_{ec} = Euler buckling strength of a concrete-filled hollow structural section

C_f = compressive force in a member or component under factored load; factored axial load

C_r = factored compressive resistance of a member or component; factored compressive resistance of steel acting at the centroid of that part of the steel area in compression

C_{rc} = factored compressive resistance of a composite column

C_{rcm} = factored compressive resistance that can coexist with M_{rc} when all of the cross-section is in compression

C_{rco} = factored compressive resistance with $\lambda = 0$

C_r' = compressive resistance of concrete acting at the centroid of the concrete area assumed to be in uniform compression; compressive resistance of concrete component of a composite column

C_w = warping torsional constant (mm^6)

C_y = axial compressive load at yield stress

c_1 = coefficient used to determine slip resistance

D = outside diameter of circular sections; diameter of rocker or roller; stiffener factor; dead load

d = depth; overall depth of a section; diameter of bolt or stud

d_b = depth of beam

E = elastic modulus of steel (200 000 MPa assumed); live load due to earthquake

E_c = elastic modulus of concrete

E_{ct} = effective modulus of concrete in tension

e	=	end distance; lever arm between the compressive resistance, C_r, and the tensile resistance, T_r
e'	=	lever arm between the compressive resistance, C'_r, of concrete and tensile resistance, T_r, of steel
F	=	strength or stress
F_{cr}	=	critical plate-buckling stress in compression, flexure, or shear
F_{cre}	=	elastic critical plate-buckling stress in shear
F_{cri}	=	inelastic critical plate-buckling stress in shear
F_s	=	ultimate shear stress
F_{sr}	=	allowable stress range in fatigue
F_{srt}	=	constant amplitude threshold stress range
F_{st}	=	factored axial force in the stiffener
F_t	=	tension-field post-buckling stress
F_u	=	specified minimum tensile strength
F_y	=	specified minimum yield stress, yield point, or yield strength
F'_y	=	yield level including effect of cold-working
F_{yr}	=	specified yield strength of reinforcing steel
f'_c	=	specified compressive strength of concrete at 28 days
G	=	shear modulus of steel (77 000 MPa assumed)
g	=	transverse spacing between fastener gauge lines (gauge distance)
h	=	clear depth of web between flanges; height of stud; storey height
h_d	=	depth of steel deck
I	=	moment of inertia of cover-plated section
I_e	=	effective moment of inertia of composite beam
I_t	=	transformed moment of inertia of composite beam
J	=	St. Venant torsion constant
K	=	effective length factor
KL	=	effective length
k	=	distance from outer face of flange to web-toe of fillet of I-shaped sections
k_b	=	buckling coefficient; required stiffness of the bracing assembly
k_s	=	mean slip coefficient
k_v	=	shear buckling coefficient
L	=	length; length of longitudinal weld; live load; length of connection in direction of loading
L_c	=	length of channel shear connector
L_{cr}	=	maximum unbraced length adjacent to a plastic hinge

L_n	=	net length (that is, gross length less design allowance for holes within the length)
M	=	bending moment in a member or component under specified load
M_f	=	bending moment in a member or component under factored load
M_{fc}	=	bending moment in a girder, under factored load, at theoretical cut-off point
M_{fg}	=	first-order moment under factored gravity loads determined assuming that there is no lateral translation of the frame
M_{ft}	=	first-order translational moment under factored lateral loads, or the moment resulting from lateral translation of an unsymmetrical frame, or the moment resulting in an unsymmetrically loaded frame under factored gravity loading (see Clause 8.6.2)
M_{f1}	=	smaller factored end moment of a beam-column; factored bending moment at a point of concentrated load
M_{f2}	=	larger factored end moment of a beam-column
M_p	=	plastic moment = ZF_y
M_r	=	factored moment resistance of a member or component
M_{rc}	=	factored moment resistance of a composite beam; factored moment resistance of a column reduced for the presence of axial load
M_u	=	critical elastic moment of a laterally unbraced beam
M_y	=	yield moment = SF_y
m	=	number of faying surfaces or shear planes in a bolted joint, equal to 1.0 for bolts in single shear and 2.0 for bolts in double shear
N	=	length of bearing of an applied load; number of passages of moving load
n	=	number of bolts; number of shear connectors required between the point of maximum positive bending moment and the adjacent point of zero moment; parameter for compressive resistance; number of threads per inch; number of stress range cycles at a given detail for each passage of the moving load
n′	=	number of shear connectors required between any concentrated load and nearest point of zero moment in a region of positive bending moment
n_t	=	modular ratio, E/E_{cl}
P	=	force to be developed in a cover plate; pitch of threads, mm
p	=	fraction of full shear connection
Q_r	=	sum of the factored resistances of all shear connectors between points of maximum and zero moment
q_r	=	factored resistance of a shear connector
q_{rr}	=	factored resistance of a shear connector in a ribbed slab
q_{rs}	=	factored resistance of a shear connector in a solid slab
R	=	end reaction or concentrated transverse load applied to a flexural member; nominal resistance of a member, connection, or structure; transition radius
r	=	radius of gyration

r_y	=	radius of gyration of a member about its weak axis
S	=	elastic section modulus of a steel section
s	=	centre-to-centre longitudinal spacing (pitch) of any two successive fastener holes
T	=	tensile force in a member or component under specified load; load due to temperature change, etc (see Clause 7.1.1)
T_f	=	tensile force in a member or component under factored load
T_{fg}	=	first order brace force under factored gravity loads determined assuming there is no translation of the braced frame
T_{ft}	=	first order translational brace force under factored gravity loads or translation of an unsymmetrical braced frame or resulting from unsymmetrical loading under gravity loading
T_r	=	factored tensile resistance of a member or component; in composite construction, factored tensile resistance of the steel acting at the centroid of that part of the steel area in tension
T_y	=	axial tensile load at yield stress
t	=	thickness; thickness of flange; average flange thickness of channel shear connector
t_c	=	concrete or cover slab thickness
U_1	=	factor to account for moment gradient and for second-order effects of axial force acting on the deformed member
U_2	=	amplification factor to account for second-order effects of gravity loads acting on the laterally displaced storey
V	=	shear force in a member or component under specified load
V_f	=	shear force in a member or component under factored load
V_h	=	total horizontal shear to be resisted at the junction of the steel section or joist and the slab or steel deck
V_r	=	factored shear resistance of a member or component
V_s	=	slip resistance of a bolted joint
V_{st}	=	factored shear force in column web to be resisted by stiffener
W	=	live load due to wind
w	=	web thickness; width of plate
w_d	=	average width of flute of steel deck
w'	=	sum of thickness of column web plus doubler plates
w_n	=	net width (that is, gross width less design allowance for holes within the width)
X_u	=	ultimate strength as rated by the electrode classification number
x	=	subscript relating to strong axis of a member
$\bar{x}$	=	eccentricity of the weld with respect to centroid of the element
y	=	subscript relating to weak axis of a member, distance from centroid of cover plate to neutral axis of cover-plated section; distance from the centroid of the effective area of concrete slab to elastic neutral axis

Z = plastic section modulus of a steel section
α = load factor; angle of inclination
β = value used to determine bracing stiffness
γ = importance factor; fatigue life constant
λ = nondimensional slenderness parameter in column formula
Δ_b = displacement of braced member and bracing assembly at the point of support under force C_f and which may be taken to Δ_o
Δ_f = relative first-order lateral (translational) displacement of the storey due to factored loads (coincident with M_{ft})
Δ_o = initial misalignment of the braced member at the point of support. This misalignment may be taken as the tolerance specified in Clause 28 for sweep or camber over the total braced length, or portion thereof, as appropriate for the method of construction
ρ = density of concrete; slenderness ratio
ρ_e = equivalent slenderness ratio of built-up member
ρ_i = maximum slenderness ratio of component part of a built-up member between interconnectors
ρ_o = slenderness ratio of built-up member acting as an integral unit
ΣC_f = sum of factored axial compressive loads of all columns in the storey
ΣV_f = sum of factored lateral loads above the storey; the total first-order storey shear
ε_f = free shrinkage strain of concrete
κ = ratio of the smaller factored moment to the larger factored moment at opposite ends of the unbraced length, positive for double curvature and negative for single curvature
ϕ = resistance factor (see definition under Factors, in Clause 2.1)
ψ = load combination factor
ω_1 = coefficient used to determine equivalent uniform bending effect in beam-columns
ω_2 = coefficient to account for increased moment resistance of a laterally unsupported beam segment when subject to a moment gradient

2.3 Units
Equations and expressions appearing in this Standard are compatible with the following SI (metric) units:
force: N (newtons)
length: mm (millimetres)
moment: N•mm
strength or stress: MPa (megapascals)

3. Reference Publications

3.1
This Standard refers to the following publications and where such reference is made it shall be to the latest edition and revisions thereto, unless otherwise specified:

CSA Standards

A23.1-94,
Concrete Materials and Methods of Concrete Construction;

A23.3-94,
Design of Concrete Structures;

B95-1962,
Surface Texture (Roughness, Waviness, and Lay);

CAN/CSA-G40.20/G40.21-92,
General Requirements for Rolled or Welded Structural Quality Steel/Structural Quality Steels;

CAN/CSA-G164-92,
Hot Dip Galvanizing of Irregularly Shaped Articles;

S37-94,
Antennas, Towers, and Antenna-Supporting Structures;

S136-94,
Cold Formed Steel Structural Members;

S304.1-94,
Masonry Design for Buildings (Limit States Design);

W47.1-92,
Certification of Companies for Fusion Welding of Steel Structures;

W48.1-M1991,
Carbon Steel Covered Electrodes for Shielded Metal Arc Welding;

W48.3-93,
Low-Alloy Steel Covered Electrodes for Shielded Metal Arc Welding;

W48.4-M1980,
Solid Mild Steel Filler Metals for Gas Shielded Arc Welding;

W48.5-M1990,
Carbon Steel Electrodes for Flux- and Metal-Cored Arc Welding;

W48.6-M1980,
Bare Mild Steel Electrodes and Fluxes for Submerged Arc Welding;

W55.3-1965,
Resistance Welding Qualification Code for Fabricators of Structural Members Used in Buildings;

W59-M1989,
Welded Steel Construction (Metal Arc Welding).

ASTM* Standards
A27/A27M-93,
Specification for Steel Castings, Carbon, for General Application;

A108-93,
Specification for Steel Bars, Carbon, Cold-Finished, Standard Quality;

A148/A148M-93b,
Specification for Steel Castings, High-Strength, for Structural Purposes;

A307-94,
Specification for Carbon Steel Bolts and Studs, 60 000 psi Tensile Strength;

A325-94,
Specification for Structural Bolts, Steel, Heat Treated 120/105 ksi Minimum Tensile Strength:

A325M-93,
Specification for High-Strength Bolts for Structural Steel Joints [Metric];

A490-93,
Specification for Heat-Treated Steel Structural Bolts, 150 ksi MinimumTensile Strength;

A490M-93,
Specification for High-Strength Steel Bolts, Classes 10.9 and 10.9.3, for Structural Steel Joints [Metric];

A514/A514M-94a,
Specification for High-Yield-Strength, Quenched and Tempered Alloy Steel Plate, Suitable for Welding;

A521-76 (R1992),
Specification for Steel, Closed-Impression Die Forgings for General Industrial Use;

A570/A570M-92 (R1993),
Specification for Steel, Sheet and Strip, Carbon, Hot-Rolled, Structural Quality;

A668-93,
Specification for Steel Forgings, Carbon and Alloy, for General Industrial Use.

CGSB‡ Standards
CAN/CGSB-1.40-M89,
Primer, Structural Steel, Oil Alkyd Type;

CAN/CGSB-1.140-M89,
Oil-Alkyd Type Red Lead, Iron Oxide Primer;

CAN/CGSB-166-M90,
Basic Lead Silicochromate Primer, Oil Alkyd Type.

CISC/CPMA‡ Standards
1-73a 1975,
A Quick-Drying One-Coat Paint for Use on Structural Steel;

2-75 1975,
A Quick-Drying Primer for Use on Structural Steel.

CISC
Code of Standard Practice for Structural Steel, 1991.

National Research Council of Canada
National Building Code of Canada, 1995.

SSPC§ Specifications
SP 2-89,
Hand Tool Cleaning;

SP 3-89,
Power Tool Cleaning;

SP 5-91,
White Metal Blast Cleaning;

SP 6-91,
Commercial Blast Cleaning;

SP 7-91,
Brush-Off Blast Cleaning;

SP 10-91,
Near-White Blast Cleaning;

SP 11-91,
Power Tool Cleaning to Bare Metal.

Structural Stability Research Council
Guide to Stability Design Criteria for Metal Structures, Fourth Edition.**

Research Council on Structural Connections
Guide to Design Criteria for Bolted and Riveted Joints, Second Edition.**

**American Society for Testing and Materials*
†*Canadian General Standards Board.*
‡*Canadian Institute of Steel Construction/Canadian Paint Manufacturers' Association.*
§*Steel Structures Painting Council.*
***Wiley Interscience.*

4. Drawings

4.1 Design Drawings

4.1.1
Design drawings shall be drawn to a scale adequate to convey the required information. The drawings shall show a complete design of the structure with members suitably designated and located, including such dimensions and detailed description as necessary to permit the preparation of fabrication and erection documents. Floor levels, column centres, and offsets shall be dimensioned. The term "drawings" may include computer output and other data.

4.1.2
Design drawings shall designate the design standards used, shall show clearly the type or types of construction (as defined in Clause 8) to be employed, shall show the category of the structural system used for seismic design (see Clause 27), and shall designate the material or product standards applicable to the members and details depicted (see Clause 5). Drawings shall give the governing combinations of shears, moments, and axial forces to be resisted by the connections.

4.1.3
Where high-strength bolted joints are required to resist shear between connected parts, design drawings shall indicate the type of joint, slip-critical or bearing-type, to be provided (see Clause 23).

4.1.4
The size and location of stiffeners, reinforcement, and bracing required to stabilize compression elements, and the camber of beams, girders, and trusses shall be shown on design drawings.

4.2 Fabrication and Erection Documents

4.2.1 Connection Design Details
Connection design details shall be prepared in advance of preparing shop details and submitted to the designer for confirmation that the intent of the design is met. Connection design details shall provide details of typical and special types of connections, and other data necessary for the preparation of shop details. Connection design details shall be referenced to the design drawings.

4.2.2 Shop Details
Shop details shall be prepared in advance of fabrication and submitted to the designer for review. Shop details shall provide complete information for the fabrication of various members and components of the structure, including the required material and product standards; the location, type, and size of all mechanical fasteners; bolt installation requirements; and welds.

4.2.3 Erection Diagrams
Erection diagrams shall be submitted to the designer for review. Erection diagrams are general arrangement drawings showing the principal dimensions of the structure, piece marks, sizes of the members, size and type of bolts, field welds, bolt installation requirements, elevations of column bases, all necessary dimensions and details for setting anchor bolts, and all other information necessary for the assembly of the structure.

4.2.4 Erection Procedures
Erection procedures shall outline the construction methods, erection sequence, temporary bracing requirements, and other engineering details necessary for shipping, erecting, and maintaining the stability of the steel frame. Erection procedures shall be supplemented by drawings and sketches to identify the location of stabilizing elements. Erection procedures shall be submitted for review when so specified.

4.2.5 Field Work Details
Field work details shall be submitted to the designer for review. Field work details shall provide complete information for modifying fabricated members in the shop or on the job site. All operations required to modify the member shall be shown on the field work details. If extra materials are necessary to make modifications, shop details shall be required.

5. Material: Standards and Identification

5.1 Standards

5.1.1 General
Acceptable material and product standards and specifications (latest editions) for use under this Standard are listed in Clauses 5.1.3 to 5.1.9, inclusive. Materials and products other than those listed also may be used if approved. Approval shall be based on published specifications that

establish the properties, characteristics, and suitability of the material or product to the extent and in the manner of those covered in listed standards.

5.1.2 Strength Levels
The yield strength, F_y, and the tensile strength, F_u, used as the basis for design shall be the specified minimum values as given in the material and product standards and specifications. The levels reported on mill test certificates shall not be used as the basis for design.

5.1.3 Structural Steel
CAN/CSA G40.21-M,
Structural Quality Steels.

5.1.4 Sheet Steel
ASTM A570,
Specification for Steel, Sheet and Strip, Carbon, Hot-Rolled, Structural Quality.

Other standards for structural sheet are listed in Clause 2.4 of CSA Standard S136. Only structural-quality sheet standards that specify chemical composition and mechanical properties will be acceptable for use with this Standard. Mill test certificates that list the chemical composition and the mechanical properties shall be available, upon request, in accordance with Clause 5.2.1(a).

5.1.5 Cast Steel
ASTM A27,
Specification for Steel Castings, Carbon, for General Application;

ASTM A148,
Specification for Steel Castings, High Strength, for Structural Purposes.

5.1.6 Forged Steel
ASTM A521,
Specification for Steel, Closed-Impression Die Forgings for General Industrial Use;

ASTM A668,
Specification for Steel Forgings, Carbon and Alloy, for General Industrial Use.

5.1.7 Bolts
ASTM A307,
Specification for Carbon Steel Bolts and Studs, 60 000 psi Tensile Strength;

ASTM A325,
Specification for Structural Bolts, Steel, Heat Treated, 120/105 ksi Minimum Tensile Strength;

ASTM A325M,
Specification for High-Strength Bolts for Structural Steel Joints [Metric];

ASTM A490,
Specification for Heat-Treated Steel Structural Bolts, 150 ksi Minimum Tensile Strength;

ASTM A490M,
Specification for High-Strength Steel Bolts, Classes 10.9 and 10.9.3, for Structural Steel Joints [Metric].

Note: *Before specifying metric bolts, the designer should check on their current availability in the quantities required.*

5.1.8 Welding Electrodes
CSA W48.1,
Mild Steel Covered Arc Welding Electrodes;

CSA W48.3,
Low-Alloy Steel Covered Arc Welding Electrodes;

CSA W48.4,
Solid Mild Steel Filler Metals for Gas Shielded Arc Welding;

CSA W48.5,
Carbon Steel Electrodes for Flux and Metal Cored Arc Welding;

CSA W48.6,
Bare Mild Steel Electrodes and Fluxes for Submerged-Arc Welding.

5.1.9 Studs
ASTM A108,
Specification for Steel Bars, Carbon, Cold Finished, Standard Quality [Grades 1015 and 1018].

5.2 Identification

5.2.1 Methods
The materials and products used shall be identified as to specification, including type or grade, if applicable, by one of the following means, except as provided in Clauses 5.2.2 and 5.2.3:
(a) mill test certificates or producer's certificates satisfactorily correlated to the materials or products to which they pertain; and
(b) legible markings on the material or product made by its producer in accordance with the applicable material or product standard.

5.2.2 Unidentified Structural Steel
Unidentified structural steel shall not be used unless approved by the building designer. If the use of unidentified steel is authorized, F_y shall be taken as 210 MPa and F_u shall be taken as 380 MPa.

5.2.3 Tests to Establish Identification
Unidentified structural steel may be tested to establish identification when permitted by the building designer. Testing shall be done by an approved testing agency in accordance with CSA Standard G40.20. The test results, taking into account both mechanical properties and chemical composition, shall form the basis for classifying the steel as to specification. Once classified, the specified minimum values for steel of that specification grade shall be used as the basis for design (see Clause 5.1.2).

5.2.4 Affidavit
The fabricator, if requested, shall provide an affidavit stating that the materials and products that have been used in fabrication conform to the applicable material or product standards called for by the design drawings or specifications.

6. Design Requirements

6.1 General

6.1.1 Limit States

As set out in this Standard, steel structures shall be designed to be serviceable during the useful life of the structure and safe from collapse during construction and during the useful life of the structure. Limit states define the various types of collapse and unserviceability that are to be avoided; those concerning safety are called the ultimate limit states (strength, overturning, sliding, and fatigue) and those concerning serviceability are called the serviceability limit states (deflections, vibration, and permanent deformation). The object of limit states design calculations is to keep the probability of a limit state being reached below a certain value previously established for the given type of structure. This is achieved in this Standard by the use of load factors applied to the specified loads (see Clause 7) and resistance factors applied to the specified resistances (see Clause 13).

The various limit states are set out in this Clause. Some of these relate to the specified loads and others to the factored loads. Camber, provisions for expansion and contraction, and corrosion protection are further design requirements related to serviceability and durability. All limit states shall be considered in the design.

6.1.2 Structural Integrity

The general arrangement of the structural system and the connection of its members shall be designed to provide resistance to widespread collapse as a consequence of local failure. The requirements of this Standard generally provide a satisfactory level of structural integrity for steel structures. Supplementary provisions may be required for structures where accidental loads such as vehicle impact or explosion are likely to occur (see Clause 1.3). (Further guidance is contained in Chapter 4, Commentary C of the Supplement to the *National Building Code of Canada*.)

6.2 Requirements Under Specified Loads

6.2.1 Deflection

6.2.1.1

Steel members and frames shall be proportioned so that deflections are within acceptable limits for the nature of the materials to be supported and for the intended use and occupancy.

6.2.1.2

In the absence of a more detailed evaluation, see Appendix I for recommended values for deflections.

6.2.1.3

Roofs shall be designed to withstand any additional loads likely to occur as a result of ponding. (See Clause 7.1.1.) (Further guidance is contained in Chapter 4, Commentary I of the Supplement to the *National Building Code of Canada*.)

6.2.2 Camber

6.2.2.1

Camber of beams, trusses, or girders, if required, shall be called for on the design drawings. Generally, trusses and crane girders of 25 m or greater span should be cambered for approximately the dead-plus-half-live-load deflection. (See Clause 16 for requirements for open-web joists and Clause 28.9.5 for fabrication tolerances.)

6.2.2.2
Any special camber requirements necessary to bring a loaded member into proper relation with the work of other trades shall be stipulated on the design drawings.

6.2.3 Dynamic Effects

6.2.3.1
Suitable provision shall be made in the design for the effect of live load that induces impact or vibration, or both. In severe cases, such as structural supports for heavy machinery that causes substantial impact or vibration when in operation, the possibility of harmonic resonance, fatigue, or unacceptable vibration shall be investigated.

6.2.3.2
Special consideration shall be given to floor systems susceptible to vibration, such as large open floor areas free of partitions, to ensure that such vibration is acceptable for the intended use and occupancy. (Guidance regarding floor vibrations is contained in Appendix G.)

6.2.3.3
Unusually flexible structures (generally those whose ratio of height to effective resisting width exceeds 4:1) shall be investigated for lateral vibrations under dynamic wind load. Lateral accelerations of the structure shall be checked to ensure that such accelerations are acceptable for the intended use and occupancy. (Information on lateral accelerations under dynamic wind loads can be found in Appendix H.)

6.2.4 Resistance to Fatigue
Structural steelwork shall be designed to resist the effects of fatigue under the specified loads in accordance with Clause 14.

6.2.5 Prevention of Permanent Deformation

6.2.5.1
For composite beams that are unshored during construction, the stress in the steel beam induced by the specified loads shall not exceed F_y. The stress shall be calculated as the sum of the stresses due to any specified loads applied before the concrete strength reaches $0.75f'_c$, calculated on the steel section alone, plus the stresses at the same location, due to the remaining specified loads calculated on the composite section.

6.2.5.2
Slip-critical joints, in which the design load is assumed to be transferred by the slip resistance of the clamped faying surfaces, shall be proportioned using the provisions of Clause 13.12 to resist without slipping the moments and forces induced by the specified loads (see Clause 23).

6.3 Requirements Under Factored Loads

6.3.1 Strength
Structural steelwork shall be proportioned to resist moments and forces resulting from the application of the factored loads acting in the most critical combination, taking into account the resistance factors as specified in the appropriate clauses of this Standard.

6.3.2 Overturning
The building or structure shall be designed to resist overturning resulting from the application of the factored loads acting in the most critical combination, taking into account the importance of the building as specified in Clause 7, and taking into account the resistance factors as specified in the appropriate clauses of this Standard.

6.4 Expansion and Contraction
Suitable provision shall be made for expansion and contraction commensurate with the service and erection conditions of the structure.

6.5 Corrosion Protection

6.5.1
Steelwork shall have sufficient corrosion protection to minimize any corrosion likely to occur in the service environment.

6.5.2
Interiors of buildings conditioned for human comfort may be generally assumed to be noncorrosive environments; however, the need for corrosion protection shall be assessed and protection shall be furnished in those buildings where it is deemed to be necessary.

6.5.3
Corrosion protection of the inside surfaces of enclosed spaces permanently sealed from any external source of oxygen is unnecessary.

6.5.4
The minimum required thickness of steelwork situated in a noncorrosive environment and therefore not requiring corrosion protection is governed by the provisions of Clause 11.

6.5.5
Corrosion protection shall be provided by means of suitable alloying elements in the steel, by protective coatings, or by other effective means, either singly or in combination.

6.5.6
Localized corrosion likely to occur from entrapped water, excessive condensation, or other factors shall be minimized by suitable design and detail. Where necessary, positive means of drainage shall be provided.

6.5.7
If the corrosion protection, specified for steelwork exposed to the weather or to other environments in which progressive corrosion can occur, is likely to require maintenance or renewal during the service life of the structure, the steelwork so protected, exclusive of fill plates and shims, shall have a minimum thickness of 4.5 mm.

7. Loads and Safety Criterion

7.1 Specified Loads

7.1.1

Except as provided for in Clauses 7.1.2 and 7.1.3, the following loads and influences as specified by the regulatory authority shall be considered in the design of structural steelwork:

D — Dead loads, including the weight of steelwork and all permanent materials of construction, partitions, stationary equipment, and additional weight of concrete and finishes resulting from deflections of supporting members, and the forces due to prestressing;

E — Live load due to earthquake;

L — Live loads, including loads due to intended use and occupancy of structures, moveable equipment, snow, rain, soil, or hydrostatic pressure, impact, and any other live loads stipulated by the regulatory authority;

T — Influences resulting from temperature changes, shrinkage, or creep of component materials or from differential settlement;

W — Live load due to wind.

7.1.2

If it can be shown by engineering principles or if it is known from experience that neglect of some or all of the effects due to T does not affect the structural safety or serviceability, they need not be considered in the calculations.

7.1.3

Suitable provision shall be made for loads imposed on the steel structure during its erection. During subsequent construction, suitable provision shall be made to support the construction loads on the steel structure with an adequate margin of safety.

7.2 Safety Criterion and Effect of Factored Loads

7.2.1

The structural steelwork shall be designed to have sufficient strength or stability, or both, such that

Factored resistance ≥ Effect of factored loads

where the factored resistance is determined in accordance with other clauses of this Standard and the effect of factored loads is determined in accordance with Clauses 7.2.2 to 7.2.5. In cases of overturning, uplift, and stress reversal, no positive anchorage is required if the stabilizing effect of dead load multiplied by a load factor of less than 1.00 given in Clause 7.2.3 is greater than the effect of loads tending to cause overturning, uplift, and stress reversal multiplied by load factors greater than 1.00 given in Clause 7.2.3.

7.2.2

For load combinations not including earthquake, the effect of factored loads, in force units, is the structural effect due to the specified loads multiplied by load factors, α, defined in Clause 7.2.3; a load combination factor, ψ, defined in Clause 7.2.4; and an importance factor, γ, defined in Clause 7.2.5. The factored load combinations shall be taken as follows:

$\alpha_D D + \gamma\psi(\alpha_L L + \alpha_W W + \alpha_T T)$

7.2.3
Load factors, α, shall be taken as follows:
α_D = 1.25 except that
α_D = 0.85 when the dead load resists overturning, uplift, or load reversal effects;
α_L = 1.50;
α_W = 1.50 for wind; and
α_T = 1.25.

7.2.4
The load combination factor, ψ, shall be taken as follows:
ψ = 1.00 when only one of L, W, and T acts;
ψ = 0.70 when two of L, W, and T act; and
ψ = 0.60 when all of L, W, and T act.
  The most unfavourable effect shall be determined by considering L, W, and T acting alone with ψ = 1.00, or in combination with ψ = 0.70 or 0.60.

7.2.5
The importance factor, γ, shall be not less than 1.00, except for those structures where it can be shown that collapse is not likely to cause injury or other serious consequences, it shall be not less than 0.80.

7.2.6
For load combinations including earthquake, the effect of factored loads, in force units, is the structural effect due to the factored load combinations taken as follows:
(a) $1.0D + \gamma(1.0E)$; and either
(b) $1.0D + \gamma(1.0L + 1.0E)$ for storage and assembly occupancies; or
(c) $1.0D + \gamma(0.5L + 1.0E)$ for all other occupancies.

8. Analysis of Structure
8.1 General
8.1.1
In proportioning the structure to meet the various design requirements of Clause 6, the methods of analysis given in this Clause shall be used. The distribution of internal forces and bending moments shall be determined both under the specified loads to satisfy the requirements of serviceability and fatigue in Clause 6 and under the factored loads as required to satisfy strength and overturning requirements in Clause 7.

8.1.2
Two basic types of construction and associated design assumptions, designated "continuous" and "simple", are permitted for all or part of a structure under this Standard. The distribution of internal forces and bending moments throughout the structure will depend on the type or types of construction chosen and the forces to be resisted.

8.2 Continuous Construction
In continuous construction the beams, girders, and trusses are rigidly framed or are continuous over supports. Connections are generally designed to resist the bending moments and internal forces calculated by assuming that the original angles between intersecting members remain unchanged as the structure is loaded.

8.3 Simple Construction

8.3.1
Simple construction assumes that the ends of beams, girders, and trusses are free to rotate under load in the plane of loading. Resistance to lateral loads, including sway effects, shall be ensured by a suitable system of bracing or shear walls or by the design of part of the structure as continuous construction, except as provided in Clause 8.3.2.

8.3.2
A frame designed to support gravity loads on the basis of simple construction may be proportioned to resist lateral loads, including sway effects, by distributing the moments resulting from such loading among selected joints of a frame by a recognized empirical method provided that
(a) the connection and connected members are proportioned to resist the moments and forces caused by lateral loads;
(b) the connected members have solid webs;
(c) the beam or girder can support the full gravity load when assumed to act as a simple beam;
(d) the connection has adequate capacity for inelastic rotation when subjected to the factored gravity and lateral loads;
(e) the mechanical fasteners or welds of the connection are proportioned to resist 1.5 times the moments and forces produced by the factored gravity and lateral loads; and
(f) in assessing the stability of the structure, in accordance with Clause 8.6, the effect of the flexibility of the connection is taken into account.

8.4 Elastic Analysis
Under a particular loading combination, the forces and moments throughout all or part of the structure may be determined by an analysis that assumes that individual members behave elastically.

8.5 Plastic Analysis
Under a particular loading combination, the forces and moments throughout all or part of the structure may be determined by a plastic analysis provided that
(a) the steel used has $F_y \leq 0.80 F_u$ and exhibits the load-strain characteristics necessary to achieve moment redistribution;
(b) the width-thickness ratios meet the requirements of Class 1 sections as given in Clause 11.2;
(c) the members are braced laterally in accordance with the requirements of Clause 13.7;
(d) web stiffeners are supplied on a member at a point of load application where a plastic hinge would form;
(e) splices in beams or columns are designed to transmit 1.1 times the maximum calculated moment under factored loads at the splice location or $0.25 M_p$, whichever is greater;
(f) members are not subject to repeated heavy impact or fatigue; and
(g) the influence of inelastic deformation on the strength of the structure is taken into account. (See also Clause 8.6.)

8.6 Stability Effects

8.6.1
The analyses referred to in Clauses 8.4 and 8.5 shall include the sway effects in each storey produced by the vertical loads acting on the structure in its displaced configuration. These second-order effects due to the relative translational displacement (sway) of the ends of a member shall preferably be determined from a second-order analysis. Alternatively, the elastic

second-order effects may be accounted for by
(a) amplifying translational load effects obtained from a first-order elastic analysis by the factor

$$U_2 = \frac{1}{1 - \left[\dfrac{\Sigma C_f \Delta_f}{\Sigma V_f h}\right]}$$

thus, eg, $M_f = M_{fg} + U_2 M_{ft}$ and $T_f = T_{fg} + U_2 T_{ft}$

(b) performing an analysis in accordance with Appendix J.

Notes:
(1) The amplification factor given in Clause 8.6.1(a) is the limit of the iterative procedure given in Appendix J.
(2) When the elastic second-order (P Δ) effects exceed 40% of the primary effects, ie, $U_2 > 1.4$, either increase the stiffness of the frame to reduce Δ_f, or perform a second-order elasto-plastic analysis, unless it can be shown that the stresses at the critical section, taking residual stresses into account, do not exceed F_y.

8.6.2
For load combinations consisting of gravity loads only, the translational load effects due to asymmetry of the loading or frame or both shall be considered. However, for all gravity load combinations, the translational load effects shall be taken as not less than the load effects produced by notional lateral loads, applied at each storey, equal to 0.005 times the factored gravity loads contributed by that storey.

9. Design Lengths of Members

9.1 Simple Span Flexural Members
Beams, girders, and trusses may be designed on the basis of simple spans, whose length may be taken as the distance between the centres of gravity of supporting members. Alternatively, the span length of beams and girders may be taken as the actual length of such members measured between centres of end connections. The length of trusses designed as simple spans may be taken as the distance between the extreme working points of the system of triangulation employed. In all cases the design of columns or other supporting members shall provide for the effect of any significant moment or eccentricity arising from the manner in which a beam, girder, or truss may actually be connected or supported.

9.2 Continuous Span Flexural Members
Beams, girders, or trusses having full or partial end restraint due to continuity or cantilever action shall be proportioned to carry all moments, shears, and other forces at any section assuming the span, in general, to be the distance between the centres of gravity of the supporting members. Supporting members shall be proportioned to carry all moments, shears, and other forces induced by the continuity of the supported beam, girder, or truss.

9.3 Members in Compression

9.3.1 General
A member in compression shall be designed on the basis of its effective length, KL (the product of the effective length factor, K, and the unbraced length, L).

Unless otherwise specified in this Standard the unbraced length, L, shall be taken as the length of the compression member between the centres of restraining members. The unbraced length may differ for different cross-sectional axes of a compression member. At the bottom storey of a multi-storey structure or for a single-storey structure, L shall be taken as the length from the top of the base plate to the centre of restraining members at the next higher level.

The effective length factor, K, depends on the potential failure modes, whether by bending in-plane or bucking as given in Clauses 9.3.2, 9.3.3, and 9.3.4.

9.3.2 Failure Mode Involving Bending In-plane
The effective length shall be taken as the actual length (K = 1.0) for beam-columns that would fail by in-plane bending provided only that, when applicable, the sway effects are included in the analysis of the structure to determine the end moments and forces acting on the beam-columns.

9.3.3 Failure Mode Involving Buckling
The effective length for axially loaded columns that would fail by buckling and for beam-columns that would fail by out-of-plane (lateral-torsional) buckling shall be based on the rotational and translational restraint afforded at the ends of the unbraced length (see Appendices B and C).

9.3.4 Compression Members in Trusses
For members that would fail by in-plane bending, the effective length shall be taken as the actual unbraced length (K=1.0) (see Clause 9.3.2).

For members that would fail by buckling, the effective length shall be based on the rotational and translational restraint afforded at the ends of the unbraced length (see Clause 9.3.3).

10. Slenderness Ratios

10.1 General
The slenderness ratio of a member in compression shall be taken as the ratio of the effective length, KL, to the corresponding radius of gyration, r. The slenderness ratio of a member in tension shall be taken as the ratio of the unbraced length, L, to the corresponding radius of gyration.

10.2 Maximum Slenderness Ratio

10.2.1
The slenderness ratio of a member in compression shall not exceed 200.

10.2.2
The slenderness ratio of a member in tension shall not exceed 300. This limit may be waived if other means are provided to control flexibility, sag, vibration, and slack in a manner commensurate with the service conditions of the structure, or if it can be shown that such factors are not detrimental to the performance of the structure or of the assembly of which the member is a part.

11. Width-Thickness Ratios: Elements in Compression

11.1 Classification of Sections

11.1.1
For the purposes of this Standard, structural sections shall be designated as Class 1, 2, 3, or 4, depending on the maximum width-thickness ratios of their elements subject to compression, and as otherwise specified in Clause 11.1.2 and 11.1.3. The Classes are defined as follows:
(a) **Class 1** sections (plastic design sections) will permit attainment of the plastic moment and subsequent redistribution of the bending moment;
(b) **Class 2** sections (compact sections) will permit attainment of the plastic moment but need not allow for subsequent moment redistribution;

(c) **Class 3** sections (noncompact sections) will permit attainment of the yield moment; and
(d) **Class 4** sections will generally have local buckling of elements in compression as the limit state of structural capacity.

11.1.2
Class 1 sections, when subject to flexure, shall have an axis of symmetry in the plane of loading and, when subject to axial compression, shall be doubly symmetric.

11.1.3
Class 2 sections, when subject to flexure, shall have an axis of symmetry in the plane of loading unless the effects of asymmetry of the section are included in the analysis.

11.2 Maximum Width-Thickness Ratios of Elements Subject to Compression
The width-thickness ratio of elements subject to compression shall not exceed the limits given in Table 1 for the specified section classification.

11.3 Width and Thickness

11.3.1
For elements supported along only one edge parallel to the direction of compressive force, the width shall be taken as follows:
(a) for plates, the width, b, is the distance from the free edge to the first row of fasteners or line of welds;
(b) for legs of angles, flanges of channels and zees, and stems of tees, the width, b, is the full nominal dimension; and
(c) for flanges of beams and tees, the width, b, is one-half of the full nominal dimension.

11.3.2
For elements supported along two edges parallel to the direction of compressive force, the width shall be taken as follows:
(a) for flange or diaphragm plates in built-up sections, the width, b, is the distance between adjacent lines of fasteners or lines of welds;
(b) for flanges of rectangular hollow structural sections, the width, b, is the clear distance between webs less the inside corner radius on each side;
(c) for webs of built-up sections, the width, h, is the distance between adjacent lines of fasteners or the clear distance between flanges when welds are used; and
(d) for webs of hot-rolled sections, the width, h, is the clear distance between flanges.

11.3.3
The thickness of elements is the nominal thickness. For tapered flanges of rolled sections, the thickness is the nominal thickness halfway between a free edge and the corresponding face of the web.

Table 1
Maximum Width-Thickness Ratios: Elements in Compression

Description of element	Section Classification*		
	Class 1 plastic design	Class 2 compact	Class 3 noncompact
Legs of angles and elements supported along one edge except as noted	—	—	$\frac{b}{t} \leq \frac{200}{\sqrt{F_y}}$
Angles in continuous contact with other elements; plate-girder stiffeners	—	—	$\frac{b}{t} \leq \frac{200}{\sqrt{F_y}}$
Stems of T-sections	$\frac{b}{t} \leq \frac{145}{\sqrt{F_y}}$ †	$\frac{b}{t} \leq \frac{170}{\sqrt{F_y}}$ †	$\frac{b}{t} \leq \frac{340}{\sqrt{F_y}}$
Flanges of I- or T-sections; plates projecting from compressive elements; outstanding legs of pairs of angles in continuous contact ‡	$\frac{b}{t} \leq \frac{145}{\sqrt{F_y}}$	$\frac{b}{t} \leq \frac{170}{\sqrt{F_y}}$	$\frac{b}{t} \leq \frac{200}{\sqrt{F_y}}$
Flanges of channels	—	—	$\frac{b}{t} \leq \frac{200}{\sqrt{F_y}}$
Flanges of rectangular hollow structural sections	$\frac{b}{t} \leq \frac{420}{\sqrt{F_y}}$	$\frac{b}{t} \leq \frac{525}{\sqrt{F_y}}$	$\frac{b}{t} \leq \frac{670}{\sqrt{F_y}}$
Flanges of box sections, flange cover plates and diaphragm plates, between lines of fasteners or welds	$\frac{b}{t} \leq \frac{525}{\sqrt{F_y}}$	$\frac{b}{t} \leq \frac{525}{\sqrt{F_y}}$	$\frac{b}{t} \leq \frac{670}{\sqrt{F_y}}$
Perforated cover plates	—	—	$\frac{b}{t} \leq \frac{840}{\sqrt{F_y}}$
Webs	$\frac{h}{w} \leq \frac{1100}{\sqrt{F_y}} \left(1 - 0.39 \frac{C_f}{C_y}\right)$	$\frac{h}{w} \leq \frac{1700}{\sqrt{F_y}} \left(1 - 0.61 \frac{C_f}{C_y}\right)$	$\frac{h}{w} \leq \frac{1900}{\sqrt{F_y}} \left(1 - 0.65 \frac{C_f}{C_y}\right)$
Circular hollow sections in axial compression	—	—	$\frac{D}{t} \leq \frac{23\,000}{F_y}$
Circular hollow sections in flexural compression	$\frac{D}{t} \leq \frac{13\,000}{F_y}$	$\frac{D}{t} \leq \frac{18\,000}{F_y}$	$\frac{D}{t} \leq \frac{66\,000}{F_y}$

*For Class 4 (slender) sections, see Clauses 13.3 or 13.5, as applicable.
†See Clause 11.1.3.
‡ Can be considered as Class 1 or Class 2 sections if angles are continuously connected by adequate mechanical fasteners or welds, and if there is an axis of symmetry in the plane of loading.

12. Gross and Net Areas

12.1 Application
Members in tension shall be proportioned on the basis of the areas associated with the potential failure modes. Members in compression shall be proportioned on the basis of the gross area. (For beams and girders, see Clause 15.)

12.2 Gross Area
Gross area shall be calculated by summing the products of the thickness and the gross width of each element (flange, web, leg, plate), as measured normal to the axis of the member.

12.3 Effective Net Area

12.3.1
The effective net area, A_{ne}, shall be determined by summing the critical net areas, A_n, of each segment along a potential path of minimum resistance. The net areas shall be calculated as follows:

(a) for a segment normal to the force (ie, in direct tension)
$A_n = w_n t$

(b) for a segment parallel to the force (ie, in shear)
$A_n = 0.6 L_n t$

(c) for a segment inclined to the force
$A_n = w_n t + s^2 t / 4g$

12.3.2
In calculating w_n and L_n, the width of bolt holes shall be taken as 2 mm larger than the specified hole diameter. Where it is known that drilled holes will be used, this allowance may be waived.

12.3.3 Effective Net Area Reduction — Shear Lag

12.3.3.1
When fasteners transmit load to each of the cross-sectional elements of a member in tension in proportion to their respective areas, the reduced effective net area is equal to the effective net area:
$A'_{ne} = A_{ne}$

12.3.3.2
When bolts transmit load to some but not all of the cross-sectional elements and only when the critical net area includes the net area of unconnected elements, the reduced effective net area shall be taken as follows:

(a) for WWF, W, M, or S shapes with flange widths not less than two-thirds the depth, and for structural tees cut from these shapes, when only the flanges are connected with three or more transverse lines of fasteners,
$A'_{ne} = 0.90 A_{ne}$

(b) for angles connected by only one leg with
   (i) four or more transverse lines of fasteners,
$A'_{ne} = 0.80 A_{ne}$
   (ii) fewer than four transverse lines of fasteners,
$A'_{ne} = 0.60 A_{ne}$

(c) for all other structural shapes connected with
  (i) three or more transverse lines of fasteners:
$A'_{ne} = 0.85 A_{ne}$ or
  (ii) with two transverse lines of fasteners:
$A'_{ne} = 0.75 A_{ne}$

where A_n is calculated in accordance with Clause 12.3.1(a) and (c).

12.3.3.3

When a tension load is transmitted by welds, the reduced effective net area shall be computed as:
$A'_{ne} = A_{ne1} + A_{ne2} + A_{ne3}$

where A_{ne1}, A_{ne2}, and A_{ne3} are the effective net areas of the connected plate elements subject to one of the following methods of load transfer:

(a) Elements connected by transverse welds, A_{ne1}
$A_{ne1} = wt$

(b) Elements connected by longitudinal welds along two parallel edges, A_{ne2}

  (i) when $L \geq 2w$, $A_{ne2} = 1.00\,wt$
  (ii) when $2w > L \geq 1.5w$, $A_{ne2} = 0.87\,wt$
  (iii) when $1.5w > L \geq w$, $A_{ne2} = 0.75\,wt$

where
L = average length of welds on the two edges
w = plate width (distance between welds)

(c) Elements connected by a single line of weld, A_{ne3}
$A_{ne3} = \left(1 - \dfrac{\bar{x}}{L}\right) wt$
where:
$\bar{x}$ = eccentricity of the weld with respect to centroid of the element
L = length of connection in the direction of the loading
Note: *The outstanding leg of an angle is considered connected by the (single) line of weld along the heel.*

12.3.3.4

Larger values of the reduced effective net area may be used if justified by test or rational analysis.

12.3.4

For angles, the gross width shall be the sum of the widths of the legs minus the thickness. The gauge for holes in opposite legs shall be the sum of the gauges from the heel of the angle minus the thickness.

12.3.5

In calculating the net area across plug or slot welds, the weld metal shall not be taken as adding to the net area.

12.4 Pin-Connected Members in Tension

12.4.1
In pin-connected members in tension, the net area, An, across the pin hole, normal to the axis of the member, shall be at least 1.33 times the cross-sectional area of the body of the member. The net area of any section on either side of the axis of the member, measured at an angle of 45° or less to the axis of the member, shall be not less than 0.9 times the cross-sectional area of the body of the member.

12.4.2
The distance from the edge of the pin hole to the edge of the member, measured transverse to the axis of the member, shall not exceed four times the thickness of the material at the pin hole.

12.4.3
The diameter of a pin hole shall be not more than 1 mm larger than the diameter of the pin.

13. Member and Connection Resistance

13.1 General
To meet the strength requirements of this Standard all factored resistances, as determined in this Clause, shall be greater than or equal to the effect of factored loads determined in accordance with Clause 7.2, and ϕ shall be taken as 0.90 unless otherwise specified.

13.2 Axial Tension
The factored tensile resistance, T_r, developed by a member subjected to an axial tensile force shall be taken as
(a) the least of
  (i) $T_r = \phi A_g F_y$
  (ii) $T_r = 0.85 \phi A_{ne} F_u$
  (iii) $T_r = 0.85 \phi A'_{ne} F_u$
(b) for pin connections,
$T_r = 0.75 \phi A_n F_y$

13.3 Axial Compression

13.3.1
The factored axial compressive resistance, C_r, of a member conforming to the requirements of Clause 11 for Class 1, 2 or 3 sections shall be taken as

$$C_r = \phi A F_y \left(1 + \lambda^{2n}\right)^{-1/n}$$

where
n = 1.34 for W shapes of Group 1, 2, and 3 of Table 1 of CSA Standard G40.20, fabricated I-shapes, fabricated box shapes, and hollow structural sections manufactured according to CSA Standard G40.20, Class C (cold-formed non-stress-relieved);

n = 2.24 for WWF shapes with flange edges flame-cut produced in accordance with CSA Standard G40.20 and hollow structural sections manufactured according to CSA Standard G40.20, Class H (hot-formed or cold-formed stress-relieved);

$$\lambda = \frac{KL}{r} \sqrt{\frac{F_y}{\pi^2 E}}$$

13.3.2
In addition to meeting the requirements of Clause 13.3.1, singly symmetric, asymmetric, or cruciform sections shall meet the requirements of Appendix D.

13.3.3
The factored compressive resistance, C_r, developed by a member subject to an axial compressive force and designated as a Class 4 section according to Clause 11 shall be determined as follows:
(a) For sections that are less than or equal to 4.5 mm thick and that are not hollow structural sections, the compressive resistance shall be calculated in accordance with CSA Standard S136.
(b) For hollow structural sections and sections greater than 4.5 mm thick that can be shown to be not critical in torsional buckling or not subject to torsional-flexural buckling, the compressive resistance shall be calculated in accordance with the requirements of Clause 13.3.1. The area, A, shall be taken as the effective area determined in accordance with CSA Standard S136. The slenderness ratio shall be calculated using gross section properties.

13.4 Shear

13.4.1 Webs of Flexural Members with Two Flanges

13.4.1.1 Elastic Analysis
Except as noted in Clause 13.4.1.2, the factored shear resistance, V_r, developed by the web of a flexural member shall be taken as
$$V_r = \phi A_w F_s$$
where
A_w = shear area (dw for rolled shapes and hw for girders); and F_s is as follows:

(a) $\dfrac{h}{w} \leq 439\sqrt{\dfrac{k_v}{F_y}}$ $\qquad F_s = 0.66 F_y$

(b) $439\sqrt{\dfrac{k_v}{F_y}} < \dfrac{h}{w} \leq 502\sqrt{\dfrac{k_v}{F_y}}$ $\qquad F_s = F_{cri} = 290 \dfrac{\sqrt{F_y k_v}}{(h/w)}$

(c) $502\sqrt{\dfrac{k_v}{F_y}} < \dfrac{h}{w} \leq 621\sqrt{\dfrac{k_v}{F_y}}$ $\qquad F_s = F_{cri} + F_t$

$$F_{cri} = \dfrac{290 \sqrt{F_y k_v}}{(h/w)}$$

$$F_t = (0.50 F_y - 0.866 F_{cri})\left(\dfrac{1}{\sqrt{1 + (a/h)^2}}\right)$$

(d) $621\sqrt{\dfrac{k_v}{F_y}} < \dfrac{h}{w}$ $\qquad F_s = F_{cre} + F_t$

$$F_{cre} = \dfrac{180\,000 k_v}{(h/w)^2}$$

$$F_t = (0.50 F_y - 0.866 F_{cre})\left(\dfrac{1}{\sqrt{1 + (a/h)^2}}\right)$$

where
k_v = shear buckling coefficient

$k_v = 4 + \dfrac{5.34}{(a/h)^2}$ when $a/h < 1$

$k_v = 5.34 + \dfrac{4}{(a/h)^2}$ when $a/h \geq 1$

a/h = aspect ratio, the ratio of the distance between stiffeners to web depth.

13.4.1.2 Plastic Analysis
In structures designed on the basis of a plastic analysis as defined in Clause 8.5, the factored shear resistance, V_r, developed by the web of a flexural member subjected to shear shall be taken as
$V_r = 0.55\phi w d F_y$

13.4.1.3 Maximum Slenderness
The slenderness ratio (h/w) of a web shall not exceed $83\,000/\sqrt{F_y}$

where
F_y = specified minimum yield point of the compressive flange steel.

This limit may be waived if analysis indicates that buckling of the compressive flange into the web will not occur at factored load levels.

13.4.1.4 Combined Shear and Moment in Girders
Transversely stiffened girders depending on tension-field action to carry shear (ie, with $h/w > 502\sqrt{k_v/F_y}$) shall be proportioned such that the following limits are met:

$\dfrac{V_f}{V_r} \leq 1.0$

$\dfrac{M_f}{M_r} \leq 1.0$

$0.727\dfrac{M_f}{M_r} + 0.455\dfrac{V_f}{V_r} \leq 1.0$

where
V_r is established according to Clause 13.4 and M_r is established according to Clause 13.5 or 13.6, as applicable.

13.4.2 Webs of Flexural Members not Having Two Flanges
The factored shear resistance for cross sections not having two flanges (eg, solid rectangles, rounds, tees) shall be determined by rational analysis. The factored shear stress at any location in the cross section shall be taken as not greater than $0.66\phi F_y$ and shall be reduced where shear buckling is a consideration.

13.4.3 Gusset Plates
The factored tensile resistance or combined tension and shear resistance of gusset plates shall be calculated in accordance with Clause 13.2(a) based on an effective net area as given in Clause 12.3.

13.4.4 Connecting Elements
The factored shear resistance of connecting elements such as gusset plates, framing angles, and shear tabs shall be taken as
$V_r = 0.50\phi L_n t F_u$

13.4.5 Pins
The total factored shear resistance of the nominal area of pins shall be taken as
$V_r = 0.66\phi A F_y$

13.5 Bending — Laterally Supported Members
The factored moment resistance, M_r, developed by a member subjected to uniaxial bending moments about a principal axis and where continuous lateral support is provided to the compressive flange shall be taken as

(a) for Class 1 and Class 2 sections:
$M_r = \phi Z F_y = \phi M_p$

(b) for Class 3 sections:
$M_r = \phi S F_y = \phi M_y$

(c) for Class 4 sections:

(i) when both the web and the compressive flange fall within Class 4 of Table 1, the value of M_r shall be determined in accordance with CSA Standard S136. The calculated value, F_y', applicable to cold-formed members shall be determined by using only the values for F_y and F_u that are specified in the relevant structural steel material standard;

(ii) for beams or girders whose flanges meet the requirements of Class 3 and whose webs exceed the limits for Class 3, see Clause 15;

(iii) for beams or girders whose webs meet the requirements of Class 3 and whose flanges exceed the limits for Class 3, the moment resistance shall be calculated in accordance with CSA Standard S136. Alternatively, the moment resistance may be calculated as
$M_r = \phi S_e F_y$

where
S_e = the effective section modulus determined using an effective flange width of $670t/\sqrt{F_y}$ for flanges supported along two edges parallel to the direction of stress and an effective width of $200t/\sqrt{F_y}$ for flanges supported along one edge parallel to the direction of stress. For flanges supported along one edge, in no case shall b/t exceed 60.

13.6 Bending — Laterally Unsupported Members
Where continuous lateral support is not provided to the compression flange of a member subjected to uniaxial strong axis bending, the factored moment resistance, M_r, may be taken as

(a) for doubly symmetric Class 1 and 2 sections

(i) when $M_u > 0.67 M_p$

$M_r = 1.15\phi M_p \left(1 - \dfrac{0.28 M_p}{M_u}\right)$ but not greater than ϕM_p

(ii) when $M_u \leq 0.67 M_p$

$M_r = \phi M_u$

where the critical elastic moment of the unbraced member is given by

$M_u = \dfrac{\omega_2 \pi}{L} \sqrt{E I_y G J + \left(\dfrac{\pi E}{L}\right)^2 I_y C_w}$

where

L = length of unbraced portion of beam, mm

ω_2 = $1.75 + 1.05\kappa + 0.3\kappa^2 \leq 2.5$, for unbraced lengths subject to end moments

 = 1.0 when the bending moment at any point within the unbraced length is larger than the larger end moment or when there is no effective lateral support for the compression flange at one of the ends of the unsupported length

C_w = 0.0 for hollow structural sections

κ = ratio of the smaller factored moment to the larger factored moment at opposite ends of the unbraced length, positive for double curvature and negative for single curvature

(b) for doubly symmetric Class 3 and 4 sections and for channels

   (i) when $M_u > 0.67 M_y$

$$M_r = 1.15\phi M_y \left(1 - \frac{0.28 M_y}{M_u}\right)$$

but not greater than ϕM_y for Class 3 sections and the value given in Clause 13.5 (c) (iii) for Class 4 sections

   (ii) when $M_u \leq 0.67 M_y$

$$M_r = \phi M_u$$

where M_u and ω_2 are defined in Clause 13.6 (a)(ii)

(c) For cantilever beams, a rational method of analysis taking into account the lateral support conditions at the support and tip of the cantilever should be used.

(d) For monosymmetric shapes, a rational method of analysis such as that given in the Structural Stability Research Council's *Guide to Stability Design Criteria for Metal Structures* should be used.

(e) for biaxial bending, the member shall meet the following criterion:

$$\frac{M_{fx}}{M_{rx}} + \frac{M_{fy}}{M_{ry}} \leq 1.0$$

13.7 Lateral Bracing for Members in Structures Analysed Plastically

Members in structures or portions of structures in which the distributions of moments and forces have been determined by a plastic analysis shall be braced to resist lateral and torsional displacement at all hinge locations. The laterally unsupported distance, L_{cr}, from such braced hinge locations to the nearest adjacent point on the frame similarly braced shall not exceed

L_{cr} = $550 r_y / \sqrt{F_y}$ for $\kappa < -0.5$

L_{cr} = $980 r_y / \sqrt{F_y}$ for $\kappa \geq -0.5$

Both bracing requirements should be checked and the more severe shall govern the location of the braced point. Bracing is not required at the location of the last hinge to form in the failure mechanism assumed as the basis for proportioning the structure. Except for the aforementioned regions, the maximum unsupported length of members in structures analysed plastically need not be less than that permitted for the same members in structures analysed elastically.

13.8 Axial Compression and Bending

Note: *More detailed methods of determining the resistance of columns subject to biaxial bending are given in Appendix F.*

13.8.1 Member Strength and Stability — All Classes of Sections Except Class 1 Sections of I-Shaped Members

Members required to resist both bending moments and an axial compressive force shall be proportioned so that

$$\frac{C_f}{C_r} + \frac{U_{1x} M_{fx}}{M_{rx}} + \frac{U_{1y} M_{fy}}{M_{ry}} \leq 1.0$$

where

M_f is the maximum moment including stability effects as defined in Clause 8.6.

The capacity of the member shall be examined for the following cases:

(a) cross-sectional strength, in which case
C_r is as defined in Clause 13.3 with the value $\lambda = 0$
M_r is as defined in Clause 13.5 (for the appropriate class of section)
U_{1x} and U_{1y} are taken as 1.0

(b) overall member strength, in which case
C_r is as defined in Clause 13.3 with the value $K = 1.0$, and is based on the maximum slenderness ratio for biaxial bending. For uniaxial strong-axis bending, $C_r = C_{rx}$ (see also Clause 9.3.2).
M_r is as defined in Clause 13.5 (for the appropriate class of section)
U_{1x} and U_{1y} are as defined in Clause 13.8.3

(c) lateral torsional buckling strength, when applicable, in which case
C_r is as defined in Clause 13.3, and is based on weak-axis buckling (see also Clause 9.3.3)
M_{rx} is as defined in Clause 13.6 (for the appropriate class of section)
M_{ry} is as defined in Clause 13.5 (for the appropriate class of section)
U_{1x} is as defined in Clause 13.8.3, but not less than 1.0
U_{1y} is as defined in Clause 13.8.3.

13.8.2 Member Strength and Stability — Class 1 Sections of I-Shaped Members

Members required to resist both bending moments and an axial compressive force shall be proportioned so that

$$\frac{C_f}{C_r} + \frac{0.85\,U_{1x}\,M_{fx}}{M_{rx}} + \frac{0.60\,U_{1y}\,M_{fy}}{M_{ry}} \leq 1.0$$

All terms in this expression are defined in Clause 13.8.1.

The capacity of the member shall be examined for

(a) cross-sectional strength;

(b) overall member strength; and

(c) lateral torsional buckling strength.

In addition, the member shall meet the following criterion:

$$\frac{M_{fx}}{M_{rx}} + \frac{M_{fy}}{M_{ry}} \leq 1.0$$

where
M_{rx} and M_{ry} are defined in Clause 13.5 or 13.6, as appropriate.

13.8.3 Value of U_1

In lieu of a more detailed analysis, the value of U_1 for the axis under consideration, accounting for the second-order effects due to the deformation of a member between its ends, shall be taken as

$$U_1 = \left[\frac{\omega_1}{1 - \frac{C_f}{C_e}}\right]$$

where
ω_1 for the axis under consideration is defined in Clause 13.8.4 and
$C_e = \pi^2\,E\,I\,/L^2$ for the axis under consideration.

13.8.4 Values of ω_1

Unless otherwise determined by analysis, the following values shall be used for ω_1:

(a) for members not subjected to transverse loads between supports
$\omega_1 = 0.6 - 0.4\kappa \geq 0.4$

(b) for members subjected to distributed loads or a series of point loads between supports
$\omega_1 = 1.0$

(c) for members subjected to a concentrated load or moment between supports
$\omega_1 = 0.85$

For the purpose of design, members subjected to concentrated load or moment between supports (eg, crane columns) may be considered to be divided into two segments at the point of load (or moment) application. Each segment shall then be treated as a member that depends on its own flexural stiffness to prevent side-sway in the plane of bending considered, and ω_1 shall be taken as 0.85. In calculating the slenderness ratio for use in Clause 13.8, the total length of the member shall be used.

13.9 Axial Tension and Bending

Members required to resist both bending moments and an axial tensile force shall be proportioned so that

(a) $\dfrac{T_f}{T_r} + \dfrac{M_f}{M_r} \leq 1.0$

where
$M_r = \phi M_p$ for Class 1 and Class 2 sections
$M_r = \phi M_y$ for Class 3 and Class 4 sections

(b) $\dfrac{M_f}{M_r} - \dfrac{T_f Z}{M_r A} \leq 1.0$ for Class 1 and Class 2 sections;

$\dfrac{M_f}{M_r} - \dfrac{T_f S}{M_r A} \leq 1.0$ for Class 3 and Class 4 sections

where
M_r is defined in Clause 13.5 or 13.6

13.10 Load Bearing

The factored bearing resistance, B_r, developed by a member or portion of a member subjected to bearing shall be taken as follows:

(a) on the contact area of machined, accurately sawn, or fitted parts,
$B_r = 1.50 \phi F_y A$

(b) on expansion rollers or rockers,
$B_r = 0.000\,26\phi \left(\dfrac{R_1}{1 - \dfrac{R_1}{R_2}} \right) L F_y^2$

where
B_r is in Newtons
R_1 and L are the radius and length, respectively, of the roller or rocker
R_2 is the radius of the groove of the supporting plate
F_y is the specified minimum yield point of the weaker part in contact

(c) in bolted connections, $B_r = 3\phi_b t d n F_u$

where
F_u is the tensile strength of the plate

Note: *See also Clause 13.2 for tensile resistance for bolted connections and Clause 22.8 for limiting end distance/bolt diameter ratios.*

13.11 Bolts in Bearing-Type Connections

13.11.1 General
For bolts in bearing-type connections, ϕ_b shall be taken as 0.67.

13.11.2 Bolts in Shear
The factored resistance developed by a bolted joint subjected to shear shall be taken as the lesser of
(a) the factored bearing resistance, B_r, given in Clause 13.10(c); or
(b) the factored shear resistance of the bolts, which shall be taken as
$V_r = 0.60\phi_b n m A_b F_u$

When the bolt threads are intercepted by any shear plane, the factored shear resistance of any joint shall be taken as 70% of V_r.

For joints longer than 1300 mm, the shearing resistance shall be taken as 80% of the aforementioned values.

Note: *The specified minimum tensile strength, F_u, for bolts are given in the relevant ASTM Standards, ie, for A325M bolts, F_u is 830 MPa; for A490M bolts, F_u is 1040 MPa; for A325 bolts 1 inch or less in diameter, F_u is 825 MPa; for A325 bolts greater than 1 inch in diameter, F_u is 725 MPa; and for A490 bolts, F_u is 1035 MPa.*

13.11.3 Bolts in Tension
The factored tensile resistance developed by a bolted joint, T_r, subjected to tension, T_f, shall be taken as
$T_r = 0.75\phi_b n A_b F_u$

High-strength bolts subjected to tensile cyclic loading shall be pretensioned to the minimum preload given in Clause 23. Connected parts shall be arranged so that prying forces are minimized, and in no case shall the calculated prying force exceed 30% of the externally applied load.

The permissible range of stress under the specified loads, based on the shank area of the bolt, shall not exceed 215 MPa for A325 and A325M bolts or 260 MPa for A490 and A490M bolts.

13.11.4 Bolts in Combined Shear and Tension
A bolt in a joint that is required to develop resistance to both tension and shear shall be proportioned so that

$$\left(\frac{V_f}{V_r}\right)^2 + \left(\frac{T_f}{T_r}\right)^2 \leq 1$$

where V_r is given in Clause 13.11.2 and T_r is given in Clause 13.11.3.

13.12 Bolts in Slip-Critical Connections

13.12.1 General
The requirement for a slip-critical connection is that under the forces and moments produced by specified loads, slip of the assembly shall not occur. In addition, the effect of factored loads shall not exceed the resistances of the connection as given in Clause 13.11.

Limit States Design of Steel Structures

13.12.2 Shear Connections

The slip resistance, V_s, of a bolted joint, subjected to shear, V, shall be taken as:
$$V_s = 0.53 c_1 k_s m n A_b F_u$$

where k_s is the mean slip coefficient as determined from Table 2 or by tests carried out in accordance with *Method to Determine the Slip Coefficient for Coatings used in Bolted Connections* of the Research Council on Structural Connections, and c_1 is a coefficient that relates the specified initial tension and mean slip to a 5% probability of slip for bolts installed by turn-of-nut procedures. Table 2 gives c_1.

Table 2
Values of c_1 and k_s

Class	Description	k_s	c_1 A325 Bolts and A325M Bolts	c_1 A490 Bolts and A490M Bolts
A	Clean mill scale, or blast-cleaned with Class A coatings	0.33	0.82	0.78
B	Blast-cleaned or blast-cleaned with Class B coatings	0.50	0.89	0.85
C	Hot-dip galvanized with wire brushed surfaces	0.40	0.90	0.85

Notes:
(1) *Class A and Class B coatings are defined as those coatings that provide a mean slip coefficient of not less than 0.33 and 0.50, respectively.*
(2) *Values of c_1 for 5% probability of slip for values of k_s other than those listed in Table 2 may be found in the Research Council on Structural Connections'* Guide to Design Criteria for Bolted and Riveted Joints, *Second Edition, listed therein as values of the Slip Factor D.*

13.12.3 Connections in Combined Shear and Tension

A bolt in a joint that is required to develop resistance to both tension and shear shall be proportioned so that the following relationship is satisfied for the specified loads:

$$\frac{V}{V_s} + 1.9 \frac{T}{n A_b F_u} \leq 1.0$$

where
V_s = slip resistance as defined in Clause 13.12.2.

13.13 Welds

13.13.1 General

The resistance factor, ϕ_w, for welded connections shall be taken as 0.67. Matching electrode classifications for CSA G40.21-M steels are given in Table 3.

13.13.2 Shear

13.13.2.1 Complete and Partial Joint Penetration Groove Welds, Plug and Slot Welds

The factored shear resistance shall be taken as the lesser of
(a) for the base metal, $V_r = 0.67 \phi_w A_m F_u$
(b) for the weld metal, $V_r = 0.67 \phi_w A_w X_u$

where
A_m = shear area of effective fusion face
A_w = area of effective weld throat, plug or slot.

13.13.2.2 Fillet Welds

The factored resistance for tension or compression-induced shear shall be taken as the lesser of
(a) for the base metal, $V_r = 0.67 \phi_w A_m F_u$
(b) for the weld metal, $V_r = 0.67 \phi_w A_w X_u (1.00 + 0.50 \sin^{1.5} \theta)$

where
θ = angle of axis of weld with the line of action of force (0° for a longitudinal weld and 90° for a transverse weld), and the other terms are defined in Clause 13.13.2.1. Conservatively, $(1.00 + 0.50 \sin^{1.5} \theta)$ can be taken as 1.0.

13.13.3 Tension Normal to Axis of Weld

13.13.3.1 Complete Joint Penetration Groove Weld Made with Matching Electrodes

The factored tensile resistance shall be taken as that of the base metal.

13.13.3.2 Partial Joint Penetration Groove Weld Made with Matching Electrodes

The factored tensile resistance shall be taken as
$T_r = \phi_w A_n F_u \leq \phi A_g F_y$

where
A_n = nominal area of fusion face face normal to the tensile force.
When overall ductile behaviour is desired (member yielding before weld fracture) $A_n F_u > A_g F_y$

13.13.3.3 Partial Joint Penetration Groove Weld Combined with a Fillet Weld, Made with Matching Electrodes

The factored tensile resistance shall be taken as
$T_r = \phi_w \sqrt{(A_n F_u)^2 + (A_w X_u)^2} \leq \phi A_g F_y$

where
A_g is the gross area of the components of the tension member connected by the welds.

Table 3
Matching Electrode Classification Strengths for G40.21-M Steels

Matching electrode classification strengths* MPa	G40.21-M Grades						
	260	300	350	380	400	480	700
410	X	X†					
480	X	X	X‡	X			
550					X‡		
620						X	
820							X

*Electrode ultimate strengths are given by electrode classification strengths.
†For HSS only.
‡For unpainted applications using "A" or "AT" steels where the deposited weld metal shall have similar atmospheric corrosion resistance or similar colour characteristics to the base metal, or both, the requirements of Clauses 5.2.1.4 and 5.2.1.5 of CSA Standard W59 shall apply.
Note: For matching condition of ASTM steels, see Table 11-1 or Table 12-1 of CSA Standard W59.

13.13.4 Compression Normal to Axis of Weld

13.13.4.1 Complete and Partial Joint Penetration Groove Welds, Made with Matching Electrodes
The compressive resistance shall be taken as that of the effective area of base metal in the joint. For partial joint penetration groove welds, the effective area in compression is the nominal area of the fusion face normal to the compression plus the area of the base metal fitted in contact bearing (see Clause 28.9.7).

13.13.4.2 Cross-Sectional Properties of Continuous Longitudinal Welds
All continuous longitudinal welds, made with matching electrodes can be considered to contribute to the cross-sectional properties, A, S, Z, I of the cross-section.

13.13.4.3 Welds for Hollow Structural Sections
The provisions of Appendix L of CSA Standard W59 may be used for hollow structural sections.

13.14 Steel Plate Shear Walls
Steel plate shear walls shall be proportioned according to the requirements of Appendix M.

14. Fatigue

14.1 General

14.1.1
In addition to meeting the requirements of Clause 14 for fatigue, any member or connection shall also meet the requirements for the static load conditions using the factored loads.

14.1.2
Members and connections subjected to fatigue loading shall be designed, detailed, and fabricated so as to minimize stress concentrations and abrupt changes in cross-section.

14.1.3
Specified loads for the design of members or connections shall be used for all fatigue calculations.

14.1.4
A specified load less than the maximum specified load but acting with a greater number of cycles may govern and shall be considered.

14.1.5
Plate girders with $h/w > 3150/\sqrt{F_y}$ shall not be used under fatigue conditions.

14.2 Fatigue Limit State
The fatigue limit state, which is the limiting case of the slow propagation of a crack within a structural element, can result from either live load effects or as the consequence of local distortion within the structure. Herein, these will be referred to as load-induced or distortion-induced fatigue effects.

For guidance in determining the number of cycles, the life of the structure shall be assumed to be not less than 50 years unless otherwise stated.

When a load is expected to be applied such that not more than 50 000 stress cycles take place at a given detail during the life of the structure, no special considerations beyond those in Clause 14.1.2 need apply. In the event that more than 50 000 stress cycles take place, the loaded members, connections, and fastening elements shall be proportioned so that the probability of fatigue failure is acceptably small. In such cases the design shall be based on the best available information on the fatigue characteristics of the materials and components to be used. In the absence of more specific information, which is subject to the approval of the owner, the requirements of Clause 14 in its entirety provide guidance in proportioning members and parts. Fatigue resistance shall be provided only for those loads considered to be repetitive. Often, the magnitude of a repeated load is less than the maximum static load for which the member or part is designed.

14.3 Live Load-Induced Fatigue

14.3.1 Calculation of Stress Range
The controlling stress feature in load-induced fatigue is the range of stress to which the element is subjected. This is calculated using ordinary elastic analysis and the principles of mechanics of materials. More sophisticated analysis is required only in cases not covered in Table 4(c), eg, major access holes and cut-outs. Stress range is the algebraic difference between the maximum stress and minimum stress at a given location: thus, only the stress due to live load need be calculated.

The load-induced fatigue provisions need be applied only at locations that undergo a net applied tensile stress. At locations where stresses resulting from the permanent loads are compressive, load-induced fatigue can be disregarded only when the compressive stress is at least twice the maximum tensile live load stress.

14.3.2 Design Criteria
For load-induced fatigue, each detail shall satisfy the requirement that

$f_{sr} \leq F_{sr}$

where f_{sr} = calculated stress range at the detail due to passage of the fatigue load
F_{sr} = fatigue resistance, including adjustment for the number of stress range cycles for each passage of load.

14.3.3 Fatigue Resistance

Fatigue resistance of a member or a detail, F_{sr}, shall be calculated as

$$F_{sr} = \left(\frac{\gamma}{nN}\right)^{1/3} \geq \frac{F_{srt}}{2}$$

where
γ = fatigue life constant pertaining to the detail category established under Clause 14.3.4. Values of γ are given in Table 4(a)
n = number of stress range cycles at given detail for each passage of the moving load, as given in Table 4(b)
N = number of passages of the moving load
F_{srt} = constant amplitude threshold stress range, tabulated in Table 4(a).

14.3.4 Detail Categories

The detail categories and their respective fatigue resistance, F_{sr}, shall be obtained from Table 4(c), with the aid of Figure K2.

The fatigue of high-strength bolts is given in Clause 13.11.3.

14.4 Distortion-Induced Fatigue

14.4.1 General

When members designed according to Clause 14.3 are provided with interconnection elements such as diaphragms, cross-bracing, lateral bracing, and the like, in addition to the deck slab, then both the members and the interconnection elements must also be examined for distortion-induced fatigue. The best protection against distortion-induced fatigue is proper detailing so that the regions of high strain do not result when transverse forces, whether calculated or uncalculated, or differential displacements are transmitted from one member to another. Whenever practicable, all the components that make up the cross-section of the primary member should be fastened to the interconnection member. Provisions intended to control web buckling and significant elastic flexing of a girder web have been noted in Clause 14.1.5.

14.4.2 Connection of Diaphragms, Cross-Frames, Lateral Bracing, or Beams

14.4.2.1 Connection to Transverse Elements

Whenever diaphragms (including internal diaphragms), cross-frames, lateral bracing, floor beams, or the like are to be connected to main members, it is preferable that the connection be made using transverse connection plates that are welded or bolted to both the tension and compression flanges of the main member. If it is necessary to use transverse stiffeners present in the main members as the connection element, such stiffeners should likewise be connected to both the tension and compression flanges of the main member.

14.4.2.2 Connection to Lateral Elements

If connection of diaphragms (including internal diaphragms), cross-frames, lateral bracing, floor beams, or the like is to be made to elements located parallel to the longitudinal axis of the main member, then it is preferable that these lateral connection plates be attached to both the tension and compression flanges of the main member. If this is not practicable, then lateral connection plates can be located as follows:

(a) Transversely Stiffened Girders – Lateral connection plates may be fastened to a transversely stiffened girder provided the attachment is located a vertical distance not less than one-half the flange width above or below the flange. If located within the depth of the web, the lateral connection plate shall be centred with respect to the transverse stiffener, even when the stiffener and connection plate are on opposite sides of the web. If the lateral connection plate and transverse stiffener are located on the same side of the web, then the plate must be attached to the stiffener. The transverse stiffener at locations where lateral connection plates are attached must be continuous between the flanges and must be fastened to them. Bracing members attached to the lateral connection plates must be located such that their ends are at least 100mm from the face of the girder web and the transverse stiffener.

(b) Transversely Unstiffened Girders – Lateral connection plates may be fastened to a transversely unstiffened girder provided the attachment is located a vertical distance not less than one-half the flange width or 150mm above or below the flange. Bracing members attached to the lateral connection plates must be located such that their ends are at least 100mm from the face of the girder web.

Table 4(a)
Fatigue Constants and F_{srt} for Various Detail Categories

Detail category	Fatigue life constant, γ	Constant amplitude threshold stress range, F_{srt} (MPa)
A	819×10^{10}	165
B	393×10^{10}	110
B1	200×10^{10}	83
C	144×10^{10}	69
C1	144×10^{10}	83
D	72.1×10^{10}	48
E	36.1×10^{10}	31
E1	12.8×10^{10}	18

Table 4(b)
Values of n

Longitudinal members	Span length, L ≥ 12 m	Span length, L < 12 m
Simple-span girders	1.0	2.0
Continuous girders:		
1. Near interior support (within 0.01L on either side)	1.5	2.0
2. All other locations	1.0	2.0
Cantilever girders	5.0	5.0
Trusses	1.0	1.0

Transverse members	Spacing ≥ 6 m	Spacing < 6 m
All cases	1.0	2.0

Table 4(c)
Detail Categories for Load-Induced Fatigue

General condition	Situation	Detail category	Illustrative example, see Figure K2
Plain members	Base metal:		1, 2
	• with rolled or cleaned surfaces. Flame-cut edges with a surface roughness not exceeding 1000 (25 μm) as defined by CSA Standard B95	A	
	• of unpainted weathering steel	B	
	• at net section of eyebar heads and pin plates	E	
Built-up members	Base metal and weld metal in components, without attachments, connected by:		3, 4, 5, 7
	• continuous full-penetration groove welds with backing bars removed, or	B	
	• continuous fillet welds parallel to the direction of applied stress	B	
	• continuous full-penetration groove welds with backing bars in place, or	B1	
	• continuous partial-penetration groove welds parallel to the direction of applied stress	B1	

(Continued)

Table 4 (c) (Continued)

General condition	Situation	Detail category	Illustrative example, see Figure K2
	Base metal at ends of partial-length cover plates:		
	• narrower than the flange, with or without end welds, or wider than the flange with end welds		7
	• flange thickness ≤ 20 mm	E	
	• flange thickness > 20 mm	E1	
	• wider than the flange without end welds	E1	
Groove-welded splice connections with weld soundness established by NDT and all required grinding in the direction of the applied stresses	Base metal and weld metal at full-penetration groove-welded splices:		
	• of plates of similar cross-sections with welds ground flush	B	8, 9
	• with 600 mm radius transitions in width with welds ground flush	B	11
	• with transitions in width or thickness with welds ground to provide slopes no steeper than 1.0 to 2.5		10, 10A
	• G40.21M-700Q and 700QT base metal	B1	
	• other base metal grades	B	
	• with or without transitions having slopes no greater than 1.0 to 2.5, when weld reinforcement is not removed	C	8, 9, 10, 10A
Longitudinally loaded groove-welded attachments	Base metal at details attached by full- or partial-penetration groove welds:		
	• when the detail length in the direction of applied stress is:		
	• less than 50 mm	C	6, 18
	• between 50 mm and 12 times the detail thickness, but less than 100 mm	D	18

(continued)

Table 4(c) (Continued)

General condition	Situation	Detail category	Illustrative example, see Figure K2
	• greater than either 12 times the detail thickness or 100 mm		
	• detail thickness < 25 mm	E	18
	• detail thickness ≥ 25 mm	E1	18
	• with a transition radius with the end welds ground smooth, regardless of detail length:		12
	• transition radius ≥ 600 mm	B	
	• 600 mm > transition radius ≥ 150 mm	C	
	• 150 mm > transition radius ≥ 50 mm	D	
	• transition radius < 50 mm	E	
	• with a transition radius with end welds not ground smooth	E	12
Transversely loaded groove-welded attachments with weld soundness established by NDT and all required grinding transverse to the direction of stress	Base metal at detail attached by full-penetration groove welds with a transition radius:		12
	• with equal plate thickness and weld reinforcement removed:		
	• transition radius ≥ 600 mm	B	
	• 600 mm > transition radius ≥ 150 mm	C	
	• 150 mm > transition radius ≥ 50 mm	D	
	• transition radius < 50 mm	E	
	• with equal plate thickness and weld reinforcement not removed:		
	• transition radius ≥ 150 mm	C	
	• 150 mm > transition radius ≥ 50 mm	D	
	• transition radius < 50 mm	E	
	• with unequal plate thickness and weld reinforcement removed:		
	• transition radius ≥ 50 mm	D	
	• transition radius < 50 mm	E	
	• for any transition radius with unequal plate thickness and weld reinforcement not removed	E	

(Continued)

Table 4(c) (Continued)

General condition	Situation	Detail category	Illustrative example, see Figure K2
Fillet-welded connections with welds normal to the direction of stress	Base metal:		
	• at details other than transverse stiffener-to-flange or transverse stiffener-to-web connections	C (see Note)	19
	• at the toe of transverse stiffener-to-flange and transverse stiffener-to-web welds	C1	6
Fillet-welded connections with welds normal and/or parallel to the direction of stress	Shear stress on weld throat	E	16
Longitudinally loaded fillet-welded attachments	Base metal at details attached by fillet welds:		
	• when the detail length in the direction of applied stress is:		
	• less than 50 mm or stud-type shear connectors	C	13, 15, 18, 20
	• between 50 mm and 12 times the detail thickness, but less than 100 mm	D	18, 20
	• greater than either 12 times the detail thickness or 100 mm		7, 16, 18, 20
	• detail thickness < 25 mm	E	
	• detail thickness ≥ 25 mm	E1	
	• with a transition radius with the end welds ground smooth, regardless of detail length		12
	• transition radius ≥ 50 mm	D	
	• transition radius < 50 mm	E	
	• with a transition radius with end welds not ground smooth	E	12

(Continued)

Table 4(c) (Concluded)

General condition	Situation	Detail category	Illustrative example, see Figure K2
Transversely loaded fillet-welded attachments with welds parallel to the direction of primary stress	Base metal at details attached by fillet welds: • with a transition radius with end welds ground smooth: 　• transition radius ≥ 50 mm 　• transition radius < 50 mm • with any transition radius with end welds not ground smooth	 D E E	12
Mechanically fastened connections	Base metal: • at gross section of high-strength bolted slip-critical connections, except axially loaded joints in which out-of-plane bending is induced in connected materials • at net section of high-strength bolted non-slip-critical connections • at net section of riveted connections	 B B D	17
Anchor bolts and threaded parts	Tensile stress range on the tensile stress area of the threaded part, including effects of bending	E	
Fillet-welded HSS to base plate	Shear stress on fillet weld	E1	21

Note: *The fatigue resistance of fillet welds transversely loaded is a function of the effective throat and plate thickness. (Ref. Frank and Fisher, Journal of the Structural Division, ASCE, Vol. 105, No. ST9, September 1979.)*

$$F_{sr} = F_{sr}^C \left[(0.06 + 0.79 H/t_p)/(0.64\, t_p^{1/6}) \right]$$

where

F_{sr}^C = *the fatigue resistance for Category C as determined in accordance with Clause 14.3.3. This assumes no penetration at the weld root.*
t_p = *plate thickness*
H = *weld leg size*

15. Beams and Girders

15.1 Proportioning
Beams and girders consisting of rolled shapes (with or without cover plates), hollow structural sections, or fabricated sections shall be proportioned on the basis of the properties of the gross section or the modified gross section. No deduction need be made for fastener holes in webs or flanges unless the reduction of flange area by such holes exceeds 15% of the gross flange area, in which case the excess shall be deducted. The effect of openings other than holes for fasteners shall be considered in accordance with Clause 15.10.

15.2 Rotational Restraint at Points of Support
Beams and girders shall be restrained against rotation about their longitudinal axes at points of support.

15.3 Copes

15.3.1
The effect of copes on the lateral torsional buckling resistance of a beam or girder shall be taken into account.

15.3.2
The effect of copes in reducing the net area of the web available to resist transverse shear and the effective net area of potential paths of minimum resistance shall be taken into account (see Clause 12.3.1).

15.4 Reduced Moment Resistance of Girders with Thin Webs
When the web slenderness ratio, h/w, exceeds $1900/\sqrt{M_f/\phi S}$, the flange shall meet the width-thickness ratios of Class 3 sections in accordance with Clause 11 and the factored moment resistance of the beam or girder, M'_r, shall be determined by

$$M'_r = M_r \left[1 - 0.0005 \frac{A_w}{A_f} \left(\frac{h}{w} - 1900/\sqrt{M_f/\phi S} \right) \right]$$

where
M_r = factored moment resistance as determined by Clause 13.5 or 13.6 but not to exceed ϕM_y. When an axial compressive force acts on the girder in addition to the moment, the constant 1900 in the expression for M'_r shall be reduced by the factor $(1 - 0.65\ C_f/C_y)$. See also Clause 11.2 and Table 1.

15.5 Flanges

15.5.1
Flanges of welded girders preferably shall consist of a single plate or a series of plates joined end-to-end by complete penetration groove welds.

15.5.2
Flanges of bolted girders shall be proportioned so that the total cross-sectional area of cover plates does not exceed 70% of the total flange area.

15.5.3
Fasteners or welds connecting flanges to webs shall be proportioned to resist horizontal shear forces due to bending combined with any loads that are transmitted from the flange to the web

other than by direct bearing. Spacing of fasteners or intermittent welds in general shall be in proportion to the intensity of the shear force and shall not exceed the maximum for compression or tension members as applicable, in accordance with Clause 19.

15.5.4
Partial-length flange cover plates shall be extended beyond the theoretical cut-off point and the extended portion shall be connected with sufficient fasteners or welds to develop a force in the cover plate at the theoretical cut-off point not less than

$$P = \frac{AM_{fc}y}{I_g}$$

where
P = required force to be developed in cover plate
A = area of cover plate
M_{fc} = moment due to factored loads at theoretical cut-off point
y = distance from centroid of cover plate to neutral axis of cover-plated section
I_g = moment of inertia of cover-plated section

Additionally, for welded cover plates the welds connecting the cover-plate termination to the beam or girder shall be designed to develop the force, P, within a length, a′, measured from the actual end of the cover plate, determined as follows:
(a) a′ = the width of cover plate when there is a continuous weld equal to or larger than three-fourths of the cover-plate thickness across the end of the plate and along both edges in the length a′;
(b) a′ = 1.5 times the width of cover plate when there is a continuous weld smaller than three-fourths of the cover-plate thickness across the end of the plate and along both edges in the length a′; and
(c) a′ = 2 times the width of cover plate when there is no weld across the end of the plate but there are continuous welds along both edges in the length a′.

15.6 Bearing Stiffeners

15.6.1
Pairs of bearing stiffeners on the webs of single-web beams and girders shall be required at points of concentrated loads and reactions wherever the bearing resistance of the web is exceeded (see Clause 15.9). Bearing stiffeners shall also be required at unframed ends of single-web girders having web depth-thickness ratios greater than $1100/\sqrt{F_y}$. Box girders may employ diaphragms designed to act as bearing stiffeners.

15.6.2
Bearing stiffeners shall bear against the flange or flanges through which they receive their loads and shall extend approximately to the edge of the flange plates or flange angles. They shall be designed as columns in accordance with Clause 13.3, assuming the column section to consist of the pair of stiffeners and a centrally located strip of the web equal to not more than 25 times its thickness at interior stiffeners, or a strip equal to not more than 12 times its thickness when the stiffeners are located at the end of the web. The effective column length, KL, shall be taken as not less than three-fourths of the length of the stiffeners in calculating the ratio KL/r. Only that portion of the stiffeners outside of the angle fillet or the flange-to-web welds shall be considered effective in bearing. Angle bearing stiffeners shall not be crimped. Bearing stiffeners shall be connected to the web so as to develop the full force required to be carried by the stiffener into the web or vice versa.

15.7 Intermediate Transverse Stiffeners

15.7.1
Intermediate transverse stiffeners, when used, shall be spaced to suit the shear resistance determined in accordance with Clause 13.4, except that at girder end panels or at panels adjacent to large openings, the tension-field component shall be taken as zero unless means are provided to anchor the tension field.

15.7.2
The maximum distance between stiffeners, when stiffeners are required, shall not exceed the values shown in Table 5. Closer spacing may be required in accordance with Clause 15.7.1.

Table 5
Maximum Intermediate Transverse Stiffener Spacing

Web depth-thickness ratio (h/w)	Maximum distance between stiffeners, a, in terms of clear web depth, h
Up to 150	3h
More than 150	$\dfrac{67\,500h}{(h/w)^2}$

15.7.3
Intermediate transverse stiffeners may be furnished singly or in pairs. Width-thickness ratios shall conform to Clause 11. The moment of inertia of the stiffener, or pair of stiffeners if so furnished, shall be not less than $(h/50)^4$ taken about an axis in the plane of the web. The gross area of intermediate stiffeners, or pairs of stiffeners if so furnished, shall be given by the expression

$$A_s \geq \frac{aw}{2}\left[1 - \frac{a/h}{\sqrt{1+(a/h)^2}}\right]CYD$$

where
a = centre-to-centre distance of adjacent stiffeners (ie, panel length)
w = web thickness
h = web depth
$C = \left[1 - \dfrac{310\,000 k_v}{F_y(h/w)^2}\right]$ but not less than 0.10
Y = ratio of specified minimum yield point of web steel to specified minimum yield point of stiffener steel
D = stiffener factor
  = 1.0 for stiffeners furnished in pairs
  = 1.8 for single angle stiffeners
  = 2.4 for single plate stiffeners
k_v = shear buckling coefficient (see Clause 13.4.1)
F_y = specified minimum yield point of web steel

When the greatest shear, V_f, in an adjacent panel is less than that permitted by Clause 13.4.1, this gross area requirement may be reduced in like proportion by multiplying by the ratio V_f/V_r.

15.7.4
Intermediate transverse stiffeners shall be connected to the web for a shear transfer per pair of stiffeners (or per single stiffener when so furnished), in newtons per millimetre of web depth, h, not less than $1 \times 10^{-4} h F_y^{1.5}$, except that when the largest calculated shear, V_f, in the adjacent panels is less than V_r as calculated by Clause 13.4.1, this shear transfer may be reduced in the same proportion. However, the total shear transfer shall in no case be less than the value of any concentrated load or reaction required to be transmitted to the web through the stiffener. Fasteners connecting intermediate transverse stiffeners to the web shall be spaced not more than 300 mm from centre to centre. If intermittent fillet welds are used, the clear distance between welds shall not exceed 16 times the web thickness or 4 times the weld length.

15.7.5
When intermediate stiffeners are used on only one side of the web, the stiffeners shall be attached to the compression flange. Intermediate stiffeners used in pairs shall have at least a snug fit against the compression flange. When stiffeners are cut short of the tension flange, the distance cut short shall be equal to or greater than four times but not greater than six times the girder web thickness. Stiffeners preferably shall be clipped to clear girder flange-to-web welds.

15.8 Lateral Forces
The flanges of beams and girders supporting cranes or other moving loads shall be proportioned to resist any lateral forces produced by such loads.

15.9 Web Crippling and Yielding
Bearing stiffeners shall be provided where the factored concentrated loads or reactions exceed the factored compressive resistances of webs of rolled beams and welded plate girders. The factored compressive resistance of the web shall be calculated as follows:

(a) for interior loads (concentrated load applied a distance from the member end greater than the member depth), the smaller of
  (i) $B_r = 1.10 \phi w (N + 5k) F_y$
  (ii) $B_r = 300 \phi w^2 \left[1 + 3 \left(\frac{N}{d} \right) \left(\frac{w}{t} \right)^{1.5} \right] \sqrt{F_y t / w}$

(b) for end reactions, the smaller of
  (i) $B_r = 1.10 \phi w (N + 2.5k) F_y$
  (ii) $B_r = 150 \phi w^2 \left[1 + 3 \left(\frac{N}{d} \right) \left(\frac{w}{t} \right)^{1.5} \right] \sqrt{F_y t / w}$

where
w = web thickness
N = length of bearing (N shall be not less than k for end reactions)
k = distance from outer face of flange to web toe of flange-to-web fillet
t = flange thickness

15.10 Openings

15.10.1
Except as provided in Clause 15.1, the effect of all openings in beams and girders shall be considered in the design. At all points where the factored shear or moments at the net section would exceed the capacity of the member, adequate reinforcement shall be added to the member at that point to provide the required strength and stability.

15.10.2
Unreinforced circular openings may be located in the web of unstiffened prismatic compact beams or girders without considering net section properties provided that
(a) the specified design load for the member is uniformly distributed;
(b) the section has an axis of symmetry in the plane of bending;
(c) the openings are located within the middle third of the depth and the middle half of the span of the member;
(d) the spacing between the centres of any two adjacent openings, measured parallel to the longitudinal axis of the member, is a minimum of 2.5 times the diameter of the larger opening; and
(e) the factored maximum shear at the support does not exceed 50% of the factored shear resistance of the section.

15.10.3
If the forces at openings are determined by an elastic analysis, the procedure adopted shall be in accordance with published, recognized principles. The forces determined by such elastic analysis shall not exceed the factored resistances given in Clause 13 and, if applicable, Clause 14.

15.10.4
The strength and stability of the member in the vicinity of openings may be determined on the basis of assumed locations of plastic hinges, such that the resulting force distributions satisfy the requirements of equilibrium, provided that the analysis is carried out in accordance with Clauses 8.5(a), (b), and (f). However, for I-type members the width-thickness ratio of the flanges may meet the requirements of Class 2 sections, provided that the webs meet the width-thickness limit of Class 1 sections.

15.11 Torsion

15.11.1
Beams and girders subjected to torsion shall have sufficient strength and rigidity to resist the torsional moment and forces in addition to other moments or forces. The connections and bracing of such members shall be adequate to transfer the reactions to the supports.

15.11.2
The factored resistance of I-shaped members subject to combined flexure and torsion may be determined from moment-torque interaction diagrams that take into account the normal stress distribution due to flexure and warping torsion and the St. Venant torsion. Assumed normal stress distributions shall be consistent with the class of section.

15.11.3
Members subject to torsional deformations required to maintain compatibility of the structure need not be designed to resist the associated torsional moments provided that the structure satisfies the requirements of equilibrium.

15.11.4
For all members subject to loads causing torsion, the torsional deformations under specified loads shall be limited in accordance with the requirements of Clause 6.2.1.1. For members subject to torsion or to combined flexure and torsion, the maximum combined normal stress, as determined by an elastic analysis, arising from warping torsion and bending due to the specified loads shall not exceed F_y.

16. Open-Web Steel Joists

16.1 Scope
Clause 16 provides requirements for the design, manufacture, transportation, and erection of open-web steel joists used in the construction of buildings. Joists intended to act compositely with the deck shall be designed using the requirements of Clause 17 in conjunction with the requirements of this Clause. This Clause shall not be used for the design of joists not having an axis of symmetry in the plane of the joist.

16.2 General
Joists are steel trusses of relatively low mass with parallel or slightly pitched chords and triangulated web systems proportioned to span between masonry walls or structural supporting members, or both, and to provide direct support for floor or roof deck. In general joists are manufactured on a production-line basis employing jigs, with certain details of the members being standardized by the individual manufacturer.

16.3 Definitions
The following definitions apply to Clause 16:

Deck or decking — the structural floor or roof element spanning between adjacent joists and directly supported thereby. The terms deck and decking include cast-in-place or precast concrete slabs, profiled metal deck, wood plank or plywood, and other relatively rigid elements suitable for floor or roof construction.

Span of an open-web steel joist — the centre-to-centre distance of joist bearings.

Special open-web steel joists or special joists —
(a) joists subjected to the loads stipulated in Clause 16.6.2;
(b) cantilever joists, continuous joists, and joists having special support conditions; and
(c) joists having other special requirements such as lateral-force-resisting braces/ties for columns and beams.

Standard open-web joists — simply supported flexural members whose design is governed by the loading given in Clause 16.6.1. This definition does not include primary trusses that support joists, other secondary members, and special joists.

Tie joists — joists that have at least one end connected to a column to facilitate erection and that are designed to resist gravity loads only, unless otherwise specified.

16.4 Materials
Steel for joists shall be of a structural quality, suitable for welding, meeting the requirements of Clause 5.1.1. Structural members cold-formed to shape may use the effect of cold-forming in accordance with Clause 5.2 of CSA Standard S136. The calculated value of F'_y shall be determined using only the values for F_y and F_u that are specified in the relevant structural steel material standard. Yield levels reported on mill test certificates or determined according to Clause 9.3 of CSA Standard S136 shall not be used as the basis for design.

16.5 Drawings

16.5.1 Building Design Drawings
The building design drawings prepared by the building designer shall show
(a) the uniformly distributed specified live and dead gravity loads, the unbalanced loading condition and the concentrated load conditions given in Clause 16.6.1 or 16.6.2, and any special

loading conditions such as nonuniform snow loads, horizontal loads, end moments, net uplift, and allowances for mechanical equipment;
(b) maximum joist spacing and, where necessary, camber, maximum joist depth, and shoe depth;
(c) where joists are not supported on steel members, maximum bearing pressures, or sizes of bearing plates;
(d) anchorage requirements in excess of the requirements of Clause 16.6.13;
(e) bracing as may be required by Clauses 16.6.7.2 or 16.9.1; and
(f) method and spacing of attachments of steel deck to the top chord.

Note: *It is recommended that the building design drawings include a note warning that attachments for mechanical, electrical, and other services shall be made by using approved clamping devices or u-bolt-type connectors and that no drilling or cutting shall be done unless approved by the building designer.*

16.5.2 Joist Design Drawings

Joist design drawings prepared by the joist manufacturer shall show, as a minimum, the specified loading, factored member loads, material specification, member sizes, dimensions, spacers, welds, shoes, anchorages, bracing, bearings, field splices, bridging locations, and camber.

16.6 Design

16.6.1 Loading for Standard Open-Web Steel Joists

Unless otherwise specified by the building designer (in accordance with Clause 16.6.2), the factored moment and shear resistances of an open-web steel joist at every section shall be not less than the moment and shear due to the following factored load conditions, considered separately:
(a) a uniformly distributed load equal to the total dead and live loads;
(b) an unbalanced load with 100% of the total dead and live loads on any continuous portion of the joist and 25% of total dead and live loads on the remainder to produce the most critical effect on any component; and
(c) a concentrated factored load applied at any panel point of 13.5 kN for floor joists for office or similar occupancy or 2 kN for roof joists.

16.6.2 Loading for Special Open-Web Steel Joists

The factored moment and shear resistances of special open-web steel joists at every section shall be not less than the moment and shear due to the loading conditions specified by the building designer in Clause 16.5.1(a) or those due to the factored dead load plus the following factored live load conditions, (a), (b), (c), or (d), considered separately:
(a) for floor joists, an unbalanced live load applied on any continuous portion of the joist to produce the most critical effect on any component;
(b) for roof joists, an unbalanced loading condition with 100% of the snow load plus other live loads applied on any continuous portion of the joist and 50% of the snow load on the remainder of the joist to produce the most critical effect on any component;
(c) for roof joists, wind uplift; and
(d) the appropriate factored concentrated load from Table 4.1.6.B of the *National Building Code of Canada*, applied at any one panel point to produce the most critical effect on any component.

16.6.3 Design Assumptions

Open-web steel joists shall be designed for loads acting in the plane of the joist applied to the compression chord, which is assumed to be prevented from lateral buckling by the deck.

For the purpose of determining axial forces in all members, the loads may be replaced by statically equivalent loads applied at the panel points.

16.6.4 Verification of Joist Manufacturer's Design
When the adequacy of the design of a joist cannot be readily demonstrated by a rational analysis based on accepted theory and engineering practice, the joist manufacturer may elect to verify the design by test. The test shall be carried out to the satisfaction of the building designer. The test loading shall be 1.10/0.90 times the factored loads used in the design.

16.6.5 Member and Connection Resistance
Member and connection resistance shall be calculated in accordance with the requirements of Clause 13 except as otherwise specified in Clause 16.

16.6.6 Width-Thickness Ratios
16.6.6.1
Width-thickness ratios of compressive elements of hot-formed sections shall be governed by Clause 11. Width-thickness ratios of compressive elements of cold-formed sections shall be governed by CSA Standard S136.

16.6.6.2
For purposes of determining the appropriate width-thickness ratio of compressive elements supported along one edge, any stiffening effect of the deck or the joist web shall be neglected.

16.6.7 Tension Chord
16.6.7.1
The tension chord shall be continuous and may be designed as an axially loaded tension member unless subject to eccentricities in excess of those permitted under Clause 16.6.11.4 or to applied load between panel points. The governing radius of gyration of the tension chord or any component thereof shall be not less than 1/240 of the corresponding unsupported length. For joists with the web in the y-plane, the unsupported length of chord for computing L_x/r_x shall be taken as the panel length centre-to-centre of panel points, and the unsupported length of chord for calculating L_y/r_y shall be taken as the distance between bridging lines connected to the tension chord. Joist shoes, when anchored, may be assumed to be equivalent to bridging lines. Moments due to concentrated loads shall be included in the design.

16.6.7.2
The chord shall be designed for the resulting compressive forces when net uplift is specified, when joists are made continuous or cantilevered, or when end moments are specified. Bracing, when required, shall be provided in accordance with the requirements of Clause 20.2.

When net uplift due to wind forces is a design requirement, a single line of bottom-chord bracing shall also be provided at each end of the joists near the first bottom-chord panel points.

16.6.8 Compression Chord
16.6.8.1
The compression chord shall be continuous and may be designed for axial compressive force alone when the panel length does not exceed 610 mm, when concentrated loads are not applied between the panel points, and when not subject to eccentricities in excess of those permitted under Clause 16.6.11.4. When the panel length exceeds 610 mm, the compression chord shall be designed as a continuous member subject to combined axial and bending forces.

16.6.8.2
The slenderness ratio, KL/r, of the compression chord or of its components shall not exceed 90 for interior panels nor 120 for end panels, where the governing KL/r shall be the maximum value determined by the following:
(a) for x-x (horizontal) axis, L_x shall be the centre-to-centre distance of panel points, and K = 0.9;
(b) for y-y (vertical) axis, L_y shall be the centre-to-centre distance of the attachments of the deck. The spacing of attachments shall be not more than the design slenderness ratio of the top chord times the radius of gyration of the top chord about its vertical axis and not more than 1000 mm, and K = 1.0;
(c) for z-z (skew) axis of individual components, L_z shall be the centre-to-centre distance of panel points or spacers, or both and K = 0.9. Decking shall not be considered to fulfil the function of batten plates or spacers for top chords consisting of two separated components;

and where
r = the appropriate radius of gyration.

16.6.8.3
Compression chords of joists in panel lengths exceeding 610 mm shall be proportioned such that
$$\frac{C_f}{C_r} + \frac{M_f}{M_r} \leq 1.0$$

where
M_r is given in Clause 13.5 and C_r is given in Clause 13.3.

At the panel point, C_r may be taken as ϕAF_y and Clause 13.5(a) may be used to determine M_r provided that the chord meets the requirements of a Class 2 section and $M_f/M_p < 0.25$.

The chord shall be assumed to be pinned at the joist supports.

16.6.8.4
When welding is used to attach steel deck to the chord of a joist for the transfer of load due to uplift or diaphragm action, any flat width of the chord component to be in contact with the deck shall be at least 5 mm larger than the nominal design dimensions of the welds, measured transverse to the longitudinal axis of the chord.

16.6.9 Webs

16.6.9.1
Webs shall be designed in accordance with the requirements of Clause 13 to resist the shear at any point due to the factored loads given in Clause 16.6.1 or 16.6.2. Particular attention shall be paid to possible reversals of shear.

16.6.9.2
The length of a web member shall be taken as the distance between the intersections of the axes of the web and the chords. For buckling in the plane of the web, the effective length factor shall be taken as 0.9 if the web consists of individual members. For all other cases, the effective length factor shall be taken as 1.0.

16.6.9.3
The slenderness ratio of a web member in tension need not be limited.

16.6.9.4
The slenderness ratio of a web member in compression shall not exceed 200.

16.6.10 Spacers and Battens
Compression members consisting of two or more sections shall be interconnected so that the slenderness ratio of each section calculated using its least radius of gyration is less than or equal to the design slenderness ratio of the built-up member. Spacers or battens shall be an integral part of the joist.

16.6.11 Connections and Splices

16.6.11.1
Component members of joists shall be connected by welding, bolting, or other approved means.

16.6.11.2
Connections and splices shall develop the factored loads without exceeding the factored member resistances given in Clause 16. Butt-welded splices shall develop the factored tensile resistance, T_r, of the member.

16.6.11.3
Splices may occur at any point in chord or web members.

16.6.11.4
Members connected at a joint preferably shall have their centroidal axes meet at a point. Where this is impractical and eccentricities are introduced, such eccentricities may be neglected if they do not exceed
(a) for continuous web members, the greater of the two distances measured from the neutral axis of the chord member to the extreme fibres of the chord member; and
(b) for noncontinuous web members, the distance measured from the neutral axis to the back (outside face) of the chord member.

When the eccentricity exceeds these limits, provision shall be made for the effects of total eccentricity.

Eccentricities assumed in design shall be those at maximum fabrication tolerances, which shall be stated on the shop details.

16.6.12 Bearings

16.6.12.1
Bearings at ends of joists shall be proportioned so that the factored bearing resistance of the supporting material is not exceeded.

16.6.12.2
Where a joist bears, with or without a bearing plate, on solid masonry or concrete support, the bearing shall meet the requirements of CSA Standards S304.1 and A23.3, respectively.

16.6.12.3
Where a joist bears on a member of the structural steel frame, the end of the bearing shall extend at least 65 mm beyond the face of the support, except that when the available bearing area is restricted this distance may be reduced provided that the bearing is adequately anchored to the support. In any case the factored bearing resistance shall not be exceeded.

16.6.12.4
The bearing detail and the end panels of the joist shall be proportioned to include the effect of the eccentricity between the centre of the bearing and the intersection of the axes of the chord and the end diagonal.

16.6.13 Anchorage
16.6.13.1
Joist ends shall be properly anchored to withstand the effect of factored loads, including net uplift, as follows:
(a) in no case shall the anchorage to masonry be less than the following:
 (i) for floor joists, a 10 mm diameter rod at least 300 mm long embedded horizontally;
 (ii) for roof joists, a 20 mm diameter anchor bolt 300 mm long embedded vertically with a 50 mm, 90° hook;
(b) the anchorage to steel shall be a connection capable of withstanding a horizontal load of not less than 10% of the end reaction of the joist but not less than one 20 mm diameter bolt or a pair of fillet welds satisfying the minimum size and length requirements of CSA Standard W59.

16.6.13.2
Tie joists may have their top and bottom chords connected to a column. Unless otherwise specified, tie joists shall have top and bottom chord connections each at least equivalent to those required by Clause 16.6.13.1. Either the top or bottom connection shall utilize a mechanical fastener.

16.6.13.3
Where joists are used as a part of a frame, the joist-to-column connections shall be designed to carry the moments and forces due to the factored loads (see Clause 7.2).

16.6.14 Deflection
16.6.14.1
Steel joists shall be proportioned so that deflection due to specified loads is within acceptable limits for the nature of the materials to be supported and the intended use and occupancy. Such deflection limits shall be as given in Clause 6.2.1 unless otherwise specified by the building designer.

16.6.14.2
The deflection may be established by test or may be calculated by assuming a moment of inertia equal to the gross moment of inertia of the chords about the centroidal axis of the joist and multiplying the calculated deflection derived on this basis by 1.10.

16.6.15 Camber
Unless otherwise specified by the building designer, the nominal camber, in millimetres, shall be equal to 0.07 times the square of the span, expressed in metres. For tolerances see Clause 16.11.9.

16.6.16 Vibration
The building designer shall give special consideration to floor systems where unacceptable vibration may occur. When requested, the joist manufacturer shall supply joist properties and details to the building designer. (See Appendix G.)

16.6.17 Welding

16.6.17.1
Arc welding design and practice shall conform to CSA Standard W59.

16.6.17.2
The resistance of resistance-welded joints shall be taken as established in CSA Standard W55.3, and the related welding practice shall conform to welding standards approved by the Canadian Welding Bureau under the same CSA Standard.

16.6.17.3
Fabricators and erectors of welded construction covered by this Standard shall be certified by the Canadian Welding Bureau in Division 1 or Division 2.1 to the requirements of CSA Standard W47.1 or W55.3, or both, as applicable. Specific welding procedures for joist fabrication shall be approved by the Canadian Welding Bureau.

16.6.17.4
The factored resistances of welds shall be equal to those given in Table 3.

16.6.17.5
When field welding joists to supporting members, surfaces to be welded shall be free of coatings that are detrimental to achieving an adequate weldment.

16.6.17.6
Flux and slag shall be removed from all welds.

16.7 Stability During Construction
Means shall be provided to support joist chords against lateral movement and to hold the joist in the vertical or specified plane during construction.

16.8 Bridging

16.8.1 General
Bridging transverse to the span of joists may be used to meet the requirements of Clause 16.7 and also to meet the slenderness ratio requirements for chords. Bridging is not to be considered "bracing" as defined in Clause 20.2.1.

16.8.2 Installation
All bridging and bridging anchors shall be completely installed before any construction loads, except for the weight of the workers necessary to install the bridging, are placed on the joists.

16.8.3 Types
Unless otherwise specified or approved by the building designer, the joist manufacturer shall supply bridging that may be either of the diagonal or horizontal type.

16.8.4 Diagonal Bridging
Diagonal bridging consisting of crossed members running from top chord to bottom chord of adjacent joists shall have a slenderness ratio, L/r, of not more than 200, where L is the length of the diagonal bridging member or one-half of this length when crossed members are connected at their point of intersection, and r is the least radius of gyration. All diagonal bridging shall be connected adequately to the joists by bolts or welds.

16.8.5 Horizontal Bridging
A line of horizontal bridging shall consist of a continuous member attached to either the top chord or the bottom chord. Horizontal bridging members shall have a slenderness ratio of not more than 300.

16.8.6 Attachment of Bridging
Attachment of diagonal and horizontal bridging to joist chords shall be by welding or mechanical means capable of resisting an axial load of at least 3 kN in the attached bridging member. These welds should meet the minimum length requirements stipulated in CSA Standard W59.

16.8.7 Anchorage of Bridging
Each line of bridging shall be adequately anchored at each end to sturdy walls or to main components of the structural frame, if practicable. If not practicable, diagonal and horizontal bridging shall be provided in combination between adjacent joists near the ends of bridging lines.

The ends of joists designed to bear on their bottom chords shall be held adequately in position by attachments to the walls or to the structural frame or by lines of bridging located at the ends except where such ends are built into masonry or concrete walls.

16.8.8 Bridging Systems
Bridging systems, including sizes of bridging members and all necessary details, shall be shown on the erection diagrams. If a specific bridging system is required by the design, the design drawings shall show all information necessary for the preparation of shop details and erection diagrams.

16.8.9 Spacing of Bridging
Diagonal and horizontal bridging, whichever is furnished, shall be spaced so that the unsupported length of the chord between bridging lines or between laterally supported ends of the joist and adjacent bridging lines does not exceed
(a) for compression chords, 170r; and
(b) for tension chords, 240r;

where
r = the applicable chord radius of gyration about its axis in the plane of the web.

Ends of joists anchored to supports may be assumed to be equivalent to bridging lines. If ends of joists are not so anchored before deck is installed, the distance from the face of the support to the nearest bridging member in the plane of the bottom chord shall not exceed 120r. In no case shall there be less than one line of horizontal or diagonal bridging attached to each joist spanning 4 m or more. If only a single line of bridging is required, it shall be placed at the centre of the joist span. If bridging is not used on joists less than 4 m in span, the ends of such joists shall be anchored to the supports so as to prevent overturning of the joist during placement of the deck.

16.9 Decking

16.9.1 Decking to Provide Lateral Support
Decking shall bear directly on the top chord of the joist and shall be sufficiently rigid to provide lateral support to the compression chord of the joist. In special cases where the decking is incapable of furnishing the required lateral support, the compression chord of the joist shall be braced laterally in accordance with the requirements of Clause 20.2.

16.9.2 Attachments
Attachments of decking considered to provide lateral support shall be capable of staying the top chords laterally. Attachments shall be deemed to fulfil this requirement when the attachments as a whole are adequate to resist a force in the plane of the decking of not less than 5% of the maximum force in the top chord and assumed to be uniformly distributed along the length of the top chord. The spacing of attachments shall be not more than the design slenderness ratio of the top chord times the radius of gyration of the top chord about its vertical axis, and not more than 1 m.

16.9.3 Diaphragm Action
Where decking is used in combination with joists to form a diaphragm for the purpose of transferring lateral applied loads to vertical bracing systems, special attachment requirements shall be fully specified on the building design drawings.

16.9.4 Cast-In-Place Slabs
Cast-in-place slabs used as decking shall have a minimum thickness of 50 mm. Forms for cast-in-place slabs shall not cause lateral displacement of the top chords of joists during installation of the forms or the placing of the concrete. Nonremovable forms shall be positively attached to top chords by means of clips, ties, wedges, fasteners, or other suitable means at intervals not exceeding 1 m; however, there shall be at least two attachments in the width of each form at each joist. Forms and their method of attachment shall be such that the cast-in-place slab, after hardening, is capable of furnishing lateral support to the joist chords.

16.9.5 Installation of Steel Deck

16.9.5.1
To facilitate installation of the steel deck, the location of the top chord of the joist shall be confirmed by marking the deck at suitable intervals, or by other suitable means.

16.9.5.2
(a) The installer of steel deck to be fastened to joists by arc spot welding shall be certified by the Canadian Welding Bureau to the requirements of CSA Standard W47.1.
(b) The installation welding procedures shall be approved by the Canadian Welding Bureau.
(c) The welders shall have current qualifications for arc spot welding issued by the Canadian Welding Bureau.

16.10 Shop Painting
Joists shall have one shop coat of protective paint of a type standard with the manufacturer unless otherwise specified.

16.11 Manufacturing Tolerances

16.11.1
The tolerance on the specified depth of the manufactured joist shall be ±7 mm.

16.11.2
The maximum deviation from the design location of a panel point measured along the length of a chord shall be 13 mm. In joists in which an individual end diagonal is attached to the bottom chord or in which the end diagonal is a continuation of an upturned bottom chord, the gravity axes of the members in such a joint should meet at a point (see Clause 16.6.11.4).

16.11.3
The maximum deviation from the design location of a panel point measured perpendicular to the longitudinal axis of the chord and in the plane of the joist shall be 7 mm.

16.11.4
The connections of web members to chords shall not deviate laterally more than 3 mm from that assumed in the design.

16.11.5
The maximum sweep of a joist or any portion of the length of the joist upon completion of manufacture shall be not greater than 1/500 of the length on which the sweep is measured.

16.11.6
The maximum tilt of bearing shoes shall be 1 in 50 measured from a plane perpendicular to the plane of the web and parallel to the longitudinal axis of the joist.

16.11.7
The tolerance on the specified shoe depth shall be ±3 mm.

16.11.8
The tolerance on the specified length of the joist shall be ±7 mm. The connection holes in a joist shall not vary from the detailed location by more than 2 mm for members 10 m or less in length or by more than 3 mm for members more than 10 m in length.

16.11.9
The tolerance in millimetres on the nominal or specified camber shall be

$$\pm \left(6 + \frac{\text{span, in metres}}{4} \right)$$

The resulting actual minimum camber in a joist shall be +3 mm, except that the maximum range in camber for joists of the same span shall be limited to 20 mm.

16.12 Inspection and Quality Control

16.12.1 Inspection
Material and workmanship shall be accessible for inspection at all times by qualified inspectors representing the building designer. Random in-process inspection shall be carried out by the manufacturer, and all joists shall be thoroughly inspected by the manufacturer before shipping.
Third-party welding inspection shall be in accordance with Clause 31.5.

16.12.2 Identification and Control of Steel
Steel used in the manufacture of joists shall, at all times in the manufacturer's plant, be marked to identify its specification (and grade, where applicable). This shall be done by suitable markings or by recognized colour-coding or by any system devised by the manufacturer that will ensure to the satisfaction of the building designer that the correct material is being used.

16.12.3 Quality Control
Upon request by the building designer, the manufacturer shall provide evidence of having suitable quality control measures to ensure that the joists meet all specified requirements. When testing is part of the manufacturer's normal quality control program, the loading criteria shall be 1.0/0.9 times the factored loads for the materials used in the joists.

For resistance welding, the quality control procedures outlined in CSA Standard W55.3 shall be met. For arc welding, the quality control requirements of CSA Standard W59 shall be met.

16.13 Handling and Erection

16.13.1 General
Care shall be exercised to avoid damage during strapping, transport, unloading, site storage and piling, and erection. Dropping of joists shall not be permitted. Special precautions shall be taken when erecting long, slender joists and hoisting cables shall preferably not be released until the member is stayed laterally by at least one line of bridging. Joists shall have all bridging attached and be permanently fastened into place before the application of any loads. Heavy construction loads shall be adequately distributed so as not to exceed the capacity of any joist. Field welding shall not cause damage to joists, bridging, deck, and supporting steel members.

16.13.2 Erection Tolerances

16.13.2.1
The maximum sweep of a joist or a portion of the length of a joist upon completion of erection shall not exceed the limit given in Clause 16.11.5 and shall be in accordance with the general requirements of Clause 30.

16.13.2.2
All members shall be free from twists, sharp kinks, and bends.

16.13.2.3
When joists are finally fastened in position in the field, the maximum deviation from the location shown on the erection diagrams shall be 15 mm.

16.13.2.4
The deviation normal to the specified plane of the web of a joist shall not exceed 1/50 of the depth of the joist.

17. Composite Beams

17.1 Application
The provisions of Clause 17 apply to composite beams consisting of steel sections, trusses, or joists interconnected with either a reinforced concrete slab or a steel deck with a concrete cover slab.

17.2 Definitions
The following definitions apply in this Clause:

Cover slab — the concrete above the flutes of the steel deck. All flutes shall be filled with concrete so as to form a ribbed slab.

Effective cover slab thickness, t — the minimum thickness of concrete measured from the top of the cover slab to the top of the steel deck. This thickness shall not be less than 65 mm unless the adequacy of a lesser thickness has been established by appropriate tests.

Effective slab thickness, t — the overall slab thickness, provided that
(a) the slab is cast with a flat underside;
(b) the slab is cast on corrugated steel forms having a height of corrugation not greater than 0.25 times the overall slab thickness; or

(c) the slab is cast on fluted steel forms whose profile meets the following requirements. The minimum concrete rib width shall be 125 mm; the maximum rib height shall be 40 mm but not more than 0.4 times the overall slab thickness; the average width between ribs shall not exceed 0.25 times the overall slab thickness nor 0.2 times the minimum width of concrete ribs.

In all other cases, effective slab thickness means the overall slab thickness minus the height of flute corrugation.

Flute — the portion of the steel deck that forms a valley.

Rib — the portion of the concrete slab that is formed by the steel deck flute.

Slab — a reinforced cast-in-place concrete slab at least 65 mm in effective thickness. The area equal to the effective width times the effective slab thickness shall be free of voids or hollows except for those specifically permitted in the definition of effective slab thickness.

Steel deck — a load-carrying steel deck, consisting of either
(a) a single fluted element (noncellular deck); or
(b) a two-element section consisting of a fluted element in conjunction with a flat sheet (cellular deck).

The maximum depth of the deck shall be 80 mm and the average width of the minimum flute shall be 50 mm. A steel deck may be of a type intended to act compositely with the cover slab in supporting applied load.

Steel joist — an open-web steel joist suitable for composite design.

Steel section — a steel structural section with a solid web or webs suitable for composite design. Web openings are permissible only on condition that their effects are fully investigated and accounted for in the design.

17.3 General

17.3.1 Deflections

Calculation of deflections shall take into account the effects of creep of concrete, shrinkage of concrete, and increased flexibility resulting from partial shear connection and from interfacial slip. These effects shall be established by test or analysis, where practicable. Consideration shall also be given to the effects of full or partial continuity in the steel beams and concrete slabs in reducing calculated deflections.

In lieu of tests or analysis, the effects of partial shear connection and interfacial slip, creep, and shrinkage may be assessed as follows:

(a) for increased flexibility resulting from partial shear connection and interfacial slip, calculate the deflections using an effective moment of inertia given by

$$I_e = I_s + 0.85 \, (p)^{0.25} \, (I_t - I_s)$$

where
I_s = moment of inertia of steel beam
I_t = transformed moment of inertia of composite beam
p = fraction of full shear connection (use $p = 1.00$ for full shear connection);

(b) for creep, increase elastic deflections caused by dead loads and long-term live loads, as calculated in Item (a), by 15%; and

(c) for shrinkage of concrete, calculate deflection using a selected free shrinkage strain, strain compatibility between the steel and concrete, and a time-dependent modulus of elasticity of the concrete in tension, E_{ct}, (See Appendix L) as it dries, shrinks, and creeps from

$$\Delta_s = \frac{\varepsilon_f \, A_c \, L^2 \, y}{8 \, n_t \, I_t}$$

where
- ε_f = free shrinkage strain of the concrete
- A_c = effective area of the concrete slab
- E_{ct} = effective modulus of concrete in tension
- L = span of beam
- n_t = modular ratio, E/E_{ct}
- y = distance from the centroid of effective area of concrete slab to the elastic neutral axis
- I_t = transformed moment of inertia of the composite beam but based on the modular ratio n_t

17.3.2 Vertical Shear
The web area of steel sections or the web system of steel trusses and joists shall be proportioned to carry the total vertical shear, V_f.

17.3.3 End Connections
End connections of steel sections, trusses, and joists shall be proportioned to transmit the total end reaction of the composite beam.

17.4 Design Effective Width of Concrete

17.4.1
Slabs or cover slabs extending on both sides of the steel section or joist shall be deemed to have a design effective width, b, equal to the lesser of
(a) 0.25 times the composite beam span; or
(b) the average distance from the centre of the steel section, truss, or joist to the centres of adjacent parallel supports.

17.4.2
Slabs or cover slabs extending on one side only of the supporting section or joist shall be deemed to have a design effective width, b, not greater than the width of the top flange of the steel section or top chord of the steel joist or truss plus the lesser of
(a) 0.1 times the composite beam span; or
(b) 0.5 times the clear distance between the steel section, truss, or joist and the adjacent parallel support.

17.5 Slab Reinforcement

17.5.1 General
Slabs shall be adequately reinforced to support all loads and to control both cracking transverse to the composite beam span and longitudinal cracking over the steel section or joist. Reinforcement shall not be less than that required by the specified fire-resistance design of the assembly.

17.5.2 Parallel Reinforcement
Reinforcement parallel to the span of the beam in regions of negative bending moment of the composite beam shall be anchored by embedment in concrete that is in compression. The reinforcement of slabs that are to be continuous over the end support of steel sections or joists fitted with flexible end connections shall be given special attention. In no case shall such reinforcement at the ends of beams supporting ribbed slabs perpendicular to the beam be less than two 15 M bars.

17.5.3 Transverse Reinforcement, Solid Slabs

Unless it is known from experience that longitudinal cracking caused by composite action directly over the steel section or joist is unlikely, additional transverse reinforcement or other effective means shall be provided. Such additional reinforcement shall be placed in the lower part of the slab and anchored so as to develop the yield strength of the reinforcement. The area of such reinforcement shall be not less than 0.002 times the concrete area being reinforced and shall be uniformly distributed.

17.5.4 Transverse Reinforcement, Ribbed Slabs

17.5.4.1
Where the ribs are parallel to the beam span, the area of transverse reinforcement shall be not less than 0.002 times the concrete cover slab area being reinforced and shall be uniformly distributed.

17.5.4.2
Where the ribs are perpendicular to the beam span, the area of transverse reinforcement shall not be less than 0.001 times the concrete cover slab area being reinforced and shall be uniformly distributed.

17.6 Interconnection

17.6.1
Except as permitted by Clauses 17.6.2 and 17.6.4, interconnection between steel sections, trusses, or joists and slabs or steel decks with cover slabs shall be attained by the use of shear connectors as prescribed in Clause 17.7.

17.6.2
Unpainted steel sections, trusses, or joists that support slabs and are totally encased in concrete do not require interconnection by means of shear connectors provided that
(a) a minimum of 50 mm of concrete covers all portions of the steel section, truss, or joist except as noted in Item (c);
(b) the cover in Item (a) is reinforced to prevent spalling; and
(c) the top of the steel section, truss, or joist is at least 40 mm below the top and 50 mm above the bottom of the slab.

17.6.3
Studs may be welded through a maximum of two steel sheets in contact, each not more than 1.71 mm in overall thickness including coatings (1.52 mm in nominal base steel thickness plus zinc coating not greater than nominal 275 g/m^2). Otherwise, holes for placing studs shall be made through the sheets as necessary. Welded studs shall meet the requirements of CSA Standard W59.

17.6.4
Other methods of interconnection that have been adequately demonstrated by test and verified by analysis may be used to effect the transfer of forces between the steel section, truss, or joist and the slab or steel deck with cover slab. In such cases the design of the composite member shall conform to the design of a similar member employing shear connectors, insofar as practicable.

17.6.5
The diameter of a welded stud shall not exceed 2.5 times the thickness of the part to which it is welded, unless test data satisfactory to the designer are provided to establish the capacity of the stud as a shear connector.

17.7 Shear Connectors

17.7.1 General
The resistance factor, ϕ_{sc}, to be used with the shear resistances given in this Clause shall be taken as 0.80. The factored shear resistance, q_r, of other shear connectors shall be established by tests acceptable to the designer.

17.7.2 End-Welded Studs
End-welded studs shall be headed or hooked with $h/d \geq 4$. The projection of a stud in a ribbed slab, based on its length prior to welding, shall be at least two stud diameters above the top surface of the steel deck.

17.7.2.1
In solid slabs,
$$q_{rs} = 0.50 \phi_{sc} A_{sc} \sqrt{f'_c E_c} \leq \phi_{sc} A_{sc} F_u$$

where F_u for commonly available studs is 415 MPa and q_{rs} is in newtons.

17.7.2.2
In ribbed slabs with ribs parallel to the beam,
(a) when $w_d/h_d \geq 1.50$
$$q_{rr} = q_{rs}$$
(b) when $w_d/h_d < 1.50$
$$q_{rr} = [0.6 (w_d/h_d)(h/h_d - 1)]q_{rs} \leq 1.0\, q_{rs}$$

17.7.2.3
In ribbed slabs with ribs perpendicular to the beam,
(a) when $h_d = 75$ mm
$$q_{rr} = 0.35\, \phi_{sc}\, \rho\, A_p\, \sqrt{f'_c} \leq q_{rs}$$
(b) when $h_d = 38$ mm
$$q_{rr} = 0.61\, \phi_{sc}\, \rho\, A_p\, \sqrt{f'_c} \leq q_{rs}$$

where
  A_p is the concrete pull-out area taking the deck profile and stud burn-off into account. For a single stud, the apex of the pyramidal pull-out area, with four sides sloping at 45°, is taken as the centre of the top surface of the head of the stud. For a pair of studs, the pull-out area has a ridge extending from stud to stud.
  ρ = 1.0 for normal-density concrete (2150 to 2500 kg/m^3)
     = 0.85 for semi-low-density concrete (1850 to 2150 kg/m^3)

17.7.2.4
The longitudinal spacing of stud connectors in both solid slabs and in ribbed slabs when ribs of formed steel deck are parallel to the beam shall be not less than six stud diameters. The maximum spacing of studs shall not exceed 1000 mm. See also Clause 17.8.
  The transverse spacing of stud connectors shall not be less than four stud diameters.

17.7.3 Channel Connectors
In solid slabs of normal-density concrete with $f'_c \geq 20$ MPa and a density of at least 2300 kg/m^3,
$$q_{rs} = 36.5\, \phi_{sc}\, (t + 0.5w) L_c\, \sqrt{f'_c}$$

17.8 Ties
Mechanical ties shall be provided between the steel section, truss, or joist and the slab or steel deck to prevent separation. Shear connectors may serve as mechanical ties if suitably proportioned. The maximum spacing of ties shall not exceed 1000 mm, and the average spacing in a span shall not exceed 600 mm or be greater than that required to achieve any specified fire-resistance rating of the composite assembly.

17.9 Design of Composite Beams with Shear Connectors

17.9.1
The composite beam shall consist of steel section, truss or joist, shear connectors, ties, and slab or steel deck with cover slab.

17.9.2
The properties of the composite section shall be calculated neglecting any concrete area that is in tension within the maximum effective area (equal to effective width times effective thickness). If a steel truss or joist is used, the area of its top chord shall be neglected in determining the properties of the composite section and only Clause 17.9.3(a) is applicable.

17.9.3
The factored moment resistance, M_{rc}, of the composite section with the slab or cover slab in compression shall be calculated as follows, where $\phi = 0.90$ and ϕ_c, the resistance factor for concrete, $= 0.60$:

(a) **Case 1** — Full shear connection and plastic neutral axis in the slab; that is, $Q_r \geq \phi A_s F_y$ and $\phi A_s F_y \leq 0.85 \phi_c b t f'_c$, where Q_r equals the sum of the factored resistances of all shear connectors between points of maximum and zero moment

$$M_{rc} = T_r e' = \phi A_s F_y e'$$

where e' is the lever arm and is calculated using

$$a = \frac{\phi A_s F_y}{0.85 \phi_c b f'_c}$$

(b) **Case 2** — Full shear connection and plastic neutral axis in the steel section; that is, $Q_r \geq 0.85 \phi_c b t f'_c$ and $0.85 \phi_c b t f'_c < \phi A_s F_y$

$$M_{rc} = C_r e + C'_r e'$$
$$C'_r = 0.85 \phi_c b t f'_c$$
$$C_r = \frac{\phi A_s F_y - C'_r}{2}$$

(c) **Case 3** — Partial shear connection; that is, $Q_r < 0.85 \phi_c b t f'_c$ and $< \phi A_s F_y$

$$M_{rc} = C_r e + C'_r e'$$
$$C'_r = Q_r$$
$$C_r = \frac{\phi A_s F_y - C'_r}{2}$$

where e' is the lever arm and is calculated using

$$a = \frac{C'_r}{0.85 \phi_c b f'_c}$$

17.9.4
No composite action shall be assumed in calculating flexural strength when Q_r is less than 0.4 times the lesser of $0.85\phi_c\, bt f'_c$ and $\phi A_s F_y$. No composite action shall be assumed in calculating deflections when Q_r is less than 0.25 times the lesser of $0.85\phi_c\, bt f'_c$ and $\phi A_s F_y$.

17.9.5
For full shear connection, the total horizontal shear, V_h, at the junction of the steel section, truss, or joist and the concrete slab or steel deck, to be resisted by shear connectors distributed between the point of maximum bending moment and each adjacent point of zero moment shall be

$V_h = \phi A_s F_y$

$V_h = 0.85\phi_c\, bt f'_c$

for Cases 1 and 2 as defined in Clause 17.9.3(a) and (b), respectively, and $Q_r \geq V_h$.

17.9.6
For partial shear connection the total horizontal shear, V_h, as defined in Clause 17.9.3(c) shall be

$V_h = Q_r$

17.9.7
Composite beams employing steel sections and concrete slabs may be designed as continuous members. The factored moment resistance of the composite section, with the concrete slab in the tension area of the composite section, shall be the factored moment resistance of the steel section alone, except that when sufficient shear connectors are placed in the negative moment region, suitably anchored concrete slab reinforcement parallel to the steel sections and within the design effective width of the concrete slab may be included in calculating the properties of the composite section. The total horizontal shear, V_h, to be resisted by shear connectors between the point of maximum negative bending moment and each adjacent point of zero moment shall be taken as $\phi A_r F_{yr}$.

17.9.8
The number of shear connectors to be located on each side of the point of maximum bending moment (positive or negative, as applicable), distributed between that point and the adjacent point of zero moment, shall be not less than

$n = \dfrac{V_h}{q_r}$

Shear connectors may be spaced uniformly, except that in a region of positive bending the number of shear connectors, n', required between any concentrated load applied in that region and the nearest point of zero moment shall be not less than

$n' = n \left(\dfrac{M_{f1} - M_r}{M_f - M_r} \right)$

where
M_{f1} = positive bending moment under factored load at concentrated load point
M_r = factored moment resistance of the steel section alone
M_f = maximum positive bending moment under factored load

17.9.9 Longitudinal Shear
The longitudinal shear of composite beams with solid slabs or with cover slabs and steel deck parallel to the beam shall be taken as

$V_u = \Sigma q_r - 0.85\phi_c f'_c A_c - \phi A_r F_{yr}$

where A_r is the area of longitudinal reinforcement within the concrete area, A_c

For normal-weight concrete, the factored shear resistance along any potential longitudinal shear surfaces in the concrete slab shall be taken as

$$V_r = (0.80 \phi A_r F_{yr} + 2.76 \phi_c A_{cv}) \leq 0.50 \phi_c f'_c A_{cv}$$

where A_r is the area of transverse reinforcement crossing shear planes, A_{cv}

17.10 Design of Composite Beams without Shear Connectors

17.10.1
Unpainted steel sections or joists supporting concrete slabs and encased in concrete in accordance with Clause 17.6.2 may be proportioned on the basis that the composite section supports the total load.

17.10.2
The properties of the composite section for determination of load carrying capacity shall be calculated by ultimate strength methods, neglecting any area of concrete in tension.

17.10.3
As an alternative method of design, encased simple-span steel sections or joists may be proportioned on the basis that the steel section, truss, or joist alone supports 0.90 times the total load.

17.11 Unshored Beams
For composite beams that are unshored during construction, the stresses in the tension flange of the steel section, truss, or joist due to the loads applied before the concrete strength reaches $0.75f'_c$ plus the stresses at the same location due to the remaining specified loads considered to act on the composite section shall not exceed F_y.

17.12 Beams During Construction
The steel section, truss, or joist alone shall be proportioned to support all factored loads applied prior to hardening of the concrete without exceeding its calculated capacity under the conditions of lateral support or shoring, or both, to be furnished during construction.

18. Concrete-Filled Hollow Structural Sections

18.1 Scope
The provisions of Clause 18 apply to composite members consisting of steel hollow structural sections completely filled with concrete.

18.2 Application
Hollow structural sections designated as Class 1, 2, or 3 sections that are completely filled with concrete may be assumed to carry compressive load as composite columns. Class 4 hollow structural sections that are completely filled with concrete may also be designed as composite columns provided that the width-thickness ratios of the walls of rectangular sections do not exceed $1350/\sqrt{F_y}$, and the outside diameter-to-thickness ratios of circular sections do not exceed $28\,000/F_y$.

Limit States Design of Steel Structures

18.3 Axial Load on Concrete
The axial load assumed to be carried by the concrete at the top level of a column shall be only that portion applied by direct bearing on the concrete. Similarly a base plate or other means shall be provided for load transfer at the bottom of a column. At intermediate floor levels, direct bearing on the concrete is not necessary.

18.4 Compressive Resistance
The factored compressive resistance of a composite column shall be taken as
$$C_{rc} = \tau C_r + \tau' C_r'$$
where
$$C_r' = 0.85 \phi_c f_c' A_c \lambda_c^{-2} \left[\sqrt{1 + 0.25 \lambda_c^{-4}} - 0.5 \lambda_c^{-2} \right]$$

in which $\lambda_c = \dfrac{KL}{r_c} \sqrt{\dfrac{f_c'}{\pi^2 E_c}}$

r_c = radius of gyration of the concrete area, A_c
E_c = initial elastic modulus for concrete, considering the effects of long-term loading. For normal-weight concrete, with f_c' expressed in megapascals, this may be taken as
$(1 + S/T)\, 2500\, \sqrt{f_c'}$

where
S is the short-term load and T is the total load on the column.

For all rectangular hollow structural sections and for circular hollow structural sections with a height-to-diameter ratio of 25 or greater, $\tau = \tau' = 1.0$

Otherwise $\tau = \dfrac{1}{\sqrt{1 + \rho + \rho^2}}$

and $\tau' = 1 + \left(\dfrac{25 \rho^2 \tau}{(D/t)} \right) \left(\dfrac{F_y}{0.85 f_c'} \right)$

where
$\rho = 0.02\,(25 - L/D)$

18.5 Bending
The factored moment resistance of a composite rectangular section shall be calculated as follows, where $\phi = 0.90$ and $\phi_c = 0.60$.

$M_{rc} = C_r e + C_r' e'$

$C_r' = \phi_c\, a(b-2t)\, f_c'$

$C_r = \dfrac{\phi A_s F_y - C_r'}{2}$

$C_r + C_r' = T_r = \phi A_{st} F_y$

Note: *The concrete in compression is taken to have a rectangular stress block of intensity f_c' over a depth of $a = 0.85c$ where c is the depth of the concrete in compression.*

18.6 Axial Compression and Bending

18.6.1 Method 1: Bending Resisted by Composite Section

Members required to resist both bending moments and axial compression shall be proportioned analogously with Clause 13.8.1 so that

$$\frac{C_f}{C_{rc}} + \frac{B \, \omega_1 \, M_f}{M_{rc}\left(1 - \frac{C_f}{C_{ec}}\right)} \leq 1.0 \text{ and}$$

$$\frac{M_f}{M_{rc}} \leq 1.0$$

where
M_{rc} is as defined in Clause 18.5

$$B = \frac{C_{rco} - C_{rcm}}{C_{rco}}$$

C_{rco} = factored compressive resistance with $\lambda = 0$
C_{rcm} = factored compressive resistance that can coexist with M_{rc} when all of the cross-section is in compression

Conservatively, B may be taken as 1.0.

18.6.2 Method 2: Bending Assumed to be Resisted by the Steel Section Alone

For members required to resist both bending moments and axial compression, under this assumption, the steel section shall be proportioned as a beam-column in parallel with Clause 13.8.1, to carry the total bending plus axial compression equal to the difference between the total axial compression and that portion that can be sustained by the concrete:
$M_f \leq \tau \, M_r$

and if $C_f > \tau' \, C_f'$

$$\frac{C_f - \tau' \, C_f'}{\tau \, C_r} + \frac{\omega_1 \, M_f}{\tau \, M_r \left(1 - \frac{C_f - \tau' \, C_f'}{C_e}\right)} \leq 1.0$$

19. General Requirements for Built-Up Members

19.1 Members in Compression

19.1.1
All components of built-up compression members and the transverse spacing of their lines of connecting bolts or welds shall meet the requirements of Clauses 10 and 11.

19.1.2
All component parts that are in contact with one another at the ends of built-up compression members shall be connected by bolts spaced longitudinally not more than four diameters apart for a distance equal to 1.5 times the width of the member or by continuous welds having a length of not less than the width of the member.

19.1.3
Unless closer spacing is required for transfer of load or for sealing inaccessible surfaces, the longitudinal spacing in-line between intermediate bolts or the clear longitudinal spacing between Intermittent welds in built-up compression members shall not exceed the following, as applicable:

(a) $330t/\sqrt{F_y}$, but not more than 300 mm for the outside component of the section consisting of a plate when the bolts on all gauge lines or intermittent welds along the component edges are not staggered, where t = thickness of the outside plate;

(b) $525t/\sqrt{F_y}$, but not more than 450 mm for the outside component of the section consisting of a plate when the bolts or intermittent welds are staggered on adjacent lines, where t = thickness of the outside plate.

19.1.4

Compression members composed of two or more rolled shapes in contact or separated from one another shall be interconnected such that the slenderness ratio of any component, based on its least radius of gyration and the distance between interconnections, shall not exceed that of the built-up member. The compressive resistance of the built-up member shall be based on:

(a) the slenderness ratio of the built-up member with respect to the appropriate axis when the buckling mode does not involve relative deformation that produces shear forces in the interconnectors;

(b) an equivalent slenderness ratio, with respect to the axis orthogonal to that in (a), when the buckling mode involves relative deformation that produces shear forces in the interconnectors, taken as

$$\rho_e = \sqrt{\rho_o^2 + \rho_i^2}$$

ρ_e = equivalent slenderness ratio of built-up member
ρ_o = slenderness ratio of built-up member acting as an integral unit
ρ_i = maximum slenderness ratio of component part of a built-up member between interconnectors;

(c) for built-up members composed of two interconnected rolled shapes, in contact or separated only by filler plates, such as back-to-back angles or channels, the maximum slenderness ratio of component parts between fasteners or welds shall be based on an effective length factor of 1.0 when the fasteners are snug-tight bolts and 0.65 when welds or pretensioned bolts are used;

(d) for built-up members composed of two interconnected rolled shapes separated by lacing or batten plates, the maximum slenderness ratio of component parts between fasteners or welds shall be based on an effective length factor of 1.0 for both snug-tight and pretensioned bolts and for welds.

19.1.5

For starred angle compression members interconnected at least at the one-third points, Clause 19.1.4 need not apply.

19.1.6

The fasteners and interconnecting parts, if any, of members defined in Clause 19.1.4(c) shall be proportioned to resist a force equal to one per cent of the total force in the built-up member.

19.1.7

The spacing requirements of Clauses 19.1.3, 19.2.3, and 19.2.4 may not always provide a continuous tight fit between components in contact. When the environment is such that corrosion could be a serious problem, the spacing of bolts or welds may need to be less than the specified maximum.

19.1.8

Open sides of compression members built up from plates or shapes shall be connected to each other by lacing, batten plates, or perforated cover plates.

19.1.9
Lacing shall provide a complete triangulated shear system and may consist of bars, rods, or shapes. Lacing shall be proportioned to resist a shear normal to the longitudinal axis of the member of not less than 2.5% of the total axial load on the member plus the shear from transverse loads, if any.

19.1.10
The slenderness ratio of lacing members shall not exceed 140. The effective length for single lacing shall be the distance between connections to the main components; for double lacing connected at the intersections, the effective length shall be 70% of that distance.

19.1.11
Lacing members shall preferably be inclined to the longitudinal axis of the built-up member at an angle of not less than 45°.

19.1.12
Lacing systems shall have diaphragms in the plane of the lacing and as near to the ends as practicable, and at intermediate points where lacing is interrupted. Such diaphragms may be plates (tie plates) or shapes.

19.1.13
End tie plates used as diaphragms shall have a length not less than the distance between the lines of bolts or welds connecting them to the main components of the member. Intermediate tie plates shall have a length of not less than one-half of that prescribed for end tie plates. The thickness of tie plates shall be at least 1/60 of the width between lines of bolts or welds connecting them to the main components, and the longitudinal spacing of the bolts or clear longitudinal spacing between welds shall not exceed 150 mm. At least three bolts shall connect the tie plate to each main component or, alternatively, a total length of weld not less than one-third the length of tie plate shall be used.

19.1.14
Shapes used as diaphragms shall be proportioned and connected to transmit from one main component to the other a longitudinal shear equal to 5% of the axial compression in the member.

19.1.15
Perforated cover plates may be used in lieu of lacing and tie plates on open sides of built-up compressive members. The net width of such plates at access holes shall be assumed to be available to resist axial load, provided that

(a) the width-thickness ratio conforms to Clause 11;

(b) the length of the access hole does not exceed twice its width;

(c) the clear distance between access holes in the direction of load is not less than the transverse distance between lines of bolts or welds connecting the perforated plate to the main components of the built-up member; and

(d) the periphery of the access hole at all points has a minimum radius of 40 mm.

19.1.16
Battens consisting of plates or shapes may be used on open sides of built-up compression members that do not carry primary bending in addition to axial load. Battens shall be provided at the ends of the member, at locations where the member is laterally supported along its length, and elsewhere as determined by Clause 19.1.4.

19.1.17

Battens shall have a length of not less than the distance between lines of bolts or welds connecting them to the main components of the member, and shall have a thickness of not less than 1/60 of this distance, if the batten consists of a flat plate. Battens and their connections shall be proportioned to resist, simultaneously, a longitudinal shear force,

$$V_f = \frac{0.025 C_f d}{na}$$

and a moment,

$$M_f = \frac{0.025 C_f d}{2n}$$

where
d = longitudinal centre-to-centre distance between battens, mm
a = distance between lines of bolts or welds connecting the batten to each main component, mm
n = number of parallel planes of battens

19.2 Members in Tension

19.2.1

Members in tension composed of two or more shapes, plates, or bars separated from one another by intermittent fillers shall have the components interconnected at fillers spaced so that the slenderness ratio of any component between points of interconnection shall not exceed 300.

19.2.2

Members in tension composed of two plate components in contact or a shape and a plate component in contact shall have the components interconnected so that the spacing between connecting bolts or clear spacing between welds does not exceed 36 times the thickness of the thinner plate nor 450 mm (see Clause 19.1.3).

19.2.3

Members in tension composed of two or more shapes in contact shall have the components interconnected so that the spacing between connecting bolts or the clear spacing between welds does not exceed 600 mm, except where it can be determined that a greater spacing would not affect the satisfactory performance of the member (see Clause 19.1.3).

19.2.4

Members in tension composed of two separated main components may have either perforated cover plates or tie plates on the open sides of the built-up member. Tie plates, including end tie plates, shall have a length of not less than two-thirds of the transverse distance between bolts or welds connecting them to the main components of the member and shall be spaced so that the slenderness ratio of any component between the tie plates does not exceed 300. The thickness of tie plates shall be at least 1/60 of the transverse distance between the bolts or welds connecting them to the main components and the longitudinal spacing of the bolts or welds shall not exceed 150 mm. Perforated cover plates shall comply with the requirements of Clause 19.1.15(b), (c), and (d).

19.3 Open Box-Type Beams and Grillages

Two or more rolled beams or channels used side-by-side to form a flexural member shall be connected together at intervals of not more than 1500 mm. Through-bolts and separators may be used provided that, in beams having a depth of 300 mm or more, no fewer than two bolts shall be used at each separator location. When concentrated loads are carried from one beam to the other or distributed between the beams, diaphragms having sufficient stiffness to distribute the load shall be bolted or welded between the beams. The design of members shall provide for torsion resulting from any unequal distribution of loads. Where beams are exposed, they shall be sealed against corrosion of interior surfaces or spaced sufficiently far apart to permit cleaning and painting.

20. Stability of Structures and Members

20.1 Structures

20.1.1
In the design of a steel structure, care shall be taken to ensure that the structural system is adequate to resist the forces caused by the factored loads and to ensure that a complete structural system is provided to transfer the factored loads to the foundations, particularly when there is a dependence on walls, floors, or roofs acting as shear-resisting elements or diaphragms. (See also Clause 8.6.)

Note: *The structure should also be checked to ensure that adequate resistance to torsional deformations has been provided.*

20.1.2
Design drawings shall indicate all load-resisting elements essential to the integrity of the completed structure and shall show details necessary to ensure the effectiveness of the load-resisting system. Design drawings shall also indicate the requirements for roofs and floors used as diaphragms.

20.1.3
Erection diagrams shall indicate all load-resisting elements essential to the integrity of the completed structure. Permanent and temporary load-resisting elements essential to the integrity of the partially completed structure shall be clearly specified in the erection procedures.

20.1.4
Where the portion of the structure under consideration does not provide adequate resistance to lateral forces, provision shall be made for transferring the forces to adjacent lateral-load-resisting elements.

20.2 Members

20.2.1
Bracing assemblies assumed to provide lateral support to columns or to the compression flange of beams and girders, or to the compression chord of trusses, and the connections of such bracing members, shall be proportioned at each point of support:
(a) to have a stiffness in the direction perpendicular to the longitudinal axis of the braced member, in the plane of buckling, at least equal to

$$k_b = \frac{\beta C_f}{L}\left(1 + \frac{\Delta_0}{\Delta_b}\right)$$

where
k_b = required stiffness of the bracing assembly
Δ_0 = initial misalignment of the braced member at the point of support. This misalignment may be taken as the tolerance specified in Clause 28 for sweep or camber over the total braced length, or portion thereof, as appropriate for the method of construction.

Δ_b = displacement of braced member and bracing assembly at the point of support under force C_f and which may be taken equal to Δ_0

β = 2, 3, 3.41, or 3.63 for 1, 2, 3, or 4 equally spaced braces, respectively

C_f = force in a column, the compressed portion of a flexural member, or the compression chord of a truss, under factored loads

L = length between brace points

(b) to have a strength perpendicular to the longitudinal axis of the braced member, in the plane of buckling, at least equal to $P_b = k_b \Delta_b$ where P_b is the force in the bracing assembly under factored loads.

In determining the actual stiffness provided by bracing assemblies, due consideration should be given to the flexibility of the brace, the brace support, and the brace connection.

20.2.2
Bracing assemblies for columns shall have adequate stiffness and strength at each floor level to sustain the lateral forces produced by possible out-of-plumbness as specified in Clause 30.7.2 (see also Clause 8.6.2).

20.2.3
When bracing of the compression flange or chord is effected by a slab or deck, the slab or deck and the means by which the calculated bracing forces are transmitted between the flange or chord and the slab or deck shall be adequate to resist a force in the plane of the slab or deck. This force shall be considered to be uniformly distributed along the length of the compression flange or chord and shall be taken as at least 5% of the maximum force in the flange or chord, unless a lesser amount can be justified by analysis.

20.2.4
Consideration shall be given to the probable accumulation of forces when a bracing member must transfer forces from one braced member to another. In such cases, the initial misalignment shall be taken as $\left(0.2 + \dfrac{0.8}{\sqrt{n}}\right)\Delta_0$, where n is the number of braced members.

20.2.5
Bracing assemblies for beams, girders, and columns designed to resist loads causing torsion shall be proportioned according to the requirements of Clause 15.11. Special consideration shall be given to the connection of asymmetric sections such as channels, angles, and zees.

21. Connections

21.1 Alignment of Members
Axially loaded members that meet at a joint shall have their gravity axes intersect at a common point if practicable; otherwise, the results of bending due to the joint eccentricity shall be provided for.

21.2 Unrestrained Members

Except as otherwise indicated on the design drawings, all connections of beams, girders, and trusses shall be designed and detailed as flexible and ordinarily may be proportioned for the reaction shears only. Flexible beam connections shall accommodate end rotations of unrestrained (simple) beams. To accomplish this, inelastic action at the specified load levels in the connection is permitted.

21.3 Restrained Members

When beams, girders, or trusses are subject to both reaction shear and end moment due to full or partial end restraint or to continuous or cantilever construction, their connections shall be designed for the combined effect of shear, bending, and axial load.

When beams are rigidly framed to the flange of an I-shaped column, stiffeners shall be provided on the column web if the following bearing and tensile resistances of the column flange are exceeded:

(a) opposite the compression flange of the beam when

$$B_r = \phi w_c (t_b + 5k) F_{yc} < \frac{M_f}{d_b}$$

except that for members with Class 3 or 4 webs,

$$B_r = \phi \frac{640\,000}{(h_c/w_c)^2} w_c (t_b + 5k)$$

(b) opposite the tension flange of the beam when

$$T_r = 7\phi(t_c)^2 F_{yc} < \frac{M_f}{d_b}$$

where
w_c = thickness of column web
t_b = thickness of beam flange
k = distance from outer face of column flange to web-toe of fillet, or to web-toe of flange-to-web weld in a welded column
F_{yc} = specified yield point of column
d_b = depth of beam
h_c = clear depth of column web
t_c = thickness of column flange

The stiffener or pair of stiffeners opposite either beam flange must develop a force equal to

$$F_{st} = \left(\frac{M_f}{d_b}\right) - B_r$$

Stiffeners shall also be provided on the web of columns, beams, or girders if V_r calculated from Clause 13.4.1.2 is exceeded, in which case the stiffener or stiffeners must transfer a shear force equal to

$$V_{st} = V_f - 0.55\phi w d F_y$$

In all cases, the stiffeners shall be connected so that the force in the stiffener is transferred through the stiffener connection. When beams frame to one side of the column only, the stiffeners need not be longer than one-half of the depth of the column.

When an axial tension or compression force is acting on the beam, their effects (additive only) shall be considered in the design of the stiffeners.

21.4 Connections of Tension or Compression Members

The connections at ends of tension or compression members not finished to bear shall develop the force due to factored loads. However, the connection shall be designed for not less than 50% of the resistance of the member based on the condition (tension or compression) that governs the selection of the member.

21.5 Bearing Joints in Compression Members

21.5.1

Where columns bear on bearing plates or are finished to bear at splices, there shall be sufficient fasteners or welds to hold all parts securely in place. At splices of I-shaped members, the flanges shall be connected.

21.5.2

Where other compression members are finished to bear, the splice material and connecting fasteners or welds shall be arranged to hold all parts in place and shall be proportioned for 50% of the calculated load.

21.6 Lamellar Tearing

Corner or T-joint details of rolled structural members or plates involving transfer of tensile forces in the through-thickness direction resulting from shrinkage due to welding executed under conditions of restraint shall be avoided where possible. If this type of connection cannot be avoided, measures shall be taken to minimize the possibility of lamellar tearing.

21.7 Placement of Fasteners and Welds

Except in members subject to repeated loads (as defined in Clause 14), disposition of fillet welds to balance the forces about the neutral axis or axes for end connections of single-angle, double-angle, or similar types of axially loaded members is not required. Eccentricity between the gravity axes of such members and the gauge lines of bolted end connections may also be neglected. In axially loaded members subject to repeated loads, the fasteners or welds in end connections shall have their centre of gravity on the gravity axis of the member unless provision is made for the effect of the resulting eccentricity.

21.8 Fillers

21.8.1

When load-carrying fasteners pass through fillers with a total thickness greater than 6 mm, the fillers shall be extended beyond the splice material and the filler extension shall be secured by sufficient fasteners to distribute the total force in the member at the ultimate limit state uniformly over the combined section of the member and the filler. Alternatively, an equivalent number of fasteners shall be included in the connection.

21.8.2

In welded construction, any filler with a total thickness greater than 6 mm shall extend beyond the edges of the splice plate and shall be welded to the part on which it is fitted with sufficient weld to transmit the splice plate load, applied at the surface of the filler, as an eccentric load. Welds that connect the splice plate to the filler shall be sufficient to transmit the splice plate load and shall be long enough to avoid overloading the filler along the toe of the weld. Any filler that is 6 mm or less in thickness shall have its edges made flush with the edges of the splice plate and the required weld size shall be equal to the thickness of the filler plate plus the size necessary to transmit the splice plate load.

21.9 Welds in Combination
If two or more of the general types of weld (groove, fillet, plug, or slot) are combined in a single connection, the effective capacity of each shall be calculated separately with reference to the axis of the group in order to determine the factored resistance of the combination.

21.10 Fasteners and Welds in Combination

21.10.1
When approved by the designer, high-strength bolts in slip-critical connections may be considered as sharing the specified load with welds in new work, provided that the factored resistance either of the high-strength bolts, or of the welds, is equal to or greater than the effect of the factored loads. At the specified load level, the load sharing shall be on the basis of the proportional capacities of the bolts in the slip-critical connection and 0.70 times the factored resistance of the welds.

21.10.2
In making alterations to structures, existing rivets and high-strength bolts may be used to carry forces resulting from existing dead loads and welding may be proportioned to carry all additional loads.

21.11 High-Strength Bolts (in Slip-Critical Joints) and Rivets in Combination
In making alterations, rivets and high-strength bolts in slip-critical joints may be considered as sharing forces due to specified dead and live loads.

21.12 Connections Using Bolts

21.12.1 Connections Using Snug-Tightened High-Strength Bolts
Snug-tightened high-strength bolts may be used in all connections except those specified in Clause 21.12.2.

21.12.2 Connections Using Pretensioned High-Strength Bolts
Pretensioned high-strength bolts (ASTM A325M, A490M, A325, A490) shall be used in
(a) slip-critical connections where slippage cannot be tolerated (such connections include those subject to fatigue or to frequent load reversal, or those in structures sensitive to deflection);
(b) connections proportioned in accordance with the requirements of Clause 27;
(c) all elements resisting crane loads;
(d) connections for supports of running machines or other live loads that produce impact or cyclic load;
(e) connections where the bolts are subject to tensile loadings (see Clause 23.1.4); and
(f) connections using oversize or slotted holes unless specifically designed to accommodate movement.

21.13 Welds
The use of welded connections is not restricted.

21.14 Special Fasteners
Fasteners of special types may be used when approved by the designer.

22. Bolting Details

22.1 High-Strength Bolts
A325M, A490M, A325, and A490 high-strength bolts and their usage shall conform to Clause 22.

Limit States Design of Steel Structures

22.2 A307 Bolts
Nuts on A307 bolts shall be tightened to an amount corresponding to the full effort of a person using a spud wrench. When so specified, nuts shall be prevented from working loose by the use of lock washers, lock nuts, jam nuts, thread burring, welding, or other methods approved by the designer.

22.3 Effective Bearing Area
The effective bearing area of bolts shall be the nominal diameter multiplied by the length in bearing. For countersunk bolts, half of the depth of the countersink shall be deducted from the bearing length.

22.4 Long Grips
A307 bolts that carry calculated loads and for which the grip exceeds five diameters shall have their number increased by 0.6% for each additional 1 mm in the grip.

22.5 Minimum Pitch
The minimum distance between centres of bolt holes should not be less than 3 bolt diameters and shall in no case be less than 2.7 diameters.

22.6 Minimum Edge Distance
The minimum distance from the centre of a bolt hole to any edge shall be that given in Table 6.

Table 6
Minimum Edge Distance for Bolt Holes

Bolt diameter		Minimum edge distance, mm	
inch*	mm	At sheared edge	At rolled, sawn, or gas-cut edge †
5/8	—	28	22
—	16	28	22
3/4	—	32	25
—	20	34	26
7/8	—	38‡	28
—	22	38	28
—	24	42	30
1	—	44‡	32
—	27	48	34
1-1/8	—	51	38
—	30	52	38
1-1/4	—	57	41
—	36	64	46
Over 1-1/4	Over 36	1.75 × diameter	1.25 × diameter

*ASTM Standards A325 and A490 are written in inch-pound units. Accordingly, bolt diameters are shown in inches for these bolts only.
†Gas-cut edges shall be smooth and free from notches. Edge distance in this column may be decreased by 3 mm when the hole is at a point where calculated stress under factored loads is not more than 0.3 of the yield stress.
‡At ends of beam-framing angles, this distance may be 32 mm.

22.7 Maximum Edge Distance
The maximum distance from the centre of any bolt to the nearest edge of parts in contact shall be 12 times the thickness of the outside connected part, but not greater than 150 mm.

22.8 Minimum End Distance
In the connection of tension members having more than two bolts in a line parallel to the direction of load, the minimum end distance (from centre of end fastener to nearest end of connected part) shall be governed by the edge distance values given in Table 6. In members having either one or two bolts in the line of load, the end distance shall be not less than 1.5 bolt diameters.

22.9 Slotted or Oversize Holes
Maximum and minimum edge distance for bolts in slotted or oversize holes (as permitted in Clause 23.3.2) shall conform to the requirements given in Clauses 22.6, 22.7, and 22.8, assuming that the fastener can be placed at any extremity of the slot or hole.

23. Structural Joints Using ASTM A325M, A490M, A325, or A490 Bolts

23.1 General

23.1.1
Clause 23 deals with the design, assembly, and inspection of structural joints using ASTM A325M, A490M, A325, or A490 bolts, or equivalent fasteners. The bolts may or may not be required to be installed to a specific minimum tension, depending on the type of connection.

23.1.2
Design, fabrication, and erection drawings shall show the type(s) of bolts to be used and shall specify whether or not pretensioning is required. Joints required to resist shear between connected parts shall be designated as either bearing-type or slip-critical. For bearing-type joints, the design drawings shall indicate whether or not the bolts shall be tightened to a specified minimum tension.

23.1.3
In joints where the bolts are subject to shear, the presence or absence of threads in the shear planes of the joint shall be considered.

Determination of the strength of bearing-type joints shall take into account both the shear capacity of the fasteners and the bearing capacity of the connected materials.

23.1.4
Bolts used in joints that are required to support load by direct tension shall be proportioned so that the tensile load on the bolt area, independent of the initial tightening force, shall not exceed the factored tensile resistance as given in Clause 13.11.3. The applied load shall be taken as the sum of the external load plus any tension caused by a prying action due to deformation of the connected parts. If the connection is subject to repeated loading, prying forces shall be avoided and the bolts shall be pretensioned.

23.1.5
Joints subject to repeated loads shall be proportioned in accordance with Clause 14.

23.2 Bolts, Nuts, and Washers

23.2.1
Except as provided in Clause 23.2.4, bolts, nuts, and washers shall conform to ASTM Standards A325, A325M, A490, and A490M.
Note: *Before specifying metric bolts, the designer should check on their current availability in the quantities required.*

23.2.2
The length of bolts shall be such that the point of the bolt will be flush with or outside the face of the nut when completely installed.

23.2.3
If required, A325M and A325 bolts, nuts, and washers may be galvanized in accordance with the requirements of ASTM Standards A325M and A325. When a galvanized nut is installed on a galvanized bolt in a solid steel connection and with three to five threads in the grip, it shall be capable of producing a tensile-type fracture of the bolt and of rotating one full turn from snug before failure.

23.2.4
Other fasteners that meet the chemical and mechanical requirements of ASTM Standards A325M, A490M, A325, or A490 and that have body diameters and bearing areas under the head and nut given in those standards, may be used. Such fasteners may differ in other dimensions and their use shall be subject to the approval of the designer.

23.2.5
If necessary, washers may be clipped on one side to a point not closer than 7/8 of the bolt diameter from the centre of the washer hole.

23.3 Bolted Parts

23.3.1
Bolted parts shall fit together solidly when assembled and shall not be separated by gaskets or any other interposed compressible material.

23.3.2
Holes may be punched, sub-punched or sub-drilled and reamed, or drilled, as permitted by Clause 28.5. The nominal diameter of a hole shall be not more than 2 mm greater than the nominal bolt size, except that for pretensioned bolted joints, where shown on the design drawings, and at other locations approved by the designer, enlarged or slotted holes may be used with high-strength bolts 16 mm in diameter and larger. Joints that use enlarged or slotted holes shall be proportioned in accordance with the requirements of Clause 23 and Clauses 13.11 and 13.12, and shall meet the following conditions:
(a) Oversize holes are 4 mm larger than bolts 22 mm and less in diameter, 6 mm larger than bolts 24 mm in diameter, and 8 mm larger than bolts 27 mm and greater in diameter. Oversize holes shall not be used in bearing-type connections but may be used in any or all plies of slip-critical connections. Hardened washers shall be used under heads or nuts adjacent to the plies containing oversize holes.

(b) Short slotted holes are 2 mm wider than the bolt diameter and have a length that does not exceed the oversize diameter provisions of Item (a) by more than 2 mm. They may be used in any or all plies of slip-critical or bearing-type connections. Such slots may be used without regard to direction of loading in slip-critical connections but shall be normal to the direction of the load in bearing-type connections. Hardened washers shall be used under heads or nuts adjacent to the plies containing the slotted holes.

(c) Long slotted holes are 2 mm wider than the bolt diameter and have a length greater than that allowed in Item (b) but not more than 2.5 times the bolt diameter and may be used
　(i) in slip-critical connections without regard to direction of loading. One-third more bolts shall be provided than would be needed to satisfy the requirements of Clause 13.12.
　(ii) in bearing-type connections with the long dimension of the slot normal to the direction of loading. No increase in the number of bolts specified in Clause 13.11 is required.
　(iii) in only one of the connected parts of either a slip-critical or bearing-type connection at an individual faying surface.
　(iv) provided that structural plate washers or a continuous bar not less than 8 mm in thickness cover long slots that are in the outer plies of joints. These washers or bars shall have a size sufficient to completely cover the slot after installation.

(d) When A490 or A490M bolts greater than 26 mm in diameter are used in oversize or slotted holes, hardened washers shall be at least 16 mm in thickness.

(e) The requirement for the nominal diameter of hole may be waived to permit the use of the following bolt diameters and hole combinations in bearing-type or slip-critical connections:
　(i) either a 3/4 inch diameter bolt or an M20 bolt in a 22 mm hole;
　(ii) either a 7/8 inch diameter bolt or an M22 bolt in a 24 mm hole; and
　(iii) either a 1 inch diameter bolt or an M24 bolt in a 27 mm hole.

23.3.3
When assembled, all joint surfaces including those adjacent to bolt heads, nuts, and washers shall be free of scale (tight mill scale excepted), burrs, dirt, and foreign material that would prevent solid seating of the parts.

23.3.4
The condition of the contact surfaces for slip-critical connections as described in Table 2 shall be as follows:

(a) For clean mill scale, the surfaces shall be free of oil, paint, lacquer, or any other coating for all areas within the bolt pattern and for a distance beyond the edge of the bolt hole that is the greater of 25 mm or the bolt diameter.

(b) For Classes A and B, the blast-cleaning and the coating application must be the same as that used in the tests to determine the mean slip coefficient.

(c) For Class C, hot-dip galvanizing must be done in accordance with CSA Standard G164 and the surface subsequently roughened by means of hand wire-brushing. Power wire-brushing is not permitted.

(d) For all other coatings, the surface preparation and coating application for the joint must be the same as that used in the tests to determine the mean slip coefficient.

Coated joints shall not be assembled before the coatings have cured for the minimum time used in the tests to determine the mean slip coefficient.

23.4 Installation

23.4.1 Bolt Tension

High-strength bolts that are not required to be pretensioned shall be installed in properly aligned holes, to a snug-tight condition. This is defined as the tightness that exists when all plies in a joint are in firm contact. (For slotted holes, see Clause 23.3.2.)

Pretensioned bolts shall be tightened to at least the minimum bolt tension given in Table 7, in accordance with Clause 23.5 or 23.6.

Table 7
Bolt Tension

Bolt diameter		Minimum bolt tension* kN	
inch	mm	A325M and A325 Bolts	A490M and A490 Bolts
1/2	—	53	67
5/8	—	85	107
—	16	91	114
3/4	—	125	157
—	20	142	178
7/8	—	174	218
—	22	176	220
—	24	205	257
1	—	227	285
—	27	267	334
1-1/8	—	249	356
—	30	326	408
1-1/4	—	316	454
1-3/8	—	378	538
—	36	475	595
1-1/2	—	458	658

*Equal to 70% of the specified minimum tensile strength given in the appropriate ASTM Standard, soft-converted where appropriate and rounded to the nearest kilonewton.

23.4.2 Hardened Washers

23.4.2.1

Hardened washers are required under the head or nut when turned

(a) as required by Clause 23.7.4; and

(b) for pretensioned A490 and A490M bolts.

23.4.2.2

Hardened washers are required

(a) for oversize or slotted holes (see Clause 23.3.2).

(b) under the head and nut for A490 and A490M bolts when used with steel having a specified minimum yield point of less than 280 MPa.

(c) of at least 16 mm in thickness, when A490 and A490M bolts greater than 26 mm in diameter are used in oversize and slotted holes.

23.4.3 Bevelled Washers

Bevelled washers shall be used to compensate for lack of parallelism where, in the case of A325M and A325 bolts, an outer face of bolted parts has more than a 5% slope with respect to a plane normal to the bolt axis. In the case of A490M and A490 bolts, bevelled washers shall be used to compensate for any lack of parallelism due to slope of outer faces.

23.5 Turn-of-Nut Tightening

23.5.1

After aligning the holes in a joint, sufficient bolts shall be placed and brought to a snug-tight condition to ensure that the parts of the joint are brought into full contact with each other. "Snug-tight" is the tightness attained by a few impacts of an impact wrench or the full effort of a person using a spud wrench.

23.5.2

Following the initial snugging operation, bolts shall be placed in any remaining open holes and brought to snug-tightness. Re-snugging may be necessary in large joints.

23.5.3

When all bolts are snug-tight, each bolt in the joint shall then be tightened additionally by the applicable amount of relative rotation given in Table 8, with tightening progressing systematically from the most rigid part of the joint to its free edges. During this operation there shall be no rotation of the part not turned by the wrench, unless the bolt and nut are match-marked to enable the amount of relative rotation to be determined.

23.6 Tightening by Use of a Direct Tension Indicator

Tightening by means of a direct tension indicator is permitted provided that it can be demonstrated by an accurate direct measurement procedure that the bolt has been tightened in accordance with Table 7.

23.7 Inspection

23.7.1

The inspector shall determine whether the requirements of Clauses 23.2, 23.3, 23.4, and 23.5 or 23.6 are met. Installation of bolts shall be observed to ascertain that a proper tightening procedure is employed. For those cases in which bolt pretension is required, the turned element of all bolts shall be visually examined for evidence that they have been tightened. For snug-tight connections, the inspection need only ensure that the bolts have been tightened sufficiently to bring the connected elements into full contact.

23.7.2

Tensions in bolts exceeding those given in Table 7 shall not be cause for rejection.

23.7.3

When bolts are installed in accordance with Clause 23.6, the verification that the bolt has been properly tightened is determined by the direct tension indicator.

23.7.4

For bolts in slip-critical connections and for bolts in pretensioned bearing-type connections, when there is disagreement concerning the results of inspection of bolt tension in the turn-of-nut method, the following arbitration inspection procedure shall be used unless a different procedure has been specified:

(a) The inspector shall use an inspection wrench that is a manual or power torque wrench capable of indicating a selected torque value.

(b) Three bolts of the same grade and diameter as those under inspection and representative of the lengths and conditions of those in the structure shall be placed individually in a calibration device capable of indicating bolt tension. There shall be a washer under the part turned if washers are so used in the structure or, if no washer is used, the material abutting the part turned shall be of the same specification as that in the structure.

(c) When the inspection wrench is a manual wrench, each bolt specified in Item (b) shall be tightened in the calibration device by any convenient means to an initial tension of approximately 15% of the required fastener tension and then to the minimum tension specified for its size in Table 7. Tightening beyond the initial condition shall not produce greater nut rotation than that permitted in Table 8. The inspection wrench shall then be applied to the tightened bolt, and the torque necessary to turn the nut or head 5° in the tightening direction shall be determined. The average torque measured in the tests of three bolts shall be taken as the job inspection torque to be used in the manner specified in Item (e). The job inspection torque shall be established at least once each working day.

(d) When the inspection wrench is a power wrench, it shall first be applied to produce an initial tension of approximately 15% of the required fastener tension and then adjusted so that it will tighten each bolt specified in Item (b) to a tension of at least 5% but not more than 10% greater than the minimum tension specified for its size in Table 7. This setting of the wrench shall be taken as the job inspection torque to be used in the manner specified in Item (e). Tightening beyond the initial condition must not produce greater nut rotation than that permitted in Table 8. The job inspection torque shall be established at least once each working day.

(e) Bolts, represented by the sample prescribed in Item (b), that have been tightened in the structure shall be inspected by applying, in the tightening direction, the inspection wrench and its job inspection torque to 10% of the bolts, but not less than 2 bolts, selected at random in each connection. If no nut or bolt head is turned by this application of the job inspection torque, the connection shall be accepted as properly tightened. If any nut or bolt head is turned by the application of the job inspection torque, this torque shall be applied to all bolts in the connection and all bolts whose nut or head is turned by the job inspection torque shall be tightened and reinspected. Alternatively, the fabricator or erector, at his option, may retighten all the bolts in the connection and then resubmit the connection for the specified inspection.

Table 8
Nut Rotation* from Snug-Tight Condition

Disposition of outer faces of bolted parts	Bolt length‡	Turn
Both faces normal to bolt axis or one face normal to axis and other face sloped 1:20 max. (bevelled washer not used)†	Up to and including 4 diameters	1/3
	Over 4 diameters and not exceeding 8 diameters or 200 mm	1/2
	Exceeding 8 diameters or 200 mm	2/3
Both faces sloped 1:20 max. from normal to bolt axis (bevelled washers not used)†	All lengths of bolts	3/4

*Nut rotation is rotation relative to a bolt regardless of whether the nut or bolt is turned. Tolerance on rotation: 30° over or under. This Table applies to coarse-thread heavy-hex structural bolts of all sizes and lengths used with heavy-hex semi-finished nuts.
†Bevelled washers are necessary when A490M or A490 bolts are used.
‡Bolt length is measured from the underside of the head to the extreme end of point.

24. Welding

24.1 Arc Welding
Arc welding design and practice shall conform with CSA Standard W59.

24.2 Resistance Welding
The resistance of resistance-welded joints shall be taken as given in CSA Standard W55.3. Related welding practice shall be in conformance with welding standards approved by the Canadian Welding Bureau as given in CSA Standard W55.3.

24.3 Fabricator and Erector Qualification
Fabricators and erectors responsible for making welds for structures fabricated or erected under this Standard shall be certified by the Canadian Welding Bureau to the requirements of CSA Standard W47.1 (Division 1 or Division 2.1), or CSA Standard W55.3, or both, as applicable. Part of the work may be sublet to a Division 2.2 or Division 3 fabricator; however, the Division 1 or Division 2.1 fabricator or erector shall retain responsibility for the sublet work.

25. Column Bases

25.1 Loads
Suitable provision shall be made to transfer column loads and moments to footings and foundations.

25.2 Resistance

25.2.1 Compressive Resistance of Concrete
The compressive resistance of concrete shall be determined in accordance with Clause 10.15 of CSA Standard A23.3. When compression exists over the entire base plate area, the bearing pressure on the concrete may be assumed to be uniform over an area equal to the width of the base plate multiplied by a depth equal to d – 2e, where e is the eccentricity of the column load.

25.2.2 Resistance to Pull-Out
Anchor bolts subject to tensile forces shall be anchored to the foundation unit in such a manner that the required factored tensile force can be developed. Full anchorage is obtained when the factored pull-out resistance of the concrete is equal to or larger than the factored tensile resistance of the bolts. For methods of transferring the tensile forces from the anchors to the concrete, see CSA Standard A23.3.

25.2.3 Resistance to Transverse Loads

25.2.3.1
Shear resistance may be developed by friction between the base plate and the foundation unit or by bearing of the anchor bolts or shear lugs against the concrete. When shear acts toward a free edge, the requirements of CSA Standard A23.3 shall be met.

25.2.3.2
When loads are transferred by friction, the requirements of CSA Standard A23.3 shall be met.

25.2.3.3
When shear is transmitted by bearing of the anchor bolts on the concrete, the factored bearing resistance shall be taken as

$B_r = 1.4 \phi_c n A f'_c$

where
$\phi_c = 0.60$
n = number of anchor bolts in shear
A = bearing area, taken as the product of the bolt diameter, d, and an assumed depth of 5d

25.2.3.4
For methods of transmitting shear by bearing of shear lugs in the concrete, see CSA Standard A23.3.

25.2.4 Moment Resistance
Moment resistance shall be taken as the couple formed by the tensile resistance of the anchor bolts determined in accordance with Clause 26.2.1 or 26.2.3, as applicable, and by the concrete compressive resistance determined in accordance with Clause 10.15 of CSA Standard A23.3.

25.3 Finishing

Column bases shall be finished in accordance with the following requirements:

(a) The bottom surfaces of bearing plates and column bases that rest on masonry or concrete foundations and are grouted to ensure full bearing need not be planed.

(b) Steel-to-steel contact bearing surfaces of rolled steel bearing plates shall be finished in such a manner that the requirements of Clauses 28.8, 28.9.7, and 30.7.6 are satisfied. In general, rolled steel bearing plates 50 mm or less in thickness may be used without machining, provided that satisfactory contact is obtained. Rolled steel bearing plates more than 50 mm but not more than 100 mm thick may be straightened by pressing, or may be machined at bearing locations to obtain a satisfactory contact. Rolled steel bearing plates more than 100 mm thick, and other column bases, shall be machined at bearing locations.

26. Anchor Bolts

26.1 General

Anchor bolts shall be designed to resist the effect of factored uplift forces, bending moments, and shears determined in accordance with Clause 7.2. The anchorage of the anchor bolts in the foundation unit shall be such that the required load capacity can be developed. Forces present during construction as well as those present in the finished structure shall be resisted.

26.2 Bolt Resistance

26.2.1 Tension

The factored tensile resistance of an anchor bolt shall be taken as

$T_r = \phi_b A_n F_u$

where

$\phi_b = 0.67$
A_n = the tensile stress area of the bolts
 $= \frac{\pi}{4}(D - 0.938P)^2$ for metric bolts
 $= \frac{\pi}{4}\left(D - \frac{0.974}{n}\right)^2$ for imperial bolts

where

P = the pitch of thread, mm, and
n = number of threads per inch

26.2.2 Shear

The factored shear resistance of the anchor bolts shall be taken as

$V_r = 0.60 \phi_b n\, A_b F_u$

but not greater than the lateral bearing resistance given in Clause 25.2.3.3.

When the bolt threads are intercepted by the shear plane, the factored shear resistance shall be taken as 70% of V_r.

26.2.3 Shear and Tension

An anchor bolt required to develop resistance to both tension and shear shall be proportioned so that

$$\left(\frac{V_f}{V_r}\right)^2 + \left(\frac{T_f}{T_r}\right)^2 \leq 1$$

where
V_f is the portion of the total shear per bolt transmitted by bearing of the anchor bolts on the concrete (see Clause 25.2.3.3).

26.2.4 Tension and Bending
An anchor bolt required to develop resistance to both tension and bending shall be proportioned to meet the requirements of Clause 13.9(a). The tensile and moment resistances, T_r and M_r, shall be based on the properties of the cross section at the critical section; M_r shall be taken as $\phi_b S F_y$.

27. Seismic Design Requirements

27.1 General

27.1.1
Clause 27 provides requirements for the design of members and connections in the lateral-load-resisting systems of steel-framed buildings for which ductile response is required under seismic loading. This Clause is to be applied in conjunction with the requirements of Clause 4.1.9 of the *National Building Code of Canada, 1995*.

27.1.2
In calculating P-Delta effects at ultimate limit states under seismic load, the provisions of *Commentary J* to the *National Building Code of Canada, 1995* may be applied. If the provisions of *Commentary J* are applied, the value of U_2 in Clause 8.6.1 may be calculated from

$$U_2 = 1 + \left(\frac{\Sigma C_f \Delta_f}{\Sigma V_f h}\right)$$

27.1.3
If structural or nonstructural elements not considered to form part of the lateral-load-resisting system have a significant effect on the structural response to earthquake motions, they shall be considered in the analysis.

27.1.4
Structural members and their connections that are not considered to form part of the lateral-load-resisting system shall be capable of maintaining their resistance when subject to seismically induced deformations.

27.1.5
Steel used in lateral-load-resisting systems shall conform to Clauses 5.1.3 and 8.5(a). Material other than this may be used if approved by the regulatory authority.

27.2 Ductile Moment-Resisting Frames

27.2.1 General

27.2.1.1
Ductile moment-resisting frames have the capacity to form plastic hinges, where necessary, in flexural members and in columns or in joints, and to maintain the member resistance at these hinges. They shall meet the specific requirements of this section.

27.2.1.2

Beams, columns, and joint panel zones shall be proportioned and braced to enable them to undergo large plastic deformations unless it can be demonstrated that the element being considered will remain elastic while one or more of the other elements at the joint is undergoing large plastic deformations.

The element or elements at a joint that may undergo large plastic deformations are termed critical elements, and they shall be identified.

An element undergoing large plastic deformations shall be assumed to apply relevant loads to the other elements at the joint equal to 1.2 times its unfactored yield resistance. An element may be considered as responding elastically if the resulting loading on it does not exceed its factored resistance.

The resistance at the ends of beams shall be determined from Clause 13.5 or 13.6. At the ends of columns, the resistance shall be determined from Clause 13.8.1 or 13.8.2, and the resistance of the joint panel zone shall be determined from Clause 27.2.4.1.

27.2.1.3

For the purposes of Clauses 27.2.2, 27.2.3, and 27.2.4, the contribution of a composite slab shall be estimated in calculating the flexural resistance of, and the loading produced by, the girders.

27.2.2 Beams

27.2.2.1

When a beam is a critical element at a joint, it shall
(a) conform to Class 1 section requirements, and
(b) be laterally braced according to the requirements of Clause 13.7.
When a beam is not critical, it shall have
(c) flanges conforming to Class 1 section requirements, and
(d) webs conforming to Classes 1 or 2.

27.2.2.2

Abrupt changes in beam flange cross sections shall be avoided in regions where plastic hinges may occur.

27.2.2.3

When horizontal framing members are trusses, the columns shall be the critical element at the joint and shall conform to Clause 27.2.3.1. Where such buildings exceed one storey, the ratio of unbraced column height to least radius of gyration shall not exceed 60. The connections of the truss chord to the column shall have factored resistances at least equal to the lesser of
(a) the factored yield resistances of the chord sections; and
(b) the forces required to impose column moments equal to 1.2 times the column factored flexural resistance.

27.2.3 Columns (Including Beam-Columns)

27.2.3.1

Columns shall be Class 1 or 2. When a column is a critical element at a joint, it shall be Class 1 and
(a) be laterally braced according to the requirements of Clause 13.7; and
(b) if in a velocity related seismic zone 4 or higher, have a factored axial load not exceeding $0.30AF_y$ for all seismic load combinations.

Note: *A column need not be treated as a critical member if*
(a) it is not designed to participate in resisting seismic shears (although it may be designed to resist seismic overturning moments), or

(b) it is in a storey which has a ratio of total storey lateral shear resistance to design shear force 1.5 or more times that of the storey above.

27.2.3.2
Splices that incorporate partial joint penetration groove welds shall be located at least one-fourth of the clear distance between beams but not less than one metre from the beam-to-column joints.

27.2.3.3
The forces arising from yielding elements shall be considered in the design of columns.

27.2.4 Column Joint Panel Zone

27.2.4.1
The horizontal shearing resistance of the column joint panel zone shall be taken as

$$V_r = 0.55 \phi d_c w' F_{yc} \left[1 + \frac{3 b_c t_c^2}{d_c d_b w'} \right]$$

where the subscripts b and c denote the beam and the column, respectively, and b_c is the width of the column flange.

27.2.4.2
When the joint panel zone is one of the critical elements at a joint, the following shall apply:
(a) In velocity related or acceleration related seismic zones of 2 or higher, the ratio of the sum of panel zone depth and width to panel zone thickness shall not exceed 90. Doubler plate thickness may be included with web thickness in calculating this ratio only if the doubler plate is plug-welded or bolted to the column web.
(b) Doubler plates, if present, shall be placed against the column web. They shall be groove- or fillet-welded to develop their full shear yield resistance.
(c) The column flange shall be braced, either directly or indirectly, through the column web. This shall be at the level of both beam flanges when the column is critical, but otherwise need only be at the level of the beam bottom flange. Bracing resistance shall correspond to the requirements of Clause 20.2.1 where β is taken as 3 and C_f is taken as the beam flange yield force.
  When the joint panel zone is not critical, the panel zone details shall be selected according to Clauses 21.3 and 27.2.4.3.

27.2.4.3
When beams are rigidly framed to the flange of an H-shaped column, the capacity of the column web to resist the tensile load due to seismic forces shall be taken as 0.6 times the value of T_r given in Clause 21.3.

27.2.5 Beam-to-Column Connections

27.2.5.1
The moment-resisting connection of a beam to a column shall have factored flexural resistance at least equal to the lesser of
(a) the factored bending resistance of the beam; and
(b) the moment that will induce in the column joint panel zone a shearing force equal to the factored resistance given in Clause 27.2.4.1.
  Partial joint penetration groove welds shall not be used in these connections.

27.2.5.2
The beam web connection shall have a resistance adequate to carry shears induced by yielding of critical members at adjacent joints.

27.2.5.3
Flanges and connection plates in bolted connections of beam-to-column joints shall have a factored net section ultimate resistance, Clause 13.2(a) (ii) or (iii), at least equal to the factored gross area yield resistance, Clause 13.2(a)(i).

27.2.5.4
Plastic hinges shall be avoided at locations where the flange area has been reduced, such as at bolt holes, unless $F_y/F_u \leq 0.67$.

27.2.6 Bracing
Beams, columns, and beam-to-column connections shall be braced by members proportioned according to Clause 20.2.1. At beam-to-column connections, the compression force to be stabilized shall be that associated with complete yielding of the critical element at the connection. The possibility of complete load reversals shall be considered.

27.2.7 Fasteners
Fasteners connecting the separate elements of built-up flexural members that are critical shall have resistance adequate to support full yielding at potential plastic hinge locations.

27.3 Moment-Resisting Frames with Nominal Ductility

27.3.1
Moment-resisting frames with nominal ductility can sustain limited amounts of inelastic deformation through flexural action, joint panel zone shearing, or connection deformations. They shall meet the specific requirements of this Clause.

27.3.2
Members shall be Class 1 or 2, unless the connection components are the critical elements in the joints. If the joint panel zone is critical, the width-thickness ratio of both the panel zone web and the doubler plate, if any, shall permit the attainment of the shear yield stress. If connection plates are critical, they shall exhibit ductile behaviour that allows significant rotation of the connection.

27.3.3
Either the beam-to-column moment-resisting connections shall satisfy the requirements of Clause 27.2.5.1, or the frame shall be designed to resist the factored load effects specified in Clause 7.2, with the earthquake loads multiplied by 1.25, and the resistance of the connections shall be associated with a ductile mode of failure.

27.3.4
The forces arising from yielding elements shall be considered in the design of columns.

27.4 Ductile Concentrically Braced Frames

27.4.1 General
Ductile braced frames with concentric bracing have the capacity to absorb energy through yielding of braces. They shall meet the specific requirements of this Clause.

27.4.2 Bracing Systems

27.4.2.1
Diagonal braces shall be oriented such that, at each level in any planar frame, at least 30% of the horizontal shear carried by the bracing system shall be carried by tension braces and at least 30% shall be carried by compression braces.

27.4.2.2
Frames in which seismic load resistance is provided by any of the following shall not be considered as frames in this category:
(a) V- or chevron bracing, in which pairs of braces are located either above or below a beam and meet the beam at a single point within the middle half of the span;
(b) K- bracing, in which pairs of braces meet a column on one side near its mid-height between floors; or
(c) systems that do not meet the requirements of Clause 27.4.2.1.

27.4.3 Diagonal Bracing Members

27.4.3.1
Bracing members shall have a slenderness ratio, L/r, less than $1900/\sqrt{F_y}$. In built-up bracing members, the slenderness ratio of the individual parts shall be not greater than 0.5 times the slenderness ratio of the member as a whole. In velocity and acceleration zone 1 or less, symmetrical open sections shall be Class 1 or 2, HSS shall be Class 1, and the width-thickness ratios for angles, tees, and flanges of channels shall not exceed $170/\sqrt{F_y}$. In all other velocity or acceleration zones, symmetrical open sections shall be Class 1, and the width-thickness ratios for angles, tees, and flanges of channels shall not exceed $145/\sqrt{F_y}$ or $330/\sqrt{F_y}$ for rectangular and square HSS and $13\,000/F_y$ for circular HSS.

27.4.3.2
The factored compressive resistance of a brace shall be determined as the product of C_r, given in Clause 13.3, and a reduction factor equal to $1/(1 + 0.35\lambda)$. This factor need not be applied if the tension braces acting at the same level and in the same plane as the compression brace have sufficient reserve capacity to compensate for the reduction.

27.4.4 Bracing Connections

27.4.4.1
Eccentricities in bracing connections shall be minimized.

27.4.4.2
In velocity or acceleration related seismic zones of 3 or higher, brace connections shall have a factored resistance at least equal to the axial tensile yield strength of the brace ($A_g F_y$) unless the engineer can show that a lower resistance is adequate. In zones 1 and 2, the connection shall resist the greater of the factored load effect and the nominal compressive resistance of the brace, but need not exceed the combined effect of 2.0 times the seismic load in the brace and the gravity load.

27.4.4.3
Brace connections including gusset plates shall be detailed to avoid brittle failures due to rotation of the brace when it buckles. This ductile rotational behaviour shall be allowed for, either in the plane of the frame or out of it, depending on the slenderness ratios.

27.4.4.4
Fasteners that connect the separate elements of built-up bracing members shall, if the overall buckling mode induces shear in the fastener, have resistances adequate to support one-half of the yield load of the smaller component being joined, with this force assumed to act at the centroid of the smaller member.

27.4.5 Other Connections

27.4.5.1
Beam-to-column connections and column splices that participate in the lateral-load-resisting system shall have factored resistances adequate to support the effects of the bracing connection loads given in Clause 27.4.4.2 while also supporting the gravity loads. They shall also support the effects of the factored load combinations detailed in Clause 7.2.

27.4.5.2
Column splices made with partial penetration groove welds and subject to net tension forces due to overturning effects shall have factored resistances equal to 150% of the calculated factored load effects, but not less than 50% of the flange yield load of the smaller column.

27.4.5.3
Connections shall resist forces arising from load redistribution following brace buckling or yielding.

27.4.6 Columns and Beams
Columns and beams shall be proportioned to resist the gravity loads together with the forces induced by the brace connection loads given in Clause 27.4.4.2. Redistributed loads due to brace buckling or yielding shall be considered. The brace compressive resistance shall not include the reduction factor defined in Clause 27.4.3.2 unless this creates a more critical condition.

27.5 Concentrically Braced Frames with Nominal Ductility

27.5.1
Concentrically braced frames with nominal ductility have the capacity to absorb limited amounts of energy through inelastic bending or extension of bracing members. They shall meet the specific requirements of this Clause. K-braced frames, in which pairs of braces meet a column near mid-height between floors, shall not be considered in this category.

27.5.2
Diagonal compression bracing members shall be Class 2 sections, or shall have cross-section elements that can undergo limited straining while sustaining the yield stress. Columns, beams, and connections in the lateral-load-resisting system shall resist the gravity load together with the forces induced by the brace connection loads given in Clause 27.5.3.

27.5.3
In velocity or acceleration related seismic zones of 2 and higher, the brace connections shall have a factored resistance at least equal to the axial tensile yield strength of the brace (A_gF_y) unless the designer can show that a lower resistance is adequate. In zone 1, the connection shall resist the greater of the factored load effect and the nominal compressive resistance of the brace, but need not exceed the combined effect of 1.33 times the seismic load in the brace and the gravity load.

For tension-only bracing the load selected shall be multiplied by an additional factor of 1.10.

27.5.4
The beam attached to chevron or V-braces shall be continuous between columns and its top and bottom flanges shall be designed to resist a lateral load of 1.5% of the flange yield force at the point of intersection with the braces.

27.5.5
When a beam is supported from below by chevron braces, it shall be a Class 1 section and shall have adequate nominal resistance to support its tributary gravity loads without the support provided by the braces. The beam connections at the columns shall resist forces corresponding to plastic bending at the brace intersection point.

Braces in chevron braced frames in velocity related seismic zones 4 and higher shall conform to the requirements of Clause 27.4.3.1.

27.6 Ductile Eccentrically Braced Frames
Members in the braced bays of eccentrically braced frames shall be designed in accordance with the following requirements.

27.6.1 Link Beam

27.6.1.1
The link beam in an eccentrically braced frame is a beam containing a segment (link) that is designed to yield, either in flexure or in shear, prior to yield of other parts of the structure. A link shall be provided at least at one end of each brace. The section used for a link beam shall be Class 1, and its yield strength, F_y, shall not exceed 350 MPa.

27.6.1.2
Axial forces in link beams due to forces from the braces and due to transfer of seismic force to the end of the frames shall be considered in the design.

27.6.2 Link Resistance
The shear resistance of the link shall be taken as the lesser of V_r' and $2M_r'/e$
where

$$V_r' = V_r \sqrt{1 - \left(\frac{P_f}{AF_y}\right)^2} \text{ and}$$

$$M_r' = 1.18 \phi M_p \left(1 - \frac{P_f}{AF_y}\right) \leq \phi M_p$$

V_r is given in Clause 13.4.1.2
P_f is the factored axial tensile or compressive force in the link, and
e is the length of the link.

When $\frac{P_f}{AF_y} \leq 0.15$, the effect of P_f on the link resistance may be neglected.

27.6.3 Length of Link
When $P_f/AF_y > 0.15$, the length of link shall not exceed:

for $\frac{A_w}{A} \geq 0.3 \frac{V_f}{P_f'}$: $\left[1.15 - 0.5 \frac{P_f}{V_f} \frac{A_w}{A}\right]\left(\frac{1.6 M_r}{V_r}\right)$

for $\frac{A_w}{A} < 0.3 \frac{V_f}{P_f'}$: $\frac{1.6 M_r}{V_r}$

27.6.4 Link Rotation

The rotation of the link segment relative to the rest of the beam, at a total frame drift of 0.5R times the drift determined for factored loading, shall not exceed the following:
(a) 0.09 radians for links having a clear length of $1.6M_r/V_r$ or less;
(b) 0.03 radians for links having a clear length of $2.6M_r/V_r$ or greater; and
(c) a value obtained by linear interpolation between the above limits for links having clear lengths between the above limits.

Note: *R is defined in Clause 4.1.9 of the* National Building Code of Canada, *1995.*

27.6.5 Link Stiffeners

27.6.5.1

Full-depth web stiffeners shall be provided on both sides of the beam web at the brace end of the link. The stiffeners shall have a combined width of not less than $b - 2w$ and a thickness of not less than $0.75w$ or 10 mm.

27.6.5.2

Intermediate link web stiffeners shall be full depth and shall be provided as follows:

(a) when $e < \dfrac{1.6 M_r}{V_r}$ stiffeners shall be spaced at intervals not exceeding $(30w - 0.2d)$ when the link rotation angle is 0.09 radians, or $(52w - 0.2d)$ when the rotation is 0.03 radians, or less. Linear interpolation shall be used for values between 0.09 and 0.03.

(b) When $\dfrac{2.6 M_r}{V_r} < e < \dfrac{5 M_r}{V_r}$ stiffeners shall be placed at a distance of 1.5b from each end of the link.

(c) When $\dfrac{1.6 M_r}{V_r} < e < \dfrac{2.6 M_r}{V_r}$ stiffeners shall be provided as in (a) and (b).

(d) When $e > \dfrac{5 M_r}{V_r}$ no intermediate stiffeners are required.

27.6.5.3

Full-depth intermediate web stiffeners are required on only one side of the web for link beams less than 650 mm in depth and on both sides of the web for beams 650 mm or greater in depth. The thickness of one-side stiffeners shall not be less than w or 10 mm whichever is larger, and the width shall not be less than $0.5b - w$.

27.6.5.4

Fillet welds connecting the stiffener to the beam web shall develop a stiffener force of A_sF_y. Fillet welds connecting the stiffener to the flanges shall develop a stiffener force of $0.25A_sF_y$.

27.6.6 Lateral Support for Link

Lateral support shall be provided to both top and bottom flanges at the ends of a link. These lateral supports shall have a resistance at least equal to $0.06btF_y$.

27.6.7 Link Beam-to-Column Connection

27.6.7.1

Links connected to columns shall not exceed a length of $1.6M_r/V_r$, unless it can be demonstrated that the link-to-column connection is adequate to undergo the required inelastic link rotation.

27.6.7.2
Where a link is adjacent to the column, the following requirements shall be met:

(a) The beam flanges shall have complete joint penetration groove welds to the column.

(b) The web connection shall be welded to develop the nominal axial, flexural and shear resistances of the beam web.

(c) The capacity of the column to resist the flange yield load shall be determined from Clause 27.2.4.3.

27.6.7.3
Where the link is connected to the column web, the beam flanges shall have complete joint penetration groove welds to the connection plates and the web connection shall be welded to develop the factored axial, flexural and shear resistance of the beam web. The rotation between the link beam and the column shall not exceed 0.015 radians at 0.5R times the drift due to factored loading.

27.6.7.4
Link beam connections to columns may be designed to resist transverse shear only if the link is not adjacent to the column. Such connections must have capacity to resist a torsional moment of $0.015 \, btdF_y$.

27.6.8 Brace-to-Link Beam Connections
Brace-to-link beam connections shall develop the nominal resistance of the brace and transfer this force to the beam web. If the brace is designed to resist a portion of the link end moment, full end restraint shall be provided. No part of the brace-to-beam connection shall extend into the web area of a link beam. The intersection of the brace and beam centre-lines shall be at or within the link. The beam shall not be spliced within or adjacent to the connection between beam and brace.

27.6.9 Link Beam Resistance

27.6.9.1
The beam outside the link shall have nominal axial, bending, and shear resistance which equals or exceeds the forces corresponding to 1.5 times the controlling resistance of the link.

27.6.9.2
The beam outside of the link shall be provided with sufficient lateral support to maintain stability of the beam under forces corresponding to 1.5 times the controlling resistance of the link. Lateral bracing shall be provided to both top and bottom flanges and shall have a resistance at least equal to $0.015 \, btF_y$.

27.6.10 Diagonal Braces
Each diagonal brace shall have a nominal resistance to support axial force and moment corresponding to 1.5 times the controlling resistance of the link beam (Clause 27.6.2). Sections shall be Class 1 or 2.

27.6.11 Columns
Moments and axial loads introduced into a column at the connection with a link or brace shall not be less than those generated by 1.25 times the controlling resistance of the link.

27.6.12 Roof Link Beam
A link beam is not required in roof beams of frames over five storeys in height.

27.6.13 Concentric Brace in Combination
The first storey of a frame over five storeys in height may be concentrically braced if this storey can be shown to have a resistance of at least 1.5 times the loading associated with yielding of any other storey of the structure.

27.7 Special Framing Systems

27.7.1 Steel Plate Shear Walls
Steel plate shear walls shall meet the requirements of Appendix M.

27.7.2 Other Framing Systems
Other framing systems and frames that incorporate special bracing, base isolation, or other energy-absorbing devices shall be designed on the basis of published research results, observed performance in past earthquakes, or special investigation.

28. Fabrication

28.1 General
Unless otherwise specified, the provisions of Clause 28 shall apply to both shop and field fabrication.

28.2 Straightness of Material
Prior to layout or fabrication, rolled material shall be straight within established rolling mill tolerances. If straightening is necessary, it shall be done by means that will not injure the material. When heat is applied locally, the temperature of the heated area shall not exceed the limits given in CSA Standard W59. Sharp kinks and bends shall be cause for rejection.

28.3 Gas Cutting
Gas cutting shall be done by machine where practicable. Gas-cut edges shall conform to CSA Standard W59. Re-entrant corners shall be free from notches and shall have the largest practical radii, with a minimum radius of 14 mm.

28.4 Sheared or Gas-Cut Edge Finish

28.4.1
Planing or finishing of sheared or gas-cut edges of plates or shapes shall not be required, unless specifically noted on the drawings or included in a stipulated edge preparation for welding.

28.4.2
The use of sheared edges in the tension area shall be avoided in locations subject to plastic hinge rotation at factored loading. If used, such edges shall be finished smooth by grinding, chipping, or planing. These requirements shall be noted on design drawings and shop details where applicable.

28.4.3
Burrs shall be removed
(a) as required in Clause 23.3.3;
(b) when required for proper fit-up for welding; and
(c) when they create a hazard during or after construction.

Limit States Design of Steel Structures

28.5 Holes for Bolts or Other Mechanical Fasteners

28.5.1
Unless otherwise shown on design drawings or as specified in Clause 23.3.2, holes shall be made 2 mm larger than the nominal diameter of the fastener. Holes may be punched when the thickness of the material is not greater than the nominal fastener diameter plus 4 mm. For greater thicknesses, holes shall be either drilled from the solid or sub-punched or sub-drilled and reamed. The die for all sub-punched holes or the drill for all sub-drilled holes shall be at least 4 mm smaller than the required diameter of the finished hole. Holes in CSA Standard G40.21 (Type 700Q) or ASTM Standard A514 steels more than 13 mm thick shall be drilled.

28.5.2
In locations subject to plastic hinge rotation at factored loading, fastener holes in the tension area shall be either sub-punched and reamed or drilled full size.

28.5.3
The requirements of Clause 28.5.2 shall be noted on design drawings and shop details where applicable.

28.6 Bolted Construction

28.6.1
Drifting done during assembly to align holes shall not distort the metal or enlarge the holes. Holes in adjacent parts shall match sufficiently well to permit easy entry of bolts. If necessary, holes, except oversize or slotted holes, may be enlarged to admit bolts by a moderate amount of reaming; however, gross mismatch of holes shall be cause for rejection.

28.6.2
Assembly of high-strength bolted joints shall be in accordance with Clause 23.

28.7 Welded Construction
Workmanship and technique in arc-welded fabrication shall conform to those prescribed by CSA Standard W59. The welding practice in resistance-welded fabrication shall conform to that required by CSA Standard W55.3, and shall be approved by the Canadian Welding Bureau.

28.8 Finishing of Bearing Surfaces
Joints in compression that depend on contact bearing shall have the bearing surfaces prepared to a common plane by milling, sawing, or other suitable means. Surface roughness shall have a roughness height rating not exceeding 500 (12.5 µm) as defined in CSA Standard B95, unless otherwise specified.

28.9 Tolerances

28.9.1
Structural members consisting primarily of a single rolled shape shall be straight within the tolerances allowed by CSA Standard G40.20, except as specified in Clause 28.9.4.

28.9.2
Built-up bolted structural members shall be straight within the tolerances allowed for rolled wide-flange shapes by CSA Standard G40.20, except as specified in Clause 28.9.4.

28.9.3
Dimensional tolerances of welded structural members shall be those prescribed by CSA Standard W59, unless otherwise specified.

28.9.4
Fabricated compression members shall not have a deviation from straightness of more than one-thousandth of the axial length between points that are to be laterally supported.

28.9.5
Beams with bow within straightness tolerances shall be fabricated so that, after erection, the bow due to rolling or fabrication shall be upward.

28.9.6
All completed members shall be free from twists, bends, and open joints. Sharp kinks or bends shall be cause for rejection.

28.9.7
Joints in compression that depend on contact bearing when assembled during fabrication shall have at least 75% of the entire contact area in full bearing. (Full bearing shall be defined as a separation not exceeding 0.5 mm.) The separation of any remaining portion shall not exceed 1 mm. A gap of up to 3 mm may be packed with nontapered steel shims in order to meet the requirements of this Clause. Shims need not be other than mild steel, regardless of the grade of the main material.

28.9.8
A variation of 1 mm is permissible in the overall length of members with both ends finished for contact bearing.

28.9.9
Members without ends finished for contact bearing that are to be framed to other steel parts of the structure may have a variation from the detailed length not greater than 2 mm for members 10 m or less in length and not greater than 4 mm for members more than 10 m in length.

29. Cleaning, Surface Preparation, and Priming

29.1 General Requirements

29.1.1
All steelwork, except as exempted in Clauses 29.1.2, 29.1.3, and 29.2, or unless otherwise noted on design drawings or in the job specifications, shall be given one coat of primer or one-coat paint (see Clause 29.5) applied in the shop. The primer or one-coat paint shall be applied thoroughly and evenly to dry, clean surfaces by suitable means.

29.1.2
Steelwork that will subsequently be concealed by interior building finish need not be given a coat of primer, unless otherwise specified (see Clause 6.5).

29.1.3
Steelwork that will be encased in concrete need not be given a coat of primer. Steelwork that is designed to act compositely with reinforced concrete and that depends on natural bond for interconnection shall not be given a coat of primer.

29.1.4
Steelwork that will be shop-primed shall be cleaned of all loose mill scale, loose rust, weld slag and flux deposit, dirt, other foreign matter, and excessive weld spatter prior to application of the primer. Oil and grease shall be removed with a solvent. The fabricator shall be free to use any satisfactory method to clean the steel and prepare the surface for painting, unless a particular method of surface preparation is specified.

29.1.5
Primer shall be dry before loading primed steelwork for shipment.

29.1.6
Steelwork that will not be shop-primed after fabrication shall be cleaned of oil and grease with solvent cleaners and shall be cleaned of dirt and other foreign matter.

29.2 Requirements for Special Surfaces

29.2.1
Surfaces that will be inaccessible after assembly shall be cleaned or cleaned and primed, as required by Clause 29.1, prior to assembly. Inside surfaces of enclosed spaces that will be entirely sealed off from any external source of oxygen need not be primed.

29.2.2
In members in compression, surfaces that are finished to bear and assembled during fabrication shall be cleaned before assembly but shall not be primed unless otherwise specified.

29.2.3
Surfaces that are finished to bear and not assembled during fabrication shall be protected by a corrosion-inhibiting coating. The coating shall be of a type that can be readily removed prior to assembly or shall be of a type that makes such removal unnecessary.

29.2.4
Faying surfaces of high-strength bolted slip-critical joints shall not be primed or otherwise coated except as permitted by Clause 23.

29.2.5
Joints to be field-welded and surfaces to which shear connections are to be welded shall be kept free of primer and/or any other coating that could be detrimental to achieving a sound weldment, except that sheet steel decks may be welded to clean, primed steelwork.

29.3 Surface Preparation
Unless otherwise specified or approved, surface preparation shall be in conformance with one of the following applicable specifications of the Steel Structures Painting Council:
SP 2; SP 3; SP 5; SP 6; SP 7; SP 10; or SP 11.

29.4 Primer
Unless otherwise specified or approved, shop primer shall conform to one of the following Standards of the Canadian General Standards Board:
1.40; 1.140; or 1-GP-166; or to CISC/CPMA Standard 2-75.

29.5 One-Coat Paint
Unless otherwise specified or approved, one-coat paint intended to withstand exposure to an essentially noncorrosive atmosphere for a period of time not exceeding six months shall conform to CISC/CPMA Standard 1-73a.

30. Erection

30.1 General
The steel framework shall be erected true and plumb within the specified tolerances. Temporary bracing shall be employed wherever necessary to withstand all loads to which the structure may be subject during erection and subsequent construction, including loads due to wind, equipment, and equipment operation. Temporary bracing shall be left in place undisturbed as long as required for the safety and integrity of the structure (see also Clause 26). The erector shall ensure during erection that an adequate margin of safety exists in the uncompleted structure and members using the factored member resistances calculated in accordance with Clause 13. (See also Clause 20.1.3.)

30.2 Temporary Loads
Wherever piles of material, erection equipment, or other loads are carried during erection, suitable provision shall be made to ensure that the loads can be safely sustained during their duration and without permanent deformation or other damage to any member of the steel frame and other building components supported thereby.

30.3 Adequacy of Temporary Connections
As erection progresses, the work shall be securely bolted or welded to take care of all dead, wind, and erection loads and to provide structural integrity as required.

30.4 Alignment
No permanent welding or bolting shall be done until as much of the structure as will be stiffened thereby has been suitably aligned.

30.5 Surface Preparation for Field Welding
The portions of surfaces that are to receive welds shall be thoroughly cleaned of all foreign matter, including paint film.

30.6 Field Painting
Unless otherwise specified, the cleaning of steelwork in preparation for field painting, touch-up of shop primer, spot-painting of field fasteners, and general field painting shall not be considered to be a part of the erection work.

30.7 Erection Tolerances

30.7.1 Elevation of Base Plates
Column base plates shall be considered to be at their proper elevation if the following tolerances are not exceeded:

(a) for single and multi-storey buildings designed as simple construction as provided in Clause 8.3, ± 5 mm from the specified elevation; and

(b) for single and multi-storey buildings designed as continuous construction as provided in Clause 8.2, ± 3 mm from the specified elevation.

30.7.2 Plumbness of Columns
Unless otherwise specified, columns shall be considered plumb if their verticality does not exceed the following tolerances:

(a) for exterior columns of multi-storey buildings,
1 to 1000, but not more than 25 mm toward or 50 mm away from the building line in the first 20 storeys plus 2 mm for each additional storey, up to a maximum of 50 mm toward or 75 mm away from the building line over the full height of the building;

(b) for columns adjacent to elevator shafts,
1 to 1000, but not more than 25 mm in the first 20 storeys plus 1 mm for each additional storey, up to a maximum of 50 mm over the full height of the elevator shaft; and

(c) for all other columns,
1 to 500.

Column plumbness is measured from the actual column centreline at the base of the column. The location of the centreline of the column at the base shall fall within a 10 mm diameter circle of the location established on the drawings.

30.7.3 Horizontal Alignment of Members
Unless otherwise specified, spandrel beams shall be considered aligned when the offset of one end relative to the other from the alignment shown on the drawings does not exceed L/1000; however, the offset need not be less than 3 mm and shall not exceed 6 mm. For all other members, the corresponding offsets are L/500, 3 mm, and 12 mm.

30.7.4 Elevations of Members
Elevations of the ends of members shall be within 10 mm of the specified member elevation. Allowances shall be made for initial base elevation, column shortening, differential deflections, temperature effects, and other special conditions, but the maximum deviation from the specified slope shall not exceed L/500. The difference from the specified elevation between member ends that meet at a joint shall not exceed 6 mm.

30.7.5 Members with Adjustable Connections
Members specified to have adjustable connections (such as shelf angles, sash angles, and lintels) shall be considered to be within tolerances when the following conditions are met:

(a) Each piece is level within L/1000; however, the difference in elevation of the ends need not be less than 3 mm and shall not exceed 6 mm.

(b) Adjoining ends of members are aligned vertically and horizontally within 2 mm.

(c) The location of these members both vertically and horizontally is within 10 mm of the location established by the dimensions on the drawings.

30.7.6 Column Splices
Column splices and other compression joints that depend on contact bearing as part of the splice resistance shall, after alignment, have a maximum allowable separation of 6 mm. Any gap exceeding 1.5 mm shall be packed with nontapered steel shims. Shims need not be other than mild steel, regardless of the grade of the main material.

30.7.7 Joint Fit-Up
The fit-up of joints that are to be field-welded shall be within the tolerances shown on the erection diagrams and shall in no case, before welding is begun, exceed the tolerances specified in CSA Standard W59.

30.7.8 Special Tolerances
Unless otherwise required by operational characteristics of the crane, crane girders and monorail beams shall be erected within the following tolerances:

(a) The slope of a member shall not exceed L/1000. However the difference in elevation of the ends need not be less than 3 mm and shall not exceed 6 mm. The difference in elevation of opposite points on two parallel girders shall not exceed 1/1000 of the distance between the girders, and shall not exceed 6 mm.

(b) The alignment of a horizontal member shall not exceed L/500. However, the difference in alignment of the ends need not be less than 3 mm and shall not exceed 8 mm.

(c) The distance between the ends of two parallel girders shall not deviate by more than 1 in 500 of the span of the girder. However the difference in the distances between the girder ends need not be less than 3 mm and shall not exceed 10 mm.

31. Inspection

31.1 General
Material and workmanship shall at all times be subject to inspection by qualified inspectors representing and responsible to the designer. The inspection shall cover shop work and field erection work to ensure compliance with this Standard.

31.2 Cooperation
Insofar as possible, all inspections shall be made in the fabricator's shop and the fabricator shall cooperate with the inspector, permitting access for inspection to all places where work is being done. The inspector shall cooperate in avoiding undue delay in the fabrication or erection of the steelwork.

31.3 Rejection
Material or workmanship not conforming to the provisions of this Standard may be rejected at any time during the progress of work when nonconformance to these provisions is established.

31.4 Inspection of High-Strength Bolted Joints
The inspection of high-strength bolted joints shall be performed in accordance with the procedures prescribed in Clause 23.7.

31.5 Third-Party Welding Inspection
When third-party welding inspection is specified, welding inspection shall be performed by firms certified to CSA Standard W178.1, except that visual inspection may be performed by persons certified to Level 2 or 3 of CSA Standard W178.2.

31.6 Identification of Steel by Marking
In the fabricator's plant, steel used for main components shall at all times be marked to identify its specification (and grade, if applicable). This shall be done by suitable markings or by recognized colour coding, except that cut pieces identified by piece mark and contract number need not continue to carry specification identification markings when it has been satisfactorily established that such cut pieces conform to the required material specifications.

Appendix A
Standard Practice for Structural Steel

Note: *This Appendix is not a mandatory part of this Standard.*

A1.
Matters concerning standard practice not covered by this Standard but pertinent to the fabrication and erection of structural steel (such as a definition of structural steel items, the computation of weights, etc) are to be in accordance with the Code of Standard Practice for Structural Steel published by the Canadian Institute of Steel Construction, unless otherwise clearly specified in the plans and specifications issued to the bidders.

Appendix B
Effective Lengths of Columns

Note: *This Appendix is not a mandatory part of this Standard.*

B1.
The slenderness ratio of a column is defined as the ratio of the effective length to the applicable radius of gyration.

The effective length, KL, may be thought of as the actual unbraced length, L, multiplied by a factor, K, such that the product, KL, is equal to the length of a pin-ended column of equal capacity to the actual member. The effective length factor, K, of a column of finite unbraced length therefore depends on the conditions of restraint afforded to the column at its braced locations.

B2.
A variation in K between 0.65 and 2.0 would apply to the majority of cases likely to be encountered in actual structures. Figure B1 illustrates six idealized cases in which joint rotation and translation are either fully realized or nonexistent.

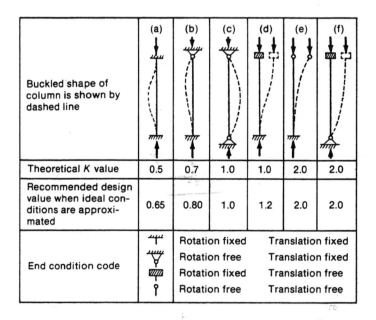

Figure B1

Appendix C
Criteria for Estimating Effective Column Lengths in Continuous Frames

Note: *This Appendix is not a mandatory part of this Standard.*

C1.
Because the Standard requires that the in-plane behaviour of beam-columns be based on their actual lengths (provided only that, when applicable, the sway effects are included in the analysis of the structure (See Clause 8.6)), this Appendix applies only to cases related to buckling; that is to axially loaded columns and beam-columns failing by out-of-plane buckling.

C2.
Figure C1 is a nomograph applicable to cases in which the equivalent I/L of adjacent girders that are rigidly attached to the columns is known; it is based on the assumption that all columns in the portion of the framework considered reach their individual critical load simultaneously.

In the usual building frame, not all columns would be loaded so as to simultaneously reach their buckling loads; thus some conservatism is introduced in the interest of simplification.

C3.
The equation on which this nomograph is based is as follows:
$$\frac{G_U G_L}{4}(\pi/K)^2 + \frac{G_U + G_L}{2}\left(1 - \frac{\pi/K}{\tan\pi/K}\right) + 2\left[\frac{\tan\pi/2K}{\pi/K}\right] = 1$$

C4.
Subscripts U and L refer to the joints at the two ends of the column section being considered. G is defined as
$$G = \frac{\Sigma I_c / L_c}{\Sigma I_g / L_g}$$

where Σ indicates a summation for all members rigidly connected to that joint and lying in the plane in which buckling of the column is being considered,

where
I_c is the moment of inertia and
L_c is the unsupported length of a column section;
I_g is the moment of inertia and
L_g is the unsupported length of a girder or other restraining member.
I_c and I_g are taken about axes perpendicular to the plane of buckling being considered.

C5.
For column ends supported by, but not rigidly connected to, a footing or foundation, G may be taken as 10 for practical designs. If the column end is rigidly attached to a properly designed footing, G may be taken as 1.0. Smaller values may be used if justified by analysis.

C6.
Refinements in girder I_g/L_g may be made when conditions at the far end of any particular girder are known definitely or when a conservative estimate can be made. For the case with no side-sway, multiply girder stiffnesses by the following factors:
(a) 1.5 if the far end of the girder is hinged; and
(b) 2.0 if the far end of the girder is fixed against rotation (ie, rigidly attached to a support that is itself relatively rigid).

C7.
Having determined G_U and G_L for a column section, the effective length factor, K, is determined at the intersection of the straight line between the appropriate points on the scales for G_U and G_L with the scale for K.

C8.
The nomograph may be used to determine the effective length factors for the in-plane behaviour of compression members of trusses designed as axially loaded members even though the joints are rigid. In this case, there should be no in-plane eccentricities and all the members of the truss meeting at the joint must not reach their ultimate load simultaneously. If it cannot be shown that all members at the joint do not reach their ultimate load simultaneously, then the effective length factor of the compression members shall be taken as 1.0.

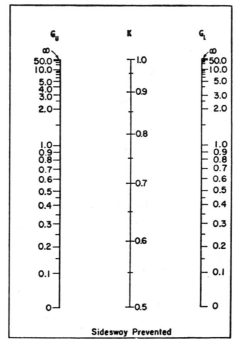

Figure C1
Nomograph for Effective Lengths of Columns in Continuous Frames

Appendix D
Torsional-Flexural Buckling of Compression Members

Note: This Appendix is a mandatory part of this Standard.

D1.
The factored compressive resistance, C_r, of asymmetric, singly symmetric, and cruciform or other bisymmetric sections not covered under Clause 13.3 shall be computed by taking torsional-flexural buckling into account. This is done using the equations given in Clause 13.3 except that λ shall be replaced by λ_e, calculated as follows:

$$\lambda_e = \sqrt{\frac{F_y}{F_e}}$$

where
(a) for doubly-symmetric sections,

$$F_e = \left[\frac{\pi^2 E C_w}{(K_z L)^2} + GJ\right]\frac{1}{I_x + I_y}$$

(b) for singly-symmetric sections with y-y taken as the axis of symmetry,

$$F_e = \frac{F_{ey} + F_{ez}}{2\beta}\left[1 - \sqrt{1 - \frac{4 F_{ey} F_{ez} \beta}{(F_{ey} + F_{ez})^2}}\right]$$

and F_{ey} is calculated for the axis of symmetry; and

(c) for asymmetric sections, F_e is the smallest root of,

$$(F_e - F_{ex})(F_e - F_{ey})(F_e - F_{ez}) - F_e^2(F_e - F_{ey})\left(\frac{x_0}{\bar{r}_0}\right)^2 - F_e^2(F_e - F_{ex})\left(\frac{y_0}{\bar{r}_0}\right)^2 = 0$$

and F_{ex}, F_{ey} are calculated with respect to the principal axes.

In these expressions
K_z = the torsional effective length factor, conservatively taken as 1.0
J = the St. Venant torsion constant
x_0, y_0 = the principal coordinates of the shear centre with respect to the centroid of the cross-section

$$\bar{r}_0^2 = x_0^2 + y_0^2 + r_x^2 + r_y^2$$

$$\beta = 1 - \left[\frac{x_0^2 + y_0^2}{\bar{r}_0^2}\right]$$

$$F_{ex} = \frac{\pi^2 E}{\left(\dfrac{K_x L}{r_x}\right)^2}$$

$$F_{ey} = \frac{\pi^2 E}{\left(\dfrac{K_y L}{r_y}\right)^2}$$

$$F_{ez} = \left(\frac{\pi^2 E C_w}{(K_z L)^2} + G J\right) \frac{1}{A \bar{r}_0^2}$$

Appendix E
Margins of Safety

Note: *This Appendix is not a mandatory part of this Standard.*

E1.
An advantage of limit states design is that the probability of failure for different loading conditions is made more consistent, by the use of distinct load factors for the different loads to which the structure is subject, than in working stress design, where a single factor of safety is used. Furthermore, different resistance factors can, in a parallel manner, be applied to determine member resistances with a uniform reliability. The combination of the load factor and the inverse of the resistance factor gives a number comparable to the traditional factor of safety. In this Standard, a resistance factor of 0.90 is generally used.

E2.
For live loads, the load factor of 1.50 multiplied by the inverse of the resistance factor (1/0.90) equals 1.67, which is comparable to the safety factor contained in former working stress design standards. By using a load factor of 1.25 for dead load, probabilistic studies indicate that consistent probabilities of failure are determined over all ranges of dead-to-live-load ratios. The same probabilistic studies also show that load combination factors of 0.70 and 0.60 (depending on the number of loads taken in combination) applied only to live, wind, and temperature loads, and a factor of 0.85 applied to dead load when it is counteractive to live loads, also result in a consistent probability of failure. The extreme value of earthquake loads (the 500 year earthquake) is used with a load factor of 1.00.

E3.
Resistance factors (see Clause 2.1) generally allow for the variation in the member or connection resistance as compared to that predicted. This variation arises from the variability in material properties, dimensions, and workmanship, as well as from simplifications in the mathematical derivation of the resistance equations.

For simplicity in some cases in this Standard, uncertainty in the formulation of the theoretical member resistance has been incorporated directly into the expression for member resistance rather than using a lower value for the resistance factor. This is the case for the column curve, where the curve predicting the ultimate strengths as a function of slenderness ratio has been derived statistically taking into account residual stresses and initial out-of-straightness.

E4.
For bolts, a resistance factor of 0.67 is used to ensure that connector failures will not occur before general failure of the member as a whole. For long bolted joints and for cases in which shear planes intersect the threads, reduction factors are applied to the resistance formulations. As with bolts, a resistance factor of 0.67 is used for welds.

Appendix F
Columns Subject to Biaxial Bending

Notes:
(1) *This Appendix is not a mandatory part of this Standard.*
(2) *The precise design of beam-columns to resist biaxial bending is extremely complex. More refined design expressions than those given in Clause 13 are available, but these are shape-dependent. The Clauses in this Appendix provide design expressions for wide-flange shapes and square hollow structural sections.*

F1.
The bending moments under factored loads to be used in Clauses F2 and F3, M_{fx} and M_{fy}, are as defined in Clause 8.6.

F2.
Class 1 and 2 wide-flange shapes may be proportioned so that

(a) $\left(\dfrac{M_{fx}}{M_{rcx}}\right)^{\zeta} + \left(\dfrac{M_{fy}}{M_{rcy}}\right)^{\zeta} \leq 1.0$

(b) $\left(\dfrac{U_{1x}M_{fx}}{M_{ox}}\right)^{\eta} + \left(\dfrac{U_{1y}M_{fy}}{M_{oy}}\right)^{\eta} \leq 1.0$

where
U_{1x} and U_{1y} are as defined in Clause 13.8.3; however, U_{1x} shall not be less than 1.0

M_{rcx} and M_{rcy} are the factored moment resistances of the section, reduced for the presence of axial load, and may be taken as

Class 1:
$M_{rcx} = 1.18\, M_{rx} \left[1 - \dfrac{C_f}{C_y}\right] \leq M_{rx}$

$M_{rcy} = 1.19 M_{ry} \left[1 - \left(\dfrac{C_f}{C_y}\right)^2\right] \leq M_{ry}$

Class 2:
$M_{rcx} = M_{rx}\left(1 - \dfrac{C_f}{C_y}\right)$

$M_{rcy} = M_{ry}\left[1 - \left(\dfrac{C_f}{C_y}\right)^2\right]$

where M_{rx} and M_{ry} are defined in Clause 13.5(a)

M_{ox} and M_{oy} = maximum factored moment resistances of the column in the presence of the axial load but in the absence of the other orthogonal moment and may be taken as

$M_{ox} = M_{rx}\left(1 - \dfrac{C_f}{C_r}\right)$

$$M_{oy} = M_{ry}\left(1 - \frac{C_f}{C_r}\right)$$

$$\zeta = 1.6 - \frac{C_f/C_y}{2\log_e(C_f/C_y)}$$

For $b/d \geq 0.3$, $\eta = 0.4 + \dfrac{C_f}{C_y} + \dfrac{b}{d}$

For $b/d < 0.3$, $\eta = 1.0$

For M_{rx}, use M_r as defined in Clause 13.6(a); for M_{ry}, use M_r as defined in Clause 13.5(a) and C_r as defined in Clause 13.3.1 with $K = 1.0$ (see Clause 9.3).

Note: *For values of $C_f/C_y < 0.3$, the value of ζ may be taken as equal to the value of η.*

F3.

Class 1 and Class 2 square hollow structural sections (rolled* or fabricated) may be proportioned so that

(a) $\dfrac{M_{fx}}{M_{rx}} + 0.5 \dfrac{M_{fy}}{M_{ry}} \leq 1.0$

(b) $\dfrac{C_f}{C_r} + 0.85\left(\dfrac{M_{fx}}{M_{rx}} + 0.5\dfrac{M_{fy}}{M_{ry}}\right) \leq 1.0$

where

M_{fx} = the numerically larger moment

M_{rx} and M_{ry} are as defined in Clause 13.5(a)

$C_r = \phi A F_y$

*Hot-rolled or stress-relieved such that residual stresses do not exceed $0.3F_y$.

(c) $\dfrac{C_f}{C_r} + \upsilon\left(\dfrac{U_{1x} M_{fx}}{M_{rx}} + \dfrac{U_{1y} M_{fy}}{M_{ry}}\right) \leq 1.0$

where $\upsilon = \dfrac{\sqrt{(\omega_{1x}M_{fx})^2 + (\omega_{1y}M_{fy})^2}}{\omega_{1x}M_{fx} + \omega_{1y}M_{fy}}$

C_r is as defined in Clause 13.3.2 with $K = 1.0$ (see Clause 9.3)

M_{rx} and M_{ry} are as defined in Clause 13.5(a)

ω_1 is as defined in Clause 13.8.4

U_{1x} and U_{1y} are as defined in Clause 13.8.3

Note: *Design of Class 1 and Class 2 sections in accordance with the above requirements takes advantage of the redistribution of stress after initiation of yielding under the specified loads. Consideration should be given to this aspect of design if yielding under the specified loads would induce undesirable lateral deformations of a structure.*

CAN/CSA-S16.1-94

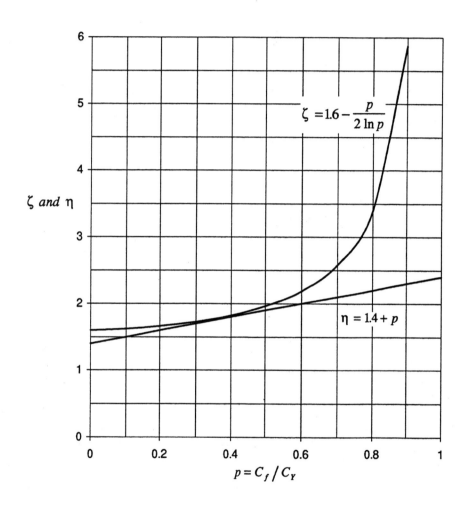

**Figure F1
Plot of ζ and η Versus C_f/C_y for b/d = 1**

Limit States Design of Steel Structures

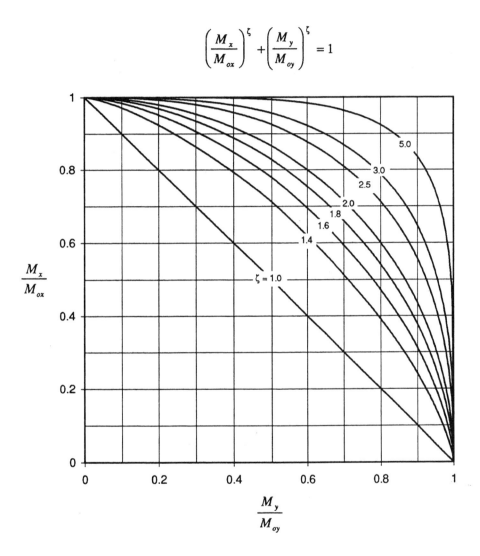

**Figure F2
Plot of the Interaction Equation**

Appendix G
Guide for Floor Vibrations

Note: *This Appendix is not a mandatory part of this Standard.*

G1.
The development of floors of lighter construction, longer spans, and less inherent damping has sometimes resulted in objectionable floor vibrations during normal human activity. Fatigue or overloading of floor structures due to vibration is not covered in this Appendix, nor are floors subject to vibrations caused by rhythmic activities such as aerobic classes.

G2.
Two types of vibration problems arise in floor construction. Continuous vibrations arise due to the periodic forces of machinery, vehicles, or certain human activities such as dancing. These vibrations can be considerably amplified when the periodic forces are in resonance with a floor frequency. Transient vibrations, which decay as shown in Figure G1, arise due to footsteps or other impact.

G3.
The most important floor characteristics affecting vibration problems are the natural frequency in hertz (cycles per second), which usually corresponds to the lowest mode of vibration, and damping. The relation between damping, expressed as a percentage of critical damping (Smith (1988)), and decay of free vibration is shown in Figure G2. Other characteristics affecting transient vibration problems are mass, especially for heavy, long-span floors, and stiffness under point load, especially for light, short-span floors.

G4. Annoyance Criteria

G4.1
Generally people do not like floors to vibrate. For a continuous sinusoidal vibration lasting more than about ten cycles, an average threshold of definite perception is shown in Figure G3 in terms of peak acceleration. The threshold levels for different people range from about one-half to twice the level shown. In the frequency range 2–8 Hz, where people are most sensitive to vibration, the threshold corresponds to 0.5% g approximately, where g is the acceleration due to gravity. The threshold of definite perception shown in Figure G3 can be used to approximate a design criterion for residential, school, and office occupancies; the design level will be lower for sensitive occupancies (eg, operating rooms and special laboratories) and greater for industrial occupancies.

G4.2
For transient vibrations, as shown in Figure G1, the design threshold in terms of initial peak acceleration of a decaying vibration increases with an increase in damping as shown in Figure G3. People find a continuous vibration much more annoying than a vibration that dies out quickly.

G4.3

Design criteria for walking (footstep) vibrations equivalent to that for continuous vibration are shown in Figure G3 for different levels of damping (Allen and Rainer (1976)). The criteria are related to acceleration from a heel impact test. Allen and Murray (1993) give a new design criterion for walking vibrations with broader application than Figure G3.

G5. Continuous Vibrations — Resonance

G5.1

Continuous vibrations caused by machines can be reduced by special design provisions (Smith (1988) and Bachmann and Ammann (1987)) such as vibration isolation. Care should be taken at the planning stage to locate such machinery away from sensitive occupancies such as offices.

G5.2

Floor vibrations can also arise from heavy street traffic on bumpy pavement over a soft subgrade. The annoyance increases considerably when repetitive vehicles such as buses create ground vibrations that synchronize with the floor frequency.

G5.3

Continuous vibrations caused by human activities may be a problem for light residential floors or for long-span floors used for special purposes such as dancing, concerts, or gymnastics. People alone or in unison can create periodic forces in the frequency range of approximately 1–4 Hz; therefore, for such occupancies natural frequencies lower than 5 Hz should be avoided. For very repetitive activities such as aerobics, it is recommended that the frequency of floors be 8 Hz or higher to avoid very noticeable vibration. More specific guidance on vibrations due to rhythmic human activities is given in the Supplement to the *National Building Code of Canada* (1995).

G6. Walking Vibrations

G6.1

Objectionable vibration due to footstep impact can occur in floor systems with light damping in residential, school, office, and similar occupancies. Since this is the most common source of annoyance, the remainder of this guide will be concerned with this problem. Types of construction that may give walking-vibration problems include open-web steel joists or steel beams with concrete deck and light wood deck floors using steel joists.

G7. Performance Test for Floor Vibration

G7.1

The vibration acceptability of a floor system to human activity can be evaluated by a performance test. Partitions, rugs, and furnishings, finishes, etc, contribute to reducing vibration annoyance and should therefore be considered in setting up the test floor. A measuring device that filters out frequencies greater than approximately 1.5 times the fundamental frequency should be located near midspan. A person who will give a subjective evaluation of the floor should also be sitting close to the measuring device.

G7.2
Figure G3 criteria apply only for a limited range of natural frequency, ie: 5 Hz < f_1 < 8Hz. The performance test involves checking floor comfort when different persons walk on the floor; the average peak acceleration can then be compared with the criterion for steady motion given in Figure G3.

G8. Long-Span Steel Floors with Concrete Deck

G8.1
Walking vibrations may be a problem for open-web steel joists or steel beams with concrete deck, composite or noncomposite, generally of spans 7–20 m with frequencies in the range 3–10 Hz. For such floors, partitions, if properly located, provide more than enough damping to avoid excessive vibrations. On the other hand, walking vibrations may be serious for bare floors with very low inherent damping. Figure G3 shows that the criterion is roughly 10 times greater for 12% damping than for 3% damping.

G8.2
To assess vibration acceptability, a knowledge of frequency, damping, and the peak acceleration from heel impact is required. If design by performance testing is not feasible, these parameters should be estimated by calculation as follows:

(a) The frequency can be estimated by assuming full composite action, even for noncomposite construction. For a simply supported one-way system, the frequency f_1 is given by

$$f_1 = 156\sqrt{\frac{EI_T}{wL^4}} \tag{1}$$

where
E is the modulus of elasticity of steel (200 000 MPa)
I_T is the moment of inertia (mm^4) of the transformed T-section (concrete transformed to steel) assuming a concrete flange equal in width to the spacing of steel joists or beams
L is the span in mm, and
w is the dead load of the T-section in N/mm of span.

When one-way systems are supported on steel girders, the frequency may be reduced. In this case the frequency can be approximated by

$$\frac{1}{f^2} = \frac{1}{f_1^2} + \frac{1}{f_2^2} \tag{2}$$

where f_2 is the frequency of floor supported on steel girders, perpendicular to joists.

A continuous beam of equal spans on flexible supports should be treated as simply supported, since adjacent spans vibrate in opposite directions. For other conditions of span and restraint, the dynamically equivalent simply supported span is less than the full span and can be estimated from the fundamental mode shape.

(b) Damping is generally more difficult to estimate than frequency. A bare steel and concrete deck floor has a damping of approximately 3–4% critical for noncomposite construction and about 2% for fully composite construction. The addition of such components as floor finishing, rugs, furnishings, ceiling, fire-proofing, and ducts increases the damping by about 3% or more. Partitions, either above or below the floor, provide the most effective damping, especially when they are located in both directions. Even light partitions that do not extend to the ceiling provide considerable damping. Partitions along support lines or parallel to the floor joists and farther apart than approximately 6 m may not be effective, however, because the nodal lines of vibration form under the partitions.

Human beings also provide damping, but this is less effective for heavy, long-span floors than for lighter, short-span floors. The following values are suggested for design calculation (see Allen (1974)):

	Damping in per cent critical
Bare floor	3
Finished floor — ceiling, ducts, flooring, and furniture	6
Finished floor with partitions	12

(c) The peak acceleration from heel impact for floors of greater than 7 m span with frequencies lower than about 10 Hz can be estimated by assuming an impulse of 70 N•s suddenly applied to a simple spring and mass system whose mass gives the same response as that of the floor system represented as a simply supported beam vibrating in the fundamental mode. The peak acceleration, a_0, as a percentage of g, the acceleration due to gravity, can be approximated by (Allen and Rainer (1976))

$$a_0 = 0.9 \left[\frac{2\pi f \times \text{impulse}}{\text{equivalent mass}} \right] \left[\frac{100}{g} \right] = \frac{60f}{wBL} \tag{3}$$

where
f is the frequency, in hertz
w is the weight of the floor plus contents, in kilopascals
L is the span, in metres
B is the width of the equivalent beam, in metres

For steel joist or beam and concrete deck systems on stiff supports, L is the joist span and B can be approximated as 40 t_c (20 t_c for edge panel adjacent to interior openings) where t_c is the thickness of the concrete deck as determined from the average mass of concrete, including ribs. For joists or beams and concrete deck supported on flexible girders, where the girder frequency is much lower than the joist frequency and therefore girder vibration predominates, L is the girder span and B can be approximated as the width of floor supported by the girder. For cases where the larger of the two frequencies is less than 1.5 times the smaller, or where the combined frequency (Equation (2)) is less than 5 Hz, it is recommended to use the design criterion for walking vibration in Allen and Murray (1993).

G8.3
For floor spans of less than 7 m, the deflection limits given in Clause 6.2.1.2 in this Standard are recommended, where, for noncomposite construction, stiffness should be based on noncomposite action. In any case, care should be taken to avoid low damping.

G9. Light Wood Deck Floors Using Steel Joists

G9.1
Walking vibrations may be objectionable for light wood deck floors using steel joists with small rolled or cold-formed sections, generally with frequencies in the range 10 – 25 Hz. Although the same principles applying to long-span floors can be used for lighter floors with higher frequencies, the motion can no longer be represented by a simple impulse applied to the floor system. This is because the persons involved — those causing and those receiving the motion — interact with the floor to damp out the motion of the floor.

G9.2

Research carried out so far on steel joist floors with wood deck indicates that, in general, their characteristics for vibration acceptability are similar to those for wood joist floors. Evaluation tests of wood floors indicate that stiffness under point loading (0.5 to 1 mm maximum deflection under 1 kN) is the most important parameter affecting vibration comfort (see Onysko (1970)). Such a stiffness requirement also helps prevent cabinet swaying, china rattling, etc. A joist deflection limit of L/360 under 2 kPa loading provides sufficient stiffness if sufficient lateral stiffness is provided either in the deck or by cross-bridging.

G9.3

Floor damping is less important for light floors than for long-span floors, since the main source of damping is provided by the persons on the floor. Also, adding mass does not improve vibration comfort, since an increase in mass corresponds to a decrease in effective damping. Spans continuous over a support that is a party wall between housing units should be avoided, since people are more annoyed by vibrations originating outside their units than from within. For cold-formed C-joists, ceiling boards or straps should be attached to the bottom flange to prevent annoying high-frequency torsional vibrations in the joists.

G10. Corrective Measures for Unacceptable Floors

G10.1

Measures for correcting floors with annoying vibrations depend on whether the vibrations are continuous or transient.

G10.2

For transient vibrations, an often effective measure is to increase the damping. This can be done either by adding partitions above or by means of posts or stiffeners attached to the underside of beams or floor joists. If these methods are not suitable, special devices such as vibration absorbers may be effective (see Allen and Pernica (1984)). For light floors, a rug is effective in reducing walking impact as well as in cushioning the sway of china cabinets.

G10.3

Corrective measures for continuous vibrations include vibration isolation, smoothing of road surface, and alteration of floor frequency to reduce resonance.

G11. References

Supplement to the National Building Code of Canada, 1995, *Commentary on Serviceability Criteria for Deflections and Vibrations.* , National Research Council of Canada, Ottawa.

Allen, D.E. and Rainer, H. 1976, *Vibration Criteria for Long Span Steel Floors.* Canadian Journal of Civil Engineering, Vol. 3, No. 2.

Allen, D.E. and Murray, T.M. 1993, *Design Criterion for Vibrations Due to Walking.* Engineering Journal, American Institute of Steel Construction, Vol. 30, No. 4, pp 117-129.

Allen, D.E. and Pernica, G. 1984, *A Simple Absorber for Walking Vibrations.* Canadian Journal of Civil Engineering, Vol. 11, No. 1.

Allen, D.L. 1974, *Vibrational Behaviour of Long-Span Floor Slabs.* Proceedings of the Canadian Structural Engineering Conference.

Bachmann, H. and Ammann, W. 1987, *Vibrations in Structures Induced by Man and Machines*, Structural Engineering Document 3e, International Association for Bridge and Structural Engineering, Zurich.

Lenzen, K.H. 1966, *Vibration of Steel Joists-Concrete Slab Floors.* AISC Engineering Journal, Vol. 3, No. 3, p. 133.

Nelson, F.C. 1968, *The Use of Viscoelastic Material to Damp Vibrations in Buildings and Large Structures.* AISC Engineering Journal, Vol. 5, No. 2, p. 72.

Onysko, D.M. 1970, *Performance of Wood-Joist Floor Systems — A Literature Review.* Forest Products Laboratory Information Report OP-X-24, Canadian Forestry Service, Department of Fisheries and Forestry.

Smith, J.W. 1988, *Vibration of Structures: Applications in Civil Engineering Design,* Chapman and Hall, New York.

Wright, D.T. and Green, R. 1959, *Human Sensitivity to Vibrations.* Department of Civil Engineering, Report No. 7, Queen's University, Kingston.

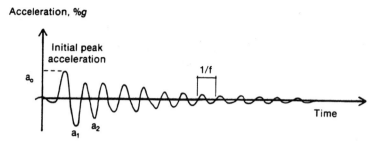

**Figure G1
Typical Transient Vibration from Heel Drop
(High Frequencies Filtered Out)**

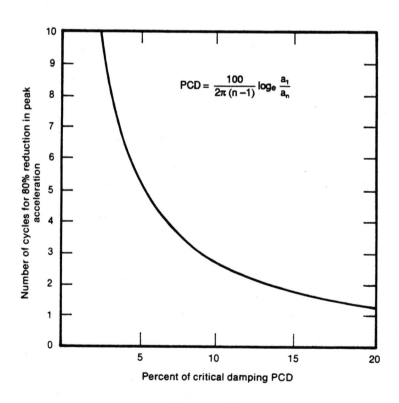

**Figure G2
Relation Between Damping and Decay**

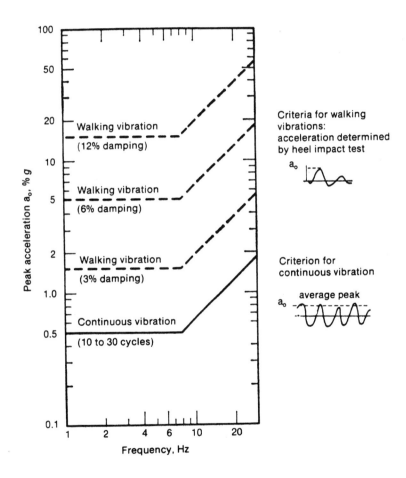

**Figure G3
Annoyance Criteria for Floor Vibrations
(Residential, School, Office Occupancies)**

Appendix H
Wind Sway Vibrations

Note: *This Appendix is not a mandatory part of this Standard.*

H1.
Wind motion of tall buildings or other flexible structures may create annoyance for human occupants unless measures are taken at the design stage. The main source of annoyance is lateral acceleration, although noise (grinding and wind howl) and visual effects can also cause concern.

H2.
For a given wind speed and direction, the motion of a building, which includes vibration parallel and perpendicular to the wind direction and twist, is best predicted by a wind tunnel test. Approximate calculation rules are, however, given in the *Supplement to the National Building Code of Canada* (1995).

H3.
In cases where wind motion is significant in design, the following should be considered:
(a) education of occupants that, although high winds may occasionally cause motion, the building is safe;
(b) minimization of noises through, for example detailing of building joints to avoid grinding and design of elevator guides to avoid scraping due to sway;
(c) minimization of twist through symmetry of layout, bracing, or outer walls (tube concept). Twist vibration also creates a magnified visual effect of relative motion of adjacent buildings;
(d) possible introduction of mechanical damping to reduce wind vibration.

H4. References

Chen, P.W. and Robertson L.E., 1972, *Human Perception Thresholds of Horizontal Motion*. Journal of the Structural Division, ASCE, Vol. 98, No. ST8, pp. 1681–1695.

Council on Tall Buildings and Urban Habitat, Monograph on the Planning and Design of Tall Buildings 1981. Volumes PC and SB. American Society of Civil Engineers.

Hansen, R.J., Reed, J.W. 1973, and Van Marcke, E.H., *Human Response to Wind-Induced Motion of Buildings*. Journal of the Structural Division, ASCE, Vol. 99, No. ST7, pp. 1589–1605.

Hogan, M. 1971, *The Influence of Wind on Tall Building Design*. Faculty of Engineering Science Research Report BLWT-4-71, University of Western Ontario.

Reed, J.W. 1971, *Wind-Induced Motion and Human Discomfort in Tall Buildings*. Department of Civil Engineering Research Report R71-42. Massachusetts Institute of Technology.

Supplement to the National Building Code of Canada, 1995, Chapter 4, Commentary B on Wind Loads, National Research Council of Canada, Ottawa.

Appendix I
Recommended Maximum Values for Deflections for Specified Design Live and Wind Loads

Note: *This Appendix is not a mandatory part of this Standard.*

I1.

Although the deflection criteria set forth in this Appendix refer to specified design live and wind loads, the designer should consider the inclusion of specified dead loads in some instances. For example, nonpermanent partitions, which are classified by the *National Building Code* as dead load, should be part of the loading considered under Appendix I if they are likely to be applied to the structure after the completion of finishes susceptible to cracking. Because some building materials augment the rigidity provided by the steelwork, the wind load assumed to be carried by the steelwork for calculating deflections can be somewhat reduced from the design wind load used in strength and stability calculations. The more common structural elements that contribute to the stiffness of a building are masonry walls, certain types of curtain walls, masonry partitions, and concrete around steel members. The maximum suggested amount of this reduction is 15%. In tall and slender structures (height greater than 4 times the width), it is recommended that the wind effects be determined by means of dynamic analysis or wind tunnel tests.

		Design load	Application	Maximum
Industrial type buildings	Vertical deflection	Live load	Simple span members supporting inelastic roof coverings	1/240 of span
		Live load	Simple span members supporting elastic roof coverings	1/180 of span
		Live load	Simple span members supporting floors	1/300 of span
		Maximum wheel loads (no impact)	Simple span crane runway girders for crane capacity of 225 kN and over	1/800 of span
		Maximum wheel loads (no impact)	Simple span crane runway girders for crane capacity under 225 kN	1/600 of span
	Lateral deflection	Crane lateral force	Simple span crane runway girders	1/600 of span
		Crane lateral force or wind	Building column sway*	1/400 to 1/200 of height

		Design load	Application	Maximum
All other buildings	**Vertical deflection**	Live load	Simple span members of floors and roofs supporting construction and finishes susceptible to cracking	1/360 of span
		Live load	Simple span members of floors and roofs supporting construction and finishes not susceptible to cracking	1/300 of span
	Lateral deflection	Wind	Building sway, due to all effects	1/400 of building height
		Wind	Storey drift (relative horizontal movement of any two consecutive floors due to shear effects) in buildings in cladding and partitions without special provision to accommodate building frame deformation	1/500 of storey height
		Wind	Storey drift (as above), with special provision to accommodate building frame deformation	1/400 of height

*Permissible sway of industrial buildings varies considerably and depends on such factors as wall construction, building height, effect of deflection on the operation of crane, etc. Where the operation of the crane is sensitive to lateral deflections, a permissible lateral deflection of less than 1/400 of the height may be required.

Appendix J
Guide to Calculation of Stability Effects

Note: *This Appendix is not a mandatory part of this Standard.*

J1. General
This Appendix gives one approach to the calculation of the additional bending moments and forces generated by the vertical loads acting through the deflected shape of the structure. In this approach, the moments and forces are incorporated into the results of the analysis of the structure. Alternatively, a second-order analysis, which formulates equilibrium on the deformed structure, may be used to include the stability effects.

J2. Combined Loading Case
Step 1. Apply the factored load combination to the structure (see Clause 7.2.2).

Step 2. Calculate the lateral deflections at each floor level, Δ_i, by first-order elastic analysis.

Step 3. Calculate the equivalent storey shears, V_i', due to the sway forces.

From Figure J1

$$V_i' = \frac{\Sigma P_i}{h_i}\left[\Delta_{i+1} - \Delta_i\right] = \text{equivalent shear in storey i due to the sway forces}$$

ΣP_i = sum of the column axial loads in storey i
h_i = height of storey i
Δ_{i+1}, Δ_i = displacements of storey i + 1 and i, respectively

Step 4. Calculate the artificial lateral loads, H_i', as follows:
$H_i' = V_{i-1}' - V_i'$

Step 5. Repeat Step 1, applying the artificial lateral loads, H_i', in addition to the factored load combination.

Step 6. Repeat Steps 2 through 5 until satisfactory convergence is achieved. Lack of convergence within 5 cycles may indicate an excessively flexible structure.

J3. Gravity Load Case
For the gravity load case, translational moments shall be taken as not less than the moment produced by notional lateral loads, applied at each storey, equal to 0.005 times the factored gravity loads acting at each storey.

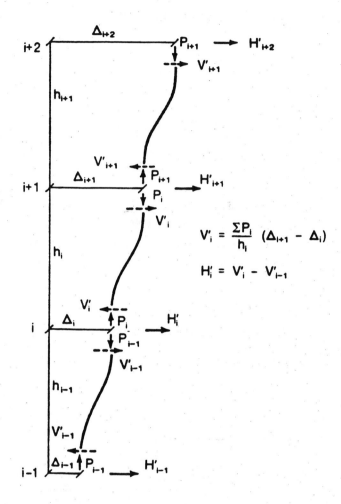

**Figure J1
Sway Forces Due to Vertical Loads**

Appendix K
Fatigue

Note: *This Appendix is not a mandatory part of this Standard.*

K1.
Figure K1 is a plot of the design curves for the fatigue resistances for categories A to E1 in Tables 4(a) and 4(c) (Clause 14).

K2.
Figure K2 gives illustrative examples of the various fatigue categories described in Table 4(c).

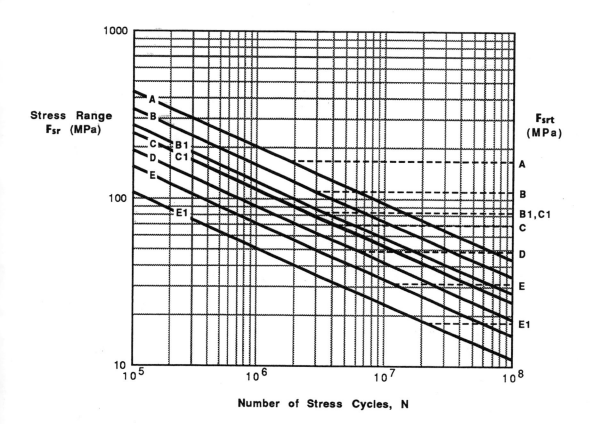

**Figure K1
Fatigue Resistances for Categories A to E1**

(Continued)

Limit States Design of Steel Structures

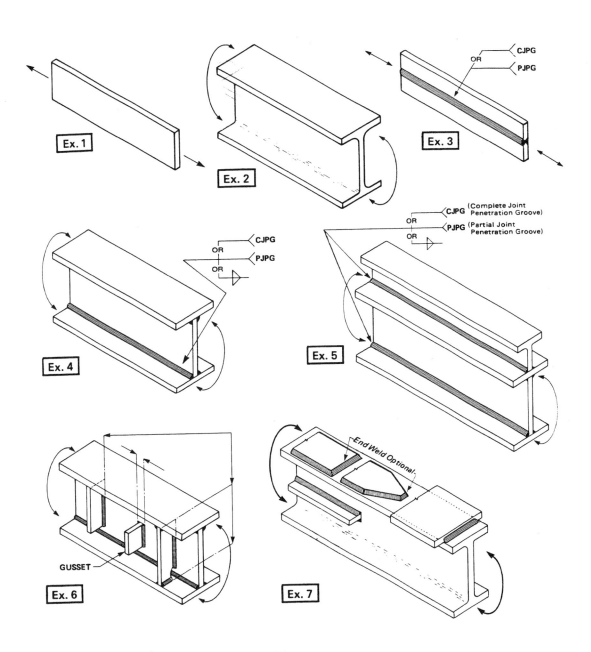

Note: *The example numbers are referenced in Table 4(C).*

**Figure K2
Illustrative Examples of Various Details
Representing Stress Range Categories**

(Continued)

CAN/CSA-S16.1-94

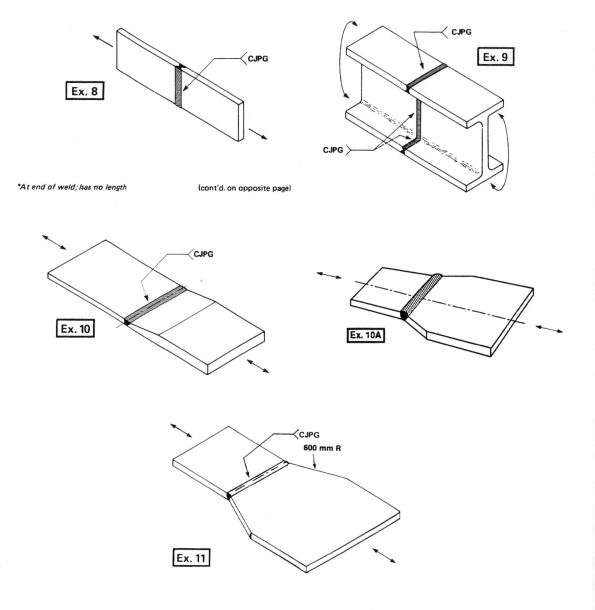

*At end of weld; has no length

(cont'd. on opposite page)

Note: *The example numbers are referenced in Table 4(C).*

Figure K2 (Continued)

(Continued)

Limit States Design of Steel Structures

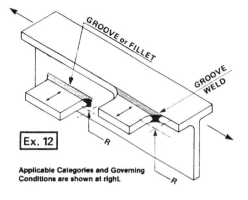

Ex. 12

Applicable Categories and Governing Conditions are shown at right.

TRANSITION RADIUS "R"	Fillet Connections	Groove Connections				
	To Web	To Web or to Flange	To Web (3)	To Flange (3)		
	Longitudinal Loading/Transverse Loading	Longitudinal Loading	Transverse Loading			
	Stress Range Category			Stress Range Category based on Condition of Joint (2)		
				1	2	3,4
50 mm > R ≥ 0 mm	E	E (1)	E	E	E	E
150 mm > R ≥ 50 mm (4)	D	D	D	D	D	E
600 mm > R ≥ 150 mm (4)	D	C	C	C	C	E
R ≥ 600 mm (4)	D	B	C	B	C	E

(1) For longitudinal loading only, use Category D if detail length is between 50 mm and 12 times the plate thickness, but less than 100 mm.
(2) Condition of Joint:
  (1) Equal thickness of parts joined — reinforcement removed.
  (2) Equal thickness of parts joined — reinforcement not removed.
  (3) Unequal thickness of parts joined — reinforcement removed.
  (4) Unequal thickness of parts joined — reinforcement not removed.
(3) Weld soundness to be established by nondestructive examination.
(4) Terminal ends of welded joints to be ground smooth.

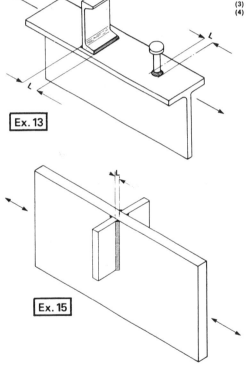

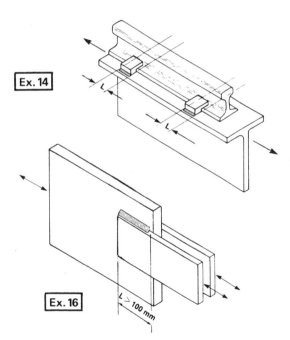

Note: The example numbers are referenced in Table 4(C).

Figure K2 (Continued)

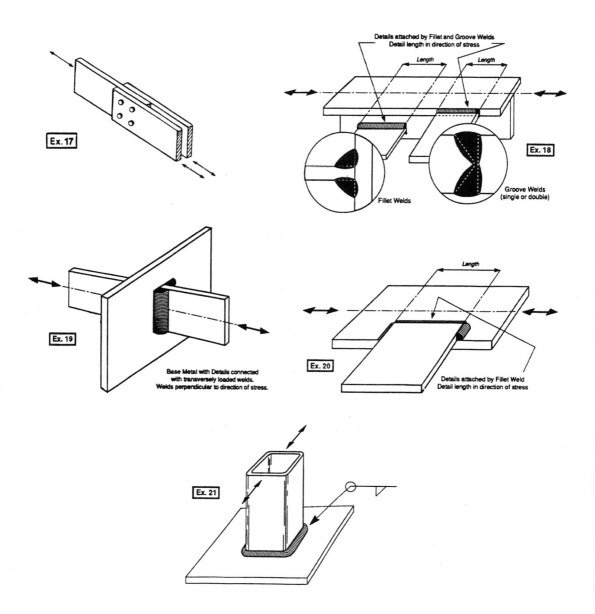

Note: *The example numbers are referenced in Table 4(C).*

Figure K2 (Concluded)

Appendix L
Deflections of Composite Beams Due to Shrinkage of Concrete

Note: *This Appendix is not a mandatory part of this Standard.*

L1.
Shrinkage induced deflections result from the following process. Concrete decreases in volume as it cures, at first rapidly and then at a decreasing rate. When restrained, tensile strains and therefore tensile stresses are developed in the concrete. (It may even crack if the tensile strength is reached.)
   A curing slab is restrained by the steel shape to which it is connected.

L2.
Figures L1 and L2 show the shrinkage strains that develop through the depth for a composite beam and truss respectively and the corresponding equilibrium conditions for unshored construction. It is evident that unshored composite members will deflect downward. (Shoring reduces the shrinkage deflection substantially, especially in the early stages when the rate of shrinkage is the greatest.)
In these figures

ε_f	=	free shrinkage strain of the concrete
ε_s	=	tensile strain in the concrete
ε_r	=	resulting restrained shrinkage strain
$\varepsilon_t, \varepsilon_{tc}$	=	compressive strain at top of steel beam or top chord of truss
$\varepsilon_b, \varepsilon_{bc}$	=	tensile strain at bottom of steel beam or bottom chord of truss
T	=	tensile force in concrete
C, C_{tc}	=	compressive force in beam or in top chord of truss
T_{bc}	=	tensile force in bottom chord of truss
M	=	moment in steel beam

L3.
Based on Figures L1 and L2, Kennedy and Brattland (1992) proposed an equilibrium method to determine shrinkage deflections. It is iterative because the concrete response is non-linear. Branson's (1964) method, given in the Standard, also based on equilibrium and strain compatibility, is equivalent when the same values are used for the free shrinkage strain and the modulus of elasticity of the concrete. It is easier to use than the equilibrium method; however, the tensile stress-strain relationship of the concrete is not necessarily satisfied. In spite of this, it gives reasonable results when appropriate values are assumed for the free shrinkage strain and modular ratio.

L4.
The shrinkage deflection is directly proportional to the assumed free shrinkage strain. The free shrinkage strain depends on the concrete properties such as water/cement ratio, percent fines, entrained air, cement content and the curing conditions. A value of 800 $\mu\varepsilon$ may be used (see ACI 209R-92) if no other data are available.

L5.

The shrinkage deflection is not sensitive to the modular ratio because both the effective moment of inertia and the distance, y, vary with it. Shaker and Kennedy (1991) show that the effective modulus of elasticity, E_{ct}, decreases with increased tensile strain, ε_t, due to increased creep of the concrete. An approximate relationship, for 30-40 MPa concrete, from Shaker and Kennedy, is:

$E_{ct} = 8300 - 4800\, \sigma_{ct}; \quad 0.3 \leq \sigma_{ct} \leq 1.2$

which could be used with Figures L1 and L2 to compute the shrinkage deflection (see Kennedy and Brattland (1992)). (At the maximum tensile stress of 1.2 MPa, reached without cracking, the effective modulus is only about 2500 MPa or about 1/9 of the 28-day modulus in compression and results in a modular ratio of 80.) Changing the modular ratio from 20 to 80, for beams of usual proportions, decreases the shrinkage deflection in the order of 30% for a given free shrinkage strain. Modular ratios of 40-60 (see Ferguson (1958)) are considered appropriate and over this range the decrease in the shrinkage deflection is only about 15%.

L6.

Montgomery et al. (1983) give an example where the shrinkage deflections were excessive. Jent (1989) provides information on shrinkage effects on continuous composite beams.

L7. References

ACI 209R-92 American Concrete Institute (ACI), 1992, *Designing for the effects of creep, shrinkage, and temperature in concrete structures,* American Concrete Institute, Detroit, Michigan.

Branson, D.E. 1964, *Time-dependent effects on composite concrete beams,* Proceedings, Journal of the American Concrete Institute, 61:212-229.

Ferguson, P.M. 1958, *Discussion of Miller, L.A., Warping of reinforced concrete due to shrinkage,* American Concrete Institute Journal, Vol 30, No. 6, Part 2, 939-950.

Kennedy, D.J.L. and Brattland, A. 1992, *Shrinkage tests of two full-scale composite trusses,* Canadian Journal of Civil Engineering, (19)2:296-309.

Montgomery, C.J., Kulak, G.L., and Shwartsburd, G.,1983, *Deflection of a composite floor system,* Canadian Journal of Civil Engineering, (10)2:192-204.

Shaker, A.F. and Kennedy, D.J.L., 1991, *The effective modulus of elasticity of concrete in tension,* Structural Engineering Report 172, Department of Civil Engineering, The University of Alberta, Edmonton, Alberta.

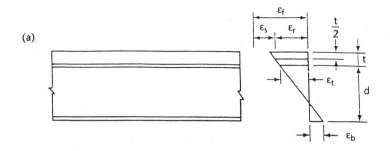

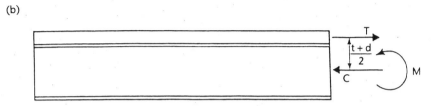

Figure L1
Composite Beam Subject to Shrinkage Forces:
(a) Shrinkage Strain; (b) Free-Body Diagram

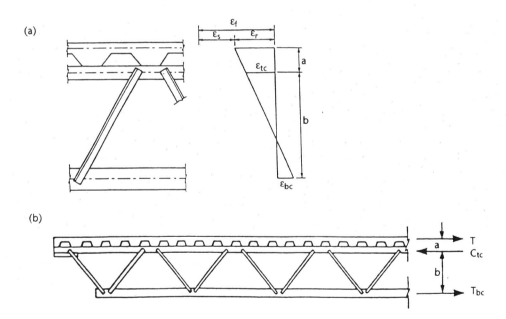

Figure L2
Composite Truss Subject to Shrinkage Forces:
(a) Shrinkage Strain; (b) Free-Body Diagram

Appendix M
Design Requirements for Steel Plate Shear Walls

Note: *This Appendix is not a mandatory part of this Standard.*

M1. General

M1.1
A steel plate shear wall in a multi-storey building (see Figure M1) consists of steel web plates connected to the beams and columns that are part of the frame of the building. The behaviour of the shear wall formed in this way is similar to that of a plate girder: the building columns form the flanges of the girder and the beams act as horizontal stiffeners.

M1.2
The web plates of the shear walls considered are assumed to be relatively thin. Lateral shears are carried by tension fields that develop in the web plates parallel to the directions of principal tensile stresses.

M2. Design Forces and Moments

M2.1
The design forces and moments for the members and connections for a shear wall may be determined by analyzing the wall, a panel at a time, using a plane frame computer program provided that:
(a) pin-ended connections are used to attach the ends of beams to columns; and
(b) the magnitudes of the vertical components of the tension fields do not vary by more than 25 percent between adjacent stories.

M2.2
Figure M2 shows a plane frame model for a typical interior panel of a shear wall. The web of the panel is represented by a series of inclined strips, which are modelled as pin-ended elements. A minimum of 10 strips, each of width equal to the strip spacing, is suggested. The beams and columns of each storey are considered to be pin-ended.

M2.3
The horizontal beams of an interior panel may be assumed to be infinitely rigid under flexural loads.

M2.4
The columns are assumed to have their actual stiffnesses.

M2.5
When moment connections are used between beams and columns or when the vertical components of the tension fields vary by more than 25 percent between adjacent stories, the entire shear wall shall be modelled. The web of each panel is represented by a series of inclined, pin-ended elements. The horizontal beams and columns are assumed to have their actual stiffnesses. Continuity is considered at the connections between beams and columns, and at column splices, as appropriate.

M3. Angle of Inclination

M3.1
In the analytical model shown in Figure M2, which assumes that the beams have pinned ends, the angle of inclination, α, for the inclined truss members may be determined from

$$\tan^4\alpha = \frac{\dfrac{2}{wL} + \dfrac{1}{A_c}}{\dfrac{2}{wL} + \dfrac{2h}{A_bL} + \dfrac{h^4}{180\,I_cL^2}}$$

where
L and h are the panel dimensions shown in Figure M2
w = thickness of web plate
A_c = cross-sectional area of a column
A_b = cross-sectional area of a beam
I_c = moment of inertia of a column

M3.2
As an approximation when the entire shear wall is modelled in the analysis, the angle of inclination, α, for the inclined truss members in each panel may be determined as indicated in Clause M3.1.

M4. Steel Web Plates
The factored tensile resistance of the inclined web plate strips shall be calculated in accordance with the requirements of Clause 13.2.

M5. Horizontal Beams

M5.1
When the vertical component of the tension fields varies by more than 25% between adjacent stories, the bending moment in the beam due to the tension fields shall be calculated. The effect is additional to other bending moments and axial forces acting on the beam.

M5.2
The horizontal beams shall be proportioned to resist both bending moments and axial compressive forces in accordance with the requirements of Clause 13.8. Class 1 or 2 sections shall be used for horizontal beams.

M6. Columns

M6.1
In addition to other bending moments and axial forces acting on the columns resulting from their overall behavior in the structure, bending moments and axial forces resulting from the tension field action shall be considered.

M6.2
The columns shall be proportioned to resist both bending moments and axial forces in accordance with the requirements of Clause 13.8 or 13.9, as appropriate. Class 1 sections shall be used for columns.

M7. Anchorage of Steel Web Plates

M7.1
The vertical components of the tension fields shall be anchored at the extreme top and bottom panels of a shear wall.

M7.2
At the top panel of the shear wall, the tension field may be anchored internally by providing a stiff horizontal beam.

M7.3
At the bottom panel of the shear wall, the tension field may be anchored by attaching the web panel directly to the substructure, or by providing a stiff horizontal beam.

M7.4
The horizontal components of the tension field in the extreme bottom web panel shall be transferred into the substructure by appropriate means.

M8. Connections
Web plate panels shall be connected to the surrounding beams and columns in accordance with Clause 13.11 or 13.13. The factored ultimate tensile strength of the web plate strips shall be developed.

M9. Seismic Design Requirements

M9.1
Steel plate shear walls used in seismic regions shall be designed according to one of the following three categories.

M9.2
For ductile steel plate shear walls ($R = 4.0$ in Clause 4.1.9 of the *National Building Code of Canada, 1995*), moment connections shall be provided between the beams and columns, and these shall be proportioned in accordance with the requirements of Clause 27.2.

M9.3
For nominally ductile steel plate shear walls ($R = 3.0$), moment connections shall be provided between the beams and columns, and these shall be proportioned in accordance with the requirements of Clause 27.3.

M9.4
Ordinary steel plate shear walls ($R = 2.0$) may be proportioned in accordance with the provisions of this Appendix, without any other special requirements.

M10. References

Kulak, G.L. 1985, *Behaviour of Steel Plate Shear Walls,* Proceedings, The 1985 International Engineering Symposium on Structural Steel, Chicago.

Kulak, G.L. 1986, *Unstiffened Steel Plate Shear Walls: Static and Seismic Behaviour, in Steel Structures,* Recent Research Advances and Their Applications to Design, Edited by M.N. Pavlovic, Elsevier Applied Science Publishers, London.

Kulak, G.L. 1991, *Structures Subjected to Repeated Loading — Stability and Strength,* Chapter 9 (Unstiffened Steel Plate Shear Walls), Elsevier Applied Science Publishers, London.

Tromposch, E.W., and Kulak, G.L. 1987, *Cyclic and Static Behaviour of Thin Panel Steel Plate Shear Walls,* Structural Engineering Report No. 145, Department of Civil Engineering, University of Alberta.

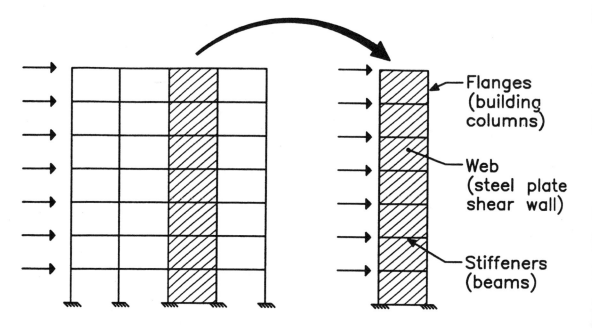

**Figure M1
Steel Plate Shear Wall**

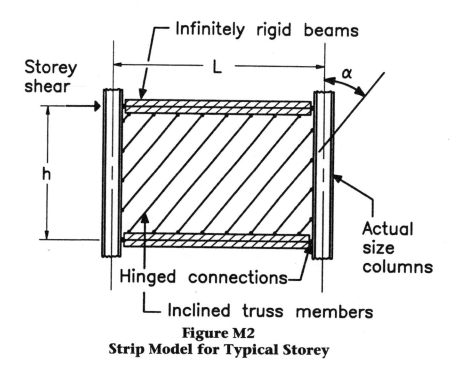

**Figure M2
Strip Model for Typical Storey**

PART TWO
CISC COMMENTARY ON CAN/CSA-S16.1-94

Preface

This Commentary has been prepared by the Canadian Institute of Steel Construction in order to provide guidance as to the intent of various provisions of the National Standard of Canada, written by CSA, CAN/CSA-S16.1-94, "Limit States Design of Steel Structures". This Commentary and the information contained in the references cited provide an extensive background to the development of the Standard, its technical requirements, and the new or revised sections of the 1994 edition.

The Institute gratefully acknowledges the efforts of the various members of the S16 Technical Committee for their valuable contributions to the commentary. D. J. L. Kennedy, Chairman of the S16 Committee, worked with the Editor on a clause-by-clause basis in preparation of this Commentary.

The information contained in the Commentary is provided by the Institute. It is not to be considered the opinion of the CSA Committee nor to detract from that Committee's responsibility and authority insofar as interpretation and revision of the Standard is concerned. For information on requesting interpretations, see Note (5) to the Preface of CAN/CSA-S16.1-94.

This Commentary is provided by the Institute as a part of its commitment to the education of those interested in the use of steel in construction. The Institute assumes no responsibility for errors or oversights resulting from the use of the information contained herein. Anyone making use of the contents of this Commentary assumes all liability arising from such use. All suggestions for improvements of this Commentary will receive full consideration for future printings.

M. I. Gilmor
Editor

Introduction

Since 1974, when the Canadian Standards Association first introduced the limit states design standard for structural steel, S16.1-1974 "Steel Structures for Buildings—Limit States Design", the Standard has undergone a number of technical improvements but its major requirements have remained virtually unchanged. However, with the introduction of the 1989 edition, a number of more significant changes were introduced, in part reflecting the maturing of the Standard but also the acquisition of more detailed information on behaviour during the intervening years. The 1994 edition continues this process with the refining of some the requirements and the addition of a new lateral load resisting system, the steel plate shear wall.

Specifically the following changes have been adopted for CAN/CSA-S16.1-94:

1) The use of other standards for design is specifically prohibited (Clause 1.4);

2) Load combinations involving earthquake have been changed significantly to be consistent with the new requirements of the 1995 National Building Code of Canada;

3) Clause 8.6.2 requires the asymmetry of load, of geometry, or both to be considered;

4) The effective length of a compression member is now clearly related to the failure mode;

5) The diameter-to-thickness ratio limits for circular hollow sections have been expanded to cover both members in compression and members in flexure;

6) Specific shear lag reduction factors for angles connected by one leg are introduced;

7) Shear lag requirements for welded members have been expanded;

8) A double exponential expression is used for the column curves replacing the series of polynomial expressions;

9) WWF columns have been assigned to the higher column curve;

10) Torsional-flexural buckling of compression members is now to be checked using a new Appendix D;

11) Biaxial bending has been added for flexural members;

12) For beam-columns, M_f is defined as the maximum moment;

13) The resistance expressions for bearing have been revised in that (a) for rollers and rockers, grooved supporting plates are accommodated and, (b) for bolted joints, the end distance rule has been removed with the user directed to the clauses on tension and net section;

14) The shear-tension interaction equation for bolts has been simplified to the basic elliptical expression;

15) The clauses on welding have been entirely rewritten. Resistance of the base metal is defined in terms of the ultimate tensile strength of the plate but with the resistance factor for welds;

16) The resistance of fillet welds is written as a function of the angle of the load to the axis of the weld;

17) Tee joints made with PJPG welds and reinforcing fillet welds are addressed;

18) Clause 13.14 refers to the static and seismic requirements for steel plate shear walls introduced in Appendix M;

19) The fatigue clauses, stress ranges and categories have been revised to be consistent with those for bridges in the new Canadian Highway Bridge Design Codes;

20) A higher stress cycle limit of 50 000 cycles has been introduced;

21) The moment resistance of girders with thin webs is reduced further when an axial compressive load is present;

22) The model for computing shrinkage deflections of composite beams has been revised;

23) For composite beams, a clause on stud spacing has been added and, for unshored beams, the serviceability limitation on stress in the bottom flange is raised from $0.90F_y$ to F_y;

24) For composite columns, a clause on the bending resistance of rectangular sections has been added;

25) For built-up compression members, the effective slenderness ratio has been extended to battened members replacing the historical slenderness limits and spacing requirements;

26) Bracing assemblies are required to have both sufficient strength and stiffness dependent, in part, on the number of braced points. An attenuation rule is also provided for the accumulation of bracing forces;

27) In Clause 27, Seismic Design Requirements, some relaxation of the previous requirements for structures in zones of low seismic risk has occurred, the requirements for eccentric braced frames have been moved into the body of the Standard. For some zones, the brace connection design force has changed to the full yield load;

28) Appendix B and C have been revised to be consistent with Clause 9;

29) Appendix D now deals with torsional-flexural buckling;

30) Appendix G has been substantially updated based on new information of Allen and Murray;

31) Appendices K and L reflects the changes made in the body of the Standard; and,

32) Appendix M is a new appendix dealing with steel plate shear walls.

Background

Standards based on working (allowable) stress, with the exception of a few clauses on plastic design, required the calculation of stresses at various points in the structure and a comparison of these stresses with allowable stresses. The allowable stresses were usually established as some portion of the yield point of the material. Elements or members subjected to compression were examined for stability, usually expressed as an allowable compressive stress. A limit states standard takes a different approach. To serve their intended purposes, all building structures must meet the requirement that the occurrence of various types of collapse or unserviceability are limited to a sufficiently small value. Limit states are those conditions of the structure corresponding to the onset of the various types of collapse or unserviceability. The conditions associated with collapse are the ultimate limit states, those associated with unserviceability are the serviceability limit states and that associated with fatigue is the fatigue limit state.

In limit states design, the capacity or performance of the structure or its components is checked against the various limit states at certain load levels. For the ultimate limit states of strength and stability, for example, the structure must retain its load carrying capacity up to factored load levels. For serviceability limit states, the performance of the structure at specified load levels must be satisfactory. (Specified loads are those prescribed by the Regulatory Authority. A factored load is the product of a specified load and its load factor.) Examples of the serviceability requirement include prevention of damage to non-structural elements, and restrictions on deflections, permanent deformations, slip in slip-critical connections, and acceleration under vibratory motion. For fatigue limit states, the stress ranges for the loads applied to the structure over its useful life must not exceed the prescribed stress ranges.

Because both the loads acting on a structure and the resistance of a member can only be defined statistically, the "factor of safety", commonly used in working stress standards, is divided, in limit states standards, into two parts—a load factor and a resistance factor. A load factor (α) is applied to the specified load to take into account the fact that loads higher than those anticipated may exist and also to take into account approximations in the analysis of the load effects. A resistance factor (ϕ) is applied to the nominal member strengths, or resistances (R), to take into account that the resistance of the member due to variability of the material properties, dimensions and workmanship may be different than anticipated, and also to take into account the type of failure and uncertainty in the prediction of the resistance. An advantage, therefore, of limit states design is that the factors assigned to loads arising from different sources can be related to their uncertainty of prediction, and the factors assigned to different members can be related to their reliability and to the different types of failure. Thus, a greater degree of consistency against failure can be obtained (Kennedy 1974; Allen 1975; Kennedy et al. 1976).

For the failure of structural steel members by yielding, the resistance factor is taken to be 0.90 (Kennedy and Gad Aly 1980). To maintain simplicity in design, the resistance formulas for buckling or other types of member failure have been adjusted so that a uniform resistance factor, $\phi = 0.90$, can be used and yet provide the necessary safety required in the definition of the resistance factor. For example, the resistance formulas for tension, clauses 13.2(a)(ii) and (iii), provide a higher safety factor against fracture across the net section than against yielding in the gross section. The only exceptions to $\phi = 0.90$ are:

- Bolts in bearing-type connections, $\phi_b = 0.67$, to ensure that the connectors will be stronger than the members being joined;

- Crushing resistance of concrete (for composite construction), $\phi_c = 0.60$, which is consistent with CAN/CSA-A23.3-94 and takes into account the greater strength variability and the brittle type of failure associated with concrete;

- Shear connectors (for composite construction), $\phi_{sc} = 0.80$; and,

- Welds, to ensure that the weld is stronger than the member joined, $\phi_w = 0.67$.

Probabilistic studies (Allen 1975) show that consistent probabilities of failure are determined for all dead-to-live load ratios when a dead load factor of 1.25 and a live load factor of 1.50 are used. For certain types of structures, if there is a high degree of uncertainty in the loads, the designer may elect to use larger load factors. Those given in S16.1 (taken from the NBCC, 1995) are minima. However, in situations where the dead load and the live loads are counteractive it is important that α_D be taken as 0.85, or less as appropriate.

Kennedy (1974) and Allen (1975) provide considerably more information on the type of probabilistic, calibration, and design studies that were performed while developing the limit states standard. The National Building Code of Canada (NBCC 1995a) contains a more extensive discussion on limit states design. Kennedy and Gad Aly (1980) and Baker and Kennedy (1984) provide information on the statistical determination of the resistance factors (ϕ).

In the Commentary clauses that follow, the numbers and headings used refer to the relevant clause numbers and headings of National Standard of Canada CAN/CSA-S16.1-94. This will be referred to simply as S16.1-94.

1. SCOPE AND APPLICATION

This Standard applies generally to steel structures, and structural steel components in other structures. The analysis, design, detailing, fabrication, and erection requirements contained in the Standard normally provide a satisfactory level of structural integrity for most steel structures.

Clause 1.2 states that requirements for some specific types of structures and members are given in other CSA Standards. Situations where additional requirements may be necessary are given in Clause 1.3. Commentary C of the "Commentaries on Part 4 of the National Building Code of Canada" provides numerous references to the technical literature on the topic of structural integrity.

Clause 1.4 prohibits the substitution of any other structural steel design standard (e.g.. CSA S16 1969 or AISC) for S16.1-94. The treatment of a number of important technical issues relating to safety, such as P–Δ effects, beam-columns, ductility of members and connections for earthquake loads, is either not covered or is treated in a manner inconsistent with the intent of S16.1-94.

Clause 1.4 notes that the designer has the freedom (subject to approval from the Regulatory Authority) to use methods of design or analyses in lieu of the formulas given in the Standard. It is required, of course, that the structural reliability provided by the alternative (as measured by the reliability index, for example) be equal to, or greater, than those in the Standard. An example of such a rational method would be the design of stub-girders using the method set out by Chien and Ritchie (1984) based on tests (Bjorhovde and Zimmerman 1980; Kullman and Hosain 1985; Ahamd et al. 1990).

2. DEFINITIONS AND SYMBOLS

2.3 Units

All coefficients appearing in equations and expressions in this Standard are consistent with forces measured in Newtons and lengths in millimetres. While most coefficients are themselves non-dimensional, in Clause 15.9, the coefficients 150 and 300 have units of $\sqrt{MPa}$ and, in Clause 17.9.9, the coefficient 2.76 has units of megapascals.

3. REFERENCE PUBLICATIONS

The Standards listed are the latest editions at the time of printing. When reference is made to undated publications in specific clauses of this Standard, it is intended that the latest edition and revisions of these publications be used.

4. DRAWINGS

4.1 Design Drawings

4.1.1 Structural steel design drawings, by themselves, should show all member designations, axis orientations, and dimensions needed to describe the complete steel structure. It should not be necessary, in order to ascertain information on structural steel components, to refer to drawings produced for the use of other trades.

4.1.2 The development of adequate connections for structural members requires that the design engineer determine the shears, moments and axial forces resulting from the governing load combinations for which the connection must be designed. For complex combinations, a useful presentation of this information may be to list the maximum value of each (e.g..

shear, moment, and axial force), along with the values of the others which coincide with that maximum. The principle is to provide co-existent sets of forces so that free body diagrams can be identified to ensure that governing forces are transmitted through connections and panels.

4.1.4 Structural stability, a fundamental consideration of design, extends to the behaviour of elements within a member as well as to the functioning of members in total. Stabilizing components are needed to achieve both the correct local behaviour and the correct overall behaviour anticipated by the design. Therefore, the design engineer must define bracing, stiffeners, and reinforcement that are required to prevent failure due to instability. An example is web reinforcement in moment connections to prevent local instability.

4.2 Fabrication and Erection Documents

Although five types of documents are identified in the Standard, many structures which use pre-engineered connections from company or industry sources require only *shop details* and *erection diagrams*.

4.2.1 Connection Design Details

Connection design details, which often take the form of design brief sheets, typically show the configuration and details of non-standard connections developed for specific situations. They are submitted to the design engineer for review to confirm that the structural intent has been understood and met, and they may be stamped by a professional engineer when appropriate. Drafting technicians use *connection design details* to prepare *shop details*.

4.2.2 Shop Details

Shop details frequently take the form of traditional shop drawings and are used to provide the fabrication shop with all the specific information required to produce the member. They are submitted to the design engineer for review to confirm that the structural intent has been understood and met. *Shop details* are not stamped by a professional engineer because they generally do not contain original engineering.

4.2.3 Erection Diagrams

Erection diagrams convey information about the permanent structure that is required by field personnel in order to assemble it. They are submitted to the design engineer for review, but are not stamped by a professional engineer because original engineering is generally not added by the fabricator.

4.2.4 Erection Procedures

Erection procedures outline methods and equipment, such as falsework and temporary guying cables, employed by the steel erector to assemble the structure safely. They may be submitted to the design engineer for review, and may be stamped by a professional engineer when appropriate.

4.2.5 Field Work Details

Field work details are drawings which describe modifications required to fabricated members. The work may be done either in the shop or at the job site depending on circumstances. When extra material is involved, *field work details* effectively become *shop details*. They are submitted to the design engineer for review.

5. MATERIALS: STANDARDS AND IDENTIFICATION

The design requirements contained in S16.1-94 have been developed on the assumption that the materials and products which will be used are those listed in Clause 5. These materials and products are all covered by standards prepared by the Canadian Standards Association (CSA) or the American Society for Testing and Materials (ASTM).

The standards listed provide controls over manufacture and delivery of the materials and products which are necessary to ensure that the materials and products will have the characteristics assumed when the design provisions of S16.1-94 were prepared. The use of materials and products other than those listed is permitted, provided that approval, based on published specifications, is obtained. In this case, the designer should assure himself that the materials and products have the characteristics required to perform satisfactorily in the structure. In particular, ductility is often as important as the strength of the material. Weldability and toughness may also be required in many structures.

The values for yield and tensile strength reported on mill test reports are not to be used for design. Only the specified minimum values published in product standards and specifications may be used. This has always been implicit in the requirements of the Standard by definition of the terms F_y and F_u but is now explicitly stated. Furthermore, when tests are done to identify steel, the *specified minimum* values of the steel, once classified, shall be used as the basis for design.

When, however, sufficient representative tests are done on the steel of an existing structure to be statistically significant, those statistical data on the variation of the material and geometric properties may be combined with that for test/predicted ratios available in the literature to develop appropriate resistance factors. This is by no means equivalent, for example, to substituting a new mean yield stress for a specified minimum value as the new reference value and the bias coefficient must be established. It could well be that, although a higher mean value of the yield stress is established, the bias coefficient, depending as it does on the reference value, would be less. It would be expected that the coefficient of variation for the material properties in particular, derived for the steel in a single structure, would be less than for steel in general.

6. DESIGN REQUIREMENTS

This clause clearly distinguishes between those requirements which must be checked using specified loads (the fatigue and serviceability limit states) and those which must be checked using factored loads (the ultimate limit states). Many of the serviceability requirements (deflections, vibrations, etc.) are stipulated qualitatively and guidance, in quantitative form, is provided in Appendices. Thus, the designer is permitted to use the best information available to him in order to satisfy the serviceability requirements, but is also provided with information that the S16.1 Committee considers to be generally suitable, when used with competent engineering judgement.

6.1.2 A clause on structural integrity acts as a reminder that measures may be necessary to guard against progressive collapse as a result of a local incident. Being inherently ductile, steel structures have generally had an excellent record of behaviour when subjected to unusual or unexpected loadings. However, connection details are particularly important in achieving this ductile behaviour. Details which rely solely on friction due to gravity to provide nominal lateral force resistance may have little or no resistance to unanticipated lateral loads if subjected to abnormal uplift conditions and should be carefully evaluated for such an eventuality or completely avoided.

6.2.1.3 Even though deflections are checked under the actions of specified loads, additional loading may result from ponding of rain on roofs, or the ponding of finishes or concrete, while in the fluid state, on floors or roofs. Such additional loads are to be included in the design of the supporting members under ultimate limit states as required by Clause 7. More information on ponding is available in the National Building Code of Canada (NBCC 1995c).

6.2.3.2 Additional information on vibrations of floor systems may be found in Allen (1974), Murray (1975), Allen and Rainer (1976), Rainer (1980), Allen et al. (1985), Allen and Murray (1993).

7. LOADS AND SAFETY CRITERION

This clause sets forth the fundamental safety criterion that must be met, namely:

Factored Resistance ≥ Effect of Factored Loads,

or

$$\phi R \geq \alpha_D D + \gamma \psi (\alpha_L L + \alpha_W W + \alpha_T T)$$

For all load combinations including those involving earthquake, the expression for the *effect of factored loads* is identical with that given in Part 4 of the National Building Code of Canada (NBCC 1995c) as are the values given for the various load factors (α), load combination factors (ψ) and importance factors (γ). This information has been included in Clause 7 for the convenience of designers using S16.1-94 and to emphasize the fact that the load and resistance factors are inextricably linked and must be considered or developed in conjunction with one another.

Dead loads are to include the additional mass of construction materials that will be built into a structure as a result of deflections of supporting members, such as a concrete floor slab placed to a level plane but supported by members that were not cambered and that deflect under the weight of the concrete.

The *factored resistance* is given by the product ϕR where ϕ is the resistance factor and R is the nominal member strength, or resistance. The *factored resistances* of various types of members are given in Clauses 13, 15, 16, 17, 18, 21, 26 and 27.

8. ANALYSIS OF STRUCTURE

Clause 8 permits the use of two basic types of construction—"continuous" and "simple"—both of which are defined. In recognition of previous successful practice, a special form of "simple" construction is permitted. In this form of construction, a building frame may be designed to support gravity loads on the basis of "simple" construction and to support lateral loads due to wind or earthquake through the provision of moment-resisting joints. A number of limitations are imposed in Clause 8.3.2 if this method is used. In particular, the requirement for solid webs, (8.3.2(b)), eliminates the use of open-web steel joists as connected members of the frame (Nixon 1981).

The limitations are intended to ensure that the moment-resisting joints designed nominally for wind or earthquake moments alone have both the strength and ductility necessary to accommodate the "overload" which will result if factored gravity and lateral loads act concurrently. It is assumed that, if the connection has adequate capacity for inelastic rotation when subjected to the first application of factored gravity and lateral loading, under subsequent loading cycles the connection will behave elastically, although it will have a permanent inelastic deformation (Sourochnikoff 1950, Disque 1964). Such an assumption is valid except in joints where load fluctuation would create alternating plasticity in the connection (Popov and Pinkney 1969).

Clause 8 also permits the use of the two general methods of analysis—elastic and plastic analysis. Methods of elastic analysis are familiar to most designers.

8.5 Plastic Analysis

The use of plastic analyses at the factored load levels to determine the forces and moments throughout a structure implies that the structure achieves its limiting load capacity when sufficient plastic hinges have developed to transform the frame into a mechanism. As successive plastic hinges form, the load-carrying capacity of the structure increases above that corresponding to the formation of the initial plastic hinge until a mechanism develops. To achieve this, the members in which the hinges form before the mechanism develops must be sufficiently stocky (Class 1 sections) and well braced so that inelastic rotations can occur without loss of moment capacity.

Deflections at the specified load level are, of course, limited in accordance with Clause 6.2.1. Plastically designed structures are usually "elastic" at specified load levels i.e. no plastic hinges have formed. Therefore, the deflections would generally be computed on the basis of an elastic analysis.

8.5(a) Material

The plastic method relies on certain basic assumptions for its validity (ASCE 1971). Therefore, restrictions are imposed to preserve the applicability of the plastic theory. The basic restriction (Clause 8.5(a)) that the steel exhibits significant amounts of strain-hardening is required to ensure that satisfactory moment redistribution will occur (Adams and Galambos 1969). This behaviour should exist at the temperatures to which the structure will be subjected in service. Also, although not explicitly stated, plastically designed structures usually entail welded fabrication, and therefore the steel specified should also be weldable. At normal temperatures all the steels referred to in Clause 5.1.2 should be satisfactory except for CSA G40.21-M, 700 Q and 700 QT steels, for which $F_y > 0.80\ F_u$.

8.5(b) Width–Thickness Ratios

In order to preclude premature local buckling, and thus ensure adequate hinge rotation, compression elements in regions of plastic moment must have width–thickness ratios no greater than those specified for Class 1 (plastic design) sections in Clause 11.2. Although both Class 1 and Class 2 sections can attain the fully plastic moment, only Class 1 sections will maintain this moment through the rotation necessary for redistribution of moments implicit in the plastic method of analysis.

8.5(c) Lateral Bracing.

The lateral bracing requirements are considerably more severe than those for structures designed on the basis of an elastic moment distribution because of the rotation needed at the location of the plastic hinges. Two values of the critical length between braces, L_{cr}, are specified, one for the case where the moment gradient is pronounced, the other (more stringent) for the case of uniform, or near uniform moment. The moment gradient dividing the two cases has been selected as $\kappa = -0.5$ based on the test results of Lay and Galambos (1967). Both criteria should be applied with the more severe requirement governing the determination of L_{cr}.

Since the final hinge in the failure mechanism does not require rotation capacity, the bracing spacing limitations of this clause do not apply, and the elastic bracing requirements of Clause 13.6(a) may be used.

Lateral bracing is required to prevent both lateral movement and twisting at a braced point. Lateral bracing is usually provided by floor beams or purlins which frame into the beam to be braced. These bracing members must have adequate axial strength and axial stiffness to resist the tendency to lateral deflection. These requirements are given in Clause 20.2. Further information on the design of bracing members is given in Lay and Galambos (1966). When the bracing member is connected to the compression flange of the braced member, the brace should possess bending stiffness to resist twisting of the braced member. Some information on the bending stiffness of braces is given in Essa and Kennedy (1995).

A concrete slab into which the compression flange is embedded or to which the compression flange is mechanically connected, as in composite construction, or metal decks welded to the top flange of the beam in the positive moment region, generally provide sufficient restraint to lateral and torsional displacements. When the lateral brace is connected to the tension flange, provision must be made for maintaining the shape of the cross-section and for preventing lateral movement of the compression flange. This can be accomplished with either diagonal struts to the compression flange or adequately designed web stiffeners.

8.5(d) Web Crippling.

Web stiffeners are required on a member at a point of load application where a plastic hinge would form. Stiffeners are also required at beam-to-column connections where the forces developed in the beam flanges would either cripple the column web or, in the case of tension loads, distort the column flange with incipient weld fracture. The rules for stiffener design are given in Clause 21.3. See ASCE (1971) for further details of stiffeners and Fisher *et al.* (1963) for special requirements pertaining to tapered and curved haunches.

When the shear force is excessive, additional stiffening may be required to limit shear deformations. The capacity of an unreinforced web to resist shear is taken to be that related to an average shear yield stress based on the Huber-Henckey-von Mises criterion of $F_y/\sqrt{3}$. For an effective depth of the web of a rolled shape of about 95% of the section depth, Clause 13.4.1.2 gives

$$V_r = 0.95 \phi \, wd \, F_y/\sqrt{3} = 0.55 \, \phi \, wd \, F_y$$

At beam-to-column connections, when the shear force exceeds that permitted above, the excess may be carried by providing doubler plates to increase the web thickness or by providing diagonal stiffeners (Figure 2–1). The force in the beam flange that is transferred into the web as a shear is closely

$$V = M/d_b$$

Equating this to the shear resistance as given in Clause 13.4.1.2 (where now, $w = w_c$ and $d = d_c$), and solving for the required web thickness,

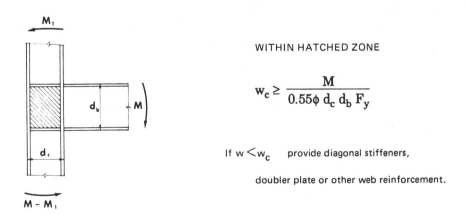

Figure 2–1
Web Thickness At Beam-to-Column Connections

$$w_c \geq \frac{M}{0.55\phi \, d_c \, d_b \, F_y}$$

If the actual web thickness is less than w_c, the required area of diagonal stiffeners may be obtained by considering the equilibrium of forces at the point where the top flange of the beam frames into the column. Using a lower bound approach, the total force to be transmitted ($V = M/d_b$) is assumed to be taken by the web and the horizontal component of the force in the diagonal stiffener:

$$V = M/d_b = 0.55 \, \phi \, w_c \, d_c \, F_y + \phi \, F_y \, A_s \cos \theta$$

where

A_s = cross sectional area of diagonal stiffeners

$\theta = \tan^{-1}(d_b/d_c)$

The required stiffener area is therefore

$$A_s = \frac{1}{\cos \theta}\left[\frac{M}{\phi \, F_y \, d_b} - 0.55 \, w_c \, d_c\right]$$

8.5(e) Splices

The bending moment diagram corresponding to the failure mechanism is the result of moment redistribution that occurred during the plastic hinging process. For example, points of inflection in the final bending moment distribution may have been required to resist significant moments to enable the failure mechanism to have developed (Hart and Milek 1965). To ensure that splices have sufficient capacity to enable the structure to reach its ultimate load capacity, a minimum connection requirement of $0.25 \, M_p$ is specified in Clause

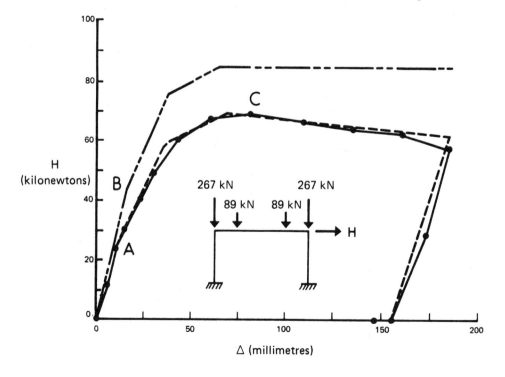

Figure 2–2
Observed And Predicted Load–Deflection Relationships

8.5(e). Also, at any splice location, the moments corresponding to various factored loading conditions must be increased by 10% above the computed value. The splice is then designed either for the larger of the moments so increased or for the minimum requirement of 0.25 M_p.

8.5(f) Impact and Fatigue

The use of moment redistribution to develop the strength of the structure corresponding to a failure mechanism implies ductile behaviour. Members which may be repeatedly subjected to heavy impact and members which may be subject to fatigue should not be designed on the basis of a plastic analysis because ductile behavior cannot be anticipated under these conditions. Such members, at least for the present, are best proportioned on the basis of an elastic bending moment distribution.

8.5(g) Inelastic Deformations

For continuous beams, inelastic deformations may have a negligible effect on the strength of the structure. For other types of structures, in particular multi-storey frames, these secondary effects may have a significant influence on the strength of the structure (ASCE 1971).

In the structure shown inset in Figure 2–2, the secondary effects have reduced the lateral load carrying capacity (while maintaining the same vertical load) by approximately 25% (ASCE 1971; Adams 1974). The first plastic hinge formed at stage A in this structure while the ultimate strength (considering moment redistribution) was not attained until stage C. The inelastic deformations between these two stages have reduced the overall strength of the structure. Clause 8.6 requires that the sway effects produced by the vertical loads be accounted for in design. Therefore Clause 8.5(g) requires that, in a structure analyzed on the basis of a plastic moment distribution, the additional effects produced by inelastic sway deformations are accommodated. In most cases the actual strength of the structure can only be predicted by tracing the complete load–deflection relationship for the structure or for selected portions (Beedle et al. 1969). Methods are available to perform this type of design. For braced multi-storey frames, however, simpler techniques have also been developed (AISI 1968).

8.6 Stability Effects

Clause 8.6 recognizes that all building structures are subjected to sway deformations. The vertical loads acting on the deformed structure produce secondary bending moments in the case of a moment resisting frame, or additional forces, in a vertical bracing system. These additional moments or forces (the stability effects) reduce the strength of the structure, as shown for a moment-resistant frame in Figure 2–2. In addition, bending moments and deflections, which exceed those predicted by a first order analysis, are produced at all stages of loading (Adams 1974). Similar effects are produced in structures containing a vertical bracing system, as shown in Figure 2–3 where the steel frame is linked to a shear wall (Adams 1974).

8.6.1

Again in the 1994 edition as was the case in the 1989 edition, the designer must account for the sway effects directly. This is done by (1) performing a second-order geometric analysis for the moments and forces, or (2) accounting for these effects by (a) amplifying the first order elastic translational moments by the factor U_2, or (b) performing an analysis according to Appendix J. In all cases, the lateral deflections of the structure under the translational loads must be determined. It is not acceptable to base the design of a moment resisting frame on a portal or cantilever type analysis to determine the moments without regard to the lateral deflections, although such methods are valuable for preliminary design.

Computer programs are available to perform analyses based on equilibrium of the deformed structure (Logcher et al. 1969; Galambos 1968). With this type of program, the additional moments or forces generated by the vertical loads acting on the displaced struc-

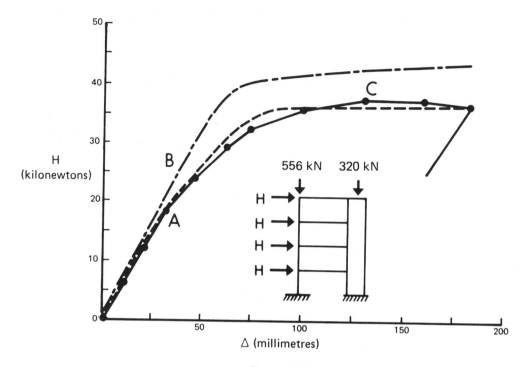

**Figure 2-3
Load-Deflection Relationships**

ture (the so-called P∆ effect) are taken into account. In addition, most second-order programs also account for the reduction in column stiffnesses, caused by their axial loads (Galambos 1968).

The second approach is simply to amplify the results of a first-order analysis to include the P∆ effects. It is implied in this approach that the change in individual member stiffnesses will be negligible. Adams (1974) suggested a simple check to ensure that this assumption is justified. The Standard, in general, limits this approach to structures for which the elastic second-order effects are less than 40% of the first-order effects. If the amplification factor exceeds this value, the designer may either increase the lateral stiffness of the structure to limit the amplification factor to 40%, or carry out a second-order elasto-plastic analysis. The limiting value of 1.4 for U_2 for an elastic second-order analysis may, however, be exceeded provided that the structure, taking residual stresses into account, remains elastic. The magnitude and pattern of residual stress are shape dependent (Galambos, 1988). For rolled I-shaped sections used as columns, a residual compressive stress of 90 MPa may conservatively be used.

In amplifying the elastic, first-order effects, two equivalent methods are given. In the first method, the first-order moments or forces are amplified by the factor U_2. The value of

$$U_2 = \frac{1}{1 - \frac{\Sigma C_f \Delta}{\Sigma V_f h}}$$

is, in fact, the limit of the so called P∆/h shears as determined by the iterative procedure, the second method given in Appendix J of the Standard (Kennedy *et al.* 1990).

As a variation on the method of Appendix J, estimated deflections may be used to calculate initial sway forces (Wood *et al.* 1976). The structure is then analyzed under the lateral load caused by wind or earthquake plus the sway forces. If the resulting deflections

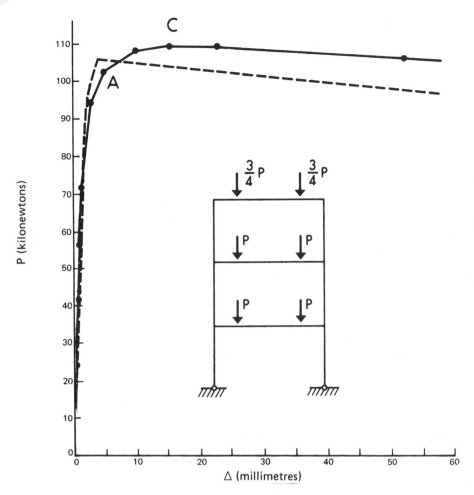

Figure 2–4
Load-Deflection Relationship—Vertical Load Only

are less than those assumed for the initial estimate of the PΔ effects then these effects have been over-estimated and (if the designer is satisfied that the situation is acceptable) the iterative process is not required.

It has been observed (Galambos 1988) that most regular structures meeting reasonable deflection limits under the specified loads will be adequately designed using the PΔ approach.

Because of its inherent simplicity, it is expected that most designers will follow the first method, using the amplification factor, U_2, to amplify the translational moments or forces. The moments and forces due to gravity loads are, of course, not amplified.

Because the designer has now included the sway effects in the analysis, either by doing a second-order analysis or by one of the two approximate methods, no further consideration of the sway effects is necessary.

8.6.2

This clause, although rewritten, is technically equivalent to the same clause in the 1989 edition. The term "pseudo-lateral load" in the latter has been replaced with the term "notional lateral load" as both this term and concept are now gaining international acceptance.

Although the function of the notional lateral load is to transform the bifurcation problem of sway buckling into a bending strength problem and therefore should really be applied to all load combinations when the potential for sway buckling exists and not just gravity load combinations, it is applied, at least for the time being, and as it was in the 1989 edition, only to combinations of gravity loads. The major reason for not applying it universally is that a smaller lateral load than $0.005\Sigma P$ should be used to limit the degree of conservatism under these circumstances. This in turn arises from the fact that the value of $0.005\Sigma P$ was obtained by calibrating the buckling strength of a flagpole column with effective length of 2L to the bending strength of a pin-ended column of actual length L when acted upon by the axial load and the notional load simultaneously. The compressive resistance in the latter case is based on the actual length, L.

The flagpole column is bent in single curvature whereas many columns in actual structures have some degree of double curvature. Consider now a sway column with complete fixity at both ends. It has very significant double curvature and an effective length of L. The sway buckling strength is now equal to the bending strength of a pin-ended column of the actual length with no notional lateral load because the effective length for buckling is equal to the actual length, L. These two cases show that the notional load required to transform the bifurcation problem of sway buckling into a bending strength problem depends on the end conditions in the actual structure and is greater when the degree of restraint is less. On the average, therefore, the notional load should be less than $0.005\Sigma P$ and should be applied to all load combinations.

By applying the notional load only to gravity load combinations in the 1994 edition, the degrees of conservatism and non-conservatism are more balanced than would be the case were a notional lateral load of $0.005\Sigma P$ (however much conceptually correct but of too large a value) applied universally. (The non-conservatism exists chiefly for beam-columns with high axial loads and relatively low moments in frames for load combinations with lateral loads and when no notional load is applied.)

The use of the notional lateral load remains of particular importance for structures subject to gravity loads only that may have insignificant lateral deflections and may only fail by elastic or inelastic sway buckling. Figure 2.4 shows a frame subject to vertical loads only. As the loads are increased, the effects of the vertical loads acting on whatever initial

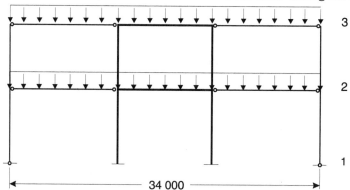

Loads	Specified		Factored Gravity			Notional Lateral Load
	DL	LL	DL	LL	Total	
Level 3	10.8	22.5	13.5	33.75	47.25	0.005(47.25 x 34) = 8.03 kN
Level 2	18.0	18.0	22.5	27.00	49.50	0.005(49.5 x 34) = 8.42 kN

Note: For complete analysis of this frame, see Kennedy, *et al*, 1990.

Figure 2–5

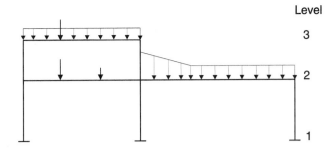

a) Asymmetrical frame with gravity loading

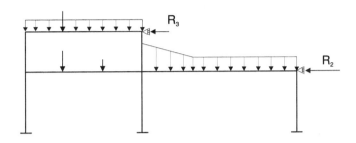

b) Computation of M_{fg}

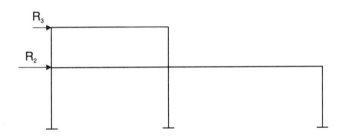

c) Computation of M_{ft} and Δ_f

Figure 2–6

imperfections exist due to fabrication and erection, lead to failure through instability, much the same as for the combined load case shown in Figures 2-2 and 2-3. The notional lateral loads of 0.005 times the factored gravity loads acting at each storey, as required by clause 8.6.2, simulate this condition. Cheong-Sait-Moy (1991) has shown that a notional lateral load of 0.0075 times the factored gravity load is adequate for frames with a U_2 factor of up to 3. In this Standard with a value of 0.005 the U_2 factor is limited to 1.4.

Figure 2-5 shows, for a frame subject to gravity loads only, the notional lateral loads that would be used to calculate the translational moments and forces for this load combination.

Clause 8.6.2 also requires that the asymmetry of the loading or structure or both should be considered. When either the gravity loads or the structure or both are asymmetric, horizontal reactions at floor levels are obtained when computing M_{fg}, defined as the first-order moment under factored gravity loads determined assuming that there is no lateral translation of the frame as shown in Figure 2-6. These horizontal reactions, when released by applying sway forces in the opposite direction, produce translational effects and must be considered for all valid load combinations, in addition to the notional lateral loads or the actual lateral loads as appropriate.

9. DESIGN LENGTH OF MEMBERS

9.1 Simple Span Flexural Members

For design purposes, it is usually convenient to consider the length of a member as equal to the distance between centres of gravity of supporting members. In most instances the difference resulting from considering a member to be that length rather than its actual length, centre-to-centre of end connections, is small. In some cases, however, there is sufficient difference to merit computing the actual length. Regardless of the length used for design, the actual connection detail may cause an eccentric load, or moment, to act on the supporting member and this effect must be taken into account.

9.3 Members in Compression

9.3.1 General

The unbraced length and the effective length factors may be different for different axes of buckling. Information about effective lengths is given in Galambos (1988) and Tall *et al.*(1974). Further guidance is provided in Appendices B and C of the Standard. The CSP, a column selection computer program available to designers (CISC 1987), contains routines for computing effective lengths, based on the principles outlined in Appendix C.

The last sentence of Clause 9.3.1 introduces the concept that effective length factors depend on the potential failure mode – how the member would fail if the forces (and moments) were increased sufficiently – as discussed in subsequent clauses.

9.3.2 Failure Modes Involving Bending In-Plane

Clause 9.3.2 now extends the case, given in the 1989 edition, when the effective length was taken as the actual length for members of sway frames (provided that the sway effects were considered in determining the stress resultants acting on the ends of the members) to members of braced frames as well. Therefore the approach has been made more consistent.

When the end moments and forces acting on a beam-column have been determined for the displaced configuration of the structure, that is to say, the sway effects have been included as required by Clause 8.6, the in-plane bending strength of the beam-column can be determined by analyzing a free-body of the member isolated from the remainder of the structure. In-plane displacements between the ends which contribute to failure arise from the end-moments and forces acting on the actual length. When the actual member length and the actual (or at least approximate) deflected shape are used, the analysis of the free-body will yield close to the correct member strength. Recourse to effective length factors is neither necessary nor appropriate.

When the actual member length is used together with the interaction expressions of Clause 13.8, the analysis is approximate and the in-plane member bending strength obtained will tend to be conservative. This simply arises because the value of the compressive resistance inherent in the interaction expression by using a length equal to the actual length (a K factor of 1.00) is that corresponding to single curvature buckling. For any other deflected shape, *having accounted for sway effects*, the compressive resistance is greater because the points of inflection of the deflected member shape are *less* than the member length apart. Under these circumstances, a better estimate of the strength, as is indeed permitted under Clause 1.4, can be obtained when the compressive resistance is based on the actual distance between points of inflection. Inelastic action of the member in the structure, however, may make this determination odious. Therefore the relatively simple but sometimes conservative approach given in the Standard which obviates the use of effective length factors is presented as the usual procedure.

9.3.3 Failure Modes Involving Buckling

The compressive resistance of an axially loaded column depends on its end restraints as does the out-of-plane buckling resistance of a beam-column under uniaxial strong axis bending. The failure is a bifurcation mechanism.

9.3.4 Compression Members in Trusses

The potential failure modes of compression members in trusses are either in-plane bending or buckling modes. The effective length factors are, therefore, either taken to be equal to one or are based on the restraint at the ends. Thus the following situations arise for in-plane and out-of-plane behaviour.

(a) In-plane behaviour

A compression member with bolted or welded end connections and with in-plane joint eccentricities acts in-plane as a beam-column with axial forces and end moments that can be established. It can be isolated from the structure and is designed as a beam-column based on its actual length, that is, with an effective length factor of 1.0.

A compression member with bolted or welded end connections and without in-plane joint eccentricities, designed as an axially loaded member, has end restraints provided that all members meeting at the two end joints do not reach their ultimate loads (yielding in tension or buckling in compression) simultaneously. The effective length factor depends on the degree of restraint. This typically occurs for trusses in which some members are oversize, for example, trusses with constant size chords. All members do not fail simultaneously and the effective length factors may be less than one.

If, however, all members reach their ultimate loads simultaneously and none restrain others, the effective length factor should be taken as 1.0.

(b) Out-of-plane behaviour

Unless members out-of-plane of the truss exist at the end joints under consideration, the restraint to out-of-plane buckling is small and should be neglected. Provided no out-of-plane displacement of the members ends occurs, an effective length factor of 1.0 is therefore appropriate.

10. SLENDERNESS RATIOS

The maximum slenderness ratio of 200 for compression members, stipulated as long ago as the 1974 Standard, has been retained in S16.1-94 for the reason that strength, or resistance, of a compression member becomes quite small as the slenderness ratio increases and the member becomes relatively inefficient.

For considerations of strength, no limiting slenderness ratio is required for a tension member and, indeed, none is applied to wire ropes and cables. However, a slenderness ratio limit of 300 is given with permission to waive this limit under specified conditions. The limit does assist in the handling of members and may help prevent flutter under oscillating loads such as those induced in wind bracing designed for tension loads only.

Members whose design is governed by earthquake loadings may be subject to more stringent slenderness ratios depending on the ductility requirements of the lateral load resisting system. See commentary on Clause 27.

11. WIDTH–THICKNESS RATIOS: ELEMENTS IN COMPRESSION

Clause 11.1 identifies four categories of cross-sections, Class 1 through Class 4, based upon the width-thickness ratios of the elements of the cross-section in compression that are needed to develop the desired flexural behaviour. With the ratios given in Table 1 of Clause 11 for Classes 1, 2, or 3, the respective ultimate limit states will be attained prior to local buckling of the plate elements. These ultimate limit states are: Class 1—maintenance of the plastic moment capacity (beams), or the plastic moment capacity reduced for the presence of axial load (beam-columns), through sufficient rotation to fulfill the assumption of plastic analysis; Class 2—attainment of the plastic moment capacity for beams, and the reduced plastic moment capacity for beam-columns, but with no requirement for rotational capacity; Class 3—attainment of the yield moment for beams, or the yield moment reduced for the presence of axial load for beam–columns. Class 4—have plate elements which buckle locally before the yield strength is reached.

Elements in flexural compression

The requirements given in Figure 2–7 for elements of Class 1, 2, and 3 sections in flexural compression (and also for axial compression), particularly those for W-shapes are based on both experimental and theoretical studies. For example, the limits on flanges have both a theoretical basis (Kulak *et al.* 1990; ASCE 1971; Galambos 1988) and an extensive experimental background (Haaijer and Thurlimann 1958; Lay 1965; Lukey and Adams 1969). For webs in flexural compression the limits $1100/\sqrt{F_y}$, $1700/\sqrt{F_y}$, and $1900/\sqrt{F_y}$ for Class 1, 2 and 3 respectively when C_f/C_y is zero come from both theory and tests on Class 1 sections (Haaijer and Thurlimann 1958) but mostly from test results for Class 2 and 3 sections (Holtz and Kulak 1973 and 1975).

For circular hollow sections in flexure, see Stelco (1973) for the requirements for Class 1 and Class 2 sections and Sherman and Tanavde (1984) for Class 3.

Elements in axial compression

The distinction between classes based on moment capacity does not apply to axially loaded members as the plate elements need only reach a strain sufficient for the plate elements to develop the yield stress. This strain is affected by the presence of residual stresses but there is no strain gradient across the cross-section as there is for members subject to flexure. Thus for webs, in Table 1 for each of Classes 1, 2 and 3 when $C_f/C_y = 1.0$ the limit on h/w is the same value of about $670/\sqrt{F_y}$ as given in Figure 2–7. The width-thickness limit for the flanges of axially loaded columns, based on the same argument, is the same as for Class 3 beam flanges, i.e., $200/\sqrt{F_y}$ (Dawe and Kulak 1984). As well the limit on the D/t ratio of $23\,000/F_y$ (Winter 1970) for circular hollow sections in axial compression is the same irrespective of the Class.

Elements in compression due to bending and axial load

In Figure 2–8, the requirements for webs in compression ranging from compression due to pure bending to that due to pure compression are plotted. Because the amount of web under compression varies from complete (columns) to one-half (beams), the depth-to-thickness limits vary as a function of the amount of axial load. The results presented here reflect the latest research results, particularly those of Dawe and Kulak (1986), and are significantly more liberal than previous limits (Perlynn and Kulak 1974; Nask and Kulak 1976).

Class 4 Sections

Sections used for columns, beams, or beam–columns may be composed of elements whose width-to-thickness ratios exceed those prescribed for Class 3 provided that the resistance equations are adjusted accordingly. These sections, called Class 4, should be evaluated according to the rules given in Clause 13.3 or 13.5 as applicable.

Detail		Class 1	Class 2	Class 3
L's connected continuously		$\dfrac{b}{t} \le \dfrac{145}{\sqrt{F_y}}$ †	$\dfrac{b}{t} \le \dfrac{170}{\sqrt{F_y}}$ †	$\dfrac{b}{t} \le \dfrac{200}{\sqrt{F_y}}$
	Flanges of I's or T's	† 2-L's or cover PLs. symmetric abt. plane of bending - abt. x-x axis.	† 2-L's or cover PLs. symmetric abt. plane of bending - abt. x-x axis.	
		—	—	$\dfrac{b}{t} \le \dfrac{200}{\sqrt{F_y}}$ L's not continuously connected, flange of C's, asymmetric cover PL, plate gdr. stiffeners.
Stems of T's	‡ symmetric abt. plane of bending, or incl. asymmetry effect in analysis.	$\dfrac{b}{t} \le \dfrac{145}{\sqrt{F_y}}$ ‡	$\dfrac{b}{t} \le \dfrac{170}{\sqrt{F_y}}$ ‡	$\dfrac{b}{t} \le \dfrac{340}{\sqrt{F_y}}$
		bending only $\dfrac{h}{w} \le \dfrac{1100}{\sqrt{F_y}}$ axial compression —	bending only $\dfrac{h}{w} \le \dfrac{1700}{\sqrt{F_y}}$ axial compression —	bending only $\dfrac{h}{w} \le \dfrac{1900}{\sqrt{F_y}}$ axial compression $\dfrac{h}{w} \le \dfrac{670}{\sqrt{F_y}}$
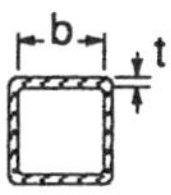 HSS		$\dfrac{b}{t} \le \dfrac{420}{\sqrt{F_y}}$	$\dfrac{b}{t} \le \dfrac{525}{\sqrt{F_y}}$	$\dfrac{b}{t} \le \dfrac{670}{\sqrt{F_y}}$
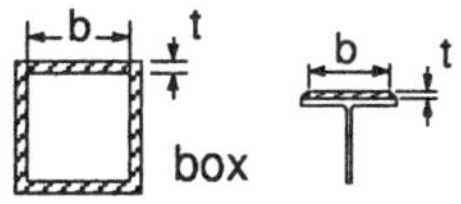 box		$\dfrac{b}{t} \le \dfrac{525}{\sqrt{F_y}}$	$\dfrac{b}{t} \le \dfrac{525}{\sqrt{F_y}}$	$\dfrac{b}{t} \le \dfrac{670}{\sqrt{F_y}}$
		—	—	$\dfrac{b}{t} \le \dfrac{840}{\sqrt{F_y}}$
		bending only $\dfrac{D}{t} \le \dfrac{13\,000}{F_y}$ axial compression —	bending only $\dfrac{D}{t} \le \dfrac{18\,000}{F_y}$ axial compression —	bending only $\dfrac{D}{t} \le \dfrac{66\,000}{F_y}$ axial compression $\dfrac{D}{t} \le \dfrac{23\,000}{F_y}$

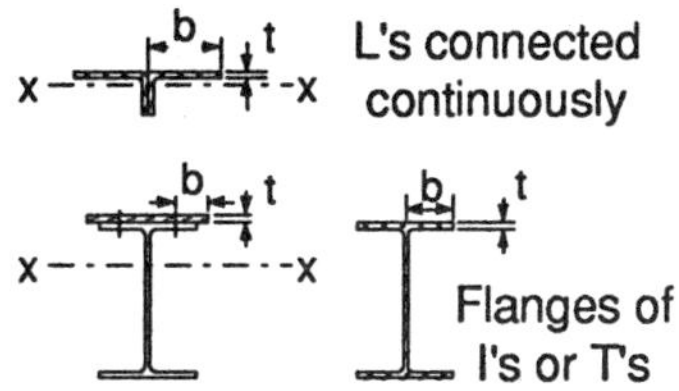

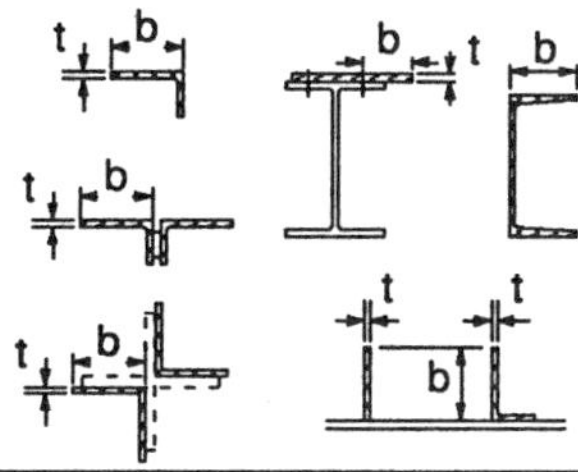

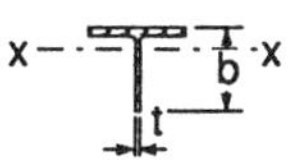

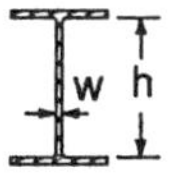

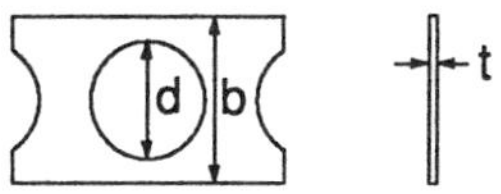

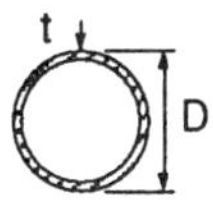

**Figure 2–7
Width–Thickness Ratios For Compression Elements**

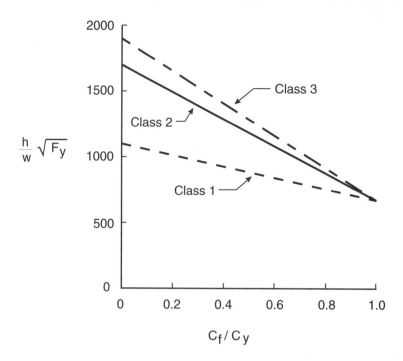

**Figure 2–8
Width-Thickness Ratios for Webs**

12. GROSS AND NET AREAS

12.1 Application

The design and behaviour of tension members is integrally related to the proportioning and detailing of connections. Consequently, Clauses 12 and 13.2 are related. Two possible overall failure modes exist: unrestricted plastic flow of the gross section and fracture of a net section. The second of these consists itself of three modes depending on the failure path and the degree of ductility available. Therefore, the four specific failure modes are:

1. *Unrestricted plastic flow of the gross section* when the deformations at yield are excessive. This represents a limit state for which the failure is gradual. A β of 3.0 is considered acceptable and thus the tensile resistance is

$$T_r = \phi A_g F_y$$

with $\phi = 0.90$.

2. *Fracture of the net section when there is sufficient ductility* to provide a reasonably uniform stress distribution. This fracture occurs with little deformation and an increased value of β is appropriate for this and the other cases of fracture of the net section. In S16.1 the tensile resistance for this mode is written as

$$T_r = 0.85 \, \phi \, A_{ne} F_u$$

where A_{ne} is the effective net area

$$A_{ne} = w_n t + \Sigma \left(\frac{s^2 t}{4g}\right)$$

The combination $0.85\phi = 0.85 \times 0.90 = 0.765$ is in effect a reduced resistance factor that results in an increased value of β of about 4.5.

3. *Fracture of a section with block tear out* when, at the end of a member, a block of material tears out with some of the failure surface in tension and some in shear.

In S16.1, the effective net area for the portion in shear is given by

$$A_{ne} = 0.60 L_n t$$

where the factor 0.60, when multiplied by F_u, gives the ultimate shear resistance, $\tau_u = 0.60 F_u$, and the effective net area for the portion in tension is as given previously.

4. *Fracture of a net section when the strain distribution is non-uniform.* This is the shear lag problem and in S16.1 equations for the effective net area are based on writing

$$A_{ne} = A_n (1 - \bar{x}/L)$$

where $\bar{x}$ is the distance from the centroid of the connected part to the face of the connection and L is the length of the connection.

The effective net area calculations are discussed in greater detail following.

12.3 Effective Net Area

This clause defines areas used to determine tension member resistances and incorporates the results of new research. The requirements apply to both bolted and welded connections.

12.3.1 The fracture path may consist of a series of segments, some loaded primarily in tension, including inclined paths, others primarily in shear. The critical path must be identified and evaluated. By determining the effective net area by summing the net area of each segment along the critical path, it is assumed, as has been demonstrated, (Birkemoe and Gilmor 1978; Ricles and Yura 1980; Hardash and Bjorhovde, 1985) that all segments fracture simultaneously.

The net area of segments normal to the force, in direct tension, and segments inclined to the force are determined as in previous standards. For the latter, the net area is increased by $s^2 / 4g$ for each inclined segment. Segments parallel to the force are in shear.

The Standard, therefore, now considers the failure mode of block tear-out that occurs, for example, when a block of material at the end of a tension member tears out through the fastener holes, as shown in Figure 2–9. According to Birkemoe and Gilmor (1978), the strength of the failure surface in shear is taken as 0.60 of that in tension. Thus, the load at

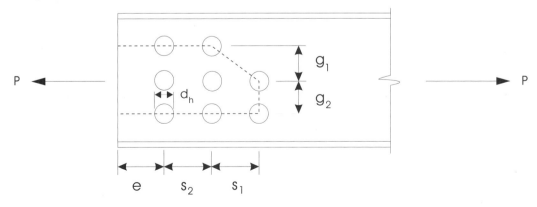

Figure 29
Block Tear-out at End Of Tension Member

which tear-out occurs is equal to the ultimate tensile strength times the net area in tension (including the $s^2/4g$ allowance for staggered holes) plus 0.60 times the ultimate tensile strength times the net area in shear. This load is equal to the ultimate tensile strength times the net area in tension plus 0.60 of the area in shear.

Block tear-out may also occur at the end of a coped beam as shown in Figure 2–10. Here the applied force is the end shear on the beam. This force causes tension on horizontal planes and shear on vertical ones. Tests by Yura *et al.* (1982) on coped beams indicate that the Birkemoe-Gilmor model derived originally for this case is somewhat liberal, but those of Hardash and Bjorhovde (1985) on gusset plate connections indicate the reverse when the net areas in shear and tension are used.

The tear-out resistance, based on these net areas and the requirements of Clause 13.2, is therefore

$$0.85 \, \phi \, F_u \, t \, [(g_2 - d_h) + \left(g_1 - d_h + \frac{s_1^2}{4 \, g_1}\right) + 0.60 \, (2s_2 + s_1 + 2e - 4d_h)]$$

for the paths shown in Figure 2–9 and

$$0.85 \, \phi \, F_u \, t \, [(e_2 - 0.5d_h) + 0.60(2g + e_1 - 2.5d_h)]$$

for the paths shown in Figure 2–10.

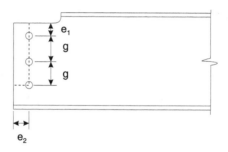

**Figure 2–10
Block Tear-out at End Of Coped Beam**

12.3.2 This allows for distortion or local material damage that may occur in forming the hole.

The limit on net areas, depending on the ratio of F_y/F_u, was deleted in 1989 because the current requirements, recognizing yielding on the gross section as a failure mode, greatly restricted its applicability.

12.3.3 Effective Net Area Reduction – Shear Lag When the critical net section fracture path crosses unconnected cross-sectional elements, the directly connected elements tend to reach their ultimate strength before the complete net section strength is reached due to shear lag. When all cross-sectional elements are directly connected, shear lag does not occur and the effective net area is the total net area.

The loss in efficiency due to shear lag can be expressed as a reduction in the net area. Munse and Chesson (1963) suggested that this reduction could be taken as $1 - \bar{x}/L$ where $\bar{x}$ is the distance from the shear plane to the centroid of that portion of the cross-section being developed and L is the connected length.

Because the connected length is usually not known at the time of tension member design, reduction factors have been derived for specific cases, as given in Clause 12.3.3.2, based on an extensive examination of the results of over 1000 tests (Kulak *et al* 1987). The reduction factor depends on the cross-sectional shape and the number of bolts (2, 3 or more) in the direction of the tensile load.

More severe reductions for shear lag are provided for angles connected by one leg based on work by Wu and Kulak (1993) who examined the results of 72 tests, including 24 of their own, on angles in tension connected with mechanical fasteners.

When block tear-out occurs in those elements which are directly connected, shear lag is not a factor. Shear lag need only be considered when a potential failure path crosses unconnected elements.

12.3.3.3 Similar behaviour has been observed in welded connections (Kulak *et al.* 1987) when welds parallel to the tensile load in the member only are used. The requirements in the 1994 Edition have been expanded to distinguish among many possible cases. For welded connections with matching electrodes and material of G40.21–300W grade steel, shear lag will be critical for cases where $A'_{ne} \leq 0.78\ A_g$. For angles, this generally occurs when the length of weld along the toe exceeds the length of weld along the heel.

When the weld length is less than the distance between welds, it is likely that the weld is critical.

The direct use of the Munse and Chesson reduction factor, $1 - \bar{x}/L$, is permitted by this clause. Figure 2–11 gives examples of dimensions used in calculating this shear lag reduction factor.

12.4 Pin-Connected Tension Members

The net area increase across the pin hole is to allow for the significant non-uniform distribution of stress (Johnston 1939). The minimum net area within the 45° arc each side of the longitudinal axis is to avoid end splitting.

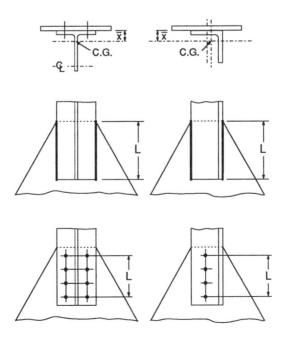

Figure 2–11
Dimensions Used For Shear Lag Calculations

13. MEMBER AND CONNECTION RESISTANCE

13.1 General

The value of ϕ of 0.90 provides consistent and adequate values of the reliability index when used with the load factors of Clause 7.2 (Kennedy and Gad Aly 1980; Baker and Kennedy 1984).

13.2 Axial Tension

The two potential failure modes for tension members are yielding of the gross section and fracture of a net section. See Clause 12 for net section area calculations. Yielding on the gross section is a ductile mode of failure for which a resistance factor of 0.90 is appropriate. Because net section fracture occurs with little deformation and therefore little warning of failure, the resistance factor is reduced by multiplying by 0.85 to about 0.76 thereby increasing the reliability index. This philosophy is consistent with the reduced resistance factor used for connectors (bolts, welds, and shear connectors).

The reduction factor of 0.75 for pin connectors recognizes the greater non-uniformity of stress that occurs around a hole that is large relative to the material in which it is formed.

13.3 Axial Compression

Steel columns are conveniently classified as short, intermediate, or long members, and each category has an associated characteristic type of behaviour. A short column is one which can resist a load equal to the yield load ($C_y = AF_y$). A long column fails by elastic buckling. The maximum load depends only on the bending stiffness (EI) and length of the member. Columns in the intermediate range are most common in steel buildings. Failure is characterized by inelastic buckling and is greatly influenced by the magnitude and pattern of residual stresses that are present and the magnitude and shape of the initial imperfections or out-of-straightness. These effects lessen for both shorter and longer columns. The expressions in this clause account for these effects which are dependent on the cross-section (Bjorhovde, 1972).

Figure 2–12 indicates the variations in strengths for columns of three different values of the slenderness parameter, λ, and with the same out-of-straightness patterns and different residual stress patterns.

The compressive resistance expressions of Clause 13.3.1 are now expressed in double exponential form (Loov 1995) rather than as a sequence of polynomials. The new expressions are easier to use with pocket calculators and computers and do not require the selection of a polynomial based on the value of the slenderness parameter λ. With values of the new parameter n of 1.34 and 2.24 for cases (a) and (b) of Clause 13.3.1, the new expressions are always within 3% and generally within 1% of Curves 2 and 1 of the Structural Stability Research Council (SSRC) (Galambos 1988) which they replace.

Steel shapes, unless explicitly stated, are assigned to SSRC Curve 2 (n = 1.34) which is therefore used for W-shapes rolled in Canada (Groups 1, 2 or 3 of CSA G40.20), fabricated boxes and I-shapes, and, based on Bjorhovde and Birkemoe (1979), cold-formed non-stress relieved, Class C, hollow structural sections.

Because of a more favourable residual stress pattern and out-of-straightness, hot-formed or cold-formed stress relieved (Class H) hollow structural sections (Kennedy and Gad Aly, 1980) are assigned to SSRC Curve 1 or its equivalent curve here with a value of n = 2.24. For the same reasons (Chernenko and Kennedy, 1991), WWF produced in Canada from plate with flame-cut edges are also assigned to the curve with n = 2.24.

For heavy sections (Groups 4 and 5 of CSA Standard G40.20) made of ASTM A36 steel and welded sections fabricated from universal mill plate, a resistance less than that corre-

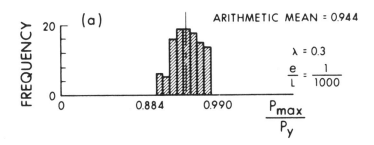

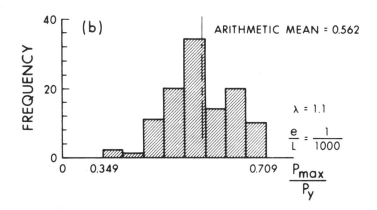

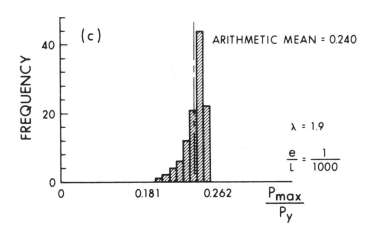

**Figure 2–12
Typical Frequency Distribution Histograms For The Maximum Strength Of 112 Column Curves (e/l = 1/1000)**

sponding to n = 1.34 (SSRC Curve 2) is appropriate and it is recommended that a value of n = 0.98, corresponding to Column Curve 3 (Galambos, 1988), be used.

Because column strengths are influenced by the magnitude and distribution of residual stresses, care should be exercised in the use of the expressions in this Standard, for example, in determining the capacity of an existing column which is being reinforced in such a manner that there is an increase in compressive residual stresses in the fibres most remote from the centroid. In such a situation, adding material to reduce the slenderness ratio may be advantageous.

13.3.2 Because the expressions for column resistance are for flexural buckling about the x or y axes only, singly symmetric, asymmetric, or cruciform sections which may fail by lateral-torsional or torsional buckling are to be checked in accordance with the requirements of Appendix D, in addition to the requirements of Clause 13.3.1 and the lesser resistance be used in design.

13.3.3

This requirement for using an effective area is equivalent to the notional removal of the material in excess of the Class 3 limit to determine the reduced area. The slenderness parameter λ is however based on the gross cross-section. For hot-rolled members, the compressive resistance of Clause 13.3 is deemed more appropriate than that of CSA S136.

13.4 Shear

13.4.1.1 Elastic Analysis

The expressions for shear strength are given for stiffened plate girders. Unstiffened

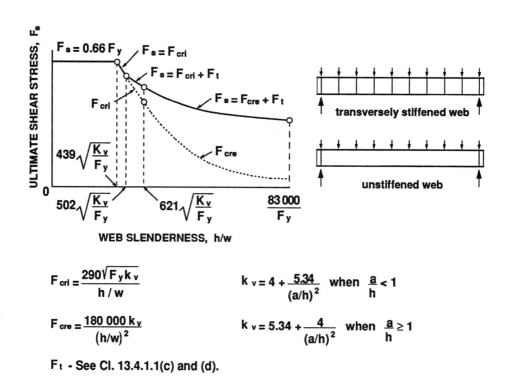

Figure 2–13
Ultimate Shear Stress — Webs Of Flexural Members

plate girders and rolled beams are simply special cases for which the shear buckling coefficient, $k_v = 5.34$.

The four ranges of resistance based on Basler (1961), correspond to the following modes of behaviour and are illustrated in Figure 2–13:

(a) Full yielding followed by strain hardening and large deformation. The limiting stress of $0.66\ F_y$ corresponds to shear deformation into the strain hardening range and is higher than that derived from von Mises criterion ($0.577\ F_y$), which forms the basis of Clause 13.4.1.2 for plastic analysis.

(b) A transition curve between strain hardening and inelastic buckling at full shear yielding. ($F_s = 0.577 F_y$);

(c) Inelastic buckling, F_{cri}, accompanied by post-buckling strength, F_t, due to tension field action, if the web is stiffened; and,

(d) Elastic buckling, F_{cre}, accompanied by post-buckling strength, F_t, due to tension field action, if the web is stiffened.

For ranges (c) and (d), the expressions were reformulated in the 1989 edition to separate buckling and tension field contributions.

In computing the shear resistance, it is assumed that the shear stress is distributed uniformly over the depth of the web. The web area (A_w) is the product of web thickness (w) and web depth (h) except for rolled shapes where it is customary to use the overall beam depth (d) in place of the web depth (h).

In panel zones and locations where strain hardening develops quickly after the onset of shear yielding, the use of $0.66 F_y$ is valid.

13.4.1.2 Plastic Analysis

For structures analysed plastically, high shears and moments may occur simultaneously at a hinge location. For this reason, the maximum shear stress is limited to the von Mises value, and when the web is assumed to be 95% of the depth of the beam, the coefficient 0.55

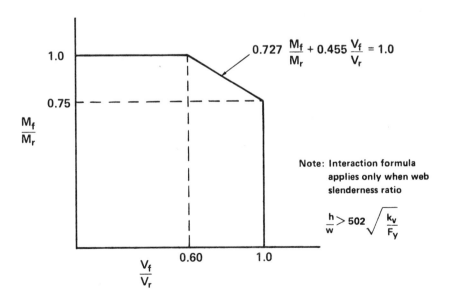

Figure 2-14
Combined Shear and Moment in Webs Of Transversly Stiffened Thin Web Plate Girders

is obtained. Experimental evidence shows that this shear can be carried without reducing the flexural resistance below M_p (Yang and Beedle 1951).

13.4.1.3 Maximum Slenderness

This limit prevents the web from buckling under the action of the vertical components of the flange force arising as a result of the curvature of the girder, (Kulak *et al*, 1995).

13.4.1.4 Combined Shear and Moment in Girders

This requirement recognizes the limit state of the web yielding by the combined action of flexural stress and the post buckling components of the tension field development in the web near the flange (Basler, 1961b).

Figure 2–14 illustrates the interaction expression provided in Clause 13.4.1.4. When Clause 15.4 applies, M'_r replaces M_r in the interaction expression.

13.4.2 Webs Of Flexural Members Not Having Two Flanges

When cross-sections do not have two flanges, the shear stress distribution can no longer be assumed to be uniform. For W-shapes with one flange coped, the elastic shear stress distribution may be determined from $\tau = VQ/It$. Limiting the maximum value to 0.66 F_y is conservative as it does not allow for any plastification as shear yielding spreads from the most heavily stressed region. For W-shapes with two flanges coped, a parabolic shear stress distribution results from this procedure with a maximum shear stress equal to 1.5 times the average. The maximum shear stress can be based on strain-hardening provided shear buckling does not occur. (Not applicable to framed webs of beams–use Clause 13.4.4.)

13.4.3 Gusset Plates

This clause applies where gusset plate failure is the result of block tear-out, (Hardash and Bjorhovde 1985).

13.4.4 Connecting Elements

This clause applies when the elements are loaded primarily in shear and is somewhat conservative when compared with segments in shear in a block tear-out mode.

13.4.5 Pins

Additional information for pins in combined shear and moment is given in the Canadian Highway Bridge Design Code 1996.

13.5 Bending: Laterally Supported Members

The factored moment resistances are consistent with the classification of cross-sections given in Clause 11, as illustrated by moment-rotation curves given in Figure 2–15.

The fully plastic moment, M_p, attained by Class 1 and 2 sections, implies that all fibres of the section are completely yielded. Any additional resistance that develops due to strain-hardening has been accounted for in the test/predicted ratio statistics used in developing resistance factors (Kennedy and Gad Aly 1980).

The stress distribution for Class 3 sections is linear with a maximum stress equal to the yield stress.

Class 4 sections reach their maximum moment resistance when a flange or web plate element buckles locally. Class 4 sections are divided into three categories.

The first consists of those sections with Class 4 flanges and webs. This type of section is designed to the requirements of CSA Standard S136 (CSA, 1994) using the material properties appropriate to the structural steel specified.

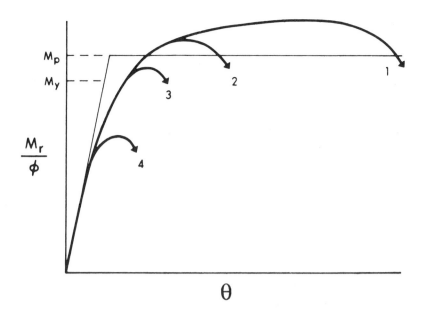

**Figure 2–15
Moment-rotation Curves**

The second category consists of those sections with Class 3 flanges and Class 4 webs. Clause 13.5(c)(ii) requires that these sections be designed in accordance with Clause 15.

The third category, with Class 4 flanges and Class 3 webs, has moment resistances, governed by local buckling of the compression elements, as given by CSA Standard S136. Alternatively, the designer can ignore the portion of the flange in excess of the b/t limit and compute a reduced, effective section modulus (Kalyanaraman *et al.*, 1977).

13.6 Bending: Laterally Unsupported Members

Laterally unsupported beams may fail by lateral torsional buckling at applied moments significantly less than the full cross-section strength (M_p or M_y). Even when the top flange is laterally supported, under some circumstances, for example, a roof beam subject to uplift, the laterally unsupported bottom flange may be in compression.

Two factors are considered to contribute to the lateral torsional buckling resistance, the St. Venant torsional resistance (related to the product GJ) developed by torsional shear stresses in the cross-section and the warping resistance (related to C_w) developed by the cross-bending of the flanges.

Analogous to columns, the lateral torsional moment resistance depends on the unsupported length. Beams may be considered to be short, intermediate, or long depending on whether the moment resistance developed is the full strength, the inelastic lateral torsional buckling strength, or the elastic lateral torsional buckling strength, respectively, as shown in Figure 2–16, for Class 1 and 2 shapes capable of attaining M_p on the cross-section. The curve for Class 3 sections is similar except that the maximum moment resistance is ϕM_y while for Class 4 sections, the maximum resistance is limited by local buckling.

Without the factor ω_2, the expression given for M_u is that for a doubly symmetric beam subject to uniform moment. The factor ω_2, expressed as a ratio of the end moments, κ, ranges from 1.0 to 2.5, and takes into account the fact that a varying moment is less critical than a uniform moment. The use of $\omega_2 = 1.0$ applied to the maximum moment in an unbraced length is somewhat conservative. Figure 2-17 illustrates ω_2 for beams with various loading

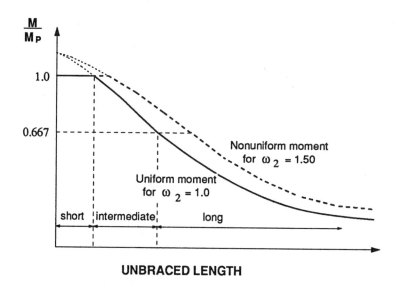

**Figure 2–16
Variation Of Uniform And Non-uniform Moment Resistance**

and lateral constraint conditions. Also plotted in Figure 2-16 is the moment resistance for a beam for which $\omega_2 = 1.5$. It is seen that in the elastic region ($M_r < \frac{2}{3} M_p$) the full value of ω_2 is realized. In the inelastic region, the increase in M_r due to non-uniform moments gradually decreases to zero as the moment approaches ϕM_p. For laterally unsupported Class 2 and 3 beams, Baker and Kennedy (1984) give resistance factors ranging from 0.93 to 0.96 and therefore the value of 0.90 used in this Standard is conservative. The length, L, is generally taken as the distance between lateral supports. When beams are continuous through a series of lateral supports, interaction buckling (Trahair, 1968) occurs. The segment which tends to buckle laterally first is restrained by the adjoining segments. Nethercot and Trahair (1976a, 1976b), Kirby and Nethercot (1978) and Schmitke and Kennedy (1985) give methods of computing effective lengths under these circumstances.

Significant information on lateral torsional buckling is summarized in Chen and Lui (1987).

The expression for M_u assumes that the beam is loaded at the shear centre. For other positions of the load, for monosymmetric beams, unusual loading, and other support conditions, the SSRC Guide (Galambos 1988) may be consulted.

For structural systems utilizing cantilever suspended span construction (Gerber girders), see Albert *et al.* (1992), Essa and Kennedy (1994(a)(b), 1995).

For members bent about both principle axes, it should be remembered that M_{ry} is either M_{yy} or M_{yp} as there is no reduction for lateral torsional buckling for weak axis buckling.

13.7 Lateral Bracing for Members in Structures Analysed Plastically

See the commentary on Clause 8.5(c).

13.8 Axial Compression and Bending

For a general discussion of all aspects of these provisions and worked examples, see Kennedy *et al* (1990).

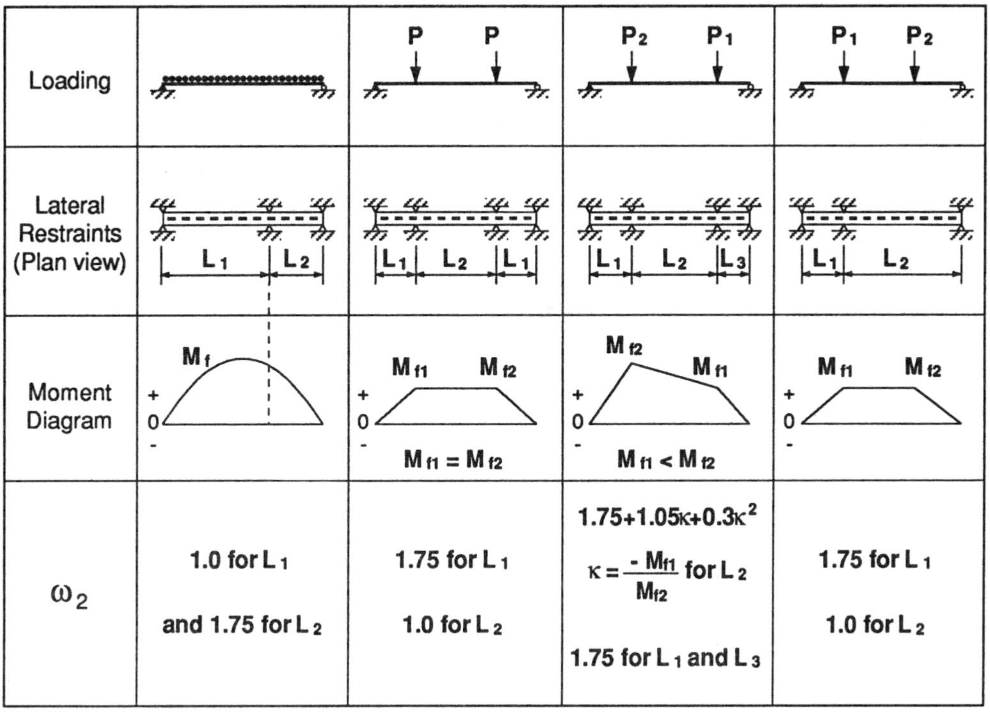

Loading	(uniform distributed load)	P, P (two point loads)	P_2, P_1	P_1, P_2
Lateral Restraints (Plan view)	L_1, L_2	L_1, L_2, L_1	L_1, L_2, L_3	L_1, L_2
Moment Diagram	M_f	M_{f1} M_{f2} ; $M_{f1} = M_{f2}$	M_{f2} M_{f1} ; $M_{f1} < M_{f2}$	M_{f1} M_{f2}
ω_2	1.0 for L_1 and 1.75 for L_2	1.75 for L_1 1.0 for L_2	$1.75+1.05\kappa+0.3\kappa^2$ $\kappa = \dfrac{-M_{f1}}{M_{f2}}$ for L_2 1.75 for L_1 and L_3	1.75 for L_1 1.0 for L_2

Figure 2–17
Various Cases Of ω_2 For Beams

In S16.1-94, the term "I-shaped" means any doubly-symmetric shape having two flanges and one web. This includes, for instance, both W-shapes and S-shapes.

The requirements in the 1994 edition are essentially the same as in the 1989 edition. In two cases, where introduced in Clause 13.8.1, terms in the interaction equations have been better or more clearly defined as follows:

(a) M_f has been more clearly defined in "M_f is the maximum moment including stability effects as defined in Clause 8.6,"

(b) the reference in the definition of C_r "(see also Clause 9.3.2)" has been changed to "(see also Clause 9.3.3)" consistent with the fact that for out-of plane buckling the effective length depends on the end restraints, e.g., the base plate anchorage may provide some fixity for buckling about the weak axis.

The value each term in the interaction equation takes is prescribed in the three subclauses (a), (b), and (c) depending on the particular mode of failure: cross-sectional strength, overall member strength, and lateral torsional buckling strength, respectively. Clause 13.8.1 is applicable to all classes of section while Clause 13.8.2 is applicable only to Class 1 sections of I-shaped members.

The interaction expressions must account for the following:

- a laterally supported member fails when it reaches its in-plane moment strength reduced for the presence of axial load;
- a laterally unsupported member may fail by lateral torsional buckling or a combination of weak axis buckling and lateral buckling;

- a relatively short member can reach its full cross-sectional strength whether it is laterally supported or not;
- when subjected to axial load only, the axial compressive resistance, C_r, depends on the maximum slenderness ratio;
- members bent about the weak axis do not exhibit out-of-plane behaviour,
- a constant moment has the most severe effect on in-plane behaviour. Other moment diagrams can be replaced by equivalent moment diagrams of reduced but uniform intensity;
- a constant moment has the most severe effect on the lateral torsional buckling behaviour. (See commentary on Clause 13.6). This effect disappears if the member is short enough, in which case, cross-sectional strength controls; and
- moments may be amplified by axial loads increasing the deflections.

All four modes of failure, addressed as follows, are to be checked in design, as appropriate:

1) local buckling of an element

According to Clause 11.2 of the Standard, the compression elements of the cross-section must be proportioned such that local buckling does not occur prior to attainment of the cross-sectional strength. (See Commentary for Clause 11).

2) strength of the cross-section

As in previous standards, the cross-sectional strength of a shape used as a beam-column cannot be exceeded. Clause 13.8.2(a) gives the cross-sectional strength requirements for Class 1 sections of I-shaped members and Clause 13.8.1(a) for all other members. The cross-sectional strength is also the limiting strength of short members.

The cross-sectional strength of a Class 1 I-shaped section comprised of relatively stocky plate elements can be derived from the fully plastic stress distribution of the cross-section as shown in Figure 2–18. For uniaxial bending about the x-x axis and the y-y axis, expressions are respectively, using the limit states notation of this Standard:

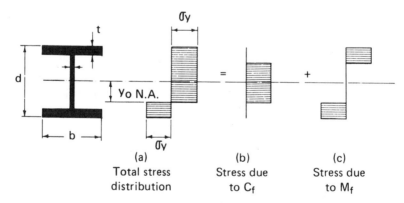

**Figure 2–18
Idealized Stress Distributions In Plastified Section Of Beam–column**

$$M_{fx} = 1.18\,\phi\,M_{px}\,(1 - C_f/\phi\,C_y) \le \phi\,M_{px}$$

$$M_{fy} = 1.19\,\phi\,M_{py}\,[1 - (C_f/\phi\,C_y)^2] \le \phi\,M_{py}$$

Transposing the terms in the first expression and using a linear approximation for the second expression gives

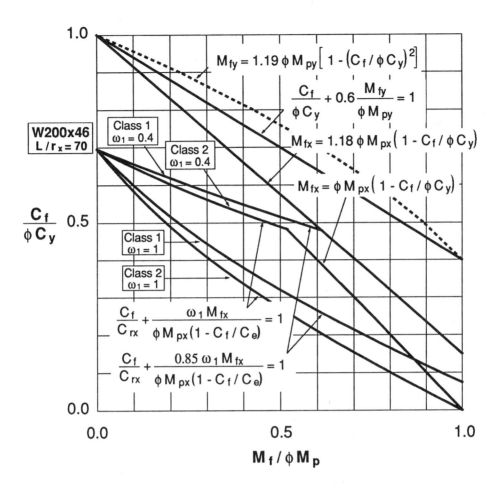

**Figure 2–19
Interaction Expressions For Class 1 And 2 W-shapes**

$$\frac{C_f}{\phi C_y} + 0.85 \frac{M_{fx}}{\phi M_{px}} \leq 1.0 \; ; \quad \frac{M_{fx}}{M_{px}} \leq 1.0$$

$$\frac{C_f}{\phi C_y} + 0.6 \frac{M_{fy}}{\phi M_{py}} \leq 1.0 \; ; \quad \frac{M_{fy}}{M_{py}} \leq 1.0$$

as shown in Figure 2–19. For biaxial bending it is conservative to combine these expressions linearly to give, using the limit states notation of this Standard:

$$\frac{C_f}{\phi C_y} + 0.85 \frac{M_{fx}}{\phi M_{px}} + 0.6 \frac{M_{fy}}{\phi M_{py}} \leq 1.0 \; ; \quad \frac{M_{fx}}{\phi M_{px}} + \frac{M_{fy}}{\phi M_{py}} \leq 1.0$$

This is identical to the two expressions in Clause 13.8.2 when, in the latter, U_{1x} and U_{1y} are set equal to 1.0, as appropriate for the cross-sectional strength, C_r for $\lambda = 0$ equals $\phi A F_y$, and M_{rx} and M_{ry} equal ϕM_{px} and ϕM_{py} respectively.

For uniaxial bending of Class 2 sections (Kulak and Dawe 1991; Dawe and Lee 1993), the appropriate interaction expression is

$$\frac{C_f}{\phi C_y} + \frac{M_{fx}}{\phi M_{px}} \leq 1.0$$

Extending this linear expression to bending about the weak axis gives, for biaxial bending

$$\frac{C_f}{\phi C_y} + \frac{M_{fx}}{\phi M_{px}} + \frac{M_{fy}}{\phi M_{py}} \leq 1.0$$

which is identical to the expression given in Clause 13.8.1 when, in the latter, the appropriate values of the various cross-sectional resistance quantities are used.

The expression of Clause 13.8.1 is also appropriate for Class 3 sections when the moment resistances are expressed in terms of M_y and for Class 4 sections when those resistances, C_r, M_{rx}, and M_{ry}, are based on local buckling.

3) overall member strength

The overall strength of a member depends on its slenderness. As an actual beam–column has length, the axial compressive resistance, C_r, depends on its slenderness ratio and will

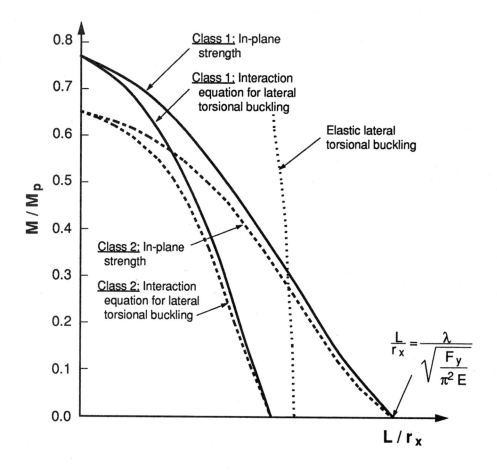

Figure 2–20
Variations Of Moment Resistance With Slenderness Ratio

be less than or equal to the yield load. For any particular beam–column this fraction of the yield load can be established and is illustrated in Figure 2–19 for a W200x46 shape.

In Figure 2–20 the variation in moment resistance in terms of M/M_p as a function of the slenderness L/r_x is plotted schematically as a solid line for a particular laterally supported Class 1 section subject to a uniform moment about the x axis and carrying an axial load of 0.35 C_y. An appropriate interaction expression for the in-plane strength of a Class 1 section is

$$\frac{C_f}{C_{rx}} + \frac{0.85 \, \omega_1 \, M_f}{\phi \, M_p \left(1 - \frac{C_f}{C_e}\right)} = 1$$

which can be deduced from Clause 13.8.2 when the terms in that expression are appropriately defined. Note that if the member is short the expression reduces to that for the cross-sectional strength. The compressive resistance, C_{rx}, is a function of the slenderness ratio L/r_x.

The term

$$\omega_1 = 0.6 - 0.4 \, \kappa \geq 0.4$$

multiplied by the maximum non-uniform moment, M_f, gives an equivalent uniform moment, $\omega_1 M_f$, having the same effect on the in-plane member strength as the non-uniform moment (Ketter 1961).

In order to account for the Pδ effects, the amplification of the moments caused by the axial loads acting on the deformed shape, the equivalent uniform moment, $\omega_1 M_f$, is amplified by the factor $\frac{1}{1 - C_f/C_e}$. The in-plane strength of a Class 1 section is shown in Figure 2–19 for a particular section with $L/r_x = 70$. When $L/r_x = 0$ and $\omega_1 = 1$, the in-plane

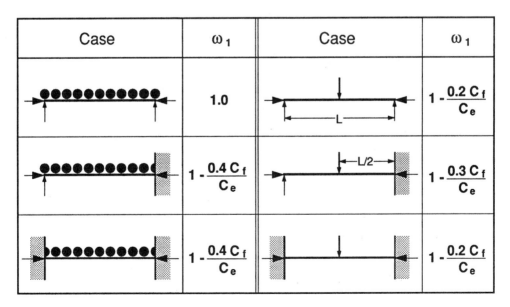

Figure 2–21
Values Of ω_1 For Special Cases Of Laterally Loaded Beam-columns

Conditions **	Design Criteria
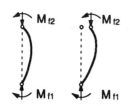 single curvature bending $M_{f2} \geq M_{f1}$ $\omega_1 = 0.6 + 0.4 \dfrac{M_{f1}}{M_{f2}}$	**Beam-Column Design Expressions** P-Δ (frame sway effects), if any, are included in analysis. (I) Class 1 Sections of I-Shapes $\dfrac{C_f}{C_r} + \dfrac{0.85\,U_{1x}M_{fx}}{M_{rx}} + \dfrac{0.6\,U_{1y}M_{fy}}{M_{ry}} \leq 1.0$ $\dfrac{M_{fx}}{M^{*}_{rx}} + \dfrac{M_{fy}}{M^{*}_{ry}} \leq 1.0$ (II) All Classes Except Class 1 Sections of I-Shapes $\dfrac{C_f}{C_r} + \dfrac{U_{1x}M_{fx}}{M_{rx}} + \dfrac{U_{1y}M_{fy}}{M_{ry}} \leq 1.0$
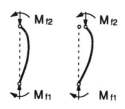 single curvature bending $M_{f1} = 0$ $\omega_1 = 0.6$	**Three Member Strength Checks** (a) Cross-sectional strength (use actual M_f at each location) $C_r = \phi A F_y$ $M_r = \phi Z F_y$ for class 1 and class 2 sections $\quad\ = \phi S F_y$ for class 3 sections $\quad\ =$ see Cl. 13.5 (c) of S16.1-94. for class 4 sections M^{*}_{rx} and M^{*}_{ry} as for M_r in Cl. 13.5 or 13.6, as appropriate U_{1x} and U_{1y} are taken as 1.0 (b) Overall strength (use M_{f2} for M_f) C_r = Factored compressive resistance (K=1) as in Cl.13.3 $\quad$ use C_{rx} for uniaxial strong-axis bending and $\quad$ max. slenderness for biaxial bending, or $\quad$ Interpolated between C_{rx} and C_{ry} on the basis of $\quad$ the proportion of interaction fractions for bending $\quad$ about two axes. $M_r = \phi Z F_y$ for class 1 and class 2 sections $\quad\ = \phi S F_y$ for class 3 sections $\quad\ =$ see Cl. 13.5 (c) of S16.1-94. for class 4 sections $U_1 = \omega_1 / (1 - C_f / C_e)$
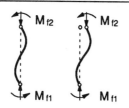 double curvature bending $M_{f2} \geq M_{f1}$ $\omega_1 = 0.6 - 0.4 \dfrac{M_{f1}}{M_{f2}} \geq 0.4$	(c) Lateral torsional buckling strength (use M_{f2} for M_f) $C_r = C_{ry}$ = Factored comp. resistance (K=1) as in Cl.13.3 M_{rx} = value given by Cl. 13.6 of S16.1-94 M_{ry} = as noted for M_r in (b) above U_{1x} = as noted for U_1 in (b) above, but not less than 1.0 U_{1y} = as noted for U_1 in (b) above
C_f = Factored compressive load C_r = Factored compressive resistance M_f = Fac. bending moment (x-x or y-y axis) M_r or M^{*}_r = Fac. moment resistance $\quad$ (x-x or y-y axis)	ω_1 = Coefficient used to determine equivalent uniform column bending effect (x-x or y-y axis) U_1 = factor to account for moment gradient & member curvature second order effects

** Moments M_{f1} and M_{f2} may be applied about one or both axes.

**Figure 2–22
Prismatic Beam-Columns–Moment At Ends–No Transverse Loads**

strength expressions 13.8.1(b) and 13.8.2(b) become the cross-sectional strength expressions 13.8.1(a) and 13.8.2(a) respectively.

In Figure 2–20, the curve of moment resistance versus slenderness ratio for the in-plane strength of a Class 2 section of equivalent cross-sectional strength to the Class 1 section is also given. It is similar to that for a Class 1 section except that, because the cross-sectional strength expression for Class 2 sections does not have the 0.85 factor that is appropriate for Class 1, the curve for Class 2 for zero slenderness ratio reaches only 0.65 of M_p and not 0.765 of M_p as for the Class 1 section.

For biaxial bending C_r is based conservatively on the maximum slenderness ratio. It could be argued that for biaxial bending the value used for C_r be interpolated between C_{rx} and C_{ry} on the basis of the proportion of the interaction fractions for bending about two axes. In other words, if a beam-column carries only a small portion of bending about the x axis, the decrease in C_r from C_{rx} toward C_{ry} should likewise be small.

In Figure 2–19, the in-plane strength interaction expressions are shown for $\omega_1 = 1$. When $\omega_1 < 1$, the limiting strength for low ratios of axial load is the cross-sectional strength expression.

4) lateral torsional buckling strength

Building beam-columns are usually laterally unsupported for their full length and, even though they are subject to strong axis bending moments, failure may occur when the column, after bending about the strong axis, buckles about the weak axis and twists simultaneously. For such columns, the lateral torsional buckling strength is likely to be the least of the cross-sectional strength, the overall member strength, and the lateral torsional buckling strength.

The curves in Figure 2–20 for a beam-column subject to uniform moment for Class 1 and 2 sections marked *interaction equation for lateral torsional buckling* demonstrate this effect. They are much below those for in-plane strength and only reach the full cross-sectional strength when the slenderness ratio is zero. The moment resistance is zero for laterally unsupported beam-columns when weak axis buckling occurs. Thus, for these members the axial compressive resistance is based on L/r_y and M_{rx} is based on the resistance of a laterally unsupported beam. When subjected to weak axis bending, members do not exhibit out-of-plane buckling behaviour and therefore the weak axis moment resistance is based on the full cross-sectional strength, the plastic moment or yield moment capacity as appropriate for the Class of the section.

In computing M_{rx} from Clause 13.6, the effect of non-uniform moments is included. Therefore, in the interaction expressions when lateral torsional buckling is being investigated, the factored moment, M_{fx}, must also be a non-uniform moment, and not be replaced by an *equivalent* lesser moment. It is for this reason that the value of U_{1x} cannot be less than 1.

13.8.4 This clause gives generally conservative values of ω_1, the factor by which the maximum value of the non-uniform moment is multiplied to give an equivalent uniform moment having the same effect as the applied non-uniform moment on the overall strength of the member. For further discussion on ω_1, see Galambos (1988) where it is called C_m. Figure 2–21, based on AISC (1994a), gives values of ω_1 for some special cases of transverse bending.

Figures 2–22 and 2–23 give guidance for the design of beam-columns subjected to various bending moment effects.

13.9 Axial Tension and Bending

The linear interaction expression of Clause 13.9(a) is a cross-sectional strength check. Conservatively, it does not take into account the fact that the bending resistance for Class

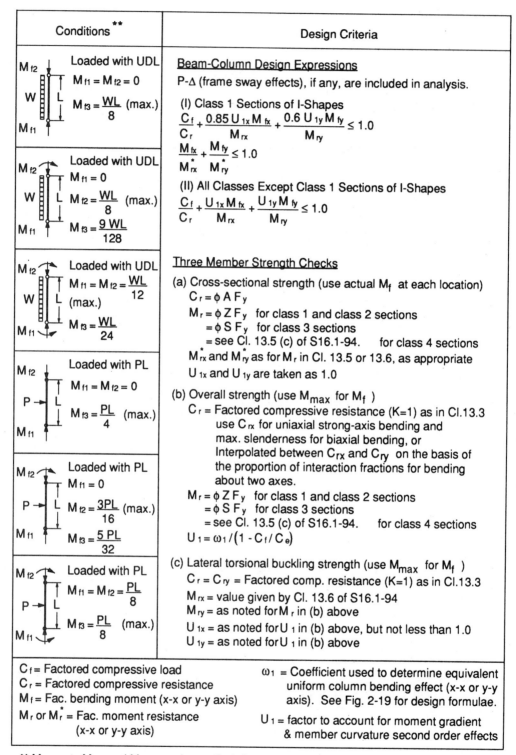

**Figure 2-23
Prismatic Beam-Columns with Transverse Loads**

1 sections does not vary linearly with axial force, for which case a factor of 0.85 multiplying the moment term would appear to be appropriate (see Clause 13.8.2).

For members subjected predominantly to bending, i.e. when the tensile force is relatively small, failure may still occur by lateral torsional buckling. The expressions of Clause 13.9(b) result from that of Clause 13.9(a) when a negative sign is assigned to the tension interaction component and when M_r is based on the overall member behaviour taking lateral torsional buckling into account.

13.10 Load Bearing

The bearing resistance given for machined, accurately sawn, or fitted parts in contact, Clause 13.10.1(a), reflects the fact that a triaxial compressive stress state, restricting yielding of the parts in contact, generally exists. The value given is based on earlier working stress design standards, which have given satisfactory results.

For a cylindrical roller or rocker, Clause 13.10.1(b), a more generalized expression has been adopted for the 1994 edition which recognizes that the roller or rocker may rest in a cylindrical groove in the supporting plate. This results in a supporting area larger than that for the case of a flat supporting plate.

In the case of a cylindrical groove in the supporting plate, the maximum shearing stress developed due to a line load of q kN/mm, (Seeley and Smith, 1957) is,

$$\tau_{max} = 0.27 \sqrt{\frac{qE}{2\pi(1-v^2)}\left(\frac{R_2 - R_1}{R_1 R_2}\right)}$$

where v is Poisson's ratio. From this, the bearing resistance is then

$$\frac{B_r}{\phi} = qL = \frac{2\pi L(1-v^2)(\tau_{max})^2}{0.27^2 E}\left(\frac{R_1 R_2}{R_2 - R_1}\right)$$

Calibrating this resistance to that given in S16-1969 at F_y = 300 MPa gives $\tau_{max} = 0.77\, F_y$, and

$$\frac{B_r}{\phi} = 0.000\,26 \left(\frac{R_1}{1 - \frac{R_1}{R_2}}\right) LF_y^2$$

For a flat plate, the "Hertz" solution, as reported by Manniche and Ward-Hall (1975), gives the allowable load as

$$2.86\, DL\, \frac{(2.7\, F_y)^2}{E} = 0.000\,20\, R_1 LF_y^2$$

and indicates that the value of $0.000\,26\, R_1$ obtained by calibration with the existing standard for a yield stress of about 300 MPa is somewhat conservative.

This is confirmed by Kennedy and Kennedy (1987) who reported that at this load no permanent deformation resulted and recommended that this value be used as a serviceability limit. They also reported that the rolling resistance of rollers varied as the fourth power of the unit normal load in kN/mm.

In bearing-type connections (Clause 13.10.1(c)) excessive deformation in front of the loaded edge of the bolt hole may occur. Tests have shown (Munse 1959; Jones, 1958; de Back and de Jong 1968; Hirano 1970) that the ratio of the bearing stress (B_r/dt) to the ultimate tensile strength of the plate (F_u) is in the same ratio as the end distance of the bolt (e) to its diameter (d). Thus,

$$\frac{B_r}{\phi\,dt} = \frac{e}{d} F_u$$

Or, for n fasteners, $B_r = \phi tneF_u$

Because the test results do not provide data for e/d greater than 3, an upper limit of $e = 3d$ is imposed. That is,

$$B_r \leq 3\,\phi\,tdnF_u$$

As for connections in general, the value of ϕ in Clause 13.10.1(c) is to be taken as 0.67.

The note directs designers to Clause 13.2 and, by implication to Clause 12.3, to investigate any potential for block tear-out when the end distance, e, is small.

13.11 Bolts In Bearing-Type Connections

13.11.2 Bolts in Shear

Based on extensive testing, it has been established that the shear strength of high strength bolts is approximately 0.60 times the tensile strength of the bolt material. However, if the bolt thread is intercepted by a shear plane, there is less shear area available. The ratio of the area through the threads of a bolt to its shank area is about 0.70 for the usual structural sizes. Moreover, in long joints, the average resistance per fastener is less. The step reduction for joints longer than 1300 mm is a reasonable approximation of the true behaviour. In this context, "joint length" refers to an axially loaded connection, such as a lap splice, whose length is measured parallel to the direction of applied force. This clause does not apply to a shear connection at the end of a girder web where the load is distributed reasonably uniformly to the fasteners.

13.11.3 Bolts in Tension

The ultimate resistance of a single high-strength bolt loaded in tension is equal to the product of its tensile stress area (a value between the gross bolt area and the area at the root of the thread) and the ultimate tensile strength of the bolt. The tensile stress area is taken as 0.75 of the gross area of the bolt.

In calculating the tensile force acting on a bolt, prying action shall be taken into account. See Clause 23.1.4.

For bolts subjected to tensile cyclic loading, such as in a T-type connection, prying must not exceed 30%. In addition, the range of stress under specified loads is limited based on recommendations of Kulak *et al*, (1987). In such circumstances, special attention must be given to the bolt installation procedures to ensure that the prescribed pretensions are attained.

13.11.4 Bolts in Combined Shear and Tension

The expression for the ultimate strength interaction between tension and shear applied to a fastener has been shown to model empirically the results of tests on single fasteners loaded simultaneously in shear and tension. The values of V_r and T_r are the full resistances in shear and tension respectively which would be used in the absence of the other loading. For small components of factored load relative to the resistance in one direction, the resistance in the other direction is reduced only a small amount; e.g., for a factored tension equal to 20 percent of the full tensile resistance, the resistance available for shear is only reduced by 2 percent of the full value which would be present in the absence of tension.

13.12 Bolts in Slip-Critical Connections

13.12.2 Different installation procedures may result in different probabilities of slip, see Kulak *et al*, (1987).

Both the slip coefficient and the initial clamping force have considerable variation about their mean values. The coefficients of friction for coatings can vary as a function of the specific coating constituents and, therefore, values of k_s may differ from one coating specification to another. The value of k_s intended for use on a project should be specified.

The resistance expression given in this edition of S16.1 replaces that used in previous editions in that the mean slip coefficient, k_s is specifically included. The clamping force is due to the pretensioning of the bolts to an initial tension, T_i, which is a minimum of 70% of the tensile strength (0.70 $A_s F_u$) where $A_s = 0.75 A_b$. Thus, the clamping force per bolt is

$$0.70 \times 0.75\, A_b\, F_u \quad \text{or} \quad 0.53\, A_b\, F_u$$

The value of c_1 establishes the probability level of slip for specific grades of bolts and the installation method. Table 2 of S16.1-94 gives values of c_1 for bolts installed by turn-of-nut procedures and for a probability level of 5%. It also provides mean slip coefficients for the three most common cases of surface conditions. Values of k_s and c_1 for many other common situations are given by Kulak *et al.* (1987).

The use of slip-critical connections should be the *exception* rather than the rule. They are the preferred solution *only where cyclic loads or frequent load reversals are present*, or where the use of the structure is such that the small one-time slips that may occur cannot be tolerated. See also Commentary Clause 21.12.2.

13.12.3 The resistance to slip is reduced as tensile load is applied and reaches zero when the parts are on the verge of separation, as no clamping force then remains. The interaction relationship is linear.

The term $1.9/(n\, A_b\, F_u)$ is the reciprocal of the initial bolt tension, $0.53\, n\, A_b\, F_u$.

13.13 Welds

13.13.1 General

This section has been revised and simplified very considerably. Resistances are now presented, in a format paralleling those for other elements, for welds in shear and in tension. A resistance factor of 0.67 is used universally in this section, recognizing that a larger value of the reliability index is used for connection resistances. Moreover the base metal resistance is written in terms of the fracture strength, F_u, rather than its yield strength. The greater strength of fillet welds loaded at an angle to their axes is formally recognized.

When electrodes with ultimate strengths equal to or greater than that of the base metal are used, they are termed "matching electrodes" and, for G40.21-M steels, are classified in Table 3. When atmospheric resisting steel grades are used in the uncoated condition, additional requirements for corrosion resistance or colour are also required for matching electrodes. In the resistance expressions, ϕ_w is taken as 0.67 so that, as for bolts, the weld has a larger reliability index and reduced probability of failure, as compared to the member itself.

13.13.2 Shear

The shear resistance of a weld is evaluated on the basis of both the resistance of the weld itself and of the base metal adjacent to the weld: the latter rarely governs. Thus, CJPG, PJPG, plug, and slot welds loaded in shear have resistances equal to the lesser of weld throat or fusion face shear strength.

The resistance of the base metal for all welds is now expressed in terms of the ultimate tensile strength of the base metal but uses the resistance factor for welds. For G40.21-M 300W steel this represents an increase of 12 percent over the resistance as determined in the 1989 edition ($0.67F_u/0.90F_y$) for the base metal.

For fillet welds in shear, the resistance of the weld metal is given as a function of the angle between the axis of the weld and the line of action of the force. Thus, this clause provides the formal basis for the ultimate strength analysis for welds.

Using the instantaneous shear centre concept, the resistance expression in 13.13.2.2(b) forms the basis of the eccentric load tables given in Part 3 (Butler and Kulak 1971, Butler et al 1972, Miazga and Kennedy 1989, Lesik and Kennedy 1990, Kennedy et al. 1990). This ultimate strength analysis, recognizing the true behaviour of the weldments, results in much more consistent strength predictions than the traditional approach (i.e., taking the quantity $1.00 + 0.50 \sin^{1.5}\theta$ as 1.0).

In the expression for the shear strength of the weld, the factor 0.67 relates the shear strength of the weld to the electrode tensile strength, as given by the electrode classification number. Lesik and Kennedy (1990) give 0.75 for this factor, based on 126 tests reported in the literature. The coefficient 0.50 in the quantity $1.00 + 0.50 \sin^{1.5}\theta$ is for tension induced shear and is slightly more liberal than the average value of tension and compression induced shear of 1.42 reported by Lesik and Kennedy. In addition, the factor 1.50 is the correct value for Clause 13.13.3.3 in which tension is the critical case. The value of 0.50 has also been adopted by AWS and AISC.

13.13.3 Tension Normal to Axis of Weld

Gagnon and Kennedy (1989) established that the net area tensile resistance, i.e. on a unit area basis, transverse to the axis of a PJPG weld, is the same as for the base metal when matching electrodes are used. The previous conservative practice of assigning shear resistances to these welds was replaced in 1989 edition with tensile resistances, consistent with the tensile resistance of complete penetration welds equaling the full tensile resistance of the member.

For T-type joints consisting of PJPG weld and a reinforcing fillet weld, Clause 13.13.3.3 provides a conservative estimate of the tensile resistance by taking the vector sum of the individual component resistances of the PJPG and fillet welds.

13.14 Steel Plate Shear Walls

See Commentary for Appendix M.

14. FATIGUE

Clause 14 provides the requirements for the design of members and connections subject to fatigue. The limit state of fatigue is one for which the requirements are checked at specified load levels. In addition, Clause 14.1.1 requires that factored resistances be sufficient for the factored static loads.

A substantial amount of experimental data, developed on steel beams since 1967 under the sponsorship of the National Co-operative Highway Research Program (NCHRP 1970, 1974; Fisher 1974) of the U.S.A., has shown that the most important factors governing fatigue resistance are the stress range and the type of detail.

The provisions of this clause are those adopted by the Canadian Highway Bridge Design

Code (CSA 1996) and AASHTO (AASHTO 1994) with few exceptions. While fatigue is generally not a design consideration for buildings such as those for commercial or residential occupancies, industrial buildings may have many members, such as crane girders, for which fatigue is a design consideration. Other instances where fatigue is a design consideration are amusement rides, wave guides, sign supports and beams supporting reciprocating machinery. However, when fatigue is a design consideration, all design, detailing, fabrication, and inspection aspects of Clause 14 must receive careful attention.

Details are assigned to one of eight stress range categories ("A" to "E1") instead of the nine categories of the 1989 edition.

The stress range categories, and details assigned to them, are identical to those in the Canadian Highway Bridge Design Code (CSA 1996), with the exception that S16.1-94 also includes a detail of a hollow structural section fillet welded to a base plate. An additional detail, based on work by Frank (1980), has been included for the fatigue strength of anchor bolts. A number of bridge details have been omitted from S16.1-94. Kulak and Smith (1993) provide an extensive background to the development of the stress range concept and the categorization of individual details.

Appendix K illustrates various details assigned to the stress range categories in Table 4(c).

A detail is considered satisfactory in fatigue provided that the stress range is less than that corresponding to the expected number of cycles of loading for the design life of the structure.

14.1.5 The maximum web slenderness permitted ($3150/\sqrt{F_y}$) is based on research conducted by Toprac and Natarajan (1971).

14.2 Fatigue Limit State

With the 1994 edition equal emphasis is placed on fatigue cracks that result from live load effects and on those that result from local distortions. For some structural elements, such as crane girders and their attachments, local distortions can be a significant source of potential fatigue cracks (Griggs 1976).

In assessing the number of cycles to which a detail may be subjected, it is important to determine the numbers of cycles that occur during each passage of the load. The 50 000 cycle demarcation value is significantly greater than the 10 000 cycle limit of previous editions.

14.3 Live Load-Induced Fatigue

Only variable loads, live loads and impact, contribute to the stress range. The situations given in Table 4(c) have been assigned to their respective detail categories on the basis of test results, field experience, and analytical methods which have included the effects of residual stress and any stress raisers. Thus, the calculation of stress range for the situations covered in Table 4(c) need only be accomplished using normal analytical means, i.e. without consideration of stress concentration factors.

Fatigue design is considered only for regions subjected to tensile stresses or stress reversal. A crack may appear in a region subject only to compression stresses if a zone of tensile residual stress exists there. Such a crack may propagate but usually stops once it leaves the tensile residual stress zone and enters the region with net compressive stresses.

Since any excursion into a tensile stress zone may result in a significant stress range, compressive load-induced fatigue may be ignored only when the compressive stresses are twice the maximum tensile live load stresses.

The fatigue life is inversely proportional to the cube of the stress range for values above the constant amplitude threshold stress range, e.g., the fatigue life increases by a factor of

2^3 if the stress range is reduced by a factor of 2. Expressing the fatigue resistance in equation form rather than in tables of discrete values for various cycles simplifies the design process.

14.4 Distortion-Induced Fatigue

Secondary stresses due to deformations and out-of-plane movements can also be a source of fatigue failures (Fisher 1978 and 1984). Crane girders, their attachments and supports, require careful design and attention to details to minimize fatigue cracks (Griggs 1976).

15. BEAMS AND GIRDERS

15.1 Proportioning

Lilley and Carpenter (1940) have shown that reductions of flange area up to 15% can be disregarded due to the limited inelastic behaviour near the holes.

15.2 Rotational Restraint at Points of Support

A severe stability problem may exist when a beam or girder is continuous over the top of a column. The compression flange of the beam tends to buckle sideways and simultaneously, the beam-column junction tends to buckle sideways because of the compression in the column. Three mechanisms exist for providing lateral restraint: direct acting bracing, such as provided by bottom chord extensions of joists, beam web stiffeners welded to the bottom flange, or the distortional stiffness of the web. In the latter two cases, the connection of the beam flange to the column cap plate must have strength and stiffness (Chien, 1989). The restraint offered by the distortion of the web requires very careful assessment. See also the commentaries on Clause 13.6 and Clause 20.2.

15.3 Copes

Flanges are coped to permit beams to be connected to girder webs with simple connections while maintaining the tops of the flanges at the same elevation. Long copes may seriously effect the lateral torsional buckling resistance of a beam (Cheng and Yura, 1986). The reduced shear and moment resistance at the coped cross-section should be examined.

15.4 Reduced Moment Resistance of Girders with Thin Webs

A plate girder with Class 3 flanges and Class 4 webs has a maximum moment resistance less than ϕM_y because the Class 4 web buckles prematurely due to the compressive bending stresses. The reduction in moment resistance is based on Basler and Thurlimann (1961). Figure 2-24 shows an approximate stress distribution in a girder with a buckled web. The reduction in moment resistance is generally small, as shown in Figure 2-25.

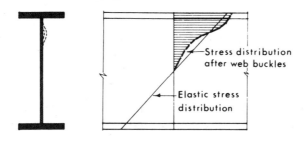

Figure 2–24
Approximate Stress Distribution In Girders With Buckled Web

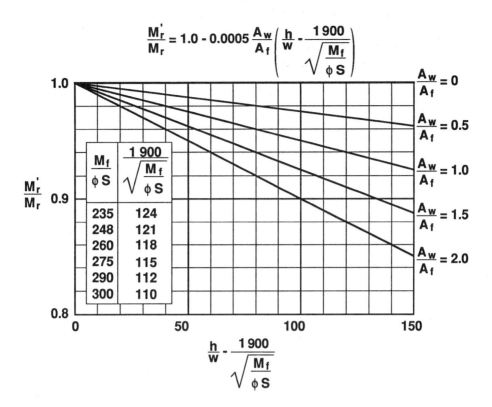

**Figure 2–25
Reduced Moment Resistance In Girders With Thin Webs**

The limit of $1900/\sqrt{F_y}$ for the slenderness of a Class 3 web is replaced in this clause by $1900/\sqrt{M_f/(\phi S)}$ to account for the possibility that the factored moment may be less than $M_r = \phi SF_y$, thereby reducing the propensity for web buckling.

In some circumstances, a plate girder may be subjected to an axial compressive force in addition to the bending moment (e.g. rafters in a heavy industrial gable frame, beams in a braced frame). The constant 1900 is then multiplied by the factor $(1.0 - 0.65\, C_f/C_y)$ to account for the increased tendency for the web to buckle. The compressive stresses due to the axial load are additive to the compressive stress due to bending thus increasing the depth of web in compression (see also commentary to Clause 11).

15.5 Flanges.

The theoretical cut-off point is the location where the moment resistance of the beam without cover plates equals the factored moment (Figure 2–26). The distance a' increases as shear lag becomes more significant, as is the case when the weld size is smaller, or when there is no weld across the end of the plate. Theoretical and experimental studies of girders with welded cover plates (ASCE, 1967) shows that the cover plate load can be developed within length (a'). Clause 15.5.4 limits the length of a' for welded cover plates and may therefore necessitate an increase in weld size or an extension of the cover plate so that the force at a distance a' from its end equals that which the terminal welds will support.

15.6 Bearing Stiffeners

The inclusion of a portion of the web in the column section resisting the direct load, and the assumption of an effective length of 0.75 times the stiffener length, are approximations

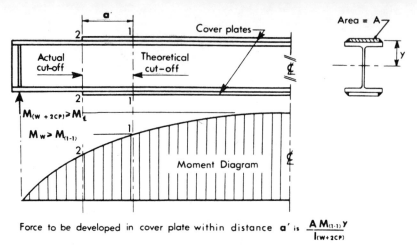

Force to be developed in cover plate within distance a' is $\dfrac{A M_{(1-1)} y}{I_{(w+2CP)}}$

For welded cover plates a' is not to exceed lengths specified in clause 15.5.4 CAN/CSA-S16.1-94. In certain cases this will necessitate an increase in weld size or an extension of cover plate so that force at distance a' from its end equals that which terminal welds will support.

**Figure 2–26
Cover Plate Development**

to the behaviour of the web under edge loading that have proved satisfactory in many years of use.

15.7 Intermediate Transverse Stiffeners

15.7.1 Figure 2–27 illustrates the action of a thin girder web under load. Tension fields are developed in the interior panels but cannot develop in the unanchored end panels, for which the maximum shear stress is, therefore, either the elastic or inelastic critical plate buckling stress in shear.

15.7.2 The limits on stiffener spacing are based on practical considerations. When a/h > 3,

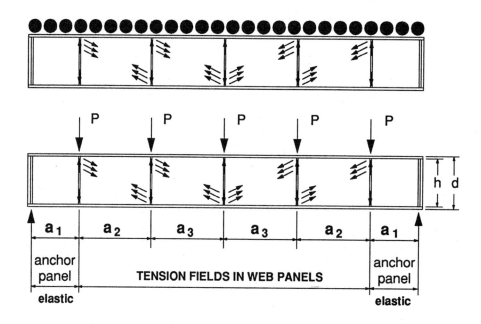

**Figure 2–27
Action Of A Thin Web Plate Girder Under Load**

the tension field contribution is reduced. When h/w > 150 the maximum stiffener spacing is reduced for ease in fabrication and handling.

15.7.3 Clause 15.7.3 requires that intermediate transverse stiffeners have both a minimum moment of inertia and a minimum area. The former provides the required stiffness when web panels are behaving in an elastic manner; the latter ensures that the stiffener can sustain the compression, to which it is subjected, when the web panel develops a tension field. Because stiffeners subject to compression act as columns, stiffeners placed only on one side of the web are loaded eccentrically and are less efficient. The stiffener factor (D) is included in the formula for stiffener area to account for the lowered efficiency of stiffeners furnished singly, rather than in pairs.

15.7.4 The minimum shear to be transferred between the stiffener and the web is based on Basler (1961c).

15.7.5 The requirement of attaching single intermediate stiffeners to the compression flange is to prevent tipping of the flange under loading.

15.9 Web Crippling and Yielding

Unstiffened webs of beams and girders carrying concentrated loads applied normal to

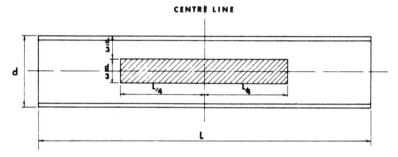

Unreinforced circular holes may be placed anywhere within the hatched zone without affecting the strength of the beam for design purposes, provided:

1. Beam supports uniformly distributed load only.
2. Beam section has an axis of symmetry in plane of bending.
3. Spacing of holes meets the requirements shown below.

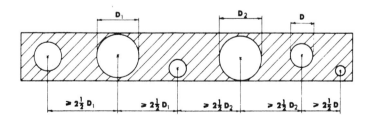

Spacing must be ⩾ 2½ times diameter of the larger opening of any two adjacent openings.

Figure 2–28
Unreinforced Circular Web Openings In Beams

one flange can fail either from web yielding or from web crippling (a localized wrinkling, folding, or buckling of the web).

If the web is relatively thick, yielding will occur prior to crippling and expressions 15.9(a)(i) and 15.9(b)(i) govern web resistance. These expressions are based on test results (Graham et al. 1959) which show that a conservative estimate of strength is obtained by assuming the web yields over a length of N + 5k. The coefficient 1.10 was obtained by calibration to the previous working stress standard (CSA 1969).

Relatively thin webs, cripple before yielding and expressions 15.9(a)(ii) and 15.9(b)(ii) govern the web resistance. The tests of Roberts (1981) and Roberts and Chong (1981) indicate that the important parameters governing the crippling load are the square of the web thickness, the length of the loaded area, the flange stiffness, and the yield strength of the web. The resistance expressions are empirically derived, and the coefficients 300 and 150 are in units of $\sqrt{MPa}$.

For unstiffened portions of webs, when concentrated compressive loads are applied opposite one another to *both* flanges, the compressive resistance of the web acting as a column should also be investigated.

15.10 Openings

The conditions under which unreinforced circular openings may be used are based on work by Redwood and McCutcheon (1968) and are illustrated in Figure 2–28.

Elastic and plastic analysis to determine the effect of openings in a member are given in Bower et al. (1971) and Redwood (1971, 1972, 1973), respectively. See Part 5 for worked examples.

A combination of vertical and horizontal intersecting stiffeners (particularly on both sides of a web) is seldom justified, and is quite expensive to fabricate. Generally, horizontal stiffeners alone are adequate. When both vertical and horizontal stiffeners are necessary, the horizontal stiffeners should be on one side of the web, and vertical stiffeners on the other, in order to achieve economy.

15.11 Torsion

In many cases, beams are not subject to torsion because of the restraint provided by slabs, bracing or other framing members. The torsional resistance of open sections having two flanges consists of the St. Venant torsional resistance and the warping torsional resistance.

Information on moment-torque interaction diagrams for I-shaped members is given in Driver and Kennedy (1989). Serviceability criteria will often govern the design of a beam subject to torsion. Limiting the maximum stress due to bending and warping, at the specified load level, to the yield strength guards against inelastic deformation. For inelastic torsion of steel I-beams, see Yong Lin Pi and Trahair (1995). For elastic analyses, see Bethlehem Steel Corporation (1967), and Brockenbrough and Johnston (1974). For methods of predicting the angle of twist in a W-shape beam, see Englekirk (1994).

16. OPEN-WEB STEEL JOISTS

16.1 Scope

Open-web steel joists (OWSJ or joists), as described in Clause 16.2, are generally proprietary products whose design, manufacture, transport, and erection are covered by the requirements of Clause 16. The Standard clarifies the information to be provided by the building designer (user-purchaser) and the joist manufacturer (joist designer–fabricator).

16.3 Definitions

There are many variations of the simply supported joist which has given rise to the definition for *special open-web steel joists or special joists*. These joists are those subjected to specific loading conditions, cantilever joists, continuous joists, joists having special support conditions, and joists used as part of the frame (see Clause 16.6.13.3). As defined, tie joists are designed to resist gravity loads only and any connection to a column is to facilitate erection.

16.4 Materials

The use of yield strength levels reported on mill test certificates for the purposes of design is prohibited here as throughout the Standard. This practice could significantly lower the margin of safety by not properly accounting for the statistical distribution of yield levels. Historically, all design rules have been, and still are, based on the use of the *specified* minimum yield point or yield strength. For structural members cold-formed to shape, the increase in yield strength due to cold forming, as given in Clause 5.2 of CAN/CSA-S136, may be taken into account provided that the increase is based on the specified minimum values in the relevant structural steel material standard.

16.5.1 Building Design Drawings

The Standard recognizes that the building designer may not be the joist designer; therefore, building design drawings are required to provide specific information for the design of the joists.

Loads such as unbalanced, non-uniform, concentrated, and net uplift, are to be shown by the building designer. Figure 2–29 shows a joist schedule that could be used to record all loads on joists.

All heavy concentrated loads such as those resulting from partitions, large pipes, mechanical, and other equipment to be supported by OWSJ, should be shown on the structural drawings. *Small* concentrated loads may be allowed for in the uniform dead load.

When the importance factor, γ, (see Clause 7.2.5) is not equal to 1.0, it should be specified by the building designer.

Options, such as attachments for deck when used as a diaphragm, special camber and any other special requirements should also be provided. Where vibration of a floor system

Mark	Depth (mm)	Spacing (mm)	Specified Dead Load	Specified Live Load	Specified Snow Load	Specified Wind Load	Remarks
J1	600	1 300	2.4 kPa	2.6 kPa			$\Delta_{live} = \dfrac{span}{320}$ Suggested I_x for vibration = _____
J2	700	2 000	8.9 kN, 1.5 kN/m, 3 m, 12 000		4.38 kN/m, 10.2 kN/m, 3 m, 12 000	-2.4 kN/m, 12 000	$\Delta_{live} = \dfrac{span}{240}$

**Figure 2–29
Joist Schedule**

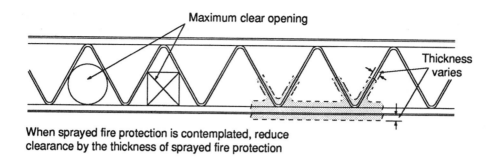

**Figure 2–30
Sizes Of Openings For Electrical And Mechanical Equipment**

is a consideration, it is recommended that the building designer give a suggested moment of inertia I_x. Because the depth of joists supplied from different joist manufacturers may vary slightly from nominal values, the depth, when it is critical, should be specified.

Although steel joist manufacturers may indicate the maximum clear openings for ducts, etc. which can be accommodated through the web openings of each depth of their OWSJs, building designers should, in general, show on the building design drawings the size, location and elevation of openings required through the OWSJs (Figure 2–30). Large ducts may be accommodated by special design. Ducts which require open panels and corresponding reinforcement of the joist should, where possible, be located within the middle half of the joist to minimize shear problems. This information is required prior to the time of tendering to permit appropriate costing.

Specific joist designations from a manufacturer's catalogue or from the AISC and Steel Joist Institute of the U.S.A. are not appropriate and should *not* be specified.

16.5.2 Joist Design Drawings

The design information of a joist manufacturer may come in varying forms such as: design sheets, computer printout, and tables. Not all joist manufacturers make "traditional" detail drawings.

16.6.1 Loading for Standard Open-Web Steel Joists

To accommodate load tables in standard OWSJ catalogues, for such purposes as estimating, the Standard provides for specific unbalanced and concentrated loading conditions. The concentrated factored loads given for floor and roof joists are total loads to be applied at a panel point. In such load tables, only the total factored load is shown because the ratio of dead-to-live load of the floor system is not known.

Three loading conditions, each considered separately, are specified to guard against overloading of joist components due to unbalanced or concentrated loads.

16.6.2 Loading for Special Open-Web Steel Joists

The four factored live load combinations are consistent with Section 4.1 of the National Building Code of Canada (1995c). In particular, as required by the National Building Code of Canada, roofs and the joists supporting them may be subject to uplift loads due to wind.

16.6.3 Design Assumptions

The loads may be replaced by statically equivalent loads applied at the panel points for the purpose of determining axial forces in all members. It is assumed that any moments induced in the joist chord by direct loading do not influence the magnitude of the axial forces

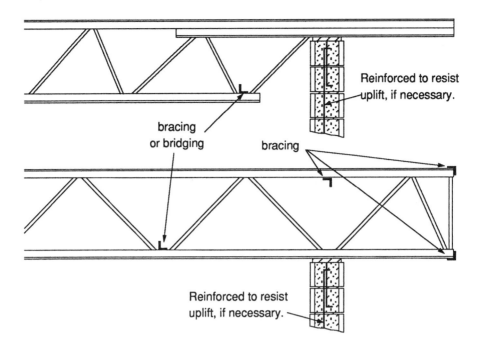

**Figure 2–31
Bracing And Bridging Of Cantilever Joists**

in the members. Tests on trusses (Aziz 1972) have shown that the secondary moments induced at rigid joints due to joint rotations do not affect the ultimate axial forces determined by a pin-jointed truss analysis.

16.6.4 Verification of Joist Manufacturer's Design

When there is difficulty in analyzing the effect of certain specific conditions, for example a particular web-chord connection, or a geometric configuration of a cold formed chord, a joist manufacturer may elect to verify the design assumption by a test. In the numerical factor of 1.10/0.90, stipulated as a multiplier for the factored loads, the factor of 1.10 provides that the results of limited number of tests bear a similar statistical relationship to the entire series of joists that the average yield strength has to the specified minimum yield strength, F_y, and the factor 0.90 is the reciprocal of the resistance factor.

16.6.7 Tension Chord

A minimum radius of gyration is specified for tension chord members to provide a minimum stiffness for handling and erection.

Under certain loading conditions, net compression forces may occur in segments of tension chords and must be considered. Bracing of the chord, for compression, may be provided by regular bridging only if the bridging meets requirements of Clause 20.2. As a minimum, lines of bracing are specifically required near the ends of tension chords in order to enhance stability when the wind causes a net uplift.

Bottom chord bracing may be required for continuous and cantilever joists as shown in Figure 2–31.

In those cases, where the bottom chord has little or no net compression, bracing is not required for cantilever joists. However, it is generally considered good practice to install a line of bridging at the first bottom chord panel point as shown in Figure 2–31.

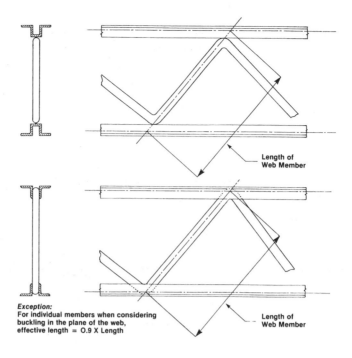

**Figure 2–32
Length Of Joist Web Members**

16.6.8 Compression Chord

When the conditions set out in Clause 16.6.8.1 are fulfilled, only axial force need be considered when the panel length is less than 610 mm (Kennedy and Rowan 1964). In these cases, the stiffness of the floor or roof structure tends to help transfer loads to the panel points of the joist, thus offsetting the reduction in chord capacity due to local bending. When the panel length exceeds 610 mm, a simplified form of the beam–column interaction formula is used, based on the larger shape factor common to joist chords. When calculating bending moments in the end panel, it is customary to assume the end of the chord to be pinned, even though the joist bearing is welded to its support. The stiffening effect of supported deck or of the web is to be neglected when determining the appropriate width–thickness ratio (Clause 16.6.6.1) of the compression top chord.

The requirement in Clause 16.6.8.4, that the flat width of the chord component be at least 5 mm larger than the nominal dimension of the weld, should be considered an absolute minimum. Increasing the dimension may improve workmanship. See Clauses 16.9.5.1 and 16.9.5.2 regarding workmanship requirements when laying and attaching deck to joists.

16.6.9 Webs

The length of web members for design purposes are shown in Figure 2–32. With the exception of web members made of individual members, the effective length factor is always taken as 1.0. For individual members this factor is 0.9 for buckling in the plane of the web (see Clause C8 of Appendix C), but is 1.0 for buckling perpendicular to the plane of the web.

Web members in tension are not required to meet a limiting slenderness ratio. This is significant when flats are used as tension members; however, attention should be paid to those loading cases where the possibility of shear reversal exists. Under these circumstances, it is likely that diagonals (except for end diagonals) may have to resist compression forces.

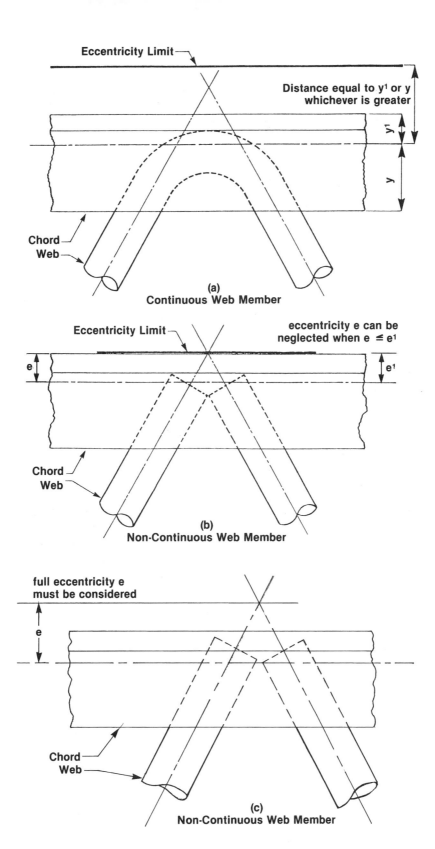

Figure 2–33
Eccentricity Limits At Panel Points Of Joists

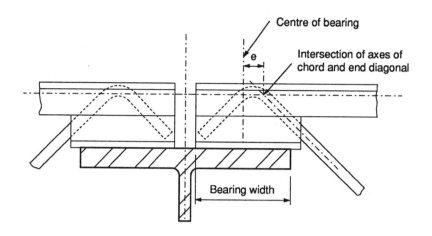

**Figure 2–34
Joist End Bearing Eccentricity**

16.6.10 Spacers and Battens

Spacers and battens must be an integral part of the joist and (see Clause 16.6.8.2(c)) steel deck cannot be considered to act as spacers or battens.

16.6.11 Connections and Splices

Although splices are permitted at any point in chord or web members, the splices must be capable of carrying the factored loads without exceeding the factored resistances of the members. Butt-welded splices are permitted provided they develop the factored tensile resistance of the member.

As a general rule, it is preferable to have the gravity axes of members meet at a common point within a joint. However, when this is not practical, eccentricities may be neglected if they do not exceed those described in Clause 16.6.11.4; see Figure 2–33. Kaliandasani, et al (1977) have shown that the effect of small eccentricities is of minor consequence, except for eccentricities at the end bearing and the intersection of the end diagonal and bottom chord. (See also Clause 16.6.12.4.)

16.6.12 Bearings

16.6.12.1 As required by Clause 16.5.1(c), the factored bearing resistance of the supporting material or the size of the bearing plates must be given on the building design drawings.

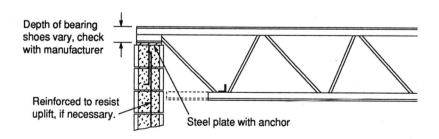

**Figure 2–35
Joists Bearing On Steel Plate Anchored To Concrete And Masonry**

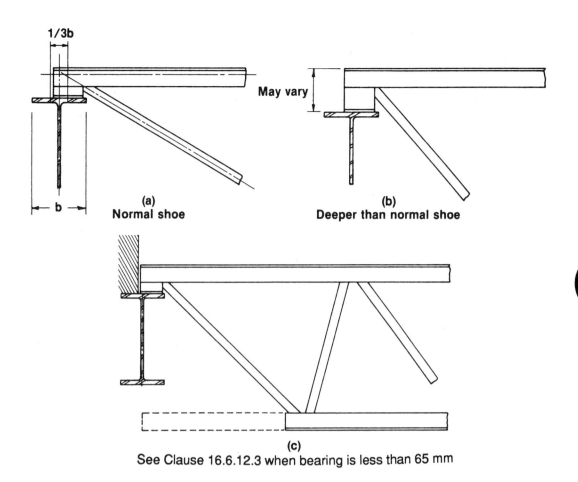

**Figure 2–36
Joists Bearing On Steel**

16.6.12.2 It is likely that the centre of bearing will be eccentric with respect to the intersection of the axes of the chord and the end diagonal as shown in Figure 2–34. Because the location of the centre of bearing is dependent on the field support conditions, and their construction tolerances, it may be wise to assume a maximum eccentricity when designing the bearing detail. In lieu of specific information, a reasonable assumption is to use a minimum eccentricity of one half the minimum bearing on a steel support of 65 mm. When detailing joists, care must be taken to provide clearance between the end diagonal and the supporting member or wall. See Figure 2–35. A maximum clearance of 25 mm is suggested to minimize eccentricities. One solution, to obtain proper bearing, is to increase the depth of the bearing shoe.

For spandrel beams and other beams on which joists frame from one side only, good practice suggests that the centre of the bearing shoe be located within the middle third of the flange of the supporting beam (Figure 2–36(a)). As the depth of bearing shoes vary, the building designer should check with the manufacturer in setting "top of steel" elevations. By using a deep shoe, interference between the support and the end diagonal will be avoided as shown in Figure 2–36(b).

If the support is found to be improperly located, such that the span of the joist is increased, the resulting eccentricity may be greater than that assumed. Increasing the length of the bearing shoe to obtain proper bearing may create the more serious problem of increasing the amount of eccentricity.

16.6.13 Anchorage

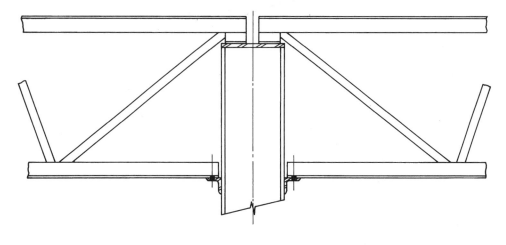

**Figure 2–37
Tie Joists**

16.6.13.1 When a joist is subject to net uplift, not only must the anchorage be sufficient to transmit the net uplift to the supporting structure but the supporting structure should be capable of accepting that force.

The anchorage of joist ends to supporting steel beams provide both lateral restraint and torsional restraint to the top flange of the supporting steel beam (Albert et al. 1992). When the supporting beam is simply supported, the restraint provided to the compression flange likely means that the full cross-sectional bending resistance can be realized. In cantilever-suspended span construction, the restraint provided by the joists is applied in part to the tension flange and is, therefore, less effective in restraining the compression (lower) flange from buckling.

Albert et al. (1992) and Essa and Kennedy (1993) show that, while the increase in moment resistance due to lateral restraint is substantial, in cantilever-suspended span construction, the further increase when torsional restraint is considered is even more substantial. The torsional restraint develops when the compression flange tends to buckle sideways distorting the web and twisting the top flange which is restrained by bending of the joists about the strong axis. The anchorage must therefore be capable of transmitting the moment that develops. For welds, a pair of 5 mm fillet welds 50 mm long coupled with the bearing of the joist seat would develop a factored moment resistance of about 1.8 kN·m

16.6.13.2 The function of tie joists is to assist in the erection and plumbing of the steel frame. Either the top or bottom chord is connected by bolting and, after plumbing the columns, the other chord is usually welded (Figure 2–37). In most buildings, tie joists remain as installed with both top and bottom chords connected; however, current practices vary throughout Canada with, in some cases, the bottom chord connections to the columns being made with slotted holes. Shrivastava et al. (1979) studied the behaviour of tie joist connections and concluded that they may be insufficient to carry lateral loads which could result from rigid bolting.

The designation *tie joist* is not intended to be used for joists participating in frame action.

16.6.13.3 When joists are used as part of a frame to brace columns, or to resist lateral forces on the finished structure, the appropriate moments and forces are to be shown on the building design drawings to enable the joists and the joist-to-column connections to be designed by the joist manufacturer.

In cantilever suspended span roof framing, joists may also be used to provide stability for girders passing over columns. See also the commentary on Clauses 16.6.13.1, and 13.6.

16.6.14 Deflection

The method of computing deflections given in Clause 16.6.14.2 has been verified by tests (Kennedy and Rowan 1964).

16.6.15 Camber

The nominal camber based on Clause 16.6.15 is tabulated in Table 2–1 rounded to the nearest millimetre. Manufacturing tolerances are covered in Clause 16.11.9. The maximum difference in camber of 20 mm for joists of the same span, set to limit the difference between two adjacent joists, is reached at a span of 17 000 mm.

Table 2–1. Camber for Joists

Span	Camber (mm)		
	Nominal Camber	Minimum Camber	Maximum Camber
Up to 6 000	–	3	10
7 000	3	3	11
8 000	4	3	12
9 000	6	3	14
10 000	7	3	16
11 000	8	3	17
12 000	10	3	19
13 000	12	3	21
14 000	14	4	23
15 000	16	6	26
16 000	18	8	28

16.6.16 Vibration

Appendix G of S16.1-94, *Guide for Floor Vibrations*, contains recommendations for floors supported on steel joists. By increasing the floor thickness (mass), both the frequency and the peak acceleration are reduced, thus reducing the annoyance more efficiently than by increasing the moment of inertia (I_x) of the joists.

16.6.17 Welding

16.6.17.3 Many welded joints used in joists are not prequalified under CSA W59, therefore

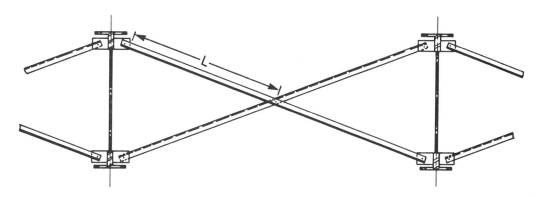

Figure 2–38
Diagonal Bridging Of Joists

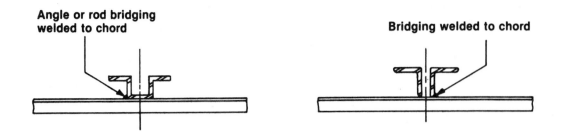

**Figure 2–39
Horizontal Bridging Connections To The Joist's Top Chord**

the certified fabricator must have all these welded joints qualified by the Canadian Welding Bureau.

16.6.17.6 Flux and slag are removed from all welds to assist in the inspection of the welds, as well as to increase the life of the protective coatings applied to the joists.

16.7 Stability During Construction

A distinction is made between bridging, put in to meet the slenderness ratio requirements for top and bottom chords, and the temporary support required by Clause 16.7 to hold joists against movement during construction. Permanent bridging, of course, can be used for both purposes.

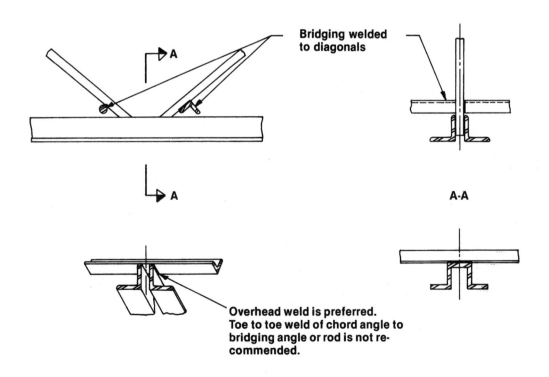

**Figure 2–40
Horizontal Bridging Connections To Joist Bottom Chord**

16.8 Bridging

Figures 2–38 to 2–40 provide illustrations of bridging and details of bridging connections.

16.8.7 Anchorage of Bridging

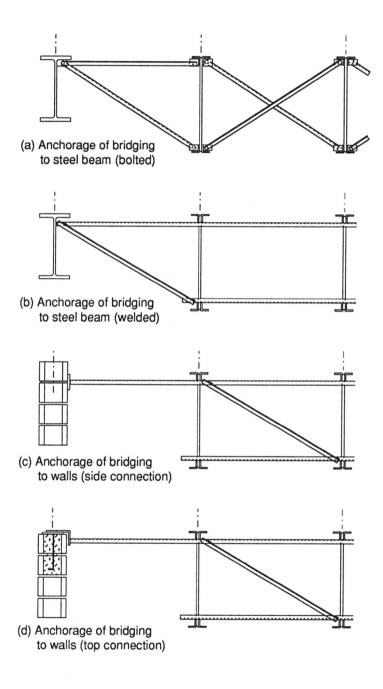

**Figure 2–41
Anchorage Of Joist Bridging**

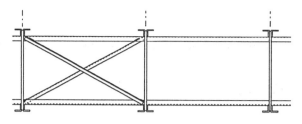

(a) diagonal bridging with horizontal bridging

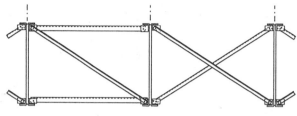

(b) horizontal bridging with diagonal bridging

**Figure 2–42
Bracing Of Joist Bridging**

Ends of bridging lines may be anchored to the adjacent steel frame or adjacent concrete or masonry walls as shown in Figure 2–41.

Where attachment to the adjacent steel frame or walls is not practicable, diagonal and horizontal bridging shall be provided in combination between adjacent joists near the ends of bridging lines as shown in Figure 2–42. Joists bearing on the bottom chord will require bridging at the ends of the top chord.

16.8.9 Spacing of Bridging

Either horizontal or diagonal bridging is acceptable, although horizontal bridging is generally recommended for shorter spans, up to about 15 m, and is usually attached by welding. Diagonal bridging is recommended for longer spans and is usually attached by bolting. Bridging need not be attached at panel points and may be fastened at any point along the length of the joists. When horizontal bridging is used, bridging lines will not necessarily appear in pairs as the requirements for support of tension chords are not the same as those for compression chords. Because the ends of joists are anchored, the supports may be assumed to be equivalent to bridging lines.

16.9.1 Decking to Provide Lateral Support

When the decking complies with Clause 16.9 and is sufficiently rigid to provide lateral support to the top (compression) chord, the top chord bridging may be removed when it is no longer required. Bottom (tension) chord bridging is permanently required such that the unsupported length of the chord does not exceed 240r, as defined in Clause 16.8.9.

16.9.5 Installation of Steel Deck

16.9.5.1 Workmanship is of concern when decking is to be attached by arc-spot welding to top chords of joists. When the joist location is marked on the deck as the deck is positioned, the welders will be more likely to position the arc-spot welds correctly.

16.9.5.2 Arc-spot welds for attaching the deck to joists are structural welds and require proper welding procedures.

16.10 Shop Painting

Interiors of buildings conditioned for human comfort are generally assumed to be of a non-corrosive environment and therefore do not require corrosion protection.

Joists normally receive one coat of paint suitable for a production line application. This paint is generally adequate for three months of exposure, which should be ample time to enclose, or paint, the joists.

Special coatings, and paints with special preparations, are expensive because they have to be applied individually to each joist by spraying or other means.

16.11 Manufacturing Tolerances

Figure 2–43 illustrates many of the manufacturing tolerance requirements.

16.12.3 Quality Control

When testing forms part of the manufacturers normal quality control programme, the test shall follow steps 1 to 4 of the loading procedure given in Part 5 of *Steel Joist Facts* (CISC 1980).

16.13.1 Erection Tolerances

Figure 2–44 illustrates many of the erection tolerance requirements.

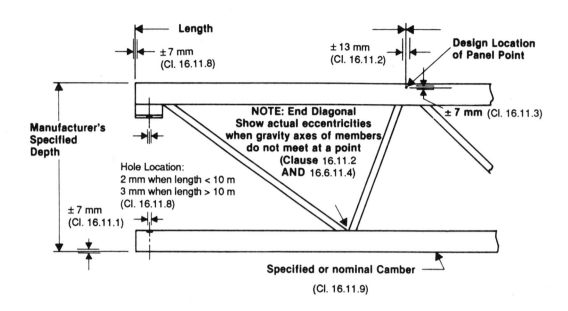

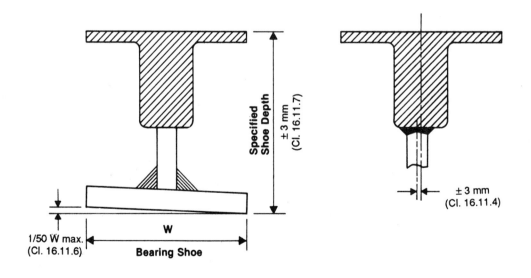

**Figure 2–43
Joist Manufacturing Tolerances**

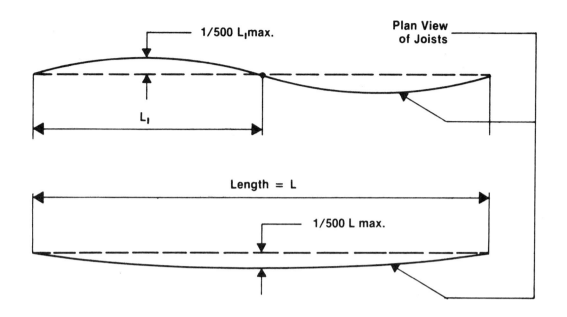

Sweep
(Cl. 16.11.5, 16.13.2.1)

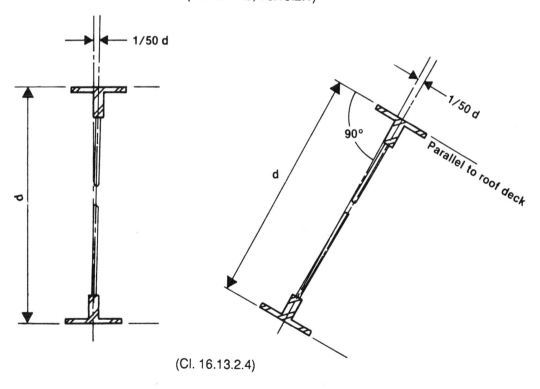

(Cl. 16.13.2.4)

**Figure 2–44
Joist Erection Tolerances**

17. COMPOSITE BEAMS

17.2 Definitions

Figure 2-45 illustrates various cases of effective slab and cover slab thickness.

17.3.1 Deflections

The moment of inertia is reduced from the transformed value to account for the increased flexibility resulting from partial shear connection, and for interfacial slip, similar to that coefficient proposed by Grant *et al.* (1977). The factor 0.85 accounts for the loss in stiffness due to interfacial slip, even with full shear connection.

The increase of the elastic deflection of 15% for creep is an arbitrary but reasonable value.

Please refer to Appendix L of the Standard for a detailed discussion of shrinkage deflections. There it is emphasized that appropriate values of the free shrinkage strain and the modulus of concrete in tension should be used in calculating these deflections. Maurer and Kennedy (1994) show that interfacial slip and, for composite joists or trusses, the open web system increase the flexibility of the system and lead to an "apparent" modulus of concrete in tension that is even less than that proposed by Shaker and Kennedy (1991) whose tests were on reinforced concrete prisms where these effects did not occur. The "apparent" value is more in line with that found by Kennedy and Brattland (1992) also based on tests on full scale trusses.

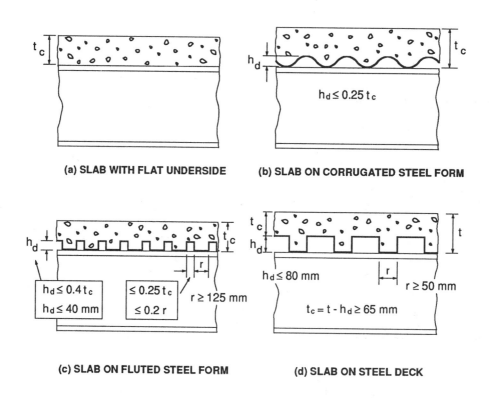

Figure 2-45
Effective Slab Thickness For Composite Beams

17.3.2 Vertical Shear

17.3.3 End Connections

These clauses follow from the assumption that the concrete does not carry any vertical shear.

17.4 Design Effective Width of Concrete

As there was no basis in fact, the effective width limit based on slab thickness has been removed, as was the case with the AISC (1993) and Eurocode No. 4 (1987) specifications. Although the effective width rules were formulated on the basis of elastic conditions (Robinson and Wallace, 1973, Adekola, 1968), the differences at ultimate load do not significantly affect the moment resistance of the composite beam (Elkelish and Robinson, 1986; Hagood et al., 1968; Johnson, 1975; Heins and Fan, 1976).

17.5 Slab Reinforcement

17.5.2 The effectiveness of the minimum requirement of two 15M bars at the ends of beams supporting ribbed slabs perpendicular to the beam proposed by Ritchie and Chien (1980) has been verified experimentally by Jent (1989).

17.5.3 The longitudinal shear forces generated by interconnecting solid concrete slabs to steel sections, trusses, or joists by means of shear connectors may cause longitudinal cracking of the slab directly over the steel. This effect is independent of any flexural cracking which may occur due to the slab spanning continuously over supports, although the two effects may combine. Longitudinal shear cracking is more apt to start from the underside of the solid slab, whereas flexural cracking is more apt to start at the top surface of the slab. Investigations by Johnson (1970), El-Ghazzi et al. (1976), and Davies (1969) have shown that a minimum area of transverse reinforcing steel is required to improve the longitudinal shear capacity of a solid slab composite beam. The minimum reinforcement ratio is the same as that specified in CSA Standard A23.3 (CSA 1994) for temperature and shrinkage reinforcement in reinforced concrete slabs.

17.5.4 For the same reasons as for Clause 17.5.3, a minimum transverse reinforcement ratio of 0.002 has also been specified for composite beams with ribbed slab when the ribs are parallel to the beam span. This ratio is reduced to 0.001 when the ribs are perpendicular to the beam span, because the steel deck provides a measure of transverse reinforcement. Reinforcement of the cover slab may also be necessary for flexure, fire resistance, shrinkage, or temperature effects.

17.6 Interconnection

When unpainted sections, trusses, or joists are totally encased in concrete as specified, effective interconnection is obtained and no shear connectors are required.

The total sheet thickness and the total amount of zinc coating are limited in order to achieve sound welds.

Tests have shown that a shear connector is not fully effective if welded to a support which is too thin or flexible (Gobel 1968). For this reason the stud diameter is limited.

17.7 Shear Connectors

The factored resistance of end welded studs in a solid slab is different from those in a ribbed slab which depend upon the orientation of the deck ribs.

For end welded studs in a solid slab, the values given in Clause 17.7.2.1 are based on work by Olgaard et al. (1971) in both normal and lightweight solid concrete slabs. The limiting value of $\phi_{sc} A_{sc} F_u$ represents the tensile strength of the stud as the stud eventually bends over and finally fails in tension.

According to Clause 17.7.2.2(a), the factored shear resistance for studs in solid slabs is applicable to ribbed slabs when the flute is wide enough (Johnson 1975). The stud resistance is reduced in narrow flutes based on Grant et al. (1977).

The provisions for ribbed slabs with ribs perpendicular to the beam are based on work by Jayas and Hosain (1988 and 1989). Push-out tests, as well as full size beam tests, indicated that failure in this type of composite beam would likely occur due to concrete pull-out. The equations of Clause 17.7.2.3, similar to those suggested by Hawkins and Mitchell (1984), provide better correlation to test results than those using the reduction factor method adopted by AISC (1993). Figure 2–46 gives diagrams of the pullout surface area. Pullout area for specific deck profiles and studs are given in Part 5.

In order to minimize excessive localized stresses in concrete, it is required that the lateral spacing centre-to-centre of studs used in pairs be not less than four stud diameters. The minimum longitudinal spacing of connectors, in both solid slabs and ribbed slabs with ribs parallel to the beam, is based on Olgaard et al. (1971). The maximum spacing limits specified for mechanical ties in Clause 17.8 has been applicable to headed studs, as they function in this capacity.

Further information on end welded studs can be found in Johnson (1970), Chien and Ritchie (1984), and Robinson (1988).

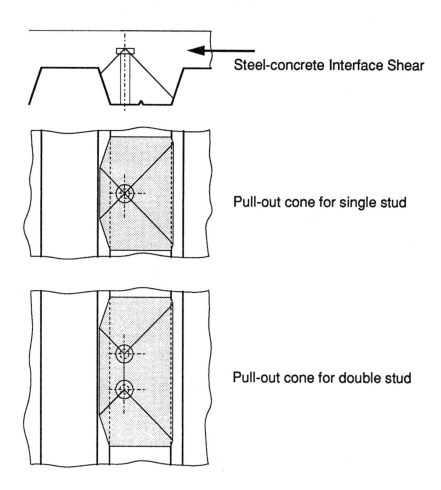

Figure 2–46
Pullout Surface Areas With Ribbed Metal Deck

17.7.3 The shear value of channel connectors is based on Slutter and Driscoll (1965).

17.9 Design of Composite Beams with Shear Connectors.

The factored moment resistance of a composite flexural member is computed based on the ultimate capacity of the cross-section (Robinson, 1969; Vincent, 1969; Hansell and Viest, 1971; Robinson and Wallace, 1973; Tall *et al.*, 1974) where the following assumptions are made:

- concrete in tension is neglected;
- only the lower chord of a steel joist or truss is considered effective when computing the moment resistance;
- the internal couple consists of equal tension and compression forces;
- the forces are obtained as the product of a limit states stress (ϕF_y for steel and 0.85 ϕ_c f'$_c$ for concrete) times the respective effective areas; and,
- to take into account the greater variability of concrete elements, the resistance factor is taken as 0.60 for concrete as compared to 0.90 for steel.

Three design cases are considered:

- Case 1 representing full shear connection with the plastic neutral axis in the slab;
- Case 2 representing full shear connection with the plastic neutral axis in the steel section; and,
- Case 3 representing partial shear connection for which the plastic neutral axis is always in the steel section.

Only Case 1 is permitted when joists or trusses are used to prevent buckling of top chord and overloading of the shear connectors. For Case 3, the depth of the concrete in compression is determined by the expression for "a" (Robinson 1969).

17.9.4 Robinson (1988) and Jayas and Hosain (1989) show that a lower limit of 40% of full shear connection is acceptable for strength calculations. Below this value, the interfacial slip is such that integral composite action cannot be assured. A lower limit is given for deflection designs as deflections are computed at specified load levels. This latter provision is used where the flexural strength is based on the bare steel beam, but the increased stiffness due to the concrete is considered for deflection calculations.

17.9.5 Between the point of zero and maximum moment, a horizontal force associated with the internal resisting couple must be transmitted across the steel-concrete interface.

17.9.8 Uniform spacing of shear connectors is generally satisfactory because the flexibility of the connectors provides a redistribution of the interface shear among them. However, to ensure that sufficient moment capacity is achieved at points of concentrated load, the second provision of this clause is invoked. As the moment capacity of the steel section does not depend on shear connectors, this capacity is subtracted from both M_f and M_{f1}.

17.9.9 Longitudinal Shear

In order to develop the compressive force in the portion of the concrete slab outside the potential shear plane shown in Figure 2–47, net shear forces, totaling V_u, must be developed on these planes. The expressions for shear resistance are based on Mattock (1974). Values for semi-low density and low density concrete are given by Mattock *et al.* and Chien and Ritchie (1984).

17.10 Design of Composite Beams Without Shear Connectors

This conservative approach assumes that the composite section is about 10% stronger

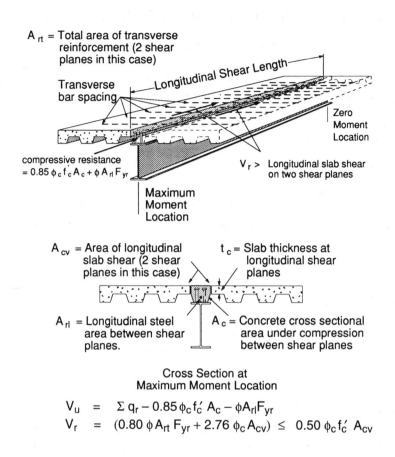

**Figure 2-47
Potential Longitudinal Shear Planes**

than the bare steel member, although the moment resistance computed according to Clause 17.10.2 typically gives a larger value.

17.11 Unshored Beams

This provision guards against permanent deformations under specified loads by limiting the total stress in the bottom fibre of the steel section. This limit has been shown (Kemp and Trinchero 1992) to be conservative. The ultimate strength of the composite beam, which exhibits ductile behaviour is not affected by the stress state at the specified load level.

18 COMPOSITE COLUMNS
(Concrete-Filled Hollow Structural Sections)

18.3 Axial Load on Concrete

Kennedy and MacGregor (1984) showed that direct bearing of the load on the concrete was not necessary for either axially loaded columns or beam-columns. When loads are applied to the steel shell, pinching between the steel and concrete quickly transfers loads to the concrete core. The Standard conservatively retains the requirement of direct bearing for the uppermost level.

18.4 Axial Compression

CIDECT (1970), Knowles and Park (1970), Wakabayashi (1977), STELCO (1981), and

Budijgnto (1983) have demonstrated that the compression resistance of composite columns, consisting of hollow structural sections (HSS) completely filled with concrete, arises from both the steel and the concrete core. The contribution of the concrete core is greatest for stocky members and decreases with increasing slenderness. The hollow steel section carries the same axial load as if it were not filled with concrete, and the concrete core is assumed to carry additional axial load, C'_r. This method of superposition and the contribution of the concrete are based on work by Knowles and Park (1970). The resistances given are in close agreement with the CIDECT solution (Stelco, 1981).

The value of E_c is modified to account for creep and leads to a reduced compressive resistance of the concrete core by increasing the value of λ and hence decreasing C'_r.

The triaxial load effect on the concrete due to the confining effect of the walls of circular HSS is based on work by Virdi & Dowling (1976). The triaxial effects increase the failure load of the concrete ($\tau' > 1.0$) and decrease the capacity of the steel section ($\tau < 1.0$) because the steel is in a biaxial stress state.

18.5 Bending

Lu and Kennedy (1994) show, for rectangular hollow sections with measured flange b/t ratios up to $720/\sqrt{F_y}$, that fully plastic stress blocks are developed in the steel and in the concrete. They got excellent agreement between the test results and their proposed model, based on such stress blocks, when the steel stress level was taken to be equal to the yield value, F_y, and the concrete stress level was taken to be equal to the concrete strength, f'_c, at the time of testing. The two components support each other. The steel restrains or confines the concrete increasing its compressive resistance to the full value rather than 0.85 of it, as used in reinforced concrete theory, while the concrete prevents inward buckling of the steel wall thus increasing the steel strain at which local buckling occurs. Therefore, sections not even meeting the requirements of Class 3 sections in bending develop fully plastic stress blocks

Tension Members	Requirements	Tension Members	Requirements
(diagram with d_{max})	TWO ROLLED SHAPES NOT IN CONTACT d_{max} = 300 x Least radius of gyration of one component.	(diagram with d_{max})	TWO ROLLED SHAPES NOT IN CONTACT d_{max} = 600 mm d_{max} may be increased, when justified.
(diagram with d_{max}, b, t)	SHAPE AND PLATE IN CONTACT d_{max} = 36 t or 450 mm whichever is lesser.	(diagram with d_1, d_2, b, t)	BATTENS $b \leq 60 t$ $d_2 \geq \dfrac{2b}{3}$ d_1 = 300 x Least radius of gyration of one component. * For intermittent welds or fasteners maximum longl. pitch = 150 mm

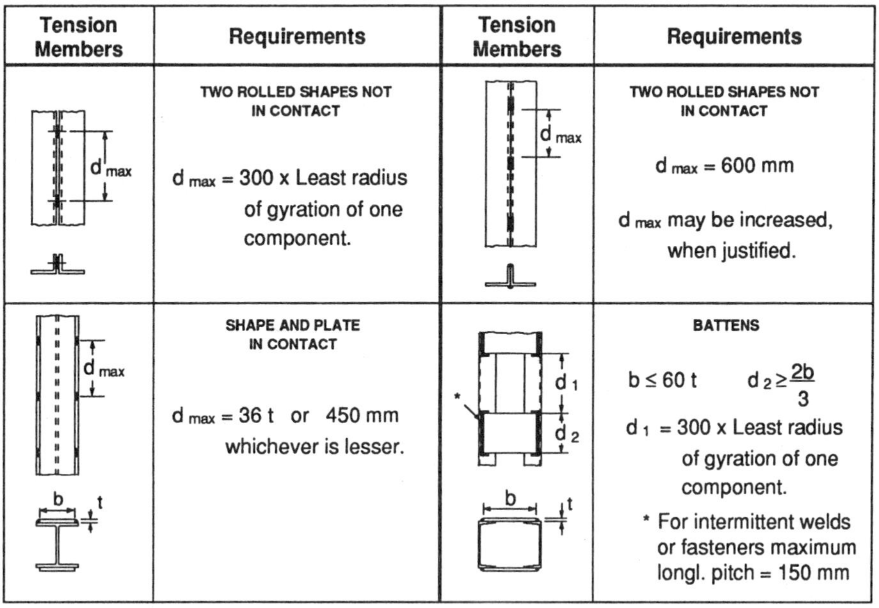

Figure 2–48a
Built-up Tension Member Details

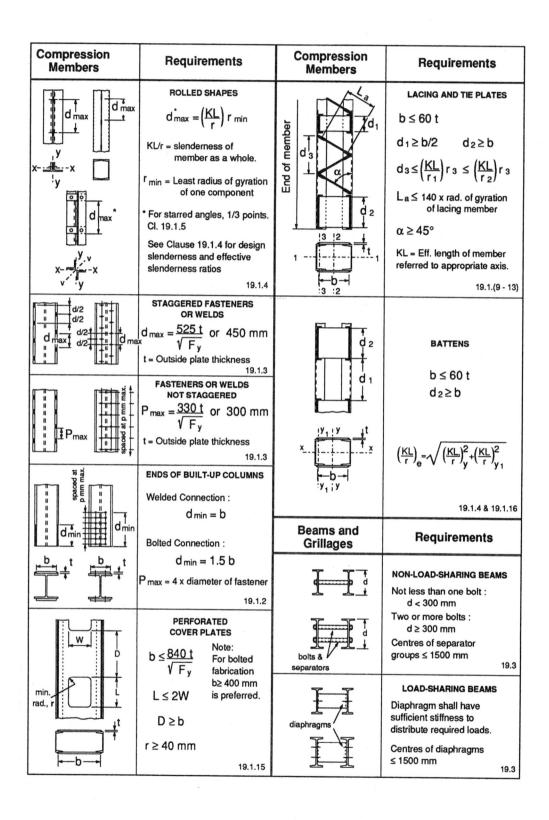

Figure 2–48b
Built-up Compression Member Details

18.6 Axial Compression and Bending

18.6.1 This clause is analogous to the expression in Clause 13.8.1 for I-shaped beam-columns. Extending the analogy, the cross-sectional resistance, in-plane strength and, if applicable, the lateral-torsional buckling strength should be checked for sections bent about their strong axis. Because of the very large torsional resistance of closed shapes, the latter is very unlikely to be a factor.

18.6.2 This section, based on the lower bound solution, is retained, principally for the design of circular hollow structural sections filled with concrete, until expressions such as those for rectangular hollow structural sections are developed.

19. GENERAL REQUIREMENTS FOR BUILT-UP MEMBERS

The term *built-up* member refers to any structural member assembled from two or more components. Such members may be used to resist compression, tension, or bending and the requirements for fastening together the various components vary accordingly.

The diagrams of Figure 2-48 illustrate the main provision of Clause 19. Many of the provisions are based on long established practice and have proven satisfactory.

Tension members are stitched together sufficiently to work in unison and to minimize vibration. For exposed members, components in contact should be fitted tightly together to minimize corrosion problems (Brockenbrough 1983).

When a built-up column buckles, shear is introduced in lacing bars (Clause 19.1.9) and battens and their connections (Clause 19.1.17), in addition to any transverse shears (Bleich, 1952).

Further discussion on columns with lacing and battens is given in Galambos (1988).

For compression members composed of two or more rolled shapes connected at intervals, Clause 19.1.4 requires the use of an equivalent slenderness ratio, increased to take into account the flexibility of the interconnector. This increase is applied to the axis of buckling where the buckling mode of the member involves relative deformation that produces shear forces (see Clause 19.1.6) in the interconnectors between the individual shapes (Duan & Chen, 1988).

The requirements for starred angles are based on work by Temple *et al.* (1986), who showed that with fewer interconnectors the buckling strength was reduced.

20. STABILITY OF STRUCTURES AND INDIVIDUAL MEMBERS

This clause has been reorganized and revised to require that braces provide both strength and stiffness.

20.1 General

Emphasis continues to be placed in the 1994 edition on the designer's responsibility to ensure stability of the structure and of the individual members. Clause 8.6 requires the structure as a whole to resist the PΔ effects.

The stability of the column-girder assembly and the girder web, when a girder is continuous over a column, requires careful assessment. The column, girder web, and the girder flange are all in compression, creating a condition of inherent instability. Stability can be achieved by providing lateral support to the girder-column joint or by properly designed web stiffeners restraining the rotation of the joint. See also the commentary on Clauses 16.6.13.1, 13.6 and the references cited therein.

20.2 Members

This clause applies equally to columns and to the compressed portion of beams. For the latter, it is only necessary to compute the factored compressive force in that portion. The basic equation for the stiffness of the brace (Winter 1958) is derived on the premise that the brace or braces force the member to buckle into a series of half-sine waves of length, L, the distance between bracing points, with nodes at the bracing points. For this to occur, the braces must provide both strength and stiffness.

The equation shows that the required stiffness of the brace, k_b, increases with the force in the braced member at the bracing location, C_f, and its initial out-of-straightness, Δ_0, and with decreased brace displacement, Δ_b. From the equation for the force in the brace, $P_b = k_b \Delta_b$, it is seen, of course, that the force in the brace increases with the product, $k_b \Delta_b$. The value of β increases as the number of braces, which are equally spaced, increases. It is 2 for one brace at the centre and reaches 4 in the limit for a large number of braces.

The factor $(1 + \Delta_0/\Delta_b)$ gives the amount by which the required bracing stiffness is multiplied when the member is initially out of alignment. It is usual to take the initial out-of straightness as the fabrication tolerance over the total braced length or some portion of it. When the former is too large a construction technique of pulling the structure into better alignment at the brace locations, say equal to the fabrication tolerance for the distance between braces, before installing the permanent braces, is suggested. This reduces Δ_0 proportionately and thereby both the strength and stiffness requirements for the bracing and, more importantly, results in a better structure.

As a starting point for brace design, having established the value of Δ_0, the brace displacement, Δ_b, can be taken equal to the initial out of alignment, Δ_0 to determine a preliminary brace stiffness and strength. It is likely that the brace system selected, taking into account minimum slenderness ratios and the like, will be considerably stiffer and an iteration may be in order as the strength required would then be less. It is good practice to keep the final value of the quantity $(1 + \Delta_0/\Delta_b)$ high and, a rule of thumb is to exceed 2, i.e., keep the brace stiff and thereby reduce Δ_b. Stiff braces are better. In determining Δ_b any movement of the remote end of the brace must be taken into account.

Massey (1962) examined lateral bracing forces for beams while Zuk (1956) and Lay and Galambos (1966) considered requirements for structures analysed plastically. Galambos (1988) has summarized many of the design requirements for bracing assemblies.

20.2.4

When an element in a structure must resist the bracing forces from more than one member, the average maximum out-of-straightness of the members should be used to compute the bracing forces. It can be shown statistically that the average maximum out-of-straightness is a function of the maximum out-of-straightness of one member divided by the square root of the number of members (Kennedy and Neville 1986). The expression given in the standard is a conservative empirical equation which applies the statistical reduction to only 0.80 of the initial misalignment. In the design of such bracing systems it must be recognized that the (axial) displacement of the in-line brace increases from the location where the brace is affixed or restrained to the most remote member and the force in the in-line brace increases in the opposite direction. Beaulieu and Adams (1980) provide more guidance in selected cases.

Worked examples are given in Section 4 of the Handbook.

20.2.5 Because the shear centre of an asymmetric section does not coincide with the centroid, this section may be unintentionally loaded so as to produce torsion and biaxial bending. Both the connections and the members providing reactions should be checked.

21. CONNECTIONS

21.3 Restrained Members

When the compressive or tensile force transmitted by a beam flange to a column (approximately equal to the factored moment divided by the depth of the beam) exceeds the factored bearing or tensile resistance of the web of the column, stiffeners are required to develop the load in excess of the bearing or tensile resistance.

Taking the length of the column web resisting the compressive force as the thickness of the beam flange plus 5k (Graham *et al.* 1959) results in the first equation given in Clause 21.3 for the bearing resistance of columns with Class 1 and 2 webs.

For members with Class 3 and 4 webs, the bearing resistance of the web is limited by its buckling strength. The second equation given in Clause 21.3(a) for these members results in a bearing resistance, when the same loaded length (t_b + 5k) is used as in the first equation, about equal to the minimum of the LRFD Specification (AISC 1993) where the loaded length is taken as the clear depth of the column. Although not stated, the bearing resistance computed from the second equation should not exceed the first. In both expressions, if the compression flange is applied at the end of a column, the loaded length should be reduced to t_b + 2.5k.

Graham *et al.* (1959) also show, based on a yield line analysis, that the column flange bending resistance, when subject to a tensile load from the beam flange, can be taken conservatively to be $7 t_c^2 F_{yc}$. Tests have shown that connections proportioned in accordance with this equation have carried the plastic moment of the beam satisfactorily.

When moment connections are made between beams and columns with relatively thick flanges (greater than about 50 mm) prudent fabrication practice suggests that the column flanges be inspected (such as radiographically) in the region surrounding the proposed weld locations to detect and thereby avoid any possible laminations that might be detrimental to the through-thickness behaviour of the column flange.

Huang *et al.* (1973) demonstrated that beam-column connections designed such that the web was connected only for the shear force were capable of reaching the plastic capacity of the beam even though in some tests the webs were connected with bolts based on bearing-type connections in round or slotted holes. The slips that occurred were not detrimental to the static ultimate load capacity. For joints in zones of high seismicity, see Commentary on Clause 27.

21.4 Connections of Tension or Compression Members.

Obviously, the end connections must transmit the factored loads. The 50% rule, which has existed in structural design standards for many years, guards against providing a connection inconsistent with the member it connects when the member size has been selected for some criterion other than strength. This provision should not be viewed as a default in the absence of specific forces, nor as a "load" to be added to the load combinations set out in Clause 7, nor accumulated throughout the structural system. The rule is concerned solely with sizing of the end connection.

21.6 Lamellar Tearing

In cases where shrinkage results as a consequence of welding under highly restrained conditions, very large tensile strains may be set up. If these are transferred across the through-thickness direction of rolled structural members or plates, lamellar tearing may result. Thornton (1973) and AISC (1973) give methods of minimizing lamellar tearing. Figure 2–49 illustrates one such case.

21.7 Placement of Fasteners and Welds

Gibson and Wake (1942) have shown that, except for cases of repeated loads, end welds

**Figure 2–49
Details To Minimize Lamellar Tearing**

on tension angles, and other similar members need not be placed so as to balance the forces about the neutral axis of the member.

21.8 Fillers

In bearing-type shear connections development of the filler before the splice material diminishes bending of the bolt. In slip-critical joints, tests with fillers up to 1 inch in thickness and with surface conditions comparable to other joint components show that the fillers act integrally with the remainder of the joint and they need not be developed before the splice material (Kulak *et al.* 1987).

21.10 Fasteners and Welds in Combination

21.10.1 Because the relative movement in a slip-critical connection at the specified load level is small, the welds and the bolts will share in carrying the load. However, as the bolt capacity is based on a slip criterion at the specified load level and the welds are usually proportioned for factored loads, the weld strengths are discounted to the specified load level by multiplying by the factor 0.70. Moreover, the factored resistance of the connection must be equal to or greater than the effect of the factored loads. As an accurate prediction of the factored resistance involves the use of the load vs. deformation response of the weld group and of the bolts (Holtz and Kulak 1970), the Standard adopts the conservative approach of using the larger of the bolt or weld group factored resistance.

21.12 Connections Using Bolts

Connections using bolts may either have the bolts pretensioned or not depending on the joint type, loads to which it is subject and, the type of holes used, as detailed in Clause 21.12.2.

Bolts which are not pretensioned must be installed to a snug-tightened condition. These may be A307, A325, or A490 bolts. Previous editions of the Standard permitted the use of A307 bolts snug-tightened but implied that all A325 and A490 bolts were pretensioned. Because the ultimate limit states of shear through the bolt and bearing on the plate material are not significantly affected by the level of pretension (Kulak *et al.* 1987), it is only logical to permit bolts of higher strength than the A307 bolt to also be installed snug-tight in similar connections. This was recognized, in part, as early as the 1984 edition in Clause 22.7.1 which relaxed the inspection requirements of certain bearing-type joints.

Only high-strength bolts, such as A325 and A490 and their metric equivalents, may be used in joints requiring pretensioned high-strength bolts.

21.12.2 Connections Using Pretensioned High-Strength Bolts

Snug-tightened bolts may be used, except for the specific cases given in Clause 21.12.2 where the use of pretensioned high-strength bolts is required.

(a) Pretensioning of the bolts provides the clamping force in slip-critical connections and hence the slip resistance at the specified load level appropriate to the condition of the faying surfaces.

(b) Pretensioning of the bolts provides energy dissipation under cyclic earthquake loading in connections proportioned in accordance with the seismic requirements given in Clause 27, although these connections are proportioned as bearing-type connections for the ultimate limit state. The contact surfaces should be Class A or better for such joints.

(c) and (d) Pretensioning in both these connections, ensures that the bolts don't work loose and, of course, is necessary to ensure adequate fatigue behaviour.

(e) An example of such a connection is a tee hanger connection. Pretensioning reduces the prying action and the stress range.

(f) In connections with oversize or slotted holes pretensioning prevents gross movement within the joint. See also Clause 23.3.2 to determine for which cases slip-critical connections are required.

For the usual building structure, full wind loads and earthquake loads are too infrequent to warrant design for fatigue as the number of stress cycles are less than the 50 000 cycles given in Clause 14.2. Therefore, slip-critical connections are not normally required in buildings for wind or seismic load combinations. However, connections of a member subject to flutter where the number of cycles is likely higher is an exception. Popov and Stephen (1972) observed that the bolted web connections of welded-bolted moment connections slipped early in the cyclic process.

Slip-critical connections are required in connections involving oversized holes, certain slotted holes, fatigue loading, or crane runways and bridges. In assessing whether or not the joint slip is detrimental at service level loads, Popov and Stephen (1972) and Kulak, *et al.* (1987) have shown that in joints with standard holes the average slip is much less than a millimetre. Bolts of joints in statically loaded structures are most likely in direct bearing after removal of the drift pins due to the member's self weight and are thus incapable of further slip (RCSC 1994).

21.13 Welds

Although the use of welded connections is not restricted, the use of certain weld details may not be permitted in fatigue or seismic loading. See Clauses 14 and 27.

Welding of high strength bolts adversely affects their strength and should not be done.

22. BOLTING DETAILS

The details given reflect good practice and, for the most part, have remained unchanged over several editions of this Standard. Kulak *et al.* (1987) contains a comprehensive summary of bolt requirements.

23. STRUCTURAL JOINTS USING ASTM A325M, A490M, A325, OR A490 BOLTS

23.1 General

A325M, A490M, A325 and A490 bolts are produced by quenching and tempering (ASTM 1995a, 1995b, 1995c, 1995d). A325 bolts are not as strong as A490 bolts but have greater ductility. For this reason and reasons of availability, the use of A490 bolts is subject to restrictions as discussed subsequently.

The behaviour of the joint depends both on how the bolts are loaded and installed. In the 1984 edition, for the first time, the use of snug-tightened high strength bolts was permitted. Their use has proved successful. As there are four basic types of connections, three with bolts in shear and one with bolts in tension, it is absolutely essential that the drawings specify the type of connections used.

Bolts subject to tension must be pretensioned. Bolts subject to shear in slip-critical connections must also be pretensioned. Bolts subject to shear in bearing-type connections may either be pretensioned or snug-tightened.

Kulak *et al.* (1987) shows that the ultimate shear and bearing resistance of a bolted connection are not dependent on the pretension in the bolt. As the number of situations (Clause 21.12.2) where pretensioning is required is very limited, the norm for building construction is that *snug-tightened bearing-type connections* are used. Departures from the norm are only to be made with due consideration. Few joints in building construction are subject to frequent load reversal nor are there many situations where a one-time slip into bearing cannot be tolerated.

As a result of normal fabrication practice, minor misalignment of bolt holes may occur in connections with two or more bolts. Such misalignment, if anything, has a beneficial effect (Kulak *et al.* 1987) resulting in a stiffer joint, improved slip resistance and decreased rigid body motion.

23.1.4 In Figure 2-50 it is assumed that the connected material is rigid as an approximation of it's high stiffness in compression relative to the bolt stiffness in tension. Thus, when the connected parts are loaded to separate them, only small changes in clamping force and bolt tension will occur prior to reaching an external load equal to the preload. (In the idealization, Fig 2-50, the assumption demonstrates zero change.) Measurement of actual bolt forces in connections of practical sizes has shown that there is an increase in the bolt force due to the flexibility of the connection but it is usually only about 5 to 10%. Thus, the bolt needs to be proportioned only for the factored tensile load.

The effect of prying action on bolt tension, Figure 2–51, is an important design consideration. Kulak *et al.* (1987) summarize the criteria for the design of tension connections when

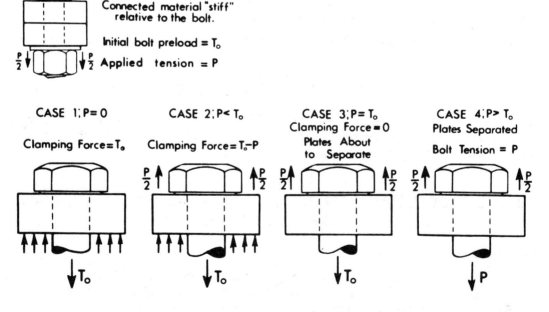

Figure 2–50
Effect Of Applied Tension On Tightened High-strength Bolts

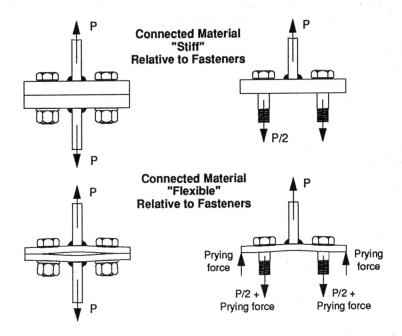

**Figure 2–51
Effect Of Prying Action On Bolt Tension**

prying action is present. For load reversal or repeated load situations these connections must be proportioned so that prying is avoided.

23.2 Bolts, Nuts and Washers

The normal bolt assembly consists of a A325 or A490 bolt, with a heavy hex head, restricted thread length, and coarse threads, and a heavy hex nut. Proprietary versions are available which differ from the normal in various aspects, and, in some cases, may offer one or more advantages. Their use is permissible under the conditions set forth in Clause 23.2.4.

Designers are cautioned to ascertain the availability of A325M and A490M bolts before specifying their use. Alternatively, by basing the design on the bolt with the lower resistance, either the Imperial series or the metric series bolt could be used depending upon availability.

Galvanized A325 bolts are permitted; however, metallic coated A490 bolts are not permitted (ASTM 1995b and 1995d) as they are especially susceptible to stress corrosion and hydrogen stress cracking (Kulak *et al.* 1987). The rotation requirement of this clause provides a means of testing the galvanized assembly for proper fit and for proper thread lubrication.

23.3 Bolted Parts

Details on the sizes and types of holes (standard, oversize, or slotted) permitted for bearing-type and slip-critical connections are given. While the Standard permits several hole making methods, punching and drilling are the most common. Incremental punching is sometimes used in fabricating slotted holes, especially long slots. Clause 23.3.2(e) allows selected Imperial bolts in metric holes as the industry phases from one series to the other. A hardened washer, when required, is intended to cover the hole (or bridge the slot) if it occurs in an outer ply. One-third more bolts are required in slip-critical connections using long slotted holes to account for the reduced clamping force that otherwise would be present (Kulak *et al.* 1987)

The treatment of the faying surfaces within the plies of slip-critical joints is to be consistent with the mean slip coefficient chosen for design (Clause 13.12). For clean mill scale, the surfaces must be free of substances which would reduce the slip coefficient. For other coatings, the surface preparation, coating application and curing should be similar to those used in the tests to obtain the slip coefficient. The Steel Structures Painting Council provides specifications (SSPC 1989) for cleaning and coating of steel structures. Kulak *et al.* (1987) provide information on slip for various surface conditions and coating types.

23.4 Installation

Bolts required to be pretensioned must be tightened to tensions of at least 70% of their specified minimum tensile strength. All other bolts need only be tightened sufficiently to ensure that all the plies are in firm contact, i.e., snug-tightened.

Except when galvanized, A325 bolts may be reused once or twice, providing that proper control on the number of reuses can be established (Kulak *et al.* 1987; RCSC, 1994). A490 bolts should not be reused.

23.5 "Turn-of-Nut" Tightening.

Any installation procedure used for pretensioning high strength bolts involves elongating the bolt to produce the desired tension. Although the shank of the bolt probably remains elastic, the threaded portion behaves plastically. Because the bolt as a whole is tightened into the inelastic range (the flat portion of the load-deformation curve) the exact location of "snug-tight" is not critical. For the same reason, application of the specified amount of nut rotation results in pretensions that are not greatly variable and that are greater than those prescribed in Table 7. Although there is a reasonable margin against twist-off, the tolerance on nut rotation prescribed in the footnote to Table 8 is good practice, particularly when galvanized A325 bolts or black A490 bolts are used.

23.6 Tightening by Use of a Direct Tension Indicator

The Standard permits use of direct tension indicator bolting systems. All of these are proprietary in nature relying on a physical change in a part of the bolt system. This change attempts to indicate that the minimum bolt tension has been achieved. Systems that rely on irreversible deformations or fracture of a part serve only to indicate that during installation, a force or torque sufficient to deform or fracture the part had been reached. Even with such a system, reliable results are unlikely to be achieved unless the installation follows procedures for snugging of the joint and patterned tightening operations as given in that for "turn-of-nut" tightening (RCSC, 1994).

23.7 Inspection

Bolts, nuts and washers are normally received with a light residual coating of oil. This coating is not detrimental, in fact it is desirable, and should not be removed. Galvanized bolts and/or nuts may be coated with a special lubricant to facilitate tightening. Obviously, this should not be removed.

The inspection procedures used depend on whether the bolts are specified to be snug-tightened or pretensioned. In all cases, the inspector shall observe that the procedure for the installation of the bolts conforms with the requirements of this Standard.

When snug-tightening is specified, the tightening is deemed satisfactory when all of the connected elements are in full contact. Inadvertent pretensioning of snug-tightened bolts is normally not a cause for concern.

When pretensioning is specified, the tightening is deemed satisfactory when all of the elements are in full contact and observation of the sides of the nuts or bolt heads shows that they have been slightly peened by the wrench.

When bolts are tightened by the turn-of-nut method and when there is rotation of the part not turned by the wrench, the outer face of the nut may be match-marked with the bolt point before final tightening, thus affording the inspector visual means of noting nut rotation. Such marks may be made with crayon or paint by the wrench operator after the bolts have been snugged.

Should disagreement arise concerning the results of inspection of bolt tension of bolts specified to be pretensioned, arbitration procedures as given in Clause 23.7.4 are to be followed. The use of inspection torque values other than those established according to the requirements of Clause 23.7.4 is invalid because of the variability of the torque-tension relationship. The inspection procedure given in Clause 23.7 is the same as that recommended by the Research Council on Structural Connections (RCSC 1994) and places its emphasis on the need to observe the installation for the proper tightening procedures.

Regardless of the installation procedure or the type of direct tension indicator used, it is important to have all of the plies drawn up tight before starting the specific tightening procedure. This is particularly so for stiff joints that require pattern tightening (RCSC 1994).

24. WELDING

Consistent with CSA policy that the requirements of one standard are not repeated in another, the user of this Standard is referred to CSA Standards W59 and W55.3 for the requirements for arc and resistance welding respectively.

24.3 Fabricator and Erector Qualifications

The intent of Clause 24.3 is simply that the responsibility for structural welding shall lie with the fabricators and erectors certified by the Canadian Welding Bureau as stated specifically in the clause. Such certification should ensure that the fabricators and erectors have the capability to make structural welds of the quality assumed by S16.1-94.

25. COLUMN BASES

The designer is referred to appropriate clauses of CSA-A23.3 (CSA 1994) for the various resistances of the reinforced concrete elements. The compressive resistance of concrete is based on a rectangular stress block as given in CSA A23.3.

In general, the use of base plates bearing directly on grout is preferred to the use of levelling plates interposed between the base plate and the grout. The latter condition may lead to uneven bearing.

26. ANCHOR BOLTS

In general, an anchor bolt is not a high-strength bolt, but rather an item which may be fabricated from a reinforcing bar or a plain bar of A36 or G40.21-300W steel, or other steel bar stock. The expressions for the tensile, shear, and combined shear and tensile resistance of anchor bolts are similar to those for high strength bolts. The tensile resistance is based directly on the tensile stress area, A_n, rather than 0.75 times the shank area as given in Clause 13.11. The basic elliptical interaction diagram is used for combined shear and tension. For tension and bending, the factored moment resistance is limited to the factored yield moment because the ductility of the steel used may be limited.

27. SEISMIC DESIGN REQUIREMENTS

Specific seismic design requirements were introduced into a steel design standard in

Canada for the first time in 1989. While the requirements represent the best available knowledge, designers should be alert to new information leading to possible changes.

27.1 General

The NBCC (1995) assigns force modification (i.e., load reduction) factors, R, to various structural systems in relation to their capacity to absorb energy by undergoing inelastic deformations. The greater the ability of the structure to absorb energy, the higher is the assigned value of R, which is used as a divisor to reduce the magnitude of the seismic forces. Values of R greater than 1.0 can be justified only if the structure has the ability to undergo inelastic deformations without loss of resistance. The degree of redundancy is also considered because alternative load paths reduce the possibility of serious consequences arising from failure of a member and enhance the energy dissipation.

The objective of Clause 27 is to provide details with ductility consistent with the R values assumed in the analysis. The clause applies to all steel structures in Canada for which an $R \geq 2$ is used. Ordinary steel structures (not meeting the special seismic requirements of this Clause) have sufficient inherent ductility to be assigned an R value of 1.5. They are especially appropriate for zones of low seismicity, and in certain cases their use in zones of high seismicity is proscribed.

Clause 27 defines the requirements for five classes of frames:

- ductile moment resisting frames,
- moment resisting frames with nominal ductility,
- ductile braced frames
- braced frames with nominal ductility, and
- ductile eccentrically braced frames.

In addition, requirements for steel plate shear walls are given in Appendix M, and other special framing systems are permitted under Clause 27.7.

Properly detailed moment resisting frames exhibit highly ductile behaviour and are highly redundant. Two categories of moment resisting frames are recognized: first, ductile moment resisting frames, in which members and connections are selected and braced to ensure that severe inelastic straining can take place; second, moment resisting frames with nominal ductility in which the proportions of members are adequate to provide the more limited inelastic straining demanded under a greater design load and the connections are adequate to resist extreme forces.

Concentrically braced frames are those in which the centrelines of diagonal braces, beams, and columns are approximately concurrent with little or no joint eccentricity. Such frames usually have limited redundancy. Inelastic straining must take place in bracing members subjected principally to axial load. Compression members can absorb considerable energy by inelastic bending after buckling and in subsequent straightening after load reversal but the amount is small for slender members. Local buckling or buckling of components of built-up members also limits energy absorption.

Two categories of concentrically braced frames are considered: first, ductile braced frames, which exclude some bracing systems that have not behaved well under seismic load; second, braced frames with nominal ductility, designed for higher loads and in which less inelastic behaviour is expected. When braces are selected to satisfy drift or slenderness requirements, and have resistances greater than required to carry the seismic loads, the connections are required to be designed for increased loads commensurate with the capacity of the braces.

The load combinations for design under this clause are given in Clause 7.2.6 and represent loads due to the extreme event earthquake combined with the expected gravity loads. In a number of locations in this clause, in particular with reference to connections, the seismic load is multiplied by a factor before combining it with the gravity loads. The intent is to thereby return the seismic load, in that instance only, to the intensity represented by R=1.5 (elastic behaviour) regardless of the R value used for the structure in general.

Because the behaviour of connections will often be critical for good performance under severe earthquake loading, the engineer's responsibility for a seismically critical structure includes not only provision of connection design loads but also specification of connection type and details. SEAOC (1988), Krawinkler and Popov (1982), and Astaneh *et al.* (1986) indicate preferred types of connections. The behaviour of welded beam-to-column connections in moment resisting frames is currently under close scrutiny as a result of the damage sustained by some structures in the Northridge, California earthquake of 1994. Designers and fabricators of moment frames similar to those being investigated should stay abreast of the results of such studies (AISC 1994b).

Recent research (Ghobarah *et al.* 1992; Osman *et al.* 1991; Tsai and Popov 1990) has demonstrated that beam-column moment connections made with extended end-plates, and having 4 bolts above and below the top and bottom flanges respectively, and that are adequately designed and detailed exhibit good performance under cyclic loads. They may be viable alternatives to welded moment joints for light and medium beams.

Yielding elements of a frame will induce forces on other members and connections. Deformations of yielding elements must not be constrained, and load paths must be identified and provided with adequate resistance.

27.1.2 In the computation of second-order effects, a linear amplification is given following the procedure outline in Appendix J of the Supplement to the National Building Code of Canada, 1995. This method is less severe than that given in Clause 8.6 as the deflections are already increased by the R factor.

27.1.4 A simple end connection maintaining its shear capacity in the displaced configuration is an example of this situation.

27.2 Ductile Moment Resisting Frames

The NBCC force modification factor, R, is 4.0.

27.2.1.1 Under severe lateral load, a ductile moment resisting frame responds by inelastic deformation at beam-to-column joints and plastic hinges. The key locations are the beam-column joints, where inelasticity may develop in the girder, the column or, in the case of H-shaped columns (bent about the strong axis) the joint panel zone consisting of the column web and flanges, and column web stiffeners, if any. It is required that these elements maintain their ultimate strength while undergoing significant deformations. The panel zone provides excellent ability to absorb energy by means of cyclic plastic shearing deformations (Popov *et al.* 1986).

At plastic hinges, members absorb energy by undergoing inelastic cyclic bending while maintaining their flexural and shear resistances. The member should be prismatic in the region of the hinge in order to avoid excessively large strains and to ensure that as large a volume of steel as practicable is strained plastically.

27.2.1.2 The requirement that non-critical elements withstand 1.2 times the unfactored resistance of critical elements ensures that the former remain elastic. Thus, the non-critical elements need only meet the bracing and width-thickness requirements for members analyzed elastically, as provided by Clauses 27.2.2, 27.2.3 and 27.2.4.

To determine the critical element at a joint, the resistances of the elements are ranked. Then 1.2 times the unfactored resistance of the single most critical element is applied as a load to the joint. Any of the remaining elements whose resistance exceeds the effect of the applied load is considered to be non-critical. It may be possible for one, two, or all three of the elements to be critical. All such elements must be braced to enable them to undergo large plastic deformations.

The Standard gives provisions for any of the three elements to be critical. To avoid a column sidesway mode of failure that demands high ductility in a multi-storey structure, it is preferable that the columns not be critical. It is advantageous to have both the beams and panel zones critical so that both absorb energy.

27.2.1.3 In evaluating the relative strengths of the structural components at the joint, an estimate should be made of the contribution of the slab. Under positive bending moment composite slabs bear against the column flange (in the case of a column bending about its strong axis) and, due to confinement of the concrete, ultimate compressive resistance of the material can reach values of 1.3 f_c'.

27.2.3 Columns (Including Beam-Columns)

27.2.3.1 The width–thickness requirements here follow from Clause 27.2.1.2. The axial load in the column is also restricted because of the rapid deterioration of beam-column flexural strength when high axial loads are acting limits the ductility.

27.2.3.2 This requirement is to avoid plastic hinges near a splice in which there are partial penetration welds. The moments in a column when the structure is responding inelastically will not, in general, be known. Conservative estimates of the moment at a splice should be made, based on the possible bending strengths at each end of the column.

27.2.3.3 Columns may accumulate forces from several yielding elements and these must be considered.

27.2.4 Column Joint Panel Zone

27.2.4.1 The column panel zone has a shear strength greater than the von Mises shear yield value on the web because of the contribution of the flexure of the column flanges necessary for the panel to yield in shear (Krawinkler and Popov 1982).

27.2.4.2 These requirements ensure that the panel zone can undergo cyclic plastic straining. The entire perimeter of the doubler plates shall be welded to contiguous elements, unless the doubler plate extends beyond the stiffeners. Lateral bracing to the column flange may be provided through transverse beams connected to the column web.

27.2.4.3. Note that the column-to-beam flange welds are subject to forces induced by both beam bending and local bending of the column and its flanges at the corners of the panel zone (Popov *et al.* 1986). The tensile resistance to a normal load on the column flange is reduced to 60% of the value given in Clause 21.3 to account for the highly non-uniform stresses in a beam flange when welded to a column with unstiffened flanges. Minor imperfections in these welds have been found to initiate premature failure of the welds and of the column flanges, especially near the girder bottom flange. Preliminary recommendations based on studies of these details are available (AISC, 1994b).

27.2.5 Beam-to-column Connections

A number of field welded beam-to-column connections of the type previously recommended (Section 8.2c, AISC 1992) did not perform well in the 1994 Northridge earthquake (AISC, 1994b). Based on the preliminary evaluation of early test results, reinforcement may be required to shift the location of the plastic hinge in the beam away from the face of the column. For new work designers should be aware of the latest recommendations available (at press time–AISC, 1994b) and alternative connection types that have demonstrated good response to seismic actions (Ghobarah *et al.* 1992; Osman *et al.* 1991; Tsai and Popov 1990).

27.2.5.1 To ensure yielding over as large a volume of material as possible, partial joint penetration groove welds are prohibited as they lead to confined zones of yielding.

27.2.5.3 and 27.2.5.4 These requirements ensure that extensive yielding on the gross section takes place before fracture on the net section and that extensive yielding takes place at plastic hinge locations.

27.2.6 Bracing of both top and bottom beam flanges as well as column flanges shall be considered.

27.2.7 Consideration should be given to the fact that plastic hinge locations will not be predicted by an elastic analysis of the frame. A few welded wide flange column shapes, with webs thicker than 20 mm, do not have complete penetration welds between their web and flanges. These are identified in the tables of properties and dimensions, Part 6.

27.3 Moment Resisting Frames With Nominal Ductility

The NBCC force modification factor, R, is 3.0.

27.3.2 These requirements are to ensure that the critical elements will undergo sufficient inelastic deformation.

27.3.3 Unless the design of the beam-column connections meets the requirements for ductile moment resisting frames, albeit with R = 3, both the members and connections shall be sized for seismic forces increased by 1.25 times (effective R = 2.4). Although a ductile mode of failure is still required of the connections, the overall ductility is less than that for a ductile moment-resisting frame.

27.4 Ductile Concentrically Braced Frames

The NBCC force modification factor, R, is 3.0.

27.4.2.1 This requirement ensures some redundancy and also similarity between the load-deflection characteristics in the two opposite directions. A significant proportion of the shear is carried by tension braces so that compression brace buckling will not cause a catastrophic loss in shear capacity. This is required in each planar braced frame because a loss of capacity in any one frame would change the torsional resistance of the structure.

27.4.2.2 The use of V- or chevron bracing is prohibited in this category because such bracing has been found to perform rather poorly under severe seismic shaking. V- or chevron bracing is allowed, with some restrictions, in braced frames with nominal ductility (See Clause 27.5) where the lateral load resisting system is designed for increased seismic forces. V-bracing alternating with chevron bracing in adjacent floors, with bracing forces transmitted directly between the braces on adjacent floors is, in effect, X-bracing and may be designed under this clause.

K-bracing is proscribed for the same reasons as V- or chevron bracing. Tension only bracing is not acceptable in view of its severely limited ability to absorb energy and the lack of redundancy.

27.4.3 Diagonal Bracing Members

27.4.3.1 The slenderness ratio is restricted because the energy absorbed by plastic bending of the braces diminishes with increased slenderness. Because of high curvatures in a member undergoing overall buckling, local buckling is exacerbated, with concomitant loss in energy absorbing capacity. To control local buckling, the limits on width-thickness ratios of HSS members are more severe than Class 1 limits in all but zones of low seismicity. Filling tubes with concrete (Liu and Goel 1988) is one way of inhibiting local buckling in critical regions.

Built-up members may deteriorate rapidly under cyclic loading and therefore the spacing of stitch fasteners is restricted.

27.4.3.2 The reduction factor takes into account the fact that, under cyclic loading, the compressive resistance diminishes with slenderness ratio. This reduction stabilizes after a few cycles. It is sufficient that the sum of the resistances of the tension and compression braces acting at the same level and in the same plane exceed the factored load effect.

27.4.4 Bracing Connections

27.4.4.1 Eccentricities which are normally considered negligible (for example at the ends of bolted or welded angle members) may influence the failure mode of connections subjected to cyclic load (Astaneh *et al.* 1986).

27.4.4.2 In all but zones of low seismicity, a connection must be designed to ensure that the member is capable of yielding on the gross section. For bolted connections of G40.21–300W steel, the net section must be at least $0.87A_g$ in oder to meet this requirement. In zones 1 and 2, it is acceptable to limit the connection capacity to 2.0 times the member force due to the seismic load plus the gravity load effect, thus providing an upper limit when the bracing members are oversized to facilitate connection, to limit deflections or, to meet a slenderness or a b/t limit. Even in these zones, however, it is recommended that in any case the bracing for the upper two or three stories be designed for nominal yield load of the brace because the dynamic response associated with higher modes of the inelastic structure may cause severe shear loading in the upper storeys.

These connection resistance requirements ensure that the connection can carry maximum compressive load to which the brace is subjected during the initial cycles.

27.4.4.3 A brace which buckles out-of-plane will form a plastic hinge at mid-length and hinges in the gusset plate at each end. When braces attached to a single gusset plate buckle out-of-plane, there is a tendency for the plate to tear if it is restrained by its attachment to the adjacent frame members (Astaneh *et al.* 1986). Provision of a clear distance, approximately twice the plate thickness, between the end of the brace and the adjacent members allows the plastic hinge to form in the plate and eliminates the restraint. When in-plane buckling of the brace may occur, ductile rotational behaviour should be possible either in the brace or in the joint.

27.4.4.4 Buckling of double angle braces (legs back-to-back) about the axis of symmetry leads to transfer of load from one angle to the other, thus imposing significant loading on the stitch fastener (Astaneh *et al.* 1986).

27.4.5.1 This ensures that other connections in the lateral load resisting system sustain, at the factored resistance level, the nominal brace resistance force in addition to the gravity load.

27.4.6 By using a resistance factor of 1.0 for the brace resistance, the other components can sustain, at the factored resistance level, the nominal brace resistance force in addition to the gravity load.

For multi-storey structures, the accumulation of the brace force in the columns is attenuated using the "modified square root of the sum of the squares" (SRSS) method (Redwood and Channagiri 1991; Redwood *et al.* 1991) as a means of accounting for the reduced probability that the forces in all braces above reach their nominal resistance simultaneously.

The reduced brace compressive resistance is not used as in the first cycles of load it does not apply. However, redistributed loads resulting from buckled compressive brace loads being transferred to the tension brace must be considered in beams and columns as well as in connections. The reduced brace resistance may affect these loads.

27.5 Concentrically Braced Frames With Nominal Ductility

The NBCC force modification factor, R, is 2.0.

27.5.1 K-braced frames are excluded from this category of framing systems because of the severe effect of columns being overloaded by lateral loads within their span.

27.5.2 The first requirement is to control local buckling in compression braces under the seismic loads. See also the Commentary on Clause 27.4.6. Whereas in Clause 27.4.4.2 the factor used is 2.0 times the member force, for the braced frame with nominal ductility the factor is taken as 1.33. These factors raise the seismic load effect in the brace to the level for which a brace would be designed in a frame for which no special provisions for ductility are imposed (i.e. when the force modification factor, R = 1.5).

27.5.3 See the Commentary on Clause 27.5.2. The additional factor of 1.10 for tension-only bracing systems is to ensure, for the slender members used in this case, that the impact resulting when slack is taken up, does not cause connection failure. Details leading to limited zones of yielding, such as occur at partial joint penetration groove welds should be avoided.

27.5.4 and 27.5.5 When the compression member of a chevron-braced system buckles, the tension brace can carry no additional load unless the beam resists the additional load in bending. This clause attempts to ensure that should one of the braces buckle, the beam can at least support the gravity load without collapse. Lateral bracing resistance is required to be higher than normal. This check involves comparing the beam nominal resistance ($\phi = 1.0$) with the gravity load effect.

Therefore the overall approach of the designer would include:

(1) designing the beam and bracing to carry the required factored combinations of dead, live, and earthquake loads. Under these circumstances the beam is supported by chevron bracing and its nominal factored resistance with $\phi = 0.90$ is used.

(2) in order to meet Clause 27.5.5, design the beam, unpropped (i.e. not supported by the chevron bracing), to carry the gravity loads. The resistance factor of the beam is to be taken as $\phi = 1.0$.

In zones of high seismicity, the braces in Chevron braced frames must be detailed as for braces of ductile concentrically braced frames because of the potentially high ductility demand.

27.6 Ductile Eccentrically Braced Frames

The design requirements of this clause are largely based on the provisions of AISC (1992) with modifications to provide consistency with this Standard and with the National Building Code of Canada. AISC (1992) contains a commentary on these requirements.

27.7 Special Framing Systems

Numerous other systems have been proposed to enhance the seismic response of structural steel frames, e.g., Steimer (1986).

28. FABRICATION

This clause and the clauses on erection and inspection serve to show that design cannot be considered in isolation, but are part of the design and construction sequence. The resistance factors used in this Standard and the methods of analysis are related to fabrication, erection, and inspection tolerances and practices.

28.4.2

The use of sheared edges is restricted because the micro-cracking induced may reduce the ductility.

28.5.1 The thickness of 700Q steels that can be punched is restricted because of the excessive damage that occurs at the edge of the hole.

28.5.2 The restriction of this clause is similar to that of Clause 28.4.2

28.9 Tolerances

The resistance factors given in this Standard are consistent with the distribution of out-of-straightness of members produced to the straightness tolerances given here (Kennedy and Gad Aly 1980; Chernenko and Kennedy 1991).

28.9.7 Milling techniques will realistically result in some measurable deviation. Tests by Popov and Stephen (1977a) on columns with intentionally introduced gaps at milled splice joints indicated that the compressive resistance of spliced columns is similar to unspliced columns. Local yielding reduces the gap. While in these tests column splice gaps of 1.6 mm were left unshimmed, the Standard is more restrictive and defines full contact as a separation not exceeding 0.5 mm. Because shims will be subjected to either biaxial or triaxial stress fields, mild steel shims may be used regardless of the grade of main material.

29. CLEANING, SURFACE PREPARATION, AND PRIMING

There are four instances where steelwork need not be or should not be coated:

a) steelwork concealed by an interior building finish;

b) steelwork encased in concrete;

c) faying surfaces of slip-critical joints, except as permitted by Clause 23; and,

d) surfaces finished to bear unless otherwise specified.

30. ERECTION

Clauses 30.7.3, 30.7.4, 30.7.5 and 30.7.8 are written in a parallel manner, in that the offset of one end relative to the other, or the elevation of one end relative to the other, is expressed as a function of the length but with upper and lower limits. The lower limit represents a realistic assessment of adequate positioning and the upper limit is a maximum not to be exceeded by the largest members, as illustrated in Figure 2–52 for horizontal alignment of spandrel beams.

30.7 Erection Tolerances

This clause has been re-written and expanded from previous editions to provide a more helpful definition of tolerances for the location of the ends of members with respect to their theoretical locations. Tolerances have now been given for column base plates, for alignment and elevations of horizontal or sloping members, and for crane or monorail beams and girders. Column splice tolerances have been redefined based on new research (see Commentary on Clause 28.9.7).

APPENDIX D
Torsional-Flexural Buckling of Compression Members

Clause 13.3 deals with the flexural buckling of shapes that are doubly symmetric such

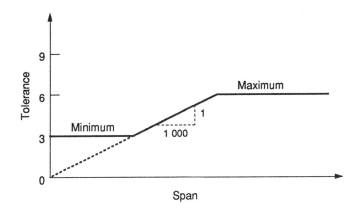

**Figure 2–52
Horizontal Alignment Tolerance Of Spandrel Beams**

as I-shaped members and HSSs which dominate in steel construction. Appendix D deals with the problems of torsional-flexural buckling. Torsional buckling (twisting about the shear centre) is a possible failure mode for point symmetric sections, e.g., a cruciform section. Torsional-flexural buckling (a combination of torsion and flexure) must be checked for open sections that are singly symmetric or asymmetric such as T's and angles. Because the shear centre and centroid of open sections that are doubly symmetric or point symmetric coincide, these sections generally are not subject to torsional-flexural buckling. Cruciform sections are an obvious exception. Thus, for sections with coincident shear centre and centroid, three potential compressive buckling modes exist (two flexural and one torsional), while for singly symmetric sections two potential compressive buckling modes (one flexural and one torsional-flexural) exist and, for a non-symmetric section, only one mode (torsional-flexural) exists. Closed sections, strong torsionally, also do not fail by torsional-flexural buckling. (See Galambos 1988). For the theory of elastic flexural-torsional buckling see Goodier (1942), Timoshenko and Gere (1961), Vlasov (1959) and Galambos (1968). The equations given here are developed in the latter among others.

As the problem of inelastic torsional-flexural buckling is quite complex and is amenable generally only to inelastic finite element analyses, the approach given here is to compute the elastic buckling stress, F_e, from the equations given for doubly symmetric, singly symmetric or asymmetric sections and then calculate an equivalent slenderness ratio from $\lambda_e = \sqrt{F_y/F_e}$ to be used in the equations of Clause 13.3. This comes from the fact that an elastic buckling curve, when non-dimensionalized by dividing by F_y can be written as $F_e/F_y = 1/\lambda^2$. When the inelastic equations of 13.3 are entered with the equivalent slenderness ratio an inelastic compressive resistance, of course, results.

The equations given here are equivalent to those in CSA Standard S-136. There however, for singly symmetric sections, the x-x axis is taken as the axis of symmetry because cold-formed channel sections are frequently used. In Appendix D, because singly symmetric sections are likely to occur for sections with one flange larger than the other, the y-y axis is taken as the axis of symmetry.

APPENDIX F
Columns Subject to Biaxial Bending

This appendix provides non-linear interaction expressions for Class 1 and 2 sections of W-shapes and HSS members, as a design alternative to the conservative linear interaction expressions of Clause 13.8. See also the commentary to Clause 13.8.

The simplest demonstration of non-linearity is that used by Pillai and Ellis (1971) for HSSs. Consider a circular tube subject to eccentric loading about the x and y axes. Since the section properties are uniform about the polar axis, the interaction expressions of clause 13.8 would sum the moment effects as if a moment of $P(e_x + e_y)$ were applied, whereas the actual moment is $P\sqrt{(e_x^2 + e_y^2)}$. Thus, the interaction curve for a circular tube is a quadrant of a circle. Finding a similar effect applied to square HSSs, Pillai determined that the available test data fitted the expression of clause F3(c) if the sum of the moment terms was modified by the expression

$$\gamma = \frac{e_x^2 + e_y^2}{e_x + e_y}$$

The value of γ is equivalent to ν in clause F3(c).

It has been found (Chen and Atsuta 1972 and 1973; Tebedge and Chen 1974, and Ross and Chen 1976) that a similar non-linear effect applied to W-shapes (Clause F1) where the exponents ζ and η have been evaluated from rigorously calculated interaction curves.

The expressions for M_{ox} and M_{oy} have been determined from Clause 13.8, transposed and applied for uniaxial bending only.

Essentially, the interaction expressions of Clause F1 define an interaction surface relating the boundary conditions on the three perpendicular axes (i.e. axial resistance and moment resistance about each axis). Therefore, the better these values are determined, the better will be the interaction surface.

Unlike W-shapes, significant biaxial bending is imposed on HSSs whenever beams frame into the column about two perpendicular axes. The design expressions given in Clause F3 for square HSSs were developed by Pillai (1970 and 1974) and Pillai and Kurian (1977).

Limitations

Significant economy may result in designs according to the expressions of this Appendix, but before adopting these, some thought should be given to the extent to which yielding under service load is likely, and whether further restrictions on the design are necessary. Springfield (1975) has proposed additional precautions.

Appendix M
Design Requirements for Steel Plate Shear Walls

M1 General

Recent research work at the University of Alberta (Kulak 1991) has demonstrated that the steel plate shear wall system is an attractive alternative for resisting lateral wind and seismic loads. The system has the advantage that it is stiff enough to minimize displacements under extreme loading conditions, and has a high degree of redundancy. The system can be used for both new construction and the upgrading of existing structures.

The steel plate shear walls considered by Appendix M imply unstiffened webs constructed from relatively thin plates. Under lateral loads, it is assumed that the buckling strength of the web plates is negligible. Lateral loads are assumed to be resisted by tensile stresses developing in the web plate panels along inclined lines (see Fig. M1).

M2 Design Forces and Moments

Thornburn, *et al.* (1983) have demonstrated that the strip model shown in Fig. M2 can be used to predict the response of steel plate shear walls to lateral loads. The expression for the angle of inclination of the tension field strips was determined by minimizing the work in one panel owing to the tension field action in the web, flexure and axial forces in the boundary columns and axial force in one boundary beam.

With the angle of inclination determined, it is possible to analyze a steel plate shear wall under lateral loads using a plane frame structural analysis program one panel at a time. From this analysis, the distribution of forces and moments can be determined within the panel. These include the tensile forces in the web plate, the forces imposed between the web plate and the boundary beams and columns, and the forces and moments in the boundary beams and columns.

For most situations, it is sufficient to analyze steel plate shear wall structures a panel at a time. If it is desirable to account for column shortening, continuity of connections between beams and columns, large variations in the magnitudes of the vertical components of the tension fields between panels and other related effects, the complete shear wall can be analyzed. This is accomplished by extending the model shown in Fig. M2 over all stories using a plane frame structural analysis program.

For preliminary design, the overall behavior of a shear wall can be approximated in a plane frame analysis by representing each web panel by a single diagonal tension brace extending from the column to beam connection below on one side of the panel to the column to beam connection above on the other side. Thornburn, *et al.* (1983) have derived the following expression for determining the equivalent area, A, of the diagonal tension brace

$$A = \frac{w\,L\,\sin^2 2\alpha}{2\,\sin\theta\,\sin 2\theta}$$

where θ is the angle between the vertical and longitudinal axis of the equivalent diagonal brace. The beams and columns are assumed to have their actual stiffness properties in the analysis.

M3 Angle of Inclination

The equation presented for determining the angle of inclination of the truss members is based on the following assumptions:

(a) The storey shear is approximately the same in the panels above and below the storey under consideration.

(b) The beams are attached to the columns with pin connections.

(c) The columns are continuous.

(d) The storey heights are approximately equal.

When these assumptions are not valid, Appendix A of Timler and Kulak (1983) can be consulted so that the least work derivation can be appropriately applied to non-standard cases.

M5 Horizontal Beams and M6 Columns

Under high lateral loads, plastic hinges can develop in the beams and columns of steel plate shear walls. To avoid premature failure, beams shall be constructed from Class 1 or 2 sections, and columns from Class 1 sections.

M8 Connections

To avoid fit-up difficulties in the field, it is recommended that continuous field welds be used to connect the web plates to fish plates that have been attached to the boundary beams and columns. A few bolts or tack welds can be used to hold web plates in place temporarily during erection.

M9 Seismic Design Requirements

Kulak (1991) has demonstrated that the hysteretic behavior of steel plate shear walls under cyclic lateral loading is stable. Much of the energy imparted to a shear wall is dissipated by the yielding of web panels in tension along inclined lines. When the direction of the lateral shear reverses, the tension field in a given panel dissipates, the panel buckles under low load and a tension field develops consistent with shear in the opposite direction.

If the beams of a shear wall panel are attached to the columns using pin-ended connections, the hysteretic behavior is pinched or S-shaped. This behavior is characteristic of any steel framing system which contains elements that buckle.

Under extreme earthquake loading, the pinched hysteretic behaviour is undesirable. Kulak (1991) has demonstrated analytically that the behaviour can be improved if moment connections are made between the beams and columns surrounding the web panels. With moment connections, a portion of the lateral load is resisted by the frame action of the boundary members. Some of the energy from seismic ground motion can be dissipated through the formation of plastic hinges in the boundary members or the yielding of the column joint panel zones. For this reason, higher R-factors are assigned to steel plate shear walls when moment connections are used.

When moment connections are provided between the beams and columns, the hierarchy of yield levels and connection requirements are the same as those specified in Clause 27 for the type of frame used for the boundary members. For ductile and nominally ductile shear walls, the design requirements for the boundary members are the same as those used for proportioning ductile moment resisting frames and moment resisting frames with nominal ductility, respectively.

REFERENCES

AASHTO. 1994. LRFD bridge design specifications, SI Units, First Edition. American Association of State Highway and Transportation Officials, Washington, DC.

ADAMS, P. F. 1974. The design of steel beam–columns. Canadian Steel Industries Construction Council, Willowdale, Ont.

ADAMS, P. F., and GALAMBOS, T. V. 1969. Material considerations in plastic design. International Association for Bridge and Structural Engineering, **29-II**.

ADEKOLA, A. O. 1968. Effective widths of composite beams of steel and concrete. The Structural Engineer, **46**(9): 285–289.

AHAMD, M.; CHIEN, E. Y. L., and HOSAIN, M. U. 1990. Modified stub-girder floor system: full scale tests. ASCE Structures Congress. Baltimore, MA.

AISC. 1973. Commentary on highly restrained welded connections. Engineering Journal, American Institute of Steel Construction, Third Quarter.

——— 1992. Seismic provisions for structural steel buildings. American Institute of Steel Construction, Chicago, IL.

——— 1993. Load and resistance factor design specification for structural steel buildings. American Institute of Steel Construction. Chicago, IL.

——— 1994a. Manual of steel construction. Load & resistance factor design, Second Edition. American Institute of Steel Construction. Chicago, IL.

——— 1994b. Interim observations and recommendation on steel moment resisting frames, AISC Northridge Technical Bulletin No. 2, American Institute of Steel Construction. Chicago, IL., October.

AISI. 1968. Plastic design of braced multi-storey steel frames. American Iron and Steel Institute, Washington, DC.

ALBERT, C., ESSA, H.S. and KENNEDY, D.J.L., 1992, Distortional buckling of steel beams in cantilever suspended span construction. Canadian Journal of Civil Engineering, **19**(5): 767-780.

ALLEN, D. L. 1974. Vibrational behaviour of long-span floor slabs. Canadian Journal of Civil Engineering, **1**(1).

ALLEN, D. E. 1975. Limit states design—a probabilistic study. Canadian Journal of Civil Engineering, **2**(1).

ALLEN, D. E., and RAINER, J. H. 1976. Vibration criteria for long-span floors. Canadian Journal of Civil Engineering, **3**(2).

ALLEN, D. E., RAINER, J. H., and PERNICA, G. 1985. Vibration criteria for assembly occumancies. Canadian Journal of Civil Engineering, **12**(3).

ALLEN, D. E., and MURRAY, T.M. 1993. Design criterion for vibrations due to walking. Engineering Journal, American Institute of Steel Construction, **30**(4): 117-129.

ASCE. 1967. Commentary on welded cover-plated beams. Subcommittee on Cover Plates, Task Committee on Flexural Members, ASCE Journal of the Structural Division, **93**(ST4).

——— 1971. Commentary on plastic design in steel. Manual of Engineering Practice, No. 41, American Society of Civil Engineers.

——— 1979. Structural design of tall steel buildings. pp. 624–628.

ASTANEH, A., GOEL, S.C., and HANSON, R.D. 1986. Earthquake-resistant design of double-angle bracings. Engineering Journal, AISC, **23**(4), 133–147.

ASTM. 1995a. Standard specification for high-strength bolts for structural steel joints [metric]. Standard A325M-93, American Society for Testing and Material.

——— 1995b. Standard specification for high-strength steel bolts, Class 10.9 and 10.9.3, for structural steel joints [metric]. Standard A490M-93, American Society for Testing and Material.

——— 1995c. Standard specification for structural bolts, steel, heat-treated, 120/105 ksi minimum tensile strength. Standard A325-94, American Society for Testing and Material.

——— 1995d. Standard specification for heat-treated, steel structural bolts, 150 ksi minimum tensile strength. Standard A490-93, American Society for Testing and Material.

AZIZ, T.S.A. 1972. Inelastic nonlinear behaviour of steel triangulated planar frames. Thesis, Carleton University, Ottawa, Ont.

de BACK, J., and de JONG, A. 1968. Measurement on connections with high strength bolts, particularly in view of the permissible arithmetical bearing stress. Report 6-68-3, Stevin Laboratory, Delft University of Technology, The Netherlands.

BAKER, K. A. and KENNEDY, D.J.L.. 1984. Resistance factors for laterally unsupported steel beams and biaxially loaded steel beam-columns. Canadian Journal of Civil Engineering, **11**(4): p. 1008-1019.

BASLER, K. 1961a. New provisions for plate girder design. Proceedings, AISC National Engineering Conference.

——— 1961b. Strength of plate girders under combined bending and shear. ASCE Journal of the Structural Division, **87**(ST7).

——— 1961c. Strength of plate girders in shear. ASCE Journal of the Structural Division, **87**(ST7).

BASLER, K., and THURLIMANN, B. 1961. Strength of plate girders in bending. ASCE Journal of the Structural Division, **87**(ST6).

BEAULIEU, D., and ADAMS, P. F. 1980. Significance of structural out-of-plumb forces and recommendations for design. Canadian Journal of Civil Engineering, **7**(1).

BEEDLE, L. S., LU, L. W., and LIM, L. C. 1969. Recent developments in plastic design practice. ASCE Journal of the Structural Division, **95**(ST9).

BETHLEHEM STEEL CORPORATION. 1967. Torsional analysis of rolled steel sections. Handbook 1963C, Bethlehem Steel Corp., Bethlehem, PA.

BIRKEMOE, P. C., and GILMOR, M. I. 1978. Behaviour of bearing critical, double-angle beam connections. Engineering Journal, AISC, Fourth Quarter.

BJORHOVDE, R. A. 1972. A probabilistic approach to maximum column strength. Procedings, ASCE Conference on Safety and Reliability of Metal Structures.

BJORHOVDE, R., and BIRKEMOE, P. C. 1979. Limit States design of HSS columns. Canadian Journal of Civil Engineering, **6**(2).

BJORHOVDE, R., and ZIMMERMAN, T. J.1980. Some aspects of stub-girder design. Proceedings of the Canadian Structural Engineering Conference. Canadian Steel Construction Council. Willowdale, Ont.

BLEICH, F. 1952. Buckling strength of metal structures. McGraw-Hill. New York., NY.

BOWER, J. E., et al. 1971. Suggested design guide for beams with web holes. ASCE Journal of the Structural Division, **97**(11).

BROCKENBROUGH, R. L. 1983. Considerations in the design of bolted steel joints for weathering steel. AISC Engineering Journal, **20**(1).

BROCKENBROUGH, R. L., and JOHNSTON, B. G. 1974. Steel design manual.United States Steel Corp., Pittsburgh, PA.

BUDIJGNTO, P. 1983. Design methods for composite columns. Project Report G83-6, Department of Civil Engineering, McGill University.

BUTLER, L.J. and KULAK, G.L. 1971. Strength of fillet welds as a function of direction of load. Welding Research Supplement, Welding Journal. Welding Research Council, **36**(5), 2315-2345

BUTLER, L.J., PAL, S. and KULAK, G.L. 1972. Eccentrically loaded welded connections. ASCE Journal of the Structural Division, **98**(ST5), 989-1005

CHEN, W. F., and ATSUTA, T. 1972. Interaction equations for biaxially loaded sections. ASCE Journal of the Structural Division, **98**(ST5).

——— 1973. Ultimate strength of biaxially loaded steel H-columns. ASCE Journal of the Structural Division, **99**(ST3).

CHEN, W. F., and LUI, E.M..1987. Structural stability, theory and implementation. Elsevier. New York, NY.

CHEONG SAIT-MOY, F. 1991. Column design in gravity loaded frames. ASCE Journal of Structural Engineering, **117**(ST5)

CHENG, J-J. R., and YURA, J. A. 1986. Local web buckling of coped beams. ASCE Journal of Structural Engineering, **112**(10).

CHERNENKO, D.E. and KENNEDY, D.J.L. 1991. An analysis of the performance of welded wide flange columns. Canadian Journal of Civil Engineering, **18**(4).

CHIEN, E.Y.L., and RITCHIE, J. K. 1984. Design and construction of composite floor systems. Canadian Institute of Steel Construction, Willowdale, Ont.

CHIEN, E.Y.L. 1989. Roof framing with cantilever (Gerber) girders & open web steel joists. Canadian Institute of Steel Construction, Willowdale, Ont.

CIDECT. 1970. Concrete filled hollow section steel columns design manual. Monograph No. 1, International Committee for the Study and Development of Tubular Structures. Whitefriars Press Ltd., London.

CISC. 1987. Column selection program. Canadian Institute of Steel Construction, Willowdale, Ont.

——— 1980. Steel Joist Facts, 2nd Edition. Canadian Institute of Steel Construction, Willowdale, Ont.

CSA. 1969. Steel structures for buildings. CSA Standard S16-1969, Canadian Standards Association, Rexdale, Ont.

——— 1994. Design of concrete structures. CSA Standard A23.3-94, Canadian Standards Association, Rexdale, Ont.

——— 1994. Cold formed steel structural members. CSA Standard S136-94. Canadian Standards Association, Rexdale, Ont.

——— 1996 Canadian highway bridge design code. CSA Standard S6-96. Canadian Standards Association, Rexdale, Ont.

DAVIES, C. 1969. Tests on half-scale steel concrete composite beams with welded stud connectors. The Structural Engineer, **47**(1): 29–40.

DAWE, J. L., and KULAK, G. L. 1984. Local buckling of W shape columns and beams. ASCE Journal of Structural Engineering, **110**(6).

——— 1986. Local buckling behavior of beam–columns. ASCE Journal of Structural Engineering, **112**(11).

DAWE, J.L., and LEE, T.S., 1993 Local buckling of Class 2 beam-column flanges. Canadian Journal of Civil Engineering, **20**(6): 931-939

DISQUE, R. O. 1964. Wind connections with simple framing. Engineering Journal, AISC, July.

DRIVER, R. G., and KENNEDY, D. J. L. 1989. Combined flexure and torsion of I-shaped steel beams. Canadian Journal of Civil Engineering, **16**(2).

DUAN, L. and CHEN, W-F. 1988. Design rules of built-up members in Load and Resistance Factor design. ASCE Journal of Structural Engineering, **114**(11), 2544-2554.

EL-GHAZZI, M. N., ROBINSON, H., and ELKHOLY, I.A.S. 1976. Longitudinal shear capacity of slabs of composite beams. Canadian Journal of Civil Engineering, **3**(4).

ELKELISH, S., and ROBINSON, H. 1986. Effective widths of composite beams with ribbed metal deck. Canadian Journal of Civil Engineering, **13**(5): 575–582.

ENGLEKIRK, R. 1994. Steel structures–controlling behavior through design. John Wiley & Sons. New York, NY.

ESSA, H.S. and KENNEDY, D.J.L., 1994a, Station Square revisited: distortional buckling collapse. Canadian Journal of Civil Engineering, **21**(3): 377-381.

——— 1994b, Design of cantilever beams: a refined approach. Journal of Structural Engineering, American Society of Civil Engineers. **120**(ST9), pp. 2623-2636.

ESSA, H.S. and KENNEDY, D.J.L., 1995, Design of steel beams in cantilever-suspended span construction. Journal of Structural Engineering, American Society of Civil Engineers. **121**(ST11).

EUROCODE No. 4. 1987. Common unified rules for composite steel and concrete structures. Report EUR 9886 EN, Commission of the European Communities, Luxembourg.

FISHER, J. W. 1974. Guide to 1974 AASHTO fatigue specifications. AISC, Chicago, IL.

——— 1978. Fatigue cracking in bridges from out-of-plane displacements. Canadian Journal of Civil Engineering, **5**(4).

——— 1984. Fatigue and fracture in steel bridges—case studies. John Wiley & Sons.

FISHER, J, W., LEE, G. C, YURA, J. A., and DRISCOLL, G. C. 1963. Plastic analysis and tests of haunched corner connections. Bulletin No. 91, Welding Research Council.

FRANK, K. H. 1980. Fatigue strength of anchor bolts. ASCE. Journal of the Structural Division, **106**(ST6).

GAGNON, D. G., and KENNEDY, D. J. L. 1989. Behaviour and ultimate tensile strength of partial joint penetration groove welds. Canadian Journal of Civil Engineering, **16**(3).

GALAMBOS, T. V. 1968. Structural members and frames. Prentice-Hall Inc., Englewood Cliffs, NJ.

GALAMBOS, T. V. 1988. Guide to stability design criteria for metal structures (4th ed.). Structural Stability Research Council, John Wiley & Sons, Inc., New York, NY.

GHOBARAH, A., KOROL, R.M., and OSMAN, A. 1992. Cyclic behaviour of extended end plate joints. ASCE. Journal of the Structural Division, **118**(ST5)

GIBSON, G. T., and WAKE, B. T. 1942. An investigation of welded connections for angle tension members. Welding Journal, American Welding Society.

GOBEL, G. 1968. Shear strength of thin flange composite specimens. Engineering Journal, AISC, April.

GOODIER, J.N., 1942, Flexural-torsional buckling of bars of open section . Cornell University Engineering Experiment Station Bulletin No. 28, January.

GRAHAM, J. D., SHERBOURNE, A. N., KHABBAZ, R. N., and JENSEN, C. D. 1959. Welded interior beam-to-column connections. American Institute of Steel Construction.

GRANT, J. A., FISHER, J. W., and SLUTTER, R. G. 1977. Composite beams with formed steel deck. Engineering Journal, AISC, First Quarter.

GRIGGS, P. H. 1976. Mill Building structures. Proceedings, Canadian Structural Engineering Conference, Canadian Steel Construction Council, Willowdale, Ont.

HAAIJER, G., and THURLIMANN, B. 1958. On inelastic buckling in steel. ASCE Journal of the Engineering Mechanics Division, April.

HAGOOD, T. A., Jr., GUTHRIE, L., and HOADLEY, G. 1968. An investigation of the effective concrete slab width for composite construction. Engineering Journal, AISC, **5**(1): 20–25.

HANSELL, W. C., and VIEST, I. M. 1971. Load factor design for steel highway bridges. Engineering Journal, AISC, **8**(4).

HARDASH, S. and BJORHOVDE, R. 1985. New design criteria for gusset plates in tension. Engineering Journal, AISC, **21**(2), 77-94

HART, W. H., and MILEK, W. A. 1965. Splices in plastically designed continuous structures. Engineering Journal, AISC, April.

HAWKINS, N. M., and MITCHELL, D. 1984. Seismic response of composite shear connections. ASCE Journal of Structural Engineering, **110**(9), 2120-2136.

HEINS, C. P., and FAN, H. M.. 1976. Effective composite beam width at ultimate load. ASCE Journal of the Structural Division, **102**(ST11).

HIRANO, N. 1970. Bearing stresses in bolted joints. Society of Steel Construction of Japan, **6**(58), Tokyo.

HOLTZ, N. M., and KULAK, G. L. 1970. High strength bolts and welds in load-sharing systems. Studies in Structural Engineering, No. 8, Technical University of Nova Scotia.

——— 1973. Web slenderness limits for compact beams. SER 43, Department of Civil Engineering, University of Alberta.

——— 1975. Web slenderness limits for non-compact beams. SER 51, Department of Civil Engineering, University of Alberta.

HUANG, J.S., CHEN, W.F., and BEEDLE, L.S. 1973. Behavour and design of steel beam-to-column moment connections. Welding Research Council Bulletin, 188, pp 1-23.

JAYAS, B. S., and HOSAIN, M. U. 1988. Behaviour of headed studs in composite beams: push-out tests. Canadian Journal of Civil Engineering, **15**(2), 240-253.

———— 1989. Behaviour of headed studs in composite beams: full-size tests. Canadian Journal of Civil Engineering, **16**(5), 712-724.

JENT, K. A. 1989. Effects of shrinkage, creep and applied loads on continuous deck-slab composite beams. M.Sc. thesis, Queen's University, Kingston, Ont.

JOHNSTON, B. G. 1939. Pin connected plate links. Transactions, American Society of Civil Engineers.

JOHNSON, R. P. 1970. Longitudinal shear strength of composite beams. ACI Journal Procedings, **67**.

————. 1975. Composite structures of steel and concrete. Volume 1: beams, columns, frames and applications in buildings. Crosby Lookwood Staples, London, England. p. 210.

JONES, J. 1958. Bearing-ratio effect on strength of riveted joints. Transactions, American Society of Civil Engineers, **123**: 964–972.

KALIANDASANI, R. A., SIMMONDS, S. H., and MURRAY, D. W. 1977. Behaviour of open web steel joists. Report No. 62, Department of Civil Engineering, University of Alberta.

KALYANARAMAN, V., PEKOZ, T., and WINTER, G. 1977. Unstiffened compression elements. ASCE. Journal of the Structural Division, **103**(ST9).

KEMP, A.R. and TRINCHERO, P. 1992. Serviceability stress limits for composite beams. Composite Construction and Engineering Foundation Conference, June 15-19, Potosi, MO.

KENNEDY, D.J.L. 1974. Limit states design—an innovation in design standards for steel structures. Canadian Journal of Civil Engineering, **1**(1).

KENNEDY, D.J.L., and ROWAN, W.H.D. 1964. Behaviour of compression chords of open web steel joists. Report to CISC.

KENNEDY, D.J.L., ALLEN, D. E., ADAMS, P. F., KULAK, G. L., TURNER, D. K., and TARLTON, D. L. 1976. Limit states design. Procedings of the Canadian Structural Engineering Conference, Canadian Steel Industries Construction Council, Willowdale, Ont.

KENNEDY, D.J.L., and BRATTLAND, A. 1992. Shrinkage tests of two full-scale composite trusses. Canadian Journal of Civil Engineering, **19**(2).

KENNEDY, D.J.L., and GAD ALY, M. 1980. Limit states design of steel structures—performance factors. Canadian Journal of Civil Engineering, **7**(1).

KENNEDY, S.J., and MacGREGOR, J.G. 1984. End connection effects on the strength of concrete filled HSS beam-columns. Structural Engineering Report 115, Department of Civil Engineering, University of Alberta, Edmonton, Alberta.

KENNEDY, D.J.L., MIAZGA, G.S. and LESIK, D.F. 1990. Discussion of Evaluation of fillet weld shear strength of FCAW electrodes by McClellan. R.W. Welding Journal, August 1989, Welding Journal Reference, 44-46

KENNEDY, D.J.L., PICARD, A., and BEAULIEU, D. 1990. New Canadian provisions for the design of steel beam_columns. Canadian Journal of Civil Engineering, **17**(6)

KENNEDY, J.B., and NEVILLE, A.M., 1986, Basic statistical methods for engineers and scientists. Third edition, Harper and Row, New York, NY

KENNEDY, S. J., and KENNEDY, D. J. L. 1987. The performance and strength of hardened steel test roller assemblies. Proceedings, CSEC Centenial Conference, May 19-22. Montreal, Que.: 513-531.

KETTER, R. L. 1961. Further studies of the strength of beam-columns. ASCE, Journal of the Structural Division, **87**(ST6): 135–152

KIRBY, P. A. and NETHERCOT, D. A. 1978. Design for structural stability. Granada. London, England.

KNOWLES, R. B., and PARK, R. 1970. Axial load design for concrete filled steel tubes. ASCE, Journal of the Structural Division, **96**(ST10).

KRAWINKLER, H., and POPOV, E.P. 1982. Seismic behaviour of moment connections and joints. ASCE, Journal of the Structural Division, **108**(ST2), 373–391.

KULAK, G. L. 1991. Unstiffened steel plate shear walls. Structures subjected to repeated loading: stability and strength, R. Narayanan and T. M. Roberts, eds., Elsevier Applied Science, New York, N. Y.

KULAK, G. L., ADAMS, P. F., and GILMOR, 1995. Limit states design in structural steel. Canadian Institute of Steel Construction, Willowdale, Ont.

KULAK, G. L., and DAWE, J. L. 1991. Discussion of design interaction equations for steel members. Sohal, I. S., Duan, L. and Chen W-F. ASCE, Journal of the Structural Division, **117**(ST7): 2191-2193.

KULAK, G. L., FISHER, J. W., and STRUIK, J.H.A. 1987. Guide to design criteria for bolted and riveted joints (2nd. ed.). John Wiley & Sons, New York, NY.

KULAK, G.L., and SMITH, I.F.C. 1993. Analysis and design of fabricated steel structures for fatigue: A primer for civil engineers. Dept. of Civil Eng. SER 190. University of Alberta, Edmonton. AB.

KULLMAN R. B. and HOSAIN, M. U. 1985. Shear capacity of stub-girders: full scale tests. ASCE, Journal of the Structural Division, **111**(ST1): 56-75.

LAY, M. G. 1965. Flange local buckling in wide-flange shapes. ASCE Journal of the Structural Division, **91**(ST6).

LAY, M. G., and GALAMBOS, T. V. 1966. Bracing requirements for inelastic steel beams. ASCE Journal of the Structural Division, **92**(ST2).

———— 1967. Inelastic beams under moment gradient. ASCE Journal of the Structural Division, **93**(ST1)

LESIK, D.F. and KENNEDY, D.J.L. 1990. Ultimate strength of fillet welded connections loaded in plane. Canadian Journal of Civil Engineering **17**(1), 55-67

LILLEY, S. B., and CARPENTER, S. T. 1940. Effective moment of inertia of a riveted plate girder. Transactions, American Society of Civil Engineers.

LIU, Z., and GOEL, S.C. 1988. Cyclic load behavior of concrete-filled tubular braces. ASCE Journal of Structural Engineering, **114**(ST7), 1488–1506.

LOGCHER, R. D., et al. 1969. ICES STRUDL II engineering users manual. Department of Civil Engineering, Massachusetts Institute of Technology, Cambridge, Mass.

LOOV, R. 1996. A simple equation for axially loaded steel colun design curves. Canadian Journal of Civil Engineering, **23**(1), 272-276.

LU, Y.Q. and KENNEDY, D.J.L. 1994. The flexural behaviour of concrete-filled hollow structural sections. Canadian Journal of Civil Engineering, **21**(1), 111-130.

LUKEY, A. F., and ADAMS, P. F. 1969. Rotation capacity of beams under moment gradient. ASCE Journal of the Structural Division, **95**(ST6).

MANNICHE, K., and WARD-HALL, G. 1975. Mission bridge—design and construction of the steel box girder. Canadian Journal of Civil Engineering, **2**(2).

MASSEY, C. 1962. Lateral bracing forces of steel I-beams. ASCE Engineering Mechanics Division, **88**(EM6).

MATTOCK, A. H. 1974. Shear transfer on concrete having reinforcement at an angle to the shear plane. Special Publication 42, Shear in Reinforced Concrete, American Concrete Institute, pp 17–42.

MATTOCK, A. H., LI, W. K., and WANG, T.C. Shear transfer in lightweight reinforced concrete. PCI Journal, **21**(1), 20-39.

MAURER, M.B. and KENNEDY D.J.L., 1994, Shrinkage and flexural tests of a full-scale composite truss. Structural Engineering Report 206, Department of Civil Engineering, The University of Alberta.

MIAZGA, G.S. and KENNEDY, D.J.L. 1989. Behaviour of fillet welds as a function of the angle of loading. Canadian Journal of Civil Engineering Vol 16 pp 583-599

MUNSE, W. H. 1959. The effect of bearing pressure on the static strength of riveted connections. Bulletin No. 454, Engineering Experimental Station, University of Illinois, Urbana, IL.

MUNSE, W. H. and CHESSON, E. 1963. Riveted and bolted joints: net section design. ASCE. Journal of the Structural Division, **89**(ST1), Part 1.

MURRAY, T. M. 1975. Design to prevent floor vibrations. Engineering Journal, AISC, third quarter.

NCRHP. 1970. Effect of weldments on the fatigue strength of steel beams. Report 102, National Cooperative Highway Research Program, Transportation Research Board, National Academy of Sciences, Washington, DC.

――― 1974. Fatigue strength of steel beams with welded stiffeners and attachments. Report 147, National Cooperative Highway Research Program, Transportation Research Board, National Academy of Sciences, Washington, DC.

NASH, D. S., and KULAK, G. L. 1976. Web slenderness limits for non-compact beam-columns. Report No. 53, Department of Civil Engineering, University of Alberta.

NBCC. 1995a. Commentary F, Commentaries on Part 4 of the National building code of Canada. National Research Council of Canada, Ottawa, Ont.

――― 1995b. Commentary I, Commentaries on Part 4 of the National building code of Canada. National Research Council of Canada, Ottawa, Ont.

――― 1995c. National building code of Canada. Clause 4.1.3. National Research Council of Cananda, Ottawa, Ont.

――― 1995d. National building code of Canada. Clause 4.1.6. National Research Council of Cananda, Ottawa, Ont.

NETHERCOT, D. A. and TRAHAIR, N. S. 1976a. Inelastic lateral buckling of determinate beams. ASCE Journal of the Structural Division, **102**(ST4):701-717.

――― 1976b. Lateral buckling approximations for elastic beams. ISE. The Stuctural Engineer. **54**(6): 197-204.

NIXON, D. 1981. The use of frame action to resist lateral loads in simple construction. Canadian Journal of Civil Engineering, **8**(4).

OLGAARD, J. G,, SLUTTER, R. G., and FISHER, J. W. 1971. Shear strength of stud connectors in light-weight and normal-weight concrete. Engineering Journal, AISC, April

OSMAN, A., KOROL, R.M., and GHOBARAH, A. 1991. Bolted beam-to-column subassemblages under repeated loading. Proceedings: Sixth Candian Earthquake Engineering Conference, University of Toronto Press, June.

PERLYNN, M. J. and KULAK, G. L. 1974. Web slenderness limits for compact beam-columns. Structural Engineering Report 50, Department of Civil Engineering, University of Alberta.

PILLAI, U. S. 1970. Review of Recent Research on the Behaviour of Beam-Columns under Biaxial Bending. Civil Engineering Research Report No. CE 70-1, Royal Military College of Canada, Kingston, Ont.

――― 1974. Beam-columns of hollow structural sections. Canadian Journal of Civil Engineering, **1**(2).

PILLAI, U. S., and ELLIS, J. S. 1971. Hollow Tubular Beam-Columns in Biaxial Bending. ASCE Journal of the Structural Division, **97**(ST5).

PILLAI, U. S., and KURIAN, V. J. 1977. Tests on hollow structural section beam-columns. Canadian Journal of Civil Engineering, **4**(2).

Popov, E. P., and Pinkney, R. B. 1969. Cyclic yield reversal in steel building connections. ASCE Journal of the Structural Division, **95**(ST3).

Popov, E. P., and Stephen, R. M. 1972. Cyclic loading of full-size steel connections. Steel Research for Construction, Bulletin No.21, AISI.

——— 1977a. Capacity of columns with splice imperfections. Engineering Journal, AISC, **14**(1).

——— 1977b. Tensile capacity of partial penetration groove welds. ASCE. Journal of the Structural Division, **103**(ST9).

Popov, E.P., Amin, N.R., Louie, J.J.C., and Stephen, R.M. 1986. Cyclic behaviour of large beam–column assemblies. Engineering Journal, AISC, **23**(1), 9–23.

Rainer, J. H. 1980. Dynamic tests on a steel-joist concrete-slab floor. Canadian Journal of Civil Engineering, **7**(2).

RCSC. 1994. Specification for structural joints using ASTM A325 or A490 bolts. Research Council on Riveted and Bolted Structural Joints.

Redwood, R. G. 1971. Simplified plastic analysis for reinforced web holes. Engineering Journal, AISC, **8**(4).

——— 1972. Tables for plastic design of beams with rectangular holes. Engineering Journal, AISC, **9**(1).

——— 1973. Design of beams with web holes. Canadian Steel Industries Construction Council, Willowdale, Ont.

Redwood, R. G., and Channagiri, V.S. 1991. Earthquake resistant design of concentricity braced steel frames. Canadian Journal of Civil Engineering, **18**(5).

Redwood, R. G., Feng Lu, Bouchard, G., and Paultre, P. 1991. Seismic response of concentrically braced steel frames. Canadian Journal of Civil Engineering, **18**(6).

Redwood, R., and McCutcheon, J. 1968. Beam tests with unreinforced web openings. ASCE Journal of the Structural Division, **94**(ST1).

Ricles, J.M., and Yura, J.A. 1980. The behavior and analysis of double row bolted shear web connections. PMFSEL Thesis no. 80-1, Department of Civil Engineering, University of Texas at Austin.

Ritchie, J.K., and Chien, E.Y.L. 1980. Composite structural systems—design, construction and cost considerations. Proceedings, Canadian Structural Engineering Conference, C.I.S.C.

Roberts, T.M. 1981. Slender plate girders subjected of edge loading. Proceedings of Institute of Civil Engineers, Part 2, 71, September.

Roberts, T.M., and Chong, C.K. 1981. Collapse of plate girders under edge loading. ASCE Journal of the Structural Division, **107**(8).

Robinson, H. 1969. Composite beam incorporating cellular steel decking. ASCE Journal of the Structural Division, **95**(ST3).

——— 1988. Multiple stud shear connections in deep ribbed metal deck. Canadian Journal of Civil Engineering, **15**(4).

Robinson, H., and Wallace, I. W. 1973. Composite beams with 1-1/2 inch metal deck and partial and full shear connection. Transactions, Canadian Society for Civil Engineering, **16**(A-8), published in the Engineering Journal, Engineering Institute of Canada.

Ross, D. A., and Chen, W. F. 1976. Design criteria for steel I-columns under axial load and biaxial bending. Canadian Journal of Civil Engineering, **3**(3).

Schmitke, C.D. and Kennedy, D.J.L. 1985. Effective lengths of laterally continuous, laterally unsupported steel beams. Canadian Journal of Civil Engineering. **12**(3), 603-616

SEAOC. 1988. Recommended lateral force requirements and tentative commentary. Structural Engineers Association of California, Seismology Committee.

Seeley, F. B., and Smith, J. O. 1957. Advanced mechanics of materials (2nd ed.). John Wiley & Sons, Inc., New York, NY. pp. 365–367.

Shaker, A.F. and Kennedy, D.J.L. 1991 The effective modulus of elasticity of concrete in tension. Structural Engineering Report 172, Department of Civil Engineering, University of Alberta, Edmonton, Alberta

Sherman, D. R., and Tanavde, A. S. 1984. Comparative study of flexural capacity of pipes. Department of Civil Engineering, University of Wisconsin-Milwaukee.

Shrivastava, S. C., Redwood, R. G., Harris, P. J., and Ettehadieh, A. A. 1979. End moments in open web steel tie joists. McGill University, June.

Slutter, R. G., and Driscoll, G. C. 1965. Flexural strength of steelconcrete composite beams. ASCE Journal of the Structural Division, **95**(ST2).

Sourochnikoff, B. 1950. Wind-stresses in semi-rigid connections of steel framework. Transactions, American Society of Civil Engineers.

Springfield, J. 1975. Design of columns subject to biaxial bending. Engineering Journal, AISC, **12**(3), Third Quarter.

Stelco. 1973. Hollow structural sections—design manual for columns and beams. The Steel Company of Canada Limited, Hamilton, Ont.

——— 1981. Hollow structural sections—design manual for concrete-filled HSS columns. The Steel Company of Canada Limited, Hamilton, Ont.

Stiemer, S. F. 1986. Innovative earthquake resistant steel design. Proceedings, Canadian Structural Engineering Conference. Canadian Steel Construction Council. Willowdale, Ont.

Tall, L., et al. 1974. Structural steel design (2nd ed.). The Ronald Press Company, New York, NY.

TEBEDGE, N., and CHEN, W. F. 1974. Design criteria for H-columns under biaxial loading. ASCE Journal of the Structural Division, **100**(ST3).

TEMPLE, M.C., SCHEPERS, J.A., and KENNEDY, D.J.L.1986. Interconnection of starred angle compression members. Canadian Journal of Civil Engineering, **13**(6).

THORNBURN, L. J., KULAK, G. L. and MONTGOMERY, C. J. 1983. Analysis of steel plate shear walls. Structural Engrg. Report No. 107, Department of Civil Engineering, Univ. of Alberta, Edmonton, Alberta.

TIMLER, P. A. and KULAK, G. L. 1983. Experimental study of steel plate shear walls. Structural Engrg. Report No. 114, Department of Civil Engineering, Univ. of Alberta, Edmonton, Alberta.

TIMOSHENKO, S.P. and GERE, J.M., 1961, Theory of elastic stability, Second edition, McGraw-Hill, New York, NY

THORNTON, C. H. 1973. Quality control in design and supervision can eliminate lamellar tearing. Engineering Journal, AISC, Fourth Quarter.

TOPRAC, A., and NATARAJAN, M. 1971. Fatigue strength of hybrid plate girders. ASCE Journal of the Structural Division, **97**(ST4).

TRAHAIR, N.S. 1968. Interaction buckling of narrow rectangular continuous beams. Civil Engineering Trans. Inst. of Eng. Australia.

TSAI, K., AND POPOV, E.P. 1990. Cyclic behaviour of end-plate moment connections. ASCE Journal of the Structural Division, **116**(ST11).

VINCENT, G. S. 1969. Tentative criteria for load factor design of steel highway bridges. Steel Research Construction Bulletin No. 15, AISI, Washington, DC.

VIRDI, K. S., and DOWLING, P. J. 1976. A unified design method for composite columns. Memoires, IABSE, No. 36-11.

VLASOZ, V.Z., 1959, Thin-walled elastic beams, Second edition, (translation from the Russian), available from the Office of Technical Services, U.S. Department of Commerce, Washington, DC.

WAKABAYASHI, M. 1977 A new design method of long composite beam-columns. Proceedings, ASCE International Colloquium on Stability of Structures under Static and Dynamic Loads, Washington, DC., May.

WINTER, G. 1958. Lateral bracing of columns and beams. ASCE Journal of the Structural Division, **84**(ST2).

——— 1970. Commentary on the 1968 edition of the specification for the design of cold-formed steel structural members. American Iron and Steel Institute, Washington, DC.

WOOD, B. R., BEAULIEU, D., and ADAMS, P. F. 1976. Column design by P-delta method. ASCE Journal of the Structural Division, **102**(ST2).

——— 1976. Further aspects of design by P-delta method. ASCE Journal of the Structural Division, **102**(ST3).

WU Y, and KULAK, G. L. 1993. Shear lag in bolted single and double angle tension members. Structural Engineering Report, Department of Civil Engineering University of Alberta, SER 187

YANG, C.H. and BEEDLE, L.S. 1951. Behavior of I and WF beams in shear. Fritz Engineering Lab. Report No. 205B21, Lehigh Univ., Bethlehem, Pa.

YONG Lin Pi and TRAHAIR, N.S. 1995. Inelastic torsion of steel I-beams. ASCE Journal of the Structural Division, **121**(ST4).

YURA, J.A., BIRKEMOE, P.C., and RICLES, J.M. Web shear connections: an experimental study. ASCE Journal of the Structural Division, **108**(ST2).

ZUK, W. 1956. Lateral bracing forces on beams and columns. ASCE Engineering Mechanics Division, **82**(EM3).

PART THREE
CONNECTIONS AND TENSION MEMBERS

General Information	3-3
Connection Loads	3-4
Bolts	
Bolt Data — Metric and Imperial Series	3-5
Bolts in Bearing-Type Connections	3-7
Bolts in Slip-Critical Connections	3-14
Bolts in Tension and Prying Action	3-19
Eccentric Loads on Bolt Groups	3-26
Welds	
Weld Data	3-38
Eccentric Loads on Weld Groups	3-42
Framed Beam Shear Connections	3-56
Double Angle Beam Connections	3-58
End Plate Connections	3-66
Single Angle Beam Connections	3-68
Shear Tab Beam Connections	3-70
Tee-Type Beam Connections	3-72
Seated Beam Shear Connections	3-74
Unstiffened Angle Seat Connections	3-74
Stiffened Seated Beam Connections	3-78
Moment Connections	3-80
Hollow Structural Section Connections	3-89
Tension Members	3-99

GENERAL INFORMATION

While the basic steel grade for W and HP shapes produced by Algoma as of May 1997 is CSA-G40.21 350W, at the time of printing, detail material (angles, plates, bars) is still G40.21 300W. Since both of these grades have identical specified minimum tensile strengths F_u of 450 MPa, those tabulated factored resistances based on F_u are unchanged. Other tables still refer to G40.21 300W, as the resistances so tabulated are generally conservative on that basis. The section on Prying Action has been revised for G40.21 350W, as F_y rather than F_u determines the behaviour. The previous version based on 300W is now in Part 8.

Part Three contains tables, examples, dimensions and general information of assistance to designers, detailers and others concerned with the design and detailing of connections and tension members according to the requirements of Clauses 12, 13.2, 13.10(c), 13.11, 13.12, 13.13, 21.0, 22.0 and 23.0 of CAN/CSA-S16.1-94. Information is included for both metric series and Imperial series bolts, though all design data is given in SI units. For convenience, Part Three is divided into seven main sections.

Bolt Data

Pages 3-5 to 3-37 contain information on diameter, area and strength of bolts, including bolt resistances and unit resistances, for evaluating bolts in bearing type connections, slip-critical connections, and bolts subjected to tension and prying action. Tables are also included for evaluating eccentric loads on various bolt groups.

Weld Data

Pages 3-38 to 3-55 contain information on factored resistance of welds, including values for various sizes of fillet welds with a comparison between Imperial and metric sized fillet welds. Tables are included for evaluating eccentric loads on various weld groups and configurations.

Framed Beam Shear Connections

Pages 3-56 to 3-73 contain information on common types of beam shear connections traditionally considered standard in the industry. Included are double angle beam connections, simple end plate connections, single angle beam connections, shear tab beam connections and tee-type beam connections.

Seated Beam Shear Connections

Pages 3-74 to 3-79 contain information on unstiffened and stiffened seated beam shear connections of a type commonly used in practice, where direct framing of the supported beam is either not desirable or possible.

Moment Connections

Pages 3-80 to 3-88 contain examples of welded and welded/bolted moment connections, and information for the design of stiffeners on supporting columns.

Hollow Structural Section Connections

Pages 3-89 to 3-98 contain information regarding the connecting of HSS sections.

Tension Members

Pages 3-99 to 3-107 contain tables and examples for calculating net effective areas and for evaluating the unit tensile resistance of bolted and welded tension members for various grades of steel.

CONNECTION LOADS

Connections are designed and detailed for the member reactions and loads given on the structural steel design drawings by the designer. Most connections are designed for factored loads, however specified loads are used for calculating the slip resistance of slip-critical connections, such as those subjected to dynamic loadings.

In evaluating member loads and forces it is preferable to keep different types of load separate to facilitate application of the different load factors and load combination factors specified in Clause 7.2 of CAN/CSA-S16.1-94 and other governing codes. However, this may not always be convenient, and Fig. 3-1 is included to permit an approximate evaluation of either the total specified load or the total factored load when either one is known, and the ratio of the specified dead load to specified live load is known or the unit specified dead load and unit specified live load are both known. The curve is based on load factors α given in Clause 7.2.3 and makes no allowance for possible live load reductions permitted by the applicable building codes.

Example

Given:

The total factored dead and live load reaction at the end of a beam to be designed as a slip-critical connection is 235 kN and the specified unit dead and live loads are 4.35 kPa and 2.16 kPa respectively. What is the reaction under specified load.

Solution:

Ratio $D/L = 4.35/2.16 = 2.0$

From Fig. 3-1, for $D/L = 2.0$, $(L + D) / (\alpha_L L + \alpha_D D) = 0.75$

Therefore specified load reaction is $0.75 \times 235 = 176$ kN

Figure 3-1

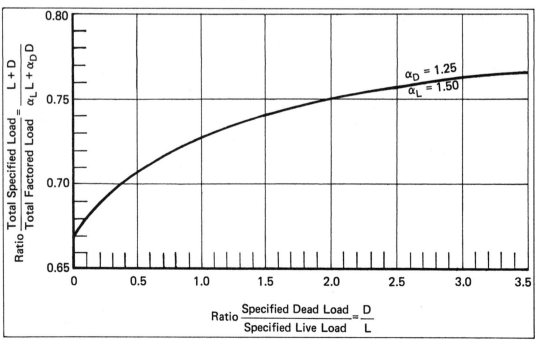

BOLT DATA

General

Tables in this section are based on CAN/CSA-S16.1-94, and include information for both metric series and Imperial series bolts. Data for the metric series bolts is based on ASTM Specifications A325M and A490M and for the Imperial series bolts on ASTM Specifications A325, A490 and A307. Values are tabulated in ascending order of nominal cross-sectional area A_b (mm^2) to facilitate comparison between metric and Imperial sizes. Bold type is used for metric series bolts when both metric sizes and Imperial sizes appear on the same table. This section includes the following:

Bolt Data Metric and Imperial Series

Table 3-1 on page 3-6 lists the size, nominal diameter (mm), nominal area (mm^2) and values of $A_b F_u$ for bolt sizes from M16 to M36 and ½ inch to 1½ inch diameter.

Bolts in Bearing-Type Connections

Tables 3-2 to 3-7 on pages 3-7 to 3-11 list values of bearing and bolt resistances computed in accordance with Clauses 13.10(c) and 13.11. Tables 3-8 and 3-9 on page 3-13 assist in evaluating combined shear and tension on bolts.

Bolts in Slip-Critical Connections

Tables 3-10 and 3-11 on page 3-15 list resistances, for use with bolts in slip-critical connections, computed in accordance with Clause 13.12.

Bolts in Tension and Prying Action

Tables and design aids on pages 3-19 to 3-25 assist in evaluating the effects of prying action on bolts loaded in tension.

Eccentric Loads on Bolt Groups

Tables for evaluating eccentric loads on bolts in bearing-type and slip-critical connections for various bolt group configurations are given on pages 3-29 to 3-37.

Availability

For more information on the range of commonly used fasteners, refer to page 6-142.

Before specifying metric bolts, the designer should check on their current availability in the quantities required. To facilitate substitution, use the lower of the metric or Imperial values for similar sizes.

BOLT DATA
Metric and Imperial Series

Table 3-1

Bolt Size		Nominal Diameter of Bolt (mm)	Nominal Area (A_b) (mm²)	$A_b F_u$** (kN)				
Metric*	Imperial			A325M	A490M	A325	A490	A307
	1/2	12.70	127			104	131	
	5/8	15.88	198			163	205	81.9
M16		16.00	201	167	209			
	3/4	19.05	285			235	295	118
M20		20.00	314	261	327			
M22		22.00	380	316	395			
	7/8	22.23	388			320	402	161
M24		24.00	452	375	470			
	1	25.40	507			418	524	
M27		27.00	573	475	595			
	1–1/8	28.58	641			465	664	
M30		30.00	707	587	735			
	1–1/4	31.75	792			574	819	
M36		36.00	1018	845	1060			
	1–1/2	38.10	1140			827	1180	

*The number following the letter M is the nominal bolt diameter in millimetres.
**See Table 3–3 page 3–8 for specified minimum tensile strengths, F_u.

BOLTS IN BEARING-TYPE CONNECTIONS

General

Connections are generally detailed as bearing-type, unless the designer has specified that the connection is "slip-critical". Bearing-type connections are designed for factored loads, and Tables 3-2 to 3-9 inclusive on the following pages assist in evaluating the requirements of Clause 13.11 of CAN/CSA-S16.1-94. Clause 23.3.2 lists the size and type of holes permitted with bearing type connections.

Table 3-2 on page 3-7 summarises the requirements of Clause 13.11.2 for bolts in shear and Clause 13.11.3 for bolts in tension, and lists expressions for factored resistance and unit factored resistance of bolts in bearing-type connections.

Table 3-3 on page 3-8 lists values of the specified minimum tension F_u values of unit factored shear resistances, $0.4\,F_u$ and $0.28\,F_u$, and values of unit factored tensile resistances, $0.5\,F_u$, for A325M, A490M, A325, A490 and A307 bolts.

Table 3-4 on page 3-8 lists factored shear and tensile resistances in kN/bolt for both metric series and Imperial series bolts.

Table 3-5 on page 3-9 lists values of the specified minimum tensile strength F_u for the common grades of structural steel, and values of unit factored bearing resistances, $0.67\,F_u$ and $3\,\phi_b\,F_u$.

Tables 3-6 and 3-7 on pages 3-9 to 3-11 list factored bearing resistance in kN/bolt for five different values of F_u for the connected material. Bearing resistances in these tables are given in terms of the material thickness t and the bolt size and grade.

The former expression for bearing on main material as a function of the bolt end distance e does not appear in S16.1-94. The requirements of Clause 12.3 (Effective Net Area) as applied in Clause 13.2 (Axial Tension) govern instead. Some failures of a shear connection by the bolts pulling through the material are a form of "block tear-out" as illustrated by examples in Tension Members starting on page 3-99.

Tables 3-8 and 3-9 on page 3-13 assist in evaluating bolts in combined shear and tension according to Clause 13.11.4.

Table 3-2
CAN/CSA-S16.1-94 SUMMARY
Bearing-Type Connections

Bolt Situation In Joint	Factored Resistance	Unit Factored Resistance ($n = m = A_b = 1$)	Clause Reference
BOLTS IN SHEAR			13.11.2
Shear on bolts with threads excluded from shear plane	$V_r = 0.60\,\phi_b\,n\,m\,A_b\,F_u$	$V_r = 0.40\,F_u$	13.11.2(b)
Shear on bolts with threads intercepted by shear plane	$V_r = 0.42\,\phi_b\,n\,m\,A_b\,F_u$	$V_r = 0.28\,F_u$	
For joints longer than 1300 mm	Shear = $0.8\,V_r$		
Bearing on main material	$B_r = 3\,\phi\,t\,d\,n\,F_u$	$B_r = 2.0\,t\,d\,F_u$	13.10(c)
BOLTS IN TENSION	$T_r = 0.75\,\phi_b\,n\,A_b\,F_u$	$T_r = 0.50\,F_u$	13.11.3

Notes: Oversize holes are not permitted in bearing-type connections (see Clause 23.3.2(a) of CAN/CSA-S16.1-94).
See Clause 23.3.2(b), (c) of CAN/CSA-S16.1-94 re use of slotted holes in bearing-type connections.
See Clause 23.3.2(e) of CAN/CSA-S16.1-94 for hole diameters permitted with M20 or 3/4-inch diameter, M22 or 7/8-inch diameter, and M24 or 1-inch diameter bolts.

UNIT FACTORED SHEAR AND TENSILE RESISTANCES**

$\phi_b = 0.67$

Table 3-3

Bolt Grade	Specified Minimum Tensile Strength, F_u (MPa)	Unit Factored Shear Resistance		Unit Factored Tensile Resistance 0.50 F_u (MPa)
		Threads Excluded 0.40 F_u (MPa)	Threads Intercepted 0.28 F_u (MPa)	
A325M	830	334	234	417
A490M	1040	418	293	523
A325 (d ≤ 1")	825	332	232	415
A325 (d ≥ 1⅛")	725	291	204	364
A490	1035	416	291	520
A307*	414	166	117	208

* Use of A307 bolts in connections is covered in Clauses 22.2 and 22.4 of CAN/CSA-S16.1-94.
**Values for Imperial series bolts are based on ASTM specifications A325 and A490 soft converted to SI units.

FACTORED SHEAR AND TENSILE RESISTANCES
(kN PER BOLT)

Table 3-4

Bolt Size		Nominal Area A_b (mm²)	Factored Shear Resistance† — Single Shear**(kN/bolt)						Factored Tensile Resistance, T_r (kN/bolt)		
			Threads Excluded			Threads Intercepted ††					
Metric*	Imperial		A325 A325M	A490 A490M	A307	A325 A325M	A490 A490M	A307	A325 A325M	A490 A490M	A307
	1/2	127	42.2	52.8		29.5	37.0		52.7	66.0	
	5/8	198	65.7	82.4	32.9	45.9	57.6	23.2	82.2	103	41.2
M16		201	67.1	84.0		47.3	58.9		83.8	105	
	3/4	285	94.6	119	47.3	66.1	82.9	33.4	118	148	59.3
M20		314	105	131		73.5	92.0		131	164	
M22		380	127	159		88.9	111		158	199	
	7/8	388	129	161	64.4	90.0	113	45.4	161	202	80.7
M24		452	151	189		106	132		188	238	
	1	507	168	211		118	148		210	264	
M27		573	191	240		134	168		239	300	
	1–1/8	641	187	267		131	187		233	333	
M30		707	236	296		165	207		295	370	
	1–1/4	792	230	329		162	230		288	412	
M36		1018	340	426		238	298		425	532	
	1–1/2	1140	332	474		233	322		415	593	

* The number following the letter M is the nominal bolt diameter in millimetres.
** For double shear, multiply tabulated values by 2.0.
† For joints longer than 1 300 mm, use 80% of the factored shear resistance.
†† Threads are intercepted if thin material next to the nut is combined with detailing for minimum bolt stick-through (the nut).

Table 3-5 UNIT FACTORED BEARING RESISTANCE

Material Standard and Grade		Specified Minimum Tensile Strength, F_u (MPa)	0.67 F_u (MPa)	3 ϕ_b F_u (MPa)
CSA-G40.21	260W, 260WT	410	275	820
	300W for HSS only	410	275	820
	300W, 350W, 300WT	450	302	900
	350G, 350WT, 350R, 350A, 350AT	480	322	960
	380W, and 380WT for HSS only	480	322	960
	400W, 400WT, 400A, 400AT	520	348	1040
	480W, 480WT, 480A, 480AT	590	395	1180
	550W, 550WT, 550A, 550AT	620	415	1250
ASTM	A36	400	268	800
	A572 Grade 42	415	278	830
	Grade 50	450	302	900
	A588 F_y = 42	435	291	870
	F_y = 46	460	308	920
	F_y = 50	485	325	970

Table 3-6 FACTORED BEARING RESISTANCE, B_r^* (kN/bolt)

CSA-G40.21 300W, 350W, 300WT (F_u = 450 MPa)

t (mm)	½	⅝	M16	¾	M20	M22	⅞	M24	1	M27	1⅛	M30	1¼	M36	1½
4	45.9	57.4	57.9	68.9	72.4	79.6	80.4	86.8	91.9	97.7	103	109	115	130	138
4.5	51.7	64.6	65.1	77.5	81.4	89.5	90.5	97.7	103	110	116	122	129	147	155
5	57.4	71.8	72.4	86.2	90.5	99.5	101	109	115	122	129	136	144	163	172
6	68.9	86.2	86.8	103	109	119	121	130	138	147	155	163	172	195	207
7	80.4	101	101	121	127	139	141	152	161	171	181	190	201	228	241
8	91.9	115	116	138	145	159	161	174	184	195	207	217	230	260	276
9	103	129	130	155	163	179	181	195	207	220	233	244	258	293	310
10	115	144	145	172	181	199	201	217	230	244	258	271	287	326	345
11		158	159	190	199	219	221	239	253	269	284	298	316	358	379
12		172	174	207	217	239	241	260	276	293	310	326	345	391	414
13				224	235	259	261	282	299	317	336	353	373	423	448
14				241	253	279	281	304	322	342	362	380	402	456	482
15					271	298	302	326	345	366	388	407	431	488	517
16						318	322	347	368	391	414	434	459	521	551
17						338	342	369	391	415	439	461	488	554	586
18								391	414	440	465	488	517	586	620
19									437	464	491	516	546	619	655
20										488	517	543	574	651	689
21											543	570	603	684	724
22												597	632	716	758
23													661	749	793
24													689	781	827
25														814	862
26														847	896
27														879	930

* $B_r = 3 \phi_b t d F_u$ for one bolt.
Shear resistance of the bolt may govern.
Block tear-out may govern when end distance (from bolt centre to material edge) is less than three bolt diameters.

FACTORED BEARING RESISTANCE, Br* (kN/bolt)

Table 3-7

ASTM A36 (F$_u$ = 400 MPa)

t (mm)	½	⅝	M16	¾	M20	M22	⅞	M24	1	M27	1⅛	M30	1¼	M36	1½
4	40.8	51.1	51.5	61.3	64.3	70.8	71.5	77.2	81.7	86.8	91.9	96.5	102	116	123
4.5	45.9	57.4	57.9	68.9	72.4	79.6	80.4	86.8	91.9	97.7	103	109	115	130	138
5	51.1	63.8	64.3	76.6	80.4	88.4	89.3	96.5	102	109	115	121	128	145	153
6	61.3	76.6	77.2	91.9	96.5	106	107	116	123	130	138	145	153	174	184
7	71.5	89.3	90.0	107	113	124	125	135	143	152	161	169	179	203	214
8	81.7	102	103	123	129	142	143	154	163	174	184	193	204	232	245
9	91.9	115	116	138	145	159	161	174	184	195	207	217	230	260	276
10	102	128	129	153	161	177	179	193	204	217	230	241	255	289	306
11		140	142	168	177	195	197	212	225	239	253	265	281	318	337
12		153	154	184	193	212	214	232	245	260	276	289	306	347	368
13				199	209	230	232	251	265	282	299	314	332	376	398
14				214	225	248	250	270	286	304	322	338	357	405	429
15					241	265	268	289	306	326	345	362	383	434	459
16						283	286	309	327	347	368	386	408	463	490
17						301	304	328	347	369	391	410	434	492	521
18								347	368	391	414	434	459	521	551
19									388	412	437	458	485	550	582
20										434	459	482	511	579	613
21											482	507	536	608	643
22												531	562	637	674
23													587	666	705
24													613	695	735
25														724	766
26														753	796

CSA-G40.21 260W, 260WT (F$_u$ = 410 MPa)

t (mm)	½	⅝	M16	¾	M20	M22	⅞	M24	1	M27	1⅛	M30	1¼	M36	1½
4	41.9	52.3	52.7	62.8	65.9	72.5	73.3	79.1	83.7	89.0	94.2	98.9	105	119	126
4.5	47.1	58.9	59.3	70.6	74.2	81.6	82.4	89.0	94.2	100	106	111	118	134	141
5	52.3	65.4	65.9	78.5	82.4	90.7	91.6	98.9	105	111	118	124	131	148	157
6	62.8	78.5	79.1	94.2	98.9	109	110	119	126	134	141	148	157	178	188
7	73.3	91.6	92.3	110	115	127	128	138	147	156	165	173	183	208	220
8	83.7	105	105	126	132	145	147	158	167	178	188	198	209	237	251
9	94.2	118	119	141	148	163	165	178	188	200	212	223	235	267	283
10	105	131	132	157	165	181	183	198	209	223	235	247	262	297	314
11		144	145	173	181	199	201	218	230	245	259	272	288	326	345
12		157	158	188	198	218	220	237	251	267	283	297	314	356	377
13				204	214	236	238	257	272	289	306	321	340	386	408
14				220	231	254	256	277	293	312	330	346	366	415	440
15					247	272	275	297	314	334	353	371	392	445	471
16						290	293	316	335	356	377	396	419	475	502
17						308	311	336	356	378	400	420	445	504	534
18								356	377	401	424	445	471	534	565
19									398	423	447	470	497	564	597
20										445	471	494	523	593	628
21											495	519	549	623	659
22												544	576	653	691
23													602	682	722
24													628	712	754
25														742	785
26														771	816

* B$_r$ = 3 φ$_b$ t d F$_u$ for one bolt.
Shear resistance of the bolt may govern.
Block tear-out may govern when end distance (from bolt centre to material edge) is less than three bolt diameters.

Table 3-7   **FACTORED BEARING RESISTANCE, B_r* (kN/bolt)**

| t (mm) | \multicolumn{14}{c}{CSA-G40.21 350G, 350WT, 350R, 350A, 380W, 380WT (F_u = 480 MPa) Bolt Size} |

t (mm)	½	⅝	M16	¾	M20	M22	⅞	M24	1	M27	1⅛	M30	1¼	M36	1½
4	49.0	61.3	61.7	73.5	77.2	84.9	85.8	92.6	98.0	104	110	116	123	139	147
4.5	55.1	68.9	69.5	82.7	86.8	95.5	96.5	104	110	117	124	130	138	156	165
5	61.3	76.6	77.2	91.9	96.5	106	107	116	123	130	138	145	153	174	184
6	73.5	91.9	92.6	110	116	127	129	139	147	156	165	174	184	208	221
7	85.8	107	108	129	135	149	150	162	172	182	193	203	214	243	257
8	98.0	123	123	147	154	170	172	185	196	208	221	232	245	278	294
9	110	138	139	165	174	191	193	208	221	234	248	260	276	313	331
10	123	153	154	184	193	212	214	232	245	260	276	289	306	347	368
11		168	170	202	212	233	236	255	270	287	303	318	337	382	404
12		184	185	221	232	255	257	278	294	313	331	347	368	417	441
13				239	251	276	279	301	319	339	358	376	398	452	478
14				257	270	297	300	324	343	365	386	405	429	486	515
15					289	318	322	347	368	391	414	434	459	521	551
16						340	343	370	392	417	441	463	490	556	588
17						361	365	394	417	443	469	492	521	590	625
18								417	441	469	496	521	551	625	662
19									466	495	524	550	582	660	698
20										521	551	579	613	695	735
21											579	608	643	729	772
22												637	674	764	809
23													705	799	845
24													735	834	882
25														868	919
26														903	956

| t (mm) | \multicolumn{14}{c}{CSA-G40.21 400W, 400WT, 400A, 400AT (F_u = 520 MPa) Bolt Size} |

t (mm)	½	⅝	M16	¾	M20	M22	⅞	M24	1	M27	1⅛	M30	1¼	M36	1½
4	53.1	66.4	66.9	79.6	83.6	92.0	92.9	100	106	113	119	125	133	151	159
4.5	59.7	74.7	75.3	89.6	94.1	103	105	113	119	127	134	141	149	169	179
5	66.4	83.0	83.6	99.6	105	115	116	125	133	141	149	157	166	188	199
6	79.6	99.6	100	119	125	138	139	151	159	169	179	188	199	226	239
7	92.9	116	117	139	146	161	163	176	186	198	209	219	232	263	279
8	106	133	134	159	167	184	186	201	212	226	239	251	265	301	319
9	119	149	151	179	188	207	209	226	239	254	269	282	299	339	358
10	133	166	167	199	209	230	232	251	265	282	299	314	332	376	398
11		183	184	219	230	253	256	276	292	310	329	345	365	414	438
12		199	201	239	251	276	279	301	319	339	358	376	398	452	478
13				259	272	299	302	326	345	367	388	408	431	489	518
14				279	293	322	325	351	372	395	418	439	465	527	558
15					314	345	348	376	398	423	448	470	498	564	597
16						368	372	401	425	452	478	502	531	602	637
17						391	395	426	451	480	508	533	564	640	677
18								452	478	508	538	564	597	677	717
19									504	536	567	596	631	715	757
20										564	597	627	664	753	796
21											627	658	697	790	836
22												690	730	828	876
23													763	865	916
24													796	903	956
25														941	996
26														978	1040

* $B_r = 3 \phi_b t d F_u$ for one bolt.
Shear resistance of the bolt may govern.
Block tear-out may govern when end distance (from bolt centre to material edge) is less than three bolt diameters.

Bolts in Combined Shear and Tension Bearing-Type Connections

Clause 13.11.4 of CAN/CSA-S16.1-94 requires bolts subjected to shear and tension to satisfy the expression $\left(\dfrac{V_f}{V_r}\right)^2 + \left(\dfrac{T_f}{T_r}\right)^2 \leq 1$, where V_f is the factored shear load on the bolt and T_f is the factored tensile load including prying effects. If the shear-tension ratio V_f/T_f is X, solving for V_f and T_f gives $V_f = X T_f$, and $T_f = \left(\dfrac{V_r^2 T_r^2}{X^2 T_r^2 + V_r^2}\right)^{1/2}$.

Combined shear and tension usually occurs for the threads-excluded case, since a plate or flange thin enough to include threads in the shear plane (about 10 mm) has little capacity to transmit tension. Table 3-8 gives values of V_f and T_f for various shear-tension ratios X for ¾, M20, M22, ⅞, M24 and 1-inch A325M or A325 bolts, with threads excluded from the shear plane. Table 3-9 gives values for A490M or A490 bolts.

Example

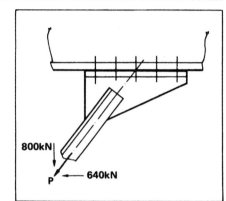

Given: A bracing connection for an inclined factored load P with a tension component T_f of 800 kN and a shear component V_f of 640 kN uses a tee section cut from a W410x74. Check the number of M20, A325M bolts required, assuming G40.21 350W steel with a 90 mm gauge on the tee flange and a 90 mm bolt pitch.

Solution: Ignore prying action for initial trial.

Shear-tension ratio is 640/800 = 0.80.

From Table 3-8, permitted V_f = 74.2 kN and permitted T_f = 92.7 kN per bolt.

Therefore, number of bolts required = 640/74.2 or 800/92.7 = 8.63.

Try 10 bolts. Check the effects of prying action upon the bolt tension, and the connection capacity (see pages 3-19 to 3-25).

Applied factored tensile load per bolt is 800/10 = 80 kN = P_f.

For W410x74, t = 16.0 mm, w = 9.7 mm,   and flange width = 180 mm.

b = (90 – 9.7)/2 = 40.2; b' = 40.2 – 10 = 30.2 mm;   1.25 b = 50.3

a = (180 – 90)/2 = 45.0 < 1.25 b; a' = 45.0 + 10 = 55.0

$K = (4 \times 30.2 \times 10^3) / (0.9 \times 90 \times 350) = 4.26$ (Eqn. 1)

$\delta = 1 - 22/90 = 0.756$ (Eqn. 2)

Use maximum allowed T_f in Eqn. 4, rather than T_r

$\alpha = \left(\dfrac{4.26 \times 92.7}{16.0^2} - 1\right) \times \dfrac{55.0}{0.756 \, (55.0 + 30.2)} = 0.463$ (for bolt T_f); $\delta\alpha = 0.350$ (Eqn. 4)

Connection capacity = $(16.0^2/4.26)(1.350)10 = 811$ kN > 800   — OK   (Eqn. 5)

To find actual bolt load (including prying):

$\alpha = \left(\dfrac{4.26 \times 80}{16.0^2} - 1\right) \times \dfrac{1}{0.756} = 0.438$ (for applied load); $\delta\alpha = 0.331$ (Eqn. 6)

$T_f \approx 80 \left[1 + \left(\dfrac{30.2}{55.0} \times \dfrac{0.331}{1.331}\right)\right] = 90.9$ kN   < 92.7   — OK   (Eqn. 7)

SHEAR AND TENSION
Bearing-Type Connections

Table 3-8 — A325M and A325 Bolts (threads excluded)

Shear-Tension Ratio		Bolt Size										
		3/4		M20		M22		7/8		M24		1	
$X = V_f/T_f$	1/X	V_f	T_f	V_f	T_f	V_f	T_f	V_f	T_f	V_f	T_f	V_f	T_f
0	T_r	0	118	0	131	0	158	0	161	0	188	0	210
0.10	10.00	11.7	117	13.0	130	15.7	157	16.0	160	18.7	187	20.8	208
0.20	5.00	22.9	115	25.4	127	30.7	153	31.2	156	36.5	182	40.7	204
0.30	3.33	33.2	111	36.8	123	44.4	148	45.2	151	52.8	176	59.0	197
0.40	2.50	42.2	106	46.9	117	56.6	142	57.6	144	67.3	168	75.1	188
0.50	2.00	50.1	100	55.6	111	67.1	134	68.3	137	79.8	160	89.0	178
0.60	1.67	56.7	94.5	62.9	105	76.0	127	77.3	129	90.4	151	101	168
0.70	1.43	62.2	88.9	69.1	98.7	83.4	119	84.9	121	99.2	142	111	158
0.80	1.25	66.8	83.5	74.2	92.7	89.6	112	91.1	114	107	133	119	149
0.90	1.11	70.6	78.5	78.4	87.1	94.7	105	96.3	107	113	125	126	140
1.00	1.00	73.8	73.8	81.9	81.9	99.0	99.0	101	101	118	118	131	131
1.11	0.90	76.7	69.0	85.2	76.6	103	92.6	105	94.2	122	110	136	123
1.25	0.80	79.6	63.7	88.4	70.7	107	85.5	109	86.9	127	102	142	113
1.43	0.70	82.5	57.7	91.6	64.1	111	77.5	113	78.8	132	92.1	147	103
1.67	0.60	85.3	51.2	94.6	56.8	114	68.6	116	69.8	136	81.6	152	90.9
2.00	0.50	87.8	43.9	97.5	48.7	118	58.9	120	59.9	140	70.1	156	78.0
2.50	0.40	90.1	36.0	100	40.0	121	48.4	123	49.1	144	57.5	160	64.0
3.33	0.30	92.0	27.6	102	30.6	124	37.0	125	37.6	147	44.0	163	49.0
5.00	0.20	93.4	18.7	104	20.7	125	25.1	127	25.5	149	29.8	166	33.2
10.00	0.10	94.3	9.4	105	10.5	127	12.7	129	12.9	151	15.1	168	16.7
V_r	0	94.6	0	105	0	127	0	129	0	151	0	168	0

Table 3-9 — A490M and A490 Bolts (threads excluded)

Shear-Tension Ratio		Bolt Size										
		3/4		M20		M22		7/8		M24		1	
$X = V_f/T_f$	1/X	V_f	T_f	V_f	T_f	V_f	T_f	V_f	T_f	V_f	T_f	V_f	T_f
0	T_r	0	148	0	164	0	199	0	202	0	238	0	264
0.10	10.00	14.7	147	16.3	163	19.7	198	20.0	200	23.6	236	26.2	262
0.20	5.00	28.7	144	31.8	159	38.6	193	39.2	196	46.2	231	51.2	256
0.30	3.33	41.6	139	46.1	154	55.9	186	56.7	189	66.8	223	74.1	247
0.40	2.50	53.0	133	58.7	147	71.2	178	72.2	181	85.0	213	94.4	236
0.50	2.00	62.8	126	69.5	139	84.3	169	85.6	171	101	201	112	224
0.60	1.67	71.2	119	78.7	131	95.5	159	96.8	161	114	190	127	211
0.70	1.43	78.1	112	86.3	123	105	150	106	152	125	179	139	199
0.80	1.25	83.9	105	92.7	116	113	141	114	143	134	168	149	187
0.90	1.11	88.7	98.6	98.0	109	119	132	121	134	142	158	158	175
1.00	1.00	92.7	92.7	102	102	124	124	126	126	148	148	165	165
1.11	0.90	96.4	86.8	106	95.7	129	116	131	118	154	138	171	154
1.25	0.80	100	80.1	110	88.3	134	107	136	109	160	128	178	142
1.43	0.70	104	72.6	114	80.0	139	97.1	141	98.4	165	116	184	129
1.67	0.60	107	64.3	118	70.9	143	86.0	145	87.1	171	102	190	114
2.00	0.50	110	55.2	122	60.8	148	73.8	150	74.8	176	87.8	196	98.0
2.50	0.40	113	45.3	125	49.9	152	60.6	153	61.4	180	72.1	201	80.4
3.33	0.30	116	34.7	127	38.2	155	46.4	157	47.0	184	55.2	205	61.6
5.00	0.20	118	23.5	129	25.9	157	31.4	159	31.8	187	37.3	208	41.7
10.00	0.10	119	11.9	131	13.1	159	15.8	161	16.0	188	18.8	210	21.0
V_r	0	119	0	131	0	159	0	161	0	189	0	211	0

BOLTS IN SLIP-CRITICAL CONNECTIONS

General

The name slip-critical emphasizes that this type of connection is required only when the consequences of slip are critical to the performance of the structure. Clause 21.12.2(a) of CAN/CSA-S16.1-94 requires slip-critical connections where slippage into bearing cannot be tolerated, such as structures sensitive to deflection, or subject to fatigue or frequent load reversals. Slip-critical shear joints transfer the specified loads by the slip resistance (friction) of the clamped faying surfaces in accordance with Clause 13.12.

In addition to the slip resistance, the strength (factored resistance) of the joint under factored loads must also be checked.

Tables

Tables 3-10 and 3-11 on page 3-15 are based on Clause 13.12.2 of S16.1-94 for bolts in slip-critical connections.

Table 3-10 lists values c_1 for a 5% probability of slip, and values of unit slip resistance ($0.53\, c_1\, k_s\, F_u$) for A325M, A490M, A325 and A490 bolts for the contact surfaces (Class A, Class B and Class C) given in Table 2 of S16.1-94.

Table 3-11 lists slip-resistance values ($V_s = 0.53\, c_1\, k_s\, m\, n\, A_b\, F_u$) for bolted joints with a single faying surface ($m = 1$) for Class A and Class B contact surface for M16 to M36 A325M and A490M bolts, and ½ to 1½ inch A325 and A490 bolts.

Example

Given:

A single shear connection is subject to 370 kN at specified load level and 550 kN at factored load level. Select the number of M20 A325M bolts required for a slip-critical connection. Steel is G40.21 350W 6mm thick, and the surface is clean mill scale (Class A). Assume 80 mm bolt pitch and 30 mm bolt end distance.

Solution:

(a) For specified loads

From Table 3-11, $V_s = 37.4$ kN (M20 A325M bolt for clean mill scale). Number of bolts required is $370/37.4 = 9.9$. Use 10 (say 2 lines of 5, parallel to the force).

(b) Confirm connection at factored loads. This includes checking bolts for shear resistance, checking material for bolt bearing, and checking material for block tear-out.

From Table 3-4 (page 3-8), $V_r = 73.5$ kN (M20 A325M, threads intercepted).

∴ factored shear resistance of bolts is $10 \times 73.5 = 735$ kN > 550 — OK

From Table 3-6 (page 3-9), the factored bearing resistance at one M20 bolt in 6 mm thick 350W material is 109 kN.

∴ 10 bolts give resistance of $109 \times 10 = 1090$ kN > 550 — OK

From S16.1-94, Clause 12.3.1, $A_{ne} = w_n\, t + 0.6\, L_n\, t$,
and from Clause 13.2(a)(ii), $T_r = 0.85\, \phi_b\, A_{ne}\, F_u$.

∴ $A_{ne} = (80-24)\, 6 + 0.6\, [8(80-24) + 2(30-12)]\, 6 = 2080\text{ mm}^2$
and $T_r = 0.85 \times 0.9 \times 2080 \times 0.450 = 716$ kN > 550 — OK

Table 3-10 — UNIT SLIP RESISTANCE, $0.53\, c_1\, k_s\, F_u$* (MPa)

For 5% Probability of Slip

Contact Surface of Bolted Parts	Class / Description	A — Clean mill scale, or blast-cleaned with Class A coatings $k_s = 0.33$	B — Blast-cleaned, or blast-cleaned with Class B coatings $k_s = 0.50$	C — Hot-dip galvanized with wire brushed surfaces $k_s = 0.40$
A325M $F_u = 830$ MPa	c_1	0.82	0.89	0.90
	$0.53\, c_1\, k_s\, F_u$ (MPa)	119	196	158
A490M $F_u = 1040$ MPa	c_1	0.78	0.85	0.85
	$0.53\, c_1\, k_s\, F_u$ (MPa)	142	234	187
A325 (d ≤ 1 inch) $F_u = 825$ MPa	c_1	0.82	0.89	0.90
	$0.53\, c_1\, k_s\, F_u$ (MPa)	118	195	157
A325 (d ≥ 1−1/8 inch) $F_u = 725$ MPa	c_1	0.82	0.89	0.90
	$0.53\, c_1\, k_s\, F_u$ (MPa)	104	171	138
A490 $F_u = 1035$ MPa	c_1	0.78	0.85	0.85
	$0.53\, c_1\, k_s\, F_u$ (MPa)	141	233	187

See CAN/CSA-S16.1-94, Clause 13.12.3 for values of c_1 and k_s.

Table 3-11 — SLIP RESISTANCE, V_s,* (kN/BOLT)

For Single Shear ** (m = 1)

Bolt Size		Nominal Area	Class A Surfaces		Class B Surfaces	
Metric[+]	Imperial	A_b (mm²)	A325M, A325	A490M, A490	A325M, A325	A490M, A490
	½	127	15.0	17.9	24.7	29.6
	⅝	198	23.4	28.0	38.5	46.2
M16		201	23.9	28.5	39.3	47.1
	¾	285	33.7	40.2	55.5	66.4
M20		314	37.4	44.5	61.5	73.6
M22		380	45.2	53.9	74.4	89.0
	⅞	388	45.9	54.8	75.5	90.5
M24		452	53.8	64.1	88.5	106
	1	507	60.0	71.6	98.7	118
M27		573	68.2	81.3	112	134
	1⅛	641	66.6	90.5	110	149
M30		707	84.2	100	138	166
	1¼	792	82.4	112	135	185
M36		1018	121	144	199	238
	1½	1140	119	161	195	266

* These resistances are for use with specified loads in accordance with Clause 13.12 of CAN/CSA-S16.1-94.
** For double shear (m = 2) multiply tabulated values by 2.0.
[+] The number following the letter M is the nominal bolt diameter in millimetres.

Bolts in Combined Shear and Tension — Slip-Critical Connections

Clause 13.12.3 of CAN/CSA-S16.1-94 requires that bolts subjected to both shear and tension in a slip-critical connection satisfy the following relationship for specified loads:

$$\frac{V}{V_s} + 1.9\frac{T}{n\,A_b\,F_u} \leq 1.0$$

The above relationship can conservatively be expressed (see Commentary page 2-43) as

$$\frac{V}{V_s} + \frac{T}{T_i} \leq 1.0 \quad \text{where } T_i \text{ is the specified installed tension.}$$

If the shear-tension ratio V/T on the bolts is X, solving for V and T gives $V = XT$, and $T = V_s/(X + V_s/T_i)$.

Table 3-12 lists values of V and T for various shear-tension ratios X for Class A contact surfaces (clean mill scale or blast cleaned with Class A coatings, $k_s = 0.33$) using A325M or A325 bolts in single shear. This table can be used to establish directly the number of bolts required to satisfy the interaction equation for slip-critical connections subjected to a combination of shear and tension.

Clause 13.12.1 requires that the factored tension T_f, including any prying forces, not exceed the bolt tensile resistance T_r. See page 3-19 for a method of evaluating the prying effects.

Example

Given:

Find the number of M20 A325M bolts required in a slip-critical connection to resist a specified tension force of 320 kN and a specified shear force of 400 kN. The single faying surface consists of clean mill scale.

Solution:

Prying is not a factor when making the specified shear vs. specified tension interaction check. Within permitted loadings, prying is only a redistribution of the contact forces between the material surfaces, having no significant affect upon resistance to slipping.

Shear-tension ratio is 400/320 = 1.25

From Table 3-12, for M20 bolts and $V/T = 1.25$,

permitted V and T are 30.9 kN and 24.7 kN, respectively, per bolt

Therefore, number of bolts required is 400/30.9 or 320/24.7 = 12.9

Try 14 bolts

The connection also has to be confirmed for strength, including bolt prying and flange bending, as a bearing-type connection at factored loads.

Use equations from page 3-20, as applied for the example on page 3-12.

Table 3-12

$c_1 = 0.82$
$k_s = 0.33$

SPECIFIED SHEAR AND TENSION (kN/BOLT)
Slip-Critical Connections, Class A Surfaces
A325M and A325 Bolts

Shear/Tension Ratio X = V/T		¾		M20		M22		⅞		M24		1	
X	1/X	V	T	V	T	V	T	V	T	V	T	V	T
0.5		21.9	43.8	24.5	49.0	29.9	59.7	30.0	60.1	35.3	70.6	39.3	78.5
0.6		23.3	38.8	26.0	43.3	31.7	52.8	31.9	53.1	37.4	62.4	41.7	69.4
0.7		24.3	34.8	27.2	38.8	33.1	47.2	33.3	47.6	39.1	55.9	43.6	62.2
0.8		25.2	31.5	28.1	35.2	34.2	42.8	34.5	43.1	40.5	50.6	45.1	56.4
0.9		25.9	28.8	28.9	32.1	35.2	39.1	35.5	39.4	41.7	46.3	46.4	51.5
1.0	1.0	26.5	26.5	29.6	29.6	36.0	36.0	36.3	36.3	42.6	42.6	47.5	47.5
1.11	0.9	27.1	24.4	30.2	27.2	36.7	33.0	37.1	33.4	43.5	39.2	48.5	43.6
1.25	0.8	27.7	22.2	30.9	24.7	37.5	30.0	37.9	30.3	44.5	35.6	49.5	39.6
1.43	0.7	28.3	19.8	31.6	22.1	38.3	26.8	38.7	27.1	45.5	31.8	50.6	35.4
1.67	0.6	29.0	17.4	32.3	19.4	39.2	23.5	39.6	23.8	46.5	27.9	51.8	31.1
2.00	0.5	29.7	14.8	33.0	16.5	40.1	20.0	40.6	20.3	47.6	23.8	53.0	26.5
2.50	0.4	30.4	12.2	33.8	13.5	41.0	16.4	41.5	16.6	48.7	19.5	54.3	21.7
3.33	0.3	31.2	9.4	34.7	10.4	42.0	12.6	42.5	12.8	49.9	15.0	55.6	16.7
5.00	0.2	32.0	6.4	35.5	7.1	43.0	8.6	43.6	8.7	51.1	10.2	57.0	11.4
10.00	0.1	32.8	3.3	36.4	3.6	44.1	4.4	44.7	4.5	52.4	5.2	58.5	5.8
Vs	0	33.7	0	37.4	0	45.2	0	45.9	0	53.8	0	60.0	0

$V = XT, \quad T = \dfrac{V_s}{X + V_s/T_i}$

NOTES

BOLTS IN TENSION AND PRYING ACTION

General

Connections with fasteners loaded in tension occur in many common situations, such as hanger and bracing connections with tee-type gussets, and end plate moment connections. Clause 23.1.4 of CAN/CSA-S16.1-94 requires that when bolts are loaded in direct tension, the effects of prying action be taken into account in proportioning the bolts and connected parts. This clause also requires that the connection be proportioned to avoid prying forces when subjected to repeated loading.

The actual stress distribution in the flange of a tee-type connection is extremely complex as it depends upon the bolt size and arrangement, and upon the strength and dimensions of the connecting flange. Consequently, various design methods have been proposed in the technical literature for proportioning such connections. The procedures given in this section are based on the recommendations contained on page 285 of *Guide to Design Criteria for Bolted and Riveted Joints*, second edition, by Kulak, Fisher and Struik.

The procedures include a set of seven equations for selecting a trial section and for evaluating the bolt forces and flange capacity. Equilibrium Eqn. 4 uses the full tensile resistance T_r of the bolts to determine α for use in the connection capacity Eqn. 5. This provides a value for the maximum capacity of the connection. Similarly, Eqn. 6 uses the applied factored tensile load per bolt P_f to determine α for use in the amplified bolt force Eqn. 7. This provides a value for the factored load per bolt (including prying) T_f of the bolts.

Based on these equations, Table 3-13 on page 3-24 and Figure 3-2 on page 3-25 provide aids for preliminary design and checking purposes. They demonstrate the effect of applied factored tensile load per bolt and flange geometry upon flange thicknesses for various bolt sizes, assuming static loads.

In general, prying effects can be minimized by dimensioning for minimum practical gauge distance and for maximum permissible edge distance. For repeated loading the flange must be made thick enough and stiff enough so that deformation of the flange is virtually eliminated. In addition, special attention must be paid to bolt installation to ensure that the bolts are properly pretensioned to provide the required clamping force.

The expressions for prying effects are based on tests carried out on tees. For angles, assuming the distribution of moment shown, the moment equilibrium equation can be derived from statics as

$$P_f b = Q a \quad \text{Therefore } Q/P_f = b/a$$

References

DOUTY, R. T., and MCGUIRE, W. 1965. High strength bolted moment connections. ASCE Journal of the Structural Division, April.

KULAK, G.L., FISHER, J.W., and STRUIK, J.H.A. 1987. Guide to design criteria for bolted and riveted joints, 2nd edition. John Wiley & Sons, New York, N.Y.

NAIR, R. S., BIRKEMOE, P. C., and MUNSE, W. H. 1969. High strength bolts subject to tension and prying. Structural Research Series 353, September, Department of Civil Engineering, U. of Ill., Urbana.

Equations

$$K = 4 b' 10^3 / (\phi p F_y) \quad (1)$$

$$\delta = 1 - d'/p \quad (2)$$

$$\text{Range of } t = \left(\frac{K P_f}{(1 + \delta \alpha)}\right)^{1/2} \begin{array}{l} t_{\min} \text{ when } \alpha = 1.0 \\ t_{\max} \text{ when } \alpha = 0.0 \end{array} \quad (3)$$

$$\alpha = \left(\frac{K T_r}{t^2} - 1\right) \times \frac{a'}{\delta (a' + b')}, \quad 0 \le \alpha \le 1.0 \quad (4)$$

$$\text{Connection capacity} = (t^2/K)(1 + \delta \alpha) n \quad (5)$$

$$\alpha = \left(\frac{K P_f}{t^2} - 1\right) \times \frac{1}{\delta} \quad \text{(for use in Eqn. 7)} \quad (6)$$

$$T_f \approx P_f \left[1 + \left(\frac{b'}{a'} \times \frac{\delta \alpha}{1 + \delta \alpha}\right)\right] \le T_r \quad (7)$$

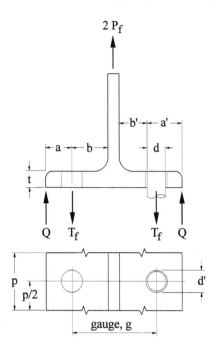

Nomenclature

- K = Parameter as defined in Eqn. 1
- P_f = Applied factored tensile load per bolt, (kN)
- Q = Prying force per bolt at factored load, $Q = T_f - P_f$, (kN)
- T_f = Factored load per bolt including prying (amplified bolt force), (kN)
- T_r = Factored tensile resistance of bolts, $\phi_\beta \, 0.75 \, A_b \, F_u$, (kN)
- F_y = Yield strength of flange material, (MPa)
- a = Distance from bolt line to edge of tee flange, not more than $1.25 b$, (mm)
- a' = $a + d/2$, (mm)
- b = Distance from bolt line (gauge line) to face of tee stem, (mm)
- b' = $b - d/2$, (mm)
- d = Bolt diameter, (mm)
- d' = Nominal hole diameter, (mm)
- n = Number of flange bolts in tension
- p = Length of flange tributary to each bolt, or bolt pitch, (mm)
- t = Thickness of flange, (mm)
- α = Ratio of sagging moment at bolt line to hogging moment at stem of tee
- δ = Ratio of net to gross flange area along a longitudinal line of bolts (see Eqn. 2)
- ϕ = Resistance factor for the tee material, (0.9)

ᑐn Tables

ᵃ 3-24 lists the maximum and minimum values of flange thickness
ᵗ using $\alpha = 0.0$ and $\alpha = 1.0$ for a range of values of P_f. Results
ᵗlange bolt patterns and bolt sizes.

The maximum and minimum values of t indicate a range of flange thickness within which the bolts and flange are in equilibrium for the particular flange geometry, and in which the effects of flange flexure and prying reduce the effective tension capacity of the bolts. When the flange thickness is greater than the larger value of t, ($\alpha = 0.0$), the flange is generally sufficiently thick and stiff to virtually eliminate prying action, and the connection capacity will be limited by the tensile resistance of the bolts. When the flange thickness is less than the smaller value of t, ($\alpha = 1.0$), the flange thickness will govern the connection capacity and the bolts will usually have excess capacity to resist the applied tension load in spite of prying effects.

Within the range of flange thickness for $0.0 \leq \alpha \leq 1.0$, with the bolts and flange in equilibrium, the ratio T_f/P_f will increase from unity for t_{max} to a maximum value for t_{min}. In this range, the bolts control the capacity with the flange strength being increasingly consumed as the flange thickness decreases. (It can be helpful to note that the typical ratio of maximum to minimum flange thickness is about 1.33, and that at the mimimum flange thickness, the prying ratio T_f/P_f is about the same. When the maximum flange thickness is used, there is essentially no prying and the ratios t_{max}/t_{used} and T_f/P_f are both 1.0. Thus, a rough guide to the ratio t_{max}/t_{reqd} is the available prying ratio T_r/P_f of the desired bolt size.)

The bolt pitch p should be approximately 4 to 5 times the bolt size ($4d \leq p \leq 5d$) and the gauge g should be kept as small as practicable. Also, dimension a for design purposes must not exceed $1.25\ b$.

Figure 3-2 on page 3-25 graphs the amplified bolt force T_f for various applied loads P_f, flange thicknesses t, and four different values of b (40mm, 45mm, 50mm and 55mm) with M20, M22 and M24 A325M bolts. These graphs can be used to evaluate the effects of flange thickness, gauge distance and bolt size on the amplified bolt force, and to establish reasonable trial connection parameters.

Design Procedure

Trial Section

1) Select an intended number and size of bolts as a function of the applied factored tensile load per bolt P_f and the anticipated prying ratio.

2) With P_f, the bolt size, and trial values of b' and p, use Eqns. 1, 2 and 3 (with $\alpha = 0.0$ and $\alpha = 1.0$) to identify a range of acceptable flange thicknesses. (Alternatively, use Table 3-13 on page 3-24.)

3) Identify an intended flange thickness.

Figure 3-2 (page 3-25) may also be used to identify an intended bolt size and flange geometry based on the amplified bolt force being less than the bolt tensile resistance.

Design Check

1) Recalculate K, if necessary, and use Eqn. 4 to determine α for use in Eqn. 5.

2) Calculate the connection capacity with Eqn. 5. (If α from Eqn. 4 < 0.0, use $\alpha = 0.0$, and if $\alpha > 1.0$, use $\alpha = 1.0$.)

3) Equations 6 and 7 can be used if desired to determine the total bolt tension, including prying (amplified bolt force), that results from the applied load.

Note:

CAN/CSA-S16.1-94, Clause 21.12.2(e) requires that all bolts subject to tensile loadings be pretensioned when installed, and Clause 23.1.4 requires that connections with repeated tension loads on bolts be proportioned to avoid prying forces.

Example 1

Given:

Design a tension tee connection with 4 ASTM A325M bolts in tension for a factored static load of 500 kN assuming the bolts are on a 100 mm gauge, at a pitch of 110 mm, with the tee connected to rigid supports. Use G40.21 350W steel.

Solution:

Trial Section

Applied load per bolt = 500/4 = 125 kN = P_f

Assume M22 bolts, 24 mm hole diameter, 15 mm web. T_r = 158 kN

Available prying ratio T_r/P_f is 158/125 = 1.26 — OK for near-minimum flange

$b = (100 - 15) / 2 = 42.5$ mm $b' = 42.5 - 11 = 31.5$

$K = 4 \times 31.5 \times 10^3 / (0.9 \times 110 \times 350) = 3.64$ (Eqn. 1)

$\delta = 1 - (24/110) = 0.782$ (Eqn. 2)

$$t_{min} = \left(\frac{3.64 \times 125}{1.782}\right)^{1/2} = 16.0 \text{ mm}; \quad t_{max} = \left(\frac{3.64 \times 125}{1.0}\right)^{1/2} = 21.3 \text{ mm} \quad \text{(Eqn. 3)}$$

(Alternatively, the range of t, by rough interpolation from Table 3-13 on page 3-24, could be seen to be about 16 to 21 mm.)

Two possible solutions are W460x89 ($t = 17.7$ mm) and W530x123 ($t = 21.2$ mm), the choice being determined by the maximum available prying ratio T_r/P_f for the bolt size finally selected.

Design Check

Try W460x89 with M22 bolts: $d = 22$ mm, $d' = 24$ mm

$t = 17.7$ mm, $w = 10.5$ mm, flange width = 192 mm

$b = (100 - 10.5)/2 = 44.8$; $b' = 44.8 - 11 = 33.8$; $1.25 b = 56.0$

$a = (192 - 100)/2 = 46.0 < 1.25 b$; $a' = 46.0 + 11 = 57.0$; $a' + b' = 90.8$

$K = 4 \times 33.8 \times 10^3 / (0.9 \times 110 \times 350) = 3.90$ (Eqn. 1)

$\delta = 0.782$ (as above) (Eqn. 2)

$$\alpha = \left(\frac{3.90 \times 158}{17.7^2} - 1\right) \times \frac{57.0}{0.782 \times 90.8} = 0.776 \quad \text{(Eqn. 4)}$$

$\delta \alpha = 0.607$

Connection capacity = $(17.7^2 / 3.90)(1.607) 4 = 516$ kN > 500   — OK (Eqn. 5)

To find actual bolt load (including prying), if desired:

$$\alpha = \left(\frac{3.90 \times 125}{17.7^2} - 1\right) \times \frac{1}{0.782} = 0.711 \quad \text{(Eqn. 6)}$$

$\delta \alpha = 0.556$

$$T_f \approx 125\left[1 + \left(\frac{33.8}{57.0} \times \frac{0.556}{1 + 0.556}\right)\right] = 151 \text{ kN} \quad < 158 \quad \text{— OK} \quad \text{(Eqn. 7)}$$

Tee stem capacity is $0.9 (2 \times 110) 10.5 \times 350 / 1\,000 = 728$ kN > 500   — OK

(With M20 bolts, $T_r/P_f = 131/125 = 1.05$, and the heavier W530x123 is required.)

Example 2

Given:

Use each of Table 3-13 and Figure 3-2 to select the bolt size and trial dimensions for a tee cut from a W460x97 section (G40.21 350W). The factored tensile load is 480 kN and a bolt gauge of 90 mm is preferred. Confirm the trial design.

Solution:

Since Table 3-13 and Figure 3-2 are intended only for the selection of a trial section that must be checked with Eqns. 1, 2, 4 and 5 on page 3-20 (illustrated in the previous example), precise interpolation is not necessary. As discussed on page 3-21, a comparison of the available ratio T_r/P_f (for the bolts chosen) to the ratio of t_{max} to a trial flange thickness (for the value of t_{max} selected from Table 3-13) will indicate whether a combination of bolt size, flange thickness and geometry may be a suitable choice.

For W460x97: $t = 19.0$ mm, $w = 11.4$ mm, flange width = 193 mm

For $g = 90$ mm, $b = (90 - 11.4)/2 = 39.3$ (Use $b = 40$ for Table 3-13)

With 4 bolts, $P_f = 480/4 = 120$ kN, and $T_r/P_f = 131/120 = 1.09$ for M20 bolts, and $159/120 = 1.32$ for M22 bolts.

Table 3-13, with $b = 40$, M20 bolts and $P_f = 120$, $p = 100$ gives $t_{max} = 21.4$ mm

$t_{max}/t = 21.4/19.0 = 1.13$ $> T_r/P_f = 1.09$ — likely not OK

Table 3-13, with $b = 40$, M22 bolts and $P_f = 120$, $p = 90$ gives $t_{max} = 22.2$ mm

$t_{max}/t = 22.2/19.0 = 1.17$ $< T_r/P_f = 1.32$ — OK, check the design

Alternatively, Figure 3-2 can be used to select the bolt size based on the flange thickness and the amplified bolt force.

Use graph for $b = 40$ mm ($b = 39.3$, see above)

Enter graph at applied load per bolt of 120 kN and flange thickness $t = 19.0$ mm

With M20 bolts, amplified bolt force ≈ 137 kN $> T_r = 131$ kN — no good

With M 22 bolts, amplified bolt force ≈ 126 kN $< T_r = 158$ kN — OK

Proceed with the design check using M22 bolts; $p = 5d = 110$ mm.

$b = 39.3$ mm; $b' = 39.3 - 11 = 28.3$; $1.25\,b = 49.1$

$a = (193 - 90)/2 = 51.5$ mm $> 1.25\,b$

Use $a = 1.25\,b = 49.1$; $a' = 49.1 + 11 = 60.1$; $a' + b' = 88.4$

$K = 4 \times 28.3 \times 10^3 / (0.9 \times 110 \times 350) = 3.27$ (Eqn. 1)

$\delta = 1 - (24/110) = 0.782$ (Eqn. 2)

$\alpha = \left(\dfrac{3.27 \times 158}{19.0^2} - 1\right) \times \dfrac{60.1}{0.782 \times 88.4} = 0.375$ $\delta\alpha = 0.293$ (Eqn. 4)

Connection capacity = $(19.0^2 / 3.27)(1.293)\,4 = 571$ kN > 480 — OK   (Eqn. 5)

Check total bolt load (amplified bolt force):

$\alpha = \left(\dfrac{3.27 \times 120}{19.0^2} - 1\right) \times \dfrac{1}{0.782} = 0.111$ $\delta\alpha = 0.0868$ (Eqn. 6)

$T_f \approx 120\left[1 + \left(\dfrac{28.3}{60.1} \times \dfrac{0.0868}{1 + 0.0868}\right)\right] = 125$ kN < 158 — OK   (Eqn. 7)

RANGE OF t

$$t = \sqrt{\frac{K P_f}{(1+\delta\alpha)}}$$

Table 3-13

t_{min} when $\alpha = 1.0$, t_{max} when $\alpha = 0.0$

b (mm)	Bolt size	P_f = 60 kN pitch p (mm)			P_f = 80 kN pitch p (mm)			P_f = 100 kN pitch p (mm)			P_f = 120 kN pitch p (mm)		
		80	90	100	80	90	100	80	90	100	80	90	100
35	3/4	11.8 / 15.6	11.1 / 14.7	10.4 / 13.9	13.6 / 18.0	12.8 / 17.0	12.0 / 16.1	15.3 / 20.1	14.3 / 19.0	13.4 / 18.0			
	M20	11.7 / 15.4	11.0 / 14.5	10.3 / 13.8	13.6 / 17.8	12.7 / 16.8	11.9 / 15.9	15.2 / 19.9	14.2 / 18.8	13.4 / 17.8	16.6 / 21.8	15.5 / 20.6	14.6 / 19.5
40	3/4	12.9 / 17.0	12.1 / 16.1	11.4 / 15.2	14.9 / 19.7	14.0 / 18.5	13.2 / 17.6	16.7 / 22.0	15.6 / 20.7	14.7 / 19.7			
	M20	12.9 / 16.9	12.0 / 15.9	11.3 / 15.1	14.9 / 19.5	13.9 / 18.4	13.1 / 17.5	16.6 / 21.8	15.5 / 20.6	14.6 / 19.5	18.2 / 23.9	17.0 / 22.5	16.0 / 21.4
45	3/4	13.9 / 18.4	13.0 / 17.3	12.3 / 16.4	16.1 / 21.2	15.1 / 20.0	14.2 / 19.0	18.0 / 23.7	16.8 / 22.4	15.9 / 21.2			
	M20	13.9 / 18.3	13.0 / 17.2	12.2 / 16.3	16.1 / 21.1	15.0 / 19.9	14.1 / 18.9	17.9 / 23.6	16.8 / 22.2	15.8 / 21.1	19.7 / 25.8	18.4 / 24.3	17.3 / 23.1
50	3/4	14.9 / 19.6	13.9 / 18.5	13.1 / 17.6	17.2 / 22.7	16.1 / 21.4	15.2 / 20.3	19.2 / 25.3	18.0 / 23.9	16.9 / 22.7			
	M20	14.9 / 19.5	13.9 / 18.4	13.1 / 17.5	17.2 / 22.5	16.0 / 21.2	15.1 / 20.2	19.2 / 25.2	17.9 / 23.8	16.9 / 22.5	21.0 / 27.6	19.6 / 26.0	18.5 / 24.7
55	3/4	15.8 / 20.8	14.8 / 19.6	13.9 / 18.6	18.2 / 24.0	17.0 / 22.7	16.1 / 21.5	20.4 / 26.9	19.1 / 25.3	18.0 / 24.0			
	M20	15.8 / 20.7	14.7 / 19.5	13.9 / 18.5	18.2 / 23.9	17.0 / 22.5	16.0 / 21.4	20.3 / 26.7	19.0 / 25.2	17.9 / 23.9	22.3 / 29.3	20.8 / 27.6	19.6 / 26.2

b (mm)	Bolt size	P_f = 100 kN pitch p (mm)			P_f = 120 kN pitch p (mm)			P_f = 140 kN pitch p (mm)			P_f = 160 kN pitch p (mm)		
		90	100	110	90	100	110	90	100	110	90	100	110
40	M22	15.4 / 20.2	14.5 / 19.2	13.7 / 18.3	16.8 / 22.2	15.8 / 21.0	15.0 / 20.0	18.2 / 23.9	17.1 / 22.7	16.2 / 21.6			
	7/8	15.3 / 20.2	14.4 / 19.2	13.7 / 18.3	16.8 / 22.1	15.8 / 21.0	15.0 / 20.0	18.2 / 23.9	17.1 / 22.7	16.2 / 21.6	19.4 / 25.5	18.3 / 24.2	17.3 / 23.1
45	M22	16.6 / 21.9	15.7 / 20.8	14.8 / 19.8	18.2 / 24.0	17.2 / 22.8	16.3 / 21.7	19.7 / 25.9	18.5 / 24.6	17.6 / 23.4			
	7/8	16.6 / 21.9	15.6 / 20.7	14.8 / 19.8	18.2 / 24.0	17.1 / 22.7	16.2 / 21.7	19.7 / 25.9	18.5 / 24.5	17.5 / 23.4	21.0 / 27.7	19.8 / 26.2	18.8 / 25.0
50	M22	17.8 / 23.5	16.8 / 22.3	15.9 / 21.2	19.5 / 25.7	18.4 / 24.4	17.4 / 23.2	21.1 / 27.8	19.8 / 26.3	18.8 / 25.1			
	7/8	17.8 / 23.4	16.8 / 22.2	15.9 / 21.2	19.5 / 25.7	18.4 / 24.3	17.4 / 23.2	21.1 / 27.7	19.8 / 26.3	18.8 / 25.1	22.5 / 29.6	21.2 / 28.1	20.1 / 26.8
55	M22	18.9 / 24.9	17.8 / 23.6	16.9 / 22.5	20.7 / 27.3	19.5 / 25.9	18.5 / 24.7	22.4 / 29.5	21.1 / 28.0	20.0 / 26.7			
	7/8	18.9 / 24.9	17.8 / 23.6	16.9 / 22.5	20.7 / 27.3	19.5 / 25.9	18.5 / 24.7	22.4 / 29.4	21.1 / 27.9	20.0 / 26.6	23.9 / 31.5	22.5 / 29.9	21.3 / 28.5

b (mm)	Bolt size	P_f = 140 kN pitch p (mm)			P_f = 160 kN pitch p (mm)			P_f = 180 kN pitch p (mm)			P_f = 200 kN pitch p (mm)		
		100	110	120	100	110	120	100	110	120	100	110	120
40	M24	16.9 / 22.3	16.0 / 21.3	15.3 / 20.4	18.1 / 23.9	17.1 / 22.7	16.3 / 21.8	19.2 / 25.3	18.2 / 24.1	17.3 / 23.1			
	1	16.8 / 22.0	15.9 / 21.0	15.1 / 20.1	17.9 / 23.6	17.0 / 22.5	16.2 / 21.5	19.0 / 25.0	18.0 / 23.8	17.1 / 22.8	20.0 / 26.3	19.0 / 25.1	18.1 / 24.0
45	M24	18.4 / 24.2	17.4 / 23.1	16.6 / 22.1	19.6 / 25.9	18.6 / 24.7	17.7 / 23.6	20.8 / 27.5	19.7 / 26.2	18.8 / 25.1			
	1	18.2 / 24.0	17.3 / 22.8	16.4 / 21.9	19.5 / 25.6	18.5 / 24.4	17.6 / 23.4	20.7 / 27.2	19.6 / 25.9	18.6 / 24.8	21.8 / 28.6	20.6 / 27.3	19.6 / 26.1
50	M24	19.7 / 26.0	18.7 / 24.8	17.8 / 23.7	21.1 / 27.8	19.9 / 26.5	19.0 / 25.4	22.3 / 29.5	21.2 / 28.1	20.1 / 26.9			
	1	19.6 / 25.8	18.6 / 24.6	17.7 / 23.5	21.0 / 27.5	19.8 / 26.2	18.9 / 25.1	22.2 / 29.2	21.0 / 27.8	20.0 / 26.7	23.4 / 30.8	22.2 / 29.3	21.1 / 28.1
55	M24	21.0 / 27.6	19.9 / 26.4	18.9 / 25.2	22.4 / 29.6	21.2 / 28.2	20.2 / 27.0	23.8 / 31.4	22.5 / 29.9	21.4 / 28.6			
	1	20.9 / 27.4	19.8 / 26.1	18.8 / 25.0	22.3 / 29.3	21.1 / 28.0	20.1 / 26.8	23.7 / 31.1	22.4 / 29.6	21.3 / 28.4	24.9 / 32.8	23.6 / 31.3	22.5 / 29.9

$K = 4 b' 10^3 / (\phi p F_y)$ where $\phi = 0.90$ and $F_y = 350$ MPa

Figure 3-2 — AMPLIFIED BOLT FORCE, T_f (kN)

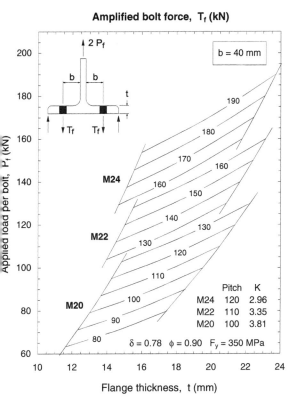
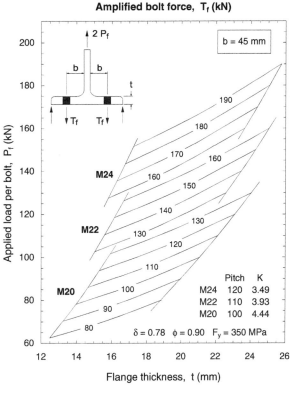
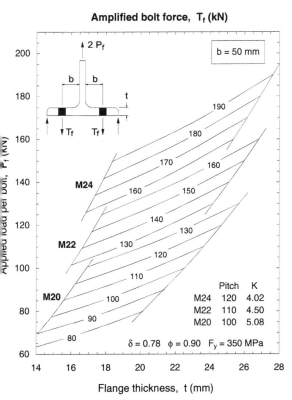
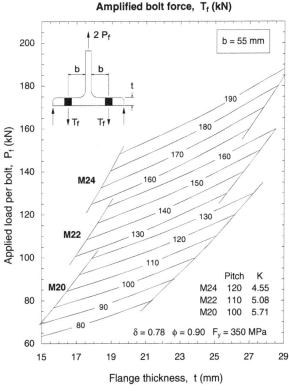

ECCENTRIC LOADS ON BOLT GROUPS

General

A bolted connection is eccentrically loaded when the line of action of the applied load passes outside the centroid of the bolt group. When the bolts are subjected to shear forces only, the effect of this eccentricity is to cause rotation about a single point called the instantaneous centre of rotation. The location of the instantaneous centre is obtained when the connection satisfies the three equilibrium equations for statics, $\Sigma F_x = 0$, $\Sigma F_y = 0$ and $\Sigma M = 0$ about the instantaneous centre.

Calculation of the instantaneous centre described in the references is a trial and error process, and the tables included in this section permit rapid evaluation of common bolt groups subjected to various eccentricities. All tables are based on symmetrical arrangements of bolts.

Bearing-Type Connections

For bearing type connections, a method of analysis is that described by Kulak et al. (1987). At the time the ultimate load is reached, it is assumed that the bolt furthest from the instantaneous centre will just reach its failure load. The resistance of each bolt is assumed to act on a line perpendicular to the radius joining the bolt to the instantaneous centre, and, Δ is assumed to vary linearly with the length of the radius. The resistance of each bolt is calculated according to the load deformation relationship:

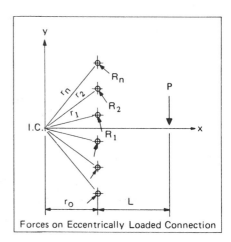

Forces on Eccentrically Loaded Connection

$$R = R_u (1 - e^{-\mu \Delta})^\lambda \quad \ldots\ldots \text{Kulak et al. (1987)}$$

and the ultimate load is reached when $\Delta = \Delta_{max}$ for the bolt furthest from the instantaneous centre, where

- R = bolt load at any given deformation
- R_u = ultimate bolt load
- Δ = shearing, bending and bearing deformation of the bolt, and local deformation of the connecting material
- μ, λ = regression coefficients
- e = base of natural logarithms

Slip-Critical Connections

For slip-critical connections, the method of anlaysis is essentially the same as that for bearing-type, except that the limiting slip resistance of the joint is reached when the maximum slip resistance of each individual bolt is reached as expressed by the relationship, $R = V_s = 0.53 c_1 k_s m n A_b F_u$ and the slip resistance of each bolt is assumed to be equal.

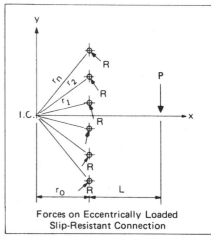

Forces on Eccentrically Loaded Slip-Resistant Connection

Tables

Tables 3-14 to 3-20 have been developed using the method described for bearing-type connections. Values tabulated are non-dimensional coefficients C and may be used for bolts of any diameter. In determining the coefficients C, the following values were used. R_u = 74 kips (329 kN), μ = 10.0, λ = 0.55, Δ_{max} = 0.34 inches (8.64 mm). These values were obtained experimentally for ¾ inch diameter A325 bolts and are reported by Crawford and Kulak (1971).

The ultimate load P for each bolt group and eccentricity was computed and then divided by the maximum value of R (when $\Delta = \Delta \text{max}$) to obtain the values of C.

The tables may thus be used to obtain the factored resistance, expressed as a vertical load P, of a connection by multiplying the coefficient C, for any particular bolt group and eccentricity, by the factored shear resistance of a single bolt. i.e. $P_f = CV_r$.

Coefficients were developed in a similar way for slip-critical connections, except that the individual bolt resistances for all bolts in the group were assumed to be equal. The coefficients calculated in this way were from 5% to 10% higher than those for bearing-type connections. Thus only one set of tables, based on the bearing-type connections is provided for use with both bearing-type and slip-critical connections.

Use of Tables

Bearing-Type Connections

1) To obtain the coefficient C required for a given geometry of bolts and eccentricity of load, divide the factored load P_f by the factored shear resistance V_r of a single bolt for the appropriate shear condition. i.e. $C = P_f / V_r$.

2) To determine the capacity of a given connection, multiply the coefficient C for the bolt group and eccentricity, by the appropriate bolt shear resistance value V_r of a single bolt. i.e. $P_f = CV_r$.

V_r is the factored shear resistance of the bolt from Table 3-4. Used in this way these tables provide a margin of safety which is consistant with bolts in joints less than 1300 mm long and subjected to shear produced by concentric loads only.

Slip-Critical Connections

Although developed using the method for bearing-type connections, these tables can also be used for slip-resistant connections using the *specified* load P and the appropriate slip resistance value V_s for the bolt size and condition of the faying surface.

1) Required $C = P / V_s$

2) Capacity $P = CV_s$

V_s is the slip resistance determined from Tables 3-10 and 3-11.

References

CRAWFORD, S.F., and KULAK, G.L. 1971. Eccentrically loaded bolted connections. ASCE Journal of the Structural Division, **97**(ST3), March.

KULAK, G.L. 1975. Eccentrically loaded slip resistant connections. AISC Engineering Journal, **12**(2), Second Quarter.

KULAK, G.L., ADAMS, P.F., and GILMOR, M.I. 1990. Limit states design in structural steel, CISC.

KULAK, G.L., FISHER, J.W., and STRUIK, J.H.A. 1987. Guide to design criteria for bolted and riveted joints (2nd. ed.). John Wiley and Sons.

SHERMER, C.L. 1971. Plastic behaviour of eccentrically loaded connections. AISC Enginecring Journal, **8**(2), April.

Example

1. Given:

A double column bracket must be designed to support a factored load of 700 kN at an eccentricity of 400 mm. Find the number of M20 A325M bolts per flange required for a gauge dimension of 120 mm and a pitch of 80 mm assuming a bearing type connection.

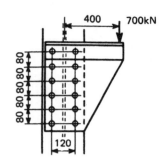

Solution:

P_f = 700/2 = 350 kN L = 400 mm

V_r = 105 kN (Table 3-4 page 3-8, single shear, threads excluded)

Required C = 350/105 = 3.33

From Tables 3-15 and 3-16, for 2 lines of bolts

   6 rows at 80 mm gauge, C = 3.49

       at 320 mm gauge, C = 4.77

Interpolating for 120 mm gauge C = 3.49 + (4.77 − 3.49) × 40/240 = 3.70

Use 6 rows of bolts (total 12 bolts)

Capacity is   3.70 × 105 = 389 kN per side

The connected material should be thick enough to provide bearing capacity for the 105 kN resistance of the bolts in accordance with CAN/CSA-S16.1-94, Clause 13.10(c). Minimum edge distances must conform with Clause 22.6.

2. Given:

Find the number of M22 bolts required for a similar bracket assuming a slip-critical connection with clean mill scale and a specified load of 500 kN.

Solution:

P = 500/2 = 250 kN L = 400 mm

V_s = 45.2 kN (Table 3-11, page 3-15)

Required C = 250/45.2 = 5.53

From Tables 3-15 and 3-16, for 2 lines of bolts

   8 rows at 80 mm gauge, C = 5.89

       at 320 mm gauge, C = 7.16

Interpolating for 120 mm gauge, C = 5.89 + (7.16 − 5.80) × 40/240 = 6.10

Use 8 rows of bolts (total 16 bolts)

Capacity is   6.10 × 45.2 = 276 kN per side

The ultimate strength of the joint would also be checked in bearing and shear for factored loads.

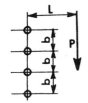

ECCENTRIC LOADS ON BOLT GROUPS
Coefficients C*

$$C = \frac{P_f}{V_r}, \text{ or } \frac{P}{V_s}$$

Table 3-14

Moment Arm, L, Millimetres												Number of Bolts	Pitch b mm
75	100	125	150	175	200	225	250	300	400	500	600		
0.94	0.74	0.60	0.51	0.44	0.39	0.35	0.31	0.26	0.20	0.16	0.13	2	
1.86	1.49	1.24	1.05	0.90	0.79	0.71	0.64	0.53	0.40	0.32	0.27	3	
2.95	2.50	2.15	1.85	1.62	1.44	1.30	1.18	0.99	0.75	0.60	0.51	4	
4.07	3.56	3.12	2.76	2.44	2.19	1.97	1.79	1.51	1.15	0.92	0.77	5	
5.17	4.67	4.21	3.77	3.37	3.06	2.77	2.53	2.16	1.65	1.34	1.12	6	80
6.25	5.79	5.30	4.82	4.38	4.01	3.66	3.37	2.89	2.23	1.80	1.51	7	
7.34	6.89	6.39	5.91	5.43	5.01	4.62	4.28	3.69	2.88	2.34	1.97	8	
8.40	7.99	7.51	7.02	6.54	6.07	5.64	5.24	4.57	3.60	2.93	2.48	9	
9.45	9.06	8.61	8.12	7.62	7.15	6.69	6.26	5.50	4.37	3.59	3.03	10	
10.5	10.1	9.71	9.23	8.73	8.25	7.75	7.30	6.49	5.20	4.30	3.65	11	
11.5	11.2	10.8	10.3	9.85	9.36	8.85	8.40	7.51	6.09	5.06	4.30	12	
1.01	0.82	0.67	0.57	0.49	0.43	0.39	0.35	0.29	0.22	0.18	0.15	2	
2.00	1.65	1.37	1.17	1.01	0.89	0.79	0.71	0.60	0.45	0.36	0.30	3	
3.11	2.68	2.33	2.04	1.80	1.60	1.44	1.31	1.11	0.84	0.68	0.57	4	
4.22	3.79	3.36	3.00	2.69	2.41	2.19	2.00	1.69	1.29	1.04	0.87	5	
5.31	4.89	4.45	4.06	3.66	3.34	3.05	2.81	2.40	1.85	1.50	1.26	6	90
6.40	5.99	5.58	5.13	4.73	4.35	4.00	3.70	3.20	2.48	2.02	1.69	7	
7.46	7.10	6.68	6.24	5.82	5.40	5.02	4.66	4.06	3.20	2.61	2.20	8	
8.51	8.18	7.77	7.34	6.90	6.48	6.07	5.69	5.00	3.98	3.27	2.76	9	
9.56	9.24	8.87	8.46	8.02	7.57	7.15	6.74	6.00	4.83	3.99	3.39	10	
10.6	10.3	9.95	9.54	9.12	8.69	8.25	7.81	7.02	5.73	4.76	4.06	11	
11.6	11.3	11.0	10.6	10.2	9.80	9.34	8.93	8.07	6.66	5.59	4.78	12	
1.09	0.89	0.74	0.63	0.55	0.48	0.43	0.39	0.33	0.25	0.20	0.16	2	
2.14	1.79	1.50	1.28	1.12	0.98	0.88	0.79	0.66	0.50	0.40	0.33	3	
3.25	2.85	2.50	2.21	1.95	1.76	1.59	1.45	1.22	0.94	0.75	0.63	4	
4.34	3.97	3.57	3.21	2.90	2.63	2.39	2.19	1.87	1.43	1.15	0.96	5	
5.44	5.07	4.69	4.29	3.92	3.60	3.32	3.05	2.63	2.04	1.66	1.39	6	100
6.51	6.16	5.78	5.40	5.01	4.64	4.31	4.01	3.49	2.73	2.23	1.88	7	
7.56	7.25	6.89	6.51	6.11	5.72	5.36	5.01	4.41	3.50	2.88	2.43	8	
8.60	8.32	7.98	7.61	7.22	6.81	6.44	6.07	5.40	4.35	3.59	3.05	9	
9.64	9.37	9.06	8.71	8.31	7.93	7.52	7.15	6.43	5.25	4.37	3.73	10	
10.7	10.4	10.1	9.79	9.42	9.03	8.64	8.24	7.50	6.20	5.21	4.46	11	
11.7	11.5	11.2	10.9	10.5	10.1	9.74	9.35	8.58	7.19	6.09	5.24	12	
1.25	1.02	0.85	0.74	0.65	0.57	0.51	0.46	0.39	0.29	0.24	0.20	2	
2.32	2.00	1.72	1.50	1.31	1.16	1.05	0.95	0.79	0.60	0.48	0.40	3	
3.44	3.11	2.80	2.51	2.25	2.04	1.85	1.69	1.45	1.11	0.90	0.75	4	
4.53	4.22	3.91	3.57	3.27	3.00	2.76	2.54	2.19	1.69	1.37	1.15	5	
5.60	5.32	5.02	4.68	4.37	4.05	3.77	3.50	3.05	2.41	1.96	1.66	6	120
6.65	6.40	6.11	5.78	5.45	5.13	4.82	4.53	4.01	3.20	2.63	2.23	7	
7.69	7.46	7.19	6.89	6.57	6.25	5.91	5.60	5.02	4.07	3.38	2.88	8	
8.72	8.51	8.26	7.98	7.68	7.36	7.01	6.69	6.06	5.00	4.20	3.59	9	
9.75	9.56	9.33	9.06	8.77	8.45	8.12	7.80	7.16	6.00	5.09	4.37	10	
10.8	10.6	10.4	10.1	9.85	9.55	9.24	8.91	8.25	7.02	6.01	5.21	11	
11.8	11.6	11.4	11.2	10.9	10.6	10.3	10.0	9.35	8.09	6.99	6.09	12	
1.45	1.24	1.07	0.94	0.83	0.74	0.66	0.60	0.51	0.39	0.31	0.26	2	
2.57	2.33	2.08	1.86	1.67	1.50	1.36	1.24	1.05	0.79	0.64	0.53	3	
3.66	3.44	3.21	2.95	2.73	2.51	2.32	2.15	1.85	1.45	1.18	0.99	4	
4.72	4.53	4.31	4.07	3.83	3.58	3.35	3.14	2.77	2.19	1.80	1.52	5	
5.77	5.60	5.40	5.17	4.93	4.68	4.44	4.21	3.77	3.06	2.54	2.16	6	160
6.80	6.65	6.47	6.26	6.03	5.79	5.54	5.30	4.83	4.01	3.37	2.89	7	
7.82	7.69	7.52	7.33	7.12	6.89	6.66	6.40	5.92	5.02	4.28	3.70	8	
8.84	8.72	8.57	8.40	8.20	7.98	7.75	7.51	7.02	6.07	5.25	4.58	9	
9.86	9.75	9.61	9.45	9.26	9.06	8.84	8.61	8.12	7.15	6.27	5.51	10	
10.9	10.8	10.6	10.5	10.3	10.1	9.92	9.70	9.23	8.25	7.31	6.50	11	
11.9	11.8	11.7	11.5	11.4	11.2	11.0	10.8	10.3	9.36	8.40	7.51	12	

* See page 3-26 for more details regarding these tables.

ECCENTRIC LOADS ON BOLT GROUPS
Coefficients C*

$$C = \frac{P_f}{V_r}, \text{ or } \frac{P}{V_s}$$

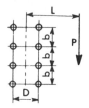

Table 3-15

Pitch b mm	Number of Bolts in Each Row	D = 80 mm Moment Arm, L, Millimetres										
		75	100	125	150	175	200	225	250	300	400	500	600
80	1	0.69	0.56	0.47	0.41	0.36	0.32	0.29	0.28	0.23	0.18	0.15	0.12
	2	2.15	1.76	1.49	1.28	1.12	1.03	0.93	0.84	0.71	0.54	0.43	0.36
	3	3.86	3.20	2.76	2.40	2.11	1.89	1.71	1.56	1.32	1.03	0.83	0.69
	4	5.89	5.09	4.46	3.90	3.51	3.14	2.84	2.59	2.20	1.70	1.38	1.15
	5	8.06	7.18	6.36	5.62	5.06	4.55	4.12	3.80	3.24	2.49	2.02	1.71
	6	10.2	9.34	8.40	7.57	6.87	6.22	5.72	5.25	4.49	3.49	2.83	2.38
	7	12.4	11.5	10.5	9.68	8.85	8.06	7.44	6.85	5.92	4.62	3.75	3.16
	8	14.6	13.7	12.8	11.8	10.9	10.1	9.33	8.63	7.50	5.89	4.81	4.07
	9	16.7	15.9	14.9	14.0	13.0	12.1	11.3	10.5	9.22	7.29	5.99	5.06
	10	18.8	18.1	17.1	16.2	15.2	14.3	13.4	12.6	11.1	8.85	7.28	6.18
	11	20.9	20.2	19.3	18.4	17.4	16.5	15.5	14.6	13.0	10.5	8.70	7.40
	12	23.0	22.3	21.5	20.6	19.6	18.6	17.7	16.7	15.0	12.2	10.2	8.70
90	2	2.21	1.87	1.59	1.37	1.20	1.07	0.99	0.90	0.76	0.58	0.46	0.39
	3	4.05	3.44	2.97	2.58	2.28	2.04	1.84	1.68	1.43	1.11	0.90	0.75
	4	6.18	5.46	4.76	4.23	3.76	3.38	3.10	2.83	2.41	1.85	1.51	1.26
	5	8.36	7.58	6.79	6.09	5.46	4.97	4.52	4.14	3.56	2.74	2.23	1.87
	6	10.6	9.75	8.91	8.11	7.40	6.79	6.21	5.71	4.94	3.85	3.13	2.64
	7	12.7	12.0	11.1	10.2	9.44	8.71	8.08	7.46	6.49	5.08	4.16	3.51
	8	14.9	14.1	13.3	12.5	11.6	10.8	10.0	9.37	8.21	6.49	5.34	4.51
	9	17.0	16.3	15.5	14.7	13.8	12.9	12.2	11.4	10.1	8.05	6.63	5.62
	10	19.1	18.4	17.7	16.8	16.0	15.1	14.3	13.5	12.0	9.72	8.07	6.85
	11	21.1	20.5	19.8	19.0	18.2	17.3	16.4	15.6	14.1	11.5	9.60	8.20
	12	23.2	22.6	22.0	21.2	20.4	19.5	18.7	17.8	16.1	13.3	11.2	9.63
100	2	2.34	1.99	1.68	1.46	1.28	1.14	1.05	0.96	0.81	0.61	0.49	0.41
	3	4.29	3.67	3.14	2.77	2.44	2.19	1.98	1.81	1.54	1.18	0.96	0.81
	4	6.45	5.74	5.09	4.50	4.06	3.66	3.32	3.07	2.62	2.01	1.63	1.38
	5	8.65	7.89	7.16	6.48	5.89	5.34	4.90	4.50	3.85	2.99	2.43	2.05
	6	10.8	10.1	9.32	8.58	7.89	7.27	6.68	6.20	5.35	4.19	3.43	2.89
	7	13.0	12.3	11.5	10.7	10.0	9.30	8.66	8.03	7.03	5.56	4.55	3.85
	8	15.1	14.4	13.7	12.9	12.2	11.4	10.7	10.1	8.87	7.09	5.83	4.96
	9	17.2	16.6	15.9	15.2	14.4	13.6	12.9	12.1	10.8	8.75	7.26	6.17
	10	19.2	18.7	18.1	17.4	16.6	15.8	15.0	14.3	12.9	10.5	8.81	7.52
	11	21.3	20.8	20.2	19.5	18.8	18.0	17.2	16.4	15.0	12.4	10.5	8.98
	12	23.3	22.9	22.3	21.7	21.0	20.2	19.5	18.6	17.1	14.4	12.2	10.5
120	2	2.57	2.15	1.88	1.63	1.44	1.28	1.16	1.05	0.91	0.69	0.56	0.47
	3	4.61	4.06	3.55	3.11	2.75	2.49	2.25	2.06	1.75	1.35	1.09	0.93
	4	6.84	6.21	5.59	5.04	4.58	4.15	3.82	3.51	3.00	2.34	1.90	1.59
	5	9.02	8.39	7.75	7.16	6.57	6.05	5.56	5.16	4.47	3.47	2.84	2.39
	6	11.1	10.6	9.98	9.35	8.68	8.11	7.53	7.04	6.16	4.88	4.01	3.39
	7	13.2	12.7	12.2	11.5	10.9	10.3	9.63	9.06	8.02	6.44	5.33	4.53
	8	15.3	14.9	14.3	13.7	13.1	12.4	11.8	11.2	10.0	8.18	6.81	5.82
	9	17.4	17.0	16.5	15.9	15.3	14.7	14.0	13.4	12.1	10.0	8.44	7.23
	10	19.5	19.1	18.6	18.1	17.5	16.9	16.2	15.5	14.3	12.0	10.2	8.79
	11	21.5	21.1	20.7	20.2	19.6	19.0	18.4	17.8	16.5	14.0	12.0	10.5
	12	23.5	23.2	22.8	22.3	21.8	21.2	20.6	20.0	18.7	16.2	14.0	12.2
160	2	2.89	2.51	2.22	1.95	1.76	1.58	1.43	1.30	1.11	0.86	0.69	0.58
	3	5.11	4.63	4.19	3.77	3.38	3.07	2.79	2.55	2.19	1.69	1.38	1.16
	4	7.29	6.85	6.35	5.89	5.44	5.03	4.67	4.32	3.77	2.96	2.42	2.05
	5	9.41	9.02	8.58	8.08	7.60	7.15	6.70	6.29	5.54	4.43	3.65	3.09
	6	11.5	11.2	10.8	10.3	9.83	9.35	8.84	8.41	7.55	6.14	5.12	4.37
	7	13.6	13.3	12.9	12.5	12.0	11.5	11.1	10.6	9.63	8.03	6.78	5.82
	8	15.6	15.3	15.0	14.6	14.2	13.7	13.3	12.8	11.8	10.0	8.57	7.42
	9	17.7	17.4	17.1	16.8	16.4	15.9	15.5	15.0	14.0	12.1	10.5	9.16
	10	19.7	19.5	19.2	18.9	18.5	18.1	17.6	17.2	16.2	14.3	12.5	11.0
	11	21.7	21.5	21.3	20.9	20.6	20.2	19.8	19.4	18.4	16.5	14.6	13.0
	12	23.7	23.5	23.3	23.0	22.7	22.3	21.9	21.5	20.6	18.7	16.8	15.0

* See page 3-26 for more details regarding these tables.

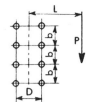

ECCENTRIC LOADS ON BOLT GROUPS
Coefficients C*

$$C = \frac{P_f}{V_r}, \text{ or } \frac{P}{V_s}$$

Table 3-16

D = 320 mm													Number of Bolts in Each Row	Pitch b mm
Moment Arm, L, Millimetres														
75	100	125	150	175	200	225	250	300	400	500	600		
1.36	1.23	1.12	1.03	0.95	0.88	0.83	0.78	0.69	0.57	0.48	0.42	1	
2.89	2.62	2.41	2.22	2.05	1.91	1.78	1.68	1.49	1.22	1.04	0.89	2	
4.56	4.17	3.81	3.53	3.26	3.03	2.84	2.66	2.38	1.95	1.65	1.43	3	
6.35	5.87	5.38	5.00	4.63	4.34	4.06	3.81	3.41	2.78	2.36	2.04	4	
8.31	7.67	7.10	6.60	6.16	5.74	5.41	5.07	4.52	3.72	3.16	2.72	5	
10.3	9.64	8.95	8.34	7.81	7.34	6.87	6.46	5.80	4.77	4.02	3.49	6	80
12.4	11.6	10.9	10.2	9.58	9.02	8.46	8.00	7.18	5.89	5.00	4.34	7	
14.5	13.7	12.9	12.2	11.5	10.8	10.2	9.64	8.67	7.16	6.08	5.25	8	
16.6	15.8	15.0	14.2	13.4	12.7	12.0	11.4	10.3	8.53	7.22	6.27	9	
18.6	17.9	17.1	16.3	15.4	14.7	13.9	13.2	12.0	9.97	8.49	7.38	10	
20.7	20.0	19.2	18.4	17.5	16.7	15.9	15.2	13.8	11.5	9.85	8.54	11	
22.8	22.1	21.3	20.5	19.7	18.8	18.0	17.1	15.6	13.2	11.3	9.82	12	
2.91	2.66	2.43	2.25	2.08	1.93	1.81	1.71	1.51	1.23	1.05	0.90	2	
4.63	4.23	3.89	3.61	3.34	3.10	2.90	2.72	2.44	1.99	1.69	1.46	3	
6.52	5.99	5.54	5.15	4.78	4.48	4.19	3.94	3.53	2.88	2.44	2.11	4	
8.48	7.87	7.34	6.85	6.36	5.97	5.58	5.28	4.70	3.87	3.26	2.84	5	
10.6	9.88	9.25	8.65	8.10	7.62	7.19	6.76	6.07	5.00	4.22	3.67	6	90
12.6	12.0	11.3	10.6	9.99	9.41	8.89	8.37	7.52	6.21	5.28	4.56	7	
14.7	14.1	13.3	12.6	11.9	11.3	10.7	10.1	9.12	7.59	6.42	5.58	8	
16.8	16.2	15.4	14.7	14.0	13.3	12.6	12.0	10.8	9.03	7.69	6.69	9	
18.9	18.3	17.6	16.8	16.1	15.3	14.6	13.9	12.7	10.6	9.05	7.87	10	
21.0	20.4	19.7	19.0	18.2	17.4	16.6	15.9	14.6	12.3	10.5	9.17	11	
23.0	22.5	21.8	21.1	20.3	19.5	18.8	18.0	16.5	14.0	12.1	10.5	12	
2.95	2.69	2.46	2.28	2.11	1.96	1.83	1.73	1.54	1.25	1.06	0.92	2	
4.71	4.32	3.98	3.66	3.42	3.18	2.97	2.79	2.50	2.04	1.73	1.50	3	
6.64	6.13	5.70	5.27	4.93	4.59	4.33	4.07	3.62	2.97	2.53	2.18	4	
8.66	8.08	7.56	7.04	6.60	6.20	5.80	5.45	4.89	4.03	3.40	2.95	5	
10.7	10.1	9.55	8.93	8.39	7.90	7.46	7.03	6.31	5.21	4.43	3.84	6	100
12.8	12.2	11.6	10.9	10.4	9.78	9.26	8.74	7.87	6.54	5.53	4.81	7	
14.9	14.3	13.7	13.0	12.4	11.8	11.1	10.6	9.57	7.98	6.80	5.91	8	
17.0	16.4	15.8	15.1	14.5	13.8	13.1	12.5	11.4	9.54	8.17	7.08	9	
19.1	18.5	17.9	17.3	16.6	15.9	15.2	14.5	13.3	11.2	9.64	8.40	10	
21.2	20.6	20.1	19.4	18.7	18.0	17.3	16.6	15.3	13.0	11.2	9.78	11	
23.2	22.7	22.2	21.5	20.9	20.2	19.4	18.7	17.3	14.9	12.9	11.3	12	
3.02	2.76	2.53	2.35	2.17	2.02	1.89	1.78	1.58	1.29	1.10	0.94	2	
4.87	4.47	4.13	3.84	3.55	3.33	3.11	2.92	2.60	2.14	1.82	1.57	3	
6.86	6.40	5.95	5.55	5.20	4.88	4.58	4.30	3.86	3.18	2.68	2.33	4	
8.95	8.44	7.94	7.44	7.02	6.58	6.21	5.88	5.25	4.33	3.68	3.18	5	120
11.0	10.5	10.0	9.48	8.93	8.48	8.00	7.59	6.84	5.70	4.82	4.19	6	
13.1	12.7	12.1	11.5	11.0	10.5	9.93	9.45	8.58	7.17	6.11	5.29	7	
15.2	14.8	14.2	13.7	13.1	12.5	12.0	11.4	10.4	8.81	7.53	6.56	8	
17.3	16.9	16.4	15.8	15.3	14.7	14.1	13.5	12.4	10.5	9.07	7.91	9	
19.3	18.9	18.5	17.9	17.4	16.8	16.2	15.6	14.5	12.4	10.7	9.40	10	
21.4	21.0	20.6	20.1	19.5	18.9	18.4	17.8	16.6	14.4	12.5	11.0	11	
23.4	23.1	22.7	22.2	21.7	21.1	20.5	19.9	18.7	16.3	14.4	12.7	12	
3.16	2.90	2.66	2.47	2.29	2.15	2.01	1.89	1.69	1.38	1.17	1.01	2	
5.14	4.78	4.46	4.16	3.86	3.62	3.39	3.20	2.85	2.35	1.99	1.73	3	
7.22	6.84	6.44	6.09	5.73	5.40	5.10	4.81	4.33	3.58	3.04	2.63	4	
9.32	8.96	8.56	8.14	7.74	7.33	6.96	6.61	6.00	4.97	4.23	3.69	5	160
11.4	11.1	10.7	10.3	9.82	9.40	8.97	8.57	7.82	6.61	5.65	4.93	6	
13.5	13.2	12.8	12.4	12.0	11.5	11.1	10.7	9.81	8.37	7.22	6.30	7	
15.5	15.3	14.9	14.5	14.1	13.7	13.2	12.8	11.9	10.3	8.94	7.84	8	
17.6	17.3	17.0	16.7	16.3	15.8	15.4	15.0	14.0	12.3	10.8	9.53	9	
19.6	19.4	19.1	18.8	18.4	18.0	17.6	17.1	16.2	14.4	12.7	11.3	10	
21.7	21.4	21.2	20.8	20.5	20.1	19.7	19.3	18.4	16.5	14.7	13.2	11	
23.7	23.5	23.2	22.9	22.6	22.2	21.8	21.4	20.5	18.7	16.8	15.2	12	

* See page 3-26 for more details regarding these tables.

ECCENTRIC LOADS ON BOLT GROUPS
Coefficients C*

$$C = \frac{P_f}{V_r}, \text{ or } \frac{P}{V_s}$$

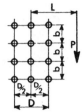

Table 3–17

Pitch b mm	Number of Bolts in Each Row	D = 160 mm											
		Moment Arm, L, Millimetres											
		75	100	125	150	175	200	225	250	300	400	500	600
	1	1.49	1.26	1.11	0.96	0.84	0.77	0.69	0.62	0.52	0.39	0.31	0.26
80	2	3.52	3.05	2.63	2.30	2.09	1.87	1.69	1.54	1.30	1.02	0.82	0.69
	3	6.06	5.18	4.61	4.07	3.63	3.27	3.03	2.77	2.35	1.84	1.48	1.24
	4	8.94	7.86	7.00	6.22	5.65	5.11	4.65	4.26	3.70	2.85	2.31	1.97
	5	12.1	10.9	9.72	8.77	7.90	7.17	6.63	6.10	5.25	4.11	3.34	2.81
	6	15.3	14.0	12.6	11.5	10.5	9.68	8.88	8.19	7.13	5.56	4.57	3.85
	7	18.5	17.2	15.9	14.6	13.4	12.4	11.4	10.6	9.18	7.25	5.94	5.05
	8	21.8	20.5	19.0	17.7	16.4	15.2	14.1	13.2	11.5	9.12	7.53	6.37
	9	25.0	23.7	22.4	20.9	19.6	18.3	17.1	15.9	14.1	11.2	9.25	7.88
	10	28.1	27.0	25.6	24.2	22.8	21.4	20.1	18.9	16.8	13.5	11.2	9.55
	11	31.3	30.2	28.9	27.5	26.0	24.6	23.3	22.0	19.6	15.9	13.2	11.3
	12	34.4	33.4	32.1	30.8	29.4	27.9	26.5	25.1	22.6	18.5	15.5	13.3
90	2	3.64	3.16	2.73	2.39	2.18	1.95	1.76	1.61	1.36	1.07	0.86	0.72
	3	6.26	5.46	4.77	4.29	3.84	3.46	3.15	2.93	2.50	1.92	1.58	1.32
	4	9.30	8.31	7.35	6.63	5.96	5.47	4.99	4.58	3.92	3.07	2.49	2.09
	5	12.6	11.4	10.3	9.30	8.48	7.72	7.07	6.52	5.67	4.41	3.63	3.06
	6	15.7	14.6	13.4	12.2	11.2	10.4	9.52	8.87	7.68	6.06	4.96	4.21
	7	19.0	17.8	16.6	15.4	14.3	13.2	12.2	11.4	10.0	7.88	6.51	5.50
	8	22.2	21.1	19.9	18.6	17.4	16.3	15.2	14.2	12.5	9.99	8.23	7.01
	9	25.4	24.3	23.1	21.9	20.6	19.4	18.2	17.1	15.2	12.3	10.2	8.65
	10	28.5	27.5	26.4	25.1	23.9	22.7	21.4	20.2	18.1	14.7	12.3	10.5
	11	31.6	30.7	29.6	28.5	27.2	25.9	24.6	23.4	21.1	17.4	14.5	12.5
	12	34.7	33.9	32.9	31.7	30.5	29.2	27.9	26.6	24.3	20.1	17.0	14.6
100	2	3.77	3.28	2.84	2.49	2.27	2.03	1.84	1.68	1.42	1.12	0.90	0.76
	3	6.54	5.72	5.00	4.44	4.04	3.66	3.33	3.05	2.65	2.03	1.67	1.40
	4	9.62	8.66	7.78	7.03	6.34	5.75	5.32	4.89	4.20	3.29	2.68	2.25
	5	12.9	11.8	10.8	9.81	8.98	8.19	7.58	6.99	6.10	4.76	3.92	3.30
	6	16.1	15.1	14.0	12.9	11.9	11.0	10.2	9.48	8.29	6.56	5.38	4.57
	7	19.3	18.3	17.2	16.1	15.0	14.0	13.0	12.2	10.7	8.54	7.07	5.99
	8	22.5	21.6	20.5	19.4	18.3	17.1	16.1	15.1	13.5	10.8	8.97	7.65
	9	25.7	24.8	23.8	22.6	21.5	20.4	19.3	18.2	16.3	13.3	11.0	9.44
	10	28.8	28.0	27.0	26.0	24.8	23.7	22.5	21.4	19.3	15.9	13.3	11.4
	11	31.9	31.1	30.2	29.2	28.1	27.0	25.8	24.7	22.4	18.7	15.8	13.6
	12	35.0	34.2	33.4	32.4	31.4	30.3	29.1	27.9	25.6	21.6	18.4	15.9
120	2	4.02	3.45	3.06	2.70	2.40	2.21	2.00	1.83	1.56	1.19	0.99	0.83
	3	6.99	6.18	5.50	4.88	4.39	4.04	3.69	3.39	2.90	2.27	1.84	1.57
	4	10.2	9.29	8.48	7.66	7.01	6.45	5.92	5.46	4.76	3.71	3.05	2.57
	5	13.4	12.5	11.7	10.7	9.89	9.16	8.50	7.87	6.88	5.45	4.47	3.80
	6	16.6	15.8	14.9	14.0	13.1	12.2	11.4	10.7	9.41	7.51	6.22	5.27
	7	19.8	19.1	18.2	17.2	16.3	15.3	14.5	13.7	12.1	9.81	8.16	6.94
	8	22.9	22.2	21.4	20.5	19.6	18.6	17.7	16.8	15.1	12.3	10.4	8.88
	9	26.0	25.4	24.6	23.8	22.9	21.9	21.0	20.0	18.2	15.1	12.8	11.0
	10	29.1	28.5	27.8	27.0	26.1	25.2	24.3	23.3	21.4	18.0	15.3	13.3
	11	32.2	31.7	31.0	30.2	29.4	28.5	27.5	26.6	24.6	21.1	18.1	15.7
	12	35.3	34.8	34.1	33.4	32.6	31.7	30.8	29.9	27.9	24.2	21.0	18.4
160	2	4.38	3.87	3.45	3.07	2.79	2.52	2.30	2.11	1.83	1.41	1.15	0.98
	3	7.60	6.92	6.30	5.72	5.17	4.75	4.35	4.01	3.50	2.73	2.25	1.89
	4	10.9	10.2	9.54	8.82	8.22	7.59	7.06	6.61	5.76	4.58	3.78	3.20
	5	14.1	13.5	12.8	12.1	11.4	10.7	10.0	9.45	8.41	6.74	5.59	4.77
	6	17.2	16.7	16.1	15.4	14.7	14.0	13.3	12.6	11.3	9.27	7.79	6.68
	7	20.3	19.8	19.3	18.7	18.0	17.2	16.6	15.8	14.5	12.1	10.2	8.79
	8	23.4	23.0	22.5	21.9	21.2	20.6	19.9	19.1	17.7	15.1	12.9	11.2
	9	26.4	26.1	25.6	25.1	24.5	23.8	23.2	22.4	21.0	18.2	15.8	13.8
	10	29.5	29.2	28.7	28.2	27.7	27.1	26.4	25.7	24.3	21.4	18.8	16.6
	11	32.5	32.2	31.8	31.4	30.9	30.3	29.7	29.0	27.6	24.7	21.9	19.5
	12	35.6	35.3	34.9	34.5	34.0	33.5	32.9	32.2	30.9	28.0	25.1	22.5

* See page 3–26 for more details regarding these tables.

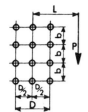

ECCENTRIC LOADS ON BOLT GROUPS
Coefficients C*

$$C = \frac{P_f}{V_r}, \text{ or } \frac{P}{V_s}$$

Table 3-18

D = 320 mm												Number of Bolts in Each Row	Pitch b mm
Moment Arm, L, Millimetres													
75	100	125	150	175	200	225	250	300	400	500	600		
1.97	1.78	1.63	1.49	1.37	1.28	1.18	1.11	0.98	0.77	0.62	0.53	1	
4.22	3.79	3.47	3.17	2.90	2.71	2.50	2.32	2.05	1.63	1.35	1.14	2	
6.66	6.04	5.49	5.07	4.66	4.30	4.03	3.75	3.32	2.66	2.18	1.87	3	
9.41	8.57	7.86	7.20	6.69	6.19	5.74	5.41	4.74	3.81	3.18	2.69	4	
12.3	11.3	10.4	9.63	8.88	8.24	7.74	7.22	6.42	5.18	4.33	3.67	5	
15.4	14.2	13.1	12.2	11.4	10.6	9.89	9.33	8.23	6.67	5.58	4.79	6	80
18.5	17.3	16.1	15.0	14.0	13.1	12.3	11.6	10.3	8.39	7.04	6.00	7	
21.7	20.4	19.2	18.0	16.9	15.9	14.9	14.0	12.5	10.2	8.59	7.40	8	
24.8	23.6	22.4	21.1	19.8	18.8	17.6	16.7	14.9	12.3	10.4	8.88	9	
27.9	26.8	25.6	24.3	23.0	21.7	20.5	19.5	17.5	14.4	12.2	10.5	10	
31.1	30.0	28.8	27.5	26.1	24.8	23.6	22.4	20.2	16.8	14.2	12.3	11	
34.2	33.2	32.0	30.7	29.4	28.0	26.7	25.4	23.1	19.3	16.4	14.2	12	
4.24	3.85	3.52	3.21	2.94	2.74	2.54	2.36	2.08	1.66	1.38	1.16	2	
6.81	6.18	5.61	5.14	4.78	4.42	4.14	3.85	3.41	2.74	2.25	1.93	3	
9.61	8.77	8.06	7.46	6.87	6.42	5.96	5.56	4.93	3.97	3.31	2.81	4	
12.6	11.7	10.8	9.97	9.27	8.60	8.02	7.56	6.66	5.44	4.51	3.87	5	
15.7	14.7	13.6	12.8	11.9	11.1	10.4	9.75	8.69	7.06	5.92	5.05	6	90
18.9	17.8	16.7	15.7	14.7	13.8	13.0	12.2	10.9	8.87	7.45	6.42	7	
22.1	21.0	19.9	18.8	17.6	16.7	15.7	14.8	13.3	10.9	9.21	7.89	8	
25.2	24.2	23.1	21.9	20.7	19.7	18.6	17.6	15.8	13.1	11.1	9.57	9	
28.3	27.4	26.3	25.1	24.0	22.8	21.6	20.5	18.6	15.5	13.1	11.3	10	
31.5	30.5	29.5	28.4	27.2	25.9	24.7	23.7	21.6	18.0	15.3	13.3	11	
34.6	33.7	32.7	31.5	30.4	29.2	28.0	26.8	24.5	20.7	17.7	15.4	12	
4.29	3.90	3.53	3.26	2.99	2.79	2.58	2.39	2.12	1.69	1.40	1.18	2	
6.91	6.28	5.75	5.27	4.90	4.53	4.21	3.96	3.47	2.79	2.32	1.99	3	
9.83	9.04	8.33	7.65	7.11	6.59	6.18	5.77	5.13	4.14	3.46	2.93	4	
12.9	12.0	11.1	10.3	9.60	8.97	8.36	7.83	6.98	5.66	4.74	4.07	5	
16.1	15.1	14.1	13.2	12.4	11.5	10.9	10.2	9.09	7.46	6.22	5.35	6	100
19.2	18.3	17.2	16.2	15.3	14.4	13.6	12.8	11.5	9.41	7.93	6.79	7	
22.4	21.4	20.5	19.4	18.4	17.4	16.4	15.6	14.0	11.6	9.78	8.44	8	
25.5	24.6	23.7	22.6	21.6	20.5	19.5	18.5	16.8	13.9	11.8	10.2	9	
28.6	27.8	26.8	25.8	24.8	23.6	22.6	21.5	19.7	16.5	14.1	12.2	10	
31.7	30.9	30.0	29.0	28.0	26.9	25.8	24.7	22.7	19.2	16.4	14.3	11	
34.8	34.1	33.2	32.3	31.2	30.1	29.0	27.9	25.8	22.0	19.0	16.5	12	
4.41	4.02	3.64	3.36	3.08	2.88	2.66	2.47	2.19	1.75	1.44	1.23	2	
7.17	6.58	5.99	5.53	5.11	4.77	4.44	4.14	3.68	2.96	2.47	2.09	3	
10.2	9.46	8.79	8.16	7.55	7.07	6.58	6.21	5.48	4.44	3.71	3.19	4	
13.4	12.6	11.7	11.0	10.3	9.66	9.03	8.51	7.56	6.21	5.18	4.45	5	
16.5	15.7	14.9	14.1	13.2	12.5	11.8	11.1	9.95	8.20	6.91	5.92	6	120
19.7	18.9	18.1	17.2	16.4	15.5	14.7	13.9	12.6	10.4	8.82	7.63	7	
22.8	22.1	21.3	20.4	19.5	18.7	17.8	17.0	15.4	12.9	11.0	9.49	8	
25.9	25.3	24.5	23.7	22.8	21.9	21.0	20.1	18.4	15.5	13.3	11.5	9	
29.0	28.4	27.7	26.9	26.0	25.1	24.2	23.3	21.5	18.3	15.8	13.8	10	
32.1	31.5	30.9	30.1	29.3	28.4	27.5	26.5	24.7	21.3	18.5	16.2	11	
35.2	34.6	34.0	33.3	32.5	31.6	30.7	29.8	27.9	24.3	21.3	18.7	12	
4.64	4.23	3.89	3.55	3.30	3.05	2.86	2.66	2.33	1.87	1.56	1.32	2	
7.62	7.05	6.55	6.06	5.64	5.23	4.88	4.61	4.07	3.30	2.77	2.38	3	
10.8	10.2	9.57	8.99	8.42	7.92	7.46	7.00	6.26	5.11	4.30	3.70	4	
14.0	13.4	12.7	12.1	11.5	10.9	10.3	9.72	8.77	7.21	6.10	5.27	5	
17.1	16.6	16.0	15.3	14.7	14.0	13.3	12.7	11.5	9.68	8.23	7.13	6	160
20.2	19.7	19.2	18.6	17.9	17.2	16.6	15.9	14.6	12.4	10.6	9.21	7	
23.3	22.9	22.4	21.8	21.2	20.5	19.8	19.1	17.7	15.2	13.2	11.5	8	
26.4	26.0	25.5	25.0	24.4	23.7	23.1	22.4	21.0	18.3	16.0	14.1	9	
29.4	29.1	28.6	28.1	27.6	27.0	26.3	25.6	24.2	21.4	18.9	16.8	10	
32.5	32.2	31.7	31.3	30.7	30.2	29.5	28.9	27.5	24.7	22.0	19.6	11	
35.5	35.2	34.8	34.4	33.9	33.4	32.7	32.1	30.8	27.9	25.1	22.6	12	

* See page 3–26 for more details regarding these tables.

ECCENTRIC LOADS ON BOLT GROUPS
Coefficients C*

$$C = \frac{P_f}{V_r}, \text{ or } \frac{P}{V_s}$$

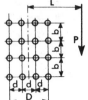

$d = D/3$

Table 3-19

Pitch b mm	Number of Bolts in Each Row	D = 240 mm Moment Arm, L, Millimetres											
		75	100	125	150	175	200	225	250	300	400	500	600
80	1	2.31	2.02	1.79	1.56	1.41	1.26	1.14	1.05	0.89	0.69	0.56	0.47
80	2	5.12	4.52	3.97	3.58	3.20	2.89	2.63	2.46	2.10	1.62	1.31	1.13
80	3	8.33	7.41	6.57	5.96	5.36	4.93	4.50	4.13	3.60	2.78	2.26	1.90
80	4	12.1	10.8	9.77	8.78	8.05	7.33	6.71	6.27	5.40	4.26	3.47	2.92
80	5	16.1	14.6	13.3	12.1	11.0	10.1	9.37	8.66	7.57	5.91	4.88	4.12
80	6	20.4	18.7	17.0	15.6	14.4	13.3	12.3	11.4	10.0	7.94	6.50	5.55
80	7	24.6	22.9	21.2	19.5	18.1	16.8	15.5	14.5	12.7	10.1	8.40	7.11
80	8	28.9	27.2	25.4	23.6	22.1	20.6	19.1	17.9	15.8	12.7	10.5	8.94
80	9	33.1	31.5	29.7	27.9	26.2	24.5	22.9	21.6	19.1	15.4	12.8	11.0
80	10	37.4	35.8	34.0	32.1	30.4	28.6	27.0	25.4	22.7	18.3	15.3	13.1
80	11	41.6	40.1	38.4	36.5	34.7	32.9	31.1	29.4	26.4	21.6	18.1	15.5
80	12	45.7	44.3	42.7	40.9	39.0	37.1	35.3	33.5	30.2	25.0	21.0	18.1
90	2	5.23	4.55	4.07	3.61	3.29	2.98	2.71	2.48	2.17	1.67	1.36	1.16
90	3	8.63	7.67	6.83	6.20	5.59	5.07	4.71	4.32	3.70	2.91	2.37	2.00
90	4	12.5	11.2	10.1	9.24	8.38	7.74	7.10	6.55	5.73	4.46	3.69	3.11
90	5	16.7	15.3	13.9	12.7	11.7	10.7	9.88	9.24	8.01	6.35	5.19	4.42
90	6	20.9	19.3	17.9	16.4	15.2	14.1	13.0	12.2	10.7	8.49	7.03	5.95
90	7	25.2	23.7	22.1	20.6	19.2	17.8	16.6	15.5	13.7	11.0	9.06	7.73
90	8	29.5	28.0	26.4	24.7	23.3	21.8	20.4	19.2	17.0	13.7	11.4	9.70
90	9	33.7	32.3	30.8	29.1	27.5	25.9	24.4	23.0	20.6	16.6	13.9	11.9
90	10	37.9	36.6	35.1	33.5	31.8	30.1	28.6	27.1	24.3	19.9	16.7	14.3
90	11	42.0	40.8	39.4	37.8	36.1	34.5	32.8	31.2	28.3	23.3	19.7	17.0
90	12	46.2	45.0	43.7	42.1	40.6	38.9	37.2	35.5	32.4	27.0	22.9	19.8
100	2	5.35	4.66	4.17	3.71	3.39	3.06	2.79	2.56	2.24	1.73	1.40	1.18
100	3	8.83	7.84	7.09	6.36	5.83	5.30	4.92	4.52	3.88	3.06	2.49	2.10
100	4	12.9	11.7	10.6	9.59	8.82	8.06	7.49	6.92	6.06	4.74	3.92	3.30
100	5	17.2	15.8	14.5	13.3	12.2	11.3	10.5	9.71	8.53	6.78	5.55	4.74
100	6	21.4	20.0	18.7	17.2	16.0	14.9	13.9	12.9	11.4	9.11	7.56	6.41
100	7	25.7	24.4	22.9	21.5	20.0	18.8	17.5	16.5	14.6	11.7	9.78	8.35
100	8	29.9	28.7	27.2	25.7	24.3	22.9	21.6	20.4	18.1	14.7	12.3	10.5
100	9	34.1	32.9	31.6	30.1	28.6	27.1	25.8	24.4	21.8	18.0	15.1	12.9
100	10	38.3	37.2	35.9	34.5	33.0	31.5	30.0	28.6	25.9	21.4	18.1	15.6
100	11	42.4	41.4	40.2	38.8	37.4	35.8	34.3	32.8	30.0	25.1	21.3	18.4
100	12	46.5	45.6	44.4	43.1	41.7	40.2	38.7	37.2	34.2	28.9	24.7	21.4
120	2	5.52	4.90	4.40	3.92	3.53	3.25	2.97	2.73	2.38	1.84	1.50	1.26
120	3	9.31	8.42	7.59	6.85	6.22	5.75	5.28	4.87	4.25	3.35	2.73	2.30
120	4	13.6	12.4	11.4	10.4	9.57	8.88	8.19	7.59	6.66	5.29	4.34	3.70
120	5	17.9	16.7	15.5	14.4	13.3	12.4	11.5	10.7	9.48	7.58	6.29	5.33
120	6	22.1	21.0	19.8	18.6	17.4	16.3	15.3	14.4	12.8	10.3	8.56	7.33
120	7	26.3	25.3	24.2	23.0	21.7	20.5	19.4	18.3	16.3	13.3	11.1	9.55
120	8	30.5	29.6	28.5	27.3	26.0	24.8	23.6	22.4	20.2	16.7	14.0	12.1
120	9	34.6	33.8	32.8	31.6	30.4	29.2	27.9	26.7	24.3	20.3	17.2	14.8
120	10	38.8	38.0	37.0	35.9	34.7	33.5	32.2	31.0	28.5	24.1	20.6	17.9
120	11	42.9	42.1	41.2	40.2	39.1	37.9	36.7	35.4	32.8	28.1	24.3	21.2
120	12	47.0	46.3	45.4	44.5	43.4	42.2	41.0	39.8	37.2	32.3	28.0	24.6
160	2	5.95	5.32	4.80	4.31	3.96	3.60	3.30	3.08	2.66	2.10	1.71	1.44
160	3	10.1	9.25	8.49	7.76	7.07	6.56	6.06	5.62	4.93	3.91	3.20	2.73
160	4	14.4	13.6	12.7	11.8	11.0	10.3	9.55	8.97	7.92	6.35	5.24	4.48
160	5	18.7	17.9	17.0	16.1	15.2	14.3	13.5	12.7	11.3	9.20	7.66	6.57
160	6	22.9	22.2	21.4	20.5	19.5	18.6	17.7	16.8	15.2	12.5	10.5	9.07
160	7	27.0	26.4	25.6	24.8	23.9	23.0	22.0	21.1	19.3	16.2	13.8	11.9
160	8	31.1	30.6	29.9	29.1	28.3	27.3	26.4	25.4	23.6	20.1	17.3	15.0
160	9	35.2	34.7	34.1	33.4	32.5	31.7	30.8	29.8	27.9	24.2	21.1	18.5
160	10	39.3	38.8	38.2	37.6	36.8	36.0	35.1	34.2	32.3	28.5	25.1	22.1
160	11	43.4	42.9	42.4	41.8	41.1	40.3	39.5	38.6	36.7	32.8	29.2	26.0
160	12	47.4	47.0	46.5	45.9	45.3	44.5	43.7	42.9	41.1	37.3	33.4	30.0

* See page 3-26 for more details regarding these tables.   * See p. 3-24 for more details regarding these tables.

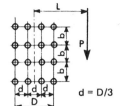

ECCENTRIC LOADS ON BOLT GROUPS
Coefficients C*

$$C = \frac{P_f}{V_r}, \text{ or } \frac{P}{V_s}$$

$d = D/3$

Table 3–20

D = 480 mm												Number of Bolts in Each Row	Pitch b mm
Moment Arm, L, Millimetres													
75	100	125	150	175	200	225	250	300	400	500	600		
2.91	2.69	2.51	2.33	2.18	2.04	1.90	1.79	1.58	1.26	1.05	0.89	1	
6.00	5.56	5.17	4.79	4.49	4.18	3.93	3.71	3.27	2.67	2.21	1.89	2	
9.32	8.62	7.99	7.48	6.96	6.54	6.17	5.78	5.15	4.18	3.50	2.99	3	
12.8	11.9	11.1	10.3	9.72	9.08	8.58	8.12	7.20	5.92	4.98	4.26	4	
16.6	15.5	14.5	13.5	12.7	11.9	11.2	10.6	9.54	7.80	6.57	5.67	5	
20.5	19.2	18.1	16.9	15.9	15.0	14.2	13.4	12.0	9.95	8.41	7.21	6	80
24.6	23.1	21.8	20.6	19.4	18.2	17.3	16.4	14.8	12.2	10.4	8.95	7	
28.7	27.3	25.8	24.3	23.0	21.7	20.6	19.6	17.7	14.8	12.6	10.9	8	
32.9	31.4	29.8	28.3	26.8	25.4	24.1	23.0	20.8	17.5	14.9	12.9	9	
37.0	35.6	34.0	32.4	30.8	29.3	27.9	26.6	24.1	20.3	17.4	15.2	10	
41.2	39.8	38.2	36.6	34.9	33.3	31.8	30.4	27.6	23.3	20.1	17.5	11	
45.4	44.0	42.4	40.8	39.1	37.4	35.8	34.2	31.4	26.6	22.9	20.0	12	
6.05	5.59	5.21	4.83	4.52	4.21	3.97	3.71	3.30	2.67	2.24	1.91	2	
9.41	8.70	8.10	7.59	7.07	6.65	6.22	5.88	5.26	4.27	3.58	3.05	3	
13.1	12.1	11.3	10.6	9.95	9.31	8.80	8.26	7.41	6.10	5.09	4.39	4	
16.9	15.9	14.8	13.8	13.0	12.2	11.6	11.0	9.80	8.10	6.84	5.90	5	
20.9	19.7	18.6	17.4	16.4	15.5	14.6	13.9	12.5	10.3	8.74	7.56	6	90
25.1	23.8	22.5	21.2	20.1	19.0	17.9	17.0	15.4	12.8	10.9	9.45	7	
29.2	27.9	26.5	25.1	23.8	22.6	21.5	20.4	18.5	15.5	13.2	11.5	8	
33.4	32.1	30.7	29.2	27.8	26.5	25.2	24.0	21.8	18.4	15.7	13.7	9	
37.6	36.3	34.9	33.4	32.0	30.6	29.1	27.8	25.4	21.4	18.5	16.1	10	
41.7	40.5	39.1	37.7	36.1	34.6	33.2	31.8	29.2	24.7	21.3	18.6	11	
45.9	44.7	43.4	41.9	40.4	38.9	37.4	35.9	33.0	28.2	24.4	21.4	12	
6.09	5.64	5.21	4.87	4.56	4.25	4.00	3.74	3.33	2.70	2.26	1.93	2	
9.53	8.83	8.22	7.65	7.19	6.77	6.33	5.99	5.36	4.36	3.66	3.12	3	
13.3	12.4	11.6	10.8	10.1	9.54	8.96	8.49	7.61	6.23	5.25	4.54	4	
17.3	16.2	15.1	14.3	13.3	12.6	11.9	11.3	10.1	8.41	7.11	6.09	5	
21.3	20.2	19.0	17.9	16.9	16.0	15.2	14.3	12.9	10.8	9.14	7.92	6	100
25.5	24.3	23.1	21.8	20.7	19.7	18.6	17.6	16.0	13.4	11.4	9.89	7	
29.6	28.5	27.2	25.9	24.7	23.5	22.3	21.3	19.3	16.2	13.9	12.1	8	
33.8	32.7	31.4	30.1	28.8	27.5	26.2	25.0	22.9	19.3	16.6	14.5	9	
38.0	36.9	35.6	34.3	33.0	31.6	30.2	29.0	26.7	22.6	19.5	17.1	10	
42.1	41.1	39.9	38.6	37.2	35.8	34.5	33.1	30.5	26.1	22.6	19.8	11	
46.3	45.2	44.1	42.8	41.5	40.1	38.7	37.3	34.6	29.8	25.9	22.7	12	
6.19	5.74	5.30	4.95	4.64	4.33	4.08	3.82	3.41	2.77	2.32	1.98	2	
9.79	9.06	8.43	7.89	7.43	6.95	6.57	6.17	5.53	4.55	3.83	3.27	3	
13.7	12.8	12.0	11.3	10.6	9.95	9.43	8.88	7.99	6.62	5.59	4.80	4	
17.7	16.8	15.8	14.9	14.1	13.3	12.6	11.9	10.8	8.98	7.61	6.59	5	
21.9	20.9	19.9	18.9	17.9	16.9	16.1	15.3	13.9	11.6	9.91	8.60	6	120
26.1	25.1	24.1	23.0	21.9	20.9	19.9	18.9	17.3	14.6	12.5	10.8	7	
30.3	29.3	28.3	27.2	26.0	24.9	23.9	22.8	20.9	17.8	15.3	13.3	8	
34.4	33.5	32.5	31.4	30.3	29.2	28.0	26.9	24.8	21.2	18.3	16.0	9	
38.5	37.7	36.8	35.7	34.6	33.4	32.3	31.1	28.8	24.9	21.6	19.0	10	
42.7	41.9	41.0	40.0	38.9	37.7	36.5	35.3	32.9	28.7	25.1	22.1	11	
46.8	46.0	45.2	44.2	43.1	42.0	40.8	39.6	37.3	32.7	28.7	25.5	12	
6.38	5.92	5.51	5.12	4.80	4.52	4.23	4.00	3.58	2.92	2.46	2.10	2	
10.2	9.54	8.94	8.39	7.88	7.41	7.02	6.62	5.96	4.93	4.16	3.60	3	
14.3	13.6	12.9	12.1	11.5	10.9	10.3	9.77	8.84	7.38	6.27	5.43	4	
18.5	17.8	17.0	16.2	15.4	14.6	13.9	13.3	12.1	10.1	8.71	7.57	5	
22.7	22.0	21.2	20.4	19.5	18.7	17.9	17.1	15.7	13.3	11.5	10.0	6	160
26.8	26.2	25.4	24.7	23.8	22.9	22.0	21.2	19.6	16.8	14.5	12.8	7	
31.0	30.4	29.7	28.9	28.1	27.2	26.4	25.4	23.7	20.6	17.9	15.8	8	
35.1	34.5	33.9	33.1	32.3	31.5	30.6	29.7	27.9	24.5	21.6	19.1	9	
39.2	38.7	38.1	37.4	36.6	35.8	34.9	34.1	32.2	28.6	25.4	22.7	10	
43.2	42.8	42.2	41.6	40.8	40.1	39.3	38.4	36.6	32.9	29.4	26.4	11	
47.3	46.9	46.3	45.7	45.1	44.3	43.5	42.7	40.9	37.2	33.6	30.3	12	

* See page 3–26 for more details regarding these tables.

ECCENTRIC LOAD ON BOLT GROUPS — SPECIAL CASE

High Strength Bolts

For connections where the eccentric load causes both shear and tension in the bolts, the following design method may be used when the fasteners are high strength bolts that have been tightened to the specified minimum initial tension.

A bracket connected by means of bolts with an initial tension T_i is shown below. Both simple and unwieldy methods are available[1] for determining tension that is applied to the upper bolts by the load on the bracket. Generally, the simpler solutions are considerably more conservative than the more accurate but unwieldy ones. The straight-forward solution (known to be extra conservative) presented in the previous edition of this Handbook is modified here to be more economical. This solution is particularly easy to use, while still conservative.

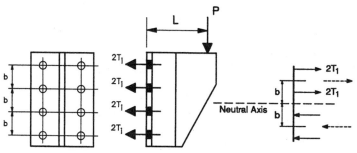

A neutral axis is assumed through the centre of gravity of the bolt group. Those bolts above the axis are said to carry the tension while those below are considered to be in "compression", so that the applied moment is resisted by a couple applied at the resultants of the upper and the lower bolts. The upper bolts are all taken to be equally loaded; this plastic stress distribution is justified by results that are still conservative compared to more precise methods.

Bolt tension from the applied moment is therefore:

$$T_1 = \frac{PL}{n' d_m}$$

where

n' = number of bolts above the neutral axis

d_m = moment arm between resultants of the tensile and compressive forces.

Bolt shear from the applied load is:

$$V = \frac{P}{n}$$

Fasteners in the top half of the connection are subjected to tension, from both the applied moment and from prying (if any), and to shear. Bolts in the bottom half are subjected to shear only, top and bottom bolts participating equally.

The connection should be proportioned so that the bolt tension T_1 due to the moment PL (plus bolt tension due to prying), when combined with the bolt shear, meets the requirements of CAN/CSA-S16.1-94 for bolts subjected to combined shear and tension. That is to say, Clause 13.11.4 for bearing-type connections or Clause 13.12.3 for slip-critical connections.

---

1. See Part Eight of the AISC LRFD Manual, second edition.

Example 1

Given:

Check the adequacy of eight M20, A325M bolts (2 rows of 4, at 80 mm pitch) for the connection shown on page 3-36 for a factored load P of 300 kN at an eccentricity L of 150 mm. Assume the material thickness is adequate so that prying action on the bolts is not significant.

Solution:

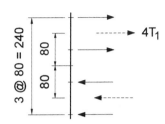

Factored tension in one bolt:

$$T_1 = \frac{P_f L}{n' d_m} = \frac{300 \times 150}{4 (2 \times 80)} = 70.3 \text{ kN}$$

$$< 131 \text{ kN} \quad - \text{OK} \quad (\text{Table 3-4})$$

Factored shear in one bolt:

$$V_f = \frac{P_f}{n} = \frac{300}{8} = 37.5 \text{ kN}$$

$$< 105 \text{ kN} \quad - \text{OK} \quad (\text{Table 3-4})$$

Check combined shear and tension for $V_f / T_f = 37.5/70.3 = 0.53$

From Table 3-8 on page 3-13, for bearing-type connections,

permissible $V_f = 57.8$ kN (by interpolation) > 37.5 — OK

and permissible $T_f = 109$ kN (by interpolation) > 70.3 — OK

Example 2

Given:

Determine the number of M20, A325M bolts required to design the connection in Example 1 as a slip-critical connection for a specified load of 200 kN. Assume clean mill scale faying surfaces.

Solution:

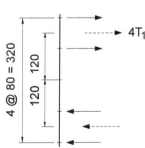

Try 10 bolts (2 rows of 5, at 80 mm pitch)

Specified tension in one bolt:

$$T_1 = \frac{PL}{n' d_m} = \frac{200 \times 150}{4 (2 \times 120)} = 31.3 \text{ kN}$$

$$T_f = 1.5 \times 31.3 = 47.0 \text{ kN} < 131 \quad - \text{OK} \quad (\text{Table 3-4})$$

Specified shear in one bolt:

$$V = \frac{P}{n} = \frac{200}{10} = 20.0 \text{ kN} < 37.4 \quad - \text{OK} \quad (\text{Table 3-11})$$

Check combined shear and tension for $V/T = 20/31.3 = 0.64$

From Table 3-12 on page 3-17,

for V/T of 0.60, permissible $V = 26.0$ kN > 20 — OK

and for V/T of 0.7, permissible $T = 38.8$ kN > 31.3 — OK

3-37

WELD DATA

General

Tables in this section are based on CAN/CSA-S16.1-94, which includes a number of significant changes from earlier editions of the Standard.

Two changes affect the resistance for shear on the base metal of all welds. The resistance factor for welds, ϕ_w, is now used rather than that for members, ϕ. This means values of 0.67, rather than 0.9. Also, ultimate base metal strength F_u is used rather than yield strength F_y.

The greater strength of fillet welds oriented transverse to the direction of force compared to those that are parallel to the force is now recognized. An amplification factor as a function of the angle between the axis of the weld and the direction of the force has been added to the previous expression for strength of the weld metal.

The strength of partial joint penetration groove welds has been capped so that the strength cannot be taken as greater than the yield strength of the member.

An expression is now provided for the tensile resistance of a partial joint penetration groove weld that has been reinforced with a fillet weld.

Tables

Table 3-21 on page 3-39 summarizes weld resistances as a function of type of load and type of weld. A sketch is included to illustrate the application of the new expression for the tensile resistance of a partial joint penetration groove weld reinforced with a fillet weld.

Table 3-22 on page 3-40 provides information on matching electrodes and gives unit factored weld resistances for various electrodes.

Table 3-23 also on page 3-40 gives factored shear resistance for a range of effective throats per millimetre of weld length, for various electrodes.

Table 3-24 on page 3-41 lists factored shear resistances of a range of fillet weld sizes per millimetre of weld length, for various electrodes.

Table 3-25 also on page 3-41 shows the increased fillet weld resistance that is now recognized in S16.1-94 as a function of the angle between the axis of the weld and the direction of the load.

Tables 3-26 to 3-33 on pages 3-44 to 3-51 present the resistance of various weld configurations when they are loaded eccentrically in the plane of the welds.

Table 3-34 on page 3-55 presents weld resistances when the eccentric load is in a plane perpendicular to the plane of the welds.

Table 3-21 FACTORED RESISTANCE OF WELDS*

Type of Load	Type of Weld	Factored Resistance
Shear (including tension or compression-induced shear in fillet welds)	Complete and partial joint penetration groove welds, and plug and slot welds	Lesser of: base metal, $V_r = 0.67\phi_w A_m F_u$ weld metal, $V_r = 0.67\phi_w A_w X_u$
	Fillet welds	Lesser of: base metal, $V_r = 0.67\phi_w A_m F_u$ weld metal, $V_r = 0.67\phi_w A_w X_u (1.00 + 0.50\sin^{1.5}\theta)$ [1]
Tension (normal to axis of load)	Complete joint penetration groove weld (made with matching electrodes)[2]	Same as the base metal
	Partial joint penetration groove weld (made with matching electrodes)[2]	$T_r = \phi_w A_n F_u \leq \phi A_g F_y$ [3]
	Partial joint penetration groove weld combined with a fillet weld (made with matching electrodes)[2]	$T_r = \phi_w \sqrt{(A_n F_u)^2 + (A_w X_u)^2} \leq \phi A_g F_y$
Compression (normal to axis of load)	Complete joint penetration groove weld (made with matching electrodes)[2]	Same as the base metal
	Partial joint penetration groove weld (made with matching electrodes)[2]	Same as the base metal, for the nominal area of the fusion face normal to the compression plus the area of the base metal fitted in contact bearing.[4]

* The detail design of welded joints is to conform to the requirements of CSA Standard W59.

A_m = shear area of effective fusion face.

A_w = area of effective weld throat, plug or slot.

A_n = nominal area of fusion face normal to the tensile force.

θ = angle of axis of weld with the line of action of force (0° for a longitudinal weld and 90° for a transverse weld).

(1) Conservatively, $(1.00 + 0.50\sin^{1.5}\theta)$ can be taken as 1.0.
(2) Summary information on matching electrodes is included in Table 3-22.
(3) When overall ductile behaviour is desired (member yielding before weld fracture) $A_n F_u > A_g F_y$.
(4) See CAN/CSA-S16.1-94, Clause 28.9.7.

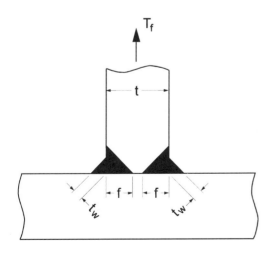

f: to be used for A_n (see CSA Standard W59 for effective fusion face)

t_w: to be used for A_w (see CSA Standard W59 for effective weld throat)

t: to be used for A_g

Application of expression $T_r = \phi_w \sqrt{(A_n F_u)^2 + (A_w X_u)^2} \leq \phi A_g F_y$

ELECTRODE CLASSIFICATION[1]
AND UNIT FACTORED WELD RESISTANCE[2]
$\phi_w = 0.67$

Table 3-22

Metric Electrode Classification Number	Imperial Electrode Classification Number	Unit Factored Resistance on Weld Metal (MPa)		Specified Minimum Tensile Strength of Base Metal F_u (MPa)	Base Metal Specification and Grade CSA-G40.21 or ASTM	F_y (MPa)	Unit Factored Resistance on Base Metal $0.67 \phi_w F_u$
		Shear on Effective Throat, A_w $0.67 \phi_w X_u$	Shear per Millimetre of Fillet Weld Size $0.67 \phi_w X_u /\sqrt{2}$				
E410XX	E60XX	184	130	410	260W, 260WT	260	184
				400	A36	250	180
E480XX	E70XX	215	152	450[3]	300W, 300WT	300	202
				450	350W	350	202
				480	350WT 350R, 350A	350	215
				480	380W, 380WT	380	215
				415	A572 Gr 42	290	186
				450	A572 Gr 50	345	202
				435	A588 F_y = 42	290	195
				460	A588 F_y = 46	315	206
				485	A588 F_y = 50	345	218
E550XX	E80XX	247	175	520	400W, 400WT 400A, 400AT	400	233
E620XX	E90XX	278	197	590	480W, 480WT 480A, 480AT	480	265
				620	550W, 550WT 550A, 550AT	550	278
E830XX	E120XX	373	263	800	700Q, 700QT	700	359

1. For complete information concerning electrode classification and strength matching of base metals, refer to CSA W59.
2. Factored weld resistance (kN) = tabulated unit resistances × (A_w or A_m)/10³.
3. F_u = 410 for 300W HSS.

FACTORED SHEAR RESISTANCE
ON EFFECTIVE THROAT PER MILLIMETRE
OF WELD LENGTH (kN)

Table 3-23

Electrode Classification	Unit Shear Resist. (MPa)	Effective Throat Thickness (mm)											
		5	6	7	8	10	12	16	20	25	30	35	40	50
E410XX	184	0.920	1.10	1.29	1.47	1.84	2.21	2.94	3.68	4.60	5.52	6.44	7.36	9.20
E480XX	215	1.08	1.29	1.51	1.72	2.15	2.59	3.45	4.31	5.39	6.46	7.54	8.62	10.8
E550XX	247	1.23	1.48	1.73	1.98	2.47	2.96	3.95	4.94	6.17	7.41	8.64	9.88	12.3
E620XX	278	1.39	1.67	1.95	2.23	2.78	3.34	4.45	5.57	6.96	8.35	9.74	11.1	13.9
E830XX	373	1.86	2.24	2.61	2.98	3.73	4.47	5.96	7.45	9.31	11.2	13.0	14.9	18.6

Table 3-24

FACTORED SHEAR RESISTANCE* OF FILLET WELDS PER MILLIMETRE OF WELD LENGTH WHEN ANGLE $\theta^+ = 0°$ (kN)

$t_w = D/\sqrt{2}$

Metric Size Fillet Welds				Fillet Weld Size, D		Imperial Size Fillet Welds			
Electrode Classification				mm	inches	Electrode Classification			
E410XX	E480XX	E550XX	E620XX			E410XX	E480XX	E550XX	E620XX
0.651	0.762	0.873	0.984	5	3/16	0.620	0.726	0.831	0.937
0.781	0.914	1.05	1.18	6	1/4	0.826	0.967	1.11	1.25
1.04	1.22	1.40	1.57	8	5/16	1.03	1.21	1.39	1.56
1.30	1.52	1.75	1.97	10	3/8	1.24	1.45	1.66	1.88
1.56	1.83	2.10	2.36	12	7/16	1.45	1.69	1.94	2.19
1.82	2.13	2.44	2.76	14	1/2	1.65	1.93	2.22	2.50
2.08	2.44	2.79	3.15	16	5/8	2.07	2.42	2.77	3.12
2.34	2.74	3.14	3.54	18	3/4	2.48	2.90	3.33	3.75
2.60	3.05	3.49	3.94	20					
184	215	247	278	Unit factored shear resistance on effective throat (MPa)					
130	152	175	197	Unit factored shear resistance per millimetre of fillet weld size (MPa)					

*Tabulated resistances for both metric and Imperial size fillet welds are based on X_u for the metric electrode classification.

+CAN/CSA-S16.1-94, Clause 13.13.2.2: $V_r = 0.67 \phi_w A_w X_u (1.0 + 0.5 \sin^{1.5} \theta)$

Table 3-25
$F_u = 450$
E480XX Electrodes

FACTORED SHEAR RESISTANCE FOR FILLET WELDS PER MILLIMETRE OF WELD LENGTH, FOR ANGLE θ^+ (kN)

Weld size (mm)	Angle θ between weld axis and direction of force						Weld size (in.)	Angle θ between weld axis and direction of force					
	0°	15°	30°	45°	50°	90°		0°	15°	30°	45°	50°	90°
5	0.762	0.812	0.896	0.988	1.01	1.01	3/16	0.726	0.773	0.854	0.941	0.962	0.962
6	0.914	0.974	1.08	1.19	1.21	1.21	1/4	0.967	1.03	1.14	1.26	1.28	1.28
8	1.22	1.30	1.43	1.58	1.62	1.62	5/16	1.21	1.29	1.42	1.57	1.60	1.60
10	1.52	1.62	1.79	1.98	2.02	2.02	3/8	1.45	1.55	1.71	1.88	1.92	1.92
12	1.83	1.95	2.15	2.37	2.42	2.42	7/16	1.69	1.80	1.99	2.20	2.24	2.24
14	2.13	2.27	2.51	2.77	2.83	2.83	1/2	1.93	2.06	2.28	2.51	2.57	2.57
16	2.44	2.60	2.87	3.16	3.23	3.23	5/8	2.42	2.58	2.85	3.14	3.21	3.21
18	2.74	2.92	3.23	3.56	3.64	3.64	3/4	2.90	3.09	3.42	3.77	3.85	3.85
20	3.05	3.25	3.59	3.95	4.04	4.04							

+S16.1-94, Clause 13.13.2.2: V_r = the lesser of $0.67 \phi_w A_m F_u$ or $0.67 \phi_w A_w X_u (1.0 + 0.5 \sin^{1.5} \theta)$

Note: For loads on specific weld patterns, use Tables 3-26 to 3-33, as appropriate.

ECCENTRIC LOADS ON WELD GROUPS

When the line of action of a load on a weld group does not pass through the centre of gravity of the group, the connection is eccentrically loaded. The traditional method of analysis of these weld groups has been elastic. Work reported by Butler et al. (1972) showed that the margins of safety for eccentrically loaded weld groups analysed elastically were both high and variable. They suggested a method of analysis based upon the load-deformation characteristics of the weld and the instantaneous centre of rotation analogy similar to that for eccentrically loaded bolt groups. For this method of analysis, the weld group is considered to be divided into a discrete number of finite weld elements. The resistance of the weld group to the external eccentric load is provided by the combined resistances of the weld elements.

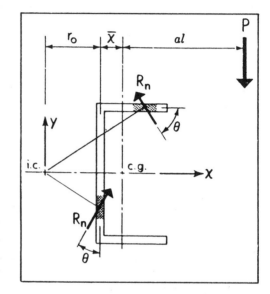

The resistance of each weld element can then be assumed to act on a line perpendicular to the radius, that radius being from the instantaneous centre of rotation to the centroid of the weld element, and can be expressed as follows:

$$R_n = R_{ULT}(1 - e^{-\mu \Delta})^\lambda \quad \text{(Butler et al.)}$$

where

R_{ULT}, μ, Δ and λ will depend on θ, the angle between the orientation of the weld and the resistance R_n, and have been determined from test specimens. (For definition of symbols see page 3-26, but applied to a weld element).

The ultimate load is obtained when the ultimate strength and deformation Δ of some weld element is reached. The resistance of the remaining weld elements is then computed from the above expression in which Δ is assumed to vary linearly with the distance from the instantaneous centre.

The correct location of the instantaneous centre is assured when the connection is in equilibrium, that is, when the three equations of statics, $\Sigma F_x = 0$, $\Sigma F_y = 0$ and $\Sigma M = 0$ are simultaneously satisfied.

Analyses by Lesik and Kennedy (1990), the basis of Clause 13.13.2.2(b) of CSA/CAN-S16.1-94, resulted in values of coefficients which are generally within 10% of those in Tables 3-26 to 3-33 which are based on the work of Butler et al. (1972).

References:

BUTLER, L.J., PAL, S., and KULAK, G.L. 1972. Eccentrically loaded welded connections. ASCE Journal of the Structural Division, **98**(ST5), May.

KULAK, G.L., and TIMLER, P.A. 1984. Tests on eccentrically loaded fillet welds. SER No 124, December, University of Alberta.

LESIK, D.F., and KENNEDY, D.J.L. 1990. Ultimate strength of fillet welded connections loaded in plane. Canadian Journal of Civil Engineering, **17**(1), February.

SWANNELL, P., and SKEWES, I.C. 1977. Design of welded brackets loaded in-plane: general theoretical ultimate load techniques and experimental programme. Australian WRA, RC #46, December, University of Queensland.

Tables

1. General

The values listed in Tables 3-26 to 3-33 inclusive were computed using the instantaneous centre of rotation method outlined above in the following manner.

For the various values of a, the eccentricity parameter, a weld length and corresponding eccentricity were selected for the appropriate weld configuration. The ultimate capacity was then computed based on the weld length and eccentricity for ¼ inch (6.35 mm) weld size and E70XX (E480XX) electrode. Butler et al. (1972) give, for the ¼ inch weld, 1 inch (25.4 mm) long, made with E60XX (E410XX) electrode, the following empirical equation for R_{ULT}, μ, λ and Δ_{max}:

$$R_{ULT} = \frac{10 + \theta}{0.92 + 0.0603\,\theta}$$

$$\mu = 75e^{0.0114\,\theta}$$

$$\lambda = 0.4e^{0.0146\,\theta}$$

$$\Delta_{max} = 0.225\,(\theta + 5)^{-0.47}$$

where θ is the angle (degrees) between the resultant force and the axis of the weld.

R_{ULT} was modified by the ratio of the electrode strengths because E410XX and E480XX have nearly the same specified ultimate elongations.

The ultimate capacities were reduced for a base weld leg size of 1 mm by division and then multiplied by the product of $0.67\,\phi$ ($\phi = 0.67$). The tabulated coefficient C is this reduced capacity divided by the weld length l in millimetres.

Values of the coefficient C above the horizontal line in each table are conservatively based on the resistance of the weld metal given in Clause 13.13 of CSA/CAN-S16.1-94 for the weld group concentrically loaded.

2. Use of Tables

The coefficients listed in Tables 3-26 to 3-33 inclusive are based on the use of E480XX electrodes. For E410XX electrodes, use 0.85 of the tabulated values.

(a) To determine the capacity P of the eccentrically loaded weld group in kilonewtons, multiply the appropriate coefficient C by the number of millimetres of weld size D and the length of the weld l, in millimetres.

(b) To determine the required number of millimetres of weld size D, divide the factored load P, in kilonewtons, by the appropriate coefficient C and the length of the weld l, in millimetres.

Values of C above the solid line are governed by the factored resistance of the weld group considered to be loaded concentrically.

3. Other Weld Configurations

For situations not covered by the tables of Eccentric Loads on Weld Groups, the method of analysis in which the vector sum of the factored longitudinal and transverse shear loads does not exceed the factored resistances of the weld is recommended as being convenient to use as it can readily be computed. Alternatively, interpolating between weld configurations in the tables which "bracket" the situation being evaluated will often be sufficient to confirm adequacy.

Example

For an example on the use of these tables, turn to page 3-52.

ECCENTRIC LOADS ON WELD GROUPS

Coefficients C*

Table 3-26

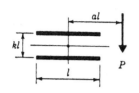

P = Factored eccentric load in kilonewtons.
l = Length of each weld in millimetres.
D = Number of millimetres in fillet weld size.
C = Coefficients tabulated below.

$$P = CDl$$

Required Minimum $C = \dfrac{P}{Dl}$

Required Minimum $D = \dfrac{P}{Cl}$

Required Minimum $l = \dfrac{P}{CD}$

a	k														
	0.0	0.1	0.2	0.3	0.4	0.5	0.6	0.7	0.8	0.9	1.0	1.2	1.4	1.6	1.8	2.0
0.2	0.278	0.282	0.294	0.305	0.305	0.305	0.305	0.305	0.305	0.305	0.305	0.305	0.305	0.305	0.305	0.305
0.3	0.230	0.234	0.245	0.260	0.277	0.294	0.305	0.305	0.305	0.305	0.305	0.305	0.305	0.305	0.305	0.305
0.4	0.194	0.198	0.208	0.222	0.238	0.255	0.272	0.291	0.305	0.305	0.305	0.305	0.305	0.305	0.305	0.305
0.5	0.168	0.171	0.180	0.193	0.207	0.224	0.239	0.256	0.273	0.290	0.305	0.305	0.305	0.305	0.305	0.305
0.6	0.147	0.150	0.158	0.170	0.183	0.198	0.214	0.228	0.244	0.260	0.276	0.305	0.305	0.305	0.305	0.305
0.7	0.130	0.133	0.140	0.151	0.164	0.177	0.192	0.206	0.220	0.235	0.250	0.280	0.304	0.305	0.305	0.305
0.8	0.117	0.119	0.126	0.136	0.148	0.161	0.174	0.187	0.201	0.214	0.228	0.257	0.282	0.304	0.305	0.305
0.9	0.106	0.108	0.114	0.123	0.134	0.147	0.159	0.171	0.184	0.197	0.209	0.236	0.263	0.285	0.304	0.305
1.0	0.096	0.098	0.105	0.113	0.123	0.134	0.146	0.158	0.169	0.182	0.193	0.217	0.245	0.267	0.287	0.305
1.2	0.082	0.084	0.089	0.097	0.105	0.115	0.125	0.136	0.146	0.156	0.168	0.189	0.213	0.235	0.255	0.274
1.4	0.071	0.073	0.077	0.084	0.092	0.100	0.109	0.119	0.129	0.138	0.147	0.167	0.187	0.208	0.229	0.247
1.6	0.063	0.064	0.068	0.074	0.081	0.089	0.097	0.105	0.114	0.123	0.132	0.149	0.167	0.185	0.206	0.224
1.8	0.056	0.058	0.061	0.067	0.073	0.080	0.087	0.095	0.102	0.111	0.119	0.134	0.151	0.167	0.185	0.203
2.0	0.051	0.052	0.055	0.060	0.066	0.072	0.079	0.086	0.093	0.100	0.108	0.123	0.138	0.152	0.168	0.185
2.2	0.047	0.048	0.051	0.055	0.060	0.066	0.072	0.079	0.085	0.092	0.099	0.113	0.126	0.140	0.154	0.170
2.4	0.043	0.044	0.046	0.051	0.055	0.061	0.067	0.073	0.079	0.085	0.091	0.104	0.116	0.129	0.142	0.156
2.6	0.040	0.040	0.043	0.047	0.051	0.056	0.062	0.067	0.073	0.079	0.085	0.097	0.108	0.120	0.132	0.145
2.8	0.037	0.038	0.040	0.043	0.048	0.052	0.058	0.063	0.068	0.074	0.079	0.090	0.101	0.112	0.124	0.135
3.0	0.034	0.035	0.037	0.041	0.045	0.049	0.054	0.059	0.064	0.069	0.074	0.084	0.095	0.105	0.116	0.126

*Coefficients in table are for E480XX electrodes.
The effect of eccentricity has been neglected for cases above the solid horizontal line.

ECCENTRIC LOADS ON WELD GROUPS

Table 3-27   Coefficients C*

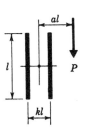

P = Factored eccentric load in kilonewtons.
l = Length of each weld in millimetres.
D = Number of millimetres in fillet weld size.
C = Coefficients tabulated below.

$$P = CDl$$

Required Minimum $C = \dfrac{P}{Dl}$

Required Minimum $D = \dfrac{P}{Cl}$

Required Minimum $l = \dfrac{P}{CD}$

a	k														
	0.0	0.1	0.2	0.3	0.4	0.5	0.6	0.7	0.8	0.9	1.0	1.2	1.4	1.6	1.8	2.0
0.2	.305	.305	.305	.305	.305	.305	.305	.305	.305	.305	.305	.305	.305	.305	.305	.305
0.3	.276	.276	.278	.281	.284	.287	.289	.291	.293	.295	.296	.298	.300	.301	.302	.303
0.4	.229	.230	.234	.240	.245	.250	.255	.259	.263	.266	.268	.273	.277	.280	.282	.284
0.5	.192	.194	.200	.207	.214	.220	.226	.232	.237	.241	.245	.251	.256	.261	.264	.267
0.6	.165	.167	.173	.181	.188	.196	.203	.209	.214	.219	.224	.232	.238	.243	.248	.252
0.7	.144	.146	.151	.160	.168	.176	.183	.190	.196	.201	.206	.215	.222	.228	.233	.238
0.8	.127	.129	.135	.143	.151	.159	.167	.174	.180	.186	.191	.200	.208	.214	.220	.225
0.9	.114	.116	.121	.129	.137	.145	.153	.160	.166	.172	.177	.187	.195	.202	.209	.214
1.0	.103	.105	.110	.118	.126	.133	.141	.148	.154	.160	.166	.176	.184	.191	.198	.204
1.2	.086	.088	.093	.100	.107	.115	.122	.128	.135	.141	.146	.156	.165	.173	.179	.186
1.4	.074	.076	.080	.086	.093	.100	.107	.113	.119	.125	.131	.140	.149	.157	.164	.170
1.6	.065	.067	.071	.076	.083	.089	.095	.101	.107	.113	.118	.128	.136	.144	.151	.157
1.8	.058	.059	.063	.068	.074	.080	.086	.092	.097	.102	.107	.117	.125	.133	.140	.146
2.0	.052	.053	.057	.062	.067	.073	.078	.084	.089	.094	.099	.108	.116	.123	.130	.136
2.2	.048	.049	.052	.056	.061	.066	.072	.077	.082	.087	.091	.100	.108	.115	.122	.128
2.4	.044	.045	.048	.052	.056	.061	.066	.071	.076	.080	.085	.093	.101	.108	.114	.120
2.6	.040	.041	.044	.048	.052	.057	.061	.066	.071	.075	.079	.087	.094	.101	.108	.113
2.8	.037	.038	.041	.044	.048	.053	.057	.062	.066	.070	.074	.082	.089	.095	.102	.107
3.0	.035	.036	.038	.041	.045	.049	.054	.058	.062	.066	.070	.077	.084	.090	.096	.102

*Coefficients in table are for E480XX electrodes.
The effect of eccentricity has been neglected for cases above the solid horizontal line.

ECCENTRIC LOADS ON WELD GROUPS

Coefficients C*

Table 3-28

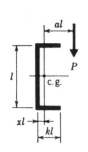

P = Factored eccentric load in kilonewtons.
l = Length of weld parallel to load P in millimetres.
D = Number of millimetres in fillet weld size.
C = Coefficients tabulated below.
xl = Distance from vertical weld to center of gravity of weld group.

$P = CDl$

Required Minimum $C = \dfrac{P}{Dl}$

Required Minimum $D = \dfrac{P}{Cl}$

Required Minimum $l = \dfrac{P}{CD}$

a	k														
	0.0	0.1	0.2	0.3	0.4	0.5	0.6	0.7	0.8	0.9	1.0	1.2	1.4	1.6	1.8	2.0
0.2	.153	.183	.214	.244	.275	.305	.336	.366	.397	.428	.458	.519	.580	.641	.702	.763
0.3	.138	.180	.214	.244	.275	.305	.336	.366	.397	.428	.458	.519	.580	.641	.702	.763
0.4	.114	.152	.188	.221	.254	.286	.320	.353	.387	.422	.455	.519	.580	.641	.702	.763
0.5	.096	.129	.161	.192	.222	.252	.282	.313	.345	.376	.409	.474	.544	.615	.687	.761
0.6	.082	.111	.140	.169	.196	.223	.251	.280	.308	.338	.368	.431	.496	.563	.632	.704
0.7	.072	.097	.122	.149	.174	.199	.225	.251	.278	.306	.334	.393	.455	.519	.586	.654
0.8	.063	.085	.109	.133	.156	.179	.203	.227	.252	.278	.305	.360	.419	.480	.545	.610
0.9	.057	.077	.098	.119	.141	.163	.184	.207	.231	.255	.280	.333	.388	.447	.508	.572
1.0	.051	.069	.088	.108	.129	.149	.169	.190	.212	.235	.259	.308	.360	.417	.476	.537
1.2	.043	.058	.074	.091	.109	.126	.144	.163	.182	.203	.223	.268	.317	.368	.422	.479
1.4	.037	.050	.064	.079	.094	.110	.126	.142	.159	.177	.196	.238	.282	.329	.378	.430
1.6	.033	.044	.056	.069	.083	.097	.111	.126	.141	.158	.176	.213	.253	.296	.342	.391
1.8	.029	.039	.050	.062	.074	.087	.100	.113	.127	.142	.159	.192	.230	.269	.312	.356
2.0	.026	.035	.045	.055	.067	.078	.090	.102	.115	.130	.144	.176	.210	.247	.286	.328
2.2	.024	.032	.041	.050	.061	.072	.082	.093	.106	.119	.132	.162	.193	.228	.264	.303
2.4	.022	.029	.037	.046	.056	.066	.076	.086	.098	.110	.122	.150	.179	.211	.246	.282
2.6	.020	.027	.035	.043	.052	.061	.070	.080	.091	.102	.113	.139	.167	.197	.229	.263
2.8	.019	.025	.032	.040	.048	.057	.065	.074	.085	.095	.106	.130	.156	.184	.215	.247
3.0	.017	.023	.030	.037	.045	.053	.061	.070	.079	.089	.099	.122	.147	.173	.202	.232
x	0	.008	.029	.056	.089	.125	.164	.204	.246	.289	.333	.424	.516	.610	.704	.800

*Coefficients in table are for E480XX electrodes.
The effect of eccentricity has been neglected for cases above the solid line.

ECCENTRIC LOADS ON WELD GROUPS

Table 3-29 — Coefficients C*

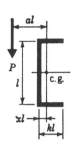

P = Factored eccentric load in kilonewtons.
l = Length of weld parallel to load P in millimetres.
D = Number of millimetres in fillet weld size.
C = Coefficients tabulated below.
xl = Distance from vertical weld to center of gravity of weld group.

$P = CDl$

Required Minimum $C = \dfrac{P}{Dl}$

Required Minimum $D = \dfrac{P}{Cl}$

Required Minimum $l = \dfrac{P}{CD}$

a	\multicolumn{16}{c}{k}
	0.0	0.1	0.2	0.3	0.4	0.5	0.6	0.7	0.8	0.9	1.0	1.2	1.4	1.6	1.8	2.0
0.2	.153	.183	.214	.244	.275	.305	.336	.366	.397	.428	.458	.519	.580	.641	.702	.763
0.3	.138	.180	.214	.244	.275	.305	.336	.366	.397	.428	.458	.519	.580	.641	.702	.763
0.4	.114	.152	.189	.226	.263	.297	.331	.365	.397	.428	.458	.519	.580	.641	.702	.763
0.5	.096	.128	.161	.195	.229	.264	.296	.328	.360	.392	.424	.488	.555	.622	.691	.761
0.6	.082	.110	.138	.166	.199	.231	.264	.295	.325	.356	.387	.449	.513	.578	.644	.711
0.7	.072	.095	.118	.147	.177	.207	.237	.266	.296	.326	.355	.415	.475	.537	.601	.666
0.8	.063	.084	.105	.131	.157	.186	.215	.244	.272	.299	.327	.384	.443	.502	.563	.626
0.9	.057	.075	.095	.117	.143	.169	.195	.222	.249	.276	.303	.357	.412	.470	.529	.589
1.0	.051	.067	.086	.106	.130	.154	.179	.205	.231	.256	.281	.333	.387	.441	.497	.556
1.2	.043	.056	.072	.090	.110	.131	.153	.176	.199	.223	.246	.293	.341	.392	.444	.498
1.4	.037	.048	.062	.078	.094	.114	.133	.153	.175	.196	.217	.260	.305	.351	.399	.449
1.6	.033	.042	.055	.069	.083	.100	.117	.136	.155	.174	.194	.234	.274	.318	.362	.409
1.8	.029	.038	.048	.061	.074	.089	.105	.122	.138	.157	.175	.211	.250	.289	.331	.374
2.0	.026	.034	.044	.055	.067	.080	.095	.110	.125	.142	.159	.193	.229	.266	.304	.345
2.2	.024	.031	.040	.050	.061	.073	.086	.100	.114	.129	.144	.178	.210	.245	.282	.320
2.4	.022	.029	.036	.046	.056	.067	.079	.091	.105	.119	.133	.163	.195	.228	.262	.297
2.6	.020	.026	.034	.042	.052	.062	.073	.085	.097	.109	.123	.151	.182	.212	.245	.278
2.8	.019	.025	.031	.039	.048	.058	.068	.079	.090	.101	.114	.140	.169	.199	.229	.261
3.0	.017	.023	.029	.037	.045	.054	.064	.073	.084	.095	.107	.131	.158	.187	.215	.246
x	0	.008	.029	.056	.089	.125	.164	.204	.246	.289	.333	.424	.516	.610	.704	.800

*Coefficients in table are for E480XX electrodes.
The effect of eccentricity has been neglected for cases above the solid horizontal line.

ECCENTRIC LOADS ON WELD GROUPS

Coefficients C*

Table 3-30

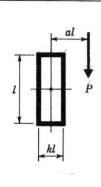

P = Factored eccentric load in kilonewtons.
l = Length of longer welds in millimetres.
D = Number of millimetres in fillet weld size.
C = Coefficients tabulated below.
Note: When load P is perpendicular to longer side l use table on facing page.

$P = CDl$

Required Minimum $C = \dfrac{P}{Dl}$

Required Minimum $D = \dfrac{P}{Cl}$

Required Minimum $l = \dfrac{P}{CD}$

a	k										
	0.0	0.1	0.2	0.3	0.4	0.5	0.6	0.7	0.8	0.9	1.0
0.2	.305	.336	.366	.397	.428	.458	.489	.519	.550	.580	.611
0.3	.276	.318	.360	.397	.428	.458	.489	.519	.550	.580	.611
0.4	.229	.267	.307	.346	.383	.417	.451	.486	.520	.555	.590
0.5	.192	.226	.263	.300	.337	.370	.403	.436	.469	.503	.536
0.6	.165	.194	.228	.262	.297	.330	.362	.394	.425	.457	.490
0.7	.144	.170	.200	.232	.265	.297	.327	.357	.388	.418	.449
0.8	.127	.150	.178	.208	.238	.269	.297	.326	.355	.384	.414
0.9	.114	.135	.159	.187	.216	.245	.272	.299	.327	.355	.383
1.0	.103	.122	.145	.170	.197	.224	.250	.276	.302	.329	.357
1.2	.086	.102	.122	.144	.167	.191	.216	.239	.264	.288	.313
1.4	.074	.088	.105	.124	.145	.167	.190	.211	.233	.255	.278
1.6	.065	.077	.092	.109	.128	.148	.169	.188	.208	.229	.250
1.8	.058	.069	.082	.098	.115	.133	.152	.170	.188	.208	.227
2.0	.052	.062	.074	.088	.104	.120	.137	.155	.172	.190	.207
2.2	.048	.056	.067	.081	.095	.110	.125	.142	.158	.174	.191
2.4	.044	.051	.062	.074	.087	.101	.116	.131	.146	.161	.177
2.6	.040	.047	.057	.069	.081	.093	.107	.122	.136	.150	.165
2.8	.037	.044	.053	.064	.075	.087	.100	.114	.127	.140	.154
3.0	.035	.041	.049	.060	.070	.081	.094	.106	.119	.132	.145

*Coefficients in table are for E480XX electrodes.
The effect of eccentricity has been neglected for cases above the solid horizontal line.

ECCENTRIC LOADS ON WELD GROUPS

Table 3-31 — Coefficients C*

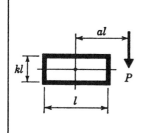

P = Factored eccentric load in kilonewtons.
l = Length of longer welds in millimetres.
D = Number of millimetres in fillet weld size.
C = Coefficients tabulated below.
Note: When load P is parallel to longer side l use table on facing page.

$P = CDl$

Required Minimum $C = \dfrac{P}{Dl}$

Required Minimum $D = \dfrac{P}{Cl}$

Required Minimum $l = \dfrac{P}{CD}$

a	k=0.0	0.1	0.2	0.3	0.4	0.5	0.6	0.7	0.8	0.9	1.0
0.2	.278	.309	.349	.388	.428	.458	.489	.519	.550	.580	.611
0.3	.230	.258	.297	.335	.375	.417	.463	.508	.550	.580	.611
0.4	.194	.220	.256	.291	.330	.367	.410	.453	.499	.543	.590
0.5	.168	.191	.223	.256	.293	.329	.366	.406	.449	.492	.536
0.6	.147	.168	.198	.228	.263	.296	.332	.367	.407	.448	.490
0.7	.130	.149	.177	.205	.237	.268	.302	.336	.371	.409	.449
0.8	.117	.134	.160	.186	.216	.245	.276	.308	.343	.376	.414
0.9	.106	.122	.145	.170	.197	.225	.255	.284	.317	.349	.383
1.0	.096	.111	.133	.156	.182	.207	.236	.264	.294	.324	.357
1.2	.082	.095	.114	.133	.157	.180	.205	.230	.257	.283	.313
1.4	.071	.083	.099	.117	.138	.158	.181	.203	.228	.252	.278
1.6	.063	.073	.088	.104	.123	.141	.162	.182	.204	.226	.250
1.8	.056	.065	.079	.093	.110	.127	.146	.165	.185	.205	.227
2.0	.051	.059	.072	.085	.100	.116	.133	.150	.169	.187	.207
2.2	.047	.054	.065	.077	.092	.106	.122	.138	.155	.173	.191
2.4	.043	.050	.060	.071	.085	.098	.113	.127	.144	.160	.177
2.6	.040	.046	.056	.066	.079	.091	.105	.118	.134	.149	.165
2.8	.037	.043	.052	.062	.073	.085	.098	.111	.125	.139	.154
3.0	.034	.040	.049	.058	.069	.079	.092	.104	.117	.130	.145

*Coefficients in table are for E480XX electrodes.
The effect of eccentricity has been neglected for cases above the solid horizontal line.

ECCENTRIC LOADS ON WELD GROUPS

Coefficients C*

Table 3-32

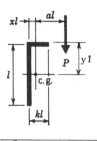

P = Factored eccentric load in kilonewtons.
l = Length of weld parallel to load P in millimetres.
D = Number of millimetres in fillet weld size.
C = Coefficients tabulated below.
xl = Distance from vertical weld to center of gravity of weld group.

$P = CDl$

Required Minimum $C = \dfrac{P}{Dl}$

Required Minimum $D = \dfrac{P}{Cl}$

Required Minimum $l = \dfrac{P}{CD}$

a	k														
	0.0	0.1	0.2	0.3	0.4	0.5	0.6	0.7	0.8	0.9	1.0	1.2	1.4	1.6	1.8	2.0
0.2	.153	.168	.183	.199	.214	.229	.244	.260	.275	.290	.305	.336	.366	.397	.428	.458
0.3	.138	.156	.175	.192	.208	.223	.238	.254	.270	.287	.304	.336	.366	.397	.428	.458
0.4	.114	.130	.147	.162	.177	.192	.206	.220	.235	.250	.266	.299	.335	.373	.411	.451
0.5	.096	.110	.124	.137	.151	.166	.179	.192	.205	.220	.234	.266	.300	.336	.373	.412
0.6	.082	.094	.107	.118	.130	.143	.157	.169	.181	.195	.209	.239	.271	.305	.341	.378
0.7	.072	.082	.093	.103	.114	.126	.138	.150	.162	.174	.188	.216	.246	.279	.313	.349
0.8	.063	.073	.082	.092	.101	.112	.123	.135	.146	.158	.170	.197	.226	.257	.290	.324
0.9	.057	.065	.074	.082	.091	.101	.111	.123	.133	.144	.155	.180	.208	.238	.269	.302
1.0	.051	.059	.067	.074	.082	.091	.101	.112	.122	.132	.143	.166	.193	.221	.251	.283
1.2	.043	.050	.056	.062	.069	.077	.085	.095	.104	.113	.122	.144	.168	.194	.221	.250
1.4	.037	.043	.048	.054	.060	.067	.074	.082	.090	.098	.107	.127	.148	.172	.197	.224
1.6	.033	.037	.042	.047	.053	.058	.065	.072	.080	.087	.095	.113	.133	.154	.177	.202
1.8	.029	.033	.038	.042	.047	.052	.058	.065	.072	.079	.086	.102	.120	.140	.161	.184
2.0	.026	.030	.034	.038	.042	.047	.052	.058	.065	.071	.078	.093	.109	.127	.147	.168
2.2	.024	.027	.031	.035	.038	.043	.048	.053	.059	.065	.071	.085	.100	.117	.136	.155
2.4	.022	.025	.028	.032	.035	.039	.044	.049	.055	.060	.066	.079	.093	.108	.126	.144
2.6	.020	.023	.026	.029	.033	.036	.041	.045	.051	.056	.061	.073	.086	.101	.117	.134
2.8	.019	.021	.024	.027	.030	.034	.038	.042	.047	.052	.057	.068	.081	.094	.109	.126
3.0	.017	.020	.023	.025	.028	.032	.035	.039	.044	.049	.053	.064	.075	.088	.103	.118
x	0	.005	.017	.035	.057	.083	.112	.144	.178	.213	.250	.327	.408	.492	.579	.667
y	.500	.455	.417	.385	.357	.333	.312	.294	.278	.263	.250	.227	.208	.192	.179	.167

*Coefficients in table are for E480XX electrodes.
The effect of eccentricity has been neglected for cases above the solid horizontal line.

ECCENTRIC LOADS ON WELD GROUPS

Table 3-33 — Coefficients C*

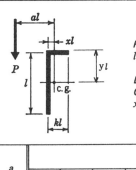

P = Factored eccentric load in kilonewtons.
l = Length of weld parallel to load P in millimetres.
D = Number of millimetres in fillet weld size.
C = Coefficients tabulated below.
xl = Distance from vertical weld to center of gravity of weld group.

$P = CDl$

Required Minimum $C = \dfrac{P}{Dl}$

Required Minimum $D = \dfrac{P}{Cl}$

Required Minimum $l = \dfrac{P}{CD}$

a	\\ k	0.0	0.1	0.2	0.3	0.4	0.5	0.6	0.7	0.8	0.9	1.0	1.2	1.4	1.6	1.8	2.0
0.2		.153	.168	.183	.199	.214	.229	.244	.260	.275	.290	.305	.336	.366	.397	.428	.458
0.3		.138	.156	.174	.189	.205	.219	.234	.250	.265	.282	.300	.336	.366	.397	.428	.458
0.4		.114	.130	.146	.159	.173	.188	.202	.217	.233	.249	.266	.303	.338	.372	.407	.443
0.5		.096	.110	.122	.135	.149	.162	.176	.191	.206	.221	.237	.273	.310	.343	.377	.413
0.6		.082	.094	.105	.117	.129	.142	.154	.169	.183	.198	.213	.248	.284	.318	.351	.386
0.7		.072	.082	.092	.102	.114	.125	.138	.151	.164	.179	.193	.226	.260	.295	.327	.360
0.8		.063	.072	.081	.091	.101	.112	.124	.136	.149	.162	.176	.207	.240	.275	.305	.338
0.9		.057	.064	.073	.082	.091	.102	.112	.123	.135	.148	.161	.191	.222	.255	.286	.318
1.0		.051	.058	.066	.074	.082	.092	.102	.113	.124	.136	.149	.176	.207	.238	.269	.300
1.2		.043	.049	.055	.062	.070	.078	.087	.096	.106	.117	.128	.153	.180	.209	.239	.268
1.4		.037	.042	.048	.054	.060	.067	.075	.083	.092	.102	.112	.134	.159	.186	.215	.242
1.6		.033	.037	.042	.047	.053	.059	.066	.073	.082	.090	.099	.120	.142	.167	.193	.220
1.8		.029	.033	.037	.042	.047	.053	.059	.066	.073	.081	.089	.108	.129	.151	.176	.201
2.0		.026	.030	.034	.038	.042	.048	.053	.059	.066	.073	.081	.098	.117	.138	.161	.185
2.2		.024	.027	.031	.034	.039	.043	.048	.054	.060	.067	.074	.089	.107	.127	.148	.170
2.4		.022	.025	.028	.031	.035	.040	.044	.050	.055	.062	.068	.082	.099	.117	.137	.157
2.6		.020	.023	.026	.029	.033	.037	.041	.046	.051	.057	.063	.076	.092	.109	.127	.146
2.8		.019	.021	.024	.027	.030	.034	.038	.043	.048	.053	.059	.071	.086	.101	.118	.136
3.0		.017	.020	.022	.025	.028	.032	.036	.040	.044	.049	.055	.067	.080	.095	.110	.127
x		0	.005	.017	.035	.057	.083	.112	.144	.178	.213	.250	.327	.408	.492	.579	.667
y		.500	.455	.417	.385	.357	.333	.312	.294	.278	.263	.250	.227	.208	.192	.179	.167

*Coefficients in table are for E480XX electrodes.
The effect of eccentricity has been neglected for cases above the solid horizontal line.

3-51

Example

Given:

A column bracket of G40.21 300W steel supports a factored load of 500 kN. The width of the bracket is 300 mm. Welds are made using E480XX electrodes. For the weld configuration shown, find the required weld size.

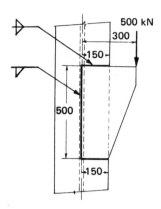

Solution:

Referring to Table 3-28, page 3-46,

$$D = \frac{P}{Cl}$$

= number of millimetres of fillet weld leg size

$k = 150/500 = 0.3$

From the bottom line of Table 3-28, for $k = 0.3$ $x = 0.056$

Referring to the figure in Table 3-28, $al + xl = 300$

For $l = 500$, $500\,a + 0.056(500) = 300$, $a = 0.544$

For $a = 0.544$ and $k = 0.3$, $C = 0.18$ by interpolation

Therefore, $D = \dfrac{500}{0.18 \times 500} = 5.5$ say 6 mm

Use 6 mm fillet welds made with E480XX electrodes.

Notes:

1. The final choice of the fillet weld size to be used in an actual connection will also be dependent upon the minimum and maximum sizes required by a) the physical thickness of the parts joined and b) the requirements of Standard CSA W59.

2. The strength of an actual connection will also be dependent upon the resistances of the connected parts.

ECCENTRIC LOADS ON WELD GROUPS
SHEAR AND MOMENT

For the case of the eccentrically loaded fillet weld group shown below in Figure 3-3(a), an analysis similar to that for the in-plane eccentricity is used but modified to account for the bearing of the plate at ultimate load. The magnitude and location of the horizontal force H_B due to plate bearing, is derived assuming a triangular stress block at ultimate load (see Figure 3-3(b)). According to Dawe and Kulak (1974), the triangular stress block assumption provides a better prediction of the ultimate load than either a rectangular or parabolic distribution. The vertical force V_B below the neutral axis is taken to be the shear resistance of that portion of the weld below the neutral axis. This method of analysis accounts for the load-deformation characteristics of the weld.

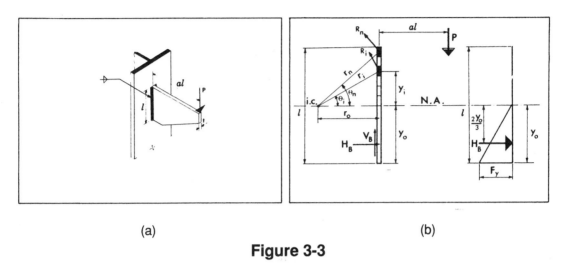

(a)　　　　　　　　　　　　　　　　(b)

Figure 3-3

Table

Table 3-34 has been computed based upon the use of E480XX electrodes and $F_y = 300$ MPa following the method outlined by Dawe and Kulak. The coefficient $C'\ (P/l)$ is tabulated for various plate thicknesses t and fillet weld sizes. For joints using lower strength materials the value of C' may be multiplied by $F_y/300$.

The tabulated values are conservative when used with plate material in excess of 300 MPa yield. The maximum value of C' is determined by shear in the plate and is accounted for in the tabulated values.

Reference:

Dawe, J.L., and Kulak, G.L. 1974. Welded connections under combined shear and moment. ASCE Journal of the Structural Division, **100**(ST4), April.

Example

Given:

A 12 mm plate carrying a 180 kN factored load is welded to a column with a pair of fillet welds 250 mm long. Find the fillet weld size required if the 180 kN load acts at a 130 mm eccentricity.

Solution:

$l = 250$ mm

$al = 130$ mm; therefore $a = 130/250 = 0.52$

C' required is $P/l = 180/250 = 0.72$

Try 6 mm fillet weld

From Table 3-34, for $t = 12$ mm, and 6 mm weld size:

$C' = 0.754$ for $a = 0.50$, and 0.629 for $a = 0.60$

Therefore, for $a = 0.52$, $C' = 0.73$ (by interpolation) > 0.72 — OK

The minimum weld size based upon the thickness of the materials joined and the resistance of the connected parts must also be checked.

ECCENTRIC LOADS ON WELD GROUPS
Coefficients C′*
Table 3-34

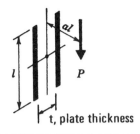

P = Factored eccentric load (kN)
l = Length of fillet welds parallel to load P (mm)
C′ = Coefficients tabulated below

$P = C' \, l$
Required minimum $C' = P/l$
Required minimum $l = P/C'$

Plate Thickness		8 mm	10 mm		12 mm			16 mm		
Weld Size		5	5	6	5	6	8	6	8	10
	0.2	1.30	1.47	1.60	1.55	1.76	2.00	1.84	2.34	2.60
	0.3	0.889	1.04	1.10	1.16	1.24	1.36	1.47	1.66	1.78
	0.4	0.671	0.784	0.828	0.885	0.941	1.02	1.13	1.25	1.34
	0.5	0.537	0.630	0.665	0.710	0.754	0.818	0.911	1.01	1.07
	0.6	0.448	0.526	0.555	0.594	0.629	0.683	0.762	0.839	0.895
	0.7	0.385	0.450	0.475	0.509	0.540	0.586	0.653	0.720	0.767
	0.8	0.336	0.394	0.416	0.446	0.473	0.512	0.573	0.631	0.673
	0.9	0.298	0.350	0.369	0.396	0.420	0.455	0.509	0.560	0.598
	1.0	0.269	0.316	0.332	0.357	0.379	0.410	0.458	0.506	0.537
a	1.2	0.223	0.263	0.276	0.297	0.316	0.342	0.382	0.421	0.448
	1.4	0.191	0.225	0.238	0.255	0.270	0.292	0.327	0.360	0.384
	1.6	0.168	0.197	0.207	0.224	0.236	0.256	0.286	0.315	0.336
	1.8	0.150	0.175	0.184	0.198	0.210	0.228	0.255	0.280	0.299
	2.0	0.134	0.157	0.166	0.178	0.188	0.205	0.229	0.251	0.269
	2.2	0.122	0.143	0.151	0.162	0.172	0.186	0.208	0.229	0.244
	2.4	0.111	0.132	0.137	0.149	0.158	0.170	0.191	0.210	0.224
	2.6	0.103	0.121	0.127	0.136	0.145	0.157	0.176	0.194	0.206
	2.8	0.096	0.113	0.118	0.127	0.135	0.146	0.162	0.180	0.192
	3.0	0.089	0.105	0.111	0.119	0.126	0.137	0.152	0.168	0.179

Plate Thickness		20 mm			25 mm				40 mm		
Weld Size		8	10	12	8	10	12	14	10	12	14	16
	0.2	2.44	2.93	3.21	2.44	3.06	3.56	3.89	3.06	3.67	4.28	4.90
	0.3	1.89	2.07	2.20	2.10	2.37	2.55	2.69	2.85	3.23	3.54	3.79
	0.4	1.45	1.57	1.66	1.65	1.81	1.93	2.03	2.31	2.55	2.74	2.90
	0.5	1.17	1.26	1.33	1.33	1.46	1.55	1.63	1.89	2.07	2.21	2.33
	0.6	0.974	1.05	1.11	1.12	1.22	1.29	1.36	1.59	1.74	1.85	1.95
	0.7	0.836	0.900	0.949	0.959	1.04	1.11	1.16	1.37	1.49	1.59	1.67
	0.8	0.732	0.789	0.833	0.840	0.916	0.973	1.02	1.20	1.31	1.39	1.46
	0.9	0.651	0.701	0.739	0.747	0.813	0.865	0.906	1.07	1.16	1.24	1.30
	1.0	0.586	0.632	0.664	0.673	0.733	0.779	0.815	0.965	1.05	1.11	1.17
a	1.2	0.488	0.526	0.554	0.562	0.610	0.649	0.679	0.806	0.873	0.930	0.977
	1.4	0.418	0.451	0.475	0.481	0.523	0.557	0.582	0.690	0.749	0.797	0.836
	1.6	0.367	0.394	0.415	0.421	0.458	0.487	0.510	0.604	0.655	0.696	0.733
	1.8	0.326	0.351	0.369	0.373	0.407	0.432	0.452	0.537	0.581	0.620	0.652
	2.0	0.293	0.316	0.333	0.336	0.366	0.390	0.408	0.484	0.528	0.558	0.586
	2.2	0.266	0.287	0.302	0.306	0.333	0.353	0.370	0.440	0.476	0.507	0.533
	2.4	0.244	0.263	0.277	0.281	0.305	0.324	0.339	0.403	0.437	0.465	0.489
	2.6	0.226	0.242	0.256	0.259	0.282	0.300	0.313	0.371	0.404	0.429	0.451
	2.8	0.209	0.225	0.237	0.240	0.261	0.278	0.291	0.345	0.374	0.399	0.419
	3.0	0.195	0.210	0.221	0.224	0.244	0.260	0.272	0.322	0.349	0.372	0.390

*Coefficients in table are for E480XX electrodes and plate with F_y = 300 MPa.

FRAMED BEAM SHEAR CONNECTIONS

General

This section of Part Three contains information on five common types of beam shear connections traditionally considered standard in the industry. Double angle, simple end plate, single angle, shear tab, and tee connections are included.

Connections of these types are generally designed for strength requirements under factored loads. The capacities of welds and of bolts in bearing-type connections are based on their factored resistances.

The CISC booklet *Standardized Shear Connections* includes tables and charts for shear connections. The tables list suitable connections, their capacities, and governing parameters for individual beams. Charts identify the range of connections that will carry the UDL capacity of various lengths of individual beams of 300W steel.

Tabulated bolt capacities for bearing-type connections are based on threads being excluded from the shear planes. When threads intercept a shear plane, these capacities must be reduced to 70% of the tabulated values (CAN/CSA-S16.1-94, Clause 13.11.2). Starting with the 1989 edition, S16.1 no longer implies that threads are excluded from the shear plane when the material thickness adjacent to the nut is equal to 10 mm. Without special precautions, such a thickness may allow threads to be intercepted. For practical reasons, it is suggested that bearing-type shear connections be designed on the assumption of intercepted threads when combinations of thin material and detailing for minimum bolt stick-through (the nuts) are expected.

Slip-critical bolt capacities are included for double angle and end plate connections for use with connections such as those subjected to fatigue or frequent load reversal. Values are based on Class A (clean mill scale or blast cleaned with Class A coatings) contact surfaces (k_s = 0.33). The capacities are to be used with *specified loads only*.

Tables of bolt and weld capacities are based on M20 and M22 A325M bolts and 3/4 and 7/8-inch diameter A325 bolts, and on E480XX electrodes. They assume the use of detail material with a specified minimum yield strength F_y of 300 MPa.

Although based on specific arrangements of bolts and welds, the tables are general in nature and are intended to facilitate the design of any shear connection without precluding types not shown. These tables can be used by steel fabricators to prepare drawing office and shop standards, by design authorities to check fabricator standards, and by educational institutions to teach structural steel design and detailing.

The standard connections of individual fabricators will depend upon fabrication methods and material sources. They may differ from those shown in the tables regarding length and size of angles and other detail material, and gauge and pitch of bolts.

Minimum Material Thickness

Associated with the tables listing capacities of welds and of bolts in bearing-type connections, is information concerning the minimum required thickness of supporting and supported material to develop the full connector capacities. These minimum thicknesses have been developed for three different values of minimum specified yield strength (F_y = 250, 300 and 350 MPa), and were determined in the following manner.

For welded connections, for each weld size and for each value of F_u, the two equations for weld V_r were equated in turn to the factored shear resistance of the web of the supported beam, and solved for the web material thickness t. The equations for V_r (S16.1-94, Clause 13.13.2.2) are $V_r = 0.67\phi_w A_m F_u$ for the base metal and $V_r = 0.67\phi_w A_w X_u$ for the weld metal (with X_u = 480 MPa, and A_w = weld throat area), and the web resistance equation (Clause 13.4.1.1) is $V_r = \phi A_w F_s$ (with $F_s = 0.66 F_y$, and A_w = web area).

For bolts in bearing-type connections, the minimum material thickness was derived by equating the bearing capacity of the material to the shear capacity of the bolts while assuming that supported beams are not coped. For webs of beams (both supporting and supported) and for webs and flanges of columns (supporting) the maximum factored bearing resistance has been assumed, namely $B_r = 3 \phi_b t d n F_u$ (Clause 13.10(c)).

At the time the tables were generated, the expression $B_r = \phi_b t n e F_u$ was available in S16.1 to determine the thickness of framing angles, end plates and tees based on the end distance e of the last bolt, which was taken as 1.5 bolt diameters. This was accomplished by using the shear value of the bolt for the required value of B_r. The method had the advantage of producing the simple rules for thicknesses reiterated below. However, it was also highly conservative in that it implicitly assigned the bearing value at the last bolt to all the bolts in the connection; that is, only half the bearing resistance at the remaining bolts was utilized.

S16.1-94 does not use an end distance check for bearing at an "edge" bolt. Block tear-out, which takes a form such as that shown in this sketch, more accurately describes the associated limit state (discussed on page 3-59). Generally though, the governing limit state is shear along the net section of the bolt line. Material bearing at the "interior" bolts should also be considered. Minimum edge and end distances must still be maintained in accordance with S16.1-94 Clauses 22.6 and 22.8.

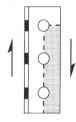

For the sake of practical simplicity, however, it has been decided to continue with thickness requirements given in this section of previous editions of the Handbook, rather than to incorporate the above S16.1-94 bolt bearing refinement.

Under the previous methodology, the factored bearing resistance at bolts for the framing angles, end plates and tees (based on $e = 1\frac{1}{2} d$) was one-half the factored bearing resistance of the beam webs, column webs, or column flanges (based on $e = 3 d$). Bolt shear values were either double or single shear as appropriate. Thus for bearing type connections, the following conditions arose for the tabulated values of minimum thickness.

(a) For double angle connections the minimum thickness is that required for the web of the supported beam, for the connection angle, and for the supporting material when beams frame from two sides. When the beams frame from one side only, one-half the value listed is required for the thickness of supporting material.

(b) For end plate connections, the minimum thickness is that required for the supporting material when beams frame in from one side only. Double the value listed is required for the end plate and also for the supporting material when beams frame from two sides.

(c) For single angle connections the minimum thickness is that required for the web of the supported beam and for the supporting material when beams frame from one side only. Double the value listed is required for the connection angle and for the supporting material when beams frame from two sides.

(d) For tee connections the minimum thickness is that required for the web of the supported beam, for the flange of the tee and for the supporting material when beams frame from two sides. Double the value listed is required for the web of the tee. One-half the value listed is required for the supporting material when beams frame from one side.

Reference

HENDERSON, J.E. 1994. Standardized shear connections. Canadian Institute of Steel Construction, Willowdale, Ont.

DOUBLE ANGLE BEAM CONNECTIONS

Tables 3-37 and 3-38 on pages 3-64 and 3-65, respectively, list capacities of bolted and welded double angle beam connections. At the bottom of each table are values for minimum material thickness required to develop the connector capacities listed in the corresponding columns. For material thickness less than those listed, the corresponding connector capacities must be reduced by the ratio of the thickness of material supplied to the thickness of material listed. Any combination of welded or bolted legs can be selected from the tables.

Bolt capacities are based on concentric loading as tests have shown that eccentricity does not influence the ultimate strength of the bolts in connections using a single line of bolts in the web framing leg. Weld capacities include the effect of eccentricity for connection angles up to 310 mm in length. For longer connection angles the weld capacity is not reduced by the effect of eccentricity.

The connection angle length L has been based on a bolt pitch of 80 mm assuming an end distance of 35 mm. For connection angles with both legs welded, the angle lengths can be adjusted and capacities interpolated in accordance with the length used. Nominal minimum and maximum depths of supported beams appropriate to each length of connection angle are included. The suggested maximum depth assumes a connection length not less than half the beam depth to provide some measure of stiffness and stability. It should be recognized that these depths may not always be appropriate for a particular structure.

Table 3-37 lists bolt capacities for both bearing-type and slip-critical connections for three sizes of angle (width of leg and gauge dimension) and includes values for 2 to 13 bolts per vertical line based on a bolt pitch of 80 mm. For web framing legs, bolt capacities are based on the "double shear" condition and for outstanding legs on the "single shear" condition. Thus two vertical lines of bolts in the outstanding legs (one line in each angle leg) have the same capacity as one vertical line in the web framing leg. When beams are connected to both sides of the supporting material, the total bolt capacity in the outstanding legs is double that listed, provided the thickness of the supporting material is equal to or greater than that listed for the web of the supported beam.

For connection angles, the minimum required thickness to develop the bolt capacities (based on an assumed end distance of 1.5 bolt diameters[1]) is the same as that listed for the web of the supported beam. This thickness will also provide adequate shear capacity for the angles. To ensure connection flexibility the angle thickness selected should not be greater than necessary, with a minimum thickness of 6 mm for practical reasons. When connection angles are less than about 12 mm thick, bolt threads may intercept a shear plane, in which case the bolt capacity must be reduced to 70% of the tabulated value. (See Clauses 13.11.2 and 23.1.3 of CSA/CAN-S16.1-94).

Table 3-38 lists weld capacities for weld configurations shown, for web framing legs with welds, and for outstanding legs with welds. Values are tabulated for four sizes of fillet weld and are based on the length and size (angle width W) of connection angles listed. The weld capacities for the outstanding leg assumes an angle thickness equal to the weld size plus 1 mm in determining the capacity under eccentric load.

Design of bearing-type connections for types and sizes of bolts other than those shown in Table 3-37 will be facilitated by the resistance tables on pages 3-7 to 3-11 inclusive, and for slip-critical connections by the tables on page 3-15.

---

1. See discussion on page 3–57.

Encroachment by Framing Angles Upon Beam Fillets

The maximum length of framing angles needs to be compatible with the clear distance T between the flange fillets of a beam. In compact situations, it is customary to tolerate a modest amount of encroachment by the angles on to the toes of the fillets. Encroachments that create a gap no more than 1 mm under the end of an angle are listed, as a function of the fillet radius, in Table 3-35 on page 60.

Supported Beams with Copes.

When copes are required at the ends of supported beams to avoid interference with the supporting material, either or both the capacity of the beam in the vicinity of the connection and the capacity of the connection may be reduced. When selecting the beam size, the designer should consider the effect of copes on the load carrying capacity of the beam, and the detailer should be aware that copes often reduce the capacity of connections on beams with thin webs.

With reference to the beam, Appendix B of the *Manual of Steel Construction*, Volume II, *Connections* (AISC 1992) provides guidance for a variety of situations that include shear at the reduced section, flexural yielding of the coped section due to bending, and web buckling in the vicinity of the cope due to shear and bending.

With reference to the connection, "block tear-out" is generally the failure mode when copes govern. This parameter more accurately describes the connection resistance than does the former "end distance" at a bolt. Therefore, Effective Net Area (Clause 12.3) as applied for block tear-out supersedes end distance (Clause 13.10.1 (c) (ii) of CAN/CSA-S16.1-M89), which is not in the 1994 edition of the Standard. See Commentary on S16.1-94, Clause 12.3 in Part Two of this Handbook.

Block tear-out takes a different pattern when connection material is bolted to the supported beam than it does when connection material is welded to the beam. In the former case, the pattern is a tension tear along a horizontal line from the end of the beam to the bottom bolt hole of the connection combined with a vertical shearing through the line of bolts to the cope. For welded connection angles, there are corresponding tension and shear lines, but along the toes of the welds. The vertical shearing extends all the way to the cope with the result that the weld across the top of the angles does not participate in the connection resistance for the block tear-out parameter.

A detailing aid for evaluating the block tear-out resistance of a bolted connection on a coped beam is presented in Table 3-36 on page 3-61 where S16.1-94 Clauses 12.3 and 13.2(a)(ii) are applied so that two coefficients have been defined. Coefficient C_1 is a function of the horizontal and vertical edge distances l_h and l_v to the beam end and the cope respectively. Coefficient C_2 is a function of the bolt diameter and the number of bolts. The sum of the coefficients multiplied by both the ultimate tensile strength F_u and the web thickness t gives the block tear-out resistance in newtons.

It is relatively easy to determine the block tear-out resistance of a welded connection on a coped beam because an assumption that the failure lines occur along the edges of the connection angles provides a conservative value. The length of the tension failure w_g is the length of the horizontal overlap of the angle on the beam web, and the length of the shear failure L_g is the vertical length of the angle plus any extra vertical distance on the web to the edge of the cope. The effective total gross length of the failure is then w_g plus $0.6\,L_g$ and the block tear-out resistance is $0.9\,(w_g + 0.6L_g)\,t\,F_y$ as implied by S16.1-94 Clauses 13.2(a)(i).

Tests cited by Yura *et al.* (1980) have shown that the capacity of single line bolted connections computed assuming failure along the "block tear-out" line are conservative, but when two lines of bolts are used in the web framing leg, the effects of eccentricity should be taken into account.

References

BIRKEMOE, P.C. and GILMOR, M.I. 1978. Behaviour of bearing critical double angle beam connections. Engineering Journal, Fourth Quarter, AISC.

YURA, J.A., BIRKEMOE, P.E. and RICLES, J.M. 1980. Beam web shear connections – an experimental study. Beam-to-Column Building Connections: State of the Art, Preprint 80-179, April, ASCE.

Fillet Encroachment

Table 3-35

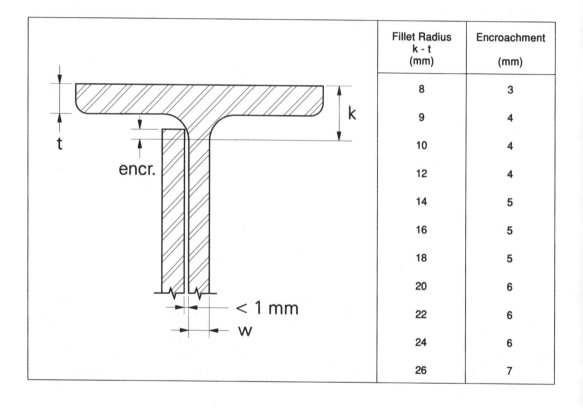

Fillet Radius k - t (mm)	Encroachment (mm)
8	3
9	4
10	4
12	4
14	5
16	5
18	5
20	6
22	6
24	6
26	7

Table 3-36

COEFFICIENTS FOR BLOCK TEAR-OUT
Based on bolt pitch of 80 mm and standard holes*

l_v (mm)	\multicolumn{12}{c}{Coefficient C_1 — l_h (mm)}
	25	26	28	30	32	34	38	45	52	59	66	73	80
25	30.6	31.4	32.9	34.4	36.0	37.5	40.5	45.9	51.3	56.6	62.0	67.3	72.7
26	31.1	31.8	33.4	34.9	36.4	37.9	41.0	46.4	51.7	57.1	62.4	67.8	73.1
28	32.0	32.7	34.3	35.8	37.3	38.9	41.9	47.3	52.6	58.0	63.3	68.7	74.1
30	32.9	33.7	35.2	36.7	38.3	39.8	42.8	48.2	53.6	58.9	64.3	69.6	75.0
32	33.8	34.6	36.1	37.6	39.2	40.7	43.8	49.1	54.5	59.8	65.2	70.5	75.9
34	34.7	35.5	37.0	38.6	40.1	41.6	44.7	50.0	55.4	60.7	66.1	71.5	76.8
38	36.6	37.3	38.9	40.4	41.9	43.5	46.5	51.9	57.2	62.6	67.9	73.3	78.6
45	39.8	40.5	42.1	43.6	45.1	46.7	49.7	55.1	60.4	65.8	71.1	76.5	81.9
52	43.0	43.8	45.3	46.8	48.3	49.9	52.9	58.3	63.6	69.0	74.4	79.7	85.1
59	46.2	47.0	48.5	50.0	51.6	53.1	56.2	61.5	66.9	72.2	77.6	82.9	88.3
66	49.4	50.2	51.7	53.2	54.8	56.3	59.4	64.7	70.1	75.4	80.8	86.1	91.5
73	52.6	53.4	54.9	56.5	58.0	59.5	62.6	67.9	73.3	78.6	84.0	89.4	94.7
80	55.8	56.6	58.1	59.7	61.2	62.7	65.8	71.1	76.5	81.9	87.2	92.6	97.9

Coefficient C_2

n	¾ in.	M20	1 in.
2	12.1	11.0	5.7
3	38.2	36.7	29.1
4	64.4	62.4	52.5
5	90.6	88.1	75.9
6	116.7	113.8	99.3
7	142.9	139.5	122.7
8	169.1	165.2	146.1
9	195.2	190.9	169.5
10	221.4	216.6	192.9

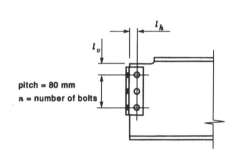

pitch = 80 mm
n = number of bolts

Block tear-out

$A_{ne} = A_n \text{ (tension)} + A_n \text{ (shear)} = w_n t + 0.6 L_n t$ (CAN/CSA-S16.1-94, Clause 12.3)

$T_r = 0.85 \, \phi \, A_{ne} F_u$ (S16.1-94, Clause 13.2(a)(ii))

$\quad = 0.85 \, (0.9) \, (w_n + 0.6 L_n) \, t \, F_u$

$\quad = 0.765 \, \{(l_h - d_h/2) + 0.6 \, [(n-1)(p - d_h) + (l_v - d_h/2)]\} \, t \, F_u$

$\quad = 0.765 \, \{(0.6 \, l_v + l_h) + 0.6 \, [(n-1)(p - d_h) - d_h/2] - d_h/2\} \, t \, F_u$

$\quad = \{[0.459 \, l_v + 0.765 \, l_h] + [0.459 \, ((n-1)(p - d_h) - d_h/2) - d_h/2.614]\} \, t \, F_u$

$\quad = (C_1 + C_2) \, t \, F_u$

where

A_{ne} = effective net area
w_n = net width in tension
L_n = net length in shear
t = thickness
T_r = factored tensile resistance (for block tear-out)
F_u = specified minimum tensile strength

l_h = distance, centre of hole to beam end
l_v = distance, centre of hole to cope
d_h = design allowance for hole diameter*
n = number of bolts
p = bolt pitch (80 mm)

*The design allowance for punched holes is 2 mm larger than the hole diameter (S16.1-94, Clause 12.3.2). Coefficient C_2 was calculated using d_h = 23 mm for ¾ in. bolts, 24 mm for M20 bolts, and 29 mm for 1 in. bolts.

Example 1

Bolted to beam web, bearing-type, welded to column flange.

Given:

W530x92 beam connected to flange of W250x73 column both G40.21 350W steel.

Reaction due to factored loads = 580 kN.

Beam web thickness = 10.2 mm; column flange thickness = 14.2 mm.

Detail material G40.21 300W steel, M20 A325M bolts, E480XX electrodes.

Solution:

Web framing legs — bolted (Table 3-37)

Vertical line with four bolts provides a nominal capacity of 840 kN.

Web thickness, based on bearing, required for steels with F_u = 450 MPa

$= 11.7 \times 580 / 840 = 8.1$ mm   < 10.2   — OK

Angle thickness required is same as required web of supported beam (see p. 3-57).

$= 8.1$ mm   — use 9.5

Minimum angle length required = 310 mm.

Actual bearing capacity of the web is $840 \times 10.2 / 11.7 = 732$ kN

Threads may be included in a shear plane; therefore apply 30% reduction in bolt shear capacity.

Bolt shear capacity is $0.7 \times 840 = 588$ kN   > 580   — OK

Outstanding legs — welded (Table 3-38)

With L = 310 mm, W = 76 or 89 mm,

8 mm fillet welds provide a capacity of 754 kN   > 580   — OK

9.5 mm thick angle required for bolting is also good for 8 mm fillet welds.

Use:

76x76x9.5 connection angles 310 mm long, four M20 A325M bolts in web framing leg and 8 mm fillet welds on outstanding legs.

Example 2

Welded to beam web, bolted to column flange, bearing-type.

Given:

Same as example 1

Solution:

Web framing legs — welded (Table 3-38)

6 mm fillet welds provide a capacity of 806 kN with angle length L = 310 mm and W = 76 mm

Web thickness required for 6 mm fillet welds, L = 310 mm

$= 10.3 \times 580 / 806 = 7.4$ mm   < 10.2   — OK

Outstanding legs — bolted (Table 3-37)

    Try 9.5 mm angle thickness (threads may be included in the shear plane).

    For $L = 310$ mm, $W = 76$ or 89 mm, four bolts per vertical line,

    Bearing capacity of angles is $840 \times 9.5 / 11.7 = 682$ kN > 580 — OK

    Bolt shear capacity is $0.7 \times 840 = 588$ kN > 580 — OK

    Required thickness of supporting material, beams framing from one side, is one half thickness of connection angles; therefore, column flange thickness of 14.2 mm is good.

Use:

    89x76x9.5 connection angles, 310 mm long, 89 mm outstanding legs, g = 130 mm with eight M20 A325M bolts (2 rows of 4) and 6 mm fillet welds.

Example 3

    Bolted to beam web and bolted to both sides of supporting member, supported beams not coped, bearing-type.

Given:

    W530x92 beam, factored reaction 580 kN, framing to both sides of 11.0 mm web of WWF800x154 girder, both of G40.21 350W steel.

    Detail material–G40.21 300W steel, M20 A325M bolts.

Solution:

Web framing legs — same as example 1

Outstanding legs — bolted to both sides of supporting member (Table 3-37)

    Total reaction on girder web is $2 \times 580 = 1\,160$ kN

    For beams connected to both sides of supporting member, bolt capacities are double those listed in table, and required web thickness of supporting member, based on bearing, is the same as that given for web thickness of supported beam.

    For angle $L = 310$ mm, $W = 76$ or 89 mm, web thickness of 11.0 mm, four M20 A325M bolts per vertical line, bearing capacity of web

        $= 2 \times 840 \times 11.0 / 11.7 = 1\,580$ kN $> 1\,160$ — OK

Use:

    89x89x9.5 connection angles, 310 mm long, four M20 A325M bolts per vertical line in both leg framing and outstanding legs.

BOLTED DOUBLE ANGLE[1] BEAM CONNECTIONS
Table 3-37

M20, M22 A325M Bolts
3/4, 7/8 A325 Bolts

BOLT CAPACITY — EITHER LEG WITH BOLTS

Angle Width and Gauge		
W	g	g_1
102	140	65
89	130	60
76	100	45

Nominal depth of supported beam (mm)		Conn. Angle Length L (mm)	Bolts per Vertical Line	BEARING-TYPE CONNECTION[2] Factored Load Resistance (kN)				SLIP-CRITICAL CONNECTION[4] Specified Load Resistance (kN)			
				Bolt Size				Bolt Size			
min.	max.			3/4	M20	M22	7/8	3/4	M20	M22	7/8
				W = 89 mm		W = 102 mm		W = 89 mm		W = 102 mm	
200	310	150	2	378	420	508	516	135	150	183	184
310	460	230	3	568	630	762	774	202	224	271	275
380	610	310	4	757	840	1020	1030	270	299	362	367
460	760	390	5	946	1050	1270	1290	337	374	452	459
530	920	470	6	1140	1260	1520	1550	404	449	542	551
610	1100	550	7	1320	1470	1780	1810	472	524	633	643
690	1200	630	8	1510	1680	2030	2060	539	598	723	734
800		710	9	1700	1890	2290	2320	607	673	814	826
900		790	10	1890	2100	2540	2580	674	748	904	918
920		870	11	2080	2310	2790	2840	741	823	994	1010
1100		950	12	2270	2520	3050	3100	809	898	1090	1100
1200		1030	13	2460	2730	3300	3350	876	972	1180	1190
				W = 76 mm		W = 89 mm		W = 76 mm		W = 89 mm	
200	310	150	2	378	420	508	516	135	150	183	184
310	460	230	3	568	630	762	774	202	224	271	275
380	610	310	4	757	840	1020	1030	270	299	362	367
460	760	390	5	946	1050	1270	1290	337	374	452	459
530	920	470	6	1140	1260	1520	1550	404	449	542	551
610	1100	550	7	1320	1470	1780	1810	472	524	633	643
690	1200	630	8	1510	1680	2030	2060	539	598	723	734
800		710	9	1700	1890	2290	2320	607	673	814	826
900		790	10	1890	2100	2540	2580	674	748	904	918
920		870	11	2080	2310	2790	2840	741	823	994	1010
1100		950	12	2270	2520	3050	3100	809	898	1090	1100
1200		1030	13	2460	2730	3300	3350	876	972	1180	1190

Specified Minimum Yield Strength of Material (MPa)		Minimum Required Web Thickness of Supported Beam[3] (mm)			
F_y = 250	(F_u = 400 MPa)	12.4	13.1	14.4	14.5
F_y = 300	(F_u = 450 MPa)	11.0	11.7	12.8	12.9
F_y = 350	(F_u = 480 MPa)	10.3	10.9	12.0	12.1

(G40.21 350W steel has F_u = 450 MPa)

1. Connection angles are assumed to be material with F_y = 300 MPa.
2. When threads intercept a shear plane, use 70% of the values tabulated for bearing-type connections.
3. For connection angles, and for supporting material with beams framing from both sides, minimum required thickness is equal to tabulated values for web thickness of supported beam, and for supporting material with beams framing from one side is one-half the tabulated values.
4. Tabulated values for slip-critical connections assume Class A contact surfaces with k_s = 0.33.

WELDED DOUBLE ANGLE BEAM CONNECTIONS [a]
Table 3-38

E480XX
Fillet Welds

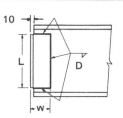

WEB FRAMING LEG WITH WELDS				OUTSTANDING LEG WITH WELDS				Conn. Angle Length L (mm)	Nominal Depth of Supported Beam (mm)	
WELD CAPACITY Factored Load Resistance (kN)				WELD CAPACITY Factored Load Resistance (kN)						
Fillet Size D (mm)				Fillet Size D (mm)						
5	6	8	10	5	6	8	10		min.	max.
Angle Width W = 76 mm				Angle Width W = 89 mm						
394	473	631	789	135	157	202	246	150	200	310
550	660	880	1100	303	356	460	563	230	310	460
672	806	1080	1340	473	568	754	933	310	380	610
794	953	1270	1590	596	715	953	1190	390	460	760
916	1100	1470	1830	718	861	1150	1440	470	530	920
1040	1250	1660	2080	840	1010	1340	1680	550	610	1100
1160	1390	1860	2320	962	1150	1540	1920	630	690	1200
1280	1540	2050	2560	1080	1300	1740	2170	710	800	
1400	1690	2250	2810	1210	1450	1930	2410	790	900	
1530	1830	2440	3050	1330	1600	2130	2660	870	920	
1650	1980	2640	3300	1450	1740	2320	2900	950	1100	
1770	2130	2830	3540	1570	1890	2520	3150	1030	1200	
Angle Width W = 64 mm				Angle Width W = 76 mm						
382	459	612	765	146	170	218	265	150	200	310
519	623	831	1040	322	380	491	602	230	310	460
641	770	1030	1280	473	568	758	947	310	380	610
764	916	1220	1530	596	715	953	1190	390	460	760
886	1060	1420	1770	718	861	1150	1440	470	530	920
1010	1210	1610	2020	840	1010	1340	1680	550	610	1100
1130	1360	1810	2260	962	1150	1540	1920	630	690	1200
1250	1500	2000	2500	1080	1300	1740	2170	710	800	
1370	1650	2200	2750	1210	1450	1930	2410	790	900	
1500	1800	2390	2990	1330	1600	2130	2660	870	920	
1620	1940	2590	3240	1450	1740	2320	2900	950	1100	
1740	2090	2780	3480	1570	1890	2520	3150	1030	1200	
Minimum Required Web Thickness of Supported Beam (mm)				Minimum Thickness of Supporting Material with Beam Attached One Side [b]				Specified Minimum Yield Strength of Material (MPa)		
10.0	12.0	16.0	20.0	5.0	6.0	8.0	10.0	F_y = 250 (MPa)		
8.6	10.3	13.8	17.2	4.3	5.2	6.9	8.6	F_y = 300 (MPa)		
7.4	8.8	11.8	14.7	3.7	4.4	5.9	7.4	F_y = 350 (MPa)		

a. Connection angles are assumed to be material with F_y = 300 MPa.
b. For supporting material with beams framing from both sides, use double the tabulated value.

END PLATE CONNECTIONS

End plate connections with the connection plate welded to the supported beam and bolted to the supporting member are commonly used because of their economy, ease of fabrication, and performance. When beams are saw cut to length, the use of simple jigging procedures to locate and support end plates during assembly and welding makes it possible to meet the tighter fabrication tolerances required without difficulty.

Research on simple beam end plate shear connections has shown that their strength and flexibility compare favourably with double angle shear connections for similar material thickness, depth of connection, and arrangement of bolts (gauge and pitch). For practical reasons it is suggested that the minimum thickness of end plate be 6 mm, and for adequate flexibility that the maximum thickness be limited to 10 mm. The gauge dimension g should preferably be between 100 mm and 150 mm for plates up to 10 mm thick, but may be as low as 80 mm for minimum thickness plates with F_y not greater than 300 MPa.

Table 3-39 lists the capacities of bolts and welds for typical end plate connections with from 2 to 8 bolts per vertical line, together with the minimum thickness of supporting and supported material to develop the full capacity of the bolts and welds respectively. The table also includes reduction factors for bolt capacity in a bearing-type connection for end plate thicknesses of 6, 7, 8 and 10 mm, (assuming an end distance of 1.5 bolt diameters[1]).

For added safety during erection, clipped end plates with one upper corner of the end plate removed may be used. Tests at Queen's University demonstrated that clipped end plate connections have similar moment-rotation characteristics to unclipped end plate connections. Therefore, weld capacities in Table 3-39 may be used directly for design, but tabulated bolt values must be reduced by the value of a single bolt.

Table 3-39 includes bolt capacities for slip-critical joints for those situations where bearing-type connections are not suitable.

1. See discussion on page 3–57.

References

VAN DALEN, K., and MACINTYRE, J.R. 1988. The rotational behaviour of clipped end plate connections. Canadian Journal of Civil Engineering, **15**(1), February.

Example

Given:

W410x60 beam framing into web of WWF700x152 girder, both G40.21 350W steel.

Factored reaction 325 kN

Beam web thickness = 7.7 mm, girder web thickness = 11.0 mm. G40.21 grade 300W steel plate detail material, M20 A325M bolts, and E480XX electrodes

Solution:

Assume 7 mm thick end plate; effective bolt capacity (bolts bearing on end plate) for 3 bolts per vertical line is $630 \times 0.60 = 378$ kN > 325 — OK (see p. 3-57)

For 230 mm long end plate, weld capacity for 5 mm fillet welds made with E480XX electrodes is 350 kN > 325 — OK

Minimum length of end plate required for 7.7 mm web thickness (350W web)

$= 230 \times 7.4 / 7.7 = 221$ mm. Use 230 mm

If beams were framing from both sides, the required web thickness for the girder would be twice the listed value, pro-rated for the actual load:

$$2 \times 5.8 \times 325 / 630 = 6.0 \text{ mm} \quad < 11.0 \quad - \text{OK}$$

Use:

End plate 160x7x230 mm connected to web of the supported beam with 5 mm E480XX fillet welds, and six M20 A325 bolts (2 rows of 3 at 100 mm gauge) with increased end distance and/or increased pitch.

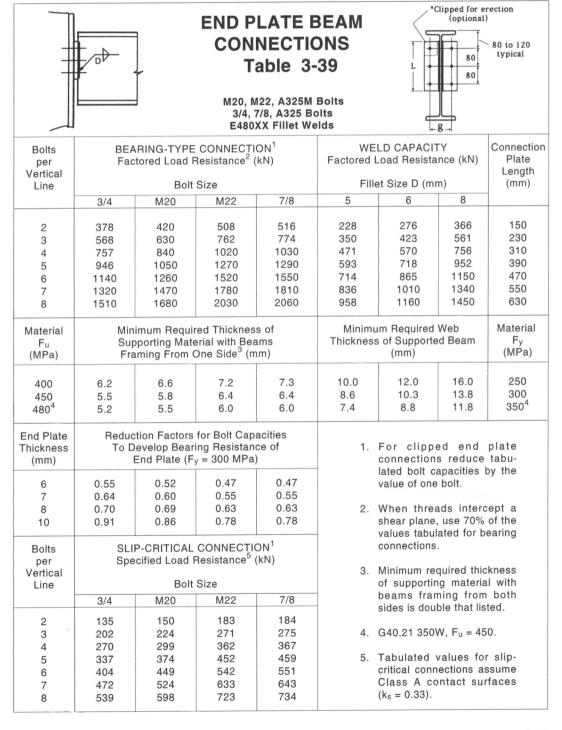

END PLATE BEAM CONNECTIONS
Table 3-39

M20, M22, A325M Bolts
3/4, 7/8, A325 Bolts
E480XX Fillet Welds

Bolts per Vertical Line	BEARING-TYPE CONNECTION[1] Factored Load Resistance[2] (kN) Bolt Size				WELD CAPACITY Factored Load Resistance (kN) Fillet Size D (mm)			Connection Plate Length (mm)
	3/4	M20	M22	7/8	5	6	8	
2	378	420	508	516	228	276	366	150
3	568	630	762	774	350	423	561	230
4	757	840	1020	1030	471	570	756	310
5	946	1050	1270	1290	593	718	952	390
6	1140	1260	1520	1550	714	865	1150	470
7	1320	1470	1780	1810	836	1010	1340	550
8	1510	1680	2030	2060	958	1160	1450	630
Material F_u (MPa)	Minimum Required Thickness of Supporting Material with Beams Framing From One Side[3] (mm)				Minimum Required Web Thickness of Supported Beam (mm)			Material F_y (MPa)
400	6.2	6.6	7.2	7.3	10.0	12.0	16.0	250
450	5.5	5.8	6.4	6.4	8.6	10.3	13.8	300
480[4]	5.2	5.5	6.0	6.0	7.4	8.8	11.8	350[4]
End Plate Thickness (mm)	Reduction Factors for Bolt Capacities To Develop Bearing Resistance of End Plate (F_y = 300 MPa)							
6	0.55	0.52	0.47	0.47				
7	0.64	0.60	0.55	0.55				
8	0.70	0.69	0.63	0.63				
10	0.91	0.86	0.78	0.78				
Bolts per Vertical Line	SLIP-CRITICAL CONNECTION[1] Specified Load Resistance[5] (kN) Bolt Size							
	3/4	M20	M22	7/8				
2	135	150	183	184				
3	202	224	271	275				
4	270	299	362	367				
5	337	374	452	459				
6	404	449	542	551				
7	472	524	633	643				
8	539	598	723	734				

1. For clipped end plate connections reduce tabulated bolt capacities by the value of one bolt.

2. When threads intercept a shear plane, use 70% of the values tabulated for bearing connections.

3. Minimum required thickness of supporting material with beams framing from both sides is double that listed.

4. G40.21 350W, F_u = 450.

5. Tabulated values for slip-critical connections assume Class A contact surfaces (k_s = 0.33).

SINGLE ANGLE BEAM CONNECTIONS

For some applications, single angle connections provide a satisfactory alternative to double angle or end plate connections. They are particularly suitable where limited access prevents the erection of beams with double angle or end plate connections, and where speed of erection is a primary consideration.

The connection angle may be either bolted or welded to the supporting and supported members; however, usual practice involves shop fillet welding to the supporting member and field bolting to the web of the supported beam.

Tests carried out at the University of British Columbia (Lipson 1968, 1977, 1980) using 4x3x3/8 inch angles with the 4 inch leg bolted to the beam web with 3/4 inch diameter A325 bolts and the 3 inch leg welded to the supporting member with 1/4 inch E70XX fillet welds, demonstrated that welded-bolted single angle connections with from 2 to 12 bolts per vertical line possess adequate rotational capacity, and that in those connections loaded to ultimate capacity (2 to 8 bolts per vertical line) the failure occurred in the bolts when the weld pattern included welding along the heel and ends of the connection angle. The tests also demonstrated that the use of horizontal slotted holes in the connection angle reduced the moment at the bolts without affecting the ultimate capacity of the connection.

Table 3-40 is based on this research, and assumes the use of 102x76x9.5 connection angles with the 76 mm leg welded to the supporting member and the 102 mm leg bolted to the supported web. Bolt capacities for bearing-type connections are provided for M20 and M22 A325M bolts and 3/4 and 7/8 inch A325 bolts based on their factored shear resistance for the appropriate number of bolts. The weld capacities have been established by assuming that a connection with 1/4 inch fillet welds has the same shear capacity as the 3/4 inch A325 bolts and then pro-rating for the three sizes of fillet welds shown in the table (i.e., weld capacity = V_r (3/4 inch bolts) x $D/6.35$, where D is the fillet weld size in mm).

References

LIPSON, S.L. 1980. Single-angle welded-bolted beam connections. Canadian Journal of Civil Engineering, **7**(2), June.

LIPSON, S.L. 1977. Single-angle welded-bolted connections. Structural Division Journal, ASCE, March.

LIPSON, S.L. 1968. Single-angle and single-plate beam framing connections. Proceedings, Canadian Structural Engineering Conference, Canadian Institute of Steel Construction, Willowdale, Ontario, February: 141–162.

Example

Single angle welded-bolted beam connection (Table 3-40)

Given:

W410x60 beam of G40.21 350W steel, factored reaction = 290 kN, web = 7.7 mm

102x76x9.5 connection angle of G40.21 300W steel.

M20 A325M bolts, E480XX electrodes.

Solution:

With threads included in the shear plane, bolt capacity with four M20 A325M bolts

is $0.7 \times 420 = 294$ kN   > 290   — OK

Web thickness required is $5.8 \times 290 / 420 = 4.0$ mm < 7.7 — OK

Angle thickness required is $2 \times 4.0 = 8.0$ mm < 10 — OK

Angle length required for 4 bolts is 310 mm, and weld capacity using 5 mm E480XX fillet welds is 298 kN > 290 — OK

Use:

102x76x9.5 connection angle, 310 mm long, 76 mm leg welded to supporting member with 5 mm E480XX fillet welds; 102 mm leg bolted to web of supported beam with four M20 A325M bolts.

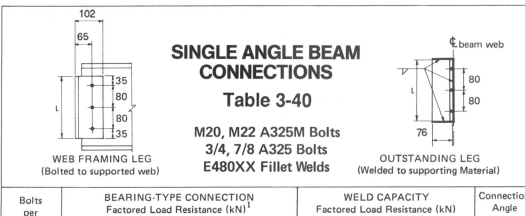

SINGLE ANGLE BEAM CONNECTIONS
Table 3-40

M20, M22 A325M Bolts
3/4, 7/8 A325 Bolts
E480XX Fillet Welds

WEB FRAMING LEG (Bolted to supported web)

OUTSTANDING LEG (Welded to supporting Material)

Bolts per Vertical Line	BEARING-TYPE CONNECTION Factored Load Resistance (kN)[1]				WELD CAPACITY Factored Load Resistance (kN)			Connection Angle Length L (mm)
	Bolt Size				Fillet Size D (mm)			
	3/4	M20	M22	7/8	5	6	8	
2	189	210	254	258	149	179	238	150
3	284	315	381	387	224	268	358	230
4	378	420	508	516	298	357	476	310
5	473	525	635	645	372	447	596	390
6	568	630	762	774	447	537	716	470
7	662	735	889	903	521	626	834	550
8	756	840	1020	1030	595	714	952	630
Material F_u (MPa)	Minimum Required Web Thickness of Supported Beam[2] (mm)				Minimum Required Thickness of Supporting Material With Beams Framing From One Side (mm)			Material F_y (MPa)
400	6.2	6.6	7.2	7.3	5.0	6.0	8.0	250
450	5.5	5.8	6.4	6.4	4.3	5.2	6.9	300
480[3]	5.2	5.5	6.0	6.0	3.7	4.4	5.9	350

1. When threads intercept a shear plane, use 70% of the values tabulated.
2. For connection angles minimum required thickness is double the tabulated values.
3. G40.21 350W steel has $F_u = 450$ MPa.

SHEAR TAB BEAM CONNECTIONS

A simple and economical connection when the loading does not require the strength of bolts in double shear is a single plate welded vertically on to a supporting member with the supported member bolted to the plate. Shear tabs as they are commonly known were studied by Astaneh et al (1989) in an experimental program to define a suitable design method for proportioning and rating them. Table 3-41 was prepared by following recommendations in that paper.

Astaneh identified that the strength of shear tabs is a function of several variables. The first is the stiffness of the supporting member. A shear tab on a column flange is restrained against following the end rotation of the supported member, whereas a shear tab on one side of a supporting beam is more free to rotate in its own plane. This results in different effective eccentricities upon the bolts. The eccentricities are also a function of the number of bolts in the connection. Generally, shear tabs on flexible supports have larger bolt eccentricities, and therefore lower resistances, than do those on rigid supports. Although for shear tabs with seven bolts, it is the same for both.

For rigid supports, efficiency in terms of capacity per bolt is a maximum for four bolts because the eccentricity varies from negative to positive as the number of bolts is increased.

The test program used only standard size holes, and the results are considered to be conservative for short slotted holes. Oversize and long slotted holes are not applicable. Holes may be either punched or drilled.

Shear tabs should be at least 6 mm thick, but no thicker than half the bolt diameter plus 2 millimetres in order to provide the potential for minor bolt hole deformation. High strength material should not be used, for the same reason.

The test specimens all measured 75 mm from the plate edge at the weld to the bolt line. A minimum edge distance of $1\frac{1}{2}$ times the bolt diameter is suggested, with 35 mm at the bottom of the lowest bolt.

Bolts may be either pretensioned or snug tight.

The design methodology used for Table 3-41 consisted of determining the effective eccentricity on the bolts according to Astaneh; finding the single shear, threads included resistance of the bolts from Table 3-14 in this Handbook; calculating the required thickness of the shear tab from CAN/CSA-S16.1-94, Clause 13.4.4; and selecting welds that will develop the shear tab material in shear as recommended by Astaneh.

Reference

ASTANEH, A., CALL, S.M., and MCMULLIN, K.M. 1989. Design of single plate shear connections. Engineering Journal, first quarter, American Institute of Steel Construction, Chicago, Illinois.

SHEAR TAB BEAM CONNECTIONS
Table 3-41

BEARING-TYPE CONNECTIONS
Factored Load Resistance (kN)
M20, M22 A325M and 3/4, 7/8 A325 Bolts
G40.21 300W Steel, E480XX Fillet Welds

RIGID SUPPORTING MATERIAL

Number of Bolts	Connection Plate Length (mm)	3/4 Bolts			M20 Bolts		
		Resistance (kN)	Plate Thickness (mm)	Weld Size D (mm)	Resistance (kN)	Plate Thickness (mm)	Weld Size D (mm)
2	150	79.3	6	5	88.2	6	5
3	230	173	6	5	193	6	5
4	310	264	6	5	294	8	6
5	390	310	6	5	345	8	6
6	470	360	8	6	401	8	6
7	550	413	8	6	459	8	6

Number of Bolts	Connection Plate Length (mm)	M22 Bolts			7/8 Bolts		
		Resistance (kN)	Plate Thickness (mm)	Weld Size D (mm)	Resistance (kN)	Plate Thickness (mm)	Weld Size D (mm)
2	150	107	6	5	108	6	5
3	230	233	8	6	236	8	6
4	310	356	10	8	360	10	8
5	390	417	10	8	422	10	8
6	470	485	10	8	491	10	8
7	550	556	10	8	563	10	8

FLEXIBLE SUPPORTING MATERIAL

Number of Bolts	Connection Plate Length (mm)	3/4 Bolts			M20 Bolts		
		Resistance (kN)	Plate Thickness (mm)	Weld Size D (mm)	Resistance (kN)	Plate Thickness (mm)	Weld Size D (mm)
2	150	62.1	6	5	69.1	6	5
3	230	123	6	5	137	6	5
4	310	195	6	5	217	6	5
5	390	269	6	5	299	6	5
6	470	342	8	6	380	8	6
7	550	413	8	6	459	8	6

Number of Bolts	Connection Plate Length (mm)	M22 Bolts			7/8 Bolts		
		Resistance (kN)	Plate Thickness (mm)	Weld Size D (mm)	Resistance (kN)	Plate Thickness (mm)	Weld Size D (mm)
2	150	83.6	6	5	84.6	6	5
3	230	165	6	5	167	6	5
4	310	262	8	6	266	8	6
5	390	362	8	6	366	8	6
6	470	460	8	6	465	8	6
7	550	556	8	6	563	8	6

TEE-TYPE BEAM CONNECTIONS

Tee-type beam connections combine some of the characteristics of single angle connections with the web-framing leg bolted in single shear, and of double-angle connections with the outstanding legs welded to the supporting member.

Their main advantage is speed and ease of erection. They are also commonly used where hole making in the supporting member is undesirable (connections to HSS columns), and to avoid coping the bottom flange of the supported beam for erection purposes.

Costs are generally higher than for other types of simple beam connections because of the higher costs of fabricating the tee-sections.

Table 3-42 lists bolt capacities for bearing-type web-framing connections, and weld capacities for connections of outstanding legs. The bolt capacities are the same as those listed in Table 3-40 for the web-framing legs of single angle connections, and the weld capacities are the same as those listed in Table 3-37 for the outstanding legs of welded double-angle connections.

Example

Tee-type welded-bolted beam connections (Table 3-42)

Given:

W460x61 beam, Factored reaction 325 kN.

Column — HSS 254x254x11

Beam web thickness 8.1 mm

G40.21 350W steel, M20 A325M bolts, E480XX electrodes.

Solution:

Try tee cut from W200x59 beam; web thickness = 9.1 mm.

These thicknesses of beam web and tee web will result in threads intercepting the shear plane.

Five bolts per vertical line provide a capacity of $0.7 \times 525 = 367$ kN.

Beam web thickness required for F_u of 450 MPa is $5.8 \times 325 / 525$

$= 3.6$ mm < 8.1 — OK

Tee web thickness required is $2 \times 3.6 = 7.2$ mm < 9.1 mm   — OK (See p. 3-57)

Length of tee required for 5 bolts is 390 mm, and weld capacity for 5 mm fillet welds is 596 kN > 325 — OK

Clear depth of beam web between fillets, T = 403 mm > 390 — OK

Use:

Tee cut from W200x59, 390 mm long, five M20 A325M bolts connecting webs of beam and tee, and 5 mm E480XX fillet welds to supporting material.

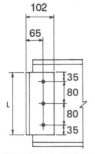

WEB FRAMING LEG
(Bolted to supported Web)

TEE-TYPE BEAM CONNECTIONS
Table 3-42

M20, M22 A325M Bolts
3/4, 7/8 A325 Bolts
E480XX Fillet Welds

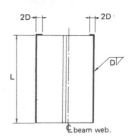

OUTSTANDING LEG
(Welded to supporting Material)

Bolts per Vertical Line	BEARING-TYPE CONNECTIONS Factored Load Resistance (kN)[1]				WELD CAPACITY Factored Load Resistance (kN)			Connection Tee Length L (mm)
	Bolt Size				Fillet Size D (mm)			
	3/4	M20	M22	7/8	5	6	8	
2	189	210	254	258	135	157	202	150
3	284	315	381	387	303	356	460	230
4	378	420	508	516	473	568	754	310
5	473	525	635	645	596	715	953	390
6	568	630	762	774	718	861	1150	470
7	662	735	889	903	840	1010	1340	550
8	756	840	1020	1030	962	1150	1540	630
Material F_u (MPa)	Minimum Required Web Thickness of Supported Beam[2] (mm)				Minimum Required Thickness of Supporting Material With Beams Framing from One Side (mm)			Material F_y (MPa)
400	6.2	6.6	7.2	7.3	5.0	6.0	8.0	250
450	5.5	5.8	6.4	6.4	4.3	5.2	6.9	300
480[3]	5.2	5.5	6.0	6.0	3.7	4.4	5.9	350

1. When threads intercept a shear plane, use 70% of the values tabulated.
2. Minimum required thickness of Tee-web is double the tabulated values.
3. G40.21 350W steel has F_u = 450 MPa.

SEATED BEAM SHEAR CONNECTIONS

General

This section of the Handbook deals with the unstiffened angle seat and the tee-type stiffened seat designed to provide a simple beam shear connection to a supporting member. Although seated beam shear connections are designed to support vertical loads only, eccentricities produced by these connections may be greater than for simple framed beam shear connections and can influence the design of supporting members.

Seated beam shear connections are most commonly used at beam-to-column supports. If used at beam-to-girder supports, the girder web must be checked for adequate local stability and resistance. Economy with seated beam shear connections results from simple shop fabrication, together with ease and speed of field erection.

The unstiffened angle seat consists of a relatively thick angle either shop welded or bolted to the supporting member. When the supporting member is a column web, access for welding may be restricted. Load capacity of an unstiffened angle seat is limited by the angle thickness. This capacity can be increased by stiffening the angle; however, stiffened angle seats are more expensive to fabricate, and stiffened seats using tee-stubs built up from plate are usually more economical. Stiffened seats designed for large loads are generally referred to as brackets and are beyond the scope of this section.

A seated beam must be stabilized with a flexible clip angle attached either to the top flange of the beam or to the beam web near the top of the beam. The clip angle must be thin enough to permit end rotation of the beam. Either welds or bolts can be used to connect the clip angle to the beam and supporting member. When welds are used, the fillet welds should be located along the toes of the angle.

Unstiffened Angle Seats

The capacity of unstiffened angle seats depends on the bending capacity of the seat angle, and is governed by the web thickness and effective bearing length of the supported beam. When the vertical leg of the seat angle is welded to the supporting member, the top of the angle is restrained by the welds, so that the capacity of the angle seat is assumed to be limited by the bending capacity of the outstanding leg. When the vertical leg is bolted to the supporting member, the top of the angle is not restrained by the bolts, so it is assumed that bending in the vertical leg, rather than the outstanding leg, controls the bending capacity of the angle seat.

Tables 3-43 and 3-44 list capacities for welded and bolted unstiffened angle seats of various thickness for seat lengths of 180 mm and 230 mm, assuming beams and seat angles of G40.21 300W material (F_y = 300 MPa) and welds made with E480XX electrodes. Capacities are based on the design models illustrated in the tables, with no allowance made for possible restraint provided by any connection between the seat and the bottom flange of the supported beam.

Tabulated values assume a seat angle fillet radius of 10 mm, and are based on a design clearance of 10 mm between the end of the supported beam and the face of the supporting member for seat angles of 10 mm or less in thickness, and a design clearance of 20 mm for seat angles greater than 10 mm in thickness.

It should be noted that beam web bearing capacities in Tables 3-43 and 3-44 are based on yielding capacity of the beam webs, Clause 15.9(b)(i) of CAN/CSA-S16.1-94, which considers web thickness but not depth of the section. Web crippling of individual beams should be checked according to Clause 15.9(b)(ii) as this could govern the beam bearing capacity.

Vertical Leg Welded to Supporting Member

Table 3-43 for welded seats, lists the beam web bearing resistance for various angle thicknesses and beam web thicknesses, and the vertical leg weld resistance for various weld sizes and angle vertical leg lengths. The beam web bearing rersistance was calculated with Clause 15.9(b)(i) by assuming $k = 3w$ and equating the value to the plastic bending resistance of the angle outstanding leg. The vertical weld capacity was determined with the model illustrated at the bottom of the table. The welds are assumed to carry the vertical load and a tension force computed from the conservative average eccentricities listed near the bottom of the table. Angle thicknesses in the top row apply throughout the table. Compression from the eccentric moment is carried by contact at the lower middle of the angle. Vertical bending resistance of the angle is also checked.

Vertical Leg Bolted to Supporting Member

Table 3-44 for bolted seats lists the beam web bearing resistance and the seat angle bending resistance based on the model illustrated at the top of the table. Values below the lines are governed by the seat angle bending resistance. Table 3-44 also gives the bolt capacity for four sizes of bolt, two or four bolts per seat angle, threads excluded and threads intercepted. These bolt capacities are based on rough assumptions thought to be conservative.

Example

Given:

W530x82 beam, factored reaction 185 kN, G40.21 300W steel.

Beam web thickness 9.5 mm, flange width 209 mm.

Solution:

(a) *Unstiffened angle seat welded to supporting member*

Seat angle thickness for beam web bearing capacity:

Assume the end clearance is 20 mm (therefore angle thickness required is more than 10 mm). Interpolate in Table 3-43 for beam web thickness of 9.5 mm with $L = 230$ (to permit the 209 mm flange to be welded to the seat). A 12.7 mm thick angle provides a beam web bearing capacity of

$(270 + 316) / 2 = 293$ kN > 185 — OK

Vertical leg connection:

For an angle thickness of 12.7 mm with a vertical leg of 127 mm and a conservatively assumed eccentricity of 40 mm, 8 mm fillet welds provide connection capacity of

190 kN > 185 — OK

Use 127x89x12.7 seat angle 230 mm long with 127 mm leg welded to supporting member with 8 mm E480XX fillet welds on each side of the vertical leg.

(b) *Unstiffened angle seat bolted to supporting member*

Seat angle thickness and beam web bearing capacity:

From Table 3-44, a 15.9 mm seat angle 230 mm long provides an angle leg flexural capacity of 188 kN, and a beam web bearing capacity of 185 kN (for an 8 mm web).

Therefore, OK for 9.5 mm web.

Four M20 bolts will provide capacity of 328 kN, (threads excluded).

Use 152x102x15.9 seat angle 230 mm long with 152 mm leg bolted to supporting member with 4 – M20 A325M bolts.

WELDED UNSTIFFENED ANGLE SEATS
Factored Resistances

Table 3-43

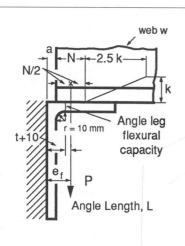

G40.21 300W Steel
E480XX Fillet Welds

Web bearing resistance
$$P = 1.1\,\phi\,w\,(N + 2.5\,k)\,F_y$$

Angle leg flexural resistance
$$P = \frac{(L\,t^2/4)\,\phi\,F_y}{N/2 + a - t - r}$$

Above expressions were equated and solved for N, which was used to calculate P for top half of table.

$t \le 10$ when $a = 10$ $\quad k = 3w$
$t > 10$ when $a = 20$ $\quad r = 10$ mm

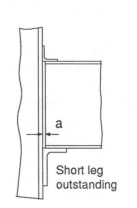

Short leg outstanding

	Angle t (mm)		7.9		9.5		12.7		15.9		19.1	
	Angle L (mm)		180	230	180	230	180	230	180	230	180	230
Beam web factored bearing resistance (kN)	Beam w (mm)	5	104	109	114	120	122	130	142	153	164	178
		6	133	138	143	149	150	159	172	184	195	210
		7	167	171	176	182	183	192	205	217	229	244
		8	205	209	214	220	220	229	242	255	267	283
		9			256	262	262	270	284	296	308	324
		10			303	309	308	316	329	342	354	370
		11					359	367	380	392	404	420
		12					414	422	435	446	459	474
	Fillet Weld D (mm)		6		6		8		8		10	
	Angle L (mm)		180	230	180	230	180	230	180	230	180	230
Seat vertical leg factored weld resistance (kN)	Seat angle	89x76	93.6	73.3	97.2	86.9	114	112				
		102x76	122	95.8	117	113	139	139				
		127x89	167	149	158	158	190	190	173	173		
		152x89[1]	209	209	199	199	243	243	223	223	259	259
		203x102[2]					355	355	332	332	389	389
	Eccentricity e_f used		25		30		40		50		60	

1. 152x102 for $t = 19.1$ (availability)
2. 203x152 for $t = 15.9$ (availability)

Tensile resistance of welds:
$$T_r = 2\left\{0.67\,\phi_w\left[\frac{(l-a)\,D}{\sqrt{2}}\right]0.480\right\} = \frac{P\,e_f}{l/2}$$

Shear resistance of welds:
$$V_r = 2\left\{0.67\,\phi_w\left[\frac{a\,D}{\sqrt{2}}\right]0.480\right\} = P$$

Above expressions were solved for P by eliminating a to obtain:
$$P = \frac{0.3047\,l^2\,D}{2\,e_f + l}$$

but P cannot exceed vertical bending resistance of total angle, taken as:
$$P = \frac{4\,\phi\,S_x\,F_y}{L}$$

Variables in the expressions are in kN and mm units.

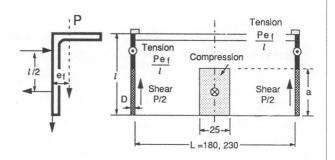

BOLTED UNSTIFFENED ANGLE SEATS
Factored Resistances
Table 3-44

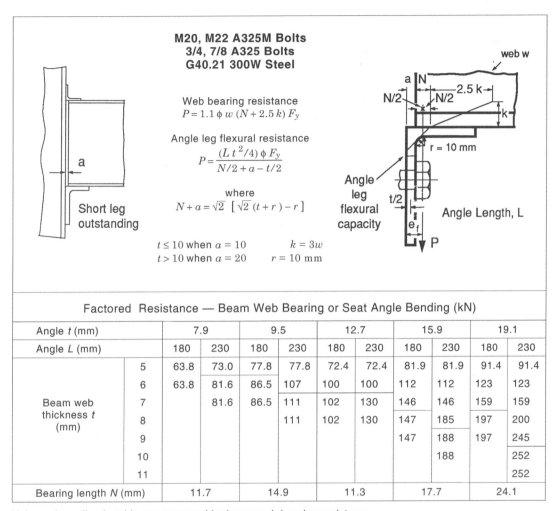

M20, M22 A325M Bolts
3/4, 7/8 A325 Bolts
G40.21 300W Steel

Web bearing resistance
$$P = 1.1\,\phi\,w\,(N + 2.5\,k)\,F_y$$

Angle leg flexural resistance
$$P = \frac{(L\,t^2/4)\,\phi\,F_y}{N/2 + a - t/2}$$

where
$$N + a = \sqrt{2}\,[\sqrt{2}\,(t+r) - r]$$

$t \le 10$ when $a = 10$ $k = 3w$
$t > 10$ when $a = 20$ $r = 10$ mm

Factored Resistance — Beam Web Bearing or Seat Angle Bending (kN)

Angle t (mm)		7.9		9.5		12.7		15.9		19.1	
Angle L (mm)		180	230	180	230	180	230	180	230	180	230
Beam web thickness t (mm)	5	63.8	73.0	77.8	77.8	72.4	72.4	81.9	81.9	91.4	91.4
	6	63.8	81.6	86.5	107	100	100	112	112	123	123
	7		81.6	86.5	111	102	130	146	146	159	159
	8				111	102	130	147	185	197	200
	9							147	188	197	245
	10								188		252
	11										252
Bearing length N (mm)		11.7		14.9		11.3		17.7		24.1	

Values above line in table are governed by beam web bearing resistance.
Values below line in table are governed by seat angle bending resistance.

Factored Bolt Capacity

Bolt size	3/4		M20		M22		7/8	
No. of bolts	2	4	2	4	2	4	2	4
Threads excluded	148	295	164	328	196	396	201	403
Threads intercepted	115	231	128	256	155	310	157	314

Angle length, L

L	180	230
g	100	130

60 min.

Bolt capacities were calculated by $V_f = n\,\sqrt{\dfrac{T_r^2\,V_r^2}{T_r^2 + V_r^2}}$

(from Clause 13.11.4 of CAN/CSA-S16.1-94), when n is number of bolts and shear-tension ratio is assumed to be 1.0.

STIFFENED SEATED BEAM CONNECTIONS

Table 3-45 lists factored resistances of stiffened seats for the tee-shaped weld configuration shown. Capacities are based on the use of E480XX electrodes, steel $F_y = 300$ MPa, and a minimum stiffener thickness t equal to 1.7 times the fillet weld leg size to ensure that the shear resistance of the stiffener is not exceeded. Factored resistances tabulated are computed using the theory outlined on page 3-53 for "Eccentric Loads on Weld Groups — Shear and Moment", modified for the tee-shaped weld configuration, with the length of the horizontal weld connecting the seat plate to the support equal to 0.4 times the length of the vertical weld connecting the stiffener plate to the support.

Stiffened seats must be proportioned so that the stiffener thickness t is not less than the web thickness w of the supported beam. If the beam has a higher specified yield strength than the stiffener, the relationship, $t \times F_y$ (stiffener) $= w \times F_y$ (beam) shall be satisfied.

When the stiffener is fitted to bear against the seat, the capacity of the welds connecting the seat plate to the stiffener shall be at least equal to the capacity of the welds connecting the seat plate to the supporting member. Welds or bolts may be used to connect the supported beam to the seat and for attachment of the clip angle required to stabilize the beam. See the sketches in Table 3-45 for the general arrangement.

When stiffened seats are in line on opposite sides of a column web, the size of the vertical fillet welds (for E480XX electrodes and $F_y = 300$ MPa material) shall not exceed $F_y/515$ times the thickness of the column web, so as not to exceed the shear resistance of the column web. As an alternative to limiting the weld size, a longer seat may be used to reduce the shear stresses in the column web.

Example

Given:

W530x82 beam with a factored reaction of 450 kN

Web thickness = 9.5 mm, flange width = 209 mm, k distance = 29 mm

Connected to web of W310x118 column, web thickness = 11.9 mm

Design stiffened welded seat for beams connected to both sides of column web

G40.21 300W steel   E480XX electrodes

Solution:

(a) *Vertical stiffener*

Required length of bearing: $B_r = 1.10 \phi w (N + 2.5 k) F_y$ (Clause 15.9(b)(i))

$\therefore N = (B_r / 1.10 \phi w F_y) - 2.5 k$

$= (450 / 1.10 \times 0.9 \times 9.5 \times 0.300) - (2.5 \times 29) = 87$ mm

For 10 mm clearance, minimum stiffener width = 87 + 10 = 97 mm

Try 100 mm stiffener width

For stiffeners both sides of column web, maximum effective weld size so that shear resistance of column web is not exceeded is

$11.9 \times 300 / 515 = 6.93$ mm

Minimum stiffener thickness for shear is $1.7 \times 6.93 = 11.9$ mm

Try 12 mm stiffener with 6 mm fillet welds.

From Table 3-45, with $w = 100$ mm, 6 mm fillet welds and $L = 275$ mm, capacity provided is 500 kN   > 450   — OK

Check $b/t = 100/12 = 8.3$ $< 200 / \sqrt{F_y} = 11.6$ — OK   (S16.1-94, Table 1)

Use 12x100 stiffener x 275 mm long welded to column web with 6 mm fillet welds.

(b) *Horizontal Seat Plate*

Try 12 mm plate and 6 mm fillet welds (same as vertical stiffener)

Minimum length of weld required to attach seat plate to column web is

$0.4\,L = 0.4 \times 275 = 110$ mm

Minimum seat plate length, including clearances for horizontal stiffener welds is

$110 + (2 \times 6) + 12 = 134$ mm

Minimum length of seat plate assuming beam bolted to seat is

beam flange width = 209 min

Minimum length of seat plate assuming beam welded to seat is

$209 + 2\,(2 \times 6) = 233$ mm

Use 12x100 seat plate x 210 mm long welded to the column web with 6 mm fillet welds on underside of seat, and bolted to bottom flange of the beam with two M20 A325M bolts.

(c) *Weld between stiffener and seat plate*

Minimum length of weld required = 110 mm (same as between seat plate and column web) for 6 mm fillets

Length available is $2 \times 100 = 200$ mm   — OK

STIFFENED SEATED BEAM CONNECTIONS
Table 3-45
E480XX Electrodes

FACTORED RESISTANCE OF WELDS (kN)

Seat Width W (mm)	Fillet Size* D (mm)	\multicolumn{11}{c}{Length of Stiffener L (mm)}										
		150	175	200	225	250	275	300	325	350	400	450
100 $e_f = 60$	6	161	220	286	361	444	500	545	591	636	727	818
	8	212	288	378	480	590	666	727	788	848	969	1090
	10	260	356	468	593	731	833	909	985	1060	1210	1360
	12	307	422	556	704	869	1000	1090	1180	1270	1450	1640
125 $e_f = 72.5$	6	133	182	239	301	372	449	533	591	636	727	818
	8	175	240	316	397	490	594	707	788	848	969	1090
	10	215	297	388	493	611	738	878	985	1060	1210	1360
	12	254	351	461	586	727	879	1050	1180	1270	1450	1640
150 $e_f = 85$	6	114	155	203	258	320	385	458	537	621	727	818
	8	149	205	269	342	421	509	607	710	820	969	1090
	10	183	252	332	422	521	631	752	880	1020	1210	1360
	12	217	299	395	501	621	755	900	1050	1220	1450	1640
200 $e_f = 110$	6	88	120	158	200	247	299	357	417	484	632	796
	8	115	158	208	264	327	397	473	553	642	838	1060
	10	142	195	256	327	406	492	585	687	798	1040	1310
	12	168	231	305	389	483	585	697	820	950	1240	1570

*Minimum plate thickness, t = 1.7D.   Material F_y = 300 MPa

MOMENT CONNECTIONS

General

Continuous construction requires moment resisting beam-to-column connections that will maintain, virtually unchanged, the original angles between intersecting members, at specified loads. Rigid moment connections can be provided by using welds, bolts or combinations of welds and bolts. Numerous configurations and details are possible; Figure 3-4 shows four of the possible arrangements.

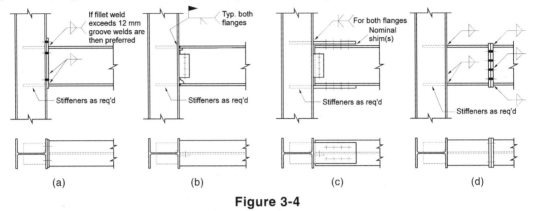

Figure 3-4

Figure 3-4 (a) illustrates a heavy plate shop welded to the end of the beam and field bolted to the column. The end plate distributes flange forces over a greater length of column web than does a fully welded joint, but prying action must be considered.

Figure 3-4 (b) illustrates beam flanges field welded directly to the column with groove welds. Shear capacity is developed by a seat angle, web framing angle or plate, or by welding the beam web directly to the column. Backing bars and run off tabs for the welds may be required.

Figure 3-4 (c) illustrates the use of moment plates shop welded to the column with groove or fillet welds and fillet-welded, or preferably bolted, to the flanges of the beam. The moment plates are spaced to accommodate rolling tolerances for beam depth and flange tilt, and nominal shims are provided to fill any significant gap. Minor gaps are closed by the action of bolting. Shear capacity is usually provided by a web plate welded to the column and field bolted to the beam.

Figure 3-4 (d) illustrates the use of short beam sections shop welded to the column, and field bolted to the beam near a point of contraflexure. An end plate connection is shown but lapping splice plates for the flanges and web may be more economical depending on the forces to be transmitted and the relative ease of achieving field fit-up.

Seismic design criteria (CAN/CSA-S16.1-94, Clause 27) impose restrictions (as a function of the level of ductility assumed in the analysis) on the type and arrangement of details permitted for moment connections in seismic environments. The behaviour of connections such as those in (b) continue to come under intense scrutiny as a result of the Northridge earthquake in California. It is known that alternatives like (a) or (c) provide desirable greater ductility. Comprehensive recommendations are expected, but were not available for inclusion in this edition of the Handbook. See the Commentary to Clause 27 in Part 2 and the FEMA reference on the next page.

Generally, in designing moment connections it is important to ensure that the connection provided will have adequate strength and stiffness, combined with sufficient rotation capacity to permit inelastic deformations to occur, as assumed or inherent in the analysis and design procedure used to proportion the structural frame. Since the behaviour of moment connections is highly complex, careful design is necessary to accommodate possible stress concentrations resulting from restrained shrinkage due to welding. In particular, it is important to provide details with adequate ductility when framing perpendicular to the strong axis of a column.

When applying the governing moments, shears and axial forces at a moment connection, a simplified approach may be used for design (except in seismic zones 3 and 4). Research (Chen and Lui 1988) has shown that the end moment and axial force in the beam can be assumed to be resisted by the beam flanges alone, with the end shear resisted by the beam web alone.

To ensure that the connection provided is consistent with the design assumptions used to proportion members of a structure, it is important that the designer provide the fabricator with governing maximum and coincident moments, shears and axial forces to be developed at the connection. See Clause 4.1.2 in S16.1-94 and the Commentary.

Column Stiffeners

Where rigid connections are required, the resistance of a column section to local deformation is important. With relatively small beams connected to heavy columns, the columns will provide the degree of fixity assumed in the design of beams. With large beams, however, the columns will usually have to be strengthened locally by means of stiffeners, doubler plates or both.

Column stiffeners are provided opposite tension flanges of the connected beams to minimize curling of the column flanges with resultant overstressing of the central portion of the weld connecting the beam flange (or moment plate) to the column. Opposite compression flanges of the beams, column stiffeners are provided to prevent buckling of the column web. The most commonly used stiffeners are horizontal plates. When different depth beams frame into opposite flanges of the column, either inclined stiffeners or horizontal plate stiffeners opposite the flange of each beam may be used. If shear generated in the column web at the moment connection exceeds the column shear capacity, "doubler" plates or diagonal plate stiffeners are used to locally increase the column web shear capacity. Clause 21.3 of S16.1-94 specifies requirements for web stiffeners on H-type columns when a beam is rigidly framed to the column flange.

References

The following references contain more detailed information on the design of moment connections. Some refer to allowable stress rules and must be interpreted for limit states applications.

ASCE. 1971. Commentary on plastic design in steel. American Society of Civil Engineers, New York, N.Y.

BLODGETT, O.W. 1966. Design of welded structures. The James F. Lincoln Arc Welding Foundation, Cleveland, Ohio.

CHEN, W.F., and LUI, E.M. 1988. Static flange moment connections. Journal of Constructional Steel Research, Elsevier Science Publishers, New York, N.Y., Vol. 10: 65–66.

CHEN, HAUNG and BEEDLE. 1974. Recent results on connection research at Lehigh. Regional Conference on Tall Buildings, Bangkok, Thailand, pages 799–813.

FEMA. 1995. Interim guidelines: evaluation, repair, modification and design of welded steel moment frame structures. Report FEMA-267, Federal Emergency Management Agency, Jessup, Maryland.

GOEL, S.C., STOJADINOVIC, B. and LEE, K-H. 1997. Truss analogy for steel moment connections. Engineering Journal, American Institute of Steel Construction, **34**(2).

GRAHAM, SHERBOURNE and KHABBAZ. 1959. Welded interior beam-to-column connections. The American Institute of Steel Construction, Chicago, Illinois.

HAUNG, CHEN and BEEDLE. 1973. Behavior and design of steel beam-to-column moment connections. Welding Research Council Bulletin.

KRISHNAMURTHY, N. 1978. A fresh look at bolted end plate behavior and design. Engineering Journal, American Institute of Steel Construction, **15**(2).

MODULAR LEARNING SYSTEM. 1996. Principles of welding design, module 32 — moment connections. Gooderham Centre for Industrial Learning (a division of the Canadian Welding Bureau), Mississauga, Ontario.

PACKER, J.A. 1977. A limit state design method for the tension region of bolted beam-column connections. The Structural Engineer, **5**(10), October.

REGEE, HAUNG and CHEN. 1973. Test of a fully-welded beam-to-column connection. Welding Research Council Bulletin.

TALL L. (editor). 1964. Structural steel design (2nd. ed.). The Ronald Press Company, New York, N.Y.

Examples

Note: In the following examples, the solution chosen in each case is intended to illustrate only one of several satisfactory solutions that could be used. In any given situation, the design will be influenced by the individual fabricator's experience, fabrication methods and erection procedures.

Example 1

Given:

Design an interior beam-to-column connection for the following coincident forces and moments due to factored loads, assuming the unit specified dead and live loads are 3.8 kPa and 4.8 kPa respectively.

Factored beam moments = 240 kN·m and 320 kN·m

Factored beam shears = 110 kN and 130 kN

Steel: CSA-G40.21 350W (W shapes), 300W (plates)   E480XX electrodes

    W310x86 Column   W410x60 Beam

    t_c = 16.3 mm t = 12.8 mm

    w_c = 9.1 mm w = 7.7 mm

    k_c = 33 mm d = 407 mm

    b = 254 mm b = 178 mm

    d = 310 mm   Class 1 (in bending)

    k_1 = 20 mm

    T = 244 mm

Solution:

(a) Web Connection

The design of the connection for the beam web to the column flange need only account for the vertical shear, neglecting eccentricity. (Design for 130 kN shear.)

Two alternatives are shown to illustrate a field welded and a field bolted condition.

Alternative 1

   Single plate field welded to beam web, shop welded to column flange, holes for 2 – M20 erection bolts

To resist factored shear, try 5 mm fillet on 6 mm plate.

Required weld length is $130/0.762 = 171$ mm   (Table 3-24, page 3-41)

Use 230 mm length of plate   — for a 410 mm beam   (Table 3-38, page 3-65)

Check plate for factored shear capacity.   (Clause 13.4.4, S16.1-94)

Net length is $230 - 2(20 + 2 + 2) = 182$ mm

V_r of plate is $0.5 \times 0.9 \times 182 \times 6 \times 0.450 = 221$ kN > 130 — OK

Use 6x75 x 230 plate with 5 mm E480XX fillet weld.

Alternative 2

   Single plate shop welded to column flange, field bolted to beam web with M20 A325M bolts at bearing values

From Table 3-4, page 3-8, factored shear resistance, single shear, threads intercepted, for M20 A325M bolts = 73.5 kN

∴ for 2 bolts, $V_r = 2 \times 73.5 = 147$ kN > 130 — OK

Check factored bearing resistance on beam web, $w = 7.7$ mm

From Table 3-6 on page 3-9,
bearing on both 300W and 350W steel for $t =$ say, 6 mm is 109 kN per bolt

∴ $w = 7.7$ is OK for 73.5 kN

Try 6 mm plate, 230 mm long, 2 bolts at 160 mm pitch, and check plate thickness for shear.

Net length = 182 mm, as for alternative 1

Required thickness of plate is

$130/(0.5 \times 0.9 \times 182 \times 0.450) = 3.5$ mm < 6 — OK

Use 6x80 x 230 plate and two M20 A325M bolts at 160 mm pitch.

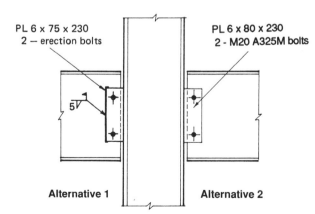

Alternative 2 replaces the two erection bolts with permanent high strength bolts, and eliminates vertical field welding (likely a better solution).

(b) Flange Connection

   Two alternatives are shown to illustrate field bolted and field welded conditions.

Alternative 1

> Top and bottom moment plates shop welded to column, field bolted to beam flanges with A325M bolts in slip-critical connection

The number of bolts required is determined on the basis of specified loads, and all other strength checks are based on factored loads and factored resistances.

Determine the specified load moment using Figure 3-1 on page 3-4.

Ratio $D/L = 3.8 / 4.8 = 0.79$ giving $(L + D) / (\alpha_L L + \alpha_D D) = 0.72$

Therefore, specified load moment is $0.72 \times 320 = 230$ kN·m

Flange force due to specified loads is $230 \times 1\,000 / 407 = 565$ kN

From Table 3-11, page 3-15, assuming M22 A325M bolts and Class A contact surfaces, number of bolts required is

$565 / 45.2 = 12.5$ — use 14 bolts (2 rows of 7)

Assuming 80 mm pitch, 40 mm end distance, 130 mm clear to column flange, plate length required is

$(6 \times 80) + 40 + 130 = 650$ mm

Flange force due to factored loads is $320 \times 1\,000 / 407 = 786$ kN

From S16.1-94, Clause 13.2(a), gross area of required plate is

$786 / (0.9 \times 0.300) = 2\,910$ mm^2

and required net effective area is

$786 / (0.85 \times 0.9 \times 0.450) = 2\,280$ mm^2

Try 200x16 plate (first preference thickness), and check areas.

Gross area is $200 \times 16 = 3\,200$ mm^2 $> 2\,910$ — OK

Net area is $(200 - 52)16 = 2\,370$ mm^2 $> 2\,280$ — OK

Ultimate strength for bolt shear and bearing on material thicknesses is OK by inspection.

Alternative 2

> Moment plate field welded to column flange and top flange of beam, bottom flange of beam welded directly to column flange with groove weld

As in alternative 1, the moment plate is designed to transmit the factored beam flange force of 786 kN.

Plate area required (gross) is $786 / (0.9 \times 0.300) = 2\,910$ mm^2

Select plate width narrower than beam flange width to permit downhand welding.

Try 150 mm plate. (Maximum weld size would be $(178 - 4 - 150) / 2 = 12$ mm.)

Plate thickness required is $2\,910 / 150 = 19.4$ mm

Use 20 mm plate (first preference thickness)

From Table 3-24, for E480XX electrode, 12 mm fillet weld = 1.83 kN/mm

Weld length required is $786 / 1.83 = 430$ mm

End weld length is 150 mm, therefore length each side is $(430 - 150)/2 = 140$ mm

It is generally recommended that an unwelded length of plate equal to at least 1.2 times the plate width be provided.

Therefore minimum plate length is $140 + (1.2 \times 150) = 320$ mm

Use 20x150 x 320 plate welded to column flange with full penetration groove weld and welded to top flange of beam with 430 mm of 12 mm fillet welds.

A possible third alternative would be to field weld the top and bottom flanges of the beam directly to the column flange with full penetration groove welds using backing bars fitted against the column flange.

(c) Column Shear Capacity

The column will be subject to a shear force due to the unbalanced moment. S16.1-94, Clause 21.3 requires stiffening of the column web if this shear exceeds

$V_r = 0.55 \, \phi \, w \, d \, F_y$
$= 0.55 \times 0.9 \times 9.1 \times 310 \times 0.350 = 489$ kN

Shear force is $(320 - 240) \times 1\,000 / 407 = 197$ kN   < 489   — OK

Thus, no reinforcing of the web is required for shear. (Shear forces from the column, above and below the moment connections, are ignored for simplicity.)

(d) Column Stiffeners

Design the column stiffeners to S16.1-94, Clause 21.3.

Clause 21.3(a): $B_r = 0.9 \times 9.1 \, (12.8 + (5 \times 33)) \, 0.350 = 510$ kN   < 786

Therefore, stiffeners are required opposite the compression flange for capacity of $786 - 510 = 276$ kN

Clause 21.3(b): $T_r = 7 \times 0.9 \times 16.3^2 \times 0.350 = 586$ kN   < 786

Stiffeners are also required opposite the tension flange for capacity of $786 - 586 = 200$ kN

Total stiffener area required at compression flange is

$276 / (0.9 \times 0.300) = 1\,020$ mm^2

Maximum b/t ratio is $145 / \sqrt{300} = 8.37$ (to match the Class 1 beam)

Try 90 mm wide stiffener each side of column web (beam flange is 178 mm wide).

Minimum $t = 90 / 8.37 = 10.8$ mm   — try 12 mm

Effective stiffener width to clear column k_1 distance is

$(178 / 2) - 20 = 69$ mm

Effective stiffener area is $2 \times 69 \times 12 = 1\,660$ mm^2 > 1 020 mm^2 — OK

Use 12x90 stiffener each side of column web opposite compression flange.

Use same stiffeners opposite tension flange.

(e) Stiffener Welds

Welds connecting stiffeners to column flange must be sufficient to develop a total force in the two stiffeners of 276 kN.

For double fillet welds at stiffener ends (length 69 mm), weld resistance required is

$276 / (2 \times 69) = 2.00$ kN/mm

From Table 3-24, page 3-41, 8 mm E480XX fillet welds provide

$2 \times 1.22 = 2.44$ kN/mm — OK

Welds connecting stiffeners to column web must transfer shear forces due to unbalanced beam moment of $197 / 2 = 98.5$ kN per side.

Approximate weld length available is T distance of 244 mm — assume 230 mm

Weld resistance required is $98.5 / 230 = 0.43$ kN/mm (one-sided weld will do).

Use single 5 mm fillet weld on each stiffener for 0.76 kN/mm. (Table 3-24)

Example 2

Given:

Design an exterior beam-to-column connection for an elastically analysed frame, in which the column size is the same as example 1 and the beam is a W460x74 having a factored end moment of 320 kN·m and a factored end shear of 130 kN.

W310x86 column W460x74 beam
See example 1 $t = 14.5$ mm
for dimensions $w = 9.0$ mm
  $d = 457$ mm
  $b = 190$ mm
  Class 1 (in bending)

Solution

This example is basically an extension of Example 1 and the solutions given are intended only to provide information on other possibilities.

(a) Web Connection

Use an unstiffened seat angle shop welded to the column to carry the beam shear and to support the beam during erection.

From Table 3-43, page 3-76, for a beam web of 9 mm and a seat length of 230 mm, a 9.5 mm thick angle will provide a beam web bearing capacity of 262 kN, > 130 kN. Also a vertical leg of 127 mm with 6 mm fillet welds provides a vertical leg connection capacity of 158 kN, > 130 kN.

Use 127x89x9.5 angle x 230 mm long with 127 mm leg vertical, welded to column flange with 6 mm E480XX fillet welds.

(b) Flange Connection

Assume field welded connection with full penetration groove welds connecting top and bottom flanges of the beam directly to the column flange (suggested alternative 3 in Example 1). The seat angle would serve as backing for the bottom flange weld.

(c) Column Shear Capacity

Shear force is $320 \times 1\,000 / 457 = 700$ kN (shears from column ignored)

Diagonal stiffeners will be used to carry shear in excess of the 489 kN shear capacity of the column web (see Example 1).

Horizontal component of stiffener force is $700 - 489 = 211$ kN

If θ is angle between stiffener and horizontal plane,

$$\cos\theta = 310 / (310^2 + 457^2)^{1/2} = 0.561$$

Force in stiffener is $211/\cos\theta = 211 / 0.561 = 376$ kN

Total stiffener area required is $376 / (0.9 \times 0.300) = 1\,390$ mm^2

For 90 mm wide stiffener, effective width is 69 mm (as in Example 1).

Stiffener thickness required is $1\,390 / (2 \times 69) = 10.1$ mm — try 12 mm

b/t is $90/12 = 7.5$ < 8.37 maximum — OK (see Example 1)

Use one 12x90 diagonal stiffener each side of column web.

(d) Horizontal Column Web Stiffeners

Calculations similar to those for Example 1 show that two 12x90 stiffeners are OK.

(e) Stiffener Welds

Diagonal Stiffeners.

Welds connecting the stiffeners to the column flanges must be sufficient to develop a total force in the two stiffeners of 376 kN (see above).

For double fillet welds at ends of stiffeners (length = 69 mm) weld resistance required is $376 / (2 \times 69) = 2.72$ kN/mm

From Table 3-24, 10 mm E480XX fillet welds provide $2 \times 1.52 = 3.04$ kN/mm

Use 10 mm E480XX fillet welds top and bottom at each end of stiffeners, and nominal 5 mm stitch welds between stiffener and column web.

Note: *The angles formed by the diagonal stiffener with the column flange and the horizontal stiffener often are not suitable for fillet welds ($60° \leq \theta \leq 120°$). However, by providing the size of a fillet weld, the designer indicates the required weld resistance. Shop detail drawings used by the fabricator must show an appropriate configuration of fillet and groove welds that will develop that resistance.*

Horizontal Stiffeners

The methods that were used for Example 1 show that the end welds must develop total forces in the stiffeners of 185 kN, for which double 6 mm fillet welds are OK.

Welds connecting the horizontal stiffeners to the column web need transfer only a portion of the stiffener load to the column web, as most of that load proceeds down the diagonal stiffeners. However, it is conservative to size these welds to transfer the total load in the stiffeners. For an approximate weld length of 230 mm (see Example 1), weld resistance required is

$185 / (2 \times 230) = 0.40$ kN/mm, for which a single 5 mm weld on each stiffener provides 0.76 kN/mm.

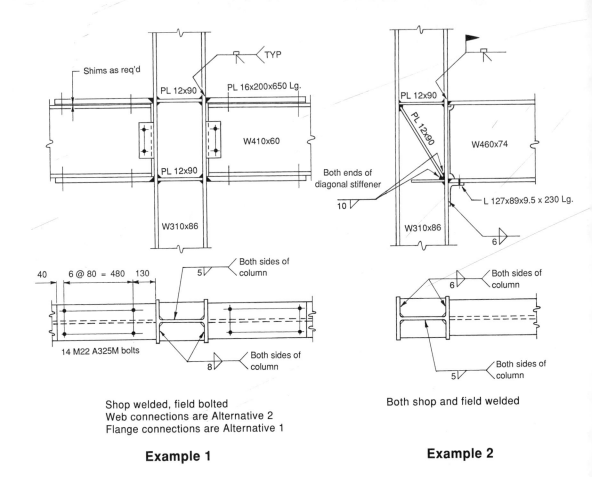

Example 1

Shop welded, field bolted
Web connections are Alternative 2
Flange connections are Alternative 1

Example 2

Both shop and field welded

Note: The angles formed by the diagonal stiffener with the column flange and the horizontal stiffener often are not suitable for fillet welds ($60° \leq \theta \leq 120°$). However, by providing the size of a fillet weld, the designer indicates the required weld resistance. Shop detail drawings used by the fabricator must show an appropriate configuration of fillet and groove welds that will develop that resistance.

HOLLOW STRUCTURAL SECTION CONNECTIONS

General

Hollow structural sections are frequently used for columns, trusses and space structures due to aesthetics, reduced weight for compression members and other reasons. This section of the Handbook presents sketches of some commonly used connections (Figures 3-5 to 3-9), and information for HSS welds (Figure 3-10, and Tables 3-46 and 3-47). Since the behaviour and resistance of welded HSS connections are not always intuitive, their detail design should be undertaken only by engineers who are familiar with current literature on the subject.

The connections illustrated in Figures 3-5 and 3-6 are simple shear connections designed in a conventional manner. The recommended width to thickness ratio of the tee flange is 13 or more in order to ensure suitable rotational flexibility.

The International Committee for the Study and Development of Tubular Structures (CIDECT) has played a major role in sponsoring international research that has resulted in the International Institute of Welding (IIW) making comprehensive design recommendations for HSS connections. Subsequently, a series of "state-of-the-art" design guides edited by CIDECT has been produced (see references). Based on this research, CISC has published *Hollow Structural Section Connections and Trusses—a Design Guide* (1997), which is presented as a practical and comprehensive 464 page book dedicated to the Canadian market with design examples that generally meet the requirements of CAN/CSA-S16.1-94.

Basic Considerations for Welded HSS Connections

A prime application of HSS members is in architecturally exposed areas where careful attention must be given to aesthetics of the connections. Simple welded connections without the use of reinforcing material often present the most pleasing and economical solutions. The following fundamentals should be kept in mind.

1. HSS members should not be selected on the basis of minimum mass. That implies that the members will need to be connected for their full capacity, which often is not possible without detail reinforcing material.

2. The force that can be transmitted from one HSS member to another is known as the "connection resistance" and is a function of the relative dimensions and wall thicknesses of the members. It is frequently less than the capacity of the connected member. Therefore, it is necessary to establish that the contemplated members have sufficient connection resistance before the member sizes can be confirmed.

3. Furthermore, design documents that specify "connect for member capacity" often have the effect of causing HSS connections to be reinforced, even if that was not the intent.

4. Square and rectangular HSS are much easier to fabricate than are round HSS because of the complexities of the connection profiles.

5. Try to avoid connections whose members are the same width. Welding is simpler and less expensive if fillet welds can be used along the sides of the connected member. On the other hand, connection resistance increases as the width of branch members approaches the width of main members, and is a maximum when the widths are the same. Therefore, to obtain optimum strength and economy with a square or rectangular HSS connection, the branch member should be as wide as possible, but not wider than the main member minus about five or six times the wall thickness of the main member (since the outer corner radius is generally between two and three times the wall thickness).

6. Connection resistance is improved when branch members have thin walls relative to the main member. A smaller size main member with a thicker wall may not be much heavier than a larger one with a thinner wall.

7. Full penetration welds are seldom justified (other than for member splices). They are not advantageous where connection resistance is less than the member capacity. In addition, they are not prequalified for HSS, and the certification for welders is more difficult. Inspection is much more difficult.

8. Ultrasonic inspection has limited application to HSS connections, and radiographic inspection is often only applicable to full strength splicing of members.

Additional Considerations for HSS Trusses

1. Optimum economy can often be achieved by reducing the number of different size members that are used in a truss. It is less expensive to procure and handle a relatively large amount each of just of few sizes than a small amount each of many sizes.

2. Simple gap connections are usually the most economical when connecting pairs of web members to a truss chord. Overlap connections require additional profiling of members, more precise fitting, and sometimes interrupted fitting to perform concealed welding. Reinforced connections are generally the most expensive.

3. If fatigue is a design consideration, careful attention should be paid to the connection details. It is suggested that overlap connections of at least 50% be used for trusses subjected to fatigue loading.

4. Primary bending moments due to eccentricity e (Figure 3-7) may be ignored, with regard to connection design, provided the intersection of the centre lines of the web members lies within the following range measured from the centre line of the chord: 25% of the chord depth towards the outside of the truss, and 55% of the chord depth towards the inside of the truss.

5. Secondary bending moments (due to local connection deformations) may be neglected provided dimensional parameters of the connected members fall within ranges presented in Packer and Henderson (1997).

6. Since the effectiveness of load transfer from one HSS section to another is more a function of dimensional parameters of the members connected than it is of the amount of welding, Packer and Henderson (1997) outline methods to calculate connection efficiency and weld effectiveness.

7. Research by Frater and Packer (1992a, 1992b) and by Packer and Cassidy (1995) has established that welding for HSS truss connections can be related to the loads rather than the member resistances. However, when web members of gap K or N connections are inclined at 60° or more to the chord, welds along the heel of the web members should be considered ineffective; when the webs are inclined at 50° or less, welds on all four faces of the webs are effective. When web members of T, Y and X connections are inclined at 60° or more to the chord, welds along both the toe and heel of the web members should be considered ineffective; when the webs are inclined at 50° or less, welds along the toes of the web members should be considered ineffective.

8. Profiling of round members is generally required when they are joined to other members. If aesthetics allow the web members to have the ends flattened instead of profiled, cost savings may be achieved.

E480XX electrodes are normally used for HSS supplied from 350W material with 450 MPa minimum ultimate stress. Table 3-46 on page 3-97 gives the size of fillet welds necessary to develop the strength of the parent material in either shear or tension in accordance with Clause 13.13.2 of S16.1-94.

In HSS connections, members are usually welded all around. Table 3-47 on page 3-98 gives the length of welds for square and rectangular web members connected to chord members at various angles θ, calculated in accordance with AWS D1.1 (1994).

References:

AWS. 1994. Structural welding code – steel, D1.1, ANSI/AWS. American Welding Society, Miami, Florida.

FRATER, G.S., and PACKER, J.A. 1992a. Weldment design for RHS truss connections. I: Applications. Journal of Structural Engineering, American Society of Civil Engineers, **118**(10), pp. 2784–2803.

FRATER, G.S., and PACKER, J.A. 1992b. Weldment design for RHS truss connections. II: Experimentation. Journal of Structural Engineering, American Society of Civil Engineers, **118**(10), pp. 2804–2820.

PACKER, J.A., and CASSIDY, C.E. 1995. Effective weld lengths for HSS T, Y and X connections. Journal of Structural Engineering, American Society of Civil Engineers, **121**(10).

PACKER, J.A., and HENDERSON, J.E. 1997. Hollow structural section connections and trusses—a design guide. Canadian Institute of Steel Construction, Willowdale, Ontario.

WARDENIER, J., KUROBANE, Y., PACKER, J.A., DUTTA, D., and YEOMANS, N. 1991. Design guide for circular hollow section (CHS) joints under predominantly static loading. CIDECT (ed.) and Verlag TÜV Rheinland GmbH, Köln, Germany.

PACKER, J.A., WARDENIER, J., KUROBANE, Y., DUTTA, D., and YEOMANS, N. 1992. Design guide for rectangular hollow section (RHS) joints under predominantly static loading. CIDECT (ed.) and Verlag TÜV Rheinland GmbH, Köln, Germany.

Figure 3-5

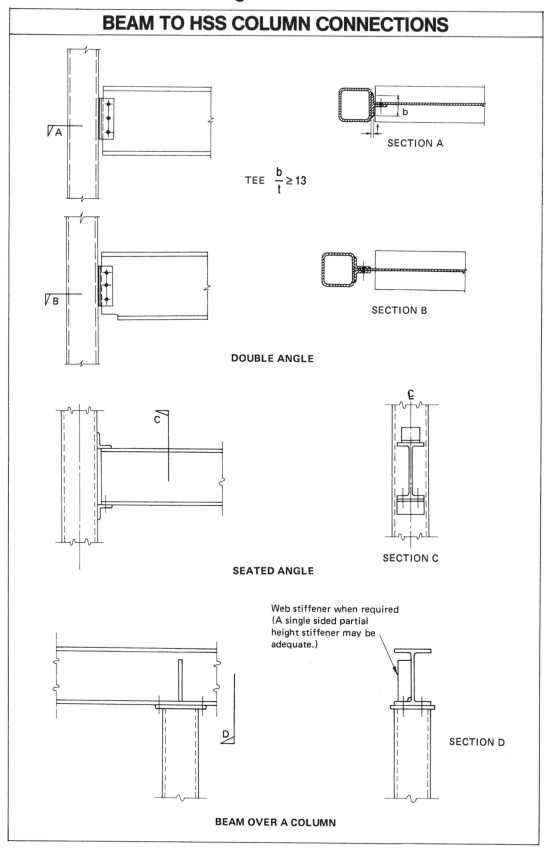

Figure 3-6
TRUSS TO COLUMN AND GIRDER CONNECTIONS

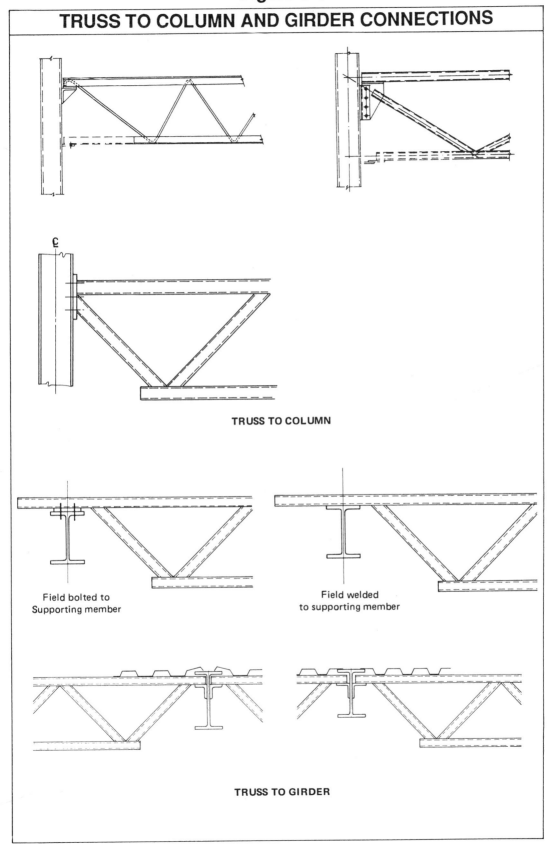

Figure 3-7
HSS TRUSS CONNECTIONS

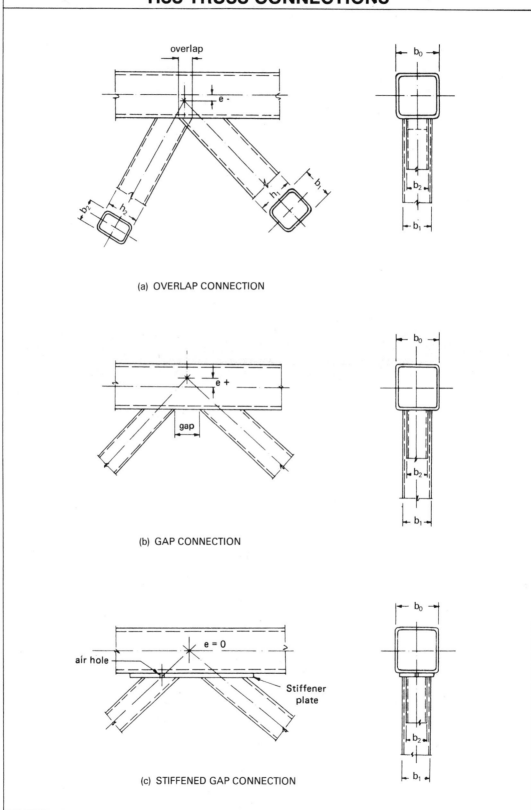

(a) OVERLAP CONNECTION

(b) GAP CONNECTION

(c) STIFFENED GAP CONNECTION

Figure 3-8
CONNECTIONS FOR MOMENT AND SHEAR

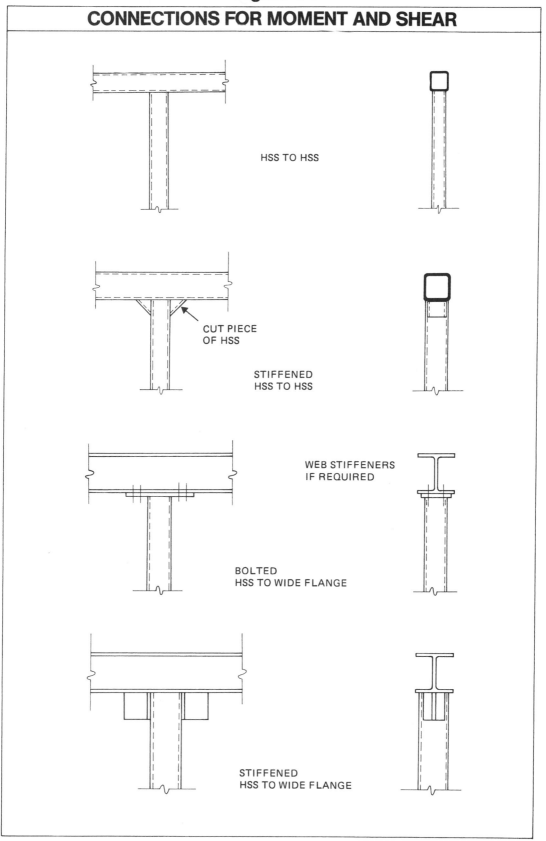

Figure 3-9

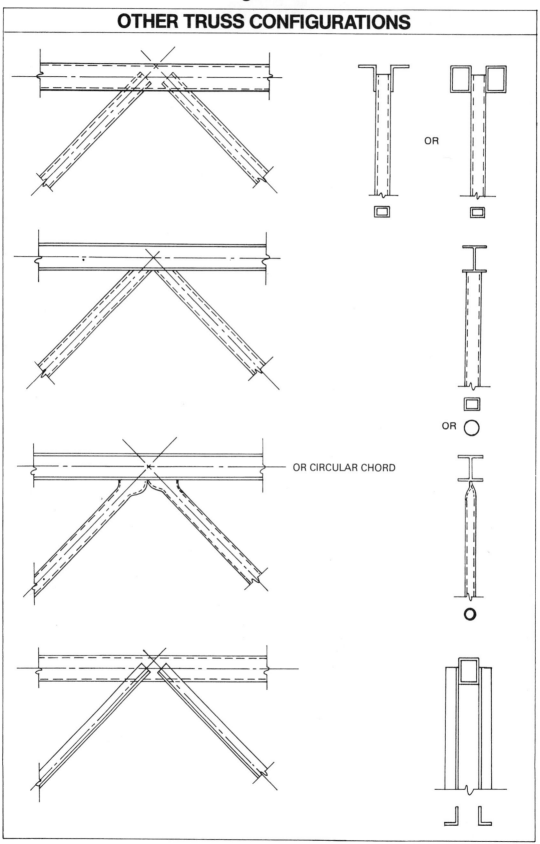

Figure 3-10

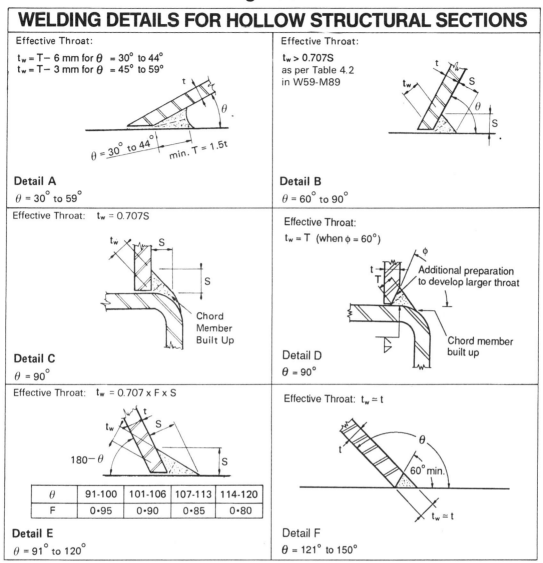

TABLE 3-46
F_u = 450 MPa
E480XX Electrodes

HSS CONNECTIONS
90° Fillet Size to Develop Wall Strength

Wall Thickness (mm)	Fillet Leg Size (mm)						
	Wall in Shear	Wall in Tension: angle between weld axis and direction of force*					
		0°	15°	30°	45°	50°	90°
3.81	6	8	8	8	8	6	6
4.78	8	10	10	10	8	8	8
6.35	10	14	14	12	10	10	10
7.95	12	18	16	14	14	14	14
9.53	14	20	20	18	16	16	16
11.13	16	24	22	20	18	18	18
12.70	18	26	26	22	20	20	20

*See CAN/CSA-S16.1-94, Clause 13.13.2.2.

LENGTH OF WELD IN MILLIMETRES
HSS Web Members

Table 3-47

HSS b x h x t (mm)	| Angle θ Between Web and Chord Member									
	30°	35°	40°	45°	50°	55°	60°	65°	70°	90°
25 x 25 x 3.8	133	121	113	107	102	98	95	93	91	89
32 x 32 x 3.8	171	157	146	138	132	127	123	120	118	114
38 x 38 x 4.8	204	187	174	164	157	151	147	143	140	136
51 x 51 x 6.4	272	249	232	219	209	201	195	191	187	181
64 x 64 x 6.4	348	319	297	280	268	258	250	244	240	232
76 x 76 x 8.0	416	381	355	335	320	308	299	292	286	278
89 x 89 x 9.5	484	443	413	390	372	359	348	340	333	323
102 x 102 x 13	544	498	464	438	418	403	391	382	374	363
127 x 127 x 13	697	637	593	561	535	516	500	488	479	464
152 x 152 x 13	849	776	723	683	652	628	610	595	584	566
178 x 178 x 13	1 000	916	853	806	770	741	719	702	689	668
203 x 203 x 13	1 150	1 055	983	928	887	854	829	809	794	769
254 x 254 x 13	1 460	1 330	1 240	1 170	1 120	1 080	1 050	1 020	1 000	972
305 x 305 x 13	1 760	1 610	1 500	1 420	1 360	1 310	1 270	1 240	1 210	1 180
51 x 25 x 4.8	181	170	161	155	150	146	143	141	139	136
25 x 51 x 4.8	227	203	186	174	164	156	150	145	142	136
76 x 51 x 8.0	317	294	277	264	254	247	241	236	233	227
51 x 76 x 8.0	363	328	302	283	268	257	248	241	235	227
102 x 51 x 8.0	370	346	329	316	306	298	292	287	283	278
51 x 102 x 8.0	463	415	380	354	334	318	306	297	289	278
89 x 64 x 8.0	393	363	342	325	313	303	295	289	285	278
64 x 89 x 8.0	439	398	367	345	327	313	303	294	288	278
102 x 76 x 9.5	461	426	400	380	365	353	344	337	332	323
76 x 102 x 9.5	507	460	425	399	379	364	351	342	335	323
127 x 51 x 9.5	415	391	374	361	351	343	337	332	329	323
51 x 127 x 9.5	554	494	451	418	393	374	359	347	338	323
127 x 64 x 9.5	464	435	413	396	384	374	366	360	356	348
64 x 127 x 9.5	580	521	477	444	419	400	384	372	363	348
127 x 76 x 13	499	464	438	419	404	393	384	377	372	363
76 x 127 x 13	590	531	489	457	432	413	398	386	377	363
152 x 102 x 13	650	602	568	541	521	505	493	484	476	464
102 x 152 x 13	743	672	619	580	549	526	507	493	482	464
178 x 127 x 13	802	741	697	664	638	618	602	590	581	566
127 x 178 x 13	896	811	749	703	667	639	617	600	587	566
203 x 102 x 13	755	706	671	644	624	608	595	585	578	566
102 x 203 x 13	943	847	776	722	681	649	624	605	590	566
203 x 152 x 13	954	880	827	786	755	731	712	697	686	668
152 x 203 x 13	1 050	951	880	826	784	752	727	707	692	668
254 x 152 x 13	1 060	984	929	889	857	833	814	799	788	769
152 x 254 x 13	1 250	1 130	1 040	968	916	875	844	819	800	769
305 x 203 x 13	1 360	1 260	1 190	1 130	1 090	1 060	1 030	1 010	997	972
203 x 305 x 13	1 560	1 410	1 300	1 210	1 150	1 100	1 060	1 030	1 010	972

Notes:
1. Outside corner radius assumed equal to $2t$.
2. Perimeters shown in table are for the thickest wall HSS of each size; therefore perimeters are conservative for smaller wall thicknesses.
3. Perimeters calculated by: $K_a [4 \pi t + 2(b - 4t) + 2(h - 4t)]$, where $K_a = ((h/\sin \theta) + b) / (h + b)$.

TENSION MEMBERS

General

Members subject to axial tension (i.e., when the resultant tensile load on the member is coincident with the longitudinal centroidal axis of the member) can be proportioned assuming a uniform stress distribution. The factored tensile resistance is calculated on the basis of yielding on the gross area and fracture on the effective net area (or effective net area reduced for shear lag) as per Clause 13.2 of CAN/CSA-S16.1-94. Effective net area and effective net area reduced for shear lag are defined in Clause 12.3, which has been expanded from the 1989 edition to included angles bolted along one leg only. Also, the clauses regarding shear lag for welded connections have been clarified and now formally include the $1 - \bar{x}/L$ expression for outstanding elements of the member that are not connected.

Fracture on the net section, rather than yielding, is the limit state because, with designs based on factored loads, yielding is unlikely under the expressions of Clause 13.2(a)(ii) or 13.2(a)(iii). Further, any deformation that might occur will be minimal since the volume of material affected is small and the beneficial effects of strainhardening will be present.

Effective Net Area

Tables 3-48 to 3-51 are intended to simplify the calculation of effective net area according to the requirements of Clause 12.3.

Hole Diameters for Effective Net Area

Table 3-48 on page 3-106 lists the specified hole diameter for various bolt sizes according to Clause 23.3.2, and the diameter of holes for calculating effective net area according to Clause 12.3.2.

Reduction of Area for Holes

Table 3-49 on page 3-106 lists values for the reduction of area for holes of different diameter in material of various thicknesses.

Staggered Holes in Tension Members

Table 3-50 on page 3-107 lists values of $s^2/4g$ required to calculate the net width of any diagonal or zig-zag line of holes according to the requirements of Clause 12.3.1(c) for various pitches from 25 to 240 mm and for various gauges from 25 to 320 mm. Values of $s^2/4g$ for pitches and gauges between those listed can be interpolated.

Effective Net Area – Reduced for Shear Lag

Clause 12.3.3 of S16.1-94 contains provisions for determining the loss of efficiency due to shear lag when tension members are not connected by all their elements.

Shear Lag Values of $1 - \bar{x}/L$

Table 3-51 on page 3-107 lists values of $1 - \bar{x}/L$ as a function of $\bar{x}$ and L for use with Clause 12.3.3.3(c) when computing effective net area reduced for shear lag of section elements projecting from a welded connection.

Example 1

Given:

The hole pattern shown occurs at a bolted splice in a built-up tension member consisting of two C310x45* sections. Determine the net area and the tensile resistance of the member for M20 A325M bolts in material conforming to G40.21 grades 300W, 350W and 480W. (* no longer available from Canadian sources)

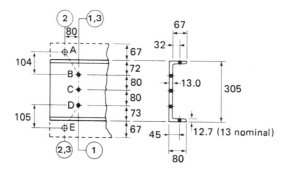

Solution:

Gross area

  Gross Area A_g for 2 – C310x45 is $2 \times 5\,690 = 11\,380$ mm^2

  Gross width is $305 + 2(80 - 13) = 439$ mm (S16.1-94, Clause 12.3.4)

Section 1—1

  Web width deduction for M20 bolt is 24 mm (Table 3-48)

  Less 3 holes, 3×24 $= -\ 72$ mm

Section 2—2

  Less 5 holes, 5×24 $= -\ 120$

  Plus $s^2/4g$ *for AB (s = 80, g = 104)* $= +\ 16$ (Table 3-50)

  Plus $s^2/4g$ *for DE (s = 80, g = 105)* $= +\ 16$

  $-\ 88$ mm (governs)

Values of $s^2/4g$ were obtained from Table 3-50 for $s = 80$ and interpolating by inspection for $g = 104$ and 105.

Section 3—3

  Less 4 holes, 4×24 $= -\ 96$

  Plus $s^2/4g$ $= +\ 16$

  $-\ 80$ mm

  Net width is $439 - 88 = 351$ mm

  Net area $A_{ne} = 2 \times 351 \times 13 = 9\,130$ mm^2

Tensile Resistances

  300W — ($F_u = 450$ MPa, from Table 3-5 on page 3-9)

    $T_r = \phi\, A_g\, F_y$ (Clause 13.2(a)(i))

      $= 0.9 \times 11\,380 \times 0.300 = 3\,070$ kN   (governs)

3-100

$T_r = 0.85 \, \phi \, A_{ne} \, F_u$ (Clause 13.2(a)(ii))

$= 0.85 \times 0.9 \times 9\,130 \times 0.450 = 3\,140$ kN

350W — ($F_u = 450$ MPa)

$T_r = 0.9 \times 11\,380 \times 0.350 \qquad = 3\,580$ kN

$T_r = 0.85 \times 0.9 \times 9\,130 \times 0.450 = 3\,140$ kN (governs)

480W — ($F_u = 590$ MPa)

$T_r = 0.9 \times 11\,380 \times 0.480 \qquad = 4\,920$ kN

$T_r = 0.85 \times 0.9 \times 9\,130 \times 0.590 = 4\,120$ kN (governs)

(More bolts than shown would be needed to develop these resistances.)

Example 2

Given:

A C380x50 tension member carries factored axial tension of 850 kN, and has been connected through only the web with ten M22 A325M bolts as illustrated. Confirm the adequacy of the connection for 300W material.

$d = 381$ mm

$w = 10.2$ mm

$A = 6\,430$ mm^2

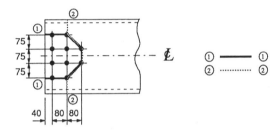

Solution:

Bolt shear

Bolt $V_r = 88.9$ kN (Table 3-4, page 3-8, threads intercepted)

∴ connection bolts $V_r = 10 \times 88.9 = 889$ kN > 850 — OK

Gross section yield

$T_r = \phi \, A_g \, F_y$ (Clause 13.2(a)(i))

$= 0.9 \times 6\,430 \times 0.300 = 1\,740$ kN > 850 — OK

Net section 1—1 rupture

Web width deduction for each bolt hole is 26 mm (Table 3-48)

Tension element $(A_n = w_n \, t)$ (Clause 12.3.1(a))

$= (75 - 26)10.2 \qquad\qquad = 500$ mm^2

Shear elements $(A_n = 0.6 \, L_n \, t)$ (Clause 12.3.1(b))

$= 2 \times 0.6 \, (40 + 80 - (1.5 \times 26)) \, 10.2 \quad = 991$

Inclined elements $(A_n = (w_n + s^2/4g) \, t)$ (Clause 12.3.1(c))

$= 2\,(75 - 26 + 21.3)\,10.2 \qquad\qquad = \underline{1\,430}$ ($s = 80, g = 75$)

$A_{ne} = 2\,920$ mm^2

$T_r = 0.85 \phi A_{ne} F_u$ (Clause 13.2(a)(ii))

$= 0.85 \times 0.9 \times 2\,920 \times 0.450 = 1\,010$ kN > 850 — OK

Net section 2—2 rupture

Gross cross section $= +\,6\,430$ mm^2

Less 4 holes, $\quad 4 \times 26 \times 10.2 \qquad = -\,1\,060$

Plus $2\,(s^2 t /4g) = 2\,(80^2 \times 10.2/(4 \times 75)) = +\quad 435$

$\qquad\qquad\qquad\qquad\qquad A_{ne} = \overline{5\,810}$ mm^2

Since section 2—2 passes through elements that are not connected (the two flanges), the net area must be reduced for shear lag.

$A'_{ne} = 0.85 \times 5\,810 = 4\,940$ mm^2 (Clause 12.3.3.2(c))

$T_r = 0.85 \phi A'_{ne} F_u$ (Clause 13.2(a)(iii))

$= 0.85 \times 0.90 \times 4\,940 \times 0.450 = 1\,700$ kN > 850 — OK

Therefore, the connection is adequate and is governed by the bolt shear resistance, 889 kN, and the net-section resistance through section 1—1, 1 010 kN.

Example 3

Given:

The flanges of a W200x52 tension member are connected to 16 mm gusset plates by 12 – M22 A325M bolts as shown. Material is 350W. Determine the connection resistance.

$t = 12.6$ mm

$A = 6\,660$ mm^2

$d = 206$ mm

$b = 204$ mm

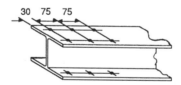

Solution:

Bolt shear

Bolt $V_r = 127$ kN (Table 3-4, page 3-8, threads excluded)

∴ connection bolts $V_r = 12 \times 127 = 1\,520$ kN

Note that the full shear resistance of the end row of bolts is available in spite of the 30 mm end distance. By removing the earlier criterion for reduced bearing resistance at bolts near an edge, S16.1-94 recognizes that the end bolts cannot tear out in isolation, but that the entire connection must fail.

Gross area yield

$T_r = \phi A_g F_y$ (Clause 13.2(a)(i))

$= 0.9 \times 6\,660 \times 0.350 = 2\,100$ kN

Net area rupture

Flange width deduction for each bolt hole is 26 mm (Table 3-48)

$$\begin{aligned}
\text{Gross area} &= 6\,660 \text{ mm}^2 \\
\text{Less 4 holes,} \quad 4 \times 26 \times 12.6 &= -1\,310 \\
A_{ne} &= 5\,350 \text{ mm}^2
\end{aligned}$$

Area reduction for shear lag is required because the web is not connected, and it will form part of the critical section.

As per S16.1-94, Clause 12.3.3.2(a), the reduction coefficient is 0.90 when there are at least 3 transverse rows of bolts, and the flange width is not less than 2/3 the member depth.

$$\therefore A'_{ne} = 0.90 \times 5\,350 = 4\,820 \text{ mm}^2$$

$T_r = 0.85\,\phi\,A'_{ne}\,F_u$ (Clause 13.2(a)(iii))

$ = 0.85 \times 0.9 \times 4\,820 \times 0.450 = 1\,660 \text{ kN}$

Therefore, the connection resistance is 1 520 kN (bolt shear governs).

Example 4

Given:

What is the tension resistance of a 300W C150x12 bolted with a single line of M20 A325M bolts in single shear as illustrated?

$A = 1\,540 \text{ mm}^2$

$w = 5.1 \text{ mm}$

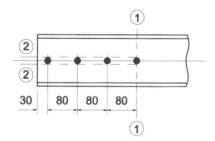

Solution:

Bolt shear

Bolt $V_r = 73.5$ kN (Table 3-4, page 3-8, threads intercepted)

$\therefore$ connection bolts $V_r = 4 \times 73.5 = 294$ kN

Material bearing at bolts

$B_r = 3\,\phi_b\,tdn\,F_u$ (Clause 13.10(c))

$ = 3 \times 0.67 \times 5.1 \times 20 \times 1 \times 0.450 = 92.3 \text{ kN} \quad > 73.5 \quad -\text{O.K.}$

Gross area yield

$T_r = \phi\,A_g\,F_y$ (Clause 13.2(a)(i))

$ = 0.9 \times 1\,540 \times 0.300 = 416 \text{ kN}$

Net section 1—1 rupture

Width deduction for one bolt hole is 24 mm (Table 3-48)

$$\therefore A_{ne} = 1\,540 - (24 \times 5.1) = 1\,420 \text{ mm}^2$$

Net area, reduced for shear lag; coefficient is 0.85 (Clause 12.3.3.2(c)(i))

$$\therefore A'_{ne} = 0.85 \times 1\,420 = 1\,210 \text{ mm}^2$$

$$T_r = 0.85\,\phi\,A'_{ne}\,F_u \qquad \text{(Clause 13.2(a)(iii))}$$

$$= 0.85 \times 0.9 \times 1\,210 \times 0.450 = 417 \text{ kN}$$

Net section 2—2 rupture

This failure is a form of block tear-out and consists of a pair of shear lines, one along each side of the bolts. A conservative measure of the length of each shear segment between the bolt holes is the clear distance between the bolts. That is, 80 – 20 = 60 mm for the interior segments and 30 – 20/2 = 20 mm for the end segment.

$$\therefore \text{ net shear length } L_n = (6 \times 60) + (2 \times 20) = 400 \text{ mm}$$

$$A_{ne} = 0.6\,L_n\,t \qquad \text{(Clause 12.3.1(b))}$$

$$= 0.6 \times 400 \times 5.1 = 1\,220 \text{ mm}^2$$

$$T_r = 0.85\,\phi\,A_{ne}\,F_u \qquad \text{(Clause 13.2(a)(ii))}$$

$$= 0.85 \times 0.9 \times 1\,220 \times 0.450 = 420 \text{ kN}$$

Therefore, the connection resistance is 294 kN (bolt shear governs).

Example 5

Given:

An HSS152x152x6.4 of 350W material carries 500 kN tension and is connected by a single plate welded into slots in the HSS walls as shown. The plate will be bolted between a pair of splice plates. Design the plate using 300W steel.

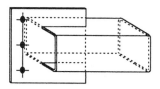

Solution:

Try a 12x230 mm plate

Bolt shear

Try three M22 A325M bolts; use 45 mm edge and end distances

Bolt $V_r = 88.9$ kN   (Table 3-4, page 3-8, threads intercepted)

$\therefore$ connection bolts $V_r = 3 \times 2 \times 88.9 = 533$ kN > 500 — OK

Gross area yield

$$T_r = \phi\,A_g\,F_y \qquad \text{(S16.1-94, Clause 13.2(a)(i))}$$

$$= 0.9 \times 12 \times 230 \times 0.300 = 745 \text{ kN} \quad > 500 \quad - \text{OK}$$

Net area rupture (through the bolt line)

If holes are known to be drilled rather than punched, the specified hole diameter can be used for the net section.   (Clause 12.3.2)

$$\therefore A_{ne} = (230 - (3 \times 24))\, 12 = 1\,900 \text{ mm}^2$$

$$T_r = 0.85\, \phi\, A_{ne}\, F_u \qquad \text{(Clause 13.2(a)(ii))}$$

$$= 0.85 \times 0.9 \times 1\,900 \times 0.450 = 654 \text{ kN} \quad > 500 \quad - \text{OK}$$

Net area rupture (by bolts pulling out the end of the plate)

When the end distance is not less than 1½ bolt diameters (as required by Clause 22.8) from the centre of the bolt to the end of the member, the failure mode is a pair of shear lines at the sides of the bolt.

A conservative measure of the length of such a shear segment is the clear distance from the bolt (centered in the hole) to the edge of the material. That is, $45 - 22/2 = 34$ mm.

$$\therefore \text{net shear length } L_n = 3 \times 2 \times 34 = 204 \text{ mm}$$

$$A_{ne} = 0.6\, L_n\, t \qquad \text{(Clause 12.3.1(b))}$$

$$= 0.6 \times 204 \times 12 = 1\,470 \text{ mm}^2$$

$$T_r = 0.85\, \phi\, A_{ne}\, F_u \qquad \text{(Clause 13.2(a)(ii))}$$

$$= 0.85 \times 0.9 \times 1\,470 \times 0.450 = 506 \text{ kN} \quad > 500 \quad - \text{OK}$$

Shear Lag for the Plate

Clause 12.3.3.3(b) in S16.1-94 provides for shear lag in plates that are connected by a pair of welds parallel to the load. The effective net area is reduced if the length of the welds is less than $2w$.

Distance between welds, $w = 152$ mm; try weld length $= w$

$$\therefore A'_{ne} = 0.75 \times 152 \times 12 = 1\,370 \text{ mm}^2 \qquad \text{(Clause 12.3.3.3(b)(iii))}$$

Plate area beyond the welds is $(230 - 152)\, 12 = 936 \text{ mm}^2$

Tensile resistance of the plate at the welds

$$T_r = 0.85\, \phi\, A'_{ne}\, F_u \qquad \text{(Clause 13.2(a)(iii))}$$

$$= 0.85 \times 0.9\, (1\,370 + 936)\, 0.450 = 794 \text{ kN} \quad > 500 \quad - \text{OK}$$

Welds

Try weld length $L = 150$ mm

Fillet weld size is $500 \text{ kN} / (4 \times 150 \times 0.153) = 5.4$ mm — use 6

Shear Lag for the HSS

The distance between the welds measured along the developed wall of the HSS is about 300 mm. This is beyond the range of L/w listed in Clause 12.3.3.3(b), but one can very conservatively ignore the additional width and deem w to be 150 mm. Then,

$$T_r = 0.85\, \phi\, A'_{ne}\, F_u$$

$$= 0.85 \times 0.9 \times 2\, (150 \times 6.35)\, 0.450 = 656 \text{ kN} \quad > 500 \quad - \text{OK}$$

Therefore a PL12x230 x 250 mm long, slotted 150 mm into the HSS, meets the requirements of the standard.

A less conservative treatment for shear lag reduced net area is possible for values of L/w less than 1.0 by extrapolating the values of the coefficients in Clause 12.3.3.3(b). However, for L/w less than approximately 0.5, the resistance of the welds (or the shear resistance of the plate along the welds) governs instead of shear lag.

HOLE DIAMETERS FOR EFFECTIVE NET AREA

Table 3-48

Bolt Size		Specified Hole Diameter Clause 23.3.2		Diameter for Net Area** Clause 12.3.2	
Metric	Imperial	Nominal Size (mm)	Oversize (mm)	Nominal Size (mm)	Oversize (mm)
M16		18	20	20	22
	3/4	22*	–	24	–
M20		22*	24	24	26
M22		24*	26	26	28
	7/8	24*	–	26	–
M24		27*	30	29	32
	1	27*	–	29	–
M27		29	35	31	37
M30		32	38	34	40
M36		38	44	40	46

* For nominal diameter of hole, see Clause 23.3.2(e).
** Diameter for net area is same as specified hole diameter if hole is drilled (Clause 12.3.2, S16.1-94).

REDUCTION OF AREA FOR HOLES*

Table 3-49

Material Thickness (mm)	Diameter of Hole (mm)											
	20	22	24	26	28	29	31	32	34	37	40	44
5	100	110	120	130	140	145	155	160	170	185	200	220
6	120	132	144	156	168	174	186	192	204	222	240	264
7	140	154	168	182	196	203	217	224	238	259	280	308
8	160	176	192	208	224	232	248	256	272	296	320	352
9	180	198	216	234	252	261	279	288	306	333	360	396
10	200	220	240	260	280	290	310	320	340	370	400	440
11	220	242	264	286	308	319	341	352	374	407	440	484
12	240	264	288	312	336	348	372	384	408	444	480	528
13	260	286	312	338	364	377	403	416	442	481	520	572
14	280	308	336	364	392	406	434	448	476	518	560	616
16	320	352	384	416	448	464	496	512	544	592	640	704
18	360	396	432	468	504	522	558	576	612	666	720	792
20	400	440	480	520	560	580	620	640	680	740	800	880
22	440	484	528	572	616	638	682	704	748	814	880	968
25	500	550	600	650	700	725	775	800	850	925	1 000	1 100
28	560	616	672	728	784	812	868	896	952	1 040	1 120	1 230
30	600	660	720	780	840	870	930	960	1 020	1 110	1 200	1 320
32	640	704	768	832	896	928	992	1 020	1 090	1 180	1 280	1 410
35	700	770	840	910	980	1 020	1 080	1 120	1 190	1 300	1 400	1 540
38	760	836	912	988	1 060	1 100	1 180	1 220	1 290	1 410	1 520	1 670
40	800	880	960	1 040	1 120	1 160	1 240	1 280	1 360	1 480	1 600	1 760
45	900	990	1 080	1 170	1 260	1 300	1 400	1 440	1 530	1 660	1 800	1 980
50	1 000	1 100	1 200	1 300	1 400	1 450	1 550	1 600	1 700	1 850	2 000	2 200
55	1 100	1 210	1 320	1 430	1 540	1 600	1 700	1 760	1 870	2 040	2 200	2 420
60	1 200	1 320	1 440	1 560	1 680	1 740	1 860	1 920	2 040	2 220	2 400	2 640
70	1 400	1 540	1 680	1 820	1 960	2 030	2 170	2 240	2 380	2 590	2 800	3 080
80	1 600	1 760	1 920	2 080	2 240	2 320	2 480	2 560	2 720	2 960	3 200	3 520
90	1 800	1 980	2 160	2 340	2 520	2 610	2 790	2 880	3 060	3 330	3 600	3 960
100	2 000	2 200	2 400	2 600	2 800	2 900	3 100	3 200	3 400	3 700	4 000	4 400

*Area (mm^2) = Diameter of hole (mm) times material thickness (mm).

Table 3-50

STAGGERED HOLES IN TENSION MEMBERS
Values of $s^2/4g$

Pitch "s" (mm)	Gauge "g" (mm)														
	25	30	35	40	45	50	60	70	80	100	120	160	200	240	280	320
25					3	3	3	2	2	2	1	1	1	1	1	0
30				6	5	4	4	3	3	2	2	1	1	1	1	1
35			9	8	7	6	5	5	4	3	3	2	2	1	1	1
40		13	11	10	9	8	7	6	5	4	3	2	2	2	1	1
45	20	17	14	13	11	10	8	7	6	5	4	3	3	2	2	2
50	25	21	18	16	14	12	10	9	8	6	4	4	3	3	2	2
55	30	25	22	19	17	15	13	11	9	8	6	5	4	3	3	2
60	36	30	26	22	20	18	15	13	11	9	7	6	4	4	3	3
65	42	35	30	26	23	21	18	15	13	11	9	7	5	4	4	3
70	49	41	35	31	27	24	20	17	15	12	10	8	6	5	4	4
75		47	40	35	31	28	23	20	18	14	12	9	7	6	5	4
80			46	40	36	32	27	23	20	16	13	10	8	7	6	5
90				51	45	40	34	29	25	20	17	13	10	8	7	6
100						50	42	36	31	25	21	16	12	10	9	8
110							50	43	38	30	25	19	15	13	11	9
120									45	36	30	22	18	15	13	11
130										42	35	26	21	18	15	13
140										49	41	31	24	20	17	15
150											47	35	28	23	20	18
160												40	32	27	23	20
170												45	36	30	26	23
180												51	40	34	29	25
190													45	38	32	28
200													50	42	36	31
210														46	39	34
220														50	43	38
230															47	41
240																45

Table 3-51

SHEAR LAG
Values of $1 - \bar{x}/L$

L (mm)	$1 - \bar{x}/L$ — Distance $\bar{x}$ (mm)														
	10	15	20	25	30	35	40	45	50	55	60	65	70	80	90	100
40	0.75	0.63	0.50	0.38	0.25	0.13										
80	0.88	0.81	0.75	0.69	0.63	0.56	0.50	0.44	0.38	0.31	0.25	0.19	0.13			
120	0.92	0.88	0.83	0.79	0.75	0.71	0.67	0.63	0.58	0.54	0.50	0.46	0.42	0.33	0.25	0.17
160	0.94	0.91	0.88	0.84	0.81	0.78	0.75	0.72	0.69	0.66	0.63	0.59	0.56	0.50	0.44	0.38
200	0.95	0.93	0.90	0.88	0.85	0.83	0.80	0.78	0.75	0.73	0.70	0.68	0.65	0.60	0.55	0.50
240	0.96	0.94	0.92	0.90	0.88	0.85	0.83	0.81	0.79	0.77	0.75	0.73	0.71	0.67	0.63	0.58
280	0.96	0.95	0.93	0.91	0.89	0.88	0.86	0.84	0.82	0.80	0.79	0.77	0.75	0.71	0.68	0.64
320	0.97	0.95	0.94	0.92	0.91	0.89	0.88	0.86	0.84	0.83	0.81	0.80	0.78	0.75	0.72	0.69
360		0.96	0.94	0.93	0.92	0.90	0.89	0.88	0.86	0.85	0.83	0.82	0.81	0.78	0.75	0.72
400		0.96	0.95	0.94	0.93	0.91	0.90	0.89	0.88	0.86	0.85	0.84	0.83	0.80	0.78	0.75
440			0.95	0.94	0.93	0.92	0.91	0.90	0.89	0.88	0.86	0.85	0.84	0.82	0.80	0.77
480			0.96	0.95	0.94	0.93	0.92	0.91	0.90	0.89	0.88	0.86	0.85	0.83	0.81	0.79

PART FOUR
COMPRESSION MEMBERS

General Information... 4–3
Limits on Width-Thickness Ratios............................... 4–5
    Width-Thickness Ratios: Compression Elements.............. 4–6
    Class of Sections for Beam-Columns......................... 4–7
Unit Factored Compressive Resistances for Compression Members... 4–12
Stability Effects... 4–18
Bending Factors for Beam-Columns.............................. 4–23
ω_1 — Equivalent Uniform Bending Coefficients........... 4–26
C_e / A — Euler Buckling Load per Unit Area................. 4–27
Amplification Factor U.. 4–28
Factored Axial Compressive Resistances of Columns............. 4–29
    WWF Shapes – CSA-G40.21 350W Steel........................ 4–32
    W Shapes – CSA-G40.21 350W Steel.......................... 4–39
    HSS — Class C – CSA-G40.21 350W Steel..................... 4–46
    HSS — Class H – CSA-G40.21 350W Steel..................... 4–70
Design of Beam-Columns.. 4–94
Factored Moment Resistances of Columns........................ 4–105
Double Angle Struts – CSA-G40.21 300W Steel................... 4–109
Bracing Assemblies.. 4–128
Column Base Plates.. 4–130
Anchor Rods... 4–135

GENERAL INFORMATION

Limits on Width-Thickness Ratios

See page 4-5.

Unit Factored Compressive Resistances for Compression Members, C_r/A

Pages 4-13 to 4-17 provide tables of unit factored compressive resistance for slenderness ratios from 1 to 200 for various yields of steel and values of n of 1.34 and 2.24. See page 4-12 for more information.

Stability Effects

Pages 4-18 to 4-22 describe the procedures given in Clause 8 of CAN/CSA-S16.1-94 to account for stability effects.

Bending Factors for Beam-Columns

Bending factors provide a simple means of converting the bending moment of a beam-column to an approximately equivalent axial load to permit easy use of the column load tables, Factored Axial Compressive Resistances of Columns. See page 4-23 for more information.

ω_1 — Equivalent Uniform Bending Coefficients

Table 4-7, page 4-26, lists values of ω_1 for various ratios M_{f1}/M_{f2} of factored end bending moments applied to beam-columns. The values of ω_1 are computed in accordance with the requirements of Clause 13.8.4, S16.1-94. See the Design of Beam-Columns, page 4-94, for more information.

C_e/A — Euler Buckling Load per Unit Area

Table 4-8, page 4-27, lists values of C_e/A for KL/r ratios varying from 1 to 200. The values of C_e/A have been computed in accordance with the definition in Clause 2.2, S16.1-94.

Amplification Factor U

Table 4-9, page 4-28, has been prepared to facilitate the design of beam-columns in accordance with the requirements of Clause 13.8, S16.1-94 which incorporates variable U in the factor U_1. Values of the amplification factor corresponding to various combinations of C_f/C_e are listed.

Factored Axial Compressive Resistances of Columns

These are the tables often referred to as "column load tables". See page 4-29 for description of contents and examples of use.

Design of Beam-Columns

For an explanation and illustrative examples on the design of beam-columns, see page 4-94.

Factored Moment Resistances of Columns

For the factored moment resistances M_{rx} and M'_{rx} for various unbraced lengths for sections not listed in the Beam Selection Table, see page 4-105.

Double Angle Struts

The tables of factored axial compressive resistances for double angle struts, Imperial series, starting on page 4-109 are based on the requirements of Clause 13.3.1, S16.1-94 with $n= 1.34$ for axis X-X and Clause 13.3.2 for axis Y-Y.

For Class 4 angles less than or equal to 4.5 mm thick, the resistances are computed based on the requirements of Clause 13.3.3(a). For Class 4 angles greater than 4.5 mm thick, the resistances are computed based on the requirements of Clauses 13.3.1 and 13.3.2 for axes X-X and Y-Y, respectively, with the area, A, taken as the effective area in accordance with CSA Standard S136. For simplicity, the effective area is computed based on the total leg width rather than the flat width.

Factored axial compressive resistances with respect to various effective lengths (in millimetres) relative to both the X-X and Y-Y axes, and the U-U and V-V axes for starred angles, are listed for angles made from CSA G40.21 300W. Yield stress F_y for G40.21 300W steel angles is 300 MPa for all thicknesses listed.

The resistances listed in the tables for axis Y-Y are based on closely spaced interconnectors; however, the resistances for axis Y-Y of double angle struts in practical configurations are generally lower than the listed values because of the additional slenderness of the component angles between interconnectors. The actual number of interconnectors and method of interconnection should therefore be taken into account in accordance with Clause 19.1.4, S16.1-94. Consult the design example on page 4-124. For starred angles, these requirements may be waived provided interconnectors are spaced no further than at the one-third points, in accordance with Clause 19.1.5.

The factored axial compressive resistances pertaining to effective lengths based on the Y-Y axis have been computed for angles spaced 10 mm back-to-back. Consult the design example on page 4-124 to obtain factored compressive resistances for different spacings.

The value r_z tabulated by itself and as part of the ratios r_z/r_x, r_z/r_y, r_z/r_u and r_z/r_v appearing with the properties of double angle struts, is the minimum radius of gyration of a single angle about its minor principal axis. Values for r_x, r_y, r_u and r_v are those for a double angle strut. See Part Six of this Handbook for a more comprehensive list of angle properties.

Column Base Plates

See page 4-130.

Anchor Bolts

See page 4-135.

LIMITS ON WIDTH-THICKNESS RATIOS

Table 4-1 below lists the particular width-thickness (b/t, h/w or D/t) ratio limits for various material yield strengths, for each general value given in Table 4-2.

Table 4-2 (page 4-6) which is taken from Clause 11 of CAN/CSA-S16.1-94 lists the width-thickness ratios for Class 1, 2, and 3 sections for various elements in compression. All sections not meeting these requirements are Class 4.

The class for webs in combined flexural and axial compression is a function of the ratio of the factored axial load to the axial compressive load at yield stress C_f/C_y as is shown in Table 4-2. Values of C_f/C_y at which the webs change class are tabulated in Table 4-3. Table 4-3 lists, on pages 4-8 to 4-11, WWF and W members in G40.21 Grade 350W, when used as beam-columns. Some members with webs that are always Class 1 are controlled by flanges that are not Class 1. Therefore, these members and their flange classification are also included in the table.

Table 4-1
WIDTH-THICKNESS LIMITS

General Value	F_y (MPa)									
	248	280	290	300	320	330	**350**	380	400	480
$145/\sqrt{F_y}$	9.2	8.7	8.5	8.4	8.1	8.0	**7.8**	7.4	7.2	6.6
$170/\sqrt{F_y}$	10.8	10.2	10.0	9.8	9.5	9.4	**9.1**	8.7	8.5	7.8
$200/\sqrt{F_y}$	12.7	12.0	11.7	11.5	11.2	11.0	**10.7**	10.2	10.0	9.1
$340/\sqrt{F_y}$	21.6	20.3	20.0	19.6	19.0	18.7	**18.2**	17.4	17.0	15.5
$420/\sqrt{F_y}$	26.7	25.1	24.7	24.2	23.5	23.1	**22.4**	21.5	21.0	19.2
$525/\sqrt{F_y}$	33.3	31.4	30.8	30.3	29.3	28.9	**28.1**	26.9	26.2	24.0
$670/\sqrt{F_y}$*	42.5	40.0	39.3	38.7	37.4	36.9	**35.8**	34.4	33.5	30.6
$840/\sqrt{F_y}$	53.3	50.2	49.3	48.5	47.0	46.2	**44.9**	43.1	42.0	38.3
$1100/\sqrt{F_y}$	69.8	65.7	64.6	63.5	61.5	60.6	**58.8**	56.4	55.0	50.2
$1700/\sqrt{F_y}$	108.0	101.6	99.8	98.1	95.0	93.6	**90.9**	87.2	85.0	77.6
$1900/\sqrt{F_y}$	120.7	113.5	111.6	109.7	106.2	104.6	**101.6**	97.5	95.0	86.7
$13\,000/F_y$	52.4	46.4	44.8	43.3	40.6	39.4	**37.1**	34.2	32.5	27.1
$18\,000/F_y$	72.6	64.3	62.1	60.0	56.2	54.5	**51.4**	47.4	45.0	37.5
$23\,000/F_y$	92.7	82.1	79.3	76.7	71.9	69.7	**65.7**	60.5	57.5	47.9
$66\,000/F_y$	266	236	228	220	206	200	**189**	174	165	138

* h/w limit for webs in pure compression (C_f/C_y = 1.0)

WIDTH-THICKNESS RATIOS: COMPRESSION ELEMENTS

Table 4-2

Description of Element	Section Classification		
	Class 1 Plastic Design	Class 2 Compact	Class 3 Non-compact
Legs of angles and elements supported along one edge, except as otherwise listed	—	—	$\dfrac{b}{t} \leq \dfrac{200}{\sqrt{F_y}}$
Angles in continuous contact with other elements; plate girder stiffeners	—	—	$\dfrac{b}{t} \leq \dfrac{200}{\sqrt{F_y}}$
Stems of T-sections	$\dfrac{b}{t} \leq \dfrac{145}{\sqrt{F_y}}$ *	$\dfrac{b}{t} \leq \dfrac{170}{\sqrt{F_y}}$ *	$\dfrac{b}{t} \leq \dfrac{340}{\sqrt{F_y}}$
Flanges of I or T sections; plates projecting from compression elements; outstanding legs of pairs of angles in continuous contact **	$\dfrac{b}{t} \leq \dfrac{145}{\sqrt{F_y}}$	$\dfrac{b}{t} \leq \dfrac{170}{\sqrt{F_y}}$	$\dfrac{b}{t} \leq \dfrac{200}{\sqrt{F_y}}$
Flanges of channels	—	—	$\dfrac{b}{t} \leq \dfrac{200}{\sqrt{F_y}}$
Flanges of rectangular hollow structural sections	$\dfrac{b}{t} \leq \dfrac{420}{\sqrt{F_y}}$	$\dfrac{b}{t} \leq \dfrac{525}{\sqrt{F_y}}$	$\dfrac{b}{t} \leq \dfrac{670}{\sqrt{F_y}}$
Flanges of box sections, flange cover plates and diaphragm plates, between lines of fasteners or welds	$\dfrac{b}{t} \leq \dfrac{525}{\sqrt{F_y}}$	$\dfrac{b}{t} \leq \dfrac{525}{\sqrt{F_y}}$	$\dfrac{b}{t} \leq \dfrac{670}{\sqrt{F_y}}$
Perforated cover plates	—	—	$\dfrac{b}{t} \leq \dfrac{840}{\sqrt{F_y}}$
Webs	$\dfrac{h}{w} \leq \dfrac{1100}{\sqrt{F_y}}\left(1 - 0.39\dfrac{C_f}{C_y}\right)$	$\dfrac{h}{w} \leq \dfrac{1700}{\sqrt{F_y}}\left(1 - 0.61\dfrac{C_f}{C_y}\right)$	$\dfrac{h}{w} \leq \dfrac{1900}{\sqrt{F_y}}\left(1 - 0.65\dfrac{C_f}{C_y}\right)$
Circular hollow sections in axial compression	—	—	$\dfrac{D}{t} \leq \dfrac{23\,000}{F_y}$
Circular hollow sections in flexural compression	$\dfrac{D}{t} \leq \dfrac{13\,000}{F_y}$	$\dfrac{D}{t} \leq \dfrac{18\,000}{F_y}$	$\dfrac{D}{t} \leq \dfrac{66\,000}{F_y}$

* See Clauses 11.1.2 and 11.1.3 of CAN/CSA-S16.1-94.
**Can be considered as Class 1 or Class 2 sections if angles are continuously connected by adequate mechanical fasteners or welds, and there is an axis of symmetry in the plane of loading.
For Class 4 sections, see Clause 11 of CAN/CSA-S16.1-94.

Table 4-3 CLASS OF SECTIONS FOR BEAM-COLUMNS
G40.21 350W

Designation	Web 1 $C_f/C_y \leq$	Web 2 $C_f/C_y \leq$	Web 3 $C_f/C_y \leq$	Flange	Designation	Web 1 $C_f/C_y \leq$	Web 2 $C_f/C_y \leq$	Web 3 $C_f/C_y \leq$	Flange
WWF2000x732	—	—	0.099	1	WWF700x245	0.027	0.590	0.657	1
x648	—	—	0.084	1	x214	—	0.573	0.643	2
x607	—	—	0.077	2	x196	—	0.563	0.635	3
x542	—	—	0.069	2	x175	—	0.573	0.643	1
					x152	—	0.557	0.630	1
WWF1800x700	—	0.106	0.251	1					
x659	—	0.097	0.243	1	WWF650x864	1.0	—	—	1
x617	—	0.088	0.236	1	x739	1.0	—	—	1
x575	—	0.079	0.228	2	x598	1.0	—	—	1
x510	—	0.070	0.221	2	x499	1.0	—	—	2
					x400	1.0	—	—	4
WWF1600x622	—	—	0.118	1					
x580	—	—	0.109	1	WWF600x793	1.0	—	—	1
x538	—	—	0.099	1	x680	1.0	—	—	1
x496	—	—	0.090	2	x551	1.0	—	—	1
x431	—	—	0.080	2	x460	1.0	—	—	1
					x369	1.0	—	—	3
WWF1400x597	—	0.174	0.308	1					
x513	—	0.151	0.289	1	WWF550x721	1.0	—	—	1
x471	—	0.140	0.279	2	x620	1.0	—	—	1
x405	—	0.128	0.270	2	x503	1.0	—	—	1
x358	—	0.128	0.270	1	x420	1.0	—	—	1
					x280	1.0	—	—	4
WWF1200x487	—	0.376	0.478	1					
x418	—	0.365	0.469	1	WWF500x651	1.0	—	—	1
x380	—	0.354	0.459	2	x561	1.0	—	—	1
x333	—	0.354	0.459	1	x456	1.0	—	—	1
x302	—	0.343	0.450	2	x381	1.0	—	—	1
x263	—	0.343	0.450	1	x343	1.0	—	—	1
					x306	1.0	—	—	2
WWF1100x458	—	0.325	0.435	1	x276	1.0	—	—	2
x388	—	0.312	0.424	1	x254	1.0	—	—	3
x351	—	0.299	0.413	2	x223	1.0	—	—	4
x304	—	0.299	0.413	1	x197	0.740	0.885	0.905	4
x273	—	0.286	0.402	2					
x234	—	0.286	0.402	1	WWF450x503	1.0	—	—	1
					x409	1.0	—	—	1
WWF1000x447	—	0.454	0.543	1	x342	1.0	—	—	1
x377	—	0.441	0.532	1	x308	1.0	—	—	1
x340	—	0.428	0.521	2	x274	1.0	—	—	1
x293	—	0.428	0.521	1	x248	1.0	—	—	2
x262	—	0.415	0.511	2	x228	1.0	—	—	2
x223	—	0.415	0.511	1	x201	1.0	—	—	3
x200	—	0.402	0.500	1	x177	0.939	0.967	0.974	4
WWF900x417	—	0.294	0.409	1	WWF400x444	1.0	—	—	1
x347	—	0.278	0.395	1	x362	1.0	—	—	1
x309	—	0.262	0.382	2	x303	1.0	—	—	1
x262	—	0.262	0.382	1	x273	1.0	—	—	1
x231	—	0.245	0.368	2	x243	1.0	—	—	1
x192	—	0.245	0.368	1	x220	1.0	—	—	1
x169	—	0.229	0.354	1	x202	1.0	—	—	2
					x178	1.0	—	—	3
WWF800x339	—	0.442	0.533	1	x157	1.0	—	—	3
x300	—	0.426	0.519	2					
x253	—	0.426	0.519	1	WWF350x315	1.0	—	—	1
x223	—	0.409	0.506	2	x263	1.0	—	—	1
x184	—	0.409	0.506	1	x238	1.0	—	—	1
x161	—	0.393	0.492	1	x212	1.0	—	—	1
					x192	1.0	—	—	1
					x176	1.0	—	—	1
					x155	1.0	—	—	2
					x137	1.0	—	—	2

See Table 4-2, page 4-6, for width-thickness criteria.
— Indicates web is never that class.

CLASS OF SECTIONS FOR BEAM-COLUMNS
G40.21 350W

Table 4-3

Designation	Web 1 $C_f/C_y \leq$	Web 2 $C_f/C_y \leq$	Web 3 $C_f/C_y \leq$	Flange	Designation	Web 1 $C_f/C_y \leq$	Web 2 $C_f/C_y \leq$	Web 3 $C_f/C_y \leq$	Flange
W1100x499*	0.846	0.928	0.942	1	W840x251*	0.534	0.800	0.833	1
x432*	0.551	0.806	0.839	1	x226*	0.420	0.752	0.794	1
x390*	0.350	0.723	0.769	1	x210*	0.323	0.712	0.760	1
x342*	0.091	0.616	0.680	1	x193*	0.218	0.669	0.723	1
					x176*	0.098	0.619	0.682	1
W1000x883*	1.0	—	—	1					
x749*	1.0	—	—	1	W760x582*	1.0	—	—	1
x641*	1.0	—	—	1	x531*	1.0	—	—	1
x591*	1.0	—	—	1	x484*	1.0	—	—	1
x554*	1.0	—	—	1	x434*	1.0	—	—	1
x539*	1.0	—	—	1	x389*	1.0	—	—	1
x483*	0.982	0.985	0.989	1	x350*	1.0	—	—	1
x443*	0.861	0.935	0.947	1	x314*	0.983	0.985	0.989	1
x412*	0.660	0.852	0.877	1	x284*	0.835	0.924	0.938	1
x371*	0.450	0.765	0.804	1	x257*	0.689	0.864	0.887	1
x321*	0.129	0.632	0.693	1					
x296*	0.130	0.632	0.693	1	W760x220*	0.677	0.859	0.883	1
					x196*	0.568	0.814	0.845	1
W1000x583*	1.0	—	—	1	x185*	0.475	0.775	0.813	1
x493*	1.0	—	—	1	x173*	0.403	0.745	0.788	1
x486*	1.0	—	—	1	x161*	0.307	0.706	0.754	1
x414*	1.0	—	—	1	x147*	0.206	0.664	0.719	1
x393*	0.917	0.958	0.966	1	x134*	—	0.557	0.630	2
x350*	0.660	0.852	0.877	1					
x314*	0.460	0.769	0.808	1	W690x802*	1.0	—	—	1
x272*	0.129	0.632	0.693	1	x548*	1.0	—	—	1
x249*	0.129	0.632	0.693	1	x500*	1.0	—	—	1
x222*	0.053	0.601	0.666	1	x457*	1.0	—	—	1
					x419*	1.0	—	—	1
W920x1188*	1.0	—	—	1	x384*	1.0	—	—	1
x967*	1.0	—	—	1	x350*	1.0	—	—	1
x784*	1.0	—	—	1	x323*	1.0	—	—	1
x653*	1.0	—	—	1	x289*	1.0	—	—	1
x585*	1.0	—	—	1	x265*	1.0	—	—	1
x534*	1.0	—	—	1	x240*	0.899	0.950	0.960	1
x488*	1.0	—	—	1	x217*	0.750	0.889	0.908	1
x446*	1.0	—	—	1					
x417*	0.932	0.964	0.971	1	W690x192*	0.759	0.893	0.911	1
x387*	0.841	0.926	0.940	1	x170*	0.636	0.842	0.869	1
x365*	0.757	0.892	0.911	1	x152*	0.430	0.756	0.797	1
x342*	0.662	0.852	0.878	1	x140*	0.308	0.706	0.755	1
					x125*	0.176	0.651	0.709	1
W920x381*	1.0	—	—	1					
x345*	0.873	0.940	0.951	1	W610x551*	1.0	—	—	1
x313*	0.793	0.907	0.923	1	x498*	1.0	—	—	1
x289*	0.638	0.843	0.869	1	x455*	1.0	—	—	1
x271*	0.533	0.799	0.833	1	x415*	1.0	—	—	1
x253*	0.404	0.746	0.788	1	x372*	1.0	—	—	1
x238*	0.299	0.702	0.752	1	x341*	1.0	—	—	1
x223*	0.214	0.667	0.722	1	x307*	1.0	—	—	1
x201*	0.106	0.623	0.685	1	x285*	1.0	—	—	1
					x262*	1.0	—	—	1
W840x576*	1.0	—	—	1	x241	1.0	—	—	1
x527*	1.0	—	—	1	x217	1.0	—	—	1
x473*	1.0	—	—	1	x195	0.941	0.968	0.975	1
x433*	1.0	—	—	1	x174	0.780	0.901	0.919	1
x392*	1.0	—	—	1	x155	0.597	0.825	0.855	2
x359*	0.929	0.963	0.971	1					
x329*	0.812	0.915	0.930	1					
x299*	0.669	0.855	0.880	1					

* Not available from Canadian mills; class of section based on ASTM A572 Grade 50 with F_y = 345 MPa.
— Indicates web is never that class.

Table 4-3   CLASS OF SECTIONS FOR BEAM-COLUMNS
G40.21 350W

Designation	Web 1 $C_f/C_y \leq$	Web 2 $C_f/C_y \leq$	Web 3 $C_f/C_y \leq$	Flange	Designation	Web 1 $C_f/C_y \leq$	Web 2 $C_f/C_y \leq$	Web 3 $C_f/C_y \leq$	Flange
W610x153*	0.791	0.906	0.923	1	W410x85	1.0	—	—	1
x140	0.658	0.851	0.876	1	x74	0.851	0.931	0.943	1
x125	0.465	0.771	0.809	1	x67	0.675	0.858	0.882	1
x113	0.331	0.716	0.763	1	x60	0.404	0.746	0.788	1
x101	0.183	0.654	0.712	1	x54	0.348	0.722	0.769	2
x91	—	0.574	0.644	2					
x84	—	0.492	0.575	3	W410x46	0.193	0.658	0.715	1
					x39	—	0.564	0.636	2
W610x92*	0.288	0.698	0.748	1					
x82*	0.081	0.612	0.676	1	W360x1086*	1.0	—	—	1
					x990*	1.0	—	—	1
W530x300*	1.0	—	—	1	x900*	1.0	—	—	1
x272*	1.0	—	—	1	x818*	1.0	—	—	1
x248*	1.0	—	—	1	x744*	1.0	—	—	1
x219*	1.0	—	—	1	x677*	1.0	—	—	1
x196*	1.0	—	—	1					
x182*	1.0	—	—	1	W360x634*	1.0	—	—	1
x165*	1.0	—	—	1	x592*	1.0	—	—	1
x150*	0.851	0.931	0.944	1	x551*	1.0	—	—	1
					x509*	1.0	—	—	1
W530x138	1.0	—	—	1	x463*	1.0	—	—	1
x123	0.894	0.949	0.958	1	x421*	1.0	—	—	1
x109	0.679	0.860	0.884	1	x382*	1.0	—	—	1
x101	0.555	0.808	0.841	1	x347*	1.0	—	—	1
x92	0.419	0.752	0.793	1	x314*	1.0	—	—	1
x82	0.262	0.687	0.739	2	x287*	1.0	—	—	1
x72	0.103	0.621	0.684	3	x262*	1.0	—	—	1
					x237*	1.0	—	—	1
W530x85*	0.454	0.766	0.805	1	x216*	1.0	—	—	1
x74*	0.324	0.713	0.760	1					
x66*	0.121	0.629	0.690	1	W360x196*	1.0	—	—	1
					x179*	1.0	—	—	1
W460x260*	1.0	—	—	1	x162*	1.0	—	—	2
x235*	1.0	—	—	1	x147*	1.0	—	—	3
x213*	1.0	—	—	1	x134*	1.0	—	—	3
x193*	1.0	—	—	1					
x177*	1.0	—	—	1	W360x122*	1.0	—	—	1
x158*	1.0	—	—	1	x110*	1.0	—	—	1
x144*	1.0	—	—	1	x101*	1.0	—	—	1
x128*	1.0	—	—	1	x91*	1.0	—	—	1
x113*	0.847	0.929	0.942	2					
					W360x79	1.0	—	—	1
W460x106	1.0	—	—	1	x72	0.942	0.968	0.975	1
x97	0.927	0.962	0.970	1	x64	0.752	0.890	0.909	1
x89	0.788	0.905	0.922	1					
x82	0.679	0.859	0.884	1	W360x57	0.733	0.882	0.902	1
x74	0.490	0.781	0.818	1	x51	0.554	0.808	0.840	1
x67	0.365	0.730	0.775	1	x45	0.463	0.770	0.809	2
x61	0.258	0.685	0.737	2					
					W360x39	0.339	0.719	0.766	1
W460x68*	0.527	0.797	0.831	1	x33	0.068	0.607	0.671	1
x60*	0.246	0.680	0.733	1					
x52*	0.124	0.630	0.691	1	W310x500*	1.0	—	—	1
					x454*	1.0	—	—	1
W410x149*	1.0	—	—	1	x415*	1.0	—	—	1
x132*	1.0	—	—	1	x375*	1.0	—	—	1
x114*	1.0	—	—	1	x342*	1.0	—	—	1
x100*	0.914	0.957	0.965	1	x313*	1.0	—	—	1

* Not available from Canadian mills; class of section based on ASTM A572 Grade 50 with F_y = 345 MPa.
— Indicates web is never that class.

CLASS OF SECTIONS FOR BEAM-COLUMNS
G40.21 350W

Table 4-3

Designation	Web 1 $C_f/C_y \leq$	Web 2 $C_f/C_y \leq$	Web 3 $C_f/C_y \leq$	Flange	Designation	Web 1 $C_f/C_y \leq$	Web 2 $C_f/C_y \leq$	Web 3 $C_f/C_y \leq$	Flange
W310x283*	1.0	—	—	1	W200x31	1.0	—	—	1
x253*	1.0	—	—	1	x27	1.0	—	—	2
x226	1.0	—	—	1	x21	0.905	0.953	0.962	3
x202	1.0	—	—	1					
x179	1.0	—	—	1	W200x22*	1.0	—	—	1
x158	1.0	—	—	1	x19*	1.0	—	—	2
x143	1.0	—	—	1	x15*	0.655	0.850	0.875	3
x129	1.0	—	—	1					
x118	1.0	—	—	2	W150x37	1.0	—	—	1
x107	1.0	—	—	2	x30	1.0	—	—	2
x97	1.0	—	—	3	x22	1.0	—	—	4
W310x86	1.0	—	—	2	W150x24*	1.0	—	—	1
x79	1.0	—	—	2	x18*	1.0	—	—	1
					x14*	1.0	—	—	2
W310x74	1.0	—	—	1	x13*	1.0	—	—	3
x67	1.0	—	—	1					
x60	0.955	0.974	0.979	1	W130x28*	1.0	—	—	1
					x24*	1.0	—	—	1
W310x52	0.891	0.947	0.957	1					
x45	0.644	0.845	0.871	1	W100x19*	1.0	—	—	1
x39	0.379	0.735	0.779	2					
x31	0.024	0.589	0.656	4					
W310x33*	0.652	0.849	0.874	1					
x28*	0.463	0.770	0.809	1					
x24*	0.310	0.707	0.755	1					
x21*	0.089	0.615	0.679	2					
W250x167	1.0	—	—	1					
x149	1.0	—	—	1					
x131	1.0	—	—	1					
x115	1.0	—	—	1					
x101	1.0	—	—	1					
x89	1.0	—	—	1					
x80	1.0	—	—	2					
x73	1.0	—	—	2					
W250x67	1.0	—	—	1					
x58	1.0	—	—	1					
x49	1.0	—	—	3					
W250x45	1.0	—	—	1					
x39	0.981	0.984	0.989	1					
x33	0.850	0.930	0.943	2					
x24	0.469	0.773	0.811	4					
W250x28*	0.940	0.968	0.974	1					
x25*	0.859	0.934	0.946	1					
x22*	0.771	0.898	0.916	1					
x18*	0.396	0.742	0.785	3					
W200x100	1.0	—	—	1					
x86	1.0	—	—	1					
x71	1.0	—	—	1					
x59	1.0	—	—	1					
x52	1.0	—	—	2					
x46	1.0	—	—	3					
W200x42	1.0	—	—	1					
x36	1.0	—	—	2					

* Not available from Canadian mills; class of section based on ASTM A572 Grade 50 with F_y = 345 MPa.
— Indicates web is never that class.

NOTES

UNIT FACTORED COMPRESSIVE RESISTANCES FOR COMPRESSION MEMBERS, C_r/A

General

Table 4-4 on pages 4-13 to 4-16 lists in MPa the unit factored compressive resistance C_r/A calculated in accordance with the requirements of Clause 13.3.1, CAN/CSA-S16.1-94 for members with F_y varying from 248 to 700 MPa, for values of KL/r from 1 to 200 with $n = 1.34$. The values for $F_y = 350$ MPa are printed in boldface type. For hollow structural sections manufactured according to G40.20, Class C with $F_y = 350$ MPa, use values given in Table 4-4 for $F_y = 350$.

Table 4-5 on page 4-17 lists the unit factored compressive resistance C_r/A for compression members consisting of WWF shapes with $F_y = 300$ and 350 MPa and HSS manufactured according to G40.20, Class H with $F_y = 350$ MPa. The resistances have been calculated, for values of KL/r from 1 to 200, in accordance with the requirements of Clause 13.3.1, S16.1-94 with $n = 2.24$.

Use

To obtain the factored compressive resistance C_r for doubly symmetric Class 1, 2 or 3 sections, multiply the unit factored compressive resistance C_r/A for the appropriate F_y and KL/r ratio, by the cross-sectional area A of the column section.

Examples

1. Given:

    Find the factored compressive resistance of a W250x131 column of CSA-G40.21 Grade 350W steel ($F_y = 350$ MPa) for a KL/r ratio of 89.

Solution:

   From page 6-52, for W250x131, $A = 16\ 700$ mm^2

   From Table 4-4, page 4-14, with $KL/r = 89$ and $F_y = 350$ MPa, $C_r/A = 155$ MPa

   Therefore, $C_r = 155$ MPa $\times\ 16\ 700$ mm^2 $= 2\ 590 \times 10^3$ N $= 2\ 590$ kN

2. Given:

    Find the factored compressive resistance of an HSS 254x152x11 Class H column for $F_y = 350$ MPa and $KL/r = 89$.

Solution:

   From page 6-98 for HSS 254x152x11, $A = 8\ 230$ mm^2

   From Table 4-5, page 4-17, with $KL/r = 89$ and $F_y = 350$ MPa, $C_r/A = 189$ MPa

   Therefore, $C_r = 189$ MPa $\times\ 8\ 230$ mm^2 $= 1\ 560 \times 10^3$ N $= 1\ 560$ kN

Notes:

1. Tables of C_r, factored axial compressive resistance, for columns in CSA-G40.21 Grade 350W steels are given on pages 4-32 to 4-44.

2. Tables of C_r for HSS Class C columns in G40.21 Grade 350W steel are given on pages 4-46 to 4-69. Tables of C_r for HSS Class H columns in G40.21 Grade 350W steel are given on pages 4-70 to 4-93.

3. For columns not manufactured in Canada, heavy sections, and built-up sections, see Clause 13.3 of the CISC Commentary in Part Two of this Handbook for more information on compressive resistance.

UNIT FACTORED COMPRESSIVE RESISTANCES, C_r / A (MPa)*
For compression members
$\phi = 0.90$ $n = 1.34$

$\dfrac{KL}{r} = 1$ to 50

Table 4-4

$\dfrac{KL}{r}$	F_y (MPa)										
	248	260	280	290	300	350	380	400	480	550	700
1	223	234	252	261	270	315	342	360	432	495	630
2	223	234	252	261	270	315	342	360	432	495	630
3	223	234	252	261	270	315	342	360	432	495	630
4	223	234	252	261	270	315	342	360	432	495	630
5	223	234	252	261	270	315	342	360	432	495	629
6	223	234	252	261	270	315	342	360	431	494	629
7	223	234	252	261	270	315	342	359	431	494	628
8	223	234	252	261	270	314	341	359	431	493	627
9	223	234	252	260	269	314	341	359	430	493	626
10	223	233	251	260	269	314	341	359	430	492	625
11	223	233	251	260	269	314	340	358	429	491	623
12	222	233	251	260	269	313	340	358	428	490	621
13	222	233	251	260	269	313	339	357	428	489	619
14	222	233	250	259	268	312	339	356	427	488	617
15	222	232	250	259	268	312	338	356	426	486	615
16	222	232	250	259	267	311	338	355	424	485	612
17	221	232	249	258	267	311	337	354	423	483	609
18	221	231	249	258	266	310	336	353	422	481	605
19	221	231	249	257	266	309	335	352	420	479	602
20	220	231	248	257	265	308	334	351	418	476	598
21	220	230	248	256	265	307	333	350	416	474	594
22	219	230	247	256	264	307	332	348	415	471	589
23	219	229	246	255	263	305	331	347	412	468	584
24	218	229	246	254	263	304	329	346	410	465	579
25	218	228	245	253	262	303	328	344	408	462	574
26	217	227	244	253	261	302	326	342	405	459	569
27	217	227	243	252	260	301	325	341	403	455	563
28	216	226	243	251	259	299	323	339	400	452	557
29	215	225	242	250	258	298	321	337	397	448	551
30	215	224	241	249	257	296	320	335	394	444	544
31	214	224	240	248	256	295	318	333	391	440	538
32	213	223	239	247	254	293	316	330	388	436	531
33	212	222	238	245	253	291	314	328	385	431	524
34	211	221	237	244	252	290	311	326	381	427	517
35	211	220	235	243	251	288	309	323	378	422	510
36	210	219	234	242	249	286	307	321	374	418	503
37	209	218	233	240	248	284	305	318	370	413	495
38	208	217	232	239	246	282	302	316	367	408	488
39	207	216	230	238	245	280	300	313	363	403	481
40	206	214	229	236	243	278	297	310	359	398	473
41	204	213	228	235	242	275	295	307	355	393	466
42	203	212	226	233	240	273	292	304	351	388	458
43	202	211	225	231	238	271	289	301	347	383	450
44	201	209	223	230	237	269	287	298	343	378	443
45	200	208	222	228	235	266	284	295	338	372	435
46	199	207	220	227	233	264	281	292	334	367	428
47	197	205	218	225	231	261	278	289	330	362	420
48	196	204	217	223	229	259	275	286	326	357	413
49	195	202	215	221	227	256	273	283	321	351	405
50	193	201	213	219	225	254	270	280	317	346	398

* Calculated in accordance with S16.1-94 Clause 13.3.1
For WWF sections and Class H hollow structural sections, see Table 4-5, page 4-17.

$\dfrac{KL}{r} = 51$ to 100

Table 4-4

UNIT FACTORED COMPRESSIVE RESISTANCES, C_r / A (MPa)*
For compression members
$\phi = 0.90 \quad n = 1.34$

KL/r	\multicolumn{11}{c}{F_y (MPa)}										
	248	260	280	290	300	350	380	400	480	550	700
51	192	200	212	218	223	**251**	267	277	313	341	391
52	191	198	210	216	222	**249**	264	273	308	335	383
53	189	196	208	214	220	**246**	261	270	304	330	376
54	188	195	206	212	218	**243**	258	267	300	325	369
55	186	193	205	210	215	**241**	255	264	296	320	362
56	185	192	203	208	213	**238**	252	260	291	315	355
57	183	190	201	206	211	**235**	249	257	287	310	349
58	182	189	199	204	209	**233**	246	254	283	305	342
59	180	187	197	202	207	**230**	243	250	279	300	336
60	179	185	195	200	205	**227**	240	247	274	295	329
61	177	184	193	198	203	**225**	236	244	270	290	323
62	176	182	192	196	201	**222**	233	241	266	285	317
63	174	180	190	194	199	**219**	230	237	262	280	310
64	173	178	188	192	197	**217**	227	234	258	275	305
65	171	177	186	190	194	**214**	224	231	254	271	299
66	170	175	184	188	192	**211**	221	228	250	266	293
67	168	173	182	186	190	**209**	219	225	246	262	287
68	166	172	180	184	188	**206**	216	222	242	257	282
69	165	170	178	182	186	**203**	213	218	238	253	276
70	163	168	176	180	184	**201**	210	215	235	249	271
71	162	167	174	178	182	**198**	207	212	231	244	266
72	160	165	172	176	180	**196**	204	209	227	240	261
73	158	163	171	174	178	**193**	201	206	224	236	256
74	157	161	169	172	175	**191**	198	203	220	232	251
75	155	160	167	170	173	**188**	196	200	217	228	246
76	154	158	165	168	171	**186**	193	198	213	224	242
77	152	156	163	166	169	**183**	190	195	210	220	237
78	150	155	161	164	167	**181**	188	192	206	217	233
79	149	153	159	162	165	**178**	185	189	203	213	228
80	147	151	157	160	163	**176**	182	186	200	209	224
81	146	150	156	158	161	**173**	180	184	197	206	220
82	144	148	154	157	159	**171**	177	181	194	202	216
83	143	146	152	155	157	**169**	175	178	190	199	212
84	141	145	150	153	155	**166**	172	176	187	195	208
85	139	143	148	151	153	**164**	170	173	184	192	204
86	138	141	147	149	151	**162**	167	171	182	189	200
87	136	140	145	147	150	**160**	165	168	179	186	197
88	135	138	143	145	148	**158**	163	166	176	183	193
89	133	137	141	144	146	**155**	160	163	173	180	190
90	132	135	140	142	144	**153**	158	161	170	177	186
91	130	133	138	140	142	**151**	156	159	168	174	183
92	129	132	136	138	140	**149**	154	156	165	171	180
93	127	130	135	137	139	**147**	151	154	162	168	177
94	126	129	133	135	137	**145**	149	152	160	165	174
95	125	127	131	133	135	**143**	147	150	157	163	171
96	123	126	130	132	133	**141**	145	147	155	160	168
97	122	124	128	130	132	**139**	143	145	153	158	165
98	120	123	127	128	130	**137**	141	143	150	155	162
99	119	121	125	127	128	**135**	139	141	148	153	159
100	118	120	124	125	127	**134**	137	139	146	150	157

* Calculated in accordance with S16.1-94 Clause 13.3.1

For WWF sections and Class H hollow structural sections, see Table 4-5, page 4-17.

UNIT FACTORED COMPRESSIVE RESISTANCES, C_r / A (MPa)*
For compression members
$\phi = 0.90$ $n = 1.34$

$\dfrac{KL}{r} = 101$ to 150

Table 4-4

KL/r	F_y (MPa)										
	248	260	280	290	300	350	380	400	480	550	700
101	116	119	122	124	125	**132**	135	137	144	148	154
102	115	117	121	122	124	**130**	133	135	141	145	151
103	114	116	119	121	122	**128**	131	133	139	143	149
104	112	114	118	119	121	**127**	130	131	137	141	147
105	111	113	116	118	119	**125**	128	129	135	139	144
106	110	112	115	116	118	**123**	126	128	133	137	142
107	109	110	113	115	116	**122**	124	126	131	134	140
108	107	109	112	113	115	**120**	123	124	129	132	137
109	106	108	111	112	113	**118**	121	122	127	130	135
110	105	107	109	111	112	**117**	119	121	125	128	133
111	104	105	108	109	110	**115**	118	119	124	127	131
112	102	104	107	108	109	**114**	116	117	122	125	129
113	101	103	105	107	108	**112**	114	116	120	123	127
114	100	102	104	105	106	**111**	113	114	118	121	125
115	98.9	100	103	104	105	**109**	111	113	117	119	123
116	97.8	99.3	102	103	104	**108**	110	111	115	117	121
117	96.7	98.2	100	101	102	**106**	108	110	113	116	119
118	95.6	97.0	99.2	100	101	**105**	107	108	112	114	117
119	94.5	95.9	98.0	98.9	99.8	**104**	106	107	110	112	116
120	93.4	94.8	96.8	97.7	98.6	**102**	104	105	109	111	114
121	92.3	93.7	95.6	96.5	97.4	**101**	103	104	107	109	112
122	91.3	92.6	94.5	95.4	96.2	**99.7**	101	102	106	108	111
123	90.3	91.5	93.4	94.2	95.0	**98.4**	100	101	104	106	109
124	89.2	90.4	92.3	93.1	93.9	**97.2**	98.8	99.7	103	105	107
125	88.2	89.4	91.2	92.0	92.7	**96.0**	97.5	98.4	101	103	106
126	87.2	88.4	90.1	90.9	91.6	**94.7**	96.2	97.1	99.9	102	104
127	86.2	87.4	89.0	89.8	90.5	**93.5**	95.0	95.9	98.6	100	103
128	85.3	86.4	88.0	88.7	89.4	**92.4**	93.8	94.6	97.3	98.9	101
129	84.3	85.4	87.0	87.7	88.4	**91.2**	92.6	93.4	96.0	97.6	99.9
130	83.4	84.4	85.9	86.6	87.3	**90.1**	91.4	92.2	94.7	96.2	98.5
131	82.4	83.4	84.9	85.6	86.3	**89.0**	90.3	91.0	93.4	94.9	97.1
132	81.5	82.5	84.0	84.6	85.2	**87.9**	89.1	89.9	92.2	93.7	95.8
133	80.6	81.5	83.0	83.6	84.2	**86.8**	88.0	88.7	91.0	92.4	94.5
134	79.7	80.6	82.0	82.6	83.2	**85.7**	86.9	87.6	89.8	91.2	93.2
135	78.8	79.7	81.1	81.7	82.3	**84.7**	85.8	86.5	88.6	90.0	91.9
136	77.9	78.8	80.1	80.7	81.3	**83.6**	84.8	85.4	87.5	88.8	90.7
137	77.1	77.9	79.2	79.8	80.4	**82.6**	83.7	84.4	86.4	87.6	89.4
138	76.2	77.1	78.3	78.9	79.4	**81.6**	82.7	83.3	85.3	86.5	88.2
139	75.4	76.2	77.4	78.0	78.5	**80.7**	81.7	82.3	84.2	85.4	87.1
140	74.6	75.4	76.6	77.1	77.6	**79.7**	80.7	81.3	83.1	84.3	85.9
141	73.8	74.5	75.7	76.2	76.7	**78.7**	79.7	80.3	82.1	83.2	84.8
142	73.0	73.7	74.8	75.3	75.8	**77.8**	78.7	79.3	81.0	82.1	83.7
143	72.2	72.9	74.0	74.5	74.9	**76.9**	77.8	78.3	80.0	81.1	82.6
144	71.4	72.1	73.2	73.6	74.1	**76.0**	76.9	77.4	79.0	80.0	81.5
145	70.6	71.3	72.3	72.8	73.3	**75.1**	76.0	76.5	78.0	79.0	80.5
146	69.8	70.5	71.5	72.0	72.4	**74.2**	75.0	75.5	77.1	78.1	79.4
147	69.1	69.8	70.7	71.2	71.6	**73.3**	74.2	74.6	76.1	77.1	78.4
148	68.4	69.0	70.0	70.4	70.8	**72.5**	73.3	73.8	75.2	76.1	77.4
149	67.6	68.3	69.2	69.6	70.0	**71.7**	72.4	72.9	74.3	75.2	76.5
150	66.9	67.5	68.4	68.8	69.2	**70.8**	71.6	72.0	73.4	74.3	75.5

* Calculated in accordance with S16.1-94 Clause 13.3.1
For WWF sections and Class H hollow structural sections, see Table 4-5, page 4-17.

Table 4-4

$\frac{KL}{r}$ = 151 to 200

UNIT FACTORED COMPRESSIVE RESISTANCES, C_r / A (MPa)*
For compression members
$\phi = 0.90 \quad n = 1.34$

KL/r	F_y (MPa)										
	248	260	280	290	300	350	380	400	480	550	700
151	66.2	66.8	67.7	68.1	68.5	70.0	70.8	71.2	72.5	73.4	74.6
152	65.5	66.1	67.0	67.3	67.7	69.2	69.9	70.4	71.7	72.5	73.6
153	64.8	65.4	66.2	66.6	67.0	68.4	69.1	69.5	70.8	71.6	72.7
154	64.1	64.7	65.5	65.9	66.2	67.7	68.3	68.7	70.0	70.7	71.8
155	63.5	64.0	64.8	65.2	65.5	66.9	67.6	68.0	69.2	69.9	71.0
156	62.8	63.3	64.1	64.5	64.8	66.2	66.8	67.2	68.3	69.1	70.1
157	62.1	62.7	63.4	63.8	64.1	65.4	66.0	66.4	67.5	68.3	69.3
158	61.5	62.0	62.7	63.1	63.4	64.7	65.3	65.7	66.8	67.5	68.4
159	60.9	61.4	62.1	62.4	62.7	64.0	64.6	64.9	66.0	66.7	67.6
160	60.2	60.7	61.4	61.7	62.0	63.3	63.9	64.2	65.2	65.9	66.8
161	59.6	60.1	60.8	61.1	61.4	62.6	63.1	63.5	64.5	65.1	66.0
162	59.0	59.5	60.1	60.4	60.7	61.9	62.4	62.8	63.8	64.4	65.3
163	58.4	58.8	59.5	59.8	60.1	61.2	61.8	62.1	63.0	63.6	64.5
164	57.8	58.2	58.9	59.2	59.4	60.6	61.1	61.4	62.3	62.9	63.7
165	57.2	57.6	58.3	58.6	58.8	59.9	60.4	60.7	61.6	62.2	63.0
166	56.6	57.1	57.7	58.0	58.2	59.3	59.8	60.1	60.9	61.5	62.3
167	56.1	56.5	57.1	57.4	57.6	58.6	59.1	59.4	60.3	60.8	61.6
168	55.5	55.9	56.5	56.8	57.0	58.0	58.5	58.8	59.6	60.1	60.9
169	55.0	55.3	55.9	56.2	56.4	57.4	57.9	58.1	59.0	59.5	60.2
170	54.4	54.8	55.3	55.6	55.8	56.8	57.2	57.5	58.3	58.8	59.5
171	53.9	54.2	54.8	55.0	55.3	56.2	56.6	56.9	57.7	58.2	58.9
172	53.3	53.7	54.2	54.5	54.7	55.6	56.0	56.3	57.1	57.5	58.2
173	52.8	53.2	53.7	53.9	54.1	55.0	55.4	55.7	56.4	56.9	57.6
174	52.3	52.6	53.1	53.4	53.6	54.5	54.9	55.1	55.8	56.3	56.9
175	51.8	52.1	52.6	52.8	53.0	53.9	54.3	54.5	55.2	55.7	56.3
176	51.3	51.6	52.1	52.3	52.5	53.3	53.7	54.0	54.6	55.1	55.7
177	50.8	51.1	51.6	51.8	52.0	52.8	53.2	53.4	54.1	54.5	55.1
178	50.3	50.6	51.1	51.3	51.5	52.3	52.6	52.8	53.5	53.9	54.5
179	49.8	50.1	50.6	50.8	51.0	51.7	52.1	52.3	52.9	53.3	53.9
180	49.3	49.6	50.1	50.3	50.4	51.2	51.6	51.8	52.4	52.8	53.3
181	48.9	49.1	49.6	49.8	50.0	50.7	51.0	51.2	51.8	52.2	52.8
182	48.4	48.7	49.1	49.3	49.5	50.2	50.5	50.7	51.3	51.7	52.2
183	47.9	48.2	48.6	48.8	49.0	49.7	50.0	50.2	50.8	51.1	51.7
184	47.5	47.7	48.1	48.3	48.5	49.2	49.5	49.7	50.3	50.6	51.1
185	47.0	47.3	47.7	47.9	48.0	48.7	49.0	49.2	49.8	50.1	50.6
186	46.6	46.8	47.2	47.4	47.6	48.2	48.5	48.7	49.2	49.6	50.1
187	46.1	46.4	46.8	46.9	47.1	47.7	48.0	48.2	48.8	49.1	49.5
188	45.7	46.0	46.3	46.5	46.7	47.3	47.6	47.7	48.3	48.6	49.0
189	45.3	45.5	45.9	46.1	46.2	46.8	47.1	47.3	47.8	48.1	48.5
190	44.9	45.1	45.5	45.6	45.8	46.4	46.6	46.8	47.3	47.6	48.0
191	44.5	44.7	45.0	45.2	45.3	45.9	46.2	46.3	46.8	47.1	47.6
192	44.0	44.3	44.6	44.8	44.9	45.5	45.7	45.9	46.4	46.7	47.1
193	43.6	43.9	44.2	44.3	44.5	45.0	45.3	45.5	45.9	46.2	46.6
194	43.2	43.5	43.8	43.9	44.1	44.6	44.9	45.0	45.5	45.7	46.1
195	42.8	43.1	43.4	43.5	43.7	44.2	44.4	44.6	45.0	45.3	45.7
196	42.5	42.7	43.0	43.1	43.2	43.8	44.0	44.2	44.6	44.9	45.2
197	42.1	42.3	42.6	42.7	42.8	43.4	43.6	43.7	44.2	44.4	44.8
198	41.7	41.9	42.2	42.3	42.4	43.0	43.2	43.3	43.7	44.0	44.3
199	41.3	41.5	41.8	41.9	42.1	42.6	42.8	42.9	43.3	43.6	43.9
200	41.0	41.1	41.4	41.6	41.7	42.2	42.4	42.5	42.9	43.1	43.5

* Calculated in accordance with S16.1-94 Clause 13.3.1

For WWF sections and Class H hollow structural sections, see Table 4-5, page 4-17.

UNIT FACTORED COMPRESSIVE RESISTANCES, C_r/A (MPa)*
For WWF and HSS Class H
$\phi = 0.90$ $n = 2.24$

WWF HSS CLASS H

Table 4-5

KL/r	F_y (MPa) WWF 300	F_y (MPa) WWF HSS 350	KL/r	F_y (MPa) WWF 300	F_y (MPa) WWF HSS 350	KL/r	F_y (MPa) WWF 300	F_y (MPa) WWF HSS 350	KL/r	F_y (MPa) WWF 300	F_y (MPa) WWF HSS 350
1	270	315	51	256	293	101	151	157	151	75.9	76.4
2	270	315	52	255	291	102	149	154	152	74.9	75.5
3	270	315	53	254	289	103	147	152	153	74.0	74.5
4	270	315	54	253	287	104	145	150	154	73.1	73.6
5	270	315	55	251	285	105	143	147	155	72.2	72.7
6	270	315	56	250	283	106	141	145	156	71.3	71.8
7	270	315	57	248	281	107	139	143	157	70.5	70.9
8	270	315	58	247	279	108	137	141	158	69.6	70.1
9	270	315	59	245	276	109	135	138	159	68.8	69.2
10	270	315	60	244	274	110	133	136	160	68.0	68.4
11	270	315	61	242	272	111	131	134	161	67.2	67.6
12	270	315	62	240	269	112	129	132	162	66.4	66.7
13	270	315	63	238	266	113	127	130	163	65.6	66.0
14	270	315	64	236	264	114	125	128	164	64.8	65.2
15	270	315	65	235	261	115	123	126	165	64.1	64.4
16	270	315	66	233	258	116	122	124	166	63.3	63.7
17	270	315	67	231	255	117	120	123	167	62.6	62.9
18	270	315	68	228	253	118	118	121	168	61.9	62.2
19	270	315	69	226	250	119	117	119	169	61.2	61.5
20	270	315	70	224	247	120	115	117	170	60.5	60.8
21	270	315	71	222	244	121	113	115	171	59.8	60.1
22	270	314	72	220	241	122	112	114	172	59.1	59.4
23	270	314	73	217	238	123	110	112	173	58.5	58.7
24	269	314	74	215	235	124	109	110	174	57.8	58.1
25	269	314	75	213	231	125	107	109	175	57.2	57.4
26	269	314	76	210	228	126	106	107	176	56.6	56.8
27	269	314	77	208	225	127	104	106	177	56.0	56.2
28	269	313	78	206	222	128	103	104	178	55.3	55.6
29	269	313	79	203	219	129	101	103	179	54.7	54.9
30	269	313	80	201	216	130	99.9	101	180	54.2	54.4
31	268	312	81	198	213	131	98.5	99.9	181	53.6	53.8
32	268	312	82	196	210	132	97.2	98.5	182	53.0	53.2
33	268	312	83	194	207	133	95.9	97.1	183	52.4	52.6
34	268	311	84	191	204	134	94.6	95.8	184	51.9	52.1
35	267	311	85	189	201	135	93.3	94.5	185	51.3	51.5
36	267	310	86	186	198	136	92.1	93.2	186	50.8	51.0
37	266	309	87	184	195	137	90.9	91.9	187	50.3	50.4
38	266	309	88	181	192	138	89.7	90.7	188	49.8	49.9
39	266	308	89	179	189	139	88.5	89.5	189	49.2	49.4
40	265	307	90	176	186	140	87.3	88.3	190	48.7	48.9
41	265	306	91	174	183	141	86.2	87.1	191	48.2	48.4
42	264	305	92	172	180	142	85.1	85.9	192	47.7	47.9
43	263	304	93	169	178	143	84.0	84.8	193	47.3	47.4
44	263	303	94	167	175	144	82.9	83.7	194	46.8	46.9
45	262	302	95	165	172	145	81.8	82.6	195	46.3	46.4
46	261	301	96	162	170	146	80.8	81.5	196	45.9	46.0
47	260	299	97	160	167	147	79.8	80.5	197	45.4	45.5
48	259	298	98	158	164	148	78.8	79.4	198	44.9	45.1
49	258	296	99	156	162	149	77.8	78.4	199	44.5	44.6
50	257	295	100	153	159	150	76.8	77.4	200	44.1	44.2

* Calculated in accordance with S16.1-94 Clause 13.3.1

For Class C hollow structural sections, see Table 4-4, pages 4-13 to 4-16.

STABILITY EFFECTS

General

Clause 8.6 of CSA Standard CAN/CSA-S16.1-94 stipulates that the sway effects produced by the vertical loads acting on the structure in its displaced configuration be accounted for directly in the analysis for forces and moments. This can be done by 1) performing a second-order geometric analysis or 2) accounting for the sway effects by a) amplifying the first-order elastic translational moments by a factor U_2, or b) performing a $P\Delta$ analysis according to Appendix J. The approximate method of accounting for the sway effects by using K factors larger than 1.0 in determining the factored resistance of a beam-column is no longer an acceptable alternative.

While some software exists for a second-order geometric analysis, most designers will probably be more accustomed to using a first-order elastic analysis. Thus, the methods of the second option given by S16.1-94 to account for sway effects will be preferred, and are illustrated here.

Amplification Factor U_2

Clause 8.6.1 of S16.1 provides the designer with the option of multiplying the first-order translational moments, i.e., the moments due to the lateral loads, by the factor U_2 and adding these amplified moments to those caused by the gravity loads to obtain the moments used in the design of the beam-column. The amplification factor U_2 is taken as

$$U_2 = \frac{1}{1 - \left(\dfrac{\Sigma\, C_f \Delta_f}{\Sigma\, V_f\, h}\right)}$$

For situations where gravity loads predominate, a minimum notional lateral load equal to 0.005 times the gravity loads acting at each storey is required by Clause 8.6.2.

$P\Delta$ Method of Analysis

Appendix J, Guide to Calculation of Stability Effects, of S16.1-94 describes an iterative procedure whereby artificial storey shears are computed and added to the primary horizontal loads in computing the moments and forces in the structure. This method accounts directly for the sway effects, the stiffness and the actual deflected shape of the lateral load resisting system.

Since, for hand computation and preliminary analysis the iterative method can become onerous, several one-step $P\Delta$ methods of analysis can be found in the literature. One such method is the maximum deflection method in which $P\Delta$ artificial storey shears are computed assuming that sway deflections will be limited to a known value. Thus the $P\Delta$ effects will be estimated based upon a maximum deflection. The designer can then control the deflections by the appropriate selection of member stiffness.

References

Additional information on stability effects can be obtained from the following references:

PART TWO of this Handbook.

KENNEDY, D.J.L., PICARD, A., and BEAULIEU, D. 1990. New Canadian provisions for the design of steel beam-columns. Canadian Journal of Civil Engineering, **17**(6), December.

SSRC. 1988. Guide to stability design criteria for metal structures (4th. edition). Structural Stability Research Council, John Wiley & Sons, Toronto, Ontario.

KULAK, G.L., ADAMS, P.F., and GILMOR, M.I. 1995. Limit states design in structural steel, chapter 9, overall stability. Canadian Institute of Steel Construction, Willowdale, Ontario.

PICARD, A., and BEAULIEU, D. 1991. Calcul des charpentes d'acier (2nd. edition). Canadian Institute of Steel Construction, Willowdale, Ontario.

WOOD, B.R., BEAULIEU, D., and ADAMS, P.F. 1976. Further aspects of design by P-delta method. Proceedings ASCE Journal of the Structural Division, **102**(ST3).

Example

1. **Given:**

    For the three storey, three bay structure shown below, the roof and the floor systems are relatively rigid diaphragms and hence the stability of the entire building is provided by the exterior rigid frames. Determine the preliminary design parameters for columns 1C2 and 2C2 with sway effects included in the analysis. Use Appendix J, CAN/CSA-S16.1-94, assuming the load combination of deal plus full wind governs the design of the rigid frame column, 1C2.

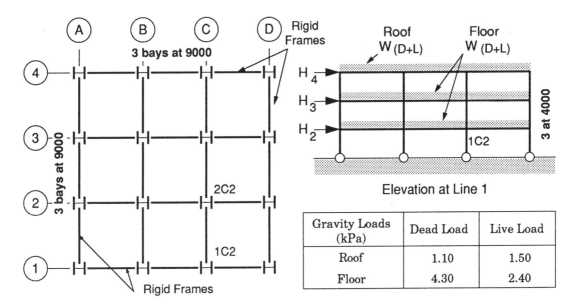

Spandrel load = 7.5 kN/m

Wind loads: 1/10 year wind, q_{10} = 0.36 kPa; 1/30 year wind, q_{30} = 0.43 kPa.
Factored 1/30 wind loads, H_4, H_3, H_2 are 43.4, 83.6 and 83.6 kN respectively.

Solution:

a) *Deflections*

The deflections due to the factored 1/30 wind plus dead load combination are 26, 21 and 14 mm respectively for Δ_4, Δ_3 and Δ_2 respectively and when determined from an elastic first order frame analysis based on assumed member stiffness.

b) *Gravity Loads*

Since four of the interior columns "lean" (for stability) upon each rigid frame, the gravity loads used in the $P\Delta$ analysis must include the loads supported by these interior columns in addition to the gravity loads supported directly by the columns of the rigid frame.

Tributary area = 13.5 m × 27.0 m = 365 m²; one-half perimeter = 27 + 27 = 54 m, therefore, for the third storey, $\Sigma P_3 = 1.25 (365 \times 1.1 + 54.0 \times 7.5) = 1\,010$ kN, for the second storey, $\Sigma P_2 = \Sigma P_3 + 1.25 (365 \times 4.3 + 54.0 \times 7.5) = \Sigma P_3 + 2\,465 = 3\,470$ kN, for the first storey, $\Sigma P_1 = \Sigma P_2 + 2\,465 = 5\,940$ kN.

c) *P∆ Analysis*

Using the first order deflections due to the factored 1/30 year wind, the P∆ shears, V', artificial forces, H', and the P∆ forces, $H + H'$, are computed and tabulated below.

$$V'_2 = \frac{(\Delta_{f3} - \Delta_{f2})\, \Sigma P_2}{h} \qquad \text{(from Appendix J, CAN/CSA-S16.1-94)}$$

Therefore, $V'_2 = \dfrac{(21 - 14)\, 3470}{4000} = 6.07$ kN

$V'_3 = V'_2 - H'_2 = 6.07 - 1.26 = 4.81$ kN

The trial frame is sufficiently stiff that convergence is achieved in 2 cycles.

Iteration	Floor Level	Factored Wind Load, H, kN	Sum of Col. Factored Axial Load, ΣP, kN	Deflection, ∆, mm	P∆ Shear, V' kN	P∆ Force, H' kN	H + H' kN
	4	43.4	1010	26	1.26	1.26	44.7
	3	83.6		21		4.81	88.4
	2	83.6	3470	14	6.07	14.72	98.3
	1		5940		20.79		
First	4			28	1.26	1.26	44.7
	3			23		4.81	88.4
	2			16	6.07	17.69	101.3
	1				23.76		
Second	4			28			
	3			24			
	2			16			
	1						

d) *Preliminary Design Parameters*

i) Interior Column, 2C2

Tributary area = 9 × 9 = 81 m²
Live load reduction factor = $0.3 + \sqrt{9.8/81} = 0.648$ (See Part 4, NBCC, 1995)
Total factored axial load, $C_f = 81 \times [1.25 (1.1 + 4.3) + 1.5 (1.5 + 2.4 \times 0.648)] = 918$ kN
Since column 2C2 has simple connections, assume $K_x = K_y = 1.0$.

ii) Rigid Frame Column 1C2

Using the forces determined from P∆ analysis, after convergence, the design moments for the beams and columns can be determined from an elastic frame analysis.

For column 1C2, they are taken as:
$C_f = 450$ kN, $M_{f1} = -47.3$ kN·m, $M_{f2} = -163$ kN·m, $\kappa = (-47.3)/(-163) = 0.290$
$\omega_1 = 0.48$ (Table 4-7, double curvature)

Since $P\Delta$ forces are included in M_{f1} and M_{f2}, K may be less than or equal to 1.0. Use $K_x = 0.9$ and $K_y = 1.0$ for trial design.

2. **Given:**

Same as Example 1, except that the storey deflection limit, h/500, is used to estimate the $P\Delta$ forces.

Solution:

Since the wind load used to compute the deflection under specified load is the 1/10 year wind, the deflection must be increased to that due to the 1/30 year wind before computing the $P\Delta$ effects. Therefore the lateral deflection under factored load is:

$$\Delta_f = \Delta_s \times \alpha_w \times \frac{q_{30}}{q_{10}} \quad \text{where } \alpha_w = 1.5 \quad \text{(Clause 7.2.3, S16.1-94)}$$

$$= \Delta_s \times 1.5 \times \frac{0.43}{0.36} = 1.79 \, \Delta_s$$

The following table summarizes the $P\Delta$ forces.

Floor Level	Factored Wind Load, H, kN	Sum of Col. Factored Axial Load, ΣP, kN	Deflection, Δ, mm	$P\Delta$ Shear, V', kN	$P\Delta$ Force, H', kN	H + H' kN
4	43.4	1010	43	3.54	3.54	46.9
3	83.6	3470	29	13.01	9.47	93.1
2	83.6	5940	14	20.79	7.78	91.4
1						

In this example, the $P\Delta$ forces are over-estimated for levels 4 and 3 and under-estimated for level 2 when compared to the more accurate iterative method.

Preliminary Design Parameters

  i) Interior Column, 2C2 – same as Example 1.

  ii) Rigid Frame Column 1C2
  For the above $P\Delta$ analysis and frame analysis,

  $C_f = 450$kN, $M_{f1} = -52.9$ kN·m, $M_{f2} = -167$ kN·m, $\kappa = (-52.9)/(-167) = 0.317$
  $\omega_1 = 0.47$ (Table 4-7, double curvature)

3. **Given:**

Using gravity loads, 1/30 year wind and first order frame deflections as given in Example 1 and assuming the load combination of dead plus full wind governs the design of the rigid frame column, 1C2, determine the preliminary design loads by using U_2 expression as in Clause 8.6.1 of CAN/CSA-S16.1-94, for column 1C2 with sway effects included in the analysis.

Solution:

The amplification factor, U_2, given in Clause 8.6.1 may be expressed as

$$U_2 = \frac{1}{1 - \left(\dfrac{\Delta_f \Sigma P_f}{h \Sigma V_f}\right)} \quad \text{where}$$

Δ_f = interstorey deflection due to factored 1/30 year wind.
$\Sigma P_f = \Sigma$ factored axial loads (all columns) in the storey
ΣV_f = factored total first order storey shear

Floor Level	Factored Wind Load, H, kN	Factored 1st. Order Storey Shear, $\Sigma V'_f$, kN	Sum of Col. Factored Axial Load, ΣP, kN	Total Deflection, Δ, mm	Storey Δ_f, mm	U_2	Total Factored Storey Shear $\Sigma V'_f$, kN	$H + H'$ kN
4	43.4	43.4	1010	26	5	1.03	44.7	44.7
3	83.6	127	3470	21	7	1.05	133	88.3
2	83.6	211	5940	14	14	1.11	234	101
1				14				

In this example, the $P\Delta$ forces are as accurate as the iterative method results in Example 1.

Preliminary Design Parameters

i) Interior Column, 2C2 – Same as in Example 1.

ii) Rigid Frame Column, 1C2 – Same as in Example 1.

While for the sake of comparison, the factored storey shears were amplified by U_2, for which an analysis for moments may then be performed, the factor U_2 can be used to amplify the translational moments and shears directly.

BENDING FACTORS FOR BEAM-COLUMNS

The initial selection of a suitable steel section subject to axial load and moment can be simplified by replacing the applied axial load and bending moments with an equivalent total axial load. This procedure is discussed in more detail on page 4-94.

To simulate the effect of the bending moments, the applied moment (larger of the two end moments for columns with unequal end moments) can be multiplied by a "bending factor" that is determined for each structural steel shape from the properties of that shape. Table 4-6 lists average bending factors for each nominal size group of WWF column shapes, W shapes and hollow structural sections. This average bending factor is sufficiently accurate for an initial shape selection.

The bending factors are computed as follows:

for Class 1 sections, $B_x = 0.85 \times 10^3 \, A/Z_x$

for Class 2 sections, $B_x = 10^3 \, A/Z_x$

for Class 3 and Class 4 sections, $B_x = 10^3 \, A/S_x$

for Class 1 sections, $B_y = 0.60 \times 10^3 \, A/Z_y$

for Class 2 sections, $B_y = 10^3 \, A/Z_y$

for Class 3 and Class 4 sections, $B_y = 10^3 \, A/S_y$

where

    A = total cross sectional area of the structural shape

    S = elastic section modulus, about the X-X or Y-Y axis, as appropriate

    Z = plastic section modulus, about the X-X or Y-Y axis, as appropriate.

The bending factors are in m^{-1} units, and multiplying a bending moment in kilonewton metres by a bending factor from Table 4-6 results in an equivalent axial load in kilonewtons.

In cases where the factored moment resistance must be reduced when the unsupported member length exceeds L_u, the bending factors may be used to estimate the size of column required (which usually will require revision), or the bending factors can be increased to compensate for the reduction in M_r.

APPROXIMATE BENDING FACTORS
WWF and W Shapes

Table 4–6

Nominal Size	B_x (m^{-1})			B_y (m^{-1})		
	Class 1 Sections	Class 2 Sections	Class 3 & Class 4 Sections	Class 1 Sections	Class 2 Sections	Class 3 & Class 4 Sections
WWF650	3.18	3.74	4.21	4.60	7.67	11.7
WWF600	3.46	4.07	4.61	4.97	8.28	12.6
WWF550	3.79	4.46	5.05	5.36	8.93	13.6
WWF500	4.14	4.88	5.47	5.98	9.97	15.1
WWF450	4.59	5.40	6.04	6.57	11.0	16.6
WWF400	5.21	6.13	6.93	7.37	12.3	18.6
WWF350	6.04	7.10	8.12	8.36	13.9	21.1
920 x 420	2.37	2.79	3.15	8.86	14.8	22.8
920 x 305	2.52	2.97	3.42	14.3	23.9	37.7
840 x 400	2.54	2.99	3.39	9.35	15.6	24.1
840 x 290	2.73	3.21	3.70	15.0	25.1	39.3
760 x 380	2.79	3.28	3.69	9.62	16.0	24.7
760 x 265	3.04	3.57	4.12	16.6	27.7	43.5
690 x 355	3.10	3.65	4.10	10.3	17.1	26.4
690 x 255	3.32	3.91	4.47	16.7	27.9	43.5
610 x 325	3.48	4.09	4.61	11.2	18.6	28.6
610 x 230	3.72	4.38	5.00	18.3	30.5	47.5
530 x 315	3.91	4.60	5.18	11.2	18.7	28.7
530 x 210	4.21	4.95	5.65	19.5	32.4	50.6
460 x 285	4.54	5.34	5.99	12.2	20.4	31.3
460 x 190	4.87	5.73	6.52	21.0	35.0	54.4
410 x 260	5.02	5.91	6.63	12.9	21.5	32.9
410 x 180	5.41	6.37	7.21	21.8	36.4	56.3
410 x 140	5.73	6.74	7.74	31.9	53.1	83.3
360 x 430	4.56	5.36	6.81	6.50	10.8	16.7
360 x 405	5.01	5.90	7.08	7.03	11.7	18.0
360 x 395	5.38	6.33	7.23	7.47	12.4	18.9
360 x 370	5.60	6.58	7.29	8.19	13.6	20.7
360 x 255	5.82	6.85	7.67	12.8	21.3	32.6
360 x 205	6.02	7.09	7.90	17.0	28.3	43.4
360 x 170	6.16	7.25	8.15	22.5	37.4	57.6
310 x 330	5.73	6.74	8.22	8.86	14.8	22.7
310 x 310	6.34	7.46	8.48	9.78	16.3	24.9
310 x 255	6.64	7.82	8.65	12.5	20.8	31.6
310 x 200	6.83	8.04	8.96	16.4	27.3	41.9
310 x 165	6.81	8.02	8.96	21.7	36.1	55.4
250 x 255	7.77	9.14	10.4	11.7	19.5	29.6
250 x 200	8.23	9.69	10.8	15.9	26.5	40.3
250 x 145	8.19	9.64	10.8	24.2	40.4	62.1
200 x 200	9.74	11.5	12.9	14.8	24.6	37.5
200 x 165	10.2	12.0	13.4	19.4	32.3	49.2
150 x 150	13.3	15.6	17.5	21.0	35.0	53.3
130 x 125	16.2	19.1	21.6	24.0	40.0	61.1
100 x 100	20.4	24.0	27.8	31.0	51.6	79.4

Table 4-6

APPROXIMATE BENDING FACTORS
Hollow Structural Sections

Nominal Size	$B_x (m^{-1})$		$B_y (m^{-1})$	
	Class 1 & Class 2 Sections	Class 3 & Class 4 Sections	Class 1 & Class 2 Sections	Class 3 & Class 4 Sections
305 x 203	9.81	11.9	13.0	14.9
354 x 152	12.2	15.1	17.3	20.1
203 x 152	14.7	17.9	17.9	20.9
203 x 102	15.9	20.4	26.0	30.5
178 x 127	17.1	21.1	21.6	25.5
152 x 102	20.3	25.4	27.0	32.1
127 x 76	25.1	32.0	36.1	43.2
127 x 64	26.1	34.1	42.9	51.8
127 x 51	27.1	36.2	53.0	64.3
102 x 76	30.3	38.0	37.1	44.9
102 x 51	32.4	42.0	53.1	63.6
89 x 64	34.8	43.5	44.0	52.8
76 x 51	42.3	54.7	56.4	69.8
51 x 25	66.1	87.1	109	133
305 x 305	9.10	10.6	9.10	10.6
254 x 254	11.0	12.9	11.0	12.9
203 x 203	13.9	16.6	13.9	16.6
178 x 178	16.0	19.0	16.0	19.0
152 x 152	18.9	22.7	18.9	22.7
127 x 127	22.8	27.6	22.8	27.6
102 x 102	28.8	35.2	28.8	35.2
89 x 89	33.1	40.6	33.1	40.6
76 x 76	38.4	46.9	38.4	46.9
64 x 64	46.2	56.5	46.2	56.5
51 x 51	58.8	73.1	58.8	73.1
38 x 38	79.7	100	79.7	100
32 x 32	96.3	122	96.3	122
25 x 25	126	165	126	165

VALUES OF ω_1 * Table 4-7

Single Curvature				Double Curvature			
$\dfrac{M_{f1}}{M_{f2}}$	ω_1	$\dfrac{M_{f1}}{M_{f2}}$	ω_1	$\dfrac{M_{f1}}{M_{f2}}$	ω_1	$\dfrac{M_{f1}}{M_{f2}}$	ω_1
1.00	1.00	0.50	0.80	0.00	0.60	0.55	0.40
0.95	0.98	0.45	0.78	0.05	0.58	0.60	0.40
0.90	0.96	0.40	0.76	0.10	0.56	0.65	0.40
0.85	0.94	0.35	0.74	0.15	0.54	0.70	0.40
0.80	0.92	0.30	0.72	0.20	0.52	0.75	0.40
0.75	0.90	0.25	0.70	0.25	0.50	0.80	0.40
0.70	0.88	0.20	0.68	0.30	0.48	0.85	0.40
0.65	0.86	0.15	0.66	0.35	0.46	0.90	0.40
0.60	0.84	0.10	0.64	0.40	0.44	0.95	0.40
0.55	0.82	0.05	0.62	0.45	0.42	1.00	0.40
		0.00	0.60	0.50	0.40		

* See Clause 13.8.4, CAN/CSA-S16.1-94.

The value of ω_1 is used to modify the bending term in the beam-column interaction expression to account for various end moment and transverse loading conditions of the columns.

For columns of a frame not subject to transverse loads between supports, use the values of ω_1 shown in Table 4-7.

For members subjected to distributed loads or a series of point loads between supports, $\omega_1 = 1.0$, and for members subjected to a concentrated load or moment between supports, $\omega_1 = 0.85$.

The values of ω_1 given in Table 4-7 are derived from

$$\omega_1 = 0.6 - 0.4\kappa \geq 0.4$$

where

κ = M_{f1}/M_{f2} for moments at opposite ends of the unbraced column length, positive for double curvature, and negative for single curvature in which,

M_{f1} = the smaller factored end moment, and

M_{f2} = the larger factored end moment.

C_e/A EULER BUCKLING LOAD
PER UNIT OF AREA, MPa

Table 4-8

$\frac{KL}{r}$	$\frac{C_e}{A}$ (MPa)	$\frac{KL}{r}$	$\frac{C_e}{A}$ (MPa)	$\frac{KL}{r}$	$\frac{C_e}{A}$ (MPa)	$\frac{KL}{r}$	$\frac{C_e}{A}$ (MPa)	$\frac{KL}{r}$	$\frac{C_e}{A}$ (MPa)
1	1 970 000	41	1 172	81	300	121	135	161	76.0
2	492 500	42	1 117	82	293	122	132	162	75.1
3	218 889	43	1 065	83	286	123	130	163	74.1
4	123 125	44	1 018	84	279	124	128	164	73.2
5	78 800	45	973	85	273	125	126	165	72.4
6	54 722	46	931	86	266	126	124	166	71.5
7	40 204	47	892	87	260	127	122	167	70.6
8	30 781	48	855	88	254	128	120	168	69.8
9	24 321	49	820	89	249	129	118	169	69.0
10	19 700	50	788	90	243	130	117	170	68.2
11	16 281	51	757	91	238	131	115	171	67.4
12	13 681	52	729	92	233	132	113	172	66.6
13	11 657	53	701	93	228	133	111	173	65.8
14	10 051	54	676	94	223	134	110	174	65.1
15	8 756	55	651	95	218	135	108	175	64.3
16	7 695	56	628	96	214	136	107	176	63.6
17	6 817	57	606	97	209	137	105	177	62.9
18	6 080	58	586	98	205	138	103	178	62.2
19	5 457	59	566	99	201	139	102	179	61.5
20	4 925	60	547	100	197	140	101	180	60.8
21	4 467	61	529	101	193	141	99.1	181	60.1
22	4 070	62	512	102	189	142	97.7	182	59.5
23	3 724	63	496	103	186	143	96.3	183	58.8
24	3 420	64	481	104	182	144	95.0	184	58.2
25	3 152	65	466	105	179	145	93.7	185	57.6
26	2 914	66	452	106	175	146	92.4	186	56.9
27	2 702	67	439	107	172	147	91.2	187	56.3
28	2 513	68	426	108	169	148	89.9	188	55.7
29	2 342	69	414	109	166	149	88.7	189	55.1
30	2 189	70	402	110	163	150	87.6	190	54.6
31	2 050	71	391	111	160	151	86.4	191	54.0
32	1 924	72	380	112	157	152	85.3	192	53.4
33	1 809	73	370	113	154	153	84.2	193	52.9
34	1 704	74	360	114	152	154	83.1	194	52.3
35	1 608	75	350	115	149	155	82.0	195	51.8
36	1 520	76	341	116	146	156	81.0	196	51.3
37	1 439	77	332	117	144	157	79.9	197	50.8
38	1 364	78	324	118	141	158	78.9	198	50.2
39	1 295	79	316	119	139	159	77.9	199	49.7
40	1 231	80	308	120	137	160	77.0	200	49.3

To obtain C_e, in kilonewtons, multiply the tabular value by the cross sectional area, **A**, in square millimetres, and divide by 1 000.

AMPLIFICATION FACTOR *

Table 4–9

$$U = \frac{1}{1 - \dfrac{C_f}{C_e}}$$

$\dfrac{C_f}{C_e}$	U	$\dfrac{C_f}{C_e}$	U	$\dfrac{C_f}{C_e}$	U	$\dfrac{C_f}{C_e}$	U
.01	1.01	.26	1.35	.51	2.04	.76	4.17
.02	1.02	.27	1.37	.52	2.08	.77	4.35
.03	1.03	.28	1.39	.53	2.13	.78	4.55
.04	1.04	.29	1.41	.54	2.17	.79	4.76
.05	1.05	.30	1.43	.55	2.22	.80	5.00
.06	1.06	.31	1.45	.56	2.27	.81	5.26
.07	1.08	.32	1.47	.57	2.33	.82	5.56
.08	1.09	.33	1.49	.58	2.38	.83	5.88
.09	1.10	.34	1.52	.59	2.44	.84	6.25
.10	1.11	.35	1.54	.60	2.50	.85	6.67
.11	1.12	.36	1.56	.61	2.56	.86	7.14
.12	1.14	.37	1.59	.62	2.63	.87	7.69
.13	1.15	.38	1.61	.63	2.70	.88	8.33
.14	1.16	.39	1.64	.64	2.78	.89	9.09
.15	1.18	.40	1.67	.65	2.86	.90	10.0
.16	1.19	.41	1.69	.66	2.94	.91	11.1
.17	1.20	.42	1.72	.67	3.03	.92	12.5
.18	1.22	.43	1.75	.68	3.13	.93	14.3
.19	1.23	.44	1.79	.69	3.23	.94	16.7
.20	1.25	.45	1.82	.70	3.33	.95	20.0
.21	1.27	.46	1.85	.71	3.45	.96	25.0
.22	1.28	.47	1.89	.72	3.57	.97	33.3
.23	1.30	.48	1.92	.73	3.70	.98	50.0
.24	1.32	.49	1.96	.74	3.85	.99	100.0
.25	1.33	.50	2.00	.75	4.00		

* See Clause 13.8.3, CAN/CSA-S16.1-94.

FACTORED AXIAL COMPRESSIVE RESISTANCES OF COLUMNS

Tables

The tables on pages 4-32 to 4-69 inclusive list the factored axial compressive resistances C_r in kilonewtons for WWF and W shapes and Class C HSS, produced to the requirements of CSA Standard G40.20. The resistances have been computed for effective lengths with respect to the least radius of gyration varying from 0 mm to 12 000 mm in accordance with the requirements of Clauses 13.3.1 and 13.3.3, CAN/CSA-S16.1-94 with $n = 1.34$ for W shapes and Class C HSS and $n = 2.24$ for WWF shapes.

The tables on pages 4-70 to 4-93 inclusive list the factored axial compressive resistances C_r in kilonewtons for Class H HSS produced to the requirements of CSA-G40.20, in accordance with Clauses 13.3.1 and 13.3.3, S16.1-94 with $n = 2.24$.

Factored axial compressive resistances of HSS produced to the requirements of ASTM A500 Grade C and calculated in accordance with CAN/CSA S16.1-94 are given in a publication entitled: "Hollow Structural Sections to ASTM A500 Grade C", available from CISC.

In all, four sets of tables are provided:

Set 1 — WWF shapes conforming to CSA-G40.20
— CSA-G40.21 Grade 350W

Set 2 — W shapes conforming to CSA-G40.20
— CSA-G40.21 Grade 350W, ASTM A992, ASTM A572 grade 50

Set 3 — HSS conforming to CSA-G40.20, Class C
— CSA-G40.21 Grade 350W

Set 4 — HSS conforming to CSA G40.20, Class H
— CSA-G40.21 Grade 350W

In each set of tables, sections which are either Class 3 or 4, in the grade of steel for which the loads have been computed, are identified. In Set 1, those welded sections in which the flange-to-web welds do not develop the entire web strength are identified. In Set 2, the minimum specified yield stress has been taken as $F_y = 345$ MPa, corresponding to the least value among the 3 grades represented (CSA-G40.21 350W, ASTM A992 and A572 grade 50).

The factored axial compressive resistances for Class 4 sections have been computed in accordance with the requirements of Clause 13.3.3 of S16.1-94 and are so identified in the tables.

The applicable steel grade is listed at the top of each table, and the metric designation of each shape is given at the top of the columns, while the equivalent imperial size and mass are listed at the bottom of the tables. Properties and design data are included at the bottom of the tables as follows:

Area = Total cross-sectional area, mm^2

Z_x = Plastic section modulus for bending about X-X axis, $10^3 \, mm^3$

S_x = Elastic section modulus for bending about X-X axis, $10^3 \, mm^3$. (S_e is given for Class 4 sections; see Clause 13.5(c), CSA-S16.1-94.)

r_x = Radius of gyration about the strong, X-X, axis, mm

Z_y = Plastic section modulus for bending about Y-Y axis, $10^3 \, mm^3$

S_y = Elastic section modulus for bending about Y-Y axis, 10^3 mm^3. (S_e is given for Class 4 sections; see Clause 13.5(c), CSA-S16.1-94.)

r_y = Radius of gyration about the weak, Y-Y, axis, mm

r_x / r_y = Ratio of radius of gyration of X-X axis to that of Y-Y axis

M_{rx} = Factored moment resistance for bending about the X-X axis, computed considering $L \le L_u$, using the Class of the section considering bending about the X-X axis only and the value of F_y shown, for Class 1 and 2 sections, $\phi Z_x F_y \times 10^{-6}$; for Class 3 sections, $\phi S_x F_y \times 10^{-6}$; and for Class 4 sections, Clause 13.5(c) of S16.1-94, kN·m.

M_{ry} = Factored moment resistance for bending about the Y-Y axis computed using the Class of the section considering bending about the Y-Y axis only and the value of F_y shown, for Class 1 and 2 section, $\phi Z_y F_y \times 10^{-6}$; for Class 3 sections, $\phi S_y F_y \times 10^{-6}$; and for class 4 sections, Clause 13.5(c) of S16.1-94, kN·m.

For hollow structural sections which have identical X-X and Y-Y axis properties, only one value of the relevant properties is shown.

The following additional constants are tabulated for the tables containing WWF and W shapes:

J = St. Venant torsional constant, 10^3 mm^4

C_w = Warping torsional constant, 10^9 mm^6

L_u = Maximum unsupported length of compression flange for which no reduction in M_r is required, mm

F_y = Specified minimum yield strength of the section, MPa

Design of Axially Loaded Columns

The design of axially loaded columns (columns theoretically not subjected to combined bending and compression) involves the determination of the governing effective length and the selection of a section with the required resistance at that effective length. Factored axial compressive resistance tables for columns enable a designer to select a suitable section directly, without following a trial-and-error procedure.

Since the factored axial compressive resistances C_r (listed in the tables supplied) have been computed on the basis of the least radius of gyration r_y for each section, the tables apply directly only to columns unbraced about the Y-Y axis. In certain cases, however, it is necessary to investigate the capacity of a column with reference to both the X-X axis and the Y-Y axis, or with reference only to the X-X axis. The ratio r_x/r_y included in the table of properties at the bottom of each resistance table provides a convenient means of investigating the strength of a column with respect to the X-X axis.

In general, a column having an effective length $K_x L_x$ with respect to the X-X axis will be able to carry a factored load equal to the tabulated factored axial compressive resistance based upon the effective length $K_y L_y$ with respect to the Y-Y axis if $K_x L_x \le K_y L_y (r_x / r_y)$.

Examples

1. **Given:**

    A W310 column is required to carry a factored axial load of 3 600 kN. The effective length $K_y L_y$ along the weak axis is 4 500 mm. The effective length $K_x L_x$ along the strong axis is 7 600 mm. Use CSA-G40.21 Grade 350W steel.

    Solution:

    With $K_y L_y$ = 4 500, the lightest W310 section with sufficient factored axial compressive resistance is W310x129. C_r = 3 820 kN; r_x/r_y = 1.76.

    $K_x L_x$ = 7 600 mm (required)

    $K_y L_y (r_x / r_y) = 4\,500 \times 1.76$

    $\qquad = 7\,920$ mm $\quad > 7\,600 \qquad$ —OK

    The W310x129 has a factored compressive resistance of 3 820 kN with an effective length of $K_x L_x$ = 7 920 mm, and hence the section is adequate. Use W310x129.

2. **Given:**

    Same as example 1, except $K_x L_x$ = 9 500 mm

    Solution:

    $K_y L_y$ = 4 500 mm

    $K_x L_x$ = 9 500 mm

    Equivalent $K_y L_y$, for $K_x L_x$ of 9 500 mm = $K_x L_x$ / (r_x/r_y)

    Assuming that a heavy W310 section will be adequate, r_x/r_y = 1.76.

    Equivalent $K_y L_y$ = 9 500 / 1.76

    $\qquad = 5\,400$ mm $\quad > 4\,500$

    Therefore, $K_x L_x$ governs, and the effective $K_y L_y$ is 5 400 mm.

    With $K_y L_y$ = 5 400 mm, a W310x143 is the lightest W310 that has a factored axial compressive resistance greater than the factored axial load of 3 600 kN (C_r for 5 500 mm = 3 630 kN; r_x/r_y = 1.76).

    Use W310x143.

Other Steel Grades

For WWF and W shapes of CSA G40.21 Grade 300W steel used as columns, see the appropriate tables in Part Eight.

For columns of steels other than G40.21 350W and 300W, ASTM A992 and A572 grade 50, see the examples on page 4-12 for a method of determining the factored axial compressive resistance of the section. For steel grades with higher specified minimum yield points, these tables are conservative.

WWF COLUMNS
Factored Axial Compressive Resistances, C_r, in kN

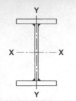

G40.21   350W
$\phi = 0.90$

Designation		WWF650			
Mass (kg/m)	864 ‡	739 ‡	598	499	400 **
Effective length (KL) in millimetres with respect to least radius of gyration					
0	34 700	29 600	24 000	20 000	15 600
2 000	34 600	29 600	24 000	20 000	15 600
2 250	34 600	29 600	24 000	20 000	15 600
2 500	34 600	29 600	24 000	20 000	15 600
2 750	34 600	29 600	24 000	20 000	15 600
3 000	34 600	29 600	24 000	20 000	15 600
3 250	34 600	29 600	24 000	20 000	15 600
3 500	34 600	29 600	24 000	20 000	15 600
3 750	34 600	29 600	24 000	20 000	15 600
4 000	34 500	29 600	23 900	20 000	15 500
4 250	34 500	29 500	23 900	20 000	15 500
4 500	34 500	29 500	23 900	20 000	15 500
4 750	34 400	29 500	23 900	19 900	15 500
5 000	34 300	29 500	23 900	19 900	15 500
5 250	34 300	29 400	23 800	19 900	15 500
5 500	34 200	29 300	23 800	19 800	15 400
6 000	33 900	29 200	23 700	19 700	15 400
6 500	33 700	29 000	23 500	19 600	15 300
7 000	33 300	28 800	23 400	19 500	15 200
7 500	32 800	28 500	23 100	19 300	15 000
8 000	32 300	28 200	22 900	19 000	14 900
8 500	31 700	27 800	22 600	18 700	14 700
9 000	30 900	27 300	22 200	18 400	14 400
9 500	30 100	26 700	21 800	18 000	14 200
10 000	29 200	26 100	21 300	17 600	13 900
10 500	28 300	25 500	20 800	17 100	13 400
11 000	27 300	24 700	20 200	16 700	13 000
11 500	26 200	24 000	19 600	16 100	12 600
12 000	25 100	23 200	18 900	15 600	12 100
PROPERTIES AND DESIGN DATA					
Area (mm^2)	110 000	94 100	76 200	63 600	51 000
Z_x (10^3 mm^3)	27 300	25 200	21 100	17 500	13 900
S_x (10^3 mm^3)	23 300	22 100	18 900	15 900	12 500
r_x (mm)	262	277	284	285	284
Z_y (10^3 mm^3)	13 200	12 800	10 600	8 510	6 400
S_y (10^3 mm^3)	8 480	8 450	7 040	5 630	4 100
r_y (mm)	158	171	173	170	164
r_x/r_y	1.66	1.62	1.64	1.68	1.73
M_{rx} (kN·m) (L < L_u)	8 600	7 940	6 650	5 510	3 940
M_{ry} (kN·m)	4 160	4 030	3 340	2 680	1 290
J (10^3 mm^4)	132 000	98 400	55 600	29 300	13 300
C_w (10^9 mm^6)	240 000	239 000	206 000	170 000	132 000
L_u (mm)	13 000	12 600	11 500	10 500	10 300
F_y (MPa)	350	350	350	350	350
IMPERIAL SIZE AND MASS					
Mass (lb./ft.)	580	497	402	336	269
Nominal Depth and Width (in.)			26 x 26		

‡ Welding does not fully develop web strength.
** Class 4: C_r calculated according to CAN/CSA-S16.1-94 Clause 13.3.3; S_x, S_y, M_{rx}, M_{ry} according to Clause 13.5(c)(iii).

G40.21 350W
$\phi = 0.90$

WWF COLUMNS
Factored Axial Compressive Resistances, C_r, in kN

Designation	WWF600				
Mass (kg/m)	793 ‡	680 ‡	551	460	369 *
Effective length (KL) in millimetres with respect to least radius of gyration					
0	31 800	27 300	22 100	18 500	14 800
2 000	31 800	27 300	22 100	18 500	14 800
2 250	31 800	27 300	22 100	18 500	14 800
2 500	31 800	27 300	22 100	18 500	14 800
2 750	31 800	27 300	22 100	18 400	14 800
3 000	31 800	27 300	22 100	18 400	14 800
3 250	31 800	27 200	22 100	18 400	14 800
3 500	31 700	27 200	22 100	18 400	14 800
3 750	31 700	27 200	22 100	18 400	14 800
4 000	31 700	27 200	22 000	18 400	14 700
4 250	31 600	27 200	22 000	18 400	14 700
4 500	31 600	27 100	22 000	18 300	14 700
4 750	31 500	27 100	22 000	18 300	14 700
5 000	31 400	27 000	21 900	18 300	14 600
5 250	31 300	27 000	21 900	18 200	14 600
5 500	31 200	26 900	21 800	18 200	14 600
6 000	30 900	26 700	21 700	18 100	14 400
6 500	30 600	26 500	21 500	17 900	14 300
7 000	30 100	26 200	21 300	17 700	14 100
7 500	29 600	25 900	21 000	17 500	13 900
8 000	28 900	25 400	20 700	17 200	13 600
8 500	28 200	24 900	20 300	16 800	13 300
9 000	27 400	24 300	19 800	16 400	13 000
9 500	26 500	23 700	19 300	16 000	12 600
10 000	25 500	23 000	18 800	15 500	12 200
10 500	24 500	22 300	18 200	15 000	11 700
11 000	23 400	21 500	17 600	14 400	11 300
11 500	22 400	20 600	16 900	13 900	10 800
12 000	21 300	19 800	16 200	13 300	10 300
PROPERTIES AND DESIGN DATA					
Area (mm^2)	101 000	86 600	70 200	58 600	47 000
Z_x (10^3 mm^3)	22 900	21 200	17 800	14 800	11 800
S_x (10^3 mm^3)	19 400	18 500	15 900	13 400	10 700
r_x (mm)	240	253	261	262	261
Z_y (10^3 mm^3)	11 200	10 900	9 050	7 250	5 460
S_y (10^3 mm^3)	7 230	7 200	6 000	4 800	3 600
r_y (mm)	147	158	160	157	152
r_x/r_y	1.63	1.60	1.63	1.67	1.72
M_{rx} (kN·m) (L < L_u)	7 210	6 680	5 610	4 660	3 370
M_{ry} (kN·m)	3 530	3 430	2 850	2 280	1 130
J (10^3 mm^4)	121 000	90 700	51 300	27 000	12 200
C_w (10^9 mm^6)	158 000	158 000	136 000	113 000	87 800
L_u (mm)	12 800	12 300	11 000	9 970	9 630
F_y (MPa)	350	350	350	350	350
IMPERIAL SIZE AND MASS					
Mass (lb./ft.)	531	456	371	309	248
Nominal Depth and Width (in.)	24 x 24				

‡ Welding does not fully develop web strength.
* Class 3 flanges.

WWF COLUMNS
Factored Axial Compressive Resistances, C_r, in kN

G40.21 350W

$\phi = 0.90$

Designation		WWF550					WWF500		
Mass (kg/m)		721 ‡	620 ‡	503	420	280 **	651 ‡	561 ‡	456
Effective length (KL) in millimetres with respect to least radius of gyration	0	29 000	24 900	20 200	16 900	10 800	26 100	22 600	18 300
	2 000	29 000	24 900	20 200	16 900	10 800	26 100	22 500	18 300
	2 250	29 000	24 900	20 200	16 900	10 800	26 100	22 500	18 300
	2 500	29 000	24 900	20 200	16 900	10 800	26 100	22 500	18 300
	2 750	28 900	24 900	20 200	16 900	10 800	26 100	22 500	18 300
	3 000	28 900	24 900	20 200	16 900	10 800	26 100	22 500	18 300
	3 250	28 900	24 900	20 200	16 800	10 800	26 000	22 500	18 300
	3 500	28 900	24 800	20 200	16 800	10 800	26 000	22 500	18 300
	3 750	28 800	24 800	20 200	16 800	10 700	25 900	22 400	18 200
	4 000	28 800	24 800	20 100	16 800	10 700	25 900	22 400	18 200
	4 250	28 700	24 800	20 100	16 800	10 700	25 800	22 300	18 200
	4 500	28 600	24 700	20 100	16 700	10 700	25 700	22 300	18 100
	4 750	28 600	24 700	20 000	16 700	10 700	25 600	22 200	18 100
	5 000	28 500	24 600	20 000	16 700	10 700	25 400	22 100	18 000
	5 250	28 300	24 500	19 900	16 600	10 600	25 300	22 000	17 900
	5 500	28 200	24 400	19 800	16 500	10 600	25 100	21 900	17 800
	6 000	27 800	24 200	19 700	16 400	10 500	24 600	21 600	17 600
	6 500	27 400	23 900	19 400	16 200	10 400	24 000	21 200	17 300
	7 000	26 800	23 500	19 100	15 900	10 300	23 300	20 700	16 900
	7 500	26 100	23 100	18 800	15 600	10 100	22 500	20 200	16 500
	8 000	25 400	22 500	18 400	15 200	9 870	21 700	19 500	16 000
	8 500	24 500	21 900	17 900	14 800	9 630	20 700	18 800	15 400
	9 000	23 600	21 300	17 400	14 400	9 360	19 700	18 000	14 800
	9 500	22 600	20 500	16 800	13 800	9 000	18 600	17 200	14 200
	10 000	21 600	19 800	16 200	13 300	8 630	17 600	16 400	13 500
	10 500	20 500	18 900	15 600	12 700	8 240	16 600	15 600	12 900
	11 000	19 500	18 100	14 900	12 200	7 850	15 600	14 700	12 200
	11 500	18 400	17 300	14 200	11 600	7 460	14 600	13 900	11 500
	12 000	17 400	16 400	13 500	11 000	7 070	13 700	13 100	10 900
PROPERTIES AND DESIGN DATA									
Area (mm^2)		92 000	79 100	64 200	53 600	35 600	83 000	71 600	58 200
Z_x (10^3 mm^3)		19 000	17 600	14 800	12 400	8 250	15 400	14 300	12 100
S_x (10^3 mm^3)		16 000	15 200	13 100	11 100	7 330	12 800	12 300	10 600
r_x (mm)		218	230	237	239	241	196	207	214
Z_y (10^3 mm^3)		9 470	9 180	7 610	6 100	3 810	7 850	7 590	6 290
S_y (10^3 mm^3)		6 080	6 050	5 040	4 030	2 380	5 030	5 000	4 170
r_y (mm)		135	145	147	144	140	123	132	134
r_x/r_y		1.61	1.59	1.61	1.66	1.72	1.59	1.57	1.60
M_{rx} (kN·m) (L < L$_u$)		5 990	5 540	4 660	3 910	2 310	4 850	4 500	3 810
M_{ry} (kN·m)		2 980	2 890	2 400	1 920	750	2 470	2 390	1 980
J (10^3 mm^4)		110 000	83 100	47 000	24 700	6 410	99 400	75 400	42 700
C_w (10^9 mm^6)		100 000	99 900	86 700	72 100	47 800	60 800	60 500	52 700
L_u (mm)		12 500	12 100	10 700	9 400	8 790	12 500	11 900	10 300
F_y (MPa)		350	350	350	350	350	350	350	350
IMPERIAL SIZE AND MASS									
Mass (lb./ft.)		484	416	338	282	188	437	377	306
Nominal Depth and Width (in.)		22 x 22					20 x 20		

‡ Welding does not fully develop web strength.
** Class 4: C_r calculated according to CAN/CSA-S16.1-94 Clause 13.3.3; S_x, S_y, M_{rx}, M_{ry} according to Clause 13.5(c)(iii).

G40.21 350W
$\phi = 0.90$

WWF COLUMNS
Factored Axial Compressive Resistances, C_r, in kN

Designation		WWF500						
Mass (kg/m)		381	343	306	276	254 *	223 **	197 **
Effective length (KL) in millimetres with respect to least radius of gyration	0	15 300	13 800	12 300	11 100	10 200	8 460	6 910
	2 000	15 300	13 800	12 300	11 100	10 200	8 460	6 910
	2 250	15 300	13 800	12 300	11 100	10 200	8 460	6 910
	2 500	15 300	13 800	12 300	11 100	10 200	8 450	6 910
	2 750	15 300	13 800	12 300	11 100	10 200	8 450	6 900
	3 000	15 300	13 800	12 300	11 100	10 100	8 450	6 900
	3 250	15 300	13 800	12 200	11 100	10 100	8 440	6 900
	3 500	15 200	13 700	12 200	11 000	10 100	8 430	6 890
	3 750	15 200	13 700	12 200	11 000	10 100	8 420	6 880
	4 000	15 200	13 700	12 200	11 000	10 100	8 410	6 870
	4 250	15 200	13 600	12 100	11 000	10 100	8 390	6 860
	4 500	15 100	13 600	12 100	10 900	10 000	8 370	6 850
	4 750	15 100	13 600	12 100	10 900	9 980	8 350	6 830
	5 000	15 000	13 500	12 000	10 800	9 930	8 320	6 810
	5 250	14 900	13 400	11 900	10 800	9 880	8 280	6 780
	5 500	14 800	13 300	11 800	10 700	9 810	8 240	6 750
	6 000	14 600	13 100	11 700	10 600	9 650	8 150	6 680
	6 500	14 300	12 900	11 400	10 300	9 450	8 030	6 590
	7 000	14 000	12 600	11 100	10 100	9 210	7 880	6 490
	7 500	13 600	12 200	10 800	9 800	8 930	7 700	6 350
	8 000	13 200	11 800	10 400	9 470	8 610	7 480	6 200
	8 500	12 700	11 300	9 970	9 100	8 260	7 240	6 030
	9 000	12 200	10 800	9 520	8 710	7 880	6 960	5 850
	9 500	11 600	10 300	9 050	8 290	7 500	6 610	5 650
	10 000	11 000	9 800	8 580	7 870	7 100	6 270	5 450
	10 500	10 500	9 270	8 110	7 450	6 710	5 920	5 230
	11 000	9 910	8 760	7 640	7 040	6 330	5 590	5 000
	11 500	9 360	8 260	7 200	6 640	5 960	5 260	4 730
	12 000	8 820	7 780	6 770	6 250	5 610	4 950	4 470

PROPERTIES AND DESIGN DATA								
Area (mm^2)		48 600	43 800	39 000	35 200	32 300	28 500	25 200
Z_x (10^3 mm^3)		10 100	9 100	8 060	7 420	6 780	6 010	5 410
S_x (10^3 mm^3)		9 010	8 140	7 240	6 740	6 160	5 180	4 330
r_x (mm)		215	216	215	218	218	219	223
Z_y (10^3 mm^3)		5 040	4 420	3 800	3 530	3 160	2 770	2 510
S_y (10^3 mm^3)		3 330	2 920	2 500	2 330	2 080	1 620	1 220
r_y (mm)		131	129	127	129	127	127	129
r_x/r_y		1.64	1.67	1.69	1.69	1.72	1.72	1.73
M_{rx} (kN·m) (L < L_u)		3 180	2 870	2 540	2 340	1 940	1 630	1 360
M_{ry} (kN·m)		1 590	1 390	1 200	1 110	655	510	384
J (10^3 mm^4)		22 500	15 400	10 200	7 920	5 820	3 970	2 870
C_w (10^9 mm^6)		44 100	39 400	34 500	32 500	29 400	26 200	24 000
L_u (mm)		8 970	8 390	7 930	7 830	8 040	8 090	8 370
F_y (MPa)		350	350	350	350	350	350	350

IMPERIAL SIZE AND MASS								
Mass (lb./ft.)		256	230	205	185	170	150	132
Nominal Depth and Width (in.)		20 x 20						

* Class 3 flanges.
** Class 4: C_r calculated according to CAN/CSA-S16.1-94 Clause 13.3.3; S_x, S_y, M_{rx}, M_{ry} according to Clause 13.5(c)(iii).

WWF COLUMNS
Factored Axial Compressive Resistances, C_r, in kN

G40.21 350W

$\phi = 0.90$

	Designation	WWF450								
	Mass (kg/m)	503 ‡	409	342	308	274	248	228	201 *	177 **
Effective length (KL) in millimetres with respect to least radius of gyration	0	20 200	16 400	13 700	12 400	11 000	9 950	9 140	8 060	6 630
	2 000	20 200	16 400	13 700	12 400	11 000	9 950	9 130	8 060	6 620
	2 250	20 200	16 400	13 700	12 400	11 000	9 940	9 120	8 050	6 620
	2 500	20 200	16 400	13 700	12 400	11 000	9 940	9 120	8 050	6 620
	2 750	20 100	16 400	13 700	12 300	11 000	9 930	9 110	8 040	6 620
	3 000	20 100	16 400	13 700	12 300	11 000	9 920	9 100	8 030	6 610
	3 250	20 100	16 400	13 700	12 300	11 000	9 900	9 080	8 020	6 600
	3 500	20 100	16 300	13 600	12 300	10 900	9 880	9 060	8 000	6 590
	3 750	20 000	16 300	13 600	12 300	10 900	9 850	9 040	7 980	6 580
	4 000	20 000	16 300	13 600	12 200	10 900	9 820	9 000	7 950	6 570
	4 250	19 900	16 200	13 500	12 200	10 800	9 780	8 960	7 910	6 550
	4 500	19 800	16 100	13 400	12 100	10 800	9 730	8 920	7 870	6 520
	4 750	19 700	16 100	13 400	12 000	10 700	9 670	8 860	7 820	6 500
	5 000	19 600	16 000	13 300	11 900	10 600	9 610	8 790	7 760	6 460
	5 250	19 400	15 900	13 200	11 800	10 500	9 530	8 710	7 690	6 430
	5 500	19 200	15 700	13 100	11 700	10 400	9 440	8 620	7 610	6 380
	6 000	18 800	15 400	12 800	11 500	10 200	9 220	8 410	7 420	6 280
	6 500	18 300	15 000	12 400	11 100	9 840	8 950	8 150	7 190	6 150
	7 000	17 700	14 600	12 000	10 700	9 470	8 630	7 850	6 930	5 990
	7 500	17 100	14 000	11 600	10 300	9 060	8 280	7 510	6 630	5 810
	8 000	16 300	13 500	11 000	9 810	8 610	7 890	7 140	6 300	5 600
	8 500	15 500	12 800	10 500	9 300	8 150	7 480	6 750	5 960	5 320
	9 000	14 700	12 200	9 920	8 780	7 670	7 060	6 360	5 610	5 040
	9 500	13 900	11 500	9 350	8 260	7 200	6 640	5 970	5 270	4 750
	10 000	13 100	10 900	8 790	7 750	6 750	6 230	5 590	4 930	4 460
	10 500	12 300	10 200	8 240	7 260	6 310	5 830	5 220	4 610	4 170
	11 000	11 500	9 580	7 720	6 790	5 890	5 460	4 880	4 310	3 900
	11 500	10 800	8 980	7 230	6 340	5 500	5 100	4 560	4 020	3 650
	12 000	10 100	8 420	6 760	5 930	5 130	4 770	4 250	3 760	3 410
	PROPERTIES AND DESIGN DATA									
	Area (mm^2)	64 100	52 200	43 600	39 300	35 000	31 600	29 000	25 600	22 600
	Z_x (10^3 mm^3)	11 400	9 640	8 100	7 290	6 470	5 960	5 450	4 840	4 360
	S_x (10^3 mm^3)	9 620	8 380	7 150	6 480	5 770	5 380	4 920	4 400	3 820
	r_x (mm)	184	190	192	193	193	196	196	197	200
	Z_y (10^3 mm^3)	6 150	5 100	4 090	3 580	3 080	2 860	2 560	2 250	2 040
	S_y (10^3 mm^3)	4 050	3 380	2 700	2 360	2 030	1 890	1 690	1 490	1 220
	r_y (mm)	119	121	118	116	114	116	114	114	116
	r_x/r_y	1.55	1.57	1.63	1.66	1.69	1.69	1.72	1.73	1.72
	M_{rx} (kN·m) (L < L_u)	3 590	3 040	2 550	2 300	2 040	1 880	1 720	1 390	1 200
	M_{ry} (kN·m)	1 940	1 610	1 290	1 130	970	901	806	469	384
	J (10^3 mm^4)	67 800	38 400	20 200	13 900	9 140	7 120	5 230	3 570	2 580
	C_w (10^9 mm^6)	34 700	30 400	25 500	22 900	20 100	18 900	17 200	15 300	14 000
	L_u (mm)	11 800	10 100	8 540	7 910	7 380	7 250	6 990	7 180	7 270
	F_y (MPa)	350	350	350	350	350	350	350	350	350
	IMPERIAL SIZE AND MASS									
	Mass (lb./ft.)	337	275	229	207	184	166	152	134	119
	Nominal Depth and Width (in.)	18 x 18								

‡ Welding does not fully develop web strength.   * Class 3 flanges.
** Class 4: C_r calculated according to CAN/CSA-S16.1-94 Clause 13.3.3; S_x, S_y, M_{rx}, M_{ry} according to Clause 13.5(c)(iii).

G40.21 350W
$\phi = 0.90$

WWF COLUMNS
Factored Axial Compressive Resistances, C_r, in kN

Designation	WWF400								
Mass (kg/m)	444 ‡	362	303	273	243	220	202	178 *	157 *
Effective length (KL) in millimetres with respect to least radius of gyration									
0	17 800	14 600	12 200	11 000	9 770	8 820	8 100	7 150	6 330
2 000	17 800	14 500	12 100	11 000	9 750	8 810	8 090	7 140	6 320
2 250	17 800	14 500	12 100	10 900	9 750	8 800	8 080	7 140	6 320
2 500	17 800	14 500	12 100	10 900	9 740	8 800	8 070	7 130	6 310
2 750	17 800	14 500	12 100	10 900	9 720	8 780	8 060	7 120	6 300
3 000	17 700	14 500	12 100	10 900	9 700	8 760	8 040	7 100	6 290
3 250	17 700	14 400	12 100	10 900	9 670	8 740	8 020	7 080	6 270
3 500	17 600	14 400	12 000	10 800	9 640	8 710	7 990	7 060	6 250
3 750	17 600	14 400	12 000	10 800	9 590	8 670	7 950	7 020	6 220
4 000	17 500	14 300	11 900	10 700	9 540	8 620	7 910	6 980	6 190
4 250	17 400	14 200	11 800	10 700	9 470	8 560	7 850	6 930	6 150
4 500	17 200	14 100	11 700	10 600	9 390	8 490	7 780	6 870	6 100
4 750	17 100	14 000	11 600	10 500	9 290	8 410	7 700	6 800	6 040
5 000	16 900	13 900	11 500	10 400	9 180	8 310	7 610	6 720	5 970
5 250	16 700	13 700	11 400	10 200	9 060	8 200	7 510	6 630	5 890
5 500	16 500	13 500	11 200	10 100	8 910	8 080	7 390	6 530	5 800
6 000	16 000	13 100	10 800	9 730	8 580	7 790	7 120	6 290	5 590
6 500	15 300	12 600	10 400	9 320	8 200	7 450	6 800	6 000	5 350
7 000	14 600	12 100	9 890	8 860	7 770	7 070	6 440	5 690	5 080
7 500	13 800	11 500	9 360	8 360	7 320	6 670	6 060	5 360	4 790
8 000	13 000	10 800	8 790	7 850	6 850	6 250	5 680	5 010	4 490
8 500	12 200	10 200	8 230	7 330	6 380	5 830	5 290	4 670	4 190
9 000	11 400	9 510	7 670	6 830	5 930	5 420	4 910	4 340	3 890
9 500	10 600	8 870	7 130	6 340	5 500	5 030	4 560	4 020	3 610
10 000	9 850	8 260	6 620	5 890	5 090	4 670	4 220	3 730	3 350
10 500	9 150	7 690	6 150	5 460	4 720	4 330	3 910	3 450	3 110
11 000	8 500	7 150	5 710	5 060	4 370	4 010	3 620	3 200	2 880
11 500	7 900	6 650	5 300	4 700	4 050	3 720	3 360	2 970	2 670
12 000	7 350	6 190	4 930	4 370	3 760	3 460	3 120	2 750	2 480
PROPERTIES AND DESIGN DATA									
Area (mm^2)	56 600	46 200	38 600	34 800	31 000	28 000	25 700	22 700	20 100
Z_x (10^3 mm^3)	8 770	7 480	6 300	5 680	5 050	4 660	4 260	3 790	3 420
S_x (10^3 mm^3)	7 300	6 410	5 500	5 000	4 470	4 170	3 830	3 430	3 120
r_x (mm)	161	166	169	170	170	173	173	174	176
Z_y (10^3 mm^3)	4 870	4 030	3 230	2 840	2 440	2 260	2 020	1 780	1 610
S_y (10^3 mm^3)	3 200	2 670	2 130	1 870	1 600	1 490	1 330	1 170	1 070
r_y (mm)	106	108	105	104	102	103	102	102	103
r_x/r_y	1.52	1.54	1.61	1.63	1.67	1.68	1.70	1.71	1.71
M_{rx} (kN·m) (L < L_u)	2 760	2 360	1 980	1 790	1 590	1 470	1 340	1 080	983
M_{ry} (kN·m)	1 530	1 270	1 020	895	769	712	636	369	337
J (10^3 mm^4)	60 100	34 100	17 900	12 300	8 110	6 320	4 640	3 170	2 290
C_w (10^9 mm^6)	18 500	16 300	13 800	12 400	11 000	10 300	9 380	8 390	7 700
L_u (mm)	11 900	9 910	8 210	7 490	6 890	6 710	6 440	6 600	6 440
F_y (MPa)	350	350	350	350	350	350	350	350	350
IMPERIAL SIZE AND MASS									
Mass (lb./ft.)	298	243	203	183	163	147	135	119	105
Nominal Depth and Width (in.)	16 x 16								

‡ Welding does not fully develop web strength.
* Class 3 flanges.

WWF COLUMNS
Factored Axial Compressive Resistances, C_r, in kN

G40.21 350W

$\phi = 0.90$

Designation	WWF350							
Mass (kg/m)	315	263	238	212	192	176	155	137
Effective length (KL) in millimetres with respect to least radius of gyration								
0	12 700	10 600	9 540	8 510	7 690	7 060	6 240	5 510
2 000	12 600	10 600	9 530	8 490	7 670	7 040	6 220	5 500
2 250	12 600	10 600	9 510	8 480	7 660	7 030	6 220	5 500
2 500	12 600	10 500	9 500	8 460	7 650	7 020	6 200	5 480
2 750	12 600	10 500	9 470	8 440	7 630	7 000	6 190	5 470
3 000	12 500	10 500	9 440	8 400	7 600	6 970	6 160	5 450
3 250	12 500	10 400	9 390	8 360	7 560	6 940	6 130	5 420
3 500	12 400	10 400	9 340	8 310	7 520	6 890	6 090	5 390
3 750	12 300	10 300	9 270	8 240	7 460	6 840	6 040	5 350
4 000	12 200	10 200	9 180	8 150	7 390	6 770	5 980	5 300
4 250	12 100	10 100	9 070	8 050	7 300	6 680	5 900	5 230
4 500	12 000	9 970	8 950	7 930	7 200	6 590	5 810	5 160
4 750	11 800	9 820	8 810	7 800	7 080	6 480	5 710	5 080
5 000	11 600	9 640	8 640	7 640	6 950	6 350	5 600	4 980
5 250	11 400	9 450	8 460	7 470	6 800	6 210	5 470	4 880
5 500	11 200	9 240	8 260	7 290	6 640	6 050	5 340	4 760
6 000	10 600	8 770	7 820	6 880	6 280	5 710	5 030	4 500
6 500	10 000	8 240	7 330	6 430	5 880	5 340	4 700	4 220
7 000	9 390	7 690	6 820	5 960	5 470	4 960	4 360	3 920
7 500	8 730	7 130	6 300	5 500	5 060	4 570	4 020	3 620
8 000	8 080	6 580	5 810	5 050	4 650	4 200	3 690	3 330
8 500	7 460	6 050	5 330	4 630	4 270	3 850	3 380	3 060
9 000	6 870	5 560	4 890	4 240	3 920	3 530	3 100	2 810
9 500	6 320	5 110	4 490	3 890	3 600	3 240	2 840	2 570
10 000	5 820	4 700	4 120	3 560	3 300	2 970	2 600	2 360
10 500	5 360	4 320	3 790	3 270	3 030	2 730	2 390	2 170
11 000	4 940	3 990	3 490	3 010	2 790	2 510	2 200	2 000
11 500	4 570	3 680	3 220	2 780	2 580	2 310	2 030	1 850
12 000	4 230	3 400	2 980	2 570	2 380	2 140	1 880	1 710
PROPERTIES AND DESIGN DATA								
Area (mm^2)	40 200	33 600	30 300	27 000	24 400	22 400	19 800	17 500
Z_x (10^3 mm^3)	5 580	4 730	4 280	3 810	3 520	3 220	2 870	2 590
S_x (10^3 mm^3)	4 710	4 070	3 720	3 330	3 120	2 870	2 580	2 350
r_x (mm)	143	146	146	147	150	150	151	153
Z_y (10^3 mm^3)	3 090	2 480	2 170	1 870	1 740	1 550	1 360	1 240
S_y (10^3 mm^3)	2 040	1 630	1 430	1 230	1 140	1 020	899	817
r_y (mm)	94.2	92.3	90.8	89.2	90.5	89.4	89.0	90.4
r_x/r_y	1.52	1.58	1.61	1.65	1.66	1.68	1.70	1.69
M_{rx} (kN·m) (L < L_u)	1 760	1 490	1 350	1 200	1 110	1 010	904	816
M_{ry} (kN·m)	973	781	684	589	548	488	428	391
J (10^3 mm^4)	29 800	15 700	10 800	7 070	5 520	4 060	2 760	2 000
C_w (10^9 mm^6)	8 040	6 870	6 210	5 490	5 190	4 720	4 230	3 890
L_u (mm)	9 900	7 970	7 140	6 460	6 230	5 930	5 630	5 510
F_y (MPa)	350	350	350	350	350	350	350	350
IMPERIAL SIZE AND MASS								
Mass (lb./ft.)	211	177	159	142	128	118	104	92
Nominal Depth and Width (in.)	14 x 14							

W COLUMNS
Factored Axial Compressive Resistances, C_r (kN)

CSA G40.21 350W
ASTM A992, A572 grade 50
$\phi = 0.90$

	Designation		W360			W310	
	Mass (kg/m)	79	72 **	64 **	226	202	179
Effective length (KL) in millimetres with respect to the least radius of gyration	0	3 140	2 740	2 400	8 970	8 010	7 080
	2 000	2 750	2 420	2 120	8 650	7 720	6 820
	2 250	2 630	2 330	2 040	8 540	7 620	6 720
	2 500	2 510	2 220	1 940	8 410	7 500	6 620
	2 750	2 370	2 110	1 850	8 260	7 360	6 490
	3 000	2 240	2 000	1 750	8 100	7 210	6 360
	3 250	2 100	1 880	1 650	7 920	7 050	6 210
	3 500	1 970	1 760	1 550	7 720	6 870	6 050
	3 750	1 840	1 650	1 460	7 520	6 680	5 880
	4 000	1 720	1 540	1 360	7 300	6 480	5 710
	4 250	1 610	1 430	1 270	7 070	6 280	5 520
	4 500	1 500	1 340	1 180	6 850	6 070	5 340
	4 750	1 400	1 250	1 100	6 610	5 860	5 150
	5 000	1 300	1 160	1 030	6 380	5 650	4 960
	5 250	1 220	1 080	957	6 140	5 440	4 780
	5 500	1 140	1 010	894	5 910	5 230	4 590
	6 000	995	885	781	5 460	4 830	4 230
	6 500	875	778	686	5 030	4 440	3 890
	7 000	773	687	606	4 630	4 080	3 570
	7 500	687	610	538	4 250	3 750	3 280
	8 000	613	545	480	3 910	3 440	3 010
	8 500	550	488	430	3 600	3 170	2 760
	9 000	496	440	388	3 310	2 910	2 540
	9 500	449	398	351	3 050	2 680	2 340
	10 000				2 820	2 480	2 160
	10 500				2 610	2 290	2 000
	11 000				2 420	2 120	1 850
	11 500				2 240	1 970	1 720
	12 000				2 090	1 830	1 590
PROPERTIES AND DESIGN DATA							
	Area (mm^2)	10 100	9 100	8 140	28 900	25 800	22 800
	Z_x (10^3 mm^3)	1 430	1 280	1 140	3 970	3 510	3 050
	S_x (10^3 mm^3)	1 280	1 150	1 030	3 420	3 050	2 670
	r_x (mm)	150	149	148	144	142	140
	Z_y (10^3 mm^3)	362	322	284	1 830	1 610	1 400
	S_y (10^3 mm^3)	236	210	186	1 190	1 050	919
	r_y (mm)	48.9	48.5	48.1	81.0	80.2	79.5
	r_x / r_y	3.07	3.07	3.08	1.78	1.77	1.76
	M_{rx} (kN·m) (L < L$_u$)	444	397	354	1 230	1 090	947
	M_{ry} (kN·m)	112	100	88.2	568	500	435
	J (10^3 mm^4)	811	601	436	10 800	7 730	5 370
	C_w (10^9 mm^6)	687	600	524	4 620	3 960	3 340
	L_u (mm)	3 010	2 940	2 870	6 680	6 220	5 820
	F_y (MPa)	345	345	345	345	345	345
IMPERIAL SIZE AND MASS							
	Mass (lb./ft.)	53	48	43	152	136	120
	Nominal Depth and Width (in.)		14 x 8			12 x 12	

** Class 4: C_r calculated according to S16.1-94 Clause 13.3.3

F_y taken as 345 MPa.

W COLUMNS
Factored Axial Compressive Resistances, C_r (kN)

CSA G40.21 350W
ASTM A992, A572 grade 50
$\phi = 0.90$

Designation		W310						W310	
Mass (kg/m)		158	143	129	118	107	97*	86	79
Effective length (KL) in millimetres with respect to the least radius of gyration	0	6 210	5 650	5 120	4 660	4 220	3 820	3 420	3 110
	2 000	5 970	5 430	4 920	4 470	4 050	3 660	3 190	2 900
	2 250	5 890	5 360	4 850	4 410	3 990	3 610	3 120	2 830
	2 500	5 800	5 270	4 770	4 330	3 930	3 550	3 030	2 750
	2 750	5 690	5 170	4 680	4 250	3 850	3 480	2 940	2 660
	3 000	5 570	5 060	4 580	4 150	3 760	3 400	2 840	2 570
	3 250	5 430	4 940	4 470	4 050	3 670	3 310	2 730	2 470
	3 500	5 290	4 810	4 350	3 940	3 570	3 220	2 620	2 370
	3 750	5 140	4 670	4 220	3 830	3 460	3 130	2 510	2 260
	4 000	4 990	4 530	4 090	3 710	3 350	3 030	2 390	2 160
	4 250	4 820	4 380	3 950	3 580	3 240	2 920	2 280	2 060
	4 500	4 660	4 230	3 820	3 460	3 120	2 820	2 170	1 960
	4 750	4 490	4 080	3 680	3 330	3 010	2 710	2 060	1 860
	5 000	4 330	3 930	3 540	3 200	2 890	2 610	1 960	1 760
	5 250	4 160	3 780	3 400	3 080	2 780	2 500	1 860	1 670
	5 500	4 000	3 630	3 270	2 950	2 670	2 400	1 760	1 580
	6 000	3 680	3 340	3 000	2 720	2 450	2 210	1 580	1 420
	6 500	3 380	3 070	2 750	2 490	2 240	2 020	1 420	1 270
	7 000	3 100	2 810	2 520	2 280	2 050	1 850	1 280	1 150
	7 500	2 850	2 580	2 310	2 090	1 880	1 690	1 150	1 030
	8 000	2 610	2 360	2 120	1 910	1 720	1 550	1 040	933
	8 500	2 400	2 170	1 950	1 760	1 580	1 420	945	845
	9 000	2 200	1 990	1 790	1 610	1 450	1 300	859	768
	9 500	2 030	1 840	1 640	1 480	1 330	1 200	783	701
	10 000	1 870	1 690	1 520	1 370	1 230	1 100	716	641
	10 500	1 730	1 560	1 400	1 260	1 130	1 020	657	588
	11 000	1 600	1 450	1 300	1 170	1 050	943	605	540
	11 500	1 490	1 340	1 200	1 080	973	874	558	498
	12 000	1 380	1 250	1 120	1 010	904	812	516	461
PROPERTIES AND DESIGN DATA									
Area (mm^2)		20 000	18 200	16 500	15 000	13 600	12 300	11 000	10 000
Z_x (10^3 mm^3)		2 670	2 420	2 160	1 950	1 760	1 590	1 420	1 280
S_x (10^3 mm^3)		2 360	2 150	1 940	1 750	1 590	1 440	1 280	1 150
r_x (mm)		139	138	137	136	135	134	134	133
Z_y (10^3 mm^3)		1 220	1 110	991	893	806	725	533	478
S_y (10^3 mm^3)		805	729	652	588	531	478	351	314
r_y (mm)		78.9	78.6	78.0	77.6	77.2	76.9	63.6	63.0
r_x / r_y		1.76	1.76	1.76	1.75	1.75	1.74	2.11	2.11
M_{rx} (kN·m) (L < L$_u$)		829	751	671	605	546	447	441	397
M_{ry} (kN·m)		379	345	308	277	250	148	166	148
J (10^3 mm^4)		3 770	2 860	2 130	1 600	1 210	909	874	655
C_w (10^9 mm^6)		2 840	2 540	2 220	1 970	1 760	1 560	961	847
L_u (mm)		5 480	5 280	5 070	4 920	4 800	4 970	3 900	3 810
F_y (MPa)		345	345	345	345	345	345	345	345
IMPERIAL SIZE AND MASS									
Mass (lb./ft.)		106	96	87	79	72	65	58	53
Nominal Depth and Width (in.)		12 x 12						12 x 10	

* Class 3 in bending

F_y taken as 345 MPa.

W COLUMNS

CSA G40.21 350W
ASTM A992, A572 grade 50
$\phi = 0.90$

Factored Axial Compressive Resistances, C_r (kN)

Designation		W310			W250				
Mass (kg/m)		74	67	60 **	167	149	131	115	101
Effective length (KL) in millimetres with respect to the least radius of gyration	0	2 940	2 640	2 290	6 610	5 900	5 190	4 530	4 010
	2 000	2 590	2 320	2 040	6 250	5 560	4 880	4 260	3 760
	2 250	2 490	2 220	1 960	6 120	5 450	4 780	4 170	3 680
	2 500	2 370	2 120	1 870	5 980	5 320	4 670	4 070	3 590
	2 750	2 250	2 010	1 780	5 820	5 180	4 540	3 960	3 480
	3 000	2 130	1 900	1 690	5 650	5 020	4 400	3 830	3 370
	3 250	2 000	1 780	1 590	5 470	4 850	4 250	3 700	3 250
	3 500	1 880	1 670	1 490	5 270	4 680	4 090	3 560	3 130
	3 750	1 760	1 560	1 390	5 070	4 500	3 930	3 420	3 000
	4 000	1 640	1 460	1 300	4 870	4 310	3 770	3 270	2 870
	4 250	1 540	1 360	1 210	4 660	4 130	3 600	3 130	2 740
	4 500	1 430	1 270	1 130	4 460	3 940	3 440	2 980	2 620
	4 750	1 340	1 190	1 060	4 260	3 760	3 280	2 840	2 490
	5 000	1 250	1 110	985	4 060	3 590	3 120	2 710	2 370
	5 250	1 170	1 040	920	3 870	3 410	2 970	2 570	2 250
	5 500	1 090	968	859	3 680	3 250	2 830	2 450	2 140
	6 000	958	848	753	3 330	2 940	2 550	2 210	1 930
	6 500	843	746	662	3 020	2 650	2 300	1 990	1 740
	7 000	746	660	585	2 730	2 400	2 080	1 800	1 570
	7 500	663	586	520	2 470	2 170	1 880	1 630	1 420
	8 000	593	524	464	2 240	1 970	1 710	1 470	1 280
	8 500	532	470	417	2 040	1 790	1 550	1 340	1 170
	9 000	480	424	375	1 860	1 630	1 420	1 220	1 060
	9 500	434	384	340	1 700	1 490	1 290	1 110	969
	10 000				1 560	1 370	1 180	1 020	887
	10 500				1 440	1 260	1 090	937	814
	11 000				1 320	1 160	1 000	862	750
	11 500				1 220	1 070	926	796	692
	12 000				1 130	991	857	737	641
PROPERTIES AND DESIGN DATA									
Area (mm^2)		9 480	8 500	7 590	21 300	19 000	16 700	14 600	12 900
Z_x (10^3 mm^3)		1 190	1 060	941	2 430	2 130	1 850	1 600	1 400
S_x (10^3 mm^3)		1 060	948	849	2 080	1 840	1 610	1 410	1 240
r_x (mm)		132	131	130	119	117	115	114	113
Z_y (10^3 mm^3)		350	310	275	1 140	1 000	870	753	656
S_y (10^3 mm^3)		229	203	180	746	656	571	495	432
r_y (mm)		49.7	49.3	49.1	68.1	67.4	66.8	66.2	65.6
r_x / r_y		2.66	2.66	2.65	1.75	1.74	1.72	1.72	1.72
M_{rx} (kN·m) (L < L$_u$)		369	329	292	755	661	574	497	435
M_{ry} (kN·m)		109	96.3	85.4	354	311	270	234	204
J (10^3 mm^4)		743	543	396	6 310	4 510	3 120	2 130	1 490
C_w (10^9 mm^6)		505	439	384	1 630	1 390	1 160	976	829
L_u (mm)		3 100	3 020	2 960	5 900	5 480	5 080	4 740	4 470
F_y (MPa)		345	345	345	345	345	345	345	345
IMPERIAL SIZE AND MASS									
Mass (lb./ft.)		50	45	40	112	100	88	77	68
Nominal Depth and Width (in.)		12 x 8			10 x 10				

** Class 4: C_r calculated according to S16.1-94 Clause 13.3.3

F_y taken as 345 MPa.

W COLUMNS
Factored Axial Compressive Resistances, C_r (kN)

CSA G40.21 350W
ASTM A992, A572 grade 50
$\phi = 0.90$

Designation		W250			W250		
Mass (kg/m)		89	80	73	67	58	49 *
Effective length (KL) in millimetres with respect to the least radius of gyration	0	3 540	3 170	2 880	2 650	2 300	1 940
	2 000	3 320	2 970	2 700	2 360	2 040	1 710
	2 250	3 250	2 900	2 640	2 270	1 960	1 630
	2 500	3 160	2 830	2 570	2 170	1 870	1 560
	2 750	3 070	2 750	2 490	2 060	1 780	1 470
	3 000	2 970	2 660	2 410	1 950	1 680	1 390
	3 250	2 860	2 560	2 320	1 840	1 580	1 310
	3 500	2 750	2 460	2 230	1 730	1 490	1 230
	3 750	2 640	2 360	2 140	1 630	1 400	1 150
	4 000	2 520	2 260	2 040	1 520	1 310	1 070
	4 250	2 410	2 150	1 950	1 430	1 220	1 000
	4 500	2 300	2 050	1 860	1 340	1 140	934
	4 750	2 190	1 950	1 770	1 250	1 070	872
	5 000	2 080	1 860	1 680	1 170	998	813
	5 250	1 970	1 760	1 590	1 090	933	760
	5 500	1 870	1 670	1 510	1 020	873	710
	6 000	1 690	1 510	1 360	899	766	622
	6 500	1 520	1 360	1 220	793	675	547
	7 000	1 370	1 220	1 100	703	598	484
	7 500	1 240	1 100	996	625	532	430
	8 000	1 120	1 000	901	559	475	384
	8 500	1 020	908	818	502	427	344
	9 000	926	826	744	453	385	310
	9 500	845	754	679	411	349	281
	10 000	773	690	621	374	317	
	10 500	710	634	570			
	11 000	654	583	525			
	11 500	603	538	484			
	12 000	558	498	448			
PROPERTIES AND DESIGN DATA							
Area (mm^2)		11 400	10 200	9 280	8 550	7 420	6 250
Z_x (10^3 mm^3)		1 230	1 090	985	901	770	633
S_x (10^3 mm^3)		1 100	982	891	806	693	572
r_x (mm)		112	111	110	110	108	106
Z_y (10^3 mm^3)		574	513	463	332	283	228
S_y (10^3 mm^3)		378	338	306	218	186	150
r_y (mm)		65.1	65.0	64.6	51.0	50.4	49.2
r_x / r_y		1.72	1.71	1.70	2.16	2.14	2.15
M_{rx} (kN·m) (L < L_u)		382	338	306	280	239	178
M_{ry} (kN·m)		178	159	144	103	87.9	46.6
J (10^3 mm^4)		1040	757	575	625	409	241
C_w (10^9 mm^6)		713	623	553	324	268	211
L_u (mm)		4 260	4 130	4 010	3 260	3 130	3 160
F_y (MPa)		345	345	345	345	345	345
IMPERIAL SIZE AND MASS							
Mass (lb./ft.)		60	54	49	45	39	33
Nominal Depth and Width (in.)		10 x 10			10 x 8		

* Class 3 in bending

F_y taken as 345 MPa.

CSA G40.21 350W
ASTM A992, A572 grade 50
$\phi = 0.90$

W COLUMNS
Factored Axial Compressive Resistances, C_r (kN)

	Designation			W200				W200	
	Mass (kg/m)	100	86	71	59	52	46 *	42	36
Effective length (KL) in millimetres with respect to the least radius of gyration	0	3 910	3 420	2 820	2 340	2 060	1 810	1 640	1 410
	2 000	3 530	3 070	2 530	2 090	1 840	1 610	1 350	1 150
	2 250	3 410	2 970	2 440	2 010	1 770	1 550	1 270	1 080
	2 500	3 270	2 850	2 340	1 930	1 690	1 480	1 180	1 010
	2 750	3 130	2 720	2 230	1 840	1 610	1 410	1 100	939
	3 000	2 980	2 590	2 120	1 740	1 530	1 330	1 020	868
	3 250	2 820	2 450	2 010	1 650	1 450	1 260	938	801
	3 500	2 670	2 320	1 900	1 550	1 360	1 190	865	737
	3 750	2 520	2 190	1 790	1 460	1 280	1 110	796	679
	4 000	2 370	2 060	1 680	1 370	1 200	1 040	733	624
	4 250	2 230	1 930	1 580	1 290	1 130	978	675	575
	4 500	2 090	1 810	1 480	1 210	1 060	915	623	530
	4 750	1 970	1 700	1 390	1 130	989	857	575	489
	5 000	1 850	1 600	1 300	1 060	926	801	531	452
	5 250	1 730	1 500	1 220	991	867	750	492	418
	5 500	1 630	1 410	1 140	929	812	703	456	387
	6 000	1 440	1 240	1 010	818	715	618	394	335
	6 500	1 270	1 100	892	723	631	545	343	291
	7 000	1 130	975	792	641	560	483	301	255
	7 500	1 010	870	706	571	499	430	266	225
	8 000	904	779	632	511	446	385	236	200
	8 500	814	701	569	460	401	346		
	9 000	735	634	514	415	362	312		
	9 500	667	575	466	376	328	283		
	10 000	608	523	424	342	299	257		
	10 500	555	478	387					
	11 000								
	11 500								
	12 000								
PROPERTIES AND DESIGN DATA									
Area (mm²)		12 600	11 000	9 070	7 530	6 620	5 820	5 280	4 540
Z_x (10³ mm³)		1 150	978	800	650	566	492	442	376
S_x (10³ mm³)		987	851	707	580	509	445	396	340
r_x (mm)		94.6	92.6	91.7	89.9	89.0	88.1	87.7	86.7
Z_y (10³ mm³)		533	458	374	302	265	229	165	141
S_y (10³ mm³)		349	300	246	199	175	151	108	92.6
r_y (mm)		53.8	53.4	52.9	52.1	51.9	51.3	41.3	41.0
r_x / r_y		1.76	1.73	1.73	1.73	1.71	1.72	2.12	2.11
M_{rx} (kN·m) (L < L_u)		357	304	248	202	176	138	137	117
M_{ry} (kN·m)		166	142	116	93.8	82.3	46.9	51.2	43.8
J (10³ mm⁴)		2 060	1 370	801	452	314	213	215	139
C_w (10⁹ mm⁶)		386	318	250	196	167	141	84.0	69.5
L_u (mm)		4 430	4 100	3 730	3 430	3 300	3 370	2 610	2 510
F_y (MPa)		345	345	345	345	345	345	345	345
IMPERIAL SIZE AND MASS									
Mass (lb./ft.)		67	58	48	40	35	31	28	24
Nominal Depth and Width (in.)		8 x 8						8 x 6½	

* Class 3 in bending
F_y taken as 345 MPa.

W COLUMNS
Factored Axial Compressive Resistances, C_r (kN)

CSA G40.21 350W
ASTM A992, A572 grade 50
$\phi = 0.90$

Designation		W200		W150		
Mass (kg/m)		31	27	37	30	22 **
Effective length (KL) in millimetres with respect to the least radius of gyration	0	1 240	1 050	1 470	1 180	839
	2 000	875	727	1 170	930	683
	2 250	794	657	1 090	866	632
	2 500	716	591	1 010	801	582
	2 750	645	530	930	738	533
	3 000	579	475	854	677	487
	3 250	521	426	783	620	445
	3 500	469	383	717	567	405
	3 750	423	345	656	519	370
	4 000	383	311	601	475	337
	4 250	347	282	551	435	308
	4 500	316	256	506	399	282
	4 750	288	234	466	367	259
	5 000	263	214	429	338	238
	5 250	242	196	396	312	220
	5 500	222	180	367	289	203
	6 000	190	154	316	248	174
	6 500			274	216	151
	7 000			240	188	132
	7 500			211	166	
	8 000					
	8 500					
	9 000					
	9 500					
	10 000					
	10 500					
	11 000					
	11 500					
	12 000					
PROPERTIES AND DESIGN DATA						
Area (mm²)		4 000	3 390	4 730	3 790	2 840
Z_x (10^3 mm³)		335	279	310	244	176
S_x (10^3 mm³)		299	249	274	218	149
r_x (mm)		88.6	87.3	68.5	67.3	65.1
Z_y (10^3 mm³)		93.8	76.1	140	111	77.5
S_y (10^3 mm³)		61.1	49.6	91.8	72.6	44.5
r_y (mm)		32.0	31.2	38.7	38.3	36.9
r_x / r_y		2.77	2.80	1.77	1.76	1.76
M_{rx} (kN·m) (L < L_u)		104	86.6	96.3	75.8	46.2
M_{ry} (kN·m)		29.1	23.6	43.5	34.5	13.8
J (10^3 mm⁴)		119	71.3	192	100	41.5
C_w (10^9 mm⁶)		40.9	32.5	40.0	30.3	20.4
L_u (mm)		1 980	1 890	2 630	2 440	2 480
F_y (MPa)		345	345	345	345	345
IMPERIAL SIZE AND MASS						
Mass (lb./ft.)		21	18	25	20	15
Nominal Depth and Width (in.)		8 x 5¼		6 x 6		

** Class 4: C_r calculated according to S16.1-94 Clause 13.3.3; S_x, S_y, M_{rx} and M_{ry} according to Clause 13.5(c)(iii).

F_y taken as 345 MPa.

FACTORED AXIAL COMPRESSIVE RESISTANCES
Hollow Structural Sections

Class C

    For rectangular hollow sections, see page 4-46.

    For square hollow sections, see page 4-54.

    For round hollow sections, see page 4-62.

Class H

    For rectangular hollow sections, see page 4-70.

    For square hollow sections, see page 4-78.

    For round hollow sections, see page 4-86.

RECTANGULAR HOLLOW SECTIONS
Factored Axial Compressive Resistances, C_r (kN)

G40.21 350W

$\phi = 0.90$

Designation (mm x mm x mm)		HSS 305 x 203					HSS 254 x 152		
		13	11	9.5	8.0 *	6.4 **	13	11	9.5
Mass (kg/m)		93.0	82.4	71.3	60.1	48.6	72.7	64.6	56.1
Effective length (KL) in millimetres with respect to the least radius of gyration	0	3 720	3 310	2 860	2 410	1 730	2 920	2 590	2 250
	1 000	3 700	3 290	2 850	2 400	1 720	2 880	2 560	2 230
	1 250	3 680	3 270	2 830	2 390	1 710	2 850	2 530	2 200
	1 500	3 650	3 250	2 820	2 370	1 700	2 810	2 500	2 170
	1 750	3 620	3 230	2 790	2 360	1 690	2 760	2 450	2 140
	2 000	3 580	3 190	2 760	2 330	1 680	2 690	2 400	2 090
	2 250	3 540	3 150	2 730	2 300	1 660	2 620	2 340	2 040
	2 500	3 480	3 100	2 690	2 270	1 640	2 540	2 270	1 980
	2 750	3 420	3 050	2 640	2 230	1 620	2 450	2 190	1 910
	3 000	3 350	2 990	2 590	2 190	1 600	2 360	2 110	1 840
	3 250	3 270	2 920	2 540	2 140	1 570	2 260	2 020	1 760
	3 500	3 190	2 850	2 480	2 090	1 540	2 150	1 930	1 690
	3 750	3 110	2 780	2 410	2 040	1 510	2 050	1 840	1 610
	4 000	3 020	2 700	2 340	1 980	1 470	1 950	1 750	1 530
	4 250	2 920	2 610	2 270	1 920	1 440	1 850	1 660	1 460
	4 500	2 830	2 530	2 200	1 860	1 400	1 750	1 570	1 380
	4 750	2 730	2 450	2 130	1 800	1 360	1 660	1 490	1 310
	5 000	2 630	2 360	2 060	1 740	1 330	1 560	1 410	1 240
	5 250	2 540	2 270	1 980	1 680	1 290	1 480	1 330	1 170
	5 500	2 440	2 190	1 910	1 620	1 250	1 400	1 260	1 110
	5 750	2 340	2 110	1 840	1 560	1 210	1 320	1 190	1 050
	6 000	2 250	2 020	1 770	1 500	1 170	1 250	1 120	992
	6 500	2 070	1 860	1 630	1 380	1 100	1 110	1 010	889
	7 000	1 910	1 720	1 500	1 280	1 030	997	902	797
	7 500	1 750	1 580	1 380	1 180	958	896	811	717
	8 000	1 610	1 450	1 270	1 080	884	807	731	647
	8 500	1 480	1 340	1 170	997	815	729	661	585
	9 000	1 360	1 230	1 080	920	752	662	600	531
	9 500	1 260	1 130	995	849	695	602	546	484
	10 000	1 160	1 050	919	785	642	550	499	442
	10 500	1 070	969	851	727	595	504	457	405
PROPERTIES AND DESIGN DATA									
Area (mm^2)		11 800	10 500	9 090	7 660	6 190	9 260	8 230	7 150
Z_x (10^3 mm^3)		1 190	1 060	926	787	640	747	671	589
S_x (10^3 mm^3)		964	867	762	652	535	592	537	475
r_x (mm)		111	112	113	114	115	90.1	91.0	91.9
Z_y (10^3 mm^3)		897	802	701	596	486	522	470	413
S_y (10^3 mm^3)		769	694	611	525	368	442	402	357
r_y (mm)		81.2	82.0	82.7	83.4	84.1	60.3	61.0	61.7
r_x / r_y		1.37	1.37	1.37	1.37	1.37	1.49	1.49	1.49
M_{rx} (kN·m)		375	334	292	248	202	235	211	186
M_{ry} (kN·m)		283	253	221	165	116	164	148	130
IMPERIAL SIZE AND MASS									
Mass (lb./ft.)		62.5	55.4	47.9	40.4	32.6	48.9	43.4	37.7
Thickness (in.)		0.500	0.438	0.375	0.313	0.250	0.500	0.438	0.375
Size (in.)		12 x 8					10 x 6		

* Class 3 in bending about Y-Y axis

** Class 4: C_r calculated according to S16.1-94 Clause 13.3.3; S_y and M_{ry} according to Clause 13.5(c)(iii).

G40.21 350W
$\phi = 0.90$

RECTANGULAR HOLLOW SECTIONS
Factored Axial Compressive Resistances, C_r (kN)

Designation (mm x mm x mm)	HSS 254 x 152		HSS 203 x 152					
	8.0	6.4 **	13	11	9.5	8.0	6.4	4.8 **
Mass (kg/m)	47.5	38.4	62.6	55.7	48.5	41.1	33.4	25.5
Effective length (KL) in millimetres with respect to the least radius of gyration								
0	1 910	1 460	2 510	2 240	1 950	1 650	1 340	954
1 000	1 880	1 450	2 480	2 210	1 920	1 630	1 320	945
1 250	1 870	1 440	2 450	2 180	1 900	1 610	1 310	937
1 500	1 840	1 420	2 410	2 150	1 870	1 590	1 290	927
1 750	1 810	1 400	2 360	2 110	1 840	1 560	1 270	914
2 000	1 770	1 380	2 310	2 060	1 800	1 530	1 240	898
2 250	1 730	1 350	2 240	2 000	1 750	1 490	1 210	879
2 500	1 680	1 320	2 160	1 940	1 690	1 440	1 170	857
2 750	1 620	1 290	2 080	1 870	1 630	1 390	1 130	834
3 000	1 560	1 250	2 000	1 790	1 570	1 340	1 090	808
3 250	1 500	1 210	1 910	1 710	1 500	1 280	1 050	781
3 500	1 440	1 170	1 820	1 630	1 430	1 230	1 000	752
3 750	1 370	1 120	1 730	1 550	1 360	1 170	955	723
4 000	1 310	1 070	1 640	1 470	1 290	1 110	909	694
4 250	1 240	1 020	1 550	1 400	1 230	1 050	863	665
4 500	1 180	967	1 460	1 320	1 160	998	819	633
4 750	1 120	918	1 380	1 250	1 100	945	776	600
5 000	1 060	870	1 300	1 180	1 040	894	734	568
5 250	1 010	825	1 230	1 110	981	845	694	538
5 500	952	781	1 160	1 050	926	798	657	509
5 750	901	740	1 090	989	875	754	621	482
6 000	852	701	1 030	934	826	713	587	456
6 500	764	629	919	833	738	638	526	409
7 000	686	565	821	745	660	571	472	367
7 500	618	509	736	669	593	513	424	330
8 000	557	460	662	602	534	463	382	298
8 500	505	416	598	544	483	419	346	270
9 000	458	378	542	493	438	380	314	245
9 500	418	345	493	448	398	346	286	223
10 000	382	315	450	409	364	316	261	204
10 500	350	289	412	375	333	289	240	187
PROPERTIES AND DESIGN DATA								
Area (mm²)	6 050	4 900	7 970	7 100	6 180	5 240	4 250	3 250
Z_x (10³ mm³)	503	411	528	476	420	360	295	228
S_x (10³ mm³)	410	338	423	385	343	297	246	192
r_x (mm)	92.7	93.6	73.4	74.3	75.1	75.9	76.7	77.5
Z_y (10³ mm³)	354	290	432	390	344	295	243	188
S_y (10³ mm³)	309	255	359	327	292	254	211	157
r_y (mm)	62.4	63.1	58.6	59.3	60.0	60.8	61.5	62.2
r_x / r_y	1.49	1.48	1.25	1.25	1.25	1.25	1.25	1.25
M_{rx} (kN·m)	158	129	166	150	132	113	92.9	71.8
M_{ry} (kN·m)	112	80.3	136	123	108	92.9	76.5	49.3
IMPERIAL SIZE AND MASS								
Mass (lb./ft.)	31.9	25.8	42.1	37.5	32.6	27.6	22.4	17.1
Thickness (in.)	0.313	0.250	0.500	0.438	0.375	0.313	0.250	0.188
Size (in.)	10 x 6		8 x 6					

** Class 4: C_r calculated according to S16.1-94 Clause 13.3.3; S_y and M_{ry} according to Clause 13.5(c)(iii).

RECTANGULAR HOLLOW SECTIONS
Factored Axial Compressive Resistances, C_r (kN)

G40.21 350W

$\phi = 0.90$

Designation (mm x mm x mm)	HSS 203 x 102						HSS 178 x 127		
	13	11	9.5	8.0	6.4	4.8 **	13	11	9.5
Mass (kg/m)	52.4	46.9	40.9	34.8	28.3	21.7	52.4	46.9	40.9
Effective length (KL) in millimetres with respect to the least radius of gyration									
0	2 100	1 880	1 640	1 400	1 140	800	2 100	1 880	1 640
1 000	2 020	1 810	1 580	1 350	1 100	780	2 060	1 840	1 610
1 250	1 960	1 760	1 540	1 310	1 070	764	2 020	1 810	1 580
1 500	1 880	1 690	1 480	1 260	1 030	743	1 970	1 760	1 540
1 750	1 780	1 600	1 410	1 210	989	718	1 900	1 710	1 500
2 000	1 680	1 510	1 330	1 140	938	689	1 830	1 650	1 440
2 250	1 560	1 410	1 250	1 070	883	657	1 750	1 570	1 380
2 500	1 450	1 310	1 160	1 000	825	623	1 660	1 500	1 320
2 750	1 340	1 210	1 070	929	768	588	1 570	1 420	1 250
3 000	1 230	1 120	992	859	712	552	1 480	1 340	1 180
3 250	1 130	1 030	913	793	658	511	1 380	1 250	1 110
3 500	1 030	942	839	730	607	472	1 290	1 170	1 040
3 750	944	864	771	672	559	436	1 210	1 100	972
4 000	865	792	708	618	515	402	1 120	1 020	908
4 250	794	728	651	569	475	371	1 050	955	848
4 500	729	669	599	524	438	343	975	890	791
4 750	671	616	552	484	404	317	908	830	738
5 000	618	568	510	447	374	293	846	774	689
5 250	571	525	472	414	346	272	789	722	643
5 500	528	486	437	384	321	252	736	674	601
5 750	490	451	405	356	299	235	688	630	562
6 000	455	419	377	332	278	219	643	590	526
6 500	395	364	328	289	242	191	565	519	463
7 000	345	319	287	253	212	167	498	458	410
7 500	304	281	253	223	188	148	442	407	364
8 000			225	198	167	131	395	363	325
8 500						117	354	326	292
9 000							319	293	263
9 500							288	266	238
10 000									
10 500									
PROPERTIES AND DESIGN DATA									
Area (mm^2)	6 680	5 970	5 210	4 430	3 610	2 760	6 680	5 970	5 210
Z_x (10^3 mm^3)	405	368	326	281	232	180	378	343	303
S_x (10^3 mm^3)	308	283	254	221	185	145	297	273	244
r_x (mm)	68.4	69.4	70.3	71.2	72.2	73.1	62.9	63.7	64.6
Z_y (10^3 mm^3)	246	224	199	172	143	111	298	271	240
S_y (10^3 mm^3)	201	187	169	148	125	93.3	244	225	202
r_y (mm)	39.1	39.8	40.5	41.3	42.0	42.7	48.1	48.9	49.6
r_x / r_y	1.75	1.74	1.74	1.72	1.72	1.71	1.31	1.30	1.30
M_{rx} (kN·m)	128	116	103	88.5	73.1	56.7	119	108	95.4
M_{ry} (kN·m)	77.5	70.6	62.7	54.2	45.0	29.4	93.9	85.4	75.6
IMPERIAL SIZE AND MASS									
Mass (lb./ft.)	35.2	31.5	27.5	23.4	19.0	14.6	35.2	31.5	27.5
Thickness (in.)	0.500	0.438	0.375	0.313	0.250	0.188	0.500	0.438	0.375
Size (in.)	8 x 4						7 x 5		

** Class 4: C_r calculated according to S16.1-94 Clause 13.3.3; S_y and M_{ry} according to Clause 13.5(c)(iii).

G40.21 350W
$\phi = 0.90$

RECTANGULAR HOLLOW SECTIONS
Factored Axial Compressive Resistances, C_r (kN)

Designation (mm x mm x mm)		HSS 178 x 127			HSS 152 x 102					
		8.0	6.4	4.8 *	13	11	9.5	8.0	6.4	4.8
Mass (kg/m)		34.8	28.3	21.7	42.3	38.0	33.3	28.4	23.2	17.9
Effective length (KL) in millimetres with respect to the least radius of gyration	0	1 400	1 140	869	1 700	1 520	1 340	1 140	932	718
	1 000	1 370	1 110	853	1 620	1 460	1 280	1 100	899	693
	1 250	1 340	1 100	840	1 570	1 410	1 240	1 060	873	675
	1 500	1 310	1 070	822	1 500	1 350	1 190	1 020	840	650
	1 750	1 280	1 040	800	1 420	1 280	1 130	973	801	621
	2 000	1 230	1 010	774	1 330	1 200	1 060	917	757	588
	2 250	1 180	969	745	1 230	1 120	994	858	709	552
	2 500	1 130	926	712	1 130	1 030	921	797	660	515
	2 750	1 070	880	678	1 040	952	850	737	612	478
	3 000	1 010	833	643	953	873	781	679	565	442
	3 250	952	786	607	871	799	717	624	520	408
	3 500	894	739	572	795	730	657	573	478	376
	3 750	838	693	537	726	668	601	525	439	346
	4 000	784	650	504	663	611	551	482	404	318
	4 250	733	608	472	607	560	506	443	371	293
	4 500	684	568	442	556	514	464	407	342	270
	4 750	639	531	413	511	472	427	375	315	249
	5 000	597	497	387	470	435	394	346	291	230
	5 250	558	465	362	433	401	364	320	269	213
	5 500	522	435	339	400	371	337	296	249	197
	5 750	488	408	318	371	344	312	275	231	183
	6 000	458	382	298	344	319	290	255	215	171
	6 500	403	337	263	298	277	252	222	187	149
	7 000	357	298	233	261	242	220	194	164	130
	7 500	317	265	208	229	213	194	171	145	115
	8 000	283	237	186					128	102
	8 500	254	213	167						
	9 000	229	192	151						
	9 500	208	174	137						
	10 000	189	158	124						
	10 500									
PROPERTIES AND DESIGN DATA										
Area (mm^2)		4 430	3 610	2 760	5 390	4 840	4 240	3 620	2 960	2 280
Z_x (10^3 mm^3)		261	216	168	252	230	206	179	148	116
S_x (10^3 mm^3)		213	178	140	193	179	162	143	121	95.6
r_x (mm)		65.4	66.2	67.1	52.2	53.1	54.0	54.8	55.7	56.5
Z_y (10^3 mm^3)		207	171	133	189	173	155	135	112	87.8
S_y (10^3 mm^3)		177	148	117	151	141	128	113	96.2	76.6
r_y (mm)		50.3	51.1	51.8	37.7	38.4	39.2	39.9	40.6	41.3
r_x / r_y		1.30	1.30	1.30	1.38	1.38	1.38	1.37	1.37	1.37
M_{rx} (kN·m)		82.2	68.0	52.9	79.4	72.5	64.9	56.4	46.6	36.5
M_{ry} (kN·m)		65.2	53.9	36.9	59.5	54.5	48.8	42.5	35.3	27.7
IMPERIAL SIZE AND MASS										
Mass (lb./ft.)		23.4	19.0	14.6	28.4	25.5	22.4	19.1	15.6	12.0
Thickness (in.)		0.313	0.250	0.188	0.500	0.438	0.375	0.313	0.250	0.188
Size (in.)		7 x 5			6 x 4					

* Class 3 in bending about Y-Y axis

RECTANGULAR HOLLOW SECTIONS
Factored Axial Compressive Resistances, C_r (kN)

G40.21 350W

$\phi = 0.90$

Designation (mm x mm x mm)		HSS 152 x 76				HSS 127 x 76				
		9.5	8.0	6.4	4.8	9.5	8.0	6.4	4.8	3.8 *
Mass (kg/m)		29.5	25.3	20.7	16.0	25.7	22.1	18.2	14.1	11.4
Effective length (KL) in millimetres with respect to the least radius of gyration	0	1 180	1 010	832	643	1 030	888	731	564	457
	1 000	1 090	937	772	599	945	816	675	523	425
	1 250	1 020	883	729	568	884	767	636	495	403
	1 500	944	818	679	530	813	708	590	460	376
	1 750	860	748	623	488	737	645	539	422	346
	2 000	774	677	565	445	661	581	488	383	315
	2 250	693	607	509	402	589	520	438	345	284
	2 500	617	543	457	362	523	463	391	310	255
	2 750	549	485	409	325	464	412	349	277	229
	3 000	489	432	366	291	412	367	311	248	205
	3 250	436	386	327	261	367	327	278	222	184
	3 500	390	346	294	235	327	292	249	199	165
	3 750	349	311	264	211	293	262	224	179	149
	4 000	314	280	238	191	263	236	202	162	134
	4 250	284	253	216	173	238	213	182	146	122
	4 500	257	229	196	157	215	193	165	133	110
	4 750	234	209	178	143	195	175	150	121	101
	5 000	213	191	163	131	178	160	137	110	92
	5 250	195	175	149	120	163	147	126	101	84
	5 500	179	160	137	110	150	135	116	93	78
	5 750	165	148	126	102		124	106	86	72
	6 000		137	117	94			98	79	66
	6 500									
	7 000									
	7 500									
	8 000									
	8 500									
	9 000									
	9 500									
	10 000									
	10 500									
PROPERTIES AND DESIGN DATA										
Area (mm^2)		3 760	3 220	2 640	2 040	3 280	2 820	2 320	1 790	1 450
Z_x (10^3 mm^3)		171	149	125	98.1	126	111	93.4	73.8	60.6
S_x (10^3 mm^3)		130	115	98.0	78.2	96.5	86.5	74.1	59.6	49.4
r_x (mm)		51.3	52.3	53.2	54.1	43.3	44.2	45.1	45.9	46.5
Z_y (10^3 mm^3)		104	91.2	76.6	60.5	87.8	77.4	65.3	51.8	42.6
S_y (10^3 mm^3)		85.0	76.3	65.5	52.9	70.8	63.9	55.2	44.8	37.3
r_y (mm)		29.4	30.1	30.8	31.5	28.7	29.4	30.1	30.8	31.3
r_x / r_y		1.74	1.74	1.73	1.72	1.51	1.50	1.50	1.49	1.49
M_{rx} (kN·m)		53.9	46.9	39.4	30.9	39.7	35.0	29.4	23.2	19.1
M_{ry} (kN·m)		32.8	28.7	24.1	19.1	27.7	24.4	20.6	16.3	11.7
IMPERIAL SIZE AND MASS										
Mass (lb./ft.)		19.8	17.0	13.9	10.7	17.3	14.9	12.2	9.46	7.66
Thickness (in.)		0.375	0.313	0.250	0.188	0.375	0.313	0.250	0.188	0.150
Size (in.)		6 x 3				5 x 3				

* Class 3 in bending about Y-Y axis

G40.21 350W
$\phi = 0.90$

RECTANGULAR HOLLOW SECTIONS
Factored Axial Compressive Resistances, C_r (kN)

Designation (mm x mm x mm)		HSS 102 x 76						HSS 102 x 51		
		9.5	8.0	6.4	4.8	3.8	3.2	8.0	6.4	4.8
Mass (kg/m)		21.9	18.9	15.6	12.2	9.89	8.35	15.8	13.1	10.3
Effective length (KL) in millimetres with respect to the least radius of gyration	0	879	759	627	488	397	334	633	526	413
	1 000	797	693	576	451	368	310	495	419	335
	1 250	743	648	541	424	347	293	424	363	292
	1 500	679	595	499	393	322	272	356	308	250
	1 750	612	539	454	359	295	249	297	259	212
	2 000	546	483	409	325	268	226	248	217	179
	2 250	484	430	365	291	241	204	208	183	151
	2 500	428	381	325	260	216	183	176	155	129
	2 750	378	338	289	232	193	163	150	133	110
	3 000	335	300	257	207	172	146	129	114	95
	3 250	297	267	229	185	154	131	112	99	83
	3 500	265	238	205	166	138	117	98	87	73
	3 750	237	213	184	149	124	105	86	76	64
	4 000	212	191	165	134	112	95			57
	4 250	191	172	149	121	101	86			
	4 500	173	156	135	110	92	78			
	4 750	157	142	123	100	84	71			
	5 000	143	129	112	91	76	65			
	5 250	131	118	103	84	70	60			
	5 500	120	109	94	77	64	55			
	5 750			87	71	59	50			
	6 000				65	55	47			
	6 500									
	7 000									
	7 500									
	8 000									
	8 500									
	9 000									
	9 500									
	10 000									
	10 500									
PROPERTIES AND DESIGN DATA										
Area (mm^2)		2 790	2 410	1 990	1 550	1 260	1 060	2 010	1 670	1 310
Z_x (10^3 mm^3)		87.9	77.9	66.0	52.6	43.3	37.0	59.0	50.7	40.8
S_x (10^3 mm^3)		67.4	61.1	52.9	43.0	35.9	30.8	43.6	38.5	31.8
r_x (mm)		35.0	35.9	36.7	37.5	38.0	38.4	33.2	34.2	35.1
Z_y (10^3 mm^3)		71.6	63.6	54.0	43.1	35.6	30.4	35.6	30.8	25.0
S_y (10^3 mm^3)		56.6	51.5	44.8	36.6	30.7	26.4	28.1	25.2	21.1
r_y (mm)		27.8	28.5	29.3	30.0	30.5	30.7	18.9	19.6	20.3
r_x / r_y		1.26	1.26	1.25	1.25	1.25	1.25	1.76	1.74	1.73
M_{rx} (kN·m)		27.7	24.5	20.8	16.6	13.6	11.7	18.6	16.0	12.9
M_{ry} (kN·m)		22.6	20.0	17.0	13.6	11.2	9.58	11.2	9.70	7.88
IMPERIAL SIZE AND MASS										
Mass (lb./ft.)		14.7	12.7	10.5	8.17	6.64	5.61	10.6	8.81	6.89
Thickness (in.)		0.375	0.313	0.250	0.188	0.150	0.125	0.313	0.250	0.188
Size (in.)		4 x 3						4 x 2		

RECTANGULAR HOLLOW SECTIONS
Factored Axial Compressive Resistances, C_r (kN)

G40.21 350W

$\phi = 0.90$

Designation (mm x mm x mm)		HSS 102 x 51		HSS 89 x 64				
		3.8	3.2	8.0	6.4	4.8	3.8	3.2
Mass (kg/m)		8.37	7.09	15.8	13.1	10.3	8.37	7.09
Effective length (KL) in millimetres with respect to the least radius of gyration	0	337	284	633	526	413	337	284
	1 000	276	235	545	458	363	298	252
	1 250	242	206	491	416	331	273	231
	1 500	208	178	433	370	296	245	208
	1 750	177	152	377	324	261	216	185
	2 000	150	129	326	282	228	190	162
	2 250	127	110	282	244	199	166	142
	2 500	108	94	243	212	173	145	124
	2 750	93	80	211	185	151	126	109
	3 000	80	70	184	161	132	111	95
	3 250	70	61	161	142	116	98	84
	3 500	61	53	142	125	103	86	74
	3 750	54	47	126	111	92	77	66
	4 000	48	42	112	99	82	69	59
	4 250			100	89	73	62	53
	4 500			90	80	66	56	48
	4 750				72	60	50	43
	5 000						46	39
	5 250							
	5 500							
	5 750							
	6 000							
	6 500							
	7 000							
	7 500							
	8 000							
	8 500							
	9 000							
	9 500							
	10 000							
	10 500							
PROPERTIES AND DESIGN DATA								
Area (mm^2)		1 070	903	2 010	1 670	1 310	1 070	903
Z_x (10^3 mm^3)		33.9	29.0	55.1	47.2	38.0	31.5	27.0
S_x (10^3 mm^3)		26.8	23.1	42.2	37.1	30.6	25.8	22.3
r_x (mm)		35.7	36.1	30.6	31.4	32.3	32.8	33.1
Z_y (10^3 mm^3)		20.8	17.9	43.3	37.3	30.1	25.0	21.4
S_y (10^3 mm^3)		18.0	15.6	34.4	30.5	25.3	21.4	18.5
r_y (mm)		20.7	21.0	23.3	24.1	24.8	25.2	25.5
r_x / r_y		1.72	1.72	1.31	1.30	1.30	1.30	1.30
M_{rx} (kN·m)		10.7	9.14	17.4	14.9	12.0	9.92	8.51
M_{ry} (kN·m)		6.55	5.64	13.6	11.7	9.48	7.88	6.74
IMPERIAL SIZE AND MASS								
Mass (lb./ft.)		5.62	4.76	10.6	8.81	6.89	5.62	4.76
Thickness (in.)		0.150	0.125	0.313	0.250	0.188	0.150	0.125
Size (in.)		4 x 2		3½ x 2½				

G40.21 350W
φ = 0.90

RECTANGULAR HOLLOW SECTIONS
Factored Axial Compressive Resistances, C_r (kN)

Designation (mm x mm x mm)		HSS 76 x 51					HSS 51 x 25
		8.0	6.4	4.8	3.8	3.2	3.2
Mass (kg/m)		12.6	10.6	8.35	6.85	5.82	3.28
Effective length (KL) in millimetres with respect to the least radius of gyration	0	504	425	334	275	233	132
	1 000	384	332	266	221	189	54
	1 250	325	285	230	192	165	39
	1 500	271	239	195	164	141	28
	1 750	224	200	164	138	120	22
	2 000	186	167	138	117	101	
	2 250	155	140	116	98	86	
	2 500	130	118	98	84	73	
	2 750	111	101	84	72	62	
	3 000	95	87	72	62	54	
	3 250	82	75	63	54	47	
	3 500	72	66	55	47	41	
	3 750		58	49	41	36	
	4 000				37	32	
	4 250						
	4 500						
	4 750						
	5 000						
	5 250						
	5 500						
	5 750						
	6 000						
	6 500						
	7 000						
	7 500						
	8 000						
	8 500						
	9 000						
	9 500						
	10 000						
	10 500						
PROPERTIES AND DESIGN DATA							
Area (mm^2)		1 600	1 350	1 060	872	741	418
Z_x (10^3 mm^3)		36.1	31.5	25.8	21.6	18.6	6.34
S_x (10^3 mm^3)		26.7	24.1	20.3	17.3	15.1	4.81
r_x (mm)		25.2	26.1	27.0	27.5	27.8	17.1
Z_y (10^3 mm^3)		26.9	23.6	19.4	16.3	14.0	3.85
S_y (10^3 mm^3)		20.7	18.9	16.1	13.8	12.0	3.15
r_y (mm)		18.1	18.9	19.6	20.0	20.3	9.78
r_x / r_y		1.39	1.38	1.38	1.38	1.37	1.75
M_{rx} (kN·m)		11.4	9.92	8.13	6.80	5.86	2.00
M_{ry} (kN·m)		8.47	7.43	6.11	5.13	4.41	1.21
IMPERIAL SIZE AND MASS							
Mass (lb./ft.)		8.46	7.11	5.61	4.60	3.91	2.21
Thickness (in.)		0.313	0.250	0.188	0.150	0.125	0.125
Size (in.)		3 x 2					2 x 1

SQUARE HOLLOW SECTIONS
Factored Axial Compressive Resistances, C_r (kN)

G40.21 350W

$\phi = 0.90$

Designation (mm x mm x mm)	HSS 305 x 305					HSS 254 x 254		
	13	11	9.5	8.0 *	6.4 **	13	11	9.5
Mass (kg/m)	113	100	86.5	72.8	58.7	93.0	82.4	71.3
0	4 540	4 030	3 470	2 920	1 910	3 720	3 310	2 860
1 000	4 530	4 020	3 460	2 920	1 900	3 700	3 300	2 850
1 250	4 520	4 020	3 450	2 910	1 900	3 690	3 290	2 850
1 500	4 510	4 010	3 440	2 910	1 900	3 680	3 270	2 830
1 750	4 490	3 990	3 430	2 900	1 900	3 660	3 260	2 820
2 000	4 470	3 980	3 420	2 890	1 890	3 630	3 240	2 800
2 250	4 450	3 960	3 400	2 870	1 880	3 600	3 210	2 780
2 500	4 430	3 940	3 380	2 860	1 880	3 570	3 180	2 750
2 750	4 390	3 910	3 360	2 840	1 870	3 530	3 140	2 720
3 000	4 360	3 880	3 340	2 820	1 860	3 480	3 100	2 690
3 250	4 320	3 840	3 310	2 790	1 850	3 430	3 060	2 650
3 500	4 270	3 800	3 270	2 760	1 840	3 380	3 010	2 610
3 750	4 230	3 760	3 240	2 730	1 820	3 310	2 960	2 560
4 000	4 170	3 710	3 200	2 700	1 810	3 250	2 900	2 520
4 250	4 110	3 660	3 160	2 670	1 790	3 180	2 840	2 460
4 500	4 050	3 610	3 110	2 630	1 780	3 110	2 780	2 410
4 750	3 990	3 550	3 060	2 590	1 760	3 030	2 710	2 350
5 000	3 920	3 500	3 010	2 550	1 740	2 960	2 640	2 300
5 250	3 850	3 430	2 960	2 500	1 720	2 880	2 570	2 240
5 500	3 780	3 370	2 910	2 460	1 700	2 800	2 500	2 180
5 750	3 700	3 300	2 850	2 410	1 670	2 720	2 430	2 110
6 000	3 620	3 230	2 790	2 360	1 650	2 640	2 360	2 050
6 500	3 460	3 090	2 670	2 270	1 600	2 470	2 220	1 930
7 000	3 300	2 950	2 550	2 160	1 550	2 320	2 080	1 810
7 500	3 140	2 810	2 430	2 060	1 500	2 160	1 940	1 690
8 000	2 970	2 660	2 300	1 960	1 450	2 020	1 810	1 580
8 500	2 820	2 520	2 190	1 860	1 390	1 880	1 690	1 480
9 000	2 660	2 390	2 070	1 760	1 340	1 750	1 580	1 380
9 500	2 510	2 260	1 960	1 660	1 290	1 640	1 470	1 290
10 000	2 370	2 130	1 850	1 570	1 240	1 530	1 370	1 200
10 500	2 240	2 010	1 750	1 490	1 190	1 420	1 280	1 120
PROPERTIES AND DESIGN DATA								
Area (mm^2)	14 400	12 800	11 000	9 280	7 480	11 800	10 500	9 090
Z (10^3 mm^3)	1 560	1 390	1 210	1 030	833	1 060	946	825
S (10^3 mm^3)	1 330	1 190	1 040	886	625	889	800	703
r (mm)	118	119	120	121	121	97.6	98.4	99.1
M_r (kN·m)	491	438	381	279	197	334	298	260
IMPERIAL SIZE AND MASS								
Mass (lb./ft.)	76.1	67.3	58.1	48.9	39.4	62.5	55.4	47.9
Thickness (in.)	0.500	0.438	0.375	0.313	0.250	0.500	0.438	0.375
Size (in.)	12 x 12					10 x 10		

Effective length (KL) in millimetres with respect to the least radius of gyration

* Class 3 in bending
** Class 4: C_r calculated according to S16.1-94 Clause 13.3.3; S and M_r according to Clause 13.5(c)(iii).

G40.21 350W
$\phi = 0.90$

SQUARE HOLLOW SECTIONS
Factored Axial Compressive Resistances, C_r (kN)

Designation (mm x mm x mm)	HSS 254 x 254		HSS 203 x 203				
	8.0	6.4 **	13	11	9.5	8.0	6.4
Mass (kg/m)	60.1	48.6	72.7	64.6	56.1	47.5	38.4

Effective length (KL) in millimetres with respect to the least radius of gyration

KL							
0	2 410	1 790	2 920	2 590	2 250	1 910	1 540
1 000	2 400	1 790	2 900	2 580	2 240	1 890	1 530
1 250	2 400	1 780	2 880	2 560	2 230	1 880	1 530
1 500	2 390	1 780	2 860	2 540	2 210	1 870	1 520
1 750	2 380	1 770	2 830	2 520	2 190	1 850	1 500
2 000	2 360	1 760	2 800	2 490	2 160	1 830	1 490
2 250	2 340	1 760	2 750	2 450	2 130	1 810	1 470
2 500	2 320	1 740	2 710	2 410	2 100	1 780	1 440
2 750	2 300	1 730	2 650	2 360	2 060	1 740	1 410
3 000	2 270	1 720	2 590	2 310	2 010	1 710	1 390
3 250	2 240	1 700	2 520	2 250	1 960	1 670	1 350
3 500	2 200	1 680	2 450	2 190	1 910	1 620	1 320
3 750	2 170	1 660	2 380	2 120	1 850	1 580	1 280
4 000	2 130	1 640	2 300	2 060	1 800	1 530	1 240
4 250	2 080	1 620	2 220	1 990	1 740	1 480	1 200
4 500	2 040	1 590	2 140	1 920	1 680	1 430	1 160
4 750	1 990	1 560	2 060	1 840	1 610	1 380	1 120
5 000	1 940	1 540	1 980	1 770	1 550	1 320	1 080
5 250	1 890	1 510	1 900	1 700	1 490	1 270	1 040
5 500	1 840	1 480	1 820	1 630	1 430	1 220	999
5 750	1 790	1 450	1 750	1 570	1 370	1 170	959
6 000	1 740	1 420	1 670	1 500	1 320	1 130	920
6 500	1 640	1 340	1 530	1 370	1 210	1 030	846
7 000	1 540	1 250	1 400	1 260	1 110	948	776
7 500	1 440	1 180	1 280	1 150	1 010	870	712
8 000	1 350	1 100	1 170	1 050	929	798	654
8 500	1 260	1 030	1 070	967	853	732	601
9 000	1 170	961	985	888	783	673	552
9 500	1 090	898	905	816	721	619	509
10 000	1 020	839	834	752	664	571	469
10 500	955	784	769	694	613	528	434

PROPERTIES AND DESIGN DATA

Area (mm^2)	7 660	6 190	9 260	8 230	7 150	6 050	4 900
Z (10^3 mm^3)	702	571	651	585	513	439	359
S (10^3 mm^3)	602	492	538	488	432	373	308
r (mm)	99.9	101	76.9	77.6	78.4	79.2	79.9
M$_r$ (kN·m)	221	155	205	184	162	138	113

IMPERIAL SIZE AND MASS

Mass (lb./ft.)	40.4	32.6	48.9	43.4	37.7	31.9	25.8
Thickness (in.)	0.313	0.250	0.500	0.438	0.375	0.313	0.250
Size (in.)	10 x 10		8 x 8				

** Class 4: C_r calculated according to S16.1-94 Clause 13.3.3; S and M$_r$ according to Clause 13.5(c)(iii).

SQUARE HOLLOW SECTIONS
Factored Axial Compressive Resistances, C_r (kN)

G40.21 350W

$\phi = 0.90$

Designation (mm x mm x mm)	HSS 178 x 178						HSS 152 x 152		
	13	11	9.5	8.0	6.4	4.8 *	13	11	9.5
Mass (kg/m)	62.6	55.7	48.5	41.1	33.4	25.5	52.4	46.9	40.9
Effective length (KL) in millimetres with respect to the least radius of gyration									
0	2 510	2 240	1 950	1 650	1 340	1 020	2 100	1 880	1 640
1 000	2 490	2 210	1 930	1 640	1 330	1 010	2 070	1 850	1 620
1 250	2 470	2 200	1 910	1 620	1 320	1 010	2 050	1 830	1 600
1 500	2 440	2 170	1 890	1 610	1 300	998	2 010	1 800	1 570
1 750	2 400	2 140	1 870	1 590	1 290	986	1 970	1 760	1 540
2 000	2 360	2 110	1 840	1 560	1 270	971	1 910	1 720	1 500
2 250	2 310	2 060	1 800	1 530	1 240	952	1 850	1 660	1 460
2 500	2 250	2 010	1 760	1 490	1 220	932	1 780	1 600	1 410
2 750	2 190	1 960	1 710	1 450	1 180	908	1 710	1 540	1 350
3 000	2 120	1 900	1 660	1 410	1 150	883	1 630	1 470	1 290
3 250	2 050	1 830	1 600	1 370	1 110	855	1 560	1 400	1 230
3 500	1 970	1 760	1 540	1 320	1 080	826	1 480	1 330	1 170
3 750	1 890	1 690	1 490	1 270	1 040	796	1 400	1 260	1 110
4 000	1 810	1 620	1 420	1 220	995	766	1 320	1 190	1 050
4 250	1 730	1 550	1 360	1 170	954	735	1 240	1 130	996
4 500	1 650	1 480	1 300	1 120	913	704	1 170	1 060	939
4 750	1 570	1 410	1 240	1 070	873	673	1 100	1 000	886
5 000	1 490	1 350	1 180	1 020	833	643	1 040	942	835
5 250	1 420	1 280	1 130	968	795	614	975	887	786
5 500	1 350	1 220	1 070	922	757	585	917	835	741
5 750	1 280	1 160	1 020	877	721	558	863	787	698
6 000	1 220	1 100	971	835	686	531	812	741	658
6 500	1 100	992	877	755	622	482	721	659	586
7 000	992	897	794	684	564	437	643	588	523
7 500	897	811	719	620	511	397	575	526	468
8 000	812	736	652	563	465	361	516	472	421
8 500	738	669	593	512	423	329	465	426	380
9 000	672	609	541	467	386	300	421	385	344
9 500	614	557	494	427	353	275	382	350	312
10 000	562	510	453	392	324	252	348	319	285
10 500	516	468	416	360	298	232	318	292	261
PROPERTIES AND DESIGN DATA									
Area (mm^2)	7 970	7 100	6 180	5 240	4 250	3 250	6 680	5 970	5 210
Z (10^3 mm^3)	484	437	385	330	271	210	342	310	275
S (10^3 mm^3)	396	361	322	279	231	181	276	253	227
r (mm)	66.5	67.2	68.0	68.8	69.6	70.3	56.1	56.9	57.6
M_r (kN·m)	152	138	121	104	85.4	57.0	108	97.7	86.6
IMPERIAL SIZE AND MASS									
Mass (lb./ft.)	42.1	37.5	32.6	27.6	22.4	17.1	35.2	31.5	27.5
Thickness (in.)	0.500	0.438	0.375	0.313	0.250	0.188	0.500	0.438	0.375
Size (in.)	7 x 7						6 x 6		

* Class 3 in bending

4-56

G40.21 350W
$\phi = 0.90$

SQUARE HOLLOW SECTIONS
Factored Axial Compressive Resistances, C_r (kN)

Designation (mm x mm x mm)	HSS 152 x 152			HSS 127 x 127					
	8.0	6.4	4.8	13	11	9.5	8.0	6.4	4.8
Mass (kg/m)	34.8	28.3	21.7	42.3	38.0	33.3	28.4	23.2	17.9
0	1 400	1 140	869	1 700	1 520	1 340	1 140	932	718
1 000	1 380	1 120	858	1 650	1 490	1 300	1 110	912	703
1 250	1 360	1 110	849	1 620	1 460	1 280	1 090	895	691
1 500	1 340	1 090	837	1 570	1 420	1 240	1 070	874	675
1 750	1 310	1 070	821	1 520	1 370	1 200	1 030	847	655
2 000	1 280	1 050	802	1 450	1 310	1 160	991	815	631
2 250	1 240	1 020	780	1 380	1 250	1 100	947	780	605
2 500	1 200	984	755	1 300	1 180	1 040	899	742	576
2 750	1 160	948	728	1 220	1 110	985	849	702	546
3 000	1 110	910	700	1 140	1 040	925	799	661	515
3 250	1 060	870	670	1 070	973	866	749	621	485
3 500	1 010	829	639	992	907	808	700	581	454
3 750	957	788	608	921	843	753	653	543	425
4 000	907	748	578	855	784	700	608	507	397
4 250	858	708	547	793	728	651	566	472	371
4 500	810	670	518	736	676	606	527	440	346
4 750	765	633	490	683	628	563	491	410	323
5 000	721	597	463	634	584	524	457	383	301
5 250	680	564	437	590	544	488	426	357	281
5 500	641	532	413	549	506	455	398	333	263
5 750	605	502	390	512	472	425	371	312	246
6 000	570	474	368	478	441	397	347	292	230
6 500	508	423	329	418	387	349	305	256	203
7 000	454	378	294	368	341	307	269	226	179
7 500	407	339	264	326	302	273	239	201	159
8 000	366	305	238	290	269	243	213	179	142
8 500	330	276	215	260	241	218	191	161	128
9 000	299	250	195	234	217	196	172	145	115
9 500	272	227	177				156	131	104
10 000	248	207	162						
10 500	227	190	148						

Effective length (KL) in millimetres with respect to the least radius of gyration

PROPERTIES AND DESIGN DATA

Area (mm^2)	4 430	3 610	2 760	5 390	4 840	4 240	3 620	2 960	2 280
Z (10^3 mm^3)	237	196	152	225	205	183	159	132	103
S (10^3 mm^3)	198	166	130	177	165	149	132	111	88.1
r (mm)	58.4	59.2	59.9	45.7	46.5	47.3	48.0	48.8	49.6
M$_r$ (kN·m)	74.7	61.7	47.9	70.9	64.6	57.6	50.1	41.6	32.4

IMPERIAL SIZE AND MASS

Mass (lb./ft.)	23.4	19.0	14.6	28.4	25.5	22.4	19.1	15.6	12.0
Thickness (in.)	0.313	0.250	0.188	0.500	0.438	0.375	0.313	0.250	0.188
Size (in.)	6 x 6			5 x 5					

4-57

SQUARE HOLLOW SECTIONS
Factored Axial Compressive Resistances, C_r (kN)

G40.21 350W

$\phi = 0.90$

Designation (mm x mm x mm)	HSS 102 x 102						HSS 89 x 89		
	9.5	8.0	6.4	4.8	3.8	3.2	9.5	8.0	6.4
Mass (kg/m)	25.7	22.1	18.2	14.1	11.4	9.62	21.9	18.9	15.6
Effective length (KL) in millimetres with respect to the least radius of gyration									
0	1 030	888	731	564	457	387	879	759	627
1 000	986	849	700	542	439	373	820	711	589
1 250	950	820	678	525	426	362	778	676	562
1 500	906	783	649	503	409	348	727	634	529
1 750	854	740	614	478	389	331	670	586	491
2 000	796	692	576	450	367	312	611	537	451
2 250	737	642	536	419	343	292	553	487	411
2 500	678	592	496	389	318	271	498	440	373
2 750	621	543	456	359	294	251	447	396	337
3 000	567	497	418	330	271	231	401	357	304
3 250	516	454	383	302	249	213	360	321	274
3 500	470	414	350	277	228	195	324	289	247
3 750	429	378	320	254	209	179	292	261	223
4 000	391	346	293	233	192	164	264	236	202
4 250	357	316	268	213	176	151	239	214	184
4 500	327	290	246	196	162	139	217	195	167
4 750	300	266	226	180	149	128	198	178	153
5 000	276	245	208	166	137	118	181	162	140
5 250	254	226	192	154	127	109	166	149	129
5 500	235	209	178	142	118	101	153	137	118
5 750	217	193	165	132	109	94	141	127	109
6 000	202	179	153	122	101	87	130	117	101
6 500	175	155	133	106	88	76			87
7 000	153	136	116	93	77	66			
7 500		119	102	82	68	58			
8 000						52			
8 500									
9 000									
9 500									
10 000									
10 500									
PROPERTIES AND DESIGN DATA									
Area (mm^2)	3 280	2 820	2 320	1 790	1 450	1 230	2 790	2 410	1 990
Z (10^3 mm^3)	110	96.8	81.4	64.3	52.8	44.9	80.5	71.4	60.5
S (10^3 mm^3)	87.6	78.5	67.3	54.2	45.0	38.5	63.0	57.1	49.5
r (mm)	36.9	37.6	38.4	39.2	39.7	40.0	31.7	32.4	33.2
M$_r$ (kN·m)	34.7	30.5	25.6	20.3	16.6	14.1	25.4	22.5	19.1
IMPERIAL SIZE AND MASS									
Mass (lb./ft.)	17.3	14.9	12.2	9.46	7.66	6.47	14.7	12.7	10.5
Thickness (in.)	0.375	0.313	0.250	0.188	0.150	0.125	0.375	0.313	0.250
Size (in.)	4 x 4						3½ x 3½		

G40.21 350W
$\phi = 0.90$

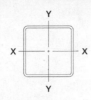

SQUARE HOLLOW SECTIONS
Factored Axial Compressive Resistances, C_r (kN)

Designation (mm x mm x mm)		HSS 89 x 89			HSS 76 x 76					
		4.8	3.8	3.2	9.5	8.0	6.4	4.8	3.8	3.2
Mass (kg/m)		12.2	9.89	8.35	18.1	15.8	13.1	10.3	8.37	7.09
Effective length (KL) in millimetres with respect to the least radius of gyration	0	488	397	334	728	633	526	413	337	284
	1 000	461	375	316	652	571	478	378	310	262
	1 250	441	360	303	603	530	446	354	291	246
	1 500	416	340	287	546	483	408	325	268	228
	1 750	387	317	268	488	434	368	295	244	207
	2 000	357	293	248	432	385	329	265	220	187
	2 250	327	269	228	380	341	292	236	196	168
	2 500	297	245	208	334	300	259	210	175	149
	2 750	269	222	189	293	265	229	186	156	133
	3 000	243	201	171	258	234	203	165	138	118
	3 250	220	182	155	228	207	180	147	123	106
	3 500	199	165	141	203	184	160	131	110	95
	3 750	180	150	128	181	164	143	118	99	85
	4 000	164	136	116	162	147	129	106	89	76
	4 250	149	124	106	146	133	116	95	80	69
	4 500	136	113	96	131	120	105	86	73	62
	4 750	124	103	88	119	109	95	79	66	57
	5 000	114	95	81	108	99	87	72	60	52
	5 250	104	87	74	99	91	79	66	55	47
	5 500	96	80	69			73	60	51	44
	5 750	89	74	63				55	47	40
	6 000	82	69	59						
	6 500	71	59	51						
	7 000									
	7 500									
	8 000									
	8 500									
	9 000									
	9 500									
	10 000									
	10 500									
PROPERTIES AND DESIGN DATA										
Area (mm^2)		1 550	1 260	1 060	2 310	2 010	1 670	1 310	1 070	903
Z (10^3 mm^3)		48.2	39.8	33.9	55.5	49.8	42.8	34.4	28.6	24.5
S (10^3 mm^3)		40.3	33.7	29.0	42.4	39.1	34.5	28.5	24.0	20.7
r (mm)		34.0	34.5	34.8	26.5	27.2	28.0	28.8	29.3	29.6
M_r (kN·m)		15.2	12.5	10.7	17.5	15.7	13.5	10.8	9.01	7.72
IMPERIAL SIZE AND MASS										
Mass (lb./ft.)		8.17	6.64	5.61	12.2	10.6	8.81	6.89	5.62	4.76
Thickness (in.)		0.188	0.150	0.125	0.375	0.313	0.250	0.188	0.150	0.125
Size (in.)		3½ x 3½			3 x 3					

SQUARE HOLLOW SECTIONS
Factored Axial Compressive Resistances, C_r (kN)

G40.21 350W

$\phi = 0.90$

Designation (mm x mm x mm)		HSS 64 x 64				HSS 51 x 51	
		6.4	4.8	3.8	3.2	6.4	4.8
Mass (kg/m)		10.6	8.35	6.85	5.82	8.05	6.45
Effective length (KL) in millimetres with respect to the least radius of gyration	0	425	334	275	233	324	259
	1 000	363	289	239	204	243	199
	1 250	326	261	217	186	204	169
	1 500	286	231	193	166	169	141
	1 750	248	202	169	145	139	117
	2 000	214	175	147	127	115	98
	2 250	184	151	128	110	95	82
	2 500	158	131	111	96	80	69
	2 750	137	113	96	83	68	59
	3 000	119	99	84	73	58	50
	3 250	104	87	74	64	50	44
	3 500	92	77	65	57	44	38
	3 750	81	68	58	50		
	4 000	72	61	52	45		
	4 250	65	54	46	40		
	4 500	58	49	42	36		
	4 750			38	33		
	5 000						
	5 250						
	5 500						
	5 750						
	6 000						
	6 500						
	7 000						
	7 500						
	8 000						
	8 500						
	9 000						
	9 500						
	10 000						
	10 500						
PROPERTIES AND DESIGN DATA							
Area (mm^2)		1 350	1 060	872	741	1 030	821
Z (10^3 mm^3)		28.1	23.0	19.2	16.6	16.4	13.8
S (10^3 mm^3)		22.2	18.7	16.0	13.9	12.6	11.0
r (mm)		22.8	23.6	24.1	24.4	17.6	18.4
M_r (kN·m)		8.85	7.25	6.05	5.23	5.17	4.35
IMPERIAL SIZE AND MASS							
Mass (lb./ft.)		7.11	5.61	4.60	3.91	5.41	4.33
Thickness (in.)		0.250	0.188	0.150	0.125	0.250	0.188
Size (in.)		2½ x 2½				2 x 2	

G40.21 350W
$\phi = 0.90$

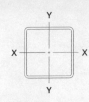

SQUARE HOLLOW SECTIONS
Factored Axial Compressive Resistances, C_r (kN)

Designation (mm x mm x mm)		HSS 51 x 51		HSS 38 x 38		
		3.8	3.2	4.8	3.8	3.2
Mass (kg/m)		5.33	4.55	4.54	3.81	3.28
Effective length (KL) in millimetres with respect to the least radius of gyration	0	214	183	182	153	132
	1 000	167	144	108	94	82
	1 250	143	124	83	73	65
	1 500	120	105	64	57	51
	1 750	100	88	50	45	40
	2 000	84	73	40	36	32
	2 250	70	62	33	29	26
	2 500	59	52	27	24	22
	2 750	51	44			18
	3 000	44	38			
	3 250	38	33			
	3 500	33	29			
	3 750	29	26			
	4 000					
	4 250					
	4 500					
	4 750					
	5 000					
	5 250					
	5 500					
	5 750					
	6 000					
	6 500					
	7 000					
	7 500					
	8 000					
	8 500					
	9 000					
	9 500					
	10 000					
	10 500					
PROPERTIES AND DESIGN DATA						
Area (mm^2)		679	580	578	485	418
Z (10^3 mm^3)		11.7	10.2	6.95	6.06	5.35
S (10^3 mm^3)		9.55	8.42	5.30	4.79	4.31
r (mm)		18.9	19.2	13.2	13.7	14.0
M_r (kN·m)		3.69	3.21	2.19	1.91	1.69
IMPERIAL SIZE AND MASS						
Mass (lb./ft.)		3.58	3.06	3.05	2.56	2.21
Thickness (in.)		0.150	0.125	0.188	0.150	0.125
Size (in.)		2 x 2		1½ x 1½		

ROUND HOLLOW SECTIONS
Factored Axial Compressive Resistances, C_r (kN)

G40.21 350W
$\phi = 0.90$

Designation (mm x mm)		HSS 610			HSS 559			HSS 508			
		13	11 *	9.5 *	13	11	9.5 *	13	11	9.5 *	8.0 *
Mass (kg/m)		187	164	141	171	150	129	155	136	117	98.0
Effective length (KL) in millimetres with respect to the least radius of gyration	0	7 500	6 580	5 670	6 870	6 020	5 170	6 240	5 480	4 690	3 940
	1 000	7 490	6 580	5 670	6 860	6 010	5 160	6 230	5 480	4 690	3 930
	1 250	7 490	6 580	5 670	6 860	6 010	5 160	6 230	5 470	4 690	3 930
	1 500	7 490	6 570	5 660	6 860	6 010	5 160	6 220	5 470	4 680	3 930
	1 750	7 480	6 570	5 660	6 850	6 000	5 150	6 220	5 460	4 680	3 920
	2 000	7 480	6 560	5 650	6 840	5 990	5 150	6 210	5 460	4 670	3 920
	2 250	7 470	6 560	5 650	6 830	5 990	5 140	6 200	5 450	4 660	3 910
	2 500	7 460	6 550	5 640	6 820	5 980	5 130	6 180	5 430	4 650	3 900
	2 750	7 450	6 540	5 630	6 810	5 970	5 120	6 170	5 420	4 640	3 890
	3 000	7 430	6 530	5 620	6 790	5 950	5 110	6 150	5 410	4 630	3 880
	3 250	7 420	6 510	5 610	6 780	5 940	5 100	6 130	5 390	4 610	3 870
	3 500	7 400	6 500	5 600	6 760	5 920	5 080	6 110	5 370	4 600	3 860
	3 750	7 380	6 480	5 580	6 730	5 900	5 070	6 080	5 350	4 580	3 840
	4 000	7 360	6 460	5 570	6 710	5 880	5 050	6 050	5 320	4 560	3 820
	4 250	7 340	6 440	5 550	6 680	5 860	5 030	6 020	5 290	4 530	3 800
	4 500	7 310	6 420	5 530	6 650	5 830	5 010	5 990	5 260	4 510	3 780
	4 750	7 280	6 400	5 510	6 620	5 800	4 980	5 950	5 230	4 480	3 760
	5 000	7 250	6 370	5 490	6 590	5 770	4 960	5 910	5 200	4 450	3 740
	5 250	7 220	6 340	5 460	6 550	5 740	4 930	5 870	5 160	4 420	3 710
	5 500	7 180	6 310	5 440	6 510	5 710	4 900	5 820	5 120	4 380	3 680
	5 750	7 150	6 280	5 410	6 470	5 670	4 870	5 770	5 080	4 350	3 650
	6 000	7 110	6 250	5 380	6 420	5 630	4 840	5 720	5 040	4 310	3 620
	6 500	7 020	6 170	5 310	6 320	5 550	4 760	5 610	4 940	4 230	3 550
	7 000	6 930	6 090	5 240	6 220	5 460	4 680	5 500	4 840	4 140	3 480
	7 500	6 820	6 000	5 170	6 100	5 360	4 600	5 370	4 730	4 050	3 400
	8 000	6 710	5 900	5 080	5 980	5 250	4 510	5 240	4 610	3 950	3 320
	8 500	6 590	5 800	4 990	5 850	5 140	4 410	5 100	4 490	3 840	3 230
	9 000	6 460	5 690	4 900	5 720	5 020	4 310	4 950	4 360	3 740	3 140
	9 500	6 330	5 570	4 800	5 570	4 900	4 200	4 800	4 240	3 630	3 050
	10 000	6 200	5 450	4 700	5 430	4 770	4 100	4 650	4 100	3 510	2 960
	10 500	6 050	5 330	4 590	5 280	4 640	3 990	4 500	3 970	3 400	2 860
PROPERTIES AND DESIGN DATA											
Area (mm^2)		23 800	20 900	18 000	21 800	19 100	16 400	19 800	17 400	14 900	12 500
Z (10^3 mm^3)		4 530	3 990	3 430	3 790	3 340	2 880	3 120	2 750	2 370	1 990
S (10^3 mm^3)		3 480	3 070	2 650	2 910	2 570	2 220	2 390	2 110	1 830	1 540
r (mm)		211	212	212	193	194	194	175	176	176	177
M_r (kN·m)		1 430	967	835	1 190	1 050	699	983	866	576	485
IMPERIAL SIZE AND MASS											
Mass (lb./ft.)		126	110	94.8	115	101	86.7	104	91.6	78.7	65.9
Thickness (in.)		0.500	0.438	0.375	0.500	0.438	0.375	0.500	0.438	0.375	0.313
Size (in.)		24 OD			22 OD			20 OD			

* Class 3 in bending

G40.21 350W
$\phi = 0.90$

ROUND HOLLOW SECTIONS
Factored Axial Compressive Resistances, C_r (kN)

Designation (mm x mm)		HSS 406					HSS 356				
		13	11	9.5	8.0	6.4 *	13	11	9.5	8.0	6.4 *
Mass (kg/m)		123	108	93.3	78.1	62.6	107	94.6	81.3	68.2	54.7
Effective length (KL) in millimetres with respect to the least radius of gyration	0	4 950	4 350	3 750	3 130	2 510	4 320	3 780	3 280	2 730	2 200
	1 000	4 940	4 340	3 740	3 130	2 510	4 310	3 770	3 270	2 730	2 190
	1 250	4 930	4 340	3 740	3 130	2 510	4 300	3 770	3 260	2 720	2 190
	1 500	4 930	4 330	3 730	3 120	2 500	4 290	3 760	3 260	2 720	2 180
	1 750	4 910	4 320	3 730	3 120	2 500	4 280	3 750	3 250	2 710	2 180
	2 000	4 900	4 310	3 720	3 110	2 490	4 260	3 730	3 240	2 700	2 170
	2 250	4 890	4 300	3 700	3 100	2 480	4 240	3 720	3 220	2 690	2 160
	2 500	4 870	4 280	3 690	3 090	2 480	4 220	3 700	3 200	2 670	2 150
	2 750	4 840	4 260	3 670	3 070	2 460	4 190	3 670	3 180	2 660	2 130
	3 000	4 820	4 240	3 650	3 060	2 450	4 160	3 640	3 160	2 640	2 120
	3 250	4 790	4 210	3 630	3 040	2 440	4 120	3 610	3 130	2 620	2 100
	3 500	4 760	4 180	3 610	3 020	2 420	4 080	3 580	3 100	2 590	2 080
	3 750	4 720	4 150	3 580	3 000	2 400	4 040	3 540	3 070	2 570	2 060
	4 000	4 680	4 120	3 550	2 970	2 380	3 990	3 500	3 030	2 540	2 040
	4 250	4 640	4 080	3 520	2 950	2 360	3 940	3 460	3 000	2 500	2 010
	4 500	4 590	4 040	3 480	2 920	2 340	3 880	3 410	2 950	2 470	1 980
	4 750	4 540	4 000	3 450	2 890	2 310	3 820	3 360	2 910	2 430	1 960
	5 000	4 490	3 950	3 410	2 850	2 290	3 760	3 300	2 860	2 400	1 920
	5 250	4 430	3 900	3 370	2 820	2 260	3 700	3 250	2 820	2 360	1 890
	5 500	4 370	3 850	3 320	2 780	2 230	3 630	3 190	2 770	2 320	1 860
	5 750	4 310	3 800	3 270	2 740	2 200	3 560	3 130	2 710	2 270	1 830
	6 000	4 250	3 740	3 230	2 700	2 170	3 490	3 070	2 660	2 230	1 790
	6 500	4 110	3 620	3 130	2 620	2 100	3 340	2 940	2 550	2 140	1 720
	7 000	3 970	3 500	3 020	2 530	2 030	3 190	2 810	2 440	2 050	1 640
	7 500	3 820	3 370	2 910	2 440	1 960	3 040	2 680	2 320	1 950	1 570
	8 000	3 670	3 240	2 800	2 350	1 880	2 890	2 550	2 210	1 860	1 490
	8 500	3 520	3 110	2 680	2 250	1 810	2 740	2 420	2 100	1 760	1 420
	9 000	3 370	2 980	2 570	2 160	1 730	2 600	2 290	1 990	1 670	1 340
	9 500	3 220	2 850	2 460	2 070	1 660	2 460	2 170	1 880	1 580	1 270
	10 000	3 070	2 720	2 350	1 970	1 580	2 320	2 050	1 780	1 500	1 200
	10 500	2 930	2 590	2 240	1 880	1 510	2 200	1 940	1 680	1 420	1 140
PROPERTIES AND DESIGN DATA											
Area (mm^2)		15 700	13 800	11 900	9 950	7 980	13 700	12 000	10 400	8 680	6 970
Z (10^3 mm^3)		1 970	1 740	1 500	1 260	1 020	1 490	1 320	1 140	961	775
S (10^3 mm^3)		1 500	1 330	1 150	972	786	1 130	1 010	873	738	598
r (mm)		139	140	140	141	141	121	122	122	123	123
M$_r$ (kN·m)		621	548	473	397	248	469	416	359	303	188
IMPERIAL SIZE AND MASS											
Mass (lb./ft.)		82.9	72.9	62.7	52.5	42.1	72.2	63.5	54.7	45.8	36.8
Thickness (in.)		0.500	0.438	0.375	0.313	0.250	0.500	0.438	0.375	0.313	0.250
Size (in.)		16 OD					14 OD				

* Class 3 in bending

ROUND HOLLOW SECTIONS
Factored Axial Compressive Resistances, C_r (kN)

G40.21 350W

$\phi = 0.90$

Designation (mm x mm)		HSS 324					HSS 273				
		13	11	9.5	8.0	6.4	13	11	9.5	8.0	6.4
Mass (kg/m)		97.5	85.8	73.9	61.9	49.7	81.6	71.9	61.9	52.0	41.8
Effective length (KL) in millimetres with respect to the least radius of gyration	0	3 910	3 430	2 960	2 490	1 990	3 280	2 890	2 490	2 090	1 680
	1 000	3 900	3 420	2 960	2 480	1 990	3 260	2 870	2 480	2 080	1 670
	1 250	3 890	3 420	2 950	2 470	1 990	3 250	2 860	2 470	2 070	1 660
	1 500	3 880	3 410	2 940	2 470	1 980	3 240	2 850	2 460	2 060	1 660
	1 750	3 860	3 390	2 930	2 460	1 970	3 220	2 830	2 440	2 050	1 650
	2 000	3 840	3 380	2 920	2 450	1 960	3 190	2 810	2 420	2 030	1 630
	2 250	3 820	3 360	2 900	2 430	1 950	3 160	2 790	2 400	2 020	1 620
	2 500	3 790	3 340	2 880	2 420	1 940	3 130	2 750	2 370	1 990	1 600
	2 750	3 760	3 310	2 860	2 400	1 920	3 080	2 720	2 340	1 970	1 580
	3 000	3 720	3 280	2 830	2 370	1 900	3 040	2 680	2 310	1 940	1 560
	3 250	3 680	3 240	2 800	2 350	1 880	2 990	2 630	2 270	1 910	1 540
	3 500	3 640	3 200	2 760	2 320	1 860	2 930	2 590	2 230	1 870	1 510
	3 750	3 590	3 160	2 730	2 290	1 840	2 870	2 530	2 190	1 840	1 480
	4 000	3 530	3 110	2 690	2 260	1 810	2 810	2 480	2 140	1 800	1 450
	4 250	3 480	3 060	2 650	2 220	1 780	2 740	2 420	2 090	1 760	1 410
	4 500	3 420	3 010	2 600	2 190	1 750	2 670	2 360	2 040	1 710	1 380
	4 750	3 350	2 960	2 550	2 150	1 720	2 600	2 300	1 980	1 670	1 340
	5 000	3 290	2 900	2 500	2 110	1 690	2 520	2 230	1 930	1 620	1 310
	5 250	3 220	2 840	2 450	2 060	1 660	2 450	2 170	1 870	1 580	1 270
	5 500	3 150	2 780	2 400	2 020	1 620	2 370	2 100	1 810	1 530	1 230
	5 750	3 070	2 720	2 340	1 970	1 580	2 300	2 030	1 760	1 480	1 200
	6 000	3 000	2 650	2 290	1 930	1 550	2 220	1 970	1 700	1 430	1 160
	6 500	2 850	2 520	2 180	1 840	1 470	2 070	1 840	1 590	1 340	1 080
	7 000	2 700	2 390	2 060	1 740	1 400	1 930	1 710	1 480	1 250	1 010
	7 500	2 550	2 260	1 950	1 650	1 320	1 800	1 590	1 380	1 160	941
	8 000	2 400	2 130	1 840	1 560	1 250	1 670	1 480	1 280	1 080	876
	8 500	2 260	2 010	1 730	1 470	1 180	1 550	1 370	1 190	1 010	814
	9 000	2 130	1 890	1 630	1 380	1 110	1 440	1 270	1 110	935	757
	9 500	2 000	1 780	1 530	1 300	1 040	1 330	1 180	1 030	870	704
	10 000	1 880	1 670	1 440	1 220	980	1 240	1 100	955	809	655
	10 500	1 760	1 570	1 350	1 150	922	1 150	1 020	889	753	610
PROPERTIES AND DESIGN DATA											
Area (mm^2)		12 400	10 900	9 410	7 890	6 330	10 400	9 160	7 890	6 620	5 320
Z (10^3 mm^3)		1 230	1 090	942	794	640	862	764	662	559	452
S (10^3 mm^3)		930	827	719	608	493	646	577	502	427	347
r (mm)		110	111	111	112	112	92.2	92.7	93.2	93.8	94.3
M_r (kN·m)		387	343	297	250	202	272	241	209	176	142
IMPERIAL SIZE AND MASS											
Mass (lb./ft.)		65.5	57.7	49.7	41.6	33.4	54.8	48.3	41.6	34.9	28.1
Thickness (in.)		0.500	0.438	0.375	0.313	0.250	0.500	0.438	0.375	0.313	0.250
Size (in.)		12.75 OD					10.75 OD				

G40.21 350W
$\phi = 0.90$

ROUND HOLLOW SECTIONS
Factored Axial Compressive Resistances, C_r (kN)

Designation (mm x mm)		HSS 219					
		13	11	9.5	8.0	6.4	4.8
Mass (kg/m)		64.6	57.1	49.3	41.4	33.3	25.3
Effective length (KL) in millimetres with respect to the least radius of gyration	0	2 590	2 290	1 980	1 660	1 340	1 010
	1 000	2 570	2 270	1 960	1 650	1 330	1 010
	1 250	2 560	2 260	1 950	1 640	1 320	1 000
	1 500	2 530	2 240	1 930	1 620	1 310	994
	1 750	2 510	2 210	1 910	1 610	1 290	983
	2 000	2 470	2 180	1 890	1 590	1 280	971
	2 250	2 430	2 150	1 850	1 560	1 260	956
	2 500	2 380	2 110	1 820	1 530	1 230	938
	2 750	2 330	2 060	1 780	1 500	1 210	918
	3 000	2 270	2 010	1 730	1 460	1 180	897
	3 250	2 200	1 950	1 690	1 420	1 150	873
	3 500	2 130	1 890	1 640	1 380	1 110	848
	3 750	2 060	1 830	1 580	1 330	1 080	821
	4 000	1 990	1 760	1 530	1 290	1 040	794
	4 250	1 910	1 700	1 470	1 240	1 000	766
	4 500	1 840	1 630	1 410	1 190	967	738
	4 750	1 760	1 560	1 360	1 150	929	709
	5 000	1 690	1 500	1 300	1 100	891	681
	5 250	1 610	1 430	1 250	1 050	854	653
	5 500	1 540	1 370	1 190	1 010	818	625
	5 750	1 470	1 310	1 140	965	783	599
	6 000	1 410	1 250	1 090	922	749	573
	6 500	1 280	1 140	994	842	684	523
	7 000	1 170	1 040	906	768	624	478
	7 500	1 060	947	826	701	570	437
	8 000	968	864	754	640	521	399
	8 500	884	789	689	585	476	365
	9 000	809	722	631	536	437	335
	9 500	742	663	579	492	401	308
	10 000	682	609	532	452	369	283
	10 500	628	561	491	417	340	261
PROPERTIES AND DESIGN DATA							
Area (mm²)		8 230	7 270	6 270	5 270	4 240	3 220
Z (10^3 mm³)		542	482	419	355	288	220
S (10^3 mm³)		402	360	315	269	219	169
r (mm)		73.1	73.6	74.2	74.7	75.3	75.8
M_r (kN·m)		171	152	132	112	90.7	69.3
IMPERIAL SIZE AND MASS							
Mass (lb./ft.)		43.4	38.4	33.1	27.8	22.4	17.0
Thickness (in.)		0.500	0.438	0.375	0.313	0.250	0.188
Size (in.)		8.625 OD					

ROUND HOLLOW SECTIONS
Factored Axial Compressive Resistances, C_r (kN)

G40.21 350W

$\phi = 0.90$

Designation (mm x mm)		HSS 168				HSS 141			
		9.5	8.0	6.4	4.8	9.5	8.0	6.4	4.8
Mass (kg/m)		37.3	31.4	25.4	19.3	31.0	26.1	21.1	16.1
Effective length (KL) in millimetres with respect to the least radius of gyration	0	1 500	1 260	1 020	775	1 240	1 050	847	646
	1 000	1 470	1 240	1 000	764	1 210	1 020	827	631
	1 250	1 450	1 230	991	755	1 190	1 000	812	619
	1 500	1 430	1 210	975	743	1 160	977	791	604
	1 750	1 400	1 180	954	728	1 120	945	766	585
	2 000	1 360	1 150	930	710	1 070	907	736	563
	2 250	1 320	1 110	902	688	1 020	865	702	538
	2 500	1 270	1 070	870	665	966	819	667	511
	2 750	1 220	1 030	836	639	910	773	629	483
	3 000	1 160	986	800	612	853	725	592	454
	3 250	1 110	939	763	584	797	679	554	426
	3 500	1 050	892	725	556	743	633	518	399
	3 750	995	845	687	527	692	590	483	372
	4 000	939	799	650	499	643	549	450	347
	4 250	886	754	614	472	597	510	419	323
	4 500	834	711	579	445	555	474	390	301
	4 750	785	669	546	420	516	441	363	280
	5 000	739	630	514	396	480	411	338	261
	5 250	695	593	484	373	447	382	315	244
	5 500	654	558	456	351	416	357	294	227
	5 750	615	526	430	331	388	333	274	212
	6 000	579	495	405	312	363	311	256	199
	6 500	514	440	360	278	318	273	225	174
	7 000	458	393	321	248	280	241	199	154
	7 500	410	351	288	222	248	213	176	137
	8 000	368	316	259	200	221	190	157	122
	8 500	332	284	233	180	198	170	141	109
	9 000	300	257	211	163	178	153	127	99
	9 500	273	234	192	148			115	89
	10 000	248	213	175	135				
	10 500	227	195	160	124				
PROPERTIES AND DESIGN DATA									
Area (mm^2)		4 750	4 000	3 230	2 460	3 950	3 330	2 690	2 050
Z (10^3 mm^3)		241	205	167	128	166	142	116	89.1
S (10^3 mm^3)		179	153	126	97.6	122	105	86.9	67.7
r (mm)		56.2	56.8	57.3	57.8	46.7	47.2	47.8	48.3
M_r (kN·m)		75.9	64.6	52.6	40.3	52.3	44.7	36.5	28.1
IMPERIAL SIZE AND MASS									
Mass (lb./ft.)		25.1	21.1	17.0	13.0	20.8	17.6	14.2	10.8
Thickness (in.)		0.375	0.313	0.250	0.188	0.375	0.313	0.250	0.188
Size (in.)		6.625 OD				5.563 OD			

G40.21 350W
$\phi = 0.90$

ROUND HOLLOW SECTIONS
Factored Axial Compressive Resistances, C_r (kN)

Designation (mm x mm)		HSS 114			HSS 102			
		8.0	6.4	4.8	8.0	6.4	4.8	3.8
Mass (kg/m)		20.9	16.9	12.9	18.4	14.9	11.4	9.19
Effective length (KL) in millimetres with respect to the least radius of gyration	0	838	677	517	737	599	457	369
	1 000	801	649	496	693	564	431	349
	1 250	774	627	480	661	539	413	334
	1 500	739	600	460	622	508	390	316
	1 750	699	568	436	577	473	364	295
	2 000	654	532	410	530	436	336	273
	2 250	607	495	382	484	399	308	250
	2 500	560	457	353	438	362	280	228
	2 750	514	421	326	396	328	254	207
	3 000	470	386	299	357	296	230	188
	3 250	430	353	274	322	268	208	170
	3 500	392	322	251	291	242	188	154
	3 750	358	295	229	263	219	171	140
	4 000	327	270	210	238	199	155	127
	4 250	299	247	193	216	181	141	115
	4 500	274	226	177	197	165	129	105
	4 750	252	208	163	180	150	118	96
	5 000	232	192	150	165	138	108	88
	5 250	214	177	138	151	127	99	81
	5 500	198	163	128	139	117	91	75
	5 750	183	151	119	129	108	84	69
	6 000	170	141	110	119	100	78	64
	6 500	147	122	96	103	86	68	55
	7 000	129	107	84				
	7 500	113	94	74				
	8 000							
	8 500							
	9 000							
	9 500							
	10 000							
	10 500							
PROPERTIES AND DESIGN DATA								
Area (mm^2)		2 660	2 150	1 640	2 340	1 900	1 450	1 170
Z (10^3 mm^3)		90.1	74.1	57.4	69.9	57.7	44.8	36.5
S (10^3 mm^3)		66.1	55.1	43.2	50.8	42.6	33.6	27.6
r (mm)		37.7	38.2	38.8	33.2	33.8	34.3	34.6
M$_r$ (kN·m)		28.4	23.3	18.1	22.0	18.2	14.1	11.5
IMPERIAL SIZE AND MASS								
Mass (lb./ft.)		14.0	11.4	8.68	12.3	10.0	7.67	6.17
Thickness (in.)		0.313	0.250	0.188	0.313	0.250	0.188	0.150
Size (in.)		4.5 OD			4 OD			

ROUND HOLLOW SECTIONS
Factored Axial Compressive Resistances, C_r (kN)

G40.21 350W

$\phi = 0.90$

Designation (mm x mm)		HSS 89				HSS 73			
		8.0	6.4	4.8	3.8	6.4	4.8	3.8	3.2
Mass (kg/m)		15.9	12.9	9.92	8.00	10.4	8.04	6.50	5.48
Effective length (KL) in millimetres with respect to the least radius of gyration	0	636	520	397	321	419	321	261	220
	1 000	582	477	366	297	363	280	228	193
	1 250	545	448	344	280	328	254	208	176
	1 500	502	414	319	259	291	226	186	157
	1 750	455	376	291	237	254	199	163	139
	2 000	409	339	263	214	220	173	142	121
	2 250	364	303	235	192	191	150	124	105
	2 500	324	270	210	172	165	130	108	92
	2 750	287	240	187	153	143	113	94	80
	3 000	255	213	167	137	125	99	82	70
	3 250	227	190	149	122	110	87	72	62
	3 500	203	170	133	110	97	77	64	54
	3 750	181	152	120	98	86	68	57	48
	4 000	163	137	108	89	77	61	51	43
	4 250	147	124	97	80	69	55	45	39
	4 500	133	112	88	73	62	49	41	35
	4 750	121	102	80	66		45	37	32
	5 000	110	93	73	60				
	5 250	101	85	67	55				
	5 500	93	78	62	51				
	5 750	85	72	57	47				
	6 000				43				
	6 500								
	7 000								
	7 500								
	8 000								
	8 500								
	9 000								
	9 500								
	10 000								
	10 500								
PROPERTIES AND DESIGN DATA									
Area (mm^2)		2 020	1 650	1 260	1 020	1 330	1 020	828	698
Z (10^3 mm^3)		52.3	43.4	33.9	27.6	28.3	22.3	18.3	15.5
S (10^3 mm^3)		37.6	31.7	25.2	20.8	20.4	16.4	13.6	11.7
r (mm)		28.8	29.3	29.8	30.1	23.7	24.2	24.5	24.7
M_r (kN·m)		16.5	13.7	10.7	8.69	8.91	7.02	5.76	4.88
IMPERIAL SIZE AND MASS									
Mass (lb./ft.)		10.7	8.69	6.66	5.37	7.01	5.40	4.37	3.68
Thickness (in.)		0.313	0.250	0.188	0.150	0.250	0.188	0.150	0.125
Size (in.)		3.5 OD				2.875 OD			

G40.21 350W
$\phi = 0.90$

ROUND HOLLOW SECTIONS
Factored Axial Compressive Resistances, C_r (kN)

Designation (mm × mm)		HSS 60				HSS 48		
		6.4	4.8	3.8	3.2	4.8	3.8	3.2
Mass (kg/m)		8.45	6.54	5.31	4.48	5.13	4.18	3.54
Effective length (KL) in millimetres with respect to the least radius of gyration	0	340	263	213	180	206	168	142
	1 000	268	210	172	146	141	116	100
	1 250	231	182	149	127	114	95	81
	1 500	195	154	127	108	91	76	66
	1 750	163	130	107	92	73	62	53
	2 000	136	109	90	77	60	50	43
	2 250	115	92	76	65	49	41	36
	2 500	97	78	65	56	41	34	30
	2 750	83	67	56	48	34	29	25
	3 000	71	58	48	41	29	25	21
	3 250	62	50	42	36			
	3 500	54	44	36	31			
	3 750	48	39	32	28			
	4 000			28	25			
	4 250							
	4 500							
	4 750							
	5 000							
	5 250							
	5 500							
	5 750							
	6 000							
	6 500							
	7 000							
	7 500							
	8 000							
	8 500							
	9 000							
	9 500							
	10 000							
	10 500							
PROPERTIES AND DESIGN DATA								
Area (mm^2)		1 080	834	676	571	654	533	451
Z (10^3 mm^3)		18.6	14.8	12.2	10.4	9.09	7.56	6.48
S (10^3 mm^3)		13.2	10.7	8.99	7.74	6.48	5.50	4.77
r (mm)		19.2	19.7	20.0	20.2	15.5	15.8	16.0
M$_r$ (kN·m)		5.86	4.66	3.84	3.28	2.86	2.38	2.04
IMPERIAL SIZE AND MASS								
Mass (lb./ft.)		5.68	4.40	3.57	3.01	3.45	2.81	2.38
Thickness (in.)		0.250	0.188	0.150	0.125	0.188	0.150	0.125
Size (in.)		2.375 OD				1.9 OD		

RECTANGULAR HOLLOW SECTIONS
Factored Axial Compressive Resistances, C_r (kN)

G40.21 350W
CLASS H
$\phi = 0.90$

Designation (mm x mm x mm)	HSS 305 x 203					HSS 254 x 152		
	13	11	9.5	8.0 *	6.4 **	13	11	9.5
Mass (kg/m)	93.0	82.4	71.3	60.1	48.6	72.7	64.6	56.1
0	3 720	3 310	2 860	2 410	1 730	2 920	2 590	2 250
1 000	3 720	3 310	2 860	2 410	1 720	2 920	2 590	2 250
1 250	3 720	3 310	2 860	2 410	1 720	2 910	2 590	2 250
1 500	3 710	3 300	2 860	2 410	1 720	2 910	2 580	2 250
1 750	3 710	3 300	2 860	2 410	1 720	2 900	2 580	2 240
2 000	3 710	3 300	2 860	2 410	1 720	2 880	2 560	2 230
2 250	3 700	3 290	2 850	2 400	1 720	2 860	2 550	2 210
2 500	3 690	3 280	2 840	2 400	1 720	2 830	2 520	2 190
2 750	3 670	3 270	2 830	2 390	1 710	2 790	2 480	2 160
3 000	3 650	3 250	2 820	2 370	1 700	2 730	2 440	2 120
3 250	3 620	3 230	2 800	2 360	1 690	2 660	2 380	2 070
3 500	3 590	3 200	2 770	2 340	1 680	2 580	2 310	2 010
3 750	3 540	3 160	2 740	2 310	1 670	2 490	2 230	1 950
4 000	3 490	3 110	2 700	2 280	1 650	2 380	2 140	1 870
4 250	3 430	3 060	2 660	2 240	1 630	2 270	2 040	1 790
4 500	3 360	3 000	2 600	2 200	1 600	2 150	1 940	1 700
4 750	3 280	2 930	2 550	2 150	1 580	2 040	1 830	1 610
5 000	3 190	2 850	2 480	2 100	1 550	1 920	1 730	1 520
5 250	3 090	2 770	2 410	2 040	1 510	1 800	1 630	1 430
5 500	2 990	2 680	2 330	1 980	1 470	1 690	1 530	1 350
5 750	2 880	2 590	2 250	1 910	1 430	1 590	1 430	1 270
6 000	2 770	2 490	2 170	1 840	1 390	1 490	1 340	1 190
6 500	2 550	2 290	2 000	1 700	1 310	1 310	1 180	1 050
7 000	2 330	2 100	1 840	1 560	1 220	1 150	1 040	924
7 500	2 120	1 910	1 670	1 430	1 130	1 020	923	819
8 000	1 920	1 740	1 520	1 300	1 050	904	821	728
8 500	1 750	1 580	1 390	1 180	967	807	733	650
9 000	1 590	1 440	1 260	1 080	884	724	658	584
9 500	1 450	1 310	1 150	984	806	652	593	526
10 000	1 320	1 200	1 050	899	737	591	537	477
10 500	1 210	1 090	962	823	675	537	488	434
PROPERTIES AND DESIGN DATA								
Area (mm²)	11 800	10 500	9 090	7 660	6 190	9 260	8 230	7 150
Z_x (10^3 mm³)	1 190	1 060	926	787	640	747	671	589
S_x (10^3 mm³)	964	867	762	652	535	592	537	475
r_x (mm)	111	112	113	114	115	90.1	91.0	91.9
Z_y (10^3 mm³)	897	802	701	596	486	522	470	413
S_y (10^3 mm³)	769	694	611	525	368	442	402	357
r_y (mm)	81.2	82.0	82.7	83.4	84.1	60.3	61.0	61.7
r_x / r_y	1.37	1.37	1.37	1.37	1.37	1.49	1.49	1.49
M_{rx} (kN·m)	375	334	292	248	202	235	211	186
M_{ry} (kN·m)	283	253	221	165	116	164	148	130
IMPERIAL SIZE AND MASS								
Mass (lb./ft.)	62.5	55.4	47.9	40.4	32.6	48.9	43.4	37.7
Thickness (in.)	0.500	0.438	0.375	0.313	0.250	0.500	0.438	0.375
Size (in.)	12 x 8					10 x 6		

Effective length (KL) in millimetres with respect to the least radius of gyration

* Class 3 in bending about Y-Y axis
** Class 4: C_r calculated according to S16.1-94 Clause 13.3.3; S_y and M_{ry} according to Clause 13.5(c)(iii).

G40.21 350W
CLASS H
$\phi = 0.90$

RECTANGULAR HOLLOW SECTIONS
Factored Axial Compressive Resistances, C_r (kN)

Designation (mm x mm x mm)		HSS 254 x 152		HSS 203 x 152					
		8.0	6.4 **	13	11	9.5	8.0	6.4	4.8 **
Mass (kg/m)		47.5	38.4	62.6	55.7	48.5	41.1	33.4	25.5
Effective length (KL) in millimetres with respect to the least radius of gyration	0	1 910	1 460	2 510	2 240	1 950	1 650	1 340	954
	1 000	1 900	1 460	2 510	2 240	1 950	1 650	1 340	954
	1 250	1 900	1 460	2 510	2 230	1 940	1 650	1 340	953
	1 500	1 900	1 460	2 500	2 230	1 940	1 650	1 330	952
	1 750	1 900	1 460	2 490	2 220	1 930	1 640	1 330	950
	2 000	1 890	1 450	2 480	2 210	1 920	1 630	1 320	947
	2 250	1 870	1 450	2 460	2 190	1 910	1 620	1 320	941
	2 500	1 860	1 430	2 430	2 170	1 890	1 600	1 300	934
	2 750	1 830	1 420	2 390	2 130	1 860	1 580	1 280	923
	3 000	1 800	1 400	2 330	2 080	1 820	1 550	1 260	910
	3 250	1 760	1 380	2 270	2 030	1 770	1 510	1 230	892
	3 500	1 710	1 350	2 190	1 960	1 720	1 470	1 200	872
	3 750	1 660	1 310	2 100	1 890	1 650	1 410	1 150	848
	4 000	1 590	1 270	2 010	1 800	1 580	1 360	1 110	820
	4 250	1 530	1 230	1 910	1 720	1 510	1 290	1 060	790
	4 500	1 450	1 180	1 800	1 620	1 430	1 230	1 010	758
	4 750	1 380	1 130	1 700	1 530	1 350	1 160	954	725
	5 000	1 300	1 070	1 590	1 440	1 270	1 100	901	691
	5 250	1 230	1 010	1 490	1 350	1 200	1 030	849	657
	5 500	1 160	952	1 400	1 270	1 120	968	798	620
	5 750	1 090	896	1 310	1 190	1 050	909	749	582
	6 000	1 020	843	1 220	1 110	984	852	703	547
	6 500	903	745	1 070	974	864	749	619	482
	7 000	798	658	941	857	761	660	546	426
	7 500	707	584	831	757	673	584	484	378
	8 000	629	520	737	672	598	519	430	336
	8 500	562	465	658	599	533	464	384	300
	9 000	505	418	590	537	478	416	345	269
	9 500	455	377	531	484	431	375	311	243
	10 000	412	341	481	438	390	340	282	220
	10 500	375	310	437	399	355	309	256	200
PROPERTIES AND DESIGN DATA									
Area (mm^2)		6 050	4 900	7 970	7 100	6 180	5 240	4 250	3 250
Z_x (10^3 mm^3)		503	411	528	476	420	360	295	228
S_x (10^3 mm^3)		410	338	423	385	343	297	246	192
r_x (mm)		92.7	93.6	73.4	74.3	75.1	75.9	76.7	77.5
Z_y (10^3 mm^3)		354	290	432	390	344	295	243	188
S_y (10^3 mm^3)		309	255	359	327	292	254	211	157
r_y (mm)		62.4	63.1	58.6	59.3	60.0	60.8	61.5	62.2
r_x / r_y		1.49	1.48	1.25	1.25	1.25	1.25	1.25	1.25
M_{rx} (kN·m)		158	129	166	150	132	113	92.9	71.8
M_{ry} (kN·m)		112	80.3	136	123	108	92.9	76.5	49.3
IMPERIAL SIZE AND MASS									
Mass (lb./ft.)		31.9	25.8	42.1	37.5	32.6	27.6	22.4	17.1
Thickness (in.)		0.313	0.250	0.500	0.438	0.375	0.313	0.250	0.188
Size (in.)		10 x 6		8 x 6					

** Class 4: C_r calculated according to S16.1-94 Clause 13.3.3; S_y and M_{ry} according to Clause 13.5(c)(iii).

RECTANGULAR HOLLOW SECTIONS
Factored Axial Compressive Resistances, C_r (kN)

G40.21 350W
CLASS H
$\phi = 0.90$

Designation (mm x mm x mm)		HSS 203 x 102						HSS 178 x 127		
		13	11	9.5	8.0	6.4	4.8 **	13	11	9.5
Mass (kg/m)		52.4	46.9	40.9	34.8	28.3	21.7	52.4	46.9	40.9
Effective length (KL) in millimetres with respect to the least radius of gyration	0	2 100	1 880	1 640	1 400	1 140	800	2 100	1 880	1 640
	1 000	2 100	1 870	1 640	1 390	1 130	798	2 100	1 880	1 640
	1 250	2 080	1 860	1 630	1 380	1 130	796	2 100	1 870	1 640
	1 500	2 060	1 840	1 610	1 370	1 120	791	2 090	1 870	1 630
	1 750	2 020	1 810	1 580	1 350	1 100	782	2 070	1 850	1 620
	2 000	1 960	1 760	1 540	1 320	1 080	768	2 040	1 830	1 600
	2 250	1 870	1 680	1 480	1 270	1 040	748	2 000	1 790	1 570
	2 500	1 760	1 590	1 410	1 210	994	722	1 950	1 750	1 530
	2 750	1 640	1 490	1 320	1 140	938	691	1 880	1 690	1 480
	3 000	1 510	1 370	1 220	1 060	875	655	1 790	1 620	1 420
	3 250	1 380	1 260	1 120	975	809	616	1 700	1 530	1 350
	3 500	1 250	1 150	1 020	893	743	575	1 590	1 440	1 280
	3 750	1 130	1 040	932	814	679	531	1 490	1 350	1 200
	4 000	1 030	943	846	741	619	485	1 380	1 260	1 120
	4 250	929	855	768	674	564	443	1 270	1 160	1 040
	4 500	842	776	698	614	514	404	1 180	1 080	959
	4 750	766	706	636	559	469	369	1 080	994	887
	5 000	698	644	580	511	429	338	1 000	918	819
	5 250	638	589	531	468	393	310	923	848	757
	5 500	584	540	487	429	361	284	852	784	701
	5 750	537	496	448	395	332	262	788	725	649
	6 000	495	458	413	365	307	242	730	672	602
	6 500	424	392	354	313	263	208	630	581	520
	7 000	367	340	307	271	228	180	548	505	453
	7 500	320	297	268	237	199	157	480	443	397
	8 000			236	208	176	139	424	391	351
	8 500						123	376	347	312
	9 000							336	311	279
	9 500							302	279	251
	10 000									
	10 500									
PROPERTIES AND DESIGN DATA										
Area (mm²)		6 680	5 970	5 210	4 430	3 610	2 760	6 680	5 970	5 210
Z_x (10³ mm³)		405	368	326	281	232	180	378	343	303
S_x (10³ mm³)		308	283	254	221	185	145	297	273	244
r_x (mm)		68.4	69.4	70.3	71.2	72.2	73.1	62.9	63.7	64.6
Z_y (10³ mm³)		246	224	199	172	143	111	298	271	240
S_y (10³ mm³)		201	187	169	148	125	93.3	244	225	202
r_y (mm)		39.1	39.8	40.5	41.3	42.0	42.7	48.1	48.9	49.6
r_x / r_y		1.75	1.74	1.74	1.72	1.72	1.71	1.31	1.30	1.30
M_{rx} (kN·m)		128	116	103	88.5	73.1	56.7	119	108	95.4
M_{ry} (kN·m)		77.5	70.6	62.7	54.2	45.0	29.4	93.9	85.4	75.6
IMPERIAL SIZE AND MASS										
Mass (lb./ft.)		35.2	31.5	27.5	23.4	19.0	14.6	35.2	31.5	27.5
Thickness (in.)		0.500	0.438	0.375	0.313	0.250	0.188	0.500	0.438	0.375
Size (in.)		8 x 4						7 x 5		

** Class 4: C_r calculated according to S16.1-94 Clause 13.3.3; S_y and M_{ry} according to Clause 13.5(c)(iii).

G40.21 350W
CLASS H
$\phi = 0.90$

RECTANGULAR HOLLOW SECTIONS
Factored Axial Compressive Resistances, C_r (kN)

Designation (mm x mm x mm)		HSS 178 x 127			HSS 152 x 102					
		8.0	6.4	4.8 *	13	11	9.5	8.0	6.4	4.8
Mass (kg/m)		34.8	28.3	21.7	42.3	38.0	33.3	28.4	23.2	17.9
Effective length (KL) in millimetres with respect to the least radius of gyration	0	1 400	1 140	869	1 700	1 520	1 340	1 140	932	718
	1 000	1 390	1 140	869	1 690	1 520	1 330	1 140	930	716
	1 250	1 390	1 130	867	1 680	1 510	1 320	1 130	925	713
	1 500	1 390	1 130	864	1 660	1 490	1 310	1 120	916	706
	1 750	1 380	1 120	859	1 620	1 460	1 280	1 100	900	695
	2 000	1 360	1 110	850	1 560	1 410	1 240	1 070	876	677
	2 250	1 340	1 090	838	1 480	1 340	1 190	1 020	842	653
	2 500	1 310	1 070	821	1 390	1 260	1 120	968	800	622
	2 750	1 270	1 040	798	1 280	1 170	1 040	904	750	585
	3 000	1 220	1 000	770	1 170	1 070	961	836	695	544
	3 250	1 160	956	737	1 060	977	878	766	639	502
	3 500	1 100	906	700	960	885	798	698	584	460
	3 750	1 030	853	661	866	800	723	633	531	419
	4 000	964	799	620	780	722	654	574	482	381
	4 250	896	745	579	705	653	592	521	438	347
	4 500	831	692	539	638	591	537	473	398	316
	4 750	770	642	500	578	537	488	430	363	288
	5 000	712	595	464	526	489	445	392	331	263
	5 250	659	551	430	481	447	407	359	303	241
	5 500	610	510	399	440	409	373	329	278	221
	5 750	565	473	370	404	376	343	302	256	203
	6 000	525	440	344	372	346	316	279	236	188
	6 500	454	381	298	319	297	271	239	202	161
	7 000	395	332	260	276	257	234	207	175	139
	7 500	347	291	229	241	224	204	181	153	122
	8 000	306	257	202					135	107
	8 500	272	229	180						
	9 000	244	205	161						
	9 500	219	184	145						
	10 000	198	166	131						
	10 500									
PROPERTIES AND DESIGN DATA										
Area (mm²)		4 430	3 610	2 760	5 390	4 840	4 240	3 620	2 960	2 280
Z_x (10³ mm³)		261	216	168	252	230	206	179	148	116
S_x (10³ mm³)		213	178	140	193	179	162	143	121	95.6
r_x (mm)		65.4	66.2	67.1	52.2	53.1	54.0	54.8	55.7	56.5
Z_y (10³ mm³)		207	171	133	189	173	155	135	112	87.8
S_y (10³ mm³)		177	148	117	151	141	128	113	96.2	76.6
r_y (mm)		50.3	51.1	51.8	37.7	38.4	39.2	39.9	40.6	41.3
r_x / r_y		1.30	1.30	1.30	1.38	1.38	1.38	1.37	1.37	1.37
M_{rx} (kN·m)		82.2	68.0	52.9	79.4	72.5	64.9	56.4	46.6	36.5
M_{ry} (kN·m)		65.2	53.9	36.9	59.5	54.5	48.8	42.5	35.3	27.7
IMPERIAL SIZE AND MASS										
Mass (lb./ft.)		23.4	19.0	14.6	28.4	25.5	22.4	19.1	15.6	12.0
Thickness (in.)		0.313	0.250	0.188	0.500	0.438	0.375	0.313	0.250	0.188
Size (in.)		7 x 5			6 x 4					

* Class 3 in bending about Y-Y axis

RECTANGULAR HOLLOW SECTIONS
Factored Axial Compressive Resistances, C_r (kN)

G40.21 350W
CLASS H
$\phi = 0.90$

Designation (mm x mm x mm)		HSS 152 x 76				HSS 127 x 76				
		9.5	8.0	6.4	4.8	9.5	8.0	6.4	4.8	3.8 *
Mass (kg/m)		29.5	25.3	20.7	16.0	25.7	22.1	18.2	14.1	11.4
Effective length (KL) in millimetres with respect to the least radius of gyration	0	1 180	1 010	832	643	1 030	888	731	564	457
	1 000	1 170	1 000	823	637	1 020	877	723	558	452
	1 250	1 150	984	809	627	995	859	709	549	445
	1 500	1 100	950	783	609	953	826	684	531	432
	1 750	1 030	897	743	580	891	776	646	504	411
	2 000	949	827	690	541	812	712	596	468	383
	2 250	853	748	627	494	725	639	539	425	349
	2 500	755	665	561	445	638	566	479	380	314
	2 750	663	587	497	396	558	497	423	337	279
	3 000	580	516	438	350	487	435	371	297	246
	3 250	508	453	386	309	425	381	326	262	218
	3 500	447	399	340	273	373	335	287	231	192
	3 750	395	353	301	242	329	296	254	204	170
	4 000	350	313	268	216	292	263	226	182	152
	4 250	312	280	239	193	260	234	202	162	136
	4 500	280	251	215	173	233	210	181	146	122
	4 750	252	226	194	156	210	189	163	131	110
	5 000	228	205	176	142	190	171	148	119	99
	5 250	208	186	160	129	173	156	134	108	90
	5 500	189	170	146	118	158	142	122	99	83
	5 750	174	156	134	108		130	112	91	76
	6 000		143	123	99			103	83	70
	6 500									
	7 000									
	7 500									
	8 000									
	8 500									
	9 000									
	9 500									
	10 000									
	10 500									
PROPERTIES AND DESIGN DATA										
Area (mm^2)		3 760	3 220	2 640	2 040	3 280	2 820	2 320	1 790	1 450
Z_x (10^3 mm^3)		171	149	125	98.1	126	111	93.4	73.8	60.6
S_x (10^3 mm^3)		130	115	98.0	78.2	96.5	86.5	74.1	59.6	49.4
r_x (mm)		51.3	52.3	53.2	54.1	43.3	44.2	45.1	45.9	46.5
Z_y (10^3 mm^3)		104	91.2	76.6	60.5	87.8	77.4	65.3	51.8	42.6
S_y (10^3 mm^3)		85.0	76.3	65.5	52.9	70.8	63.9	55.2	44.8	37.3
r_y (mm)		29.4	30.1	30.8	31.5	28.7	29.4	30.1	30.8	31.3
r_x / r_y		1.74	1.74	1.73	1.72	1.51	1.50	1.50	1.49	1.49
M_{rx} (kN·m)		53.9	46.9	39.4	30.9	39.7	35.0	29.4	23.2	19.1
M_{ry} (kN·m)		32.8	28.7	24.1	19.1	27.7	24.4	20.6	16.3	11.7
IMPERIAL SIZE AND MASS										
Mass (lb./ft.)		19.8	17.0	13.9	10.7	17.3	14.9	12.2	9.46	7.66
Thickness (in.)		0.375	0.313	0.250	0.188	0.375	0.313	0.250	0.188	0.150
Size (in.)		6 x 3				5 x 3				

* Class 3 in bending about Y-Y axis

4-74

RECTANGULAR HOLLOW SECTIONS
Factored Axial Compressive Resistances, C_r (kN)

G40.21 350W CLASS H
$\phi = 0.90$

Designation (mm x mm x mm)	HSS 102 x 76						HSS 102 x 51		
	9.5	8.0	6.4	4.8	3.8	3.2	8.0	6.4	4.8
Mass (kg/m)	21.9	18.9	15.6	12.2	9.89	8.35	15.8	13.1	10.3

Effective length (KL) in millimetres with respect to the least radius of gyration

KL									
0	879	759	627	488	397	334	633	526	413
1 000	865	748	619	483	393	330	582	489	388
1 250	842	731	606	473	386	325	518	441	354
1 500	802	699	582	457	373	314	438	379	308
1 750	744	652	547	431	353	298	359	314	258
2 000	672	593	501	397	327	276	292	258	214
2 250	595	528	450	359	296	251	239	212	176
2 500	520	464	398	319	265	224	197	175	147
2 750	452	406	349	281	234	199	165	147	123
3 000	393	354	305	247	206	175	140	124	104
3 250	342	309	267	217	181	154	119	107	89
3 500	300	271	235	191	160	136	103	92	77
3 750	264	239	208	169	141	120	90	81	68
4 000	234	212	184	150	126	107			60
4 250	208	189	164	134	112	96			
4 500	187	169	147	120	101	86			
4 750	168	152	133	108	91	77			
5 000	152	138	120	98	82	70			
5 250	138	125	109	89	75	64			
5 500	126	114	100	81	68	58			
5 750			91	74	63	53			
6 000				68	57	49			
6 500									
7 000									
7 500									
8 000									
8 500									
9 000									
9 500									
10 000									
10 500									

PROPERTIES AND DESIGN DATA

Area (mm^2)	2 790	2 410	1 990	1 550	1 260	1 060	2 010	1 670	1 310
Z_x (10^3 mm^3)	87.9	77.9	66.0	52.6	43.3	37.0	59.0	50.7	40.8
S_x (10^3 mm^3)	67.4	61.1	52.9	43.0	35.9	30.8	43.6	38.5	31.8
r_x (mm)	35.0	35.9	36.7	37.5	38.0	38.4	33.2	34.2	35.1
Z_y (10^3 mm^3)	71.6	63.6	54.0	43.1	35.6	30.4	35.6	30.8	25.0
S_y (10^3 mm^3)	56.6	51.5	44.8	36.6	30.7	26.4	28.1	25.2	21.1
r_y (mm)	27.8	28.5	29.3	30.0	30.5	30.7	18.9	19.6	20.3
r_x / r_y	1.26	1.26	1.25	1.25	1.25	1.25	1.76	1.74	1.73
M_{rx} (kN·m)	27.7	24.5	20.8	16.6	13.6	11.7	18.6	16.0	12.9
M_{ry} (kN·m)	22.6	20.0	17.0	13.6	11.2	9.58	11.2	9.70	7.88

IMPERIAL SIZE AND MASS

Mass (lb./ft.)	14.7	12.7	10.5	8.17	6.64	5.61	10.6	8.81	6.89
Thickness (in.)	0.375	0.313	0.250	0.188	0.150	0.125	0.313	0.250	0.188
Size (in.)	4 x 3						4 x 2		

RECTANGULAR HOLLOW SECTIONS
Factored Axial Compressive Resistances, C_r (kN)

G40.21 350W
CLASS H
$\phi = 0.90$

Designation (mm x mm x mm)		HSS 102 x 51		HSS 89 x 64				
		3.8	3.2	8.0	6.4	4.8	3.8	3.2
Mass (kg/m)		8.37	7.09	15.8	13.1	10.3	8.37	7.09
Effective length (KL) in millimetres with respect to the least radius of gyration	0	337	284	633	526	413	337	284
	1 000	318	269	611	510	402	329	278
	1 250	292	249	579	487	385	316	267
	1 500	256	219	528	448	357	294	250
	1 750	216	186	465	399	321	266	226
	2 000	180	155	399	345	280	233	199
	2 250	149	129	338	295	241	201	173
	2 500	124	107	286	251	206	173	148
	2 750	104	90	243	214	176	148	127
	3 000	88	77	207	183	151	127	110
	3 250	76	66	179	158	131	110	95
	3 500	66	57	155	138	114	96	83
	3 750	57	50	136	120	100	84	73
	4 000	51	44	120	106	88	74	64
	4 250			106	94	78	66	57
	4 500			95	84	70	59	51
	4 750				76	63	53	46
	5 000						48	41
	5 250							
	5 500							
	5 750							
	6 000							
	6 500							
	7 000							
	7 500							
	8 000							
	8 500							
	9 000							
	9 500							
	10 000							
	10 500							

PROPERTIES AND DESIGN DATA

Area (mm^2)		1 070	903	2 010	1 670	1 310	1 070	903
Z_x (10^3 mm^3)		33.9	29.0	55.1	47.2	38.0	31.5	27.0
S_x (10^3 mm^3)		26.8	23.1	42.2	37.1	30.6	25.8	22.3
r_x (mm)		35.7	36.1	30.6	31.4	32.3	32.8	33.1
Z_y (10^3 mm^3)		20.8	17.9	43.3	37.3	30.1	25.0	21.4
S_y (10^3 mm^3)		18.0	15.6	34.4	30.5	25.3	21.4	18.5
r_y (mm)		20.7	21.0	23.3	24.1	24.8	25.2	25.5
r_x / r_y		1.72	1.72	1.31	1.30	1.30	1.30	1.30
M_{rx} (kN·m)		10.7	9.14	17.4	14.9	12.0	9.92	8.51
M_{ry} (kN·m)		6.55	5.64	13.6	11.7	9.48	7.88	6.74

IMPERIAL SIZE AND MASS

Mass (lb./ft.)		5.62	4.76	10.6	8.81	6.89	5.62	4.76
Thickness (in.)		0.150	0.125	0.313	0.250	0.188	0.150	0.125
Size (in.)		4 x 2		3½ x 2½				

G40.21 350W CLASS H
$\phi = 0.90$

RECTANGULAR HOLLOW SECTIONS
Factored Axial Compressive Resistances, C_r (kN)

Designation (mm x mm x mm)		HSS 76 x 51					HSS 51 x 25
		8.0	6.4	4.8	3.8	3.2	3.2
Mass (kg/m)		12.6	10.6	8.35	6.85	5.82	3.28
Effective length (KL) in millimetres with respect to the least radius of gyration	0	504	425	334	275	233	132
	1 000	456	391	310	257	219	64
	1 250	399	348	280	233	200	44
	1 500	332	294	240	202	174	31
	1 750	268	241	199	169	146	23
	2 000	216	196	164	139	121	
	2 250	176	160	134	114	100	
	2 500	145	132	111	95	83	
	2 750	121	111	93	80	70	
	3 000	102	94	79	68	59	
	3 250	87	80	68	58	51	
	3 500	76	69	59	50	44	
	3 750		61	51	44	38	
	4 000				39	34	
	4 250						
	4 500						
	4 750						
	5 000						
	5 250						
	5 500						
	5 750						
	6 000						
	6 500						
	7 000						
	7 500						
	8 000						
	8 500						
	9 000						
	9 500						
	10 000						
	10 500						
PROPERTIES AND DESIGN DATA							
Area (mm^2)		1 600	1 350	1 060	872	741	418
Z_x (10^3 mm^3)		36.1	31.5	25.8	21.6	18.6	6.34
S_x (10^3 mm^3)		26.7	24.1	20.3	17.3	15.1	4.81
r_x (mm)		25.2	26.1	27.0	27.5	27.8	17.1
Z_y (10^3 mm^3)		26.9	23.6	19.4	16.3	14.0	3.85
S_y (10^3 mm^3)		20.7	18.9	16.1	13.8	12.0	3.15
r_y (mm)		18.1	18.9	19.6	20.0	20.3	9.78
r_x / r_y		1.39	1.38	1.38	1.38	1.37	1.75
M_{rx} (kN·m)		11.4	9.92	8.13	6.80	5.86	2.00
M_{ry} (kN·m)		8.47	7.43	6.11	5.13	4.41	1.21
IMPERIAL SIZE AND MASS							
Mass (lb./ft.)		8.46	7.11	5.61	4.60	3.91	2.21
Thickness (in.)		0.313	0.250	0.188	0.150	0.125	0.125
Size (in.)		3 x 2					2 x 1

SQUARE HOLLOW SECTIONS
Factored Axial Compressive Resistances, C_r (kN)

G40.21 350W
CLASS H
$\phi = 0.90$

Designation (mm x mm x mm)	HSS 305 x 305					HSS 254 x 254		
	13	11	9.5	8.0 *	6.4 **	13	11	9.5
Mass (kg/m)	113	100	86.5	72.8	58.7	93.0	82.4	71.3

Effective length (KL) in millimetres with respect to the least radius of gyration	HSS 305 x 305					HSS 254 x 254		
0	4 540	4 030	3 470	2 920	1 910	3 720	3 310	2 860
1 000	4 540	4 030	3 460	2 920	1 910	3 720	3 310	2 860
1 250	4 540	4 030	3 460	2 920	1 910	3 720	3 310	2 860
1 500	4 540	4 030	3 460	2 920	1 910	3 720	3 310	2 860
1 750	4 530	4 030	3 460	2 920	1 910	3 710	3 310	2 860
2 000	4 530	4 030	3 460	2 920	1 910	3 710	3 300	2 860
2 250	4 530	4 030	3 460	2 920	1 910	3 710	3 300	2 860
2 500	4 530	4 030	3 460	2 920	1 910	3 700	3 300	2 850
2 750	4 530	4 020	3 460	2 920	1 900	3 700	3 290	2 850
3 000	4 520	4 020	3 450	2 910	1 900	3 690	3 280	2 840
3 250	4 510	4 010	3 450	2 910	1 900	3 670	3 270	2 830
3 500	4 500	4 010	3 440	2 910	1 900	3 660	3 260	2 820
3 750	4 490	4 000	3 430	2 900	1 900	3 640	3 240	2 810
4 000	4 480	3 980	3 430	2 890	1 890	3 610	3 220	2 790
4 250	4 460	3 970	3 410	2 880	1 890	3 580	3 190	2 770
4 500	4 440	3 950	3 400	2 870	1 880	3 540	3 160	2 740
4 750	4 420	3 930	3 380	2 850	1 880	3 500	3 120	2 710
5 000	4 390	3 910	3 360	2 840	1 870	3 450	3 080	2 670
5 250	4 350	3 880	3 340	2 820	1 860	3 400	3 030	2 630
5 500	4 320	3 840	3 310	2 800	1 850	3 330	2 980	2 580
5 750	4 270	3 800	3 280	2 770	1 840	3 260	2 920	2 530
6 000	4 220	3 760	3 240	2 740	1 830	3 190	2 850	2 480
6 500	4 110	3 660	3 160	2 670	1 800	3 030	2 710	2 360
7 000	3 970	3 540	3 060	2 590	1 760	2 850	2 550	2 220
7 500	3 820	3 410	2 950	2 500	1 710	2 660	2 390	2 080
8 000	3 640	3 260	2 820	2 390	1 670	2 480	2 230	1 940
8 500	3 460	3 100	2 690	2 280	1 610	2 300	2 070	1 800
9 000	3 280	2 940	2 550	2 170	1 550	2 120	1 910	1 670
9 500	3 090	2 770	2 400	2 050	1 490	1 960	1 770	1 540
10 000	2 900	2 610	2 260	1 930	1 430	1 810	1 630	1 430
10 500	2 720	2 450	2 130	1 810	1 370	1 670	1 510	1 320

PROPERTIES AND DESIGN DATA								
Area (mm^2)	14 400	12 800	11 000	9 280	7 480	11 800	10 500	9 090
Z (10^3 mm^3)	1 560	1 390	1 210	1 030	833	1 060	946	825
S (10^3 mm^3)	1 330	1 190	1 040	886	625	889	800	703
r (mm)	118	119	120	121	121	97.6	98.4	99.1
M_r (kN·m)	491	438	381	279	197	334	298	260

IMPERIAL SIZE AND MASS								
Mass (lb./ft.)	76.1	67.3	58.1	48.9	39.4	62.5	55.4	47.9
Thickness (in.)	0.500	0.438	0.375	0.313	0.250	0.500	0.438	0.375
Size (in.)	12 x 12					10 x 10		

* Class 3 in bending

** Class 4: C_r calculated according to S16.1-94 Clause 13.3.3; S and M_r according to Clause 13.5(c)(iii).

G40.21 350W
CLASS H
$\phi = 0.90$

SQUARE HOLLOW SECTIONS
Factored Axial Compressive Resistances, C_r (kN)

Designation (mm x mm x mm)		HSS 254 x 254		HSS 203 x 203				
		8.0	6.4 **	13	11	9.5	8.0	6.4
Mass (kg/m)		60.1	48.6	72.7	64.6	56.1	47.5	38.4
Effective length (KL) in millimetres with respect to the least radius of gyration	0	2 410	1 790	2 920	2 590	2 250	1 910	1 540
	1 000	2 410	1 790	2 920	2 590	2 250	1 910	1 540
	1 250	2 410	1 790	2 920	2 590	2 250	1 900	1 540
	1 500	2 410	1 790	2 910	2 590	2 250	1 900	1 540
	1 750	2 410	1 790	2 910	2 590	2 250	1 900	1 540
	2 000	2 410	1 790	2 910	2 580	2 240	1 900	1 540
	2 250	2 410	1 790	2 900	2 580	2 240	1 890	1 540
	2 500	2 410	1 790	2 890	2 570	2 230	1 890	1 530
	2 750	2 400	1 780	2 870	2 550	2 220	1 880	1 520
	3 000	2 400	1 780	2 850	2 540	2 200	1 870	1 510
	3 250	2 390	1 780	2 820	2 510	2 190	1 850	1 500
	3 500	2 380	1 770	2 790	2 480	2 160	1 830	1 490
	3 750	2 370	1 770	2 750	2 450	2 130	1 810	1 470
	4 000	2 350	1 760	2 700	2 400	2 090	1 780	1 440
	4 250	2 330	1 750	2 640	2 350	2 050	1 740	1 420
	4 500	2 310	1 740	2 570	2 290	2 000	1 700	1 380
	4 750	2 290	1 730	2 500	2 230	1 950	1 660	1 350
	5 000	2 260	1 710	2 420	2 160	1 890	1 610	1 310
	5 250	2 220	1 690	2 330	2 090	1 830	1 560	1 270
	5 500	2 190	1 670	2 240	2 010	1 760	1 500	1 230
	5 750	2 140	1 650	2 150	1 930	1 690	1 440	1 180
	6 000	2 100	1 630	2 060	1 850	1 620	1 390	1 130
	6 500	2 000	1 570	1 870	1 680	1 480	1 270	1 040
	7 000	1 890	1 510	1 700	1 530	1 350	1 150	946
	7 500	1 770	1 440	1 530	1 380	1 220	1 050	859
	8 000	1 650	1 350	1 380	1 250	1 100	948	779
	8 500	1 540	1 260	1 250	1 130	999	860	706
	9 000	1 420	1 170	1 130	1 020	906	780	641
	9 500	1 320	1 080	1 030	930	823	709	584
	10 000	1 220	1 000	938	848	750	647	532
	10 500	1 130	927	857	774	686	591	487
PROPERTIES AND DESIGN DATA								
Area (mm^2)		7 660	6 190	9 260	8 230	7 150	6 050	4 900
Z (10^3 mm^3)		702	571	651	585	513	439	359
S (10^3 mm^3)		602	492	538	488	432	373	308
r (mm)		99.9	101	76.9	77.6	78.4	79.2	79.9
M$_r$ (kN·m)		221	155	205	184	162	138	113
IMPERIAL SIZE AND MASS								
Mass (lb./ft.)		40.4	32.6	48.9	43.4	37.7	31.9	25.8
Thickness (in.)		0.313	0.250	0.500	0.438	0.375	0.313	0.250
Size (in.)		10 x 10		8 x 8				

** Class 4: C_r calculated according to S16.1-94 Clause 13.3.3; S and M_r according to Clause 13.5(c)(iii).

SQUARE HOLLOW SECTIONS
Factored Axial Compressive Resistances, C_r (kN)

G40.21 350W
CLASS H
$\phi = 0.90$

Designation (mm x mm x mm)	HSS 178 x 178						HSS 152 x 152		
	13	11	9.5	8.0	6.4	4.8 *	13	11	9.5
Mass (kg/m)	62.6	55.7	48.5	41.1	33.4	25.5	52.4	46.9	40.9
Effective length (KL) in millimetres with respect to the least radius of gyration									
0	2 510	2 240	1 950	1 650	1 340	1 020	2 100	1 880	1 640
1 000	2 510	2 240	1 950	1 650	1 340	1 020	2 100	1 880	1 640
1 250	2 510	2 230	1 950	1 650	1 340	1 020	2 100	1 880	1 640
1 500	2 510	2 230	1 940	1 650	1 340	1 020	2 100	1 870	1 630
1 750	2 500	2 230	1 940	1 640	1 330	1 020	2 090	1 870	1 630
2 000	2 490	2 220	1 930	1 640	1 330	1 020	2 070	1 850	1 620
2 250	2 480	2 210	1 920	1 630	1 330	1 010	2 050	1 830	1 600
2 500	2 460	2 190	1 910	1 620	1 320	1 010	2 020	1 810	1 580
2 750	2 440	2 170	1 890	1 610	1 310	1 000	1 980	1 770	1 550
3 000	2 400	2 150	1 870	1 590	1 290	989	1 930	1 730	1 520
3 250	2 360	2 110	1 840	1 570	1 270	976	1 860	1 680	1 470
3 500	2 310	2 070	1 810	1 540	1 250	959	1 790	1 610	1 420
3 750	2 250	2 010	1 760	1 500	1 220	938	1 710	1 540	1 360
4 000	2 180	1 950	1 710	1 460	1 190	914	1 620	1 470	1 290
4 250	2 100	1 890	1 650	1 410	1 150	887	1 530	1 390	1 230
4 500	2 020	1 810	1 590	1 360	1 110	857	1 440	1 310	1 160
4 750	1 930	1 740	1 530	1 310	1 070	825	1 350	1 230	1 090
5 000	1 840	1 660	1 460	1 250	1 020	791	1 260	1 150	1 020
5 250	1 750	1 570	1 390	1 190	978	755	1 180	1 070	953
5 500	1 660	1 490	1 320	1 130	931	720	1 100	1 000	891
5 750	1 570	1 410	1 250	1 070	884	684	1 020	935	832
6 000	1 480	1 340	1 180	1 020	838	649	955	873	777
6 500	1 320	1 190	1 050	910	751	582	833	763	680
7 000	1 170	1 060	940	812	671	521	730	669	597
7 500	1 040	944	839	725	600	466	643	590	526
8 000	930	844	750	649	537	418	569	523	467
8 500	833	757	673	583	483	376	507	466	416
9 000	750	681	606	525	435	339	454	417	373
9 500	677	615	548	475	393	307	409	376	336
10 000	614	558	497	431	357	278	370	340	304
10 500	559	508	453	392	325	254	336	309	276
PROPERTIES AND DESIGN DATA									
Area (mm^2)	7 970	7 100	6 180	5 240	4 250	3 250	6 680	5 970	5 210
Z (10^3 mm^3)	484	437	385	330	271	210	342	310	275
S (10^3 mm^3)	396	361	322	279	231	181	276	253	227
r (mm)	66.5	67.2	68.0	68.8	69.6	70.3	56.1	56.9	57.6
M_r (kN·m)	152	138	121	104	85.4	57.0	108	97.7	86.6
IMPERIAL SIZE AND MASS									
Mass (lb./ft.)	42.1	37.5	32.6	27.6	22.4	17.1	35.2	31.5	27.5
Thickness (in.)	0.500	0.438	0.375	0.313	0.250	0.188	0.500	0.438	0.375
Size (in.)	7 x 7						6 x 6		

* Class 3 in bending

SQUARE HOLLOW SECTIONS
Factored Axial Compressive Resistances, C_r (kN)

G40.21 350W
CLASS H
$\phi = 0.90$

Designation (mm x mm x mm)		HSS 152 x 152			HSS 127 x 127					
		8.0	6.4	4.8	13	11	9.5	8.0	6.4	4.8
Mass (kg/m)		34.8	28.3	21.7	42.3	38	33.3	28.4	23.2	17.9
Effective length (KL) in millimetres with respect to the least radius of gyration	0	1 400	1 140	869	1 700	1 520	1 340	1 140	932	718
	1 000	1 390	1 140	869	1 690	1 520	1 330	1 140	931	717
	1 250	1 390	1 140	868	1 690	1 520	1 330	1 140	929	716
	1 500	1 390	1 130	867	1 680	1 510	1 320	1 130	925	713
	1 750	1 390	1 130	864	1 660	1 490	1 310	1 120	918	708
	2 000	1 380	1 120	859	1 630	1 470	1 290	1 110	906	699
	2 250	1 370	1 110	853	1 590	1 440	1 270	1 080	889	687
	2 500	1 350	1 100	843	1 540	1 390	1 230	1 050	866	670
	2 750	1 320	1 080	830	1 470	1 340	1 180	1 020	836	649
	3 000	1 290	1 060	813	1 400	1 270	1 120	969	800	622
	3 250	1 260	1 030	792	1 310	1 200	1 060	917	759	592
	3 500	1 210	997	767	1 220	1 120	995	861	715	558
	3 750	1 170	958	738	1 130	1 040	925	803	668	523
	4 000	1 110	916	707	1 040	957	857	745	622	488
	4 250	1 060	871	673	958	882	791	689	576	453
	4 500	997	824	638	879	811	729	636	532	420
	4 750	939	777	602	807	745	671	586	492	388
	5 000	881	731	567	741	685	618	540	454	359
	5 250	826	686	533	682	631	569	498	419	331
	5 500	773	642	499	628	582	525	460	387	307
	5 750	723	601	468	580	537	485	425	358	284
	6 000	676	563	438	536	497	449	394	332	263
	6 500	592	494	385	462	428	388	340	287	228
	7 000	520	434	339	401	372	337	296	250	198
	7 500	459	384	300	351	326	295	259	219	174
	8 000	407	340	266	309	287	260	229	193	153
	8 500	363	304	237	275	255	231	203	172	136
	9 000	326	272	213	245	228	207	182	153	122
	9 500	293	245	192				163	138	110
	10 000	265	222	174						
	10 500	241	202	158						
PROPERTIES AND DESIGN DATA										
Area (mm²)		4 430	3 610	2 760	5 390	4 840	4 240	3 620	2 960	2 280
Z (10^3 mm³)		237	196	152	225	205	183	159	132	103
S (10^3 mm³)		198	166	130	177	165	149	132	111	88.1
r (mm)		58.4	59.2	59.9	45.7	46.5	47.3	48.0	48.8	49.6
M_r (kN·m)		74.7	61.7	47.9	70.9	64.6	57.6	50.1	41.6	32.4
IMPERIAL SIZE AND MASS										
Mass (lb./ft.)		23.4	19.0	14.6	28.4	25.5	22.4	19.1	15.6	12.0
Thickness (in.)		0.313	0.250	0.188	0.500	0.438	0.375	0.313	0.250	0.188
Size (in.)		6 x 6			5 x 5					

SQUARE HOLLOW SECTIONS
Factored Axial Compressive Resistances, C_r (kN)

G40.21 350W
CLASS H
$\phi = 0.90$

Designation (mm x mm x mm)		HSS 102 x 102						HSS 89 x 89		
		9.5	8.0	6.4	4.8	3.8	3.2	9.5	8.0	6.4
Mass (kg/m)		25.7	22.1	18.2	14.1	11.4	9.62	21.9	18.9	15.6
Effective length (KL) in millimetres with respect to the least radius of gyration	0	1 030	888	731	564	457	387	879	759	627
	1 000	1 030	885	728	562	455	386	871	753	622
	1 250	1 020	878	723	559	453	384	858	743	615
	1 500	1 010	866	714	552	448	380	833	723	600
	1 750	979	845	698	541	439	373	795	693	577
	2 000	941	815	675	524	426	362	743	650	544
	2 250	891	774	644	502	409	348	680	598	504
	2 500	831	724	605	473	386	329	613	542	459
	2 750	764	669	561	441	361	308	546	485	413
	3 000	695	611	515	406	333	285	483	431	368
	3 250	629	554	468	371	305	261	427	382	328
	3 500	566	500	424	337	278	238	378	339	291
	3 750	509	451	383	305	252	216	335	301	259
	4 000	458	407	346	276	228	196	299	268	232
	4 250	413	367	313	250	207	178	267	240	207
	4 500	373	332	283	227	188	161	240	216	187
	4 750	338	301	257	206	171	147	216	195	169
	5 000	308	274	234	188	156	134	196	177	153
	5 250	281	250	214	172	142	122	178	161	139
	5 500	257	229	196	157	131	112	163	147	127
	5 750	236	210	180	145	120	103	149	135	117
	6 000	217	194	166	133	111	95	137	124	107
	6 500	186	166	142	114	95	82			92
	7 000	161	143	123	99	82	71			
	7 500		125	107	86	72	62			
	8 000						54			
	8 500									
	9 000									
	9 500									
	10 000									
	10 500									
PROPERTIES AND DESIGN DATA										
Area (mm^2)		3 280	2 820	2 320	1 790	1 450	1 230	2 790	2 410	1 990
Z (10^3 mm^3)		110	96.8	81.4	64.3	52.8	44.9	80.5	71.4	60.5
S (10^3 mm^3)		87.6	78.5	67.3	54.2	45.0	38.5	63.0	57.1	49.5
r (mm)		36.9	37.6	38.4	39.2	39.7	40.0	31.7	32.4	33.2
M_r (kN·m)		34.7	30.5	25.6	20.3	16.6	14.1	25.4	22.5	19.1
IMPERIAL SIZE AND MASS										
Mass (lb./ft.)		17.3	14.9	12.2	9.46	7.66	6.47	14.7	12.7	10.5
Thickness (in.)		0.375	0.313	0.250	0.188	0.150	0.125	0.375	0.313	0.250
Size (in.)		4 x 4						3½ x 3½		

G40.21 350W
CLASS H
$\phi = 0.90$

SQUARE HOLLOW SECTIONS
Factored Axial Compressive Resistances, C_r (kN)

Designation (mm x mm x mm)		HSS 89 x 89			HSS 76 x 76					
		4.8	3.8	3.2	9.5	8.0	6.4	4.8	3.8	3.2
Mass (kg/m)		12.2	9.89	8.35	18.1	15.8	13.1	10.3	8.37	7.09
Effective length (KL) in millimetres with respect to the least radius of gyration	0	488	397	334	728	633	526	413	337	284
	1 000	485	394	332	713	622	518	407	333	281
	1 250	480	390	329	691	604	505	398	326	275
	1 500	469	382	322	651	573	481	381	313	265
	1 750	453	370	312	596	528	447	357	294	249
	2 000	429	351	297	531	474	405	325	269	229
	2 250	399	328	278	465	417	359	291	242	206
	2 500	366	301	256	402	363	314	256	214	183
	2 750	331	273	232	347	315	274	224	188	161
	3 000	296	246	209	300	273	238	195	164	141
	3 250	265	220	187	260	237	208	171	144	123
	3 500	236	196	167	227	208	182	150	126	109
	3 750	210	175	150	200	183	160	132	112	96
	4 000	188	157	134	177	162	142	117	99	85
	4 250	169	141	120	157	144	126	105	88	76
	4 500	152	127	108	141	129	113	94	79	68
	4 750	137	115	98	127	116	102	84	71	61
	5 000	125	104	89	114	105	92	76	65	56
	5 250	114	95	81	104	95	84	69	59	51
	5 500	104	87	74			76	63	54	46
	5 750	95	80	68				58	49	42
	6 000	88	73	63						
	6 500	75	63	54						
	7 000									
	7 500									
	8 000									
	8 500									
	9 000									
	9 500									
	10 000									
	10 500									
PROPERTIES AND DESIGN DATA										
Area (mm^2)		1 550	1 260	1 060	2 310	2 010	1 670	1 310	1 070	903
Z (10^3 mm^3)		48.2	39.8	33.9	55.5	49.8	42.8	34.4	28.6	24.5
S (10^3 mm^3)		40.3	33.7	29.0	42.4	39.1	34.5	28.5	24.0	20.7
r (mm)		34.0	34.5	34.8	26.5	27.2	28.0	28.8	29.3	29.6
M_r (kN·m)		15.2	12.5	10.7	17.5	15.7	13.5	10.8	9.01	7.72
IMPERIAL SIZE AND MASS										
Mass (lb./ft.)		8.17	6.64	5.61	12.2	10.6	8.81	6.89	5.62	4.76
Thickness (in.)		0.188	0.150	0.125	0.375	0.313	0.250	0.188	0.150	0.125
Size (in.)		3½ x 3½			3 x 3					

SQUARE HOLLOW SECTIONS
Factored Axial Compressive Resistances, C_r (kN)

G40.21 350W
CLASS H
$\phi = 0.90$

Designation (mm x mm x mm)		HSS 64 x 64				HSS 51 x 51	
		6.4	4.8	3.8	3.2	6.4	4.8
Mass (kg/m)		10.6	8.35	6.85	5.82	8.05	6.45
Effective length (KL) in millimetres with respect to the least radius of gyration	0	425	334	275	233	324	259
	1 000	409	323	266	227	290	235
	1 250	386	307	254	217	251	207
	1 500	349	281	234	200	206	174
	1 750	305	248	208	179	166	141
	2 000	260	214	180	156	133	114
	2 250	219	182	154	133	108	93
	2 500	185	154	131	114	88	77
	2 750	157	131	112	97	74	64
	3 000	134	112	96	83	62	54
	3 250	115	96	83	72	53	46
	3 500	100	84	72	62	46	40
	3 750	88	73	63	55		
	4 000	77	65	56	48		
	4 250	69	58	49	43		
	4 500	61	51	44	38		
	4 750			40	35		
	5 000						
	5 250						
	5 500						
	5 750						
	6 000						
	6 500						
	7 000						
	7 500						
	8 000						
	8 500						
	9 000						
	9 500						
	10 000						
	10 500						
PROPERTIES AND DESIGN DATA							
Area (mm^2)		1 350	1 060	872	741	1 030	821
Z (10^3 mm^3)		28.1	23.0	19.2	16.6	16.4	13.8
S (10^3 mm^3)		22.2	18.7	16.0	13.9	12.6	11.0
r (mm)		22.8	23.6	24.1	24.4	17.6	18.4
M_r (kN·m)		8.85	7.25	6.05	5.23	5.17	4.35
IMPERIAL SIZE AND MASS							
Mass (lb./ft.)		7.11	5.61	4.60	3.91	5.41	4.33
Thickness (in.)		0.250	0.188	0.150	0.125	0.250	0.188
Size (in.)		2½ x 2½				2 x 2	

G40.21 350W
CLASS H
$\phi = 0.90$

SQUARE HOLLOW SECTIONS
Factored Axial Compressive Resistances, C_r (kN)

Designation (mm x mm x mm)		HSS 51 x 51		HSS 38 x 38		
		3.8	3.2	4.8	3.8	3.2
Mass (kg/m)		5.33	4.55	4.54	3.81	3.28
Effective length (KL) in millimetres with respect to the least radius of gyration	0	214	183	182	153	132
	1 000	197	169	132	115	101
	1 250	175	151	100	89	79
	1 500	148	129	75	67	60
	1 750	121	106	56	51	46
	2 000	99	87	44	40	36
	2 250	81	71	35	32	28
	2 500	67	59	28	26	23
	2 750	56	49			19
	3 000	47	42			
	3 250	40	36			
	3 500	35	31			
	3 750	30	27			
	4 000					
	4 250					
	4 500					
	4 750					
	5 000					
	5 250					
	5 500					
	5 750					
	6 000					
	6 500					
	7 000					
	7 500					
	8 000					
	8 500					
	9 000					
	9 500					
	10 000					
	10 500					
PROPERTIES AND DESIGN DATA						
Area (mm^2)		679	580	578	485	418
Z (10^3 mm^3)		11.7	10.2	6.95	6.06	5.35
S (10^3 mm^3)		9.55	8.42	5.30	4.79	4.31
r (mm)		18.9	19.2	13.2	13.7	14.0
M_r (kN·m)		3.69	3.21	2.19	1.91	1.69
IMPERIAL SIZE AND MASS						
Mass (lb./ft.)		3.58	3.06	3.05	2.56	2.21
Thickness (in.)		0.150	0.125	0.188	0.150	0.125
Size (in.)		2 x 2		1½ x 1½		

ROUND HOLLOW SECTIONS
Factored Axial Compressive Resistances, C_r (kN)

G40.21 350W
CLASS H
$\phi = 0.90$

Designation (mm x mm)		HSS 610			HSS 559			HSS 508			
		13	11 *	9.5 *	13	11	9.5 *	13	11	9.5 *	8.0 *
Mass (kg/m)		187	164	141	171	150	129	155	136	117	98.0
Effective length (KL) in millimetres with respect to the least radius of gyration	0	7 500	6 580	5 670	6 870	6 020	5 170	6 240	5 480	4 690	3 940
	1 000	7 500	6 580	5 670	6 870	6 020	5 170	6 240	5 480	4 690	3 940
	1 250	7 500	6 580	5 670	6 870	6 020	5 170	6 240	5 480	4 690	3 940
	1 500	7 500	6 580	5 670	6 870	6 020	5 170	6 240	5 480	4 690	3 940
	1 750	7 500	6 580	5 670	6 870	6 020	5 170	6 240	5 480	4 690	3 940
	2 000	7 500	6 580	5 670	6 870	6 020	5 170	6 240	5 480	4 690	3 940
	2 250	7 500	6 580	5 670	6 870	6 020	5 170	6 240	5 480	4 690	3 940
	2 500	7 500	6 580	5 670	6 870	6 020	5 170	6 240	5 480	4 690	3 940
	2 750	7 500	6 580	5 670	6 870	6 010	5 160	6 230	5 480	4 690	3 940
	3 000	7 500	6 580	5 670	6 860	6 010	5 160	6 230	5 480	4 690	3 940
	3 250	7 490	6 580	5 670	6 860	6 010	5 160	6 230	5 480	4 690	3 930
	3 500	7 490	6 580	5 670	6 860	6 010	5 160	6 230	5 470	4 690	3 930
	3 750	7 490	6 580	5 670	6 860	6 010	5 160	6 230	5 470	4 690	3 930
	4 000	7 490	6 580	5 660	6 860	6 010	5 160	6 220	5 470	4 680	3 930
	4 250	7 490	6 580	5 660	6 850	6 010	5 160	6 220	5 470	4 680	3 930
	4 500	7 490	6 570	5 660	6 850	6 000	5 150	6 210	5 460	4 680	3 920
	4 750	7 480	6 570	5 660	6 850	6 000	5 150	6 210	5 460	4 670	3 920
	5 000	7 480	6 570	5 660	6 840	5 990	5 150	6 200	5 450	4 670	3 920
	5 250	7 470	6 560	5 650	6 830	5 990	5 140	6 190	5 440	4 660	3 910
	5 500	7 470	6 560	5 650	6 830	5 980	5 140	6 180	5 430	4 650	3 900
	5 750	7 460	6 550	5 640	6 820	5 980	5 130	6 170	5 420	4 640	3 900
	6 000	7 450	6 550	5 640	6 810	5 970	5 120	6 160	5 410	4 630	3 890
	6 500	7 440	6 530	5 620	6 780	5 950	5 110	6 120	5 380	4 610	3 870
	7 000	7 410	6 510	5 610	6 750	5 920	5 080	6 080	5 340	4 580	3 840
	7 500	7 380	6 490	5 590	6 710	5 880	5 050	6 020	5 300	4 540	3 810
	8 000	7 350	6 450	5 560	6 660	5 840	5 020	5 960	5 240	4 490	3 770
	8 500	7 300	6 410	5 520	6 600	5 790	4 970	5 880	5 170	4 430	3 720
	9 000	7 250	6 370	5 480	6 530	5 730	4 920	5 790	5 090	4 360	3 660
	9 500	7 180	6 310	5 440	6 450	5 660	4 860	5 680	5 000	4 280	3 600
	10 000	7 110	6 250	5 380	6 350	5 580	4 790	5 560	4 900	4 190	3 530
	10 500	7 020	6 170	5 320	6 250	5 480	4 710	5 430	4 780	4 100	3 450

PROPERTIES AND DESIGN DATA											
Area (mm^2)		23 800	20 900	18 000	21 800	19 100	16 400	19 800	17 400	14 900	12 500
Z (10^3 mm^3)		4 530	3 990	3 430	3 790	3 340	2 880	3 120	2 750	2 370	1 990
S (10^3 mm^3)		3 480	3 070	2 650	2 910	2 570	2 220	2 390	2 110	1 830	1 540
r (mm)		211	212	212	193	194	194	175	176	176	177
M_r (kN·m)		1 430	967	835	1 190	1 050	699	983	866	576	485

IMPERIAL SIZE AND MASS											
Mass (lb./ft.)		126	110	94.8	115	101	86.7	104	91.6	78.7	65.9
Thickness (in.)		0.500	0.438	0.375	0.500	0.438	0.375	0.500	0.438	0.375	0.313
Size (in.)		24 OD			22 OD			20 OD			

* Class 3 in bending

G40.21 350W
CLASS H
$\phi = 0.90$

ROUND HOLLOW SECTIONS
Factored Axial Compressive Resistances, C_r (kN)

Designation (mm x mm)		HSS 406					HSS 356				
		13	11	9.5	8.0	6.4 *	13	11	9.5	8.0	6.4 *
Mass (kg/m)		123	108	93.3	78.1	62.6	107	94.6	81.3	68.2	54.7
Effective length (KL) in millimetres with respect to the least radius of gyration	0	4 950	4 350	3 750	3 130	2 510	4 320	3 780	3 280	2 730	2 200
	1 000	4 950	4 350	3 750	3 130	2 510	4 320	3 780	3 280	2 730	2 200
	1 250	4 950	4 350	3 750	3 130	2 510	4 320	3 780	3 280	2 730	2 200
	1 500	4 950	4 350	3 750	3 130	2 510	4 310	3 780	3 280	2 730	2 200
	1 750	4 940	4 350	3 750	3 130	2 510	4 310	3 780	3 280	2 730	2 190
	2 000	4 940	4 350	3 750	3 130	2 510	4 310	3 780	3 270	2 730	2 190
	2 250	4 940	4 350	3 750	3 130	2 510	4 310	3 780	3 270	2 730	2 190
	2 500	4 940	4 340	3 750	3 130	2 510	4 310	3 770	3 270	2 730	2 190
	2 750	4 940	4 340	3 740	3 130	2 510	4 310	3 770	3 270	2 730	2 190
	3 000	4 940	4 340	3 740	3 130	2 510	4 300	3 770	3 270	2 730	2 190
	3 250	4 930	4 340	3 740	3 130	2 510	4 300	3 760	3 260	2 720	2 190
	3 500	4 930	4 330	3 740	3 120	2 510	4 290	3 760	3 260	2 720	2 180
	3 750	4 920	4 330	3 730	3 120	2 500	4 280	3 750	3 250	2 710	2 180
	4 000	4 920	4 320	3 730	3 120	2 500	4 270	3 740	3 240	2 710	2 170
	4 250	4 910	4 310	3 720	3 110	2 500	4 250	3 730	3 230	2 700	2 170
	4 500	4 900	4 300	3 710	3 100	2 490	4 240	3 710	3 220	2 690	2 160
	4 750	4 880	4 290	3 700	3 100	2 480	4 210	3 690	3 200	2 670	2 150
	5 000	4 870	4 280	3 690	3 090	2 480	4 190	3 670	3 180	2 660	2 140
	5 250	4 850	4 260	3 680	3 080	2 470	4 160	3 650	3 160	2 640	2 120
	5 500	4 830	4 240	3 660	3 060	2 460	4 130	3 620	3 140	2 620	2 110
	5 750	4 800	4 220	3 640	3 050	2 440	4 090	3 590	3 110	2 600	2 090
	6 000	4 770	4 200	3 620	3 030	2 430	4 050	3 550	3 080	2 570	2 070
	6 500	4 700	4 140	3 570	2 990	2 400	3 940	3 470	3 000	2 510	2 020
	7 000	4 620	4 070	3 510	2 940	2 360	3 820	3 360	2 910	2 440	1 960
	7 500	4 510	3 980	3 430	2 880	2 310	3 690	3 240	2 810	2 360	1 890
	8 000	4 390	3 870	3 340	2 800	2 250	3 530	3 110	2 700	2 260	1 820
	8 500	4 260	3 760	3 240	2 720	2 180	3 370	2 970	2 580	2 160	1 740
	9 000	4 110	3 630	3 130	2 630	2 110	3 200	2 820	2 450	2 060	1 650
	9 500	3 950	3 490	3 010	2 530	2 030	3 020	2 670	2 320	1 950	1 570
	10 000	3 780	3 340	2 880	2 430	1 950	2 850	2 520	2 180	1 840	1 480
	10 500	3 610	3 190	2 750	2 320	1 860	2 680	2 370	2 060	1 730	1 390
PROPERTIES AND DESIGN DATA											
Area (mm²)		15 700	13 800	11 900	9 950	7 980	13 700	12 000	10 400	8 680	6 970
Z (10³ mm³)		1 970	1 740	1 500	1 260	1 020	1 490	1 320	1 140	961	775
S (10³ mm³)		1 500	1 330	1 150	972	786	1 130	1 010	873	738	598
r (mm)		139	140	140	141	141	121	122	122	123	123
M_r (kN·m)		621	548	473	397	248	469	416	359	303	188
IMPERIAL SIZE AND MASS											
Mass (lb./ft.)		82.9	72.9	62.7	52.5	42.1	72.2	63.5	54.7	45.8	36.8
Thickness (in.)		0.500	0.438	0.375	0.313	0.250	0.500	0.438	0.375	0.313	0.250
Size (in.)		16 OD					14 OD				

* Class 3 in bending

ROUND HOLLOW SECTIONS
Factored Axial Compressive Resistances, C_r (kN)

G40.21 350W
CLASS H
$\phi = 0.90$

Designation (mm x mm)		HSS 324					HSS 273				
		13	11	9.5	8.0	6.4	13	11	9.5	8.0	6.4
Mass (kg/m)		97.5	85.8	73.9	61.9	49.7	81.6	71.9	61.9	52.0	41.8
Effective length (KL) in millimetres with respect to the least radius of gyration	0	3 910	3 430	2 960	2 490	1 990	3 280	2 890	2 490	2 090	1 680
	1 000	3 910	3 430	2 960	2 490	1 990	3 280	2 890	2 490	2 090	1 680
	1 250	3 910	3 430	2 960	2 490	1 990	3 280	2 880	2 480	2 080	1 680
	1 500	3 910	3 430	2 960	2 480	1 990	3 270	2 880	2 480	2 080	1 680
	1 750	3 900	3 430	2 960	2 480	1 990	3 270	2 880	2 480	2 080	1 670
	2 000	3 900	3 430	2 960	2 480	1 990	3 270	2 880	2 480	2 080	1 670
	2 250	3 900	3 430	2 960	2 480	1 990	3 270	2 880	2 480	2 080	1 670
	2 500	3 900	3 430	2 960	2 480	1 990	3 260	2 870	2 470	2 080	1 670
	2 750	3 890	3 420	2 950	2 480	1 990	3 250	2 870	2 470	2 070	1 670
	3 000	3 890	3 420	2 950	2 470	1 990	3 240	2 860	2 460	2 070	1 660
	3 250	3 880	3 410	2 940	2 470	1 980	3 230	2 840	2 450	2 060	1 650
	3 500	3 870	3 400	2 940	2 460	1 980	3 210	2 830	2 440	2 050	1 640
	3 750	3 860	3 390	2 930	2 460	1 970	3 190	2 810	2 420	2 030	1 630
	4 000	3 840	3 380	2 920	2 450	1 960	3 160	2 780	2 400	2 020	1 620
	4 250	3 820	3 360	2 900	2 430	1 950	3 120	2 750	2 380	2 000	1 610
	4 500	3 800	3 340	2 880	2 420	1 940	3 080	2 720	2 350	1 970	1 590
	4 750	3 770	3 320	2 860	2 400	1 930	3 040	2 680	2 310	1 940	1 560
	5 000	3 740	3 290	2 840	2 380	1 910	2 980	2 630	2 270	1 910	1 540
	5 250	3 700	3 260	2 810	2 360	1 890	2 920	2 580	2 230	1 880	1 510
	5 500	3 650	3 220	2 780	2 340	1 870	2 860	2 530	2 180	1 840	1 480
	5 750	3 600	3 180	2 740	2 310	1 850	2 790	2 460	2 130	1 790	1 450
	6 000	3 550	3 130	2 700	2 270	1 820	2 710	2 400	2 070	1 750	1 410
	6 500	3 430	3 030	2 610	2 200	1 760	2 550	2 260	1 950	1 650	1 330
	7 000	3 280	2 900	2 510	2 110	1 690	2 380	2 110	1 820	1 540	1 240
	7 500	3 120	2 770	2 390	2 020	1 620	2 200	1 950	1 690	1 430	1 160
	8 000	2 960	2 620	2 260	1 910	1 530	2 030	1 800	1 560	1 320	1 070
	8 500	2 780	2 470	2 130	1 800	1 450	1 870	1 660	1 440	1 220	988
	9 000	2 610	2 320	2 000	1 700	1 360	1 720	1 530	1 330	1 120	910
	9 500	2 440	2 170	1 870	1 590	1 270	1 580	1 400	1 220	1 030	838
	10 000	2 280	2 030	1 750	1 480	1 190	1 450	1 290	1 120	951	771
	10 500	2 120	1 890	1 630	1 390	1 110	1 340	1 190	1 030	876	710
PROPERTIES AND DESIGN DATA											
Area (mm^2)		12 400	10 900	9 410	7 890	6 330	10 400	9 160	7 890	6 620	5 320
Z (10^3 mm^3)		1 230	1 090	942	794	640	862	764	662	559	452
S (10^3 mm^3)		930	827	719	608	493	646	577	502	427	347
r (mm)		110	111	111	112	112	92.2	92.7	93.2	93.8	94.3
M$_r$ (kN·m)		387	343	297	250	202	272	241	209	176	142
IMPERIAL SIZE AND MASS											
Mass (lb./ft.)		65.5	57.7	49.7	41.6	33.4	54.8	48.3	41.6	34.9	28.1
Thickness (in.)		0.500	0.438	0.375	0.313	0.250	0.500	0.438	0.375	0.313	0.250
Size (in.)		12.75 OD					10.75 OD				

G40.21 350W
CLASS H
$\phi = 0.90$

ROUND HOLLOW SECTIONS
Factored Axial Compressive Resistances, C_r (kN)

Designation (mm x mm)		HSS 219					
		13	11	9.5	8.0	6.4	4.8
Mass (kg/m)		64.6	57.1	49.3	41.4	33.3	25.3
Effective length (KL) in millimetres with respect to the least radius of gyration	0	2 590	2 290	1 980	1 660	1 340	1 010
	1 000	2 590	2 290	1 970	1 660	1 340	1 010
	1 250	2 590	2 290	1 970	1 660	1 330	1 010
	1 500	2 590	2 290	1 970	1 660	1 330	1 010
	1 750	2 590	2 280	1 970	1 660	1 330	1 010
	2 000	2 580	2 280	1 970	1 650	1 330	1 010
	2 250	2 570	2 270	1 960	1 650	1 330	1 010
	2 500	2 560	2 260	1 950	1 640	1 320	1 000
	2 750	2 540	2 250	1 940	1 630	1 310	997
	3 000	2 520	2 230	1 920	1 620	1 300	990
	3 250	2 490	2 200	1 900	1 600	1 290	980
	3 500	2 450	2 170	1 870	1 580	1 270	967
	3 750	2 410	2 130	1 840	1 550	1 250	951
	4 000	2 350	2 080	1 800	1 520	1 230	933
	4 250	2 290	2 030	1 760	1 480	1 200	912
	4 500	2 220	1 970	1 710	1 440	1 160	887
	4 750	2 150	1 910	1 650	1 400	1 130	861
	5 000	2 070	1 840	1 590	1 350	1 090	832
	5 250	1 980	1 760	1 530	1 300	1 050	801
	5 500	1 900	1 690	1 470	1 240	1 010	769
	5 750	1 810	1 610	1 400	1 190	964	737
	6 000	1 730	1 540	1 340	1 130	920	704
	6 500	1 560	1 390	1 210	1 030	835	639
	7 000	1 400	1 250	1 090	926	754	578
	7 500	1 260	1 120	982	834	680	521
	8 000	1 130	1 010	884	751	613	470
	8 500	1 020	911	797	678	553	425
	9 000	921	824	721	613	500	384
	9 500	834	746	653	556	454	349
	10 000	759	679	594	506	413	318
	10 500	692	619	542	462	377	290
PROPERTIES AND DESIGN DATA							
Area (mm^2)		8 230	7 270	6 270	5 270	4 240	3 220
Z (10^3 mm^3)		542	482	419	355	288	220
S (10^3 mm^3)		402	360	315	269	219	169
r (mm)		73.1	73.6	74.2	74.7	75.3	75.8
M_r (kN·m)		171	152	132	112	90.7	69.3
IMPERIAL SIZE AND MASS							
Mass (lb./ft.)		43.4	38.4	33.1	27.8	22.4	17.0
Thickness (in.)		0.500	0.438	0.375	0.313	0.250	0.188
Size (in.)		8.625 OD					

4-89

ROUND HOLLOW SECTIONS
Factored Axial Compressive Resistances, C_r (kN)

G40.21 350W
CLASS H
$\phi = 0.90$

Designation (mm x mm)		HSS 168				HSS 141			
		9.5	8.0	6.4	4.8	9.5	8.0	6.4	4.8
Mass (kg/m)		37.3	31.4	25.4	19.3	31.0	26.1	21.1	16.1
Effective length (KL) in millimetres with respect to the least radius of gyration	0	1 500	1 260	1 020	775	1 240	1 050	847	646
	1 000	1 500	1 260	1 020	774	1 240	1 050	846	645
	1 250	1 490	1 260	1 020	774	1 240	1 040	844	643
	1 500	1 490	1 250	1 010	772	1 230	1 040	840	640
	1 750	1 480	1 250	1 010	769	1 220	1 030	833	635
	2 000	1 470	1 240	1 000	764	1 200	1 010	821	627
	2 250	1 460	1 230	994	757	1 170	993	804	614
	2 500	1 440	1 210	980	747	1 140	964	782	598
	2 750	1 410	1 190	962	734	1 090	926	753	577
	3 000	1 370	1 160	938	717	1 040	882	718	551
	3 250	1 330	1 120	910	696	979	833	679	522
	3 500	1 270	1 080	877	671	915	779	637	490
	3 750	1 220	1 030	839	643	850	725	594	458
	4 000	1 150	982	798	613	785	671	551	425
	4 250	1 090	928	756	581	724	619	509	393
	4 500	1 030	874	712	548	666	570	470	363
	4 750	961	820	669	515	612	525	433	335
	5 000	898	767	627	483	563	483	399	309
	5 250	839	717	586	452	519	445	368	285
	5 500	782	669	548	423	478	411	339	263
	5 750	730	625	512	395	442	380	314	244
	6 000	681	583	478	369	409	352	291	226
	6 500	594	510	418	323	352	303	251	195
	7 000	520	447	367	284	306	263	218	169
	7 500	459	394	323	250	268	231	191	149
	8 000	406	349	287	222	237	204	169	131
	8 500	362	311	255	198	210	181	150	116
	9 000	324	279	229	177	188	162	134	104
	9 500	292	251	206	160			120	94
	10 000	264	227	187	144				
	10 500	240	206	170	131				
PROPERTIES AND DESIGN DATA									
Area (mm^2)		4 750	4 000	3 230	2 460	3 950	3 330	2 690	2 050
Z (10^3 mm^3)		241	205	167	128	166	142	116	89.1
S (10^3 mm^3)		179	153	126	97.6	122	105	86.9	67.7
r (mm)		56.2	56.8	57.3	57.8	46.7	47.2	47.8	48.3
M_r (kN·m)		75.9	64.6	52.6	40.3	52.3	44.7	36.5	28.1
IMPERIAL SIZE AND MASS									
Mass (lb./ft.)		25.1	21.1	17.0	13.0	20.8	17.6	14.2	10.8
Thickness (in.)		0.375	0.313	0.250	0.188	0.375	0.313	0.250	0.188
Size (in.)		6.625 OD				5.563 OD			

G40.21 350W CLASS H
$\phi = 0.90$

ROUND HOLLOW SECTIONS
Factored Axial Compressive Resistances, C_r (kN)

Designation (mm x mm)		HSS 114			HSS 102			
		8.0	6.4	4.8	8.0	6.4	4.8	3.8
Mass (kg/m)		20.9	16.9	12.9	18.4	14.9	11.4	9.19
Effective length (KL) in millimetres with respect to the least radius of gyration	0	838	677	517	737	599	457	369
	1 000	834	675	515	732	594	454	366
	1 250	828	670	511	723	588	449	363
	1 500	817	661	505	706	575	440	355
	1 750	798	647	495	678	554	425	344
	2 000	769	625	479	640	525	403	327
	2 250	731	595	457	592	488	376	305
	2 500	685	559	431	539	446	345	281
	2 750	632	517	400	485	403	312	255
	3 000	578	474	368	433	360	281	229
	3 250	525	431	335	385	321	251	205
	3 500	474	390	304	342	286	224	183
	3 750	427	352	275	305	255	200	164
	4 000	385	318	249	272	228	179	146
	4 250	348	287	225	244	205	160	131
	4 500	315	260	204	219	184	144	118
	4 750	285	236	185	198	166	131	107
	5 000	260	215	169	180	151	119	97
	5 250	237	196	154	164	138	108	89
	5 500	217	180	141	150	126	99	81
	5 750	199	165	130	137	115	91	74
	6 000	184	152	120	126	106	83	68
	6 500	157	130	103	108	91	71	58
	7 000	136	113	89				
	7 500	119	98	77				
	8 000							
	8 500							
	9 000							
	9 500							
	10 000							
	10 500							
PROPERTIES AND DESIGN DATA								
Area (mm^2)		2 660	2 150	1 640	2 340	1 900	1 450	1 170
Z (10^3 mm^3)		90.1	74.1	57.4	69.9	57.7	44.8	36.5
S (10^3 mm^3)		66.1	55.1	43.2	50.8	42.6	33.6	27.6
r (mm)		37.7	38.2	38.8	33.2	33.8	34.3	34.6
M_r (kN·m)		28.4	23.3	18.1	22.0	18.2	14.1	11.5
IMPERIAL SIZE AND MASS								
Mass (lb./ft.)		14.0	11.4	8.68	12.3	10.0	7.67	6.17
Thickness (in.)		0.313	0.250	0.188	0.313	0.250	0.188	0.150
Size (in.)		4.5 OD			4 OD			

ROUND HOLLOW SECTIONS
Factored Axial Compressive Resistances, C_r (kN)

G40.21 350W
CLASS H
$\phi = 0.90$

Designation (mm x mm)		HSS 89				HSS 73			
		8.0	6.4	4.8	3.8	6.4	4.8	3.8	3.2
Mass (kg/m)		15.9	12.9	9.92	8.00	10.4	8.04	6.50	5.48
Effective length (KL) in millimetres with respect to the least radius of gyration	0	636	520	397	321	419	321	261	220
	1 000	628	513	392	318	406	312	254	214
	1 250	613	502	385	312	385	298	243	205
	1 500	588	483	371	301	353	274	224	190
	1 750	550	453	349	284	313	244	201	170
	2 000	502	415	321	262	270	212	175	149
	2 250	448	373	290	237	229	181	150	128
	2 500	395	330	257	211	194	154	128	109
	2 750	345	289	226	186	165	132	109	93
	3 000	301	253	199	163	142	113	94	80
	3 250	264	222	174	143	122	97	81	69
	3 500	231	195	153	126	106	85	70	60
	3 750	204	172	136	112	93	74	62	53
	4 000	181	153	120	99	82	65	54	47
	4 250	161	136	107	89	73	58	48	41
	4 500	145	122	96	79	65	52	43	37
	4 750	130	110	87	72		47	39	33
	5 000	118	100	79	65				
	5 250	107	90	71	59				
	5 500	98	83	65	54				
	5 750	90	76	60	49				
	6 000				45				
	6 500								
	7 000								
	7 500								
	8 000								
	8 500								
	9 000								
	9 500								
	10 000								
	10 500								
PROPERTIES AND DESIGN DATA									
Area (mm^2)		2 020	1 650	1 260	1 020	1 330	1 020	828	698
Z (10^3 mm^3)		52.3	43.4	33.9	27.6	28.3	22.3	18.3	15.5
S (10^3 mm^3)		37.6	31.7	25.2	20.8	20.4	16.4	13.6	11.7
r (mm)		28.8	29.3	29.8	30.1	23.7	24.2	24.5	24.7
M$_r$ (kN·m)		16.5	13.7	10.7	8.69	8.91	7.02	5.76	4.88
IMPERIAL SIZE AND MASS									
Mass (lb./ft.)		10.7	8.69	6.66	5.37	7.01	5.40	4.37	3.68
Thickness (in.)		0.313	0.250	0.188	0.150	0.250	0.188	0.150	0.125
Size (in.)		3.5 OD				2.875 OD			

G40.21 350W
CLASS H
$\phi = 0.90$

ROUND HOLLOW SECTIONS
Factored Axial Compressive Resistances, C_r (kN)

	Designation (mm × mm)	HSS 60				HSS 48		
		6.4	4.8	3.8	3.2	4.8	3.8	3.2
	Mass (kg/m)	8.45	6.54	5.31	4.48	5.13	4.18	3.54
Effective length (KL) in millimetres with respect to the least radius of gyration	0	340	263	213	180	206	168	142
	1 000	314	245	199	169	172	142	121
	1 250	282	221	181	154	140	117	100
	1 500	240	190	156	133	110	92	79
	1 750	197	158	131	112	85	72	62
	2 000	161	130	108	92	67	57	49
	2 250	132	107	89	76	54	46	39
	2 500	109	88	74	63	44	37	32
	2 750	91	74	62	53	37	31	27
	3 000	77	63	52	45	31	26	23
	3 250	66	54	45	39			
	3 500	57	47	39	33			
	3 750	50	41	34	29			
	4 000			30	26			
	4 250							
	4 500							
	4 750							
	5 000							
	5 250							
	5 500							
	5 750							
	6 000							
	6 500							
	7 000							
	7 500							
	8 000							
	8 500							
	9 000							
	9 500							
	10 000							
	10 500							
PROPERTIES AND DESIGN DATA								
	Area (mm^2)	1 080	834	676	571	654	533	451
	Z (10^3 mm^3)	18.6	14.8	12.2	10.4	9.09	7.56	6.48
	S (10^3 mm^3)	13.2	10.7	8.99	7.74	6.48	5.50	4.77
	r (mm)	19.2	19.7	20.0	20.2	15.5	15.8	16.0
	M$_r$ (kN·m)	5.86	4.66	3.84	3.28	2.86	2.38	2.04
IMPERIAL SIZE AND MASS								
	Mass (lb./ft.)	5.68	4.40	3.57	3.01	3.45	2.81	2.38
	Thickness (in.)	0.250	0.188	0.150	0.125	0.188	0.150	0.125
	Size (in.)	2.375 OD				1.9 OD		

DESIGN OF BEAM-COLUMNS

General

For members which are subjected to both axial compression and bending (usually referred to as beam-columns), Clauses 8.6 and 13.8 of CAN/CSA-S16.1-94 *Limit States Design of Steel Structures* are applicable, along with Appendix F of that Standard.

Clause 13.8.1 applies to all sections except Class 1 of I-shaped members; and Clause 13.8.2 applies only to the Class 1 sections of I-shaped members. If a more detailed design of W or HSS shaped Class 1 or 2 sections is required than that given in Clause 13.8, Appendix F is available. The appendix takes advantage of the redistribution of stress after initiation of yielding under specified loads.

In each of Clauses 13.8.1 and 13.8.2, a single comprehensive axial-flexural interaction expression applies for checking cross-sectional strength, overall member strength and lateral torsional buckling strength of the member. The first and last of these three checks correspond to those in S16.1-M84 for strength and stability, while the check for overall member strength applies to members which have sufficient lateral bracing to be torsionally stable, or which are inherently stable like HSS. The three checks involve the use of different values for the variables in the expressions, as outlined below.

For all sections except Class 1 of I-shaped members, the following design expression applies (Clause 13.8.1):

$$\frac{C_f}{C_r} + \frac{U_{1x} M_{fx}}{M_{rx}} + \frac{U_{1y} M_{fy}}{M_{ry}} \leq 1.0 \qquad [1]$$

where

M_{fx}, M_{fy} = maximum factored moments as defined in Clause 8.6, *Stability Effects*, to include frame sway effects

C_f = factored axial load

When checking cross-sectional strength:

C_r = factored axial compressive resistance of the member as defined in Clause 13.3, *Axial Compression*, with $\lambda = 0$

M_{rx}, M_{ry} = factored moment resistances of the member as defined in Clause 13.5, *Bending — Laterally Supported Members*, (for the appropriate class of section)

U_{1x}, U_{1y} = 1.0

When checking overall member strength:

C_r = factored axial compressive resistance of the member as defined in Clause 13.3, *Axial Compression*, with $K = 1.0$, based on the greater slenderness ratio, L_x/r_x or L_y/r_y. For uniaxial strong axis bending, C_{rx} may be used

M_{rx}, M_{ry} = factored moment resistances of the member as defined in Clause 13.5, *Bending — Laterally Supported Members*, (for the appropriate class of section)

$$U_1 = \frac{\omega_1}{1 - \left(\frac{C_f}{C_e}\right)} \qquad [2]$$

and is a term that includes factors to account for 1) the effects of a varying moment along the member, and 2) the $P\delta$ moment applification, which accounts for the effects of member curvature

ω_1 = coefficient used to determine the bending effect which is equivalent to that derived from an ideal uniform bending moment that approximates the effect due to the actual design moments

C_e = Euler buckling load equal to $\pi^2 EI/L^2$ where L is the column length in the plane of buckling

When checking lateral torsional buckling strength:

C_r = factored axial compressive resistance of the member as defined in Clause 13.3, *Axial Compression*, based on weak-axis buckling

M_{rx} = factored moment resistance of the member as defined in Clause 13.6, *Bending — Laterally Unsupported Members*, (for the appropriate class of section)

M_{ry} = factored moment resistance of the member as defined in Clause 13.5, *Bending — Laterally Supported Members*, (for the appropriate class of section)

U_{1x} = as defined for the overall member strength check, but not less than 1.0

U_{1y} = as defined for the overall member strength check

For Class 1 sections of I-shaped members, the following design expression applies (Clause 13.8.2):

$$\frac{C_f}{C_r} + \frac{0.85\, U_{1x}\, M_{fx}}{M_{rx}} + \frac{0.60\, U_{1y}\, M_{fy}}{M_{ry}} \leq 1.0 \qquad [3]$$

in which the coefficients 0.85 and 0.60 reflect the increased bending resistance of Class 1 sections. Other terms are the same as for [1].

The member must also satisfy the condition

$$\frac{M_{fx}}{M_{rx}} + \frac{M_{fy}}{M_{ry}} \leq 1.0 \qquad [4]$$

The design of a beam-column may be controlled by any of the three failure modes. Often the requirement guarding against one failure mode governs for one loading condition, while the requirement against another mode governs for a different loading condition.

The value of ω_1, (ω_{1x} or ω_{1y}), depends upon the direction and magnitude of the end moments, and any presence of transverse loads. The value of ω_1 may be determined by analysis, or the following values may be used:

(a) for members not subjected to transverse loads between supports,

$\omega_1 = 0.6 - 0.4\kappa$ but not less than 0.4 (see Table 4-7, page 4-26)

where

κ = M_{f1}/M_{f2} for moments at opposite ends of the unbraced length, positive for double curvature and negative for single curvature, in which

M_{f1} = the smaller factored end moment, and

M_{f2} = the larger factored end moment.

(b) for members subjected to distributed loads or a series of point loads between supports,

$\omega_1 = 1.0$

(c) for members subjected to a concentrated load or moment between supports,

$\omega_1 = 0.85$

The member may be considered to be divided into two segments at the point of load (or moment) application. Each segment is then treated as a member which depends on its own flexural stiffness to prevent side-sway in the plane of bending considered. In computing the slenderness ratio KL/r for use in [1] or [3], the total length of the member is used.

It is important to remember that the factored axial compressive resistance C_r in [1] and [3] is determined as though only compressive loads were present. Therefore, it is always the value associated with the maximum effective slenderness ratio, regardless of the axis about which bending occurs, while C_e is always computed using the slenderness ratio in the plane of bending.

References

Additional information on beam-columns can be obtained from:

PART TWO of this Handbook.

KENNEDY, D.J.L., PICARD, A., and BEAULIEU, D. 1990. New Canadian provisions for the design of steel beam-columns. Canadian Journal of Civil Engineering, **17**(6), December.

KULAK, G.L., ADAMS, P.F., and GILMOR, M.I. 1995. Limit states design in structural steel. Canadian Institute of Steel Construction, Willowdale, Ontario.

PICARD, A., and BEAULIEU, D. 1991. Calcul des charpentes d'acier (2nd. edition). Canadian Institute of Steel Construction, Willowdale, Ontario.

SSRC. 1988. Guide to stability design criteria for metal structures (4th. edition). Structural Stability Research Council, John Wiley, Toronto, Ontario.

Material provided in this Handbook should help the designer minimize design time when a computer is not available.

Design

The selection of a suitable steel section to use as a beam-column is a trial-and-error procedure. A trial section must be selected and then its suitability checked in accordance with the appropriate design expressions [1], [3] or [4]. The procedure is then repeated if the initial choice is not suitable.

The following terms are a result of the analysis for forces and moments:

C_f = factored axial load

M_{fx}, M_{fy} = factored bending moments

The following terms can be determined from tables in this Handbook:

C_r = factored axial compressive resistance, ($\leq \phi A F_y$), or

= value in tables of Factored Axial Compressive Resistances of columns for the appropriate effective length, page 4-32*

M_{rx} = factored moment resistance for X-X axis bending

= tabulated value with properties in tables of Factored Axial Compressive

M_{ry} ⎵ Resistances of columns when the unsupported length of compression flange $L \leq L_u$, page 4-32*, or

= tabulated in Factored Moment Resistances for Columns table under the appropriate heading for length of compression flange which is unsupported, or may be interpolated from these values, page 4-105*

M_{ry} = factored moment resistance for Y-Y axis bending

= tabulated value with properties in the Column tables, page 4-32*

ω_{1x}, ω_{1y} = equivalent uniform bending factors to account for moment gradient

= tabulated in Table 4-7, page 4-26

$U = \dfrac{1}{1 - \left(\dfrac{C_f}{C_e}\right)}$, an amplification factor to account for secondary effects of axial force acting on the deformed member, $(U_1 = \omega_1 U)$

= tabulated in Table 4-9, page 4-28, values for various C_f/C_e ratios

C_e = Euler buckling load

= tabulated as C_e/A for KL/r ratios from 1 to 200 in Table 4-8, page 4-27

* First page of the series of like tables

Trial Selection by Equivalent-Loads Method

Any convenient method can be used to choose a trial section. One method is that of equivalent-loads. Basically, this method involves converting applied loads and bending moments into equivalent total axial loads computed from the "cross-sectional strength" requirement and from the "lateral torsional buckling strength" requirement. The trial section is chosen for the larger equivalent load from the Factored Axial Compressive Resistances column tables (pages 4-32 to 4-85), and is then checked using the appropriate expression [1] or [3]. This method is applicable only to prismatic members governed by [1] or [3].

The "overall member strength" requirement need not be used when making trial selections for I-shaped members because it seldom governs for those members; but that check is still made when confirming the section.

Beam-Columns Without Transverse Loads Between Brace Points

Cross-Sectional Strength Requirement

$$C_f' = C_f + B\, M_{f2} \qquad [5]$$

where

C_f' = equivalent total factored axial load for cross-sectional strength, (kN)

C_f = applied factored axial load, (kN)

B = appropriate bending factor (m^{-1}) chosen from Table 4-6 (page 4-24)

M_{f2} = larger applied factored end moment, (kN·m)

Lateral Torsional Buckling Strength Requirement

$$C_f'' = C_f + (\omega_1 U) M_{f2} \left(B \frac{C_{rL}}{C_{r0}} \right) \qquad [6]$$

where

C_f'' = equivalent total factored axial load for lateral torsional buckling, (kN)

ω_1 = equivalent uniform bending factor

C_{rL} = factored axial compressive resistance at the actual effective length for the trial section chosen, (kN)

C_{r0} = factored axial compressive resistance at an effective length of zero millimetres for the trial section chosen, (kN)

$$U = \frac{1}{1 - \left(\dfrac{C_f}{C_e}\right)}$$

an amplification factor listed in Table 4-9 for various combinations of C_f/C_e in the plane of bending, where $\omega_1 U$ must be ≥ 1.0

C_f, M_{f2} and B are the same as for [5].

Multiplying the larger end moment M_{f2} with the value of ω_1 produces a uniform moment for the unbraced segment, which would have the same effect on the beam-column as the applied end moments so far as lateral torsional buckling strength is concerned.

The expression C_{rL}/C_{r0} in [6] modifies the factor B to account for the difference between the capacity of the beam-column along the unbraced length and that at a brace location. If [6] is used with moments at the brace location, then $C_{rL} = C_{r0}$, and the ratio C_{rL}/C_{r0} can be deleted.

The other refinement which [6] provides (as compared to [5]) is the introduction of the amplification factor U. This factor takes into account the increased second-order moment produced by the applied load as the curvature of the column under load increases. Accordingly, U increases rapidly as the length of the column (and hence the effective slenderness ratio in the plane of bending increases) and/or as the applied axial load increases.

When bending about two axes is involved, the second term of [5] and [6] should be expanded to two terms using variables appropriate to each axis.

Although [6] looks cumbersome, it is easily used since all the unknowns in the equation can be picked directly from tables.

However, [6] can be simplified with a possible reduction in accuracy, by assuming that the ratio $C_{rL}/C_{r0} = 0.70$. This is an average value for 350 MPa yield steel columns with KL/r from 50 to 80.

Beam-Columns With Transverse Loads Between Brace Points

For beam-columns subject to transverse loading between the brace locations, [5] satisfies the cross-sectional strength requirement if the larger end moment is greater than the maximum bending moment along the unbraced length.

The lateral torsional buckling strength requirement is satisfied by a modified form of [6] identified as [6a]. If the maximum bending moment along the unbraced length is greater than the larger end moment, [6a] will satisfy both the cross-sectional strength and the lateral torsional buckling strength requirements for the beam-column.

$$C_f'' = C_f + U M_{f2}' \left(B \frac{C_{rL}}{C_{r0}} \right)$$ [6a]

M_{f2}' = the greater of the maximum bending moment along the unbraced length, or the larger end moment.

The other symbols are the same as for [6].

Suggested Design Procedure

A. *Trial Section Selection*

1. Choose a nominal size group (WWF500, W310, etc.) from which the trial section will be chosen and select the appropriate bending factor or factors from Table 4-6.

2. Use [5] to compute equivalent axial load for cross-sectional strength C_f'.

3. Use C_f' to select an initial trial section from the tables of Factored Axial Compressive Resistances of columns. C_f' must not be greater than C_r in the tables at zero effective length.

4. Use [6] to compute equivalent axial load for lateral torsional buckling strength, C_f''. The value of ω_1 can be chosen from Table 4-7, and the value of B will be the same value used to establish C_f'. The values of C_{rL} and C_{r0} will be those listed in the table of Factored Axial Compressive Resistances of columns for the initial trial section chosen in step 3. The value of U will be selected from Table 4-9, using the properties of the initial trial section chosen in step 3; the product of $\omega_1 U$ must not be less than 1.0. Alternatively, an assumed value of C_{rL}/C_{r0} can be chosen such as 0.70. If the column is subject to transverse loads, use [6a] for this step.

5. Use the Factored Axial Compressive Resistance tables to verify that the initial trial section is adequate for C_f'' at the appropriate effective length, or choose a new trial section to satisfy C_f'' at the appropriate KL.

B. *Suitability of the Trial Section*

1. Use the properties and resistances of the trial section to check its suitability in accordance with the requirements of Clause 13.8 of S16.1-94. Since most of the values used in these equations are tabulated this should be a quick computation.

2. If the trial section is unsuitable, choose a new trial section (which should be the next lighter or heavier section listed) and verify suitability.

Examples

1. Given:

Design a steel column for the third storey of a six storey building for the loading conditions shown. Moments are due to rigidly framed beams and gravity loading, and cause bending about the X-X axis of the column. The $P\Delta$ effects have been included in the analysis. Beams framing to the minor axis have flexible connections. Steel is G40.21 350W.

Solution:

$L = 3\,700$ mm $M_{f1} = 200$ kN·m

$C_f = 2\,000$ kN $M_{f2} = 300$ kN·m

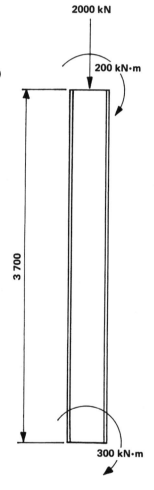

A. Trial Section Selection

Assume W310, Class 1 section

From Table 4-6 (page 4-24), for a Class 1 W shape, 310x310

$B_x = 6.34$ m^{-1}

$C_f' = C_f + B_x M_{f2}$ [5]

$= 2\,000 + 6.34\,(300) = 3\,900$ kN

Select W310x118 from table on page 4-40

C_{r0} (for $KL = 0$) $= 4\,730$ kN, comfortably $> 3\,900$ kN

$A = 15\,000$ mm^2, $r_x = 136$ mm, $r_y = 77.6$ mm, $r_x/r_y = 1.75$

$M_{rx} = 614$ kN·m, $L_u = 4\,880$ mm

$C_{rL} = 3\,890$ kN   — by interpolation ($KL = 3\,700$ mm)

$C_f'' = C_f + (\omega_1 U) M_{f2} \left(B_x \dfrac{C_{rL}}{C_{r0}} \right)$ [6]

$M_{f1}/M_{f2} = 200/300 = 0.67$ (double curvature)

From Table 4-7 (page 4-26), $\omega_1 = 0.4$

$KL_x/r_x = 1.0\,(3\,700/136) = 27.2$

From Table 4-8 (page 4-27), $C_e/A = 2\,664$ MPa

Then, $C_e = 2\,664$ MPa $\times\, 15\,000$ mm$^2 = 40\,000$ kN

and $C_f/C_e = 2\,000/40\,000 = 0.05$

From Table 4-9 (page 4-28), $U = 1.05$

Where $\omega_1 U < 1.0$, use $\omega_1 U = 1.0$

From Table 4-6 (page 4-24), $B_x = 6.34$ m^{-1} (as for C_f')

$C_f'' = 2\,000 + (1.0)\,300\,(6.34 \times 3\,890/4\,730)$

$= 2\,000 + 1\,560$

$= 3\,560$ kN $< 3\,800$ — OK   Use W310x118 as trial section.

4-100

B. *Suitability of the Trial Section*

Confirm class of section as a beam-column.

Although Table 5-1, Classes of Sections in Bending (page 5-7), lists the W310x118 as Class 2, the addition of axial load might change that class (as per "Webs in axial compression" line, Table 1 in Clause 11.2 of S16.1-94). However, an examination of Table 4-3, Class of Section for Beam-Columns (page 4-11), shows that the W310x118 is always a Class 2 section, and [1] applies:

$$\frac{C_f}{C_r} + \frac{U_{1x}\, M_{fx}}{M_{rx}} \leq 1.0$$

i) *Cross-sectional strength check*

$C_f = 2\,000$ kN

$M_{fx} = 300$ kN·m

From Table of Factored Axial Compressive Resistances, page 4-40,

$C_r = C_{r0} = 4\,730$ kN, and $M_{rx} = 614$ kN.m

U_1 is taken as 1.0 for cross-sectional strength check

Therefore,

$$\frac{2\,000}{4\,730} + \frac{1.0 \times 300}{614} = 0.423 + 0.489 = 0.912 \quad \leq 1.0 \quad - \text{OK}$$

ii) *Overall member strength check*

$KL_x/r_x = 1.0\ (3\,700/136) = 27.2$

From Table 4-4 (page 4-13), $C_{rx}/A = 301$ MPa, for $F_y = 350$ MPa

$C_r = C_{rx} = 301 \times 15\,000/10^3 = 4\,520$ kN

$U_1 = \omega_1 U = 0.4 \times 1.05 = 0.42$

Therefore,

$$\frac{2\,000}{4\,520} + \frac{0.42 \times 300}{614} = 0.442 + 0.205 = 0.647 \quad \leq 1.0 \quad - \text{OK}$$

iii) *Lateral torsional buckling strength check*

$C_r = C_{ry} = C_{rL} = 3\,890$ kN   (by interpolation, table on page 4-40)

$L = 3\,700$ mm $L_u = 4\,880$ mm

Therefore, $M_{rx} = 614$ kN·m

$U_1 = \omega_1 U$ (but ≥ 1.0) $= 0.4 \times 1.05$ $\therefore U_1 = 1.0$

Therefore,

$$\frac{2\,000}{3\,890} + \frac{1.0 \times 300}{614} = 0.514 + 0.488 = 1.00 \ = 1.0 \quad - \text{OK}$$

Use the W310x118 column section.

Comments:

1. C_r could more accurately be determined by computing the KL/r values and entering the tables of Unit Factored Compressive Resistances for the larger KL/r and multiplying that value by the area of the column.

2. When $L > L_u$ the tables on pages 4-106 to 4-108, Factored Moment Resistances of

Columns, will be more useful. (Caution: if a column section changes from Class 2 to Class 3 on account of high axial loads, M_{rx} or M'_{rx} values need to be adjusted as noted on page 4-105.)

2. **Given:**

Same as example 1, except that beams framing to Y-Y axis have rigid connections and induce moments of 100 kN·m at each end of the column. The direction of the moments is such that double curvature is induced in the column.

Solution:

$L = 3\,700$ mm $M_{fx1} = 200$ kN·m $M_{fy1} = 100$ kN·m

$C_f = 2\,000$ kN $M_{fx2} = 300$ kN·m $M_{fy2} = 100$ kN·m

A. *Trial Section Selection*

Assume W310 section, Class 1

From Table 4-6 (page 4-24), for a Class 1 W shape, 310x310

$B_x = 6.34$ and $B_y = 9.78$

$C'_f = C_f + B_x M_{fx2} + B_y M_{fy2}$ [5]

$= 2\,000 + (6.34 \times 300) + (9.78 \times 100)$

$= 4\,880$ kN

Select W310x129 from table on page 4-40, a Class 1 section in G40.21 350W.

$C_{r0} = 5\,200$ kN $> 4\,880$ — OK

$A = 16\,500$ mm^2 $M_{rx} = 680$ kN·m

$r_x = 137$ mm $M_{ry} = 312$ kN·m

$r_y = 78.0$ mm

$C_{rL} = 4\,300$ kN   — by interpolation ($KL = 3\,700$ mm)

$C''_f = C_f + (\omega_{1x}\, U_x)\, M_{fx2} \left(B_x \dfrac{C_{rLx}}{C_{r0}} \right) + (\omega_{1y}\, U_y)\, M_{fy2} \left(B_y \dfrac{C_{rLy}}{C_{r0}} \right)$ [6]

$M_{fx1}/M_{fx2} = 200/300 = 0.67$ (double curvature)

$M_{fy1}/M_{fy2} = 100/100 = 1.00$ (double curvature)

From Table 4-7 (page 4-26), $\omega_{1x} = 0.40$, $\omega_{1y} = 0.40$

$KL_x/r_x = 1.0\,(3\,700/137) = 27.0$

$KL_y/r_y = 1.0\,(3\,700/78.0) = 47.4$

From Table 4-8 (page 4-27), $(C_e/A)_x = 2\,702$ MPa ($KL/r = 27$)

$(C_e/A)_y = 877$ MPa ($KL/r = 47.4$)

$C_{ex} = 2\,702$ MPa $\times\, 16\,500$ mm^2 $= 44\,600 \times 10^3$ N

$= 44\,600$ kN

$C_{ey} = 877$ MPa $\times\, 16\,500$ mm^2 $= 14\,500 \times 10^3$ N

$= 14\,500$ kN

$C_f/C_{ex} = 2\,000/44\,600 = 0.04$

$C_f/C_{ey} = 2\,000/14\,500 = 0.14$

From Table 4-9 (page 4-28), $U_x = 1.04$, $U_y = 1.16$

 Since $\omega_{1x} U_x = U_{1x} < 1.0$, use 1.0

but $\omega_{1y} U_y = 0.4 \times 1.16 = 0.464$

From Table 4-6, $B_x = 6.34 \text{ m}^{-1}$ and $B_y = 9.78 \text{ m}^{-1}$ (as for C_f')

$C_f'' = 2\,000 + (1.0)\,300\,(6.34 \times 4\,300 / 5\,200)$

$\quad + (0.464)\,100\,(9.78 \times 4\,300 / 5\,200)$

$= 3\,950 \text{ kN}$

Since C_r for $L = 3\,700$ mm is $4\,300$ kN $> 3\,950$ kN, the W310x129 satisfies the lateral torsional buckling strength requirement as well as the cross-sectional strength requirement.

Use W310x129 as a trial section.

B. *Suitability of the Trial Section*

This section is a heavier one of the same series as the W310x118 used in Example 1, but is a Class 1 section; therefore, [3] applies:

$$\frac{C_f}{C_r} + \frac{0.85\,U_{1x}\,M_{fx}}{M_{rx}} + \frac{0.60\,U_{1y}\,M_{fy}}{M_{ry}} \le 1.0$$

i) *Cross-sectional strength check*

 $C_f = 2\,000$ kN

 $M_{fx} = 300$ kN·m

 $M_{fy} = 100$ kN·m

 From Table of Factored Axial Compressive Resistances, page 4-40,

 $C_r = C_{r0} = 5\,200$ kN, $M_{rx} = 680$ kN·m and $M_{ry} = 312$ kN·m

 U_{1x} and U_{1y} are taken as 1.0 for cross-sectional strength check

 Therefore,

 $$\frac{2\,000}{5\,200} + \frac{0.85 \times 1.0 \times 300}{680} + \frac{0.60 \times 1.0 \times 100}{312}$$

 $= 0.385 + 0.375 + 0.192 = 0.952 \quad \le 1.0 \quad -\text{OK}$

ii) *Overall member strength check*

 $KL_x/r_x = 1.0\,(3\,700/137) = 27.0$

 $KL_y/r_y = 1.0\,(3\,700/78.0) = 47.4$

 From Table 4-4 (page 4-13), $C_r/A = 260$ MPa (use $KL/r = 47.4$)

 $C_r = 260 \times 16\,500 / 10^3 = 4\,290$ kN

 $U_{1x} = \omega_{1x} U_x = 0.4 \times 1.04 = 0.42$

$U_{1y} = \omega_{1y} U_y = 0.4 \times 1.16 = 0.46$

Therefore,

$$\frac{2\,000}{4\,290} + \frac{0.85 \times 0.42 \times 300}{680} + \frac{0.60 \times 0.46 \times 100}{312}$$

$= 0.466 + 0.158 + 0.088 = 0.712 \quad \leq 1.0 \quad -\text{OK}$

iii) *Lateral torsional buckling check*

$C_r = C_{ry} = C_{rL} = 4\,290$ kN

$L = 3\,700$ mm $\quad < L_u = 5\,030$ mm

Therefore, $M_{rx} = 680$ kN·m

$U_{1x} = \omega_{1x} U_x \quad (\text{but} \geq 1.0) \quad = 0.4 \times 1.04 \quad \therefore U_{1x} = 1.0$

$U_{1y} = \omega_{1y} U_y = 0.4 \times 1.16 = 0.46$

Therefore,

$$\frac{2\,000}{4\,290} + \frac{0.85 \times 1.0 \times 300}{680} + \frac{0.60 \times 0.46 \times 100}{312}$$

$= 0.466 + 0.375 + 0.088 = 0.929 \quad \leq 1.0 \quad -\text{OK}$

Use the W310x129 column section.

Shear

Where beams with large end moments are connected to columns with thin webs, a check for shear capacity in the column web will be necessary.

FACTORED MOMENT RESISTANCES OF COLUMNS

The tables on pages 4-106 to 4-108 inclusive list: 1) the factored moment resistance for strong axis bending M_{rx} for cases where the unsupported length of compression flange L is less than L_u, and 2) the factored moment resistance M'_{rx} where L is greater than L_u. The listings in G40.21 steel are for WWF and W shapes in Grade 350W normally used as columns. Sections are ordered as in Part Six of this Handbook, with all of the sections of the same nominal dimensions listed together.

The M_{rx} and M'_{rx} values are based on the class of the section in bending about the X-X axis, without axial load. However, the class of a section used as a beam-column is a function of the ratio of the factored axial load to the axial compressive load at yield stress C_f/C_y in accordance with CAN/CSA-S16.1-94 Clause 11, width-thickness criteria. For example, a W410x39 of G40.21 Grade 350W steel becomes a Class 3 section when C_f/C_y exceeds 0.564, based upon the S16.1-94 Class 2 limit for h/w of

$$\frac{1700}{\sqrt{F_y}}\left(1 - 0.61\frac{C_f}{C_y}\right)$$

Table 4-3 on page 4-7 lists the classes for sections when they are used as beam-columns, and the values of C_f/C_y at which the class changes.

Thus, sections whose loading causes a change from Class 2 to Class 3 need to have their tabulated values of M_{rx} and M'_{rx} adjusted. A conservative method is to multiply the listed values by the factor S_x/Z_x.

FACTORED MOMENT RESISTANCES OF COLUMNS, M_{rx} and M'_{rx} (kN·m)

G40.21 350W

$\phi = 0.90$

Section	M_{rx}	\multicolumn{10}{c}{M'_{rx} for the following unsupported lengths in millimetres}
		8 000	9 000	10 000	11 000	12 000	13 000	14 000	16 000	18 000	20 000
WWF650x864	8 600	—	—	—	—	—	—	8 470	8 240	8 000	7 760
WWF650x739	7 940	—	—	—	—	—	7 900	7 770	7 540	7 300	7 060
WWF650x598	6 650	—	—	—	—	6 590	6 460	6 340	6 090	5 850	5 600
WWF650x499	5 510	—	—	—	5 450	5 320	5 200	5 070	4 810	4 540	4 280
** WWF650x400	3 940	—	—	—	3 870	3 770	3 660	3 560	3 340	3 110	2 880
WWF600x793	7 210	—	—	—	—	—	7 190	7 090	6 890	6 700	6 500
WWF600x680	6 680	—	—	—	—	—	6 610	6 510	6 320	6 120	5 930
WWF600x551	5 610	—	—	—	—	5 510	5 410	5 300	5 100	4 900	4 690
WWF600x460	4 660	—	—	—	4 550	4 440	4 330	4 220	4 000	3 780	3 560
* WWF600x369	3 370	—	—	3 340	3 250	3 150	3 060	2 960	2 770	2 570	2 370
WWF550x721	5 990	—	—	—	—	—	5 950	5 870	5 710	5 550	5 390
WWF550x620	5 540	—	—	—	—	—	5 460	5 380	5 220	5 070	4 910
WWF550x503	4 660	—	—	—	4 630	4 550	4 460	4 380	4 210	4 050	3 880
WWF550x420	3 910	—	—	3 860	3 760	3 670	3 580	3 490	3 300	3 120	2 940
** WWF550x280	2 310	—	2 300	2 230	2 150	2 080	2 000	1 920	1 770	1 610	1 420
WWF500x651	4 850	—	—	—	—	—	4 820	4 760	4 630	4 510	4 380
WWF500x561	4 500	—	—	—	—	—	4 430	4 370	4 240	4 120	3 990
WWF500x456	3 810	—	—	—	3 760	3 690	3 630	3 560	3 430	3 300	3 160
WWF500x381	3 180	—	—	3 100	3 030	2 960	2 880	2 810	2 670	2 520	2 380
WWF500x343	2 870	—	2 820	2 750	2 670	2 590	2 510	2 440	2 280	2 130	1 980
WWF500x306	2 540	2 530	2 460	2 380	2 300	2 220	2 140	2 060	1 900	1 740	1 550
WWF500x276	2 340	2 330	2 250	2 170	2 100	2 020	1 940	1 860	1 690	1 520	1 340
* WWF500x254	1 940	—	1 880	1 820	1 750	1 680	1 620	1 550	1 410	1 270	1 110
** WWF500x223	1 630	—	1 580	1 530	1 470	1 410	1 350	1 290	1 160	1 030	892
** WWF500x197	1 360	—	1 330	1 290	1 240	1 190	1 140	1 090	985	869	752
WWF450x503	3 590	—	—	—	—	3 580	3 530	3 480	3 390	3 290	3 190
WWF450x409	3 040	—	—	—	2 990	2 940	2 890	2 830	2 730	2 630	2 530
WWF450x342	2 550	—	2 520	2 460	2 410	2 350	2 290	2 230	2 120	2 010	1 900
WWF450x308	2 300	2 290	2 230	2 170	2 110	2 050	1 990	1 930	1 810	1 690	1 570
WWF450x274	2 040	2 000	1 940	1 870	1 810	1 740	1 680	1 620	1 490	1 360	1 210
WWF450x248	1 880	1 830	1 770	1 710	1 640	1 580	1 510	1 450	1 320	1 180	1 040
WWF450x228	1 720	1 660	1 590	1 530	1 460	1 400	1 330	1 270	1 130	976	861
* WWF450x201	1 390	1 350	1 300	1 240	1 190	1 130	1 080	1 020	905	779	684
** WWF450x177	1 200	1 170	1 120	1 070	1 030	976	926	875	763	653	570

Note: Moment resistances are based on class of section for X-X axis bending only, $\omega_2 = 1.0$.

* Class 3 section

** Class 4 section; M_{rx} and M'_{rx} calculated according to S16.1-94 Clause 13.5(c)(iii).

G40.21 350W

$\phi = 0.90$

FACTORED MOMENT RESISTANCES OF COLUMNS, M_{rx} and M'_{rx} (kN·m)

Section	M_{rx}	M'_{rx} for the following unsupported lengths in millimetres									
		4 000	5 000	6 000	7 000	8 000	9 000	10 000	12 000	14 000	16 000
WWF400x444	2 760	—	—	—	—	—	—	—	—	2 690	2 610
WWF400x362	2 360	—	—	—	—	—	—	2 360	2 280	2 200	2 120
WWF400x303	1 980	—	—	—	—	—	1 940	1 900	1 810	1 730	1 640
WWF400x273	1 790	—	—	—	—	1 770	1 720	1 670	1 580	1 490	1 400
WWF400x243	1 590	—	—	—	1 580	1 540	1 490	1 440	1 340	1 240	1 140
WWF400x220	1 470	—	—	—	1 460	1 410	1 360	1 310	1 210	1 110	1 010
WWF400x202	1 340	—	—	—	1 310	1 260	1 210	1 160	1 060	958	846
* WWF400x178	1 080	—	—	—	1 060	1 020	982	939	854	769	673
* WWF400x157	983	—	—	—	961	921	879	836	749	662	560
WWF350x315	1 760	—	—	—	—	—	—	—	1 700	1 640	1 590
WWF350x263	1 490	—	—	—	—	—	1 460	1 430	1 360	1 300	1 240
WWF350x238	1 350	—	—	—	—	1 320	1 290	1 250	1 180	1 120	1 050
WWF350x212	1 200	—	—	—	1 180	1 140	1 110	1 070	999	928	858
WWF350x192	1 110	—	—	—	1 080	1 040	1 010	971	899	827	756
WWF350x176	1 010	—	—	—	970	932	894	857	782	709	625
WWF350x155	904	—	—	890	852	813	774	734	656	572	490
WWF350x137	816	—	—	798	760	722	683	643	565	475	405
W310x226	1 250	—	—	—	1 240	1 200	1 170	1 140	1 070	1 010	940
W310x202	1 110	—	—	—	1 080	1 050	1 010	980	910	850	780
W310x179	961	—	—	953	920	887	854	822	758	695	629
W310x158	841	—	—	822	790	757	725	693	630	568	494
W310x143	762	—	—	737	705	673	641	609	546	476	411
W310x129	680	—	—	650	618	587	555	524	462	390	336
W310x118	614	—	610	580	549	518	486	455	388	324	279
W310x107	554	—	547	518	487	456	425	395	325	271	233
* W310x97	454	—	452	429	404	379	354	328	272	226	193
W250x167	765	—	—	761	738	716	694	671	627	584	540
W250x149	671	—	—	658	636	613	592	570	527	484	439
W250x131	583	—	—	561	539	518	496	475	433	391	341
W250x115	504	—	497	476	454	433	412	391	349	301	262
W250x101	441	—	429	407	386	365	344	324	279	236	205
W250x89	387	—	371	350	329	308	287	266	220	186	161
W250x80	343	—	325	304	284	263	243	221	179	151	130
W250x73	310	—	290	270	250	230	209	185	149	126	108
W200x100	362	—	353	338	324	309	295	281	253	220	192
W200x86	308	—	295	280	266	253	239	225	196	167	145
W200x71	252	248	234	220	206	192	179	164	136	115	100
W200x59	205	197	183	169	155	142	126	112	92.2	78.2	68.0
W200x52	178	169	155	142	129	114	100	88.4	72.3	61.3	53.2
* W200x46	140	133	123	112	101	89.1	77.4	68.5	55.8	47.2	40.9

Note: Moment resistances are based on class of section for X-X axis bending only, $\omega_2 = 1.0$.

* Class 3 section

** Class 4 section; M_{rx} and M'_{rx} calculated according to S16.1-94 Clause 13.5(c)(iii).

NOTES

G40.21-M 300W
$\phi = 0.90$

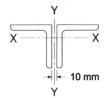

DOUBLE ANGLE STRUTS
Imperial Series – Equal Leg Angles

Factored Axial Compressive Resistances – kN
Legs 10 mm Back to Back *

Designation (mm x mm x mm)		L 152 x 152			L 127 x 127			
		16	13**	9.5**	16	13	9.5**	7.9**
Mass (kg/m)		71.9	58.1	44.1	59.4	48.1	36.6	30.7
Effective length (KL) in millimetres with respect to indicated axis	X–X Axis							
	0	2 470	1 890	1 180	2 040	1 660	1 110	821
	500	2 470	1 890	1 180	2 030	1 650	1 110	818
	1 000	2 420	1 860	1 170	1 980	1 600	1 090	805
	1 500	2 330	1 810	1 140	1 860	1 510	1 050	775
	2 000	2 190	1 740	1 090	1 680	1 370	980	731
	2 500	2 000	1 620	1 040	1 480	1 210	899	675
	3 000	1 800	1 460	968	1 270	1 040	799	613
	3 500	1 590	1 290	894	1 080	888	683	551
	4 000	1 390	1 140	819	920	755	582	492
	4 500	1 220	993	746	782	642	496	420
	5 000	1 060	868	667	667	549	424	360
	5 500	929	759	585	573	472	365	310
	6 000	814	667	514	496	408	316	268
	6 500	717	588	453	432	356	275	234
	7 000	635	520	401	379	312	242	206
	7 500	565	463	357	334	276	214	182
	8 000	505	414	320				
	8 500	453	372	287				
	9 000	409	335	259				
	9 500			235				
	Y–Y Axis							
	0	2 470	1 890	1 180	2 040	1 660	1 110	821
	500	2 140	1 560	878	1 850	1 390	884	598
	1 000	2 090	1 500	841	1 820	1 360	861	579
	1 500	2 070	1 490	832	1 800	1 350	853	573
	2 000	2 040	1 470	826	1 770	1 330	845	568
	2 500	2 010	1 450	820	1 700	1 290	833	562
	3 000	1 960	1 420	813	1 610	1 230	810	555
	3 500	1 880	1 380	803	1 490	1 150	777	544
	4 000	1 780	1 330	790	1 360	1 060	732	528
	4 500	1 660	1 260	773	1 230	965	678	508
	5 000	1 530	1 180	750	1 110	871	620	481
	5 500	1 410	1 090	723	994	784	563	444
	6 000	1 290	1 000	686	892	704	510	406
	6 500	1 180	921	640	801	633	460	370
	7 000	1 080	844	594	720	570	416	336
	7 500	985	772	549	649	514	376	305
	8 000	900	707	506	586	464	340	278
	8 500	824	648	467	531	421	309	253
	9 000	755	595	431	483	383	282	231
	9 500	694	547	397	440	349	257	211
	10 000	639	504	367	403	319	235	194
PROPERTIES OF 2 ANGLES — 10 mm BACK TO BACK								
Area (mm²)		9 160	7 400	5 610	7 570	6 130	4 660	3 910
r_x (mm)		46.7	47.1	47.6	38.7	39.1	39.5	39.8
r_y (mm)		67.6	67.1	66.6	57.5	57.0	56.4	56.2
* r_z (mm)		29.8	30.0	30.2	24.8	25.0	25.1	25.2
* r_z / r_x		0.638	0.637	0.634	0.641	0.639	0.635	0.633
* r_z / r_y		0.441	0.447	0.453	0.431	0.439	0.445	0.448
IMPERIAL SIZE AND MASS								
Mass (lb./ft.)		48.4	39.2	29.8	40.0	32.4	24.6	20.6
Thickness (in.)		5/8	1/2	3/8	5/8	1/2	3/8	5/16
Size (in.)		6 x 6			5 x 5			

* See page 4–4 for more information regarding these tables.
** Factored axial compressive resistances calculated according to Clause 13.3.3, CAN/CSA-S16.1-94.

DOUBLE ANGLE STRUTS
Imperial Series – Equal Leg Angles

Factored Axial Compressive Resistances – kN
Legs 10 mm Back to Back *

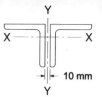

G40.21-M
300W
$\phi = 0.90$

	Designation (mm x mm x mm)		L 102 x 102				L 89 x 89			
			13	9.5	7.9**	6.4**	13	9.5	7.9	6.4**
	Mass (kg/m)		38.1	29.1	24.4	19.7	32.9	25.2	21.2	17.1
Effective length (KL) in millimetres with respect to indicated axis	X–X Axis	0	1 310	1 000	760	525	1 130	867	729	503
		500	1 300	992	755	522	1 120	855	719	499
		1 000	1 240	945	731	507	1 040	797	671	477
		1 500	1 110	853	682	476	896	692	584	435
		2 000	952	733	613	432	732	568	481	380
		2 500	789	610	515	383	581	453	384	313
		3 000	645	500	422	334	459	359	305	249
		3 500	526	409	346	281	365	286	244	199
		4 000	431	336	284	232	294	231	197	161
		4 500	357	279	236	192	241	189	161	132
		5 000	299	233	198	161	200	157	134	109
		5 500	253	198	168	137			113	92.1
		6 000	216	169	143	117				
	Y–Y Axis	0	1 310	1 000	760	525	1 130	867	729	503
		500	1 180	805	604	374	1 040	735	567	381
		1 000	1 160	788	589	365	1 030	723	555	374
		1 500	1 140	777	581	362	999	708	545	370
		2 000	1 100	757	569	358	932	674	525	363
		2 500	1 020	721	549	352	839	615	490	349
		3 000	924	666	518	343	738	545	440	324
		3 500	822	599	475	329	640	474	386	291
		4 000	725	532	427	311	551	409	334	256
		4 500	635	468	379	283	474	352	289	223
		5 000	556	411	334	253	409	304	250	194
		5 500	488	361	294	226	354	263	216	169
		6 000	429	318	260	201	308	229	189	148
		6 500	379	281	230	179	270	200	165	130
		7 000	336	249	204	160	238	176	146	115
		7 500	299	222	182	143	210	156	129	102
		8 000	267	199	163	128	187	139	115	90.8
		8 500	240	179	147	116				
		9 000	217	161	133	105				
	PROPERTIES OF 2 ANGLES — 10 mm BACK TO BACK									
	Area (mm²)		4 860	3 710	3 110	2 510	4 190	3 210	2 700	2 180
	r_x (mm)		31.1	31.5	31.7	31.9	26.9	27.3	27.5	27.7
	r_y (mm)		46.9	46.4	46.1	45.8	41.7	41.1	40.8	40.5
	* r_z (mm)		19.9	20.1	20.2	20.3	17.3	17.4	17.5	17.6
	* r_z / r_x		0.640	0.638	0.637	0.636	0.643	0.637	0.636	0.635
	* r_z / r_y		0.424	0.433	0.438	0.443	0.415	0.423	0.429	0.435
	IMPERIAL SIZE AND MASS									
	Mass (lb./ft.)		25.6	19.6	16.4	13.2	22.2	17.0	14.4	11.6
	Thickness (in.)		1/2	3/8	5/16	1/4	1/2	3/8	5/16	1/4
	Size (in.)		4 x 4				3 1/2 x 3 1/2			

* See page 4–4 for more information regarding these tables.
** Factored axial compressive resistances calculated according to Clause 13.3.3, CAN/CSA-S16.1-94.

G40.21-M
300W
$\phi = 0.90$

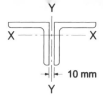

DOUBLE ANGLE STRUTS
Imperial Series – Equal Leg Angles

Factored Axial Compressive Resistances – kN
Legs 10 mm Back to Back *

Designation (mm x mm x mm)		L 76 x 76				L 64 x 64		L 51 x 51				
		13	9.5	7.9	6.4**	6.4	4.8**	9.5	7.9	6.4	4.8	3.2**
Mass (kg/m)		27.9	21.4	18.0	14.6	12.0	9.14	13.8	11.7	9.50	7.24	4.91
X–X Axis	0	959	734	618	472	413	276	473	402	327	249	110
	500	938	719	606	465	400	271	443	378	308	235	107
	1 000	841	648	547	433	341	244	336	289	237	182	100
	1 500	684	531	450	365	259	199	224	194	160	124	81.7
	2 000	526	411	349	285	188	145	148	129	107	83.1	48.8
	2 500	398	313	266	218	136	106	102	89.1	74.0	57.7	31.2
	3 000	304	239	204	167	102	79.0	73.8	64.4	53.5	41.8	21.7
	3 500	236	187	159	131	77.6	60.4					
	4 000	187	148	127	104							
	4 500	152	120	103	84.1							
Y–Y Axis	0	959	734	618	472	413	276	473	402	327	249	110
	500	904	652	515	369	334	210	448	371	285	191	87.4
	1 000	889	642	506	361	327	206	423	353	275	186	85.9
	1 500	839	616	490	352	311	199	363	304	240	170	83.3
	2 000	754	560	453	334	275	185	294	245	194	141	76.6
	2 500	656	488	399	303	231	162	231	192	153	112	61.1
	3 000	558	415	341	263	190	136	182	151	120	88.1	46.4
	3 500	471	350	288	225	155	113	144	119	94.5	69.8	35.5
	4 000	397	295	243	190	128	93.5	116	95.7	75.9	56.1	27.8
	4 500	335	249	205	161	106	77.9	94.3	78.0	61.9	45.8	22.3
	5 000	285	211	174	137	88.6	65.5	78.1	64.6	51.2	38.0	
	5 500	244	181	149	118	75.1	55.6					
	6 000	210	156	129	102	64.2	47.7					
	6 500	183	135	112	88.4							
	7 000	160	119	98.0	77.5							
PROPERTIES OF 2 ANGLES — 10 mm BACK TO BACK												
Area (mm²)		3 550	2 720	2 290	1 850	1 530	1 160	1 750	1 490	1 210	922	626
r_x (mm)		22.8	23.2	23.4	23.6	19.5	19.8	15.1	15.3	15.5	15.7	15.9
r_y (mm)		36.6	36.0	35.7	35.4	30.3	30.1	26.0	25.6	25.3	25.0	24.7
* r_z (mm)		14.8	14.9	15.0	15.0	12.5	12.6	9.89	9.90	9.93	10.0	10.1
* r_z / r_x		0.649	0.642	0.641	0.636	0.641	0.636	0.655	0.647	0.641	0.637	0.635
* r_z / r_y		0.404	0.414	0.420	0.424	0.413	0.419	0.380	0.387	0.392	0.400	0.409
IMPERIAL SIZE AND MASS												
Mass (lb./ft.)		18.8	14.4	12.2	9.80	8.20	6.14	9.40	7.84	6.38	4.88	3.30
Thickness (in.)		1/2	3/8	5/16	1/4	1/4	3/16	3/8	5/16	1/4	3/16	1/8
Size (in.)			3 x 3			2 1/2 x 2 1/2			2 x 2			

Effective length (KL) in millimetres with respect to indicated axis

* See page 4–4 for more information regarding these tables.
** Factored axial compressive resistances calculated according to Clause 13.3.3, CAN/CSA-S16.1-94.

DOUBLE ANGLE STRUTS
Imperial Series – Equal Leg Angles

Factored Axial Compressive Resistances – kN
Legs 10 mm Back to Back *

G40.21-M
300W
$\phi = 0.90$

	Designation (mm x mm x mm)		L 44 x 44			L 38 x 38	
			6.4	4.8	3.2**	4.8	3.2**
	Mass (kg/m)		8.24	6.30	4.28	5.34	3.65
Effective length (KL) in millimetres with respect to indicated axis	X–X Axis	0	284	217	105	184	98.7
		500	260	199	102	162	93.9
		1 000	183	142	91.3	103	70.2
		1 500	114	90.2	57.8	59.9	35.4
		2 000	73.3	58.1	32.5	37.0	19.9
		2 500	49.7	39.5	20.8		
		3 000					
	Y–Y Axis	0	284	217	105	184	98.7
		500	255	177	88.2	158	82.2
		1 000	239	169	86.6	145	79.5
		1 500	198	145	81.1	114	69.9
		2 000	153	113	67.1	84.5	50.3
		2 500	116	86.1	47.9	62.1	33.5
		3 000	89.1	66.0	34.6	46.6	23.7
		3 500	69.4	51.5	25.9	35.8	17.5
		4 000	55.1	40.9	20.1	28.2	
		4 500	44.6	33.1			
		5 000					

PROPERTIES OF 2 ANGLES — 10 mm BACK TO BACK

Area (mm²)		1 050	802	546	680	464
r_x (mm)		13.4	13.7	13.9	11.6	11.8
r_y (mm)		22.8	22.5	22.2	20.0	19.6
* r_z (mm)		8.68	8.73	8.82	7.45	7.52
* r_z / r_x		0.648	0.637	0.635	0.642	0.637
* r_z / r_y		0.381	0.388	0.397	0.373	0.384

IMPERIAL SIZE AND MASS

Mass (lb./ft.)		5.54	4.24	2.88	3.60	2.46
Thickness (in.)		1/4	3/16	1/8	3/16	1/8
Size (in.)			1 3/4 x 1 3/4		1 1/2 x 1 1/2	

* See page 4–4 for more information regarding these tables.
** Factored axial compressive resistances calculated according to Clause 13.3.3, CAN/CSA-S16.1-94.

G40.21-M 300W
$\phi = 0.90$

DOUBLE ANGLE STRUTS
Imperial Series – Unequal Leg Angles

Factored Axial Compressive Resistances – kN
Long Legs 10 mm Back to Back *

Designation (mm x mm x mm)		L 178 x 102		L 152 x 102				L 152 x 89			
		13**	9.5**	16	13**	9.5**	7.9**	16	13**	9.5**	7.9**
Mass (kg/m)		53.3	40.5	59.4	48.1	36.6	30.7	56.2	45.5	34.6	29.0
Effective length (KL) in millimetres with respect to indicated axis											
X–X Axis	0	1 660	1 120	2 040	1 600	1 090	811	1 930	1 510	1 020	794
	500	1 660	1 120	2 040	1 600	1 090	809	1 930	1 510	1 020	792
	1 000	1 650	1 110	2 000	1 580	1 070	802	1 900	1 490	1 010	783
	1 500	1 620	1 090	1 930	1 530	1 050	784	1 830	1 450	982	762
	2 000	1 570	1 060	1 820	1 460	998	756	1 720	1 380	939	730
	2 500	1 500	1 010	1 680	1 360	937	719	1 590	1 290	882	686
	3 000	1 420	958	1 510	1 230	865	674	1 430	1 170	816	636
	3 500	1 320	897	1 350	1 100	789	620	1 280	1 040	746	582
	4 000	1 220	833	1 190	970	714	562	1 130	920	676	529
	4 500	1 120	769	1 040	852	642	507	988	808	609	478
	5 000	1 010	706	911	747	574	457	866	709	546	431
	5 500	906	647	798	656	505	411	760	623	480	389
	6 000	811	593	702	578	445	370	668	548	423	351
	6 500	727	542	620	510	393	333	590	484	374	316
	7 000	652	497	549	452	349	295	523	430	332	281
	7 500	587	452	489	403	311	263	466	383	296	250
	8 000	529	408	438	361	278	236	417	343	265	224
	8 500	479	369	393	324	250	212	375	308	238	202
	9 000	435	335	355	293	226	192	338	278	215	182
	9 500	396	306	322	265	205	174	306	252	195	165
	10 000	362	279								
Y–Y Axis	0	1 660	1 120	2 040	1 600	1 090	811	1 930	1 510	1 020	794
	500	1 370	784	1 830	1 370	820	564	1 740	1 300	779	536
	1 000	1 310	739	1 780	1 320	786	536	1 680	1 250	744	508
	1 500	1 270	716	1 720	1 280	766	523	1 580	1 180	716	491
	2 000	1 210	689	1 620	1 210	740	508	1 430	1 080	676	469
	2 500	1 120	655	1 470	1 120	703	488	1 260	959	624	440
	3 000	1 010	613	1 310	1 000	656	462	1 080	827	562	404
	3 500	890	565	1 150	882	601	431	916	705	488	365
	4 000	776	514	996	770	535	397	776	599	420	327
	4 500	675	463	863	669	471	361	658	510	360	284
	5 000	586	407	749	582	413	325	561	435	310	247
	5 500	511	359	652	507	363	288	482	374	268	215
	6 000	447	317	569	444	319	255	417	324	233	188
	6 500	393	280	500	390	282	226	363	283	204	165
	7 000	348	249	442	345	250	202	319	248	179	146
	7 500	309	222	392	307	223	180				
	8 000	276		350	274	200	162				
	8 500			314							
PROPERTIES OF 2 ANGLES — 10 mm BACK TO BACK											
Area (mm²)		6 790	5 160	7 570	6 130	4 660	3 910	7 160	5 800	4 410	3 700
r_x (mm)		57.3	57.8	48.0	48.5	48.9	49.2	48.2	48.6	49.1	49.3
r_y (mm)		40.2	39.7	42.7	42.1	41.6	41.3	36.6	36.0	35.4	35.2
* r_z (mm)		22.2	22.4	22.0	22.2	22.4	22.5	19.1	19.3	19.5	19.6
* r_z / r_x		0.387	0.388	0.458	0.458	0.458	0.457	0.396	0.397	0.397	0.398
* r_z / r_y		0.552	0.564	0.515	0.527	0.538	0.545	0.522	0.536	0.551	0.557
IMPERIAL SIZE AND MASS											
Mass (lb./ft.)		35.8	27.2	40.0	32.4	24.6	20.6	37.8	30.6	23.4	19.6
Thickness (in.)		1/2	3/8	5/8	1/2	3/8	5/16	5/8	1/2	3/8	5/16
Size (in.)		7 x 4		6 x 4				6 x 3 1/2			

* See page 4–4 for more information regarding these tables.
** Factored axial compressive resistances calculated according to Clause 13.3.3, CAN/CSA-S16.1-94.

DOUBLE ANGLE STRUTS
Imperial Series – Unequal Leg Angles

G40.21-M
300W

Factored Axial Compressive Resistances – kN
Long Legs 10 mm Back to Back *

$\phi = 0.90$

Designation (mm x mm x mm)			L 127 x 89						L 127 x 76			
			19	16	13	9.5**	7.9**	6.4**	13	9.5**	7.9**	6.4**
Mass (kg/m)			59.0	49.9	40.5	30.9	25.9	20.9	38.0	29.0	24.3	19.6
Effective length (KL) in millimetres with respect to indicated axis	X–X Axis	0	2 030	1 720	1 390	988	773	529	1 310	923	719	514
		500	2 020	1 710	1 390	984	771	527	1 300	920	717	513
		1 000	1 970	1 660	1 350	964	755	519	1 270	902	703	505
		1 500	1 850	1 570	1 280	921	723	502	1 200	863	674	488
		2 000	1 680	1 430	1 160	857	674	475	1 090	804	629	461
		2 500	1 480	1 260	1 030	778	613	441	972	731	574	423
		3 000	1 280	1 100	897	690	548	403	845	651	515	380
		3 500	1 100	938	769	594	485	362	725	560	457	339
		4 000	932	799	657	508	426	319	620	480	402	300
		4 500	794	681	561	434	367	281	530	411	347	265
		5 000	679	583	480	373	315	248	454	353	298	234
		5 500	584	502	414	322	272	220	391	304	258	208
		6 000	505	435	359	279	236	192	339	264	224	182
		6 500	441	379	313	244	206	168	296	231	195	159
		7 000	387	333	275	214	181	147	260	203	172	140
		7 500	341	294	243	189	160	130	230	179	152	124
		8 000			216	168	143	116	204	159	135	110
	Y–Y Axis	0	2 030	1 720	1 390	988	773	529	1 310	923	719	514
		500	1 930	1 590	1 220	800	568	355	1 150	756	537	337
		1 000	1 890	1 550	1 190	775	548	340	1 100	726	515	321
		1 500	1 790	1 480	1 130	751	532	331	1 020	687	491	308
		2 000	1 630	1 350	1 040	711	510	321	894	619	455	291
		2 500	1 450	1 190	926	645	477	306	758	533	408	268
		3 000	1 250	1 030	803	568	436	286	630	448	352	242
		3 500	1 070	880	687	491	387	263	522	373	297	214
		4 000	914	749	585	422	335	239	433	311	249	186
		4 500	780	638	499	361	290	215	362	261	210	158
		5 000	668	546	427	311	251	188	305	221	178	136
		5 500	575	470	368	268	217	165	259	188	153	117
		6 000	498	407	319	233	189	145	223	162	132	101
		6 500	435	355	278	204	166	128				
		7 000	382	312	244	179	146	113				
		7 500	338	276	216							
PROPERTIES OF 2 ANGLES — 10 mm BACK TO BACK												
Area (mm²)			7 520	6 360	5 160	3 930	3 300	2 660	4 840	3 690	3 100	2 500
r_x (mm)			39.3	39.7	40.1	40.6	40.8	41.0	40.3	40.8	41.0	41.2
r_y (mm)			39.2	38.5	37.9	37.4	37.1	36.8	32.0	31.4	31.1	30.8
* r_z (mm)			19.0	19.1	19.2	19.3	19.4	19.6	16.5	16.6	16.7	16.8
* r_z / r_x			0.483	0.481	0.479	0.475	0.475	0.478	0.409	0.407	0.407	0.408
* r_z / r_y			0.485	0.496	0.507	0.516	0.523	0.533	0.516	0.529	0.537	0.545
IMPERIAL SIZE AND MASS												
Mass (lb./ft.)			39.6	33.6	27.2	20.8	17.4	14.0	25.6	19.6	16.4	13.2
Thickness (in.)			3/4	5/8	1/2	3/8	5/16	1/4	1/2	3/8	5/16	1/4
Size (in.)			5 x 3 1/2						5 x 3			

* See page 4–4 for more information regarding these tables.
** Factored axial compressive resistances calculated according to Clause 13.3.3, CAN/CSA-S16.1-94.

DOUBLE ANGLE STRUTS
Imperial Series – Unequal Leg Angles

G40.21-M 300W
$\phi = 0.90$

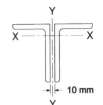

Factored Axial Compressive Resistances – kN
Long Legs 10 mm Back to Back *

Designation (mm x mm x mm)		L 102 x 89				L 102 x 76				L 89 x 76
		13	9.5	7.9**	6.4**	13	9.5	7.9**	6.4**	7.9
Mass (kg/m)		35.5	27.1	22.8	18.4	33.0	25.2	21.2	17.1	19.6

Effective length (KL) in millimetres with respect to indicated axis

X–X Axis

KL	13	9.5	7.9**	6.4**	13	9.5	7.9**	6.4**	7.9
0	1 220	934	746	513	1 130	867	689	498	675
500	1 210	926	740	510	1 120	859	684	496	666
1 000	1 150	883	712	495	1 070	820	659	481	624
1 500	1 040	800	656	465	969	744	609	452	545
2 000	895	690	580	423	836	644	540	406	451
2 500	745	576	487	374	697	539	456	352	362
3 000	610	473	401	319	573	444	376	301	288
3 500	499	388	329	267	469	365	309	253	231
4 000	410	319	271	220	386	301	255	209	187
4 500	340	265	225	183	320	250	212	174	153
5 000	285	222	189	154	268	210	178	146	127
5 500	241	188	160	130	227	178	151	124	107
6 000	206	161	137	112	195	152	129	106	
6 500							112	91.9	

Y–Y Axis

KL	13	9.5	7.9**	6.4**	13	9.5	7.9**	6.4**	7.9
0	1 220	934	746	513	1 130	867	689	498	675
500	1 110	772	585	377	1 040	730	556	364	543
1 000	1 090	755	571	369	1 010	708	540	353	530
1 500	1 050	734	559	362	943	669	518	341	508
2 000	978	694	535	353	836	602	474	323	467
2 500	875	632	496	337	716	519	415	296	410
3 000	765	557	445	315	601	438	353	263	349
3 500	659	483	390	288	501	366	297	224	294
4 000	565	416	337	254	418	306	249	190	247
4 500	485	357	291	222	350	257	210	162	208
5 000	416	307	251	193	296	217	178	138	177
5 500	360	266	218	168	253	185	152	118	151
6 000	312	231	190	147	217	159	131	102	130
6 500	273	202	166	129	188	138	114	88.8	113
7 000	240	178	146	114					
7 500	212	158	130	101					
8 000	189								

PROPERTIES OF 2 ANGLES — 10 mm BACK TO BACK

	13	9.5	7.9**	6.4**	13	9.5	7.9**	6.4**	7.9
Area (mm²)	4 530	3 460	2 910	2 340	4 200	3 210	2 700	2 180	2 500
r_x (mm)	31.5	31.9	32.1	32.3	31.8	32.2	32.4	32.7	27.9
r_y (mm)	40.2	39.7	39.4	39.1	34.0	33.4	33.1	32.8	34.3
* r_z (mm)	18.4	18.5	18.6	18.7	16.2	16.4	16.5	16.6	15.9
* r_z / r_x	0.584	0.580	0.579	0.579	0.509	0.509	0.509	0.508	0.570
* r_z / r_y	0.458	0.466	0.472	0.478	0.476	0.491	0.498	0.506	0.464

IMPERIAL SIZE AND MASS

	13	9.5	7.9**	6.4**	13	9.5	7.9**	6.4**	7.9
Mass (lb./ft.)	23.8	18.2	15.4	12.4	22.2	17.0	14.4	11.6	13.2
Thickness (in.)	1/2	3/8	5/16	1/4	1/2	3/8	5/16	1/4	5/16
Size (in.)	4 x 3 1/2				4 x 3				3 1/2 x 3

* See page 4–4 for more information regarding these tables.
** Factored axial compressive resistances calculated according to Clause 13.3.3, CAN/CSA-S16.1-94.

DOUBLE ANGLE STRUTS
Imperial Series – Unequal Leg Angles

Factored Axial Compressive Resistances – kN
Long Legs 10 mm Back to Back *

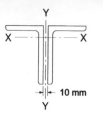

G40.21-M
300W
$\phi = 0.90$

Designation (mm x mm x mm)			L 64 x 51				L 51 x 38	
			9.5	7.9	6.4	4.8**	6.4	4.8
Mass (kg/m)			15.7	13.3	10.8	8.19	8.23	6.29
Effective length (KL) in millimetres with respect to indicated axis	X–X Axis	0	540	456	370	262	284	216
		500	522	442	358	256	268	205
		1 000	446	378	308	226	208	160
		1 500	339	289	236	181	142	110
		2 000	246	210	173	133	95.5	74.3
		2 500	178	153	126	97.2	66.4	51.8
		3 000	133	114	94.1	72.6	48.1	37.5
		3 500	101	87.3	72.0	55.7		
		4 000				43.7		
	Y–Y Axis	0	540	456	370	262	284	216
		500	504	411	311	200	252	175
		1 000	469	385	294	192	221	156
		1 500	396	327	254	173	168	121
		2 000	315	260	203	143	122	88.2
		2 500	245	202	158	113	88.4	64.3
		3 000	190	157	123	88.9	65.8	47.9
		3 500	150	123	96.5	70.3	50.3	36.7
		4 000	120	98.7	77.2	56.5		
		4 500	97.4	80.3	62.8	46.1		
PROPERTIES OF 2 ANGLES — 10 mm BACK TO BACK								
Area (mm^2)			2 000	1 690	1 370	1 040	1 050	801
r_x (mm)			19.5	19.7	19.9	20.1	15.8	16.0
r_y (mm)			24.6	24.3	23.9	23.6	19.0	18.6
* r_z (mm)			10.7	10.7	10.8	10.9	8.12	8.18
* r_z / r_x			0.549	0.543	0.543	0.542	0.514	0.511
* r_z / r_y			0.435	0.440	0.452	0.462	0.427	0.440
IMPERIAL SIZE AND MASS								
Mass (lb./ft.)			10.6	9.00	7.24	5.50	5.54	4.24
Thickness (in.)			3/8	5/16	1/4	3/16	1/4	3/16
Size (in.)			2 1/2 x 2				2 x 1 1/2	

* See page 4–4 for more information regarding these tables.
** Factored axial compressive resistances calculated according to Clause 13.3.3, CAN/CSA-S16.1-94.

G40.21-M
300W
$\phi = 0.90$

DOUBLE ANGLE STRUTS
Imperial Series – Unequal Leg Angles

Factored Axial Compressive Resistances – kN
Short Legs 10 mm Back to Back *

	Designation (mm x mm x mm)		L 178 x 102		L 152 x 102				L 152 x 89			
			13**	9.5**	16	13**	9.5**	7.9**	16	13**	9.5**	7.9**
	Mass (kg/m)		53.3	40.5	59.4	48.1	36.6	30.7	56.2	45.5	34.6	29.0
Effective length (KL) in millimetres with respect to indicated axis	X–X Axis	0	1 660	1 120	2 040	1 600	1 090	811	1 930	1 510	1 020	794
		500	1 650	1 110	2 020	1 590	1 080	805	1 900	1 490	1 010	784
		1 000	1 570	1 060	1 900	1 510	1 030	778	1 730	1 390	943	734
		1 500	1 410	958	1 680	1 370	942	722	1 440	1 180	825	644
		2 000	1 220	833	1 400	1 150	822	645	1 140	935	688	540
		2 500	1 010	706	1 140	936	699	551	875	724	561	443
		3 000	805	593	915	754	586	464	677	562	437	362
		3 500	647	497	736	609	474	391	530	441	344	293
		4 000	525	408	598	495	387	328	423	352	275	234
		4 500	431	335	492	408	319	271	343	286	224	191
		5 000	359	279	410	340	266	226			185	158
		5 500	302	235	345	287	225	191				
		6 000						163				
	Y–Y Axis	0	1 660	1 120	2 040	1 600	1 090	811	1 930	1 510	1 020	794
		500	1 340	762	1 810	1 350	802	549	1 720	1 280	759	519
		1 000	1 300	724	1 790	1 320	775	526	1 690	1 250	735	498
		1 500	1 290	716	1 780	1 310	769	521	1 690	1 240	730	494
		2 000	1 280	713	1 770	1 300	766	518	1 680	1 240	728	492
		2 500	1 280	711	1 760	1 300	764	517	1 680	1 240	726	491
		3 000	1 280	709	1 740	1 290	761	515	1 660	1 230	724	490
		3 500	1 280	708	1 690	1 270	757	513	1 620	1 220	722	489
		4 000	1 270	707	1 610	1 240	751	510	1 550	1 200	718	487
		4 500	1 260	705	1 510	1 180	741	507	1 460	1 150	712	485
		5 000	1 250	702	1 410	1 110	726	501	1 370	1 090	702	482
		5 500	1 230	699	1 310	1 040	702	494	1 270	1 010	683	477
		6 000	1 190	694	1 210	959	669	482	1 180	938	655	469
		6 500	1 140	687	1 110	884	632	466	1 090	866	620	456
		7 000	1 080	675	1 020	813	593	445	1 000	798	583	438
		7 500	1 010	657	938	747	549	422	920	735	545	416
		8 000	946	633	862	686	507	398	847	676	502	392
		8 500	882	605	792	631	467	374	780	623	463	369
		9 000	822	576	729	580	430	350	719	574	427	346
		9 500	765	546	672	535	397	326	663	529	395	325
		10 000	713	517	620	494	367	302	613	489	365	301
	PROPERTIES OF 2 ANGLES — 10 mm BACK TO BACK											
	Area (mm²)		6 790	5 160	7 570	6 130	4 660	3 910	7 160	5 800	4 410	3 700
	r_x (mm)		28.5	28.9	28.9	29.3	29.8	30.0	24.3	24.7	25.1	25.3
	r_y (mm)		87.7	87.1	74.1	73.5	72.9	72.6	76.1	75.5	74.9	74.6
	* r_z (mm)		22.2	22.4	22.0	22.2	22.4	22.5	19.1	19.3	19.5	19.6
	* r_z / r_x		0.779	0.775	0.761	0.758	0.752	0.750	0.786	0.781	0.777	0.775
	* r_z / r_y		0.253	0.257	0.297	0.302	0.307	0.310	0.251	0.256	0.260	0.263
	IMPERIAL SIZE AND MASS											
	Mass (lb./ft.)		35.8	27.2	40.0	32.4	24.6	20.6	37.8	30.6	23.4	19.6
	Thickness (in.)		1/2	3/8	5/8	1/2	3/8	5/16	5/8	1/2	3/8	5/16
	Size (in.)		7 x 4		6 x 4				6 x 3 1/2			

* See page 4–4 for more information regarding these tables.
** Factored axial compressive resistances calculated according to Clause 13.3.3, CAN/CSA-S16.1-94.

DOUBLE ANGLE STRUTS
Imperial Series – Unequal Leg Angles

Factored Axial Compressive Resistances – kN
Short Legs 10 mm Back to Back *

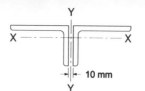

G40.21-M
300W
$\phi = 0.90$

Designation (mm x mm x mm)			L 127 x 89						L 127 x 76			
			19	16	13	9.5**	7.9**	6.4**	13	9.5**	7.9**	6.4**
Mass (kg/m)			59.0	49.9	40.5	30.9	25.9	20.9	38.0	29.0	24.3	19.6
Effective length (KL) in millimetres with respect to indicated axis	X–X Axis	0	2 030	1 720	1 390	988	773	529	1 310	923	719	514
		500	1 990	1 690	1 370	975	764	524	1 270	904	705	506
		1 000	1 830	1 550	1 260	915	718	499	1 120	817	640	468
		1 500	1 530	1 310	1 070	804	634	453	879	677	535	396
		2 000	1 220	1 050	862	666	534	395	656	511	426	318
		2 500	940	815	674	523	438	329	486	381	324	252
		3 000	730	635	527	410	348	270	366	288	245	201
		3 500	580	500	416	325	276	223	282	222	190	155
		4 000	460	400	333	261	222	181	222	175	150	123
		4 500	374	325	272	213	181	148				
		5 000		269	225	176	150	122				
	Y–Y Axis	0	2 030	1 720	1 390	988	773	529	1 310	923	719	514
		500	1 930	1 580	1 210	786	555	345	1 130	741	523	325
		1 000	1 910	1 570	1 190	771	541	333	1 120	727	510	314
		1 500	1 910	1 560	1 180	767	537	330	1 120	724	507	312
		2 000	1 880	1 550	1 180	763	535	329	1 110	722	506	311
		2 500	1 820	1 510	1 160	759	532	327	1 100	719	504	310
		3 000	1 720	1 440	1 130	751	529	326	1 080	715	502	309
		3 500	1 600	1 340	1 060	735	523	324	1 020	705	499	308
		4 000	1 480	1 240	986	707	513	321	952	684	493	306
		4 500	1 350	1 130	903	659	496	316	875	645	481	304
		5 000	1 230	1 030	822	605	471	309	798	593	459	299
		5 500	1 120	933	745	551	440	299	725	541	431	292
		6 000	1 010	844	673	499	407	285	658	491	401	280
		6 500	915	762	608	452	370	268	596	445	367	265
		7 000	828	689	550	409	336	250	540	404	334	247
		7 500	750	624	498	370	305	233	490	366	303	230
		8 000	681	566	451	336	278	216	445	333	276	214
		8 500	619	515	410	306	253	198	405	303	251	198
		9 000	565	470	374	279	231	181	370	276	229	182
		9 500	516	429	342	255	211	166	338	253	210	166
		10 000	473	393	313	234	193	153	310	232	193	153
PROPERTIES OF 2 ANGLES — 10 mm BACK TO BACK												
Area (mm²)			7 520	6 360	5 160	3 930	3 300	2 660	4 840	3 690	3 100	2 500
r_x (mm)			24.8	25.2	25.6	26.0	26.2	26.4	21.1	21.5	21.7	21.9
r_y (mm)			63.1	62.5	61.9	61.3	61.0	60.7	63.8	63.2	62.9	62.6
* r_z (mm)			19.0	19.1	19.2	19.3	19.4	19.6	16.5	16.6	16.7	16.8
* r_z / r_x			0.766	0.758	0.750	0.742	0.740	0.742	0.782	0.772	0.770	0.767
* r_z / r_y			0.301	0.306	0.310	0.315	0.318	0.323	0.259	0.263	0.266	0.268
IMPERIAL SIZE AND MASS												
Mass (lb./ft.)			39.6	33.6	27.2	20.8	17.4	14.0	25.6	19.6	16.4	13.2
Thickness (in.)			3/4	5/8	1/2	3/8	5/16	1/4	1/2	3/8	5/16	1/4
Size (in.)			5 x 3 1/2						5 x 3			

* See page 4–4 for more information regarding these tables.
** Factored axial compressive resistances calculated according to Clause 13.3.3, CAN/CSA-S16.1-94.

G40.21-M
300W
$\phi = 0.90$

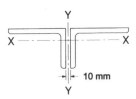

DOUBLE ANGLE STRUTS
Imperial Series – Unequal Leg Angles

Factored Axial Compressive Resistances – kN
Short Legs 10 mm Back to Back *

			L 102 x 89					L 102 x 76			L 89 x 76
Designation (mm x mm x mm)		13	9.5	7.9**	6.4**	13	9.5	7.9**	6.4**		7.9	
Mass (kg/m)		35.5	27.1	22.8	18.4	33.0	25.2	21.2	17.1		19.6	

Effective length (KL) in millimetres with respect to indicated axis	Axis	KL (mm)										
	X-X Axis	0	1 220	934	746	513	1 130	867	689	498	675	
		500	1 200	921	737	508	1 110	847	676	491	661	
		1 000	1 120	856	694	486	981	754	616	456	594	
		1 500	959	739	615	443	786	609	515	389	485	
		2 000	778	603	512	387	595	464	394	313	374	
		2 500	615	478	407	324	445	349	297	243	284	
		3 000	483	377	322	262	338	266	227	185	217	
		3 500	384	300	257	209	261	206	176	144	169	
		4 000	308	242	207	168	206	163	139	114	134	
		4 500	252	197	169	138			113	92.3	108	
		5 000	209	164	140	114						
	Y-Y Axis	0	1 220	934	746	513	1 130	867	689	498	675	
		500	1 110	767	580	373	1 030	721	548	357	539	
		1 000	1 100	755	570	366	1 030	712	540	350	531	
		1 500	1 080	747	565	363	1 020	707	536	347	524	
		2 000	1 050	732	556	359	991	698	532	345	509	
		2 500	979	702	541	354	931	675	523	342	478	
		3 000	891	650	513	345	851	629	500	335	431	
		3 500	797	587	472	331	766	569	462	324	379	
		4 000	705	522	424	310	681	508	416	306	329	
		4 500	621	460	376	285	602	450	370	283	285	
		5 000	546	405	332	255	531	397	328	255	247	
		5 500	480	356	293	227	469	350	289	227	214	
		6 000	423	314	259	202	414	309	256	201	187	
		6 500	374	278	229	180	367	274	227	179	164	
		7 000	332	247	204	160	326	244	202	160	145	
		7 500	296	220	182	143	292	218	180	143	128	
		8 000	265	197	163	129	261	195	162	128	114	
		8 500	238	177	147	116	235	176	146	116		
		9 000	215	160	133	105	213	159	132	105		
		9 500	195	145	120		193	144	120	94.9		
		10 000					176					

PROPERTIES OF 2 ANGLES — 10 mm BACK TO BACK

Area (mm²)	4 530	3 460	2 910	2 340	4 200	3 210	2 700	2 180	2 500
r_x (mm)	26.4	26.8	27.1	27.3	21.9	22.3	22.5	22.7	23.0
r_y (mm)	48.5	47.9	47.6	47.4	50.2	49.6	49.3	49.0	42.3
* r_z (mm)	18.4	18.5	18.6	18.7	16.2	16.4	16.5	16.6	15.9
* r_z / r_x	0.697	0.690	0.686	0.685	0.740	0.735	0.733	0.731	0.691
* r_z / r_y	0.379	0.386	0.391	0.395	0.323	0.331	0.335	0.339	0.376

IMPERIAL SIZE AND MASS

Mass (lb./ft.)	23.8	18.2	15.4	12.4	22.2	17.0	14.4	11.6	13.2
Thickness (in.)	1/2	3/8	5/16	1/4	1/2	3/8	5/16	1/4	5/16
Size (in.)	4 x 3 1/2				4 x 3				3 1/2 x 3

* See page 4–4 for more information regarding these tables.
** Factored axial compressive resistances calculated according to Clause 13.3.3, CAN/CSA-S16.1-94.

DOUBLE ANGLE STRUTS
Imperial Series – Unequal Leg Angles

Factored Axial Compressive Resistances – kN
Short Legs 10 mm Back to Back *

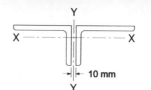

G40.21-M
300W
$\phi = 0.90$

Designation (mm x mm x mm)		L 64 x 51				L 51 x 38	
		9.5	7.9	6.4	4.8**	6.4	4.8
Mass (kg/m)		15.7	13.3	10.8	8.19	8.23	6.29
Effective length (KL) in millimetres with respect to indicated axis	X–X Axis						
	0	540	456	370	262	284	216
	500	503	426	346	249	246	189
	1 000	374	319	262	198	149	116
	1 500	245	211	174	134	85.0	66.7
	2 000	161	139	115	89.1	52.0	41.0
	2 500	110	95.4	79.2	61.5		
	3 000			57.1	44.4		
	Y–Y Axis						
	0	540	456	370	262	284	216
	500	504	409	308	197	253	173
	1 000	496	404	304	194	248	171
	1 500	459	381	294	191	221	160
	2 000	400	334	263	181	182	135
	2 500	337	281	223	160	145	108
	3 000	279	232	185	135	115	85.2
	3 500	230	191	152	111	91.6	67.9
	4 000	190	158	126	92.3	73.9	54.8
	4 500	158	131	105	77.0	60.5	44.8
	5 000	133	110	87.8	64.7	50.2	37.2
	5 500	113	93.7	74.5	55.0		
	6 000	96.7	80.3	63.9	47.1		
	6 500	83.6					
PROPERTIES OF 2 ANGLES — 10 mm BACK TO BACK							
Area (mm²)		2 000	1 690	1 370	1 040	1 050	801
r_x (mm)		14.6	14.8	15.0	15.2	11.0	11.2
r_y (mm)		32.6	32.3	32.0	31.6	27.0	26.6
* r_z (mm)		10.7	10.7	10.8	10.9	8.12	8.18
* r_z / r_x		0.733	0.723	0.720	0.717	0.738	0.730
* r_z / r_y		0.328	0.331	0.338	0.345	0.301	0.308
IMPERIAL SIZE AND MASS							
Mass (lb./ft.)		10.6	9.00	7.24	5.50	5.54	4.24
Thickness (in.)		3/8	5/16	1/4	3/16	1/4	3/16
Size (in.)		2 1/2 x 2				2 x 1 1/2	

* See page 4–4 for more information regarding these tables.
** Factored axial compressive resistances calculated according to Clause 13.3.3, CAN/CSA-S16.1-94.

G40.21-M
300W
$\phi = 0.90$

DOUBLE ANGLE STRUTS
Imperial Series – Star Shaped *
Factored Axial Compressive Resistances – kN

	Designation (mm x mm x mm)	L 152 x 152	L 127 x 127		L 102 x 102	
		16	16	13	13	9.5
	Mass (kg/m)	71.9	59.4	48.1	38.1	29.1
	Legs	12 mm apart	12 mm apart		10 mm apart	
U–U Axis	0	2 470	2 040	1 660	1 310	1 000
	500	2 080	1 800	1 340	1 150	777
	1 000	2 030	1 780	1 320	1 140	764
	1 500	2 020	1 780	1 310	1 140	761
	2 000	2 010	1 770	1 310	1 130	760
	2 500	2 010	1 770	1 310	1 130	760
	3 000	2 010	1 770	1 310	1 040	759
	3 500	2 010	1 670	1 310	942	704
	4 000	2 010	1 550	1 240	846	629
	4 500	1 900	1 430	1 140	755	559
	5 000	1 780	1 310	1 040	671	495
	5 500	1 650	1 190	943	596	438
	6 000	1 520	1 090	856	530	388
	6 500	1 410	988	776	472	345
	7 000	1 300	898	704	421	307
	7 500	1 190	816	639	377	275
	8 000	1 100	743	581	339	247
	8 500	1 010	678	529	306	222
	9 000	931	620	483	277	201
	9 500	859	568	442	252	183
	10 000	794	522	406	230	167
V–V Axis	0	2 470	2 040	1 660	1 310	1 000
	500	2 080	1 800	1 340	1 150	777
	1 000	2 030	1 780	1 320	1 140	764
	1 500	2 020	1 780	1 310	1 140	761
	2 000	2 010	1 770	1 310	1 080	760
	2 500	2 010	1 690	1 310	956	737
	3 000	2 010	1 530	1 250	825	638
	3 500	1 890	1 360	1 110	704	546
	4 000	1 720	1 200	986	599	465
	4 500	1 560	1 060	868	509	397
	5 000	1 410	929	763	435	339
	5 500	1 260	816	671	374	292
	6 000	1 130	718	591	324	253
	6 500	1 020	635	523	282	221
	7 000	916	563	464	248	194
	7 500	826	502	414	219	171
	8 000	746	449	371		
	8 500	676	404	333		
	9 000	615	365	301		
	9 500	560	331	273		
	10 000	513				
PROPERTIES OF 2 STARRED ANGLES						
Area (mm^2)		9 160	7 570	6 130	4 860	3 710
r_u (mm)		76.6	66.5	64.9	53.6	52.1
r_v (mm)		58.9	48.7	49.3	39.1	39.6
* r_z (mm)		29.8	24.8	25.0	19.9	20.1
* r_z / r_u		0.389	0.373	0.385	0.371	0.386
* r_z / r_v		0.506	0.509	0.508	0.509	0.507
IMPERIAL SIZE AND MASS						
Mass (lb./ft.)		48.4	40.0	32.4	25.6	19.6
Thickness (in.)		5/8	5/8	1/2	1/2	3/8
Size (in.)		6 x 6	5 x 5		4 x 4	

Effective length (KL) in millimetres with respect to indicated axis

* See page 4–4 for more information regarding these tables.
See Clauses 19.1.4 and 19.1.5 CAN/CSA-S16.1-94 for interconnecting requirements.

DOUBLE ANGLE STRUTS
Imperial Series – Star Shaped *

Factored Axial Compressive Resistances – kN

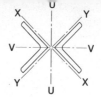

G40.21-M
300W
$\phi = 0.90$

Designation (mm x mm x mm)		L 89 x 89			L 76 x 76			L 64 x 64
		13	9.5	7.9	13	9.5	7.9	6.4
Mass (kg/m)		32.9	25.2	21.2	27.9	21.4	18.0	12.0
Legs		8 mm apart			8 mm apart			8 mm apart
Effective length (KL) in millimetres with respect to indicated axis	**U–U Axis**							
0		1 130	867	729	959	734	618	413
500		1 030	718	549	893	639	500	323
1 000		1 020	711	541	891	636	496	321
1 500		1 020	709	540	885	635	495	321
2 000		1 000	709	539	815	615	495	317
2 500		918	692	539	729	547	456	269
3 000		825	618	514	639	475	395	224
3 500		731	544	451	553	408	339	186
4 000		642	474	393	475	349	289	154
4 500		561	413	341	408	298	246	129
5 000		490	359	296	352	256	211	109
5 500		429	313	258	304	220	181	92.5
6 000		376	274	225	264	191	157	79.4
6 500		332	241	198	231	167	137	68.7
7 000		294	213	175	203	147	120	
7 500		261	189	155	180	130	106	
8 000		233	169	138	160	115		
8 500		210	151	124				
9 000		189	136					
	V–V Axis							
0		1 130	867	729	959	734	618	413
500		1 030	718	549	893	639	500	323
1 000		1 020	711	541	890	636	496	321
1 500		989	709	540	783	606	495	312
2 000		867	670	539	652	509	432	247
2 500		736	571	485	527	414	353	191
3 000		614	478	407	422	333	285	148
3 500		509	397	339	339	269	230	117
4 000		422	331	282	275	219	188	93.1
4 500		353	277	237	226	180	155	75.6
5 000		298	234	200	188	150	129	
5 500		253	199	170	159	126	109	
6 000		217	171	146				
6 500		188	148	127				
PROPERTIES OF 2 STARRED ANGLES								
Area (mm²)		4 190	3 210	2 700	3 550	2 720	2 290	1 530
r_u (mm)		47.0	45.5	44.8	41.8	40.3	39.8	33.9
r_v (mm)		33.8	34.3	34.7	28.6	29.2	29.6	24.7
* r_z (mm)		17.3	17.4	17.5	14.8	14.9	15.0	12.5
* r_z / r_u		0.368	0.383	0.390	0.354	0.369	0.377	0.369
* r_z / r_v		0.511	0.507	0.504	0.518	0.510	0.507	0.506
IMPERIAL SIZE AND MASS								
Mass (lb./ft.)		22.2	17.0	14.4	18.8	14.4	12.2	8.20
Thickness (in.)		1/2	3/8	5/16	1/2	3/8	5/16	1/4
Size (in.)		3 1/2 x 3 1/2			3 x 3			2 1/2 x 2 1/2

* See page 4–4 for more information regarding these tables.
See Clauses 19.1.4 and 19.1.5 CAN/CSA-S16.1-94 for interconnecting requirements.

G40.21-M
300W
$\phi = 0.90$

DOUBLE ANGLE STRUTS
Imperial Series – Star Shaped

Factored Axial Compressive Resistances – kN

Designation (mm x mm x mm)		L 51 x 51				L 44 x 44		L 38 x 38
		9.5	7.9	6.4	4.8	6.4	4.8	4.8
Mass (kg/m)		13.8	11.7	9.50	7.24	8.24	6.30	5.34
Legs		6 mm apart				6 mm apart		6 mm apart
Effective length (KL) in millimetres with respect to indicated axis								
U–U Axis	0	473	402	327	249	284	217	184
	500	446	369	282	188	253	174	156
	1 000	439	368	281	187	253	174	156
	1 500	388	326	261	187	214	160	126
	2 000	324	270	215	160	170	126	94.5
	2 500	263	217	172	127	132	96.7	70.4
	3 000	211	173	136	99.8	102	74.6	53.2
	3 500	170	139	109	79.2	80.4	58.4	41.1
	4 000	138	112	87.6	63.8	64.3	46.5	32.4
	4 500	114	92.0	71.7	52.1	52.2	37.7	
	5 000	94.5	76.5	59.5	43.2			
	5 500	79.7	64.4					
V–V Axis	0	473	402	327	249	284	217	184
	500	446	369	282	188	253	174	156
	1 000	385	330	270	187	217	168	127
	1 500	288	249	205	158	154	120	83.3
	2 000	206	179	149	115	106	82.8	54.6
	2 500	149	130	108	84.2	74.4	58.5	37.5
	3 000	110	96.2	80.3	62.8	54.3	42.8	
	3 500	84.1	73.4	61.4	48.0			
PROPERTIES OF 2 STARRED ANGLES								
Area (mm²)		1 750	1 490	1 210	922	1 050	802	680
r_u (mm)		28.9	28.0	27.4	26.7	24.8	24.0	21.5
r_v (mm)		18.9	19.2	19.5	19.8	16.9	17.2	14.6
* r_z (mm)		9.89	9.90	9.93	10.0	8.68	8.73	7.45
* r_z / r_u		0.343	0.353	0.362	0.375	0.350	0.363	0.346
* r_z / r_v		0.523	0.516	0.509	0.505	0.514	0.508	0.510
IMPERIAL SIZE AND MASS								
Mass (lb./ft.)		9.40	7.84	6.38	4.88	5.54	4.24	3.60
Thickness (in.)		3/8	5/16	1/4	3/16	1/4	3/16	3/16
Size (in.)		2 x 2				1 3/4 x 1 3/4		1 1/2 x 1 1/2

* See page 4–4 for more information regarding these tables.
See Clauses 19.1.4 and 19.1.5 CAN/CSA-S16.1-94 for interconnecting requirements.

Example

General

The following example illustrates the design of a double angle strut in accordance with S16.1-94 Clauses 13.3 and 19.1. Also see page 4-4 and the tables of factored axial compressive resistances starting on page 4-109.

Given:

Find the factored axial compressive resistance of a 2L102x76x6.4 double angle strut, with long legs 10 mm back-to-back. The steel grade is G40.21 300W (F_y = 300 MPa), L = 2 000 mm and there are two welded intermediate connectors.

Solution:

A. Class

Width-to-thickness ratios, S16.1-94, Clause 11.2

$$\frac{d}{t} = \frac{102}{6.35} = 16.1 > \frac{200}{\sqrt{F_y}} = 11.5$$

$$\frac{b}{t} = \frac{76.2}{6.35} = 12.0 > \frac{200}{\sqrt{F_y}} = 11.5$$

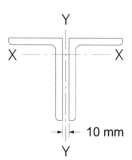

The angle is therefore a class 4 section.

B. Compressive Resistance About Axis X-X, Flexural Mode

Slenderness parameter, S16.1-94, Clause 13.3.1

$$\lambda = \left(\frac{KL}{r}\right)_x \sqrt{\frac{F_y}{\pi^2 E}} = \frac{2\ 000}{32.7} \sqrt{\frac{300}{\pi^2 \times 200 \times 10^3}} = 0.754$$

Calculated stress

$$f = F_y (1 + \lambda^{2n})^{-1/n} = 300\ (1 + 0.754^{\,2 \times 1.34})^{-1/1.34} = 225 \text{ MPa}$$

Effective leg widths, S136-94, Clause 5.6.2.1 (for simplicity, the total leg width is used instead of the flat width).

$$W_{lim} = 0.644 \sqrt{k\,E/f} = 0.644 \sqrt{0.43 \times 200 \times 10^3 / 225} = 12.6$$

$$W_d = \frac{d}{t} = 16.1 > W_{lim} = 12.6 \qquad W_b = \frac{b}{t} = 12.0 < W_{lim} = 12.6$$

Therefore, the effective width of the long leg must be reduced.

$$B_d = 0.95 \sqrt{k\,E/f} \left(1 - \frac{0.208}{W_d} \sqrt{k\,E/f}\right)$$

$$= 0.95 \sqrt{0.43 \times 200 \times 10^3 / 225} \left(1 - \frac{0.208}{16.1} \sqrt{0.43 \times 200 \times 10^3 / 225}\right) = 13.9$$

Effective area

$$A_e = A - 2\,(W_d - B_d)\,t^2 = 2\ 180 - 2\,(16.1 - 13.9)\,6.35^2 = 2\ 000 \text{ mm}^2$$

Compressive resistance

$$C_r = \phi A_e f = 0.90 \times 2\,000 \times 225 = 405 \text{ kN}$$

The table on page 4-115 indicates a compressive resistance of 406 kN.

C. *Compressive Resistance About Axis Y-Y, Torsional-Flexural Mode (Detailed Calculation)*

Shear centre location, S16.1-94, Appendix D

$$x_0 = 0 \qquad y_0 = y - \frac{t}{2} = 31.6 - \frac{6.35}{2} = 28.4 \text{ mm}$$

Torsional-flexural section properties

$$\bar{r}_0^2 = x_0^2 + y_0^2 + r_x^2 + r_y^2 = 0^2 + 28.4^2 + 32.7^2 + 32.8^2 = 2\,950 \text{ mm}^2$$

$$\beta = 1 - \left(\frac{x_0^2 + y_0^2}{\bar{r}_0^2}\right) = 1 - \left(\frac{0^2 + 28.4^2}{2\,950}\right) = 0.727$$

$$F_{ey} = \frac{\pi^2 E}{(KL/r_y)^2} = \frac{\pi^2 \times 200 \times 10^3}{(2\,000/32.8)^2} = 531 \text{ MPa}$$

$$F_{ez} = \left(\frac{\pi^2 E C_w}{(KL)^2} + GJ\right)\frac{1}{A\bar{r}_0^2}$$

$$= \left(\frac{\pi^2 \times 200 \times 10^3 \times 19.3 \times 10^6}{2\,000^2} + 77 \times 10^3 \times 29.3 \times 10^3\right)\frac{1}{2\,180 \times 2\,950} = 352 \text{ MPa}$$

Elastic torsional-flexural buckling stress

$$F_e = \frac{F_{ey} + F_{ez}}{2\beta}\left(1 - \sqrt{1 - \frac{4 F_{ey} F_{ez} \beta}{(F_{ey} + F_{ez})^2}}\right)$$

$$= \frac{531 + 352}{2 \times 0.727}\left(1 - \sqrt{1 - \frac{4 \times 531 \times 352 \times 0.727}{(531 + 352)^2}}\right) = 273 \text{ MPa}$$

Equivalent slenderness parameter

$$\lambda_e = \sqrt{\frac{F_y}{F_e}} = \sqrt{\frac{300}{273}} = 1.05$$

Slenderness ratio of the built-up member, S16.1-94, Clause 19.1.4(b)

$$\rho_o = \left(\frac{KL}{r}\right)_o = \lambda_e \sqrt{\frac{\pi^2 E}{F_y}} = 1.05 \sqrt{\frac{\pi^2 \times 200 \times 10^3}{300}} = 85.2$$

Slenderness ratio of a component angle, with two welded intermediate connectors spaced at $L/3 = 667$ mm and $K = 0.65$

$$\rho_i = \left(\frac{KL}{r}\right)_z = \frac{0.65 \times 667}{16.6} = 26.1$$

Equivalent slenderness ratio

$$\rho_e = \sqrt{\rho_o^2 + \rho_i^2} = \sqrt{85.2^2 + 26.1^2} = 89.1$$

$$\lambda = \rho_e \sqrt{\frac{F_y}{\pi^2 E}} = 89.1 \sqrt{\frac{300}{\pi^2 \times 200 \times 10^3}} = 1.10$$

Calculated stress

$$f = F_y(1 + \lambda^{2n})^{-1/n} = 300(1 + 1.10^{2 \times 1.34})^{-1/1.34} = 162 \text{ MPa}$$

Effective leg widths, S136-94, Clause 5.6.2.1

$$W_{lim} = 0.644 \sqrt{k\, E/f} = 0.644 \sqrt{0.43 \times 200 \times 10^3 / 162} = 14.8$$

$$W_d = \frac{d}{t} = 16.1 > W_{lim} = 14.8$$

$$W_b = \frac{b}{t} = 12.0 < W_{lim} = 14.8$$

Therefore, the effective width of the long leg must be reduced.

$$B_d = 0.95 \sqrt{k\, E/f} \left(1 - \frac{0.208}{W_d} \sqrt{k\, E/f}\right)$$

$$= 0.95 \sqrt{0.43 \times 200 \times 10^3 / 162} \left(1 - \frac{0.208}{16.1} \sqrt{0.43 \times 200 \times 10^3 / 162}\right) = 15.4$$

Effective area

$$A_e = A - 2(W_d - B_d)\, t^2 = 2\,180 - 2(16.1 - 15.4)\, 6.35^2 = 2\,120 \text{ mm}^2$$

Compressive resistance

$$C_r = \phi\, A_e\, f = 0.90 \times 2\,120 \times 162 = 309 \text{ kN}$$

D. *Approximate Compressive Resistance About Axis Y-Y* (Using Table on page 4-115)

The actual length $L = 2\,000$ mm is replaced by an equivalent length L_e that accounts for the slenderness of the component angles between the connectors.

$$L_e = r_y \sqrt{\left(\frac{KL}{r}\right)_y^2 + \left(\frac{KL}{r}\right)_z^2} = 32.8 \sqrt{\left(\frac{2\,000}{32.8}\right)^2 + \left(\frac{0.65 \times 667}{16.6}\right)^2} = 2\,180 \text{ mm}$$

The table indicates $C_r = 323$ kN for $L = 2\,000$ mm and $C_r = 296$ for $L = 2\,500$ mm. The compressive resistance is obtained by interpolation.

$$C_r = 313 \text{ kN}$$

By comparison, the detailed calculation in c) yielded a resistance of 309 kN. For shorter lengths, the value computed using the table is only approximate since the controlling compressive resistance includes a torsional-flexural component that also depends on section properties other than r_y.

E. *Approximate Compressive Resistance of Struts With Spacings Other Than 10 mm*

The actual length is replaced by an equivalent length based on the radius of gyration of the built-up section r'_y and the slenderness ratio of the component angles. Consider a double angle strut with long legs spaced 16 mm back-to-back ($r'_y = 35.0$ mm).

$$L_e = r_y \sqrt{\left(\frac{KL}{r'}\right)_y^2 + \left(\frac{KL}{r}\right)_z^2} = 32.8 \sqrt{\left(\frac{2\,000}{35.0}\right)^2 + \left(\frac{0.65 \times 667}{16.6}\right)^2} = 2\,060 \text{ mm}$$

By interpolation

$C_r = 320$ kN

BRACING ASSEMBLIES

General

The following example illustrates the design of a bracing assembly in accordance with S16.1-94 Clause 20.2.

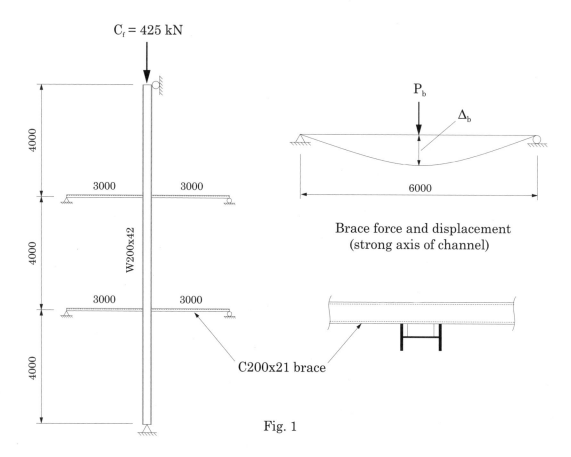

Fig. 1

Example

A W200x42 column is braced about its weak axis by channels located at the one-third points, as shown on Fig. 1. Given a factored axial load of 425 kN acting on the 12-m column, design the channel braces.

Solution

The design objective consists in sizing the channel braces with sufficient stiffness and strength to force the column into a buckling mode between bracing points.

A. Initial Imperfections

For the assumed imperfect shape shown on Fig. 2, the initial imperfection is:

$\Delta_o = 8.0$ mm

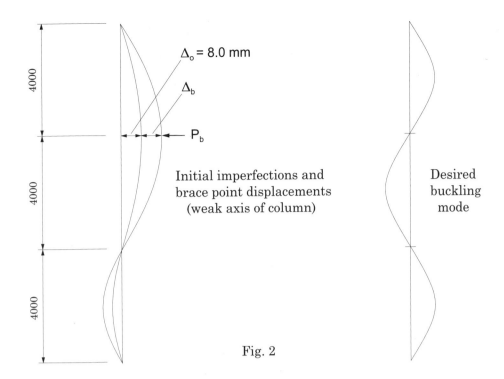

Fig. 2

B. *Stiffness Requirement*

The required flexural stiffness of the channel braces is calculated according to Clause 20.2.1(a). For two equally spaced braces, $\beta = 3$. The brace point displacement is taken equal to the initial imperfection:

$\Delta_b = \Delta_o = 8.0$ mm

The brace stiffness is given by:

$$k_b = \frac{\beta C_f}{L}\left(1 + \frac{\Delta_o}{\Delta_b}\right) = \frac{3 \times 425 \times 10^3}{4000}\left(1 + \frac{8.0}{8.0}\right) = 638 \text{ N/mm}$$

For channels of length $L = 6000$ mm, the required strong-axis moment of inertia is:

$$I_x = \frac{k_b L^3}{48 E} = \frac{638 \times 6000^3}{48 \times 200 \times 10^3} = 14.4 \times 10^6 \text{ mm}^4$$

Try C200x21 channels: $I_x = 14.9 \times 10^6$ mm^4 > 14.4×10^6 mm^4

C. *Strength Requirement*

The factored brace force is calculated according to Clause 20.2.1(b):

$P_b = k_b \Delta_b = 638 \times 8.0 = 5100$ N $= 5.1$ kN

Factored moment acting on a channel:

$M_f = P_b L / 4 = 5.1 \times 6.0 / 4 = 7.65$ kN·m

The moment resistance of a C200x21 channel with unbraced length $L/2 = 3000$ mm and $F_y = 300$ MPa may determined using the Beam Selection Table, p. 5-92:

$M_r' = 27.5$ kN·m > 7.65 kN·m

The selected channel section is adequate.

COLUMN BASE PLATES

When steel columns bear on concrete footings, steel base plates are required to distribute the column load to the footing without exceeding the bearing resistance of the concrete. In general the ends of columns are saw-cut or milled to a plane surface so as to bear evenly on the base plate. Connection of the column to the base plate and then to the footing depends on the loading conditions. For columns carrying vertical gravity loads only, this connection is required only to hold the parts in line, and two anchor bolts are generally sufficient. When practical, to facilitate erection, four anchor bolts should be considered.

For base plates subjected to vertical gravity loads only, the following assumptions and design method are recommended:

1. The factored gravity load is assumed uniformly distributed over the base plate within a rectangle of $0.95d \times 0.80b$ (see diagram).
2. The base plate exerts a uniform pressure over the footing.
3. The base plate projecting beyond the area of $0.95d \times 0.80b$ acts as a cantilever subject to the uniform bearing pressure.

C_f = total factored column load (kN)

A = $B \times C$ = area of plate (mm²)

t_p = plate thickness (mm)

F_y = specified minimum yield strength of base plate steel (MPa)

f'_c = specified 28-day strength of concrete (MPa)

ϕ = 0.90 for steel

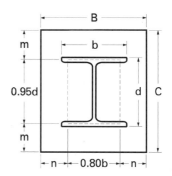

1. Determine the required area $A = C_f/B_r$ where B_r is the factored bearing resistance per unit of bearing area. For concrete, B_r is assumed to be $0.85 \phi_c f'_c$ where $\phi_c = 0.60$ in bearing. (Clause 10.8 of CSA-A23.3-94 states when B_r may be increased.)
2. Determine B and C so that the dimensions m and n (the projections of the plate beyond the area $0.95d \times 0.80b$) are approximately equal.
3. Determine m and n and solve for t_p, where

$$t_p = \sqrt{\frac{2 C_f m^2}{BC \phi F_y}} \quad \text{or} \quad \sqrt{\frac{2 C_f n^2}{BC \phi F_y}} \quad \text{whichever is greater.}$$

These formulae were derived by equating the factored moment acting on the portion of the plate taken as a cantilever to the factored moment resistance of the plate ($M_r = \phi Z F_y$) and solving for the plate thickness t_p.

To minimize deflection of the base plate, the thickness should be generally not less than about 1/5 of the overhang, m or n.

Examples

1. **Given:**

    A W310x118 column subjected to a factored axial load of 2 500 kN is supported by a concrete foundation whose 28-day specified strength is 20 MPa. Design the base plate assuming 300 MPa steel.

Solution:

For W310x118, $b = 307$ mm, $d = 314$ mm.

Area of plate required $= \dfrac{2\,500}{0.85 \times 0.60 \times 20/10^3} = 245\,000 \text{ mm}^2$

Try $B = C = 500$ mm; $A = 250\,000 \text{ mm}^2$

Determine m and n

$\quad 0.95d = 0.95 \times 314 = 298$ mm

$\quad\quad$ Therefore, $m = (500 - 298)/2 = 101$ mm

$\quad 0.80b = 0.80 \times 307 = 246$ mm

$\quad\quad$ Therefore, $n = (500 - 246)/2 = 127$ mm

Use n for design

Plate thickness required $= \sqrt{\dfrac{2 \times 2\,500 \times 127^2}{500 \times 500 \times 0.90 \times 300 /10^3}} = 34.6 \text{ mm}$

$\dfrac{n}{5} = \dfrac{127}{5} = 25 \text{ mm} \quad < 34.6 \text{ mm} \quad - \text{OK}$

Use 35 mm (first preference plate thickness).

Since the plate thickness of 35 mm is less than 65 mm, $F_y = 300$ MPa for G40.21 Grade 300W steel. For plates greater than 65 mm in thickness, $F_y = 280$ MPa for 300W steel (see Table 6-3).

Therefore, use PL 35x500x500 for the base plate.

2. **Given:**

    An HSS 203x203x9.5 column supports a factored axial load of 1 470 kN.

    Select a base plate assuming $f'_c = 20$ MPa and $F_y = 300$ MPa.

Solution:

Area required is $\dfrac{1\,470}{0.85 \times 0.60 \times 20/10^3} = 144\,000 \text{ mm}^2$

$B = C = \sqrt{A} = \sqrt{144 \times 10^3} = 379$ mm

$n = \dfrac{380 - (203 - 9)}{2} = 93$ mm

Therefore, $t_p = \sqrt{\dfrac{2 \times 1\,470 \times 93^2}{379 \times 379 \times 0.9 \times 300/10^3}} = 25.6 \text{ mm}$

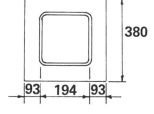

Use 30 mm (first preference plate thickness).

Therefore, use PL 30x380x380 for the base plate.

Design Chart

As an alternative to computing the plate thickness, Figure 4-1 provides a means of selecting t_p knowing the length of cantilever m or n and the unit factored bearing resistance.

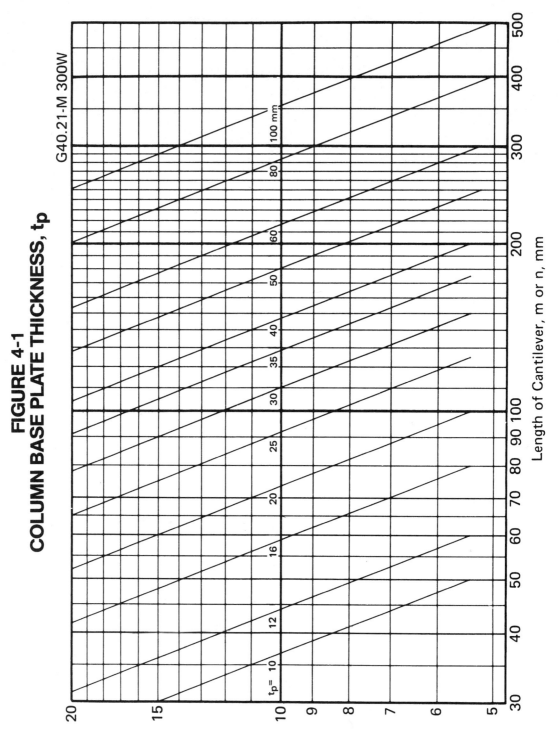

**FIGURE 4-1
COLUMN BASE PLATE THICKNESS, t_p**

Example

Given:

Same as example 1

Solution:

Unit factored bearing resistance is $0.85 \times 0.60 \times 20 = 10.2$ MPa

From Figure 4-1 for 10.2 MPa and $n = 127$, select $t_p = 35$ mm.

Base plate assemblies (including anchor bolts) that are subjected to applied bending moments, uplift tension, and shear forces must be designed to resist all such forces.

Lightly Loaded Base Plates

For lightly loaded base plates where the required bearing area is less than or about equal to the area bounded by the column dimensions b and d, the above method does not give realistic results for the base plate thickness, and other methods have been proposed in the literature. Fling (1970) uses a yield line theory to derive an equation for plate thickness. When modified for limit states design the equation becomes:

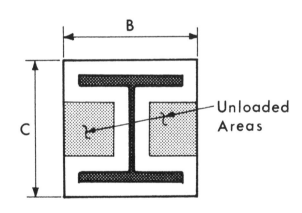

$$t_p = 0.43\, b\, \beta \sqrt{\frac{B_r}{\phi F_y (1 - \beta^2)}}$$

where

$B_r = 0.85\, \phi_c f'_c$

$\beta = \sqrt{0.75 + \dfrac{1}{4\lambda^2}} - \dfrac{1}{2\lambda}$

$\lambda = 2d/b$

b = column width (mm)

d = column depth (mm)

Stockwell (1975) assumes an effective bearing area where only an H-shaped pattern under a W column is loaded, and that the remainder of the base plate is unloaded. The assumed width of flange strips can be derived from the required bearing area, and the thickness of plate can be determined by the expression:

$$t_p = \sqrt{\frac{2\, C_f\, m^2}{A\, \phi\, F_y}}$$

where

A = the effective bearing area

m = half the width of the bearing strips

References

FLING, R.S. 1970. Design of steel bearing plates. Engineering Journal, American Institute of Steel Construction, **7**(2), April.

STOCKWELL, F.J.Jr. 1975. Preliminary base plate selection. Engineering Journal, American Institute of Steel Construction, **12**(3), third quarter.

ANCHOR RODS

Theoretically, anchor rods are not required at base plates for concentrically loaded columns carrying gravity loads only since neither end moments nor horizontal forces are present. In practice, however, anchor rods are provided to locate the column base, to provide a means for levelling the base plate, and to resist nominal end moments and horizontal forces which may occur. *Note: the expression "anchor rod" has replaced "anchor bolt" in order to avoid confusion with bolts produced to ASTM A325 and A490 (See ASTM F1554).*

Clauses 25 and 26 cover requirements for the design of column bases and anchor rods for cases where anchor rods are required to transfer end moments and horizontal forces due to lateral loads.

Fabricators normally supply anchor rods manufactured from round bar stock. The bars are threaded at one end to receive a washer and nut and may be bent at the other end to form a hook, or both ends may be threaded. The material used for most common applications is usually produced to CSA-G40.21 Grade 300W (F_y = 300 MPa) or to ASTM A36 (F_y = 248 MPa). However, ASTM A36 round bar stock is generally more readily available, and should be assumed for anchor rod design in most situations. For specialized applications, fastener suppliers or fabricators should be consulted.

The diameter of anchor rod holes in base plates should provide for possible horizontal adjustments for alignment purposes. The following table of Suggested Anchor Rod Hole Sizes can be used as a guide, although actual sizes used by fabricators may vary depending on shop and field practices. If hole diameters other than those suggested are required or if other requirements are necessary for such situations as column bases designed to transfer end moments and/or horizontal loads, they should be clearly identified in the contract documents.

References

CHIEN, E.Y.L. 1991. Low rise building design aid. Canadian Institute of Steel Construction, Willowdale, Ontario.

CHIEN, E.Y.L. 1985. Single storey building design aid. Canadian Institute of Steel Construction, Willowdale, Ontario.

K.S.M. 1974. Structural engineering aspects of headed concrete anchors and deformed bar anchors in the concrete construction industry. K.S.M. Welding Systems Division, Omark Industries.

NIXON, D. and ADAMS, P. F. 1978. Design of light industrial buildings. Proceedings, Canadian Structural Engineering Conference, C.I.S.C., Willowdale, Ontario.

STELCO Inc. 1983. Structural steels: selection and uses, part 8.

SUGGESTED ANCHOR ROD HOLE SIZES*

Rod Diameter mm	Suggested Hole Diameter mm	Rod Diameter mm	Suggested Hole Diameter mm
20	26	36	48
22	28	42	54
24	30	48	60
27	33	56	68
30	42	64	76

* Anchor rod holes in base plates which will receive anchor rods that are grouted may be flame cut.
Anchor rod hole sizes will vary with individual fabricators depending on shop and field practices.

PART FIVE
FLEXURAL MEMBERS

General Information	5–3
Class of Sections in Bending	5–4
Factored Ultimate Shear Stress in Girder Webs	5–9
Plate Girders	5–12
Composite Beams	5–20
Factored Shear Resistance of Studs	
Solid Slabs or Deck-Slabs with Ribs Parallel to Beam	5–23
Deck-slabs with Ribs Transverse to Beam	5–24
Trial Selection Tables	
75 mm Deck with 65 mm Slab, 20 MPa, 2300 kg/m^3 concrete	5–28
75 mm Deck with 75 mm Slab, 25 MPa, 2300 kg/m^3 concrete	5–36
75 mm Deck with 85 mm Slab, 25 MPa, 1850 kg/m^3 concrete	5–44
75 mm Deck with 85 mm Slab, 25 MPa, 2000 kg/m^3 concrete	5–52
75 mm Deck with 90 mm Slab, 20 MPa, 2300 kg/m^3 concrete	5–60
Deflection of Flexural Members	5–68
Factored Resistances of Beams	5–72
Beam Selection Tables	
CSA-G40.21 350W — WWF and W Shapes	5–78
CSA-G40.21 300W — C Shapes	5–92
ASTM A572 Gr. 50 — S Shapes	5–94
Beam Load Tables	
CSA-G40.21 350W — WWF Shapes	5–96
CSA-G40.21 350W — W Shapes	5–104
CSA-G40.21 350W — Rectangular HSS	5–126
Beam Diagrams and Formulae	5–132
Beam Bearing Plates	5–151
Beams with Web Holes	5–154

GENERAL INFORMATION

Class of Sections in Bending

See page 5-4

Factored Ultimate Shear Stress in Girder Webs

The tables on pages 5-9 to 5-11 list the factored ultimate shear stress ϕF_s in a girder web, computed in accordance with the requirements of Clause 13.4.1.1 of CAN/CSA-S16.1-94, and the required gross area of pairs of intermediate stiffeners, computed in accordance with the requirements of Clause 15.7.3 of S16.1-94. Values are provided for minimum specified yield strength levels F_y of 248, 300 and 350 MPa, for aspect ratios (a/h) from 0.50 to 3.00, and for web slenderness ratios (h/w) varying between 60 and 320 for F_y = 248, between 50 and 260 for F_y = 300, and between 50 and 220 for F_y = 350. The required gross area of stiffeners is provided as a percentage of web area $(h\,w)$ and is shown in italics.

Plate Girders

For design information and illustrative example, see page 5-12.

Composite Beams

Tables for Shear Resistance of Shear Studs in solid slabs and in deck-slabs are given on pages 5-23 to 5-25, and Trial Selection Tables for composite beams with various combinations of cover slab and cellular steel deck (hollow composite construction), and with solid slabs are given on pages 5-28 to 5-67. See page 5-20 for explanatory text.

Deflection of Flexural Members

See page 5-68 for design chart, table, and illustrative examples.

Factored Resistance of Beams

Beam Selection Tables, which list the factored moment resistance of beams under various conditions of lateral support, are provided on pages 5-78 to 5-95 to facilitate the design of flexural members. See page 5-72 for explanatory text.

Beam Load Tables which list total uniformly distributed factored loads for laterally supported beams of various spans are provided on pages 5-96 to 5-131. For explanatory text, see page 5-89.

Beam Diagrams and Formulae

Pages 5-132 to 5-150 contain diagrams and formulae to facilitate the design of flexural members in accordance with "elastic theory".

Beam Bearing Plates

See page 5-151 for design information, design chart and illustrative example.

Beams with Web Holes

See page 5-154 for design information, design tables and illustrative example.

Guide for Floor Vibration

See Appendix G of CAN/CSA S16.1-94, page 1-116.

CLASS OF SECTIONS IN BENDING

Table 5-1 on pages 5-5 to 5-8 lists the class of section in bending of WWF sizes and W shapes for four common grades of steel. Listed are the Canadian WWF sizes and W shapes plus the non-Canadian W shapes provided in Part Six of this Handbook. For WWF sizes and Canadian W shapes, values are given for CAN/CSA-G40.21 grades 300W and 350W, ASTM A36 and ASTM A572 grade 50. For non-Canadian W shapes, values are given for ASTM A36 and ASTM A572 grade 50. For these four steel grades all S shapes are Class 1, and all C and MC shapes are Class 3.

Table 5-1 also lists for each section size the ratios b/t and h/w, where; b = one-half the flange width, t = flange thickness, h = clear distance between flanges and w = web thickness. See also "Limits on Width-Thickness Ratios", page 4-5.

Table 5-1 — CLASS OF SECTIONS IN BENDING

Designation	CSA G40.21 300W	CSA G40.21 350W	ASTM A36	ASTM A572 Gr 50	b/t	h/w
WWF2000x732	2	3			5.50	95.0
x648	2	3			6.88	96.0
x607	2	3			7.86	96.5
x542	2	3			8.33	97.0
WWF1800x700	2	2			5.50	85.0
x659	2	2			6.11	85.5
x617	2	2			6.88	86.0
x575	2	2			7.86	86.5
x510	2	2			8.33	87.0
WWF1600x622	2	3			5.50	93.8
x580	2	3			6.11	94.4
x538	2	3			6.88	95.0
x496	2	3			7.86	95.6
x431	2	3			8.33	96.3
WWF1400x597	2	2			5.50	81.3
x513	2	2			6.88	82.5
x471	2	2			7.86	83.1
x405	2	2			8.33	83.8
x358	2	2			6.67	83.8
WWF1200x487	2	2			6.88	70.0
x418	2	2			7.14	70.6
x380	2	2			8.33	71.3
x333	2	2			6.67	71.3
x302	2	2			8.00	71.9
x263	2	2			6.00	71.9
WWF1100x458	2	2			6.88	72.9
x388	2	2			7.14	73.6
x351	2	2			8.33	74.3
x304	2	2			6.67	74.3
x273	2	2			8.00	75.0
x234	2	2			6.00	75.0
WWF1000x447	2	2			6.88	65.7
x377	2	2			7.14	66.4
x340	2	2			8.33	67.1
x293	2	2			6.67	67.1
x262	2	2			8.00	67.9
x223	2	2			6.00	67.9
x200	2	2			7.50	68.6
WWF900x417	2	2			6.88	74.5
x347	2	2			7.14	75.5
x309	2	2			8.33	76.4
x262	2	2			6.67	76.4
x231	2	2			8.00	77.3
x192	2	2			6.00	77.3
x169	2	2			7.50	78.2
WWF800x339	2	2			7.14	66.4
x300	2	2			8.33	67.3
x253	2	2			6.67	67.3
x223	2	2			8.00	68.2
x184	2	2			6.00	68.2
x161	2	2			7.50	69.1

Designation	CSA G40.21 300W	CSA G40.21 350W	ASTM A36	ASTM A572 Gr 50	b/t	h/w
WWF700x245	1	1			6.67	58.2
x214	1	2			8.00	59.1
x196	2	3			9.09	59.6
x175	1	2			6.00	59.1
x152	1	2			7.50	60.0
WWF650x864	1	1			5.42	8.83
x739	1	1			5.42	17.7
x598	1	1			6.50	27.5
x499	1	2			8.13	28.5
x400	3	4			10.8	29.5
WWF600x793	1	1			5.00	8.00
x680	1	1			5.00	16.0
x551	1	1			6.00	25.0
x460	1	1			7.50	26.0
x369	3	3			10.0	27.0
WWF550x721	1	1			4.58	7.17
x620	1	1			4.58	14.3
x503	1	1			5.50	22.5
x420	1	1			6.88	23.5
x280	3	4			11.0	31.3
WWF500x651	1	1			4.17	6.33
x561	1	1			4.17	12.7
x456	1	1			5.00	20.0
x381	1	1			6.25	21.0
x343	1	1			7.14	21.5
x306	1	2			8.33	22.0
x276	2	2			8.93	27.8
x254	3	3			10.0	28.1
x223	3	4			11.4	32.6
x197	4	4			12.5	41.8
WWF450x503	1	1			3.75	11.0
x409	1	1			4.50	17.5
x342	1	1			5.63	18.5
x308	1	1			6.43	19.0
x274	1	1			7.50	19.5
x248	1	2			8.04	24.6
x228	2	2			9.00	25.0
x201	3	3			10.2	29.0
x177	3	4			11.3	37.3
WWF400x444	1	1			3.33	9.33
x362	1	1			4.00	15.0
x303	1	1			5.00	16.0
x273	1	1			5.71	16.5
x243	1	1			6.67	17.0
x220	1	1			7.14	21.5
x202	1	2			8.00	21.9
x178	2	3			9.09	25.4
x157	3	3			10.0	32.7
WWF350x315	1	1			3.50	12.5
x263	1	1			4.38	13.5
x238	1	1			5.00	14.0
x212	1	1			5.83	14.5
x192	1	1			6.25	18.4
x176	1	1			7.00	18.8
x155	1	2			7.95	21.9
x137	2	2			8.75	28.2

CLASS OF SECTIONS IN BENDING   Table 5-1

Designation	CSA G40.21 300W	CSA G40.21 350W	ASTM A36	ASTM A572 Gr 50	b/t	h/w	Designation	CSA G40.21 300W	CSA G40.21 350W	ASTM A36	ASTM A572 Gr 50	b/t	h/w
W1100x499*			1	1	4.50	39.7	W840x251*			1	1	4.71	46.9
x432*			1	1	5.01	46.5	x226*			1	1	5.49	49.5
x390*			1	1	5.54	51.1	x210*			1	1	6.00	51.8
x342*			1	1	6.45	57.1	x193*			1	1	6.73	54.2
							x176*			1	1	7.77	57.0
W1000x883*			1	1	2.59	20.4							
x749*			1	1	2.97	23.7	W760x582*			1	1	3.19	20.8
x641*			1	1	3.44	27.3	x531*			1	1	3.45	22.8
x591*			1	1	3.66	29.9	x484*			1	1	3.74	24.8
x554*			1	1	3.92	31.5	x434*			1	1	4.12	27.8
x539*			1	1	3.98	32.7	x389*			1	1	4.59	30.5
x483*			1	1	4.39	36.5	x350*			1	1	5.01	34.1
x443*			1	1	4.80	39.3	x314*			1	1	5.75	36.5
x412*			1	1	5.03	44.0	x284*			1	1	6.35	39.9
x371*			1	1	5.54	48.8	x257*			1	1	7.03	43.3
x321*			1	1	6.45	56.2							
x296*			1	1	7.38	56.2	W760x220*			1	1	4.43	43.6
							x196*			1	1	5.28	46.1
W1000x583*			1	1	2.45	25.7	x185*			1	1	5.66	48.2
x493*			1	1	2.86	29.9	x173*			1	1	6.18	49.9
x486*			1	1	2.85	30.9	x161*			1	1	6.89	52.1
x414*			1	1	3.30	35.8	x147*			1	1	7.79	54.5
x393*			1	1	3.45	38.0	x134*			1	2	8.52	60.4
x350*			1	1	3.78	44.0							
x314*			1	1	4.18	48.6	W690x802*			1	1	2.15	12.9
x272*			1	1	4.84	56.2	x548*			1	1	2.95	18.4
x249*			1	1	5.77	56.2	x500*			1	1	3.19	20.2
x222*			1	1	7.11	58.0	x457*			1	1	3.46	21.9
							x419*			1	1	3.71	24.0
W920x1188*			1	1	2.10	14.0	x384*			1	1	4.02	25.9
x967*			1	1	2.48	17.0	x350*			1	1	4.40	28.0
x784*			1	1	2.96	20.7	x323*			1	1	4.71	30.6
x653*			1	1	3.48	24.6	x289*			1	1	5.24	34.0
x585*			1	1	3.82	27.4	x265*			1	1	5.93	35.1
x534*			1	1	4.16	29.9	x240*			1	1	6.50	38.5
x488*			1	1	4.49	32.7	x217*			1	1	7.16	41.9
x446*			1	1	4.95	35.3							
x417*			1	1	5.29	37.7	W690x192*			1	1	4.55	41.7
x387*			1	1	5.74	39.8	x170*			1	1	5.42	44.5
x365*			1	1	6.11	41.7	x152*			1	1	6.02	49.3
x342*			1	1	6.53	43.9	x140*			1	1	6.72	52.1
							x125*			1	1	7.76	55.2
W920x381*			1	1	3.53	35.4							
x345*			1	1	3.86	39.1	W610x551*			1	1	2.51	14.8
x313*			1	1	4.48	40.9	x498*			1	1	2.72	16.3
x289*			1	1	4.81	44.5	x455*			1	1	2.94	17.9
x271*			1	1	5.12	46.9	x415*			1	1	3.18	19.4
x253*			1	1	5.48	49.9	x372*			1	1	3.49	21.7
x238*			1	1	5.89	52.3	x341*			1	1	3.79	23.5
x223*			1	1	6.36	54.3	x307*			1	1	4.14	25.9
x201*			1	1	7.56	56.8	x285*			1	1	4.43	27.8
							x262*			1	1	4.81	30.2
W840x576*			1	1	3.55	24.9	x241	1	1	1	1	5.31	32.0
x527*			1	1	3.85	27.0	x217	1	1	1	1	5.92	34.7
x473*			1	1	4.23	30.2	x195	1	1	1	1	6.70	37.2
x433*			1	1	4.60	32.7	x174	1	1	1	1	7.52	40.9
x392*			1	1	5.03	36.1	x155	2	2	1	2	8.53	45.1
x359*			1	1	5.66	37.8							
x329*			1	1	6.19	40.5							
x299*			1	1	6.85	43.8							

* Not available from Canadian mills

Table 5-1 — CLASS OF SECTIONS IN BENDING

Designation	CSA G40.21 300W	CSA G40.21 350W	ASTM A36	ASTM A572 Gr 50	b/t	h/w	Designation	CSA G40.21 300W	CSA G40.21 350W	ASTM A36	ASTM A572 Gr 50	b/t	h/w
W610x153*			1	1	4.60	40.9	W410x46	1	1	1	1	6.25	54.4
x140	1	1	1	1	5.18	43.7	x39	1	2	1	2	7.95	59.6
x125	1	1	1	1	5.84	48.1							
x113	1	1	1	1	6.59	51.2	W360x1086*			1	1	1.82	4.09
x101	1	1	1	1	7.65	54.6	x990*			1	1	1.95	4.45
x91	2	2	1	2	8.94	59.0	x900*			1	1	2.08	4.84
x84	2	3	2	3	9.66	63.6	x818*			1	1	2.25	5.29
							x744*			1	1	2.43	5.76
W610x92*			1	1	5.97	52.6	x677*			1	1	2.63	6.25
x82*			1	1	6.95	57.3							
							W360x634*			1	1	2.75	6.72
W530x300*			1	1	3.85	21.7	x592*			1	1	2.91	7.12
x272*			1	1	4.23	23.8	x551*			1	1	3.09	7.61
x248*			1	1	4.57	26.4	x509*			1	1	3.32	8.20
x219*			1	1	5.45	27.4	x463*			1	1	3.59	8.94
x196*			1	1	6.01	30.4	x421*			1	1	3.89	9.75
x182*			1	1	6.45	33.0	x382*			1	1	4.23	10.7
x165*			1	1	7.05	35.8	x347*			1	1	4.62	11.8
x150*			1	1	7.68	39.6	x314*			1	1	5.06	12.8
							x287*			1	1	5.45	14.2
W530x138	1	1	1	1	4.53	34.1	x262*			1	1	5.98	15.2
x123	1	1	1	1	5.00	38.3	x237*			1	1	6.54	16.9
x109	1	1	1	1	5.61	43.2	x216*			1	1	7.11	18.5
x101	1	1	1	1	6.03	46.1							
x92	1	1	1	1	6.70	49.2	W360x196*			1	1	7.14	19.5
x82	1	2	1	2	7.86	52.8	x179*			1	1	7.80	21.3
x72	2	3	2	3	9.50	56.4	x162*			1	2	8.51	24.1
							x147*			2	3	9.34	26.0
W530x85*			1	1	5.03	48.7	x134*			2	3	10.3	28.6
x74*			1	1	6.10	51.7							
x66*			1	1	7.24	56.4	W360x122*			1	1	5.92	24.6
							x110*			1	1	6.43	28.1
W460x260*			1	1	3.58	18.9	x101*			1	1	6.97	30.5
x235*			1	1	3.92	20.8	x91*			1	1	7.74	33.7
x213*			1	1	4.25	23.1							
x193*			1	1	4.64	25.2	W360x79	1	1	1	1	6.10	34.1
x177*			1	1	5.32	25.8	x72	1	1	1	1	6.75	37.2
x158*			1	1	5.94	28.5	x64	1	1	1	1	7.52	41.6
x144*			1	1	6.40	31.5							
x128*			1	1	7.19	35.1	W360x57	1	1	1	1	6.56	42.0
x113*			1	2	8.09	39.7	x51	1	1	1	1	7.37	46.1
							x45	2	2	1	2	8.72	48.2
W460x106	1	1	1	1	4.71	34.0							
x97	1	1	1	1	5.08	37.5	W360x39	1	1	1	1	5.98	51.0
x89	1	1	1	1	5.42	40.7	x33	1	1	1	1	7.47	57.2
x82	1	1	1	1	5.97	43.2							
x74	1	1	1	1	6.55	47.6	W310x500*			1	1	2.26	6.14
x67	1	1	1	1	7.48	50.4	x454*			1	1	2.45	6.72
x61	2	2	1	2	8.75	52.9	x415*			1	1	2.66	7.14
							x375*			1	1	2.88	7.81
W460x68*			1	1	5.00	47.1	x342*			1	1	3.12	8.49
x60*			1	1	5.75	53.6	x313*			1	1	3.36	9.25
x52*			1	1	7.04	56.4							
							W310x283*			1	1	3.65	10.3
W410x149*			1	1	5.30	25.6	x253*			1	1	4.03	11.3
x132*			1	1	5.92	28.6	x226	1	1	1	1	4.45	12.5
x114*			1	1	6.76	32.9	x202	1	1	1	1	4.95	13.8
x100*			1	1	7.69	38.1	x179	1	1	1	1	5.57	15.4
							x158	1	1	1	1	6.18	17.9
W410x85	1	1	1	1	4.97	34.9	x143	1	1	1	1	6.75	19.8
x74	1	1	1	1	5.63	39.3	x129	1	1	1	1	7.48	21.1
x67	1	1	1	1	6.22	43.3	x118	1	2	1	2	8.21	23.2
x60	1	1	1	1	6.95	49.5	x107	2	2	2	2	9.00	25.4
x54	1	2	1	2	8.12	50.8	x97	3	3	2	3	9.90	28.0

* Not available from Canadian mills

CLASS OF SECTIONS IN BENDING — Table 5-1

Designation	CSA G40.21 300W	CSA G40.21 350W	ASTM A36	ASTM A572 Gr 50	$\dfrac{b}{t}$	$\dfrac{h}{w}$	Designation	CSA G40.21 300W	CSA G40.21 350W	ASTM A36	ASTM A572 Gr 50	$\dfrac{b}{t}$	$\dfrac{h}{w}$
W310x86	1	2	1	1	7.79	30.5	W150x24*			1	1	4.95	21.1
x79	2	2	1	2	8.70	31.5	x18*			1	1	7.18	23.9
							x14*			1	2	9.09	32.3
W310x74	1	1	1	1	6.29	29.5	x13*			2	3	10.2	32.1
x67	1	1	1	1	6.99	32.6							
x60	1	1	1	1	7.75	36.9	W130x28*			1	1	5.87	15.8
							x24*			1	1	6.98	17.8
W310x52	1	1	1	1	6.33	38.4							
x45	1	1	1	1	7.41	44.0	W100x19*			1	1	5.85	12.5
x39	2	2	1	2	8.51	50.1							
x31	3	4	3	4	11.1	58.2							
W310x33*			1	1	4.72	44.2							
x28*			1	1	5.73	48.5							
x24*			1	1	7.54	52.1							
x21*			1	2	8.86	57.2							
W250x167	1	1	1	1	4.17	11.7							
x149	1	1	1	1	4.63	13.0							
x131	1	1	1	1	5.20	14.6							
x115	1	1	1	1	5.86	16.7							
x101	1	1	1	1	6.56	18.9							
x89	1	1	1	1	7.40	21.1							
x80	1	2	1	2	8.17	23.9							
x73	2	2	1	2	8.94	26.1							
W250x67	1	1	1	1	6.50	25.3							
x58	1	1	1	1	7.52	28.1							
x49	2	3	2	3	9.18	30.4							
W250x45	1	1	1	1	5.69	31.6							
x39	1	1	1	1	6.56	36.3							
x33	1	2	1	2	8.02	39.3							
x24	3	4	3	4	11.3	48.0							
W250x28*			1	1	5.10	37.5							
x25*			1	1	6.07	39.4							
x22*			1	1	7.39	41.4							
x18*			2	3	9.53	50.1							
W200x100	1	1	1	1	4.43	12.5							
x86	1	1	1	1	5.07	13.9							
x71	1	1	1	1	5.92	17.8							
x59	1	1	1	1	7.22	20.0							
x52	1	2	1	2	8.10	22.9							
x46	2	3	2	3	9.23	25.1							
W200x42	1	1	1	1	7.03	25.2							
x36	1	2	1	2	8.09	29.1							
W200x31	1	1	1	1	6.57	29.6							
x27	1	2	1	2	7.92	32.8							
x21	3	3	2	3	10.4	38.0							
W200x22*			1	1	6.38	30.6							
x19*			1	2	7.85	32.8							
x15*			2	3	9.62	44.1							
W150x37	1	1	1	1	6.64	17.1							
x30	1	2	1	2	8.23	21.0							
x22	3	4	3	4	11.5	23.9							

* Not available from Canadian mills

F_y = 248 MPa
ϕ = 0.90

FACTORED ULTIMATE SHEAR STRESS, ϕF_s
GIRDER WEBS (Elastic Analysis), (MPa)

Required Gross Area of Pairs of Intermediate Stiffeners
(Percent of Web Area, hw) * +

Web Slenderness Ratio $\frac{h}{w}$	Aspect Ratio $\frac{a}{h}$: Stiffener Spacing to Web Depth										No Intermediate Stiffeners
	0.50	0.67	0.75	1.00	1.25	1.50	1.75	2.00	2.50	3.00	
60										147 / 0.77	147
70								147 / 1.06	144 / 0.89	141 / 0.77	135
80				147 / 1.46	144 / 1.37	137 / 1.26	132 / 1.15	129 / 1.06	127 / 0.89	125 / 0.77	118
90			147 / 1.50	140 / 1.46	129 / 1.37	125 / 1.26	123 / 1.15	120 / 1.06	117 / 0.89	115 / 0.83	105
100			147 / 1.50	128 / 1.46	123 / 1.37	119 / 1.39	116 / 1.95	113 / 2.19	107 / 2.26	103 / 2.13	86.5
110		147 / 1.49	137 / 1.50	123 / 1.46	118 / 2.52	111 / 3.33	106 / 3.61	102 / 3.64	95.8 / 3.42	91.5 / 3.10	71.4
120	147 / 1.38	137 / 1.49	128 / 1.50	119 / 2.77	111 / 4.30	104 / 4.81	98.0 / 4.88	93.6 / 4.75	87.1 / 4.30	82.5 / 3.83	60.0
130	147 / 1.38	128 / 1.49	125 / 1.50	114 / 4.53	104 / 5.69	97.4 / 5.96	91.7 / 5.86	87.1 / 5.61	80.3 / 4.99	75.6 / 4.40	51.1
140	147 / 1.38	126 / 1.49	122 / 2.09	109 / 5.92	99.7 / 6.80	92.5 / 6.88	86.7 / 6.64	82.0 / 6.29	75.0 / 5.53	70.0 / 4.86	44.1
150	138 / 1.38	123 / 1.74	119 / 3.76	105 / 7.05	95.8 / 7.68	88.5 / 7.62	82.7 / 7.27	77.9 / 6.84	70.7 / 5.97	65.5 / 5.22	38.4
160	129 / 1.38	121 / 3.32	116 / 5.12	102 / 7.97	92.7 / 8.41	85.3 / 8.22	79.4 / 7.79	74.5 / 7.29	67.1 / 6.33		33.7
170	127 / 1.38	118 / 4.64	113 / 6.25	99.2 / 8.73	90.0 / 9.02	82.6 / 8.72	76.6 / 8.21	71.7 / 7.66			29.9
180	126 / 1.38	115 / 5.74	110 / 7.19	97.0 / 9.37	87.8 / 9.52	80.4 / 9.14	74.3 / 8.57	69.3 / 7.97			26.7
190	124 / 1.68	113 / 6.68	108 / 7.99	95.2 / 9.91	86.0 / 9.95	78.5 / 9.49	72.4 / 8.88				23.9
200	123 / 2.87	111 / 7.48	106 / 8.67	93.6 / 10.4	84.4 / 10.3	76.9 / 9.79					21.6
220	119 / 4.77	108 / 8.76	103 / 9.77	91.0 / 11.1	81.9 / 10.9						17.8
240	116 / 6.21	105 / 9.73	101 / 10.6	89.1 / 11.7							15.0
260	114 / 7.34	103 / 10.5	99.2 / 11.3								12.8
280	112 / 8.23	102 / 11.1	97.8 / 11.8								11.0
300	110 / 8.95	101 / 11.6	96.7 / 12.2								9.61
320	109 / 9.54										8.45

\# Clause 13.4.1.1, CAN/CSA-S16.1-94
* For single stiffeners on one side of web only, multiply percentages shown by 1.8 for angle stiffeners and by 2.4 for plate stiffeners.
+ When stiffener F_y is not the same as the web F_y, multiply gross area by ratio (F_y web / F_y stiffener)

FACTORED ULTIMATE SHEAR STRESS, ϕF_s
GIRDER WEBS (Elastic Analysis), (MPa)[#]

$F_y = 300$ MPa
$\phi = 0.90$

Required Gross Area of Pairs of Intermediate Stiffeners
(Percent of Web Area, hw) * +

Web Slenderness Ratio $\frac{h}{w}$	Aspect Ratio $\frac{a}{h}$: Stiffener Spacing to Web Depth										No Intermediate Stiffeners
	0.50	0.67	0.75	1.00	1.25	1.50	1.75	2.00	2.50	3.00	
50											178
60										178 / 0.77	174
70					178 / 1.37	172 / 1.26	166 / 1.15	163 / 1.06	158 / 0.89	155 / 0.77	149
80			178 / 1.50	173 / 1.46	159 / 1.37	153 / 1.26	150 / 1.15	148 / 1.06	144 / 0.89	141 / 0.77	130
90			178 / 1.50	155 / 1.46	149 / 1.37	145 / 1.26	141 / 1.75	138 / 2.02	131 / 2.12	127 / 2.02	106
100		178 / 1.49	166 / 1.50	149 / 1.46	143 / 2.52	135 / 3.33	128 / 3.61	123 / 3.64	116 / 3.42	111 / 3.10	86.5
110	178 / 1.38	164 / 1.49	154 / 1.50	144 / 2.96	133 / 4.46	124 / 4.94	118 / 4.99	112 / 4.84	104 / 4.37	98.9 / 3.90	71.4
120	178 / 1.38	154 / 1.49	151 / 1.50	136 / 4.83	125 / 5.93	116 / 6.16	110 / 6.03	104 / 5.75	95.8 / 5.10	89.9 / 4.50	60.0
130	175 / 1.38	151 / 1.49	147 / 2.62	130 / 6.28	119 / 7.08	110 / 7.11	103 / 6.84	97.6 / 6.46	89.0 / 5.67	83.0 / 4.98	51.1
140	163 / 1.38	148 / 2.41	142 / 4.33	125 / 7.43	114 / 7.99	105 / 7.87	98.3 / 7.49	92.5 / 7.03	83.7 / 6.12	77.4 / 5.35	44.1
150	155 / 1.38	144 / 4.01	138 / 5.70	122 / 8.36	110 / 8.73	102 / 8.48	94.3 / 8.01	88.3 / 7.48	79.3 / 6.49	72.9 / 5.65	38.4
160	153 / 1.38	140 / 5.32	134 / 6.83	118 / 9.12	107 / 9.33	98.3 / 8.98	91.0 / 8.44	85.0 / 7.86	75.8 / 6.78		33.7
170	151 / 1.38	137 / 6.41	131 / 7.76	116 / 9.75	105 / 9.83	95.6 / 9.39	88.2 / 8.79	82.1 / 8.16			29.9
180	149 / 2.64	134 / 7.32	129 / 8.54	114 / 10.3	102 / 10.2	93.4 / 9.74	85.9 / 9.09	79.8 / 8.42			26.7
190	146 / 3.79	132 / 8.10	127 / 9.21	112 / 10.7	101 / 10.6	91.5 / 10.0	84.0 / 9.34				23.9
200	144 / 4.77	130 / 8.75	125 / 9.77	110 / 11.1	99.0 / 10.9	89.9 / 10.3					21.6
220	140 / 6.34	127 / 9.81	122 / 10.7	108 / 11.7	96.5 / 11.4						17.8
240	137 / 7.53	125 / 10.6	120 / 11.4	106 / 12.2							15.0
260	134 / 8.46	123 / 11.2	118 / 11.9								12.8

[#] **Clause 13.4.1.1, CAN/CSA-S16.1-94**
* For single stiffeners on one side of web only, multiply percentages shown by 1.8 for angle stiffeners and by 2.4 for plate stiffeners.
\+ When stiffener F_y is not the same as the web F_y, multiply gross area by ratio (F_y web / F_y stiffener)

$F_y = 350$ MPa
$\phi = 0.90$

FACTORED ULTIMATE SHEAR STRESS, ϕF_s
GIRDER WEBS (Elastic Analysis), (MPa)

Required Gross Area of Pairs of Intermediate Stiffeners
(Percent of Web Area, hw) * +

Web Slenderness Ratio $\frac{h}{w}$	Aspect Ratio $\frac{a}{h}$: Stiffener Spacing to Web Depth										No Intermediate Stiffeners
	0.50	0.67	0.75	1.00	1.25	1.50	1.75	2.00	2.50	3.00	
50									208 / 0.77	207	
60							208 / 1.15	205 / 1.06	199 / 0.89	196 / 0.77	188
70				208 / 1.46	196 / 1.37	186 / 1.26	181 / 1.15	178 / 1.06	174 / 0.89	172 / 0.77	161
80			208 / 1.50	187 / 1.46	177 / 1.37	172 / 1.26	168 / 1.15	165 / 1.29	160 / 1.54	156 / 1.54	135
90		208 / 1.49	199 / 1.50	176 / 1.46	168 / 1.86	161 / 2.79	154 / 3.15	148 / 3.24	140 / 3.09	134 / 2.83	106
100	208 / 1.38	195 / 1.49	181 / 1.50	169 / 2.53	157 / 4.11	147 / 4.66	140 / 4.74	133 / 4.63	124 / 4.21	118 / 3.75	86.5
110	208 / 1.38	180 / 1.49	176 / 1.50	160 / 4.63	147 / 5.78	137 / 6.03	129 / 5.92	122 / 5.66	113 / 5.03	106 / 4.44	71.4
120	205 / 1.38	176 / 1.49	172 / 2.55	152 / 6.23	139 / 7.04	129 / 7.08	121 / 6.82	114 / 6.44	104 / 5.65	97.1 / 4.96	60.0
130	189 / 1.38	173 / 2.48	166 / 4.39	146 / 7.48	133 / 8.03	123 / 7.90	114 / 7.51	108 / 7.05	97.4 / 6.14	90.1 / 5.36	51.1
140	180 / 1.38	168 / 4.18	160 / 5.85	141 / 8.46	128 / 8.81	118 / 8.54	109 / 8.07	103 / 7.53	92.0 / 6.52	84.5 / 5.69	44.1
150	178 / 1.38	163 / 5.56	156 / 7.03	137 / 9.26	124 / 9.44	114 / 9.07	105 / 8.51	98.4 / 7.92	87.7 / 6.84	80.0 / 5.94	38.4
160	176 / 1.69	159 / 6.68	152 / 8.00	134 / 9.91	121 / 9.95	111 / 9.49	102 / 8.88	95.0 / 8.24	84.2 / 7.09		33.7
170	173 / 3.08	156 / 7.62	149 / 8.80	132 / 10.5	119 / 10.4	108 / 9.85	99.4 / 9.18	92.2 / 8.51			29.9
180	169 / 4.24	153 / 8.40	147 / 9.47	129 / 10.9	117 / 10.7	106 / 10.1	97.1 / 9.43	89.9 / 8.73			26.7
190	167 / 5.22	151 / 9.06	145 / 10.0	128 / 11.3	115 / 11.0	104 / 10.4	95.2 / 9.65				23.9
200	164 / 6.06	149 / 9.63	143 / 10.5	126 / 11.6	113 / 11.3	102 / 10.6					21.6
220	160 / 7.41	146 / 10.5	140 / 11.3	123 / 12.1	111 / 11.7						17.8

\# Clause 13.4.1.1, CAN/CSA-S16.1-94
* For single stiffeners on one side of web only, multiply percentages shown by 1.8 for angle stiffeners and by 2.4 for plate stiffeners.
\+ When stiffener F_y is not the same as the web F_y, multiply gross area by ratio (F_y web / F_y stiffener)

PLATE GIRDERS

General

When the required capacity of a steel flexural member exceeds that which can be provided by the rolled or welded wide flange sections listed in Part Six in this book, a plate girder can usually provide an economical solution. A plate girder basically consists of flange plates connected to a relatively thin web plate. Transverse or longitudinal stiffeners (usually consisting of plates or small shapes) may be used to increase the strength of the web. Heavy concentrated loads and reactions are supported directly on bearing stiffeners.

Clause 15 of CAN/CSA S16.1-94 *Limit States Design of Steel Structures* contains requirements for the design of plate girders. Clause 15.1 specifies that girders shall be proportioned on the basis of the section properties, i.e., by the moment of inertia method, where $M_r = \phi(I/y)F_y$. To obtain a trial section for calculating section properties an approximate method is frequently used. One such method assumes the applied bending moment resisted by the flanges alone, so that the required area of one flange A_f is given approximately by $A_f = \dfrac{M_f}{\phi F_y d_{eff}}$, where M_f = bending moment in the girder under factored load, ϕ = the performance factor, 0.90, F_y = specified minimum yield stress and d_{eff} = an assumed effective girder depth. The moment of resistance of the trial section should then be checked using the expression $M_r = \phi(I/y)F_y > M_f$, where I = the moment of inertia of the girder (gross or net) in accordance with Clause 15.1 of S16.1-94, and y = the distance from the neutral axis to the extreme fibre of a plate girder flange. Formulae in Part Seven can be used to compute I.

Flanges

The compression flange may fail either by buckling or by yielding, and will generally govern the flange design of a symmetrical plate girder. S16.1-94 clauses intended to minimize the possibility of buckling are:

1. Local flange buckling
    - Clause 11.2, Table 1
    - $b/2t \leq 200/\sqrt{F_y}$
    - Limits on width-thickness ratios, page 4-5

2. Lateral buckling
    - Clause 13.6

3. Vertical buckling
    - Clause 13.4.1.3
    - $\dfrac{h}{w} \leq \dfrac{83\,000}{F_y}$

where: F_y = specified minimum yield point of compression flange steel (MPa)

h = clear depth of web between flanges (mm)

w = web thickness (mm)

b = flange width (mm)

t = flange thickness (mm)

The portion of a girder web subject to compressive stress tends to deflect laterally, causing a stress redistribution within the girder which results in an increase to the compression flange stress. For relatively thick webs this increase is insignificant, but when the web slenderness ratio h/w exceeds $1\,900/\sqrt{M_f/\phi S}$, Clause 15.4 of S16.1-94 limits the factored moment resistance to M'_r as follows:

$$M'_r = M_r [1.0 - (0.0005 A_w / A_f)(h/w - 1\,900/\sqrt{M_f/\phi S})]$$

where: M'_r = reduced factored moment resistance

M_r = factored moment resistance determined by Clause 13.5 or Clause 13.6 but not exceeding ϕM_y

S = elastic section modulus, I/y (mm^3)

ϕ = performance factor, 0.90

A_w = web area (mm^2)

A_f = compression flange area (mm^2)

and the other symbols are as previously defined.

The constant $1\,900$ in the expression $1\,900/\sqrt{M_f/\phi S}$ is reduced by the factor $(1 - 0.65 C_f/C_y)$ when an axial load acts upon the girder in addition to the bending moments.

When designing plate girders with long spans, economy may be achieved by reducing the area of flange plates where the bending moment is substantially less than the maximum. The cost of making flange splices must, however, be balanced against the weight savings achieved.

Webs

Plate girders are subjected to a complex combination of bending and shear stresses. The shear capacity of unstiffened plate girder webs decreases rapidly as the web slenderness ratio increases. Transverse web stiffeners increase the factored shear resistance of plate girder webs, with the amount of increase depending on the aspect ratio a/h where a = stiffener spacing and h = clear web depth.

Clause 13.4.1.1 of CAN/CSA S16.1-94 specifies factored shear resistances for both stiffened and unstiffened webs (which have been proportioned elastically) for different values of the web slenderness ratio. The tables on pages 5-9, 5-10 and 5-11 list the factored ultimate shear stress ϕF_s, for plate girder webs, from which the factored shear resistance can be computed according to Clause 13.4.1.1. The values are tabulated for various combinations of web slenderness ratio h/w stiffener aspect ratio a/h and for specified minimum yield strengths F_y of 248, 300 and 350 MPa.

The formulae in Clause 13.4.1.1 consider the post buckling strength of transversely stiffened plate girder webs in establishing the factored shear resistance. In addition to shear resistance, web crippling and yielding of thin webs subject to loads acting on the compression edge of the web plate, Clause 15.9, S16.1-94 must be investigated.

Stiffeners

Bearing stiffeners, intermediate transverse stiffeners and longitudinal stiffeners are used with thin web plate girders. Longitudinal stiffeners are seldom used in building construction but are common for bridge construction. CAN/CSA S16.1-94 provides detailed requirements for bearing stiffeners, Clause 15.6, and for intermediate transverse stiffeners, Clause 15.7. The tables on pages 5-9, 5-10 and 5-11 list also the required gross area of pairs of intermediate transverse stiffeners as a percentage of the web area.

When intermediate transverse stiffeners are required, the maximum spacing must not exceed $3h$ for web slenderness ratios h/w less than or equal to 150. For $h/w \geq 150$, the maximum spacing a_{max} is established by the formula:

$$a_{max} = \frac{67\,500\,h}{(h/w)^2}$$

Thin-web stiffened plate girders may be less economical than plate girders with thicker webs which do not require intermediate transverse stiffeners, since possible saving in weight can sometimes be more than overcome by the additional fabrication costs. Framing angles at the ends of plate girders can be assumed to provide web stability at girder ends, provided they comprise at least 2/3 of the girder depth.

References

PART TWO of this Handbook.

KULAK, ADAMS and GILMOR, 1995. Limit states design in structural steel. CISC.

PICARD, A., and BEAULIEU, D., 1991. Calcul des charpentes d'acier. CISC.

BALSER and THÜRLIMANN, 1961. Strength of plate girders in bending. Proceedings, ASCE Journal of the Structural Division, **87**(ST6), August.

BASLER, 1961. Strength of plate girders in shear. Proceedings, ASCE Journal of the Structural Division, **87**(ST7), October.

BASLER, 1961. Strength of plate girders under combined bending and shear. Proceedings, ASCE Journal of the Structural Division, **87**(ST7), October.

Example

Given:

Design a simply supported welded plate girder spanning 22 metres and loaded as shown. The compression flange is laterally supported along the entire length of the

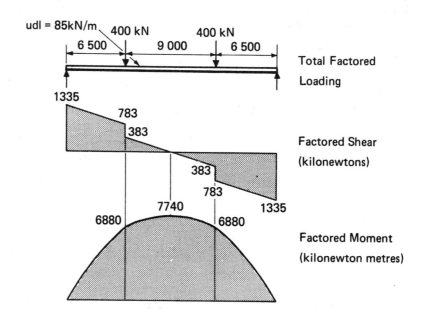

member. The uniform load includes an estimate of 3.50 kN/m for the girder dead load. Total depth is limited to 1 800 mm. Use G40.21 300W Steel.

Solution:

Trial Section – Flanges

Approximate flange area, $A_f = \dfrac{M_f}{\phi \, F_y \, d_{eff}}$

Assume $d_{eff} = 1\,770$ mm

Maximum ultimate stress $= \phi \, F_y = 0.90 \times 300 = 270$ MPa

Required $A_f = \dfrac{7\,740 \times 10^6}{250 \times 1\,770} \times 90\,\% = 14\,600 \text{ mm}^2$

where the 90% factor allows for the contribution of the web to flexural capacity

Flange slenderness ratio, $b/2t \leq 200/\sqrt{F_y} = 11.5$ \hfill (page 4-5)

Try 30 x 500 flange plate, $A_f = 500 \times 30 = 15\,000 \text{ mm}^2$

$b/2t = 500/(2 \times 30) = 8.33 < 11.5$ — OK (Class 1)

Trial Section – Web

Maximum h/w permitted $= 83\,000/F_y$ \hfill (Clause 13.4.1.3, S16.1-94)

$= 83\,000/300 = 277$

Maximum h/w permitted before the factored moment resistance is reduced

$= 1\,900/\sqrt{M_f/\phi S}$ \hfill (Clause 15.4, S16.1-94)

For $M_f = \phi \, S \, F_y$, $h/w = 1\,900/\sqrt{F_y} = 1\,900/\sqrt{300} = 110$

Try 10 x 1 740 web   (10 mm is a first preference thickness)

$h/w = 1\,740/10 = 174$

Factored ultimate shear stress in girder web

$= 1\,335 \times 10^3/(1\,740 \times 10) = 76.7$ MPa

From table on page 5-10, for $h/w = 180$, $a/h = 2.00$ (stiffened web)

$\phi \, F_s = 79.8$ MPa > 76.7 — OK

$A_w = 10 \times 1\,740 = 17\,400 \text{ mm}^2$

Check Trial Section

(a) Compute Girder Properties

Moment of inertia $I = \dfrac{bd^3 - (b-w)h^3}{12}$

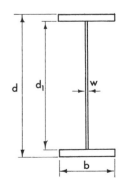

$= \dfrac{500 \times 1\,800^3 - (500-10)\,1\,740^3}{12} = 27\,900 \times 10^6 \text{ mm}^4$

Section modulus $S = I/y = 27\,900 \times 10^6/900 = 31\,000 \times 10^3 \text{ mm}^3$

Flange area $A_f = 15\,000 \text{ mm}^2$

Web area $A_w = 17\,400$ mm^2

(b) Calculate moment resistance.

Since h/w exceeds 110, check reduced moment resistance

$M'_r = M_r [1.0 - (0.0005\, A_w/A_f)(h/w - 1\,900/\sqrt{M_f/\phi S}\,)]$

$M_r = \phi M_y = \phi S F_y = 0.9 \times 31\,000 \times 10^3 \times 300 \times 10^{-6} = 8\,370$ kN·m (Clause 15.4)

$A_w/A_f = 17\,400/15\,000 = 1.16$

$1\,900/\sqrt{M_f/\phi S} = 1\,900/\sqrt{(7\,740 \times 10^6)/(0.9 \times 31\,000 \times 10^3)} = 114$

$M'_r = 8\,370\,[1.0 - 0.0005 \times 1.16\,(174 - 114)]$

$\quad\quad = 8\,080$ kN·m $> 7\,740$ — OK

Locate Web Stiffeners

(a) Establish end panel length for no tension field. (Clause 15.7.1)

By Clause 13.4.1.1(d), for $h/w = 174$ and $F_t = 0$

$V_r = 0.9 \times 17\,400 \times \dfrac{180\,000\,k_v}{174^2 \times 1000} = 93.1\,k_v$

and equating V_r to $V_f = 1\,335$ kN, gives $k_v = 14.34$

but $k_v = 4 + \dfrac{5.34}{(a/h)^2}$, therefore $a/h = \sqrt{5.34/10.34} = 0.719$

Therefore maximum end panel length is $0.719 \times 1\,740 = 1\,250$ mm

(b) Check shear resistance of unstiffened web.

Using table on page 5-10 and interpolating for $h/w = 174$

$\phi F_s = 29.9 - 0.4\,(29.9 - 26.7) = 28.6$ MPa

Therefore V_r (unstiffened web) $= 28.6 \times 17\,400 / 1000 = 498$ kN > 383

Therefore, stiffeners are not required between the two concentrated loads, but are required at the concentrated load locations.

Using an end panel length of $1\,000$ mm, and two equal panels between the end stiffeners and the stiffeners at the concentrated loads gives an intermediate stiffener spacing of

$(6\,500 - 1\,000)/2 = 2\,750$ mm,

and an aspect ratio $a/h = 2\,750/1\,740 = 1.58$

From table on page 5-10 for $a/h = 1.75$ and interpolating for $h/w = 174$

$\phi F_s = 88.2 - 0.4\,(88.2 - 85.9) = 87.3$ MPa (conservative)

V_r (stiffened web) $= 87.3 \times 17\,400 / 1000 = 1\,520$ kN $> V_f$

Stiffener Size

(a) Intermediate Stiffeners

From table on page 5-10, with $h/w = 174$ and $a/h = 1.58$, the required total area of a pair of intermediate stiffeners is approximately 9.5% of the web area.

$$= 0.095 \times 17\,400 = 1\,650 \text{ mm}^2$$

Required $I = (h/50)^4 = (1\,740/50)^4 = 1.47 \times 10^6 \text{ mm}^4$ \hfill (Clause 15.7.3)

Maximum $b/t = 200 / \sqrt{F_y} = 11.5$

Use two 10 x 110 stiffeners

$A_s = 2 \times 10 \times 110 = 2\,200 \text{ mm}^2 > 1\,650$

$I = 10 \times 230^3 / 12 = 10.14 \times 10^6 \text{ mm}^4 > 1.47 \times 10^6 \text{ mm}^4$

$b/t = 110/10 = 11.0 < 11.5$

(b) Stiffeners at concentrated loads

Assuming the concentrated load is applied to the top flange of the girder, the web must be checked for web crippling and web yielding. \hfill (Clause 15.9(a))

Assume the load is applied through a column base plate (N = 300 mm) and that k = flange thickness plus 8 mm (assumed weld size),

B_r (web yielding) is

$$1.10 \times 0.9 \times 10\,[\,300 + 5\,(30 + 8)\,]\,300 / 10^3 = 1\,460 \text{ kN} > 400$$

B_r (web crippling) is

$$300 \times 0.9 \times 10^2\,[1 + 3\,(300/1\,740)\,(10/30)^{1.5}\,]\,\sqrt{300 \times (\,30/10\,)} / 10^3$$

$$= 891 \text{ kN} > 400$$

Therefore, stiffeners are not required to bear at the top flange.

Use two 10 x 110 intermediate stiffeners as above.

Combined Shear and Moment

Combined shear and moment in girder webs is governed by Clause 13.4.1.4 of S16.1-94. The worst effect will occur under the concentrated load.

$V_f = 783$ kN, $\quad M_f = 6\,880$ kN·m

Check whether $h/w = 174$ exceeds $502\,\sqrt{k_v/F_y}$

$$k_v = 5.34 + 4\,/\,(a/h)^2 = 5.34 + 4/1.58^2 = 6.94$$

$$502\,\sqrt{k_v/F_y} = 502 \times \sqrt{6.94/300} = 76.4 < 174$$

Therefore, Clause 13.4.1.4 must be checked.

From table on page 5-10, for $h/w = 174$, and $a/h = 1.58$, $\phi F_s = 87.3$ MPa

Therefore $V_r = 87.3 \times 17\,400 / 10^3 = 1\,520$ kN

$$\frac{V_f}{V_r} = \frac{783}{1\,520} = 0.515 < 1$$

$$\frac{M_f}{M_r} = \frac{6\,880}{8\,080} = 0.851 < 1$$

$$0.727\,\frac{M_f}{M_r} + 0.455\,\frac{V_f}{V_r} = (0.727 \times 0.851) + (0.455 \times 0.515) = 0.853 < 1 \quad \text{— OK}$$

Final Selection

Use — 30 x 500   flange plates

10 x 1 740   web plate

10 x 110   intermediate stiffeners in pairs spaced as shown

Girder UDL, excluding stiffeners

Flanges — $2 \times 30 \times 500$ = 30 000

Web — $10 \times 1\,740$ = 17 400

Total area — = 47 400 mm^2

Dead Load is $47\,400 \times 7\,850 \text{ kg/m}^3 \times 9.81 / (10^6 \times 10^3)$ = 3.65 kN/m

Assumed dead load was 3.50 kN/m

Since there is reserve moment and shear capacity, this slight increase in UDL has no effect on the design. If desired, the flange area could be reduced and a thinner web with additional stiffeners might be used to obtain a lighter girder, provided the saving in weight is not offset by possible additional fabrication costs.

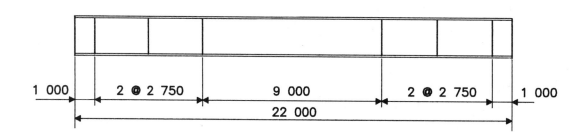

Girder-to-Column Connection

Assuming the girder ends bear on top of supporting columns, then bearing stiffeners designed according to Clause 15.6 of S16.1-94 must be investigated.

$1100 / \sqrt{F_y} = 1100 / \sqrt{300} = 63.5 < h/w = 174$

Therefore, bearing stiffeners are required.

Try two 16 x 175 stiffeners

$b/t = 175/16 = 10.9 < 200 / \sqrt{F_y} = 11.5$

Area = $(2 \times 175 \times 16) + (12 \times 10 \times 10) = 5\,600 + 1\,200 = 6\,800$ mm^2

$I = 16 \times 360^3 / 12 = 62.2 \times 10^6$ mm^4

$r = \sqrt{62.2 \times 10^6 / 6\,800} = 95.6$ mm

$$\frac{KL}{r} = 0.75 \times 1\,740 / 95.6 = 13.7 \quad \text{(for top flange laterally supported at the reaction)}$$

Using Table 4-4 on page 4-11, $\quad C_r/A = 268$ MPa

Therefore $C_r = 268 \times 6\,800 / 1\,000 = 1\,820$ kN $> 1\,335$ kN

Bearing on contact area (Clause 13.10.1(a))

Assume 150 mm bearing (stiffener clipped 25 mm to clear fillet welds)

Bearing area $= 2 \times 150 \times 16 = 4\,800$ mm^2

$B_r = 1.50 \times 0.9 \times 300$ MPa $\times 4\,800$ mm$^2 / 1000 = 1\,940$ kN $> 1\,335$ kN

If the girder ends are framed to the supporting columns, the girder-to-column connections may consist of pairs of header angles bolted or welded to the web plate and connected to the supporting column flange. These connection angles must also serve as end web stiffeners; therefore, their length must be almost equal to the clear depth of the girder web.

Note: Although not considered in this example, a deep header connection as described above can develop significant bending moment at the column, and the effects of this moment upon the members must be considered in design. A method for assessing this moment is described in "Advanced Design in Structural Steel" by J. E. Lothers, Prentice Hall, 1960.

COMPOSITE BEAMS

General

A composite beam, in general, consists of a steel beam and concrete slab so interconnected that both the steel beam and the slab act jointly to resist bending. Several combinations which effectively act as composite beams occur in practice. These include a steel beam or girder with a concrete slab inter-connected with mechanical shear connectors, a steel beam or girder with a ribbed concrete slab formed by steel deck inter-connected by mechanical shear connectors, and a steel beam or girder fully encased by the concrete in such a way that the encased beam and the concrete slab behave monolithically. Clause 17 of CAN/CSA-S16.1-94 *Limit States Design of Steel Structures* contains requirements for composite beams.

Some advantages of composite construction are:

- Reduced weight of steel members
- Reduced depth of steel members
- Reduced deflections under superimposed load
- Simplified changes to electrical services when steel deck is used.

Composite construction is most advantageous when heavy loads and long spans are involved. For this reason composite construction is widely used for bridges. For building construction, composite beams consisting of steel beams with steel deck and concrete cover slab utilizing steel stud shear connectors welded to the beam top flange are most frequently used. Other types of composite construction used in buildings include composite trusses and joists, and stub-girders.

A comprehensive examination of the design and behaviour of steel-concrete composite floor systems and their components is contained in the CISC publication *Design and Construction of Composite Floor Systems* (Chien and Ritchie 1984). This book contains design aids and examples for composite beams, girders, trusses, and stub-girders. The Windows based GFD4 computer program is also available from CISC to assist in the design of composite floor systems.

Tables

The Composite Beam Trial Selection Tables from pages 5-28 to 5-67 are based on CSA G40.21 350W steel and list composite members for the practical range of rolled W shapes from 200 mm to 610 mm nominal depth, and for eleven WWF sizes. (See also "Principal Sources of Structural Steel Sections", page 6-17, and "Rolled Structural Shapes", page 6-38 for comments regarding availability.) Tables are provided for the following combinations of deck-slab concrete strength and concrete density:

- 75 mm steel deck with 65 mm cover slab with f'_c of 20 MPa, 2 300 kg/m^3 concrete
- 75 mm steel deck with 75 mm cover slab with f'_c of 25 MPa, 2 300 kg/m^3 concrete
- 75 mm steel deck with 85 mm cover slab with f'_c of 25 MPa, 1 850 kg/m^3 concrete
- 75 mm steel deck with 85 mm cover slab with f'_c of 25 MPa, 2 000 kg/m^3 concrete
- 75 mm steel deck with 90 mm cover slab with f'_c of 20 MPa, 2 300 kg/m^3 concrete

The tables show steel shapes listed in descending order of nominal depth and mass, and include the following properties, design data and resistances:

b = flange width of steel shape (mm)

t = flange thickness of steel shape (mm)

d = overall depth of steel shape (mm)

b_1 = effective width of slab used in computing values of M_{rc}, Q_r, I_t, S_t and I_t^*, (mm). (Refer to Clause 17.4 of S16.1-94 for appropriate design effective width.)

M_{rc} = factored moment resistance of composite beam for percentage of full shear connection equal to 100%, 70% and 40% (kN·m)

Q_r = required sum of factored shear resistances between adjacent points of maximum and zero moment for 100 per cent shear connection, (kN). Q_r = lesser of $\phi A_s F_y$ or $0.85 \phi_c b_1 t_c f'_c$, where t_c = effective slab thickness or effective cover slab thickness

I_t = moment of inertia of the composite section, transformed into steel properties, computed using mass density as shown on each table (10^6 mm^4)

S_t = section modulus of the composite section related to the extreme fibre of the bottom flange of the steel beam based on the value of I_t (10^3 mm^3)

I_t^* = transformed moment of inertia for calculating shrinkage deflection, based on modular ratio $n_t = 50$ (10^6 mm^4)

M_r = factored moment resistance of laterally supported bare steel section (kN·m)

V_r = factored shear resistance of the bare steel beam (kN)

L_u = maximum unsupported length of compression flange of the steel beam alone for which no reduction in M_r is required (mm)

I_x = moment of inertia about the x-x axis of the bare steel beam (10^6 mm^4)

S_x = section modulus of the bare steel beam (10^3 mm^3)

M'_r = factored moment resistance of the bare steel beam for an unsupported length L' (kN·m).

Since the concrete slab and/or the steel deck prevent movement of the top flange, lateral buckling is not a consideration at composite action. During construction, however, the unsupported length of the compression flange may be greater than L_u, and the moment resistance for the non-composite shape for the appropriate unsupported length of compression flange must be used.

The tabulated factored shear resistance V_r is computed according to Clause 13.4.1.1 of S16.1-94 for the appropriate h/w ratio.

Shear Connectors

Clauses 17.9.5 and 17.9.6 of S16.1-94 stipulate the amount of total factored horizontal shear force that must be resisted by shear connectors.

For full (i.e., 100%) shear connection the total factored horizontal shear force V_h to be transferred between the point of maximum positive moment and adjacent points of zero moment is either

- $\phi A_s F_y$ when the plastic neutral axis is in the slab, or
- $0.85 \phi_c b_1 t_c f'_c$ when the plastic neutral axis is in the steel section.

For partial shear connection the total factored horizontal shear force V_h is the sum of the factored resistances of all the shear connectors between the point of maximum positive moment and each adjacent point of zero moment. Clause 17.9.4 of S16.1-94 limits the minimum amount of partial shear connection to 40 percent of either $\phi A_s F_y$ or $0.85 \phi_c b_1 t_c f'_c$, whichever is the lesser, when computing flexural strength.

Generally, shear connectors may be uniformly spaced in regions of positive or negative bending. However, when a concentrated load occurs within a region of positive bending, the number of shear connectors and the shear connector spacing is determined by Clause 17.9.8 of S16.1-94.

Tables 5-2, 5-3 and 5-4 provide values of factored shear resistance q_r for the most common sizes of end-welded shear studs according to the requirements of Clause 17.7 of S16.1-94 when the stud height is at least four stud diameters, and when the stud projection in a ribbed slab is at least two stud diameters above the top surface of the steel deck.

Table 5-2 on page 5-23 gives values of q_r for stud diameters of 3/4 inch (19 mm), 5/8 inch (15.9 mm), and 1/2 inch (12.7 mm) in solid slabs, or in deck-slabs with ribs parallel to the beam, based on three concrete strength levels f'_c of 20 MPa, 25 MPa, and 30 MPa for both normal density (2 300 kg/m^3) and semi-low density (1 850 kg/m^3) concrete. Values are calculated according to Clause 17.7.2.1 and Clause 17.7.2.2 of S16.1-94.

Table 5-3 on page 5-24 and Table 5-4 on page 5-25 give values of q_r for 3/4 inch (19 mm) and 5/8 inch (15.9 mm) diameter studs respectively in ribbed slabs for 75 mm or 38 mm deck, with ribs perpendicular to the beam, calculated according to Clause 17.7.2.3 of S16.1-94. Values are given for three concrete strength levels f'_c of 20 MPa, 25 MPa, and 30 MPa for both normal density (2 300 kg/m^3) and semi-low density (1 850 kg/m^3) concrete.

Deflections

Composite beams are stiffer than similar non-composite beams, and deflections are reduced when composite construction is used. Due to creep of the concrete slab over time, maximum deflections may increase, especially if the full load is sustained. Appendix L of CAN/CSA-S16.1-94 provides guidance for estimating deflections caused by shrinkage of the concrete slab. Beam deflection during construction, due to loads supported prior to hardening of the concrete while the steel beam alone supports the loads, should be checked. Cambering or the use of temporary shores will reduce the total final deflection.

For steel beams unshored during construction, Clause 17.11 of S16.1-94 limits the stress (caused by the total of specified loads applied before the concrete strength reaches $0.75 f'_c$ and, at the same location, the remaining specified loads acting on the composite section) in the tension flange to F_y.

Other Composite Members

Other composite members suitable for floor construction include composite trusses, composite open web steel joists, and stub-girders. Optimum spans for performance and economy depend on overall building considerations such as storey height restrictions, and integration of building services.

For composite trusses and joists, Clause 17.9.2 of S16.1-94 stipulates that the area of the top chord shall be neglected in determining the properties of the composite section, and that the factored moment resistance of the composite truss or joist shall be computed on the basis of full shear connection with the plastic neutral axis in the slab.

Composite stub-girders use wide flange column shapes with short W-shape stubs shop welded to the top of the girders and interconnected with the deck-slab by shear connectors to provide vierendeel girder action. Deck-slabs usually consist of 75 mm composite steel deck with 75 mm or 85 mm cover slabs.

Examples

An example to illustrate the use of the tables in this book follows on page 5-26. Further examples and design information for composite trusses, composite joists and stub girders, including design tables for stub giders, are contained in the CISC publication *Design and Construction of Composite Floor Systems*.

References

PART TWO of this Handbook. See commentary on Clause 17.

CHIEN, E.Y.L., and RITCHIE, J.K. 1984. Design and construction of composite floor systems. Canadian Institute of Steel Construction, Willowdale, ON.

KULAK, ADAMS and GILMOR. 1995. Limit states design in structural steel. Canadian Institute of Steel Construction, Willowdale, ON.

PICARD, A., BEAULIEU, D. 1991. Calcul des charpentes d'acier. Canadian Institute of Steel Construction, Willowdale, ON.

Table 5-2
Factored Shear Resistance of Shear Studs, q_r (kN), in Solid Slabs or in Deck-Slabs with Ribs Parallel to Beam

Stud Diameter	Stud Location	Factored Shear Resistance of a Stud, q_r (kN) for listed f'_c as					
		w_c = 2 300 kg/m³			w_c = 1 850 kg/m³		
		20 MPa	25 MPa	30 MPa	20 MPa	25 MPa	30 MPa
3/4" (19 mm)	Solid Slab (or) Deck-Slab with $w_d/h_d \geq 1.5$	73.9	87.3	94.1	62.7	74.2	85.0
5/8" (15.9 mm)		51.7	61.2	65.9	43.9	51.9	59.6
1/2" (12.7 mm)		33.0	39.0	42.1	28.0	33.1	38.0
3/4" (19 mm)	Deck-Slab with $w_d/h_d \approx 1.4$ $h/h_d \approx 2.0$	62.0	73.4	79.1	52.7	62.3	71.4
5/8" (15.9 mm)		43.5	51.4	55.4	36.9	43.6	50.0
1/2" (12.7 mm)		27.7	32.8	35.3	23.5	27.8	31.9
3/4" (19 mm)	Deck-Slab with $w_d/h_d \approx 1.4$ $h/h_d \geq 2.2$	73.9	87.3	94.1	62.7	74.2	85.0
5/8" (15.9 mm)		51.7	61.2	65.9	43.9	51.9	59.6
1/2" (12.7 mm)		33.0	39.0	42.1	28.0	33.1	38.0

Notes: q_r = factored shear resistance per stud connection, (kN)
= least of a) $0.5 \phi_{sc} A_{sc} \sqrt{f'_c E_c}$ b) $415 \phi_{sc} A_{sc}$ and
c) $0.6(w_d/h_d)(h/h_d - 1)$ times lesser of a) and b)
where, $\phi_{sc} = 0.8$, A_{sc} = stud cross sectional area, $E_c = w_c^{1.5} \, 0.043 \sqrt{f'_c}$

Table 5-3

Factored Shear Resistance of 3/4" Dia. Studs, q_r (kN), with Shear Transverse to Concrete Ribs of Deck-Slabs Formed by 75 mm or 38 mm Deep Steel Deck

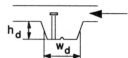

Deck		Stud Connector(s)				Pullout Cone Area A_p × 10^3 mm²	Factored Shear Resistance (kN) of Stud(s), $q_{r(rib)}$, for listed f'_c as					
							N.D. w_c = 2 300 kg/m³			S.L.D. w_c = 1 850 kg/m³		
Depth h_d mm	$\frac{w_d}{h_d}$	Diameter in. (mm)	Length mm	value of n	Edge Distance mm		20 MPa	25 MPa	30 MPa	20 MPa	25 MPa	30 MPa
75	2.4	3/4" (19)	115	1	Int.	52.2	65.4	73.1	80.1	55.6	62.1	68.0
				1	65	41.3	51.7	57.8	63.3	44.0	49.1	53.8
				1	35	34.3	43.0	48.0	52.6	36.5	40.8	44.7
				2	Int.	70.0	87.7	98.0	107	74.5	83.3	91.3
			150	1	Int.	80.9	73.9	87.3	94.1	62.7	74.2	85.0
				1	65	59.4	73.9	83.2	91.1	62.7	70.7	77.4
				1	35	50.7	63.5	71.0	77.8	54.0	60.3	66.1
				2	Int.	103	129	144	158	110	123	134
75	2.0	3/4" (19)	115	1	Int.	48.4	60.6	67.8	74.2	51.5	57.6	63.1
				1	65	38.3	48.0	53.6	58.7	40.8	45.6	49.9
				1	35	31.8	39.8	44.5	48.8	33.8	37.8	41.5
				2	Int.	64.9	81.3	90.9	99.5	69.1	77.2	84.6
			150	1	Int.	71.3	73.9	87.3	94.1	62.7	74.2	85.0
				1	65	52.6	65.9	73.6	80.7	56.0	62.6	68.6
				1	35	44.8	56.1	62.7	68.7	47.7	53.3	58.4
				2	Int.	91.2	114	128	140	97.1	109	119
38	2.5	3/4" (19)	75	1	Int.	20.5	44.7	50.0	54.8	38.0	42.5	46.6
				1	65	19.3	42.1	47.1	51.6	35.8	40.0	43.8
				1	35	15.2	33.2	37.1	40.6	28.2	31.5	34.5
				2	Int.	31.2	68.1	76.1	83.4	57.9	64.7	70.9
			100	1	Int.	32.0	69.8	78.1	85.5	59.4	66.4	72.7
				1	65	27.3	59.6	66.6	73.0	50.6	56.6	62.0
				1	35	22.1	48.2	53.9	59.1	41.0	45.8	50.2
				2	Int.	45.3	98.9	111	121	84.0	94.0	103
38	1.4	3/4" (19)	75	1	Int.	13.6	29.7	33.2	36.4	25.2	28.2	30.9
				1	65	13.0	28.4	31.7	34.7	24.1	27.0	29.5
				1	35	10.4	22.7	25.4	27.8	19.3	21.6	23.6
				2	Int.	21.6	47.1	52.7	57.7	40.1	44.8	49.1
			100	1	Int.	27.6	60.2	67.3	73.8	51.2	57.2	62.7
				1	65	24.9	54.3	60.8	66.6	46.2	51.6	56.6
				1	35	19.9	43.4	48.6	53.2	36.9	41.3	45.2
				2	Int.	40.9	89.3	99.8	109	75.9	84.8	92.9

$q_{r(rib)}$ = factored shear resistance (kN) per rib connection (Total for 1- or 2-stud connection as noted)
= least of a) $0.5 n \phi_{sc} A_{sc} \sqrt{f'_c E_c}$,   b) $415 n \phi_{sc} A_{sc}$ and c) $C \phi_{sc} \rho A_p \sqrt{f'_c}$
where, n = number of studs per rib, A_{sc} = stud cross sectional area, A_p = pullout cone area,
ρ = 1.0 for N.D. concrete or = 0.85 for S.L.D. concrete,
C = 0.35 for 75 mm deck or 0.61 for 38 mm deck, ϕ_{sc} = 0.8, $E_c = w_c^{1.5} \, 0.043 \sqrt{f'_c}$

Notes: 1. Stud length listed = length after welding.
  Minimum length prior to welding ≈ stud length listed + 10 mm as fusion allowance.
2. Double studs transversely spaced at minimum 4 stud diameters.
3. Int. = interior condition.
4. Studs placed off-centre in ribs of 75 mm deck and on-centre in ribs of 38 mm deck.

Table 5-4

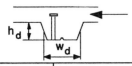

Factored Shear Resistance of 5/8" Dia. Studs, q_r (kN), with Shear Transverse to Concrete Ribs of Deck-Slabs Formed by 75 mm or 38 mm Deep Steel Deck

Deck		Stud Connector(s)				Pullout Cone Area A_p ×10³ mm²	Factored Shear Resistance (kN) of Stud(s), $q_{r(rib)}$, for listed f'_c as					
Depth h_d mm	$\frac{w_d}{h_d}$	Diameter in. (mm)	Length mm	value of n	Edge Distance mm		N.D. w_c = 2 300 kg/m³			S.L.D. w_c = 1 850 kg/m³		
							20 MPa	25 MPa	30 MPa	20 MPa	25 MPa	30 MPa
75	2.4	5/8" (15.9)	115	1	Int.	52.2	51.7	61.2	65.9	43.9	51.9	59.6
				1	65	41.3	51.7	57.8	63.3	43.9	49.1	53.8
				1	35	34.3	43.0	48.0	52.6	36.5	40.8	44.7
				2	Int.	67.1	84.0	93.9	103	71.4	79.8	87.5
			150	1	Int.	80.9	51.7	61.2	65.9	43.9	51.9	59.6
				1	65	59.4	51.7	61.2	65.9	43.9	51.9	59.6
				1	35	50.7	51.7	61.2	65.9	43.9	51.9	59.6
				2	Int.	99.5	103	122	132	87.9	104	119
75	2.0	5/8" (15.9)	115	1	Int.	48.4	51.7	61.2	65.9	43.9	51.9	59.6
				1	65	38.3	48.0	53.6	58.7	40.8	45.6	49.9
				1	35	31.8	39.8	44.5	48.8	33.8	37.8	41.5
				2	Int.	62.3	78.0	87.2	95.5	66.3	74.1	81.2
			150	1	Int.	71.3	51.7	61.2	65.9	43.9	51.9	59.6
				1	65	52.6	51.7	61.2	65.9	43.9	51.9	59.6
				1	35	44.8	51.7	61.2	65.9	43.9	51.9	58.4
				2	Int.	87.9	103	122	132	87.9	104	115
38	2.5	5/8" (15.9)	75	1	Int.	20.5	44.7	50.0	54.8	38.0	42.5	46.6
				1	65	19.3	42.1	47.1	51.6	35.8	40.0	43.8
				1	35	15.2	33.2	37.1	40.6	28.2	31.5	34.5
				2	Int.	29.4	64.2	71.7	78.6	54.5	61.0	66.8
			100	1	Int.	32.0	51.7	61.2	65.9	43.9	51.9	59.6
				1	65	27.3	51.7	61.2	65.9	43.9	51.9	59.6
				1	35	22.1	48.2	53.9	59.1	41.0	45.8	50.2
				2	Int.	43.2	94.3	105	115	80.1	89.6	98.1
38	1.4	5/8" (15.9)	75	1	Int.	13.6	29.7	33.2	36.4	25.2	28.2	30.9
				1	65	13.0	28.4	31.7	34.7	24.1	27.0	29.5
				1	35	10.4	22.7	25.4	27.8	19.3	21.6	23.6
				2	Int.	20.3	44.3	49.5	54.3	37.7	42.1	46.1
			100	1	Int.	27.6	51.7	61.2	65.9	43.9	51.9	59.6
				1	65	24.9	51.7	60.8	65.9	43.9	51.6	56.6
				1	35	19.9	43.4	48.6	53.2	36.9	41.3	45.2
				2	Int.	38.8	84.7	94.7	104	72.0	80.5	88.2

$q_{r(rib)}$ = factored shear resistance (kN) per rib connection (Total for 1- or 2-stud connection as noted)
= least of a) $0.5 n \phi_{sc} A_{sc} \sqrt{f'_c E_c}$, b) $415 n \phi_{sc} A_{sc}$ and c) $C \phi_{sc} \rho A_p \sqrt{f'_c}$
where, n = number of studs per rib, A_{sc} = stud cross sectional area, A_p = pullout cone area,
ρ = 1.0 for N.D. concrete or = 0.85 for S.L.D. concrete,
C = 0.35 for 75 mm deck or 0.61 for 38 mm deck, ϕ_{sc} = 0.8, $E_c = w_c^{1.5} \, 0.043 \sqrt{f'_c}$

Notes: 1. Stud length listed = length after welding.
Minimum length prior to welding ≈ stud length listed + 10 mm as fusion allowance.
2. Double studs transversely spaced at minimum 4 stud diameters.
3. Int. = interior condition.
4. Studs placed off-centre in ribs of 75 mm deck and on-centre in ribs of 38 mm deck.

Example

Given:

Select a simply supported composite beam to span 12 m and carry a uniformly distributed specified load of 15 kN/m live load and 9 kN/m dead load. Beams are spaced at 3 m on centre and support a 75 mm steel deck (ribs perpendicular to the beam) with a 65 mm cover slab of 20 MPa normal density concrete. Use G40.21 350W steel and limit live load deflections to $L/300$.

Solution:

Total factored load $= (1.25 \times 9) + (1.50 \times 15) = 33.8 \text{ kN/m}$

Therefore $M_f = 33.8 \times 12^2 / 8 = 608 \text{ kN·m}$

and $V_f = 33.8 \times 12 / 2 = 203 \text{ kN}$

Compute minimum I_{reqd} for deflection limit $L/300$ using Figure 5-1 and Table 5-5 on pages 5-70 and 5-71.

Total specified live load, $W = 15 \times 12 = 180 \text{ kN}$

$B_d = 1.0$ simple span udl   (Table 5-5)

$C_d = 2.8$ for 12 m span and $L/\Delta = 300$ (Figure 5-1)

$I_{reqd} = W \times C_d \times B_d$

$= (180 \times 2.8 \times 1.0)\,1.15 = 580 \times 10^6 \text{ mm}^4$ (with 15% allowance for creep)

Effective Width   (Clause 17.4.1)

a) $0.25\,L = 0.25 \times 12\,000 \text{ mm} = 3\,000 \text{ mm}$

b) beam spacing $= 3 \text{ m} = 3\,000 \text{ mm}$

Therefore, maximum effective width $= 3\,000 \text{ mm}$

Beam Selection

From composite beam selection tables for 75 mm steel deck with 65 mm cover slab and $b_1 = 3000$ mm, page 5-32, a suitable shape is a W460 x 61 with M_{rc} for 40 per cent shear connection $= 623 \text{ kN·m} > 608 \text{ kN·m}$

$V_r = 758 \text{ kN} > 203 \text{ kN}$

$I_t = 880 \times 10^6 \text{ mm}^4$

For 40% shear connection, $I_e = I_s + 0.85\,p^{0.25}(I_t - I_s)$ (Clause 17.3.1(a))

$= 255 + 0.85\,(0.4)^{0.25}(880 - 255) = 677 \times 10^6 \text{ mm}^4 > 580 \times 10^6 \text{ mm}^4$

$Q_r = 1\,990 \text{ kN}\,;\ S_t = 1\,880 \times 10^3 \text{ mm}^3\,;\ M_r = 406 \text{ kN·m}\,;\ L_u = 2\,390 \text{ mm}$

Check Steel Beam Under Construction Loads   (Clause 17.12)

Clause 17.12 requires that the steel section alone must be capable of supporting all factored loads applied before concrete hardens. In this case the steel deck will provide lateral support to the compression flange of the beam.

Thus $M_r = 406$ kN·m applies.

Assuming dead load due to deck-slab and steel beam as 7 kN/m and construction live load as 2.5 kN/m, the total factored load applied before the concrete hardens is

$$(1.25 \times 7) + (1.5 \times 2.5) = 12.5 \text{ kN/m}$$

$$M_f = 12.5 \times 12^2 / 8 = 225 \text{ kN·m} < 406 \text{ kN·m}$$

Check Unshored Beam Tension Flange (Clause 17.11)

Assume that load applied before concrete strength reaches $0.75 f'_c$ is the specified dead load (7 kN/m), and that the remaining dead load (9 − 7 = 2 kN/m) and the specified live load act on the composite section.

Stress in tension flange due to specified load acting on steel beam alone:

S_x of steel beam $= 1\,130 \times 10^3$ mm^3

$$f_1 = \frac{M_1}{S_x} = \frac{7 \times 12\,000^2}{8 \times 1\,130 \times 10^3} = 112 \text{ MPa}$$

Stress in tension flange due to specified live and superimposed dead loads acting on composite section:

$$f_2 = \frac{M_2}{S_t} = \frac{(15 + 2) \times 12\,000^2}{8 \times 1\,880 \times 10^3} = 163 \text{ MPa}$$

$$f_1 + f_2 = 112 + 163 = 275 \text{ MPa} < 350 \text{ MPa}$$

Shear Connectors

Q_r (100% connection) $= 1\,990$ kN

Assume 3/4 inch (19 mm) diameter studs.

Minimum flange thickness $= 19 / 2.5 = 7.6$ mm < 12.7 mm (Clause 17.6.5)

From Table 5-3 (page 5-24), for 3/4 inch diameter studs, $h_d = 75$, $w_d/h_d = 2.0$, 20 MPa 2 300 kg/m^3 concrete, the factored shear resistance per stud q_r is 60.6 kN

Number of studs required

$$= \frac{2 \times Q_r \times (\%\text{ shear connection} /100)}{q_r}$$

$$= \frac{2 \times 1\,990 \times 40/100}{60.6} = 26.2 \quad \text{Use 28 studs}$$

Since there are no concentrated loads, the studs can be spaced uniformly along the full length of the beam as permitted by the deck flutes.

COMPOSITE BEAMS
Trial Selection Table

75 mm Deck with 65 mm Slab
$\phi = 0.90$, $\phi_c = 0.60$

G40.21 350W
$f'_c = 20$ MPa

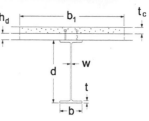

Steel section	b_1	M_{rc} (kN·m) for % shear connection			Q_r (kN)	I_t 10^6	S_t 10^3	I_t^* 10^6	Steel section data	Unbraced condition			
		Composite							**Non-composite**				
	mm	100%	70%	40%	100%	mm⁴	mm³	mm⁴		L' mm	M_r' kN·m	L' mm	M_r' kN·m
WWF1000x223	5 000	4 700	4 430	4 030	3 320	10 300	12 400	6 540	M_r 3 310	4 000	3 250	12 000	920
WWF39x150	4 000	4 520	4 260	3 900	2 650	9 760	12 200	6 210	V_r 2 210	5 000	2 960	14 000	729
b = 300	3 000	4 290	4 050	3 760	1 990	9 010	11 900	5 860	L_u 3 790	6 000	2 620	16 000	602
t = 25	2 000	4 010	3 830	3 620	1 330	8 020	11 500	5 470	I_x 4 590	8 000	1 790	18 000	512
d = 1000	1 000	3 680	3 570	3 460	663	6 640	10 700	5 050	S_x 9 190	10 000	1 230	20 000	446
WWF1000x200	5 000	4 280	4 010	3 610	3 320	9 370	11 000	5 860	M_r 2 890	4 000	2 790	12 000	699
WWF39x134	4 000	4 100	3 830	3 480	2 650	8 850	10 900	5 540	V_r 2 210	5 000	2 510	14 000	549
b = 300	3 000	3 860	3 630	3 340	1 990	8 170	10 600	5 190	L_u 3 610	6 000	2 180	16 000	449
t = 20	2 000	3 580	3 400	3 190	1 330	7 250	10 200	4 810	I_x 3 940	8 000	1 390	18 000	379
d = 1000	1 000	3 250	3 150	3 030	663	5 950	9 460	4 400	S_x 7 880	10 000	945	20 000	328
WWF900x231	5 000	4 470	4 330	4 030	3 320	9 360	12 500	6 070	M_r 3 400	4 000	3 400	12 000	1 760
WWF35x156	4 000	4 370	4 200	3 920	2 650	8 850	12 300	5 790	V_r 1 350	5 000	3 400	14 000	1 370
b = 400	3 000	4 220	4 040	3 800	1 990	8 200	12 100	5 480	L_u 5 460	6 000	3 300	16 000	1 110
t = 25	2 000	4 000	3 850	3 670	1 330	7 340	11 700	5 150	I_x 4 410	8 000	2 870	18 000	933
d = 900	1 000	3 720	3 630	3 530	663	6 150	11 100	4 800	S_x 9 810	10 000	2 360	20 000	801
WWF900x192	5 000	3 780	3 640	3 340	3 320	7 920	10 200	5 060	M_r 2 710	4 000	2 700	12 000	842
WWF35x128	4 000	3 680	3 510	3 230	2 650	7 500	10 100	4 800	V_r 1 350	5 000	2 490	14 000	670
b = 300	3 000	3 530	3 350	3 110	1 990	6 950	9 860	4 510	L_u 3 980	6 000	2 240	16 000	555
t = 25	2 000	3 310	3 160	2 980	1 330	6 210	9 540	4 190	I_x 3 460	8 000	1 630	18 000	473
d = 900	1 000	3 040	2 940	2 840	663	5 130	8 970	3 840	S_x 7 680	10 000	1 120	20 000	412
WWF900x169	5 000	3 390	3 250	2 950	3 320	7 070	8 920	4 490	M_r 2 320	4 000	2 290	12 000	625
WWF35x113	4 000	3 290	3 120	2 840	2 650	6 700	8 790	4 230	V_r 1 340	5 000	2 080	14 000	491
b = 300	3 000	3 140	2 960	2 720	1 990	6 220	8 610	3 960	L_u 3 820	6 000	1 850	16 000	402
t = 20	2 000	2 920	2 770	2 590	1 330	5 550	8 320	3 650	I_x 2 930	8 000	1 250	18 000	339
d = 900	1 000	2 650	2 550	2 450	663	4 560	7 790	3 310	S_x 6 510	10 000	845	20 000	293
WWF800x223	5 000	3 870	3 760	3 510	3 320	7 430	11 000	4 770	M_r 2 940	4 000	2 940	10 000	2 120
WWF31x150	4 000	3 780	3 650	3 410	2 650	7 020	10 800	4 540	V_r 1 370	5 000	2 940	12 000	1 630
b = 400	3 000	3 670	3 520	3 310	1 990	6 500	10 600	4 290	L_u 5 570	6 000	2 880	14 000	1 280
t = 25	2 000	3 480	3 350	3 190	1 330	5 810	10 300	4 020	I_x 3 410	7 000	2 710	16 000	1 050
d = 800	1 000	3 240	3 160	3 060	663	4 840	9 700	3 730	S_x 8 520	8 000	2 530	18 000	882
WWF800x184	5 000	3 260	3 150	2 900	3 320	6 260	8 920	3 970	M_r 2 330	4 000	2 330	10 000	1 040
WWF31x123	4 000	3 170	3 040	2 800	2 650	5 930	8 790	3 760	V_r 1 370	5 000	2 170	12 000	791
b = 300	3 000	3 060	2 910	2 700	1 990	5 490	8 610	3 520	L_u 4 060	6 000	1 970	14 000	634
t = 25	2 000	2 870	2 740	2 580	1 330	4 900	8 330	3 260	I_x 2 660	7 000	1 750	16 000	529
d = 800	1 000	2 630	2 550	2 450	663	4 030	7 830	2 980	S_x 6 640	8 000	1 490	18 000	454
WWF800x161	5 000	2 910	2 800	2 550	3 320	5 570	7 760	3 520	M_r 1 990	4 000	1 980	10 000	778
WWF31x108	4 000	2 830	2 700	2 460	2 650	5 290	7 650	3 320	V_r 1 370	5 000	1 810	12 000	581
b = 300	3 000	2 710	2 560	2 350	1 990	4 910	7 500	3 100	L_u 3 900	6 000	1 620	14 000	460
t = 20	2 000	2 530	2 400	2 240	1 330	4 380	7 250	2 850	I_x 2 250	7 000	1 400	16 000	379
d = 800	1 000	2 280	2 200	2 110	663	3 580	6 790	2 570	S_x 5 610	8 000	1 130	18 000	322

Note: Resistances are based on a concrete density of 2300 kg/m³.

Units: M_r - kN·m, V_r - kN, L_u - mm, I_x - 10^6 mm⁴, S_x - 10^3 mm³, b - mm, t - mm, d - mm

* Transformed moment of inertia for calculating shrinkage deflections, based on a modular ratio $n_t = 50$

COMPOSITE BEAMS
Trial Selection Table

G40.21 350W
f'$_c$ = 20 MPa

75 mm Deck with 65 mm Slab
φ = 0.90, φ$_c$ = 0.60

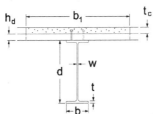

Steel section	b$_1$	Composite							Non-composite				
		M$_{rc}$ (kN·m) for % shear connection			Q$_r$	I$_t$	S$_t$	I$_t$*	Steel section data		Unbraced condition			
		100%	70%	40%	100%	10^6	10^3	10^6			L'	M$_r$'	L'	M$_r$'
	mm				(kN)	mm^4	mm^3	mm^4			mm	kN·m	mm	kN·m
WWF700x214	5 000	3 310	3 200	3 010	3 320	5 740	9 480	3 640	M$_r$	2 500	4 000	2 500	10 000	1 870
WWF28x144	4 000	3 220	3 120	2 920	2 650	5 420	9 350	3 460	V$_r$	1 370	5 000	2 500	12 000	1 500
b = 400	3 000	3 130	3 010	2 830	1 990	5 010	9 160	3 260	L$_u$	5 690	6 000	2 470	14 000	1 190
t = 25	2 000	2 980	2 870	2 730	1 330	4 460	8 860	3 040	I$_x$	2 540	7 000	2 330	16 000	982
d = 700	1 000	2 770	2 700	2 620	663	3 690	8 350	2 800	S$_x$	7 270	8 000	2 190	18 000	835
WWF700x175	5 000	2 780	2 670	2 480	3 320	4 820	7 670	3 030	M$_r$	1 970	4 000	1 970	10 000	965
WWF28x117	4 000	2 690	2 590	2 390	2 650	4 560	7 560	2 850	V$_r$	1 370	5 000	1 850	12 000	741
b = 300	3 000	2 600	2 480	2 300	1 990	4 220	7 410	2 670	L$_u$	4 160	6 000	1 700	14 000	601
t = 25	2 000	2 450	2 340	2 200	1 330	3 760	7 160	2 460	I$_x$	1 970	7 000	1 530	16 000	505
d = 700	1 000	2 240	2 160	2 080	663	3 070	6 710	2 230	S$_x$	5 640	8 000	1 350	18 000	436
WWF700x152	5 000	2 470	2 370	2 180	3 320	4 270	6 640	2 680	M$_r$	1 680	4 000	1 680	10 000	711
WWF28x102	4 000	2 390	2 290	2 090	2 650	4 050	6 550	2 520	V$_r$	1 370	5 000	1 540	12 000	537
b = 300	3 000	2 300	2 180	2 000	1 990	3 760	6 420	2 340	L$_u$	3 990	6 000	1 390	14 000	429
t = 20	2 000	2 150	2 040	1 900	1 330	3 350	6 210	2 140	I$_x$	1 660	7 000	1 230	16 000	356
d = 700	1 000	1 940	1 870	1 790	663	2 730	5 800	1 920	S$_x$	4 760	8 000	1 020	18 000	305
W610x241	4 000	3 230	3 060	2 830	2 650	4 790	9 240	2 960	M$_r$	2 420	3 000	2 420	8 000	1 950
W24x162	3 000	3 080	2 930	2 730	1 990	4 390	9 000	2 780	V$_r$	2 360	4 000	2 420	10 000	1 640
b = 329	2 000	2 890	2 770	2 630	1 330	3 880	8 630	2 580	L$_u$	4 750	5 000	2 380	12 000	1 300
t = 31	1 500	2 790	2 690	2 570	995	3 550	8 370	2 480	I$_x$	2 150	6 000	2 240	14 000	1 070
d = 635	1 000	2 670	2 600	2 520	663	3 170	8 020	2 380	S$_x$	6 780	7 000	2 100	16 000	911
W610x217	4 000	2 940	2 790	2 560	2 650	4 380	8 360	2 690	M$_r$	2 160	3 000	2 160	8 000	1 690
W24x146	3 000	2 800	2 660	2 470	1 990	4 020	8 150	2 520	V$_r$	2 150	4 000	2 160	10 000	1 370
b = 328	2 000	2 630	2 510	2 370	1 330	3 550	7 830	2 330	L$_u$	4 640	5 000	2 110	12 000	1 070
t = 27.7	1 500	2 520	2 420	2 310	995	3 250	7 590	2 230	I$_x$	1 910	6 000	1 980	14 000	878
d = 628	1 000	2 410	2 340	2 260	663	2 890	7 280	2 130	S$_x$	6 070	7 000	1 840	16 000	745
W610x195	4 000	2 660	2 520	2 310	2 650	3 980	7 500	2 440	M$_r$	1 910	3 000	1 910	8 000	1 450
W24x131	3 000	2 540	2 400	2 220	1 990	3 670	7 320	2 270	V$_r$	1 990	4 000	1 910	10 000	1 120
b = 327	2 000	2 370	2 260	2 120	1 330	3 240	7 040	2 090	L$_u$	4 530	5 000	1 860	12 000	870
t = 24.4	1 500	2 270	2 180	2 070	995	2 960	6 830	2 000	I$_x$	1 680	6 000	1 730	14 000	709
d = 622	1 000	2 160	2 090	2 010	663	2 630	6 540	1 900	S$_x$	5 400	7 000	1 590	16 000	598
W610x174	4 000	2 390	2 280	2 080	2 650	3 600	6 690	2 200	M$_r$	1 690	3 000	1 690	8 000	1 240
W24x117	3 000	2 290	2 170	1 990	1 990	3 330	6 540	2 040	V$_r$	1 790	4 000	1 690	10 000	924
b = 325	2 000	2 130	2 030	1 890	1 330	2 940	6 300	1 870	L$_u$	4 440	5 000	1 630	12 000	709
t = 21.6	1 500	2 040	1 950	1 840	995	2 690	6 120	1 780	I$_x$	1 470	6 000	1 510	14 000	574
d = 616	1 000	1 930	1 860	1 790	663	2 380	5 860	1 680	S$_x$	4 780	7 000	1 380	16 000	482
W610x155	4 000	2 150	2 050	1 870	2 650	3 260	5 970	1 990	M$_r$	1 490	3 000	1 490	8 000	1 060
W24x104	3 000	2 060	1 950	1 780	1 990	3 020	5 840	1 850	V$_r$	1 610	4 000	1 490	10 000	762
b = 324	2 000	1 920	1 820	1 690	1 330	2 680	5 630	1 680	L$_u$	4 370	5 000	1 430	12 000	579
t = 19	1 500	1 830	1 740	1 640	995	2 450	5 470	1 590	I$_x$	1 290	6 000	1 310	14 000	465
d = 611	1 000	1 730	1 660	1 590	663	2 160	5 240	1 500	S$_x$	4 220	7 000	1 190	16 000	388

Note: Resistances are based on a concrete density of 2300 kg/m^3.

Units: M$_r$ - kN·m, V$_r$ - kN, L$_u$ - mm, I$_x$ - 10^6 mm^4, S$_x$ - 10^3 mm^3, b - mm, t - mm, d - mm

* Transformed moment of inertia for calculating shrinkage deflections, based on a modular ratio n$_t$ = 50

COMPOSITE BEAMS
Trial Selection Table

75 mm Deck with 65 mm Slab
$\phi = 0.90$, $\phi_c = 0.60$

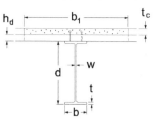

G40.21 350W
$f'_c = 20$ MPa

Steel section	b_1	M_{rc} (kN·m) for % shear connection			Q_r	I_t	S_t	I_t^*	Steel section data	Unbraced condition			
										L'	M_r'	L'	M_r'
	mm	100%	70%	40%	100% (kN)	10^6 mm⁴	10^3 mm³	10^6 mm⁴		mm	kN·m	mm	kN·m
W610x140	4 000	1 990	1 890	1 700	2 650	3 010	5 360	1 820	M_r 1 310	3 000	1 310	8 000	573
W24x94	3 000	1 900	1 780	1 610	1 990	2 790	5 250	1 680	V_r 1 680	4 000	1 180	10 000	422
b = 230	2 000	1 750	1 640	1 510	1 330	2 470	5 050	1 510	L_u 3 050	5 000	1 040	12 000	334
t = 22.2	1 500	1 660	1 570	1 460	995	2 260	4 900	1 420	I_x 1 120	6 000	878	14 000	277
d = 617	1 000	1 550	1 480	1 410	663	1 980	4 680	1 330	S_x 3 630	7 000	695	16 000	237
W610x125	4 000	1 800	1 710	1 530	2 650	2 720	4 780	1 660	M_r 1 160	3 000	1 160	8 000	470
W24x84	3 000	1 710	1 610	1 450	1 990	2 530	4 690	1 520	V_r 1 510	4 000	1 040	10 000	342
b = 229	2 000	1 580	1 480	1 360	1 330	2 250	4 520	1 370	L_u 2 990	5 000	895	12 000	269
t = 19.6	1 500	1 490	1 410	1 310	995	2 060	4 390	1 280	I_x 985	6 000	733	14 000	222
d = 612	1 000	1 390	1 330	1 260	663	1 810	4 200	1 190	S_x 3 220	7 000	575	16 000	189
W610x113	4 000	1 650	1 570	1 410	2 650	2 490	4 330	1 520	M_r 1 040	3 000	1 030	8 000	391
W24x76	3 000	1 570	1 480	1 330	1 990	2 320	4 240	1 400	V_r 1 420	4 000	915	10 000	282
b = 228	2 000	1 450	1 360	1 230	1 330	2 070	4 100	1 250	L_u 2 930	5 000	781	12 000	220
t = 17.3	1 500	1 370	1 290	1 190	995	1 900	3 990	1 170	I_x 875	6 000	617	14 000	180
d = 608	1 000	1 270	1 210	1 130	663	1 670	3 810	1 080	S_x 2 880	7 000	481	16 000	153
W610x101	4 000	1 510	1 430	1 280	2 650	2 250	3 890	1 390	M_r 914	3 000	901	8 000	320
W24x68	3 000	1 430	1 350	1 200	1 990	2 110	3 810	1 270	V_r 1 310	4 000	795	10 000	228
b = 228	2 000	1 320	1 230	1 110	1 330	1 890	3 690	1 130	L_u 2 870	5 000	668	12 000	176
t = 14.9	1 500	1 240	1 160	1 060	995	1 730	3 590	1 050	I_x 764	6 000	512	14 000	144
d = 603	1 000	1 140	1 080	1 010	663	1 520	3 430	962	S_x 2 530	7 000	396	16 000	121
W610x91	4 000	1 370	1 300	1 160	2 650	2 030	3 460	1 260	M_r 806	3 000	788	8 000	261
W24x61	3 000	1 300	1 220	1 080	1 990	1 900	3 400	1 150	V_r 1 110	4 000	687	10 000	185
b = 227	2 000	1 200	1 110	992	1 330	1 710	3 300	1 020	L_u 2 790	5 000	567	12 000	142
t = 12.7	1 500	1 120	1 040	944	995	1 570	3 210	944	I_x 667	6 000	421	14 000	115
d = 598	1 000	1 020	964	892	663	1 380	3 070	861	S_x 2 230	7 000	325	16 000	97.1
W530x138	4 000	1 780	1 680	1 500	2 650	2 440	4 800	1 450	M_r 1 140	3 000	1 130	8 000	515
W21x93	3 000	1 680	1 570	1 420	1 990	2 260	4 690	1 330	V_r 1 680	4 000	1 010	10 000	390
b = 214	2 000	1 540	1 450	1 330	1 330	1 990	4 510	1 190	L_u 2 910	5 000	891	12 000	314
t = 23.6	1 500	1 460	1 380	1 280	995	1 820	4 360	1 120	I_x 861	6 000	762	14 000	263
d = 549	1 000	1 360	1 300	1 230	663	1 590	4 150	1 040	S_x 3 140	7 000	616	16 000	227
W530x123	4 000	1 610	1 520	1 360	2 650	2 210	4 300	1 330	M_r 1 010	3 000	996	8 000	421
W21x83	3 000	1 530	1 430	1 280	1 990	2 050	4 210	1 210	V_r 1 480	4 000	888	10 000	316
b = 212	2 000	1 400	1 310	1 200	1 330	1 820	4 060	1 080	L_u 2 840	5 000	768	12 000	253
t = 21.2	1 500	1 320	1 240	1 150	995	1 660	3 930	1 010	I_x 761	6 000	631	14 000	211
d = 544	1 000	1 230	1 170	1 100	663	1 450	3 750	934	S_x 2 800	7 000	505	16 000	182
W530x109	4 000	1 450	1 370	1 230	2 650	1 990	3 830	1 210	M_r 891	3 000	872	8 000	342
W21x73	3 000	1 370	1 290	1 160	1 990	1 860	3 750	1 100	V_r 1 300	4 000	771	10 000	254
b = 211	2 000	1 270	1 190	1 070	1 330	1 660	3 620	979	L_u 2 790	5 000	656	12 000	202
t = 18.8	1 500	1 190	1 120	1 030	995	1 510	3 520	911	I_x 667	6 000	520	14 000	168
d = 539	1 000	1 100	1 050	982	663	1 320	3 360	836	S_x 2 480	7 000	413	16 000	144

Note: Resistances are based on a concrete density of 2300 kg/m³.
Units: M_r - kN·m, V_r - kN, L_u - mm, I_x - 10^6 mm⁴, S_x - 10^3 mm³, b - mm, t - mm, d - mm
* Transformed moment of inertia for calculating shrinkage deflections, based on a modular ratio $n_t = 50$

COMPOSITE BEAMS
Trial Selection Table

G40.21 350W
$f'_c = 20$ MPa

75 mm Deck with 65 mm Slab
$\phi = 0.90$, $\phi_c = 0.60$

Steel section	b_1	Composite M_{rc} (kN·m) for % shear connection			Q_r	I_t	S_t	I_t^*	Non-composite Steel section data	Unbraced condition			
		100%	70%	40%	100%	10^6	10^3	10^6		L'	M_r'	L'	M_r'
	mm				(kN)	mm⁴	mm³	mm⁴		mm	kN·m	mm	kN·m
W530×101	4 000	1 370	1 290	1 160	2 650	1 870	3 560	1 140	M_r 825	3 000	804	8 000	301
W21×68	3 000	1 290	1 220	1 090	1 990	1 750	3 490	1 040	V_r 1 220	4 000	706	10 000	222
b = 210	2 000	1 190	1 110	1 010	1 330	1 560	3 380	924	L_u 2 750	5 000	594	12 000	176
t = 17.4	1 500	1 120	1 050	962	995	1 430	3 290	857	I_x 617	6 000	462	14 000	146
d = 537	1 000	1 040	980	916	663	1 250	3 140	784	S_x 2 300	7 000	365	16 000	125
W530×92	4 000	1 270	1 200	1 070	2 650	1 720	3 250	1 060	M_r 743	3 000	719	8 000	253
W21×62	3 000	1 190	1 130	1 000	1 990	1 610	3 190	963	V_r 1 130	4 000	626	10 000	185
b = 209	2 000	1 100	1 030	924	1 330	1 450	3 090	851	L_u 2 700	5 000	519	12 000	146
t = 15.6	1 500	1 040	968	881	995	1 330	3 010	787	I_x 552	6 000	393	14 000	120
d = 533	1 000	952	898	834	663	1 160	2 880	716	S_x 2 070	7 000	309	16 000	103
W530×82	4 000	1 160	1 080	969	2 650	1 540	2 880	960	M_r 652	3 000	625	8 000	204
W21×55	3 000	1 080	1 020	904	1 990	1 450	2 830	872	V_r 1 040	4 000	538	10 000	148
b = 209	2 000	997	929	827	1 330	1 310	2 750	767	L_u 2 640	5 000	436	12 000	115
t = 13.3	1 500	935	870	785	995	1 200	2 680	706	I_x 478	6 000	321	14 000	94.8
d = 528	1 000	854	802	739	663	1 050	2 560	638	S_x 1 810	7 000	250	16 000	80.5
W460×106	4 000	1 270	1 190	1 060	2 650	1 560	3 360	929	M_r 753	3 000	727	7 000	366
W18×71	3 000	1 190	1 120	997	1 990	1 450	3 290	843	V_r 1 230	3 500	686	8 000	308
b = 194	2 000	1 090	1 020	922	1 330	1 290	3 170	744	L_u 2 670	4 000	643	10 000	235
t = 20.6	1 500	1 030	964	882	995	1 180	3 070	688	I_x 488	5 000	553	12 000	190
d = 469	1 000	948	898	839	663	1 020	2 930	627	S_x 2 080	6 000	450	14 000	160
W460×97	4 000	1 180	1 100	988	2 650	1 440	3 080	871	M_r 687	3 000	660	7 000	314
W18×65	3 000	1 100	1 040	926	1 990	1 350	3 020	790	V_r 1 100	3 500	621	8 000	264
b = 193	2 000	1 010	949	854	1 330	1 200	2 910	695	L_u 2 630	4 000	579	10 000	200
t = 19	1 500	954	893	814	995	1 100	2 830	641	I_x 445	5 000	491	12 000	161
d = 466	1 000	878	830	772	663	960	2 700	582	S_x 1 910	6 000	389	14 000	135
W460×89	4 000	1 110	1 040	929	2 650	1 350	2 850	821	M_r 633	3 000	605	7 000	276
W18×60	3 000	1 030	974	869	1 990	1 260	2 800	744	V_r 1 010	3 500	568	8 000	231
b = 192	2 000	952	891	800	1 330	1 130	2 710	653	L_u 2 600	4 000	527	10 000	174
t = 17.7	1 500	896	838	761	995	1 040	2 640	601	I_x 409	5 000	442	12 000	140
d = 463	1 000	823	776	719	663	905	2 520	544	S_x 1 770	6 000	343	14 000	117
W460×82	4 000	1 040	963	861	2 650	1 240	2 610	767	M_r 576	3 000	546	7 000	234
W18×55	3 000	960	902	806	1 990	1 170	2 560	694	V_r 947	3 500	510	8 000	195
b = 191	2 000	880	826	738	1 330	1 050	2 490	608	L_u 2 540	4 000	471	10 000	146
t = 16	1 500	829	775	701	995	964	2 420	558	I_x 370	5 000	386	12 000	116
d = 460	1 000	760	715	660	663	845	2 320	502	S_x 1 610	6 000	292	14 000	97.3
W460×74	4 000	965	893	797	2 650	1 140	2 370	713	M_r 520	3 000	489	7 000	198
W18×50	3 000	890	833	745	1 990	1 070	2 330	645	V_r 855	3 500	455	8 000	164
b = 190	2 000	811	763	680	1 330	970	2 270	563	L_u 2 510	4 000	417	10 000	122
t = 14.5	1 500	766	715	643	995	893	2 210	515	I_x 332	5 000	332	12 000	96.8
d = 457	1 000	701	658	603	663	784	2 120	461	S_x 1 460	6 000	249	14 000	80.6

Note: Resistances are based on a concrete density of 2300 kg/m³.

Units: M_r - kN·m, V_r - kN, L_u - mm, I_x - 10^6 mm⁴, S_x - 10^3 mm³, b - mm, t - mm, d - mm

* Transformed moment of inertia for calculating shrinkage deflections, based on a modular ratio $n_t = 50$

COMPOSITE BEAMS
Trial Selection Table

75 mm Deck with 65 mm Slab
$\phi = 0.90$, $\phi_c = 0.60$

G40.21 350W
$f'_c = 20$ MPa

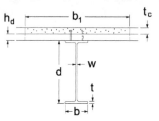

Steel section	b_1	Composite							Non-composite				
		M_{rc} (kN·m) for % shear connection			Q_r	I_t	S_t	I_t^*	Steel section data		Unbraced condition		
											L'	M_r'	L'	M_r'
	mm	100%	70%	40%	(kN)	10^6 mm^4	10^3 mm^3	10^6 mm^4			mm	kN·m	mm	kN·m
W460x67	4 000	898	827	735	2 650	1 040	2 150	660	M_r	466	3 000	435	7 000	165
W18x45	3 000	825	769	686	1 990	981	2 120	597	V_r	802	3 500	402	8 000	136
b = 190	2 000	747	703	623	1 330	892	2 060	520	L_u	2 460	4 000	366	10 000	99.8
t = 12.7	1 500	705	657	587	995	824	2 010	474	I_x	296	5 000	281	12 000	79.0
d = 454	1 000	643	601	547	663	726	1 930	423	S_x	1 300	6 000	209	14 000	65.4
W460x61	4 000	809	742	655	2 410	927	1 910	599	M_r	406	3 000	374	7 000	132
W18x41	3 000	756	702	623	1 990	880	1 880	542	V_r	758	3 500	343	8 000	108
b = 189	2 000	680	639	562	1 330	804	1 830	471	L_u	2 390	4 000	309	10 000	78.5
t = 10.8	1 500	640	595	527	995	745	1 790	428	I_x	255	5 000	228	12 000	61.7
d = 450	1 000	581	540	488	663	659	1 720	378	S_x	1 130	6 000	168	14 000	50.8
W410x85	4 000	992	917	815	2 650	1 100	2 520	667	M_r	542	3 000	512	7 000	242
W16x57	3 000	914	855	761	1 990	1 030	2 480	602	V_r	945	3 500	480	8 000	205
b = 181	2 000	833	780	697	1 330	925	2 400	525	L_u	2 510	4 000	447	10 000	157
t = 18.2	1 500	783	732	661	995	846	2 330	481	I_x	315	5 000	377	12 000	127
d = 417	1 000	718	675	623	663	738	2 220	431	S_x	1 510	6 000	297	14 000	107
W410x74	4 000	905	832	737	2 650	984	2 240	609	M_r	476	3 000	445	7 000	194
W16x50	3 000	830	772	688	1 990	926	2 200	549	V_r	833	3 500	414	8 000	163
b = 180	2 000	751	705	627	1 330	836	2 130	477	L_u	2 450	4 000	382	10 000	123
t = 16	1 500	707	660	592	995	768	2 080	435	I_x	275	5 000	312	12 000	99.7
d = 413	1 000	646	606	555	663	672	1 990	388	S_x	1 330	6 000	239	14 000	83.8
W410x67	4 000	840	770	678	2 650	895	2 020	563	M_r	428	3 000	397	7 000	161
W16x45	3 000	767	711	633	1 990	845	1 990	508	V_r	750	3 500	367	8 000	135
b = 179	2 000	689	648	575	1 330	767	1 930	441	L_u	2 400	4 000	336	10 000	102
t = 14.4	1 500	649	606	542	995	707	1 880	401	I_x	245	5 000	264	12 000	81.6
d = 410	1 000	593	554	505	663	621	1 810	356	S_x	1 200	6 000	201	14 000	68.3
W410x60	4 000	750	683	599	2 390	801	1 790	516	M_r	375	3 000	344	7 000	131
W16x40	3 000	699	645	573	1 990	760	1 760	466	V_r	652	3 500	317	8 000	109
b = 178	2 000	623	585	520	1 330	694	1 720	404	L_u	2 370	4 000	288	10 000	81.3
t = 12.8	1 500	584	548	488	995	643	1 680	367	I_x	216	5 000	218	12 000	64.9
d = 407	1 000	536	500	453	663	568	1 610	324	S_x	1 060	6 000	165	14 000	54.1
W410x54	4 000	675	612	535	2 140	717	1 590	469	M_r	331	3 000	298	7 000	104
W16x36	3 000	645	592	523	1 990	682	1 570	424	V_r	628	3 500	272	8 000	86.0
b = 177	2 000	571	534	471	1 330	626	1 530	367	L_u	2 290	4 000	243	10 000	63.5
t = 10.9	1 500	533	498	440	995	581	1 500	331	I_x	186	5 000	176	12 000	50.4
d = 403	1 000	486	452	405	663	515	1 440	291	S_x	923	6 000	132	14 000	41.8
W410x46	4 000	591	533	462	1 860	629	1 380	422	M_r	278	3 000	212	7 000	61.7
W16x31	3 000	577	526	460	1 860	601	1 360	381	V_r	585	3 500	177	8 000	51.8
b = 140	2 000	515	478	419	1 330	555	1 330	329	L_u	1 780	4 000	142	10 000	39.2
t = 11.2	1 500	477	445	389	995	518	1 300	297	I_x	156	5 000	99.9	12 000	31.6
d = 403	1 000	433	400	354	663	462	1 250	258	S_x	772	6 000	76.4	14 000	26.5

Note: Resistances are based on a concrete density of 2300 kg/m^3.

Units: M_r - kN·m, V_r - kN, L_u - mm, I_x - 10^6 mm^4, S_x - 10^3 mm^3, b - mm, t - mm, d - mm

* Transformed moment of inertia for calculating shrinkage deflections, based on a modular ratio $n_t = 50$

COMPOSITE BEAMS
Trial Selection Table

G40.21 350W
$f'_c = 20$ MPa

75 mm Deck with 65 mm Slab
$\phi = 0.90$, $\phi_c = 0.60$

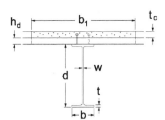

Steel section	b_1	Composite M_{rc} (kN·m) for % shear connection			Q_r (kN)	I_t 10^6 mm⁴	S_t 10^3 mm³	I_t^* 10^6 mm⁴	Non-composite Steel section data		Unbraced condition			
		100%	70%	40%	100%						L' mm	M_r' kN·m	L' mm	M_r' kN·m
W410x39	4 000	503	452	388	1 570	534	1 160	368	M_r	230	3 000	167	7 000	44.0
W16x26	3 000	493	447	387	1 570	512	1 150	334	V_r	484	3 500	132	8 000	36.5
b = 140	2 000	456	420	366	1 330	476	1 120	288	L_u	1 720	4 000	105	10 000	27.3
t = 8.8	1 500	419	390	338	995	447	1 100	259	I_x	126	5 000	73.0	12 000	21.8
d = 399	1 000	378	348	304	663	401	1 060	224	S_x	634	6 000	55.1	14 000	18.2
W360x79	3 000	771	714	638	1 990	783	2 130	455	M_r	450	2 000	450	6 000	319
W14x53	2 000	692	652	585	1 330	703	2 060	394	V_r	692	2 500	450	7 000	267
b = 205	1 500	651	613	553	995	643	2 000	359	L_u	2 980	3 000	450	8 000	225
t = 16.8	1 000	600	565	519	663	560	1 920	320	I_x	226	4 000	409	10 000	171
d = 354	500	530	508	482	332	435	1 750	276	S_x	1 280	5 000	364	12 000	139
W360x72	3 000	712	656	584	1 990	713	1 920	420	M_r	403	2 000	403	6 000	273
W14x48	2 000	635	596	534	1 330	643	1 860	363	V_r	626	2 500	403	7 000	222
b = 204	1 500	595	560	504	995	591	1 820	330	L_u	2 910	3 000	400	8 000	186
t = 15.1	1 000	548	515	470	663	516	1 740	292	I_x	201	4 000	360	10 000	141
d = 350	500	481	459	434	332	402	1 590	250	S_x	1 150	5 000	317	12 000	114
W360x64	3 000	657	603	534	1 990	646	1 720	388	M_r	359	2 000	359	6 000	228
W14x43	2 000	581	543	487	1 330	587	1 680	335	V_r	555	2 500	359	7 000	183
b = 203	1 500	542	511	458	995	541	1 640	303	L_u	2 850	3 000	354	8 000	153
t = 13.5	1 000	499	468	425	663	475	1 570	267	I_x	178	4 000	316	10 000	115
d = 347	500	436	414	389	332	370	1 440	226	S_x	1 030	5 000	275	12 000	92.2
W360x57	3 000	620	566	497	1 990	606	1 550	369	M_r	318	2 000	318	6 000	147
W14x38	2 000	545	507	450	1 330	553	1 510	318	V_r	588	2 500	312	7 000	119
b = 172	1 500	506	474	421	995	512	1 470	287	L_u	2 340	3 000	292	8 000	99.7
t = 13.1	1 000	462	431	389	663	452	1 420	251	I_x	160	4 000	246	10 000	75.6
d = 358	500	399	378	353	332	353	1 300	209	S_x	896	5 000	192	12 000	61.0
W360x51	3 000	574	521	454	1 990	547	1 390	340	M_r	281	2 000	281	6 000	121
W14x34	2 000	500	463	410	1 330	502	1 350	292	V_r	531	2 500	274	7 000	96.9
b = 171	1 500	462	433	383	995	467	1 320	263	L_u	2 300	3 000	255	8 000	80.9
t = 11.6	1 000	421	393	351	663	414	1 270	229	I_x	141	4 000	211	10 000	60.8
d = 355	500	361	341	316	332	325	1 170	189	S_x	796	5 000	159	12 000	48.8
W360x45	3 000	517	467	405	1 800	490	1 230	310	M_r	245	2 000	245	6 000	96.0
W14x30	2 000	459	423	372	1 330	452	1 200	267	V_r	505	2 500	237	7 000	76.4
b = 171	1 500	422	393	345	995	422	1 180	240	L_u	2 240	3 000	219	8 000	63.3
t = 9.8	1 000	382	355	314	663	376	1 140	208	I_x	122	4 000	177	10 000	47.1
d = 352	500	324	304	280	332	296	1 050	169	S_x	691	5 000	128	12 000	37.5
W360x39	3 000	456	410	352	1 570	433	1 070	280	M_r	209	2 000	195	6 000	54.1
W14x26	2 000	419	383	334	1 330	402	1 040	241	V_r	477	2 500	173	7 000	44.2
b = 128	1 500	382	354	308	995	377	1 020	216	L_u	1 650	3 000	149	8 000	37.4
t = 10.7	1 000	343	317	278	663	338	989	186	I_x	102	4 000	97.0	10 000	28.7
d = 353	500	287	267	243	332	268	914	149	S_x	580	5 000	69.7	12 000	23.3

Note: Resistances are based on a concrete density of 2300 kg/m³.

Units: M_r - kN·m, V_r - kN, L_u - mm, I_x - 10^6 mm⁴, S_x - 10^3 mm³, b - mm, t - mm, d - mm

* Transformed moment of inertia for calculating shrinkage deflections, based on a modular ratio $n_t = 50$

COMPOSITE BEAMS
Trial Selection Table

75 mm Deck with 65 mm Slab
$\phi = 0.90$, $\phi_c = 0.60$

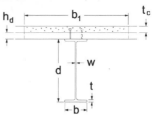

G40.21 350W
$f'_c = 20$ MPa

Steel section	b_1	Composite							Non-composite			
		M_{rc} (kN·m) for % shear connection			Q_r	I_t	S_t	I_t^*	Steel section data	Unbraced condition		
										L'	M_r'	L'	M_r'
	mm	100%	70%	40%	(kN)	10^6 mm⁴	10^3 mm³	10^6 mm⁴		mm	kN·m	mm	kN·m
W360x33	3 000	385	343	293	1 310	366	894	244	M_r 170	2 000	156	6 000	38.0
W14x22	2 000	371	336	291	1 310	342	875	211	V_r 399	2 500	136	7 000	30.8
b = 127	1 500	335	308	267	995	323	859	189	L_u 1 590	3 000	114	8 000	25.8
t = 8.5	1 000	298	275	238	663	292	833	162	I_x 82.6	4 000	70.2	10 000	19.6
d = 349	500	247	228	205	332	234	774	128	S_x 473	5 000	49.6	12 000	15.8
W310x74	3 000	673	616	545	1 990	620	1 850	357	M_r 375	2 000	375	6 000	279
W12x50	2 000	595	555	497	1 330	557	1 790	307	V_r 606	2 500	375	7 000	243
b = 205	1 500	554	521	469	995	509	1 740	277	L_u 3 070	3 000	375	8 000	206
t = 16.3	1 000	509	479	437	663	442	1 670	244	I_x 165	4 000	345	10 000	159
d = 310	500	447	427	403	332	340	1 510	207	S_x 1 060	5 000	312	12 000	129
W310x67	3 000	622	566	497	1 990	561	1 660	328	M_r 334	2 000	334	6 000	238
W12x45	2 000	545	507	453	1 330	507	1 610	281	V_r 541	2 500	334	7 000	200
b = 204	1 500	506	475	425	995	465	1 570	253	L_u 2 990	3 000	334	8 000	169
t = 14.6	1 000	463	435	395	663	406	1 500	222	I_x 145	4 000	303	10 000	129
d = 306	500	404	385	361	332	313	1 370	186	S_x 948	5 000	271	12 000	105
W310x60	3 000	575	522	453	1 990	509	1 490	303	M_r 296	2 000	296	6 000	203
W12x40	2 000	500	463	413	1 330	462	1 450	260	V_r 472	2 500	296	7 000	166
b = 203	1 500	461	433	387	995	426	1 420	234	L_u 2 930	3 000	295	8 000	139
t = 13.1	1 000	422	396	358	663	374	1 360	204	I_x 129	4 000	266	10 000	106
d = 303	500	366	347	324	332	289	1 240	169	S_x 849	5 000	235	12 000	85.2
W310x52	3 000	548	495	427	1 990	484	1 340	295	M_r 265	2 000	265	6 000	130
W12x35	2 000	474	436	386	1 330	443	1 310	253	V_r 502	2 500	260	7 000	106
b = 167	1 500	435	406	361	995	411	1 280	227	L_u 2 350	3 000	244	8 000	89.4
t = 13.2	1 000	395	370	332	663	362	1 230	197	I_x 119	4 000	208	10 000	68.4
d = 318	500	340	322	299	332	282	1 130	161	S_x 750	5 000	167	12 000	55.5
W310x45	3 000	479	429	367	1 790	417	1 150	262	M_r 223	2 000	223	6 000	98.2
W12x30	2 000	422	386	339	1 330	385	1 120	224	V_r 429	2 500	217	7 000	79.3
b = 166	1 500	384	356	316	995	359	1 100	201	L_u 2 290	3 000	202	8 000	66.5
t = 11.2	1 000	346	324	288	663	319	1 060	173	I_x 99.2	4 000	168	10 000	50.4
d = 313	500	296	279	257	332	250	976	140	S_x 634	5 000	128	12 000	40.6
W310x39	3 000	419	373	318	1 560	367	1 000	237	M_r 192	2 000	192	6 000	77.7
W12x26	2 000	383	348	303	1 330	341	978	203	V_r 374	2 500	186	7 000	62.2
b = 165	1 500	347	319	282	995	319	959	182	L_u 2 250	3 000	172	8 000	51.8
t = 9.7	1 000	309	289	256	663	286	928	156	I_x 85.1	4 000	140	10 000	38.8
d = 310	500	263	247	226	332	226	861	125	S_x 549	5 000	103	12 000	31.1
W250x67	3 000	558	503	433	1 990	448	1 520	255	M_r 284	2 000	284	5 000	246
W10x45	2 000	481	443	393	1 330	403	1 470	216	V_r 476	2 500	284	6 000	225
b = 204	1 500	442	412	370	995	368	1 430	193	L_u 3 230	3 000	284	7 000	203
t = 15.7	1 000	401	377	342	663	319	1 360	167	I_x 104	3 500	278	8 000	180
d = 257	500	350	333	312	332	242	1 230	138	S_x 806	4 000	268	10 000	139

Note: Resistances are based on a concrete density of 2300 kg/m³.

Units: M_r - kN·m, V_r - kN, L_u - mm, I_x - 10^6 mm⁴, S_x - 10^3 mm³, b - mm, t - mm, d - mm

* Transformed moment of inertia for calculating shrinkage deflections, based on a modular ratio $n_t = 50$

COMPOSITE BEAMS
Trial Selection Table

G40.21 350W
f′_c = 20 MPa

75 mm Deck with 65 mm Slab
$\phi = 0.90$, $\phi_c = 0.60$

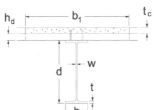

Steel section	b_1	Composite M_{rc} (kN·m) for % shear connection			Q_r	I_t	S_t	I_t^*	Non-composite Steel section data	Unbraced condition			
		100%	70%	40%	100%	10^6	10^3	10^6		L'	M_r'	L'	M_r'
	mm				(kN)	mm⁴	mm³	mm⁴		mm	kN·m	mm	kN·m
W250x58	3 000	508	454	386	1 990	392	1 320	228	M_r 243	2 000	243	5 000	204
W10x39	2 000	433	396	348	1 330	355	1 280	193	V_r 419	2 500	243	6 000	182
b = 203	1 500	394	366	326	995	326	1 250	172	L_u 3 100	3 000	243	7 000	161
t = 13.5	1 000	355	333	300	663	285	1 190	148	I_x 87.3	3 500	235	8 000	137
d = 252	500	307	291	271	332	217	1 080	120	S_x 693	4 000	225	10 000	105
W250x45	3 000	439	389	326	1 800	338	1 050	207	M_r 190	2 000	190	5 000	114
W10x30	2 000	381	344	298	1 330	310	1 020	175	V_r 420	2 500	182	6 000	90.6
b = 148	1 500	343	315	276	995	288	1 000	156	L_u 2 150	3 000	169	7 000	75.2
t = 13	1 000	304	283	250	663	254	962	133	I_x 71.1	3 500	156	8 000	64.4
d = 266	500	257	241	221	332	197	880	105	S_x 534	4 000	143	10 000	50.1
W250x39	3 000	381	335	279	1 550	293	907	185	M_r 162	2 000	162	5 000	88.0
W10x26	2 000	345	310	265	1 330	271	884	158	V_r 360	2 500	153	6 000	69.2
b = 147	1 500	308	281	245	995	253	866	140	L_u 2 090	3 000	141	7 000	57.0
t = 11.2	1 000	270	251	221	663	226	836	119	I_x 60.1	3 500	129	8 000	48.6
d = 262	500	227	212	192	332	177	770	93.0	S_x 459	4 000	116	10 000	37.6
W250x33	3 000	325	284	235	1 310	250	768	163	M_r 134	2 000	134	5 000	63.6
W10x22	2 000	311	277	233	1 310	233	750	139	V_r 327	2 500	124	6 000	49.4
b = 146	1 500	276	249	215	995	219	735	124	L_u 2 000	3 000	113	7 000	40.3
t = 9.1	1 000	238	220	192	663	197	712	105	I_x 48.9	3 500	101	8 000	34.1
d = 258	500	197	183	164	332	156	660	80.6	S_x 379	4 000	88.6	10 000	26.1
W200x42	3 000	358	310	252	1 660	233	865	141	M_r 139	2 000	139	5 000	106
W8x28	2 000	312	277	232	1 330	214	841	118	V_r 307	2 500	139	6 000	92.0
b = 166	1 500	275	248	213	995	198	821	104	L_u 2 580	3 000	134	7 000	76.5
t = 11.8	1 000	237	218	192	663	175	790	87.1	I_x 40.6	3 500	127	8 000	65.6
d = 205	500	197	184	167	332	134	721	66.4	S_x 396	4 000	120	10 000	51.2
W200x36	3 000	311	267	215	1 430	202	747	126	M_r 118	2 000	118	5 000	85.4
W8x24	2 000	286	251	207	1 330	187	727	107	V_r 259	2 500	118	6 000	70.2
b = 165	1 500	250	223	189	995	174	711	93.8	L_u 2 490	3 000	112	7 000	58.0
t = 10.2	1 000	212	194	170	663	155	686	78.3	I_x 34.1	3 500	106	8 000	49.5
d = 201	500	173	162	146	332	120	631	58.9	S_x 340	4 000	99.0	10 000	38.4
W200x31	3 000	283	242	195	1 260	190	671	122	M_r 106	2 000	105	5 000	57.0
W8x21	2 000	270	236	193	1 260	176	654	103	V_r 279	2 500	97.8	6 000	45.6
b = 134	1 500	239	212	178	995	165	640	91.3	L_u 1 960	3 000	90.2	7 000	38.0
t = 10.2	1 000	201	183	158	663	148	619	76.2	I_x 31.4	3 500	82.3	8 000	32.7
d = 210	500	162	150	133	332	116	571	56.9	S_x 299	4 000	74.4	10 000	25.6
W200x27	3 000	241	205	165	1 070	163	571	108	M_r 87.9	2 000	86.3	5 000	41.2
W8x18	2 000	232	201	163	1 070	152	556	92.2	V_r 250	2 500	79.5	6 000	32.6
b = 133	1 500	217	191	158	995	143	545	81.6	L_u 1 880	3 000	72.2	7 000	27.1
t = 8.4	1 000	181	163	139	663	129	528	68.1	I_x 25.8	3 500	64.5	8 000	23.1
d = 207	500	143	131	115	332	103	491	50.3	S_x 249	4 000	56.0	10 000	18.0

Note: Resistances are based on a concrete density of 2300 kg/m³.
Units: M_r - kN·m, V_r - kN, L_u - mm, I_x - 10^6 mm⁴, S_x - 10^3 mm³, b - mm, t - mm, d - mm
* Transformed moment of inertia for calculating shrinkage deflections, based on a modular ratio $n_t = 50$

COMPOSITE BEAMS
Trial Selection Table

75 mm Deck with 75 mm Slab
$\phi = 0.90$, $\phi_c = 0.60$

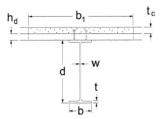

G40.21 350W
$f'_c = 25$ MPa

Steel section	b_1	Composite M_{rc} (kN·m) for % shear connection			Q_r (kN)	I_t 10^6	S_t 10^3	I_t^* 10^6	Steel section data	Non-composite Unbraced condition			
										L'	M_r'	L'	M_r'
	mm	100%	70%	40%	100%	mm⁴	mm³	mm⁴		mm	kN·m	mm	kN·m
WWF1000x223	5 000	4 970	4 760	4 310	4 780	11 100	12 700	6 820	M_r 3 310	4 000	3 250	12 000	920
WWF39x150	4 000	4 830	4 570	4 140	3 830	10 500	12 500	6 450	V_r 2 210	5 000	2 960	14 000	729
b = 300	3 000	4 600	4 330	3 960	2 870	9 760	12 300	6 050	L_u 3 790	6 000	2 620	16 000	602
t = 25	2 000	4 270	4 040	3 750	1 910	8 700	11 800	5 610	I_x 4 590	8 000	1 790	18 000	512
d = 1000	1 000	3 840	3 690	3 530	956	7 130	11 000	5 130	S_x 9 190	10 000	1 230	20 000	446
WWF1000x200	5 000	4 540	4 340	3 890	4 780	10 100	11 300	6 120	M_r 2 890	4 000	2 790	12 000	699
WWF39x134	4 000	4 400	4 150	3 720	3 830	9 540	11 100	5 770	V_r 2 210	5 000	2 510	14 000	549
b = 300	3 000	4 180	3 900	3 530	2 870	8 850	10 900	5 380	L_u 3 610	6 000	2 180	16 000	449
t = 20	2 000	3 840	3 610	3 330	1 910	7 890	10 500	4 950	I_x 3 940	8 000	1 390	18 000	379
d = 1000	1 000	3 410	3 260	3 100	956	6 420	9 760	4 470	S_x 7 880	10 000	945	20 000	328
WWF900x231	5 000	4 680	4 530	4 250	4 780	10 000	12 700	6 310	M_r 3 400	4 000	3 400	12 000	1 760
WWF35x156	4 000	4 550	4 420	4 120	3 830	9 530	12 600	5 990	V_r 1 350	5 000	3 400	14 000	1 370
b = 400	3 000	4 420	4 260	3 970	2 870	8 860	12 300	5 650	L_u 5 460	6 000	3 300	16 000	1 110
t = 25	2 000	4 210	4 030	3 790	1 910	7 930	12 000	5 270	I_x 4 410	8 000	2 870	18 000	933
d = 900	1 000	3 860	3 740	3 600	956	6 580	11 300	4 860	S_x 9 810	10 000	2 360	20 000	801
WWF900x192	5 000	3 990	3 840	3 560	4 780	8 480	10 400	5 280	M_r 2 710	4 000	2 700	12 000	842
WWF35x128	4 000	3 860	3 730	3 430	3 830	8 070	10 300	4 990	V_r 1 350	5 000	2 490	14 000	670
b = 300	3 000	3 730	3 570	3 280	2 870	7 510	10 100	4 660	L_u 3 980	6 000	2 240	16 000	555
t = 25	2 000	3 520	3 340	3 100	1 910	6 730	9 800	4 310	I_x 3 460	8 000	1 630	18 000	473
d = 900	1 000	3 170	3 050	2 910	956	5 530	9 220	3 910	S_x 7 680	10 000	1 120	20 000	412
WWF900x169	5 000	3 590	3 440	3 170	4 780	7 550	9 110	4 690	M_r 2 320	4 000	2 290	12 000	625
WWF35x113	4 000	3 470	3 340	3 040	3 830	7 200	9 000	4 420	V_r 1 340	5 000	2 080	14 000	491
b = 300	3 000	3 340	3 180	2 890	2 870	6 720	8 830	4 110	L_u 3 820	6 000	1 850	16 000	402
t = 20	2 000	3 130	2 950	2 710	1 910	6 020	8 560	3 760	I_x 2 930	8 000	1 250	18 000	339
d = 900	1 000	2 780	2 660	2 520	956	4 930	8 020	3 370	S_x 6 510	10 000	845	20 000	293
WWF800x223	5 000	4 080	3 930	3 700	4 780	7 980	11 200	4 970	M_r 2 940	4 000	2 940	10 000	2 120
WWF31x150	4 000	3 960	3 830	3 590	3 830	7 570	11 100	4 710	V_r 1 370	5 000	2 940	12 000	1 630
b = 400	3 000	3 830	3 700	3 460	2 870	7 040	10 900	4 430	L_u 5 570	6 000	2 880	14 000	1 280
t = 25	2 000	3 660	3 510	3 300	1 910	6 290	10 500	4 120	I_x 3 410	7 000	2 710	16 000	1 050
d = 800	1 000	3 360	3 250	3 130	956	5 190	9 940	3 780	S_x 8 520	8 000	2 530	18 000	882
WWF800x184	5 000	3 470	3 320	3 090	4 780	6 710	9 120	4 160	M_r 2 330	4 000	2 330	10 000	1 040
WWF31x123	4 000	3 350	3 220	2 980	3 830	6 380	9 000	3 920	V_r 1 370	5 000	2 170	12 000	791
b = 300	3 000	3 220	3 090	2 840	2 870	5 950	8 840	3 650	L_u 4 060	6 000	1 970	14 000	634
t = 25	2 000	3 050	2 900	2 690	1 910	5 320	8 570	3 360	I_x 2 660	7 000	1 750	16 000	529
d = 800	1 000	2 750	2 640	2 510	956	4 360	8 050	3 030	S_x 6 640	8 000	1 490	18 000	454
WWF800x161	5 000	3 110	2 970	2 750	4 780	5 960	7 930	3 700	M_r 1 990	4 000	1 980	10 000	778
WWF31x108	4 000	2 990	2 880	2 630	3 830	5 680	7 840	3 470	V_r 1 370	5 000	1 810	12 000	581
b = 300	3 000	2 870	2 750	2 500	2 870	5 310	7 690	3 220	L_u 3 900	6 000	1 620	14 000	460
t = 20	2 000	2 700	2 550	2 350	1 910	4 760	7 460	2 940	I_x 2 250	7 000	1 400	16 000	379
d = 800	1 000	2 400	2 300	2 170	956	3 880	7 000	2 620	S_x 5 610	8 000	1 130	18 000	322

Note: Resistances are based on a concrete density of 2300 kg/m³.

Units: M_r - kN·m, V_r - kN, L_u - mm, I_x - 10^6 mm⁴, S_x - 10^3 mm³, b - mm, t - mm, d - mm

* Transformed moment of inertia for calculating shrinkage deflections, based on a modular ratio $n_t = 50$

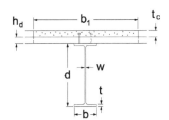

COMPOSITE BEAMS
Trial Selection Table

G40.21 350W
$f'_c = 25$ MPa

75 mm Deck with 75 mm Slab
$\phi = 0.90$, $\phi_c = 0.60$

Steel section	b_1	M_{rc} (kN·m) for % shear connection			Q_r (kN)	I_t 10^6 mm⁴	S_t 10^3 mm³	I_t^* 10^6 mm⁴	Steel section data	Unbraced condition			
		100%	70%	40%	100%					L' mm	M_r' kN·m	L' mm	M_r' kN·m
WWF700x214	5 000	3 520	3 370	3 170	4 780	6 180	9 710	3 800	M_r 2 500	4 000	2 500	10 000	1 870
WWF28x144	4 000	3 390	3 270	3 080	3 830	5 860	9 580	3 600	V_r 1 370	5 000	2 500	12 000	1 500
b = 400	3 000	3 270	3 170	2 960	2 870	5 440	9 400	3 370	L_u 5 690	6 000	2 470	14 000	1 190
t = 25	2 000	3 130	3 010	2 830	1 910	4 850	9 110	3 120	I_x 2 540	7 000	2 330	16 000	982
d = 700	1 000	2 870	2 780	2 670	956	3 980	8 580	2 840	S_x 7 270	8 000	2 190	18 000	835
WWF700x175	5 000	2 980	2 840	2 640	4 780	5 170	7 850	3 170	M_r 1 970	4 000	1 970	10 000	965
WWF28x117	4 000	2 860	2 740	2 540	3 830	4 920	7 750	2 980	V_r 1 370	5 000	1 850	12 000	741
b = 300	3 000	2 740	2 630	2 430	2 870	4 580	7 610	2 770	L_u 4 160	6 000	1 700	14 000	601
t = 25	2 000	2 590	2 470	2 290	1 910	4 090	7 380	2 540	I_x 1 970	7 000	1 530	16 000	505
d = 700	1 000	2 340	2 250	2 140	956	3 330	6 920	2 270	S_x 5 640	8 000	1 350	18 000	436
WWF700x152	5 000	2 670	2 530	2 340	4 780	4 570	6 800	2 820	M_r 1 680	4 000	1 680	10 000	711
WWF28x102	4 000	2 560	2 440	2 250	3 830	4 360	6 720	2 640	V_r 1 370	5 000	1 540	12 000	537
b = 300	3 000	2 430	2 340	2 130	2 870	4 080	6 600	2 440	L_u 3 990	6 000	1 390	14 000	429
t = 20	2 000	2 300	2 180	2 000	1 910	3 650	6 400	2 220	I_x 1 660	7 000	1 230	16 000	356
d = 700	1 000	2 040	1 950	1 840	956	2 970	5 990	1 960	S_x 4 760	8 000	1 020	18 000	305
W610x241	4 000	3 430	3 280	3 000	3 830	5 220	9 520	3 080	M_r 2 420	3 000	2 420	8 000	1 950
W24x162	3 000	3 280	3 120	2 870	2 870	4 810	9 290	2 880	V_r 2 360	4 000	2 420	10 000	1 640
b = 329	2 000	3 070	2 920	2 730	1 910	4 240	8 930	2 660	L_u 4 750	5 000	2 380	12 000	1 300
t = 31	1 500	2 930	2 810	2 650	1 430	3 880	8 670	2 540	I_x 2 150	6 000	2 240	14 000	1 070
d = 635	1 000	2 780	2 680	2 570	956	3 430	8 290	2 420	S_x 6 780	7 000	2 100	16 000	911
W610x217	4 000	3 120	2 990	2 730	3 830	4 770	8 610	2 810	M_r 2 160	3 000	2 160	8 000	1 690
W24x146	3 000	2 990	2 840	2 600	2 870	4 400	8 410	2 620	V_r 2 150	4 000	2 160	10 000	1 370
b = 328	2 000	2 790	2 650	2 460	1 910	3 890	8 100	2 400	L_u 4 640	5 000	2 110	12 000	1 070
t = 27.7	1 500	2 660	2 540	2 390	1 430	3 560	7 860	2 290	I_x 1 910	6 000	1 980	14 000	878
d = 628	1 000	2 510	2 420	2 310	956	3 140	7 520	2 170	S_x 6 070	7 000	1 840	16 000	745
W610x195	4 000	2 830	2 700	2 470	3 830	4 320	7 720	2 550	M_r 1 910	3 000	1 910	8 000	1 450
W24x131	3 000	2 700	2 570	2 350	2 870	4 000	7 550	2 370	V_r 1 990	4 000	1 910	10 000	1 120
b = 327	2 000	2 530	2 400	2 210	1 910	3 550	7 280	2 160	L_u 4 530	5 000	1 860	12 000	870
t = 24.4	1 500	2 410	2 290	2 140	1 430	3 240	7 070	2 050	I_x 1 680	6 000	1 730	14 000	709
d = 622	1 000	2 260	2 170	2 060	956	2 860	6 770	1 930	S_x 5 400	7 000	1 590	16 000	598
W610x174	4 000	2 560	2 440	2 230	3 830	3 900	6 890	2 310	M_r 1 690	3 000	1 690	8 000	1 240
W24x117	3 000	2 440	2 320	2 120	2 870	3 630	6 740	2 130	V_r 1 790	4 000	1 690	10 000	924
b = 325	2 000	2 280	2 160	1 990	1 910	3 220	6 510	1 940	L_u 4 440	5 000	1 630	12 000	709
t = 21.6	1 500	2 170	2 060	1 910	1 430	2 950	6 330	1 830	I_x 1 470	6 000	1 510	14 000	574
d = 616	1 000	2 030	1 940	1 840	956	2 600	6 060	1 720	S_x 4 780	7 000	1 380	16 000	482
W610x155	4 000	2 310	2 200	2 010	3 830	3 520	6 130	2 100	M_r 1 490	3 000	1 490	8 000	1 060
W24x104	3 000	2 190	2 100	1 900	2 870	3 280	6 010	1 930	V_r 1 610	4 000	1 490	10 000	762
b = 324	2 000	2 060	1 940	1 780	1 910	2 930	5 820	1 750	L_u 4 370	5 000	1 430	12 000	579
t = 19	1 500	1 950	1 850	1 710	1 430	2 690	5 660	1 640	I_x 1 290	6 000	1 310	14 000	465
d = 611	1 000	1 820	1 740	1 640	956	2 360	5 430	1 530	S_x 4 220	7 000	1 190	16 000	388

Note: Resistances are based on a concrete density of 2300 kg/m³.
Units: M_r - kN·m, V_r - kN, L_u - mm, I_x - 10^6 mm⁴, S_x - 10^3 mm³, b - mm, t - mm, d - mm
* Transformed moment of inertia for calculating shrinkage deflections, based on a modular ratio $n_t = 50$

COMPOSITE BEAMS
Trial Selection Table

75 mm Deck with 75 mm Slab
$\phi = 0.90$, $\phi_c = 0.60$

G40.21 350W
$f'_c = 25$ MPa

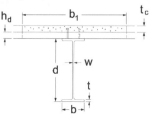

Steel section	b_1	Composite							Non-composite			
		M_{rc} (kN·m) for % shear connection			Q_r	I_t	S_t	I_t^*	Steel section data	Unbraced condition			
		100%	70%	40%	100%					L'	M_r'	L'	M_r'
	mm				(kN)	10^6 mm^4	10^3 mm^3	10^6 mm^4		mm	kN·m	mm	kN·m
W610x140 W24x94 b = 230 t = 22.2 d = 617	4 000 3 000 2 000 1 500 1 000	2 160 2 040 1 890 1 780 1 650	2 040 1 930 1 770 1 670 1 560	1 840 1 730 1 600 1 530 1 460	3 830 2 870 1 910 1 430 956	3 250 3 030 2 710 2 480 2 180	5 520 5 410 5 230 5 080 4 860	1 920 1 760 1 580 1 480 1 370	M_r 1 310 V_r 1 680 L_u 3 050 I_x 1 120 S_x 3 630	3 000 4 000 5 000 6 000 7 000	1 310 1 180 1 040 878 695	8 000 10 000 12 000 14 000 16 000	573 422 334 277 237
W610x125 W24x84 b = 229 t = 19.6 d = 612	4 000 3 000 2 000 1 500 1 000	1 960 1 840 1 710 1 610 1 490	1 850 1 750 1 610 1 510 1 410	1 670 1 570 1 450 1 380 1 310	3 830 2 870 1 910 1 430 956	2 930 2 740 2 460 2 260 1 990	4 920 4 830 4 680 4 550 4 360	1 750 1 600 1 430 1 330 1 230	M_r 1 160 V_r 1 510 L_u 2 990 I_x 985 S_x 3 220	3 000 4 000 5 000 6 000 7 000	1 160 1 040 895 733 575	8 000 10 000 12 000 14 000 16 000	470 342 269 222 189
W610x113 W24x76 b = 228 t = 17.3 d = 608	4 000 3 000 2 000 1 500 1 000	1 810 1 690 1 570 1 480 1 360	1 700 1 610 1 470 1 390 1 280	1 540 1 440 1 320 1 250 1 180	3 830 2 870 1 910 1 430 956	2 670 2 510 2 260 2 080 1 830	4 450 4 370 4 240 4 130 3 960	1 610 1 470 1 310 1 210 1 110	M_r 1 040 V_r 1 420 L_u 2 930 I_x 875 S_x 2 880	3 000 4 000 5 000 6 000 7 000	1 030 915 781 617 481	8 000 10 000 12 000 14 000 16 000	391 282 220 180 153
W610x101 W24x68 b = 228 t = 14.9 d = 603	4 000 3 000 2 000 1 500 1 000	1 660 1 550 1 430 1 350 1 240	1 560 1 470 1 350 1 260 1 160	1 410 1 310 1 200 1 130 1 060	3 830 2 870 1 910 1 430 956	2 410 2 270 2 060 1 900 1 670	3 990 3 920 3 810 3 710 3 560	1 470 1 340 1 180 1 090 995	M_r 914 V_r 1 310 L_u 2 870 I_x 764 S_x 2 530	3 000 4 000 5 000 6 000 7 000	901 795 668 512 396	8 000 10 000 12 000 14 000 16 000	320 228 176 144 121
W610x91 W24x61 b = 227 t = 12.7 d = 598	4 000 3 000 2 000 1 500 1 000	1 510 1 410 1 300 1 220 1 110	1 410 1 330 1 220 1 140 1 040	1 260 1 190 1 080 1 010 941	3 650 2 870 1 910 1 430 956	2 160 2 050 1 860 1 720 1 520	3 550 3 500 3 400 3 320 3 190	1 340 1 220 1 070 987 893	M_r 806 V_r 1 110 L_u 2 790 I_x 667 S_x 2 230	3 000 4 000 5 000 6 000 7 000	788 687 567 421 325	8 000 10 000 12 000 14 000 16 000	261 185 142 115 97.1
W530x138 W21x93 b = 214 t = 23.6 d = 549	4 000 3 000 2 000 1 500 1 000	1 940 1 820 1 680 1 580 1 450	1 820 1 720 1 570 1 480 1 370	1 630 1 530 1 410 1 350 1 280	3 830 2 870 1 910 1 430 956	2 640 2 460 2 190 2 000 1 750	4 950 4 850 4 680 4 540 4 330	1 530 1 400 1 250 1 160 1 070	M_r 1 140 V_r 1 680 L_u 2 910 I_x 861 S_x 3 140	3 000 4 000 5 000 6 000 7 000	1 130 1 010 891 762 616	8 000 10 000 12 000 14 000 16 000	515 390 314 263 227
W530x123 W21x83 b = 212 t = 21.2 d = 544	4 000 3 000 2 000 1 500 1 000	1 770 1 650 1 520 1 430 1 320	1 660 1 560 1 430 1 340 1 240	1 490 1 390 1 280 1 220 1 150	3 830 2 870 1 910 1 430 956	2 390 2 230 2 000 1 830 1 600	4 430 4 350 4 200 4 090 3 900	1 410 1 280 1 130 1 050 964	M_r 1 010 V_r 1 480 L_u 2 840 I_x 761 S_x 2 800	3 000 4 000 5 000 6 000 7 000	996 888 768 631 505	8 000 10 000 12 000 14 000 16 000	421 316 253 211 182
W530x109 W21x73 b = 211 t = 18.8 d = 539	4 000 3 000 2 000 1 500 1 000	1 610 1 490 1 370 1 300 1 190	1 500 1 410 1 290 1 210 1 120	1 350 1 260 1 150 1 090 1 030	3 830 2 870 1 910 1 430 956	2 140 2 010 1 810 1 670 1 460	3 940 3 870 3 750 3 650 3 490	1 280 1 170 1 030 951 865	M_r 891 V_r 1 300 L_u 2 790 I_x 667 S_x 2 480	3 000 4 000 5 000 6 000 7 000	872 771 656 520 413	8 000 10 000 12 000 14 000 16 000	342 254 202 168 144

Note: Resistances are based on a concrete density of 2300 kg/m^3.

Units: M_r - kN·m, V_r - kN, L_u - mm, I_x - 10^6 mm^4, S_x - 10^3 mm^3, b - mm, t - mm, d - mm

* Transformed moment of inertia for calculating shrinkage deflections, based on a modular ratio $n_t = 50$

COMPOSITE BEAMS
Trial Selection Table

G40.21 350W
$f'_c = 25$ MPa

75 mm Deck with 75 mm Slab
$\phi = 0.90$, $\phi_c = 0.60$

Steel section	b_1	Composite M_{rc} (kN·m) for % shear connection			Q_r (kN)	I_t 10^6	S_t 10^3	I_t^* 10^6	Non-composite Steel section data		Unbraced condition			
		100%	70%	40%	100%	mm⁴	mm³	mm⁴			L' mm	M_r' kN·m	L' mm	M_r' kN·m
	mm													
W530x101	4 000	1 520	1 420	1 270	3 830	2 010	3 660	1 210	M_r	825	3 000	804	8 000	301
W21x68	3 000	1 410	1 320	1 190	2 870	1 890	3 600	1 100	V_r	1 220	4 000	706	10 000	222
b = 210	2 000	1 290	1 220	1 080	1 910	1 710	3 500	972	L_u	2 750	5 000	594	12 000	176
t = 17.4	1 500	1 220	1 140	1 030	1 430	1 580	3 410	896	I_x	617	6 000	462	14 000	146
d = 537	1 000	1 120	1 050	960	956	1 390	3 270	813	S_x	2 300	7 000	365	16 000	125
W530x92	4 000	1 410	1 310	1 170	3 720	1 840	3 340	1 130	M_r	743	3 000	719	8 000	253
W21x62	3 000	1 310	1 230	1 100	2 870	1 740	3 290	1 020	V_r	1 130	4 000	626	10 000	185
b = 209	2 000	1 190	1 130	1 000	1 910	1 580	3 190	897	L_u	2 700	5 000	519	12 000	146
t = 15.6	1 500	1 130	1 050	943	1 430	1 460	3 120	824	I_x	552	6 000	393	14 000	120
d = 533	1 000	1 030	965	879	956	1 280	2 990	744	S_x	2 070	7 000	309	16 000	103
W530x82	4 000	1 260	1 160	1 040	3 310	1 640	2 960	1 020	M_r	652	3 000	625	8 000	204
W21x55	3 000	1 200	1 120	997	2 870	1 560	2 920	927	V_r	1 040	4 000	538	10 000	148
b = 209	2 000	1 080	1 020	902	1 910	1 420	2 840	810	L_u	2 640	5 000	436	12 000	115
t = 13.3	1 500	1 020	953	845	1 430	1 320	2 770	742	I_x	478	6 000	321	14 000	94.8
d = 528	1 000	931	867	782	956	1 160	2 670	665	S_x	1 810	7 000	250	16 000	80.5
W460x106	4 000	1 430	1 320	1 170	3 830	1 680	3 460	991	M_r	753	3 000	727	7 000	366
W18x71	3 000	1 310	1 230	1 090	2 870	1 580	3 400	896	V_r	1 230	3 500	686	8 000	308
b = 194	2 000	1 190	1 120	995	1 910	1 420	3 290	785	L_u	2 670	4 000	643	10 000	235
t = 20.6	1 500	1 120	1 050	940	1 430	1 300	3 200	721	I_x	488	5 000	553	12 000	190
d = 469	1 000	1 020	961	880	956	1 140	3 060	651	S_x	2 080	6 000	450	14 000	160
W460x97	4 000	1 330	1 230	1 090	3 830	1 550	3 170	929	M_r	687	3 000	660	7 000	314
W18x65	3 000	1 220	1 140	1 020	2 870	1 460	3 120	840	V_r	1 100	3 500	621	8 000	264
b = 193	2 000	1 100	1 040	924	1 910	1 320	3 020	734	L_u	2 630	4 000	579	10 000	200
t = 19	1 500	1 040	973	871	1 430	1 220	2 940	673	I_x	445	5 000	491	12 000	161
d = 466	1 000	951	891	813	956	1 070	2 820	605	S_x	1 910	6 000	389	14 000	135
W460x89	4 000	1 240	1 140	1 010	3 590	1 450	2 940	877	M_r	633	3 000	605	7 000	276
W18x60	3 000	1 150	1 070	955	2 870	1 370	2 890	793	V_r	1 010	3 500	568	8 000	231
b = 192	2 000	1 030	975	868	1 910	1 240	2 810	691	L_u	2 600	4 000	527	10 000	174
t = 17.7	1 500	973	914	816	1 430	1 140	2 740	632	I_x	409	5 000	442	12 000	140
d = 463	1 000	893	835	759	956	1 000	2 630	567	S_x	1 770	6 000	343	14 000	117
W460x82	4 000	1 140	1 040	917	3 280	1 330	2 690	819	M_r	576	3 000	546	7 000	234
W18x55	3 000	1 080	995	886	2 870	1 260	2 650	741	V_r	947	3 500	510	8 000	195
b = 191	2 000	961	904	804	1 910	1 150	2 570	645	L_u	2 540	4 000	471	10 000	146
t = 16	1 500	901	847	755	1 430	1 060	2 510	588	I_x	370	5 000	386	12 000	116
d = 460	1 000	827	773	699	956	937	2 410	524	S_x	1 610	6 000	292	14 000	97.3
W460x74	4 000	1 040	947	831	2 980	1 220	2 440	762	M_r	520	3 000	489	7 000	198
W18x50	3 000	1 000	925	820	2 870	1 160	2 410	689	V_r	855	3 500	455	8 000	164
b = 190	2 000	891	835	744	1 910	1 060	2 340	598	L_u	2 510	4 000	417	10 000	122
t = 14.5	1 500	832	783	696	1 430	981	2 290	544	I_x	332	5 000	332	12 000	96.8
d = 457	1 000	764	713	642	956	869	2 210	483	S_x	1 460	6 000	249	14 000	80.6

Note: Resistances are based on a concrete density of 2300 kg/m³.
Units: M_r - kN·m, V_r - kN, L_u - mm, I_x - 10^6 mm⁴, S_x - 10^3 mm³, b - mm, t - mm, d - mm
* Transformed moment of inertia for calculating shrinkage deflections, based on a modular ratio $n_t = 50$

COMPOSITE BEAMS
Trial Selection Table

75 mm Deck with 75 mm Slab
$\phi = 0.90$, $\phi_c = 0.60$

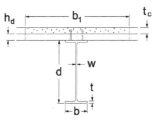

G40.21 350W
$f'_c = 25$ MPa

Steel section	b_1	M_{rc} (kN·m) for % shear connection			Q_r	I_t	S_t	I_t^*	Steel section data		Unbraced condition			
		100%	70%	40%	100%	10^6	10^3	10^6			L´	$M_r´$	L´	$M_r´$
	mm				(kN)	mm⁴	mm³	mm⁴			mm	kN·m	mm	kN·m
W460x67	4 000	946	859	751	2 700	1 110	2 210	706	M_r	466	3 000	435	7 000	165
W18x45	3 000	922	847	748	2 700	1 060	2 180	638	V_r	802	3 500	402	8 000	136
b = 190	2 000	825	771	685	1 910	971	2 130	554	L_u	2 460	4 000	366	10 000	99.8
t = 12.7	1 500	767	722	639	1 430	904	2 080	502	I_x	296	5 000	281	12 000	79.0
d = 454	1 000	703	655	585	956	803	2 010	444	S_x	1 300	6 000	209	14 000	65.4
W460x61	4 000	848	766	667	2 410	990	1 970	641	M_r	406	3 000	374	7 000	132
W18x41	3 000	829	757	664	2 410	945	1 940	580	V_r	758	3 500	343	8 000	108
b = 189	2 000	757	704	622	1 910	873	1 890	502	L_u	2 390	4 000	309	10 000	78.5
t = 10.8	1 500	700	657	577	1 430	816	1 860	454	I_x	255	5 000	228	12 000	61.7
d = 450	1 000	639	593	525	956	728	1 790	399	S_x	1 130	6 000	168	14 000	50.8
W410x85	4 000	1 110	1 010	877	3 400	1 190	2 610	715	M_r	542	3 000	512	7 000	242
W16x57	3 000	1 030	949	840	2 870	1 120	2 560	644	V_r	945	3 500	480	8 000	205
b = 181	2 000	915	856	761	1 910	1 020	2 490	558	L_u	2 510	4 000	447	10 000	157
t = 18.2	1 500	854	801	713	1 430	936	2 430	508	I_x	315	5 000	377	12 000	127
d = 417	1 000	781	730	660	956	822	2 320	451	S_x	1 510	6 000	297	14 000	107
W410x74	4 000	983	889	772	3 010	1 060	2 310	653	M_r	476	3 000	445	7 000	194
W16x50	3 000	943	865	760	2 870	1 000	2 270	588	V_r	833	3 500	414	8 000	163
b = 180	2 000	830	774	688	1 910	915	2 210	509	L_u	2 450	4 000	382	10 000	123
t = 16	1 500	771	724	642	1 430	847	2 160	461	I_x	275	5 000	312	12 000	99.7
d = 413	1 000	705	658	591	956	748	2 070	408	S_x	1 330	6 000	239	14 000	83.8
W410x67	4 000	890	802	695	2 710	960	2 080	604	M_r	428	3 000	397	7 000	161
W16x45	3 000	866	790	691	2 710	913	2 050	545	V_r	750	3 500	367	8 000	135
b = 179	2 000	768	713	633	1 910	838	2 000	471	L_u	2 400	4 000	336	10 000	102
t = 14.4	1 500	710	666	590	1 430	778	1 960	426	I_x	245	5 000	264	12 000	81.6
d = 410	1 000	648	605	541	956	690	1 880	374	S_x	1 200	6 000	201	14 000	68.3
W410x60	4 000	788	707	611	2 390	857	1 850	553	M_r	375	3 000	344	7 000	131
W16x40	3 000	770	698	608	2 390	818	1 820	500	V_r	652	3 500	317	8 000	109
b = 178	2 000	700	647	574	1 910	755	1 780	432	L_u	2 370	4 000	288	10 000	81.3
t = 12.8	1 500	643	602	535	1 430	705	1 740	390	I_x	216	5 000	218	12 000	64.9
d = 407	1 000	584	548	488	956	629	1 680	342	S_x	1 060	6 000	165	14 000	54.1
W410x54	4 000	708	633	545	2 140	766	1 650	503	M_r	331	3 000	298	7 000	104
W16x36	3 000	693	625	543	2 140	733	1 620	455	V_r	628	3 500	272	8 000	86.0
b = 177	2 000	647	594	524	1 910	679	1 580	393	L_u	2 290	4 000	243	10 000	63.5
t = 10.9	1 500	591	550	485	1 430	636	1 550	354	I_x	186	5 000	176	12 000	50.4
d = 403	1 000	533	498	439	956	570	1 500	308	S_x	923	6 000	132	14 000	41.8
W410x46	4 000	618	550	471	1 860	671	1 420	452	M_r	278	3 000	212	7 000	61.7
W16x31	3 000	607	545	469	1 860	644	1 400	410	V_r	585	3 500	177	8 000	51.8
b = 140	2 000	585	534	466	1 860	601	1 370	353	L_u	1 780	4 000	142	10 000	39.2
t = 11.2	1 500	534	494	433	1 430	565	1 350	318	I_x	156	5 000	99.9	12 000	31.6
d = 403	1 000	477	444	388	956	510	1 310	275	S_x	772	6 000	76.4	14 000	26.5

Note: Resistances are based on a concrete density of 2300 kg/m³.

Units: M_r - kN·m, V_r - kN, L_u - mm, I_x - 10^6 mm⁴, S_x - 10^3 mm³, b - mm, t - mm, d - mm

* Transformed moment of inertia for calculating shrinkage deflections, based on a modular ratio $n_t = 50$

COMPOSITE BEAMS
Trial Selection Table

G40.21 350W
f'$_c$ = 25 MPa

75 mm Deck with 75 mm Slab
φ = 0.90, φ$_c$ = 0.60

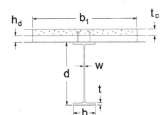

Steel section	b_1	M_{rc} (kN·m) for % shear connection			Q_r	I_t	S_t	I_t^*	Steel section data		Unbraced condition			
											L'	M_r'	L'	M_r'
	mm	100%	70%	40%	100% (kN)	10^6 mm^4	10^3 mm^3	10^6 mm^4			mm	kN·m	mm	kN·m
W410x39	4 000	525	465	396	1 570	569	1 200	394	M_r	230	3 000	167	7 000	44.0
W16x26	3 000	517	462	394	1 570	548	1 180	358	V_r	484	3 500	132	8 000	36.5
b = 140	2 000	501	454	392	1 570	514	1 160	310	L_u	1 720	4 000	105	10 000	27.3
t = 8.8	1 500	475	436	380	1 430	486	1 140	278	I_x	126	5 000	73.0	12 000	21.8
d = 399	1 000	419	390	337	956	441	1 110	239	S_x	634	6 000	55.1	14 000	18.2
W360x79	3 000	885	806	702	2 870	853	2 200	489	M_r	450	2 000	450	6 000	319
W14x53	2 000	772	716	639	1 910	774	2 140	421	V_r	692	2 500	450	7 000	267
b = 205	1 500	713	669	599	1 430	714	2 090	381	L_u	2 980	3 000	450	8 000	225
t = 16.8	1 000	651	612	553	956	627	2 000	336	I_x	226	4 000	409	10 000	171
d = 354	500	566	536	500	478	487	1 830	285	S_x	1 280	5 000	364	12 000	139
W360x72	3 000	824	747	645	2 870	774	1 990	452	M_r	403	2 000	403	6 000	273
W14x48	2 000	713	658	585	1 910	707	1 940	389	V_r	626	2 500	403	7 000	222
b = 204	1 500	655	612	548	1 430	654	1 890	351	L_u	2 910	3 000	400	8 000	186
t = 15.1	1 000	595	560	503	956	577	1 820	308	I_x	201	4 000	360	10 000	141
d = 350	500	516	487	452	478	450	1 670	259	S_x	1 150	5 000	317	12 000	114
W360x64	3 000	744	670	576	2 560	701	1 780	417	M_r	359	2 000	359	6 000	228
W14x43	2 000	658	605	535	1 910	643	1 740	359	V_r	555	2 500	359	7 000	183
b = 203	1 500	601	560	500	1 430	598	1 700	323	L_u	2 850	3 000	354	8 000	153
t = 13.5	1 000	542	511	457	956	530	1 640	282	I_x	178	4 000	316	10 000	115
d = 347	500	469	442	407	478	415	1 510	235	S_x	1 030	5 000	275	12 000	92.2
W360x57	3 000	680	610	523	2 270	655	1 600	398	M_r	318	2 000	318	6 000	147
W14x38	2 000	621	568	499	1 910	604	1 560	341	V_r	588	2 500	312	7 000	119
b = 172	1 500	565	523	463	1 430	564	1 530	307	L_u	2 340	3 000	292	8 000	99.7
t = 13.1	1 000	506	474	421	956	503	1 480	266	I_x	160	4 000	246	10 000	75.6
d = 358	500	432	405	371	478	396	1 360	218	S_x	896	5 000	192	12 000	61.0
W360x51	3 000	611	545	466	2 030	590	1 430	366	M_r	281	2 000	281	6 000	121
W14x34	2 000	575	524	456	1 910	547	1 400	315	V_r	531	2 500	274	7 000	96.9
b = 171	1 500	520	479	423	1 430	513	1 370	282	L_u	2 300	3 000	255	8 000	80.9
t = 11.6	1 000	462	433	383	956	459	1 330	244	I_x	141	4 000	211	10 000	60.8
d = 355	500	393	367	334	478	364	1 230	198	S_x	796	5 000	159	12 000	48.8
W360x45	3 000	546	485	414	1 800	527	1 270	334	M_r	245	2 000	245	6 000	96.0
W14x30	2 000	525	475	410	1 800	491	1 250	287	V_r	505	2 500	237	7 000	76.4
b = 171	1 500	478	439	385	1 430	462	1 220	258	L_u	2 240	3 000	219	8 000	63.3
t = 9.8	1 000	422	394	345	956	416	1 180	222	I_x	122	4 000	177	10 000	47.1
d = 352	500	355	330	297	478	332	1 100	178	S_x	691	5 000	128	12 000	37.5
W360x39	3 000	480	424	360	1 570	465	1 110	302	M_r	209	2 000	195	6 000	54.1
W14x26	2 000	464	417	357	1 570	435	1 080	260	V_r	477	2 500	173	7 000	44.2
b = 128	1 500	438	399	346	1 430	411	1 060	232	L_u	1 650	3 000	149	8 000	37.4
t = 10.7	1 000	382	355	308	956	373	1 030	199	I_x	102	4 000	97.0	10 000	28.7
d = 353	500	317	293	260	478	300	960	157	S_x	580	5 000	69.7	12 000	23.3

Note: Resistances are based on a concrete density of 2300 kg/m^3.
Units: M_r - kN·m, V_r - kN, L_u - mm, I_x - 10^6 mm^4, S_x - 10^3 mm^3, b - mm, t - mm, d - mm
* Transformed moment of inertia for calculating shrinkage deflections, based on a modular ratio n_t = 50

COMPOSITE BEAMS
Trial Selection Table

75 mm Deck with 75 mm Slab
$\phi = 0.90$, $\phi_c = 0.60$

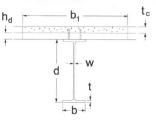

G40.21 350W
$f'_c = 25$ MPa

Steel section	b_1	Composite							Non-composite				
		M_{rc} (kN·m) for % shear connection			Q_r (kN)	I_t 10^6	S_t 10^3	I_t^* 10^6	Steel section data	Unbraced condition			
										L'	M_r'	L'	M_r'
	mm	100%	70%	40%	100%	mm⁴	mm³	mm⁴		mm	kN·m	mm	kN·m
W360x33	3 000	404	355	299	1 310	392	928	263	M_r 170	2 000	156	6 000	38.0
W14x22	2 000	392	350	297	1 310	369	907	228	V_r 399	2 500	136	7 000	30.8
b = 127	1 500	381	344	295	1 310	351	892	204	L_u 1 590	3 000	114	8 000	25.8
t = 8.5	1 000	336	309	267	956	321	867	174	I_x 82.6	4 000	70.2	10 000	19.6
d = 349	500	275	253	222	478	262	812	135	S_x 473	5 000	49.6	12 000	15.8
W310x74	3 000	786	708	605	2 870	678	1 920	385	M_r 375	2 000	375	6 000	279
W12x50	2 000	674	618	546	1 910	616	1 870	329	V_r 606	2 500	375	7 000	243
b = 205	1 500	615	572	510	1 430	568	1 820	296	L_u 3 070	3 000	375	8 000	206
t = 16.3	1 000	554	521	468	956	498	1 750	258	I_x 165	4 000	345	10 000	159
d = 310	500	479	453	420	478	384	1 600	215	S_x 1 060	5 000	312	12 000	129
W310x67	3 000	718	642	545	2 680	612	1 730	355	M_r 334	2 000	334	6 000	238
W12x45	2 000	623	568	499	1 910	559	1 680	303	V_r 541	2 500	334	7 000	200
b = 204	1 500	565	523	465	1 430	517	1 640	271	L_u 2 990	3 000	334	8 000	169
t = 14.6	1 000	506	475	425	956	456	1 580	236	I_x 145	4 000	303	10 000	129
d = 306	500	435	410	378	478	353	1 450	194	S_x 948	5 000	271	12 000	105
W310x60	3 000	646	575	486	2 390	553	1 550	328	M_r 296	2 000	296	6 000	203
W12x40	2 000	576	524	455	1 910	508	1 510	280	V_r 472	2 500	296	7 000	166
b = 203	1 500	520	479	425	1 430	472	1 480	251	L_u 2 930	3 000	295	8 000	139
t = 13.1	1 000	462	433	387	956	419	1 420	217	I_x 129	4 000	266	10 000	106
d = 303	500	396	372	341	478	327	1 310	177	S_x 849	5 000	235	12 000	85.2
W310x52	3 000	592	525	443	2 100	525	1 400	319	M_r 265	2 000	265	6 000	130
W12x35	2 000	549	497	429	1 910	485	1 360	273	V_r 502	2 500	260	7 000	106
b = 167	1 500	493	453	399	1 430	453	1 330	244	L_u 2 350	3 000	244	8 000	89.4
t = 13.2	1 000	435	407	361	956	404	1 290	210	I_x 119	4 000	208	10 000	68.4
d = 318	500	370	346	315	478	318	1 190	169	S_x 750	5 000	167	12 000	55.5
W310x45	3 000	507	447	376	1 790	451	1 190	283	M_r 223	2 000	223	6 000	98.2
W12x30	2 000	486	436	372	1 790	419	1 160	243	V_r 429	2 500	217	7 000	79.3
b = 166	1 500	441	402	351	1 430	394	1 140	217	L_u 2 290	3 000	202	8 000	66.5
t = 11.2	1 000	385	357	316	956	354	1 110	186	I_x 99.2	4 000	168	10 000	50.4
d = 313	500	324	302	273	478	282	1 030	148	S_x 634	5 000	128	12 000	40.6
W310x39	3 000	443	388	325	1 560	396	1 040	256	M_r 192	2 000	192	6 000	77.7
W12x26	2 000	427	380	323	1 560	370	1 020	220	V_r 374	2 500	186	7 000	62.2
b = 165	1 500	402	364	314	1 430	349	997	197	L_u 2 250	3 000	172	8 000	51.8
t = 9.7	1 000	347	320	283	956	317	968	168	I_x 85.1	4 000	140	10 000	38.8
d = 310	500	289	269	241	478	255	904	132	S_x 549	5 000	103	12 000	31.1
W250x67	3 000	655	580	482	2 690	493	1 580	277	M_r 284	2 000	284	5 000	246
W10x45	2 000	559	505	435	1 910	448	1 540	234	V_r 476	2 500	284	6 000	225
b = 204	1 500	501	459	405	1 430	413	1 500	209	L_u 3 230	3 000	284	7 000	203
t = 15.7	1 000	442	413	370	956	361	1 440	179	I_x 104	3 500	278	8 000	180
d = 257	500	377	356	328	478	276	1 310	144	S_x 806	4 000	268	10 000	139

Note: Resistances are based on a concrete density of 2300 kg/m³.

Units: M_r - kN·m, V_r - kN, L_u - mm, I_x - 10^6 mm⁴, S_x - 10^3 mm³, b - mm, t - mm, d - mm

* Transformed moment of inertia for calculating shrinkage deflections, based on a modular ratio $n_t = 50$

COMPOSITE BEAMS
Trial Selection Table

G40.21 350W
$f'_c = 25$ MPa

75 mm Deck with 75 mm Slab
$\phi = 0.90$, $\phi_c = 0.60$

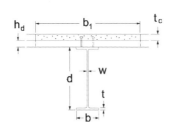

Steel section	b_1	Composite M_{rc} (kN·m) for % shear connection			Q_r	I_t	S_t	I_t^*	Non-composite Steel section data		Unbraced condition			
		100%	70%	40%	100%	10^6	10^3	10^6			L'	M_r'	L'	M_r'
	mm				(kN)	mm^4	mm^3	mm^4			mm	kN·m	mm	kN·m
W250x58	3 000	574	503	416	2 340	430	1 380	249	M_r	243	2 000	243	5 000	204
W10x39	2 000	509	456	388	1 910	393	1 340	210	V_r	419	2 500	243	6 000	182
b = 203	1 500	453	412	360	1 430	365	1 310	187	L_u	3 100	3 000	243	7 000	161
t = 13.5	1 000	395	366	326	956	322	1 260	159	I_x	87.3	3 500	235	8 000	137
d = 252	500	333	313	286	478	248	1 150	126	S_x	693	4 000	225	10 000	105
W250x45	3 000	467	406	335	1 800	367	1 100	225	M_r	190	2 000	190	5 000	114
W10x30	2 000	446	396	331	1 800	340	1 070	191	V_r	420	2 500	182	6 000	90.6
b = 148	1 500	400	360	309	1 430	319	1 050	169	L_u	2 150	3 000	169	7 000	75.2
t = 13	1 000	343	316	276	956	285	1 010	143	I_x	71.1	3 500	156	8 000	64.4
d = 266	500	283	263	236	478	224	932	111	S_x	534	4 000	143	10 000	50.1
W250x39	3 000	404	349	286	1 550	319	947	202	M_r	162	2 000	162	5 000	88.0
W10x26	2 000	388	342	284	1 550	297	923	172	V_r	360	2 500	153	6 000	69.2
b = 147	1 500	364	326	275	1 430	279	905	153	L_u	2 090	3 000	141	7 000	57.0
t = 11.2	1 000	309	282	245	956	252	876	129	I_x	60.1	3 500	129	8 000	48.6
d = 262	500	250	233	207	478	201	814	99.0	S_x	459	4 000	116	10 000	37.6
W250x33	3 000	344	295	241	1 310	272	804	177	M_r	134	2 000	134	5 000	63.6
W10x22	2 000	333	290	239	1 310	254	783	152	V_r	327	2 500	124	6 000	49.4
b = 146	1 500	321	284	238	1 310	240	768	135	L_u	2 000	3 000	113	7 000	40.3
t = 9.1	1 000	276	250	215	956	219	745	114	I_x	48.9	3 500	101	8 000	34.1
d = 258	500	219	203	178	478	176	696	86.3	S_x	379	4 000	88.6	10 000	26.1
W200x42	3 000	384	326	260	1 660	257	911	155	M_r	139	2 000	139	5 000	106
W8x28	2 000	366	317	257	1 660	237	884	130	V_r	307	2 500	139	6 000	92.0
b = 166	1 500	332	293	242	1 430	222	864	115	L_u	2 580	3 000	134	7 000	76.5
t = 11.8	1 000	276	249	214	956	198	834	95.3	I_x	40.6	3 500	127	8 000	65.6
d = 205	500	218	203	180	478	154	768	71.3	S_x	396	4 000	120	10 000	51.2
W200x36	3 000	332	280	222	1 430	223	787	139	M_r	118	2 000	118	5 000	85.4
W8x24	2 000	318	273	220	1 430	207	764	117	V_r	259	2 500	118	6 000	70.2
b = 165	1 500	305	267	217	1 430	194	748	103	L_u	2 490	3 000	112	7 000	58.0
t = 10.2	1 000	250	224	190	956	175	723	86.0	I_x	34.1	3 500	106	8 000	49.5
d = 201	500	193	179	158	478	138	671	63.7	S_x	340	4 000	99.0	10 000	38.4
W200x31	3 000	301	254	201	1 260	208	707	134	M_r	106	2 000	105	5 000	57.0
W8x21	2 000	290	249	200	1 260	194	687	114	V_r	279	2 500	97.8	6 000	45.6
b = 134	1 500	280	243	198	1 260	183	673	101	L_u	1 960	3 000	90.2	7 000	38.0
t = 10.2	1 000	239	213	179	956	166	651	83.7	I_x	31.4	3 500	82.3	8 000	32.7
d = 210	500	182	168	146	478	133	606	61.6	S_x	299	4 000	74.4	10 000	25.6
W200x27	3 000	256	215	170	1 070	178	602	118	M_r	87.9	2 000	86.3	5 000	41.2
W8x18	2 000	248	211	169	1 070	167	586	101	V_r	250	2 500	79.5	6 000	32.6
b = 133	1 500	241	207	167	1 070	158	574	90.0	L_u	1 880	3 000	72.2	7 000	27.1
t = 8.4	1 000	218	192	159	956	144	556	75.0	I_x	25.8	3 500	64.5	8 000	23.1
d = 207	500	162	149	128	478	118	520	54.8	S_x	249	4 000	56.0	10 000	18.0

Note: Resistances are based on a concrete density of 2300 kg/m^3.

Units: M_r - kN·m, V_r - kN, L_u - mm, I_x - 10^6 mm^4, S_x - 10^3 mm^3, b - mm, t - mm, d - mm

* Transformed moment of inertia for calculating shrinkage deflections, based on a modular ratio $n_t = 50$

COMPOSITE BEAMS
Trial Selection Table

75 mm Deck with 85 mm Slab
$\phi = 0.90$, $\phi_c = 0.60$

G40.21 350W
$f'_c = 25$ MPa

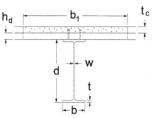

Steel section	b_1	Composite							Non-composite				
		M_{rc} (kN·m) for % shear connection			Q_r	I_t	S_t	I_t^*	Steel section data	Unbraced condition			
		100%	70%	40%	100%	10^6	10^3	10^6		L'	M_r'	L'	M_r'
	mm				(kN)	mm⁴	mm³	mm⁴		mm	kN·m	mm	kN·m
WWF1000x223	5 000	5 080	4 890	4 430	5 420	10 700	12 600	7 090	M_r 3 310	4 000	3 250	12 000	920
WWF39x150	4 000	4 930	4 700	4 250	4 340	10 100	12 400	6 690	V_r 2 210	5 000	2 960	14 000	729
b = 300	3 000	4 720	4 450	4 040	3 250	9 310	12 100	6 240	L_u 3 790	6 000	2 620	16 000	602
t = 25	2 000	4 380	4 130	3 820	2 170	8 270	11 700	5 750	I_x 4 590	8 000	1 790	18 000	512
d = 1000	1 000	3 900	3 750	3 570	1 080	6 800	10 900	5 200	S_x 9 190	10 000	1 230	20 000	446
WWF1000x200	5 000	4 650	4 460	4 010	5 420	9 690	11 200	6 380	M_r 2 890	4 000	2 790	12 000	699
WWF39x134	4 000	4 500	4 280	3 820	4 340	9 150	11 000	5 990	V_r 2 210	5 000	2 510	14 000	549
b = 300	3 000	4 300	4 020	3 620	3 250	8 440	10 800	5 560	L_u 3 610	6 000	2 180	16 000	449
t = 20	2 000	3 950	3 700	3 390	2 170	7 480	10 300	5 090	I_x 3 940	8 000	1 390	18 000	379
d = 1000	1 000	3 480	3 320	3 140	1 080	6 100	9 590	4 550	S_x 7 880	10 000	945	20 000	328
WWF900x231	5 000	4 790	4 620	4 340	5 420	9 680	12 700	6 540	M_r 3 400	4 000	3 400	12 000	1 760
WWF35x156	4 000	4 640	4 500	4 210	4 340	9 140	12 500	6 190	V_r 1 350	5 000	3 400	14 000	1 370
b = 400	3 000	4 490	4 340	4 040	3 250	8 460	12 200	5 810	L_u 5 460	6 000	3 300	16 000	1 110
t = 25	2 000	4 290	4 110	3 850	2 170	7 560	11 900	5 390	I_x 4 410	8 000	2 870	18 000	933
d = 900	1 000	3 920	3 790	3 630	1 080	6 290	11 200	4 930	S_x 9 810	10 000	2 360	20 000	801
WWF900x192	5 000	4 100	3 930	3 650	5 420	8 200	10 400	5 500	M_r 2 710	4 000	2 700	12 000	842
WWF35x128	4 000	3 950	3 810	3 520	4 340	7 760	10 200	5 180	V_r 1 350	5 000	2 490	14 000	670
b = 300	3 000	3 800	3 660	3 350	3 250	7 190	10 000	4 820	L_u 3 980	6 000	2 240	16 000	555
t = 25	2 000	3 600	3 420	3 160	2 170	6 410	9 690	4 420	I_x 3 460	8 000	1 630	18 000	473
d = 900	1 000	3 230	3 100	2 940	1 080	5 270	9 090	3 970	S_x 7 680	10 000	1 120	20 000	412
WWF900x169	5 000	3 690	3 530	3 260	5 420	7 320	9 080	4 900	M_r 2 320	4 000	2 290	12 000	625
WWF35x113	4 000	3 560	3 420	3 130	4 340	6 940	8 950	4 600	V_r 1 340	5 000	2 080	14 000	491
b = 300	3 000	3 410	3 270	2 960	3 250	6 440	8 760	4 260	L_u 3 820	6 000	1 850	16 000	402
t = 20	2 000	3 210	3 030	2 770	2 170	5 740	8 460	3 880	I_x 2 930	8 000	1 250	18 000	339
d = 900	1 000	2 840	2 710	2 550	1 080	4 690	7 900	3 440	S_x 6 510	10 000	845	20 000	293
WWF800x223	5 000	4 190	4 020	3 780	5 420	7 700	11 200	5 170	M_r 2 940	4 000	2 940	10 000	2 120
WWF31x150	4 000	4 050	3 900	3 660	4 340	7 280	11 000	4 880	V_r 1 370	5 000	2 940	12 000	1 630
b = 400	3 000	3 900	3 780	3 520	3 250	6 730	10 800	4 570	L_u 5 570	6 000	2 880	14 000	1 280
t = 25	2 000	3 730	3 580	3 350	2 170	5 990	10 400	4 230	I_x 3 410	7 000	2 710	16 000	1 050
d = 800	1 000	3 410	3 290	3 150	1 080	4 960	9 820	3 840	S_x 8 520	8 000	2 530	18 000	882
WWF800x184	5 000	3 580	3 410	3 170	5 420	6 490	9 090	4 330	M_r 2 330	4 000	2 330	10 000	1 040
WWF31x123	4 000	3 430	3 290	3 050	4 340	6 150	8 960	4 080	V_r 1 370	5 000	2 170	12 000	791
b = 300	3 000	3 290	3 170	2 910	3 250	5 700	8 770	3 780	L_u 4 060	6 000	1 970	14 000	634
t = 25	2 000	3 120	2 970	2 740	2 170	5 070	8 470	3 460	I_x 2 660	7 000	1 750	16 000	529
d = 800	1 000	2 800	2 680	2 540	1 080	4 150	7 940	3 090	S_x 6 640	8 000	1 490	18 000	454
WWF800x161	5 000	3 220	3 060	2 830	5 420	5 780	7 910	3 860	M_r 1 990	4 000	1 980	10 000	778
WWF31x108	4 000	3 080	2 950	2 710	4 340	5 490	7 800	3 620	V_r 1 370	5 000	1 810	12 000	581
b = 300	3 000	2 940	2 820	2 570	3 250	5 090	7 640	3 350	L_u 3 900	6 000	1 620	14 000	460
t = 20	2 000	2 770	2 620	2 400	2 170	4 540	7 380	3 030	I_x 2 250	7 000	1 400	16 000	379
d = 800	1 000	2 460	2 340	2 200	1 080	3 690	6 890	2 670	S_x 5 610	8 000	1 130	18 000	322

Note: Resistances are based on a concrete density of 1850 kg/m³.
Units: M_r - kN·m, V_r - kN, L_u - mm, I_x - 10^6 mm⁴, S_x - 10^3 mm³, b - mm, t - mm, d - mm
* Transformed moment of inertia for calculating shrinkage deflections, based on a modular ratio $n_t = 50$

COMPOSITE BEAMS
Trial Selection Table

G40.21 350W
$f'_c = 25$ MPa

75 mm Deck with 85 mm Slab
$\phi = 0.90$, $\phi_c = 0.60$

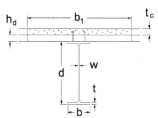

Steel section	b_1	Composite							Non-composite				
		M_{rc} (kN·m) for % shear connection			Q_r (kN)	I_t 10^6	S_t 10^3	I_t^* 10^6	Steel section data	Unbraced condition			
		100%	70%	40%	100%					L'	M_r'	L'	M_r'
	mm					mm⁴	mm³	mm⁴		mm	kN·m	mm	kN·m
WWF700x214	5 000	3 630	3 460	3 240	5 420	5 970	9 680	3 960	M_r 2 500	4 000	2 500	10 000	1 870
WWF28x144	4 000	3 480	3 340	3 140	4 340	5 640	9 530	3 730	V_r 1 370	5 000	2 500	12 000	1 500
b = 400	3 000	3 340	3 230	3 020	3 250	5 200	9 330	3 480	L_u 5 690	6 000	2 470	14 000	1 190
t = 25	2 000	3 180	3 060	2 870	2 170	4 620	9 020	3 200	I_x 2 540	7 000	2 330	16 000	982
d = 700	1 000	2 920	2 820	2 700	1 080	3 800	8 470	2 890	S_x 7 270	8 000	2 190	18 000	835
WWF700x175	5 000	3 090	2 930	2 700	5 420	5 010	7 840	3 320	M_r 1 970	4 000	1 970	10 000	965
WWF28x117	4 000	2 950	2 810	2 610	4 340	4 750	7 720	3 110	V_r 1 370	5 000	1 850	12 000	741
b = 300	3 000	2 800	2 700	2 490	3 250	4 390	7 560	2 880	L_u 4 160	6 000	1 700	14 000	601
t = 25	2 000	2 650	2 530	2 340	2 170	3 900	7 300	2 620	I_x 1 970	7 000	1 530	16 000	505
d = 700	1 000	2 390	2 290	2 160	1 080	3 170	6 820	2 320	S_x 5 640	8 000	1 350	18 000	436
WWF700x152	5 000	2 770	2 620	2 410	5 420	4 450	6 790	2 960	M_r 1 680	4 000	1 680	10 000	711
WWF28x102	4 000	2 640	2 510	2 310	4 340	4 220	6 700	2 760	V_r 1 370	5 000	1 540	12 000	537
b = 300	3 000	2 500	2 400	2 190	3 250	3 920	6 560	2 550	L_u 3 990	6 000	1 390	14 000	429
t = 20	2 000	2 350	2 230	2 040	2 170	3 490	6 330	2 290	I_x 1 660	7 000	1 230	16 000	356
d = 700	1 000	2 090	1 990	1 870	1 080	2 820	5 910	2 000	S_x 4 760	8 000	1 020	18 000	305
W610x241	4 000	3 520	3 370	3 080	4 340	5 000	9 450	3 210	M_r 2 420	3 000	2 420	8 000	1 950
W24x162	3 000	3 360	3 200	2 930	3 250	4 580	9 200	2 980	V_r 2 360	4 000	2 420	10 000	1 640
b = 329	2 000	3 150	2 990	2 770	2 170	4 030	8 810	2 730	L_u 4 750	5 000	2 380	12 000	1 300
t = 31	1 500	3 000	2 860	2 690	1 630	3 680	8 530	2 600	I_x 2 150	6 000	2 240	14 000	1 070
d = 635	1 000	2 830	2 720	2 600	1 080	3 270	8 160	2 460	S_x 6 780	7 000	2 100	16 000	911
W610x217	4 000	3 210	3 070	2 800	4 340	4 570	8 560	2 930	M_r 2 160	3 000	2 160	8 000	1 690
W24x146	3 000	3 060	2 920	2 660	3 250	4 200	8 340	2 720	V_r 2 150	4 000	2 160	10 000	1 370
b = 328	2 000	2 870	2 720	2 510	2 170	3 700	8 000	2 480	L_u 4 640	5 000	2 110	12 000	1 070
t = 27.7	1 500	2 730	2 590	2 420	1 630	3 380	7 750	2 350	I_x 1 910	6 000	1 980	14 000	878
d = 628	1 000	2 560	2 460	2 340	1 080	2 990	7 400	2 210	S_x 6 070	7 000	1 840	16 000	745
W610x195	4 000	2 920	2 780	2 540	4 340	4 160	7 680	2 660	M_r 1 910	3 000	1 910	8 000	1 450
W24x131	3 000	2 770	2 650	2 410	3 250	3 830	7 490	2 460	V_r 1 990	4 000	1 910	10 000	1 120
b = 327	2 000	2 600	2 460	2 260	2 170	3 370	7 200	2 230	L_u 4 530	5 000	1 860	12 000	870
t = 24.4	1 500	2 470	2 340	2 180	1 630	3 080	6 970	2 110	I_x 1 680	6 000	1 730	14 000	709
d = 622	1 000	2 310	2 210	2 090	1 080	2 720	6 660	1 970	S_x 5 400	7 000	1 590	16 000	598
W610x174	4 000	2 650	2 510	2 300	4 340	3 770	6 860	2 420	M_r 1 690	3 000	1 690	8 000	1 240
W24x117	3 000	2 500	2 390	2 170	3 250	3 480	6 700	2 220	V_r 1 790	4 000	1 690	10 000	924
b = 325	2 000	2 340	2 220	2 030	2 170	3 070	6 440	2 010	L_u 4 440	5 000	1 630	12 000	709
t = 21.6	1 500	2 230	2 110	1 950	1 630	2 800	6 250	1 890	I_x 1 470	6 000	1 510	14 000	574
d = 616	1 000	2 080	1 980	1 860	1 080	2 470	5 970	1 760	S_x 4 780	7 000	1 380	16 000	482
W610x155	4 000	2 400	2 270	2 070	4 340	3 410	6 110	2 200	M_r 1 490	3 000	1 490	8 000	1 060
W24x104	3 000	2 260	2 150	1 960	3 250	3 160	5 980	2 020	V_r 1 610	4 000	1 490	10 000	762
b = 324	2 000	2 110	2 000	1 820	2 170	2 800	5 760	1 810	L_u 4 370	5 000	1 430	12 000	579
t = 19	1 500	2 010	1 900	1 740	1 630	2 550	5 600	1 700	I_x 1 290	6 000	1 310	14 000	465
d = 611	1 000	1 870	1 770	1 660	1 080	2 240	5 350	1 570	S_x 4 220	7 000	1 190	16 000	388

Note: Resistances are based on a concrete density of 1850 kg/m³.

Units: M_r - kN·m, V_r - kN, L_u - mm, I_x - 10^6 mm⁴, S_x - 10^3 mm³, b - mm, t - mm, d - mm

* Transformed moment of inertia for calculating shrinkage deflections, based on a modular ratio $n_t = 50$

COMPOSITE BEAMS
Trial Selection Table

75 mm Deck with 85 mm Slab
$\phi = 0.90$, $\phi_c = 0.60$

G40.21 350W
$f'_c = 25$ MPa

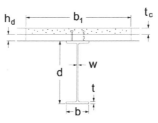

Steel section	b_1	Composite							Non-composite				
		M_{rc} (kN·m) for % shear connection			Q_r	I_t	S_t	I_t^*	Steel section data	Unbraced condition			
		100%	70%	40%	100%	10^6	10^3	10^6		L'	M_r'	L'	M_r'
	mm				(kN)	mm⁴	mm³	mm⁴		mm	kN·m	mm	kN·m
W610x140 W24x94 b = 230 t = 22.2 d = 617	4 000 3 000 2 000 1 500 1 000	2 240 2 100 1 950 1 840 1 690	2 110 2 000 1 830 1 720 1 600	1 910 1 790 1 650 1 570 1 480	4 340 3 250 2 170 1 630 1 080	3 150 2 920 2 580 2 360 2 060	5 500 5 380 5 180 5 020 4 780	2 020 1 840 1 640 1 530 1 400	M_r 1 310 V_r 1 680 L_u 3 050 I_x 1 120 S_x 3 630	3 000 4 000 5 000 6 000 7 000	1 310 1 180 1 040 878 695	8 000 10 000 12 000 14 000 16 000	573 422 334 277 237
W610x125 W24x84 b = 229 t = 19.6 d = 612	4 000 3 000 2 000 1 500 1 000	2 040 1 900 1 760 1 660 1 530	1 910 1 800 1 660 1 560 1 440	1 730 1 620 1 490 1 410 1 330	4 340 3 250 2 170 1 630 1 080	2 850 2 650 2 360 2 150 1 880	4 910 4 810 4 640 4 500 4 290	1 840 1 680 1 490 1 380 1 260	M_r 1 160 V_r 1 510 L_u 2 990 I_x 985 S_x 3 220	3 000 4 000 5 000 6 000 7 000	1 160 1 040 895 733 575	8 000 10 000 12 000 14 000 16 000	470 342 269 222 189
W610x113 W24x76 b = 228 t = 17.3 d = 608	4 000 3 000 2 000 1 500 1 000	1 890 1 760 1 610 1 530 1 400	1 770 1 660 1 530 1 430 1 320	1 590 1 490 1 360 1 290 1 210	4 340 3 250 2 170 1 630 1 080	2 600 2 430 2 170 1 980 1 740	4 440 4 350 4 210 4 090 3 900	1 700 1 550 1 370 1 260 1 150	M_r 1 040 V_r 1 420 L_u 2 930 I_x 875 S_x 2 880	3 000 4 000 5 000 6 000 7 000	1 030 915 781 617 481	8 000 10 000 12 000 14 000 16 000	391 282 220 180 153
W610x101 W24x68 b = 228 t = 14.9 d = 603	4 000 3 000 2 000 1 500 1 000	1 730 1 610 1 480 1 400 1 280	1 610 1 520 1 400 1 300 1 190	1 440 1 360 1 240 1 160 1 080	4 100 3 250 2 170 1 630 1 080	2 360 2 210 1 980 1 810 1 590	3 990 3 910 3 780 3 680 3 510	1 550 1 410 1 240 1 140 1 030	M_r 914 V_r 1 310 L_u 2 870 I_x 764 S_x 2 530	3 000 4 000 5 000 6 000 7 000	901 795 668 512 396	8 000 10 000 12 000 14 000 16 000	320 228 176 144 121
W610x91 W24x61 b = 227 t = 12.7 d = 598	4 000 3 000 2 000 1 500 1 000	1 550 1 470 1 340 1 270 1 150	1 430 1 380 1 270 1 180 1 070	1 280 1 240 1 120 1 040 964	3 650 3 250 2 170 1 630 1 080	2 120 1 990 1 800 1 650 1 450	3 560 3 490 3 380 3 290 3 150	1 410 1 280 1 130 1 030 925	M_r 806 V_r 1 110 L_u 2 790 I_x 667 S_x 2 230	3 000 4 000 5 000 6 000 7 000	788 687 567 421 325	8 000 10 000 12 000 14 000 16 000	261 185 142 115 97.1
W530x138 W21x93 b = 214 t = 23.6 d = 549	4 000 3 000 2 000 1 500 1 000	2 030 1 880 1 730 1 630 1 490	1 890 1 780 1 620 1 520 1 410	1 700 1 580 1 450 1 380 1 300	4 340 3 250 2 170 1 630 1 080	2 560 2 370 2 090 1 900 1 660	4 940 4 820 4 630 4 480 4 260	1 620 1 470 1 300 1 200 1 100	M_r 1 140 V_r 1 680 L_u 2 910 I_x 861 S_x 3 140	3 000 4 000 5 000 6 000 7 000	1 130 1 010 891 762 616	8 000 10 000 12 000 14 000 16 000	515 390 314 263 227
W530x123 W21x83 b = 212 t = 21.2 d = 544	4 000 3 000 2 000 1 500 1 000	1 850 1 720 1 570 1 480 1 360	1 730 1 610 1 480 1 380 1 270	1 550 1 440 1 320 1 250 1 170	4 340 3 250 2 170 1 630 1 080	2 330 2 160 1 920 1 740 1 520	4 430 4 330 4 170 4 040 3 840	1 480 1 350 1 190 1 100 995	M_r 1 010 V_r 1 480 L_u 2 840 I_x 761 S_x 2 800	3 000 4 000 5 000 6 000 7 000	996 888 768 631 505	8 000 10 000 12 000 14 000 16 000	421 316 253 211 182
W530x109 W21x73 b = 211 t = 18.8 d = 539	4 000 3 000 2 000 1 500 1 000	1 690 1 560 1 420 1 340 1 230	1 570 1 460 1 340 1 250 1 150	1 400 1 310 1 190 1 120 1 050	4 340 3 250 2 170 1 630 1 080	2 090 1 950 1 740 1 590 1 390	3 940 3 860 3 720 3 610 3 450	1 350 1 230 1 080 991 895	M_r 891 V_r 1 300 L_u 2 790 I_x 667 S_x 2 480	3 000 4 000 5 000 6 000 7 000	872 771 656 520 413	8 000 10 000 12 000 14 000 16 000	342 254 202 168 144

Note: Resistances are based on a concrete density of 1850 kg/m³.
Units: M_r - kN·m, V_r - kN, L_u - mm, I_x - 10^6 mm⁴, S_x - 10^3 mm³, b - mm, t - mm, d - mm
* Transformed moment of inertia for calculating shrinkage deflections, based on a modular ratio $n_t = 50$

COMPOSITE BEAMS
Trial Selection Table

G40.21 350W
f′c = 25 MPa

75 mm Deck with 85 mm Slab
$\phi = 0.90$, $\phi_c = 0.60$

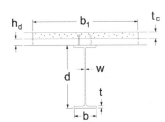

Steel section	b_1	Composite M_{rc} (kN·m) for % shear connection			Q_r	I_t	S_t	I_t^*	Non-composite Steel section data	Unbraced condition			
										L'	M_r'	L'	M_r'
	mm	100%	70%	40%	(kN)	10^6 mm⁴	10^3 mm³	10^6 mm⁴		mm	kN·m	mm	kN·m
W530x101 W21x68 b = 210 t = 17.4 d = 537	4 000 3 000 2 000 1 500 1 000	1 580 1 470 1 330 1 260 1 150	1 460 1 380 1 260 1 180 1 080	1 300 1 230 1 120 1 050 981	4 060 3 250 2 170 1 630 1 080	1 970 1 840 1 640 1 500 1 310	3 670 3 600 3 480 3 380 3 220	1 280 1 160 1 020 936 842	M_r 825 V_r 1 220 L_u 2 750 I_x 617 S_x 2 300	3 000 4 000 5 000 6 000 7 000	804 706 594 462 365	8 000 10 000 12 000 14 000 16 000	301 222 176 146 125
W530x92 W21x62 b = 209 t = 15.6 d = 533	4 000 3 000 2 000 1 500 1 000	1 450 1 370 1 240 1 160 1 070	1 340 1 280 1 170 1 090 996	1 190 1 140 1 040 972 900	3 720 3 250 2 170 1 630 1 080	1 810 1 690 1 520 1 390 1 220	3 350 3 290 3 180 3 090 2 950	1 190 1 080 943 863 772	M_r 743 V_r 1 130 L_u 2 700 I_x 552 S_x 2 070	3 000 4 000 5 000 6 000 7 000	719 626 519 393 309	8 000 10 000 12 000 14 000 16 000	253 185 146 120 103
W530x82 W21x55 b = 209 t = 13.3 d = 528	4 000 3 000 2 000 1 500 1 000	1 300 1 260 1 120 1 050 964	1 190 1 170 1 060 990 897	1 050 1 040 937 874 803	3 310 3 250 2 170 1 630 1 080	1 620 1 520 1 370 1 260 1 110	2 970 2 920 2 830 2 760 2 640	1 080 981 854 778 692	M_r 652 V_r 1 040 L_u 2 640 I_x 478 S_x 1 810	3 000 4 000 5 000 6 000 7 000	625 538 436 321 250	8 000 10 000 12 000 14 000 16 000	204 148 115 94.8 80.5
W460x106 W18x71 b = 194 t = 20.6 d = 469	4 000 3 000 2 000 1 500 1 000	1 500 1 380 1 230 1 160 1 060	1 380 1 280 1 160 1 080 990	1 210 1 140 1 030 968 900	4 250 3 250 2 170 1 630 1 080	1 650 1 530 1 360 1 240 1 080	3 470 3 400 3 270 3 170 3 010	1 050 950 827 755 676	M_r 753 V_r 1 230 L_u 2 670 I_x 488 S_x 2 080	3 000 3 500 4 000 5 000 6 000	727 686 643 553 450	7 000 8 000 10 000 12 000 14 000	366 308 235 190 160
W460x97 W18x65 b = 193 t = 19 d = 466	4 000 3 000 2 000 1 500 1 000	1 380 1 280 1 150 1 070 983	1 260 1 190 1 080 1 010 919	1 100 1 060 958 898 832	3 870 3 250 2 170 1 630 1 080	1 520 1 420 1 270 1 160 1 010	3 180 3 120 3 010 2 920 2 780	987 891 774 706 630	M_r 687 V_r 1 100 L_u 2 630 I_x 445 S_x 1 910	3 000 3 500 4 000 5 000 6 000	660 621 579 491 389	7 000 8 000 10 000 12 000 14 000	314 264 200 161 135
W460x89 W18x60 b = 192 t = 17.7 d = 463	4 000 3 000 2 000 1 500 1 000	1 280 1 210 1 080 1 010 923	1 170 1 120 1 010 948 863	1 020 994 900 843 778	3 590 3 250 2 170 1 630 1 080	1 420 1 330 1 200 1 090 954	2 950 2 900 2 800 2 720 2 600	932 841 730 664 590	M_r 633 V_r 1 010 L_u 2 600 I_x 409 S_x 1 770	3 000 3 500 4 000 5 000 6 000	605 568 527 442 343	7 000 8 000 10 000 12 000 14 000	276 231 174 140 117
W460x82 W18x55 b = 191 t = 16 d = 460	4 000 3 000 2 000 1 500 1 000	1 170 1 140 1 000 933 855	1 060 1 050 938 879 799	930 923 836 781 718	3 280 3 250 2 170 1 630 1 080	1 310 1 230 1 110 1 020 891	2 700 2 650 2 570 2 500 2 390	871 786 681 619 547	M_r 576 V_r 947 L_u 2 540 I_x 370 S_x 1 610	3 000 3 500 4 000 5 000 6 000	546 510 471 386 292	7 000 8 000 10 000 12 000 14 000	234 195 146 116 97.3
W460x74 W18x50 b = 190 t = 14.5 d = 457	4 000 3 000 2 000 1 500 1 000	1 070 1 040 932 864 790	968 953 869 813 739	843 839 774 721 660	2 980 2 980 2 170 1 630 1 080	1 200 1 130 1 030 945 828	2 460 2 410 2 340 2 280 2 190	810 732 633 574 506	M_r 520 V_r 855 L_u 2 510 I_x 332 S_x 1 460	3 000 3 500 4 000 5 000 6 000	489 455 417 332 249	7 000 8 000 10 000 12 000 14 000	198 164 122 96.8 80.6

Note: Resistances are based on a concrete density of 1850 kg/m³.

Units: M_r - kN·m, V_r - kN, L_u - mm, I_x - 10^6 mm⁴, S_x - 10^3 mm³, b - mm, t - mm, d - mm

* Transformed moment of inertia for calculating shrinkage deflections, based on a modular ratio $n_t = 50$

COMPOSITE BEAMS
Trial Selection Table

75 mm Deck with 85 mm Slab
$\phi = 0.90$, $\phi_c = 0.60$

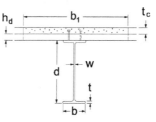

G40.21 350W
$f'_c = 25$ MPa

Steel section	b_1	Composite							Non-composite			
		M_{rc} (kN·m) for % shear connection			Q_r	I_t	S_t	I_t^*	Steel section data	Unbraced condition		
										L'	M_r'	L'	M_r'
	mm	100%	70%	40%	(kN)	10^6 mm⁴	10^3 mm³	10^6 mm⁴		mm	kN·m	mm	kN·m
W460x67	4 000	973	877	762	2 700	1 100	2 230	750	M_r 466	3 000	435	7 000	165
W18x45	3 000	949	866	758	2 700	1 040	2 190	679	V_r 802	3 500	402	8 000	136
b = 190	2 000	866	805	714	2 170	945	2 130	587	L_u 2 460	4 000	366	10 000	99.8
t = 12.7	1 500	799	751	663	1 630	872	2 080	531	I_x 296	5 000	281	12 000	79.0
d = 454	1 000	728	680	604	1 080	767	1 990	465	S_x 1 300	6 000	209	14 000	65.4
W460x61	4 000	872	783	677	2 410	984	1 980	682	M_r 406	3 000	374	7 000	132
W18x41	3 000	853	774	674	2 410	933	1 950	617	V_r 758	3 500	343	8 000	108
b = 189	2 000	797	737	651	2 170	852	1 900	533	L_u 2 390	4 000	309	10 000	78.5
t = 10.8	1 500	731	684	602	1 630	789	1 850	481	I_x 255	5 000	228	12 000	61.7
d = 450	1 000	663	617	543	1 080	697	1 780	419	S_x 1 130	6 000	168	14 000	50.8
W410x85	4 000	1 140	1 030	891	3 400	1 170	2 620	763	M_r 542	3 000	512	7 000	242
W16x57	3 000	1 090	1 000	876	3 250	1 100	2 570	686	V_r 945	3 500	480	8 000	205
b = 181	2 000	957	891	791	2 170	982	2 480	592	L_u 2 510	4 000	447	10 000	157
t = 18.2	1 500	886	833	738	1 630	898	2 410	536	I_x 315	5 000	377	12 000	127
d = 417	1 000	809	755	678	1 080	781	2 300	472	S_x 1 510	6 000	297	14 000	107
W410x74	4 000	1 010	910	784	3 010	1 050	2 320	696	M_r 476	3 000	445	7 000	194
W16x50	3 000	983	896	780	3 010	985	2 280	627	V_r 833	3 500	414	8 000	163
b = 180	2 000	872	808	716	2 170	888	2 210	540	L_u 2 450	4 000	382	10 000	123
t = 16	1 500	803	753	666	1 630	816	2 150	488	I_x 275	5 000	312	12 000	99.7
d = 413	1 000	731	683	609	1 080	713	2 060	427	S_x 1 330	6 000	239	14 000	83.8
W410x67	4 000	917	821	705	2 710	953	2 100	644	M_r 428	3 000	397	7 000	161
W16x45	3 000	893	809	702	2 710	900	2 060	581	V_r 750	3 500	367	8 000	135
b = 179	2 000	809	747	660	2 170	816	2 000	500	L_u 2 400	4 000	336	10 000	102
t = 14.4	1 500	741	693	613	1 630	752	1 950	451	I_x 245	5 000	264	12 000	81.6
d = 410	1 000	671	628	558	1 080	659	1 870	394	S_x 1 200	6 000	201	14 000	68.3
W410x60	4 000	812	724	621	2 390	853	1 860	589	M_r 375	3 000	344	7 000	131
W16x40	3 000	793	715	618	2 390	809	1 830	533	V_r 652	3 500	317	8 000	109
b = 178	2 000	740	680	599	2 170	738	1 780	460	L_u 2 370	4 000	288	10 000	81.3
t = 12.8	1 500	674	627	556	1 630	683	1 740	414	I_x 216	5 000	218	12 000	64.9
d = 407	1 000	606	569	504	1 080	603	1 670	360	S_x 1 060	6 000	165	14 000	54.1
W410x54	4 000	729	648	554	2 140	764	1 660	536	M_r 331	3 000	298	7 000	104
W16x36	3 000	714	640	551	2 140	727	1 630	486	V_r 628	3 500	272	8 000	86.0
b = 177	2 000	684	626	546	2 140	666	1 590	419	L_u 2 290	4 000	243	10 000	63.5
t = 10.9	1 500	621	575	507	1 630	619	1 550	376	I_x 186	5 000	176	12 000	50.4
d = 403	1 000	554	519	456	1 080	548	1 500	325	S_x 923	6 000	132	14 000	41.8
W410x46	4 000	637	563	478	1 860	672	1 440	481	M_r 278	3 000	212	7 000	61.7
W16x31	3 000	626	558	477	1 860	641	1 420	437	V_r 585	3 500	177	8 000	51.8
b = 140	2 000	603	547	473	1 860	591	1 380	377	L_u 1 780	4 000	142	10 000	39.3
t = 11.2	1 500	565	519	454	1 630	552	1 350	339	I_x 156	5 000	99.9	12 000	31.6
d = 403	1 000	498	465	404	1 080	492	1 300	291	S_x 772	6 000	76.4	14 000	26.5

Note: Resistances are based on a concrete density of 1850 kg/m³.
Units: M_r - kN·m, V_r - kN, L_u - mm, I_x - 10^6 mm⁴, S_x - 10^3 mm³, b - mm, t - mm, d - mm

* Transformed moment of inertia for calculating shrinkage deflections, based on a modular ratio $n_t = 50$

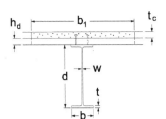

COMPOSITE BEAMS
Trial Selection Table

G40.21 350W
$f'_c = 25$ MPa

75 mm Deck with 85 mm Slab
$\phi = 0.90$, $\phi_c = 0.60$

Steel section	b_1	M_{rc} (kN·m) for % shear connection			Q_r (kN)	I_t 10^6	S_t 10^3	I_t^* 10^6	Steel section data		Unbraced condition			
											L´	$M_r´$	L´	$M_r´$
	mm	100%	70%	40%	100%	mm⁴	mm³	mm⁴			mm	kN·m	mm	kN·m
W410x39	4 000	541	477	402	1 570	572	1 220	419	M_r	230	3 000	167	7 000	44.0
W16x26	3 000	533	473	401	1 570	547	1 200	383	V_r	484	3 500	132	8 000	36.5
b = 140	2 000	517	465	398	1 570	508	1 170	331	L_u	1 720	4 000	105	10 000	27.3
t = 8.8	1 500	500	457	396	1 570	477	1 140	297	I_x	126	5 000	73.0	12 000	21.8
d = 399	1 000	440	408	353	1 080	428	1 100	254	S_x	634	6 000	55.1	14 000	18.2
W360x79	3 000	940	851	731	3 180	838	2 210	523	M_r	450	2 000	450	6 000	319
W14x53	2 000	814	750	665	2 170	752	2 140	449	V_r	692	2 500	450	7 000	267
b = 205	1 500	745	695	621	1 630	687	2 080	404	L_u	2 980	3 000	450	8 000	225
t = 16.8	1 000	673	634	569	1 080	597	1 990	353	I_x	226	4 000	409	10 000	171
d = 354	500	582	550	509	542	460	1 800	295	S_x	1 280	5 000	364	12 000	139
W360x72	3 000	853	767	656	2 870	764	2 000	484	M_r	403	2 000	403	6 000	273
W14x48	2 000	754	692	610	2 170	689	1 940	415	V_r	626	2 500	403	7 000	222
b = 204	1 500	687	638	569	1 630	632	1 890	373	L_u	2 910	3 000	400	8 000	186
t = 15.1	1 000	617	581	519	1 080	551	1 810	325	I_x	201	4 000	360	10 000	141
d = 350	500	531	501	461	542	425	1 650	268	S_x	1 150	5 000	317	12 000	114
W360x64	3 000	769	688	586	2 560	693	1 800	447	M_r	359	2 000	359	6 000	228
W14x43	2 000	699	638	558	2 170	629	1 750	383	V_r	555	2 500	359	7 000	183
b = 203	1 500	632	585	520	1 630	579	1 700	344	L_u	2 850	3 000	354	8 000	153
t = 13.5	1 000	564	530	473	1 080	508	1 630	298	I_x	178	4 000	316	10 000	115
d = 347	500	484	455	416	542	393	1 490	244	S_x	1 030	5 000	275	12 000	92.2
W360x57	3 000	702	626	533	2 270	650	1 620	426	M_r	318	2 000	318	6 000	147
W14x38	2 000	661	602	521	2 170	593	1 570	365	V_r	588	2 500	312	7 000	119
b = 172	1 500	596	549	484	1 630	548	1 530	327	L_u	2 340	3 000	292	8 000	99.7
t = 13.1	1 000	527	494	436	1 080	483	1 470	282	I_x	160	4 000	246	10 000	75.6
d = 358	500	447	418	379	542	375	1 350	227	S_x	896	5 000	192	12 000	61.0
W360x51	3 000	631	559	474	2 030	587	1 450	392	M_r	281	2 000	281	6 000	121
W14x34	2 000	604	546	470	2 030	538	1 410	337	V_r	531	2 500	274	7 000	96.9
b = 171	1 500	550	505	442	1 630	500	1 380	302	L_u	2 300	3 000	255	8 000	80.9
t = 11.6	1 000	483	451	398	1 080	443	1 330	259	I_x	141	4 000	211	10 000	60.8
d = 355	500	408	380	342	542	345	1 220	207	S_x	796	5 000	159	12 000	48.8
W360x45	3 000	564	498	421	1 800	526	1 290	358	M_r	245	2 000	245	6 000	96.0
W14x30	2 000	543	487	417	1 800	485	1 250	308	V_r	505	2 500	237	7 000	76.4
b = 171	1 500	509	464	403	1 630	452	1 230	276	L_u	2 240	3 000	219	8 000	63.3
t = 9.8	1 000	443	411	360	1 080	402	1 180	236	I_x	122	4 000	177	10 000	47.1
d = 352	500	369	343	306	542	316	1 090	186	S_x	691	5 000	128	12 000	37.5
W360x39	3 000	496	435	366	1 570	466	1 120	323	M_r	209	2 000	195	6 000	54.1
W14x26	2 000	480	428	363	1 570	431	1 090	279	V_r	477	2 500	173	7 000	44.2
b = 128	1 500	464	420	361	1 570	404	1 070	249	L_u	1 650	3 000	149	8 000	37.4
t = 10.7	1 000	403	372	322	1 080	362	1 030	213	I_x	102	4 000	97.0	10 000	28.7
d = 353	500	331	305	269	542	286	951	165	S_x	580	5 000	69.7	12 000	23.3

Note: Resistances are based on a concrete density of 1850 kg/m³.

Units: M_r - kN·m, V_r - kN, L_u - mm, I_x - 10^6 mm⁴, S_x - 10^3 mm³, b - mm, t - mm, d - mm

* Transformed moment of inertia for calculating shrinkage deflections, based on a modular ratio $n_t = 50$

COMPOSITE BEAMS
Trial Selection Table

75 mm Deck with 85 mm Slab
$\phi = 0.90$, $\phi_c = 0.60$

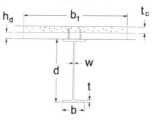

G40.21 350W
$f'_c = 25$ MPa

Steel section	b_1	Composite							Non-composite			
		M_{rc} (kN·m) for % shear connection			Q_r	I_t 10^6	S_t 10^3	I_t^* 10^6	Steel section data	Unbraced condition		
		100%	70%	40%	100%					L'	M_r'	L'	M_r'
	mm				(kN)	mm⁴	mm³	mm⁴		mm	kN·m	mm	kN·m
W360x33 W14x22 b = 127 t = 8.5 d = 349	3 000 2 000 1 500 1 000 500	417 406 394 356 288	364 359 353 326 265	304 302 301 281 230	1 310 1 310 1 310 1 080 542	394 368 346 313 250	939 916 898 869 806	281 244 219 187 143	M_r 170 V_r 399 L_u 1 590 I_x 82.6 S_x 473	2 000 2 500 3 000 4 000 5 000	156 136 114 70.2 49.6	6 000 7 000 8 000 10 000 12 000	38.0 30.8 25.8 19.6 15.8
W310x74 W12x50 b = 205 t = 16.3 d = 310	3 000 2 000 1 500 1 000 500	824 715 647 576 494	737 652 598 541 466	623 569 530 483 429	2 990 2 170 1 630 1 080 542	669 600 548 475 362	1 940 1 870 1 820 1 740 1 570	414 353 316 273 223	M_r 375 V_r 606 L_u 3 070 I_x 165 S_x 1 060	2 000 2 500 3 000 4 000 5 000	375 375 375 345 312	6 000 7 000 8 000 10 000 12 000	279 243 206 159 129
W310x67 W12x45 b = 204 t = 14.6 d = 306	3 000 2 000 1 500 1 000 500	744 663 596 527 450	661 602 549 494 423	556 521 485 440 387	2 680 2 170 1 630 1 080 542	606 546 501 436 334	1 740 1 690 1 640 1 570 1 430	381 325 290 250 202	M_r 334 V_r 541 L_u 2 990 I_x 145 S_x 948	2 000 2 500 3 000 4 000 5 000	334 334 334 303 271	6 000 7 000 8 000 10 000 12 000	238 200 169 129 105
W310x60 W12x40 b = 203 t = 13.1 d = 303	3 000 2 000 1 500 1 000 500	670 617 551 483 409	591 557 504 451 384	495 477 443 401 349	2 390 2 170 1 630 1 080 542	550 499 459 402 309	1 570 1 520 1 480 1 420 1 300	353 301 269 231 185	M_r 296 V_r 472 L_u 2 930 I_x 129 S_x 849	2 000 2 500 3 000 4 000 5 000	296 296 295 266 235	6 000 7 000 8 000 10 000 12 000	203 166 139 106 85.2
W310x52 W12x35 b = 167 t = 13.2 d = 318	3 000 2 000 1 500 1 000 500	613 584 524 456 383	539 525 478 424 358	452 447 417 375 324	2 100 2 100 1 630 1 080 542	523 477 442 389 301	1 410 1 370 1 340 1 280 1 170	343 293 262 224 177	M_r 265 V_r 502 L_u 2 350 I_x 119 S_x 750	2 000 2 500 3 000 4 000 5 000	265 260 244 208 167	6 000 7 000 8 000 10 000 12 000	130 106 89.4 68.4 55.5
W310x45 W12x30 b = 166 t = 11.2 d = 313	3 000 2 000 1 500 1 000 500	525 504 471 405 336	459 449 427 374 314	383 380 367 330 281	1 790 1 790 1 630 1 080 542	451 415 386 343 268	1 210 1 170 1 150 1 110 1 020	304 261 233 198 155	M_r 223 V_r 429 L_u 2 290 I_x 99.2 S_x 634	2 000 2 500 3 000 4 000 5 000	223 217 202 168 128	6 000 7 000 8 000 10 000 12 000	98.2 79.3 66.5 50.4 40.6
W310x39 W12x26 b = 165 t = 9.7 d = 310	3 000 2 000 1 500 1 000 500	459 443 427 367 300	399 391 383 337 280	332 329 326 295 249	1 560 1 560 1 560 1 080 542	397 368 344 308 243	1 050 1 030 1 000 971 898	275 237 212 180 139	M_r 192 V_r 374 L_u 2 250 I_x 85.1 S_x 549	2 000 2 500 3 000 4 000 5 000	192 186 172 140 103	6 000 7 000 8 000 10 000 12 000	77.7 62.2 51.8 38.8 31.1
W250x67 W10x45 b = 204 t = 15.7 d = 257	3 000 2 000 1 500 1 000 500	682 600 533 463 390	599 538 485 430 367	493 457 423 383 335	2 690 2 170 1 630 1 080 542	489 439 401 346 261	1 600 1 550 1 500 1 430 1 290	300 253 224 191 151	M_r 284 V_r 476 L_u 3 230 I_x 104 S_x 806	2 000 2 500 3 000 3 500 4 000	284 284 284 278 268	5 000 6 000 7 000 8 000 10 000	246 225 203 180 139

Note: Resistances are based on a concrete density of 1850 kg/m³.
Units: M_r - kN·m, V_r - kN, L_u - mm, I_x - 10^6 mm⁴, S_x - 10^3 mm³, b - mm, t - mm, d - mm
* Transformed moment of inertia for calculating shrinkage deflections, based on a modular ratio $n_t = 50$

COMPOSITE BEAMS
Trial Selection Table

G40.21 350W
$f'_c = 25$ MPa

75 mm Deck with 85 mm Slab
$\phi = 0.90$, $\phi_c = 0.60$

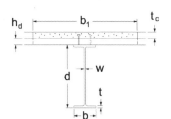

Steel section	b_1	M_{rc} (kN·m) for % shear connection			Q_r (kN)	I_t 10^6 mm⁴	S_t 10^3 mm³	I_t^* 10^6 mm⁴	Steel section data		Unbraced condition			
		100%	70%	40%	100%						L´ mm	$M_r´$ kN·m	L´ mm	$M_r´$ kN·m
W250x58	3 000	597	519	425	2 340	429	1 390	270	M_r	243	2 000	243	5 000	204
W10x39	2 000	549	489	410	2 170	387	1 350	228	V_r	419	2 500	243	6 000	182
b = 203	1 500	484	437	376	1 630	356	1 310	201	L_u	3 100	3 000	243	7 000	161
t = 13.5	1 000	416	384	339	1 080	309	1 260	170	I_x	87.3	3 500	235	8 000	137
d = 252	500	345	324	293	542	234	1 140	133	S_x	693	4 000	225	10 000	105
W250x45	3 000	485	419	342	1 800	369	1 110	243	M_r	190	2 000	190	5 000	114
W10x30	2 000	464	409	339	1 800	337	1 080	207	V_r	420	2 500	182	6 000	90.6
b = 148	1 500	430	385	326	1 630	313	1 050	183	L_u	2 150	3 000	169	7 000	75.2
t = 13	1 000	364	333	289	1 080	276	1 010	154	I_x	71.1	3 500	156	8 000	64.4
d = 266	500	295	274	243	542	213	924	118	S_x	534	4 000	143	10 000	50.1
W250x39	3 000	420	360	293	1 550	321	962	218	M_r	162	2 000	162	5 000	88.0
W10x26	2 000	404	352	290	1 550	296	934	186	V_r	360	2 500	153	6 000	69.2
b = 147	1 500	388	345	287	1 550	276	913	165	L_u	2 090	3 000	141	7 000	57.0
t = 11.2	1 000	329	298	258	1 080	246	880	139	I_x	60.1	3 500	129	8 000	48.6
d = 262	500	261	243	214	542	191	809	105	S_x	459	4 000	116	10 000	37.6
W250x33	3 000	357	305	247	1 310	275	818	191	M_r	134	2 000	134	5 000	63.6
W10x22	2 000	346	299	245	1 310	255	794	165	V_r	327	2 500	124	6 000	49.4
b = 146	1 500	335	294	243	1 310	239	777	147	L_u	2 000	3 000	113	7 000	40.3
t = 9.1	1 000	297	267	227	1 080	214	750	123	I_x	48.9	3 500	101	8 000	34.1
d = 258	500	230	213	185	542	169	694	92.3	S_x	379	4 000	88.6	10 000	26.1
W200x42	3 000	400	338	266	1 660	259	928	169	M_r	139	2 000	139	5 000	106
W8x28	2 000	382	329	263	1 660	237	897	143	V_r	307	2 500	139	6 000	92.0
b = 166	1 500	361	317	259	1 630	219	875	125	L_u	2 580	3 000	134	7 000	76.5
t = 11.8	1 000	296	265	225	1 080	193	839	104	I_x	40.6	3 500	127	8 000	65.6
d = 205	500	228	212	187	542	147	765	76.5	S_x	396	4 000	120	10 000	51.2
W200x36	3 000	346	290	227	1 430	226	803	151	M_r	118	2 000	118	5 000	85.4
W8x24	2 000	332	283	225	1 430	207	777	129	V_r	259	2 500	118	6 000	70.2
b = 165	1 500	319	277	223	1 430	193	759	113	L_u	2 490	3 000	112	7 000	58.0
t = 10.2	1 000	270	240	201	1 080	171	730	93.9	I_x	34.1	3 500	106	8 000	49.5
d = 201	500	204	188	165	542	132	670	68.6	S_x	340	4 000	99.0	10 000	38.4
W200x31	3 000	313	262	206	1 260	212	722	145	M_r	106	2 000	105	5 000	57.0
W8x21	2 000	303	257	205	1 260	195	699	124	V_r	279	2 500	97.8	6 000	45.6
b = 134	1 500	292	252	203	1 260	183	683	110	L_u	1 960	3 000	90.2	7 000	38.0
t = 10.2	1 000	259	230	190	1 080	164	658	91.5	I_x	31.4	3 500	82.3	8 000	32.7
d = 210	500	193	177	153	542	128	606	66.6	S_x	299	4 000	74.4	10 000	25.6
W200x27	3 000	266	222	174	1 070	182	616	128	M_r	87.9	2 000	86.3	5 000	41.2
W8x18	2 000	259	219	173	1 070	169	597	111	V_r	250	2 500	79.5	6 000	32.6
b = 133	1 500	252	215	172	1 070	159	583	98.4	L_u	1 880	3 000	72.2	7 000	27.1
t = 8.4	1 000	237	208	169	1 070	143	563	82.1	I_x	25.8	3 500	64.5	8 000	23.1
d = 207	500	173	157	134	542	114	522	59.5	S_x	249	4 000	56.0	10 000	18.0

Note: Resistances are based on a concrete density of 1850 kg/m³.

Units: M_r - kN·m, V_r - kN, L_u - mm, I_x - 10^6 mm⁴, S_x - 10^3 mm³, b - mm, t - mm, d - mm

* Transformed moment of inertia for calculating shrinkage deflections, based on a modular ratio $n_t = 50$

COMPOSITE BEAMS
Trial Selection Table

75 mm Deck with 85 mm Slab
$\phi = 0.90$, $\phi_c = 0.60$

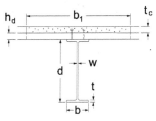

G40.21 350W
$f'_c = 25$ MPa

Steel section	b_1	Composite M_{rc} (kN·m) for % shear connection			Q_r	I_t	S_t	I_t^*	Non-composite Steel section data	Unbraced condition			
		100%	70%	40%	100%	10^6	10^3	10^6		L'	M_r	L'	M_r'
	mm				(kN)	mm⁴	mm³	mm⁴		mm	kN·m	mm	kN·m
WWF1000x223	5 000	5 080	4 890	4 430	5 420	11 000	12 700	7 090	M_r 3 310	4 000	3 250	12 000	920
WWF39x150	4 000	4 930	4 700	4 250	4 340	10 400	12 500	6 690	V_r 2 210	5 000	2 960	14 000	729
b = 300	3 000	4 720	4 450	4 040	3 250	9 620	12 200	6 240	L_u 3 790	6 000	2 620	16 000	602
t = 25	2 000	4 380	4 130	3 820	2 170	8 560	11 800	5 750	I_x 4 590	8 000	1 790	18 000	512
d = 1000	1 000	3 900	3 750	3 570	1 080	7 010	11 000	5 200	S_x 9 190	10 000	1 230	20 000	446
WWF1000x200	5 000	4 650	4 460	4 010	5 420	9 970	11 300	6 380	M_r 2 890	4 000	2 790	12 000	699
WWF39x134	4 000	4 500	4 280	3 820	4 340	9 430	11 100	5 990	V_r 2 210	5 000	2 510	14 000	549
b = 300	3 000	4 300	4 020	3 620	3 250	8 730	10 900	5 560	L_u 3 610	6 000	2 180	16 000	449
t = 20	2 000	3 950	3 700	3 390	2 170	7 750	10 500	5 090	I_x 3 940	8 000	1 390	18 000	379
d = 1000	1 000	3 480	3 320	3 140	1 080	6 310	9 710	4 550	S_x 7 880	10 000	945	20 000	328
WWF900x231	5 000	4 790	4 620	4 340	5 420	9 950	12 800	6 540	M_r 3 400	4 000	3 400	12 000	1 760
WWF35x156	4 000	4 640	4 500	4 210	4 340	9 420	12 600	6 190	V_r 1 350	5 000	3 400	14 000	1 370
b = 400	3 000	4 490	4 340	4 040	3 250	8 740	12 300	5 810	L_u 5 460	6 000	3 300	16 000	1 110
t = 25	2 000	4 290	4 110	3 850	2 170	7 810	12 000	5 390	I_x 4 410	8 000	2 870	18 000	933
d = 900	1 000	3 920	3 790	3 630	1 080	6 480	11 300	4 930	S_x 9 810	10 000	2 360	20 000	801
WWF900x192	5 000	4 100	3 930	3 650	5 420	8 420	10 500	5 500	M_r 2 710	4 000	2 700	12 000	842
WWF35x128	4 000	3 950	3 810	3 520	4 340	7 990	10 300	5 180	V_r 1 350	5 000	2 490	14 000	670
b = 300	3 000	3 800	3 660	3 350	3 250	7 420	10 100	4 820	L_u 3 980	6 000	2 240	16 000	555
t = 25	2 000	3 600	3 420	3 160	2 170	6 630	9 790	4 420	I_x 3 460	8 000	1 630	18 000	473
d = 900	1 000	3 230	3 100	2 940	1 080	5 440	9 190	3 970	S_x 7 680	10 000	1 120	20 000	412
WWF900x169	5 000	3 690	3 530	3 260	5 420	7 510	9 140	4 900	M_r 2 320	4 000	2 290	12 000	625
WWF35x113	4 000	3 560	3 420	3 130	4 340	7 140	9 020	4 600	V_r 1 340	5 000	2 080	14 000	491
b = 300	3 000	3 410	3 270	2 960	3 250	6 640	8 840	4 260	L_u 3 820	6 000	1 850	16 000	402
t = 20	2 000	3 210	3 030	2 770	2 170	5 940	8 550	3 880	I_x 2 930	8 000	1 250	18 000	339
d = 900	1 000	2 840	2 710	2 550	1 080	4 840	8 000	3 440	S_x 6 510	10 000	845	20 000	293
WWF800x223	5 000	4 190	4 020	3 780	5 420	7 920	11 200	5 170	M_r 2 940	4 000	2 940	10 000	2 120
WWF31x150	4 000	4 050	3 900	3 660	4 340	7 500	11 100	4 880	V_r 1 370	5 000	2 940	12 000	1 630
b = 400	3 000	3 900	3 780	3 520	3 250	6 950	10 900	4 570	L_u 5 570	6 000	2 880	14 000	1 280
t = 25	2 000	3 730	3 580	3 350	2 170	6 200	10 500	4 230	I_x 3 410	7 000	2 710	16 000	1 050
d = 800	1 000	3 410	3 290	3 150	1 080	5 110	9 920	3 840	S_x 8 520	8 000	2 530	18 000	882
WWF800x184	5 000	3 580	3 410	3 170	5 420	6 670	9 150	4 330	M_r 2 330	4 000	2 330	10 000	1 040
WWF31x123	4 000	3 430	3 290	3 050	4 340	6 330	9 030	4 080	V_r 1 370	5 000	2 170	12 000	791
b = 300	3 000	3 290	3 170	2 910	3 250	5 880	8 850	3 780	L_u 4 060	6 000	1 970	14 000	634
t = 25	2 000	3 120	2 970	2 740	2 170	5 250	8 560	3 460	I_x 2 660	7 000	1 750	16 000	529
d = 800	1 000	2 800	2 680	2 540	1 080	4 290	8 030	3 090	S_x 6 640	8 000	1 490	18 000	454
WWF800x161	5 000	3 220	3 060	2 830	5 420	5 930	7 960	3 860	M_r 1 990	4 000	1 980	10 000	778
WWF31x108	4 000	3 080	2 950	2 710	4 340	5 640	7 860	3 620	V_r 1 370	5 000	1 810	12 000	581
b = 300	3 000	2 940	2 820	2 570	3 250	5 250	7 710	3 350	L_u 3 900	6 000	1 620	14 000	460
t = 20	2 000	2 770	2 620	2 400	2 170	4 700	7 460	3 030	I_x 2 250	7 000	1 400	16 000	379
d = 800	1 000	2 460	2 340	2 200	1 080	3 820	6 980	2 670	S_x 5 610	8 000	1 130	18 000	322

Note: Resistances are based on a concrete density of 2000 kg/m³.

Units: M_r - kN·m, V_r - kN, L_u - mm, I_x - 10^6 mm⁴, S_x - 10^3 mm³, b - mm, t - mm, d - mm

* Transformed moment of inertia for calculating shrinkage deflections, based on a modular ratio $n_t = 50$

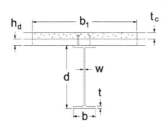

COMPOSITE BEAMS
Trial Selection Table

G40.21 350W
$f'_c = 25$ MPa

75 mm Deck with 85 mm Slab
$\phi = 0.90$, $\phi_c = 0.60$

Steel section	b_1	Composite M_{rc} (kN·m) for % shear connection			Q_r (kN)	I_t 10^6 mm⁴	S_t 10^3 mm³	I_t^* 10^6 mm⁴	Non-composite Steel section data	Unbraced condition			
		100%	70%	40%	100%					L´ mm	M_r´ kN·m	L´ mm	M_r´ kN·m
WWF700x214	5 000	3 630	3 460	3 240	5 420	6 140	9 740	3 960	M_r 2 500	4 000	2 500	10 000	1 870
WWF28x144	4 000	3 480	3 340	3 140	4 340	5 810	9 610	3 730	V_r 1 370	5 000	2 500	12 000	1 500
b = 400	3 000	3 340	3 230	3 020	3 250	5 380	9 420	3 480	L_u 5 690	6 000	2 470	14 000	1 190
t = 25	2 000	3 180	3 060	2 870	2 170	4 780	9 110	3 200	I_x 2 540	7 000	2 330	16 000	982
d = 700	1 000	2 920	2 820	2 700	1 080	3 920	8 560	2 890	S_x 7 270	8 000	2 190	18 000	835
WWF700x175	5 000	3 090	2 930	2 700	5 420	5 150	7 890	3 320	M_r 1 970	4 000	1 970	10 000	965
WWF28x117	4 000	2 950	2 810	2 610	4 340	4 890	7 780	3 110	V_r 1 370	5 000	1 850	12 000	741
b = 300	3 000	2 800	2 700	2 490	3 250	4 540	7 630	2 880	L_u 4 160	6 000	1 700	14 000	601
t = 25	2 000	2 650	2 530	2 340	2 170	4 040	7 380	2 620	I_x 1 970	7 000	1 530	16 000	505
d = 700	1 000	2 390	2 290	2 160	1 080	3 280	6 900	2 320	S_x 5 640	8 000	1 350	18 000	436
WWF700x152	5 000	2 770	2 620	2 410	5 420	4 560	6 840	2 960	M_r 1 680	4 000	1 680	10 000	711
WWF28x102	4 000	2 640	2 510	2 310	4 340	4 340	6 750	2 760	V_r 1 370	5 000	1 540	12 000	537
b = 300	3 000	2 500	2 400	2 190	3 250	4 040	6 620	2 550	L_u 3 990	6 000	1 390	14 000	429
t = 20	2 000	2 350	2 230	2 040	2 170	3 610	6 400	2 290	I_x 1 660	7 000	1 230	16 000	356
d = 700	1 000	2 090	1 990	1 870	1 080	2 920	5 980	2 000	S_x 4 760	8 000	1 020	18 000	305
W610x241	4 000	3 520	3 370	3 080	4 340	5 170	9 550	3 210	M_r 2 420	3 000	2 420	8 000	1 950
W24x162	3 000	3 360	3 200	2 930	3 250	4 750	9 300	2 980	V_r 2 360	4 000	2 420	10 000	1 640
b = 329	2 000	3 150	2 990	2 770	2 170	4 180	8 930	2 730	L_u 4 750	5 000	2 380	12 000	1 300
t = 31	1 500	3 000	2 860	2 690	1 630	3 820	8 650	2 600	I_x 2 150	6 000	2 240	14 000	1 070
d = 635	1 000	2 830	2 720	2 600	1 080	3 380	8 260	2 460	S_x 6 780	7 000	2 100	16 000	911
W610x217	4 000	3 210	3 070	2 800	4 340	4 730	8 640	2 930	M_r 2 160	3 000	2 160	8 000	1 690
W24x146	3 000	3 060	2 920	2 660	3 250	4 350	8 430	2 720	V_r 2 150	4 000	2 160	10 000	1 370
b = 328	2 000	2 870	2 720	2 510	2 170	3 840	8 100	2 480	L_u 4 640	5 000	2 110	12 000	1 070
t = 27.7	1 500	2 730	2 590	2 420	1 630	3 500	7 850	2 350	I_x 1 910	6 000	1 980	14 000	878
d = 628	1 000	2 560	2 460	2 340	1 080	3 090	7 500	2 210	S_x 6 070	7 000	1 840	16 000	745
W610x195	4 000	2 920	2 780	2 540	4 340	4 290	7 750	2 660	M_r 1 910	3 000	1 910	8 000	1 450
W24x131	3 000	2 770	2 650	2 410	3 250	3 960	7 570	2 460	V_r 1 990	4 000	1 910	10 000	1 120
b = 327	2 000	2 600	2 460	2 260	2 170	3 500	7 280	2 230	L_u 4 530	5 000	1 860	12 000	870
t = 24.4	1 500	2 470	2 340	2 180	1 630	3 200	7 060	2 110	I_x 1 680	6 000	1 730	14 000	709
d = 622	1 000	2 310	2 210	2 090	1 080	2 810	6 750	1 970	S_x 5 400	7 000	1 590	16 000	598
W610x174	4 000	2 650	2 510	2 300	4 340	3 880	6 920	2 420	M_r 1 690	3 000	1 690	8 000	1 240
W24x117	3 000	2 500	2 390	2 170	3 250	3 590	6 760	2 220	V_r 1 790	4 000	1 690	10 000	924
b = 325	2 000	2 340	2 220	2 030	2 170	3 180	6 520	2 010	L_u 4 440	5 000	1 630	12 000	709
t = 21.6	1 500	2 230	2 110	1 950	1 630	2 910	6 330	1 890	I_x 1 470	6 000	1 510	14 000	574
d = 616	1 000	2 080	1 980	1 860	1 080	2 560	6 050	1 760	S_x 4 780	7 000	1 380	16 000	482
W610x155	4 000	2 400	2 270	2 070	4 340	3 510	6 160	2 200	M_r 1 490	3 000	1 490	8 000	1 060
W24x104	3 000	2 260	2 150	1 960	3 250	3 260	6 040	2 020	V_r 1 610	4 000	1 490	10 000	762
b = 324	2 000	2 110	2 000	1 820	2 170	2 900	5 830	1 810	L_u 4 370	5 000	1 430	12 000	579
t = 19	1 500	2 010	1 900	1 740	1 630	2 650	5 660	1 700	I_x 1 290	6 000	1 310	14 000	465
d = 611	1 000	1 870	1 770	1 660	1 080	2 330	5 420	1 570	S_x 4 220	7 000	1 190	16 000	388

Note: Resistances are based on a concrete density of 2000 kg/m³.

Units: M_r - kN·m, V_r - kN, L_u - mm, I_x - 10^6 mm⁴, S_x - 10^3 mm³, b - mm, t - mm, d - mm

* Transformed moment of inertia for calculating shrinkage deflections, based on a modular ratio $n_t = 50$

COMPOSITE BEAMS
Trial Selection Table

75 mm Deck with 85 mm Slab
$\phi = 0.90$, $\phi_c = 0.60$

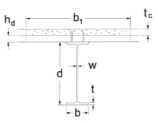

G40.21 350W
$f'_c = 25$ MPa

Steel section	b_1	Composite M_{rc} (kN·m) for % shear connection			Q_r	I_t	S_t	I_t^*	Non-composite Steel section data	Unbraced condition			
		100%	70%	40%	100%	10^6	10^3	10^6		L'	M_r'	L'	M_r'
	mm				kN	mm⁴	mm³	mm⁴		mm	kN·m	mm	kN·m
W610x140	4 000	2 240	2 110	1 910	4 340	3 240	5 550	2 020	M_r 1 310	3 000	1 310	8 000	573
W24x94	3 000	2 100	2 000	1 790	3 250	3 010	5 430	1 840	V_r 1 680	4 000	1 180	10 000	422
b = 230	2 000	1 950	1 830	1 650	2 170	2 680	5 240	1 640	L_u 3 050	5 000	1 040	12 000	334
t = 22.2	1 500	1 840	1 720	1 570	1 630	2 450	5 080	1 530	I_x 1 120	6 000	878	14 000	277
d = 617	1 000	1 690	1 600	1 480	1 080	2 140	4 850	1 400	S_x 3 630	7 000	695	16 000	237
W610x125	4 000	2 040	1 910	1 730	4 340	2 920	4 950	1 840	M_r 1 160	3 000	1 160	8 000	470
W24x84	3 000	1 900	1 800	1 620	3 250	2 730	4 850	1 680	V_r 1 510	4 000	1 040	10 000	342
b = 229	2 000	1 760	1 660	1 490	2 170	2 440	4 690	1 490	L_u 2 990	5 000	895	12 000	269
t = 19.6	1 500	1 660	1 560	1 410	1 630	2 230	4 560	1 380	I_x 985	6 000	733	14 000	222
d = 612	1 000	1 530	1 440	1 330	1 080	1 960	4 350	1 260	S_x 3 220	7 000	575	16 000	189
W610x113	4 000	1 890	1 770	1 590	4 340	2 670	4 480	1 700	M_r 1 040	3 000	1 030	8 000	391
W24x76	3 000	1 760	1 660	1 490	3 250	2 500	4 390	1 550	V_r 1 420	4 000	915	10 000	282
b = 228	2 000	1 610	1 530	1 360	2 170	2 240	4 250	1 370	L_u 2 930	5 000	781	12 000	220
t = 17.3	1 500	1 530	1 430	1 290	1 630	2 060	4 140	1 260	I_x 875	6 000	617	14 000	180
d = 608	1 000	1 400	1 320	1 210	1 080	1 810	3 960	1 150	S_x 2 880	7 000	481	16 000	153
W610x101	4 000	1 730	1 610	1 440	4 100	2 420	4 020	1 550	M_r 914	3 000	901	8 000	320
W24x68	3 000	1 610	1 520	1 360	3 250	2 270	3 950	1 410	V_r 1 310	4 000	795	10 000	228
b = 228	2 000	1 480	1 400	1 240	2 170	2 050	3 820	1 240	L_u 2 870	5 000	668	12 000	176
t = 14.9	1 500	1 400	1 300	1 160	1 630	1 880	3 720	1 140	I_x 764	6 000	512	14 000	144
d = 603	1 000	1 280	1 190	1 080	1 080	1 650	3 560	1 030	S_x 2 530	7 000	396	16 000	121
W610x91	4 000	1 550	1 430	1 280	3 650	2 170	3 580	1 410	M_r 806	3 000	788	8 000	261
W24x61	3 000	1 470	1 380	1 240	3 250	2 050	3 520	1 280	V_r 1 110	4 000	687	10 000	185
b = 227	2 000	1 340	1 270	1 120	2 170	1 850	3 420	1 130	L_u 2 790	5 000	567	12 000	142
t = 12.7	1 500	1 270	1 180	1 040	1 630	1 710	3 330	1 030	I_x 667	6 000	421	14 000	115
d = 598	1 000	1 150	1 070	964	1 080	1 500	3 190	925	S_x 2 230	7 000	325	16 000	97.1
W530x138	4 000	2 030	1 890	1 700	4 340	2 640	4 990	1 620	M_r 1 140	3 000	1 130	8 000	515
W21x93	3 000	1 880	1 780	1 580	3 250	2 450	4 870	1 470	V_r 1 680	4 000	1 010	10 000	390
b = 214	2 000	1 730	1 620	1 450	2 170	2 170	4 690	1 300	L_u 2 910	5 000	891	12 000	314
t = 23.6	1 500	1 630	1 520	1 380	1 630	1 980	4 540	1 200	I_x 861	6 000	762	14 000	263
d = 549	1 000	1 490	1 410	1 300	1 080	1 720	4 320	1 100	S_x 3 140	7 000	616	16 000	227
W530x123	4 000	1 850	1 730	1 550	4 340	2 390	4 470	1 480	M_r 1 010	3 000	996	8 000	421
W21x83	3 000	1 720	1 610	1 440	3 250	2 230	4 370	1 350	V_r 1 480	4 000	888	10 000	316
b = 212	2 000	1 570	1 480	1 320	2 170	1 990	4 220	1 190	L_u 2 840	5 000	768	12 000	253
t = 21.2	1 500	1 480	1 380	1 250	1 630	1 810	4 090	1 100	I_x 761	6 000	631	14 000	211
d = 544	1 000	1 360	1 270	1 170	1 080	1 580	3 900	995	S_x 2 800	7 000	505	16 000	182
W530x109	4 000	1 690	1 570	1 400	4 340	2 150	3 970	1 350	M_r 891	3 000	872	8 000	342
W21x73	3 000	1 560	1 460	1 310	3 250	2 010	3 890	1 230	V_r 1 300	4 000	771	10 000	254
b = 211	2 000	1 420	1 340	1 190	2 170	1 800	3 760	1 080	L_u 2 790	5 000	656	12 000	202
t = 18.8	1 500	1 340	1 250	1 120	1 630	1 650	3 660	991	I_x 667	6 000	520	14 000	168
d = 539	1 000	1 230	1 150	1 050	1 080	1 440	3 500	895	S_x 2 480	7 000	413	16 000	144

Note: Resistances are based on a concrete density of 2000 kg/m³.

Units: M_r - kN·m, V_r - kN, L_u - mm, I_x - 10^6 mm⁴, S_x - 10^3 mm³, b - mm, t - mm, d - mm

* Transformed moment of inertia for calculating shrinkage deflections, based on a modular ratio $n_t = 50$

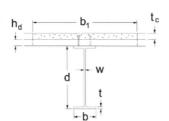

COMPOSITE BEAMS
Trial Selection Table

G40.21 350W
$f'_c = 25$ MPa

75 mm Deck with 85 mm Slab
$\phi = 0.90$, $\phi_c = 0.60$

Steel section	b_1	Composite M_{rc} (kN·m) for % shear connection			Q_r (kN)	I_t 10^6	S_t 10^3	I_t^* 10^6	Non-composite Steel section data		Unbraced condition			
		100%	70%	40%	100%	mm⁴	mm³	mm⁴			L' mm	M_r' kN·m	L' mm	M_r' kN·m
	mm													
W530x101	4 000	1 580	1 460	1 300	4 060	2 010	3 700	1 280	M_r	825	3 000	804	8 000	301
W21x68	3 000	1 470	1 380	1 230	3 250	1 890	3 630	1 160	V_r	1 220	4 000	706	10 000	222
b = 210	2 000	1 330	1 260	1 120	2 170	1 700	3 510	1 020	L_u	2 750	5 000	594	12 000	176
t = 17.4	1 500	1 260	1 180	1 050	1 630	1 560	3 420	936	I_x	617	6 000	462	14 000	146
d = 537	1 000	1 150	1 080	981	1 080	1 370	3 270	842	S_x	2 300	7 000	365	16 000	125
W530x92	4 000	1 450	1 340	1 190	3 720	1 850	3 370	1 190	M_r	743	3 000	719	8 000	253
W21x62	3 000	1 370	1 280	1 140	3 250	1 740	3 310	1 080	V_r	1 130	4 000	626	10 000	185
b = 209	2 000	1 240	1 170	1 040	2 170	1 570	3 210	943	L_u	2 700	5 000	519	12 000	146
t = 15.6	1 500	1 160	1 090	972	1 630	1 450	3 130	863	I_x	552	6 000	393	14 000	120
d = 533	1 000	1 070	996	900	1 080	1 270	3 000	772	S_x	2 070	7 000	309	16 000	103
W530x82	4 000	1 300	1 190	1 050	3 310	1 650	2 990	1 080	M_r	652	3 000	625	8 000	204
W21x55	3 000	1 260	1 170	1 040	3 250	1 560	2 940	981	V_r	1 040	4 000	538	10 000	148
b = 209	2 000	1 120	1 060	937	2 170	1 420	2 860	854	L_u	2 640	5 000	436	12 000	115
t = 13.3	1 500	1 050	990	874	1 630	1 310	2 790	778	I_x	478	6 000	321	14 000	94.8
d = 528	1 000	964	897	803	1 080	1 150	2 670	692	S_x	1 810	7 000	250	16 000	80.5
W460x106	4 000	1 500	1 380	1 210	4 250	1 690	3 500	1 050	M_r	753	3 000	727	7 000	366
W18x71	3 000	1 380	1 280	1 140	3 250	1 580	3 430	950	V_r	1 230	3 500	686	8 000	308
b = 194	2 000	1 230	1 160	1 030	2 170	1 410	3 310	827	L_u	2 670	4 000	643	10 000	235
t = 20.6	1 500	1 160	1 080	968	1 630	1 290	3 210	755	I_x	488	5 000	553	12 000	190
d = 469	1 000	1 060	990	900	1 080	1 120	3 060	676	S_x	2 080	6 000	450	14 000	160
W460x97	4 000	1 380	1 260	1 100	3 870	1 560	3 200	987	M_r	687	3 000	660	7 000	314
W18x65	3 000	1 280	1 190	1 060	3 250	1 470	3 140	891	V_r	1 100	3 500	621	8 000	264
b = 193	2 000	1 150	1 080	958	2 170	1 320	3 040	774	L_u	2 630	4 000	579	10 000	200
t = 19	1 500	1 070	1 010	898	1 630	1 210	2 960	706	I_x	445	5 000	491	12 000	161
d = 466	1 000	983	919	832	1 080	1 050	2 820	630	S_x	1 910	6 000	389	14 000	135
W460x89	4 000	1 280	1 170	1 020	3 590	1 460	2 970	932	M_r	633	3 000	605	7 000	276
W18x60	3 000	1 210	1 120	994	3 250	1 370	2 920	841	V_r	1 010	3 500	568	8 000	231
b = 192	2 000	1 080	1 010	900	2 170	1 240	2 830	730	L_u	2 600	4 000	527	10 000	174
t = 17.7	1 500	1 010	948	843	1 630	1 140	2 750	664	I_x	409	5 000	442	12 000	140
d = 463	1 000	923	863	778	1 080	994	2 630	590	S_x	1 770	6 000	343	14 000	117
W460x82	4 000	1 170	1 060	930	3 280	1 340	2 720	871	M_r	576	3 000	546	7 000	234
W18x55	3 000	1 140	1 050	923	3 250	1 270	2 670	786	V_r	947	3 500	510	8 000	195
b = 191	2 000	1 000	938	836	2 170	1 150	2 590	681	L_u	2 540	4 000	471	10 000	146
t = 16	1 500	933	879	781	1 630	1 060	2 530	619	I_x	370	5 000	386	12 000	116
d = 460	1 000	855	799	718	1 080	928	2 420	547	S_x	1 610	6 000	292	14 000	97.3
W460x74	4 000	1 070	968	843	2 980	1 230	2 470	810	M_r	520	3 000	489	7 000	198
W18x50	3 000	1 040	953	839	2 980	1 160	2 430	732	V_r	855	3 500	455	8 000	164
b = 190	2 000	932	869	774	2 170	1 060	2 360	633	L_u	2 510	4 000	417	10 000	122
t = 14.5	1 500	864	813	721	1 630	978	2 310	574	I_x	332	5 000	332	12 000	96.8
d = 457	1 000	790	739	660	1 080	862	2 220	506	S_x	1 460	6 000	249	14 000	80.6

Note: Resistances are based on a concrete density of 2000 kg/m³.

Units: M_r - kN·m, V_r - kN, L_u - mm, I_x - 10^6 mm⁴, S_x - 10^3 mm³, b - mm, t - mm, d - mm

* Transformed moment of inertia for calculating shrinkage deflections, based on a modular ratio $n_t = 50$

COMPOSITE BEAMS
Trial Selection Table

75 mm Deck with 85 mm Slab
$\phi = 0.90$, $\phi_c = 0.60$

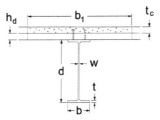

G40.21 350W

$f'_c = 25$ MPa

Steel section	b_1	M_{rc} (kN·m) for % shear connection			Q_r	I_t	S_t	I_t^*	Steel section data		Unbraced condition			
											|	|	|	|
						10^6	10^3	10^6			L'	M_r'	L'	M_r'
	mm	100%	70%	40%	100% (kN)	mm⁴	mm³	mm⁴			mm	kN·m	mm	kN·m
W460x67	4 000	973	877	762	2 700	1 120	2 240	750	M_r	466	3 000	435	7 000	165
W18x45	3 000	949	866	758	2 700	1 060	2 210	679	V_r	802	3 500	402	8 000	136
b = 190	2 000	866	805	714	2 170	973	2 150	587	L_u	2 460	4 000	366	10 000	99.8
t = 12.7	1 500	799	751	663	1 630	902	2 100	531	I_x	296	5 000	281	12 000	79.0
d = 454	1 000	728	680	604	1 080	797	2 020	465	S_x	1 300	6 000	209	14 000	65.4
W460x61	4 000	872	783	677	2 410	1 000	2 000	682	M_r	406	3 000	374	7 000	132
W18x41	3 000	853	774	674	2 410	954	1 960	617	V_r	758	3 500	343	8 000	108
b = 189	2 000	797	737	651	2 170	877	1 910	533	L_u	2 390	4 000	309	10 000	78.5
t = 10.8	1 500	731	684	602	1 630	815	1 870	481	I_x	255	5 000	228	12 000	61.7
d = 450	1 000	663	617	543	1 080	724	1 800	419	S_x	1 130	6 000	168	14 000	50.8
W410x85	4 000	1 140	1 030	891	3 400	1 200	2 640	763	M_r	542	3 000	512	7 000	242
W16x57	3 000	1 090	1 000	876	3 250	1 130	2 590	686	V_r	945	3 500	480	8 000	205
b = 181	2 000	957	891	791	2 170	1 020	2 510	592	L_u	2 510	4 000	447	10 000	157
t = 18.2	1 500	886	833	738	1 630	933	2 440	536	I_x	315	5 000	377	12 000	127
d = 417	1 000	809	755	678	1 080	814	2 330	472	S_x	1 510	6 000	297	14 000	107
W410x74	4 000	1 010	910	784	3 010	1 070	2 340	696	M_r	476	3 000	445	7 000	194
W16x50	3 000	983	896	780	3 010	1 010	2 300	627	V_r	833	3 500	414	8 000	163
b = 180	2 000	872	808	716	2 170	917	2 230	540	L_u	2 450	4 000	382	10 000	123
t = 16	1 500	803	753	666	1 630	846	2 180	488	I_x	275	5 000	312	12 000	99.7
d = 413	1 000	731	683	609	1 080	742	2 080	427	S_x	1 330	6 000	239	14 000	83.8
W410x67	4 000	917	821	705	2 710	973	2 110	644	M_r	428	3 000	397	7 000	161
W16x45	3 000	893	809	702	2 710	922	2 080	581	V_r	750	3 500	367	8 000	135
b = 179	2 000	809	747	660	2 170	841	2 020	500	L_u	2 400	4 000	336	10 000	102
t = 14.4	1 500	741	693	613	1 630	778	1 970	451	I_x	245	5 000	264	12 000	81.6
d = 410	1 000	671	628	558	1 080	686	1 890	394	S_x	1 200	6 000	201	14 000	68.3
W410x60	4 000	812	724	621	2 390	870	1 870	589	M_r	375	3 000	344	7 000	131
W16x40	3 000	793	715	618	2 390	827	1 840	533	V_r	652	3 500	317	8 000	109
b = 178	2 000	740	680	599	2 170	759	1 800	460	L_u	2 370	4 000	288	10 000	81.3
t = 12.8	1 500	674	627	556	1 630	706	1 760	414	I_x	216	5 000	218	12 000	64.9
d = 407	1 000	606	569	504	1 080	626	1 690	360	S_x	1 060	6 000	165	14 000	54.1
W410x54	4 000	729	648	554	2 140	779	1 670	536	M_r	331	3 000	298	7 000	104
W16x36	3 000	714	640	551	2 140	742	1 650	486	V_r	628	3 500	272	8 000	86.0
b = 177	2 000	684	626	546	2 140	685	1 600	419	L_u	2 290	4 000	243	10 000	63.5
t = 10.9	1 500	621	575	507	1 630	638	1 570	376	I_x	186	5 000	176	12 000	50.4
d = 403	1 000	554	519	456	1 080	569	1 510	325	S_x	923	6 000	132	14 000	41.8
W410x46	4 000	637	563	478	1 860	684	1 450	481	M_r	278	3 000	212	7 000	61.7
W16x31	3 000	626	558	477	1 860	654	1 430	437	V_r	585	3 500	177	8 000	51.8
b = 140	2 000	603	547	473	1 860	606	1 390	377	L_u	1 780	4 000	142	10 000	39.2
t = 11.2	1 500	565	519	454	1 630	568	1 360	339	I_x	156	5 000	99.9	12 000	31.6
d = 403	1 000	498	465	404	1 080	509	1 320	291	S_x	772	6 000	76.4	14 000	26.5

Note: Resistances are based on a concrete density of 2000 kg/m³.

Units: M_r - kN·m, V_r - kN, L_u - mm, I_x - 10^6 mm⁴, S_x - 10^3 mm³, b - mm, t - mm, d - mm

* Transformed moment of inertia for calculating shrinkage deflections, based on a modular ratio $n_t = 50$

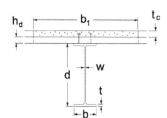

COMPOSITE BEAMS
Trial Selection Table

G40.21 350W
$f'_c = 25$ MPa

75 mm Deck with 85 mm Slab
$\phi = 0.90$, $\phi_c = 0.60$

Steel section	b_1	M_{rc} (kN·m) for % shear connection			Q_r (kN)	I_t 10^6	S_t 10^3	I_t^* 10^6	Steel section data		Unbraced condition			
											L'	M_r'	L'	M_r'
	mm	100%	70%	40%	100%	mm⁴	mm³	mm⁴			mm	kN·m	mm	kN·m
W410x39	4 000	541	477	402	1 570	581	1 220	419	M_r	230	3 000	167	7 000	44.0
W16x26	3 000	533	473	401	1 570	557	1 200	383	V_r	484	3 500	132	8 000	36.5
b = 140	2 000	517	465	398	1 570	520	1 180	331	L_u	1 720	4 000	105	10 000	27.3
t = 8.8	1 500	500	457	396	1 570	490	1 150	297	I_x	126	5 000	73.0	12 000	21.8
d = 399	1 000	440	408	353	1 080	442	1 120	254	S_x	634	6 000	55.1	14 000	18.2
W360x79	3 000	940	851	731	3 180	861	2 230	523	M_r	450	2 000	450	6 000	319
W14x53	2 000	814	750	665	2 170	777	2 160	449	V_r	692	2 500	450	7 000	267
b = 205	1 500	745	695	621	1 630	714	2 110	404	L_u	2 980	3 000	450	8 000	225
t = 16.8	1 000	673	634	569	1 080	623	2 020	353	I_x	226	4 000	409	10 000	171
d = 354	500	582	550	509	542	481	1 840	295	S_x	1 280	5 000	364	12 000	139
W360x72	3 000	853	767	656	2 870	784	2 020	484	M_r	403	2 000	403	6 000	273
W14x48	2 000	754	692	610	2 170	711	1 960	415	V_r	626	2 500	403	7 000	222
b = 204	1 500	687	638	569	1 630	655	1 910	373	L_u	2 910	3 000	400	8 000	186
t = 15.1	1 000	617	581	519	1 080	574	1 830	325	I_x	201	4 000	360	10 000	141
d = 350	500	531	501	461	542	445	1 670	268	S_x	1 150	5 000	317	12 000	114
W360x64	3 000	769	688	586	2 560	710	1 810	447	M_r	359	2 000	359	6 000	228
W14x43	2 000	699	638	558	2 170	648	1 760	383	V_r	555	2 500	359	7 000	183
b = 203	1 500	632	585	520	1 630	600	1 720	344	L_u	2 850	3 000	354	8 000	153
t = 13.5	1 000	564	530	473	1 080	528	1 650	298	I_x	178	4 000	316	10 000	115
d = 347	500	484	455	416	542	411	1 520	244	S_x	1 030	5 000	275	12 000	92.2
W360x57	3 000	702	626	533	2 270	665	1 630	426	M_r	318	2 000	318	6 000	147
W14x38	2 000	661	602	521	2 170	610	1 580	365	V_r	588	2 500	312	7 000	119
b = 172	1 500	596	549	484	1 630	566	1 550	327	L_u	2 340	3 000	292	8 000	99.7
t = 13.1	1 000	527	494	436	1 080	502	1 490	282	I_x	160	4 000	246	10 000	75.6
d = 358	500	447	418	379	542	392	1 370	227	S_x	896	5 000	192	12 000	61.0
W360x51	3 000	631	559	474	2 030	600	1 460	392	M_r	281	2 000	281	6 000	121
W14x34	2 000	604	546	470	2 030	553	1 420	337	V_r	531	2 500	274	7 000	96.9
b = 171	1 500	550	505	442	1 630	516	1 390	302	L_u	2 300	3 000	255	8 000	80.9
t = 11.6	1 000	483	451	398	1 080	459	1 340	259	I_x	141	4 000	211	10 000	60.8
d = 355	500	408	380	342	542	361	1 240	207	S_x	796	5 000	159	12 000	48.8
W360x45	3 000	564	498	421	1 800	537	1 300	358	M_r	245	2 000	245	6 000	96.0
W14x30	2 000	543	487	417	1 800	497	1 260	308	V_r	505	2 500	237	7 000	76.4
b = 171	1 500	509	464	403	1 630	466	1 240	276	L_u	2 240	3 000	219	8 000	63.3
t = 9.8	1 000	443	411	360	1 080	417	1 200	236	I_x	122	4 000	177	10 000	47.1
d = 352	500	369	343	306	542	330	1 110	186	S_x	691	5 000	128	12 000	37.5
W360x39	3 000	496	435	366	1 570	475	1 130	323	M_r	209	2 000	195	6 000	54.1
W14x26	2 000	480	428	363	1 570	442	1 100	279	V_r	477	2 500	173	7 000	44.2
b = 128	1 500	464	420	361	1 570	415	1 080	249	L_u	1 650	3 000	149	8 000	37.4
t = 10.7	1 000	403	372	322	1 080	374	1 040	213	I_x	102	4 000	97.0	10 000	28.7
d = 353	500	331	305	269	542	298	966	165	S_x	580	5 000	69.7	12 000	23.3

Note: Resistances are based on a concrete density of 2000 kg/m³.

Units: M_r - kN·m, V_r - kN, L_u - mm, I_x - 10^6 mm⁴, S_x - 10^3 mm³, b - mm, t - mm, d - mm

* Transformed moment of inertia for calculating shrinkage deflections, based on a modular ratio $n_t = 50$

COMPOSITE BEAMS
Trial Selection Table

75 mm Deck with 85 mm Slab
$\phi = 0.90$, $\phi_c = 0.60$

G40.21 350W
$f'_c = 25$ MPa

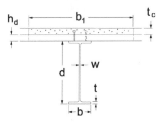

Steel section	b_1	Composite							Non-composite				
		M_{rc} (kN·m) for % shear connection			Q_r	I_t	S_t	I_t^*	Steel section data	Unbraced condition			
										L'	M_r'	L'	M_r'
	mm	100%	70%	40%	100%	10^6 mm⁴	10^3 mm³	10^6 mm⁴		mm	kN·m	mm	kN·m
W360x33	3 000	417	364	304	1 310	401	946	281	M_r 170	2 000	156	6 000	38.0
W14x22	2 000	406	359	302	1 310	376	923	244	V_r 399	2 500	136	7 000	30.8
b = 127	1 500	394	353	301	1 310	355	905	219	L_u 1 590	3 000	114	8 000	25.8
t = 8.5	1 000	356	326	281	1 080	323	878	187	I_x 82.6	4 000	70.2	10 000	19.6
d = 349	500	288	265	230	542	261	818	143	S_x 473	5 000	49.6	12 000	15.8
W310x74	3 000	824	737	623	2 990	687	1 960	414	M_r 375	2 000	375	6 000	279
W12x50	2 000	715	652	569	2 170	620	1 890	353	V_r 606	2 500	375	7 000	243
b = 205	1 500	647	598	530	1 630	569	1 840	316	L_u 3 070	3 000	375	8 000	206
t = 16.3	1 000	576	541	483	1 080	496	1 760	273	I_x 165	4 000	345	10 000	159
d = 310	500	494	466	429	542	380	1 600	223	S_x 1 060	5 000	312	12 000	129
W310x67	3 000	744	661	556	2 680	622	1 760	381	M_r 334	2 000	334	6 000	238
W12x45	2 000	663	602	521	2 170	564	1 700	325	V_r 541	2 500	334	7 000	200
b = 204	1 500	596	549	485	1 630	520	1 660	290	L_u 2 990	3 000	334	8 000	169
t = 14.6	1 000	527	494	440	1 080	455	1 590	250	I_x 145	4 000	303	10 000	129
d = 306	500	450	423	387	542	349	1 450	202	S_x 948	5 000	271	12 000	105
W310x60	3 000	670	591	495	2 390	563	1 580	353	M_r 296	2 000	296	6 000	203
W12x40	2 000	617	557	477	2 170	514	1 530	301	V_r 472	2 500	296	7 000	166
b = 203	1 500	551	504	443	1 630	476	1 500	269	L_u 2 930	3 000	295	8 000	139
t = 13.1	1 000	483	451	401	1 080	419	1 440	231	I_x 129	4 000	266	10 000	106
d = 303	500	409	384	349	542	324	1 320	185	S_x 849	5 000	235	12 000	85.2
W310x52	3 000	613	539	452	2 100	535	1 420	343	M_r 265	2 000	265	6 000	130
W12x35	2 000	584	525	447	2 100	491	1 380	293	V_r 502	2 500	260	7 000	106
b = 167	1 500	524	478	417	1 630	457	1 350	262	L_u 2 350	3 000	244	8 000	89.4
t = 13.2	1 000	456	424	375	1 080	405	1 300	224	I_x 119	4 000	208	10 000	68.4
d = 318	500	383	358	324	542	315	1 190	177	S_x 750	5 000	167	12 000	55.5
W310x45	3 000	525	459	383	1 790	461	1 220	304	M_r 223	2 000	223	6 000	98.2
W12x30	2 000	504	449	380	1 790	426	1 180	261	V_r 429	2 500	217	7 000	79.3
b = 166	1 500	471	427	367	1 630	398	1 160	233	L_u 2 290	3 000	202	8 000	66.5
t = 11.2	1 000	405	374	330	1 080	356	1 120	198	I_x 99.2	4 000	168	10 000	50.4
d = 313	500	336	314	281	542	280	1 030	155	S_x 634	5 000	128	12 000	40.6
W310x39	3 000	459	399	332	1 560	405	1 060	275	M_r 192	2 000	192	6 000	77.7
W12x26	2 000	443	391	329	1 560	377	1 030	237	V_r 374	2 500	186	7 000	62.2
b = 165	1 500	427	383	326	1 560	354	1 010	212	L_u 2 250	3 000	172	8 000	51.8
t = 9.7	1 000	367	337	295	1 080	319	981	180	I_x 85.1	4 000	140	10 000	38.8
d = 310	500	300	280	249	542	254	911	139	S_x 549	5 000	103	12 000	31.1
W250x67	3 000	682	599	493	2 690	502	1 620	300	M_r 284	2 000	284	5 000	246
W10x45	2 000	600	538	457	2 170	454	1 560	253	V_r 476	2 500	284	6 000	225
b = 204	1 500	533	485	423	1 630	416	1 520	224	L_u 3 230	3 000	284	7 000	203
t = 15.7	1 000	463	430	383	1 080	362	1 450	191	I_x 104	3 500	278	8 000	180
d = 257	500	390	367	335	542	274	1 320	151	S_x 806	4 000	268	10 000	139

Note: Resistances are based on a concrete density of 2000 kg/m³.

Units: M_r - kN·m, V_r - kN, L_u - mm, I_x - 10^6 mm⁴, S_x - 10^3 mm³, b - mm, t - mm, d - mm

* Transformed moment of inertia for calculating shrinkage deflections, based on a modular ratio $n_t = 50$

COMPOSITE BEAMS
Trial Selection Table

G40.21 350W
$f'_c = 25$ MPa

75 mm Deck with 85 mm Slab
$\phi = 0.90$, $\phi_c = 0.60$

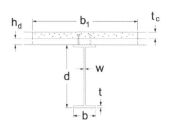

Steel section	b_1	Composite							Non-composite				
		M_{rc} (kN·m) for % shear connection			Q_r	I_t	S_t	I_t^*	Steel section data		Unbraced condition			
											L'	M_r'	L'	M_r'
	mm	100%	70%	40%	100%	10^6 mm⁴	10^3 mm³	10^6 mm⁴			mm	kN·m	mm	kN·m
W250x58	3 000	597	519	425	2 340	440	1 410	270	M_r	243	2 000	243	5 000	204
W10x39	2 000	549	489	410	2 170	400	1 360	228	V_r	419	2 500	243	6 000	182
b = 203	1 500	484	437	376	1 630	369	1 330	201	L_u	3 100	3 000	243	7 000	161
t = 13.5	1 000	416	384	339	1 080	323	1 270	170	I_x	87.3	3 500	235	8 000	137
d = 252	500	345	324	293	542	246	1 160	133	S_x	693	4 000	225	10 000	105
W250x45	3 000	485	419	342	1 800	377	1 120	243	M_r	190	2 000	190	5 000	114
W10x30	2 000	464	409	339	1 800	347	1 090	207	V_r	420	2 500	182	6 000	90.6
b = 148	1 500	430	385	326	1 630	323	1 060	183	L_u	2 150	3 000	169	7 000	75.2
t = 13	1 000	364	333	289	1 080	287	1 020	154	I_x	71.1	3 500	156	8 000	64.4
d = 266	500	295	274	243	542	223	940	118	S_x	534	4 000	143	10 000	50.1
W250x39	3 000	420	360	293	1 550	328	970	218	M_r	162	2 000	162	5 000	88.0
W10x26	2 000	404	352	290	1 550	303	942	186	V_r	360	2 500	153	6 000	69.2
b = 147	1 500	388	345	287	1 550	284	922	165	L_u	2 090	3 000	141	7 000	57.0
t = 11.2	1 000	329	298	258	1 080	255	890	139	I_x	60.1	3 500	129	8 000	48.6
d = 262	500	261	243	214	542	200	822	105	S_x	459	4 000	116	10 000	37.6
W250x33	3 000	357	305	247	1 310	280	824	191	M_r	134	2 000	134	5 000	63.6
W10x22	2 000	346	299	245	1 310	261	801	165	V_r	327	2 500	124	6 000	49.4
b = 146	1 500	335	294	243	1 310	245	784	147	L_u	2 000	3 000	113	7 000	40.3
t = 9.1	1 000	297	267	227	1 080	222	758	123	I_x	48.9	3 500	101	8 000	34.1
d = 258	500	230	213	185	542	177	704	92.3	S_x	379	4 000	88.6	10 000	26.1
W200x42	3 000	400	338	266	1 660	266	937	169	M_r	139	2 000	139	5 000	106
W8x28	2 000	382	329	263	1 660	244	906	143	V_r	307	2 500	139	6 000	92.0
b = 166	1 500	361	317	259	1 630	227	884	125	L_u	2 580	3 000	134	7 000	76.5
t = 11.8	1 000	296	265	225	1 080	201	850	104	I_x	40.6	3 500	127	8 000	65.6
d = 205	500	228	212	187	542	155	779	76.5	S_x	396	4 000	120	10 000	51.2
W200x36	3 000	346	290	227	1 430	231	810	151	M_r	118	2 000	118	5 000	85.4
W8x24	2 000	332	283	225	1 430	213	785	129	V_r	259	2 500	118	6 000	70.2
b = 165	1 500	319	277	223	1 430	199	766	113	L_u	2 490	3 000	112	7 000	58.0
t = 10.2	1 000	270	240	201	1 080	178	739	93.9	I_x	34.1	3 500	106	8 000	49.5
d = 201	500	204	188	165	542	139	681	68.6	S_x	340	4 000	99.0	10 000	38.4
W200x31	3 000	313	262	206	1 260	216	728	145	M_r	106	2 000	105	5 000	57.0
W8x21	2 000	303	257	205	1 260	200	706	124	V_r	279	2 500	97.8	6 000	45.6
b = 134	1 500	292	252	203	1 260	188	690	110	L_u	1 960	3 000	90.2	7 000	38.0
t = 10.2	1 000	259	230	190	1 080	169	665	91.5	I_x	31.4	3 500	82.3	8 000	32.7
d = 210	500	193	177	153	542	134	616	66.6	S_x	299	4 000	74.4	10 000	25.6
W200x27	3 000	266	222	174	1 070	186	620	128	M_r	87.9	2 000	86.3	5 000	41.2
W8x18	2 000	259	219	173	1 070	173	603	111	V_r	250	2 500	79.5	6 000	32.6
b = 133	1 500	252	215	172	1 070	163	589	98.4	L_u	1 880	3 000	72.2	7 000	27.1
t = 8.4	1 000	237	208	169	1 070	148	569	82.1	I_x	25.8	3 500	64.5	8 000	23.1
d = 207	500	173	157	134	542	119	529	59.5	S_x	249	4 000	56.0	10 000	18.0

Note: Resistances are based on a concrete density of 2000 kg/m³.

Units: M_r - kN·m, V_r - kN, L_u - mm, I_x - 10^6 mm⁴, S_x - 10^3 mm³, b - mm, t - mm, d - mm

* Transformed moment of inertia for calculating shrinkage deflections, based on a modular ratio $n_t = 50$

COMPOSITE BEAMS
Trial Selection Table

75 mm Deck with 90 mm Slab
$\phi = 0.90$, $\phi_c = 0.60$

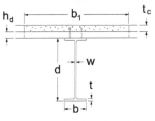

G40.21 350W
$f'_c = 20$ MPa

Steel section	b_1	Composite M_{rc} (kN·m) for % shear connection			Q_r	I_t	S_t	I_t^*	Non-composite Steel section data	Unbraced condition			
		100%	70%	40%	100%	10^6	10^3	10^6		L'	M_r'	L'	M_r'
	mm				(kN)	mm⁴	mm³	mm⁴		mm	kN·m	mm	kN·m
WWF1000x223	5 000	4 970	4 760	4 300	4 590	11 500	12 900	7 220	M_r 3 310	4 000	3 250	12 000	920
WWF39x150	4 000	4 820	4 560	4 130	3 670	10 900	12 700	6 800	V_r 2 210	5 000	2 960	14 000	729
b = 300	3 000	4 590	4 320	3 950	2 750	10 100	12 400	6 340	L_u 3 790	6 000	2 620	16 000	602
t = 25	2 000	4 250	4 020	3 750	1 840	8 990	12 000	5 820	I_x 4 590	8 000	1 790	18 000	512
d = 1000	1 000	3 820	3 680	3 530	918	7 340	11 200	5 240	S_x 9 190	10 000	1 230	20 000	446
WWF1000x200	5 000	4 550	4 340	3 880	4 590	10 400	11 500	6 510	M_r 2 890	4 000	2 790	12 000	699
WWF39x134	4 000	4 400	4 140	3 700	3 670	9 860	11 300	6 110	V_r 2 210	5 000	2 510	14 000	549
b = 300	3 000	4 160	3 890	3 520	2 750	9 160	11 100	5 660	L_u 3 610	6 000	2 180	16 000	449
t = 20	2 000	3 830	3 600	3 320	1 840	8 160	10 700	5 160	I_x 3 940	8 000	1 390	18 000	379
d = 1000	1 000	3 390	3 260	3 100	918	6 620	9 900	4 590	S_x 7 880	10 000	945	20 000	328
WWF900x231	5 000	4 690	4 540	4 250	4 590	10 400	12 900	6 660	M_r 3 400	4 000	3 400	12 000	1 760
WWF35x156	4 000	4 560	4 420	4 110	3 670	9 840	12 800	6 300	V_r 1 350	5 000	3 400	14 000	1 370
b = 400	3 000	4 420	4 250	3 960	2 750	9 150	12 500	5 900	L_u 5 460	6 000	3 300	16 000	1 110
t = 25	2 000	4 200	4 020	3 790	1 840	8 190	12 200	5 450	I_x 4 410	8 000	2 870	18 000	933
d = 900	1 000	3 850	3 730	3 590	918	6 760	11 500	4 960	S_x 9 810	10 000	2 360	20 000	801
WWF900x192	5 000	4 000	3 850	3 560	4 590	8 750	10 600	5 600	M_r 2 710	4 000	2 700	12 000	842
WWF35x128	4 000	3 870	3 730	3 420	3 670	8 330	10 500	5 270	V_r 1 350	5 000	2 490	14 000	670
b = 300	3 000	3 730	3 560	3 270	2 750	7 770	10 300	4 900	L_u 3 980	6 000	2 240	16 000	555
t = 25	2 000	3 510	3 330	3 100	1 840	6 960	9 950	4 480	I_x 3 460	8 000	1 630	18 000	473
d = 900	1 000	3 160	3 040	2 900	918	5 700	9 340	4 010	S_x 7 680	10 000	1 120	20 000	412
WWF900x169	5 000	3 600	3 460	3 170	4 590	7 790	9 250	5 000	M_r 2 320	4 000	2 290	12 000	625
WWF35x113	4 000	3 480	3 340	3 030	3 670	7 440	9 140	4 690	V_r 1 340	5 000	2 080	14 000	491
b = 300	3 000	3 340	3 170	2 880	2 750	6 950	8 970	4 340	L_u 3 820	6 000	1 850	16 000	402
t = 20	2 000	3 120	2 940	2 710	1 840	6 240	8 690	3 930	I_x 2 930	8 000	1 250	18 000	339
d = 900	1 000	2 770	2 650	2 520	918	5 090	8 140	3 470	S_x 6 510	10 000	845	20 000	293
WWF800x223	5 000	4 090	3 940	3 700	4 590	8 260	11 400	5 260	M_r 2 940	4 000	2 940	10 000	2 120
WWF31x150	4 000	3 960	3 840	3 590	3 670	7 840	11 200	4 970	V_r 1 370	5 000	2 940	12 000	1 630
b = 400	3 000	3 830	3 700	3 450	2 750	7 280	11 000	4 640	L_u 5 570	6 000	2 880	14 000	1 280
t = 25	2 000	3 650	3 500	3 300	1 840	6 510	10 700	4 280	I_x 3 410	7 000	2 710	16 000	1 050
d = 800	1 000	3 350	3 250	3 120	918	5 350	10 100	3 870	S_x 8 520	8 000	2 530	18 000	882
WWF800x184	5 000	3 480	3 330	3 090	4 590	6 940	9 260	4 420	M_r 2 330	4 000	2 330	10 000	1 040
WWF31x123	4 000	3 350	3 230	2 980	3 670	6 610	9 150	4 150	V_r 1 370	5 000	2 170	12 000	791
b = 300	3 000	3 220	3 090	2 840	2 750	6 160	8 980	3 850	L_u 4 060	6 000	1 970	14 000	634
t = 25	2 000	3 040	2 890	2 690	1 840	5 520	8 700	3 510	I_x 2 660	7 000	1 750	16 000	529
d = 800	1 000	2 740	2 640	2 510	918	4 500	8 170	3 110	S_x 6 640	8 000	1 490	18 000	454
WWF800x161	5 000	3 120	2 980	2 750	4 590	6 160	8 070	3 950	M_r 1 990	4 000	1 980	10 000	778
WWF31x108	4 000	3 000	2 890	2 630	3 670	5 880	7 970	3 700	V_r 1 370	5 000	1 810	12 000	581
b = 300	3 000	2 880	2 750	2 500	2 750	5 500	7 820	3 410	L_u 3 900	6 000	1 620	14 000	460
t = 20	2 000	2 700	2 550	2 340	1 840	4 940	7 580	3 080	I_x 2 250	7 000	1 400	16 000	379
d = 800	1 000	2 400	2 290	2 170	918	4 020	7 110	2 700	S_x 5 610	8 000	1 130	18 000	322

Note: Resistances are based on a concrete density of 2300 kg/m³.
Units: M_r - kN·m, V_r - kN, L_u - mm, I_x - 10^6 mm⁴, S_x - 10^3 mm³, b - mm, t - mm, d - mm
* Transformed moment of inertia for calculating shrinkage deflections, based on a modular ratio $n_t = 50$

COMPOSITE BEAMS
Trial Selection Table

G40.21 350W
$f'_c = 20$ MPa

75 mm Deck with 90 mm Slab
$\phi = 0.90$, $\phi_c = 0.60$

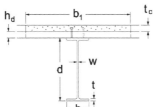

Steel section	b_1 mm	Composite M_{rc} (kN·m) for % shear connection 100%	70%	40%	Q_r (kN) 100%	I_t 10^6 mm⁴	S_t 10^3 mm³	I_t^* 10^6 mm⁴	Non-composite Steel section data	Unbraced condition L´ mm	$M_r´$ kN·m	L´ mm	$M_r´$ kN·m
WWF700x214	5 000	3 530	3 380	3 170	4 590	6 410	9 870	4 040	M_r 2 500	4 000	2 500	10 000	1 870
WWF28x144	4 000	3 400	3 280	3 080	3 670	6 080	9 750	3 800	V_r 1 370	5 000	2 500	12 000	1 500
b = 400	3 000	3 270	3 170	2 960	2 750	5 650	9 560	3 540	L_u 5 690	6 000	2 470	14 000	1 190
t = 25	2 000	3 120	3 000	2 820	1 840	5 030	9 260	3 250	I_x 2 540	7 000	2 330	16 000	982
d = 700	1 000	2 870	2 780	2 670	918	4 110	8 710	2 910	S_x 7 270	8 000	2 190	18 000	835
WWF700x175	5 000	2 990	2 850	2 640	4 590	5 360	8 000	3 390	M_r 1 970	4 000	1 970	10 000	965
WWF28x117	4 000	2 870	2 750	2 540	3 670	5 110	7 890	3 180	V_r 1 370	5 000	1 850	12 000	741
b = 300	3 000	2 740	2 640	2 430	2 750	4 760	7 750	2 930	L_u 4 160	6 000	1 700	14 000	601
t = 25	2 000	2 590	2 470	2 290	1 840	4 250	7 510	2 660	I_x 1 970	7 000	1 530	16 000	505
d = 700	1 000	2 340	2 250	2 140	918	3 450	7 030	2 340	S_x 5 640	8 000	1 350	18 000	436
WWF700x152	5 000	2 680	2 550	2 340	4 590	4 740	6 930	3 020	M_r 1 680	4 000	1 680	10 000	711
WWF28x102	4 000	2 560	2 450	2 250	3 670	4 530	6 840	2 820	V_r 1 370	5 000	1 540	12 000	537
b = 300	3 000	2 440	2 340	2 130	2 750	4 230	6 720	2 600	L_u 3 990	6 000	1 390	14 000	429
t = 20	2 000	2 290	2 170	1 990	1 840	3 800	6 510	2 330	I_x 1 660	7 000	1 230	16 000	356
d = 700	1 000	2 040	1 950	1 840	918	3 080	6 100	2 030	S_x 4 760	8 000	1 020	18 000	305
W610x241	4 000	3 430	3 280	3 000	3 670	5 440	9 720	3 270	M_r 2 420	3 000	2 420	8 000	1 950
W24x162	3 000	3 280	3 120	2 870	2 750	5 010	9 480	3 030	V_r 2 360	4 000	2 420	10 000	1 640
b = 329	2 000	3 060	2 910	2 720	1 840	4 420	9 110	2 770	L_u 4 750	5 000	2 380	12 000	1 300
t = 31	1 500	2 930	2 800	2 650	1 380	4 030	8 830	2 630	I_x 2 150	6 000	2 240	14 000	1 070
d = 635	1 000	2 770	2 680	2 570	918	3 550	8 440	2 480	S_x 6 780	7 000	2 100	16 000	911
W610x217	4 000	3 130	2 990	2 730	3 670	4 970	8 790	2 990	M_r 2 160	3 000	2 160	8 000	1 690
W24x146	3 000	2 990	2 840	2 600	2 750	4 590	8 590	2 770	V_r 2 150	4 000	2 160	10 000	1 370
b = 328	2 000	2 790	2 650	2 460	1 840	4 050	8 260	2 510	L_u 4 640	5 000	2 110	12 000	1 070
t = 27.7	1 500	2 660	2 540	2 390	1 380	3 700	8 020	2 380	I_x 1 910	6 000	1 980	14 000	878
d = 628	1 000	2 510	2 420	2 310	918	3 250	7 660	2 230	S_x 6 070	7 000	1 840	16 000	745
W610x195	4 000	2 840	2 710	2 470	3 670	4 510	7 880	2 720	M_r 1 910	3 000	1 910	8 000	1 450
W24x131	3 000	2 710	2 570	2 350	2 750	4 170	7 710	2 510	V_r 1 990	4 000	1 910	10 000	1 120
b = 327	2 000	2 520	2 390	2 210	1 840	3 700	7 430	2 270	L_u 4 530	5 000	1 860	12 000	870
t = 24.4	1 500	2 400	2 280	2 140	1 380	3 380	7 210	2 130	I_x 1 680	6 000	1 730	14 000	709
d = 622	1 000	2 260	2 170	2 060	918	2 970	6 900	1 990	S_x 5 400	7 000	1 590	16 000	598
W610x174	4 000	2 570	2 450	2 230	3 670	4 070	7 030	2 470	M_r 1 690	3 000	1 690	8 000	1 240
W24x117	3 000	2 440	2 330	2 110	2 750	3 780	6 880	2 270	V_r 1 790	4 000	1 690	10 000	924
b = 325	2 000	2 280	2 160	1 980	1 840	3 360	6 650	2 040	L_u 4 440	5 000	1 630	12 000	709
t = 21.6	1 500	2 160	2 050	1 910	1 380	3 070	6 460	1 910	I_x 1 470	6 000	1 510	14 000	574
d = 616	1 000	2 030	1 940	1 840	918	2 700	6 180	1 780	S_x 4 780	7 000	1 380	16 000	482
W610x155	4 000	2 320	2 210	2 010	3 670	3 670	6 260	2 250	M_r 1 490	3 000	1 490	8 000	1 060
W24x104	3 000	2 200	2 100	1 900	2 750	3 420	6 140	2 060	V_r 1 610	4 000	1 490	10 000	762
b = 324	2 000	2 050	1 940	1 780	1 840	3 060	5 940	1 840	L_u 4 370	5 000	1 430	12 000	579
t = 19	1 500	1 950	1 850	1 710	1 380	2 800	5 780	1 720	I_x 1 290	6 000	1 310	14 000	465
d = 611	1 000	1 820	1 740	1 640	918	2 460	5 540	1 590	S_x 4 220	7 000	1 190	16 000	388

Note: Resistances are based on a concrete density of 2300 kg/m³.

Units: M_r - kN·m, V_r - kN, L_u - mm, I_x - 10^6 mm⁴, S_x - 10^3 mm³, b - mm, t - mm, d - mm

* Transformed moment of inertia for calculating shrinkage deflections, based on a modular ratio $n_t = 50$

COMPOSITE BEAMS
Trial Selection Table

75 mm Deck with 90 mm Slab
$\phi = 0.90$, $\phi_c = 0.60$

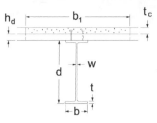

G40.21 350W
$f'_c = 20$ MPa

Steel section	b_1	M_{rc} (kN·m) for % shear connection			Q_r	I_t 10^6	S_t 10^3	I_t^* 10^6	Steel section data	Unbraced condition			
										L´	M_r´	L´	M_r´
	mm	100%	70%	40%	100%	mm⁴	mm³	mm⁴		mm	kN·m	mm	kN·m
W610x140	4 000	2 170	2 050	1 840	3 670	3 380	5 640	2 070	M_r 1 310	3 000	1 310	8 000	573
W24x94	3 000	2 040	1 940	1 730	2 750	3 160	5 530	1 890	V_r 1 680	4 000	1 180	10 000	422
b = 230	2 000	1 890	1 770	1 600	1 840	2 830	5 340	1 670	L_u 3 050	5 000	1 040	12 000	334
t = 22.2	1 500	1 780	1 670	1 530	1 380	2 590	5 190	1 550	I_x 1 120	6 000	878	14 000	277
d = 617	1 000	1 640	1 560	1 460	918	2 270	4 960	1 420	S_x 3 630	7 000	695	16 000	237
W610x125	4 000	1 970	1 860	1 670	3 670	3 050	5 020	1 890	M_r 1 160	3 000	1 160	8 000	470
W24x84	3 000	1 850	1 750	1 570	2 750	2 860	4 930	1 720	V_r 1 510	4 000	1 040	10 000	342
b = 229	2 000	1 710	1 610	1 440	1 840	2 570	4 780	1 520	L_u 2 990	5 000	895	12 000	269
t = 19.6	1 500	1 610	1 510	1 380	1 380	2 360	4 650	1 410	I_x 985	6 000	733	14 000	222
d = 612	1 000	1 480	1 400	1 300	918	2 070	4 450	1 280	S_x 3 220	7 000	575	16 000	189
W610x113	4 000	1 820	1 710	1 540	3 670	2 780	4 540	1 740	M_r 1 040	3 000	1 030	8 000	391
W24x76	3 000	1 700	1 610	1 440	2 750	2 620	4 460	1 580	V_r 1 420	4 000	915	10 000	282
b = 228	2 000	1 570	1 470	1 320	1 840	2 360	4 330	1 400	L_u 2 930	5 000	781	12 000	220
t = 17.3	1 500	1 480	1 380	1 250	1 380	2 180	4 220	1 290	I_x 875	6 000	617	14 000	180
d = 608	1 000	1 360	1 280	1 180	918	1 910	4 050	1 160	S_x 2 880	7 000	481	16 000	153
W610x101	4 000	1 670	1 570	1 410	3 670	2 510	4 080	1 590	M_r 914	3 000	901	8 000	320
W24x68	3 000	1 560	1 480	1 310	2 750	2 370	4 010	1 450	V_r 1 310	4 000	795	10 000	228
b = 228	2 000	1 440	1 350	1 200	1 840	2 150	3 890	1 270	L_u 2 870	5 000	668	12 000	176
t = 14.9	1 500	1 350	1 260	1 130	1 380	1 980	3 800	1 160	I_x 764	6 000	512	14 000	144
d = 603	1 000	1 230	1 150	1 060	918	1 750	3 640	1 050	S_x 2 530	7 000	396	16 000	121
W610x91	4 000	1 530	1 430	1 280	3 650	2 250	3 630	1 450	M_r 806	3 000	788	8 000	261
W24x61	3 000	1 420	1 340	1 190	2 750	2 140	3 580	1 320	V_r 1 110	4 000	687	10 000	185
b = 227	2 000	1 300	1 220	1 080	1 840	1 950	3 480	1 150	L_u 2 790	5 000	567	12 000	142
t = 12.7	1 500	1 220	1 140	1 010	1 380	1 800	3 400	1 050	I_x 667	6 000	421	14 000	115
d = 598	1 000	1 110	1 030	939	918	1 590	3 260	942	S_x 2 230	7 000	325	16 000	97.1
W530x138	4 000	1 950	1 830	1 640	3 670	2 760	5 070	1 660	M_r 1 140	3 000	1 130	8 000	515
W21x93	3 000	1 820	1 720	1 530	2 750	2 580	4 970	1 510	V_r 1 680	4 000	1 010	10 000	390
b = 214	2 000	1 680	1 570	1 410	1 840	2 300	4 790	1 330	L_u 2 910	5 000	891	12 000	314
t = 23.6	1 500	1 570	1 470	1 350	1 380	2 100	4 650	1 230	I_x 861	6 000	762	14 000	263
d = 549	1 000	1 450	1 370	1 280	918	1 830	4 430	1 120	S_x 3 140	7 000	616	16 000	227
W530x123	4 000	1 780	1 670	1 490	3 670	2 500	4 540	1 520	M_r 1 010	3 000	996	8 000	421
W21x83	3 000	1 660	1 570	1 390	2 750	2 340	4 450	1 380	V_r 1 480	4 000	888	10 000	316
b = 212	2 000	1 520	1 430	1 280	1 840	2 100	4 310	1 210	L_u 2 840	5 000	768	12 000	253
t = 21.2	1 500	1 430	1 340	1 210	1 380	1 920	4 190	1 120	I_x 761	6 000	631	14 000	211
d = 544	1 000	1 310	1 240	1 150	918	1 680	4 000	1 010	S_x 2 800	7 000	505	16 000	182
W530x109	4 000	1 620	1 510	1 350	3 670	2 240	4 030	1 390	M_r 891	3 000	872	8 000	342
W21x73	3 000	1 500	1 410	1 260	2 750	2 110	3 960	1 260	V_r 1 300	4 000	771	10 000	254
b = 211	2 000	1 380	1 290	1 150	1 840	1 900	3 840	1 100	L_u 2 790	5 000	656	12 000	202
t = 18.8	1 500	1 300	1 210	1 090	1 380	1 750	3 740	1 010	I_x 667	6 000	520	14 000	168
d = 539	1 000	1 190	1 110	1 030	918	1 530	3 580	910	S_x 2 480	7 000	413	16 000	144

Note: Resistances are based on a concrete density of 2300 kg/m³.

Units: M_r - kN·m, V_r - kN, L_u - mm, I_x - 10^6 mm⁴, S_x - 10^3 mm³, b - mm, t - mm, d - mm

* Transformed moment of inertia for calculating shrinkage deflections, based on a modular ratio $n_t = 50$

COMPOSITE BEAMS
Trial Selection Table

G40.21 350W
$f'_c = 20$ MPa

75 mm Deck with 90 mm Slab
$\phi = 0.90$, $\phi_c = 0.60$

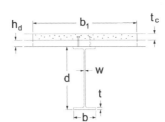

Steel section	b_1	Composite M_{rc} (kN·m) for % shear connection			Q_r	I_t	S_t	I_t^*	Non-composite Steel section data	Unbraced condition			
		100%	70%	40%	100%	10^6	10^3	10^6		L'	M_r'	L'	M_r'
	mm				(kN)	mm⁴	mm³	mm⁴		mm	kN·m	mm	kN·m
W530x101	4 000	1 530	1 430	1 280	3 670	2 100	3 750	1 320	M_r 825	3 000	804	8 000	301
W21x68	3 000	1 420	1 330	1 190	2 750	1 980	3 690	1 200	V_r 1 220	4 000	706	10 000	222
b = 210	2 000	1 290	1 220	1 080	1 840	1 790	3 580	1 040	L_u 2 750	5 000	594	12 000	176
t = 17.4	1 500	1 220	1 140	1 020	1 380	1 650	3 490	956	I_x 617	6 000	462	14 000	146
d = 537	1 000	1 110	1 050	959	918	1 450	3 350	857	S_x 2 300	7 000	365	16 000	125
W530x92	4 000	1 430	1 330	1 180	3 670	1 920	3 430	1 220	M_r 743	3 000	719	8 000	253
W21x62	3 000	1 320	1 240	1 100	2 750	1 820	3 370	1 110	V_r 1 130	4 000	626	10 000	185
b = 209	2 000	1 200	1 130	1 000	1 840	1 650	3 270	966	L_u 2 700	5 000	519	12 000	146
t = 15.6	1 500	1 130	1 050	942	1 380	1 530	3 190	882	I_x 552	6 000	393	14 000	120
d = 533	1 000	1 030	963	878	918	1 350	3 070	787	S_x 2 070	7 000	309	16 000	103
W530x82	4 000	1 280	1 190	1 050	3 310	1 720	3 040	1 110	M_r 652	3 000	625	8 000	204
W21x55	3 000	1 200	1 120	1 000	2 750	1 630	2 990	1 010	V_r 1 040	4 000	538	10 000	148
b = 209	2 000	1 090	1 020	902	1 840	1 490	2 910	876	L_u 2 640	5 000	436	12 000	115
t = 13.3	1 500	1 020	953	845	1 380	1 380	2 840	797	I_x 478	6 000	321	14 000	94.8
d = 528	1 000	929	865	782	918	1 220	2 730	706	S_x 1 810	7 000	250	16 000	80.5
W460x106	4 000	1 440	1 330	1 180	3 670	1 770	3 560	1 080	M_r 753	3 000	727	7 000	366
W18x71	3 000	1 320	1 230	1 090	2 750	1 660	3 490	976	V_r 1 230	3 500	686	8 000	308
b = 194	2 000	1 190	1 120	996	1 840	1 490	3 380	848	L_u 2 670	4 000	643	10 000	235
t = 20.6	1 500	1 120	1 050	940	1 380	1 370	3 290	773	I_x 488	5 000	553	12 000	190
d = 469	1 000	1 020	959	879	918	1 200	3 140	689	S_x 2 080	6 000	450	14 000	160
W460x97	4 000	1 340	1 240	1 090	3 670	1 630	3 260	1 020	M_r 687	3 000	660	7 000	314
W18x65	3 000	1 230	1 140	1 020	2 750	1 540	3 200	916	V_r 1 100	3 500	621	8 000	264
b = 193	2 000	1 110	1 040	925	1 840	1 390	3 100	794	L_u 2 630	4 000	579	10 000	200
t = 19	1 500	1 040	973	871	1 380	1 280	3 020	723	I_x 445	5 000	491	12 000	161
d = 466	1 000	950	890	812	918	1 120	2 900	642	S_x 1 910	6 000	389	14 000	135
W460x89	4 000	1 270	1 160	1 020	3 590	1 520	3 030	959	M_r 633	3 000	605	7 000	276
W18x60	3 000	1 160	1 080	959	2 750	1 440	2 970	865	V_r 1 010	3 500	568	8 000	231
b = 192	2 000	1 040	979	869	1 840	1 300	2 890	749	L_u 2 600	4 000	527	10 000	174
t = 17.7	1 500	976	915	816	1 380	1 200	2 820	681	I_x 409	5 000	442	12 000	140
d = 463	1 000	892	835	759	918	1 060	2 700	602	S_x 1 770	6 000	343	14 000	117
W460x82	4 000	1 160	1 060	932	3 280	1 400	2 770	896	M_r 576	3 000	546	7 000	234
W18x55	3 000	1 080	1 000	891	2 750	1 320	2 720	809	V_r 947	3 500	510	8 000	195
b = 191	2 000	965	908	806	1 840	1 210	2 640	700	L_u 2 540	4 000	471	10 000	146
t = 16	1 500	904	849	755	1 380	1 120	2 580	634	I_x 370	5 000	386	12 000	116
d = 460	1 000	826	772	699	918	986	2 480	559	S_x 1 610	6 000	292	14 000	97.3
W460x74	4 000	1 060	967	846	2 980	1 280	2 520	833	M_r 520	3 000	489	7 000	198
W18x50	3 000	1 010	933	826	2 750	1 210	2 480	753	V_r 855	3 500	455	8 000	164
b = 190	2 000	895	840	746	1 840	1 110	2 410	651	L_u 2 510	4 000	417	10 000	122
t = 14.5	1 500	835	785	697	1 380	1 030	2 360	589	I_x 332	5 000	332	12 000	96.8
d = 457	1 000	764	713	642	918	915	2 270	517	S_x 1 460	6 000	249	14 000	80.6

Note: Resistances are based on a concrete density of 2300 kg/m³.

Units: M_r - kN·m, V_r - kN, L_u - mm, I_x - 10^6 mm⁴, S_x - 10^3 mm³, b - mm, t - mm, d - mm

* Transformed moment of inertia for calculating shrinkage deflections, based on a modular ratio $n_t = 50$

COMPOSITE BEAMS
Trial Selection Table

75 mm Deck with 90 mm Slab
$\phi = 0.90$, $\phi_c = 0.60$

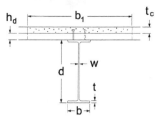

G40.21 350W
$f'_c = 20$ MPa

Steel section	b_1	Composite							Non-composite				
		M_{rc} (kN·m) for % shear connection			Q_r	I_t	S_t	I_t^*	Steel section data	Unbraced condition			
										L'	M_r'	L'	M_r'
	mm	100%	70%	40%	(kN)	10^6 mm⁴	10^3 mm³	10^6 mm⁴		mm	kN·m	mm	kN·m
W460x67	4 000	969	878	765	2 700	1 160	2 280	772	M_r 466	3 000	435	7 000	165
W18x45	3 000	939	864	760	2 700	1 110	2 250	699	V_r 802	3 500	402	8 000	136
b = 190	2 000	830	776	687	1 840	1 020	2 190	604	L_u 2 460	4 000	366	10 000	99.8
t = 12.7	1 500	771	725	639	1 380	951	2 140	545	I_x 296	5 000	281	12 000	79.0
d = 454	1 000	704	655	585	918	846	2 060	476	S_x 1 300	6 000	209	14 000	65.4
W460x61	4 000	870	784	679	2 410	1 040	2 030	702	M_r 406	3 000	374	7 000	132
W18x41	3 000	846	773	676	2 410	994	2 000	636	V_r 758	3 500	343	8 000	108
b = 189	2 000	762	709	625	1 840	919	1 950	549	L_u 2 390	4 000	309	10 000	78.5
t = 10.8	1 500	704	660	578	1 380	859	1 910	494	I_x 255	5 000	228	12 000	61.7
d = 450	1 000	640	594	525	918	767	1 840	430	S_x 1 130	6 000	168	14 000	50.8
W410x85	4 000	1 130	1 030	893	3 400	1 250	2 690	786	M_r 542	3 000	512	7 000	242
W16x57	3 000	1 040	957	845	2 750	1 180	2 640	707	V_r 945	3 500	480	8 000	205
b = 181	2 000	919	861	762	1 840	1 070	2 560	609	L_u 2 510	4 000	447	10 000	157
t = 18.2	1 500	857	803	713	1 380	989	2 500	550	I_x 315	5 000	377	12 000	127
d = 417	1 000	781	730	660	918	868	2 390	483	S_x 1 510	6 000	297	14 000	107
W410x74	4 000	1 010	910	787	3 010	1 110	2 380	717	M_r 476	3 000	445	7 000	194
W16x50	3 000	951	873	766	2 750	1 060	2 340	646	V_r 833	3 500	414	8 000	163
b = 180	2 000	835	779	690	1 840	966	2 280	556	L_u 2 450	4 000	382	10 000	123
t = 16	1 500	774	727	643	1 380	895	2 220	501	I_x 275	5 000	312	12 000	99.7
d = 413	1 000	706	658	591	918	790	2 140	438	S_x 1 330	6 000	239	14 000	83.8
W410x67	4 000	912	821	708	2 710	1 010	2 150	664	M_r 428	3 000	397	7 000	161
W16x45	3 000	882	807	703	2 710	963	2 120	599	V_r 750	3 500	367	8 000	135
b = 179	2 000	772	718	635	1 840	884	2 060	515	L_u 2 400	4 000	336	10 000	102
t = 14.4	1 500	713	669	591	1 380	822	2 010	464	I_x 245	5 000	264	12 000	81.6
d = 410	1 000	649	605	541	918	729	1 940	403	S_x 1 200	6 000	201	14 000	68.3
W410x60	4 000	810	725	623	2 390	904	1 910	607	M_r 375	3 000	344	7 000	131
W16x40	3 000	787	714	619	2 390	863	1 880	550	V_r 652	3 500	317	8 000	109
b = 178	2 000	705	652	577	1 840	797	1 830	474	L_u 2 370	4 000	288	10 000	81.3
t = 12.8	1 500	647	606	536	1 380	744	1 790	426	I_x 216	5 000	218	12 000	64.9
d = 407	1 000	586	549	488	918	665	1 730	369	S_x 1 060	6 000	165	14 000	54.1
W410x54	4 000	729	650	556	2 140	808	1 700	552	M_r 331	3 000	298	7 000	104
W16x36	3 000	710	640	553	2 140	773	1 680	501	V_r 628	3 500	272	8 000	86.0
b = 177	2 000	652	600	527	1 840	717	1 640	432	L_u 2 290	4 000	243	10 000	63.5
t = 10.9	1 500	594	554	487	1 380	672	1 600	388	I_x 186	5 000	176	12 000	50.4
d = 403	1 000	535	499	440	918	603	1 550	334	S_x 923	6 000	132	14 000	41.8
W410x46	4 000	638	566	481	1 860	709	1 480	496	M_r 278	3 000	212	7 000	61.7
W16x31	3 000	624	559	479	1 860	680	1 450	451	V_r 585	3 500	177	8 000	51.8
b = 140	2 000	594	544	473	1 840	634	1 420	389	L_u 1 780	4 000	142	10 000	39.2
t = 11.2	1 500	538	498	434	1 380	597	1 390	349	I_x 156	5 000	99.9	12 000	31.6
d = 403	1 000	479	446	388	918	539	1 350	300	S_x 772	6 000	76.4	14 000	26.5

Note: Resistances are based on a concrete density of 2300 kg/m³.

Units: M_r - kN·m, V_r - kN, L_u - mm, I_x - 10^6 mm⁴, S_x - 10^3 mm³, b - mm, t - mm, d - mm

* Transformed moment of inertia for calculating shrinkage deflections, based on a modular ratio $n_t = 50$

COMPOSITE BEAMS
Trial Selection Table

G40.21 350W
$f'_c = 20$ MPa

75 mm Deck with 90 mm Slab
$\phi = 0.90$, $\phi_c = 0.60$

Steel section	b_1	\multicolumn{4}{c}{Composite}						Non-composite				
		M_{rc} (kN·m) for % shear connection			Q_r	I_t	S_t	I_t^*	Steel section data		Unbraced condition			
		100%	70%	40%	(kN)	10^6	10^3	10^6			L'	M_r'	L'	M_r'
	mm					mm⁴	mm³	mm⁴			mm	kN·m	mm	kN·m
W410x39	4 000	543	479	404	1 570	602	1 250	432	M_r	230	3 000	167	7 000	44.0
W16x26	3 000	533	474	403	1 570	579	1 230	394	V_r	484	3 500	132	8 000	36.5
b = 140	2 000	512	464	399	1 570	543	1 200	342	L_u	1 720	4 000	105	10 000	27.3
t = 8.8	1 500	479	440	382	1 380	513	1 180	307	I_x	126	5 000	73.0	12 000	21.8
d = 399	1 000	421	392	338	918	467	1 140	262	S_x	634	6 000	55.1	14 000	18.2
W360x79	3 000	893	814	708	2 750	904	2 280	540	M_r	450	2 000	450	6 000	319
W14x53	2 000	776	721	642	1 840	822	2 210	463	V_r	692	2 500	450	7 000	267
b = 205	1 500	716	673	600	1 380	758	2 160	416	L_u	2 980	3 000	450	8 000	225
t = 16.8	1 000	653	613	553	918	665	2 070	362	I_x	226	4 000	409	10 000	171
d = 354	500	565	536	500	459	515	1 890	300	S_x	1 280	5 000	364	12 000	139
W360x72	3 000	832	756	651	2 750	821	2 060	500	M_r	403	2 000	403	6 000	273
W14x48	2 000	718	663	589	1 840	750	2 000	428	V_r	626	2 500	403	7 000	222
b = 204	1 500	658	616	549	1 380	695	1 960	384	L_u	2 910	3 000	400	8 000	186
t = 15.1	1 000	597	561	504	918	613	1 880	333	I_x	201	4 000	360	10 000	141
d = 350	500	515	487	452	459	477	1 720	273	S_x	1 150	5 000	317	12 000	114
W360x64	3 000	761	686	588	2 560	743	1 850	461	M_r	359	2 000	359	6 000	228
W14x43	2 000	663	610	539	1 840	682	1 800	395	V_r	555	2 500	359	7 000	183
b = 203	1 500	605	564	502	1 380	635	1 760	354	L_u	2 850	3 000	354	8 000	153
t = 13.5	1 000	544	512	458	918	563	1 700	306	I_x	178	4 000	316	10 000	115
d = 347	500	468	442	407	459	440	1 560	248	S_x	1 030	5 000	275	12 000	92.2
W360x57	3 000	697	625	534	2 270	694	1 660	440	M_r	318	2 000	318	6 000	147
W14x38	2 000	626	574	502	1 840	641	1 620	377	V_r	588	2 500	312	7 000	119
b = 172	1 500	568	527	465	1 380	598	1 580	337	L_u	2 340	3 000	292	8 000	99.7
t = 13.1	1 000	508	476	421	918	534	1 530	290	I_x	160	4 000	246	10 000	75.6
d = 358	500	432	405	371	459	420	1 410	232	S_x	896	5 000	192	12 000	61.0
W360x51	3 000	628	560	476	2 030	626	1 490	405	M_r	281	2 000	281	6 000	121
W14x34	2 000	580	529	460	1 840	580	1 450	348	V_r	531	2 500	274	7 000	96.9
b = 171	1 500	523	483	425	1 380	544	1 420	311	L_u	2 300	3 000	255	8 000	80.9
t = 11.6	1 000	465	435	383	918	488	1 370	267	I_x	141	4 000	211	10 000	60.8
d = 355	500	393	367	334	459	387	1 270	211	S_x	796	5 000	159	12 000	48.8
W360x45	3 000	562	499	423	1 800	560	1 320	369	M_r	245	2 000	245	6 000	96.0
W14x30	2 000	536	486	418	1 800	521	1 290	318	V_r	505	2 500	237	7 000	76.4
b = 171	1 500	482	443	387	1 380	490	1 270	285	L_u	2 240	3 000	219	8 000	63.3
t = 9.8	1 000	424	396	346	918	442	1 220	243	I_x	122	4 000	177	10 000	47.1
d = 352	500	355	330	297	459	353	1 140	191	S_x	691	5 000	128	12 000	37.5
W360x39	3 000	496	437	368	1 570	494	1 150	333	M_r	209	2 000	195	6 000	54.1
W14x26	2 000	475	427	365	1 570	462	1 120	288	V_r	477	2 500	173	7 000	44.2
b = 128	1 500	442	403	349	1 380	436	1 100	258	L_u	1 650	3 000	149	8 000	37.4
t = 10.7	1 000	384	357	309	918	396	1 070	219	I_x	102	4 000	97.0	10 000	28.7
d = 353	500	317	293	260	459	319	993	170	S_x	580	5 000	69.7	12 000	23.3

Note: Resistances are based on a concrete density of 2300 kg/m³.
Units: M_r - kN·m, V_r - kN, L_u - mm, I_x - 10^6 mm⁴, S_x - 10^3 mm³, b - mm, t - mm, d - mm
* Transformed moment of inertia for calculating shrinkage deflections, based on a modular ratio $n_t = 50$

COMPOSITE BEAMS
Trial Selection Table

75 mm Deck with 90 mm Slab
$\phi = 0.90$, $\phi_c = 0.60$

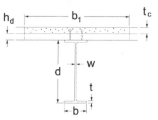

G40.21 350W
$f'_c = 20$ MPa

Steel section	b_1	Composite M_{rc} (kN·m) for % shear connection			Q_r (kN)	I_t 10^6	S_t 10^3	I_t^* 10^6	Non-composite Steel section data		Unbraced condition			
		100%	70%	40%	100%						L´	M_r'	L´	M_r'
	mm					mm⁴	mm³	mm⁴			mm	kN·m	mm	kN·m
W360x33	3 000	418	366	306	1 310	418	967	290	M_r	170	2 000	156	6 000	38.0
W14x22	2 000	404	359	304	1 310	392	943	253	V_r	399	2 500	136	7 000	30.8
b = 127	1 500	390	352	301	1 310	372	925	227	L_u	1 590	3 000	114	8 000	25.8
t = 8.5	1 000	338	312	268	918	341	898	193	I_x	82.6	4 000	70.2	10 000	19.6
d = 349	500	276	253	222	459	279	841	147	S_x	473	5 000	49.6	12 000	15.8
W310x74	3 000	793	716	611	2 750	723	2 000	429	M_r	375	2 000	375	6 000	279
W12x50	2 000	678	623	549	1 840	657	1 940	365	V_r	606	2 500	375	7 000	243
b = 205	1 500	618	576	512	1 380	606	1 890	326	L_u	3 070	3 000	375	8 000	206
t = 16.3	1 000	556	523	469	918	531	1 810	281	I_x	165	4 000	345	10 000	159
d = 310	500	479	453	420	459	408	1 650	228	S_x	1 060	5 000	312	12 000	129
W310x67	3 000	734	659	558	2 680	653	1 800	395	M_r	334	2 000	334	6 000	238
W12x45	2 000	627	574	503	1 840	596	1 740	336	V_r	541	2 500	334	7 000	200
b = 204	1 500	568	527	467	1 380	552	1 700	300	L_u	2 990	3 000	334	8 000	169
t = 14.6	1 000	508	477	426	918	487	1 640	257	I_x	145	4 000	303	10 000	129
d = 306	500	435	410	378	459	376	1 500	206	S_x	948	5 000	271	12 000	105
W310x60	3 000	663	591	497	2 390	591	1 610	365	M_r	296	2 000	296	6 000	203
W12x40	2 000	581	529	459	1 840	542	1 570	311	V_r	472	2 500	296	7 000	166
b = 203	1 500	523	483	427	1 380	504	1 530	278	L_u	2 930	3 000	295	8 000	139
t = 13.1	1 000	464	436	388	918	447	1 480	237	I_x	129	4 000	266	10 000	106
d = 303	500	396	373	341	459	348	1 360	189	S_x	849	5 000	235	12 000	85.2
W310x52	3 000	609	540	454	2 100	559	1 450	355	M_r	265	2 000	265	6 000	130
W12x35	2 000	554	502	433	1 840	517	1 410	303	V_r	502	2 500	260	7 000	106
b = 167	1 500	497	457	401	1 380	483	1 380	270	L_u	2 350	3 000	244	8 000	89.4
t = 13.2	1 000	438	409	362	918	431	1 330	231	I_x	119	4 000	208	10 000	68.4
d = 318	500	370	347	316	459	339	1 230	181	S_x	750	5 000	167	12 000	55.5
W310x45	3 000	524	460	385	1 790	481	1 240	315	M_r	223	2 000	223	6 000	98.2
W12x30	2 000	498	448	381	1 790	447	1 210	270	V_r	429	2 500	217	7 000	79.3
b = 166	1 500	445	406	354	1 380	420	1 190	241	L_u	2 290	3 000	202	8 000	66.5
t = 11.2	1 000	387	360	317	918	378	1 150	205	I_x	99.2	4 000	168	10 000	50.4
d = 313	500	324	303	273	459	301	1 060	159	S_x	634	5 000	128	12 000	40.6
W310x39	3 000	458	400	333	1 560	423	1 090	284	M_r	192	2 000	192	6 000	77.7
W12x26	2 000	439	391	330	1 560	395	1 060	245	V_r	374	2 500	186	7 000	62.2
b = 165	1 500	406	368	317	1 380	373	1 040	219	L_u	2 250	3 000	172	8 000	51.8
t = 9.7	1 000	349	323	284	918	338	1 000	186	I_x	85.1	4 000	140	10 000	38.8
d = 310	500	289	270	242	459	272	937	143	S_x	549	5 000	103	12 000	31.1
W250x67	3 000	672	597	495	2 690	530	1 660	312	M_r	284	2 000	284	5 000	246
W10x45	2 000	564	510	439	1 840	482	1 600	263	V_r	476	2 500	284	6 000	225
b = 204	1 500	505	463	408	1 380	444	1 560	233	L_u	3 230	3 000	284	7 000	203
t = 15.7	1 000	444	415	371	918	389	1 500	197	I_x	104	3 500	278	8 000	180
d = 257	500	378	356	328	459	296	1 360	155	S_x	806	4 000	268	10 000	139

Note: Resistances are based on a concrete density of 2300 kg/m³.
Units: M_r - kN·m, V_r - kN, L_u - mm, I_x - 10^6 mm⁴, S_x - 10^3 mm³, b - mm, t - mm, d - mm
* Transformed moment of inertia for calculating shrinkage deflections, based on a modular ratio $n_t = 50$

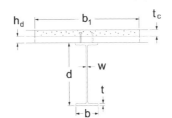

COMPOSITE BEAMS
Trial Selection Table

G40.21 350W
$f'_c = 20$ MPa

75 mm Deck with 90 mm Slab
$\phi = 0.90$, $\phi_c = 0.60$

Steel section	b_1	Composite M_{rc} (kN·m) for % shear connection			Q_r	I_t	S_t	I_t^*	Non-composite Steel section data		Unbraced condition			
		100%	70%	40%	100%	10^6	10^3	10^6			L'	M_r'	L'	M_r'
	mm				(kN)	mm⁴	mm³	mm⁴			mm	kN·m	mm	kN·m
W250x58	3 000	591	519	427	2 340	463	1 440	280	M_r	243	2 000	243	5 000	204
W10x39	2 000	514	462	392	1 840	424	1 400	236	V_r	419	2 500	243	6 000	182
b = 203	1 500	456	416	363	1 380	393	1 360	209	L_u	3 100	3 000	243	7 000	161
t = 13.5	1 000	397	369	328	918	346	1 310	176	I_x	87.3	3 500	235	8 000	137
d = 252	500	334	314	286	459	266	1 200	136	S_x	693	4 000	225	10 000	105
W250x45	3 000	484	420	344	1 800	396	1 150	253	M_r	190	2 000	190	5 000	114
W10x30	2 000	457	407	340	1 800	366	1 120	215	V_r	420	2 500	182	6 000	90.6
b = 148	1 500	404	365	312	1 380	342	1 090	190	L_u	2 150	3 000	169	7 000	75.2
t = 13	1 000	346	318	278	918	306	1 050	160	I_x	71.1	3 500	156	8 000	64.4
d = 266	500	283	264	236	459	241	971	121	S_x	534	4 000	143	10 000	50.1
W250x39	3 000	419	362	294	1 550	344	995	226	M_r	162	2 000	162	5 000	88.0
W10x26	2 000	400	352	291	1 550	319	966	194	V_r	360	2 500	153	6 000	69.2
b = 147	1 500	368	330	279	1 380	300	945	172	L_u	2 090	3 000	141	7 000	57.0
t = 11.2	1 000	311	284	247	918	271	914	144	I_x	60.1	3 500	129	8 000	48.6
d = 262	500	251	234	208	459	216	848	109	S_x	459	4 000	116	10 000	37.6
W250x33	3 000	358	307	248	1 310	294	845	198	M_r	134	2 000	134	5 000	63.6
W10x22	2 000	344	300	246	1 310	274	822	171	V_r	327	2 500	124	6 000	49.4
b = 146	1 500	330	293	244	1 310	259	804	152	L_u	2 000	3 000	113	7 000	40.3
t = 9.1	1 000	279	253	217	918	235	778	128	I_x	48.9	3 500	101	8 000	34.1
d = 258	500	221	204	179	459	190	726	95.4	S_x	379	4 000	88.6	10 000	26.1
W200x42	3 000	400	339	268	1 660	281	964	176	M_r	139	2 000	139	5 000	106
W8x28	2 000	377	328	265	1 660	258	933	149	V_r	307	2 500	139	6 000	92.0
b = 166	1 500	335	297	246	1 380	241	910	131	L_u	2 580	3 000	134	7 000	76.5
t = 11.8	1 000	278	251	216	918	215	876	108	I_x	40.6	3 500	127	8 000	65.6
d = 205	500	219	204	181	459	168	807	79.3	S_x	396	4 000	120	10 000	51.2
W200x36	3 000	346	292	229	1 430	244	834	158	M_r	118	2 000	118	5 000	85.4
W8x24	2 000	330	283	227	1 430	225	809	134	V_r	259	2 500	118	6 000	70.2
b = 165	1 500	309	271	221	1 380	211	789	118	L_u	2 490	3 000	112	7 000	58.0
t = 10.2	1 000	253	227	192	918	190	761	98.0	I_x	34.1	3 500	106	8 000	49.5
d = 201	500	194	180	159	459	150	705	71.2	S_x	340	4 000	99.0	10 000	38.4
W200x31	3 000	314	264	208	1 260	228	749	151	M_r	106	2 000	105	5 000	57.0
W8x21	2 000	301	258	206	1 260	212	727	130	V_r	279	2 500	97.8	6 000	45.6
b = 134	1 500	288	252	204	1 260	199	710	115	L_u	1 960	3 000	90.2	7 000	38.0
t = 10.2	1 000	242	216	181	918	181	685	95.5	I_x	31.4	3 500	82.3	8 000	32.7
d = 210	500	184	169	147	459	145	637	69.2	S_x	299	4 000	74.4	10 000	25.6
W200x27	3 000	268	224	176	1 070	195	638	133	M_r	87.9	2 000	86.3	5 000	41.2
W8x18	2 000	259	220	174	1 070	182	620	115	V_r	250	2 500	79.5	6 000	32.6
b = 133	1 500	249	215	173	1 070	172	606	103	L_u	1 880	3 000	72.2	7 000	27.1
t = 8.4	1 000	221	195	162	918	157	586	85.7	I_x	25.8	3 500	64.5	8 000	23.1
d = 207	500	163	150	129	459	128	547	61.9	S_x	249	4 000	56.0	10 000	18.0

Note: Resistances are based on a concrete density of 2300 kg/m³.
Units: M_r - kN·m, V_r - kN, L_u - mm, I_x - 10^6 mm⁴, S_x - 10^3 mm³, b - mm, t - mm, d - mm
* Transformed moment of inertia for calculating shrinkage deflections, based on a modular ratio $n_t = 50$

DEFLECTION OF FLEXURAL MEMBERS

CAN/CSA-S16.1-94, *Limit States Design of Steel Structures*, considers deflection to be a serviceability limit state which must be accounted for in the design of flexural members. Appendix I of the S16.1-94 Standard, *Recommended Maximum Values for Deflections for Specified Live and Wind Loads*, provides some guidance to designers. Deflections tend to be more significant with longer clear spans, shallower members and with the use of high strength steels. Deflection calculations are based on specified loads.

Three methods for dealing with deflection of prismatic beams are summarized below:

1. Compute the required minimum moment of inertia to satisfy the deflection constraint, prior to selection of the beam size.

   $I_{reqd} = W\, C_d\, B_d$

   where

   I_{reqd} = required value of moment of inertia (10^6 mm^4)

   W = specified load value as described in Table 5-5 (kN)

   C_d = value of deflection constant obtained from Figure 5-1 for the appropriate span L and span/deflection limit L/Δ (10^6 mm^4 / kN)

   B_d = a number to relate the actual load and support condition to a uniformly distributed load (udl) on a simply supported beam, Table 5-5. Values of B_d are computed for the maximum deflection within the span. For a uniformly distributed load, $B_d = 1.0$.

   The actual deflection of a beam can be computed as:

   $\Delta = (I_{reqd} / I)\, \Delta_m$

   where

   Δ = actual deflection (mm)

   I = moment of inertia of beam (10^6 mm^4)

   Δ_m = maximum deflection permitted (mm)

   I_{reqd} = moment of inertia required to meet Δ_m (10^6 mm^4).

2. Compute deflections using the formulae for deflection of beams included with the Beam Diagrams and Formulae provided on pages 5-132 to 5-144.

3. The beam load tables for WWF shapes, W shapes and rectangular HSS on pages 5-96 through 5-131 list approximate deflections for the various steel sections and spans, based on the tabulated uniformly distributed total factored loads at an assumed stress of 215 MPa for 300W steel, and 240 MPa for 350W steel. (See "Vertical Deflection" on page 5-74.) One can determine deflections (for live load only or for total load) caused by stress levels that are different from those assumed, by multiplying the tabulated deflection with the ratio of actual stress to assumed stress (either 215 or 240 MPa).

Examples

Given:

A W410x85 section has been chosen for a simply supported non-composite beam spanning 10 m and subjected to a uniformly distributed specified load of 15 kN/m live and 7 kN/m dead. Check for live load deflection assuming the beam is laterally supported, 350W steel and deflection is limited to = $L/300$ = 33 mm.

Solutions:

Method 1

Using Figure 5-1 and Table 5-5 on pages 5-70 and 5-71,

B_d = 1.0 (simple span udl, Figure 5-5)

C_d = 1.95×10^6 mm^4/kN (for L/Δ = 300 and L = 10 m, Figure 5-1)

I_{reqd} = $W \cdot C_d \cdot B_d$

$= (15 \times 10) \times 1.95 \times 10^6 \times 1.0$

$= 293 \times 10^6$ mm^4

For W410x85, I_x = 315×10^6 mm^4

Actual deflection, $\Delta = (I_{reqd} / I) \Delta_m$

$= (293 / 315)\, 33 = 31$ mm

Method 2

From Beam Diagrams and Formulae, page 5-132,

$$\Delta = \frac{5\, w\, L^4}{384\, E\, I}$$

$I = 315 \times 10^6$ mm^4

$E = 200\,000$ MPa

Therefore $\Delta = \dfrac{5 \times 15 \times (10 \times 10^3)^4}{384 \times 200\,000 \times 315 \times 10^6} = 31$ mm

Method 3

From beam load tables, page 5-120, approximate deflection for W410x85 beam, span 10 m, loaded to a stress of 240 MPa = 61 mm

Stress due to live load is $\dfrac{M}{S} = \dfrac{W L}{8 S}$

$= \dfrac{(15 \times 10 \times 10^3)(10 \times 10^3)}{8 \times 1\,510 \times 10^3}$

$= 124$ MPa

Live load deflection is $(124 / 240)\, 61 = 32$ mm

DEFLECTION CONSTANT C_d

Figure 5-1

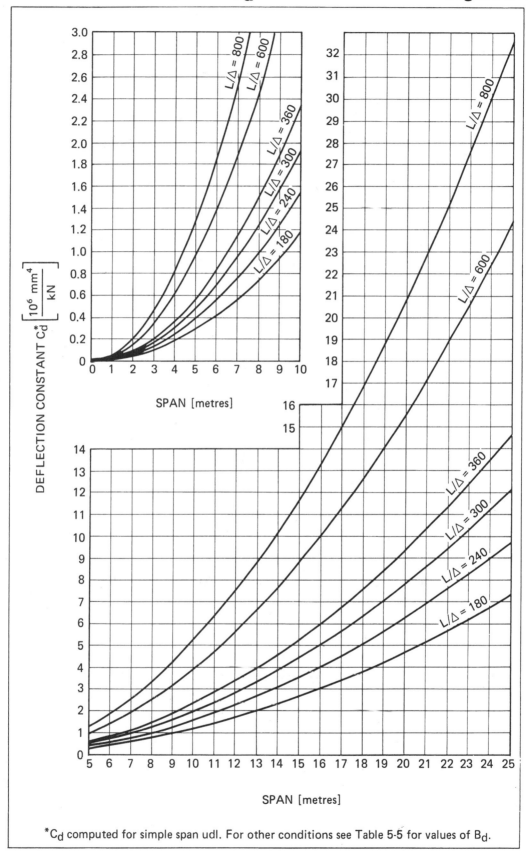

*C_d computed for simple span udl. For other conditions see Table 5-5 for values of B_d.

Table 5-5

Values of B_d for Various Loadings & Support Conditions

LOADING CONDITION	a/L	B_d	LOADING CONDITION	a/L	B_d	LOADING CONDITION	B_d	LOADING CONDITION	B_d
(cantilever point load at a, simply supported)	1. .8 .6 .5 .4 .2	0.0 .91 1.5 1.6 1.5 .91	(fixed-fixed point load)	1. .8 .6 .5 .4 .2	0.0 .16 .37 .40 .37 .16	(simple span UDL)	1.0	(fixed-fixed UDL)	.20
						(propped cantilever UDL)	.42	(cantilever UDL)	9.6
(simple, partial UDL from end)	1. .8 .6 .4 .2	1.0 1.13 1.10 .86 .48	(fixed-fixed partial UDL)	1. .8 .6 .4 .2	.20 .24 .23 .16 .057	(two point loads at L/4)	1.4	(two point loads at L/4 overhang)	2.2
						(triangular load)	1.0	(triangular peak center)	1.3
(propped cantilever partial UDL)	1. .8 .6 .4 .2	.415 .452 .390 .242 .079	(propped cantilever partial UDL other end)	1. .8 .6 .4 .2	.415 .500 .539 .463 .268	(two point loads)	.72	(single point load center)	1.2
(cantilever partial UDL)	1. .8 .6 .5 .4 .2	0.0 .51 .75 .72 .59 .22	(cantilever partial UDL from fixed)	1. .8 .6 .4 .2	25.6 17.8 10.9 5.27 1.41	(four point loads)	1.17	(four point loads overhang)	1.9
						(six point loads)	1.6	(six point loads overhang)	2.7

$I_{REQUIRED} = W \cdot C_d \cdot B_d$

Where I 10^6 mm^4
      W kN
      C_d from graph
      B_d from this table
      = 1.0 for simple span, udl.

						LOADING	B_d	LOADING	B_d
						(UDL 2 pt supports)	.42	(UDL short)	.71
						(UDL 3 pt supports)	.53	(UDL 2 short spans)	.77
						(3 point loads on 3 supports)	.89	(2 point loads on 2 supports)	1.24
LOADING CONDITION	n	B_d	LOADING CONDITION	n	B_d	6×W	1.44	2×W 2×W	2.1
n no. of spaces $(n-1) \cdot W$	2 3 4 5 6 7	1.60 2.72 3.80 4.84 5.87 6.87	n no. of spaces $(n-1) \cdot W$	2 3 4 5 6 7	.40 .59 .80 1.0 1.2 1.4	9×W	2.1	3×W 3×W	2.9

FACTORED RESISTANCE OF BEAMS

General

The blue pages contain tables which can be used to select flexural members and tables for estimating maximum beam reactions. These tables include Beam Selection Tables on pages 5-78 to 5-95 and Beam Load Tables on pages 5-96 to 5-131. The beam selection tables facilitate proportioning of flexural members subject to forces and moments determined by elastic analysis. Some values in these tables may also be used for selection of Class 1 flexural members for which the moments and forces have been determined by plastic analysis. The beam load tables list total uniformly distributed factored loads for laterally supported beams.

When using these tables, factored moments or forces must be equal to or less than the appropriate factored resistances, V_r, M_r or M_r' given in the tables.

Beam Selection Tables

Table

The beam selection tables list beam sizes in descending order of their factored moment resistance M_r (shown in bold) based on full lateral support (Clause 13.5, CAN/CSA S16.1-94). Listed beams include all WWF 2000 to WWF 700 sizes, W shapes normally used as beams, and all Canadian C shapes. Tables for WWF sizes are based on F_y = 350 MPa for G40.21 grade 350W, while those for W shapes are based on F_y = 345 MPa, corresponding to the least value among the 3 grades represented (G40.21 350W, ASTM A992 and A572 grade 50). Tables for C shapes are based on G40.21 300W with F_y = 300 MPa. Tables from earlier editions for WWF sizes and W shapes based on G40.21 300W material are now in Part Eight of this Handbook. A table for S shapes based on ASTM A572 grade 50 steel with F_y = 345 MPa is also included.

Shapes shown in bold type are generally the economy sections, based on both M_r and mass. Other shapes listed below a bold-type section are heavier sections, but may be economy sections when depth limitations require a shallower beam, or when shear resistance of a coped beam influences the beam selection. (See comments on availability of rolled shapes under "Principal Sources of Structural Steel Sections" on page 6-17.) Guidance as to factored shear resistance of coped beams is provided in the Beam Load Tables. (See comments on page 5-74.)

For each beam size, the tables list the maximum unsupported length L_u for which the factored moment resistance M_r is applicable. In addition the tables list the factored moment resistance M_r' for laterally unsupported beams (Clause 13.6, S16.1-94) for selected values of the unbraced beam length greater than L_u. For other values of unbraced length greater than L_u, M_r' can be interpolated.

The following items are included in the table:

V_r = factored shear resistance (kN)

= $\phi A_w F_s$ (Clause 13.4.1.1, S16.1-94)

I_x = moment of inertia about X-X axis (10^6 mm^4)

b = flange width (mm)

L_u = maximum unsupported beam length for which **M_r** is applicable (mm)

M_r = factored moment resistance for laterally supported member (kN·m)

= $\phi Z F_y$ for Class 1 and Class 2 sections

= $\phi S F_y$ for Class 3 sections (Clause 13.5, S16.1-94)

M_r' = factored moment resistance for tabulated unbraced beam length when greater than L_u (kN·m)

(M_r' is computed according to Clause 13.6 of S16.1-94 using $\omega_2 = 1.0$ in the expression for M_u.)

Design

Compute the maximum bending moment in the beam under factored loads M_f and the required moment of inertia I_{reqd} to meet the deflection limit using the specified loads. (For I_{reqd}, see "Deflection of Flexural Members", page 5-68)

For a laterally supported beam, proceed up the M_r column until a value of $M_r > M_f$ is obtained. Any beam above will satisfy the factored moment requirement. Check to ensure that $V_r > V_f$, the maximum factored beam shear, that $I_x > I_{reqd}$, and that L_u is greater than the maximum unsupported beam length.

For a laterally unsupported beam, proceed up the M_r column until a value of $M_r > M_f$ is reached. Then move to the right across the table to the column headed by the unsupported length (or the first listed unsupported length greater than that required) to obtain a value of M_r'. Proceed up this column comparing a few beams that have an $M_r' > M_f$ and choose the lightest section. Check $V_r > V_f$ and $I_x > I_{reqd}$ if a deflection check is necessary.

Other Steel Grades

For steel grades where F_y is less than that used in the table, reduce the tabulated values by the ratio of the yield values. For steel grades with F_y greater than that used in the table, the tabulated values are conservative.

Plastic Analysis

For beams analysed plastically (Clause 8.5, S16.1-94), the Beam Selection Tables may be used to facilitate selection of a beam size as follows:

Proceed up the M_r column until a value of $M_r > M_f$ is obtained.

Any beam above will satisfy the factored moment requirement, provided the beam is a Class 1 section. (See Table 5-1 on pages 5-5 to 5-8 for the Class of Sections in Bending for various steel grades.)

Check to ensure that $0.83\ V_r > V_f$, the maximum factored beam shear. (Clause 13.4.1.2, S16.1-94)

Provide suitable lateral bracing. (Clause 13.7, S16.1-94)

Beam Load Tables — WWF Beams and W Shapes

Loads

The beam load tables list total uniformly distributed factored loads for simply supported beams with the top flange fully supported (i.e., the unsupported length of beam is less than or equal to L_u). Tables for WWF sizes are based on $F_y = 350$ MPa for G40.21 grade 350W, while those for W shapes are based on $F_y = 345$ MPa, corresponding to the least value among the 3 grades represented (G40.21 350W, ASTM A992 and A572 grade 50). Tables from earlier editions based on G40.21 300W material with $F_y = 300$ MPa for WWF sizes and W shapes are now in Part Eight of this Handbook. Beam sizes listed include the same sections listed in the beam selection tables. To obtain the net supported load (factored), the beam factored dead load should be deducted from the total tabulated load.

For laterally supported beams the beam load tables may also be used to estimate loads for other loading conditions by dividing the tabulated values by the coefficient of the "Equivalent Tabular Load" value for the particular loading condition. (See "Beam

Diagrams and Formulae", pages 5-132 to 5-144.) Thus, for a simple beam, laterally supported and carrying equal concentrated loads at third points (loading condition 9), each factored concentrated load is 3/8 of the tabulated uniformly distributed factored load, and the total load is 3/4 of the tabulated load for the same span.

For steel grades where F_y is less than that used in the table, reduce the tabulated values by the ratio of the yield values. For steel grades with F_y greater than that used in the table, the tabulated values are conservative.

The beam load tables (sometimes referred to as "book loads") are frequently used to obtain maximum reactions for design of connections, when beam reactions are not provided on the structural design drawings. It should be noted that reactions computed from beam load tables are generally lower than beam reactions for compositely designed beams.

Vertical Deflection

The column headed "Approximate Deflection" lists the approximate theoretical mid-span deflection, at assumed bending stress levels of 215 MPa for 300W steel and 240 MPa for 350W, ASTM A992 and A572 grade 50 steels, for beams of various spans designed to support the tabulated factored loads.

The listed deflections are based on the nominal depth of the beam, and are calculated using the formula $\Delta = \dfrac{5}{384} \dfrac{W L^3}{EI}$.

For E = 200 000 MPa and an assumed bending unit stress of 215 MPa this formula reduces to $\Delta = \dfrac{224 \times 10^{-6} \times L^2}{d}$ where:

Δ = deflection (mm)

W = total uniform load including the dead load of the beam (kN)

L = beam span (mm)

E = modulus of elasticity (MPa)

I = moment of inertia of beam (mm^4)

d = depth of beam (mm)

More accurate deflections can be determined by multiplying the approximate deflection values listed by the ratio of actual bending stress to the assumed unit bending stress of 215 MPa for 300W steel or 240 MPa for 350W, ASTM A992 and A527 grade 50 steels. (See also "Deflection of Flexural Members", page 5-68.)

Web Shear

For beams with very short spans, with high end shear, and with flanges coped at the supports, the loads for beams may be limited by the shear capacity of the web rather than the bending capacity of the section. The designer should consider the effect of copes on the load carrying capacity of beams when selecting appropriate member sizes.

Both the depth and length of copes can vary considerably depending on the relative size and elevations of intersecting beams. For rolled shapes and for welded wide flange sections, the beam load tables list the factored shear resistance V_r for uncoped beams (Clause 13.4.1.1, S16.1-94). The factored shear resistance of singly and doubly coped beams should be adjusted for the depth and length of the copes, to account for possible non-uniform shear distribution across the effective web depth and to account for local web buckling.

Web Crippling & Yielding

Bearing stiffeners are required when the factored compressive resistance (Clause 15.9, S16.1-94) of the web is exceeded. For most common beam sizes the bearing resistance is governed by web crippling, except for very short bearing lengths when web yielding may govern.

The beam load tables for W shapes list values of, R (kN), the maximum factored end reaction for 100 mm of bearing per Clause 15.9(b)(ii), and values of the increment in bearing resistance, G (kN), per 10 millimetres of bearing length. For steels with a minimum specified yield stress other than F_y = 345 MPa, values of R and G can be computed by multiplying the values listed by the ratio $\sqrt{F_y / 345}$.

For WWF shapes, values of R and G are not listed, since the web slenderness of WWF shapes generally exceeds $1100 / \sqrt{F_y}$ and bearing stiffeners would be required (Clause 15.6.1, S16.1-94).

Proper lateral support must be provided for the top flanges of beams at the reaction point to ensure that the web crippling strength is not decreased.

Properties and Dimensions

The properties and dimensions listed on the beam load tables for rolled shapes include the beam depth d, the flange width width b, the flange thickness t, the web thickness w, and the beam k-distance, (all in millimetres). Dimensions d, t, w and k are required for calculating the compressive resistance of the web (yielding or crippling) per Clause 15.9, S16.1-94.

Beam Load Tables — Rectangular HSS

The beam load tables for rectangular Hollow Structural Sections (HSS) on pages 5-126 to 5-131 are based on G40.21 350W steel and list total uniformly distributed factored loads for laterally supported rectangular hollow sections in strong axis bending. As with the beam load tables for W shapes, the "Equivalent Tabular Load" values for other loading conditions may be used.

Approximate deflections listed are based on an assumed bending stress of 240 MPa.

Examples

1. **Given:**

    Design a simply supported beam spanning 8 metres to carry a uniformly distributed load of 15 kN/m specified live load and 7 kN/m specified dead load. The dead load includes an assumed beam dead load of 0.7 kN/m. Live load deflection is limited to L/300. Assume the beam frames into supporting members and that the beam is laterally supported. Use G40.21 350W steel.

   Solutions:

   (a) Using Beam Selection Tables — Elastic Analysis:

   Factored load $= a_D D + a_L L = (1.25 \times 7) + (1.50 \times 15) = 31.3$ kN/m

   M_f (factored load moment) $= \dfrac{wL^2}{8} = \dfrac{31.3 \times 8^2}{8} = 250$ kN·m

   V_f (factored end shear) $= \dfrac{wL}{2} = \dfrac{31.3 \times 8}{2} = 125$ kN

Compute I_{reqd} to meet deflection limit, see page 5-68

For udl, $B_d = 1.0$; — from Figure 5-1, for $L/\Delta = 300$, $C_d = 1.25 \times 10^6$

$I_{reqd} = W \cdot C_d \cdot B_d = (15 \times 8) \times 1.25 \times 10^6 \times 1.0 = 150 \times 10^6 \text{ mm}^4$

From table page 5-88, select a W410x46

$M_r = 274 \text{ kN} \cdot \text{m} \quad > 250 \quad$ — OK

$V_r = 578 \text{ kN} \quad > 125 \quad$ — OK

$I_x = 156 \times 10^6 \text{ mm}^4 \quad > 150 \times 10^6 \quad$ — OK

Use W410x46

(b) Using Beam Load Tables:

Total factored load, $W_f = 31.3 \times 8 = 250 \text{ kN}$

End reaction is $250/2 = 125 \text{ kN}$

From tables page 5-120, select a W410x46

$W_r \ (8\ 000) = 274 \text{ kN} \quad > 250 \quad$ — OK

$V_r = 578 \text{ kN} \quad > 125 \quad$ — (uncoped)

(If the beam were load bearing rather than framing into a girder, it would be necessary to check that R listed in the tables is greater than or equal to the factored end reaction.)

Approximate deflection listed at assumed stress of 240 MPa = 39 mm

Stress at specified live load $= \dfrac{M_{Live}}{S_x} = \dfrac{15}{31.3} \times \dfrac{250 \text{ kN} \cdot \text{m} \times 10^6}{773 \times 10^3} = 155 \text{ MPa}$

Live load deflection $= \dfrac{155}{240} \times 39 = 25 \text{ mm}$

$\dfrac{L}{300} = \dfrac{8\ 000}{300} = 27 \text{ mm} \quad > 25 \quad$ — OK

Use W410x46

2. Given:

Same as in Example (1), except assume that beam is laterally supported at quarter points, mid span and ends of beam.

Solution:

As in (1), select a W410x46

Since $L_u = 1\ 790 \text{ mm} \quad < 8\ 000/4 = 2\ 000$ must check for M_r' for actual unbraced length of 2 000 mm.

$M_r' \ (2\ 000) = 265 \text{ kN} \cdot \text{m} \quad > 250 \text{ kN} \cdot \text{m}$

Use W410x46

3. **Given:**

Same as in Example (1), except assume that beam is laterally supported at midpoint and ends of beam only.

Solution:

As in (1), select a W410x46 on basis of M_r and check for M_r' for unbraced length of $8\,000/2 = 4\,000$ mm

M_r' (4 000) = 142 kN·m   < 250 kN·m

Thus the W410x46 is not adequate; therefore check further up the table for lightest section with $M_r' > 250$ kN·m for unbraced length of 4 000 mm.

For W410x60, M_r' (4 000) = 286 kN·m   > 250 kN·m   — OK

$V_r = 642$ kN   > 125 kN   — (uncoped)

$I_x = 216 \times 10^6$ mm^4 > 150×10^6

Use W410x60

BEAM SELECTION TABLE
WWF and W Shapes

CSA G40.21 350W
ASTM A992, A572 grade 50

Designation	V_r	I_x	b	L_u	M_r	Factored moment resistance M_r' (kN·m) Unbraced length (mm)					
	kN	10^6 mm^4	mm	mm	≤ L_u	5 000	5 500	6 000	6 500	7 000	7 500
†WWF2000x732	3 640	63 900	550	7 620	20 100	—	—	—	—	—	—
WWF1800x700	4 070	50 400	550	7 300	19 800	—	—	—	—	—	19 600
WWF1800x659	4 050	46 600	550	7 170	18 300	—	—	—	—	—	18 100
W920x1188	13 200	26 100	457	8 930	18 300	—	—	—	—	—	—
†WWF2000x648	3 600	54 200	550	7 370	17 100	—	—	—	—	—	17 000
WWF1800x617	4 020	42 700	550	7 020	16 900	—	—	—	—	—	16 600
†WWF2000x607	3 590	49 300	550	7 210	15 500	—	—	—	—	—	15 300
WWF1800x575	4 000	38 800	550	6 860	15 500	—	—	—	—	15 400	15 000
†WWF1600x622	2 360	37 600	550	8 010	14 800	—	—	—	—	—	—
W920x967	10 500	20 300	446	7 890	14 500	—	—	—	—	—	—
W1000x883	10 200	21 000	424	6 850	14 100	—	—	—	—	14 000	13 800
WWF1400x597	2 730	28 100	550	7 790	13 900	—	—	—	—	—	—
†WWF1600x580	2 350	34 600	550	7 890	13 600	—	—	—	—	—	—
WWF1800x510	3 980	32 400	500	5 960	13 200	—	—	13 100	12 800	12 400	12 000
†WWF2000x542	3 570	41 400	500	6 310	13 000	—	—	—	12 900	12 600	12 300
†WWF1600x538	2 330	31 600	550	7 750	12 400	—	—	—	—	—	—
W1000x749	8 560	17 300	417	6 400	11 800	—	—	—	11 700	11 500	11 300
WWF1400x513	2 680	23 500	550	7 510	11 700	—	—	—	—	—	—
W920x784	8 350	15 900	437	7 120	11 600	—	—	—	—	—	11 400
†WWF1600x496	2 320	28 400	550	7 620	11 200	—	—	—	—	—	—
WWF1400x471	2 660	21 100	550	7 360	10 500	—	—	—	—	—	—
W1000x641	7 300	14 500	412	6 040	9 970	—	—	—	—	9 780	9 580	9 360
WWF1200x487	2 890	16 700	550	7 710	9 640	—	—	—	—	—	—
W690x802	8 460	10 600	387	7 980	9 590	—	—	—	—	—	—
W920x653	6 870	12 900	431	6 640	9 530	—	—	—	—	9 410	9 240
†WWF1600x431	2 300	23 400	500	6 720	9 230	—	—	—	—	9 110	8 900
W1000x591	6 610	13 300	409	5 920	9 160	—	—	9 130	8 930	8 730	8 530
WWF1400x405	2 640	17 300	500	6 450	8 760	—	—	—	8 740	8 530	8 310
W1000x583	7 810	12 500	314	4 530	8 730	8 500	8 270	8 020	7 780	7 540	7 290
W1000x554	6 240	12 300	408	5 820	8 540	—	—	8 470	8 280	8 080	7 880
WWF1100x458	2 210	13 600	550	7 890	8 540	—	—	—	—	—	—
W920x585	6 100	11 400	427	6 400	8 510	—	—	—	8 470	8 310	8 140
W1000x539	5 990	12 000	407	5 790	8 320	—	—	8 240	8 050	7 860	7 660
W1100x499	5 930	12 900	405	5 490	8 260	—	8 250	8 040	7 830	7 600	7 360
WWF1200x418	2 890	13 800	500	6 790	8 060	—	—	—	—	7 990	7 810
W840x576	5 990	10 100	411	6 330	7 920	—	—	—	7 870	7 720	7 570
W920x534	5 530	10 300	425	6 240	7 700	—	—	—	7 620	7 460	7 300
WWF1000x447	2 210	11 100	550	8 030	7 620	—	—	—	—	—	—

Note: The designation comprises the nominal depth in millimetres and the mass in kilograms per metre.
† Class 3 section
F_y taken as 350 MPa for WWF shapes and 345 MPa for W shapes. $\phi = 0.90$

CSA G40.21 350W
ASTM A992, A572 grade 50

BEAM SELECTION TABLE
WWF and W Shapes

Nominal mass	Factored moment resistance M_r' (kN·m)									Imperial designation
	Unbraced length (mm)									
kg/m	8 000	9 000	10 000	11 000	12 000	14 000	16 000	18 000	20 000	
732	19 800	19 000	18 100	17 100	16 100	13 800	11 100	9 060	7 610	**WWF79x490**
700	19 200	18 300	17 400	16 300	15 300	12 800	10 200	8 420	7 120	**WWF71x470**
659	17 700	16 800	15 900	14 900	13 800	11 300	8 990	7 360	6 190	**WWF71x442**
1 188	—	18 200	17 900	17 500	17 100	16 400	15 700	14 900	14 200	W36x798
648	16 600	15 900	15 000	14 100	13 100	10 800	8 510	6 910	5 760	**WWF79x436**
617	16 200	15 400	14 500	13 500	12 400	9 900	7 810	6 370	5 330	**WWF71x415**
607	15 000	14 300	13 500	12 600	11 600	9 360	7 330	5 930	4 920	**WWF79x408**
575	14 700	13 900	13 000	12 000	11 000	8 530	6 700	5 440	4 540	**WWF71x388**
622	—	14 300	13 700	13 100	12 400	11 000	9 370	7 750	6 580	WWF63x419
967	—	14 100	13 800	13 400	13 000	12 300	11 600	10 900	10 200	W36x650
883	13 600	13 100	12 700	12 200	11 800	10 900	10 000	9 070	8 080	W40x593
597	13 800	13 200	12 700	12 100	11 500	10 100	8 610	7 180	6 140	WWF55x402
580	—	13 100	12 500	11 900	11 200	9 850	8 170	6 730	5 680	WWF63x388
510	11 600	10 700	9 750	8 650	7 350	5 530	4 340	3 530	2 940	**WWF71x344**
542	11 900	11 100	10 200	9 260	8 100	6 070	4 750	3 850	3 200	WWF79x364
538	12 300	11 800	11 300	10 700	10 100	8 710	7 050	5 770	4 850	WWF63x361
749	11 100	10 600	10 200	9 760	9 320	8 450	7 500	6 560	5 830	W40x503
513	11 400	10 900	10 400	9 850	9 250	7 950	6 400	5 270	4 460	WWF55x344
784	11 300	10 900	10 600	10 200	9 840	9 140	8 440	7 750	6 900	W36x527
496	11 000	10 600	10 000	9 480	8 890	7 590	6 010	4 890	4 090	**WWF63x333**
471	10 200	9 780	9 280	8 730	8 150	6 840	5 410	4 430	3 730	**WWF55x316**
641	9 150	8 720	8 290	7 850	7 410	6 500	5 530	4 820	4 270	W40x431
487	9 530	9 160	8 750	8 320	7 870	6 910	5 770	4 800	4 090	WWF47x326
802	—	9 380	9 160	8 950	8 740	8 330	7 920	7 510	7 100	W27x539
653	9 070	8 720	8 370	8 020	7 670	6 970	6 250	5 450	4 840	W36x439
431	8 670	8 180	7 630	7 040	6 410	4 920	3 870	3 140	2 630	**WWF63x289**
591	8 320	7 900	7 470	7 040	6 610	5 650	4 790	4 160	3 680	W40x397
405	8 080	7 580	7 030	6 440	5 800	4 390	3 470	2 840	2 390	**WWF55x272**
583	7 040	6 550	6 070	5 500	4 970	4 160	3 580	3 150	2 810	W40x392
554	7 680	7 260	6 840	6 410	5 980	5 000	4 230	3 670	3 240	W40x372
458	8 500	8 190	7 850	7 500	7 120	6 340	5 450	4 550	3 890	WWF43x307
585	7 980	7 630	7 290	6 940	6 590	5 890	5 080	4 420	3 920	W36x393
539	7 460	7 040	6 620	6 200	5 770	4 790	4 050	3 510	3 100	W40x362
499	7 120	6 620	6 100	5 570	4 920	3 960	3 300	2 840	2 490	**W44x335**
418	7 630	7 230	6 790	6 330	5 850	4 700	3 780	3 130	2 670	WWF47x281
576	7 420	7 110	6 800	6 490	6 190	5 570	4 880	4 260	3 780	W33x387
534	7 140	6 800	6 460	6 120	5 780	5 070	4 290	3 720	3 290	W36x359
447	—	7 360	7 080	6 780	6 470	5 820	5 140	4 340	3 730	WWF39x300

BEAM SELECTION TABLE
WWF and W Shapes

CSA G40.21 350W
ASTM A992, A572 grade 50

Designation	V_r	I_x	b	L_u	M_r	Factored moment resistance M_r' (kN·m) Unbraced length (mm)					
	kN	10^6 mm^4	mm	mm	$\leq L_u$	4 500	5 000	5 500	6 000	6 500	7 000
WWF1400x358	2 640	14 500	400	5 000	7 470	—	—	7 240	6 990	6 730	6 450
W1000x483	5 310	10 700	404	5 650	7 420	—	—	—	7 300	7 120	6 930
W760x582	5 960	8 600	396	6 470	7 360	—	—	—	—	7 350	7 230
W1000x493	6 580	10 300	309	4 280	7 270	7 170	6 950	6 720	6 490	6 250	6 010
W840x527	5 460	9 140	409	6 150	7 230	—	—	—	—	7 130	6 990
W1100x432	5 020	11 300	402	5 400	7 200	—	—	7 170	6 980	6 770	6 560
W1000x486	6 370	10 200	308	4 270	7 200	7 100	6 880	6 660	6 420	6 190	5 950
WWF1200x380	2 890	12 300	500	6 630	7 180	—	—	—	—	—	7 060
WWF1100x388	2 210	11 200	500	6 970	7 060	—	—	—	—	—	7 050
W920x488	5 000	9 350	422	6 100	7 020	—	—	—	—	6 900	6 750
W1000x443	4 890	9 670	402	5 530	6 770	—	—	—	6 610	6 440	6 250
W760x531	5 380	7 750	393	6 240	6 680	—	—	—	—	6 610	6 490
WWF900x417	1 370	8 680	550	8 310	6 550	—	—	—	—	—	—
W1100x390	4 530	10 100	400	5 320	6 460	—	—	6 400	6 220	6 030	5 830
W840x473	4 830	8 120	406	5 970	6 460	—	—	—	6 450	6 310	6 170
W920x446	4 590	8 470	423	6 010	6 400	—	—	—	—	6 260	6 110
W1000x412	4 360	9 100	402	5 530	6 370	—	—	—	6 220	6 050	5 880
W690x548	5 550	6 710	372	6 400	6 300	—	—	—	—	6 280	6 180
WWF1000x377	2 210	9 120	500	7 080	6 300	—	—	—	—	—	—
WWF1100x351	2 210	9 930	500	6 810	6 270	—	—	—	—	—	6 220
WWF1200x333	2 890	10 200	400	5 150	6 080	—	—	5 960	5 770	5 570	5 360
W1000x414	5 410	8 520	304	4 090	6 050	5 890	5 680	5 470	5 250	5 020	4 790
W760x484	4 890	6 980	390	6 030	6 050	—	—	—	—	5 940	5 820
W920x417	4 280	7 880	422	5 940	5 960	—	—	—	5 950	5 810	5 670
W840x433	4 430	7 350	404	5 840	5 870	—	—	—	5 830	5 690	5 560
W610x551	5 620	5 580	347	6 620	5 780	—	—	—	—	—	5 710
W1000x393	5 080	8 080	303	4 050	5 740	5 570	5 370	5 160	4 940	4 710	4 480
W1000x371	3 890	8 140	400	5 440	5 710	—	—	5 700	5 550	5 390	5 220
W690x500	5 000	6 050	369	6 140	5 710	—	—	—	—	5 640	5 540
W1100x342	3 850	8 670	400	5 230	5 620	—	—	5 530	5 370	5 190	5 010
WWF1000x340	2 210	8 060	500	6 910	5 580	—	—	—	—	—	5 550
W920x387	4 020	7 180	420	5 830	5 460	—	—	—	5 420	5 290	5 150
W760x434	4 320	6 180	387	5 820	5 400	—	—	—	5 360	5 250	5 130
WWF1200x302	2 890	8 970	400	4 980	5 390	—	5 380	5 220	5 040	4 850	4 640
WWF900x347	1 370	7 100	500	7 320	5 390	—	—	—	—	—	—
W920x381	4 760	6 970	310	4 250	5 280	5 200	5 020	4 850	4 660	4 470	4 280
W840x392	3 970	6 590	401	5 710	5 280	—	—	—	5 210	5 080	4 950
WWF1100x304	2 210	8 220	400	5 310	5 260	—	—	5 210	5 060	4 890	4 720
W690x457	4 550	5 450	367	5 910	5 220	—	—	—	5 200	5 100	5 000
W610x498	5 030	4 950	343	6 230	5 190	—	—	—	—	5 140	5 060
W1000x350	4 360	7 230	302	4 010	5 150	4 980	4 780	4 580	4 370	4 160	3 940
W920x365	3 810	6 700	419	5 770	5 120	—	—	—	5 070	4 940	4 810

Note: The designation comprises the nominal depth in millimetres and the mass in kilograms per metre.
F_y taken as 350 MPa for WWF shapes and 345 MPa for W shapes. $\phi = 0.90$

CSA G40.21 350W
ASTM A992, A572 grade 50

BEAM SELECTION TABLE
WWF and W Shapes

Nominal mass	Factored moment resistance M_r' (kN·m) Unbraced length (mm)									Imperial designation
kg/m	8 000	9 000	10 000	11 000	12 000	14 000	16 000	18 000	20 000	
358	5 830	5 160	4 330	3 650	3 130	2 400	1 920	1 590	1 360	**WWF55x240**
483	6 540	6 140	5 730	5 310	4 860	3 940	3 320	2 860	2 520	W40x324
582	6 970	6 720	6 470	6 220	5 970	5 480	4 990	4 400	3 930	W30x391
493	5 530	5 050	4 480	3 980	3 580	2 980	2 560	2 240	2 000	W40x331
527	6 690	6 390	6 080	5 770	5 460	4 850	4 130	3 600	3 190	W33x354
432	6 120	5 650	5 150	4 590	4 020	3 210	2 660	2 270	1 980	**W44x290**
486	5 460	4 980	4 410	3 910	3 510	2 930	2 510	2 200	1 960	W40x327
380	6 720	6 330	5 920	5 470	5 000	3 890	3 100	2 560	2 160	WWF47x255
388	6 750	6 420	6 070	5 690	5 300	4 400	3 550	2 960	2 530	WWF43x260
488	6 430	6 100	5 770	5 430	5 090	4 320	3 640	3 150	2 780	W36x328
443	5 880	5 480	5 070	4 660	4 160	3 360	2 810	2 420	2 130	W40x297
531	6 240	5 990	5 740	5 490	5 250	4 760	4 210	3 690	3 290	W30x357
417	—	6 400	6 180	5 950	5 710	5 210	4 690	4 110	3 550	WWF35x279
390	5 400	4 950	4 480	3 900	3 400	2 700	2 220	1 890	1 640	**W44x262**
473	5 880	5 590	5 280	4 980	4 680	4 000	3 390	2 940	2 600	W33x318
446	5 810	5 490	5 160	4 830	4 490	3 720	3 120	2 690	2 360	W36x300
412	5 510	5 130	4 730	4 330	3 830	3 080	2 570	2 200	1 930	W40x277
548	5 980	5 770	5 570	5 370	5 170	4 780	4 390	3 930	3 520	W27x368
377	6 060	5 780	5 480	5 170	4 840	4 130	3 360	2 810	2 420	WWF39x253
351	5 930	5 620	5 280	4 920	4 530	3 620	2 890	2 390	2 030	WWF43x236
333	4 900	4 410	3 820	3 230	2 780	2 160	1 740	1 460	1 250	WWF47x224
414	4 320	3 780	3 280	2 890	2 590	2 140	1 830	1 600	1 420	W40x278
484	5 580	5 330	5 090	4 840	4 590	4 100	3 530	3 090	2 740	W30x326
417	5 370	5 060	4 730	4 410	4 070	3 310	2 770	2 380	2 090	W36x280
433	5 280	4 990	4 690	4 390	4 090	3 390	2 870	2 480	2 190	W33x291
551	5 550	5 390	5 230	5 070	4 910	4 590	4 280	3 970	3 600	W24x370
393	4 020	3 470	3 000	2 650	2 370	1 950	1 670	1 450	1 290	W40x264
371	4 880	4 510	4 120	3 700	3 240	2 590	2 150	1 840	1 600	**W40x249**
500	5 340	5 140	4 940	4 740	4 540	4 150	3 740	3 290	2 940	W27x336
342	4 610	4 190	3 730	3 180	2 760	2 160	1 770	1 490	1 290	**W44x230**
340	5 310	5 040	4 760	4 450	4 130	3 380	2 720	2 260	1 930	WWF39x228
387	4 870	4 560	4 250	3 930	3 580	2 870	2 390	2 050	1 790	W36x260
434	4 890	4 650	4 400	4 160	3 910	3 370	2 870	2 500	2 220	W30x292
302	4 200	3 710	3 110	2 620	2 250	1 730	1 390	1 150	978	WWF47x203
347	5 250	5 030	4 790	4 550	4 300	3 770	3 150	2 660	2 290	WWF35x233
381	3 890	3 470	3 010	2 650	2 370	1 960	1 670	1 460	1 300	W36x256
392	4 680	4 400	4 110	3 810	3 510	2 840	2 390	2 060	1 820	W33x263
304	4 350	3 950	3 530	3 000	2 590	2 010	1 630	1 370	1 180	WWF43x204
457	4 800	4 600	4 400	4 200	4 000	3 610	3 150	2 770	2 470	W27x307
498	4 900	4 740	4 580	4 420	4 260	3 950	3 640	3 290	2 940	W24x335
350	3 490	2 950	2 540	2 230	1 980	1 630	1 380	1 200	1 070	W40x235
365	4 520	4 230	3 920	3 600	3 230	2 580	2 140	1 830	1 600	W36x245

BEAM SELECTION TABLE
WWF and W Shapes

CSA G40.21 350W
ASTM A992, A572 grade 50

Designation	V_r	I_x	b	L_u	M_r	Factored moment resistance M_r' (kN·m)					
						Unbraced length (mm)					
	kN	10^6 mm^4	mm	mm	≤ L_u	4 000	4 500	5 000	5 500	6 000	6 500
W1000x321	3 250	6 960	400	5 360	4 910	—	—	—	4 870	4 730	4 590
W920x342	3 610	6 250	418	5 700	4 780	—	—	—	—	4 710	4 590
W840x359	3 750	5 910	403	5 620	4 780	—	—	—	—	4 690	4 570
W760x389	3 880	5 440	385	5 650	4 780	—	—	—	—	4 710	4 600
W920x345	4 270	6 260	308	4 170	4 750	—	4 650	4 480	4 310	4 130	3 950
W690x419	4 100	4 940	364	5 710	4 750	—	—	—	—	4 700	4 600
WWF900x309	1 370	6 240	500	7 160	4 730	—	—	—	—	—	—
WWF800x339	1 370	5 500	500	7 490	4 690	—	—	—	—	—	—
W610x455	4 520	4 450	340	5 940	4 690	—	—	—	—	4 680	4 600
WWF1000x293	2 210	6 640	400	5 400	4 660	—	—	—	4 640	4 510	4 370
WWF1100x273	2 210	7 160	400	5 150	4 630	—	—	—	4 540	4 390	4 240
W1000x314	3 910	6 440	300	3 910	4 630	4 600	4 430	4 240	4 050	3 850	3 640
WWF1200x263	2 890	7 250	300	3 560	4 470	4 300	4 090	3 850	3 600	3 330	3 040
W1000x296	3 230	6 200	400	5 230	4 440	—	—	—	4 370	4 240	4 110
W840x329	3 480	5 350	401	5 530	4 350	—	—	—	—	4 240	4 130
W690x384	3 760	4 470	362	5 560	4 320	—	—	—	—	4 230	4 140
W760x350	3 440	4 860	382	5 530	4 280	—	—	—	—	4 190	4 080
W610x415	4 100	4 000	338	5 700	4 250	—	—	—	—	4 210	4 130
W920x313	4 030	5 480	309	4 060	4 220	—	4 090	3 940	3 770	3 600	3 420
WWF800x300	1 370	4 840	500	7 280	4 130	—	—	—	—	—	—
WWF1000x262	2 210	5 780	400	5 230	4 100	—	—	—	4 030	3 910	3 780
W1000x272	3 250	5 540	300	3 870	3 970	3 940	3 780	3 620	3 440	3 260	3 060
W920x289	3 690	5 040	308	4 010	3 910	—	3 780	3 620	3 460	3 300	3 120
W840x299	3 190	4 790	400	5 450	3 910	—	—	—	3 900	3 800	3 690
W690x350	3 450	4 020	360	5 390	3 910	—	—	—	3 890	3 800	3 710
WWF900x262	1 370	5 110	400	5 630	3 910	—	—	—	—	3 830	3 730
W760x314	3 170	4 270	384	5 420	3 820	—	—	—	3 800	3 710	3 610
W610x372	3 620	3 530	335	5 450	3 790	—	—	—	3 780	3 700	3 620
WWF1100x234	2 210	5 720	300	3 710	3 780	3 690	3 520	3 340	3 140	2 930	2 710
W920x271	3 480	4 720	307	3 970	3 660	—	3 520	3 380	3 220	3 060	2 890
W690x323	3 120	3 700	359	5 310	3 600	—	—	—	3 570	3 480	3 390
W1000x249	3 220	4 810	300	3 740	3 510	3 440	3 290	3 130	2 960	2 780	2 590
W610x341	3 310	3 180	333	5 250	3 450	—	—	—	3 410	3 330	3 260
W920x253	3 260	4 370	306	3 930	3 420	3 400	3 270	3 130	2 980	2 820	2 660
W760x284	2 870	3 810	382	5 320	3 420	—	—	—	3 380	3 290	3 200
WWF900x231	1 350	4 410	400	5 460	3 400	—	—	—	—	3 300	3 200
WWF800x253	1 370	3 950	400	5 730	3 400	—	—	—	—	3 360	3 270
WWF1000x223	2 210	4 590	300	3 790	3 310	3 250	3 110	2 960	2 800	2 620	2 440
W840x251	2 990	3 860	292	3 890	3 200	3 170	3 050	2 920	2 790	2 650	2 500
W690x289	2 780	3 250	356	5 140	3 200	—	—	—	3 140	3 060	2 970
W920x238	3 090	4 060	305	3 890	3 170	3 140	3 020	2 890	2 740	2 590	2 430
W760x257	2 630	3 420	381	5 230	3 080	—	—	—	3 040	2 950	2 860
W610x307	2 960	2 840	330	5 080	3 080	—	—	—	3 020	2 950	2 870

Note: The designation comprises the nominal depth in millimetres and the mass in kilograms per metre.
F_y taken as 350 MPa for WWF shapes and 345 MPa for W shapes. $\phi = 0.90$

CSA G40.21 350W
ASTM A992, A572 grade 50

BEAM SELECTION TABLE
WWF and W Shapes

Nominal mass	Factored moment resistance M_r' (kN·m)									Imperial designation
	Unbraced length (mm)									
kg/m	7 000	8 000	9 000	10 000	11 000	12 000	14 000	16 000	18 000	
321	4 440	4 120	3 780	3 420	2 980	2 600	2 050	1 690	1 430	**W40x215**
342	4 460	4 190	3 900	3 600	3 290	2 910	2 310	1 910	1 630	W36x230
359	4 450	4 190	3 910	3 630	3 340	3 000	2 410	2 020	1 730	W33x241
389	4 480	4 250	4 010	3 770	3 530	3 290	2 730	2 320	2 010	W30x261
345	3 760	3 380	2 940	2 530	2 220	1 980	1 630	1 390	1 210	W36x232
419	4 500	4 300	4 100	3 900	3 710	3 510	3 100	2 660	2 330	W27x281
309	—	4 560	4 350	4 130	3 900	3 650	3 120	2 520	2 110	WWF35x208
339	—	4 610	4 430	4 240	4 050	3 850	3 440	2 980	2 530	WWF31x228
455	4 520	4 360	4 200	4 040	3 880	3 730	3 420	3 100	2 740	W24x306
293	4 230	3 920	3 580	3 230	2 800	2 420	1 900	1 550	1 310	WWF39x197
273	4 070	3 720	3 330	2 860	2 410	2 070	1 600	1 280	1 070	WWF43x184
314	3 420	2 940	2 460	2 110	1 840	1 640	1 340	1 130	979	**W40x211**
263	2 670	2 100	1 710	1 430	1 220	1 060	834	684	578	WWF47x176
296	3 960	3 650	3 320	2 960	2 530	2 190	1 720	1 410	1 190	**W40x199**
329	4 010	3 750	3 490	3 210	2 920	2 570	2 060	1 710	1 470	W33x221
384	4 040	3 850	3 650	3 460	3 260	3 070	2 630	2 250	1 970	W27x258
350	3 970	3 750	3 520	3 280	3 040	2 780	2 270	1 920	1 660	W30x235
415	4 050	3 890	3 730	3 580	3 420	3 270	2 960	2 610	2 300	W24x279
313	3 240	2 860	2 410	2 070	1 810	1 600	1 310	1 110	962	W36x210
300	—	4 010	3 840	3 660	3 470	3 270	2 860	2 370	1 990	WWF31x202
262	3 640	3 340	3 020	2 650	2 240	1 930	1 500	1 210	1 010	WWF39x176
272	2 860	2 390	1 990	1 690	1 470	1 300	1 050	883	761	**W40x183**
289	2 940	2 560	2 130	1 820	1 590	1 410	1 140	963	833	W36x194
299	3 580	3 340	3 080	2 810	2 510	2 190	1 740	1 440	1 230	W33x201
350	3 620	3 420	3 230	3 030	2 840	2 640	2 190	1 870	1 640	W27x235
262	3 620	3 380	3 130	2 860	2 580	2 240	1 760	1 450	1 230	WWF35x176
314	3 500	3 280	3 060	2 820	2 590	2 300	1 860	1 560	1 350	W30x211
372	3 550	3 390	3 230	3 080	2 920	2 770	2 440	2 110	1 860	W24x250
234	2 450	1 930	1 570	1 320	1 120	978	770	632	535	WWF43x157
271	2 710	2 320	1 930	1 640	1 430	1 260	1 020	859	741	**W36x182**
323	3 300	3 110	2 920	2 730	2 540	2 320	1 910	1 630	1 420	W27x217
249	2 400	1 940	1 600	1 360	1 170	1 030	831	694	596	**W40x167**
341	3 180	3 020	2 870	2 710	2 560	2 400	2 050	1 760	1 550	W24x229
253	2 480	2 090	1 730	1 470	1 270	1 120	905	758	652	W36x170
284	3 100	2 890	2 670	2 450	2 190	1 930	1 550	1 300	1 120	W30x191
231	3 100	2 870	2 630	2 360	2 040	1 760	1 370	1 110	933	WWF35x156
253	3 180	2 990	2 790	2 570	2 350	2 090	1 660	1 370	1 170	WWF31x170
223	2 250	1 790	1 460	1 230	1 050	920	729	602	512	WWF39x150
251	2 350	2 010	1 680	1 440	1 260	1 120	911	771	668	W33x169
289	2 880	2 690	2 510	2 310	2 110	1 880	1 540	1 310	1 130	W27x194
238	2 270	1 880	1 550	1 310	1 130	998	802	669	575	**W36x160**
257	2 770	2 570	2 360	2 140	1 870	1 640	1 310	1 090	931	W30x173
307	2 800	2 640	2 490	2 340	2 180	2 020	1 690	1 450	1 270	W24x207

BEAM SELECTION TABLE
WWF and W Shapes

CSA G40.21 350W
ASTM A992, A572 grade 50

Designation	V_r	I_x	b	L_u	M_r	Factored moment resistance M_r' (kN·m) Unbraced length (mm)					
	kN	10^6 mm⁴	mm	mm	≤ L_u	3 500	4 000	4 500	5 000	5 500	6 000
W1000x222	3 000	4 080	300	3 590	**3 040**	—	2 940	2 800	2 650	2 480	2 310
W920x223	2 970	3 770	304	3 830	2 960	—	2 920	2 800	2 670	2 530	2 380
WWF800x223	1 370	3 410	400	5 570	2 940	—	—	—	—	—	2 880
WWF700x245	1 370	2 950	400	5 880	2 900	—	—	—	—	—	2 890
WWF1000x200	2 210	3 940	300	3 610	2 890	—	2 790	2 660	2 510	2 350	2 180
W690x265	2 660	2 900	358	5 060	2 880	—	—	—	—	2 820	2 740
W610x285	2 730	2 610	329	4 970	2 850	—	—	—	—	2 780	2 700
W840x226	2 810	3 400	294	3 830	2 840	—	2 810	2 700	2 570	2 450	2 310
WWF900x192	1 350	3 460	300	3 980	2 710	—	2 700	2 600	2 490	2 370	2 240
W530x300	2 770	2 210	319	5 210	2 690	—	—	—	—	2 660	2 600
W840x210	2 670	3 110	293	3 770	**2 620**	—	2 570	2 460	2 350	2 220	2 090
W690x240	2 410	2 610	356	4 970	2 610	—	—	—	2 600	2 530	2 450
W920x201	2 710	3 250	304	3 720	**2 600**	—	2 540	2 430	2 300	2 170	2 030
W610x262	2 500	2 360	327	4 850	2 590	—	—	—	2 570	2 500	2 430
W760x220	2 630	2 780	266	3 570	2 540	—	2 460	2 350	2 230	2 120	1 990
WWF700x214	1 370	2 540	400	5 690	2 500	—	—	—	—	—	2 470
W530x272	2 490	1 970	318	5 040	2 430	—	—	—	—	2 370	2 320
W610x241	2 330	2 150	329	4 790	2 380	—	—	—	2 350	2 290	2 220
W840x193	2 530	2 780	292	3 680	**2 370**	—	2 310	2 210	2 100	1 980	1 850
W690x217	2 190	2 340	355	4 900	2 350	—	—	—	2 340	2 270	2 190
WWF800x184	1 370	2 660	300	4 060	2 330	—	—	2 260	2 170	2 070	1 970
WWF900x169	1 340	2 930	300	3 820	2 320	—	2 290	2 190	2 080	1 970	1 850
W760x196	2 460	2 400	268	3 500	2 230	—	2 130	2 030	1 920	1 810	1 690
W530x248	2 220	1 770	315	4 880	2 190	—	—	—	2 180	2 120	2 070
W610x217	2 120	1 910	328	4 680	2 130	—	—	—	2 090	2 030	1 960
W840x176	2 300	2 460	292	3 600	**2 120**	—	2 050	1 950	1 850	1 730	1 610
W760x185	2 340	2 230	267	3 450	2 080	2 070	1 980	1 880	1 780	1 670	1 550
†WWF700x196	1 370	2 300	400	5 860	2 070	—	—	—	—	—	2 050
W460x260	2 360	1 440	289	4 980	2 030	—	—	—	—	1 980	1 940
W690x192	2 230	1 980	254	3 440	2 010	2 000	1 910	1 830	1 730	1 640	1 540
WWF800x161	1 370	2 250	300	3 900	1 990	—	1 980	1 900	1 810	1 720	1 620
WWF700x175	1 370	1 970	300	4 160	1 970	—	—	1 930	1 850	1 780	1 700
W760x173	2 250	2 060	267	3 410	**1 930**	1 910	1 830	1 730	1 630	1 520	1 410
W530x219	2 100	1 510	318	4 720	1 900	—	—	—	1 870	1 820	1 760
W610x195	1 960	1 680	327	4 570	1 880	—	—	—	1 840	1 780	1 710
W460x235	2 120	1 270	287	4 770	1 810	—	—	—	1 790	1 750	1 710
W760x161	2 140	1 860	266	3 330	**1 760**	1 730	1 650	1 560	1 460	1 360	1 250
W690x170	2 060	1 700	256	3 380	1 750	1 730	1 650	1 570	1 480	1 390	1 290
W530x196	1 870	1 340	316	4 600	1 700	—	—	—	1 660	1 610	1 550
WWF700x152	1 370	1 660	300	3 990	1 680	—	—	1 610	1 540	1 470	1 390
W610x174	1 770	1 470	325	4 480	1 660	—	—	—	1 610	1 550	1 490
W460x213	1 880	1 140	285	4 590	1 640	—	—	—	1 600	1 560	1 520
W760x147	2 040	1 660	265	3 260	**1 580**	1 550	1 470	1 380	1 290	1 190	1 090
W530x182	1 720	1 240	315	4 530	1 560	—	—	—	1 520	1 470	1 420
W690x152	1 850	1 510	254	3 320	1 550	1 530	1 460	1 380	1 290	1 210	1 110
W460x193	1 700	1 020	283	4 440	1 470	—	—	—	1 430	1 390	1 350
W610x155	1 590	1 290	324	4 400	1 470	—	—	1 460	1 410	1 360	1 300

Note: The designation comprises the nominal depth in millimetres and the mass in kilograms per metre.
† Class 3 section
F_y taken as 350 MPa for WWF shapes and 345 MPa for W shapes. $\phi = 0.90$

CSA G40.21 350W
ASTM A992, A572 grade 50

BEAM SELECTION TABLE
WWF and W Shapes

Nominal mass kg/m	Factored moment resistance M_r' (kN·m) Unbraced length (mm)									Imperial designation
	7 000	8 000	9 000	10 000	11 000	12 000	14 000	16 000	18 000	
222	1 910	1 520	1 250	1 050	906	794	634	527	451	**W40x149**
223	2 070	1 680	1 380	1 170	1 010	884	707	589	504	W36x150
223	2 710	2 530	2 330	2 120	1 880	1 630	1 280	1 050	882	WWF31x150
245	2 750	2 600	2 440	2 280	2 110	1 950	1 560	1 310	1 120	WWF28x164
200	1 770	1 390	1 130	945	806	699	549	449	379	WWF39x134
265	2 570	2 390	2 200	2 010	1 790	1 590	1 300	1 090	946	W27x178
285	2 550	2 400	2 250	2 100	1 950	1 760	1 470	1 260	1 100	W24x192
226	2 020	1 680	1 400	1 190	1 040	915	741	623	538	W33x152
192	1 960	1 630	1 330	1 120	964	842	670	555	473	WWF35x128
300	2 480	2 370	2 250	2 130	2 020	1 900	1 650	1 420	1 250	W21x201
210	1 810	1 480	1 220	1 040	900	793	640	536	461	**W33x141**
240	2 290	2 110	1 930	1 750	1 520	1 340	1 090	913	788	W27x161
201	1 720	1 370	1 120	941	809	707	561	464	396	**W36x135**
262	2 290	2 140	1 990	1 830	1 670	1 490	1 240	1 060	924	W24x176
220	1 740	1 430	1 210	1 050	922	823	679	579	504	W30x148
214	2 330	2 190	2 040	1 870	1 710	1 500	1 190	982	835	WWF28x144
272	2 200	2 090	1 970	1 860	1 740	1 630	1 370	1 180	1 040	W21x182
241	2 080	1 930	1 780	1 630	1 450	1 300	1 070	911	795	W24x162
193	1 580	1 260	1 040	878	758	666	534	446	382	**W33x130**
217	2 040	1 870	1 700	1 500	1 300	1 140	920	769	660	W27x146
184	1 750	1 490	1 230	1 040	900	791	634	529	454	WWF31x123
169	1 570	1 250	1 010	845	721	625	491	402	339	WWF35x113
196	1 430	1 160	973	835	731	650	532	451	391	W30x132
248	1 960	1 840	1 730	1 610	1 500	1 360	1 140	979	859	W21x166
217	1 820	1 680	1 530	1 370	1 200	1 070	878	745	647	W24x146
176	1 330	1 060	868	731	629	551	439	364	311	**W33x118**
185	1 280	1 040	867	743	649	576	470	397	344	W30x124
196	1 950	1 830	1 700	1 570	1 430	1 260	992	812	686	WWF28x132
260	1 850	1 770	1 690	1 600	1 520	1 440	1 250	1 080	958	W18x175
192	1 330	1 090	924	802	708	634	525	449	392	W27x129
161	1 400	1 130	927	778	666	581	460	379	322	WWF31x108
175	1 530	1 350	1 130	965	839	741	601	505	436	WWF28x117
173	1 150	924	770	657	572	506	411	346	299	**W30x116**
219	1 650	1 540	1 430	1 310	1 180	1 060	880	754	659	W21x147
195	1 580	1 440	1 300	1 120	981	870	709	598	518	W24x131
235	1 630	1 540	1 460	1 380	1 290	1 210	1 020	887	783	W18x158
161	988	793	658	560	486	429	347	291	251	**W30x108**
170	1 070	875	736	634	557	497	408	347	302	W27x114
196	1 450	1 340	1 230	1 110	977	873	721	615	537	W21x132
152	1 230	1 020	842	711	613	537	429	356	305	WWF28x102
174	1 370	1 230	1 090	924	803	709	574	482	415	W24x117
213	1 440	1 350	1 270	1 190	1 110	1 010	848	733	646	W18x143
147	840	671	555	470	407	358	288	241	207	**W30x99**
182	1 320	1 210	1 100	969	851	759	625	531	463	W21x122
152	898	728	610	523	458	406	332	281	244	W27x102
193	1 270	1 180	1 100	1 020	925	836	702	606	533	W18x130
155	1 180	1 050	901	762	659	579	465	388	333	W24x104

BEAM SELECTION TABLE
WWF and W Shapes

CSA G40.21 350W
ASTM A992, A572 grade 50

Designation	V_r	I_x	b	L_u	M_r	Factored moment resistance M_r' (kN·m) Unbraced length (mm)					
	kN	10^6 mm^4	mm	mm	$\leq L_u$	2 500	3 000	3 500	4 000	4 500	5 000
W760x134	1 650	1 500	264	3 230	**1 440**	—	—	1 400	1 330	1 250	1 160
W610x153	1 790	1 250	229	3 110	1 430	—	—	1 380	1 310	1 240	1 160
W690x140	1 740	1 360	254	3 270	1 410	—	—	1 380	1 320	1 240	1 160
W530x165	1 570	1 110	313	4 440	1 410	—	—	—	—	—	1 360
W460x177	1 640	910	286	4 330	1 330	—	—	—	—	1 320	1 280
W610x140	1 660	1 120	230	3 070	1 290	—	—	1 240	1 170	1 100	1 030
W530x150	1 410	1 010	312	4 380	1 290	—	—	—	—	1 280	1 240
W690x125	1 610	1 180	253	3 190	**1 250**	—	—	1 210	1 140	1 070	999
W460x158	1 460	796	284	4 200	1 170	—	—	—	—	1 150	1 110
W610x125	1 490	985	229	3 020	1 140	—	—	1 090	1 020	959	889
W530x138	1 650	861	214	2 930	1 120	—	1 110	1 060	1 000	945	884
W460x144	1 320	726	283	4 130	1 070	—	—	—	—	1 050	1 010
W610x113	1 400	875	228	2 950	**1 020**	—	—	964	906	843	775
W410x149	1 320	619	265	4 080	1 010	—	—	—	—	984	953
W530x123	1 460	761	212	2 860	997	—	984	933	879	822	762
W360x162	992	515	371	5 980	975	—	—	—	—	—	—
W460x128	1 170	637	282	4 040	947	—	—	—	—	917	884
W610x101	1 300	764	228	2 890	**900**	—	891	842	787	728	664
W410x132	1 160	538	263	3 940	885	—	—	—	882	853	823
W530x109	1 280	667	211	2 810	879	—	862	815	764	709	652
W460x113	1 020	556	280	3 950	829	—	—	—	826	796	765
W530x101	1 200	617	210	2 770	814	—	794	749	699	647	591
†W360x147	907	463	370	6 190	798	—	—	—	—	—	—
W610x91	1 100	667	227	2 810	**795**	—	779	732	681	625	564
W610x92	1 350	646	179	2 180	779	744	683	614	540	448	376
W410x114	998	462	261	3 810	764	—	—	—	754	727	698
W460x106	1 210	488	194	2 690	742	—	719	679	637	594	549
W530x92	1 110	552	209	2 720	733	—	711	668	621	570	516
†W360x134	817	415	369	6 030	723	—	—	—	—	—	—
W360x122	967	365	257	4 040	705	—	—	—	—	686	664
W610x82	1 170	560	178	2 110	**683**	644	587	522	448	364	304
W460x97	1 090	445	193	2 650	677	—	652	614	574	531	488
W310x129	854	308	308	5 080	671	—	—	—	—	—	—
W410x100	850	398	260	3 730	661	—	—	—	648	623	596
W530x85	1 130	485	166	2 110	652	616	564	507	446	373	316
W530x82	1 030	478	209	2 660	643	—	618	578	533	485	434
†W610x84	944	613	226	2 980	640	—	639	605	568	527	483
W360x110	841	331	256	3 940	640	—	—	—	637	617	596
W460x89	996	409	192	2 620	624	—	598	562	523	482	439
W310x118	766	275	307	4 920	605	—	—	—	—	—	603
W360x101	768	301	255	3 860	584	—	—	—	578	559	538
W460x82	933	370	191	2 560	568	—	540	505	466	426	384
W530x74	1 050	411	166	2 040	**562**	523	474	420	357	293	247
W310x107	695	248	306	4 800	546	—	—	—	—	—	541
W410x85	931	315	181	2 530	534	—	507	475	443	409	375
W360x91	687	267	254	3 760	522	—	—	—	513	494	475
W460x74	843	332	190	2 530	512	—	484	450	414	375	332

Note: The designation comprises the nominal depth in millimetres and the mass in kilograms per metre.

† Class 3 section

F_y taken as 350 MPa for WWF shapes and 345 MPa for W shapes. $\phi = 0.90$

CSA G40.21 350W
ASTM A992, A572 grade 50

BEAM SELECTION TABLE
WWF and W Shapes

| Nominal mass | Factored moment resistance M_r' (kN·m) ||||||||| Imperial designation |
| | Unbraced length (mm) ||||||||| |
kg/m	6 000	7 000	8 000	9 000	10 000	11 000	12 000	14 000	16 000	
134	967	738	587	483	408	351	308	246	205	**W30x90**
153	1 000	815	675	576	502	445	400	333	286	W24x103
140	987	778	628	523	447	389	345	280	236	W27x94
165	1 270	1 170	1 060	954	823	721	641	525	445	W21x111
177	1 200	1 120	1 030	953	865	770	694	581	500	W18x119
140	874	695	573	486	422	373	334	277	237	W24x94
150	1 150	1 050	943	829	709	618	548	446	377	W21x101
125	834	640	513	425	362	314	277	223	187	**W27x84**
158	1 040	955	875	794	696	617	555	462	397	W18x106
125	733	575	470	396	342	301	269	222	189	W24x84
138	759	616	515	444	390	347	314	263	227	W21x93
144	936	858	779	693	602	533	478	396	339	W18x97
113	617	481	391	328	282	247	220	180	153	**W24x76**
149	889	825	761	697	622	555	502	421	364	W16x100
123	631	505	421	361	316	281	253	211	182	W21x83
162	974	935	895	855	814	773	733	653	558	W14x109
128	812	736	658	566	489	431	385	318	271	W18x86
101	512	396	320	267	228	199	176	144	121	**W24x68**
132	762	699	635	566	496	441	398	333	287	W16x89
109	520	413	342	291	254	225	202	168	144	W21x73
113	696	623	545	458	394	345	307	252	213	W18x76
101	462	365	301	255	222	196	176	146	125	W21x68
147	—	773	740	708	675	642	609	544	467	W14x99
91	421	325	261	217	185	161	142	115	97.1	**W24x61**
92	281	222	183	156	135	120	107	88.9	76.1	W24x62
114	638	576	514	439	383	339	305	254	218	W16x77
106	450	366	308	266	235	210	190	160	138	W18x71
92	393	309	253	214	185	163	146	120	103	W21x62
134	—	694	663	630	598	565	532	461	392	W14x90
122	621	577	534	491	441	395	357	301	260	W14x82
82	225	177	145	123	106	93.4	83.4	68.9	58.7	**W24x55**
97	389	314	264	227	200	178	161	135	117	W18x65
129	643	612	581	551	520	490	460	390	336	W12x87
100	539	480	412	349	302	266	238	197	168	W16x67
85	240	193	162	139	122	108	97.8	81.9	70.6	W21x57
82	321	250	204	171	148	130	115	94.8	80.5	W21x55
84	377	289	232	192	163	142	125	101	84.4	W24x56
110	553	510	467	423	372	332	300	252	217	W14x74
89	343	276	231	198	174	155	140	117	101	W18x60
118	574	543	513	482	452	422	388	324	279	W12x79
101	497	454	411	363	318	283	255	214	184	W14x68
82	292	234	195	166	146	129	116	97.3	83.6	W18x55
74	186	148	123	105	91.7	81.4	73.2	61.0	52.4	**W21x50**
107	512	483	453	423	392	362	325	271	233	W12x72
85	297	242	205	177	157	140	127	107	92.8	W16x57
91	434	392	350	299	261	232	209	174	150	W14x61
74	249	198	164	140	122	108	96.8	80.6	69.1	W18x50

BEAM SELECTION TABLE
WWF and W Shapes

CSA G40.21 350W
ASTM A992, A572 grade 50

Designation	V_r	I_x	b	L_u	M_r	Factored moment resistance M_r' (kN·m) Unbraced length (mm)					
	kN	10^6 mm⁴	mm	mm	≤ L_u	2 000	2 500	3 000	3 500	4 000	4 500
W530x66	928	351	165	1 980	**484**	483	444	398	347	284	232
†W530x72	926	401	207	2 750	475	—	—	462	434	402	368
W410x74	821	275	180	2 470	469	—	467	440	410	379	346
W460x68	856	297	154	2 010	463	—	429	390	348	301	250
W460x67	791	296	190	2 480	460	—	458	430	398	363	326
†W310x97	625	222	305	4 970	447	—	—	—	—	—	—
W360x79	682	226	205	3 010	444	—	—	—	425	404	383
W310x86	578	198	254	3 900	441	—	—	—	—	438	424
W250x101	644	164	257	4 470	435	—	—	—	—	—	434
W410x67	739	245	179	2 420	422	—	418	392	364	333	301
W460x61	747	255	189	2 410	**401**	—	396	370	340	307	271
W460x60	746	255	153	1 970	**397**	396	365	329	289	242	200
W360x72	617	201	204	2 940	397	—	—	395	377	357	336
W310x79	552	177	254	3 810	397	—	—	—	—	392	378
W250x89	570	143	256	4 260	382	—	—	—	—	—	377
W410x60	642	216	178	2 390	369	—	365	341	314	286	255
W310x74	597	165	205	3 100	369	—	—	—	357	342	326
W200x100	680	113	210	4 430	357	—	—	—	—	—	356
W360x64	548	178	203	2 870	354	—	—	350	332	313	293
W460x52	680	212	152	1 890	**338**	333	303	269	231	185	152
W250x80	493	126	255	4 130	338	—	—	—	—	—	331
W310x67	533	145	204	3 020	329	—	—	—	315	300	285
W410x54	619	186	177	2 310	326	—	318	295	269	241	210
W360x57	580	160	172	2 360	314	—	309	289	267	244	220
W250x73	446	113	254	4 010	306	—	—	—	—	—	297
W200x86	591	94.4	209	4 110	304	—	—	—	—	—	298
W310x60	466	129	203	2 950	292	—	—	291	277	263	248
W250x67	469	104	204	3 260	280	—	—	—	275	265	254
W360x51	524	141	171	2 320	**277**	—	271	252	232	210	187
W410x46	578	156	140	1 790	**274**	265	239	210	177	142	117
W310x52	495	119	167	2 370	261	—	257	241	224	206	188
W200x71	452	76.3	206	3 730	248	—	—	—	—	245	238
W360x45	498	122	171	2 260	**242**	—	234	217	197	176	152
W250x58	413	87.3	203	3 130	239	—	—	—	232	222	212
W410x39	480	126	140	1 730	**227**	216	193	166	132	105	86.5
W310x45	423	99.2	166	2 310	220	—	215	200	184	167	150
W360x39	470	102	128	1 660	206	193	172	148	120	97.0	81.2
W200x59	392	60.9	205	3 430	202	—	—	—	201	194	188
W310x39	368	85.1	165	2 260	189	—	184	170	155	139	121
W250x45	414	71.1	148	2 170	187	—	179	167	155	142	129
†W250x49	375	70.6	202	3 160	178	—	—	—	173	165	157
W200x52	334	52.5	204	3 300	176	—	—	—	173	167	160

Note: The designation comprises the nominal depth in millimetres and the mass in kilograms per metre.
† Class 3 section
F_y taken as 350 MPa for WWF shapes and 345 MPa for W shapes. $\phi = 0.90$

BEAM SELECTION TABLE
WWF and W Shapes

CSA G40.21 350W
ASTM A992, A572 grade 50

Nominal mass	Factored moment resistance M_r' (kN·m) Unbraced length (mm)									Imperial designation
kg/m	5 000	6 000	7 000	8 000	9 000	10 000	12 000	14 000	16 000	
66	195	145	115	94.9	80.6	70.0	55.5	46.0	39.4	**W21x44**
72	331	247	191	155	129	111	85.7	69.9	59.1	W21x48
74	312	239	194	163	140	123	99.7	83.8	72.3	W16x50
68	214	164	133	112	96.8	85.3	69.0	58.0	50.1	W18x46
67	281	209	165	136	115	99.8	79.0	65.4	55.9	W18x45
97	446	423	400	375	351	326	272	226	193	W12x65
79	361	317	267	225	194	171	139	117	101	W14x53
86	409	379	347	316	282	248	199	167	144	W12x58
101	424	403	382	362	342	322	279	236	205	W10x68
67	264	201	161	135	116	102	81.6	68.3	58.9	W16x45
61	228	168	132	108	90.9	78.5	61.7	50.8	43.2	**W18x41**
60	169	129	104	86.7	74.4	65.3	52.4	43.9	37.8	**W18x40**
72	315	272	222	186	160	141	114	95.6	82.5	W14x48
79	364	334	304	273	237	207	166	139	119	W12x53
89	367	346	326	306	285	265	220	186	161	W10x60
60	218	165	131	109	93.1	81.3	64.9	54.1	46.4	W16x40
74	310	277	243	206	179	159	129	109	94.5	W12x50
100	349	335	321	307	293	279	252	220	192	W8x67
64	273	228	183	153	131	115	92.2	77.2	66.5	W14x43
52	128	96.3	76.8	63.7	54.4	47.4	37.9	31.6	27.1	**W18x35**
80	321	302	282	262	242	221	179	151	130	W10x54
67	269	237	200	169	146	129	105	88.4	76.4	W12x45
54	176	132	104	86.0	73.1	63.5	50.4	41.8	35.8	W16x36
57	192	147	119	99.7	85.9	75.6	61.0	51.2	44.2	W14x38
73	287	268	248	228	209	185	149	126	108	W10x49
86	291	278	264	250	237	224	196	167	145	W8x58
60	233	202	166	139	120	106	85.2	71.6	61.8	W12x40
67	244	223	202	180	157	139	114	96.5	83.8	W10x45
51	159	121	96.9	80.9	69.4	60.8	48.8	40.8	35.2	**W14x34**
46	99.9	76.4	61.7	51.8	44.6	39.2	31.6	26.5	22.9	**W16x31**
52	167	130	106	89.4	77.5	68.4	55.5	46.8	40.5	W12x35
71	231	218	204	191	178	164	136	115	100	W8x48
45	128	96.0	76.4	63.3	54.0	47.1	37.5	31.3	26.8	**W14x30**
58	202	181	161	137	119	105	85.7	72.4	62.8	W10x39
39	73.0	55.1	44.0	36.5	31.2	27.3	21.8	18.2	15.6	**W16x26**
45	128	98.2	79.3	66.5	57.3	50.4	40.6	34.1	29.4	W12x30
39	69.7	54.1	44.2	37.4	32.5	28.7	23.3	19.7	17.0	W14x26
59	181	168	154	141	126	112	92.2	78.2	68.0	W8x40
39	103	77.7	62.2	51.8	44.3	38.8	31.1	25.9	22.3	W12x26
45	114	90.6	75.2	64.4	56.3	50.1	41.1	34.9	30.3	W10x30
49	149	133	115	97.2	84.0	74.1	60.0	50.5	43.6	W10x33
52	154	141	128	114	99.6	88.4	72.3	61.3	53.2	W8x35

BEAM SELECTION TABLE
WWF and W Shapes

CSA G40.21 350W
ASTM A992, A572 grade 50

Designation	V_r	I_x	b	L_u	M_r	Factored moment resistance M_r' (kN·m)					
						Unbraced length (mm)					
	kN	$10^6\,mm^4$	mm	mm	$\leq L_u$	1 500	2 000	2 500	3 000	3 500	4 000
W360x33	396	82.6	127	1 600	**168**	—	155	135	113	87.5	70.2
W250x39	354	60.1	147	2 110	159	—	—	151	140	128	115
W310x33	423	65.0	102	1 330	149	143	125	104	80.5	64.2	53.3
†W200x46	300	45.2	203	3 370	138	—	—	—	—	137	132
W200x42	302	40.6	166	2 610	137	—	—	—	132	126	119
‡W310x31	294	67.2	164	2 310	**133**	—	—	130	121	110	99.2
W250x33	323	48.9	146	2 020	132	—	—	122	112	100	88.4
W310x28	380	54.3	102	1 290	**126**	120	103	83.1	61.9	48.8	40.2
W200x36	255	34.1	165	2 510	117	—	—	—	111	105	98.1
W250x28	341	40.0	102	1 370	110	107	94.5	81.0	65.4	52.6	44.0
W200x31	275	31.4	134	1 980	104	—	—	96.7	89.3	81.7	74.0
W310x24	350	42.7	101	1 210	**102**	94.2	78.3	58.1	42.8	33.5	27.3
W150x37	269	22.2	154	2 630	96.3	—	—	—	93.3	89.4	85.4
W250x25	321	34.2	102	1 330	95.3	91.7	80.0	66.9	51.6	41.2	34.2
W310x21	303	37.0	101	1 190	**89.1**	81.5	66.7	47.9	35.0	27.2	22.0
W200x27	246	25.8	133	1 890	86.6	—	85.3	78.7	71.5	64.1	56.0
‡W250x24	259	34.7	145	2 080	81.7	—	—	76.8	70.1	62.6	54.7
W250x22	302	28.9	102	1 280	81.7	77.6	66.6	53.9	40.3	31.9	26.3
W150x30	212	17.1	153	2 440	75.8	—	—	75.3	71.6	67.8	63.9
W200x22	262	20.0	102	1 390	68.9	67.4	60.1	52.1	43.2	35.0	29.4
†W200x21	208	19.8	133	1 930	60.5	—	59.9	55.2	50.0	44.4	38.0
W150x24	216	13.4	102	1 630	59.3	—	56.2	51.9	47.7	43.4	39.1
W200x19	241	16.6	102	1 340	**58.1**	56.0	49.2	41.5	32.7	26.2	21.9
†W250x18	247	22.4	101	1 330	**55.6**	53.4	46.1	37.5	27.8	21.7	17.7
‡W150x22	181	12.0	152	2 480	46.2	—	—	46.1	43.7	41.1	38.5
W150x18	182	9.15	102	1 480	42.2	42.1	38.2	34.2	30.1	25.4	21.5
†**W200x15**	176	12.7	100	1 380	**39.4**	38.5	33.8	28.5	22.1	17.4	14.4
W150x14	132	6.85	100	1 400	**31.7**	31.0	27.6	23.8	19.6	15.7	13.1
†**W150x13**	130	6.11	100	1 460	**25.6**	25.4	22.8	19.9	16.8	13.5	11.2

Note: The designation comprises the nominal depth in millimetres and the mass in kilograms per metre.
† Class 3 section   ‡ Class 4 section. M_r and M_r' calculated according to CAN/CSA-S16.1-94 Clause 13.5(c)(iii).
F_y taken as 350 MPa for WWF shapes and 345 MPa for W shapes. $\phi = 0.90$

CSA G40.21 350W
ASTM A992, A572 grade 50

BEAM SELECTION TABLE
WWF and W Shapes

Nominal mass	Factored moment resistance M_r' (kN·m)									Imperial designation
	Unbraced length (mm)									
kg/m	5 000	6 000	7 000	8 000	9 000	10 000	11 000	12 000	14 000	
33	49.6	38.0	30.8	25.8	22.3	19.6	17.5	15.8	13.3	**W14x22**
39	88.0	69.2	57.0	48.6	42.3	37.6	33.8	30.7	26.0	W10x26
33	39.7	31.7	26.4	22.7	19.9	17.7	16.0	14.6	12.4	W12x22
46	121	111	100	89.1	77.4	68.5	61.5	55.8	47.2	W8x31
42	105	91.9	76.5	65.6	57.5	51.2	46.2	42.1	35.7	W8x28
31	72.7	54.5	43.2	35.7	30.4	26.4	23.4	21.0	17.4	**W12x21**
33	63.6	49.4	40.3	34.1	29.6	26.1	23.4	21.2	17.9	W10x22
28	29.5	23.4	19.4	16.5	14.5	12.8	11.6	10.5	8.93	**W12x19**
36	84.9	70.2	58.0	49.5	43.2	38.4	34.6	31.4	26.7	W8x24
28	33.1	26.6	22.3	19.2	16.9	15.1	13.6	12.4	10.6	W10x19
31	57.0	45.6	38.0	32.7	28.7	25.6	23.1	21.0	17.9	W8x21
24	19.8	15.5	12.8	10.9	9.45	8.37	7.52	6.83	5.78	**W12x16**
37	77.6	69.8	61.5	53.2	46.9	42.0	38.0	34.7	29.6	W6x25
25	25.5	20.4	17.0	14.6	12.8	11.4	10.3	9.40	7.99	W10x17
21	15.8	12.3	10.0	8.49	7.36	6.50	5.83	5.29	4.46	**W12x14**
27	41.2	32.6	27.1	23.1	20.2	18.0	16.2	14.7	12.5	W8x18
24	38.2	29.0	23.3	19.4	16.6	14.6	13.0	11.7	9.77	W10x16
22	19.5	15.4	12.8	11.0	9.60	8.54	7.70	7.01	5.95	W10x15
30	56.3	48.1	40.2	34.6	30.4	27.2	24.5	22.4	19.1	W6x20
22	22.3	18.0	15.1	13.0	11.5	10.3	9.28	8.47	7.22	W8x15
21	27.4	21.4	17.5	14.8	12.9	11.4	10.2	9.29	7.85	W8x14
24	30.4	25.0	21.2	18.4	16.3	14.6	13.3	12.1	10.4	W6x16
19	16.4	13.2	11.0	9.48	8.33	7.44	6.72	6.13	5.22	**W8x13**
18	12.8	10.0	8.24	7.00	6.09	5.40	4.85	4.40	3.73	**W10x12**
22	33.2	27.2	22.5	19.2	16.8	14.9	13.4	12.2	10.4	W6x15
18	16.5	13.4	11.3	9.80	8.64	7.74	7.01	6.41	5.47	W6x12
15	10.6	8.35	6.92	5.91	5.17	4.59	4.14	3.76	3.19	**W8x10**
14	9.91	7.97	6.67	5.75	5.06	4.51	4.08	3.72	3.17	**W6x9**
13	8.42	6.75	5.64	4.85	4.26	3.80	3.44	3.13	2.67	**W6x8.5**

BEAM SELECTION TABLE
C Shapes

G40.21 300W

$\phi = 0.90$

Designation	V_r	I_x	b	L_u	M_r	Factored moment resistance M_r' (kN·m)					
						Unbraced length (mm)					
	kN	10^6 mm^4	mm	mm	$\leq L_u$	1 500	2 000	2 500	3 000	3 500	4 000
†C250x37	607	37.9	73	1 630	**80.7**	—	77.0	71.9	66.9	61.9	57.0
†C250x30	435	32.7	69	1 460	**69.4**	69.0	63.4	57.7	52.0	46.4	39.8
†C250x23	276	27.8	65	1 340	**59.1**	57.3	51.4	45.2	38.9	32.1	27.3
†C230x22	294	21.3	63	1 330	**50.2**	48.7	44.0	39.2	34.4	29.0	24.9
†C200x28	449	18.2	64	1 540	48.6	—	45.8	42.9	39.9	37.0	34.1
†C230x20	241	19.8	61	1 270	**46.7**	44.6	39.8	35.0	29.7	24.7	21.1
†C200x21	279	14.9	59	1 290	39.7	38.2	34.6	31.1	27.5	23.5	20.3
†C200x17	203	13.5	57	1 210	**35.9**	33.8	30.1	26.3	22.2	18.6	16.0
†C180x22	336	11.3	58	1 410	34.3	33.9	31.6	29.3	27.0	24.8	22.5
†C180x18	254	10.0	55	1 250	30.5	29.2	26.6	24.1	21.5	18.7	16.2
†C180x15	168	8.86	53	1 150	**26.9**	25.0	22.2	19.4	16.4	13.8	11.9
†C150x19	301	7.12	54	1 470	25.3	25.2	23.7	22.2	20.7	19.3	17.8
†C150x16	217	6.22	51	1 230	22.1	21.2	19.4	17.7	16.0	14.2	12.3
†C150x12	138	5.36	48	1 080	**19.1**	17.4	15.5	13.6	11.4	9.67	8.38
†C130x13	188	3.66	47	1 260	15.6	15.0	13.9	12.9	11.8	10.7	9.48
†C130x10	109	3.09	44	1 050	**13.1**	11.9	10.7	9.43	8.04	6.83	5.94
†C100x11	149	1.91	43	1 330	10.1	9.88	9.25	8.63	8.02	7.40	6.79
†C100x9	115	1.77	42	1 170	**9.34**	8.88	8.19	7.51	6.84	6.14	5.35
†C100x8	85.4	1.61	40	1 040	**8.53**	7.83	7.08	6.35	5.59	4.76	4.15
†C75x9	122	0.846	40	1 670	6.02	—	5.84	5.56	5.28	5.01	4.73
†C75x7	89.4	0.749	37	1 310	**5.32**	5.19	4.87	4.55	4.23	3.91	3.59
†C75x6	58.2	0.670	35	1 070	**4.75**	4.44	4.07	3.71	3.35	2.94	2.56

Note: The designation comprises the nominal depth in millimetres and the mass in kilograms per metre.
† Class 3 section

G40.21 300W
$\phi = 0.90$

BEAM SELECTION TABLE
C Shapes

Nominal mass	Factored moment resistance M_r' (kN·m)									Imperial designation
	Unbraced length (mm)									
kg/m	4 500	5 000	6 000	7 000	8 000	9 000	10 000	11 000	12 000	
37	51.5	46.1	38.0	32.4	28.2	25.0	22.5	20.4	18.7	**C10x25**
30	34.8	31.0	25.5	21.7	18.8	16.7	15.0	13.6	12.4	C10x20
23	23.8	21.1	17.3	14.6	12.7	11.2	10.1	9.13	8.35	C10x15.3
22	21.8	19.4	16.0	13.6	11.8	10.4	9.37	8.50	7.78	**C9x15**
28	30.9	27.7	23.0	19.6	17.1	15.2	13.7	12.4	11.4	C8x18.75
20	18.5	16.5	13.5	11.5	9.98	8.83	7.92	7.18	6.57	**C9x13.4**
21	17.8	15.9	13.2	11.2	9.77	8.67	7.78	7.07	6.47	C8x13.75
17	14.0	12.5	10.3	8.77	7.64	6.77	6.08	5.51	5.05	**C8x11.5**
22	19.9	17.8	14.8	12.6	11.0	9.79	8.81	8.00	7.33	C7x14.75
18	14.3	12.8	10.6	9.04	7.89	7.00	6.29	5.71	5.23	C7x12.25
15	10.5	9.36	7.73	6.59	5.75	5.10	4.58	4.16	3.81	**C7x9.8**
19	16.2	14.6	12.1	10.4	9.05	8.04	7.23	6.57	6.02	C6x13
16	10.9	9.78	8.12	6.94	6.06	5.38	4.84	4.40	4.03	C6x10.5
12	7.41	6.63	5.50	4.69	4.10	3.64	3.27	2.97	2.72	**C6x8.2**
13	8.41	7.55	6.28	5.37	4.70	4.17	3.75	3.41	3.13	C5x9
10	5.26	4.72	3.92	3.35	2.93	2.60	2.34	2.12	1.95	**C5x6.7**
11	6.04	5.43	4.52	3.87	3.38	3.01	2.71	2.46	2.25	C4x7.25
9	4.75	4.27	3.55	3.04	2.66	2.36	2.12	1.93	1.77	**C4x6.25**
8	3.68	3.31	2.75	2.36	2.06	1.83	1.65	1.50	1.37	**C4x5.4**
9	4.45	4.18	3.54	3.03	2.65	2.36	2.12	1.93	1.77	C3x6
7	3.20	2.88	2.39	2.05	1.79	1.59	1.44	1.30	1.20	**C3x5**
6	2.28	2.05	1.70	1.46	1.28	1.13	1.02	0.928	0.850	**C3x4.1**

BEAM SELECTION TABLE
S Shapes

ASTM A572 Gr. 50
$\phi = 0.90$

Designation	V_r	I_x	b	L_u	M_r	Factored moment resistance M_r' (kN·m)					
						Unbraced length (mm)					
	kN	10^6 mm^4	mm	mm	$\leq L_u$	1 500	2 000	2 500	3 000	3 500	4 000
S610x180*	2 590	1 310	204	2 570	**1 560**	—	—	—	1 490	1 410	1 320
S610x158*	2 000	1 220	200	2 560	**1 420**	—	—	—	1 360	1 280	1 200
S610x149*	2 360	996	184	2 150	**1 220**	—	—	1 170	1 080	993	902
S610x134*	1 990	939	181	2 140	**1 130**	—	—	1 080	1 000	914	825
S610x119*	1 590	879	178	2 150	**1 040**	—	—	995	920	839	754
S510x143*	2 150	700	183	2 310	1 010	—	—	989	933	875	816
S510x128*	1 780	658	179	2 280	935	—	—	911	856	798	740
S510x112*	1 680	532	162	1 960	**776**	—	772	717	659	599	538
S510x98.2*	1 330	497	159	1 950	**711**	—	707	655	600	541	482
S460x104*	1 700	387	159	1 890	637	—	627	583	536	489	441
S460x81.4*	1 100	335	152	1 850	**534**	—	522	480	435	388	335
S380x74*	1 090	203	143	1 750	**394**	—	380	350	318	286	251
S380x64*	812	187	140	1 750	**354**	—	340	311	281	250	213
S310x74*	1 090	127	139	1 950	311	—	309	291	274	257	240
S310x60.7*	731	113	133	1 840	**270**	—	264	246	229	211	193
S310x52*	681	95.8	129	1 650	**229**	—	216	198	179	160	138
S310x47*	556	91.1	127	1 640	**214**	—	202	184	165	146	124
S250x52*	786	61.6	126	1 700	181	—	174	162	151	139	128
S250x38*	411	51.4	118	1 570	**144**	—	134	122	110	97.7	83.3
S200x34*	466	27.0	106	1 470	**98.1**	97.6	90.2	82.8	75.4	68.1	59.7
S200x27*	287	24.0	102	1 400	**84.5**	83.1	75.7	68.1	60.6	52.2	44.7
S150x26*	368	10.9	91	1 410	**53.7**	53.1	49.4	45.8	42.3	38.8	35.1
S150x19*	184	9.19	85	1 230	**43.2**	41.1	37.1	33.1	29.2	24.7	21.4
S130x15*	141	5.12	76	1 160	**28.8**	27.0	24.3	21.7	19.1	16.2	14.0
S100x14.1*	173	2.85	71	1 230	**20.6**	19.8	18.3	16.8	15.4	13.9	12.2
S100x11*	102	2.56	68	1 100	**18.0**	16.7	15.1	13.5	12.0	10.2	8.9
S75x11*	139	1.22	64	1 390	12.0	11.9	11.2	10.5	9.8	9.1	8.4
S75x8*	67.0	1.04	59	1 100	**9.9**	9.3	8.5	7.7	6.9	6.0	5.2

Note: The designation comprises the nominal depth in millimetres and the mass in kilograms per metre.

* Not available from Canadian mills; M_r and M_r' calculated using ASTM A572 Grade 50 with F_y = 345 MPa.

ASTM A572 Gr. 50
φ = 0.90

BEAM SELECTION TABLE
S Shapes

Nominal mass	Factored moment resistance M_r' (kN·m)									Imperial designation
	Unbraced length (mm)									
kg/m	5 000	6 000	7 000	8 000	9 000	10 000	12 000	14 000	16 000	
180	1 140	944	778	662	577	511	418	353	307	**S24x121**
158	1 030	832	682	578	502	444	362	305	265	**S24x106**
149	698	549	453	385	336	298	244	206	179	**S24x100**
134	625	488	401	340	295	262	213	180	156	**S24x90**
119	564	437	357	302	262	231	188	158	137	**S24x80**
143	698	566	471	405	355	316	260	221	192	S20x96
128	621	492	409	350	306	272	224	190	165	S20x86
112	406	323	269	230	202	180	148	126	109	**S20x75**
98.2	356	282	234	200	174	155	127	108	93.7	**S20x66**
104	338	271	227	196	172	154	127	108	94.2	S18x70
81.4	248	196	163	140	122	109	89.2	75.7	65.9	**S18x54.7**
74	190	153	129	111	97.8	87.4	72.2	61.5	53.6	**S15x50**
64	160	128	107	92.2	80.9	72.2	59.5	50.6	44.1	**S15x42.9**
74	206	168	143	124	110	98.5	81.8	69.9	61.0	S12x50
60.7	154	125	106	91.7	80.9	72.5	60.1	51.3	44.8	**S12x40.8**
52	105	85.4	72.0	62.3	55.0	49.2	40.7	34.7	30.3	**S12x35**
47	94.4	76.4	64.3	55.6	49.0	43.8	36.2	30.9	26.9	**S12x31.8**
52	102	83.9	71.3	62.0	54.9	49.3	40.9	35.0	30.6	S10x35
38	64.1	52.2	44.1	38.2	33.8	30.2	25.0	21.4	18.7	**S10x25.4**
34	46.8	38.6	32.9	28.6	25.4	22.8	18.9	16.2	14.2	**S8x23**
27	34.8	28.6	24.3	21.1	18.7	16.8	13.9	11.9	10.4	**S8x18.4**
26	27.8	23.1	19.7	17.2	15.3	13.7	11.4	9.79	8.57	**S6x17.25**
19	16.8	13.9	11.9	10.3	9.17	8.24	6.85	5.87	5.13	**S6x12.5**
15	11.1	9.20	7.85	6.86	6.09	5.47	4.55	3.90	3.41	**S5x10**
14.1	9.7	8.06	6.90	6.03	5.35	4.82	4.01	3.44	3.01	**S4x9.5**
11	7.0	5.84	5.00	4.37	3.88	3.49	2.90	2.49	2.18	**S4x7.7**
11	6.9	5.72	4.90	4.28	3.81	3.43	2.85	2.45	2.14	S3x7.5
8	4.2	3.46	2.97	2.59	2.30	2.07	1.73	1.48	1.29	**S3x5.7**

BEAM LOAD TABLES
WWF Shapes

Total Uniformly Distributed Factored Loads
for Laterally Supported Beams – kN

G40.21-M
350W

$\phi = 0.90$

Designation Mass (kg/m)	WWF2000				Approx. Deflect. (mm)	WWF1800					Approx. Deflect. (mm)
	732*	648*	607*	542*		700	659	617	575	510	
13 000					22					7 960	25
13 500					24					7 800	26
14 000					26					7 520	28
14 500				7 140	27					7 260	31
15 000				6 960	29				8 000	7 020	33
15 500				6 730	31				7 980	6 800	35
16 000				6 520	33				7 730	6 580	37
16 500				6 320	36			8 050	7 500	6 380	40
17 000			7 170	6 140	38			7 960	7 280	6 200	42
17 500			7 100	5 960	40			7 730	7 070	6 020	44
18 000			6 900	5 800	42		8 090	7 520	6 870	5 850	47
18 500		7 210	6 720	5 640	45		7 930	7 320	6 690	5 690	50
19 000		7 190	6 540	5 490	47	8 140	7 720	7 120	6 510	5 540	52
19 500		7 000	6 370	5 350	50	8 100	7 520	6 940	6 340	5 400	55
20 000		6 830	6 210	5 220	52	7 900	7 330	6 770	6 190	5 270	58
20 500		6 660	6 060	5 090	55	7 710	7 150	6 600	6 040	5 140	61
21 000		6 500	5 920	4 970	58	7 520	6 980	6 440	5 890	5 020	64
21 500		6 350	5 780	4 850	60	7 350	6 820	6 290	5 760	4 900	67
22 000	7 280	6 210	5 650	4 740	63	7 180	6 670	6 150	5 620	4 790	70
22 500	7 160	6 070	5 520	4 640	66	7 020	6 520	6 010	5 500	4 680	74
23 000	7 000	5 940	5 400	4 540	69	6 870	6 380	5 880	5 380	4 580	77
23 500	6 850	5 810	5 290	4 440	72	6 720	6 240	5 760	5 260	4 480	80
24 000	6 710	5 690	5 180	4 350	75	6 580	6 110	5 640	5 160	4 390	84
24 500	6 570	5 580	5 070	4 260	78	6 450	5 990	5 520	5 050	4 300	87
25 000	6 440	5 460	4 970	4 170	82	6 320	5 870	5 410	4 950	4 210	91
25 500	6 320	5 360	4 870	4 090	85	6 200	5 750	5 310	4 850	4 130	94
26 000	6 190	5 250	4 780	4 010	88	6 080	5 640	5 200	4 760	4 050	98
26 500	6 080	5 150	4 690	3 940	92	5 960	5 530	5 110	4 670	3 980	102
27 000	5 960	5 060	4 600	3 860	95	5 850	5 430	5 010	4 580	3 900	106
27 500	5 860	4 970	4 520	3 790	99	5 750	5 330	4 920	4 500	3 830	110
28 000	5 750	4 880	4 440	3 730	102	5 640	5 240	4 830	4 420	3 760	114
28 500	5 650	4 790	4 360	3 660	106	5 540	5 150	4 750	4 340	3 700	118
29 000	5 550	4 710	4 280	3 600	110	5 450	5 060	4 670	4 270	3 630	122
29 500	5 460	4 630	4 210	3 540	114	5 360	4 970	4 590	4 190	3 570	126
30 000	5 370	4 550	4 140	3 480	118	5 270	4 890	4 510	4 120	3 510	131

Span in Millimetres

DESIGN DATA AND PROPERTIES

V_r (kN)	3 640	3 600	3 590	3 570		4 070	4 050	4 020	4 000	3 980	
L_u (mm)	7 620	7 370	7 210	6 310		7 300	7 170	7 020	6 860	5 960	
d (mm)	2 000	2 000	2 000	2 000		1 800	1 800	1 800	1 800	1 800	
b (mm)	550	550	550	500		550	550	550	550	500	
t (mm)	50.0	40.0	35.0	30.0		50.0	45.0	40.0	35.0	30.0	
w (mm)	20.0	20.0	20.0	20.0		20.0	20.0	20.0	20.0	20.0	
k (mm)	61	51	46	41		61	56	51	46	41	

IMPERIAL SIZE AND MASS

Mass (lb./ft.)	490	436	408	364		470	442	415	388	344	
Nominal Depth (in.)		79						71			

* Class 3 section

G40.21-M 350W
$\phi = 0.90$

BEAM LOAD TABLES
WWF Shapes

Total Uniformly Distributed Factored Loads for Laterally Supported Beams – kN

Designation	WWF1600					Approx. Deflect. (mm)	WWF1400					Approx. Deflect. (mm)
Mass (kg/m)	622*	580*	538*	496*	431*		597	513	471	405	358	
11 000						20					5 290	23
11 500						22					5 190	25
12 000						24					4 980	27
12 500						26					4 780	29
13 000						28				5 290	4 590	32
13 500						30				5 190	4 420	34
14 000						32				5 000	4 270	37
14 500						34				4 830	4 120	39
15 000						37				4 670	3 980	42
15 500						39				5 330	4 520	3 850	45
16 000					4 600	42			5 260	4 380	3 730	48
16 500					4 480	44			5 100	4 250	3 620	51
17 000					4 340	47		5 370	4 950	4 120	3 510	54
17 500					4 220	50		5 330	4 810	4 000	3 410	57
18 000					4 100	53		5 180	4 680	3 890	3 320	60
18 500					3 990	56		5 040	4 550	3 790	3 230	64
19 000				4 630	3 890	59		4 910	4 430	3 690	3 140	67
19 500				4 590	3 790	62		4 780	4 320	3 590	3 060	71
20 000				4 470	3 690	65	5 450	4 660	4 210	3 500	2 990	75
20 500				4 360	3 600	69	5 410	4 550	4 110	3 420	2 910	78
21 000			4 660	4 260	3 520	72	5 280	4 440	4 010	3 340	2 840	82
21 500			4 630	4 160	3 430	76	5 160	4 340	3 920	3 260	2 780	86
22 000			4 520	4 070	3 360	79	5 040	4 240	3 830	3 180	2 720	90
22 500			4 420	3 980	3 280	83	4 930	4 140	3 740	3 110	2 650	95
23 000		4 690	4 330	3 890	3 210	86	4 820	4 050	3 660	3 050	2 600	99
23 500		4 640	4 240	3 810	3 140	90	4 720	3 970	3 580	2 980	2 540	103
24 000		4 550	4 150	3 730	3 080	94	4 620	3 880	3 510	2 920	2 490	108
24 500		4 450	4 060	3 650	3 010	98	4 530	3 810	3 440	2 860	2 440	112
25 000	4 720	4 360	3 980	3 580	2 950	102	4 440	3 730	3 370	2 800	2 390	117
25 500	4 660	4 280	3 900	3 510	2 900	106	4 350	3 660	3 300	2 750	2 340	121
26 000	4 560	4 200	3 830	3 440	2 840	110	4 260	3 590	3 240	2 690	2 300	126
26 500	4 480	4 120	3 760	3 380	2 790	115	4 180	3 520	3 180	2 640	2 250	131
27 000	4 400	4 040	3 690	3 310	2 740	119	4 110	3 450	3 120	2 600	2 210	136
27 500	4 320	3 970	3 620	3 250	2 680	124	4 030	3 390	3 060	2 550	2 170	141
28 000	4 240	3 900	3 560	3 200	2 640	128	3 960	3 330	3 010	2 500	2 130	146

Span in Millimetres

DESIGN DATA AND PROPERTIES

V_r (kN)	2 360	2 350	2 330	2 320	2 300		2 730	2 680	2 660	2 640	2 640	
L_u (mm)	8 010	7 890	7 750	7 620	6 720		7 790	7 510	7 360	6 450	5 000	
d (mm)	1 600	1 600	1 600	1 600	1 600		1 400	1 400	1 400	1 400	1 400	
b (mm)	550	550	550	550	500		550	550	550	500	400	
t (mm)	50.0	45.0	40.0	35.0	30.0		50.0	40.0	35.0	30.0	30.0	
w (mm)	16.0	16.0	16.0	16.0	16.0		16.0	16.0	16.0	16.0	16.0	
k (mm)	61	56	51	44	39		61	51	44	39	39	

IMPERIAL SIZE AND MASS

Mass (lb./ft.)	419	388	361	333	289		402	344	316	272	240	
Nominal Depth (in.)			63						55			

* Class 3 section

BEAM LOAD TABLES
WWF Shapes

Total Uniformly Distributed Factored Loads
for Laterally Supported Beams – kN

G40.21-M
350W

$\phi = 0.90$

Designation	WWF1200						Approximate Deflection (mm)
Mass (kg/m)	487	418	380	333	302	263	
6 000						5 780	8
6 500						5 500	9
7 000					5 780	5 110	11
7 500					5 750	4 770	12
8 000				5 780	5 390	4 470	14
8 500				5 720	5 070	4 210	16
9 000				5 400	4 790	3 980	18
9 500			5 780	5 120	4 540	3 770	20
10 000			5 750	4 860	4 310	3 580	22
10 500			5 470	4 630	4 100	3 410	24
11 000		5 780	5 220	4 420	3 920	3 250	26
11 500		5 610	5 000	4 230	3 750	3 110	29
12 000		5 380	4 790	4 050	3 590	2 980	31
12 500		5 160	4 600	3 890	3 450	2 860	34
13 000	5 780	4 960	4 420	3 740	3 320	2 750	37
13 500	5 710	4 780	4 260	3 600	3 190	2 650	40
14 000	5 510	4 610	4 100	3 470	3 080	2 560	43
14 500	5 320	4 450	3 960	3 350	2 970	2 470	46
15 000	5 140	4 300	3 830	3 240	2 870	2 390	49
15 500	4 980	4 160	3 710	3 140	2 780	2 310	52
16 000	4 820	4 030	3 590	3 040	2 690	2 240	56
16 500	4 670	3 910	3 480	2 950	2 610	2 170	59
17 000	4 540	3 800	3 380	2 860	2 540	2 100	63
17 500	4 410	3 690	3 280	2 780	2 460	2 040	67
18 000	4 280	3 580	3 190	2 700	2 390	1 990	71
18 500	4 170	3 490	3 110	2 630	2 330	1 930	75
19 000	4 060	3 400	3 020	2 560	2 270	1 880	79
19 500	3 950	3 310	2 950	2 490	2 210	1 840	83
20 000	3 860	3 230	2 870	2 430	2 160	1 790	87
20 500	3 760	3 150	2 800	2 370	2 100	1 750	92
21 000	3 670	3 070	2 740	2 320	2 050	1 700	96
21 500	3 590	3 000	2 670	2 260	2 000	1 660	101
22 000	3 500	2 930	2 610	2 210	1 960	1 630	105
22 500	3 430	2 870	2 550	2 160	1 920	1 590	110
23 000	3 350	2 800	2 500	2 120	1 870	1 560	115

Span in Millimetres

DESIGN DATA AND PROPERTIES							
V_r (kN)	2 890	2 890	2 890	2 890	2 890	2 890	
L_u (mm)	7 710	6 790	6 630	5 150	4 980	3 560	
d (mm)	1 200	1 200	1 200	1 200	1 200	1 200	
b (mm)	550	500	500	400	400	300	
t (mm)	40.0	35.0	30.0	30.0	25.0	25.0	
w (mm)	16.0	16.0	16.0	16.0	16.0	16.0	
k (mm)	51	44	39	39	34	34	

IMPERIAL SIZE AND MASS							
Mass (lb./ft.)	326	281	255	224	203	176	
Nominal Depth (in.)	47						

G40.21-M
350W
ϕ = 0.90

BEAM LOAD TABLES
WWF Shapes

Total Uniformly Distributed Factored Loads
for Laterally Supported Beams – kN

Designation	WWF1100						Approximate Deflection (mm)
Mass (kg/m)	458	388	351	304	273	234	
Span in Millimetres							
6 000							9
6 500						4 420	10
7 000						4 320	12
7 500						4 030	13
8 000					4 420	3 780	15
8 500					4 360	3 560	17
9 000					4 120	3 360	19
9 500				4 420	3 900	3 180	21
10 000				4 210	3 700	3 020	24
10 500				4 010	3 530	2 880	26
11 000			4 420	3 830	3 370	2 750	29
11 500			4 360	3 660	3 220	2 630	31
12 000			4 180	3 510	3 090	2 520	34
12 500		4 420	4 010	3 370	2 960	2 420	37
13 000		4 340	3 860	3 240	2 850	2 330	40
13 500		4 180	3 720	3 120	2 740	2 240	43
14 000		4 030	3 580	3 010	2 650	2 160	47
14 500		3 890	3 460	2 900	2 560	2 090	50
15 000	4 420	3 760	3 340	2 810	2 470	2 020	53
15 500	4 410	3 640	3 240	2 720	2 390	1 950	57
16 000	4 270	3 530	3 130	2 630	2 320	1 890	61
16 500	4 140	3 420	3 040	2 550	2 240	1 830	65
17 000	4 020	3 320	2 950	2 480	2 180	1 780	69
17 500	3 900	3 230	2 870	2 400	2 120	1 730	73
18 000	3 790	3 140	2 790	2 340	2 060	1 680	77
18 500	3 690	3 050	2 710	2 280	2 000	1 640	81
19 000	3 590	2 970	2 640	2 220	1 950	1 590	86
19 500	3 500	2 900	2 570	2 160	1 900	1 550	90
20 000	3 420	2 820	2 510	2 100	1 850	1 510	95
20 500	3 330	2 750	2 450	2 050	1 810	1 480	100
21 000	3 250	2 690	2 390	2 000	1 760	1 440	105
21 500	3 180	2 620	2 330	1 960	1 720	1 410	110
22 000	3 100	2 570	2 280	1 910	1 680	1 380	115
22 500	3 040	2 510	2 230	1 870	1 650	1 340	120
23 000	2 970	2 450	2 180	1 830	1 610	1 320	126
DESIGN DATA AND PROPERTIES							
V_r (kN)	2 210	2 210	2 210	2 210	2 210	2 210	
L_u (mm)	7 890	6 970	6 810	5 310	5 150	3 710	
d (mm)	1 100	1 100	1 100	1 100	1 100	1 100	
b (mm)	550	500	500	400	400	300	
t (mm)	40.0	35.0	30.0	30.0	25.0	25.0	
w (mm)	14.0	14.0	14.0	14.0	14.0	14.0	
k (mm)	51	44	39	39	34	34	
IMPERIAL SIZE AND MASS							
Mass (lb./ft.)	307	260	236	204	184	157	
Nominal Depth (in.)	43						

BEAM LOAD TABLES
WWF Shapes

Total Uniformly Distributed Factored Loads
for Laterally Supported Beams – kN

G40.21-M
350W
$\phi = 0.90$

Designation	\multicolumn{7}{c}{WWF1000}	Approx. Deflect. (mm)
Mass (kg/m)	447	377	340	293	262	223	200	
Span in Millimetres								
5 000							4 420	7
5 500						4 420	4 200	8
6 000						4 410	3 850	9
6 500						4 070	3 560	11
7 000					4 420	3 780	3 300	13
7 500					4 370	3 530	3 080	15
8 000				4 420	4 100	3 310	2 890	17
8 500				4 390	3 850	3 110	2 720	19
9 000				4 140	3 640	2 940	2 570	21
9 500				3 930	3 450	2 780	2 430	24
10 000			4 420	3 730	3 280	2 650	2 310	26
10 500			4 250	3 550	3 120	2 520	2 200	29
11 000		4 420	4 060	3 390	2 980	2 400	2 100	32
11 500		4 380	3 880	3 240	2 850	2 300	2 010	35
12 000		4 200	3 720	3 110	2 730	2 200	1 930	38
12 500		4 030	3 570	2 980	2 620	2 120	1 850	41
13 000		3 880	3 430	2 870	2 520	2 040	1 780	44
13 500	4 420	3 730	3 300	2 760	2 430	1 960	1 710	48
14 000	4 360	3 600	3 190	2 660	2 340	1 890	1 650	51
14 500	4 210	3 480	3 080	2 570	2 260	1 820	1 590	55
15 000	4 070	3 360	2 970	2 490	2 180	1 760	1 540	59
15 500	3 930	3 250	2 880	2 410	2 110	1 710	1 490	63
16 000	3 810	3 150	2 790	2 330	2 050	1 650	1 440	67
16 500	3 700	3 060	2 700	2 260	1 980	1 600	1 400	71
17 000	3 590	2 960	2 620	2 190	1 930	1 560	1 360	76
17 500	3 480	2 880	2 550	2 130	1 870	1 510	1 320	80
18 000	3 390	2 800	2 480	2 070	1 820	1 470	1 280	85
18 500	3 300	2 720	2 410	2 020	1 770	1 430	1 250	89
19 000	3 210	2 650	2 350	1 960	1 720	1 390	1 220	94
19 500	3 130	2 580	2 290	1 910	1 680	1 360	1 180	99
20 000	3 050	2 520	2 230	1 860	1 640	1 320	1 160	105
20 500	2 980	2 460	2 180	1 820	1 600	1 290	1 130	110
21 000	2 900	2 400	2 120	1 780	1 560	1 260	1 100	115
21 500	2 840	2 340	2 080	1 740	1 520	1 230	1 080	121
22 000	2 770	2 290	2 030	1 700	1 490	1 200	1 050	126
DESIGN DATA AND PROPERTIES								
V_r (kN)	2 210	2 210	2 210	2 210	2 210	2 210	2 210	
L_u (mm)	8 030	7 080	6 910	5 400	5 230	3 790	3 610	
d (mm)	1 000	1 000	1 000	1 000	1 000	1 000	1 000	
b (mm)	550	500	500	400	400	300	300	
t (mm)	40.0	35.0	30.0	30.0	25.0	25.0	20.0	
w (mm)	14.0	14.0	14.0	14.0	14.0	14.0	14.0	
k (mm)	51	44	39	39	34	34	29	
IMPERIAL SIZE AND MASS								
Mass (lb./ft.)	300	253	228	197	176	150	134	
Nominal Depth (in.)	\multicolumn{7}{c}{39}	

5-100

G40.21-M
350W
$\phi = 0.90$

BEAM LOAD TABLES
WWF Shapes

Total Uniformly Distributed Factored Loads
for Laterally Supported Beams – kN

Designation	WWF900							Approx. Deflect. (mm)
Mass (kg/m)	417	347	309	262	231	192	169	
Span in Millimetres								
6 000								10
6 500							2 680	12
7 000							2 650	14
7 500						2 710	2 480	16
8 000						2 710	2 320	19
8 500						2 550	2 180	21
9 000						2 410	2 060	24
9 500						2 280	1 960	26
10 000					2 710	2 170	1 860	29
10 500					2 590	2 060	1 770	32
11 000				2 730	2 470	1 970	1 690	35
11 500				2 720	2 370	1 880	1 620	38
12 000				2 600	2 270	1 810	1 550	42
12 500				2 500	2 180	1 730	1 490	45
13 000				2 400	2 090	1 670	1 430	49
13 500			2 730	2 320	2 020	1 600	1 380	53
14 000			2 700	2 230	1 940	1 550	1 330	57
14 500			2 610	2 160	1 880	1 500	1 280	61
15 000			2 520	2 080	1 810	1 440	1 240	65
15 500		2 730	2 440	2 020	1 760	1 400	1 200	70
16 000		2 690	2 360	1 950	1 700	1 360	1 160	74
16 500		2 610	2 290	1 890	1 650	1 310	1 130	79
17 000		2 540	2 220	1 840	1 600	1 280	1 090	84
17 500		2 460	2 160	1 790	1 560	1 240	1 060	89
18 000		2 390	2 100	1 740	1 510	1 200	1 030	94
18 500		2 330	2 040	1 690	1 470	1 170	1 000	99
19 000	2 730	2 270	1 990	1 640	1 430	1 140	977	105
19 500	2 690	2 210	1 940	1 600	1 400	1 110	952	110
20 000	2 620	2 160	1 890	1 560	1 360	1 080	929	116
20 500	2 560	2 100	1 840	1 520	1 330	1 060	906	122
21 000	2 500	2 050	1 800	1 490	1 300	1 030	884	128
21 500	2 440	2 000	1 760	1 450	1 270	1 010	864	134
22 000	2 380	1 960	1 720	1 420	1 240	985	844	141
22 500	2 330	1 920	1 680	1 390	1 210	963	825	147
23 000	2 280	1 870	1 640	1 360	1 180	942	807	154
DESIGN DATA AND PROPERTIES								
V_r (kN)	1 370	1 370	1 370	1 370	1 350	1 350	1 340	
L_u (mm)	8 310	7 320	7 160	5 630	5 460	3 980	3 820	
d (mm)	900	900	900	900	900	900	900	
b (mm)	550	500	500	400	400	300	300	
t (mm)	40.0	35.0	30.0	30.0	25.0	25.0	20.0	
w (mm)	11.0	11.0	11.0	11.0	11.0	11.0	11.0	
k (mm)	51	44	39	39	34	34	29	
IMPERIAL SIZE AND MASS								
Mass (lb./ft.)	279	233	208	176	156	128	113	
Nominal Depth (in.)	35							

BEAM LOAD TABLES
WWF Shapes

Total Uniformly Distributed Factored Loads
for Laterally Supported Beams – kN

G40.21-M
350W

$\phi = 0.90$

Designation				WWF800			Approximate Deflection (mm)
Mass (kg/m)	339	300	253	223	184	161	
Span in Millimetres							
5 000							8
5 500						2 730	10
6 000						2 650	12
6 500					2 730	2 450	14
7 000					2 670	2 280	16
7 500					2 490	2 120	18
8 000					2 330	1 990	21
8 500				2 730	2 200	1 870	24
9 000				2 620	2 080	1 770	26
9 500			2 730	2 480	1 970	1 680	29
10 000			2 720	2 350	1 870	1 590	33
10 500			2 590	2 240	1 780	1 520	36
11 000			2 470	2 140	1 700	1 450	40
11 500			2 370	2 050	1 620	1 380	43
12 000		2 730	2 270	1 960	1 560	1 330	47
12 500		2 640	2 180	1 880	1 490	1 270	51
13 000		2 540	2 090	1 810	1 440	1 220	55
13 500	2 730	2 440	2 020	1 740	1 380	1 180	60
14 000	2 680	2 360	1 940	1 680	1 330	1 140	64
14 500	2 590	2 280	1 880	1 620	1 290	1 100	69
15 000	2 500	2 200	1 810	1 570	1 240	1 060	74
15 500	2 420	2 130	1 760	1 520	1 200	1 030	78
16 000	2 350	2 060	1 700	1 470	1 170	995	84
16 500	2 280	2 000	1 650	1 430	1 130	965	89
17 000	2 210	1 940	1 600	1 380	1 100	937	94
17 500	2 150	1 890	1 560	1 340	1 070	910	100
18 000	2 090	1 830	1 510	1 310	1 040	885	106
18 500	2 030	1 780	1 470	1 270	1 010	861	112
19 000	1 980	1 740	1 430	1 240	983	838	118
19 500	1 930	1 690	1 400	1 210	958	817	124
20 000	1 880	1 650	1 360	1 180	934	796	131
20 500	1 830	1 610	1 330	1 150	911	777	137
21 000	1 790	1 570	1 300	1 120	889	758	144
21 500	1 750	1 540	1 270	1 100	869	741	151
22 000	1 710	1 500	1 240	1 070	849	724	158
DESIGN DATA AND PROPERTIES							
V_r (kN)	1 370	1 370	1 370	1 370	1 370	1 370	
L_u (mm)	7 490	7 280	5 730	5 570	4 060	3 900	
d (mm)	800	800	800	800	800	800	
b (mm)	500	500	400	400	300	300	
t (mm)	35.0	30.0	30.0	25.0	25.0	20.0	
w (mm)	11.0	11.0	11.0	11.0	11.0	11.0	
k (mm)	44	39	39	34	34	29	
IMPERIAL SIZE AND MASS							
Mass (lb./ft.)	228	202	170	150	123	108	
Nominal Depth (in.)				31			

BEAM LOAD TABLES
WWF Shapes

G40.21-M
350W
$\phi = 0.90$

Total Uniformly Distributed Factored Loads
for Laterally Supported Beams – kN

Designation	WWF700					Approximate Deflection (mm)
Mass (kg/m)	245	214	196	175	152	
Span in Millimetres						
4 000						6
4 500					2 730	8
5 000					2 680	9
5 500				2 730	2 440	11
6 000			2 730	2 630	2 230	13
6 500			2 540	2 430	2 060	16
7 000		2 730	2 360	2 260	1 920	18
7 500		2 670	2 200	2 110	1 790	21
8 000	2 730	2 500	2 070	1 980	1 680	24
8 500	2 730	2 360	1 940	1 860	1 580	27
9 000	2 580	2 230	1 840	1 760	1 490	30
9 500	2 440	2 110	1 740	1 660	1 410	34
10 000	2 320	2 000	1 650	1 580	1 340	37
10 500	2 210	1 910	1 570	1 500	1 280	41
11 000	2 110	1 820	1 500	1 440	1 220	45
11 500	2 020	1 740	1 440	1 370	1 170	49
12 000	1 930	1 670	1 380	1 320	1 120	54
12 500	1 860	1 600	1 320	1 260	1 070	58
13 000	1 780	1 540	1 270	1 220	1 030	63
13 500	1 720	1 480	1 220	1 170	993	68
14 000	1 660	1 430	1 180	1 130	958	73
14 500	1 600	1 380	1 140	1 090	925	78
15 000	1 550	1 340	1 100	1 050	894	84
15 500	1 500	1 290	1 070	1 020	865	90
16 000	1 450	1 250	1 030	988	838	96
16 500	1 410	1 210	1 000	958	813	102
17 000	1 360	1 180	972	929	789	108
17 500	1 330	1 140	945	903	766	114
18 000	1 290	1 110	918	878	745	121
18 500	1 260	1 080	894	854	725	128
19 000	1 220	1 050	870	832	706	135
19 500	1 190	1 030	848	810	688	142
20 000	1 160	1 000	827	790	670	149
20 500	1 130	977	806	771	654	157
21 000	1 100	954	787	752	638	165
DESIGN DATA AND PROPERTIES						
V_r (kN)	1 370	1 370	1 370	1 370	1 370	
L_u (mm)	5 880	5 690	5 860	4 160	3 990	
d (mm)	700	700	700	700	700	
b (mm)	400	400	400	300	300	
t (mm)	30.0	25.0	22.0	25.0	20.0	
w (mm)	11.0	11.0	11.0	11.0	11.0	
k (mm)	39	34	31	34	29	
IMPERIAL SIZE AND MASS						
Mass (lb./ft.)	164	144	132	117	102	
Nominal Depth (in.)	28					

BEAM LOAD TABLES
W Shapes

CSA G40.21 350W
ASTM A992, A572 grade 50

Total Uniformly Distributed Factored Loads for Laterally Supported Beams (kN)

Designation	W1100				Approx. Deflect. (mm)	W1000					Approx. Deflect. (mm)
Mass (kg/m)	499	432	390	342		883	749	641	591	554	539	
5 000					6		17 100	14 600		12 500		6
5 500	11 900	10 000	9 060	7 700	7	20 400	17 100	14 500	13 200	12 400	12 000	8
6 000	11 000	9 600	8 610	7 490	8	18 800	15 700	13 300	12 200	11 400	11 100	9
6 500	10 200	8 870	7 950	6 920	10	17 300	14 500	12 300	11 300	10 500	10 200	11
7 000	9 440	8 230	7 380	6 420	11	16 100	13 400	11 400	10 500	9 760	9 510	12
7 500	8 810	7 680	6 890	5 990	13	15 000	12 600	10 600	9 770	9 110	8 880	14
8 000	8 260	7 200	6 460	5 620	15	14 100	11 800	9 970	9 160	8 540	8 320	16
8 500	7 770	6 780	6 080	5 290	16	13 200	11 100	9 380	8 620	8 040	7 830	18
9 000	7 340	6 400	5 740	5 000	18	12 500	10 500	8 860	8 140	7 590	7 400	20
9 500	6 960	6 070	5 440	4 730	21	11 800	9 910	8 390	7 710	7 190	7 010	23
10 000	6 610	5 760	5 170	4 500	23	11 300	9 410	7 970	7 330	6 830	6 660	25
10 500	6 290	5 490	4 920	4 280	25	10 700	8 970	7 590	6 980	6 510	6 340	28
11 000	6 010	5 240	4 700	4 090	28	10 200	8 560	7 250	6 660	6 210	6 050	30
11 500	5 750	5 010	4 490	3 910	30	9 780	8 190	6 930	6 370	5 940	5 790	33
12 000	5 510	4 800	4 310	3 750	33	9 380	7 850	6 640	6 110	5 690	5 550	36
12 500	5 290	4 610	4 130	3 600	36	9 000	7 530	6 380	5 860	5 460	5 330	39
13 000	5 080	4 430	3 970	3 460	38	8 660	7 240	6 130	5 640	5 250	5 120	42
13 500	4 890	4 270	3 830	3 330	41	8 340	6 970	5 910	5 430	5 060	4 930	46
14 000	4 720	4 120	3 690	3 210	45	8 040	6 720	5 700	5 230	4 880	4 760	49
14 500	4 560	3 970	3 560	3 100	48	7 760	6 490	5 500	5 050	4 710	4 590	53
15 000	4 400	3 840	3 440	3 000	51	7 500	6 280	5 320	4 890	4 550	4 440	56
15 500	4 260	3 720	3 330	2 900	55	7 260	6 070	5 140	4 730	4 410	4 290	60
16 000	4 130	3 600	3 230	2 810	58	7 030	5 880	4 980	4 580	4 270	4 160	64
16 500	4 000	3 490	3 130	2 720	62	6 820	5 710	4 830	4 440	4 140	4 030	68
17 000	3 890	3 390	3 040	2 640	66	6 620	5 540	4 690	4 310	4 020	3 920	72
17 500	3 780	3 290	2 950	2 570	70	6 430	5 380	4 560	4 190	3 900	3 800	77
18 000	3 670	3 200	2 870	2 500	74	6 250	5 230	4 430	4 070	3 800	3 700	81
18 500	3 570	3 120	2 790	2 430	78	6 080	5 090	4 310	3 960	3 690	3 600	86
19 000	3 480	3 030	2 720	2 370	82	5 920	4 950	4 200	3 860	3 600	3 500	90
19 500	3 390	2 960	2 650	2 310	86	5 770	4 830	4 090	3 760	3 500	3 410	95
20 000	3 300	2 880	2 580	2 250	91	5 630	4 710	3 990	3 660	3 420	3 330	100
20 500	3 220	2 810	2 520	2 190	96	5 490	4 590	3 890	3 570	3 330	3 250	105
21 000	3 150	2 740	2 460	2 140	100	5 360	4 480	3 800	3 490	3 250	3 170	110
21 500	3 070	2 680	2 400	2 090	105	5 230	4 380	3 710	3 410	3 180	3 100	116
22 000	3 000	2 620	2 350	2 040	110	5 110	4 280	3 620	3 330	3 110	3 030	121

Span in Millimetres

DESIGN DATA AND PROPERTIES												
V_r (kN)	5 930	5 020	4 530	3 850		10 200	8 560	7 300	6 610	6 240	5 990	
R (kN)	2 480	1 830	1 510	1 200		7 760	5 730	4 320	3 620	3 260	3 040	
G (kN)	26.0	18.3	15.4	13.0		79.1	60.1	47.1	38.5	35.9	32.7	
L_u (mm)	5 490	5 400	5 320	5 230		6 850	6 400	6 040	5 920	5 820	5 790	
d (mm)	1 118	1 108	1 100	1 090		1 092	1 068	1 048	1 040	1 032	1 030	
b (mm)	405	402	400	400		424	417	412	409	408	407	
t (mm)	45.0	40.1	36.1	31.0		82.0	70.1	59.9	55.9	52.1	51.1	
w (mm)	25.9	22.1	20.1	18.0		45.5	39.1	34.0	31.0	29.5	28.4	
k (mm)	71	66	62	57		115	103	93	89	85	84	

IMPERIAL SIZE AND MASS												
Mass (lb./ft.)	335	290	262	230		593	503	431	397	372	362	
Nominal Depth (in.)	44					40						

Note: F_y taken as 345 MPa. $\phi = 0.90$

CSA G40.21 350W
ASTM A992, A572 grade 50

BEAM LOAD TABLES
W Shapes

Total Uniformly Distributed Factored Loads for Laterally Supported Beams (kN)

Designation						W1000						Approx. Deflect. (mm)
Mass (kg/m)	483	443	412	371	321	296	583	493	486	414	393	
Span in Millimetres												
4 000							15 600	13 200		10 800		4
4 500							15 500	12 900	12 700	10 800	10 200	5
5 000						6 460	14 000	11 600	11 500	9 690	9 190	6
5 500	10 600	9 780	8 720	7 780		6 460	12 700	10 600	10 500	8 810	8 360	8
6 000	9 890	9 030	8 490	7 620	6 500	5 920	11 600	9 690	9 600	8 070	7 660	9
6 500	9 130	8 330	7 830	7 030	6 040	5 460	10 700	8 940	8 870	7 450	7 070	11
7 000	8 480	7 740	7 270	6 530	5 610	5 070	9 970	8 300	8 230	6 920	6 560	12
7 500	7 920	7 220	6 790	6 090	5 230	4 740	9 310	7 750	7 680	6 460	6 130	14
8 000	7 420	6 770	6 370	5 710	4 910	4 440	8 730	7 270	7 200	6 050	5 740	16
8 500	6 980	6 370	5 990	5 380	4 620	4 180	8 210	6 840	6 780	5 700	5 410	18
9 000	6 600	6 020	5 660	5 080	4 360	3 950	7 760	6 460	6 400	5 380	5 110	20
9 500	6 250	5 700	5 360	4 810	4 130	3 740	7 350	6 120	6 070	5 100	4 840	23
10 000	5 940	5 420	5 090	4 570	3 920	3 550	6 980	5 810	5 760	4 840	4 600	25
10 500	5 650	5 160	4 850	4 350	3 740	3 380	6 650	5 540	5 490	4 610	4 380	28
11 000	5 400	4 920	4 630	4 160	3 570	3 230	6 350	5 280	5 240	4 400	4 180	30
11 500	5 160	4 710	4 430	3 970	3 410	3 090	6 070	5 050	5 010	4 210	4 000	33
12 000	4 950	4 510	4 240	3 810	3 270	2 960	5 820	4 840	4 800	4 040	3 830	36
12 500	4 750	4 330	4 070	3 660	3 140	2 840	5 580	4 650	4 610	3 880	3 680	39
13 000	4 570	4 170	3 920	3 520	3 020	2 730	5 370	4 470	4 430	3 730	3 530	42
13 500	4 400	4 010	3 770	3 390	2 910	2 630	5 170	4 310	4 270	3 590	3 400	46
14 000	4 240	3 870	3 640	3 260	2 800	2 540	4 990	4 150	4 120	3 460	3 280	49
14 500	4 090	3 730	3 510	3 150	2 710	2 450	4 810	4 010	3 970	3 340	3 170	53
15 000	3 960	3 610	3 390	3 050	2 620	2 370	4 650	3 880	3 840	3 230	3 060	56
15 500	3 830	3 490	3 290	2 950	2 530	2 290	4 500	3 750	3 720	3 130	2 960	60
16 000	3 710	3 380	3 180	2 860	2 450	2 220	4 360	3 630	3 600	3 030	2 870	64
16 500	3 600	3 280	3 090	2 770	2 380	2 150	4 230	3 520	3 490	2 940	2 790	68
17 000	3 490	3 190	3 000	2 690	2 310	2 090	4 110	3 420	3 390	2 850	2 700	72
17 500	3 390	3 090	2 910	2 610	2 240	2 030	3 990	3 320	3 290	2 770	2 630	77
18 000	3 300	3 010	2 830	2 540	2 180	1 970	3 880	3 230	3 200	2 690	2 550	81
18 500	3 210	2 930	2 750	2 470	2 120	1 920	3 770	3 140	3 120	2 620	2 480	86
19 000	3 120	2 850	2 680	2 410	2 070	1 870	3 670	3 060	3 030	2 550	2 420	90
19 500	3 040	2 780	2 610	2 340	2 010	1 820	3 580	2 980	2 960	2 480	2 360	95
20 000	2 970	2 710	2 550	2 290	1 960	1 780	3 490	2 910	2 880	2 420	2 300	100
20 500	2 900	2 640	2 480	2 230	1 910	1 730	3 400	2 840	2 810	2 360	2 240	105
21 000	2 830	2 580	2 420	2 180	1 870	1 690	3 320	2 770	2 740	2 310	2 190	110
DESIGN DATA AND PROPERTIES												
V_r (kN)	5 310	4 890	4 360	3 890	3 250	3 230	7 810	6 580	6 370	5 410	5 080	
R (kN)	2 440	2 090	1 710	1 390	1 050	1 000	4 870	3 580	3 390	2 520	2 250	
G (kN)	26.3	23.3	17.5	14.3	11.0	12.7	52.4	40.0	36.2	27.9	24.5	
L_u (mm)	5 650	5 530	5 530	5 440	5 360	5 230	4 530	4 280	4 270	4 090	4 050	
d (mm)	1 020	1 012	1 008	1 000	990	982	1 056	1 036	1 036	1 020	1 016	
b (mm)	404	402	402	400	400	400	314	309	308	304	303	
t (mm)	46.0	41.9	40.0	36.1	31.0	27.1	64.0	54.1	54.1	46.0	43.9	
w (mm)	25.4	23.6	21.1	19.0	16.5	16.5	36.1	31.0	30.0	25.9	24.4	
k (mm)	79	75	73	69	64	60	97	87	87	79	77	
IMPERIAL SIZE AND MASS												
Mass (lb./ft.)	324	297	277	249	215	199	392	331	327	278	264	
Nominal Depth (in.)						40						

Note: F_y taken as 345 MPa. $\phi = 0.90$

BEAM LOAD TABLES
W Shapes

CSA G40.21 350W
ASTM A992, A572 grade 50

Total Uniformly Distributed Factored Loads for Laterally Supported Beams (kN)

Designation		W1000					Approx. Deflect. (mm)	W920				Approx. Deflect. (mm)
Mass (kg/m)		350	314	272	249	222		1188	967	784	653	
Span in Millimetres	4 000				6 440	6 000	4					4
	4 500	8 720	7 820	6 500	6 240	5 410	5					6
	5 000	8 250	7 400	6 360	5 610	4 870	6					7
	5 500	7 500	6 730	5 780	5 100	4 430	8	26 400	21 000	16 700	13 700	8
	6 000	6 870	6 170	5 300	4 680	4 060	9	24 300	19 400	15 400	12 700	10
	6 500	6 340	5 690	4 890	4 320	3 750	11	22 500	17 900	14 300	11 700	11
	7 000	5 890	5 290	4 540	4 010	3 480	12	20 900	16 600	13 200	10 900	13
	7 500	5 500	4 930	4 240	3 740	3 250	14	19 500	15 500	12 400	10 200	15
	8 000	5 150	4 630	3 970	3 510	3 040	16	18 300	14 500	11 600	9 530	17
	8 500	4 850	4 350	3 740	3 300	2 860	18	17 200	13 700	10 900	8 970	20
	9 000	4 580	4 110	3 530	3 120	2 700	20	16 200	12 900	10 300	8 470	22
	9 500	4 340	3 900	3 350	2 950	2 560	23	15 400	12 200	9 750	8 030	25
	10 000	4 120	3 700	3 180	2 810	2 430	25	14 600	11 600	9 270	7 630	27
	10 500	3 930	3 520	3 030	2 670	2 320	28	13 900	11 100	8 820	7 260	30
	11 000	3 750	3 360	2 890	2 550	2 210	30	13 300	10 600	8 420	6 930	33
	11 500	3 590	3 220	2 760	2 440	2 120	33	12 700	10 100	8 060	6 630	36
	12 000	3 440	3 080	2 650	2 340	2 030	36	12 200	9 690	7 720	6 350	39
	12 500	3 300	2 960	2 540	2 250	1 950	39	11 700	9 300	7 410	6 100	42
	13 000	3 170	2 850	2 450	2 160	1 870	42	11 200	8 940	7 130	5 870	46
	13 500	3 050	2 740	2 360	2 080	1 800	46	10 800	8 610	6 860	5 650	50
	14 000	2 950	2 640	2 270	2 000	1 740	49	10 400	8 300	6 620	5 450	53
	14 500	2 840	2 550	2 190	1 940	1 680	53	10 100	8 020	6 390	5 260	57
	15 000	2 750	2 470	2 120	1 870	1 620	56	9 740	7 750	6 180	5 080	61
	15 500	2 660	2 390	2 050	1 810	1 570	60	9 420	7 500	5 980	4 920	65
	16 000	2 580	2 310	1 990	1 750	1 520	64	9 130	7 270	5 790	4 770	70
	16 500	2 500	2 240	1 930	1 700	1 480	68	8 850	7 050	5 620	4 620	74
	17 000	2 430	2 180	1 870	1 650	1 430	72	8 590	6 840	5 450	4 490	79
	17 500	2 360	2 110	1 820	1 600	1 390	77	8 350	6 640	5 290	4 360	83
	18 000	2 290	2 060	1 770	1 560	1 350	81	8 110	6 460	5 150	4 240	88
	18 500	2 230	2 000	1 720	1 520	1 320	86	7 900	6 280	5 010	4 120	93
	19 000	2 170	1 950	1 670	1 480	1 280	90	7 690	6 120	4 880	4 010	98
	19 500	2 110	1 900	1 630	1 440	1 250	95	7 490	5 960	4 750	3 910	103
	20 000	2 060	1 850	1 590	1 400	1 220	100	7 300	5 810	4 630	3 810	109
	20 500	2 010	1 810	1 550	1 370	1 190	105	7 120	5 670	4 520	3 720	114
	21 000	1 960	1 760	1 510	1 340	1 160	110	6 960	5 540	4 410	3 630	120
DESIGN DATA AND PROPERTIES												
V_r (kN)		4 360	3 910	3 250	3 220	3 000		13 200	10 500	8 350	6 870	
R (kN)		1 710	1 400	1 050	990	888		13 800	9 420	6 340	4 510	
G (kN)		17.5	14.6	11.0	13.3	15.1		143	102	69.9	51.3	
L_u (mm)		4 010	3 910	3 870	3 740	3 590		8 930	7 890	7 120	6 640	
d (mm)		1 008	1 000	990	980	970		1 066	1 028	996	972	
b (mm)		302	300	300	300	300		457	446	437	431	
t (mm)		40.0	35.9	31.0	26.0	21.1		109.0	89.9	73.9	62.0	
w (mm)		21.1	19.1	16.5	16.5	16.0		60.5	50.0	40.9	34.5	
k (mm)		73	69	64	59	54		137	117	101	90	
IMPERIAL SIZE AND MASS												
Mass (lb./ft.)		235	211	183	167	149		798	650	527	439	
Nominal Depth (in.)				40						36		

Note: F_y taken as 345 MPa. $\phi = 0.90$

CSA G40.21 350W
ASTM A992, A572 grade 50

BEAM LOAD TABLES
W Shapes

Total Uniformly Distributed Factored Loads for Laterally Supported Beams (kN)

Designation				W920						Approx. Deflect. (mm)
Mass (kg/m)	585	534	488	446	417	387	365	342		
Span in Millimetres										
5 000						8 040	7 620	7 220		7
5 500	12 200	11 100	10 000	9 180	8 560	7 950	7 450	6 960		8
6 000	11 300	10 300	9 360	8 530	7 950	7 290	6 830	6 380		10
6 500	10 500	9 480	8 640	7 870	7 340	6 730	6 310	5 890		11
7 000	9 720	8 800	8 020	7 310	6 810	6 250	5 860	5 460		13
7 500	9 070	8 210	7 490	6 820	6 360	5 830	5 460	5 100		15
8 000	8 510	7 700	7 020	6 400	5 960	5 460	5 120	4 780		17
8 500	8 010	7 250	6 600	6 020	5 610	5 140	4 820	4 500		20
9 000	7 560	6 840	6 240	5 690	5 300	4 860	4 550	4 250		22
9 500	7 160	6 480	5 910	5 390	5 020	4 600	4 310	4 030		25
10 000	6 810	6 160	5 610	5 120	4 770	4 370	4 100	3 830		27
10 500	6 480	5 870	5 350	4 870	4 540	4 160	3 900	3 640		30
11 000	6 190	5 600	5 100	4 650	4 340	3 970	3 730	3 480		33
11 500	5 920	5 360	4 880	4 450	4 150	3 800	3 560	3 330		36
12 000	5 670	5 130	4 680	4 260	3 970	3 640	3 420	3 190		39
12 500	5 440	4 930	4 490	4 090	3 820	3 500	3 280	3 060		42
13 000	5 240	4 740	4 320	3 940	3 670	3 360	3 150	2 940		46
13 500	5 040	4 560	4 160	3 790	3 530	3 240	3 040	2 830		50
14 000	4 860	4 400	4 010	3 660	3 410	3 120	2 930	2 730		53
14 500	4 690	4 250	3 870	3 530	3 290	3 020	2 830	2 640		57
15 000	4 540	4 110	3 740	3 410	3 180	2 910	2 730	2 550		61
15 500	4 390	3 970	3 620	3 300	3 080	2 820	2 640	2 470		65
16 000	4 250	3 850	3 510	3 200	2 980	2 730	2 560	2 390		70
16 500	4 120	3 730	3 400	3 100	2 890	2 650	2 480	2 320		74
17 000	4 000	3 620	3 300	3 010	2 810	2 570	2 410	2 250		79
17 500	3 890	3 520	3 210	2 920	2 730	2 500	2 340	2 190		83
18 000	3 780	3 420	3 120	2 840	2 650	2 430	2 280	2 130		88
18 500	3 680	3 330	3 030	2 770	2 580	2 360	2 220	2 070		93
19 000	3 580	3 240	2 950	2 690	2 510	2 300	2 160	2 010		98
19 500	3 490	3 160	2 880	2 620	2 450	2 240	2 100	1 960		103
20 000	3 400	3 080	2 810	2 560	2 380	2 190	2 050	1 910		109
20 500	3 320	3 010	2 740	2 500	2 330	2 130	2 000	1 870		114
21 000	3 240	2 930	2 670	2 440	2 270	2 080	1 950	1 820		120
21 500	3 170	2 870	2 610	2 380	2 220	2 030	1 910	1 780		126
22 000	3 090	2 800	2 550	2 330	2 170	1 990	1 860	1 740		132
DESIGN DATA AND PROPERTIES										
V_r (kN)	6 100	5 530	5 000	4 590	4 280	4 020	3 810	3 610		
R (kN)	3 650	3 070	2 560	2 190	1 920	1 710	1 540	1 390		
G (kN)	41.8	35.5	29.5	26.1	23.1	21.6	20.0	18.5		
L_u (mm)	6 400	6 240	6 100	6 010	5 940	5 830	5 770	5 700		
d (mm)	960	950	942	933	928	921	916	912		
b (mm)	427	425	422	423	422	420	419	418		
t (mm)	55.9	51.1	47.0	42.7	39.9	36.6	34.3	32.0		
w (mm)	31.0	28.4	25.9	24.0	22.5	21.3	20.3	19.3		
k (mm)	83	79	75	70	67	64	62	60		
IMPERIAL SIZE AND MASS										
Mass (lb./ft.)	393	359	328	300	280	260	245	230		
Nominal Depth (in.)				36						

Note: F_y taken as 345 MPa. $\phi = 0.90$

BEAM LOAD TABLES
W Shapes

CSA G40.21 350W
ASTM A992, A572 grade 50

Total Uniformly Distributed Factored Loads for Laterally Supported Beams (kN)

Designation		W920									Approx. Deflect. (mm)
Mass (kg/m)		381	345	313	289	271	253	238	223	201	
Span in Millimetres	3 000										2
	3 500								5 940	5 420	3
	4 000	9 520	8 540	8 060	7 380	6 960	6 520	6 180	5 920	5 190	4
	4 500	9 380	8 450	7 510	6 960	6 510	6 070	5 630	5 270	4 610	6
	5 000	8 450	7 600	6 760	6 260	5 860	5 460	5 070	4 740	4 150	7
	5 500	7 680	6 910	6 140	5 690	5 330	4 970	4 610	4 310	3 780	8
	6 000	7 040	6 330	5 630	5 220	4 890	4 550	4 220	3 950	3 460	10
	6 500	6 500	5 850	5 200	4 820	4 510	4 200	3 900	3 650	3 190	11
	7 000	6 030	5 430	4 830	4 470	4 190	3 900	3 620	3 390	2 970	13
	7 500	5 630	5 070	4 500	4 170	3 910	3 640	3 380	3 160	2 770	15
	8 000	5 280	4 750	4 220	3 910	3 660	3 420	3 170	2 960	2 600	17
	8 500	4 970	4 470	3 970	3 680	3 450	3 210	2 980	2 790	2 440	20
	9 000	4 690	4 220	3 750	3 480	3 260	3 040	2 820	2 630	2 310	22
	9 500	4 450	4 000	3 560	3 290	3 090	2 880	2 670	2 490	2 190	25
	10 000	4 220	3 800	3 380	3 130	2 930	2 730	2 530	2 370	2 080	27
	10 500	4 020	3 620	3 220	2 980	2 790	2 600	2 410	2 260	1 980	30
	11 000	3 840	3 460	3 070	2 850	2 660	2 480	2 300	2 150	1 890	33
	11 500	3 670	3 300	2 940	2 720	2 550	2 380	2 200	2 060	1 810	36
	12 000	3 520	3 170	2 820	2 610	2 440	2 280	2 110	1 970	1 730	39
	12 500	3 380	3 040	2 700	2 500	2 340	2 190	2 030	1 900	1 660	42
	13 000	3 250	2 920	2 600	2 410	2 250	2 100	1 950	1 820	1 600	46
	13 500	3 130	2 820	2 500	2 320	2 170	2 020	1 880	1 760	1 540	50
	14 000	3 020	2 710	2 410	2 240	2 090	1 950	1 810	1 690	1 480	53
	14 500	2 910	2 620	2 330	2 160	2 020	1 880	1 750	1 630	1 430	57
	15 000	2 820	2 530	2 250	2 090	1 950	1 820	1 690	1 580	1 380	61
	15 500	2 720	2 450	2 180	2 020	1 890	1 760	1 630	1 530	1 340	65
	16 000	2 640	2 380	2 110	1 960	1 830	1 710	1 580	1 480	1 300	70
	16 500	2 560	2 300	2 050	1 900	1 780	1 660	1 540	1 440	1 260	74
	17 000	2 480	2 240	1 990	1 840	1 720	1 610	1 490	1 390	1 220	79
	17 500	2 410	2 170	1 930	1 790	1 670	1 560	1 450	1 350	1 190	83
	18 000	2 350	2 110	1 880	1 740	1 630	1 520	1 410	1 320	1 150	88
	18 500	2 280	2 050	1 830	1 690	1 580	1 480	1 370	1 280	1 120	93
	19 000	2 220	2 000	1 780	1 650	1 540	1 440	1 330	1 250	1 090	98
	19 500	2 170	1 950	1 730	1 610	1 500	1 400	1 300	1 220	1 060	103
	20 000	2 110	1 900	1 690	1 560	1 470	1 370	1 270	1 180	1 040	109
DESIGN DATA AND PROPERTIES											
V_r (kN)		4 760	4 270	4 030	3 690	3 480	3 260	3 090	2 970	2 710	
R (kN)		2 260	1 860	1 650	1 400	1 250	1 100	998	916	812	
G (kN)		26.2	21.6	22.0	18.5	16.9	15.2	14.3	13.9	14.6	
L_u (mm)		4 250	4 170	4 060	4 010	3 970	3 930	3 890	3 830	3 720	
d (mm)		951	943	932	927	923	919	915	911	903	
b (mm)		310	308	309	308	307	306	305	304	304	
t (mm)		43.9	39.9	34.5	32.0	30.0	27.9	25.9	23.9	20.1	
w (mm)		24.4	22.1	21.1	19.4	18.4	17.3	16.5	15.9	15.2	
k (mm)		69	65	60	58	56	53	51	49	46	
IMPERIAL SIZE AND MASS											
Mass (lb./ft.)		256	232	210	194	182	170	160	150	135	
Nominal Depth (in.)		36									

Note: F_y taken as 345 MPa. $\phi = 0.90$

CSA G40.21 350W
ASTM A992, A572 grade 50

BEAM LOAD TABLES
W Shapes

Total Uniformly Distributed Factored Loads for Laterally Supported Beams (kN)

Designation	W840								Approx. Deflect. (mm)
Mass (kg/m)	576	527	473	433	392	359	329	299	
Span in Millimetres									
4 000									5
4 500							6 960	6 380	6
5 000	12 000	10 900	9 660	8 860	7 940	7 500	6 960	6 260	7
5 500	11 500	10 500	9 390	8 540	7 680	6 960	6 320	5 690	9
6 000	10 600	9 650	8 610	7 820	7 040	6 380	5 800	5 220	11
6 500	9 740	8 900	7 950	7 220	6 500	5 890	5 350	4 820	13
7 000	9 050	8 270	7 380	6 710	6 030	5 460	4 970	4 470	15
7 500	8 450	7 720	6 890	6 260	5 630	5 100	4 640	4 170	17
8 000	7 920	7 230	6 460	5 870	5 280	4 780	4 350	3 910	19
8 500	7 450	6 810	6 080	5 520	4 970	4 500	4 090	3 680	22
9 000	7 040	6 430	5 740	5 220	4 690	4 250	3 860	3 480	24
9 500	6 670	6 090	5 440	4 940	4 450	4 030	3 660	3 290	27
10 000	6 330	5 790	5 170	4 690	4 220	3 830	3 480	3 130	30
10 500	6 030	5 510	4 920	4 470	4 020	3 640	3 310	2 980	33
11 000	5 760	5 260	4 700	4 270	3 840	3 480	3 160	2 850	36
11 500	5 510	5 030	4 490	4 080	3 670	3 330	3 020	2 720	39
12 000	5 280	4 820	4 310	3 910	3 520	3 190	2 900	2 610	43
12 500	5 070	4 630	4 130	3 760	3 380	3 060	2 780	2 500	47
13 000	4 870	4 450	3 970	3 610	3 250	2 940	2 680	2 410	50
13 500	4 690	4 290	3 830	3 480	3 130	2 830	2 580	2 320	54
14 000	4 520	4 130	3 690	3 350	3 020	2 730	2 480	2 240	58
14 500	4 370	3 990	3 560	3 240	2 910	2 640	2 400	2 160	63
15 000	4 220	3 860	3 440	3 130	2 820	2 550	2 320	2 090	67
15 500	4 090	3 730	3 330	3 030	2 720	2 470	2 240	2 020	72
16 000	3 960	3 620	3 230	2 930	2 640	2 390	2 170	1 960	76
16 500	3 840	3 510	3 130	2 850	2 560	2 320	2 110	1 900	81
17 000	3 730	3 400	3 040	2 760	2 480	2 250	2 050	1 840	86
17 500	3 620	3 310	2 950	2 680	2 410	2 190	1 990	1 790	91
18 000	3 520	3 220	2 870	2 610	2 350	2 130	1 930	1 740	96
18 500	3 420	3 130	2 790	2 540	2 280	2 070	1 880	1 690	102
19 000	3 330	3 050	2 720	2 470	2 220	2 010	1 830	1 650	107
19 500	3 250	2 970	2 650	2 410	2 170	1 960	1 780	1 610	113
20 000	3 170	2 890	2 580	2 350	2 110	1 910	1 740	1 560	119
20 500	3 090	2 820	2 520	2 290	2 060	1 870	1 700	1 530	125
21 000	3 020	2 760	2 460	2 240	2 010	1 820	1 660	1 490	131
DESIGN DATA AND PROPERTIES									
V_r (kN)	5 990	5 460	4 830	4 430	3 970	3 750	3 480	3 190	
R (kN)	3 920	3 330	2 680	2 280	1 880	1 680	1 450	1 230	
G (kN)	46.6	40.3	32.3	28.1	23.2	22.9	20.6	18.2	
L_u (mm)	6 330	6 150	5 970	5 840	5 710	5 620	5 530	5 450	
d (mm)	913	903	893	885	877	868	862	855	
b (mm)	411	409	406	404	401	403	401	400	
t (mm)	57.9	53.1	48.0	43.9	39.9	35.6	32.4	29.2	
w (mm)	32.0	29.5	26.4	24.4	22.1	21.1	19.7	18.2	
k (mm)	79	75	70	65	61	57	54	51	
IMPERIAL SIZE AND MASS									
Mass (lb./ft.)	387	354	318	291	263	241	221	201	
Nominal Depth (in.)	33								

Note: F_y taken as 345 MPa. $\phi = 0.90$

BEAM LOAD TABLES
W Shapes

CSA G40.21 350W
ASTM A992, A572 grade 50

Total Uniformly Distributed Factored Loads for Laterally Supported Beams (kN)

Designation		W840				Approx. Deflect. (mm)		W760				Approx. Deflect. (mm)
Mass (kg/m)	251	226	210	193	176		582	531	484	434	389	
Span in Millimetres												
3 000						3						3
3 500			5 340	5 060	4 600	4						4
4 000	5 980	5 620	5 240	4 740	4 240	5						5
4 500	5 690	5 060	4 650	4 210	3 760	6	11 900	10 800	9 780		7 760	7
5 000	5 120	4 550	4 190	3 790	3 390	7	11 800	10 700	9 690	8 640	7 650	8
5 500	4 650	4 140	3 810	3 450	3 080	9	10 700	9 710	8 810	7 860	6 960	10
6 000	4 260	3 790	3 490	3 160	2 820	11	9 810	8 900	8 070	7 200	6 380	12
6 500	3 940	3 500	3 220	2 920	2 610	13	9 060	8 220	7 450	6 650	5 890	14
7 000	3 660	3 250	2 990	2 710	2 420	15	8 410	7 630	6 920	6 170	5 460	16
7 500	3 410	3 030	2 790	2 530	2 260	17	7 850	7 120	6 460	5 760	5 100	19
8 000	3 200	2 840	2 620	2 370	2 120	19	7 360	6 680	6 050	5 400	4 780	21
8 500	3 010	2 680	2 460	2 230	1 990	22	6 930	6 280	5 700	5 080	4 500	24
9 000	2 840	2 530	2 330	2 110	1 880	24	6 540	5 930	5 380	4 800	4 250	27
9 500	2 690	2 400	2 200	2 000	1 780	27	6 200	5 620	5 100	4 550	4 030	30
10 000	2 560	2 280	2 090	1 900	1 690	30	5 890	5 340	4 840	4 320	3 830	33
10 500	2 440	2 170	1 990	1 810	1 610	33	5 610	5 090	4 610	4 120	3 640	36
11 000	2 330	2 070	1 900	1 720	1 540	36	5 350	4 860	4 400	3 930	3 480	40
11 500	2 220	1 980	1 820	1 650	1 470	39	5 120	4 640	4 210	3 760	3 330	44
12 000	2 130	1 900	1 750	1 580	1 410	43	4 910	4 450	4 040	3 600	3 190	47
12 500	2 050	1 820	1 680	1 520	1 360	47	4 710	4 270	3 880	3 460	3 060	51
13 000	1 970	1 750	1 610	1 460	1 300	50	4 530	4 110	3 730	3 320	2 940	56
13 500	1 900	1 690	1 550	1 400	1 250	54	4 360	3 960	3 590	3 200	2 830	60
14 000	1 830	1 630	1 500	1 350	1 210	58	4 210	3 810	3 460	3 090	2 730	64
14 500	1 760	1 570	1 440	1 310	1 170	63	4 060	3 680	3 340	2 980	2 640	69
15 000	1 710	1 520	1 400	1 260	1 130	67	3 920	3 560	3 230	2 880	2 550	74
15 500	1 650	1 470	1 350	1 220	1 090	72	3 800	3 450	3 130	2 790	2 470	79
16 000	1 600	1 420	1 310	1 180	1 060	76	3 680	3 340	3 030	2 700	2 390	84
16 500	1 550	1 380	1 270	1 150	1 030	81	3 570	3 240	2 940	2 620	2 320	90
17 000	1 510	1 340	1 230	1 110	997	86	3 460	3 140	2 850	2 540	2 250	95
17 500	1 460	1 300	1 200	1 080	968	91	3 360	3 050	2 770	2 470	2 190	101
18 000	1 420	1 260	1 160	1 050	941	96	3 270	2 970	2 690	2 400	2 130	107
18 500	1 380	1 230	1 130	1 020	916	102	3 180	2 890	2 620	2 340	2 070	113
19 000	1 350	1 200	1 100	998	892	107	3 100	2 810	2 550	2 270	2 010	119
19 500	1 310	1 170	1 070	972	869	113	3 020	2 740	2 480	2 220	1 960	125
20 000	1 280	1 140	1 050	948	847	119	2 940	2 670	2 420	2 160	1 910	132

DESIGN DATA AND PROPERTIES

V_r (kN)	2 990	2 810	2 670	2 530	2 300		5 960	5 380	4 890	4 320	3 880
R (kN)	1 120	976	882	789	701		4 590	3 840	3 250	2 610	2 150
G (kN)	13.9	13.8	13.3	13.1	13.1		59.1	49.6	42.8	34.2	29.4
L_u (mm)	3 890	3 830	3 770	3 680	3 600		6 470	6 240	6 030	5 820	5 650
d (mm)	859	851	846	840	835		843	833	823	813	803
b (mm)	292	294	293	292	292		396	393	390	387	385
t (mm)	31.0	26.8	24.4	21.7	18.8		62.0	56.9	52.1	47.0	41.9
w (mm)	17.0	16.1	15.4	14.7	14.0		34.5	31.5	29.0	25.9	23.6
k (mm)	53	48	46	43	40		84	78	74	69	63

IMPERIAL SIZE AND MASS

Mass (lb./ft.)	169	152	141	130	118		391	357	326	292	261
Nominal Depth (in.)			33						30		

Note: F_y taken as 345 MPa. $\phi = 0.90$

CSA G40.21 350W
ASTM A992, A572 grade 50

BEAM LOAD TABLES
W Shapes

Total Uniformly Distributed Factored Loads for Laterally Supported Beams (kN)

Designation						W760							Approx. Deflect. (mm)
Mass (kg/m)		350	314	284	257	220	196	185	173	161	147	134	
Span in Millimetres	3 000								4 500	4 280	4 080	3 300	3
	3 500					5 260	4 920	4 680	4 410	4 020	3 620	3 290	4
	4 000					5 090	4 450	4 150	3 860	3 510	3 170	2 880	5
	4 500	6 880	6 340	5 740	5 260	4 520	3 960	3 690	3 430	3 120	2 820	2 560	7
	5 000	6 860	6 110	5 460	4 930	4 070	3 560	3 320	3 090	2 810	2 530	2 300	8
	5 500	6 230	5 560	4 970	4 480	3 700	3 240	3 020	2 800	2 560	2 300	2 090	10
	6 000	5 710	5 090	4 550	4 110	3 390	2 970	2 770	2 570	2 340	2 110	1 920	12
	6 500	5 270	4 700	4 200	3 790	3 130	2 740	2 560	2 370	2 160	1 950	1 770	14
	7 000	4 900	4 360	3 900	3 520	2 910	2 540	2 370	2 200	2 010	1 810	1 640	16
	7 500	4 570	4 070	3 640	3 290	2 710	2 370	2 220	2 060	1 870	1 690	1 530	19
	8 000	4 280	3 820	3 420	3 080	2 540	2 230	2 080	1 930	1 760	1 580	1 440	21
	8 500	4 030	3 590	3 210	2 900	2 390	2 100	1 960	1 810	1 650	1 490	1 350	24
	9 000	3 810	3 390	3 040	2 740	2 260	1 980	1 850	1 710	1 560	1 410	1 280	27
	9 500	3 610	3 220	2 880	2 600	2 140	1 870	1 750	1 620	1 480	1 330	1 210	30
	10 000	3 430	3 060	2 730	2 470	2 030	1 780	1 660	1 540	1 410	1 270	1 150	33
	10 500	3 260	2 910	2 600	2 350	1 940	1 700	1 580	1 470	1 340	1 210	1 100	36
	11 000	3 120	2 780	2 480	2 240	1 850	1 620	1 510	1 400	1 280	1 150	1 050	40
	11 500	2 980	2 660	2 380	2 140	1 770	1 550	1 450	1 340	1 220	1 100	1 000	44
	12 000	2 860	2 550	2 280	2 060	1 700	1 480	1 380	1 290	1 170	1 060	958	47
	12 500	2 740	2 440	2 190	1 970	1 630	1 420	1 330	1 230	1 120	1 010	920	51
	13 000	2 640	2 350	2 100	1 900	1 560	1 370	1 280	1 190	1 080	974	885	56
	13 500	2 540	2 260	2 020	1 830	1 510	1 320	1 230	1 140	1 040	938	852	60
	14 000	2 450	2 180	1 950	1 760	1 450	1 270	1 190	1 100	1 000	905	821	64
	14 500	2 360	2 110	1 880	1 700	1 400	1 230	1 150	1 060	970	874	793	69
	15 000	2 290	2 040	1 820	1 640	1 360	1 190	1 110	1 030	937	845	767	74
	15 500	2 210	1 970	1 760	1 590	1 310	1 150	1 070	995	907	817	742	79
	16 000	2 140	1 910	1 710	1 540	1 270	1 110	1 040	964	879	792	719	84
	16 500	2 080	1 850	1 660	1 490	1 230	1 080	1 010	935	852	768	697	90
	17 000	2 020	1 800	1 610	1 450	1 200	1 050	978	907	827	745	677	95
	17 500	1 960	1 750	1 560	1 410	1 160	1 020	950	881	803	724	657	101
	18 000	1 900	1 700	1 520	1 370	1 130	989	923	857	781	704	639	107
	18 500	1 850	1 650	1 480	1 330	1 100	963	898	834	760	685	622	113
	19 000	1 800	1 610	1 440	1 300	1 070	937	875	812	740	667	605	119
	19 500	1 760	1 570	1 400	1 260	1 040	913	852	791	721	650	590	125
	20 000	1 710	1 530	1 370	1 230	1 020	891	831	771	703	633	575	132

DESIGN DATA AND PROPERTIES													
V_r (kN)		3 440	3 170	2 870	2 630	2 630	2 460	2 340	2 250	2 140	2 040	1 650	
R (kN)		1 730	1 490	1 240	1 050	1 070	925	838	773	700	631	514	
G (kN)		23.3	21.9	18.7	16.4	14.5	14.6	13.8	13.6	13.5	13.5	10.9	
L_u (mm)		5 530	5 420	5 320	5 230	3 570	3 500	3 450	3 410	3 330	3 260	3 230	
d (mm)		795	786	779	773	779	770	766	762	758	753	750	
b (mm)		382	384	382	381	266	268	267	267	266	265	264	
t (mm)		38.1	33.4	30.1	27.1	30.0	25.4	23.6	21.6	19.3	17.0	15.5	
w (mm)		21.1	19.7	18.0	16.5	16.5	15.6	14.9	14.4	13.8	13.2	11.9	
k (mm)		60	55	52	49	52	47	45	43	41	39	37	

IMPERIAL SIZE AND MASS													
Mass (lb./ft.)		235	211	191	173	148	132	124	116	108	99	90	
Nominal Depth (in.)							30						

Note: F_y taken as 345 MPa. $\phi = 0.90$

BEAM LOAD TABLES
W Shapes

CSA G40.21 350W
ASTM A992, A572 grade 50

Total Uniformly Distributed Factored Loads for Laterally Supported Beams (kN)

Designation		W690									Approx. Deflect. (mm)
Mass (kg/m)		802	548	500	457	419	384	350	323	289	
Span in Millimetres	4 000										6
	4 500	16 900	11 100	10 000	9 100	8 200	7 520	6 900	6 240	5 560	7
	5 000	15 400	10 100	9 140	8 350	7 600	6 910	6 260	5 760	5 120	9
	5 500	14 000	9 170	8 310	7 590	6 910	6 280	5 690	5 240	4 650	11
	6 000	12 800	8 400	7 620	6 960	6 330	5 750	5 220	4 800	4 260	13
	6 500	11 800	7 760	7 030	6 420	5 850	5 310	4 820	4 430	3 940	15
	7 000	11 000	7 200	6 530	5 960	5 430	4 930	4 470	4 120	3 660	18
	7 500	10 200	6 720	6 090	5 560	5 070	4 600	4 170	3 840	3 410	20
	8 000	9 590	6 300	5 710	5 220	4 750	4 320	3 910	3 600	3 200	23
	8 500	9 030	5 930	5 380	4 910	4 470	4 060	3 680	3 390	3 010	26
	9 000	8 530	5 600	5 080	4 640	4 220	3 840	3 480	3 200	2 840	29
	9 500	8 080	5 310	4 810	4 390	4 000	3 630	3 290	3 030	2 690	33
	10 000	7 680	5 040	4 570	4 170	3 800	3 450	3 130	2 880	2 560	36
	10 500	7 310	4 800	4 350	3 970	3 620	3 290	2 980	2 740	2 440	40
	11 000	6 980	4 580	4 160	3 790	3 460	3 140	2 850	2 620	2 330	44
	11 500	6 670	4 380	3 970	3 630	3 300	3 000	2 720	2 510	2 220	48
	12 000	6 400	4 200	3 810	3 480	3 170	2 880	2 610	2 400	2 130	52
	12 500	6 140	4 030	3 660	3 340	3 040	2 760	2 500	2 310	2 050	57
	13 000	5 900	3 880	3 520	3 210	2 920	2 660	2 410	2 220	1 970	61
	13 500	5 690	3 740	3 390	3 090	2 820	2 560	2 320	2 130	1 900	66
	14 000	5 480	3 600	3 260	2 980	2 710	2 470	2 240	2 060	1 830	71
	14 500	5 290	3 480	3 150	2 880	2 620	2 380	2 160	1 990	1 760	76
	15 000	5 120	3 360	3 050	2 780	2 530	2 300	2 090	1 920	1 710	82
	15 500	4 950	3 250	2 950	2 690	2 450	2 230	2 020	1 860	1 650	87
	16 000	4 800	3 150	2 860	2 610	2 380	2 160	1 960	1 800	1 600	93
	16 500	4 650	3 060	2 770	2 530	2 300	2 090	1 900	1 750	1 550	99
	17 000	4 520	2 970	2 690	2 450	2 240	2 030	1 840	1 690	1 510	105
	17 500	4 390	2 880	2 610	2 380	2 170	1 970	1 790	1 650	1 460	111
	18 000	4 260	2 800	2 540	2 320	2 110	1 920	1 740	1 600	1 420	117
	18 500	4 150	2 730	2 470	2 260	2 050	1 870	1 690	1 560	1 380	124
	19 000	4 040	2 650	2 410	2 200	2 000	1 820	1 650	1 520	1 350	131
	19 500	3 940	2 590	2 340	2 140	1 950	1 770	1 610	1 480	1 310	138
	20 000	3 840	2 520	2 290	2 090	1 900	1 730	1 560	1 440	1 280	145
	20 500	3 740	2 460	2 230	2 040	1 850	1 680	1 530	1 410	1 250	152
	21 000	3 660	2 400	2 180	1 990	1 810	1 640	1 490	1 370	1 220	160
DESIGN DATA AND PROPERTIES											
V_r (kN)		8 460	5 550	5 000	4 550	4 100	3 760	3 450	3 120	2 780	
R (kN)		9 670	4 810	4 010	3 410	2 850	2 440	2 090	1 760	1 420	
G (kN)		127	66.9	55.9	48.4	40.2	35.1	31.1	25.7	21.3	
L_u (mm)		7 980	6 400	6 140	5 910	5 710	5 560	5 390	5 310	5 140	
d (mm)		826	772	762	752	744	736	728	722	714	
b (mm)		387	372	369	367	364	362	360	359	356	
t (mm)		89.9	63.0	57.9	53.1	49.0	45.0	40.9	38.1	34.0	
w (mm)		50.0	35.1	32.0	29.5	26.9	24.9	23.1	21.1	19.0	
k (mm)		111	85	79	75	71	67	62	60	56	
IMPERIAL SIZE AND MASS											
Mass (lb./ft.)		539	368	336	307	281	258	235	217	194	
Nominal Depth (in.)						27					

Note: F_y taken as 345 MPa. $\phi = 0.90$

CSA G40.21 350W
ASTM A992, A572 grade 50

BEAM LOAD TABLES
W Shapes

Total Uniformly Distributed Factored Loads for Laterally Supported Beams (kN)

Designation				W690					Approx. Deflect. (mm)	W610		Approx. Deflect. (mm)
Mass (kg/m)	265	240	217	192	170	152	140	125		551	498	
Span in Millimetres												
3 000					4 120	3 700	3 480	3 220	3			4
3 500				4 460	3 990	3 550	3 230	2 850	4			5
4 000	5 320	4 820	4 380	4 010	3 490	3 110	2 830	2 490	6	11 200	10 100	7
4 500	5 130	4 630	4 180	3 570	3 100	2 760	2 510	2 210	7	10 300	9 220	8
5 000	4 620	4 170	3 760	3 210	2 790	2 480	2 260	1 990	9	9 240	8 300	10
5 500	4 200	3 790	3 420	2 920	2 540	2 260	2 050	1 810	11	8 400	7 540	12
6 000	3 850	3 470	3 130	2 670	2 330	2 070	1 880	1 660	13	7 700	6 910	15
6 500	3 550	3 210	2 890	2 470	2 150	1 910	1 740	1 530	15	7 110	6 380	17
7 000	3 300	2 980	2 690	2 290	1 990	1 770	1 610	1 420	18	6 600	5 930	20
7 500	3 080	2 780	2 510	2 140	1 860	1 660	1 510	1 330	20	6 160	5 530	23
8 000	2 880	2 610	2 350	2 010	1 750	1 550	1 410	1 250	23	5 780	5 190	26
8 500	2 710	2 450	2 210	1 890	1 640	1 460	1 330	1 170	26	5 440	4 880	30
9 000	2 560	2 320	2 090	1 780	1 550	1 380	1 260	1 110	29	5 130	4 610	33
9 500	2 430	2 190	1 980	1 690	1 470	1 310	1 190	1 050	33	4 860	4 370	37
10 000	2 310	2 080	1 880	1 600	1 400	1 240	1 130	996	36	4 620	4 150	41
10 500	2 200	1 980	1 790	1 530	1 330	1 180	1 080	949	40	4 400	3 950	45
11 000	2 100	1 890	1 710	1 460	1 270	1 130	1 030	906	44	4 200	3 770	50
11 500	2 010	1 810	1 640	1 400	1 210	1 080	983	866	48	4 020	3 610	54
12 000	1 920	1 740	1 570	1 340	1 160	1 040	942	830	52	3 850	3 460	59
12 500	1 850	1 670	1 500	1 280	1 120	994	904	797	57	3 700	3 320	64
13 000	1 780	1 600	1 450	1 230	1 070	955	869	766	61	3 550	3 190	69
13 500	1 710	1 540	1 390	1 190	1 030	920	837	738	66	3 420	3 070	75
14 000	1 650	1 490	1 340	1 150	997	887	807	711	71	3 300	2 960	80
14 500	1 590	1 440	1 300	1 110	963	857	779	687	76	3 190	2 860	86
15 000	1 540	1 390	1 250	1 070	931	828	753	664	82	3 080	2 770	92
15 500	1 490	1 340	1 210	1 040	901	801	729	643	87	2 980	2 680	98
16 000	1 440	1 300	1 180	1 000	873	776	706	623	93	2 890	2 590	105
16 500	1 400	1 260	1 140	973	846	753	685	604	99	2 800	2 510	112
17 000	1 360	1 230	1 110	944	821	731	665	586	105	2 720	2 440	118
17 500	1 320	1 190	1 070	917	798	710	646	569	111	2 640	2 370	126
18 000	1 280	1 160	1 040	891	776	690	628	553	117	2 570	2 300	133
18 500	1 250	1 130	1 020	867	755	671	611	538	124	2 500	2 240	140
19 000	1 210	1 100	990	845	735	654	595	524	131	2 430	2 180	148
19 500	1 180	1 070	964	823	716	637	580	511	138	2 370	2 130	156
20 000	1 150	1 040	940	802	698	621	565	498	145	2 310	2 070	164
DESIGN DATA AND PROPERTIES												
V_r (kN)	2 660	2 410	2 190	2 230	2 060	1 850	1 740	1 610		5 620	5 030	
R (kN)	1 310	1 090	914	951	813	663	587	514		5 880	4 880	
G (kN)	22.0	18.6	15.9	14.3	14.0	11.6	11.1	10.9		88.1	73.9	
L_u (mm)	5 060	4 970	4 900	3 440	3 380	3 320	3 270	3 190		6 620	6 230	
d (mm)	706	701	695	702	693	688	684	678		711	699	
b (mm)	358	356	355	254	256	254	254	253		347	343	
t (mm)	30.2	27.4	24.8	27.9	23.6	21.1	18.9	16.3		69.1	63.0	
w (mm)	18.4	16.8	15.4	15.5	14.5	13.1	12.4	11.7		38.6	35.1	
k (mm)	52	49	46	49	45	43	40	38		91	85	
IMPERIAL SIZE AND MASS												
Mass (lb./ft.)	178	161	146	129	114	102	94	84		370	335	
Nominal Depth (in.)				27						24		

Note: F_y taken as 345 MPa. $\phi = 0.90$.

BEAM LOAD TABLES
W Shapes

CSA G40.21 350W
ASTM A992, A572 grade 50

Total Uniformly Distributed Factored Loads for Laterally Supported Beams (kN)

Designation					W610						Approx. Deflect. (mm)
Mass (kg/m)	455	415	372	341	307	285	262	241	217	195	
Span in Millimetres											
3 000											4
3 500										3 920	5
4 000	9 040	8 200	7 240	6 620	5 920	5 460	5 000	4 660	4 240	3 770	7
4 500	8 340	7 560	6 730	6 130	5 480	5 070	4 610	4 230	3 780	3 350	8
5 000	7 500	6 810	6 060	5 510	4 930	4 560	4 150	3 810	3 400	3 020	10
5 500	6 820	6 190	5 510	5 010	4 480	4 150	3 770	3 460	3 090	2 740	12
6 000	6 250	5 670	5 050	4 600	4 110	3 800	3 460	3 180	2 840	2 510	15
6 500	5 770	5 240	4 660	4 240	3 790	3 510	3 190	2 930	2 620	2 320	17
7 000	5 360	4 860	4 330	3 940	3 520	3 260	2 960	2 720	2 430	2 150	20
7 500	5 000	4 540	4 040	3 680	3 290	3 040	2 770	2 540	2 270	2 010	23
8 000	4 690	4 250	3 790	3 450	3 080	2 850	2 590	2 380	2 130	1 880	26
8 500	4 410	4 000	3 570	3 240	2 900	2 680	2 440	2 240	2 000	1 770	30
9 000	4 170	3 780	3 370	3 060	2 740	2 530	2 300	2 120	1 890	1 680	33
9 500	3 950	3 580	3 190	2 900	2 600	2 400	2 180	2 010	1 790	1 590	37
10 000	3 750	3 400	3 030	2 760	2 470	2 280	2 070	1 910	1 700	1 510	41
10 500	3 570	3 240	2 890	2 630	2 350	2 170	1 980	1 810	1 620	1 440	45
11 000	3 410	3 090	2 750	2 510	2 240	2 070	1 890	1 730	1 550	1 370	50
11 500	3 260	2 960	2 640	2 400	2 140	1 980	1 800	1 660	1 480	1 310	54
12 000	3 130	2 840	2 530	2 300	2 060	1 900	1 730	1 590	1 420	1 260	59
12 500	3 000	2 720	2 420	2 210	1 970	1 820	1 660	1 520	1 360	1 210	64
13 000	2 890	2 620	2 330	2 120	1 900	1 750	1 600	1 470	1 310	1 160	69
13 500	2 780	2 520	2 240	2 040	1 830	1 690	1 540	1 410	1 260	1 120	75
14 000	2 680	2 430	2 160	1 970	1 760	1 630	1 480	1 360	1 220	1 080	80
14 500	2 590	2 350	2 090	1 900	1 700	1 570	1 430	1 310	1 170	1 040	86
15 000	2 500	2 270	2 020	1 840	1 640	1 520	1 380	1 270	1 130	1 010	92
15 500	2 420	2 200	1 960	1 780	1 590	1 470	1 340	1 230	1 100	973	98
16 000	2 340	2 130	1 890	1 720	1 540	1 430	1 300	1 190	1 060	942	105
16 500	2 270	2 060	1 840	1 670	1 490	1 380	1 260	1 150	1 030	914	112
17 000	2 210	2 000	1 780	1 620	1 450	1 340	1 220	1 120	1 000	887	118
17 500	2 140	1 940	1 730	1 580	1 410	1 300	1 190	1 090	972	862	126
18 000	2 080	1 890	1 680	1 530	1 370	1 270	1 150	1 060	945	838	133
18 500	2 030	1 840	1 640	1 490	1 330	1 230	1 120	1 030	920	815	140
19 000	1 970	1 790	1 590	1 450	1 300	1 200	1 090	1 000	896	794	148

DESIGN DATA AND PROPERTIES

V_r (kN)	4 520	4 100	3 620	3 310	2 960	2 730	2 500	2 330	2 120	1 960	
R (kN)	4 070	3 460	2 790	2 380	1 960	1 700	1 450	1 280	1 080	930	
G (kN)	61.8	53.6	43.1	37.7	31.2	27.4	23.7	21.9	19.4	18.1	
L_u (mm)	5 940	5 700	5 450	5 250	5 080	4 970	4 850	4 790	4 680	4 570	
d (mm)	689	679	669	661	653	647	641	635	628	622	
b (mm)	340	338	335	333	330	329	327	329	328	327	
t (mm)	57.9	53.1	48.0	43.9	39.9	37.1	34.0	31.0	27.7	24.4	
w (mm)	32.0	29.5	26.4	24.4	22.1	20.6	19.0	17.9	16.5	15.4	
k (mm)	79	75	70	65	61	59	56	53	49	46	

IMPERIAL SIZE AND MASS

Mass (lb./ft.)	306	279	250	229	207	192	176	162	146	131
Nominal Depth (in.)					24					

Note: F_y taken as 345 MPa. $\phi = 0.90$

CSA G40.21 350W
ASTM A992, A572 grade 50

BEAM LOAD TABLES
W Shapes

Total Uniformly Distributed Factored Loads for Laterally Supported Beams (kN)

Designation Mass (kg/m)	W610											Approx. Deflect. (mm)
Span in Millimetres	174	155	153	140	125	113	101	91	84†	92	82	
2 000										2 700	2 340	2
2 500						2 800	2 600	2 200	1 890	2 490	2 190	3
3 000			3 580	3 320	2 980	2 720	2 400	2 120	1 710	2 080	1 820	4
3 500	3 540	3 180	3 260	2 950	2 600	2 330	2 060	1 820	1 460	1 780	1 560	5
4 000	3 330	2 940	2 860	2 580	2 280	2 040	1 800	1 590	1 280	1 560	1 370	7
4 500	2 960	2 610	2 540	2 290	2 030	1 820	1 600	1 410	1 140	1 390	1 210	8
5 000	2 660	2 350	2 290	2 060	1 820	1 630	1 440	1 270	1 020	1 250	1 090	10
5 500	2 420	2 140	2 080	1 870	1 660	1 490	1 310	1 160	930	1 130	994	12
6 000	2 220	1 960	1 900	1 720	1 520	1 360	1 200	1 060	853	1 040	911	15
6 500	2 050	1 810	1 760	1 590	1 400	1 260	1 110	978	787	959	841	17
7 000	1 900	1 680	1 630	1 470	1 300	1 170	1 030	908	731	891	781	20
7 500	1 780	1 570	1 520	1 370	1 220	1 090	960	848	682	831	729	23
8 000	1 660	1 470	1 430	1 290	1 140	1 020	900	795	640	779	683	26
8 500	1 570	1 380	1 340	1 210	1 070	961	847	748	602	734	643	30
9 000	1 480	1 310	1 270	1 150	1 010	908	800	707	569	693	607	33
9 500	1 400	1 240	1 200	1 090	960	860	758	669	539	656	575	37
10 000	1 330	1 170	1 140	1 030	912	817	720	636	512	623	546	41
10 500	1 270	1 120	1 090	982	868	778	686	606	487	594	520	45
11 000	1 210	1 070	1 040	937	829	743	655	578	465	567	497	50
11 500	1 160	1 020	994	896	793	711	626	553	445	542	475	54
12 000	1 110	979	952	859	760	681	600	530	426	520	455	59
12 500	1 070	940	914	825	729	654	576	509	409	499	437	64
13 000	1 020	904	879	793	701	629	554	489	394	480	420	69
13 500	986	870	846	764	675	605	534	471	379	462	405	75
14 000	951	839	816	736	651	584	515	454	366	445	390	80
14 500	918	810	788	711	629	564	497	439	353	430	377	86
15 000	888	783	762	687	608	545	480	424	341	416	364	92
15 500	859	758	737	665	588	527	465	410	330	402	353	98
16 000	832	734	714	644	570	511	450	397	320	390	342	105
16 500	807	712	693	625	553	495	437	385	310	378	331	112
17 000	783	691	672	606	536	481	424	374	301	367	321	118
17 500	761	671	653	589	521	467	412	363	292	356	312	126
18 000	740	653	635	573	506	454	400	353	284	346	304	133
18 500	720	635	618	557	493	442	389	344	277	337	295	140
19 000	701	618	601	543	480	430	379	335	269	328	288	148

DESIGN DATA AND PROPERTIES

V_r (kN)	1 770	1 590	1 790	1 660	1 490	1 400	1 300	1 100	944	1 350	1 170	
R (kN)	766	627	789	684	561	491	426	360	310	457	382	
G (kN)	15.5	13.3	13.3	12.3	10.6	10.0	9.69	9.04	7.86	10.8	9.81	
L_u (mm)	4 480	4 400	3 110	3 070	3 020	2 950	2 890	2 810	2 980	2 180	2 110	
d (mm)	616	611	623	617	612	608	603	598	596	603	599	
b (mm)	325	324	229	230	229	228	228	227	226	179	178	
t (mm)	21.6	19.0	24.9	22.2	19.6	17.3	14.9	12.7	11.7	15.0	12.8	
w (mm)	14.0	12.7	14.0	13.1	11.9	11.2	10.5	9.7	9.0	10.9	10.0	
k (mm)	43	41	46	44	41	39	36	34	33	37	34	

IMPERIAL SIZE AND MASS

Mass (lb./ft.)	117	104	103	94	84	76	68	61	56	62	55
Nominal Depth (in.)	24										

Note: F_y taken as 345 MPa. $\phi = 0.90$

† Class 3

BEAM LOAD TABLES
W Shapes

CSA G40.21 350W
ASTM A992, A572 grade 50

Total Uniformly Distributed Factored Loads for Laterally Supported Beams (kN)

Designation					W530						Approx. Deflect. (mm)
Mass (kg/m)	300	272	248	219	196	182	165	150	138	123	
Span in Millimetres — 2 000											2
2 500									3 300	2 920	3
3 000									2 990	2 660	4
3 500	5 540	4 980	4 440	4 200	3 740	3 440	3 140	2 820	2 560	2 280	6
4 000	5 380	4 850	4 380	3 790	3 390	3 130	2 830	2 580	2 240	1 990	8
4 500	4 790	4 310	3 900	3 370	3 010	2 780	2 510	2 290	1 990	1 770	10
5 000	4 310	3 880	3 510	3 040	2 710	2 500	2 260	2 060	1 790	1 590	12
5 500	3 920	3 530	3 190	2 760	2 470	2 280	2 050	1 870	1 630	1 450	14
6 000	3 590	3 230	2 920	2 530	2 260	2 090	1 880	1 720	1 490	1 330	17
6 500	3 310	2 980	2 700	2 330	2 090	1 930	1 740	1 590	1 380	1 230	20
7 000	3 080	2 770	2 510	2 170	1 940	1 790	1 610	1 470	1 280	1 140	23
7 500	2 870	2 590	2 340	2 020	1 810	1 670	1 510	1 370	1 200	1 060	27
8 000	2 690	2 430	2 190	1 900	1 700	1 560	1 410	1 290	1 120	997	30
8 500	2 530	2 280	2 060	1 790	1 600	1 470	1 330	1 210	1 050	938	34
9 000	2 390	2 160	1 950	1 690	1 510	1 390	1 260	1 150	996	886	38
9 500	2 270	2 040	1 850	1 600	1 430	1 320	1 190	1 090	944	839	43
10 000	2 150	1 940	1 750	1 520	1 360	1 250	1 130	1 030	897	797	47
10 500	2 050	1 850	1 670	1 450	1 290	1 190	1 080	982	854	759	52
11 000	1 960	1 760	1 590	1 380	1 230	1 140	1 030	937	815	725	57
11 500	1 870	1 690	1 520	1 320	1 180	1 090	983	896	780	693	62
12 000	1 790	1 620	1 460	1 260	1 130	1 040	942	859	747	664	68
12 500	1 720	1 550	1 400	1 210	1 090	1 000	904	825	717	638	74
13 000	1 660	1 490	1 350	1 170	1 040	963	869	793	690	613	80
13 500	1 600	1 440	1 300	1 120	1 000	927	837	764	664	591	86
14 000	1 540	1 390	1 250	1 080	969	894	807	736	641	570	92
14 500	1 490	1 340	1 210	1 050	935	863	779	711	618	550	99
15 000	1 440	1 290	1 170	1 010	904	835	753	687	598	532	106
15 500	1 390	1 250	1 130	979	875	808	729	665	579	514	113
16 000	1 350	1 210	1 100	949	848	782	706	644	560	498	121

DESIGN DATA AND PROPERTIES

V_r (kN)	2 770	2 490	2 220	2 100	1 870	1 720	1 570	1 410	1 650	1 460	
R (kN)	2 170	1 820	1 480	1 340	1 090	931	789	651	871	694	
G (kN)	38.3	32.6	26.2	28.2	23.2	19.6	17.0	14.0	18.4	14.7	
L_u (mm)	5 210	5 040	4 880	4 720	4 600	4 530	4 440	4 380	2 930	2 860	
d (mm)	585	577	571	560	554	551	546	543	549	544	
b (mm)	319	318	315	318	316	315	313	312	214	212	
t (mm)	41.4	37.6	34.5	29.2	26.3	24.4	22.2	20.3	23.6	21.2	
w (mm)	23.1	21.1	19.0	18.3	16.5	15.2	14.0	12.7	14.7	13.1	
k (mm)	59	56	53	47	44	42	40	38	42	39	

IMPERIAL SIZE AND MASS

Mass (lb./ft.)	201	182	166	147	132	122	111	101	93	83	
Nominal Depth (in.)					21						

Note: F_y taken as 345 MPa. $\phi = 0.90$

CSA G40.21 350W
ASTM A992, A572 grade 50

BEAM LOAD TABLES
W Shapes

Total Uniformly Distributed Factored Loads for Laterally Supported Beams (kN)

Designation		W530								Approx. Deflect. (mm)
Mass (kg/m)		109	101	92	82	72†	85	74	66	
Span in Millimetres	2 000				2 060	1 850	2 260	2 100	1 860	2
	2 500	2 560	2 400	2 220	2 060	1 520	2 090	1 800	1 550	3
	3 000	2 340	2 170	1 950	1 710	1 270	1 740	1 500	1 290	4
	3 500	2 010	1 860	1 670	1 470	1 090	1 490	1 280	1 110	6
	4 000	1 760	1 630	1 470	1 290	950	1 300	1 120	969	8
	4 500	1 560	1 450	1 300	1 140	845	1 160	999	861	10
	5 000	1 410	1 300	1 170	1 030	760	1 040	899	775	12
	5 500	1 280	1 180	1 070	935	691	948	817	705	14
	6 000	1 170	1 080	977	857	633	869	749	646	17
	6 500	1 080	1 000	902	791	585	803	692	596	20
	7 000	1 000	930	837	735	543	745	642	554	23
	7 500	937	868	782	686	507	696	599	517	27
	8 000	879	814	733	643	475	652	562	484	30
	8 500	827	766	690	605	447	614	529	456	34
	9 000	781	723	651	571	422	580	500	431	38
	9 500	740	685	617	541	400	549	473	408	43
	10 000	703	651	586	514	380	522	450	388	47
	10 500	669	620	558	490	362	497	428	369	52
	11 000	639	592	533	467	346	474	409	352	57
	11 500	611	566	510	447	330	454	391	337	62
	12 000	586	542	489	428	317	435	375	323	68
	12 500	562	521	469	411	304	417	360	310	74
	13 000	541	501	451	396	292	401	346	298	80
	13 500	521	482	434	381	282	386	333	287	86
	14 000	502	465	419	367	271	373	321	277	92
	14 500	485	449	404	355	262	360	310	267	99
	15 000	469	434	391	343	253	348	300	258	106
	15 500	454	420	378	332	245	337	290	250	113
	16 000	439	407	366	321	238	326	281	242	121

DESIGN DATA AND PROPERTIES										
V_r (kN)		1 280	1 200	1 110	1 030	926	1 130	1 050	928	
R (kN)		545	481	419	360	313	430	375	313	
G (kN)		11.6	10.4	9.60	9.18	9.28	9.31	9.54	8.86	
L_u (mm)		2 810	2 770	2 720	2 660	2 750	2 110	2 040	1 980	
d (mm)		539	537	533	528	524	535	529	525	
b (mm)		211	210	209	209	207	166	166	165	
t (mm)		18.8	17.4	15.6	13.3	10.9	16.5	13.6	11.4	
w (mm)		11.6	10.9	10.2	9.5	8.9	10.3	9.7	8.9	
k (mm)		37	35	34	29	27	35	32	29	

IMPERIAL SIZE AND MASS										
Mass (lb./ft.)		73	68	62	55	48	57	50	44	
Nominal Depth (in.)		21								

Note: F_y taken as 345 MPa. $\phi = 0.90$ † Class 3

BEAM LOAD TABLES
W Shapes

CSA G40.21 350W
ASTM A992, A572 grade 50

Total Uniformly Distributed Factored Loads for Laterally Supported Beams (kN)

Designation				W460						Approx. Deflect. (mm)
Mass (kg/m)	260	235	213	193	177	158	144	128	113	
3 000	4 720	4 240	3 760	3 400	3 280	2 920	2 640	2 340	2 040	5
3 500	4 630	4 140	3 740	3 370	3 040	2 680	2 450	2 160	1 890	7
4 000	4 060	3 630	3 270	2 950	2 660	2 340	2 140	1 890	1 660	9
4 500	3 600	3 220	2 910	2 620	2 360	2 080	1 900	1 680	1 470	11
5 000	3 240	2 900	2 620	2 360	2 130	1 870	1 710	1 520	1 330	14
5 500	2 950	2 640	2 380	2 150	1 930	1 700	1 560	1 380	1 210	16
6 000	2 700	2 420	2 180	1 970	1 770	1 560	1 430	1 260	1 110	20
6 500	2 500	2 230	2 010	1 820	1 640	1 440	1 320	1 170	1 020	23
7 000	2 320	2 070	1 870	1 690	1 520	1 340	1 220	1 080	947	27
7 500	2 160	1 930	1 750	1 570	1 420	1 250	1 140	1 010	884	31
8 000	2 030	1 810	1 640	1 470	1 330	1 170	1 070	947	829	35
8 500	1 910	1 710	1 540	1 390	1 250	1 100	1 010	891	780	39
9 000	1 800	1 610	1 450	1 310	1 180	1 040	952	842	737	44
9 500	1 710	1 530	1 380	1 240	1 120	986	902	797	698	49
10 000	1 620	1 450	1 310	1 180	1 060	936	857	758	663	54
10 500	1 540	1 380	1 250	1 120	1 010	892	816	722	632	60
11 000	1 470	1 320	1 190	1 070	967	851	779	689	603	66
11 500	1 410	1 260	1 140	1 030	924	814	745	659	577	72
12 000	1 350	1 210	1 090	983	886	780	714	631	553	78
12 500	1 300	1 160	1 050	944	851	749	686	606	531	85
13 000	1 250	1 120	1 010	908	818	720	659	583	510	92
13 500	1 200	1 070	970	874	788	694	635	561	491	99
14 000	1 160	1 040	935	843	759	669	612	541	474	107

Span in Millimetres

DESIGN DATA AND PROPERTIES

V_r (kN)	2 360	2 120	1 880	1 700	1 640	1 460	1 320	1 170	1 020	
R (kN)	2 130	1 780	1 440	1 220	1 140	935	773	622	488	
G (kN)	42.2	35.9	28.7	24.8	26.5	22.3	18.1	14.9	11.8	
L_u (mm)	4 980	4 770	4 590	4 440	4 330	4 200	4 130	4 040	3 950	
d (mm)	509	501	495	489	482	476	472	467	463	
b (mm)	289	287	285	283	286	284	283	282	280	
t (mm)	40.4	36.6	33.5	30.5	26.9	23.9	22.1	19.6	17.3	
w (mm)	22.6	20.6	18.5	17.0	16.6	15.0	13.6	12.2	10.8	
k (mm)	60	56	53	50	46	43	42	39	37	

IMPERIAL SIZE AND MASS

Mass (lb./ft.)	175	158	143	130	119	106	97	86	76	
Nominal Depth (in.)					18					

Note: F_y taken as 345 MPa. $\phi = 0.90$

CSA G40.21 350W
ASTM A992, A572 grade 50

BEAM LOAD TABLES
W Shapes

Total Uniformly Distributed Factored Loads for Laterally Supported Beams (kN)

Designation					W460							Approx. Deflect. (mm)
Mass (kg/m)		106	97	89	82	74	67	61	68	60	52	
Span in Millimetres	2 000	2 420	2 180	2 000	1 870	1 690	1 580	1 490	1 710	1 490	1 350	2
	2 500	2 370	2 170	2 000	1 820	1 640	1 470	1 280	1 480	1 270	1 080	3
	3 000	1 980	1 810	1 660	1 520	1 370	1 230	1 070	1 230	1 060	903	5
	3 500	1 700	1 550	1 430	1 300	1 170	1 050	916	1 060	908	774	7
	4 000	1 480	1 350	1 250	1 140	1 020	919	801	925	795	677	9
	4 500	1 320	1 200	1 110	1 010	911	817	712	822	707	602	11
	5 000	1 190	1 080	999	909	820	735	641	740	636	542	14
	5 500	1 080	985	908	826	745	668	583	673	578	492	16
	6 000	989	903	832	758	683	613	534	617	530	451	20
	6 500	913	833	768	699	631	566	493	569	489	417	23
	7 000	848	774	713	649	586	525	458	529	454	387	27
	7 500	792	722	666	606	546	490	427	493	424	361	31
	8 000	742	677	624	568	512	460	401	463	397	338	35
	8 500	698	637	587	535	482	433	377	435	374	319	39
	9 000	660	602	555	505	455	408	356	411	353	301	44
	9 500	625	570	526	478	431	387	337	390	335	285	49
	10 000	594	542	499	455	410	368	320	370	318	271	54
	10 500	565	516	476	433	390	350	305	352	303	258	60
	11 000	540	492	454	413	373	334	291	336	289	246	66
	11 500	516	471	434	395	356	320	279	322	276	235	72
	12 000	495	451	416	379	342	306	267	308	265	226	78
	12 500	475	433	399	364	328	294	256	296	254	217	85
	13 000	457	417	384	350	315	283	246	285	245	208	92
	13 500	440	401	370	337	304	272	237	274	236	201	99
	14 000	424	387	357	325	293	263	229	264	227	193	107

DESIGN DATA AND PROPERTIES												
V_r (kN)		1 210	1 090	1 000	933	843	791	747	856	746	680	
R (kN)		665	547	465	412	341	302	272	350	271	241	
G (kN)		15.6	12.6	10.6	9.9	8.28	8.01	8.23	8.02	6.36	6.79	
L_u (mm)		2 690	2 650	2 620	2 560	2 530	2 480	2 410	2 010	1 970	1 890	
d (mm)		469	466	463	460	457	454	450	459	455	450	
b (mm)		194	193	192	191	190	190	189	154	153	152	
t (mm)		20.6	19.0	17.7	16.0	14.5	12.7	10.8	15.4	13.3	10.8	
w (mm)		12.6	11.4	10.5	9.9	9.0	8.5	8.1	9.1	8.0	7.6	
k (mm)		37	36	34	33	31	25	23	32	30	28	

IMPERIAL SIZE AND MASS												
Mass (lb./ft.)		71	65	60	55	50	45	41	46	40	35	
Nominal Depth (in.)						18						

Note: F_y taken as 345 MPa. $\phi = 0.90$.

BEAM LOAD TABLES
W Shapes

CSA G40.21 350W
ASTM A992, A572 grade 50

Total Uniformly Distributed Factored Loads for Laterally Supported Beams (kN)

Designation		W410										Approx. Deflect. (mm)
Mass (kg/m)		149	132	114	100	85	74	67	60	54	46	39	
Span in Millimetres	1 000												1
	1 500										1 160	960	1
	2 000					1 860	1 640	1 480	1 280	1 240	1 100	907	2
	2 500					1 710	1 500	1 350	1 180	1 040	878	725	4
	3 000	2 640	2 320	2 000	1 700	1 420	1 250	1 130	985	869	732	604	5
	3 500	2 310	2 020	1 750	1 510	1 220	1 070	965	845	745	627	518	7
	4 000	2 020	1 770	1 530	1 320	1 070	938	845	739	652	549	453	10
	4 500	1 790	1 570	1 360	1 180	949	834	751	657	580	488	403	12
	5 000	1 610	1 420	1 220	1 060	854	750	676	591	522	439	363	15
	5 500	1 470	1 290	1 110	962	777	682	614	537	474	399	330	18
	6 000	1 350	1 180	1 020	882	712	625	563	493	435	366	302	22
	6 500	1 240	1 090	940	814	657	577	520	455	401	338	279	26
	7 000	1 150	1 010	873	756	610	536	483	422	373	314	259	30
	7 500	1 080	944	815	705	570	500	450	394	348	293	242	34
	8 000	1 010	885	764	661	534	469	422	369	326	274	227	39
	8 500	950	833	719	622	503	441	397	348	307	258	213	44
	9 000	897	787	679	588	475	417	375	328	290	244	201	49
	9 500	850	745	643	557	450	395	356	311	275	231	191	55
	10 000	807	708	611	529	427	375	338	296	261	220	181	61
	10 500	769	674	582	504	407	357	322	282	248	209	173	67
	11 000	734	644	556	481	388	341	307	269	237	200	165	74
	11 500	702	616	531	460	372	326	294	257	227	191	158	81
	12 000	673	590	509	441	356	313	282	246	217	183	151	88
	12 500	646	566	489	423	342	300	270	236	209	176	145	95
	13 000	621	545	470	407	329	289	260	227	201	169	139	103

DESIGN DATA AND PROPERTIES

V_r (kN)		1 320	1 160	998	850	931	821	739	642	619	578	480	
R (kN)		952	761	580	433	513	407	335	258	242	213	177	
G (kN)		23.1	18.8	14.5	10.7	12.8	10.4	8.68	6.59	7.22	5.72	5.62	
L_u (mm)		4 080	3 940	3 810	3 730	2 530	2 470	2 420	2 390	2 310	1 790	1 730	
d (mm)		431	425	420	415	417	413	410	407	403	403	399	
b (mm)		265	263	261	260	181	180	179	178	177	140	140	
t (mm)		25.0	22.2	19.3	16.9	18.2	16.0	14.7	12.8	10.9	11.2	8.8	
w (mm)		14.9	13.3	11.6	10.0	10.9	9.7	8.8	7.7	7.5	7.0	6.4	
k (mm)		42	39	36	34	35	33	31	30	28	28	26	

IMPERIAL SIZE AND MASS

Mass (lb./ft.)		100	89	77	67	57	50	45	40	36	31	26	
Nominal Depth (in.)		16											

Note: F_y taken as 345 MPa. $\phi = 0.90$

CSA G40.21 350W
ASTM A992, A572 grade 50

BEAM LOAD TABLES
W Shapes

Total Uniformly Distributed Factored Loads for Laterally Supported Beams (kN)

Designation	W360								Approx. Deflect. (mm)
Mass (kg/m)	122	110	101	91	79	72	64	57	
Span in Millimetres									
2 000								1 160	3
2 500	1 930				1 360	1 230	1 100	1 000	4
3 000	1 880	1 680	1 540	1 370	1 180	1 060	944	836	6
3 500	1 610	1 460	1 330	1 190	1 010	908	809	717	9
4 000	1 410	1 280	1 170	1 040	888	795	708	627	11
4 500	1 250	1 140	1 040	927	789	707	629	558	14
5 000	1 130	1 020	934	835	710	636	566	502	17
5 500	1 030	930	849	759	646	578	515	456	21
6 000	940	853	778	696	592	530	472	418	25
6 500	867	787	718	642	546	489	436	386	29
7 000	806	731	667	596	507	454	405	358	34
7 500	752	682	623	556	474	424	378	335	39
8 000	705	640	584	522	444	397	354	314	44
8 500	663	602	549	491	418	374	333	295	50
9 000	627	569	519	464	395	353	315	279	56
9 500	594	539	492	439	374	335	298	264	63
10 000	564	512	467	417	355	318	283	251	69
10 500	537	487	445	397	338	303	270	239	77
11 000	513	465	425	379	323	289	257	228	84

DESIGN DATA AND PROPERTIES									
V_r (kN)	967	841	768	687	682	617	548	580	
R (kN)	757	586	498	409	401	336	270	281	
G (kN)	21.0	15.6	13.3	11.1	10.5	9.05	7.33	7.91	
L_u (mm)	4 040	3 940	3 860	3 760	3 010	2 940	2 870	2 360	
d (mm)	363	360	357	353	354	350	347	358	
b (mm)	257	256	255	254	205	204	203	172	
t (mm)	21.7	19.9	18.3	16.4	16.8	15.1	13.5	13.1	
w (mm)	13.0	11.4	10.5	9.5	9.4	8.6	7.7	7.9	
k (mm)	41	39	38	36	36	34	33	26	

IMPERIAL SIZE AND MASS									
Mass (lb./ft.)	82	74	68	61	53	48	43	38	
Nominal Depth (in.)	14								

Note: F_y taken as 345 MPa. $\phi = 0.90$

BEAM LOAD TABLES
W Shapes

CSA G40.21 350W
ASTM A992, A572 grade 50

Total Uniformly Distributed Factored Loads for Laterally Supported Beams (kN)

Designation		W360				Approx. Deflect. (mm)	W310					Approx. Deflect. (mm)
Mass (kg/m)		51	45	39	33		86	79	74	67	60	
Span in Millimetres	1 000					1						1
	1 500		996	940	792	2						2
	2 000	1 050	966	822	672	3			1 190	1 070		3
	2 500	887	773	658	538	4		1 100	1 180	1 050	932	5
	3 000	739	644	548	448	6	1 160	1 060	985	878	779	7
	3 500	634	552	470	384	9	1 010	908	845	752	668	10
	4 000	555	483	411	336	11	882	795	739	658	584	13
	4 500	493	429	365	299	14	784	707	657	585	519	16
	5 000	444	387	329	269	17	705	636	591	527	467	20
	5 500	403	351	299	244	21	641	578	537	479	425	24
	6 000	370	322	274	224	25	588	530	493	439	390	29
	6 500	341	297	253	207	29	543	489	455	405	360	34
	7 000	317	276	235	192	34	504	454	422	376	334	40
	7 500	296	258	219	179	39	470	424	394	351	312	45
	8 000	277	242	206	168	44	441	397	369	329	292	52
	8 500	261	227	193	158	50	415	374	348	310	275	58
	9 000	246	215	183	149	56	392	353	328	293	260	65
	9 500	233	203	173	141	63	371	335	311	277	246	73
	10 000	222	193	164	134	69	353	318	296	263	234	81
	10 500	211	184	157	128	77	336	303	282	251	223	89
	11 000	202	176	149	122	84	321	289	269	239	212	98

DESIGN DATA AND PROPERTIES

V_r (kN)		524	498	470	396		578	552	597	533	466	
R (kN)		233	214	191	152		390	365	415	341	266	
G (kN)		6.82	7.16	5.47	4.95		11.2	11.5	12.4	10.3	8.00	
L_u (mm)		2 320	2 260	1 660	1 600		3 900	3 810	3 100	3 020	2 950	
d (mm)		355	352	353	349		310	306	310	306	303	
b (mm)		171	171	128	127		254	254	205	204	203	
t (mm)		11.6	9.8	10.7	8.5		16.3	14.6	16.3	14.6	13.1	
w (mm)		7.2	6.9	6.5	5.8		9.1	8.8	9.4	8.5	7.5	
k (mm)		24	22	23	21		34	32	34	32	31	

IMPERIAL SIZE AND MASS

Mass (lb./ft.)		34	30	26	22		58	53	50	45	40	
Nominal Depth (in.)			14						12			

Note: F_y taken as 345 MPa. $\phi = 0.90$

CSA G40.21 350W
ASTM A992, A572 grade 50

BEAM LOAD TABLES
W Shapes

Total Uniformly Distributed Factored Loads for Laterally Supported Beams (kN)

Designation				W310					Approx. Deflect. (mm)
Mass (kg/m)	52	45	39	31‡	33	28	24	21	
Span in Millimetres									
1 000					846	760	700	606	1
1 500				588	795	674	543	475	2
2 000	990	846	736	533	596	505	407	356	3
2 500	836	703	606	426	477	404	326	285	5
3 000	696	586	505	355	397	337	272	238	7
3 500	597	502	433	305	341	289	233	204	10
4 000	522	440	379	267	298	253	204	178	13
4 500	464	391	337	237	265	225	181	158	16
5 000	418	352	303	213	238	202	163	143	20
5 500	380	320	275	194	217	184	148	130	24
6 000	348	293	253	178	199	168	136	119	29
6 500	321	271	233	164	183	156	125	110	34
7 000	298	251	216	152	170	144	116	102	40
7 500	279	234	202	142	159	135	109	95	45
8 000	261	220	189	133	149	126	102	89	52
8 500	246	207	178	125	140	119	96	84	58
9 000	232	195	168	118	132	112	91	79	65
9 500	220	185	159	112	126	106	86	75	73
10 000	209	176	152	107	119	101	82	71	81

DESIGN DATA AND PROPERTIES

V_r (kN)	495	423	368	294	423	380	350	303	
R (kN)	270	204	158	118	204	169	151	127	
G (kN)	7.87	6.17	4.88	4.15	6.40	5.91	6.46	5.78	
L_u (mm)	2 370	2 310	2 260	2 310	1 330	1 290	1 210	1 190	
d (mm)	318	313	310	306	313	309	305	303	
b (mm)	167	166	165	164	102	102	101	101	
t (mm)	13.2	11.2	9.7	7.4	10.8	8.9	6.7	5.7	
w (mm)	7.6	6.6	5.8	5.0	6.6	6.0	5.6	5.1	
k (mm)	27	25	24	22	22	20	18	17	

IMPERIAL SIZE AND MASS

Mass (lb./ft.)	35	30	26	21	22	19	16	14	
Nominal Depth (in.)				12					

Note: F_y taken as 345 MPa. $\phi = 0.90$ ‡ Class 4

BEAM LOAD TABLES
W Shapes

CSA G40.21 350W
ASTM A992, A572 grade 50

Total Uniformly Distributed Factored Loads for Laterally Supported Beams (kN)

Designation				W250								Approx. Deflect. (mm)
Mass (kg/m)	67	58	49†	45	39	33	24‡	28	25	22	18†	
Span in Millimetres												
1 000							518	682	642	604	445	1
1 500			750	828	708	646	436	585	508	436	296	2
2 000	938	826	710	748	637	527	327	438	381	327	222	4
2 500	895	765	568	598	510	421	262	351	305	261	178	6
3 000	746	638	474	498	425	351	218	292	254	218	148	9
3 500	639	546	406	427	364	301	187	251	218	187	127	12
4 000	560	478	355	374	319	263	163	219	191	163	111	16
4 500	497	425	316	332	283	234	145	195	169	145	99	20
5 000	448	383	284	299	255	211	131	175	153	131	89	25
5 500	407	348	258	272	232	191	119	159	139	119	81	30
6 000	373	319	237	249	212	176	109	146	127	109	74	36
6 500	344	294	219	230	196	162	101	135	117	101	68	42
7 000	320	273	203	214	182	150	93	125	109	93	64	49
7 500	298	255	189	199	170	140	87	117	102	87	59	56
8 000	280	239	178	187	159	132	82	110	95	82	56	64

DESIGN DATA AND PROPERTIES

V_r (kN)	469	413	375	414	354	323	259	341	321	302	247	
R (kN)	395	322	280	285	216	187	129	204	189	176	123	
G (kN)	13.1	11.3	11.2	9.55	7.37	7.27	5.81	7.58	7.91	8.37	6.25	
L_u (mm)	3 260	3 130	3 160	2 170	2 110	2 020	2 080	1 370	1 330	1 280	1 330	
d (mm)	257	252	247	266	262	258	253	260	257	254	251	
b (mm)	204	203	202	148	147	146	145	102	102	102	101	
t (mm)	15.7	13.5	11.0	13.0	11.2	9.1	6.4	10.0	8.4	6.9	5.3	
w (mm)	8.9	8.0	7.4	7.6	6.6	6.1	5.0	6.4	6.1	5.8	4.8	
k (mm)	33	31	28	28	26	24	16	20	18	17	15	

IMPERIAL SIZE AND MASS

| Mass (lb./ft.) | 45 | 39 | 33 | 30 | 26 | 22 | 16 | 19 | 17 | 15 | 12 | |
| Nominal Depth (in.) | | | | | | 10 | | | | | | |

Note: F_y taken as 345 MPa. $\phi = 0.90$. † Class 3   ‡ Class 4

CSA G40.21 350W
ASTM A992, A572 grade 50

BEAM LOAD TABLES
W Shapes

Total Uniformly Distributed Factored Loads for Laterally Supported Beams (kN)

Designation	W200								Approx. Deflect. (mm)
Mass (kg/m)	42	36	31	27	21†	22	19	15†	
1 000				492	416	524	465	315	1
1 500	604	510	550	462	323	368	310	210	3
2 000	549	467	416	347	242	276	232	158	5
2 500	439	374	333	277	194	221	186	126	8
3 000	366	311	277	231	161	184	155	105	11
3 500	314	267	238	198	138	158	133	90	15
4 000	274	233	208	173	121	138	116	79	20
4 500	244	208	185	154	108	123	103	70	25
5 000	220	187	166	139	97	110	93	63	31
5 500	200	170	151	126	88	100	85	57	38
6 000	183	156	139	116	81	92	77	53	45
6 500	169	144	128	107	75	85	72	49	53
7 000	157	133	119	99	69	79	66	45	61

Span in Millimetres

DESIGN DATA AND PROPERTIES

V_r (kN)	302	255	275	246	208	262	241	176	
R (kN)	282	211	222	186	143	218	201	108	
G (kN)	11.6	8.74	9.21	8.44	7.24	10.9	11.1	5.75	
L_u (mm)	2 610	2 510	1 980	1 890	1 930	1 390	1 340	1 380	
d (mm)	205	201	210	207	203	206	203	200	
b (mm)	166	165	134	133	133	102	102	100	
t (mm)	11.8	10.2	10.2	8.4	6.4	8.0	6.5	5.2	
w (mm)	7.2	6.2	6.4	5.8	5.0	6.2	5.8	4.3	
k (mm)	24	22	22	20	16	18	16	15	

IMPERIAL SIZE AND MASS

| Mass (lb./ft.) | 28 | 24 | 21 | 18 | 14 | 15 | 13 | 10 | |
| Nominal Depth (in.) | 8 | | | | | | | | |

Note: F_y taken as 345 MPa. $\phi = 0.90$ † Class 3

BEAM LOAD TABLES[1]
Rectangular HSS

G40.21 350W

$\phi = 0.90$

Total Uniformly Distributed Factored Loads for Laterally Supported Beams (kN)

Designation mm x mm mm		HSS 305 x 203					Approx. Deflect. (mm)	HSS 254 x 152					Approx. Deflect. (mm)
		13	11	9.5	8.0	6.4		13	11	9.5	8.0	6.4	
Mass (kg/m)		93.0	82.4	71.3	60.1	48.6		72.7	64.6	56.1	47.5	38.4	
Span in Millimetres	500							2 150	1 940	1 710	1 470	1 210	
	1 000	2 680	2 410	2 110	1 800	1 480	1	1 880	1 690	1 480	1 270	1 040	1
	1 500	2 000	1 780	1 560	1 320	1 080	2	1 250	1 130	990	845	690	2
	2 000	1 500	1 340	1 170	992	806	3	941	845	742	634	518	4
	2 500	1 200	1 070	933	793	645	5	753	676	594	507	414	6
	3 000	1 000	890	778	661	538	7	627	564	495	423	345	9
	3 500	857	763	667	567	461	10	538	483	424	362	296	12
	4 000	750	668	583	496	403	13	471	423	371	317	259	16
	4 500	666	594	519	441	358	17	418	376	330	282	230	20
	5 000	600	534	467	397	323	21	376	338	297	254	207	25
	5 500	545	486	424	361	293	25	342	307	270	230	188	30
	6 000	500	445	389	331	269	30	314	282	247	211	173	35
	6 500	461	411	359	305	248	35	290	260	228	195	159	42
	7 000	428	382	333	283	230	40	269	242	212	181	148	48
	7 500	400	356	311	264	215	46	251	225	198	169	138	55
	8 000	375	334	292	248	202	52	235	211	186	158	129	63
	8 500	353	314	275	233	190	59	221	199	175	149	122	71
	9 000	333	297	259	220	179	66	209	188	165	141	115	80
	9 500	316	281	246	209	170	74	198	178	156	133	109	89
	10 000	300	267	233	198	161	82	188	169	148	127	104	98
DESIGN DATA AND PROPERTIES													
S_x (10^3 mm^3)		964	867	762	652	535		592	537	475	410	338	
Z_x (10^3 mm^3)		1 190	1 060	926	787	640		747	671	589	503	411	
I_x (10^6 mm^4)		147	132	116	99.4	81.5		75.2	68.2	60.4	52.0	42.9	
C_{rt} (mm^2)		6 450	5 790	5 080	4 340	3 550		5 160	4 660	4 110	3 530	2 900	
V_r (kN)		1 340	1 200	1 060	902	738		1 070	969	855	735	604	
IMPERIAL SIZE AND MASS													
Mass (lb./ft.)		62.5	55.4	47.9	40.4	32.6		48.9	43.4	37.7	31.9	25.8	
Thickness (in.)		0.500	0.438	0.375	0.313	0.250		0.500	0.438	0.375	0.313	0.250	
Size (in.)		12 x 8						10 x 6					

[1] For strong axis bending only

G40.21 350W
$\phi = 0.90$

BEAM LOAD TABLES[1]
Rectangular HSS

Total Uniformly Distributed Factored Loads for Laterally Supported Beams (kN)

Designation mm x mm mm		HSS 203 x 152						HSS 203 x 102						Approx. Deflect. (mm)
		13	11	9.5	8.0	6.4	4.8	13	11	9.5	8.0	6.4	4.8	
Mass (kg/m)		62.6	55.7	48.5	41.1	33.4	25.5	52.4	46.9	40.9	34.8	28.3	21.7	
Span in Millimetres	500	1 610	1 470	1 310	1 130	939	732	1 610	1 470	1 310	1 130	939	732	
	1 000	1 330	1 200	1 060	907	743	575	1 020	927	822	708	585	454	1
	1 500	887	800	706	605	496	383	680	618	548	472	390	302	3
	2 000	665	600	529	454	372	287	510	464	411	354	292	227	5
	2 500	532	480	423	363	297	230	408	371	329	283	234	181	8
	3 000	444	400	353	302	248	192	340	309	274	236	195	151	11
	3 500	380	343	302	259	212	164	292	265	235	202	167	130	15
	4 000	333	300	265	227	186	144	255	232	205	177	146	113	20
	4 500	296	267	235	202	165	128	227	206	183	157	130	101	25
	5 000	266	240	212	181	149	115	204	185	164	142	117	91	31
	5 500	242	218	192	165	135	104	186	169	149	129	106	83	37
	6 000	222	200	176	151	124	96	170	155	137	118	97	76	44
	6 500	205	185	163	140	114	88	157	143	126	109	90	70	52
	7 000	190	171	151	130	106	82	146	132	117	101	84	65	60

DESIGN DATA AND PROPERTIES														
S_x (10^3 mm^3)		423	385	343	297	246	192	308	283	254	221	185	145	
Z_x (10^3 mm^3)		528	476	420	360	295	228	405	368	326	281	232	180	
I_x (10^6 mm^4)		43.0	39.2	34.8	30.2	25.0	19.5	31.3	28.7	25.8	22.5	18.8	14.7	
C_{rt} (mm^2)		3 870	3 530	3 150	2 730	2 260	1 760	3 870	3 530	3 150	2 730	2 260	1 760	
V_r (kN)		805	734	654	567	469	366	805	734	654	567	469	366	

IMPERIAL SIZE AND MASS														
Mass (lb./ft.)		42.1	37.5	32.6	27.6	22.4	17.1	35.2	31.5	27.5	23.4	19.0	14.6	
Thickness (in.)		0.500	0.438	0.375	0.313	0.250	0.188	0.500	0.438	0.375	0.313	0.250	0.188	
Size (in.)		8 x 6						8 x 4						

[1] For strong axis bending only

BEAM LOAD TABLES[1]
Rectangular HSS

G40.21 350W
$\phi = 0.90$

Total Uniformly Distributed Factored Loads for Laterally Supported Beams (kN)

Designation mm x mm mm	HSS 178 x 127						Approximate Deflection (mm)
	13	11	9.5	8.0	6.4	4.8	
Mass (kg/m)	52.4	46.9	40.9	34.8	28.3	21.7	

Span in Millimetres								
500	1 340	1 230	1 110	965	805	631	
1 000	953	864	764	658	544	423	1
1 500	635	576	509	438	363	282	3
2 000	476	432	382	329	272	212	6
2 500	381	346	305	263	218	169	9
3 000	318	288	255	219	181	141	13
3 500	272	247	218	188	156	121	17
4 000	238	216	191	164	136	106	22
4 500	212	192	170	146	121	94	28
5 000	191	173	153	132	109	85	35
5 500	173	157	139	120	99	77	43
6 000	159	144	127	110	91	71	51

DESIGN DATA AND PROPERTIES

S_x (10^3 mm^3)	297	273	244	213	178	140	
Z_x (10^3 mm^3)	378	343	303	261	216	168	
I_x (10^6 mm^4)	26.4	24.2	21.7	19.0	15.8	12.4	
C_{rt} (mm^2)	3 230	2 970	2 660	2 320	1 940	1 520	
V_r (kN)	671	617	553	483	402	315	

IMPERIAL SIZE AND MASS

Mass (lb./ft.)	35.2	31.5	27.5	23.4	19.0	14.6	
Thickness (in.)	0.500	0.438	0.375	0.313	0.250	0.188	
Size (in.)	7 x 5						

[1] For strong axis bending only

G40.21 350W
$\phi = 0.90$

BEAM LOAD TABLES[1]
Rectangular HSS

Total Uniformly Distributed Factored Loads for Laterally Supported Beams (kN)

Designation mm x mm mm	HSS 152 x 102						HSS 152 x 76				Approx. Deflect. (mm)
	13	11	9.5	8.0	6.4	4.8	9.5	8.0	6.4	4.8	
Mass (kg/m)	42.3	38.0	33.3	28.4	23.2	17.9	29.5	25.3	20.7	16.0	
Span in Millimetres											
500	1 070	999	906	797	671	530	862	751	630	494	1
1 000	635	580	519	451	373	292	431	375	315	247	2
1 500	423	386	346	301	249	195	287	250	210	165	4
2 000	318	290	260	226	186	146	215	188	158	124	7
2 500	254	232	208	180	149	117	172	150	126	99	10
3 000	212	193	173	150	124	97	144	125	105	82	15
3 500	181	166	148	129	107	84	123	107	90	71	20
4 000	159	145	130	113	93	73	108	94	79	62	26
4 500	141	129	115	100	83	65	96	83	70	55	33
5 000	127	116	104	90	75	59	86	75	63	49	41
DESIGN DATA AND PROPERTIES											
S_x (10^3 mm^3)	193	179	162	143	121	95.6	130	115	98.0	78.2	
Z_x (10^3 mm^3)	252	230	206	179	148	116	171	149	125	98.1	
I_x (10^6 mm^4)	14.7	13.6	12.4	10.9	9.19	7.28	9.89	8.79	7.47	5.96	
C_{rt} (mm^2)	2 580	2 400	2 180	1 920	1 610	1 270	2 180	1 920	1 610	1 270	
V_r (kN)	537	499	453	399	335	265	453	399	335	265	
IMPERIAL SIZE AND MASS											
Mass (lb./ft.)	28.4	25.5	22.4	19.1	15.6	12.0	19.8	17.0	13.9	10.7	
Thickness (in.)	0.500	0.438	0.375	0.313	0.250	0.188	0.375	0.313	0.250	0.188	
Size (in.)	6 x 4						6 x 3				

[1] For strong axis bending only

BEAM LOAD TABLES[1]
Rectangular HSS

G40.21 350W
$\phi = 0.90$

Total Uniformly Distributed Factored Loads for Laterally Supported Beams (kN)

Designation mm x mm mm	HSS 127 x 76					Approximate Deflection (mm)
	9.5	8.0	6.4	4.8	3.8	
Mass (kg/m)	25.7	22.1	18.2	14.1	11.4	
Span in Millimetres						
500	635	559	471	372	305	1
1 000	318	280	235	186	153	2
1 500	212	186	157	124	102	4
2 000	159	140	118	93	76	8
2 500	127	112	94	74	61	12
3 000	106	93	79	62	51	18
3 500	91	80	67	53	44	24
4 000	79	70	59	47	38	31
DESIGN DATA AND PROPERTIES						
S_x (10^3 mm^3)	96.5	86.5	74.1	59.6	49.4	
Z_x (10^3 mm^3)	126	111	93.4	73.8	60.6	
I_x (10^6 mm^4)	6.13	5.49	4.70	3.78	3.14	
C_{rt} (mm^2)	1 690	1 510	1 290	1 030	852	
V_r (kN)	352	315	268	214	177	
IMPERIAL SIZE AND MASS						
Mass (lb./ft.)	17.3	14.9	12.2	9.46	7.66	
Thickness (in.)	0.375	0.313	0.250	0.188	0.150	
Size (in.)	5 x 3					

[1] For strong axis bending only

G40.21 350W
$\phi = 0.90$

BEAM LOAD TABLES[1]
Rectangular HSS

Total Uniformly Distributed Factored Loads for Laterally Supported Beams (kN)

Designation mm x mm mm		HSS 102 x 76						HSS 102 x 51					Approx. Deflect. (mm)
		9.5	8.0	6.4	4.8	3.8	3.2	8.0	6.4	4.8	3.8	3.2	
Mass (kg/m)		21.9	18.9	15.6	12.2	9.89	8.35	15.8	13.1	10.3	8.37	7.09	
Span in Millimetres	500	443	393	333	265	218	186	297	256	206	171	146	1
	1 000	222	196	166	133	109	93	149	128	103	85	73	2
	1 500	148	131	111	88	73	62	99	85	69	57	49	6
	2 000	111	98	83	66	55	47	74	64	51	43	37	10
	2 500	89	79	67	53	44	37	60	51	41	34	29	15
	3 000	74	65	55	44	36	31	50	43	34	29	24	22
	3 500	63	56	48	38	31	27	43	37	29	24	21	30
	4 000	55	49	42	33	27	23	37	32	26	21	18	39
DESIGN DATA AND PROPERTIES													
S_x (10^3 mm^3)		67.4	61.1	52.9	43.0	35.9	30.8	43.6	38.5	31.8	26.8	23.1	
Z_x (10^3 mm^3)		87.9	77.9	66.0	52.6	43.3	37.0	59.0	50.7	40.8	33.9	29.0	
I_x (10^6 mm^4)		3.42	3.10	2.69	2.18	1.82	1.57	2.21	1.95	1.61	1.36	1.17	
C_{rt} (mm^2)		1 210	1 110	968	789	658	565	1 110	968	789	658	565	
V_r (kN)		252	231	201	164	137	118	231	201	164	137	118	
IMPERIAL SIZE AND MASS													
Mass (lb./ft.)		14.7	12.7	10.5	8.17	6.64	5.61	10.6	8.81	6.89	5.62	4.76	
Thickness (in.)		0.375	0.313	0.250	0.188	0.150	0.125	0.313	0.250	0.188	0.150	0.125	
Size (in.)		4 x 3						4 x 2					

[1] For strong axis bending only

BEAM DIAGRAMS AND FORMULAE

Equivalent Tabular Load is the uniformly distributed factored load given in the Beam Load Tables

1   SIMPLE BEAM—UNIFORMLY DISTRIBUTED LOAD

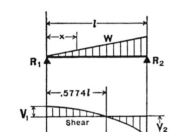

Equivalent Tabular Load . . . $= wl$

$R = V$ . . . . $= \dfrac{wl}{2}$

V_x . . . . $= w\left(\dfrac{l}{2} - x\right)$

M max. (at center) . . . . $= \dfrac{wl^2}{8}$

M_x . . . . $= \dfrac{wx}{2}(l - x)$

Δ max. (at center) . . . . $= \dfrac{5\,wl^4}{384\,EI}$

Δ_x . . . . $= \dfrac{wx}{24EI}(l^3 - 2lx^2 + x^3)$

2   SIMPLE BEAM—LOAD INCREASING UNIFORMLY TO ONE END

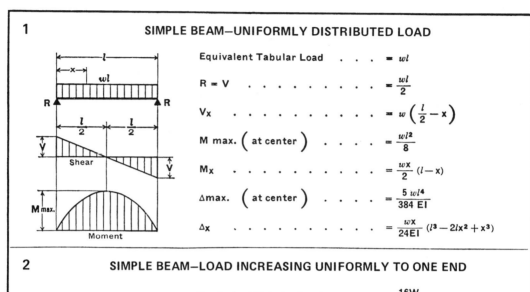

Equivalent Tabular Load . . . $= \dfrac{16W}{9\sqrt{3}} = 1.0264W$

$R_1 = V_1$ . . . . $= \dfrac{W}{3}$

$R_2 = V_2$ max. . . . . $= \dfrac{2W}{3}$

V_x . . . . $= \dfrac{W}{3} - \dfrac{Wx^2}{l^2}$

M max. $\left(\text{at } x = \dfrac{l}{\sqrt{3}} = .5774l\right)$ . . $= \dfrac{2Wl}{9\sqrt{3}} = .1283\,Wl$

M_x . . . . $= \dfrac{Wx}{3l^2}(l^2 - x^2)$

Δ max. $\left(\text{at } x = l\sqrt{1 - \sqrt{\dfrac{8}{15}}} = .5193l\right) = .01304\,\dfrac{Wl^3}{EI}$

Δ_x . . . . $= \dfrac{Wx}{180EI\,l^2}(3x^4 - 10l^2x^2 + 7l^4)$

3   SIMPLE BEAM—LOAD INCREASING UNIFORMLY TO CENTER

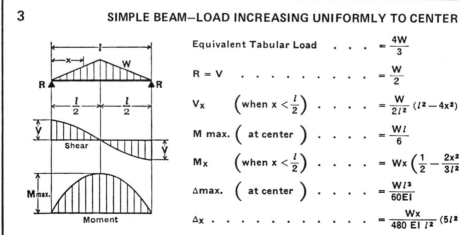

Equivalent Tabular Load . . . $= \dfrac{4W}{3}$

$R = V$ . . . . $= \dfrac{W}{2}$

V_x $\left(\text{when } x < \dfrac{l}{2}\right)$ . . . . $= \dfrac{W}{2l^2}(l^2 - 4x^2)$

M max. (at center) . . . . $= \dfrac{Wl}{6}$

M_x $\left(\text{when } x < \dfrac{l}{2}\right)$ . . . . $= Wx\left(\dfrac{1}{2} - \dfrac{2x^2}{3l^2}\right)$

Δ max. (at center) . . . . $= \dfrac{Wl^3}{60EI}$

Δ_x . . . . $= \dfrac{Wx}{480\,EI\,l^2}(5l^2 - 4x^2)^2$

Note: For deflection calculations, use specified loads.

BEAM DIAGRAMS AND FORMULAE

4. SIMPLE BEAM—UNIFORM LOAD PARTIALLY DISTRIBUTED

$R_1 = V_1$ (max. when $a < c$) $= \dfrac{wb}{2l}(2c + b)$

$R_2 = V_2$ (max. when $a > c$) $= \dfrac{wb}{2l}(2a + b)$

V_x (when $x > a$ and $< (a+b)$) $= R_1 - w(x-a)$

M max. (at $x = a + \dfrac{R_1}{w}$) $= R_1\left(a + \dfrac{R_1}{2w}\right)$

M_x (when $x < a$) $= R_1 x$

M_x (when $x > a$ and $< (a+b)$) $= R_1 x - \dfrac{w}{2}(x-a)^2$

M_x (when $x > (a+b)$) $= R_2(l-x)$

5. SIMPLE BEAM—UNIFORM LOAD PARTIALLY DISTRIBUTED AT ONE END

$R_1 = V_1$ max. $= \dfrac{wa}{2l}(2l - a)$

$R_2 = V_2$ $= \dfrac{wa^2}{2l}$

V (when $x < a$) $= R_1 - wx$

M max. (at $x = \dfrac{R_1}{w}$) $= \dfrac{R_1^2}{2w}$

M_x (when $x < a$) $= R_1 x - \dfrac{wx^2}{2}$

M_x (when $x > a$) $= R_2(l-x)$

Δ_x (when $x < a$) $= \dfrac{wx}{24EIl}\left(a^2(2l-a)^2 - 2ax^2(2l-a) + lx^3\right)$

Δ_x (when $x > a$) $= \dfrac{wa^2(l-x)}{24EIl}(4xl - 2x^2 - a^2)$

6. SIMPLE BEAM—UNIFORM LOADS PARTIALLY DISTRIBUTED AT EACH END

$R_1 = V_1$ $= \dfrac{w_1 a(2l - a) + w_2 c^2}{2l}$

$R_2 = V_2$ $= \dfrac{w_2 c(2l - c) + w_1 a^2}{2l}$

V_x (when $x < a$) $= R_1 - w_1 x$

V_x (when $x > a$ and $< (a+b)$) $= R_1 - w_1 a$

V_x (when $x > (a+b)$) $= R_2 - w_2(l-x)$

M max. (at $x = \dfrac{R_1}{w_1}$ when $R_1 < w_1 a$) $= \dfrac{R_1^2}{2w_1}$

M max. (at $x = l - \dfrac{R_2}{w_2}$ when $R_2 < w_2 c$) $= \dfrac{R_2^2}{2w_2}$

M_x (when $x < a$) $= R_1 x - \dfrac{w_1 x^2}{2}$

M_x (when $x > a$ and $< (a+b)$) $= R_1 x - \dfrac{w_1 a}{2}(2x - a)$

M_x (when $x > (a+b)$) $= R_2(l-x) - \dfrac{w_2(l-x)^2}{2}$

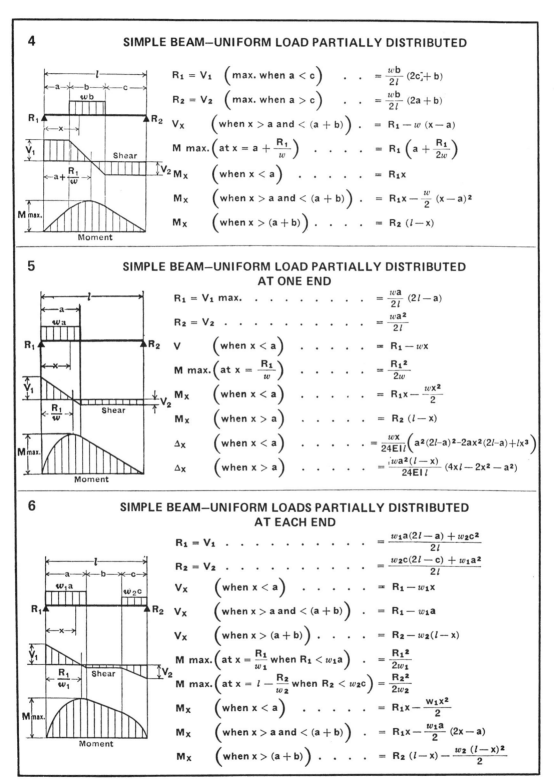

Note: For deflection calculations, use specified loads.

BEAM DIAGRAMS AND FORMULAE

Equivalent Tabular Load is the uniformly distributed factored load given in the Beam Load Tables

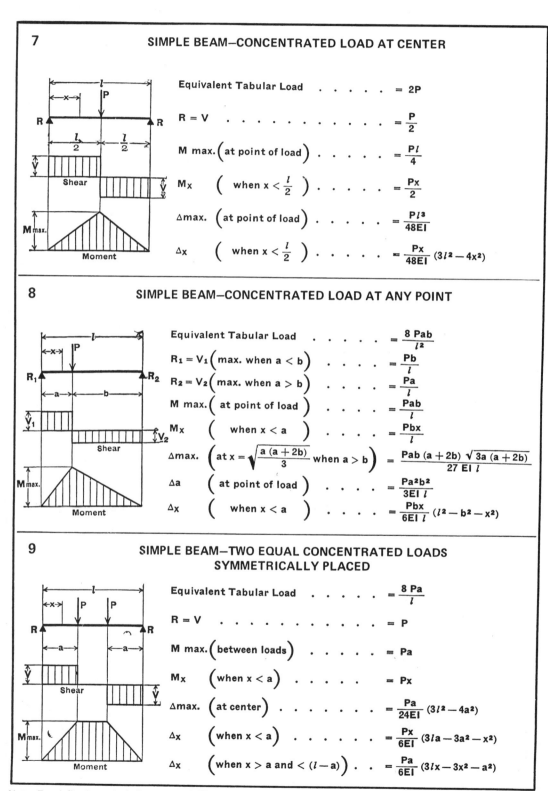

7. SIMPLE BEAM—CONCENTRATED LOAD AT CENTER

Equivalent Tabular Load $= 2P$

$R = V = \dfrac{P}{2}$

M max. (at point of load) $= \dfrac{Pl}{4}$

M_x (when $x < \dfrac{l}{2}$) $= \dfrac{Px}{2}$

Δmax. (at point of load) $= \dfrac{Pl^3}{48EI}$

Δ_x (when $x < \dfrac{l}{2}$) $= \dfrac{Px}{48EI}(3l^2 - 4x^2)$

8. SIMPLE BEAM—CONCENTRATED LOAD AT ANY POINT

Equivalent Tabular Load $= \dfrac{8\,Pab}{l^2}$

$R_1 = V_1$ (max. when $a < b$) $= \dfrac{Pb}{l}$

$R_2 = V_2$ (max. when $a > b$) $= \dfrac{Pa}{l}$

M max. (at point of load) $= \dfrac{Pab}{l}$

M_x (when $x < a$) $= \dfrac{Pbx}{l}$

Δmax. $\left(\text{at } x = \sqrt{\dfrac{a(a+2b)}{3}} \text{ when } a > b\right) = \dfrac{Pab(a+2b)\sqrt{3a(a+2b)}}{27\,EI\,l}$

Δ_a (at point of load) $= \dfrac{Pa^2b^2}{3EI\,l}$

Δ_x (when $x < a$) $= \dfrac{Pbx}{6EI\,l}(l^2 - b^2 - x^2)$

9. SIMPLE BEAM—TWO EQUAL CONCENTRATED LOADS SYMMETRICALLY PLACED

Equivalent Tabular Load $= \dfrac{8\,Pa}{l}$

$R = V = P$

M max. (between loads) $= Pa$

M_x (when $x < a$) $= Px$

Δmax. (at center) $= \dfrac{Pa}{24EI}(3l^2 - 4a^2)$

Δ_x (when $x < a$) $= \dfrac{Px}{6EI}(3la - 3a^2 - x^2)$

Δ_x (when $x > a$ and $< (l-a)$) $= \dfrac{Pa}{6EI}(3lx - 3x^2 - a^2)$

Note: For deflection calculations, use specified loads.

BEAM DIAGRAMS AND FORMULAE

Equivalent Tabular Load is the uniformly distributed factored load given in the Beam Load Tables

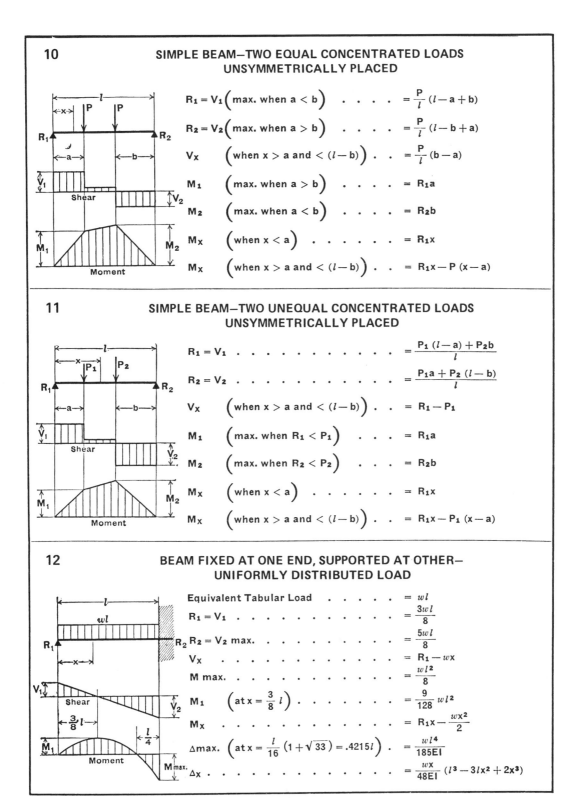

10 SIMPLE BEAM—TWO EQUAL CONCENTRATED LOADS UNSYMMETRICALLY PLACED

$R_1 = V_1$ (max. when $a < b$) $= \frac{P}{l}(l - a + b)$

$R_2 = V_2$ (max. when $a > b$) $= \frac{P}{l}(l - b + a)$

V_x (when $x > a$ and $< (l-b)$) $= \frac{P}{l}(b - a)$

M_1 (max. when $a > b$) $= R_1 a$

M_2 (max. when $a < b$) $= R_2 b$

M_x (when $x < a$) $= R_1 x$

M_x (when $x > a$ and $< (l-b)$) $= R_1 x - P(x - a)$

11 SIMPLE BEAM—TWO UNEQUAL CONCENTRATED LOADS UNSYMMETRICALLY PLACED

$R_1 = V_1$ $= \frac{P_1(l-a) + P_2 b}{l}$

$R_2 = V_2$ $= \frac{P_1 a + P_2(l-b)}{l}$

V_x (when $x > a$ and $< (l-b)$) $= R_1 - P_1$

M_1 (max. when $R_1 < P_1$) $= R_1 a$

M_2 (max. when $R_2 < P_2$) $= R_2 b$

M_x (when $x < a$) $= R_1 x$

M_x (when $x > a$ and $< (l-b)$) $= R_1 x - P_1(x-a)$

12 BEAM FIXED AT ONE END, SUPPORTED AT OTHER— UNIFORMLY DISTRIBUTED LOAD

Equivalent Tabular Load $= wl$

$R_1 = V_1$ $= \frac{3wl}{8}$

$R_2 = V_2$ max. $= \frac{5wl}{8}$

V_x $= R_1 - wx$

M max. $= \frac{wl^2}{8}$

M_1 (at $x = \frac{3}{8}l$) $= \frac{9}{128}wl^2$

M_x $= R_1 x - \frac{wx^2}{2}$

Δ max. (at $x = \frac{l}{16}(1 + \sqrt{33}) = .4215l$) $= \frac{wl^4}{185EI}$

Δ_x $= \frac{wx}{48EI}(l^3 - 3lx^2 + 2x^3)$

Note: For deflection calculations, use specified loads.

BEAM DIAGRAMS AND FORMULAE

Equivalent Tabular Load is the uniformly distributed factored load given in the Beam Load Tables

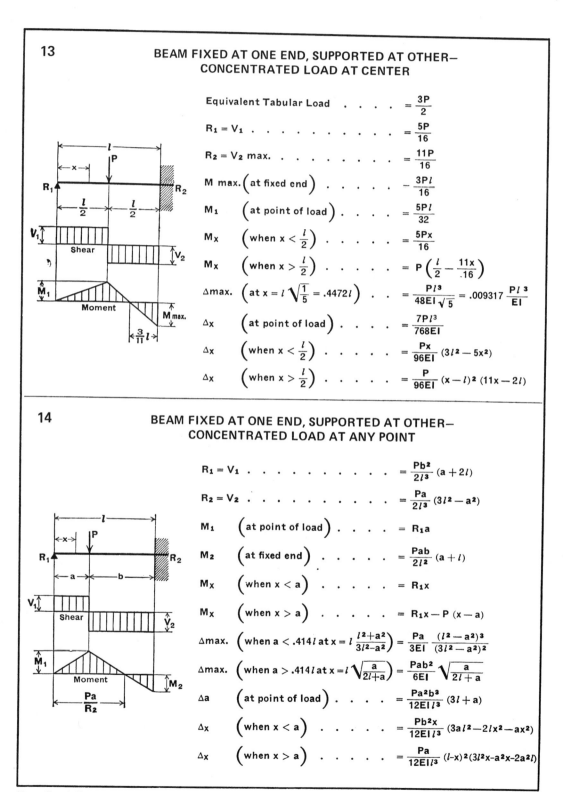

13   BEAM FIXED AT ONE END, SUPPORTED AT OTHER—CONCENTRATED LOAD AT CENTER

Equivalent Tabular Load $= \dfrac{3P}{2}$

$R_1 = V_1 = \dfrac{5P}{16}$

$R_2 = V_2$ max. $= \dfrac{11P}{16}$

M max. (at fixed end) $= \dfrac{3Pl}{16}$

M_1 (at point of load) $= \dfrac{5Pl}{32}$

M_x (when $x < \dfrac{l}{2}$) $= \dfrac{5Px}{16}$

M_x (when $x > \dfrac{l}{2}$) $= P\left(\dfrac{l}{2} - \dfrac{11x}{16}\right)$

Δmax. (at $x = l\sqrt{\dfrac{1}{5}} = .4472l$) $= \dfrac{Pl^3}{48EI\sqrt{5}} = .009317\dfrac{Pl^3}{EI}$

Δ_x (at point of load) $= \dfrac{7Pl^3}{768EI}$

Δ_x (when $x < \dfrac{l}{2}$) $= \dfrac{Px}{96EI}(3l^2 - 5x^2)$

Δ_x (when $x > \dfrac{l}{2}$) $= \dfrac{P}{96EI}(x-l)^2(11x-2l)$

14   BEAM FIXED AT ONE END, SUPPORTED AT OTHER—CONCENTRATED LOAD AT ANY POINT

$R_1 = V_1 = \dfrac{Pb^2}{2l^3}(a + 2l)$

$R_2 = V_2 = \dfrac{Pa}{2l^3}(3l^2 - a^2)$

M_1 (at point of load) $= R_1 a$

M_2 (at fixed end) $= \dfrac{Pab}{2l^2}(a + l)$

M_x (when $x < a$) $= R_1 x$

M_x (when $x > a$) $= R_1 x - P(x - a)$

Δmax. (when $a < .414l$ at $x = l\dfrac{l^2+a^2}{3l^2-a^2}$) $= \dfrac{Pa}{3EI}\dfrac{(l^2-a^2)^3}{(3l^2-a^2)^2}$

Δmax. (when $a > .414l$ at $x = l\sqrt{\dfrac{a}{2l+a}}$) $= \dfrac{Pab^2}{6EI}\sqrt{\dfrac{a}{2l+a}}$

Δa (at point of load) $= \dfrac{Pa^2b^3}{12EIl^3}(3l + a)$

Δ_x (when $x < a$) $= \dfrac{Pb^2x}{12EIl^3}(3al^2 - 2lx^2 - ax^2)$

Δ_x (when $x > a$) $= \dfrac{Pa}{12EIl^3}(l-x)^2(3l^2x - a^2x - 2a^2l)$

Note: For deflection calculations, use specified loads.

BEAM DIAGRAMS AND FORMULAE

Equivalent Tabular Load is the uniformly distributed factored load given in the Beam Load Tables

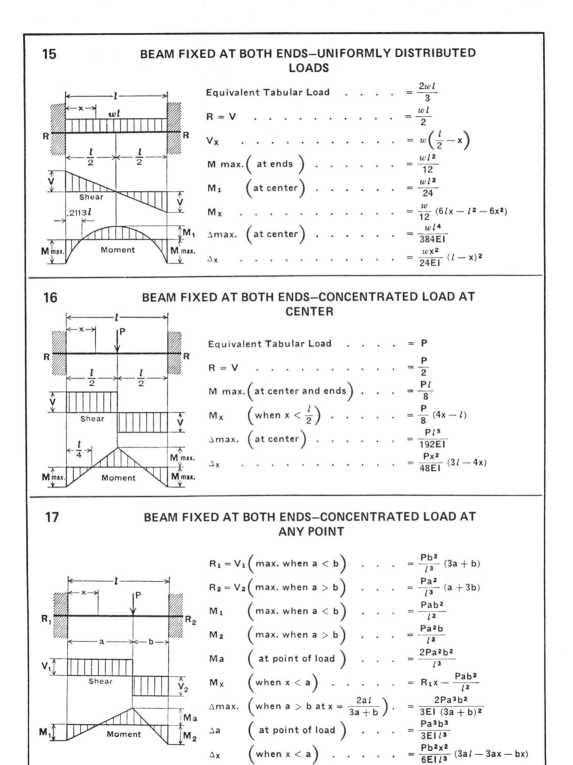

15   **BEAM FIXED AT BOTH ENDS—UNIFORMLY DISTRIBUTED LOADS**

Equivalent Tabular Load $= \dfrac{2wl}{3}$

$R = V = \dfrac{wl}{2}$

$V_x = w\left(\dfrac{l}{2} - x\right)$

M max. (at ends) $= \dfrac{wl^2}{12}$

M_1 (at center) $= \dfrac{wl^2}{24}$

$M_x = \dfrac{w}{12}(6lx - l^2 - 6x^2)$

Δ max. (at center) $= \dfrac{wl^4}{384EI}$

$\Delta_x = \dfrac{wx^2}{24EI}(l-x)^2$

16   **BEAM FIXED AT BOTH ENDS—CONCENTRATED LOAD AT CENTER**

Equivalent Tabular Load $= P$

$R = V = \dfrac{P}{2}$

M max. (at center and ends) $= \dfrac{Pl}{8}$

$M_x \left(\text{when } x < \dfrac{l}{2}\right) = \dfrac{P}{8}(4x - l)$

Δ max. (at center) $= \dfrac{Pl^3}{192EI}$

$\Delta_x = \dfrac{Px^2}{48EI}(3l - 4x)$

17   **BEAM FIXED AT BOTH ENDS—CONCENTRATED LOAD AT ANY POINT**

$R_1 = V_1$ (max. when $a < b$) $= \dfrac{Pb^2}{l^3}(3a + b)$

$R_2 = V_2$ (max. when $a > b$) $= \dfrac{Pa^2}{l^3}(a + 3b)$

M_1 (max. when $a < b$) $= \dfrac{Pab^2}{l^2}$

M_2 (max. when $a > b$) $= \dfrac{Pa^2b}{l^2}$

M_a (at point of load) $= \dfrac{2Pa^2b^2}{l^3}$

M_x (when $x < a$) $= R_1 x - \dfrac{Pab^2}{l^2}$

Δ max. $\left(\text{when } a > b \text{ at } x = \dfrac{2al}{3a + b}\right) = \dfrac{2Pa^3b^2}{3EI(3a + b)^2}$

Δ_a (at point of load) $= \dfrac{Pa^3b^3}{3EIl^3}$

Δ_x (when $x < a$) $= \dfrac{Pb^2x^2}{6EIl^3}(3al - 3ax - bx)$

Note: For deflection calculations, use specified loads.

BEAM DIAGRAMS AND FORMULAE

Equivalent Tabular Load is the uniformly distributed factored load given in the Beam Load Tables

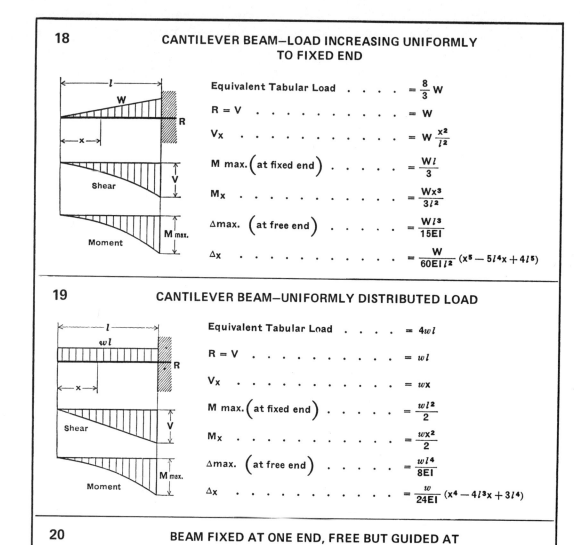

18. CANTILEVER BEAM—LOAD INCREASING UNIFORMLY TO FIXED END

Equivalent Tabular Load $= \dfrac{8}{3} W$

$R = V = W$

$V_x = W \dfrac{x^2}{l^2}$

M max. (at fixed end) $= \dfrac{Wl}{3}$

$M_x = \dfrac{Wx^3}{3l^2}$

Δ max. (at free end) $= \dfrac{Wl^3}{15EI}$

$\Delta_x = \dfrac{W}{60EIl^2}(x^5 - 5l^4x + 4l^5)$

19. CANTILEVER BEAM—UNIFORMLY DISTRIBUTED LOAD

Equivalent Tabular Load $= 4wl$

$R = V = wl$

$V_x = wx$

M max. (at fixed end) $= \dfrac{wl^2}{2}$

$M_x = \dfrac{wx^2}{2}$

Δ max. (at free end) $= \dfrac{wl^4}{8EI}$

$\Delta_x = \dfrac{w}{24EI}(x^4 - 4l^3x + 3l^4)$

20. BEAM FIXED AT ONE END, FREE BUT GUIDED AT OTHER—UNIFORMLY DISTRIBUTED LOAD

Equivalent Tabular Load $= \dfrac{8}{3} wl$

$R = V = wl$

$V_x = wx$

M max. (at fixed end) $= \dfrac{wl^2}{3}$

M_1 (at deflected end) $= \dfrac{wl^2}{6}$

$M_x = \dfrac{w}{6}(l^2 - 3x^2)$

Δ max. (at deflected end) $= \dfrac{wl^4}{24EI}$

$\Delta_x = \dfrac{w(l^2 - x^2)^2}{24EI}$

Note: For deflection calculations, use specified loads.

BEAM DIAGRAMS AND FORMULAE

Equivalent Tabular Load is the uniformly distributed factored load given in the Beam Load Tables

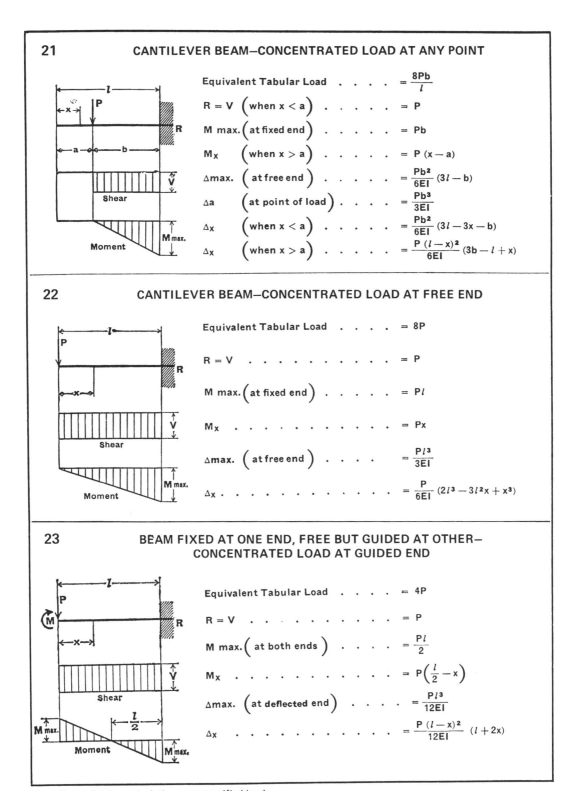

21 CANTILEVER BEAM—CONCENTRATED LOAD AT ANY POINT

Equivalent Tabular Load $= \dfrac{8Pb}{l}$

$R = V$ (when $x < a$) $= P$

M max. (at fixed end) $= Pb$

M_x (when $x > a$) $= P(x-a)$

Δ max. (at free end) $= \dfrac{Pb^2}{6EI}(3l - b)$

Δa (at point of load) $= \dfrac{Pb^3}{3EI}$

Δ_x (when $x < a$) $= \dfrac{Pb^2}{6EI}(3l - 3x - b)$

Δ_x (when $x > a$) $= \dfrac{P(l-x)^2}{6EI}(3b - l + x)$

22 CANTILEVER BEAM—CONCENTRATED LOAD AT FREE END

Equivalent Tabular Load $= 8P$

$R = V = P$

M max. (at fixed end) $= Pl$

$M_x = Px$

Δ max. (at free end) $= \dfrac{Pl^3}{3EI}$

$\Delta_x = \dfrac{P}{6EI}(2l^3 - 3l^2 x + x^3)$

23 BEAM FIXED AT ONE END, FREE BUT GUIDED AT OTHER—CONCENTRATED LOAD AT GUIDED END

Equivalent Tabular Load $= 4P$

$R = V = P$

M max. (at both ends) $= \dfrac{Pl}{2}$

$M_x = P\left(\dfrac{l}{2} - x\right)$

Δ max. (at deflected end) $= \dfrac{Pl^3}{12EI}$

$\Delta_x = \dfrac{P(l-x)^2}{12EI}(l + 2x)$

Note: For deflection calculations, use specified loads.

5-139

BEAM DIAGRAMS AND FORMULAE

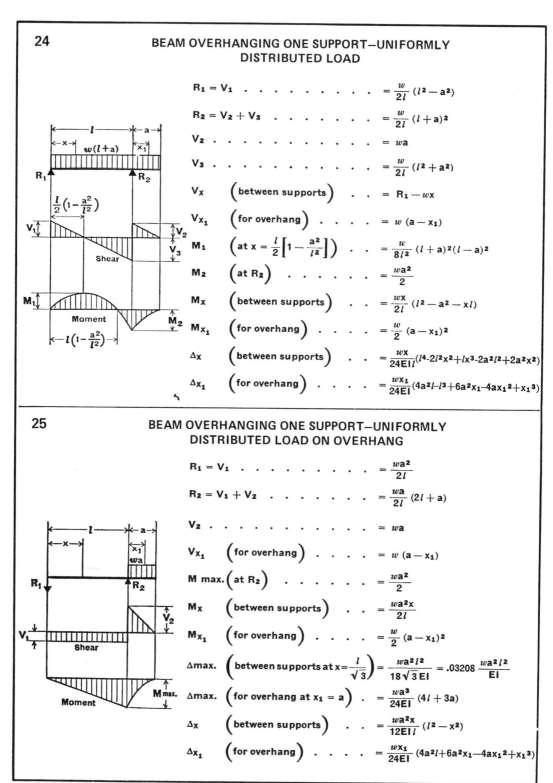

24 — BEAM OVERHANGING ONE SUPPORT—UNIFORMLY DISTRIBUTED LOAD

$R_1 = V_1 \quad = \dfrac{w}{2l}(l^2 - a^2)$

$R_2 = V_2 + V_3 \quad = \dfrac{w}{2l}(l + a)^2$

$V_2 \quad = wa$

$V_3 \quad = \dfrac{w}{2l}(l^2 + a^2)$

V_x (between supports) $= R_1 - wx$

V_{x_1} (for overhang) $= w(a - x_1)$

$M_1 \left(\text{at } x = \dfrac{l}{2}\left[1 - \dfrac{a^2}{l^2}\right]\right) = \dfrac{w}{8l^2}(l+a)^2(l-a)^2$

M_2 (at R_2) $= \dfrac{wa^2}{2}$

M_x (between supports) $= \dfrac{wx}{2l}(l^2 - a^2 - xl)$

M_{x_1} (for overhang) $= \dfrac{w}{2}(a - x_1)^2$

Δ_x (between supports) $= \dfrac{wx}{24EIl}(l^4 - 2l^2x^2 + lx^3 - 2a^2l^2 + 2a^2x^2)$

Δ_{x_1} (for overhang) $= \dfrac{wx_1}{24EI}(4a^2l - l^3 + 6a^2x_1 - 4ax_1^2 + x_1^3)$

25 — BEAM OVERHANGING ONE SUPPORT—UNIFORMLY DISTRIBUTED LOAD ON OVERHANG

$R_1 = V_1 \quad = \dfrac{wa^2}{2l}$

$R_2 = V_1 + V_2 \quad = \dfrac{wa}{2l}(2l + a)$

$V_2 \quad = wa$

V_{x_1} (for overhang) $= w(a - x_1)$

$M \text{ max.}$ (at R_2) $= \dfrac{wa^2}{2}$

M_x (between supports) $= \dfrac{wa^2 x}{2l}$

M_{x_1} (for overhang) $= \dfrac{w}{2}(a - x_1)^2$

$\Delta \text{max.}$ $\left(\text{between supports at } x = \dfrac{l}{\sqrt{3}}\right) = \dfrac{wa^2 l^2}{18\sqrt{3}\,EI} = .03208 \dfrac{wa^2 l^2}{EI}$

$\Delta \text{max.}$ (for overhang at $x_1 = a$) $= \dfrac{wa^3}{24EI}(4l + 3a)$

Δ_x (between supports) $= \dfrac{wa^2 x}{12EIl}(l^2 - x^2)$

Δ_{x_1} (for overhang) $= \dfrac{wx_1}{24EI}(4a^2 l + 6a^2 x_1 - 4ax_1^2 + x_1^3)$

Note: For deflection calculations, use specified loads.

BEAM DIAGRAMS AND FORMULAE

Equivalent Tabular Load is the uniformly distributed factored load given in the Beam Load Tables

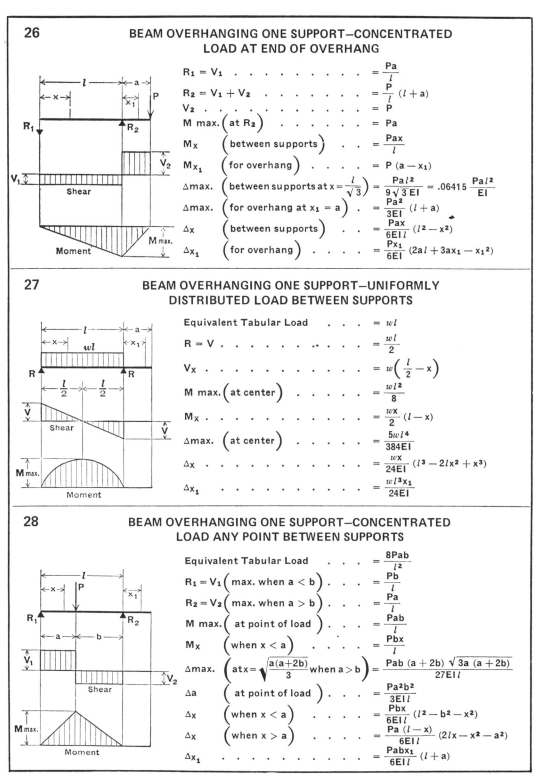

26   **BEAM OVERHANGING ONE SUPPORT—CONCENTRATED LOAD AT END OF OVERHANG**

$R_1 = V_1 = \dfrac{Pa}{l}$

$R_2 = V_1 + V_2 = \dfrac{P}{l}(l+a)$

$V_2 = P$

M max. (at R_2) $= Pa$

M_x (between supports) $= \dfrac{Pax}{l}$

M_{x_1} (for overhang) $= P(a - x_1)$

Δmax. (between supports at $x = \dfrac{l}{\sqrt{3}}$) $= \dfrac{Pal^2}{9\sqrt{3}\,EI} = .06415\dfrac{Pal^2}{EI}$

Δmax. (for overhang at $x_1 = a$) $= \dfrac{Pa^2}{3EI}(l+a)$

Δ_x (between supports) $= \dfrac{Pax}{6EIl}(l^2 - x^2)$

Δ_{x_1} (for overhang) $= \dfrac{Px_1}{6EI}(2al + 3ax_1 - x_1^2)$

27   **BEAM OVERHANGING ONE SUPPORT—UNIFORMLY DISTRIBUTED LOAD BETWEEN SUPPORTS**

Equivalent Tabular Load $= wl$

$R = V = \dfrac{wl}{2}$

$V_x = w\left(\dfrac{l}{2} - x\right)$

M max. (at center) $= \dfrac{wl^2}{8}$

$M_x = \dfrac{wx}{2}(l - x)$

Δmax. (at center) $= \dfrac{5wl^4}{384EI}$

$\Delta_x = \dfrac{wx}{24EI}(l^3 - 2lx^2 + x^3)$

$\Delta_{x_1} = \dfrac{wl^3 x_1}{24EI}$

28   **BEAM OVERHANGING ONE SUPPORT—CONCENTRATED LOAD ANY POINT BETWEEN SUPPORTS**

Equivalent Tabular Load $= \dfrac{8Pab}{l^2}$

$R_1 = V_1$ (max. when $a < b$) $= \dfrac{Pb}{l}$

$R_2 = V_2$ (max. when $a > b$) $= \dfrac{Pa}{l}$

M max. (at point of load) $= \dfrac{Pab}{l}$

M_x (when $x < a$) $= \dfrac{Pbx}{l}$

Δmax. $\left(\text{at } x = \sqrt{\dfrac{a(a+2b)}{3}} \text{ when } a > b\right) = \dfrac{Pab(a+2b)\sqrt{3a(a+2b)}}{27EIl}$

Δa (at point of load) $= \dfrac{Pa^2b^2}{3EIl}$

Δ_x (when $x < a$) $= \dfrac{Pbx}{6EIl}(l^2 - b^2 - x^2)$

Δ_x (when $x > a$) $= \dfrac{Pa(l-x)}{6EIl}(2lx - x^2 - a^2)$

$\Delta_{x_1} = \dfrac{Pabx_1}{6EIl}(l+a)$

Note: For deflection calculations, use specified loads.

BEAM DIAGRAMS AND FORMULAE

29   BEAM—UNIFORMLY DISTRIBUTED LOAD AND VARIABLE END MOMENTS

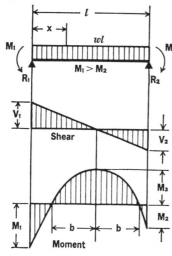

$$R_1 = V_1 = \frac{wl}{2} + \frac{M_1 - M_2}{l}$$

$$R_2 = V_2 = \frac{wl}{2} - \frac{M_1 - M_2}{l}$$

$$V_x = w\left(\frac{l}{2} - x\right) + \frac{M_1 - M_2}{l}$$

$$M_3 \left(\text{at } x = \frac{l}{2} + \frac{M_1 - M_2}{wl}\right)$$

$$= \frac{wl^2}{8} - \frac{M_1 + M_2}{2} + \frac{(M_1 - M_2)^2}{2wl^2}$$

$$M_x = \frac{wx}{2}(l - x) + \left(\frac{M_1 - M_2}{l}\right)x - M_1$$

$$b\left(\begin{matrix}\text{To locate}\\ \text{inflection points}\end{matrix}\right) = \sqrt{\frac{l^2}{4} - \left(\frac{M_1 + M_2}{w}\right) + \left(\frac{M_1 - M_2}{wl}\right)^2}$$

$$\Delta_x = \frac{wx}{24EI}\left[x^3 - \left(2l + \frac{4M_1}{wl} - \frac{4M_2}{wl}\right)x^2 + \frac{12M_1}{w}x + l^3 - \frac{8M_1 l}{w} - \frac{4M_2 l}{w}\right]$$

---

30   BEAM—CONCENTRATED LOAD AT CENTER AND VARIABLE END MOMENTS

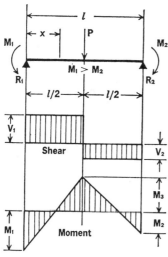

$$R_1 = V_1 = \frac{P}{2} + \frac{M_1 - M_2}{l}$$

$$R_2 = V_2 = \frac{P}{2} - \frac{M_1 - M_2}{l}$$

$$M_3 \text{ (At center)} = \frac{Pl}{4} - \frac{M_1 + M_2}{2}$$

$$M_x \left(\text{When } x < \frac{l}{2}\right) = \left(\frac{P}{2} + \frac{M_1 - M_2}{l}\right)x - M_1$$

$$M_x \left(\text{When } x > \frac{l}{2}\right) = \frac{P}{2}(l - x) + \frac{(M_1 - M_2)x}{l} - M_1$$

$$\Delta_x \left(\text{When } x < \frac{l}{2}\right) = \frac{Px}{48EI}\left(3l^2 - 4x^2 - \frac{8(l-x)}{Pl}[M_1(2l - x) + M_2(l + x)]\right)$$

Note: For deflection calculations, use specified loads.

BEAM DIAGRAMS AND FORMULAE

31 SIMPLE BEAM—ONE CONCENTRATED MOVING LOAD

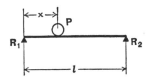

R_1 max. $= V_1$ max. $\left(\text{at } x = o\right)$ . . . . $= P$

M max. $\left(\text{at point of load, when } x = \dfrac{l}{2}\right)$. $= \dfrac{Pl}{4}$

---

32 SIMPLE BEAM—TWO EQUAL CONCENTRATED MOVING LOADS

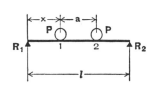

R_1 max. $= V_1$ max. $\left(\text{at } x = o\right)$ . . . . $= P\left(2 - \dfrac{a}{l}\right)$

M max. $\begin{cases} \left[\begin{array}{l}\text{when } a < (2-\sqrt{2})\,l = .586l \\ \text{under load 1 at } x = \dfrac{1}{2}\left(l - \dfrac{a}{2}\right)\end{array}\right] = \dfrac{P}{2l}\left(l - \dfrac{a}{2}\right)^2 \\[1em] \left[\begin{array}{l}\text{when } a > (2-\sqrt{2})\,l = .586l \\ \text{with one load at center of span} \\ \text{(case 31)}\end{array}\right] = \dfrac{Pl}{4} \end{cases}$

---

33 SIMPLE BEAM—TWO UNEQUAL CONCENTRATED MOVING LOADS

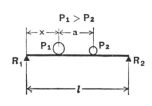

R_1 max. $= V_1$ max. $\left(\text{at } x = o\right)$ . . . . $= P_1 + P_2 \dfrac{l-a}{l}$

M max. $\begin{cases} \left[\text{under } P_1, \text{ at } x = \dfrac{1}{2}\left(l - \dfrac{P_2 a}{P_1 + P_2}\right)\right] = (P_1 + P_2)\dfrac{x^2}{l} \\[1em] \left[\begin{array}{l}\text{M max. may occur with larger} \\ \text{load at center of span and other} \\ \text{load off span (case 31)}\end{array}\right] = \dfrac{P_1 l}{4} \end{cases}$

GENERAL RULES FOR SIMPLE BEAMS CARRYING MOVING CONCENTRATED LOADS

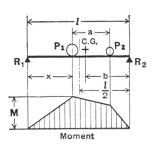

The maximum shear due to moving concentrated loads occurs at one support when one of the loads is at that support. With several moving loads, the location that will produce maximum shear must be determined by trial.

The maximum bending moment produced by moving concentrated loads occurs under one of the loads when that load is as far from one support as the center of gravity of all the moving loads on the beam is from the other support.

In the accompanying diagram, the maximum bending moment occurs under load P_1 when $x = b$. It should also be noted that this condition occurs when the center line of the span is midway between the center of gravity of loads and the nearest concentrated load.

Note: For deflection calculations, use specified loads.

BEAM DIAGRAMS AND FORMULAE

Equivalent Tabular Load is the uniformly distributed factored load given in the Beam Load Tables

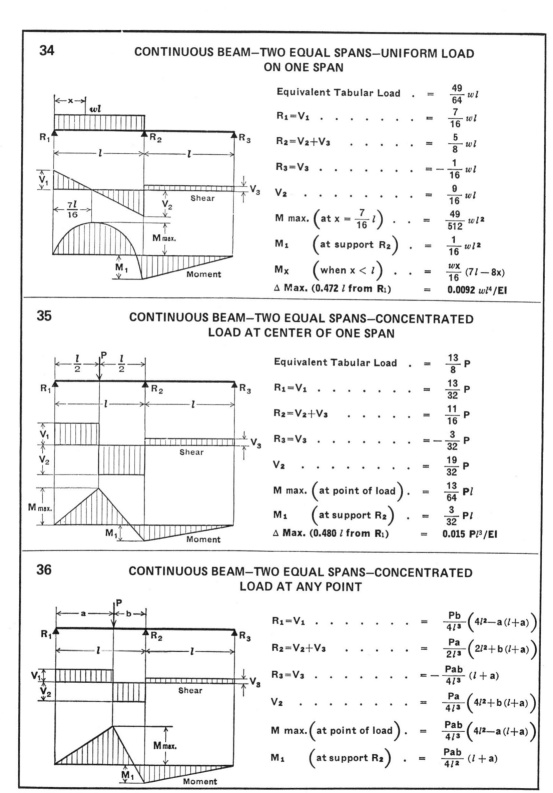

34 CONTINUOUS BEAM—TWO EQUAL SPANS—UNIFORM LOAD ON ONE SPAN

Equivalent Tabular Load $= \dfrac{49}{64} wl$

$R_1 = V_1 = \dfrac{7}{16} wl$

$R_2 = V_2 + V_3 = \dfrac{5}{8} wl$

$R_3 = V_3 = -\dfrac{1}{16} wl$

$V_2 = \dfrac{9}{16} wl$

M max. $\left(\text{at } x = \dfrac{7}{16} l\right) = \dfrac{49}{512} wl^2$

M_1 (at support R_2) $= \dfrac{1}{16} wl^2$

M_x (when $x < l$) $= \dfrac{wx}{16}(7l - 8x)$

Δ Max. (0.472 l from R_1) $= 0.0092\, wl^4/EI$

35 CONTINUOUS BEAM—TWO EQUAL SPANS—CONCENTRATED LOAD AT CENTER OF ONE SPAN

Equivalent Tabular Load $= \dfrac{13}{8} P$

$R_1 = V_1 = \dfrac{13}{32} P$

$R_2 = V_2 + V_3 = \dfrac{11}{16} P$

$R_3 = V_3 = -\dfrac{3}{32} P$

$V_2 = \dfrac{19}{32} P$

M max. (at point of load) $= \dfrac{13}{64} Pl$

M_1 (at support R_2) $= \dfrac{3}{32} Pl$

Δ Max. (0.480 l from R_1) $= 0.015\, Pl^3/EI$

36 CONTINUOUS BEAM—TWO EQUAL SPANS—CONCENTRATED LOAD AT ANY POINT

$R_1 = V_1 = \dfrac{Pb}{4l^3}\left(4l^2 - a(l+a)\right)$

$R_2 = V_2 + V_3 = \dfrac{Pa}{2l^3}\left(2l^2 + b(l+a)\right)$

$R_3 = V_3 = -\dfrac{Pab}{4l^3}(l+a)$

$V_2 = \dfrac{Pa}{4l^3}\left(4l^2 + b(l+a)\right)$

M max. (at point of load) $= \dfrac{Pab}{4l^3}\left(4l^2 - a(l+a)\right)$

M_1 (at support R_2) $= \dfrac{Pab}{4l^2}(l+a)$

Note: For deflection calculations, use specified loads.

MOMENTS, REACTIONS
Equal Span Continuous Beams
UNIFORMLY DISTRIBUTED LOADS

Moment = Coefficient x W x L
Reaction = Coefficient x W
Where: W = Total uniformly distributed load on one span
L = Length of one span

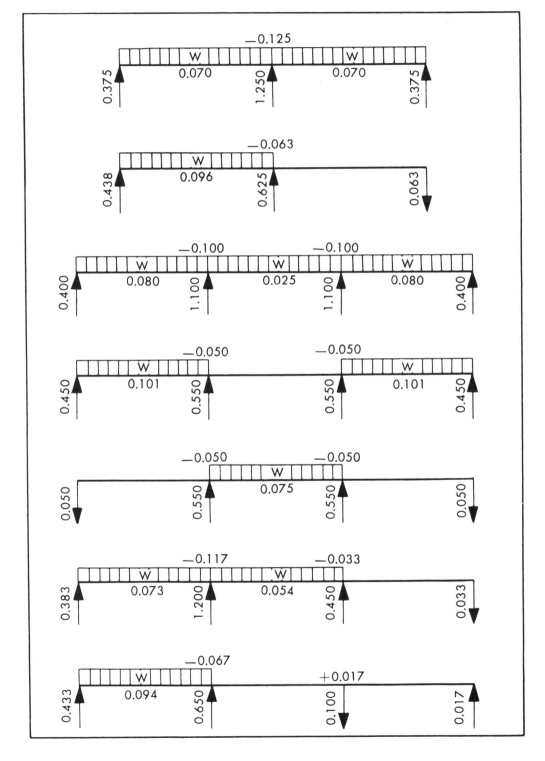

5-145

MOMENTS, REACTIONS
Equal Span Continuous Beams
UNIFORMLY DISTRIBUTED LOADS

Moment = Coefficient x W x L
Reaction = Coefficient x W
Where: W = Total uniformly distributed load on one span
L = Length of one span

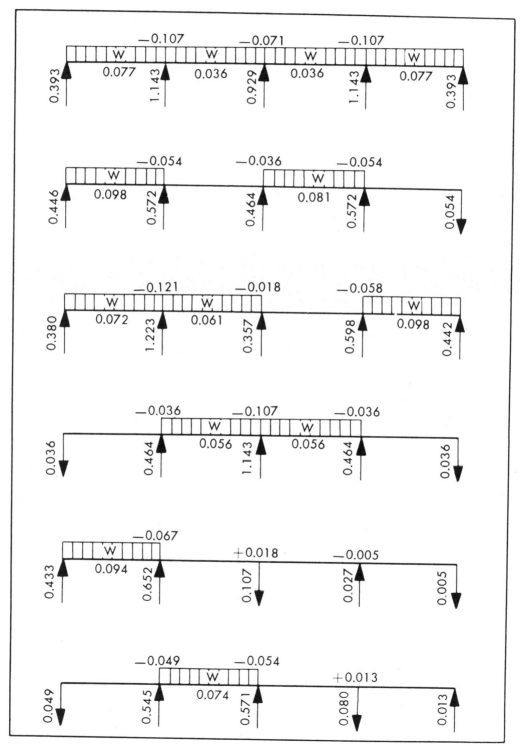

MOMENTS, REACTIONS
Equal Span Continuous Beams
CENTRAL POINT LOADS

Moment = Coefficient x W x L
Reaction = Coefficient x W
Where: W = The concentrated load on one span
L = Length of one span

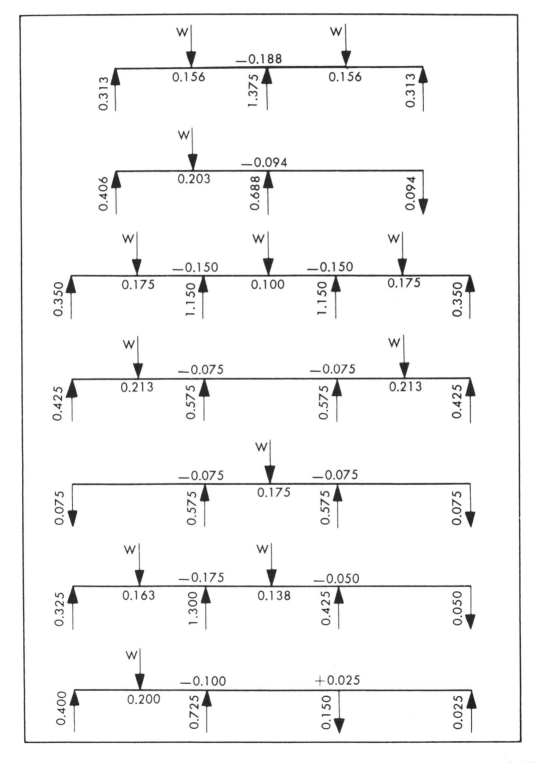

MOMENTS, REACTIONS
Equal Span Continuous Beams
CENTRAL POINT LOADS

Moment = Coefficient x W x L
Reaction = Coefficient x W
Where: W = The concentrated load on one span
L = Length of one span

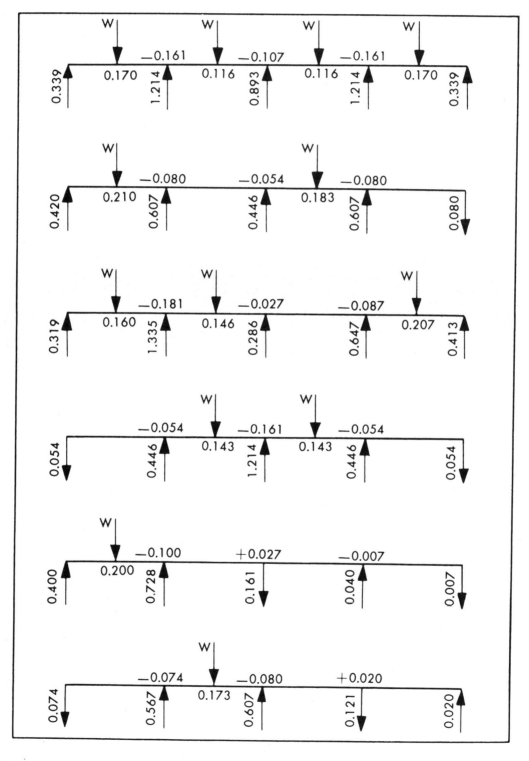

MOMENTS, REACTIONS
Equal Span Continuous Beams

POINT LOADS AT THIRD POINTS OF SPAN

Moment = Coefficient x W x L
Reaction = Coefficient x W
Where: W = The total load on one span
L = Length of one span

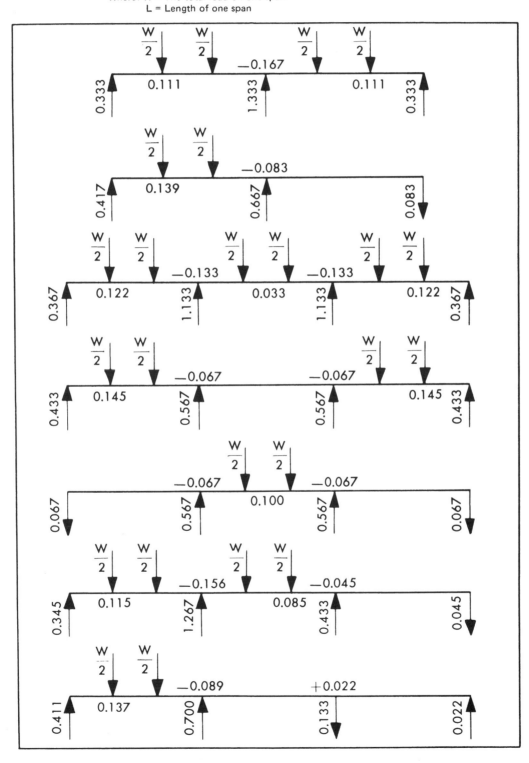

5-149

MOMENTS, REACTIONS
Equal Span Continuous Beams

POINT LOADS AT THIRD POINTS OF SPAN

Moment = Coefficient x W x L
Reaction = Coefficient x W
Where: W = The total load on one span
L = Length of one span

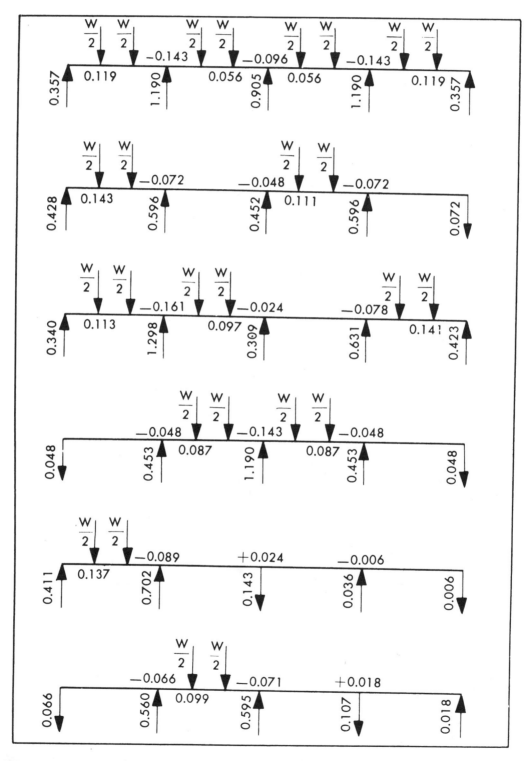

5-150

BEAM BEARING PLATES

General

When a flexural member is supported by a masonry wall or pier, the beam reaction must be distributed over sufficient area to avoid exceeding the bearing capacity of the masonry or concrete. Steel bearing plates may be used for this purpose.

Bearing plates are usually set in place and grouted level at the required elevation before positioning the beam. Thus, even though the beam flange may be able to distribute the reaction to supporting masonry or concrete, a bearing plate can be useful to facilitate erection. Some form of anchorage is required to ensure that the beam is connected to the pier or wall either longitudinally or for uplift forces.

Design Chart

Figure 5-2 on page 5-153 provides a graph to determine the thickness of bearing plates using G40.21 300W steel, for beams without bearing stiffeners, based on the following assumptions:

- The beam reaction R is uniformly distributed to the bearing plate over an effective area of width $2k$ and length C

- The bearing pressure between the effective area of the bearing plate and the concrete or masonry support is uniform over the area of the plate

- The bearing pressure under the portion of plate projecting beyond the k-distance from the centre line of the beam is ignored, since in practice the flange may be slightly "curled".

Equating the factored moment acting on the portion of the bearing plate taken as a cantilever, to the factored moment resistance of the plate, $(M_r = \phi Z F_y)$, the bearing plate thickness is calculated as:

$$t_p = \sqrt{\frac{2 B_r n^2}{A \phi F_y}}$$

where

F_y = specified minimum yield strength of the bearing plate steel (MPa)

A = $B \times C$ = area of plate (mm^2)

t_p = required thickness of bearing plate (mm)

k = beam k-distance = distance from web toe of fillet to outer face of flange (mm)

n = $B/2 - k$, (mm)

b = width of beam flange (mm)

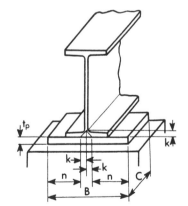

To minimize deflection of the bearing plate, the thickness generally should not be less than about one fifth of the overhang, i.e., $t_p \geq (B - b) / 10$.

Use of chart

1. Required area, A = beam reaction due to factored loads divided by the unit factored concrete bearing resistance, $(0.85 \phi_c f'_c)$, where $\phi_c = 0.60$

2. Determine C and solve for B. (C, the length of bearing, is usually governed by the available wall thickness or other structural considerations.)

3. Determine n and enter Figure 5-2 to determine t_p.

Example

Given:

A W610x140 of G40.21 350W steel beam has a factored end reaction of 850 kN and is supported on a concrete pier with 28 day compressive strength of 20 MPa. Design the bearing plate assuming G40.21 300W steel and a concrete bearing length of 300 mm.

Solution:

Unit factored bearing resistance of concrete is

$0.85 \times 0.60 \times 20 = 10.2$ MPa

Area required is $(850 \times 10^3) / 10.2 = 83\,300$ mm²

Therefore required B is $83\,300 / 300 = 278$ mm

For W610x140,

$b = 230$ mm, $t = 22.2$ mm, $k = 42$ mm, $w = 13.1$ mm

Select $B = 280$ mm   (greater than flange width, $b = 230$ mm)

$n = (B/2) - k = (280/2) - 42 = 98$ mm

From Figure 5-2, for unit factored bearing resistance of 10.2 MPa and n of 98 mm,

minimum $t_p = 27$ mm   — select 30 mm

Use plate 30 x 300 x 280

Check for web crippling and web yielding   (Clause 15.9(b), S16.1-94)

Web yielding

$B_r = 1.10 \, \phi \, w \, (N + 2.5k) \, F_y$

$= 1.10 \times 0.9 \times 13.1 \, (300 + (2.5 \times 42)) \, 350 / 10^3$ kN

$= 1\,840$ kN

Web crippling

$B_r = 150 \, \phi \, w^2 \, [\, 1 + 3 \, (N/d) \, (w/t)^{1.5} \,] \, (F_y \, t/w)^{0.5}$

$= 150 \times 0.9 \times 13.1^2 \, [1 + 3 \, (300 / 617) \, (13.1 / 22.2)^{1.5}] \, (350 \times 22.2 / 13.1)^{0.5} / 10^3$

$= 937$ kN

Therefore web crippling, Clause 15.9(b)(ii), governs and $B_r = 937$ kN

Alternatively, the bearing resistance can be established using the values of R and G from the beam load tables for a W610x140, on page 5-110.

$R = 689$ kN, $G = 12.4$ kN

Bearing capacity for bearing length of 300 mm is

$R + (200 \times G/10) = 689 + (200 \times 12.4 / 10) = 937$ kN

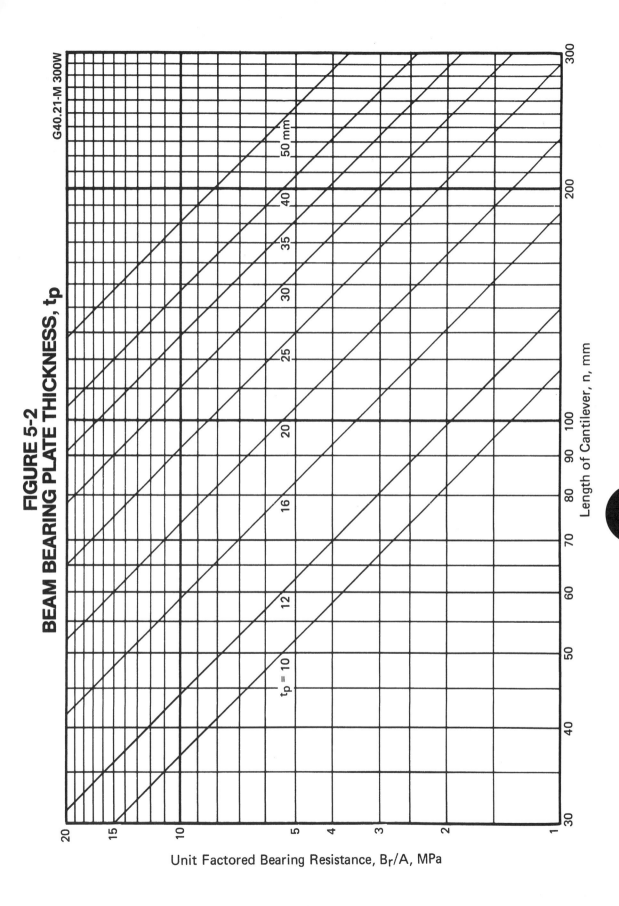

FIGURE 5-2
BEAM BEARING PLATE THICKNESS, t_p

BEAMS WITH WEB HOLES

General

Structures may support a variety of pipes, ducts, conduit and other services, and efforts to reduce floor heights have led to these items being placed in the same plane as the structural floor members. Structural systems using stub girders, trusses and open web steel joists provide openings for structural/mechanical integration; however, when beams with solid webs are used it may be necessary to cut openings through the webs. This section, based on research summarized by Redwood and Shrivastava (1980), describes a method to account for web holes during design of the member.

Special precautions may be required if it becomes necessary to cut holes in beam webs after construction is complete.

Design

The formulae are applicable for beams of Class 1 and Class 2 sections with openings between 0.3 and 0.7 times the depth of the beam, and hole lengths up to three times the hole height. The steel should meet the requirements of Clause 8.5(a) of CAN/CSA-S16.1-94 and exhibit the characteristics necessary to achieve moment redistribution, such as G40.21 grade 350W.

The hole corner should have a radius at least equal to the larger of 16 mm or twice the web thickness. Fatigue loading considerations have not been accounted for in the formulae of this section and if holes are necessary in a member subjected to fatigue, some guidance is available from Frost & Leffler (1971).

Special design considerations are required if concentrated loads are to be located within the hole length or within one beam depth from either end of a hole.

The width-to-thickness ratio of outstanding reinforcing plates should meet Class 1 requirements.

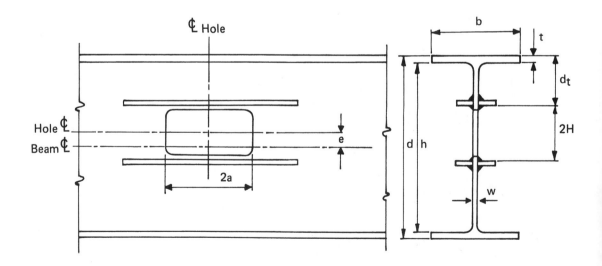

Nomenclature

A_f area of one flange ($b\ t$)

A_r area of reinforcement along top or bottom edge of the hole

A_w area of web ($d\,w$)

e eccentricity of centreline of the hole above or below beam centreline — *always positive*

M_f bending moment due to factored loads at centreline of hole

M_r factored moment resistance of an unperforated beam

M_o, M_l values of moment resistance defined in web hole formulae

R radius of a circular hole

s length of web between adjacent holes

V_f shear force at centreline of hole due to factored loads

V_r' factored shear resistance based on plastic analysis of an unperforated beam = $0.55\,\phi\,w\,d\,F_y$. This can be obtained by multiplying tabulated values of V_r (the factored shear resistance based on elastic analysis) given in the "Beam Selection Tables" or the "Beam Load Tables" by 0.833

V_o, V_l values of shear resistance defined in web hole formulae

Web Stability

This section of the Handbook is valid for the following range of values:

For Class One Sections	For Class Two Sections
$V_f \leq 0.67\,V_r'$ and in addition, for rectangular holes $a/H \leq 3.0$ $(a/H) + 6\,(2\,H/d) \leq 5.6$	$V_f \leq 0.45\,V_r'$ and in addition, for rectangular holes $a/H \leq 2.2$ $(a/H) + 6\,(2\,H/d) \leq 5.6$

If these values are exceeded, refer to Redwood and Shrivastava (1980).

Deflections

One or two small circular holes normally result in negligible additional deflections; however, deflections of beams with large holes will increase because of local deformations caused by:

(a) effect of rotation produced by change in length of the tee sections above and below the hole

(b) local bending over the length of the hole

(c) shear deformations.

Multiple Holes

To avoid effects of interaction between two adjacent holes which may occur with high shear, the length of the web between the holes should satisfy the following, where s = clear length of solid web between the holes:

Rectangular holes

$$s \geq 2H$$

$$s \geq 2a \left(\frac{V_f / V_r'}{1 - (V_f / V_r')} \right)$$

Circular holes

$$s \geq 3R$$

$$s \geq 2R \left(\frac{V_f / V_r'}{1 - (V_f / V_r')} \right)$$

where in each case the length, height or radius refers to that of the larger of the two holes.

Lateral Stability

The presence of a web hole has only a minor effect on the lateral stability of a beam, when the strength of the beam is governed by the resistance of a section remote from the hole. For members that may be susceptible to lateral buckling, refer to the paper by Redwood and Shrivastava.

Unreinforced Holes

According to Clause 15.10.2 of S16.1-94 (see Part Two), unreinforced circular openings may be used under stipulated conditions. Round holes that are not covered by Clause 15.10.2 may be checked using the unreinforced hole formulas below by equating a and H to hole radius R as follows:

$$2a = 0.9R \quad \text{and} \quad 2H = 1.8R$$

Beam Resistance — Unreinforced Holes

Web stability must always be confirmed (page 5-155), and compression zone stability of the tee section must be checked when $2a > 4\ d_t$.

The factored shear force V_f and factored moment M_f applied at the web hole centreline must satisfy:

$$V_f \leq V_l \tag{1}$$

$$M_f \leq M_o - (M_o - M_l)\ V_f / V_l \tag{2}$$

in which

$$\frac{M_o}{M_r} = 1 - \frac{\frac{A_w}{4 A_f} \left[\left(\frac{2H}{d}\right)^2 + \left(\frac{4e}{d}\right)\left(\frac{2H}{d}\right) \right]}{1 + \frac{A_w}{4 A_f}} \tag{3}$$

$$\frac{M_l}{M_r} = \frac{1 - \frac{2}{\sqrt{3}} \left(\frac{A_w}{A_f}\right)\left(\frac{a}{d}\right)\sqrt{\frac{\alpha_2}{1 + \alpha_2}}}{1 + \frac{A_w}{4 A_f}} \tag{4}$$

$$\frac{V_l}{V_r'} = \frac{2}{\sqrt{3}} \left(\frac{a}{d}\right) \left(\frac{\alpha_1}{\sqrt{1 + \alpha_1}} + \frac{\alpha_2}{\sqrt{1 + \alpha_2}} \right) \tag{5}$$

where

$$\alpha_1 = \frac{3}{16}\left(\frac{d}{a}\right)^2\left(1 - \frac{2H}{d} - \frac{2e}{d}\right)^2 \qquad [6]$$

$$\alpha_2 = \frac{3}{16}\left(\frac{d}{a}\right)^2\left(1 - \frac{2H}{d} + \frac{2e}{d}\right)^2 \qquad [7]$$

Table 5-6, page 5-164, and Table 5-7, page 5-165, provide a means of evaluating [1] and [2]. For further explanation of these tables, see page 5-160.

Reinforced Holes

Horizontal Bars Only

Equal areas of reinforcement should be placed above and below the opening, with the reinforcement as close as possible to the edges of the hole. Welds attaching the reinforcement to the beam web should be continuous and may be placed on only one side of the reinforcing bar (with a short weld at each end on the opposite side of the bar to maintain alignment). Within the length of the hole, the welds should develop twice the factored tensile resistance of the reinforcement except that the weld capacity need not exceed $1.15\, a\, w\, F_y$. The reinforcement should extend past the hole far enough for the weld to develop the factored tensile resistance of the reinforcement but not less than a distance of $a/2$.

Reinforcement may be placed on only one side of the web of Class 1 sections (for economy) providing the following conditions are satisfied:

$A_r \leq 0.333\, A_f$

$M_f \leq 20\, V_f d$ — at the hole centreline

$a/H \leq 2.5$

$d_t/w \leq 370\,/\sqrt{F_y}$

Round holes may be checked using the reinforced hole formulae by relating a and H to R as follows:

$2a = 0.9R$ and $2H = 2R$

Once it is established that hole reinforcement is required, Table 5-8 provides a means of checking the resistance of a beam with a reinforced hole for an assumed area of reinforcement. To determine minimum reinforcement requirements, a flow chart for writing programs for a programmable calculator is shown on page 5-159.

Vertical Bars

The compression zone stability of the reinforced tee should be checked by treating it as an axially loaded column with effective length equal to $2a$.

If it is determined that web instability could be a problem, vertical reinforcing at the ends of the hole will be required. Attachment of both vertical and horizontal bars is generally more economical when the horizontal bars are placed on one side of the web with the vertical bars on the other side.

Beam Resistance — Holes with Horizontal Reinforcing Bars

Web stability and compression zone stability must be checked in addition to the following strength criteria. The factored shear force V_f and factored moment M_f at the web hole centreline must satisfy the following, where A_r is less than A_f:

$$V_f \le V_l \tag{8a}$$

$$V_f/V_r' \le 1 - \frac{2H}{d} \tag{8b}$$

$$M_f \le M_o - (M_0 - M_l)\, V_f/V_l \tag{9a}$$

$$M_f \le M_r \tag{9b}$$

in which

$$\left(\frac{M_o}{M_r}\right)_a = 1 + \frac{\dfrac{A_r}{A_f}\left(\dfrac{2H}{d}\right) - \dfrac{A_w}{4A_f}\left[\left(\dfrac{2H}{d}\right)^2 + 4\left(\dfrac{2H}{d}\right)\left(\dfrac{e}{d}\right) - 4\left(\dfrac{e}{d}\right)^2\right]}{1 + \dfrac{A_w}{4A_f}} \quad \text{— for } \frac{e}{d} \le \frac{A_r}{A_w} \tag{10a}$$

or

$$\left(\frac{M_o}{M_r}\right)_b = \left(\frac{M_o}{M_r}\right)_a - \frac{\dfrac{A_w}{A_f}\left(\dfrac{e}{d} - \dfrac{A_r}{A_w}\right)^2}{1 + \dfrac{A_w}{4A_f}} \quad \text{— for } \frac{e}{d} > \frac{A_r}{A_w} \tag{10b}$$

$$\frac{M_l}{M_r} = \frac{1 - \dfrac{A_r}{A_f}}{1 + \dfrac{A_w}{4A_f}} \tag{11}$$

$$\frac{V_l}{V_r'} = \sqrt{3}\left(\frac{d}{a}\right)\left(\frac{A_r}{A_w}\right)\left(1 - \frac{2H}{d}\right) \tag{12}$$

Flow Chart

The flow chart on page 5-159 is provided as a guide in developing programs for programmable calculators and computers. The logic provided determines the minimum A_r which will satisfy [9a]. It is anticipated that the individuals implementing the program will modify it to their own needs.

References

FROST, R.W., and LEFFLER, R.E. 1971. Fatigue tests of beams with rectangular web holes. Journal of the Structural Division, ASCE, **97**(ST2): 509–527.

PART TWO of this Handbook.

REDWOOD, R.G. 1974. The influence of web holes on the design of steel beams. Proceedings, Canadian Structural Engineering Conference, Canadian Steel Construction Council, Willowdale, Ontario.

REDWOOD, R.G. 1971. Simplified plastic analysis for reinforced web holes. Engineering Journal, AISC, **8**(3): 128–131.

REDWOOD, R.G., and SHRIVASTAVA, S.C. 1980. Design recommendations for steel beams with web holes. Canadian Journal of Civil Engineering, **7**(4), December.

REDWOOD, R.G., and WONG, P. 1982. Web holes in composite beams with steel deck. Procedings, Canadian Structrual Engineering Conference, Canadian Steel Construction Council, Willowdale, Ontario.

REINFORCED HOLE PROGRAM FLOWCHART

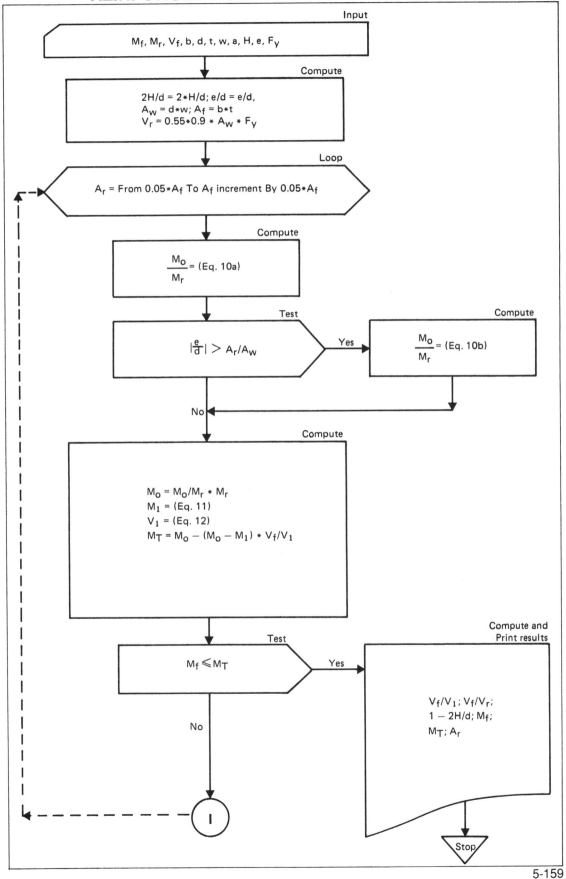

Tables

(a) *Unreinforced Holes*

Table 5-6, page 5-164 gives the values of constants C_1 and C_2 for unreinforced holes where,

$$C_1 = \frac{M_o}{M_r} \quad \text{and} \quad C_2 = \frac{M_o/M_r - M_l/M_r}{V_l/V_r'}$$

where M_o, M_l and V_l are defined in [3] to [5].

Table 5-7, page 5-165, gives the value of constant C_3 taken as V_l/V_r'.

A_w/A_f varies from 0.5 to 2.25, $2H/d$ from 0.3 to 0.6, a/H from 0.50 to 2.2, and $e/d = 0$.

Written in terms of the constants C_1 and C_2, [2] becomes

$$\frac{M_f}{M_r} \leq C_1 - C_2 \left(\frac{V_f}{V_r'} \right) \tag{13}$$

and [1] becomes

$$\frac{V_f}{V_r'} \leq C_3 \tag{14}$$

Use

For concentric ($e/d = 0$) unreinforced holes, compute A_w/A_f, $2H/d$ and a/H. Determine C_1, C_2 and C_3 with the aid of Tables 5-6 and 5-7 for use in [13] and [14].

(b) *Reinforced Holes*

Table 5-8, pages 5-166 to 5-168, gives the values of the constants C_4 and C_5 for reinforced holes where

$$C_4 = \frac{M_o}{M_r} \quad \text{and} \quad C_5 = \frac{M_o/M_r - M_l/M_r}{V_l/V_r'}$$

where M_o, M_l and V_l are defined in [10] to [12] for concentric holes ($e/d = 0$).

A_w/A_f varies from 0.5 to 2.25, $2H/d$ from 0.3 to 0.6 and a/H from 0.45 to 2.2, for the three values of A_r/A_f, 0.33, 0.67 and 1.0.

Written in terms of the constants C_4 and C_5, [9a] becomes

$$\frac{M_f}{M_r} \leq C_4 - C_5 \left(\frac{V_f}{V_r'} \right) \tag{15}$$

Use

For concentric ($e/d = 0$) reinforced holes, compute A_w/A_f, $2H/d$ and a/H. Determine C_4 and C_5, for one of the assumed values of A_r/A_f, for use in [15].

Calculate V_l from [12] for use in [8a].

Then check [8b] and [9b].

Example

Given:

A simple span W610x101 beam spanning 12 m supports a factored total uniformly distributed load of 420 kN (35 kN/m). Check the adequacy of the section for two rectangular holes located as shown. Steel is G40.21 350W. Lateral support to compression flange is provided.

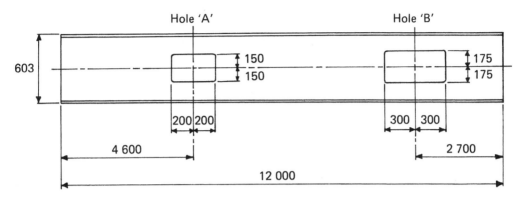

Solution for Hole 'A':

Class of beam: From page 5-7, for G40.21 350W steel, W610x101 is Class 1

From page 5-96, $M_r = 914$ kN·m

$V_r = 1\ 310$ kN

For use in formulae for holes in beam webs, V_r must be reduced to V_r' used for plastically analysed beams.

Therefore $V_r' = 0.833 \times 1\ 310 = 1\ 090$ kN

At centreline of hole

$M_f = 35$ kN/m $\times 4.6$ m $\times (12 - 4.6) / 2 = 596$ kN·m

$V_f = 35$ kN/m $\times ((12/2) - 4.6) = 49$ kN

Therefore, $\dfrac{M_f}{M_r} = \dfrac{596}{914} = 0.652$, and $\dfrac{V_f}{V_r'} = \dfrac{49}{1090} = 0.045$

Check web stability

Refer to page 5-155.

$\dfrac{V_f}{V_r'} = 0.045 \quad \leq 0.67 \quad$ — OK

$a/H = 1.33 \quad \leq 3.0$ (limit for Class 1 beam) — OK

$a/H + 6(2H/d) = 200/150 + 6(300/603) = 1.33 + 6(0.498) = 4.3 \quad \leq 5.6$ — OK

Check compression zone stability

OK, if $2a \leq 4\ d_t$

$\leq 4((603/2) - 150) = 606$ mm

$2a = 400$ mm $\ < 606 \quad$ — OK

5-161

Check for unreinforced hole

$$\frac{A_w}{A_f} = \frac{10.5 \times 603}{228 \times 14.9} = 1.86 \quad \text{and} \quad \frac{2H}{d} = 0.50$$

$C_1 = 0.92$, from Table 5-6, page 5-164

For $a/H = 1.33$ — use 1.4

$C_2 = 1.9$, from Table 5-6

$C_3 = 0.263$, from Table 5-7, page 5-165

$$\frac{M_f}{M_r} \leq C_1 - C_2 \left(\frac{V_f}{V_r'}\right) \quad [13]$$

$$\leq 0.92 - 1.9\,(0.045) = 0.83$$

$M_f/M_r = 0.65 \quad < 0.82 \qquad \text{OK}$

$$\frac{V_f}{V_r'} \leq C_3 \quad [14]$$

$0.045 < 0.263 \quad -\text{OK}$

Therefore, reinforcement not required.

Solution for Hole 'B':

At centreline of hole

$M_f = 35 \text{ kN/m} \times 2.7 \text{ m} \times (12 - 2.7)/2 = 440 \text{ kN·m}$

$V_f = 35 \text{ kN/m} \times ((12/2) - 2.7) = 116 \text{ kN}$

Therefore $\dfrac{M_f}{M_r} = \dfrac{440}{914} = 0.481,$ and $\dfrac{V_f}{V_r'} = \dfrac{116}{1090} = 0.11$

Check spacing between holes

Use $2H$ of larger hole.

OK, if $s \geq 2H$

≥ 350

$s = 12\,000 - (2\,700 + 4\,600) = 4\,700 \quad > 350 \quad -\text{OK}$

Check web stability

Refer to page 5-155.

$\dfrac{V_f}{V_r'} = 0.11 \quad \leq 0.67 \quad -\text{OK}$

$a/H = 1.7 \quad \leq 3.0$ (limit for Class 1 beam) $\quad -\text{OK}$

$a/H + 6(2H/d) = 300/175 + 6(350/603) = 1.71 + 6(0.580) = 5.2 \quad \leq 5.6 \quad -\text{OK}$

Check compression zone stability

OK, if $2a \leq 4\,d_t$ (unreinforced tee)

$\qquad \leq 4\,((603/2) - 175) = 506$ mm

$2a = 600$ mm $\quad > 506 \quad$ — not adequate

Check for unreinforced hole

From Table 5-6, page 5-164, for $A_w/A_f = 1.86 \quad$ (use 2.0)

and $2H/d = 0.58 \quad$ (use 0.60),

$C_1 = 0.88$

For $a/H = 1.71 \quad$ (use 1.8),

$C_2 = 3.83$

$$\frac{M_f}{M_r} \leq C_1 - C_2\left(\frac{V_f}{V_r'}\right) \quad [13]$$

$\qquad \leq 0.88 - 3.83\,(0.11) = 0.46$

$M_f/M_r = 0.48 \quad > 0.46 \quad$ — reinforcement required

Reinforcement

Assume $A_r/A_f = 0.33$ (maximum permitted for one-sided reinforcement)

From Table 5-8, page 5-166,

for $\dfrac{A_r}{A_f} = 0.33,\ \dfrac{A_w}{A_f} = 2.0,\ \dfrac{2H}{d} = 0.60,$

$C_4 = 1.013$

For $\dfrac{a}{H} = 1.7,\ C_5 = 2.51$ (by interpolation)

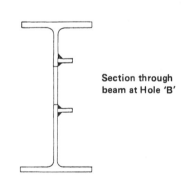

Section through beam at Hole 'B'

$$\frac{M_f}{M_r} \leq C_4 - C_5\left(\frac{V_f}{V_r'}\right) \quad [15]$$

$\qquad \leq 1.01 - 2.51\,(0.11) = 0.74$

$M_f/M_r = 0.48 \quad < 0.74 \quad$ — OK

Further refinement of A_r/A_f can be accomplished by using the expressions previously given.

Check one-sided reinforcement

$M_f \leq 20\,V_f\,d$ at hole centreline (see page 5-157)

$\qquad \leq 20 \times 116 \times 603 = 1\,400\,000$ kN·mm $= 1\,400$ kN·m

$M_f = 440$ kN·m $\quad < 1\,400$ kN·m $\quad$ — OK

$a/H = 1.7 \quad \leq 2.5 \quad$ — OK

$\dfrac{d_t}{w} \leq \dfrac{370}{\sqrt{F_y}} \leq 19.8$

VALUES OF C_1 AND C_2
For Unreinforced Concentric Holes in Beam Webs

Table 5-6

$\dfrac{A_w}{A_f}$	$\dfrac{2H}{d}$	C_1	C_2 For following $\dfrac{a}{H}$ values							
			.50	1.0	1.2	1.4	1.6	1.8	2.0	2.2
0.50	.30	.990	0.204	0.271	0.300	0.330	0.360	0.391	0.423	0.455
	.35	.986	0.226	0.315	0.353	0.392	0.433	0.474	0.516	0.558
	.40	.982	0.252	0.367	0.417	0.468	0.520	0.574	0.628	0.682
	.45	.978	0.283	0.432	0.495	0.561	0.628	0.696	0.764	0.833
	.50	.972	0.321	0.511	0.593	0.676	0.761	0.846	0.933	1.020
	.55	.966	0.368	0.612	0.715	0.820	0.927	1.035	1.143	1.252
	.60	.960	0.428	0.740	0.872	1.005	1.140	1.275	1.411	*
0.75	.30	.986	0.290	0.385	0.426	0.468	0.512	0.556	0.601	0.647
	.35	.981	0.321	0.447	0.502	0.557	0.615	0.673	0.733	0.793
	.40	.975	0.358	0.522	0.593	0.665	0.740	0.815	0.892	0.970
	.45	.968	0.402	0.613	0.704	0.797	0.892	0.989	1.086	1.184
	.50	.961	0.456	0.726	0.842	0.961	1.081	1.203	1.325	1.449
	.55	.952	0.522	0.869	1.016	1.166	1.317	1.470	1.624	1.779
	.60	.943	0.608	1.052	1.239	1.428	1.619	1.812	2.005	*
1.00	.30	.982	0.367	0.488	0.540	0.593	0.648	0.705	0.762	0.820
	.35	.975	0.407	0.567	0.635	0.706	0.779	0.853	0.928	1.005
	.40	.968	0.454	0.661	0.751	0.843	0.937	1.033	1.130	1.228
	.45	.960	0.510	0.777	0.892	1.010	1.130	1.252	1.376	1.500
	.50	.950	0.577	0.920	1.067	1.217	1.369	1.523	1.679	1.835
	.55	.939	0.662	1.101	1.287	1.477	1.669	1.862	2.057	2.253
	.60	.928	0.770	1.333	1.569	1.809	2.051	2.295	2.539	*
1.25	.30	.979	0.437	0.581	0.643	0.706	0.772	0.839	0.907	0.976
	.35	.971	0.485	0.675	0.756	0.841	0.927	1.015	1.105	1.196
	.40	.962	0.540	0.787	0.894	1.003	1.115	1.229	1.345	1.462
	.45	.952	0.607	0.925	1.062	1.202	1.346	1.491	1.638	1.786
	.50	.940	0.687	1.095	1.270	1.449	1.630	1.813	1.999	2.185
	.55	.928	0.788	1.310	1.532	1.758	1.987	2.217	2.449	2.682
	.60	.914	0.916	1.587	1.868	2.154	2.442	2.732	3.023	*
1.50	.30	.975	0.500	0.666	0.736	0.809	0.884	0.961	1.039	1.118
	.35	.967	0.555	0.773	0.866	0.963	1.062	1.163	1.266	1.370
	.40	.956	0.619	0.902	1.024	1.149	1.278	1.408	1.541	1.675
	.45	.945	0.695	1.059	1.216	1.377	1.541	1.708	1.876	2.046
	.50	.932	0.787	1.255	1.455	1.659	1.867	2.077	2.289	2.502
	.55	.917	0.902	1.501	1.755	2.014	2.276	2.540	2.805	3.072
	.60	.902	1.050	1.817	2.140	2.467	2.797	3.129	3.463	*
1.75	.30	.973	0.558	0.743	0.822	0.903	0.987	1.072	1.159	1.248
	.35	.963	0.619	0.862	0.967	1.075	1.185	1.298	1.413	1.529
	.40	.951	0.691	1.006	1.142	1.282	1.426	1.572	1.719	1.869
	.45	.938	0.776	1.182	1.357	1.537	1.720	1.906	2.094	2.283
	.50	.924	0.879	1.400	1.624	1.852	2.084	2.318	2.555	2.793
	.55	.908	1.007	1.675	1.959	2.247	2.539	2.834	3.131	3.429
	.60	.890	1.171	2.028	2.388	2.753	3.122	3.492	3.864	*
2.00	.30	.970	0.611	0.813	0.900	0.989	1.081	1.174	1.270	1.366
	.35	.959	0.678	0.944	1.059	1.177	1.298	1.421	1.547	1.674
	.40	.947	0.757	1.102	1.251	1.405	1.561	1.721	1.883	2.047
	.45	.933	0.849	1.295	1.486	1.683	1.884	2.087	2.293	2.500
	.50	.917	0.962	1.534	1.778	2.028	2.282	2.539	2.798	3.059
	.55	.899	1.103	1.835	2.145	2.461	2.781	3.104	3.429	3.755
	.60	.880	1.283	2.221	2.616	3.016	3.419	3.825	4.232	*
2.25	.30	.968	0.660	0.878	0.972	1.068	1.167	1.268	1.371	1.476
	.35	.956	0.733	1.020	1.144	1.271	1.402	1.535	1.671	1.808
	.40	.942	0.817	1.190	1.351	1.517	1.686	1.859	2.034	2.211
	.45	.927	0.917	1.398	1.605	1.818	2.034	2.254	2.476	2.700
	.50	.910	1.039	1.656	1.920	2.190	2.465	2.742	3.022	3.303
	.55	.891	1.191	1.981	2.317	2.658	3.004	3.352	3.703	4.056
	.60	.870	1.386	2.399	2.825	3.257	3.692	4.131	4.571	*

*a/H plus $6(2H/d)$ exceeds 5.6.

$d_t/w = ((603 / 2) - 175) / 10.5 = 12.0 \quad < 19.8 \quad$ — OK

Therefore, one-sided reinforcement OK

$A_r = 0.33, \quad A_f = 0.33 (228 \times 14.9) = 1\,120 \text{ mm}^2$

Check shear

$$V_l = \sqrt{3} \left(\frac{d}{a}\right)\left(\frac{A_r}{A_w}\right)\left(1 - \frac{2H}{d}\right) V_r' \quad [12]$$

$$= \sqrt{3} \left(\frac{603}{300}\right)\left(\frac{1\,120}{10.5 \times 603}\right)(1 - 0.58)\,1090 = 281$$

$V_f \leq V_l \quad$ [8a]

$\quad 116 < 281 \quad$ — OK

$V_f/V_r' \leq 1 - \dfrac{2H}{d} \quad$ [8b]

$\quad \leq 1 - 0.58 = 0.42$

$V_f/V_r' = 0.11 \quad < 0.42 \quad$ — OK

Try 16 x 70 reinforcement

$\dfrac{b}{t} \leq \dfrac{145}{\sqrt{F_y}} \quad$ (for Class 1)

$\quad \leq 7.75$

$\quad b/t = 70/16 = 4.3 \quad < 7.75 \quad$ — OK

Therefore, use 16 x 70 one-sided reinforcement.

VALUES OF C₃
For Unreinforced Concentric Holes in Beam Webs

Table 5–7

$\dfrac{2H}{d}$	a/H							
	0.5	1.0	1.2	1.4	1.6	1.8	2.0	2.2
.30	.680	.627	.602	.575	.549	.523	.498	.474
.35	.621	.552	.521	.490	.461	.433	.407	.384
.40	.560	.475	.441	.408	.378	.351	.327	.305
.45	.497	.400	.364	.332	.303	.279	.257	.238
.50	.433	.327	.293	.263	.238	.217	.199	.183
.55	.368	.260	.229	.203	.182	.165	.150	.138
.60	.302	.200	.173	.152	.136	.122	.111	.102

VALUES OF C_4 AND C_5
For Reinforced Concentric Holes in Beam Webs

Table 5-8
$A_r/A_f = 0.33$

$\dfrac{A_w}{A_f}$	$\dfrac{2H}{d}$	C_4	C_5 For following $\dfrac{a}{H}$ values							
			.45	1.0	1.2	1.4	1.6	1.8	2.0	2.2
0.50	.30	1.079	0.041	0.090	0.108	0.126	0.144	0.162	0.180	0.199
	.35	1.090	0.052	0.116	0.139	0.162	0.186	0.209	0.232	0.255
	.40	1.101	0.066	0.147	0.176	0.205	0.235	0.264	0.293	0.323
	.45	1.111	0.083	0.184	0.220	0.257	0.294	0.330	0.367	0.404
	.50	1.120	0.103	0.229	0.274	0.320	0.366	0.411	0.457	0.503
	.55	1.129	0.128	0.284	0.341	0.398	0.455	0.511	0.568	0.625
	.60	1.138	0.159	0.354	0.425	0.496	0.567	0.637	0.708	*
0.75	.30	1.070	0.064	0.142	0.170	0.198	0.227	0.255	0.283	0.311
	.35	1.079	0.081	0.181	0.217	0.253	0.290	0.326	0.362	0.398
	.40	1.087	0.102	0.228	0.273	0.319	0.364	0.410	0.455	0.501
	.45	1.094	0.127	0.283	0.340	0.397	0.453	0.510	0.566	0.623
	.50	1.101	0.158	0.350	0.420	0.491	0.561	0.631	0.701	0.771
	.55	1.107	0.195	0.433	0.519	0.606	0.693	0.779	0.866	0.952
	.60	1.112	0.241	0.536	0.643	0.750	0.858	0.965	1.072	*
1.00	.30	1.062	0.088	0.196	0.235	0.275	0.314	0.353	0.392	0.432
	.35	1.069	0.112	0.250	0.300	0.350	0.400	0.449	0.499	0.549
	.40	1.075	0.141	0.313	0.375	0.438	0.500	0.563	0.625	0.688
	.45	1.079	0.174	0.387	0.464	0.542	0.619	0.697	0.774	0.851
	.50	1.083	0.214	0.476	0.572	0.667	0.762	0.857	0.953	1.048
	.55	1.086	0.263	0.585	0.702	0.819	0.936	1.053	1.170	1.287
	.60	1.088	0.324	0.721	0.865	1.009	1.153	1.297	1.441	*
1.25	.30	1.055	0.114	0.254	0.304	0.355	0.406	0.457	0.507	0.558
	.35	1.060	0.145	0.322	0.386	0.450	0.515	0.579	0.643	0.708
	.40	1.063	0.180	0.401	0.481	0.561	0.642	0.722	0.802	0.882
	.45	1.066	0.222	0.494	0.593	0.692	0.791	0.890	0.989	1.088
	.50	1.067	0.273	0.606	0.727	0.848	0.969	1.090	1.211	1.333
	.55	1.068	0.333	0.741	0.889	1.037	1.185	1.333	1.481	1.629
	.60	1.067	0.408	0.907	1.089	1.270	1.452	1.633	1.815	*
1.50	.30	1.048	0.141	0.314	0.376	0.439	0.502	0.565	0.627	0.690
	.35	1.051	0.178	0.396	0.476	0.555	0.634	0.713	0.793	0.872
	.40	1.053	0.222	0.492	0.591	0.689	0.788	0.886	0.985	1.083
	.45	1.054	0.272	0.605	0.726	0.847	0.968	1.089	1.210	1.331
	.50	1.053	0.332	0.738	0.886	1.033	1.181	1.329	1.476	1.624
	.55	1.051	0.404	0.899	1.078	1.258	1.438	1.618	1.797	1.977
	.60	1.047	0.493	1.096	1.315	1.534	1.753	1.973	2.192	*
1.75	.30	1.042	0.169	0.376	0.451	0.526	0.601	0.676	0.751	0.827
	.35	1.044	0.213	0.473	0.568	0.663	0.757	0.852	0.947	1.041
	.40	1.044	0.264	0.586	0.704	0.821	0.938	1.055	1.173	1.290
	.45	1.043	0.323	0.718	0.861	1.005	1.149	1.292	1.436	1.579
	.50	1.040	0.393	0.873	1.048	1.222	1.397	1.572	1.746	1.921
	.55	1.035	0.477	1.059	1.271	1.483	1.694	1.906	2.118	2.330
	.60	1.030	0.579	1.286	1.543	1.801	2.058	2.315	2.572	*
2.00	.30	1.037	0.198	0.440	0.528	0.615	0.703	0.791	0.879	0.967
	.35	1.037	0.249	0.553	0.663	0.774	0.884	0.995	1.105	1.216
	.40	1.036	0.307	0.683	0.819	0.956	1.092	1.229	1.365	1.502
	.45	1.033	0.375	0.833	1.000	1.167	1.333	1.500	1.667	1.833
	.50	1.028	0.455	1.010	1.212	1.415	1.617	1.819	2.021	2.223
	.55	1.021	0.550	1.221	1.466	1.710	1.954	2.198	2.443	2.687
	.60	1.013	0.665	1.478	1.774	2.069	2.365	2.660	2.956	*
2.25	.30	1.032	0.227	0.505	0.606	0.707	0.808	0.909	1.010	1.111
	.35	1.031	0.285	0.634	0.760	0.887	1.014	1.141	1.267	1.394
	.40	1.028	0.351	0.781	0.937	1.093	1.249	1.405	1.562	1.718
	.45	1.023	0.428	0.951	1.141	1.331	1.521	1.712	1.902	2.092
	.50	1.017	0.517	1.150	1.380	1.610	1.839	2.069	2.299	2.529
	.55	1.008	0.623	1.386	1.663	1.940	2.217	2.494	2.771	3.048
	.60	0.998	0.752	1.671	2.005	2.340	2.674	3.008	3.342	*

*a/H plus $6(2H/d)$ exceeds 5.6.

VALUES OF C_4 AND C_5
For Reinforced Concentric Holes in Beam Webs

Table 5-8
$A_r/A_f = 0.67$

$\dfrac{A_w}{A_f}$	$\dfrac{2H}{d}$	C_4	C_5 For following $\dfrac{a}{H}$ values							
			.45	1.0	1.2	1.4	1.6	1.8	2.0	2.2
0.50	.30	1.168	0.036	0.081	0.097	0.113	0.129	0.146	0.162	0.178
	.35	1.194	0.047	0.105	0.126	0.146	0.167	0.188	0.209	0.230
	.40	1.219	0.060	0.133	0.160	0.187	0.213	0.240	0.266	0.293
	.45	1.244	0.076	0.168	0.201	0.235	0.269	0.302	0.336	0.369
	.50	1.269	0.095	0.210	0.253	0.295	0.337	0.379	0.421	0.463
	.55	1.292	0.119	0.264	0.316	0.369	0.422	0.474	0.527	0.580
	.60	1.316	0.149	0.331	0.397	0.463	0.530	0.596	0.662	*
0.75	.30	1.154	0.055	0.122	0.146	0.170	0.195	0.219	0.243	0.267
	.35	1.177	0.071	0.157	0.188	0.219	0.251	0.282	0.314	0.345
	.40	1.199	0.089	0.199	0.239	0.278	0.318	0.358	0.398	0.438
	.45	1.221	0.112	0.250	0.300	0.350	0.400	0.450	0.500	0.549
	.50	1.241	0.140	0.312	0.374	0.437	0.499	0.561	0.624	0.686
	.55	1.261	0.175	0.389	0.467	0.545	0.623	0.700	0.778	0.856
	.60	1.280	0.219	0.487	0.584	0.682	0.779	0.876	0.974	*
1.00	.30	1.142	0.073	0.162	0.195	0.227	0.260	0.292	0.325	0.357
	.35	1.162	0.094	0.209	0.251	0.292	0.334	0.376	0.418	0.459
	.40	1.181	0.119	0.264	0.317	0.370	0.422	0.475	0.528	0.581
	.45	1.199	0.149	0.330	0.397	0.463	0.529	0.595	0.661	0.727
	.50	1.217	0.185	0.411	0.494	0.576	0.658	0.740	0.823	0.905
	.55	1.233	0.230	0.511	0.614	0.716	0.818	0.920	1.023	1.125
	.60	1.248	0.287	0.637	0.765	0.892	1.020	1.147	1.275	*
1.25	.30	1.131	0.092	0.203	0.244	0.285	0.325	0.366	0.407	0.448
	.35	1.149	0.117	0.261	0.313	0.365	0.417	0.469	0.521	0.574
	.40	1.165	0.148	0.329	0.395	0.460	0.526	0.592	0.658	0.723
	.45	1.180	0.185	0.410	0.492	0.574	0.656	0.738	0.821	0.903
	.50	1.194	0.229	0.509	0.611	0.713	0.814	0.916	1.018	1.120
	.55	1.207	0.284	0.631	0.757	0.883	1.009	1.135	1.261	1.388
	.60	1.219	0.353	0.784	0.940	1.097	1.254	1.410	1.567	*
1.50	.30	1.121	0.110	0.245	0.293	0.342	0.391	0.440	0.489	0.538
	.35	1.136	0.141	0.313	0.375	0.438	0.500	0.563	0.625	0.688
	.40	1.150	0.177	0.393	0.472	0.550	0.629	0.708	0.786	0.865
	.45	1.163	0.220	0.489	0.587	0.685	0.783	0.881	0.978	1.076
	.50	1.174	0.272	0.605	0.726	0.847	0.968	1.089	1.210	1.332
	.55	1.184	0.336	0.748	0.897	1.047	1.196	1.346	1.495	1.645
	.60	1.193	0.417	0.926	1.111	1.296	1.481	1.667	1.852	*
1.75	.30	1.112	0.129	0.286	0.343	0.400	0.457	0.514	0.571	0.629
	.35	1.125	0.164	0.364	0.437	0.510	0.583	0.656	0.729	0.802
	.40	1.137	0.206	0.457	0.549	0.640	0.731	0.823	0.914	1.006
	.45	1.147	0.255	0.567	0.681	0.794	0.908	1.021	1.135	1.248
	.50	1.156	0.315	0.700	0.840	0.980	1.120	1.260	1.400	1.540
	.55	1.163	0.388	0.862	1.035	1.207	1.380	1.552	1.725	1.897
	.60	1.169	0.479	1.065	1.278	1.491	1.704	1.917	2.130	*
2.00	.30	1.103	0.147	0.327	0.392	0.458	0.523	0.589	0.654	0.719
	.35	1.115	0.187	0.416	0.499	0.583	0.666	0.749	0.832	0.916
	.40	1.124	0.234	0.521	0.625	0.729	0.833	0.938	1.042	1.146
	.45	1.133	0.290	0.645	0.774	0.903	1.032	1.161	1.290	1.419
	.50	1.139	0.357	0.794	0.953	1.111	1.270	1.429	1.588	1.746
	.55	1.144	0.439	0.975	1.170	1.365	1.560	1.755	1.951	2.146
	.60	1.147	0.540	1.201	1.441	1.681	1.921	2.162	2.402	*
2.25	.30	1.096	0.166	0.368	0.442	0.516	0.589	0.663	0.737	0.810
	.35	1.105	0.211	0.468	0.561	0.655	0.749	0.842	0.936	1.029
	.40	1.113	0.263	0.584	0.701	0.818	0.935	1.052	1.169	1.286
	.45	1.119	0.325	0.722	0.866	1.011	1.155	1.300	1.444	1.588
	.50	1.123	0.399	0.887	1.064	1.241	1.419	1.596	1.773	1.951
	.55	1.126	0.489	1.087	1.304	1.521	1.738	1.956	2.173	2.390
	.60	1.126	0.600	1.334	1.601	1.868	2.135	2.402	2.669	*

*a/H plus 6(2H/d) exceeds 5.6.

VALUES OF C_4 AND C_5
For Reinforced Concentric Holes in Beam Webs

Table 5-8
$A_r/A_f = 1.00$

$\frac{A_w}{A_f}$	$\frac{2H}{d}$	C_4	C_5 For following $\frac{a}{H}$ values							
			.45	1.0	1.2	1.4	1.6	1.8	2.0	2.2
0.50	.30	1.257	0.035	0.078	0.093	0.109	0.124	0.140	0.155	0.171
	.35	1.298	0.045	0.101	0.121	0.141	0.161	0.182	0.202	0.222
	.40	1.338	0.058	0.129	0.154	0.180	0.206	0.232	0.257	0.283
	.45	1.378	0.073	0.163	0.195	0.228	0.260	0.293	0.325	0.358
	.50	1.417	0.092	0.204	0.245	0.286	0.327	0.368	0.409	0.450
	.55	1.455	0.116	0.257	0.308	0.359	0.411	0.462	0.513	0.565
	.60	1.493	0.145	0.323	0.388	0.453	0.517	0.582	0.647	*
0.75	.30	1.238	0.052	0.115	0.138	0.161	0.184	0.207	0.230	0.253
	.35	1.275	0.067	0.149	0.178	0.208	0.238	0.268	0.297	0.327
	.40	1.312	0.085	0.189	0.227	0.265	0.303	0.341	0.379	0.416
	.45	1.347	0.107	0.239	0.286	0.334	0.382	0.429	0.477	0.525
	.50	1.382	0.135	0.299	0.359	0.419	0.479	0.538	0.598	0.658
	.55	1.415	0.169	0.375	0.449	0.524	0.599	0.674	0.749	0.824
	.60	1.448	0.212	0.470	0.564	0.659	0.753	0.847	0.941	*
1.00	.30	1.222	0.068	0.151	0.181	0.212	0.242	0.272	0.302	0.333
	.35	1.255	0.088	0.195	0.234	0.273	0.312	0.351	0.390	0.429
	.40	1.288	0.112	0.248	0.297	0.347	0.397	0.446	0.496	0.545
	.45	1.320	0.140	0.312	0.374	0.436	0.499	0.561	0.623	0.686
	.50	1.350	0.175	0.390	0.468	0.546	0.624	0.701	0.779	0.857
	.55	1.380	0.219	0.487	0.584	0.681	0.779	0.876	0.973	1.071
	.60	1.408	0.274	0.610	0.732	0.854	0.975	1.097	1.219	*
1.25	.30	1.207	0.084	0.187	0.224	0.261	0.299	0.336	0.373	0.411
	.35	1.238	0.108	0.240	0.289	0.337	0.385	0.433	0.481	0.529
	.40	1.267	0.137	0.305	0.366	0.427	0.488	0.548	0.609	0.670
	.45	1.295	0.172	0.382	0.459	0.535	0.612	0.688	0.764	0.841
	.50	1.321	0.215	0.477	0.572	0.668	0.763	0.858	0.954	1.049
	.55	1.347	0.267	0.594	0.713	0.832	0.951	1.069	1.188	1.307
	.60	1.371	0.334	0.742	0.891	1.039	1.188	1.336	1.485	*
1.50	.30	1.194	0.100	0.222	0.266	0.310	0.354	0.399	0.443	0.487
	.35	1.221	0.128	0.285	0.342	0.399	0.456	0.512	0.569	0.626
	.40	1.247	0.162	0.360	0.432	0.504	0.576	0.648	0.720	0.792
	.45	1.272	0.203	0.451	0.541	0.631	0.721	0.811	0.901	0.991
	.50	1.295	0.252	0.561	0.673	0.785	0.898	1.010	1.122	1.234
	.55	1.317	0.314	0.697	0.837	0.976	1.116	1.255	1.395	1.534
	.60	1.338	0.391	0.869	1.043	1.217	1.391	1.565	1.738	*
1.75	.30	1.181	0.115	0.256	0.307	0.358	0.409	0.460	0.512	0.563
	.35	1.206	0.148	0.328	0.394	0.459	0.525	0.591	0.656	0.722
	.40	1.230	0.186	0.414	0.497	0.580	0.663	0.745	0.828	0.911
	.45	1.251	0.233	0.517	0.621	0.724	0.828	0.931	1.034	1.138
	.50	1.272	0.289	0.642	0.771	0.899	1.028	1.156	1.285	1.413
	.55	1.291	0.359	0.797	0.956	1.116	1.275	1.434	1.594	1.753
	.60	1.308	0.446	0.991	1.189	1.387	1.586	1.784	1.982	*
2.00	.30	1.170	0.130	0.289	0.347	0.405	0.463	0.521	0.579	0.637
	.35	1.192	0.167	0.371	0.445	0.519	0.593	0.667	0.741	0.816
	.40	1.213	0.210	0.467	0.560	0.654	0.747	0.841	0.934	1.027
	.45	1.233	0.262	0.582	0.699	0.815	0.932	1.048	1.164	1.281
	.50	1.250	0.325	0.722	0.866	1.010	1.155	1.299	1.443	1.588
	.55	1.266	0.402	0.893	1.072	1.251	1.429	1.608	1.786	1.965
	.60	1.280	0.499	1.109	1.330	1.552	1.774	1.995	2.217	*
2.25	.30	1.160	0.145	0.323	0.387	0.452	0.516	0.581	0.646	0.710
	.35	1.180	0.186	0.413	0.495	0.578	0.660	0.743	0.825	0.908
	.40	1.198	0.234	0.519	0.623	0.726	0.830	0.934	1.038	1.142
	.45	1.215	0.291	0.646	0.775	0.904	1.033	1.162	1.291	1.421
	.50	1.230	0.360	0.799	0.959	1.118	1.278	1.438	1.598	1.758
	.55	1.243	0.444	0.987	1.184	1.382	1.579	1.776	1.974	2.171
	.60	1.254	0.550	1.222	1.467	1.711	1.955	2.200	2.444	*

*a/H plus 6(2H/d) exceeds 5.6.

PART SIX
PROPERTIES AND DIMENSIONS

Structural Steels	6–3
Historical Listing of Selected Structural Steels	6–5
Relationship Between ISO and CSA Standards for Structural Steels	6–10
Standard Mill Practice	6–11
Principal Sources of Structural Steel Sections	6–17
Metric and Imperial Designations	6–19
Welded Shapes	6–29
Welded Reduced Flange Shapes (WRF)	6–30
Welded Wide Flange Shapes (WWF)	6–32
Rolled Structural Shapes	6–38
W Shapes	6–40
HP Shapes	6–56
M Shapes	6–58
SLB Shapes (Super Light Beams)	6–60
S Shapes	6–62
Standard Channels (C)	6–66
Miscellaneous Channels (MC)	6–68
Angles (L) – Imperial Series	6–72
Structural Tees (WT)	6–82
Hollow Structural Sections (HSS)	6–96
Rectangular	6–98
Square	6–100
Round	6–102
A53 Pipe	6–104
Cold Formed Channels	6–105
Built-up Sections	6–111
Double Angles—Imperial Series	
Equal Legs Back-to-Back	6–112
Long Legs Back-to-Back	6–114
Short Legs Back-to-Back	6–117
Double Channels	6–120
W Shapes and Channels	6–122
Diagrams and formulae	6–125
Bars and Plates	6–129

Crane Rails	6–137
Fasteners	6–142
Welding	6–156
Steel Products — Record of Changes	6–163

STRUCTURAL STEELS

General

Canadian structural steels are covered by two standards prepared by the Canadian Standards Association Technical Committee on Structural Steel, G40. These are CSA G40.20 and CSA G40.21. The information provided in this section is based on the current 1998 editions of both standards, and on the SI metric values, in keeping with Canadian design standards for steel structures.

CSA G40.20, "General Requirements for Rolled or Welded Structural Quality Steel" sets out the general requirements governing the delivery of structural quality steels. These requirements include: Definitions, Chemical Composition, Variations in dimensions, Methods of Testing, Frequency of Testing, Heat Treatment, Repairs of defects, Marking, etc.

CSA G40.21, "Structural Quality Steels" governs the chemical and mechanical properties of 7 types and 8 strength levels of structural steels for general construction and engineering purposes. All strength levels are not available in all types, and selection of the proper grade (type and strength level) is important for a particular application. CSA G40.21 350A and CSA G40.21 350AT are atmospheric corrosion-resistant steels normally used in bridge construction. For HSS sections, 350W is the normal grade used when produced to CSA G40.21.

The 7 types covered in CSA G40.21 are:

(a) **Type W – Weldable Steel.** Steels of this type meet specified strength requirements and are suitable for general welded construction where notch toughness at low temperature is not a design requirement. Applications may include buildings, compression members of bridges, etc.

(b) **Type WT – Weldable Notch-Tough Steel.** Steels of this type meet specified strength and Charpy V-Notch impact requirements and are suitable for welded construction where notch toughness at low temperature is a design requirement. The purchaser, in addition to specifying the grade, must specify the category of steel required that establishes the Charpy V-Notch test temperature and energy level. Applications may include primary tension members in bridges and similar elements.

(c) **Type R – Atmospheric Corrosion-Resistant Steel.** Steels of this type meet specified strength requirements and display an atmospheric corrosion-resistance approximately four times that of plain carbon steels*. These steels may be readily welded up to the maximum thickness covered by this standard. Applications include unpainted siding, unpainted light structural members, etc, where notch toughness at low temperature is not a design requirement.

(d) **Type A – Atmospheric Corrosion-Resistant Weldable Steel.** Steels of this type meet specified strength requirements and display an atmospheric corrosion resistance approximately four times that of plain carbon steels*. These steels are suitable for welded construction where notch toughness at low temperature is not a design requiremermt and are often used in structures in the unpainted condition. Applications are similar to those for type W.

(e) **Type AT – Atmospheric Corrosion-Resistant Weldable Notch-Tough Steel.** Steels of this type meet specified strength and Charpy V-Notch impact requirements and display an atmospheric corrosion-resistance of approximately four times that of plain carbon steels*. These steels are suitable for welded construction where notch toughness at low temperature is a design requirement and are often used in structures in the unpainted condition. The purchaser, in addition to specifying the grade, must specify the category of steel required that establishes the Charpy V-Notch

---

* Copper content not exceeding 0.02%

test temperature and energy level. Applications may include primary tension members in bridges and similar elements.

(f) **Type Q – Quenched and Tempered Low Alloy Steel Plate.** Steels of this type meet specified strength requirements. While these steels may be readily welded, the welding and fabrication techniques are of fundamental importance and must not adversely affect the properties of the plate, especially the heat-affected zone. Applications may include bridges and similar structures.

(g) **Type QT – Quenched and Tempered Low Alloy Notch-Tough Steel Plate.** Steels of this type meet specified strength and Charpy V-Notch impact requirements. They provide good resistance to brittle fracture and are suitable for structures where notch toughness at low temperature is a design requirement. The purchaser, in addition to specifying the grade, must specify the category of steel required that establishes the Charpy V-Notch test temperature and energy level. While these steels may be readily welded, the welding and fabrication techniques are of fundamental importance and must not adversely affect the properties of the plate, especially the heat-affected zone. Applications may include primary tension members in bridges and similar elements.

Tables

Table 6-1, "Grades, Types, Strength Levels", page 6-6 gives the grade designation of the various types and strength levels of structural steels according to the requirements of CSA G40.21.

Availability of any grade and shape combination should be kept in mind when designing to ensure overall economy, since a specified product may not always be available in the tonnage and time frame contemplated. Local availability should always be checked.

Table 6-3, "Mechanical Properties Summary", page 6-7 provides a summary of the various grades and strength levels, tensile strength and yield strength, and the usual maximum thickness and size available for plates, bars and welded shapes, rolled shapes and sheet piling, and hollow structural sections based on CSA G40.21. Table 6-2, "Shape Size Grouping for Tensile Property Classification", page 6-6 summarizes the size groupings for rolled shapes.

Table 6-4, "Chemical Composition", page 6-8 summarizes the chemical requirements of various grades of steel covered by CSA G40.21.

Steel is identified at the mill as to type and grade according to the requirements of the G40.21 standard. Table 6-5, "Steel Marking Colour Code", page 6-9 summarizes the colour code according to the standard. Normally one end of each piece is marked with the appropriate colour code, however, where products are bundled or are shipped as secured lifts only the top or an outside piece may be marked, or a substantial tag may be used.

The particular standards, CSA G40.20 and CSA G40.21 should be consulted for more detail. Similar information about steel covered by ASTM standards should be consulted when appropriate.

Historical Remarks

When confronted with an unidentified structural steel, Clause 5.2.2 of CAN/CSA-S16.1-94 requires F_y be taken as 210 MPa and F_u as 380 MPa. This provides a minimum in lieu of more precise information, such as coupon testing. The following tables list selected dates of publication and data from various CSA and ASTM structural steel standards and specifications, many of which preceded current standards.

For more information on ASTM specifications and properties and dimensions of iron and steel beams previously produced in the U.S.A., consult "Iron and Steel Beams 1873 to 1952", published by the American Institute of Steel Construction. In that publication, the first date listed for both ASTM A7 and A9 is the year 1900. Between 1900 and 1909, medium steel in A7 and A9 had a tensile strength 5 ksi higher than that adoped in 1914. For CSA standards, consult original documents.

Historical Listing of Selected Structural Steels

CSA Standards

Designation	Date Published	Yield Strength		Tensile Strength (F_u)	
		ksi	MPa	ksi	MPa
A16	1924	½ F_u	½ F_u	55-65	380-450
S39	1935	30	210	55-65	380-450
S40	1935	33	230	60-72	410-500
G40.4	1950	33	230	60-72	410-500
G40.5	1950	33	230	60-72	410-500
G40.6	1950	45[1]	310	80-95	550-650
G40.8	1960	40[3]	280	65-85	450-590
G40.12	1964*	44[2]	300	65	450
G40.21	1973**	Replaced all previous Standards, see CISC Handbook			

* Introduced in May 1962 by the Algoma Steel Corporation as 'Algoma 44'
** In May 1997, grade 350W became the only grade for W and HP shapes produced by Algoma Steel Inc.
[1] Silicon steel
[2] Yield reduces when thickness exceeds 1½ inches (40 mm).
[3] Yield reduces when thickness exceeds ⅝ inches (16 mm).

ASTM Specifications

Designation	Date Published	Yield Strength		Tensile Strength (F_u)	
		ksi	MPa	ksi	MPa
A7 (bridges) A9 (buildings)	1914*	½ F_u	½ F_u	55-65	380-450
	1924	½ F_u ≥ 30	½ F_u ≥ 210	55-65	380-450
	1934	½ F_u ≥ 33	½ F_u ≥ 230	60-72	410-500
A373	1954	32	220	58-75	400-520
A242	1955	50[1]	350	70[1]	480
A36	1960	36	250	60-80	410-550
A440	1959	50[1]	350	70[1]	480
A441	1960	50[1]	350	70[1]	480
A572 grade 50	1966	50	345	65	450
A588	1968	50[1]	345	70[1]	485
A992	1998	50 min. to 65 max.	345 min. to 450 max.	65	450

[1] Reduces with increasing thickness
* See text, Historical Remarks, above.

GRADES, TYPES, STRENGTH LEVELS* Table 6–1

Type	Yield Strength, MPa							
	260	300	350	380	400	480	550	700
W	260W	300W	350W	380W**	400W	480W	550W	—
WT	260WT	300WT	350WT	380WT***	400WT	480WT	550WT	—
R	—	—	350R	—	—	—	—	—
A	—	—	350A	—	400A	480A	550A	—
AT	—	—	350AT	—	400AT	480AT	550AT	—
Q	—	—	—	—	—	—	—	700Q
QT	—	—	—	—	—	—	—	700QT

*See CSA-G40.20/G40.21
**This grade is available in hollow structural sections, angles and bars only.
***This grade is available in hollow structural sections only.

SHAPE SIZE GROUPINGS FOR TENSILE PROPERTY CLASSIFICATION* Table 6–2

Shape Type	Group 1	Group 2	Group 3	Group 4	Group 5
W Shapes	W610x82–92 W530x66–85 W460x52–106 W410x39–85 W360x33–79 W310x21–86 W250x18–67 W200x15–71 W150x13–37 W130x24 & 28 W100x19	W1000x222–399 W920x201–313 W840x176–226 W760x134–314 W690x125–265 W610x101–241 W530x92–219 W460x113–213 W410x100–149 W360x91–196 W310x97–158 W250x73–167 W200x86 & 100	W1000x412–488 W920x342–446 W840x299–433 W760x350–389 W690x289–384 W610x262–341 W530x248–331 W460x235–286 W360x216–314 W310x179–283	W1000x539–976 W920x488–1188 W840x473–922 W760x434–865 W690x419–802 W610x372–732 W530x370–599 W460x315–464 W360x347–818 W310x313–500	W920x1262 W360x900–1086
M Shapes and Super Light Beams	To 56 kg/m				
S Shapes	To 52 kg/m	Over 52 kg/m			
HP Shapes		To 150 kg/m	Over 152 kg/m		
C Shapes	To 30.8 kg/m	Over 30.8 kg/m			
MC Shapes	To 42.4 kg/m	Over 42.4 kg/m			
L Shapes	To 13 mm	Over 13 to 20 mm	Over 20 mm		

Note: Tees cut from W, M, and S shapes fall in the same group as the shape from which they are cut.
* See CSA-G40.20/G40.21

Table 6-3 — MECHANICAL PROPERTIES SUMMARY

Type	CSA G40.21* Grade	Tensile Strength F_u (MPa)	Plates, Sheets, Floor Plates, Bars and Welded Shapes			Rolled Shapes and Sheet Piling				Hollow Structural Sections	
			Nominal Maximum Thickness t (mm)	F_y (MPa) min. t ≤ 65	F_y (MPa) min. t > 65	Usual Maximum Shape Size Group	F_y (MPa) min. Groups 1 to 3	Group 4	Group 5	Usual Maximum Wall Thickness (mm)	F_y (MPa) min.
W	260W	410-590	200	260	250	4	260	260			
	300W	450-620[1]	200	300	280	3	300			16	300
	350W	450-650[2]	150	350	320	2	350			16	350
	380W	480-650		380		2[3]	380			16	380
	400W	520-690	20	400		1	400				400
	480W	590-790	20	480		1	480				480
	550W	620-860	14	550			550				550
WT	260WT	410-590	150	260	250	5	260	260	250		
	300WT	450-620	150	300	280	5	300	290	280		
	350WT	480-650[4]	60	350	320	4	350	330		16	350
	380WT	480-650								16	380
	400WT	520-690	20	400		2	400				400
	480WT	590-790	20			1	480				480
	550WT	620-860	14			1					550
R	350R	480-650	14	350		1	350				
A	350A	480-650	100	350	350	5	350	350	320	16	350
	400A	520-690	40	400		2	400				400
	480A	590-790	20	480							480
	550A	620-860	14	550							550
AT	350AT	480-650	100	350	350	5	350	350	320	16	350
	400AT	520-690	40	400		2	400				400
	480AT	590-790	20	480							480
	550AT	620-860	14	550							550
Q	700Q	800-950	65	700	700						
QT	700QT	800-950	65	700	700						

[1] 410-590 MPa for HSS only.
[2] Upper bound F_u = 620 MPa for HSS.
[3] For angles only.
[4] F_u = 450 - 620 MPa for HSS.
* See CSA-G40.20/G40.21.

CHEMICAL COMPOSITION[1]

Table 6-4

CSA G40.21 Grade	Chemical Composition (Heat Analysis) Per Cent[14]								
	All percentages are maxima unless otherwise indicated								
	C	Mn[16]	P	S	Si[12,13]	Other[2]	Cr	Ni	Cu[9]
260W	0.20[15]	0.50/1.50	0.04	0.05	0.40	0.10	—	—	—
300W[5]	0.22[15]	0.50/1.50[21]	0.04	0.05	0.40	0.10	—	—	—
350W	0.23	0.50/1.50[21]	0.04	0.05	0.40	0.10	—	—	—
380W[6]	0.23	0.50/1.50[21]	0.04	0.05	0.40	0.10	—	—	—
400W	0.23[17]	0.50/1.50	0.04	0.05	0.40	0.10	—	—	—
480W	0.26[17]	0.50/1.50	0.04	0.05	0.40	0.10[19]	—	—	—
550W	0.15	1.75[7]	0.04	0.05	0.40	0.15	—	—	—
260WT	0.20[15]	0.80/1.50	0.03	0.04	0.15/0.40	0.10	—	—	—
300WT	0.22[15]	0.80/1.50	0.03	0.04	0.15/0.40	0.10	—	—	—
350WT	0.22[15]	0.80/1.50[7,21]	0.03	0.04	0.15/0.40	0.10[8]	—	—	—
380WT[6]	0.22	0.80/1.50[7,21]	0.03	0.04	0.15/0.40	0.10	—	—	—
400WT	0.22[17]	0.80/1.50[7]	0.03	0.04[18]	0.15/0.40	0.10[8]	—	—	—
480WT	0.26[17]	0.80/1.50[7]	0.03	0.04[18]	0.15/0.40	0.10[8,19]	—	—	—
550WT	0.15	1.75[7]	0.03	0.04[18]	0.15/0.40	0.15	—	—	—
350R	0.16	0.75	0.05/0.15	0.04	0.75	0.10	0.30/1.25[10]	0.90[10]	0.20/0.60[10]
350A	0.20	0.75/1.35[7,21]	0.03	0.04	0.15/0.50	0.10	0.70[11]	0.90[11]	0.20/0.60
400A	0.20	0.75/1.35[7]	0.03	0.04[18]	0.15/0.50	0.10	0.70[11]	0.90[11]	0.20/0.60
480A	0.20	1.00/1.60	0.025[20]	0.035[18]	0.15/0.50	0.12	0.70[11]	0.25/0.50[11]	0.20/0.60
550A	0.15	1.75[7]	0.025[20]	0.035[18]	0.15/0.50	0.15	0.70[11]	0.25/0.50[11]	0.20/0.60
350AT	0.20	0.75/1.35[7,21]	0.03	0.04[18]	0.15/0.50	0.10	0.70[11]	0.90[11]	0.20/0.60
400AT	0.20	0.75/1.35[7]	0.03	0.04[18]	0.15/0.50	0.10	0.70[11]	0.90[11]	0.20/0.60
480AT	0.20	1.00/1.60	0.025[20]	0.035[18]	0.15/0.50	0.12	0.70[11]	0.25/0.50[11]	0.20/0.60
550AT	0.15	1.75[7]	0.025[20]	0.035[18]	0.15/0.50	0.15	0.70[11]	0.25/0.50[11]	0.20/0.60
700Q	0.20	1.50	0.03	0.04	0.15/0.40	—	Boron 0.0005/0.005		—
700QT	0.20	1.50	0.03	0.04	0.15/0.40	—	Boron 0.0005/0.005		—

Notes:
1. For full details, consult CSA Standard G40.20/G40.21. Usual deoxidation for all grades is killed.
2. Other includes grain refining elements Cb, V, Al. Elements (Cb, V) may be used singly or in combination—see G40.20/G40.21 for qualifications. Al, when used, is not included in summation.
3. May have 1.50% Mn.
4. May have 0.32% C for thicknesses over 20 mm.
5. For HSS 0.26% C and 0.30/1.20% Mn.
6. Only angles, bars, and HSS in 380W grade, and only HSS in 380WT grade.
7. Mn may be increased—see G40.20/G40.21 for qualifications.
8. 0.01/0.02% N may be used but N ≤ ¼ V.
9. Copper content of 0.20% minimum may be specified.
10. Cr + Ni + Cu ≥ 1.00%
11. Cr + Ni ≥ 0.40% and for HSS, 0.90% Ni max.
12. Si content of 0.15% to 0.40% is required for type W steel over 40 mm thickness, HSS of WT, A or AT steel, or bar diameter except as required by Note 13.
13. By purchaser's request or producer's option, no minimum Si content is required provided that 0.02% Al is used.
14. Additional alloying elements may be used when approved.
15. For thicknesses over 100 mm, C may be 0.22% for 260W and 260WT grades and 0.23% for 300W, 300WT, and 350WT grades.
16. For HSS Mn 1.65% for 400 yield, 1.75% for 480 yield and 1.85% for 550 yield steels.
17. For HSS 0.20% C.
18. For HSS 0.03% S.
19. For HSS 0.12%
20. For HSS 0.03% P
21. For HSS mininum limit for Mn shall be 0.30% provided that the ratio of Mn to C is not less than 2 to 1 and the ratio of Mn to S is not less than 20 to 1.

Table 6-5 STEEL MARKING COLOUR CODE

Steel Grade	Primary Colour	Secondary Colour
260W 300W 350W 380W 400W 480W 550W	White Green Blue Brown Black Yellow Pink	Green Green Green Green Green Green Green
260WT 300WT 350WT 380WT 400WT 480WT 550WT	White Green Blue Brown Black Yellow Pink	White White White White White White White
350R	Blue	Blue
350A 400A 480A 550A	Blue Black Yellow Pink	Yellow Yellow Yellow Yellow
350AT 400AT 480AT 550AT	Blue Black Yellow Pink	Brown Brown Brown Brown
700Q	Red	Red
700QT	Red	Purple

In this Code, the following colour system applies:

Strength Level	Primary Colour	Type	Secondary Colour
260 300 350 380 400 480 550 700	White Green Blue Brown Black Yellow Pink Red	W WT R A AT Q QT	Green White Blue Yellow Brown Red Purple

RELATIONSHIP BETWEEN ISO AND CSA STANDARDS FOR STRUCTURAL STEELS

ISO Specification	ISO Grade—Quality	Corresponding* CSA G40.21 grade	Significant Requirement in ISO Spec vs CSA
630	Fe 310-0	None	—
	Fe 360-A	230G	C 0.20 max.
	Fe 360-B	None	A and B permit batch testing
	Fe 360-C	None	—
	Fe 360-D	None	—
	Fe 430-A	260W or 300W	—
	Fe 430-B	260W or 300W	—
	Fe 430-C	260WT or 300WT Cat. 1	Charpy 27 J min.
	Fe 430-D	260WT or 300WT Cat. 2	Charpy 27 J min.
	Fe 510-B	350W	—
	Fe 510-C	350WT Cat. 1	—
	Fe 510-D	350WT Cat. 2	
4950-2	E355-CC	350WT Cat. 1	Charpy 40 J min.
	E355-DD	350WT Cat. 2	Charpy 40 J min.
	E355-E	350WT Cat. 4	Test temp. $-50°C$
	E390/420-CC	400WT Cat. 1	Charpy 40 J min.
	E390/420-DD	400WT Cat. 2	Charpy 40 J min.
	E390/420-E	400WT Cat. 4	Test temp. $-50°C$
	E460-CC	480WT Cat. 1	Charpy 40 J min.
	E460-DD	480WT Cat. 2	Charpy 40 J min.
	E460-E	480WT Cat. 4	Test temp. $-50°C$
4950-3	E420-DD/E	None	—
	E460-DD/E	None	—
	E500-DD/E	None	—
	E550-DD/E	None	—
	E620-DD/E	None	—
	E690-DD	700QT Cat. 2	Charpy 40 J min.
	E690-E	700QT Cat. 4	Charpy 27 J min. at $-50°C$
4952	Fe 235W-B/C/D	None	—
	Fe 355W-1A	350R	—
	Fe 355W-1D	None	—
	Fe 355W-2B	350AT	Charpy 27 J min. at $-50°C$
	Fe 355W-2C	350AT Cat. 1	—
	Fe 355W-2D	350AT Cat. 2	—

*Correspondence between standards is only approximate and varies with product thickness. Consult respective standards for actual requirements.
ISO 630 covers plate, strip in coils, wide flats, bars, and hot rolled sections including HSS.
ISO 4950-2 and 4950-3 cover plate, wide strip in coils, and wide flats.
ISO 4952 covers plate, wide flats and sections including HSS.

STANDARD MILL PRACTICE

General

Rolled structural shapes are produced by passing hot blooms, billets or slabs of steel through a series of grooved rolls. Wear on the rolls can cause the dimensions of the finished product to vary slightly from the theoretical, published dimensions. Standard rolling tolerances have been established to make allowance for roll wear, and other factors. These tolerances are contained in CSA G40.20 for shapes supplied according to CSA material standards, and in ASTM Standard A6 for shapes supplied according to ASTM material standards.

Letter symbols for dimensions on sketches shown in this section are in accordance with CSA G40.20, ASTM A6, and mill catalogs.

Methods of increasing area and mass by spreading rolls

Most nominal size groups of rolled shapes contain several specific shapes, each of which is slightly different in mass, area and properties from other shapes in the same size group. Methods used to increase the area and mass, from the minimum nominal size, by spreading the rolls are described below:

For W Shapes (Fig. 1), the thickness of both flange and web is increased, resulting in an increase to the overall beam depth and flange width, with the distance between inside faces of flanges being unchanged.

For S Shapes and Channels (Fig. 2 and 3), the web thickness and flange width are increased by equal amounts, all other dimensions remaining unchanged.

For angles (Fig. 4) the thickness of each leg is increased an equal amount, resulting in a corresponding increase in leg length.

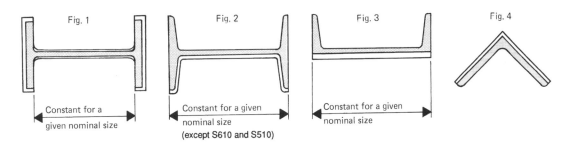

Tolerances

Tolerances are the permissible variations in the mass, cross-sectional area, length, depth, flange width, camber, sweep and other geometric properties of a rolled or welded section. A summary of the basic manufacturing tolerances, taken from CSA G40.20, are provided in the following tables. While these tables are provided for convenience, the actual Standard should be referred to for complete information.

Camber and Sweep

After a section is rolled, it is cold straightened to meet the specified sweep and camber tolerances.

Camber is a deflection, approximating a simple regular curve, measured along the depth of a section. It is usually measured half way between two specified points. The length for purposes of determining the "maximum permissible variation" is the distance between the two specified points.

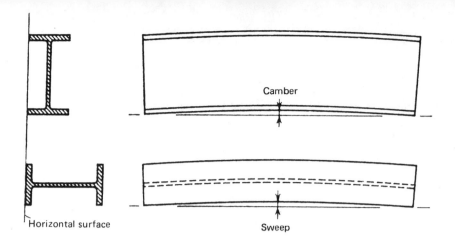

Positions for measuring camber and sweep

Sweep is a deflection, similar to camber, measured along the width of the section. The following table lists Permissible Variations in Straightness.

PERMISSIBLE VARIATIONS IN STRAIGHTNESS

Shape	Maximum Permissible Variation in Straightness, mm
W-shape beams with flange width ≥ 150 mm [1] (camber and sweep) Welded beams or girders where there is no specified camber or sweep	L / 1000
W-shape beams with flange width < 150 mm [1] (camber and sweep)	L / 500
Welded beams or girders with specified camber	6 + L / 4000

W-shapes specified as columns, with flange width approximately equal to depth [1, 2]
 	
Welded columns and compression members in trusses
	L ≤ 14 000 mm:
	    L / 1000 ≤ 10 mm
	L > 14 000 mm:
	    10 + (L − 14000) / 1000

Standard (S) shapes [1]	Camber:
	    L / 500
	Sweep:
	    Negotiable

Bars [1, 3]	L / 250 [4]

Bar-size shapes [1] (greatest cross-sectional dimension < 75 mm)	Camber:
	    L / 250
	Sweep:
	    Negotiable

Notes:

[1] See ASTM A6 / A6M.

[2] Applies only to: 200 mm-deep sections - 46 kg/m and heavier, 250 mm-deep sections - 73 kg/m and heavier, W310 mm-deep sections - 97 kg/m and heavier, and 360 mm-deep sections - 116 kg/m and heavier. For other sections specified as columns, tolerances are negotiable.

[3] Permissible variations do not apply to hot-rolled bars if any subsequent heating operation has been performed.

[4] Round to the nearest whole millimetre.

Sectional Dimensions

The permissible variations in sectional dimensions for welded shapes and rolled shapes are given in the following tables.

PERMISSIBLE VARIATIONS IN SECTIONAL DIMENSIONS OF WELDED STRUCTURAL SHAPES

Nominal Depth, mm	Depth, A, mm		Width of flange, B, mm		Combined warpage and tilt,* mm	Web off-centre, E, mm	Web flatness**	Diagram
	Over	Under	Over	Under	Maximum	Maximum	Maximum	
900 and under	5	3	6	5	Greater of B/100 or 6	6	A/150	
Over 900 to 2000 incl.	5	5	6	5		6	A/150	

* The combined warpage and tilt of the flange is measured from the toe of the flange to a line normal to the plane of the web through the intersection of the centreline of the web with the outside surface of the flange plate.
** The deviation from flatness of the web is measured in any length of the web equal to the total depth of the beam.

PERMISSIBLE VARIATIONS IN SECTIONAL DIMENSIONS OF WIDE-FLANGE SHAPES

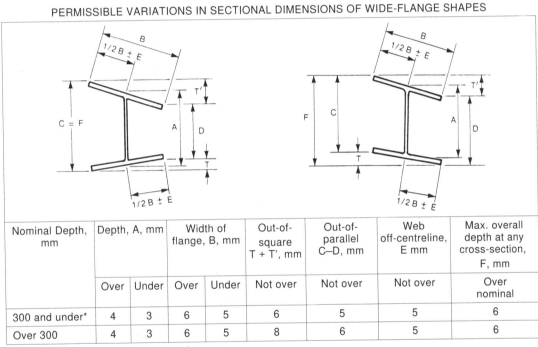

Nominal Depth, mm	Depth, A, mm		Width of flange, B, mm		Out-of-square T + T', mm	Out-of-parallel C–D, mm	Web off-centreline, E mm	Max. overall depth at any cross-section, F, mm
	Over	Under	Over	Under	Not over	Not over	Not over	Over nominal
300 and under*	4	3	6	5	6	5	5	6
Over 300	4	3	6	5	8	6	5	6

A is measured at the centreline of the web.
B is the actual flange width and is measured parallel to the flange.
F is measured parallel to the web.
* Includes all H-beams rolled on mills having vertical rolls.

PERMISSIBLE VARIATIONS IN LENGTH FOR WIDE-FLANGE BEAMS

W Shapes	Variations from Specified Length for Lengths Given, mm			
	9000 and under		Over 9000	
	Over	Under	Over	Under
Beams 610 mm and under in nominal depth	10	10	10 plus 1 for each additional 1000 mm or fraction thereof	10
Beams over 610 mm in nominal depth and all columns	13	13	13 plus 1 for each additional 1000 mm or fraction thereof	13

Note: For W shapes used as bearing piles, the length tolerance is +125 mm, -0 mm. See ASTM A6 / A6M.

PERMISSIBLE VARIATIONS IN LENGTH FOR STANDARD BEAMS

Shapes	Variations from Specified Length For Lengths Given, mm										
	1500 to 3000 excl.		3000 to 6000 excl.		6000 to 9000 excl.		9000 to 12000 excl.		12 000 to 20 000 excl.		Over 20 000	
	Over	Under	Over	Under	Over	Under	Over	Under	Over	Under	Over	Under
S Shapes	25	0	38	0	45	0	57	0	70	0	...	...

Note: Where "..." appears in this table, there is no requirement. See ASTM A6 / A6M.

PERMISSIBLE VARIATIONS IN SECTION DIMENSIONS FOR STANDARD BEAMS, SUPER LIGHT BEAMS AND CHANNELS

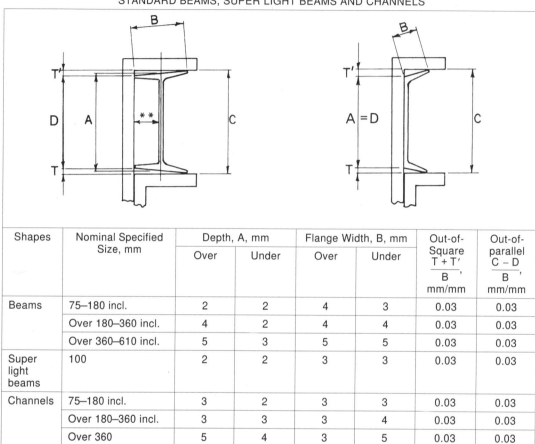

Shapes	Nominal Specified Size, mm	Depth, A, mm		Flange Width, B, mm		Out-of-Square $\frac{T + T'}{B}$, mm/mm	Out-of-parallel $\frac{C - D}{B}$, mm/mm
		Over	Under	Over	Under		
Beams	75–180 incl.	2	2	4	3	0.03	0.03
	Over 180–360 incl.	4	2	4	4	0.03	0.03
	Over 360–610 incl.	5	3	5	5	0.03	0.03
Super light beams	100	2	2	3	3	0.03	0.03
Channels	75–180 incl.	3	2	3	3	0.03	0.03
	Over 180–360 incl.	3	3	3	4	0.03	0.03
	Over 360	5	4	3	5	0.03	0.03

** Back of square and centreline of web to be parallel when measuring out-of-square.
A is measured at centreline for beams and at back of web channels.

Mass and Area Tolerances

Structural-Size Shapes–Cross-sectional area or mass— ±2.5% from theoretical.
Super Light Beams–Cross-sectional area or mass— −2.5% or +7.5% from theoretical.

Web Thickness Tolerance in CSA G40.20

For W shapes, standard beams, H-beams and Super Light Beams under 150 mm specified size, the permissible variation given in CSA Standard G40.20 is ±15% and for specified sizes 150 mm and over, ±1.0 mm.

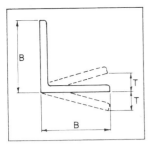

Tolerances for Angles

Permissible variations for cross-sectional dimemsions of bar-size angles, bar-size defined as rolled angles having maximum cross-sectional dimensions less than 75 mm, differ from structural size angles and both varations are given in the following table (see ASTM A6 / A6M).

Structural Size Angles				Bar Size Angles**				
Specified Size*, mm	Length of Leg, B, mm		Out-of-Square T/B	Specified Length of Leg*, mm	Variations from Thickness Given, mm			Variations from Length of Leg Over and Under, mm
	Over	Under			5 and under	Over 5 to 10 incl.	Over 10	
75 to 100 incl.	3	2	0.026	25 and Under	0.2	0.2	...	1
Over 100 to 150 incl.	3	3	0.026	Over 25 to 50 incl.	0.2	0.2	0.3	1
Over 150	5	3	0.026	Over 50	0.3	0.4	0.4	2

Note: Where "..." appears in this table, there is no requirement.
*For unequal leg angles, longer leg determines classification.
**Permissible out-of-square in either direction is 1.5 degrees.

HOLLOW STRUCTURAL SECTIONS

Class of Section

Class H means hollow sections made by:

(i) A seamless or furnace-buttwelded (continuous-welded) or automatic electric-welding process hot-formed to final shape; or

(ii) A seamless or automatic electric-welding process producing a continuous weld, and cold-formed to final shape, subsequently stress-relieved by heating to a temperature of 450°C or higher, followed by cooling in air.

Class C means hollow sections that are cold-formed from a section produced by a seamless process or by an automatic electric-welding process producing a continuous weld.

Cross-Sectional Dimensions

Outside dimensions measured across the flats or diameter at least 50 mm from either end of a piece, including an allowance for convexity or concavity, shall not vary from the specified dimensions of the section by more than the prescribed tolerances.

*Tolerance includes allowance for convexity or concavity. Tolerance may be increased by 50 per cent when applied to the smaller dimension of rectangular sections whose ratio of cross sectional dimensions is between 1.5 and 3, and by 100 per cent when this ratio exceeds 3.

Largest Outside Dimension Across Flats or Diameter, mm	Tolerance*, mm
To 65	±0.5
Over 65–90 incl.	±0.8
Over 90–140 incl.	±1.0
Over 140	±1%

Mass Variation

Based on a mass density of 7850 kg/m^3, the actual mass shall not deviate from the published mass by more than minus 3.5 or plus 10 per cent.

Wall Thickness

Not more than plus 10 or minus 5 per cent from the nominal specified wall thickness, except for the weld seam.

Maximum Outside Corner radius

Wall Thickness mm	Maximum Outside Corner Radii, mm	
	Perimeter to 700 mm Incl.	Perimeter Over 700 mm
To 3 incl.	6	–
Over 3–4 incl.	8	–
Over 4–5 incl.	15	–
Over 5–6 incl.	18	18
Over 6–8 incl.	21	24
Over 8–10 incl.	27	30
Over 10–13 incl.	36	39
Over 13	–	3 x wall thickness

Corner squareness

For rectangular sections, corners shall be square (90°) within plus or minus 1° for hot formed sections and plus or minus 2° for cold formed sections, with the average slope of the sides being the basis for determination.

Straightness Variation

Deviation from straightness in millimetres shall not exceed total length in millimetres divided by 500.

Permissible Twist

Twist of a rectangular section measured by holding down the side of one end of the section on a flat surface and noting the height above the surface of either corner at the opposite end of that side shall not exceed the prescribed tolerances.

Largest Outside Dimension mm	Maximum Twist per 1000 mm of Length mm
To 40 incl.	1.3
Over 40 – 65 incl.	1.7
Over 65 – 105 incl.	2.1
Over 105 – 155 incl.	2.4
Over 155 – 205 incl.	2.8
Over 205	3.1

Cutting Tolerances

Tolerances on ordered cold cut lengths are:
plus 12 and minus 6 millimetres for lengths 7500 mm and under;
plus 18 and minus 6 millimetres for lenghts over 7500 mm.

Tolerances on ordered hot cut lengths of hot rolled sections are:
plus or minus 25 millimetres for lengths 7500 mm and under;
plus or minus 50 millimetres for lengths over 7500 mm.

PRINCIPAL SOURCES OF STRUCTURAL STEEL SECTIONS

General

Standard Canadian and North American sections can be supplied by a number of steel mills in Canada and elsewhere. Principal sources for the various section sizes listed in this Handbook are indicated below.

In 1999, Algoma Steel Inc., the sole Canadian producer of W and HP shapes for 30 years, announced its withdrawal from the rolled shape market.

Canadian Sections

Structural sections available from Canadian mills are listed below. Canadian sizes should be specified to the CSA G40.20/G40.21 material standards. When Canadian sizes are obtained from non-Canadian sources, the material should be supplied to the CSA G40.20/G40.21 Standards. If material substitutions are offered, the sections must be designed using the minimum specified strength levels of the substitute material.

STRUCTURAL SECTIONS AVAILABLE FROM CANADIAN MILLS

WRF — All sizes listed	ANGLES — Sizes less than 203 mm in leg length—see Tables of Properties and Dimensions.
WWF — All sizes listed	
C250 x 37—23	
C230 x 30—20	
C200 x 28—17	HSS — All sizes listed
C180 x 18—15	Pipe — All Standard sizes
C150 x 19—12	
C130 x 17—10	SLB— All sizes listed
C100 x 11—8	
C75 x 9—6	

Non-Canadian Sections

Channels and angles not available from Canadian mills are identified by an asterisk (*) in the Designation Tables on pages 6-24 and 6-25, and in the Tables of Section Properties on pages 6-66 to 6-81. When non-Canadian sizes are required, generally the material should also be specified to the CSA G40.20/G40.21 Standards, and if material substitutions are offered, the sections must be designed using the minimum specified strength levels applicable to the substitute material.

W shapes most commonly produced to ASTM Standards by North American mills will be to ASTM A572 grade 50 and A992 (a more restrictive version of A572 grade 50) and may be considered equivalent to grade 350W of CSA G40.21.

Principal Sources

Some of the more common sources (for Canada) of structural sections and other products are listed below. Producers' catalogues should be consulted for more information and details about particular sections and other products produced. This list is a general guide and not necessarily complete.

Algoma Steel Inc. (WWF, WRF shapes, plate, checkered floor plate)
British Steel Canada Inc.* (shapes, plate)
Chaparral Steel * (shapes)
LTV Copperweld Canada, Sonco Steel Tube Division (HSS)
LTV Copperweld Canada, Standard Tube Division (HSS)
Dofasco (sheet steel in coils and cut lengths)
Co-Steel Lasco (angles and channels > 50 mm)
Gerdau Courtice Steel Inc. (angles and channels up to 51mm, bars)
Gerdau MRM Steel Inc. (SLB)
IPSCO Inc. (HSS, pipe, plate, coil, sheet)
Ispat Sidbec Inc. (bars, hot-rolled coil)
Northwestern Steel & Wire Co.* (shapes)
Prudential Steel (HSS)
Stelco Inc. (plate, bar, sheet steel in coil, pipe)
Trade ARBED Canada Inc.* (shapes)
Nucor-Yamato Steel Co.* (shapes)

non-Canadian sources

Availability of Rolled Shapes

Section sizes are generally produced by steel mills according to rolling schedules. Steel mills and warehouses carry various inventories, usually of the more commonly used sections, and serve as a buffer between rolling cycles to provide ready availability of material. The designer should consider material availability when specifying section sizes, particularly for the heavier mass per metre sizes in a nominal size range and for small quantities of the less commonly used sizes.

The tables of properties and dimensions highlight those section sizes which are more commonly used, and which are generally readily available. Other section sizes are available from mill rollings. Guidance as to the availability of particular sizes can be obtained from local steel fabricators and from steel mill and warehouse sources.

Availability of Angles

Tables of properties and dimensions of angles listed in this Part are based on the imperial series as the sole Canadian producer of metric angles is not currently in the angle market. At the time of writing, there are two Canadian producers of angles. Availability of angles is indicated in the tables as follows:

* = Not available from Canadian mills

+ = Check availability.

The hard metric angles available currently only from offshore producers are not necessarily those listed in previous editions of the Handbook nor CAN/CSA-G312.3-M92. Please consult producer catalogues for sections, properties and dimensions.

METRIC AND IMPERIAL DESIGNATIONS

General

In Canada, the official designation for structural steel sections for purposes of design, detailing and ordering material, is the metric (SI) designation. For WWF, W, SLB and HP shapes, angles, cold-formed channels, and hollow structural sections (HSS), this is described in the CAN/CSA-G312.3-M92 Standard. Canadian and North American sections may also be defined using imperial designations. However, all tables of properties and dimensions, and all design tables included elsewhere in this handbook provide only Canadian metric designations, metric properties and metric design information. Tables on pages 6-21 to 6-28 list Canadian (SI) Designations and corresponding Imperial Designations.

WWF Shapes

Canadian WWF shapes, page 6-21 are hard metric shapes and the Canadian (SI) designation is the total depth in millimetres times the mass in kilograms per metre. The corresponding imperial designation is expressed as inches x lb./ft.

W Shapes

Canadian (SI) designations and imperial designations for W shapes are given on pages 6-22 and 6-23. The Canadian (SI) designation is the nominal depth in millimetres times the mass in kilograms per metre. The corresponding imperial designation is generally the ASTM A6 designation and is expressed as inches x lb./ft.

HP Shapes, S Shapes, M shapes

Canadian (SI) designations and imperial designations for HP, S and M shapes are given on page 6-24. The Canadian (SI) designation is the nominal depth in millimetres times the mass in kilograms per metre and is generally soft converted from the corresponding imperial designation expressed as inches x lb./ft.

C Shapes, MC Shapes

Canadian (SI) designations and imperial designations for C and MC shapes are given on page 6-24. The Canadian (SI) designation is the nominal depth in millimetres times the mass in kilograms per metre and is soft converted from the corresponding imperial designation expressed as inches x lb./ft.

Angles (L)

While the G312.3 Standard includes a metric angle series and an imperial angle series, only the imperial series is included herein as it is the only series currently produced in Canada. The metric series is hard metric with the designation being the leg lengths and thickness in millimetres. The designation for the imperial series is a soft conversion of the imperial size, expressed as leg lengths in millimetres and thickness in millimetres to two significant figures. Designations for angles are given on page 6-25.

Hollow Structural Sections (HSS)

Revisions to the G312.3 Standard for square, rectangular and round hollow structural sections provide for Canadian (SI) designations to be expressed as the nominal outside dimensions in millimetres times the thickness in millimetres to two significant figures. The imperial designation (size) is the outside dimensions times the thickness, both in inches. Designations for rectangular HSS, square HSS and round HSS are given on pages 6-26, 6-27 and 6-28.

NOTES

DESIGNATION TABLE FOR WWF SHAPES

Canadian (SI) Designation (mm x kg/m)	Imperial Designation (in. x lb./ft.)	Canadian (SI) Designation (mm x kg/m)	Imperial Designation (in. x lb./ft.)
WWF2000x732	WWF79x490	WWF700x245	WWF28x164
x648	x436	x214	x144
x607	x408	x196	x132
x542	x364	x175	x117
		x152	x102
WWF1800x700	WWF71x470		
x659	x442	WWF650x864	WWF26x580
x617	x415	x739	x497
x575	x388	x598	x402
x510	x344	x499	x336
		x400	x269
WWF1600x622	WWF63x419		
x580	x388	WWF600x793	WWF24x531
x538	x361	x680	x456
x496	x333	x551	x371
x431	x289	x460	x309
		x369	x248
WWF1400x597	WWF55x402		
x513	x344	WWF550x721	WWF22x484
x471	x316	x620	x416
x405	x272	x503	x338
x358	x240	x420	x282
		x280	x188
WWF1200x487	WWF47x326		
x418	x281	WWF500x651	WWF20x437
x380	x255	x561	x377
x333	x224	x456	x306
x302	x203	x381	x256
x263	x176	x343	x230
		x306	x205
WWF1100x458	WWF43x307	x276	x185
x388	x260	x254	x170
x351	x236	x223	x150
x304	x204	x197	x132
x273	x184		
x234	x157	WWF450x503	WWF18x337
		x409	x275
WWF1000x447	WWF39x300	x342	x229
x377	x253	x308	x207
x340	x228	x274	x184
x293	x197	x248	x166
x262	x176	x228	x152
x223	x150	x201	x134
x200	x134	x177	x119
WWF900x417	WWF35x279	WWF400x444	WWF16x298
x347	x233	x362	x243
x309	x208	x303	x203
x262	x176	x273	x183
x231	x156	x243	x163
x192	x128	x220	x147
x169	x113	x202	x135
		x178	x119
WWF800x339	WWF31x228	x157	x105
x300	x202		
x253	x170	WWF350x315	WWF14x211
x223	x150	x263	x177
x184	x123	x238	x159
x161	x108	x212	x142
		x192	x128
		x176	x118
		x155	x104
		x137	x92

DESIGNATION TABLE FOR W SHAPES

Canadian (SI) Designation (mm x kg/m)	Imperial Designation (in. x lb./ft.)	Canadian (SI) Designation (mm x kg/m)	Imperial Designation (in. x lb./ft.)	Canadian (SI) Designation (mm x kg/m)	Imperial Designation (in. x lb./ft.)
W1100x499	W44x335	W840x251	W33x169	W610x153	W24x103
x432	x290	x226	x152	x140	x94
x390	x262	x210	x141	x125	x84
x342	x230	x193	x130	x113	x76
		x176	x118	x101	x68
W1000x883	W40x593			x91	x61
x749	x503	W760x582	W30x391	x84	x56
x641	x431	x531	x357		
x591	x397	x484	x326	W610x92	W24x62
x554	x372	x434	x292	x82	x55
x539	x362	x389	x261		
x483	x324	x350	x235	W530x300	W21x201
x443	x297	x314	x211	x272	x182
x412	x277	x284	x191	x248	x166
x371	x249	x257	x173	x219	x147
x321	x215			x196	x132
x296	x199	W760x220	W30x148	x182	x122
		x196	x132	x165	x111
W1000x583	W40x392	x185	x124	x150	x101
x493	x331	x173	x116		
x486	x327	x161	x108	W530x138	W21x93
x414	x278	x147	x99	x123	x83
x393	x264	x134	x90	x109	x73
x350	x235			x101	x68
x314	x211	W690x802	W27x539	x92	x62
x272	x183	x548	x368	x82	x55
x249	x167	x500	x336	x72	x48
x222	x149	x457	x307		
		x419	x281	W530x85	W21x57
W920x1188	W36x798	x384	x258	x74	x50
x967	x650	x350	x235	x66	x44
x784	x527	x323	x217		
x653	x439	x289	x194	W460x260	W18x175
x585	x393	x265	x178	x235	x158
x534	x359	x240	x161	x213	x143
x488	x328	x217	x146	x193	x130
x446	x300			x177	x119
x417	x280	W690x192	W27x129	x158	x106
x387	x260	x170	x114	x144	x97
x365	x245	x152	x102	x128	x86
x342	x230	x140	x94	x113	x76
		x125	x84		
W920x381	W36x256			W460x106	W18x71
x345	x232	W610x551	W24x370	x97	x65
x313	x210	x498	x335	x89	x60
x289	x194	x455	x306	x82	x55
x271	x182	x415	x279	x74	x50
x253	x170	x372	x250	x67	x45
x238	x160	x341	x229	x61	x41
x223	x150	x307	x207		
x201	x135	x285	x192	W460x68	W18x46
		x262	x176	x60	x40
W840x576	W33x387	x241	x162	x52	x35
x527	x354	x217	x146		
x473	x318	x195	x131	W410x149	W16x100
x433	x291	x174	x117	x132	x89
x392	x263	x155	x104	x114	x77
x359	x241			x100	x67
x329	x221				
x299	x201				

DESIGNATION TABLE FOR W SHAPES

Canadian (SI) Designation (mm x kg/m)	Imperial Designation (in. x lb./ft.)	Canadian (SI) Designation (mm x kg/m)	Imperial Designation (in. x lb./ft.)	Canadian (SI) Designation (mm x kg/m)	Imperial Designation (in. x lb./ft.)
W410x85	W16x57	W310x283	W12x190	W200x31	W8x21
x74	x50	x253	x170	x27	x18
x67	x45	x226	x152	x21	x14
x60	x40	x202	x136		
x54	x36	x179	x120	W200x22	W8x15
		x158	x106	x19	x13
W410x46	W16x31	x143	x96	x15	x10
x39	x26	x129	x87		
		x118	x79	W150x37	W6x25
W360x1086	W14x730	x107	x72	x30	x20
x990	x665	x97	x65	x22	x15
x900	x605				
x818	x550	W310x86	W12x58	W150x24	W6x16
x744	x500	x79	x53	x18	x12
x677	x455			x14	x9
		W310x74	W12x50	x13	x8.5
W360x634	W14x426	x67	x45		
x592	x398	x60	x40	W130x28	W5x19
x551	x370			x24	x16
x509	x342	W310x52	W12x35		
x463	x311	x45	x30	W100x19	W4x13
x421	x283	x39	x26		
x382	x257	x31	x21		
x347	x233				
x314	x211	W310x33	W12x22		
x287	x193	x28	x19		
x262	x176	x24	x16		
x237	x159	x21	x14		
x216	x145				
		W250x167	W10x112		
W360x196	W14x132	x149	x100		
x179	x120	x131	x88		
x162	x109	x115	x77		
x147	x99	x101	x68		
x134	x90	x89	x60		
		x80	x54		
W360x122	W14x82	x73	x49		
x110	x74				
x101	x68	W250x67	W10x45		
x91	x61	x58	x39		
		x49	x33		
W360x79	W14x53				
x72	x48	W250x45	W10x30		
x64	x43	x39	x26		
		x33	x22		
W360x57	W14x38	x24	x16		
x51	x34				
x45	x30	W250x28	W10x19		
		x25	x17		
W360x39	W14x26	x22	x15		
x33	x22	x18	x12		
W310x500	W12x336	W200x100	W8x67		
x454	x305	x86	x58		
x415	x279	x71	x48		
x375	x252	x59	x40		
x342	x230	x52	x35		
x313	x210	x46	x31		
		W200x42	W8x28		
		x36	x24		

DESIGNATION TABLE FOR HP, M, SLB, S, C, MC SHAPES

Canadian (SI) Designation (mm x kg/m)	Imperial Designation (in. x lb./ft.)	Canadian (SI) Designation (mm x kg/m)	Imperial Designation (in. x lb./ft.)	Canadian (SI) Designation (mm x kg/m)	Imperial Designation (in. x lb./ft.)
HP360x174	HP14x117	S250x52	S10x35	MC460x86	MC18x58
x152	x102	x38	x25.4	x77.2	x51.9
x132	x89			x68.2	x45.8
x108	x73	S200x34	S8x23	x63.5	x42.7
		x27	x18.4		
HP310x174	HP12x117			MC330x74	MC13x50
x152	x102	S150x26	S6x17.25	x60	x40
x132	x89	x19	x12.5	x52	x35
x125	x84			x47.3	x31.8
x110	x74	S130x15	S5x10		
x94	x63			MC310x74	MC12x50
x79	x53	S100x14.1	S4x9.5	x67	x45
		x11	x7.7	x60	x40
HP250x85	HP10x57			x52	x35
x62	x42	S75x11	S3x7.5	x46	x31
		x8	x5.7		
HP200x54	HP8x36			MC310x15.8	MC12x10.6
		C380x74	C15x50	MC250x61.2	MC10x41.1
M310x17.6	M12x11.8	x60	x40	x50	x33.6
x16.1	x10.8	x50	x33.9	x42.4	x28.5
M250x13.4	M10x9	C310x45	C12x30	MC250x37	MC10x25
x11.9	x8	x37	x25	x33	x22
		x31	x20.7		
M200x9.7	M8x6.5			MC250x12.5	MC10x8.4
		C250x45	C10x30		
M130x28.1	M5x18.9	x37	x25	MC230x37.8	MC9x25.4
		x30	x20	x35.6	x23.9
M100x8.9	M4x6	x23	x15.3		
				MC200x33.9	MC8x22.8
		C230x30	C9x20	x31.8	x21.4
SLB100x5.4	SLB4x3.64	x22	x15		
x4.8	x3.20	x20	x13.4	MC200x29.8	MC8x20
				x27.8	x18.7
SLB75x4.5	SLB3x3.05	C200x28	C8x18.75		
x4.3	x2.90	x21	x13.75	MC200x12.6	MC8x8.5
		x17	x11.5		
S610x180	S24x121	C180x22	C7x14.75	MC180x33.8	MC7x22.7
x158	x106	x18	x12.25	x28.4	x19.1
		x15	x9.8		
S610x149	S24x100			MC150x26.8	MC6x18
x134	x90	C150x19	C6x13		
x119	x80	x16	x10.5	MC150x24.3	MC6x16.3
		x12	x8.2	x22.5	x15.1
S510x143	S20x96				
x128	x86	C130x13	C5x9	MC150x17.9	MC6x12
		x10	x6.7		
S510x112	S20x75				
x98.2	x66	C100x11	C4x7.25		
		x9	x6.25		
S460x104	S18x70	x8	x5.4		
x81.4	x54.7				
		C75x9	C3x6		
S380x74	S15x50	x7	x5		
x64	x42.9	x6	x4.1		
S310x74	S12x50				
x60.7	x40.8				
S310x52	S12x35				
x47	x31.8				

DESIGNATION TABLE FOR ANGLES

Canadian (SI) Designation (mm x mm x mm)	Imperial Designation (in. x in. x in.)	Canadian (SI) Designation (mm x mm x mm)	Imperial Designation (in. x in. x in.)	Canadian (SI) Designation (mm x mm x mm)	Imperial Designation (in. x in. x in.)
L203x 203x 29	L8x 8x $1^1/_8$	L127x 76x 13	L5x 3x $^1/_2$	L64x 64x 13	L$2^1/_2$ x $2^1/_2$ x $^1/_2$
x 25	x 1	x 11	x $^7/_{16}$	x 9.5	x $^3/_8$
x 22	x $^7/_8$	x 9.5	x $^3/_8$	x 7.9	x $^5/_{16}$
x 19	x $^3/_4$	x 7.9	x $^5/_{16}$	x 6.4	x $^1/_4$
x 16	x $^5/_8$	x 6.4	x $^1/_4$	x 4.8	x $^3/_{16}$
x 13	x $^1/_2$	L102x 102x 19	L4x 4x $^3/_4$	L64x 51x 9.5	L$2^1/_2$ x 2x $^3/_8$
L203x 152x 25	L8x 6x 1	x 16	x $^5/_8$	x 7.9	x $^5/_{16}$
x 19	x $^3/_4$	x 13	x $^1/_2$	x 6.4	x $^1/_4$
x 13	x $^1/_2$	x 11	x $^7/_{16}$	x 4.8	x $^3/_{16}$
x 11	x $^7/_{16}$	x 9.5	x $^3/_8$	L51x 51x 9.5	L2x 2x $^3/_8$
		x 7.9	x $^5/_{16}$	x 7.9	x $^5/_{16}$
L203x 102x 25	L8x 4x 1	x 6.4	x $^1/_4$	x 6.4	x $^1/_4$
x 19	x $^3/_4$			x 4.8	x $^3/_{16}$
x 13	x $^1/_2$	L102x 89x 13	L4x $3^1/_2$ x $^1/_2$	x 3.2	x $^1/_8$
		x 9.5	x $^3/_8$		
L178x 102x 19	L7x 4x $^3/_4$	x 7.9	x $^5/_{16}$	L51x 38x 6.4	L2x $1^1/_2$ x $^1/_4$
x 16	x $^5/_8$	x 6.4	x $^1/_4$	x 4.8	x $^3/_{16}$
x 13	x $^1/_2$				
x 11	x $^7/_{16}$	L102x 76x 16	L4x 3x $^5/_8$	L44x 44x 6.4	L$1^3/_4$ x $1^3/_4$ x $^1/_4$
x 9.5	x $^3/_8$	x 13	x $^1/_2$	x 4.8	x $^3/_{16}$
		x 11	x $^7/_{16}$	x 3.2	x $^1/_8$
L152x 152x 25	L6x 6x 1	x 9.5	x $^3/_8$		
x 22	x $^7/_8$	x 7.9	x $^5/_{16}$	L38x 38x 6.4	L$1^1/_2$ x $1^1/_2$ x $^1/_4$
x 19	x $^3/_4$	x 6.4	x $^1/_4$	x 4.8	x $^3/_{16}$
x 16	x $^5/_8$			x 3.2	x $^1/_8$
x 14	x $^9/_{16}$	L89x 89x 13	L$3^1/_2$ x $3^1/_2$ x $^1/_2$		
x 13	x $^1/_2$	x 11	x $^7/_{16}$	L32x 32x 6.4	L$1^1/_4$ x $1^1/_4$ x $^1/_4$
x 11	x $^7/_{16}$	x 9.5	x $^3/_8$	x 4.8	x $^3/_{16}$
x 9.5	x $^3/_8$	x 7.9	x $^5/_{16}$	x 3.2	x $^1/_8$
x 7.9	x $^5/_{16}$	x 6.4	x $^1/_4$		
x 6.4	x $^1/_4$			L25x 25x 6.4	L1x 1x $^1/_4$
		L89x 76x 13	L$3^1/_2$ x 3x $^1/_2$	x 4.8	x $^3/_{16}$
L152x 102x 22	L6x 4x $^7/_8$	x 9.5	x $^3/_8$	x 3.2	x $^1/_8$
x 19	x $^3/_4$	x 7.9	x $^5/_{16}$		
x 16	x $^5/_8$	x 6.4	x $^1/_4$	L19x 19x 3.2	L$^3/_4$ x $^3/_4$ x $^1/_8$
x 14	x $^9/_{16}$				
x 13	x $^1/_2$	L89x 64x 13	L$3^1/_2$ x $2^1/_2$ x $^1/_2$		
x 11	x $^7/_{16}$	x 9.5	x $^3/_8$		
x 9.5	x $^3/_8$	x 7.9	x $^5/_{16}$		
x 7.9	x $^5/_{16}$	x 6.4	x $^1/_4$		
L152x 89x 16	L6x $3^1/_2$ x $^5/_8$	L76x 76x 13	L3x 3x $^1/_2$		
x 13	x $^1/_2$	x 11	x $^7/_{16}$		
x 9.5	x $^3/_8$	x 9.5	x $^3/_8$		
x 7.9	x $^5/_{16}$	x 7.9	x $^5/_{16}$		
		x 6.4	x $^1/_4$		
L127x 127x 22	L5x 5x $^7/_8$	x 4.8	x $^3/_{16}$		
x 19	x $^3/_4$				
x 16	x $^5/_8$	L76x 64x 13	L3x $2^1/_2$ x $^1/_2$		
x 13	x $^1/_2$	x 9.5	x $^3/_8$		
x 11	x $^7/_{16}$	x 7.9	x $^5/_{16}$		
x 9.5	x $^3/_8$	x 6.4	x $^1/_4$		
x 7.9	x $^5/_{16}$	x 4.8	x $^3/_{16}$		
x 6.4	x $^1/_4$				
		L76x 51x 13	L3x 2x $^1/_2$		
L127x 89x 19	L5x $3^1/_2$ x $^3/_4$	x 9.5	x $^3/_8$		
x 16	x $^5/_8$	x 7.9	x $^5/_{16}$		
x 13	x $^1/_2$	x 6.4	x $^1/_4$		
x 9.5	x $^3/_8$	x 4.8	x $^3/_{16}$		
x 7.9	x $^5/_{16}$				
x 6.4	x $^1/_4$				

DESIGNATION TABLE FOR RECTANGULAR HSS

Canadian (SI) Designation (mm x mm x mm)	Imperial Designation (in. x in. x in.)	Canadian (SI) Designation (mm x mm x mm)	Imperial Designation (in. x in. x in.)
HSS 305x 203x 13	HSS 12x 8x 0.500	HSS 102x 51x 8.0	HSS 4x 2x 0.313
x 11	x 0.438	x 6.4	x 0.250
x 9.5	x 0.375	x 4.8	x 0.188
x 8.0	x 0.313	x 3.8	x 0.150
x 6.4	x 0.250	x 3.2	x 0.125
HSS 254x 152x 13	HSS 10x 6x 0.500	HSS 89x 64x 8.0	HSS $3^1/_2$ x $2^1/_2$ x 0.313
x 11	x 0.438	x 6.4	x 0.250
x 9.5	x 0.375	x 4.8	x 0.188
x 8.0	x 0.313	x 3.8	x 0.150
x 6.4	x 0.250	x 3.2	x 0.125
HSS 203x 152x 13	HSS 8x 6x 0.500	HSS 76x 51x 8.0	HSS 3x 2x 0.313
x 11	x 0.438	x 6.4	x 0.250
x 9.5	x 0.375	x 4.8	x 0.188
x 8.0	x 0.313	x 3.8	x 0.150
x 6.4	x 0.250	x 3.2	x 0.125
x 4.8	x 0.188		
HSS 203x 102x 13	HSS 8x 4x 0.500	HSS 51x 25x 3.2	HSS 2x 1x 0.125
x 11	x 0.438		
x 9.5	x 0.375		
x 8.0	x 0.313		
x 6.4	x 0.250		
x 4.8	x 0.188		
HSS 178x 127x 13	HSS 7x 5x 0.500		
x 11	x 0.438		
x 9.5	x 0.375		
x 8.0	x 0.313		
x 6.4	x 0.250		
x 4.8	x 0.188		
HSS 152x 102x 13	HSS 6x 4x 0.500		
x 11	x 0.438		
x 9.5	x 0.375		
x 8.0	x 0.313		
x 6.4	x 0.250		
x 4.8	x 0.188		
HSS 152x 76x 9.5	HSS 6x 3x 0.375		
x 8.0	x 0.313		
x 6.4	x 0.250		
x 4.8	x 0.188		
HSS 127x 76x 9.5	HSS 5x 3x 0.375		
x 8.0	x 0.313		
x 6.4	x 0.250		
x 4.8	x 0.188		
x 3.8	x 0.150		
HSS 102x 76x 9.5	HSS 4x 3x 0.375		
x 8.0	x 0.313		
x 6.4	x 0.250		
x 4.8	x 0.188		
x 3.8	x 0.150		
x 3.2	x 0.125		

DESIGNATION TABLE FOR SQUARE HSS

Canadian (SI) Designation (mm x mm x mm)	Imperial Designation (in. x in. x in.)	Canadian (SI) Designation (mm x mm x mm)	Imperial Designation (in. x in. x in.)
HSS 305x 305x 13	HSS 12x 12x 0.500	HSS 64x 64x 6.4	HSS 2½ x 2½ x 0.250
x 11	x 0.438	x 4.8	x 0.188
x 9.5	x 0.375	x 3.8	x 0.150
x 8.0	x 0.313	x 3.2	x 0.125
x 6.4	x 0.250		
		HSS 51x 51x 6.4	HSS 2x 2x 0.250
HSS 254x 254x 13	HSS 10x 10x 0.500	x 4.8	x 0.188
x 11	x 0.438	x 3.8	x 0.150
x 9.5	x 0.375	x 3.2	x 0.125
x 8.0	x 0.313		
x 6.4	x 0.250	HSS 38x 38x 4.8	HSS 1½ x 1½ x 0.188
		x 3.8	x 0.150
HSS 203x 203x 13	HSS 8x 8x 0.500	x 3.2	x 0.125
x 11	x 0.438		
x 9.5	x 0.375		
x 8.0	x 0.313		
x 6.4	x 0.250		
HSS 178x 178x 13	HSS 7x 7x 0.500		
x 11	x 0.438		
x 9.5	x 0.375		
x 8.0	x 0.313		
x 6.4	x 0.250		
x 4.8	x 0.188		
HSS 152x 152x 13	HSS 6x 6x 0.500		
x 11	x 0.438		
x 9.5	x 0.375		
x 8.0	x 0.313		
x 6.4	x 0.250		
x 4.8	x 0.188		
HSS 127x 127x 13	HSS 5x 5x 0.500		
x 11	x 0.438		
x 9.5	x 0.375		
x 8.0	x 0.313		
x 6.4	x 0.250		
x 4.8	x 0.188		
HSS 102x 102x 9.5	HSS 4x 4x 0.375		
x 8.0	x 0.313		
x 6.4	x 0.250		
x 4.8	x 0.188		
x 3.8	x 0.150		
x 3.2	x 0.125		
HSS 89x 89x 9.5	HSS 3½ x 3½ x 0.375		
x 8.0	x 0.313		
x 6.4	x 0.250		
x 4.8	x 0.188		
x 3.8	x 0.150		
x 3.2	x 0.125		
HSS 76x 76x 9.5	HSS 3x 3x 0.375		
x 8.0	x 0.313		
x 6.4	x 0.250		
x 4.8	x 0.188		
x 3.8	x 0.150		
x 3.2	x 0.125		

DESIGNATION TABLE FOR ROUND HSS

Canadian (SI) Designation (mm x mm)	Imperial Designation (in. x in.)	Canadian (SI) Designation (mm x mm)	Imperial Designation (in. x in.)
HSS 610x 13	HSS 24x 0.500	HSS 114x 8.0	HSS 4.5x 0.313
x 11	x 0.438	x 6.4	x 0.250
x 9.5	x 0.375	x 4.8	x 0.188
x 8.0	x 0.313		
x 6.4	x 0.250	HSS 102x 8.0	HSS 4x 0.313
		x 6.4	x 0.250
HSS 559x 13	HSS 22x 0.500	x 4.8	x 0.188
x 11	x 0.438	x 3.8	x 0.150
x 9.5	x 0.375		
x 8.0	x 0.313	HSS 89x 8.0	HSS 3.5x 0.313
x 6.4	x 0.250	x 6.4	x 0.250
		x 4.8	x 0.188
HSS 508x 13	HSS 20x 0.500	x 3.8	x 0.150
x 11	x 0.438		
x 9.5	x 0.375	HSS 73x 6.4	HSS 2.875x 0.250
x 8.0	x 0.313	x 4.8	x 0.188
x 6.4	x 0.250	x 3.8	x 0.150
		x 3.2	x 0.125
HSS 406x 13	HSS 16x 0.500		
x 11	x 0.438	HSS 60x 6.4	HSS 2.375x 0.250
x 9.5	x 0.375	x 4.8	x 0.188
x 8.0	x 0.313	x 3.8	x 0.150
x 6.4	x 0.250	x 3.2	x 0.125
HSS 356x 13	HSS 14x 0.500	HSS 48x 4.8	HSS 1.9x 0.188
x 11	x 0.438	x 3.8	x 0.150
x 9.5	x 0.375	x 3.2	x 0.125
x 8.0	x 0.313		
x 6.4	x 0.250		
HSS 324x 13	HSS 12.75x 0.500		
x 11	x 0.438		
x 9.5	x 0.375		
x 8.0	x 0.313		
x 6.4	x 0.250		
HSS 273x 13	HSS 10.75x 0.500		
x 11	x 0.438		
x 9.5	x 0.375		
x 8.0	x 0.313		
x 6.4	x 0.250		
HSS 219x 13	HSS 8.625x 0.500		
x 11	x 0.438		
x 9.5	x 0.375		
x 8.0	x 0.313		
x 6.4	x 0.250		
x 4.8	x 0.188		
HSS 168x 9.5	HSS 6.625x 0.375		
x 8.0	x 0.313		
x 6.4	x 0.250		
x 4.8	x 0.188		
HSS 141x 9.5	HSS 5.563x 0.375		
x 8.0	x 0.313		
x 6.4	x 0.250		
x 4.8	x 0.188		

WELDED SHAPES

General

Welded shapes are produced to the requirements of CSA-G40.20 Standard, using plate meeting the requirements of CSA-G40.21 material Standard. Canadian welded shapes are produced by the Algoma Steel Inc. and are available in 300W and 350W grade material. If the use of other grades of steel is contemplated, the manufacturer should be consulted. Properties and dimensions of WWF (welded wide flange) shapes and WRF (welded reduced flange) shapes are provided on pages 6-30 to 6-37.

Manufacture

The G40.20 Standard defines welded shapes as I-type sections produced by automatic welding processes from three individual components. Hot rolled plates are flame-cut to required width, then assembled together on a special jig where pressure rolls hold the flange and web plates in contact for automatic submerged arc welding. Welds are made simultaneously along both flange-to-web joints in the horizontal fillet position.

For web thicknesses up to and including 20 mm, the flange-to-web welds develop the full capacity of the flange-to-web joint. For web thicknesses greater than 20 mm, which occurs in the heavier WWF column shapes, the full strength of the web is not developed by the web-to-flange welds, so that additional welding by the purchaser may be required at major connection points. Column WWF shapes with partial strength web welds are identified by asterisks (**) and the note "welding does not fully develop the web strength for these sections".

Products

WWF beams range in depth from 700 mm to 2000 mm and have equal width flanges which range in width between 300 mm and 550 mm. Flange thickness varies while the overall depth of beams in each size range is constant. The Class in bending for WWF beams for G40.21 grades 300W and 350W steels is given in Table 5-1.

WWF column shapes range in depth and width from 350 mm to 650 mm in increments of 50 mm. Overall depth and flange width in each size range is the same.

WRF shapes are asymmetrical beams in which one flange is narrower than the other. Overall depths vary from 1000 mm to 1800 mm and are constant for all beams in each size range. Flange widths are 300 mm and 550 mm.

Availability of WWF and WRF Shapes

WWF beam and column shapes, and WRF shapes are generally not carried in inventory and are produced on demand. Typical lead time to source welded shapes is eight to ten weeks.

A fabricator may request the approval of the designer to substitute a welded three-plate section, welded to the requirements of CSA Standard W59 of oxy-flame-cut plates of the same grade as that of the WWF section originally designed.

WELDED REDUCED FLANGE SHAPES
WRF1800-WRF1000

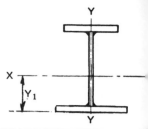

PROPERTIES

Designation ‡	Dead Load	Total Area	Axis X-X					Axis Y-Y			Torsional Constant	Warping Constant
			I_x	S_{xt}	S_{xb}	r_x	Y_1	I_y	S_y	r_y	J	C_w
	kN/m	mm²	10^6 mm⁴	10^3 mm³	10^3 mm³	mm	mm	10^6 mm⁴	10^3 mm³	mm	10^3 mm⁴	10^9 mm⁶
WRF1800												
X543	5.33	69 200	35 700	34 200	47 100	718	757	726	2 640	102	29 100	268 000
X480	4.70	61 100	30 100	29 300	38 900	702	774	565	2 050	96	15 500	211 000
X416	4.08	53 000	24 300	24 200	30 600	678	795	404	1 470	87	7 830	152 000
WRF1600												
X491	4.82	62 500	26 600	28 300	40 200	652	660	726	2 640	108	27 900	211 000
X427	4.19	54 400	22 200	24 000	32 900	639	674	565	2 050	102	14 200	166 000
X362	3.56	46 200	17 700	19 500	25 500	619	693	403	1 470	93	6 540	120 000
WRF1400												
X413	4.05	52 600	17 600	21 200	30 800	578	571	645	2 350	111	19 300	143 000
X348	3.42	44 400	14 200	17 500	24 400	566	584	484	1 760	104	8 880	109 000
X284	2.78	36 200	10 800	13 500	17 800	545	605	323	1 170	94	3 510	73 700
WRF1200												
X373	3.66	47 600	12 200	16 900	25 500	506	478	645	2 350	116	18 800	104 000
X309	3.03	39 300	9 760	13 700	20 000	498	488	484	1 760	111	8 310	79 500
X244	2.39	31 000	7 240	10 400	14 300	483	505	322	1 170	102	2 940	53 900
WRF1000												
X340	3.34	43 300	7 980	13 100	20 500	429	389	645	2 340	122	18 400	71 400
X275	2.70	35 000	6 340	10 500	16 000	426	396	484	1 760	117	7 960	54 600
X210	2.06	26 700	4 620	7 820	11 300	416	408	322	1 170	110	2 590	37 200

‡Nominal depth in millimetres and mass in kilograms per metre.

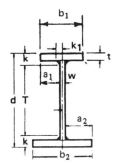

WELDED REDUCED FLANGE SHAPES
WRF1800-WRF1000

DIMENSIONS AND SURFACE AREAS

Nominal Mass kg/m	Depth d mm	Flange Width b_1 mm	Flange Width b_2 mm	Flange Thickness t mm	Web Thickness w mm	Distances a_1 mm	Distances a_2 mm	Distances T mm	Distances k mm	Distances k_1 mm	Distances d-2t mm	Surface Area (m²) per metre of length Total	Surface Area (m²) per metre of length Minus Top of Top Flange
543	1 800	300	550	45.0	18.0	141	266	1 688	56	19	1 710	5.26	4.96
480	1 800	300	550	35.0	18.0	141	266	1 708	46	19	1 730	5.26	4.96
416	1 800	300	550	25.0	18.0	141	266	1 728	36	19	1 750	5.26	4.96
491	1 600	300	550	45.0	16.0	142	267	1 492	54	16	1 510	4.87	4.57
427	1 600	300	550	35.0	16.0	142	267	1 512	44	16	1 530	4.87	4.57
362	1 600	300	550	25.0	16.0	142	267	1 532	34	16	1 550	4.87	4.57
413	1 400	300	550	40.0	14.0	143	268	1 302	49	15	1 320	4.47	4.17
348	1 400	300	550	30.0	14.0	143	268	1 322	39	15	1 340	4.47	4.17
284	1 400	300	550	20.0	14.0	143	268	1 342	29	15	1 360	4.47	4.17
373	1 200	300	550	40.0	12.0	144	269	1 102	49	14	1 120	4.08	3.78
309	1 200	300	550	30.0	12.0	144	269	1 122	39	14	1 140	4.08	3.78
244	1 200	300	550	20.0	12.0	144	269	1 142	29	14	1 160	4.08	3.78
340	1 000	300	550	40.0	10.0	145	270	902	49	13	920	3.68	3.38
275	1 000	300	550	30.0	10.0	145	270	922	39	13	940	3.68	3.38
210	1 000	300	550	20.0	10.0	145	270	942	29	13	960	3.68	3.38

WELDED WIDE FLANGE SHAPES
WWF2000-WWF1100

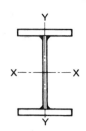

PROPERTIES

Designation‡	Dead Load	Total Area	Axis X-X				Axis Y-Y				Torsional Constant	Warping Constant
			I_x	S_x	r_x	Z_x	I_y	S_y	r_y	Z_y	J	C_w
	kN/m	mm²	10^6mm⁴	10^3mm³	mm	10^3mm³	10^6mm⁴	10^3mm³	mm	10^3mm³	10^3mm⁴	10^9mm⁶
WWF2000												
*X732	7.17	93 200	63 900	63 900	828	71 800	1 390	5 050	122	7 750	50 900	1 320 000
*X648	6.35	82 600	54 200	54 200	810	61 700	1 110	4 040	116	6 240	28 600	1 070 000
*X607	5.94	77 300	49 300	49 300	799	56 600	972	3 530	112	5 490	20 900	938 000
X542	5.31	69 000	41 400	41 400	775	48 500	626	2 510	95.2	3 950	14 200	608 000
WWF1800												
*X700	6.86	89 200	50 400	56 000	752	62 700	1 390	5 050	125	7 730	50 400	1 060 000
*X659	6.45	83 900	46 600	51 800	745	58 200	1 250	4 540	122	6 980	38 000	962 000
*X617	6.04	78 600	42 700	47 400	737	53 700	1 110	4 040	119	6 220	28 100	860 000
*X575	5.64	73 300	38 800	43 100	728	49 100	972	3 530	115	5 470	20 300	757 000
X510	5.00	65 000	32 400	36 000	706	41 800	626	2 500	98.1	3 930	13 600	490 000
WWF1600												
*X622	6.09	79 200	37 600	47 100	689	51 800	1 390	5 040	132	7 660	47 900	833 000
*X580	5.68	73 800	34 600	43 300	685	47 700	1 250	4 540	130	6 900	35 500	755 000
X538	5.27	68 500	31 600	39 500	679	43 700	1 110	4 040	127	6 150	25 500	675 000
X496	4.85	63 100	28 400	35 500	671	39 600	971	3 530	124	5 390	17 800	595 000
X431	4.21	54 800	23 400	29 300	653	33 100	626	2 500	107	3 850	11 100	385 000
WWF1400												
*X597	5.84	76 000	28 100	40 100	608	44 000	1 390	5 040	135	7 650	47 600	632 000
X513	5.02	65 300	23 500	33 600	600	37 000	1 110	4 040	130	6 140	25 300	513 000
X471	4.61	59 900	21 100	30 200	594	33 400	971	3 530	127	5 380	17 500	452 000
X405	3.97	51 600	17 300	24 800	579	27 800	625	2 500	110	3 840	10 800	293 000
X358	3.50	45 600	14 500	20 800	564	23 700	320	1 600	83.8	2 490	9 030	150 000
WWF1200												
X487	4.78	62 100	16 700	27 900	519	30 600	1 110	4 030	134	6 120	25 000	373 000
X418	4.09	53 200	13 800	23 100	509	25 600	730	2 920	117	4 450	15 800	248 000
X380	3.72	48 400	12 300	20 500	504	22 800	625	2 500	114	3 820	10 600	214 000
X333	3.26	42 400	10 200	17 100	490	19 300	320	1 600	86.9	2 470	8 760	110 000
X302	2.96	38 500	8 970	15 000	483	17 100	267	1 340	83.3	2 070	5 740	92 200
X263	2.58	33 500	7 250	12 100	465	14 200	113	753	58.1	1 200	4 700	39 000
WWF1100												
X458	4.50	58 500	13 700	24 800	484	27 100	1 110	4 030	138	6 100	24 400	312 000
X388	3.81	49 500	11 200	20 400	476	22 400	729	2 920	121	4 430	15 200	207 000
X351	3.44	44 700	9 940	18 100	472	19 900	625	2 500	118	3 800	9 950	179 000
X304	2.97	38 700	8 220	14 900	461	16 700	320	1 600	90.9	2 450	8 150	91 700
X273	2.68	34 800	7 160	13 000	454	14 700	267	1 330	87.6	2 050	5 130	77 100
X234	2.29	29 800	5 720	10 400	438	12 000	113	752	61.6	1 180	4 090	32 600

‡Nominal depth in millimetres and mass in kilograms per metre.
*Maximum piece weight may limit length.

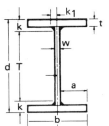

WELDED WIDE FLANGE SHAPES
WWF2000-WWF1100

DIMENSIONS AND SURFACE AREAS

Nominal Mass	Depth d	Flange		Web Thickness w	Distances					Surface Area (m²) per metre of length	
		Width b	Thickness t		a	T	k	k₁	d-2t	Total	Minus Top of Top flange
kg/m	mm	mm	mm	mm	mm	mm	mm	mm	mm		
732	2 000	550	50.0	20.0	265	1 878	61	20	1 900	6.16	5.61
648	2 000	550	40.0	20.0	265	1 898	51	20	1 920	6.16	5.61
607	2 000	550	35.0	20.0	265	1 908	46	20	1 930	6.16	5.61
542	2 000	500	30.0	20.0	240	1 918	41	20	1 940	5.96	5.46
700	1 800	550	50.0	20.0	265	1 678	61	20	1 700	5.76	5.21
659	1 800	550	45.0	20.0	265	1 688	56	20	1 710	5.76	5.21
617	1 800	550	40.0	20.0	265	1 698	51	20	1 720	5.76	5.21
575	1 800	550	35.0	20.0	265	1 708	46	20	1 730	5.76	5.21
510	1 800	500	30.0	20.0	240	1 718	41	20	1 740	5.56	5.06
622	1 600	550	50.0	16.0	267	1 478	61	18	1 500	5.37	4.82
580	1 600	550	45.0	16.0	267	1 488	56	18	1 510	5.37	4.82
538	1 600	550	40.0	16.0	267	1 498	51	18	1 520	5.37	4.82
496	1 600	550	35.0	16.0	267	1 512	44	16	1 530	5.37	4.82
431	1 600	500	30.0	16.0	242	1 522	39	16	1 540	5.17	4.67
597	1 400	550	50.0	16.0	267	1 278	61	18	1 300	4.97	4.42
513	1 400	550	40.0	16.0	267	1 298	51	18	1 320	4.97	4.42
471	1 400	550	35.0	16.0	267	1 312	44	16	1 330	4.97	4.42
405	1 400	500	30.0	16.0	242	1 322	39	16	1 340	4.77	4.27
358	1 400	400	30.0	16.0	192	1 322	39	16	1 340	4.37	3.97
487	1 200	550	40.0	16.0	267	1 098	51	18	1 120	4.57	4.02
418	1 200	500	35.0	16.0	242	1 112	44	16	1 130	4.37	3.87
380	1 200	500	30.0	16.0	242	1 122	39	16	1 140	4.37	3.87
333	1 200	400	30.0	16.0	192	1 122	39	16	1 140	3.97	3.57
302	1 200	400	25.0	16.0	192	1 132	34	16	1 150	3.97	3.57
263	1 200	300	25.0	16.0	142	1 132	34	16	1 150	3.57	3.27
458	1 100	550	40.0	14.0	268	998	51	17	1 020	4.37	3.82
388	1 100	500	35.0	14.0	243	1 012	44	15	1 030	4.17	3.67
351	1 100	500	30.0	14.0	243	1 022	39	15	1 040	4.17	3.67
304	1 100	400	30.0	14.0	193	1 022	39	15	1 040	3.77	3.37
273	1 100	400	25.0	14.0	193	1 032	34	15	1 050	3.77	3.37
234	1 100	300	25.0	14.0	143	1 032	34	15	1 050	3.37	3.07

WELDED WIDE FLANGE SHAPES
WWF1000-WWF600

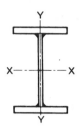

PROPERTIES

Designation‡	Dead Load	Total Area	Axis X-X				Axis Y-Y				Torsional Constant	Warping Constant
			I_x	S_x	r_x	Z_x	I_y	S_y	r_y	Z_y	J	C_w
	kN/m	mm²	10^6mm⁴	10^3mm³	mm	10^3mm³	10^6mm⁴	10^3mm³	mm	10^3mm³	10^3mm⁴	10^9mm⁶
WWF1000												
X447	4.39	57 100	11 100	22 200	441	24 200	1 110	4 030	139	6 100	24 300	256 000
X377	3.70	48 100	9 120	18 200	435	20 000	729	2 920	123	4 420	15 100	170 000
X340	3.33	43 300	8 060	16 100	431	17 700	625	2 500	120	3 800	9 860	147 000
X293	2.87	37 300	6 640	13 300	422	14 800	320	1 600	92.6	2 450	8 060	75 300
X262	2.57	33 400	5 780	11 600	416	13 000	267	1 330	89.4	2 050	5 040	63 400
X223	2.18	28 400	4 590	9 190	402	10 500	113	752	63.1	1 170	3 990	26 800
X200	1.96	25 600	3 940	7 890	392	9 170	90.2	602	59.4	948	2 480	21 700
WWF900												
X417	4.09	53 200	8 680	19 300	404	20 800	1 110	4 030	144	6 080	23 800	205 000
X347	3.40	44 300	7 100	15 800	400	17 100	729	2 920	128	4 400	14 700	136 000
X309	3.03	39 400	6 240	13 900	398	15 000	625	2 500	126	3 780	9 370	118 000
X262	2.56	33 400	5 110	11 400	391	12 400	320	1 600	97.9	2 430	7 570	60 600
X231	2.26	29 500	4 410	9 810	387	10 800	267	1 330	95.1	2 030	4 540	51 100
X192	1.88	24 500	3 460	7 680	376	8 600	113	751	67.9	1 150	3 500	21 600
X169	1.66	21 600	2 930	6 510	368	7 370	90.1	601	64.6	927	1 980	17 400
WWF800												
X339	3.32	43 200	5 500	13 700	357	14 900	729	2 920	130	4 400	14 600	107 000
X300	2.94	38 300	4 840	12 100	355	13 100	625	2 500	128	3 770	9 330	92 700
X253	2.48	32 300	3 950	9 870	350	10 800	320	1 600	99.5	2 420	7 530	47 400
X223	2.18	28 400	3 410	8 520	347	9 340	267	1 330	97.0	2 020	4 500	40 100
X184	1.80	23 400	2 660	6 640	337	7 410	113	751	69.5	1 150	3 460	16 900
X161	1.57	20 500	2 250	5 610	331	6 320	90.1	601	66.3	924	1 940	13 700
WWF700												
X245	2.39	31 200	2 950	8 420	307	9 210	320	1 600	101	2 420	7 480	35 900
X214	2.10	27 300	2 540	7 270	305	7 950	267	1 330	98.9	2 020	4 460	30 400
X196	1.92	24 900	2 300	6 560	304	7 190	235	1 170	97.1	1 780	3 130	27 000
X175	1.71	22 300	1 970	5 640	297	6 270	113	751	71.2	1 150	3 410	12 800
X152	1.49	19 400	1 660	4 760	293	5 320	90.1	601	68.1	921	1 890	10 400
WWF650												
*X864**	8.46	110 000	7 570	23 300	262	27 300	2 760	8 480	158	13 200	132 000	240 000
*X739**	7.24	94 100	7 200	22 100	277	25 200	2 750	8 450	171	12 800	98 400	239 000
*X598	5.86	76 200	6 150	18 900	284	21 100	2 290	7 040	173	10 600	55 600	206 000
X499	4.89	63 600	5 170	15 900	285	17 500	1 830	5 630	170	8 510	29 300	170 000
X400	3.92	51 000	4 110	12 600	284	13 900	1 370	4 230	164	6 400	13 300	132 000
WWF600												
*X793**	7.77	101 000	5 830	19 400	240	22 900	2 170	7 230	147	11 200	121 000	158 000
*X680**	6.66	86 600	5 560	18 500	253	21 200	2 160	7 200	158	10 900	90 700	158 000
X551	5.40	70 200	4 770	15 900	261	17 800	1 800	6 000	160	9 050	51 300	136 000
X460	4.51	58 600	4 020	13 400	262	14 800	1 440	4 800	157	7 250	27 000	113 000
X369	3.61	47 000	3 200	10 700	261	11 800	1 080	3 600	152	5 460	12 200	87 800

‡Nominal depth in millimetres and mass in kilograms per metre.
**Welding does not fully develop the web strength for these sections.
*Maximum piece weight may limit length.

WELDED WIDE FLANGE SHAPES
WWF1000-WWF600

DIMENSIONS AND SURFACE AREAS

Nominal Mass	Depth d	Flange		Web Thickness w	Distances					Surface Area (m²) per metre of length	
		Width b	Thickness t		a	T	k	k_1	d-2t	Total	Minus Top of Top flange
kg/m	mm	mm	mm	mm	mm	mm	mm	mm	mm		
447	1 000	550	40.0	14.0	268	898	51	17	920	4.17	3.62
377	1 000	500	35.0	14.0	243	912	44	15	930	3.97	3.47
340	1 000	500	30.0	14.0	243	922	39	15	940	3.97	3.47
293	1 000	400	30.0	14.0	193	922	39	15	940	3.57	3.17
262	1 000	400	25.0	14.0	193	932	34	15	950	3.57	3.17
223	1 000	300	25.0	14.0	143	932	34	15	950	3.17	2.87
200	1 000	300	20.0	14.0	143	942	29	15	960	3.17	2.87
417	900	550	40.0	11.0	270	798	51	16	820	3.98	3.43
347	900	500	35.0	11.0	245	812	44	14	830	3.78	3.28
309	900	500	30.0	11.0	245	822	39	14	840	3.78	3.28
262	900	400	30.0	11.0	195	822	39	14	840	3.38	2.98
231	900	400	25.0	11.0	195	832	34	14	850	3.38	2.98
192	900	300	25.0	11.0	145	832	34	14	850	2.98	2.68
169	900	300	20.0	11.0	145	842	29	14	860	2.98	2.68
339	800	500	35.0	11.0	245	712	44	14	730	3.58	3.08
300	800	500	30.0	11.0	245	722	39	14	740	3.58	3.08
253	800	400	30.0	11.0	195	722	39	14	740	3.18	2.78
223	800	400	25.0	11.0	195	732	34	14	750	3.18	2.78
184	800	300	25.0	11.0	145	732	34	14	750	2.78	2.48
161	800	300	20.0	11.0	145	742	29	14	760	2.78	2.48
245	700	400	30.0	11.0	195	622	39	14	640	2.98	2.58
214	700	400	25.0	11.0	195	632	34	14	650	2.98	2.58
196	700	400	22.0	11.0	195	638	31	14	656	2.98	2.58
175	700	300	25.0	11.0	145	632	34	14	650	2.58	2.28
152	700	300	20.0	11.0	145	642	29	14	660	2.58	2.28
864	650	650	60.0	60.0	295	508	71	40	530	3.78	3.13
739	650	650	60.0	30.0	310	508	71	25	530	3.84	3.19
598	650	650	50.0	20.0	315	528	61	20	550	3.86	3.21
499	650	650	40.0	20.0	315	548	51	20	570	3.86	3.21
400	650	650	30.0	20.0	315	568	41	20	590	3.86	3.21
793	600	600	60.0	60.0	270	458	71	40	480	3.48	2.88
680	600	600	60.0	30.0	285	458	71	25	480	3.54	2.94
551	600	600	50.0	20.0	290	478	61	20	500	3.56	2.96
460	600	600	40.0	20.0	290	498	51	20	520	3.56	2.96
369	600	600	30.0	20.0	290	518	41	20	540	3.56	2.96

WELDED WIDE FLANGE SHAPES
WWF550-WWF350

PROPERTIES

Designation‡	Dead Load	Total Area	Axis X-X				Axis Y-Y				Torsional Constant	Warping Constant
			I_x	S_x	r_x	Z_x	I_y	S_y	r_y	Z_y	J	C_w
	kN/m	mm²	10^6mm⁴	10^3mm³	mm	10^3mm³	10^6mm⁴	10^3mm³	mm	10^3mm³	10^3mm⁴	10^9mm⁶
WWF550												
*X721**	7.08	92 000	4 390	16 000	218	19 000	1 670	6 080	135	9 470	110 000	100 000
*X620**	6.08	79 100	4 190	15 200	230	17 600	1 660	6 050	145	9 180	83 100	99 900
X503	4.94	64 200	3 610	13 100	237	14 800	1 390	5 040	147	7 610	47 000	86 700
X420	4.12	53 600	3 050	11 100	239	12 400	1 110	4 030	144	6 100	24 700	72 100
X280	2.74	35 600	2 070	7 530	241	8 250	693	2 520	140	3 810	6 410	47 800
WWF500												
*X651**	6.38	83 000	3 200	12 800	196	15 400	1 260	5 030	123	7 850	99 400	60 800
X561**	5.51	71 600	3 070	12 300	207	14 300	1 250	5 000	132	7 590	75 400	60 500
X456	4.47	58 200	2 660	10 600	214	12 100	1 040	4 170	134	6 290	42 700	52 700
X381	3.74	48 600	2 250	9 010	215	10 100	834	3 330	131	5 040	22 500	44 100
X343	3.37	43 800	2 040	8 140	216	9 100	729	2 920	129	4 420	15 400	39 400
X306	3.00	39 000	1 810	7 240	215	8 060	625	2 500	127	3 800	10 200	34 500
X276	2.71	35 200	1 680	6 740	218	7 420	583	2 330	129	3 530	7 920	32 500
X254	2.48	32 300	1 540	6 160	218	6 780	521	2 080	127	3 160	5 820	29 400
X223	2.19	28 500	1 370	5 500	219	6 010	458	1 830	127	2 770	3 970	26 200
X197	1.93	25 200	1 250	4 990	223	5 410	417	1 670	129	2 510	2 870	24 000
WWF450												
X503**	4.93	64 100	2 160	9 620	184	11 400	912	4 050	119	6 150	67 800	34 700
X409	4.01	52 200	1 890	8 380	190	9 640	760	3 380	121	5 100	38 400	30 400
X342	3.35	43 600	1 610	7 150	192	8 100	608	2 700	118	4 090	20 200	25 500
X308	3.02	39 300	1 460	6 480	193	7 290	532	2 360	116	3 580	13 900	22 900
X274	2.69	35 000	1 300	5 770	193	6 470	456	2 030	114	3 080	9 140	20 100
X248	2.43	31 600	1 210	5 380	196	5 960	425	1 890	116	2 860	7 120	18 900
X228	2.23	29 000	1 110	4 920	196	5 450	380	1 690	114	2 560	5 230	17 200
X201	1.97	25 600	991	4 400	197	4 840	334	1 490	114	2 250	3 570	15 300
X177	1.74	22 600	901	4 000	200	4 360	304	1 350	116	2 040	2 580	14 000
WWF400												
X444**	4.35	56 600	1 460	7 300	161	8 770	641	3 200	106	4 870	60 100	18 500
X362	3.55	46 200	1 280	6 410	166	7 480	534	2 670	108	4 030	34 100	16 300
X303	2.97	38 600	1 100	5 500	169	6 300	427	2 130	105	3 230	17 900	13 800
X273	2.67	34 800	1 000	5 000	170	5 680	374	1 870	104	2 840	12 300	12 400
X243	2.38	31 000	894	4 470	170	5 050	320	1 600	102	2 440	8 110	11 000
X220	2.15	28 000	834	4 170	173	4 660	299	1 490	103	2 260	6 320	10 300
X202	1.98	25 700	765	3 830	173	4 260	267	1 330	102	2 020	4 640	9 380
X178	1.74	22 700	686	3 430	174	3 790	235	1 170	102	1 780	3 170	8 390
X157	1.54	20 100	625	3 120	176	3 420	213	1 070	103	1 610	2 290	7 700
WWF350												
X315	3.09	40 200	824	4 710	143	5 580	357	2 040	94.2	3 090	29 800	8 040
X263	2.58	33 600	712	4 070	146	4 730	286	1 630	92.3	2 480	15 700	6 870
X238	2.33	30 300	650	3 720	146	4 280	250	1 430	90.8	2 170	10 800	6 210
X212	2.07	27 000	583	3 330	147	3 810	215	1 230	89.2	1 870	7 070	5 490
X192	1.88	24 400	546	3 120	150	3 520	200	1 140	90.5	1 740	5 520	5 190
X176	1.72	22 400	502	2 870	150	3 220	179	1 020	89.4	1 550	4 060	4 720
X155	1.52	19 800	451	2 580	151	2 870	157	899	89.0	1 360	2 760	4 230
X137	1.35	17 500	412	2 350	153	2 590	143	817	90.4	1 240	2 000	3 890

‡ Nominal depth in millimetres and mass in kilograms per metre.
**Welding does not fully develop the web strength for these sections.
*Maximum piece weight may limit length.

WELDED WIDE FLANGE SHAPES
WWF550-WWF350

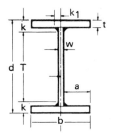

DIMENSIONS AND SURFACE AREAS

Nominal Mass	Depth d	Flange Width b	Flange Thickness t	Web Thickness w	Distances a	Distances T	Distances k	Distances k_1	Distances d-2t	Surface Area (m²) per metre of length Total	Surface Area (m²) per metre of length Minus Top of Top flange
kg/m	mm	mm	mm	mm	mm	mm	mm	mm	mm		
721	550	550	60.0	60.0	245	408	71	40	430	3.18	2.63
620	550	550	60.0	30.0	260	408	71	25	430	3.24	2.69
503	550	550	50.0	20.0	265	428	61	20	450	3.26	2.71
420	550	550	40.0	20.0	265	448	51	20	470	3.26	2.71
280	550	550	25.0	16.0	267	482	34	16	500	3.27	2.72
651	500	500	60.0	60.0	220	358	71	40	380	2.88	2.38
561	500	500	60.0	30.0	235	358	71	25	380	2.94	2.44
456	500	500	50.0	20.0	240	378	61	20	400	2.96	2.46
381	500	500	40.0	20.0	240	398	51	20	420	2.96	2.46
343	500	500	35.0	20.0	240	408	46	20	430	2.96	2.46
306	500	500	30.0	20.0	240	418	41	20	440	2.96	2.46
276	500	500	28.0	16.0	242	426	37	16	444	2.97	2.47
254	500	500	25.0	16.0	242	432	34	16	450	2.97	2.47
223	500	500	22.0	14.0	243	438	31	15	456	2.97	2.47
197	500	500	20.0	11.0	245	442	29	14	460	2.98	2.48
503	450	450	60.0	30.0	210	308	71	25	330	2.64	2.19
409	450	450	50.0	20.0	215	328	61	20	350	2.66	2.21
342	450	450	40.0	20.0	215	348	51	20	370	2.66	2.21
308	450	450	35.0	20.0	215	358	46	20	380	2.66	2.21
274	450	450	30.0	20.0	215	368	41	20	390	2.66	2.21
248	450	450	28.0	16.0	217	376	37	16	394	2.67	2.22
228	450	450	25.0	16.0	217	382	34	16	400	2.67	2.22
201	450	450	22.0	14.0	218	388	31	15	406	2.67	2.22
177	450	450	20.0	11.0	220	392	29	14	410	2.68	2.23
444	400	400	60.0	30.0	185	258	71	25	280	2.34	1.94
362	400	400	50.0	20.0	190	278	61	20	300	2.36	1.96
303	400	400	40.0	20.0	190	298	51	20	320	2.36	1.96
273	400	400	35.0	20.0	190	308	46	20	330	2.36	1.96
243	400	400	30.0	20.0	190	318	41	20	340	2.36	1.96
220	400	400	28.0	16.0	192	326	37	16	344	2.37	1.97
202	400	400	25.0	16.0	192	332	34	16	350	2.37	1.97
178	400	400	22.0	14.0	193	338	31	15	356	2.37	1.97
157	400	400	20.0	11.0	195	342	29	14	360	2.38	1.98
315	350	350	50.0	20.0	165	228	61	20	250	2.06	1.71
263	350	350	40.0	20.0	165	248	51	20	270	2.06	1.71
238	350	350	35.0	20.0	165	258	46	20	280	2.06	1.71
212	350	350	30.0	20.0	165	268	41	20	290	2.06	1.71
192	350	350	28.0	16.0	167	276	37	16	294	2.07	1.72
176	350	350	25.0	16.0	167	282	34	16	300	2.07	1.72
155	350	350	22.0	14.0	168	288	31	15	306	2.07	1.72
137	350	350	20.0	11.0	170	292	29	14	310	2.08	1.73

ROLLED STRUCTURAL SHAPES

General

The majority of rolled shapes available in Canada are produced either to CSA Standard G40.21 350W, to ASTM A572 grade 50, or ASTM A992. All of these grades have similar specified minimum values of yield. See pages 6-3 to 6-9 for more information on steel grades, and pages 6-11 to 6-17 for information on tolerances and mill practice.

The tables of properties and dimensions on pages 6-40 to 6-95 include most of the rolled shapes used in construction. Channels and angles not rolled in Canada are identified in the tables by an asterisk (*) and the note "not available from Canadian mills". (See "Principal sources of Structural Sections", page 6-17 for comments regarding Canadian and non-Canadian sections).

Special shapes, such as rolled Tees, Zees, Bulb Angles, Carbuilding and Shipbuilding Channels are produced by some mills. These shapes are generally rolled only at irregular intervals and usually by special arrangement. Their use should, therefore, be avoided unless the quantity of any one size can warrant a rolling. Properties and dimensions of these shapes may be obtained from the appropriate mill catalogues.

Properties and Dimensions

The basic metric dimensions used to compute properties of the rolled steel shapes are taken from the CAN/CSA-G312.3-M92 Standard "Metric Dimensions for Structural Steel Shapes and Hollow Structural Sections". For shapes not listed in this Standard, they are taken from the ASTM A6 Standard or information provided by the producer.

W shapes rolled by most mills supplying the Canadian market have essentially parallel flanges, although some mills offer W shapes with slightly tapered (approximately 3°) flanges. Also the web-to-flange fillet radius may vary slightly for different mills. Properties for W shapes are calculated using the smallest theoretical fillet radius while dimensions for detailing are adjusted for the largest theoretical fillet radius.

HP shapes are essentially square (equal flange width and overall depth) with parallel flange surfaces and equal thickness flanges and web.

S shapes and Standard channels (C shapes) have tapered flanges with the inside face sloping at approximately $16\frac{2}{3}\%$ (2 in 12). The tabulated thickness is the mean thickness. S shapes are not available from any Canadian producer.

SLB shapes (super light beams) are produced in Canada by MRM and the properties and dimensions are computed from the information supplied by the mill on the same basis as W shapes.

M and MC shapes are essentially shapes that cannot be classified as W, HP, S or C shapes. They are not rolled in Canada, and are usually only produced by a single mill. Availability should be checked before specifying their use. These shapes may be produced with parallel flanges or with tapered flanges of various slopes. Dimensions and properties provided in this Handbook should be suitable for general use, in spite of possible variations in actual dimensions.

Availability of W Shapes

The tables for W shapes highlight in yellow those section sizes which are commonly used and which are generally readily available from inventory sources or from mill rollings. These sizes should be selected for design when small quantities and fast delivery are required. For larger projects requiring significant tonnage of various section sizes, availability will depend on mill rolling cycles, and the most appropriate section size for the design condition should be selected. (See comments on availability of rolled shapes under "Principal sources of Structural Sections" on page 6-17). Further information on the availability of W shapes may be found on CISC's web site (www.cisc-icca.ca).

Angles

Properties and dimensions in metric units are provided for the imperial series hot-rolled angles for equal-leg angles and for unequal-leg angles. The tables include properties and dimensions for single angles and for two equal-leg angles back to back, two unequal-leg angles with short legs back to back, and two unequal-leg angles with long legs back to back. Section properties of hot-rolled angles are based on flat rectangular legs excluding the fillet and roundings.

The properties of angles produced by cold-forming may be up to 7 percent less than the properties of hot-rolled angles of similar size due to the absence of a heel. Designers encountering cold-formed angles should consult the manufacturer's catalogue for the exact dimensions and properties.

Tees cut from W shapes

Properties and dimensions of Tees are based on W shapes assuming the depth of the Tee equal to one-half the depth of the W shape. Tees are not rolled in Canada, and are usually fabricated from W shapes by splitting the web either using rotary shears or flame cutting and subsequently straightening to meet published tolerances.

Structural Section Tables (SST File)

Properties and dimensions of welded and rolled steel shapes are available on computer diskette from the Canadian Institute of Steel Construction (CISC). The SST database is in ASCII text format and contains additional information such as fillet radii, flange slopes, width-to-thickness ratios, CSA-G40.20 groupings, and special torsional and buckling constants. Further details on the SST database and other CISC software may be found on CISC's web site (www.cisc-icca.ca).

W SHAPES
W1100 - W920

PROPERTIES

Designation‡	Dead Load	Area	Axis X-X				Axis Y-Y				Torsional Constant	Warping Constant
			I_x	S_x	r_x	Z_x	I_y	S_y	r_y	Z_y	J	C_w
	kN/m	mm^2	10^6 mm^4	10^3 mm^3	mm	10^3 mm^3	10^6 mm^4	10^3 mm^3	mm	10^3 mm^3	10^3 mm^4	10^9 mm^6
W1100												
x499	4.88	63 400	12 900	23 100	452	26 600	500	2 470	88.8	3 870	31 000	144 000
x432	4.26	55 300	11 300	20 400	452	23 200	435	2 170	88.7	3 370	21 400	124 000
x390	3.84	49 900	10 100	18 300	450	20 800	386	1 930	87.9	3 000	15 700	109 000
x342	3.36	43 600	8 670	15 900	446	18 100	331	1 660	87.1	2 570	10 300	92 900
W1000												
x883	8.67	113 000	21 000	38 400	432	45 300	1 050	4 950	96.6	7 870	185 000	268 000
x749	7.36	95 500	17 300	32 500	426	37 900	852	4 090	94.5	6 470	116 000	212 000
x641	6.29	81 700	14 500	27 700	421	32 100	702	3 410	92.7	5 370	73 500	171 000
x591	5.80	75 300	13 300	25 600	421	29 500	640	3 130	92.2	4 920	59 000	155 000
x554	5.44	70 700	12 300	23 900	418	27 500	592	2 900	91.5	4 550	48 600	142 000
x539	5.29	68 700	12 000	23 400	418	26 800	576	2 830	91.6	4 440	45 300	138 000
x483	4.74	61 500	10 700	20 900	417	23 900	507	2 510	90.8	3 920	33 100	120 000
x443	4.34	56 400	9 670	19 100	414	21 800	455	2 260	89.8	3 530	25 400	107 000
x412	4.04	52 500	9 100	18 100	416	20 500	434	2 160	90.9	3 350	21 400	102 000
x371	3.64	47 300	8 140	16 300	415	18 400	386	1 930	90.3	2 980	15 900	89 600
x321	3.15	40 900	6 960	14 100	413	15 800	331	1 660	90.0	2 550	10 300	76 100
x296	2.91	37 800	6 200	12 600	405	14 300	290	1 450	87.6	2 240	7 640	66 000
W1000												
x583	5.73	74 500	12 500	23 600	409	28 100	334	2 130	67.0	3 480	71 700	82 300
x493	4.85	63 000	10 300	19 900	404	23 400	269	1 740	65.3	2 820	44 100	64 800
x486	4.77	61 900	10 200	19 700	406	23 200	266	1 730	65.5	2 790	42 900	64 100
x414	4.06	52 800	8 520	16 700	402	19 500	217	1 430	64.1	2 300	27 000	51 500
x393	3.85	50 000	8 080	15 900	402	18 500	205	1 350	64.0	2 170	23 300	48 400
x350	3.43	44 500	7 230	14 300	403	16 600	185	1 220	64.4	1 940	17 200	43 200
x314	3.08	40 000	6 440	12 900	401	14 900	162	1 080	63.7	1 710	12 600	37 700
x272	2.67	34 700	5 540	11 200	400	12 800	140	933	63.5	1 470	8 350	32 200
x249	2.44	31 700	4 810	9 820	390	11 300	118	783	60.9	1 240	5 820	26 700
x222	2.18	28 300	4 080	8 410	380	9 800	95.4	636	58.1	1 020	3 900	21 500
W920												
x1188	11.7	151 000	26 100	48 900	415	58 800	1 750	7 660	108	12 200	439 000	401 000
x967	9.48	123 000	20 300	39 500	406	46 800	1 340	6 000	104	9 490	246 000	294 000
x784	7.68	99 800	15 900	32 000	400	37 300	1 030	4 730	102	7 420	136 000	220 000
x653	6.41	83 200	12 900	26 600	394	30 700	830	3 850	99.9	6 020	80 500	172 000
x585	5.74	74 500	11 400	23 800	392	27 400	728	3 410	98.8	5 310	58 900	149 000
x534	5.24	68 000	10 300	21 700	389	24 800	656	3 090	98.2	4 800	45 100	132 000
x488	4.78	62 100	9 350	19 900	388	22 600	590	2 800	97.5	4 340	35 000	118 000
x446	4.39	57 000	8 470	18 200	386	20 600	540	2 550	97.3	3 950	26 700	107 000
x417	4.10	53 300	7 880	17 000	385	19 200	501	2 370	97.0	3 670	21 900	98 700
x387	3.80	49 300	7 180	15 600	382	17 600	453	2 160	95.8	3 330	17 300	88 500
x365	3.58	46 400	6 700	14 600	380	16 500	421	2 010	95.2	3 110	14 400	81 900
x342	3.36	43 600	6 250	13 700	379	15 400	390	1 870	94.6	2 880	11 900	75 500

‡ Nominal depth in millimetres and mass in kilograms per metre

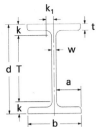

W SHAPES
W1100 - W920

DIMENSIONS AND SURFACE AREAS

Nominal Mass	Theoretical Mass	Depth d	Flange Width b	Flange Thickness t	Web Thickness w	Distances					Surface Area (m²) per metre of length		Imperial Designation
						a	T	k	k_1	d-2t	Total	Minus Top of Top Flange	
kg/m	kg/m	mm	mm	mm	mm	mm	mm	mm	mm	mm			
499	497.8	1 118	405	45.0	25.9	190	977	71	37	1 028	3.80	3.40	W44x335
432	434.1	1 108	402	40.1	22.1	190	977	66	35	1 028	3.78	3.38	W44x290
390	391.6	1 100	400	36.1	20.1	190	977	62	34	1 028	3.76	3.36	W44x262
342	342.6	1 090	400	31.0	18.0	191	977	57	33	1 028	3.74	3.34	W44x230
883	883.4	1 092	424	82.0	45.5	189	861	115	55	928	3.79	3.37	W40x593
749	749.8	1 068	417	70.1	39.1	189	861	103	51	928	3.73	3.31	W40x503
641	641.3	1 048	412	59.9	34.0	189	862	93	49	928	3.68	3.26	W40x431
591	590.9	1 040	409	55.9	31.0	189	862	89	47	928	3.65	3.25	W40x397
554	554.7	1 032	408	52.1	29.5	189	861	85	47	928	3.64	3.23	W40x372
539	539.4	1 030	407	51.1	28.4	189	861	84	46	928	3.63	3.22	W40x362
483	482.9	1 020	404	46.0	25.4	189	861	79	45	928	3.61	3.20	W40x324
443	442.5	1 012	402	41.9	23.6	189	862	75	44	928	3.58	3.18	W40x297
412	412.2	1 008	402	40.0	21.1	190	861	73	42	928	3.58	3.18	W40x277
371	371.2	1 000	400	36.1	19.0	191	861	69	41	928	3.56	3.16	W40x249
321	320.9	990	400	31.0	16.5	192	861	64	40	928	3.55	3.15	W40x215
296	296.4	982	400	27.1	16.5	192	861	60	40	928	3.53	3.13	W40x199
583	584.6	1 056	314	64.0	36.1	139	861	97	50	928	3.30	2.98	W40x392
493	494.3	1 036	309	54.1	31.0	139	861	87	47	928	3.25	2.94	W40x331
486	486.2	1 036	308	54.1	30.0	139	861	87	47	928	3.24	2.94	W40x327
414	414.3	1 020	304	46.0	25.9	139	861	79	45	928	3.20	2.90	W40x278
393	392.7	1 016	303	43.9	24.4	139	862	77	44	928	3.20	2.89	W40x264
350	349.4	1 008	302	40.0	21.1	140	861	73	42	928	3.18	2.88	W40x235
314	314.3	1 000	300	35.9	19.1	140	862	69	41	928	3.16	2.86	W40x211
272	272.3	990	300	31.0	16.5	142	861	64	40	928	3.15	2.85	W40x183
249	248.7	980	300	26.0	16.5	142	861	59	40	928	3.13	2.83	W40x167
222	222.0	970	300	21.1	16.0	142	861	54	40	928	3.11	2.81	W40x149
1 188	1 188.7	1 066	457	109.0	60.5	198	793	137	56	848	3.84	3.38	W36x798
967	966.3	1 028	446	89.9	50.0	198	793	117	51	848	3.74	3.29	W36x650
784	783.2	996	437	73.9	40.9	198	793	101	46	848	3.66	3.22	W36x527
653	653.1	972	431	62.0	34.5	198	793	90	43	848	3.60	3.17	W36x439
585	585.0	960	427	55.9	31.0	198	793	83	42	848	3.57	3.14	W36x393
534	533.9	950	425	51.1	28.4	198	793	79	40	848	3.54	3.12	W36x359
488	487.7	942	422	47.0	25.9	198	793	75	39	848	3.52	3.10	W36x328
446	447.1	933	423	42.7	24.0	200	793	70	38	848	3.51	3.09	W36x300
417	418.0	928	422	39.9	22.5	200	793	67	37	848	3.50	3.08	W36x280
387	387.0	921	420	36.6	21.3	199	793	64	37	848	3.48	3.06	W36x260
365	364.6	916	419	34.3	20.3	199	792	62	36	847	3.47	3.05	W36x245
342	342.4	912	418	32.0	19.3	199	793	60	36	848	3.46	3.04	W36x230

W SHAPES
W920 - W760

PROPERTIES

Designation[‡]	Dead Load	Area	Axis X-X				Axis Y-Y				Torsional Constant	Warping Constant
			I_x	S_x	r_x	Z_x	I_y	S_y	r_y	Z_y	J	C_w
	kN/m	mm^2	10^6 mm^4	10^3 mm^3	mm	10^3 mm^3	10^6 mm^4	10^3 mm^3	mm	10^3 mm^3	10^3 mm^4	10^9 mm^6
W920												
x381	3.74	48 600	6 970	14 700	379	17 000	219	1 410	67.2	2 240	22 000	45 100
x345	3.39	44 000	6 260	13 300	377	15 300	195	1 270	66.6	2 000	16 500	39 800
x313	3.07	39 800	5 480	11 800	371	13 600	170	1 100	65.4	1 750	11 600	34 300
x289	2.83	36 800	5 040	10 900	370	12 600	156	1 020	65.2	1 600	9 230	31 300
x271	2.67	34 600	4 720	10 200	369	11 800	145	946	64.8	1 490	7 690	28 900
x253	2.49	32 300	4 370	9 520	368	11 000	134	874	64.3	1 370	6 260	26 500
x238	2.34	30 400	4 060	8 880	366	10 200	123	806	63.6	1 270	5 140	24 300
x223	2.20	28 600	3 770	8 270	363	9 540	112	738	62.7	1 160	4 220	22 100
x201	1.97	25 600	3 250	7 200	356	8 360	94.4	621	60.7	982	2 910	18 400
W840												
x576	5.65	73 400	10 100	22 100	371	25 500	672	3 270	95.7	5 100	61 300	123 000
x527	5.18	67 200	9 140	20 200	369	23 300	607	2 970	95.1	4 620	47 500	110 000
x473	4.64	60 300	8 120	18 200	367	20 800	537	2 640	94.3	4 100	34 900	95 800
x433	4.25	55 200	7 350	16 600	365	18 900	483	2 390	93.6	3 710	26 800	85 500
x392	3.84	49 900	6 590	15 000	363	17 000	430	2 140	92.8	3 310	20 100	75 300
x359	3.53	45 800	5 910	13 600	359	15 400	389	1 930	92.2	2 980	15 000	67 400
x329	3.23	42 000	5 350	12 400	357	14 000	349	1 740	91.2	2 690	11 500	60 000
x299	2.94	38 100	4 790	11 200	355	12 600	312	1 560	90.4	2 410	8 560	53 200
W840												
x251	2.46	31 900	3 860	9 000	348	10 300	129	884	63.6	1 380	7 360	22 100
x226	2.22	28 900	3 400	7 990	343	9 160	114	774	62.8	1 210	5 150	19 300
x210	2.07	26 900	3 110	7 350	340	8 430	103	700	61.8	1 100	4 060	17 300
x193	1.90	24 700	2 780	6 630	336	7 630	90.3	618	60.5	971	3 060	15 100
x176	1.73	22 400	2 460	5 900	331	6 820	78.2	536	59.1	844	2 220	13 000
W760												
x582	5.71	74 200	8 600	20 400	341	23 700	644	3 250	93.2	5 080	71 500	98 200
x531	5.21	67 600	7 750	18 600	339	21 500	578	2 940	92.4	4 580	55 100	87 000
x484	4.75	61 700	6 980	17 000	336	19 500	517	2 650	91.5	4 120	42 400	76 800
x434	4.25	55 200	6 180	15 200	334	17 400	455	2 350	90.8	3 640	30 900	66 800
x389	3.81	49 500	5 440	13 500	332	15 400	399	2 070	89.8	3 210	22 200	57 800
x350	3.43	44 500	4 860	12 200	330	13 800	355	1 860	89.2	2 860	16 500	50 800
x314	3.08	40 100	4 270	10 900	327	12 300	316	1 640	88.8	2 540	11 600	44 700
x284	2.79	36 200	3 810	9 790	325	11 000	280	1 470	88.0	2 260	8 580	39 300
x257	2.53	32 800	3 420	8 840	323	9 930	250	1 310	87.3	2 020	6 380	34 800
W760												
x220	2.16	28 100	2 780	7 140	315	8 190	94.4	710	58.0	1 110	6 050	13 200
x196	1.93	25 100	2 400	6 240	309	7 170	81.7	610	57.1	959	4 040	11 300
x185	1.81	23 500	2 230	5 820	308	6 690	75.1	563	56.5	884	3 330	10 300
x173	1.70	22 100	2 060	5 400	305	6 210	68.7	515	55.7	810	2 690	9 420
x161	1.57	20 400	1 860	4 900	302	5 660	60.7	457	54.5	720	2 070	8 280
x147	1.44	18 700	1 660	4 410	298	5 100	52.9	399	53.1	631	1 560	7 160
x134	1.31	17 000	1 500	4 010	297	4 630	47.7	361	53.0	568	1 180	6 430

[‡] Nominal depth in millimetres and mass in kilograms per metre

W SHAPES
W920 - W760

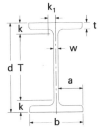

DIMENSIONS AND SURFACE AREAS

Nominal Mass	Theoretical Mass	Depth d	Flange Width b	Flange Thickness t	Web Thickness w	Distances					Surface Area (m²) per metre of length		Imperial Designation
						a	T	k	k_1	d-2t	Total	Minus Top of Top Flange	
kg/m	kg/m	mm	mm	mm	mm	mm	mm	mm	mm	mm			
381	381.4	951	310	43.9	24.4	143	812	69	36	863	3.09	2.78	W36x256
345	345.1	943	308	39.9	22.1	143	812	65	35	863	3.07	2.77	W36x232
313	312.7	932	309	34.5	21.1	144	812	60	35	863	3.06	2.75	W36x210
289	288.6	927	308	32.0	19.4	144	812	58	34	863	3.05	2.74	W36x194
271	271.7	923	307	30.0	18.4	144	812	56	33	863	3.04	2.73	W36x182
253	253.7	919	306	27.9	17.3	144	812	53	33	863	3.03	2.72	W36x170
238	238.3	915	305	25.9	16.5	144	812	51	32	863	3.02	2.71	W36x160
223	224.2	911	304	23.9	15.9	144	812	49	32	863	3.01	2.70	W36x150
201	201.3	903	304	20.1	15.2	144	812	46	32	863	2.99	2.69	W36x135
576	576.1	913	411	57.9	32.0	190	754	79	36	797	3.41	3.00	W33x387
527	527.7	903	409	53.1	29.5	190	754	75	35	797	3.38	2.97	W33x354
473	473.3	893	406	48.0	26.4	190	754	70	33	797	3.36	2.95	W33x318
433	433.3	885	404	43.9	24.4	190	754	65	32	797	3.34	2.93	W33x291
392	391.7	877	401	39.9	22.1	189	754	61	31	797	3.31	2.91	W33x263
359	359.4	868	403	35.6	21.1	191	754	57	31	797	3.31	2.90	W33x241
329	329.4	862	401	32.4	19.7	191	754	54	30	797	3.29	2.89	W33x221
299	299.4	855	400	29.2	18.2	191	754	51	29	797	3.27	2.87	W33x201
251	250.7	859	292	31.0	17.0	138	754	53	29	797	2.85	2.56	W33x169
226	226.7	851	294	26.8	16.1	139	754	48	28	797	2.85	2.55	W33x152
210	210.8	846	293	24.4	15.4	139	754	46	28	797	2.83	2.54	W33x141
193	193.6	840	292	21.7	14.7	139	754	43	27	797	2.82	2.53	W33x130
176	176.0	835	292	18.8	14.0	139	754	40	27	797	2.81	2.52	W33x118
582	582.1	843	396	62.0	34.5	181	676	84	37	719	3.20	2.81	W30x391
531	530.9	833	393	56.9	31.5	181	676	78	36	719	3.18	2.78	W30x357
484	484.6	823	390	52.1	29.0	181	676	74	35	719	3.15	2.76	W30x326
434	433.7	813	387	47.0	25.9	181	676	69	33	719	3.12	2.74	W30x292
389	388.5	803	385	41.9	23.6	181	676	63	32	719	3.10	2.71	W30x261
350	349.5	795	382	38.1	21.1	180	676	60	31	719	3.08	2.69	W30x235
314	314.4	786	384	33.4	19.7	182	676	55	30	719	3.07	2.68	W30x211
284	283.9	779	382	30.1	18.0	182	676	52	29	719	3.05	2.67	W30x191
257	257.6	773	381	27.1	16.6	182	676	49	28	719	3.04	2.66	W30x173
220	220.2	779	266	30.0	16.5	125	676	52	28	719	2.59	2.32	W30x148
196	196.8	770	268	25.4	15.6	126	676	47	28	719	2.58	2.31	W30x132
185	184.8	766	267	23.6	14.9	126	676	45	27	719	2.57	2.30	W30x124
173	173.6	762	267	21.6	14.4	126	676	43	27	719	2.56	2.30	W30x116
161	160.4	758	266	19.3	13.8	126	676	41	27	719	2.55	2.29	W30x108
147	147.1	753	265	17.0	13.2	126	676	39	27	719	2.54	2.27	W30x99
134	133.2	750	264	15.5	11.9	126	676	37	26	719	2.53	2.27	W30x90

W SHAPES
W690 - W610

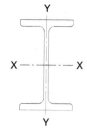

PROPERTIES

Designation‡	Dead Load	Area	Axis X-X				Axis Y-Y				Torsional Constant	Warping Constant
			I_x	S_x	r_x	Z_x	I_y	S_y	r_y	Z_y	J	C_w
	kN/m	mm²	10⁶ mm⁴	10³ mm³	mm	10³ mm³	10⁶ mm⁴	10³ mm³	mm	10³ mm³	10³ mm⁴	10⁹ mm⁶
W690												
x802	7.86	102 000	10 600	25 700	322	30 900	875	4 520	92.6	7 140	203 000	119 000
x548	5.37	69 700	6 710	17 400	310	20 300	543	2 920	88.2	4 560	69 600	68 200
x500	4.90	63 600	6 050	15 900	308	18 400	487	2 640	87.5	4 110	53 700	60 300
x457	4.48	58 200	5 450	14 500	306	16 800	439	2 390	86.8	3 720	41 600	53 600
x419	4.10	53 200	4 940	13 300	305	15 300	395	2 170	86.1	3 370	32 400	47 700
x384	3.76	48 900	4 470	12 200	303	13 900	357	1 970	85.4	3 050	25 200	42 600
x350	3.43	44 600	4 020	11 000	300	12 600	319	1 770	84.6	2 740	19 100	37 600
x323	3.17	41 200	3 700	10 200	300	11 600	294	1 640	84.5	2 530	15 300	34 400
x289	2.82	36 700	3 250	9 100	298	10 300	256	1 440	83.6	2 220	10 900	29 600
x265	2.59	33 700	2 900	8 220	294	9 290	231	1 290	82.9	1 990	8 110	26 400
x240	2.35	30 600	2 610	7 450	292	8 390	206	1 160	82.2	1 780	6 080	23 400
x217	2.14	27 700	2 340	6 740	291	7 570	185	1 040	81.7	1 600	4 560	20 800
W690												
x192	1.88	24 400	1 980	5 640	285	6 460	76.4	602	56.0	941	4 610	8 680
x170	1.67	21 600	1 700	4 900	280	5 620	66.2	517	55.3	809	3 040	7 410
x152	1.49	19 400	1 510	4 380	279	5 000	57.8	455	54.6	710	2 200	6 420
x140	1.37	17 800	1 360	3 980	276	4 550	51.7	407	53.9	636	1 670	5 720
x125	1.23	16 000	1 180	3 500	272	4 010	44.1	349	52.5	546	1 170	4 830
W610												
x551	5.41	70 200	5 580	15 700	282	18 600	484	2 790	83.0	4 380	83 900	49 900
x498	4.89	63 500	4 950	14 200	279	16 700	426	2 480	81.9	3 890	63 300	43 100
x455	4.46	57 900	4 450	12 900	277	15 100	381	2 240	81.1	3 500	48 800	37 900
x415	4.08	52 900	4 000	11 800	275	13 700	343	2 030	80.5	3 160	37 800	33 600
x372	3.65	47 400	3 530	10 600	273	12 200	302	1 800	79.8	2 800	27 700	29 100
x341	3.34	43 400	3 180	9 630	271	11 100	271	1 630	79.0	2 520	21 300	25 800
x307	3.01	39 100	2 840	8 690	269	9 930	240	1 450	78.2	2 240	15 900	22 500
x285	2.80	36 400	2 610	8 060	268	9 180	221	1 340	77.9	2 070	12 800	20 500
x262	2.56	33 300	2 360	7 360	266	8 350	198	1 210	77.2	1 870	9 910	18 300
x241	2.37	30 800	2 150	6 780	264	7 670	184	1 120	77.4	1 730	7 700	16 800
x217	2.14	27 800	1 910	6 070	262	6 850	163	995	76.7	1 530	5 600	14 700
x195	1.92	24 900	1 680	5 400	260	6 070	142	871	75.6	1 340	3 970	12 700
x174	1.71	22 200	1 470	4 780	257	5 360	124	761	74.7	1 170	2 800	10 900
x155	1.52	19 700	1 290	4 220	256	4 730	108	666	73.9	1 020	1 950	9 450
W610												
x153	1.51	19 600	1 250	4 020	253	4 600	50.0	437	50.5	682	2 950	4 470
x140	1.37	17 900	1 120	3 630	250	4 150	45.1	392	50.3	613	2 180	3 990
x125	1.23	15 900	985	3 220	249	3 670	39.3	343	49.7	535	1 540	3 450
x113	1.11	14 400	875	2 880	246	3 290	34.3	300	48.7	469	1 120	2 990
x101	0.997	13 000	764	2 530	243	2 900	29.5	259	47.7	404	781	2 550
x91	0.892	11 600	667	2 230	240	2 560	24.8	219	46.3	343	577	2 130
x84	0.824	10 700	613	2 060	239	2 360	22.6	200	45.9	313	462	1 930

‡ Nominal depth in millimetres and mass in kilograms per metre

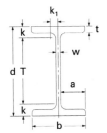

W SHAPES
W690 - W610

DIMENSIONS AND SURFACE AREAS

Nominal Mass	Theoretical Mass	Depth	Flange Width	Flange Thickness	Web Thickness	Distances					Surface Area (m²) per metre of length		Imperial Designation
		d	b	t	w	a	T	k	k_1	d-2t	Total	Minus Top of Top Flange	
kg/m	kg/m	mm	mm	mm	mm	mm	mm	mm	mm	mm			
802	801.4	826	387	89.9	50.0	169	603	111	45	646	3.10	2.71	W27x539
548	547.5	772	372	63.0	35.1	168	603	85	38	646	2.96	2.59	W27x368
500	499.3	762	369	57.9	32.0	169	603	79	36	646	2.94	2.57	W27x336
457	457.0	752	367	53.1	29.5	169	603	75	35	646	2.91	2.55	W27x307
419	418.0	744	364	49.0	26.9	169	603	71	33	646	2.89	2.53	W27x281
384	383.5	736	362	45.0	24.9	169	603	67	32	646	2.87	2.51	W27x258
350	349.9	728	360	40.9	23.1	168	603	62	32	646	2.85	2.49	W27x235
323	323.2	722	359	38.1	21.1	169	603	60	31	646	2.84	2.48	W27x217
289	287.9	714	356	34.0	19.0	169	603	56	30	646	2.81	2.46	W27x194
265	264.5	706	358	30.2	18.4	170	603	52	29	646	2.81	2.45	W27x178
240	239.9	701	356	27.4	16.8	170	603	49	28	646	2.79	2.44	W27x161
217	217.8	695	355	24.8	15.4	170	602	46	28	645	2.78	2.42	W27x146
192	191.4	702	254	27.9	15.5	119	603	49	28	646	2.39	2.14	W27x129
170	169.9	693	256	23.6	14.5	121	603	45	27	646	2.38	2.13	W27x114
152	152.1	688	254	21.1	13.1	120	603	43	27	646	2.37	2.11	W27x102
140	139.8	684	254	18.9	12.4	121	603	40	26	646	2.36	2.11	W27x94
125	125.5	678	253	16.3	11.7	121	602	38	26	645	2.34	2.09	W27x84
551	551.2	711	347	69.1	38.6	154	530	91	39	573	2.73	2.39	W24x370
498	498.3	699	343	63.0	35.1	154	530	85	38	573	2.70	2.36	W24x335
455	454.2	689	340	57.9	32.0	154	530	79	36	573	2.67	2.33	W24x306
415	415.6	679	338	53.1	29.5	154	530	75	35	573	2.65	2.31	W24x279
372	372.3	669	335	48.0	26.4	154	530	70	33	573	2.63	2.29	W24x250
341	340.4	661	333	43.9	24.4	154	530	65	32	573	2.61	2.27	W24x229
307	307.3	653	330	39.9	22.1	154	530	61	31	573	2.58	2.25	W24x207
285	285.4	647	329	37.1	20.6	154	530	59	30	573	2.57	2.24	W24x192
262	261.2	641	327	34.0	19.0	154	530	56	30	573	2.55	2.23	W24x176
241	241.7	635	329	31.0	17.9	156	530	53	29	573	2.55	2.22	W24x162
217	217.9	628	328	27.7	16.5	156	530	49	28	573	2.54	2.21	W24x146
195	195.6	622	327	24.4	15.4	156	530	46	28	573	2.52	2.19	W24x131
174	174.3	616	325	21.6	14.0	156	530	43	27	573	2.50	2.18	W24x117
155	154.9	611	324	19.0	12.7	156	530	41	26	573	2.49	2.17	W24x104
153	153.6	623	229	24.9	14.0	108	530	46	27	573	2.13	1.91	W24x103
140	140.1	617	230	22.2	13.1	108	530	44	27	573	2.13	1.90	W24x94
125	125.1	612	229	19.6	11.9	109	530	41	26	573	2.12	1.89	W24x84
113	113.4	608	228	17.3	11.2	108	530	39	26	573	2.11	1.88	W24x76
101	101.7	603	228	14.9	10.5	109	530	36	25	573	2.10	1.87	W24x68
91	90.9	598	227	12.7	9.7	109	530	34	25	573	2.08	1.86	W24x61
84	84.0	596	226	11.7	9.0	109	530	33	25	573	2.08	1.85	W24x56

6-45

W SHAPES
W610 - W460

PROPERTIES

Designation‡	Dead Load	Area	Axis X-X				Axis Y-Y				Torsional Constant	Warping Constant
			I_x	S_x	r_x	Z_x	I_y	S_y	r_y	Z_y	J	C_w
	kN/m	mm²	10^6 mm⁴	10^3 mm³	mm	10^3 mm³	10^6 mm⁴	10^3 mm³	mm	10^3 mm³	10^3 mm⁴	10^9 mm⁶
W610												
x92	0.905	11 800	646	2 140	234	2 510	14.4	161	35.0	258	710	1 250
x82	0.803	10 400	560	1 870	232	2 200	12.1	136	34.0	218	488	1 040
W530												
x300	2.94	38 200	2 210	7 550	241	8 670	225	1 410	76.7	2 180	17 000	16 600
x272	2.67	34 600	1 970	6 840	239	7 810	202	1 270	76.4	1 960	12 800	14 700
x248	2.42	31 400	1 770	6 220	238	7 060	180	1 140	75.7	1 760	9 770	13 000
x219	2.15	27 900	1 510	5 390	233	6 110	157	986	75.0	1 520	6 420	11 000
x196	1.93	25 000	1 340	4 840	231	5 460	139	877	74.4	1 350	4 700	9 640
x182	1.78	23 100	1 240	4 480	231	5 040	127	808	74.2	1 240	3 740	8 820
x165	1.62	21 100	1 110	4 060	230	4 550	114	726	73.4	1 110	2 830	7 790
x150	1.48	19 200	1 010	3 710	229	4 150	103	659	73.2	1 010	2 160	7 030
W530												
x138	1.36	17 600	861	3 140	221	3 610	38.7	362	46.9	569	2 500	2 670
x123	1.21	15 700	761	2 800	220	3 210	33.8	319	46.4	499	1 800	2 310
x109	1.07	13 900	667	2 480	219	2 830	29.5	280	46.1	437	1 260	2 000
x101	0.995	12 900	617	2 300	219	2 620	26.9	256	45.6	400	1 020	1 820
x92	0.907	11 800	552	2 070	217	2 360	23.8	228	44.9	355	762	1 590
x82	0.808	10 500	478	1 810	214	2 070	20.3	194	44.0	303	529	1 340
x72	0.705	9 150	401	1 530	209	1 760	16.2	156	42.0	245	342	1 060
W530												
x85	0.831	10 800	485	1 810	212	2 100	12.6	152	34.2	242	737	849
x74	0.733	9 520	411	1 550	208	1 810	10.4	125	33.1	200	480	692
x66	0.645	8 370	351	1 340	205	1 560	8.57	104	32.0	166	320	565
W460												
x260	2.55	33 100	1 440	5 650	208	6 530	163	1 130	70.1	1 740	14 100	8 950
x235	2.30	29 900	1 270	5 080	206	5 840	145	1 010	69.5	1 550	10 500	7 790
x213	2.09	27 100	1 140	4 620	205	5 270	129	909	69.1	1 400	7 970	6 890
x193	1.90	24 600	1 020	4 190	204	4 750	115	816	68.5	1 250	6 030	6 060
x177	1.74	22 600	910	3 780	201	4 280	105	735	68.2	1 130	4 400	5 440
x158	1.55	20 100	796	3 350	199	3 770	91.4	643	67.4	989	3 110	4 670
x144	1.42	18 400	726	3 080	199	3 450	83.6	591	67.4	906	2 440	4 230
x128	1.26	16 400	637	2 730	197	3 050	73.3	520	67.0	796	1 710	3 670
x113	1.11	14 400	556	2 400	196	2 670	63.3	452	66.3	691	1 180	3 150
W460												
x106	1.04	13 500	488	2 080	190	2 390	25.1	259	43.2	405	1 460	1 260
x97	0.947	12 300	445	1 910	190	2 180	22.8	237	43.1	368	1 130	1 140
x89	0.876	11 400	409	1 770	190	2 010	20.9	218	42.9	339	905	1 040
x82	0.804	10 400	370	1 610	188	1 830	18.6	195	42.2	303	690	918
x74	0.728	9 450	332	1 460	188	1 650	16.6	175	41.9	271	516	813
x67	0.660	8 570	296	1 300	186	1 480	14.5	153	41.2	238	377	708
x61	0.590	7 660	255	1 130	182	1 290	12.2	129	39.9	201	260	587

‡ Nominal depth in millimetres and mass in kilograms per metre

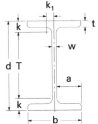

W SHAPES
W610 - W460

DIMENSIONS AND SURFACE AREAS

Nominal Mass	Theoretical Mass	Depth d	Flange Width b	Flange Thickness t	Web Thickness w	Distances					Surface Area (m²) per metre of length		Imperial Designation
						a	T	k	k_1	d-2t	Total	Minus Top of Top Flange	
kg/m	kg/m	mm	mm	mm	mm	mm	mm	mm	mm	mm			
92	92.3	603	179	15.0	10.9	84	530	37	25	573	1.90	1.72	W24x62
82	81.9	599	178	12.8	10.0	84	530	34	25	573	1.89	1.71	W24x55
300	299.5	585	319	41.4	23.1	148	466	59	28	502	2.40	2.08	W21x201
272	271.9	577	318	37.6	21.1	148	466	56	27	502	2.38	2.07	W21x182
248	246.6	571	315	34.5	19.0	148	466	53	26	502	2.36	2.05	W21x166
219	218.9	560	318	29.2	18.3	150	466	47	26	502	2.36	2.04	W21x147
196	196.5	554	316	26.3	16.5	150	465	44	25	501	2.34	2.02	W21x132
182	181.7	551	315	24.4	15.2	150	466	42	24	502	2.33	2.02	W21x122
165	165.3	546	313	22.2	14.0	150	466	40	24	502	2.32	2.00	W21x111
150	150.6	543	312	20.3	12.7	150	466	38	23	502	2.31	2.00	W21x101
138	138.3	549	214	23.6	14.7	100	466	42	24	502	1.92	1.71	W21x93
123	123.2	544	212	21.2	13.1	99	466	39	23	502	1.91	1.70	W21x83
109	109.0	539	211	18.8	11.6	100	465	37	22	501	1.90	1.69	W21x73
101	101.4	537	210	17.4	10.9	100	466	35	22	502	1.89	1.68	W21x68
92	92.5	533	209	15.6	10.2	99	466	34	22	502	1.88	1.67	W21x62
82	82.4	528	209	13.3	9.5	100	470	29	19	501	1.87	1.66	W21x55
72	71.8	524	207	10.9	8.9	99	471	27	19	502	1.86	1.65	W21x48
85	84.7	535	166	16.5	10.3	78	466	35	22	502	1.71	1.55	W21x57
74	74.7	529	166	13.6	9.7	78	466	32	21	502	1.70	1.54	W21x50
66	65.7	525	165	11.4	8.9	78	466	29	21	502	1.69	1.53	W21x44
260	259.9	509	289	40.4	22.6	133	389	60	29	428	2.13	1.84	W18x175
235	234.8	501	287	36.6	20.6	133	389	56	28	428	2.11	1.82	W18x158
213	212.7	495	285	33.5	18.5	133	389	53	27	428	2.09	1.81	W18x143
193	193.3	489	283	30.5	17.0	133	389	50	27	428	2.08	1.79	W18x130
177	177.3	482	286	26.9	16.6	135	389	46	26	428	2.07	1.79	W18x119
158	157.7	476	284	23.9	15.0	135	389	43	26	428	2.06	1.77	W18x106
144	144.5	472	283	22.1	13.6	135	389	42	25	428	2.05	1.77	W18x97
128	128.4	467	282	19.6	12.2	135	389	39	24	428	2.04	1.76	W18x86
113	113.0	463	280	17.3	10.8	135	389	37	23	428	2.02	1.74	W18x76
106	105.7	469	194	20.6	12.6	91	394	37	22	428	1.69	1.49	W18x71
97	96.5	466	193	19.0	11.4	91	395	36	21	428	1.68	1.49	W18x65
89	89.3	463	192	17.7	10.5	91	394	34	20	428	1.67	1.48	W18x60
82	81.9	460	191	16.0	9.9	91	395	33	20	428	1.66	1.47	W18x55
74	74.2	457	190	14.5	9.0	91	395	31	20	428	1.66	1.47	W18x50
67	67.3	454	190	12.7	8.5	91	403	25	15	429	1.65	1.46	W18x45
61	60.1	450	189	10.8	8.1	90	403	23	15	428	1.64	1.45	W18x41

W SHAPES
W460 - W360

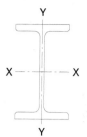

PROPERTIES

Designation‡	Dead Load	Area	Axis X-X				Axis Y-Y				Torsional Constant	Warping Constant
			I_x	S_x	r_x	Z_x	I_y	S_y	r_y	Z_y	J	C_w
	kN/m	mm²	10⁶ mm⁴	10³ mm³	mm	10³ mm³	10⁶ mm⁴	10³ mm³	mm	10³ mm³	10³ mm⁴	10⁹ mm⁶
W460												
x68	0.672	8 730	297	1 290	184	1 490	9.41	122	32.8	192	509	463
x60	0.584	7 590	255	1 120	183	1 280	7.96	104	32.4	163	335	388
x52	0.510	6 630	212	943	179	1 090	6.34	83.4	30.9	131	210	306
W410												
x149	1.46	19 000	619	2 870	180	3 250	77.7	586	63.9	900	3 220	3 200
x132	1.30	16 800	538	2 530	179	2 850	67.4	512	63.3	785	2 260	2 730
x114	1.12	14 600	462	2 200	178	2 460	57.2	439	62.6	671	1 490	2 300
x100	0.977	12 700	398	1 920	177	2 130	49.5	381	62.5	581	994	1 960
W410												
x85	0.833	10 800	315	1 510	171	1 720	18.0	199	40.8	310	924	717
x74	0.735	9 540	275	1 330	170	1 510	15.6	173	40.4	269	636	614
x67	0.662	8 600	245	1 200	169	1 360	13.8	154	40.1	239	468	540
x60	0.584	7 580	216	1 060	169	1 190	12.0	135	39.9	209	327	468
x54	0.524	6 800	186	923	165	1 050	10.1	114	38.5	177	225	388
W410												
x46	0.453	5 890	156	772	163	884	5.14	73.4	29.5	115	192	197
x39	0.384	4 990	126	634	159	730	4.04	57.6	28.4	90.6	110	154
W360												
x1086	10.7	139 000	5 960	20 900	207	27 200	1 960	8 650	119	13 400	605 000	96 700
x990	9.72	126 000	5 190	18 900	203	24 300	1 730	7 740	117	12 000	469 000	82 000
x900	8.85	115 000	4 500	17 000	198	21 600	1 530	6 940	116	10 700	364 000	69 200
x818	8.03	104 000	3 920	15 300	194	19 300	1 360	6 200	114	9 560	278 000	58 900
x744	7.30	94 800	3 420	13 700	190	17 200	1 200	5 550	112	8 550	214 000	50 200
x677	6.65	86 300	2 990	12 400	186	15 300	1 070	4 990	111	7 680	164 000	43 100
W360												
x634	6.22	80 800	2 740	11 600	184	14 200	983	4 630	110	7 120	138 000	38 700
x592	5.81	75 500	2 500	10 800	182	13 100	902	4 280	109	6 570	114 000	34 800
x551	5.40	70 100	2 260	9 940	180	12 100	825	3 950	108	6 050	92 500	31 000
x509	5.00	64 900	2 050	9 170	178	11 000	754	3 630	108	5 550	73 900	27 700
x463	4.54	59 000	1 800	8 280	175	9 880	670	3 250	107	4 980	56 500	23 900
x421	4.14	53 700	1 600	7 510	172	8 880	601	2 940	106	4 490	43 400	20 800
x382	3.75	48 700	1 410	6 790	170	7 960	536	2 640	105	4 030	32 800	18 200
x347	3.40	44 200	1 250	6 140	168	7 140	481	2 380	104	3 630	24 800	15 900
x314	3.07	39 900	1 100	5 530	166	6 370	426	2 120	103	3 240	18 500	13 800
x287	2.82	36 600	997	5 070	165	5 810	388	1 940	103	2 960	14 500	12 300
x262	2.58	33 500	894	4 620	163	5 260	350	1 760	102	2 680	11 000	11 000
x237	2.32	30 100	788	4 150	162	4 690	310	1 570	102	2 390	8 180	9 500
x216	2.12	27 500	711	3 790	161	4 260	283	1 430	101	2 180	6 320	8 520

‡ Nominal depth in millimetres and mass in kilograms per metre

When subject to tension, bolted connections are preferred for these sections.

W SHAPES
W460 - W360

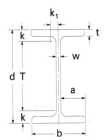

DIMENSIONS AND SURFACE AREAS

Nominal Mass	Theo-retical Mass	Depth	Flange Width	Flange Thick-ness	Web Thick-ness	Distances					Surface Area (m²) per metre of length		Imperial Designation
		d	b	t	w	a	T	k	k₁	d-2t	Total	Minus Top of Top Flange	
kg/m	kg/m	mm	mm	mm	mm	mm	mm	mm	mm	mm			
68	68.5	459	154	15.4	9.1	72	395	32	20	428	1.52	1.36	W18x46
60	59.6	455	153	13.3	8.0	73	395	30	19	428	1.51	1.35	W18x40
52	52.0	450	152	10.8	7.6	72	395	28	19	428	1.49	1.34	W18x35
149	149.3	431	265	25.0	14.9	125	348	42	23	381	1.89	1.63	W16x100
132	132.1	425	263	22.2	13.3	125	347	39	22	381	1.88	1.61	W16x89
114	114.5	420	261	19.3	11.6	125	348	36	21	381	1.86	1.60	W16x77
100	99.6	415	260	16.9	10.0	125	348	34	20	381	1.85	1.59	W16x67
85	85.0	417	181	18.2	10.9	85	347	35	21	381	1.54	1.36	W16x57
74	74.9	413	180	16.0	9.7	85	348	33	20	381	1.53	1.35	W16x50
67	67.5	410	179	14.4	8.8	85	348	31	20	381	1.52	1.34	W16x45
60	59.5	407	178	12.8	7.7	85	348	30	19	381	1.51	1.33	W16x40
54	53.4	403	177	10.9	7.5	85	348	28	19	381	1.50	1.32	W16x36
46	46.2	403	140	11.2	7.0	67	347	28	19	381	1.35	1.21	W16x31
39	39.2	399	140	8.8	6.4	67	348	26	18	381	1.35	1.21	W16x26
1 086	1 087.8	569	454	125.0	78.0	188	286	142	54	319	2.80	2.34	W14x730
990	991.0	550	448	115.0	71.9	188	287	132	51	320	2.75	2.30	W14x665
900	902.1	531	442	106.0	65.9	188	286	123	48	319	2.70	2.26	W14x605
818	819.0	514	437	97.0	60.5	188	287	114	45	320	2.66	2.22	W14x550
744	744.2	498	432	88.9	55.6	188	287	106	43	320	2.61	2.18	W14x500
677	677.8	483	428	81.5	51.2	188	287	98	41	320	2.58	2.15	W14x455
634	634.3	474	424	77.1	47.6	188	277	99	44	320	2.55	2.12	W14x426
592	592.6	465	421	72.3	45.0	188	277	94	43	320	2.52	2.10	W14x398
551	550.6	455	418	67.6	42.0	188	277	89	41	320	2.50	2.08	W14x370
509	509.4	446	416	62.7	39.1	188	278	84	40	321	2.48	2.06	W14x342
463	462.8	435	412	57.4	35.8	188	277	79	38	320	2.45	2.03	W14x311
421	421.6	425	409	52.6	32.8	188	277	74	36	320	2.42	2.01	W14x283
382	382.3	416	406	48.0	29.8	188	277	70	35	320	2.40	1.99	W14x257
347	346.9	407	404	43.7	27.2	188	277	65	34	320	2.38	1.97	W14x233
314	313.3	399	401	39.6	24.9	188	277	61	32	320	2.35	1.95	W14x211
287	287.5	393	399	36.6	22.6	188	277	58	31	320	2.34	1.94	W14x193
262	262.7	387	398	33.3	21.1	188	277	55	31	320	2.32	1.93	W14x176
237	236.2	380	395	30.2	18.9	188	277	52	29	320	2.30	1.91	W14x159
216	216.3	375	394	27.7	17.3	188	277	49	29	320	2.29	1.90	W14x145

W SHAPES
W360 - W310

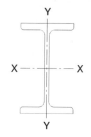

PROPERTIES

Designation‡	Dead Load	Area	Axis X-X				Axis Y-Y				Torsional Constant	Warping Constant
			I_x	S_x	r_x	Z_x	I_y	S_y	r_y	Z_y	J	C_w
	kN/m	mm²	10⁶ mm⁴	10³ mm³	mm	10³ mm³	10⁶ mm⁴	10³ mm³	mm	10³ mm³	10³ mm⁴	10⁹ mm⁶
W360												
x196	1.93	25 000	636	3 420	159	3 840	229	1 220	95.6	1 860	5 130	6 830
x179	1.76	22 800	574	3 120	159	3 480	207	1 110	95.2	1 680	3 910	6 120
x162	1.59	20 600	515	2 830	158	3 140	186	1 000	94.9	1 520	2 940	5 430
x147	1.45	18 800	463	2 570	157	2 840	167	904	94.3	1 370	2 230	4 840
x134	1.31	17 100	415	2 330	156	2 560	151	817	94.0	1 240	1 680	4 310
W360												
x122	1.19	15 500	365	2 010	154	2 270	61.5	478	63.0	732	2 110	1 790
x110	1.08	14 000	331	1 840	154	2 060	55.7	435	63.0	664	1 600	1 610
x101	0.993	12 900	301	1 690	153	1 880	50.6	397	62.7	605	1 250	1 450
x91	0.891	11 600	267	1 510	152	1 680	44.8	353	62.3	538	914	1 270
W360												
x79	0.777	10 100	226	1 280	150	1 430	24.2	236	48.9	362	811	687
x72	0.701	9 100	201	1 150	149	1 280	21.4	210	48.5	322	601	600
x64	0.627	8 140	178	1 030	148	1 140	18.8	186	48.1	284	436	524
W360												
x57	0.555	7 210	160	896	149	1 010	11.1	129	39.3	199	333	331
x51	0.496	6 440	141	796	148	893	9.68	113	38.8	174	237	285
x45	0.441	5 730	122	691	146	778	8.18	95.7	37.8	148	159	239
W360												
x39	0.384	4 980	102	580	143	662	3.75	58.6	27.4	91.6	150	110
x33	0.321	4 170	82.6	473	141	541	2.91	45.8	26.4	71.8	85.3	84.3
W310												
x500	4.91	63 700	1 690	7 910	163	9 880	494	2 910	88.0	4 490	101 000	15 300
x454	4.45	57 800	1 480	7 130	160	8 820	436	2 600	86.8	4 000	77 100	13 100
x415	4.07	52 900	1 300	6 450	157	7 900	391	2 340	86.0	3 610	59 500	11 300
x375	3.68	47 700	1 130	5 770	154	7 000	344	2 080	84.9	3 210	44 900	9 570
x342	3.37	43 700	1 010	5 260	152	6 330	310	1 890	84.2	2 910	34 900	8 420
x313	3.07	39 900	896	4 790	150	5 720	277	1 700	83.3	2 620	27 000	7 350
W310												
x283	2.78	36 000	787	4 310	148	5 100	246	1 530	82.6	2 340	20 300	6 330
x253	2.48	32 200	682	3 830	146	4 490	215	1 350	81.6	2 060	14 800	5 370
x226	2.22	28 900	596	3 420	144	3 970	189	1 190	81.0	1 830	10 800	4 620
x202	1.99	25 800	520	3 050	142	3 510	166	1 050	80.2	1 610	7 730	3 960
x179	1.75	22 800	445	2 670	140	3 050	144	919	79.5	1 400	5 370	3 340
x158	1.54	20 000	386	2 360	139	2 670	125	805	78.9	1 220	3 770	2 840
x143	1.40	18 200	348	2 150	138	2 420	113	729	78.6	1 110	2 860	2 540
x129	1.27	16 500	308	1 940	137	2 160	100	652	78.0	991	2 130	2 220
x118	1.15	15 000	275	1 750	136	1 950	90.2	588	77.6	893	1 600	1 970
x107	1.05	13 600	248	1 590	135	1 760	81.2	531	77.2	806	1 210	1 760
x97	0.950	12 300	222	1 440	134	1 590	72.9	478	76.9	725	909	1 560

‡ Nominal depth in millimetres and mass in kilograms per metre

W SHAPES
W360 - W310

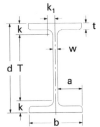

DIMENSIONS AND SURFACE AREAS

Nominal Mass	Theoretical Mass	Depth	Flange Width	Flange Thickness	Web Thickness	Distances					Surface Area (m²) per metre of length		Imperial Designation
		d	b	t	w	a	T	k	k₁	d-2t	Total	Minus Top of Top Flange	
kg/m	kg/m	mm	mm	mm	mm	mm	mm	mm	mm	mm			
196	196.5	372	374	26.2	16.4	179	277	48	28	320	2.21	1.83	W14x132
179	179.2	368	373	23.9	15.0	179	277	45	28	320	2.20	1.83	W14x120
162	161.9	364	371	21.8	13.3	179	277	43	27	320	2.19	1.81	W14x109
147	147.5	360	370	19.8	12.3	179	277	41	26	320	2.18	1.81	W14x99
134	133.9	356	369	18.0	11.2	179	277	40	26	320	2.17	1.80	W14x90
122	121.7	363	257	21.7	13.0	122	281	41	24	320	1.73	1.47	W14x82
110	110.2	360	256	19.9	11.4	122	282	39	24	320	1.72	1.47	W14x74
101	101.2	357	255	18.3	10.5	122	282	38	23	320	1.71	1.46	W14x68
91	90.8	353	254	16.4	9.5	122	282	36	23	320	1.70	1.45	W14x61
79	79.2	354	205	16.8	9.4	98	282	36	23	320	1.51	1.30	W14x53
72	71.5	350	204	15.1	8.6	98	281	34	22	320	1.50	1.29	W14x48
64	63.9	347	203	13.5	7.7	98	281	33	22	320	1.49	1.29	W14x43
57	56.6	358	172	13.1	7.9	82	307	26	15	332	1.39	1.22	W14x38
51	50.6	355	171	11.6	7.2	82	307	24	15	332	1.38	1.21	W14x34
45	45.0	352	171	9.8	6.9	82	307	22	15	332	1.37	1.20	W14x30
39	39.1	353	128	10.7	6.5	61	306	23	14	332	1.21	1.08	W14x26
33	32.7	349	127	8.5	5.8	61	307	21	14	332	1.19	1.07	W14x22
500	500.4	427	340	75.1	45.1	147	243	92	38	277	2.12	1.78	W12x336
454	453.9	415	336	68.7	41.3	147	244	85	36	278	2.09	1.76	W12x305
415	415.1	403	334	62.7	38.9	148	244	79	35	278	2.06	1.73	W12x279
375	374.7	391	330	57.2	35.4	147	243	74	33	277	2.03	1.70	W12x252
342	343.2	382	328	52.6	32.6	148	243	69	32	277	2.01	1.68	W12x230
313	313.3	374	325	48.3	30.0	148	244	65	30	277	1.99	1.66	W12x210
283	282.9	365	322	44.1	26.9	148	243	61	29	277	1.96	1.64	W12x190
253	252.9	356	319	39.6	24.4	147	243	56	27	277	1.94	1.62	W12x170
226	226.7	348	317	35.6	22.1	147	242	53	27	277	1.92	1.60	W12x152
202	202.6	341	315	31.8	20.1	147	243	49	26	277	1.90	1.59	W12x136
179	178.7	333	313	28.1	18.0	148	242	46	25	277	1.88	1.57	W12x120
158	157.4	327	310	25.1	15.5	147	242	43	24	277	1.86	1.55	W12x106
143	143.1	323	309	22.9	14.0	148	242	40	23	277	1.85	1.55	W12x96
129	129.6	318	308	20.6	13.1	147	242	38	22	277	1.84	1.53	W12x87
118	117.5	314	307	18.7	11.9	148	242	36	22	277	1.83	1.53	W12x79
107	106.9	311	306	17.0	10.9	148	242	34	21	277	1.82	1.52	W12x72
97	96.8	308	305	15.4	9.9	148	242	33	21	277	1.82	1.51	W12x65

W SHAPES
W310 - W250

PROPERTIES

Designation[‡]	Dead Load	Area	Axis X-X				Axis Y-Y				Torsional Constant	Warping Constant
			I_x	S_x	r_x	Z_x	I_y	S_y	r_y	Z_y	J	C_w
	kN/m	mm^2	10^6 mm^4	10^3 mm^3	mm	10^3 mm^3	10^6 mm^4	10^3 mm^3	mm	10^3 mm^3	10^3 mm^4	10^9 mm^6
W310												
x86	0.847	11 000	198	1 280	134	1 420	44.5	351	63.6	533	874	961
x79	0.774	10 000	177	1 150	133	1 280	39.9	314	63.0	478	655	847
W310												
x74	0.730	9 480	165	1 060	132	1 190	23.4	229	49.7	350	743	505
x67	0.655	8 500	145	948	131	1 060	20.7	203	49.3	310	543	439
x60	0.584	7 590	129	849	130	941	18.3	180	49.1	275	396	384
W310												
x52	0.514	6 670	119	750	134	841	10.3	123	39.2	189	308	238
x45	0.438	5 690	99.2	634	132	708	8.55	103	38.8	158	191	195
x39	0.380	4 940	85.1	549	131	610	7.27	88.1	38.4	135	126	164
x31	0.310	4 020	67.2	439	129	488	5.45	66.4	36.8	102	73.1	121
W310												
x33	0.322	4 180	65.0	415	125	480	1.92	37.6	21.4	59.6	122	43.8
x28	0.278	3 610	54.3	351	123	407	1.58	31.0	20.9	49.2	75.7	35.6
x24	0.234	3 040	42.7	280	119	328	1.16	22.9	19.5	36.7	42.5	25.7
x21	0.207	2 690	37.0	244	117	287	0.983	19.5	19.1	31.2	29.4	21.7
W250												
x167	1.64	21 300	300	2 080	119	2 430	98.8	746	68.1	1 140	6 310	1 630
x149	1.46	19 000	259	1 840	117	2 130	86.2	656	67.4	1 000	4 510	1 390
x131	1.29	16 700	221	1 610	115	1 850	74.5	571	66.8	870	3 120	1 160
x115	1.13	14 600	189	1 410	114	1 600	64.1	495	66.2	753	2 130	976
x101	0.992	12 900	164	1 240	113	1 400	55.5	432	65.6	656	1 490	829
x89	0.879	11 400	143	1 100	112	1 230	48.4	378	65.1	574	1 040	713
x80	0.786	10 200	126	982	111	1 090	43.1	338	65.0	513	757	623
x73	0.715	9 280	113	891	110	985	38.8	306	64.6	463	575	553
W250												
x67	0.659	8 550	104	806	110	901	22.2	218	51.0	332	625	324
x58	0.571	7 420	87.3	693	108	770	18.8	186	50.4	283	409	268
x49	0.481	6 250	70.6	572	106	633	15.1	150	49.2	228	241	211
W250												
x45	0.441	5 720	71.1	534	111	602	7.03	95.1	35.1	146	261	113
x39	0.379	4 920	60.1	459	110	513	5.94	80.8	34.7	124	169	93.4
x33	0.321	4 170	48.9	379	108	424	4.73	64.7	33.7	99.5	98.5	73.2
x24	0.240	3 110	34.7	275	106	307	3.26	44.9	32.4	69.0	40.1	49.5
W250												
x28	0.279	3 630	40.0	307	105	353	1.78	34.8	22.1	54.7	96.7	27.7
x25	0.249	3 230	34.2	266	103	307	1.49	29.2	21.5	46.2	65.2	23.0
x22	0.220	2 850	28.9	227	101	263	1.23	24.0	20.7	38.1	43.4	18.7
x18	0.175	2 270	22.4	179	99.3	207	0.913	18.1	20.0	28.6	22.4	13.8

‡ Nominal depth in millimetres and mass in kilograms per metre

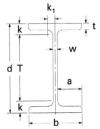

W SHAPES
W310 - W250

DIMENSIONS AND SURFACE AREAS

Nominal Mass	Theoretical Mass	Depth	Flange Width	Flange Thickness	Web Thickness	Distances					Surface Area (m^2) per metre of length		Imperial Designation
		d	b	t	w	a	T	k	k_1	d-2t	Total	Minus Top of Top Flange	
kg/m	kg/m	mm	mm	mm	mm	mm	mm	mm	mm	mm			
86	86.3	310	254	16.3	9.1	122	243	34	20	277	1.62	1.36	W12x58
79	78.9	306	254	14.6	8.8	123	242	32	20	277	1.61	1.36	W12x53
74	74.4	310	205	16.3	9.4	98	243	34	21	277	1.42	1.22	W12x50
67	66.7	306	204	14.6	8.5	98	242	32	20	277	1.41	1.21	W12x45
60	59.6	303	203	13.1	7.5	98	242	31	20	277	1.40	1.20	W12x40
52	52.4	318	167	13.2	7.6	80	263	27	17	292	1.29	1.12	W12x35
45	44.6	313	166	11.2	6.6	80	262	25	16	291	1.28	1.11	W12x30
39	38.7	310	165	9.7	5.8	80	262	24	16	291	1.27	1.10	W12x26
31	31.6	306	164	7.4	5.0	80	263	22	15	291	1.26	1.09	W12x21
33	32.8	313	102	10.8	6.6	48	270	22	13	291	1.02	0.919	W12x22
28	28.4	309	102	8.9	6.0	48	269	20	12	291	1.01	0.912	W12x19
24	23.8	305	101	6.7	5.6	48	270	18	12	292	1.00	0.902	W12x16
21	21.1	303	101	5.7	5.1	48	270	17	12	292	1.00	0.899	W12x14
167	167.4	289	265	31.8	19.2	123	191	49	26	225	1.60	1.33	W10x112
149	148.9	282	263	28.4	17.3	123	190	46	25	225	1.58	1.32	W10x100
131	131.1	275	261	25.1	15.4	123	190	43	24	225	1.56	1.30	W10x88
115	114.8	269	259	22.1	13.5	123	190	40	23	225	1.55	1.29	W10x77
101	101.2	264	257	19.6	11.9	123	190	37	22	225	1.53	1.28	W10x68
89	89.6	260	256	17.3	10.7	123	191	35	21	225	1.52	1.27	W10x60
80	80.1	256	255	15.6	9.4	123	190	33	21	225	1.51	1.26	W10x54
73	72.9	253	254	14.2	8.6	123	190	32	20	225	1.50	1.25	W10x49
67	67.1	257	204	15.7	8.9	98	191	33	20	226	1.31	1.11	W10x45
58	58.2	252	203	13.5	8.0	98	190	31	20	225	1.30	1.10	W10x39
49	49.0	247	202	11.0	7.4	97	190	28	20	225	1.29	1.09	W10x33
45	44.9	266	148	13.0	7.6	70	211	28	17	240	1.11	0.961	W10x30
39	38.7	262	147	11.2	6.6	70	211	26	16	240	1.10	0.952	W10x26
33	32.7	258	146	9.1	6.1	70	211	24	16	240	1.09	0.942	W10x22
24	24.4	253	145	6.4	5.0	70	221	16	10	240	1.08	0.931	W10x16
28	28.5	260	102	10.0	6.4	48	221	20	11	240	0.915	0.813	W10x19
25	25.3	257	102	8.4	6.1	48	221	18	11	240	0.910	0.808	W10x17
22	22.4	254	102	6.9	5.8	48	221	17	11	240	0.904	0.802	W10x15
18	17.9	251	101	5.3	4.8	48	221	15	11	240	0.896	0.795	W10x12

W SHAPES
W200 - W100

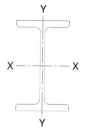

PROPERTIES

Designation[‡]	Dead Load	Area	Axis X-X				Axis Y-Y				Torsional Constant	Warping Constant
			I_x	S_x	r_x	Z_x	I_y	S_y	r_y	Z_y	J	C_w
	kN/m	mm^2	10^6 mm^4	10^3 mm^3	mm	10^3 mm^3	10^6 mm^4	10^3 mm^3	mm	10^3 mm^3	10^3 mm^4	10^9 mm^6
W200												
x100	0.973	12 600	113	987	94.6	1 150	36.6	349	53.8	533	2 060	386
x86	0.848	11 000	94.4	851	92.6	978	31.4	300	53.4	458	1 370	318
x71	0.699	9 070	76.3	707	91.7	800	25.4	246	52.9	374	801	250
x59	0.580	7 530	60.9	580	89.9	650	20.4	199	52.1	302	452	196
x52	0.510	6 620	52.5	509	89.0	566	17.8	175	51.9	265	314	167
x46	0.448	5 820	45.2	445	88.1	492	15.3	151	51.3	229	213	141
W200												
x42	0.406	5 280	40.6	396	87.7	442	9.00	108	41.3	165	215	84.0
x36	0.350	4 540	34.1	340	86.7	376	7.64	92.6	41.0	141	139	69.5
W200												
x31	0.308	4 000	31.4	299	88.6	335	4.10	61.1	32.0	93.8	119	40.9
x27	0.261	3 390	25.8	249	87.3	279	3.30	49.6	31.2	76.1	71.3	32.5
x21	0.208	2 710	19.8	195	85.5	218	2.51	37.8	30.5	58.0	35.9	24.3
W200												
x22	0.220	2 860	20.0	194	83.6	222	1.42	27.8	22.3	43.7	56.6	13.9
x19	0.191	2 480	16.6	163	81.7	187	1.15	22.6	21.6	35.6	36.2	11.1
x15	0.147	1 900	12.7	127	81.8	145	0.869	17.4	21.4	27.1	17.6	8.24
W150												
x37	0.364	4 730	22.2	274	68.5	310	7.07	91.8	38.7	140	192	40.0
x30	0.292	3 790	17.1	218	67.3	244	5.56	72.6	38.3	111	100	30.3
x22	0.219	2 840	12.0	159	65.1	176	3.87	50.9	36.9	77.5	41.5	20.4
W150												
x24	0.235	3 050	13.4	168	66.3	191	1.83	35.8	24.5	55.2	92.3	10.2
x18	0.176	2 280	9.15	120	63.3	136	1.26	24.7	23.5	38.2	36.9	6.70
x14	0.133	1 730	6.85	91.3	63.0	102	0.918	18.4	23.0	28.3	16.8	4.79
x13	0.124	1 610	6.11	82.5	61.7	92.7	0.818	16.4	22.6	25.2	13.3	4.19
W130												
x28	0.276	3 580	10.9	167	55.3	190	3.81	59.6	32.6	90.8	128	13.8
x24	0.232	3 010	8.80	139	54.1	156	3.11	49.0	32.1	74.6	76.9	10.8
W100												
x19	0.190	2 470	4.76	89.8	43.9	103	1.61	31.2	25.5	47.9	62.9	3.79

‡ Nominal depth in millimetres and mass in kilograms per metre

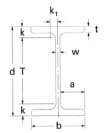

W SHAPES
W200 - W100

DIMENSIONS AND SURFACE AREAS

Nominal Mass	Theoretical Mass	Depth	Flange Width	Flange Thickness	Web Thickness	Distances					Surface Area (m²) per metre of length		Imperial Designation
		d	b	t	w	a	T	k	k_1	d-2t	Total	Minus Top of Top Flange	
kg/m	kg/m	mm	mm	mm	mm	mm	mm	mm	mm	mm			
100	99.2	229	210	23.7	14.5	98	158	35	17	182	1.27	1.06	W8x67
86	86.5	222	209	20.6	13.0	98	157	32	17	181	1.25	1.05	W8x58
71	71.2	216	206	17.4	10.2	98	158	29	15	181	1.24	1.03	W8x48
59	59.1	210	205	14.2	9.1	98	158	26	15	182	1.22	1.02	W8x40
52	52.0	206	204	12.6	7.9	98	157	24	14	181	1.21	1.01	W8x35
46	45.7	203	203	11.0	7.2	98	158	23	14	181	1.20	1.00	W8x31
42	41.4	205	166	11.8	7.2	79	158	24	14	181	1.06	0.894	W8x28
36	35.6	201	165	10.2	6.2	79	157	22	13	181	1.05	0.885	W8x24
31	31.4	210	134	10.2	6.4	64	167	22	13	190	0.943	0.809	W8x21
27	26.6	207	133	8.4	5.8	64	167	20	13	190	0.934	0.801	W8x18
21	21.2	203	133	6.4	5.0	64	171	16	10	190	0.928	0.795	W8x14
22	22.4	206	102	8.0	6.2	48	171	18	11	190	0.808	0.706	W8x15
19	19.4	203	102	6.5	5.8	48	171	16	11	190	0.802	0.700	W8x13
15	15.0	200	100	5.2	4.3	48	170	15	10	190	0.791	0.691	W8x10
37	37.1	162	154	11.6	8.1	73	121	21	12	139	0.924	0.770	W6x25
30	29.8	157	153	9.3	6.6	73	120	18	11	138	0.913	0.760	W6x20
22	22.3	152	152	6.6	5.8	73	121	16	11	139	0.900	0.748	W6x15
24	24.0	160	102	10.3	6.6	48	124	18	10	139	0.715	0.613	W6x16
18	17.9	153	102	7.1	5.8	48	123	15	9	139	0.702	0.600	W6x12
14	13.6	150	100	5.5	4.3	48	121	15	10	139	0.691	0.591	W6x9
13	12.6	148	100	4.9	4.3	48	120	14	10	138	0.687	0.587	W6x8.5
28	28.1	131	128	10.9	6.9	61	90	20	11	109	0.760	0.632	W5x19
24	23.6	127	127	9.1	6.1	60	90	19	11	109	0.750	0.623	W5x16
19	19.4	106	103	8.8	7.1	48	73	17	10	88	0.610	0.507	W4x13

HP SHAPES

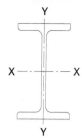

PROPERTIES

Designation[‡]	Dead Load	Area	Axis X-X				Axis Y-Y				Torsional Constant	Warping Constant
			I_x	S_x	r_x	Z_x	I_y	S_y	r_y	Z_y	J	C_w
	kN/m	mm^2	$10^6 mm^4$	$10^3 mm^3$	mm	$10^3 mm^3$	$10^6 mm^4$	$10^3 mm^3$	mm	$10^3 mm^3$	$10^3 mm^4$	$10^9 mm^6$
HP360												
x174	1.71	22 200	508	2 820	152	3 180	184	973	91.1	1 490	3 310	5 330
x152	1.49	19 400	439	2 470	150	2 770	159	845	90.5	1 290	2 240	4 540
x132	1.30	16 800	375	2 140	149	2 380	135	724	89.6	1 110	1 490	3 800
x108	1.06	13 800	303	1 750	148	1 940	108	585	88.6	891	832	3 000
HP310												
x125	1.22	15 900	270	1 730	130	1 960	88.2	566	74.5	870	1 760	1 910
x110	1.08	14 100	237	1 540	130	1 730	77.1	497	74.0	763	1 240	1 650
x94	0.916	11 900	196	1 300	129	1 450	63.9	415	73.3	635	764	1 340
x79	0.768	9 980	163	1 090	128	1 210	52.6	344	72.6	525	460	1 090
HP250												
x85	0.837	10 900	123	968	106	1 090	42.3	325	62.3	500	829	606
x62	0.614	7 980	87.5	711	105	792	30.0	234	61.3	358	339	415
HP200												
x54	0.525	6 820	49.8	488	85.5	552	16.7	162	49.5	249	321	155

[‡] Nominal depth in millimetres and mass in kilograms per metre

HP SHAPES

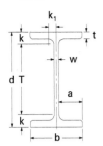

DIMENSIONS AND SURFACE AREAS

Nominal Mass	Theo-retical Mass	Depth	Flange Width	Flange Thick-ness	Web Thick-ness	Distances					Surface Area (m²) per metre of length		Imperial Designation
		d	b	t	w	a	T	k	k₁	d-2t	Total	Minus Top of Top Flange	
kg/m	kg/m	mm	mm	mm	mm	mm	mm	mm	mm	mm			
174	173.9	361	378	20.4	20.4	179	288	37	25	320	2.19	1.82	HP14x117
152	152.2	356	376	17.9	17.9	179	288	34	24	320	2.18	1.80	HP14x102
132	132.1	351	373	15.6	15.6	179	287	32	23	320	2.16	1.79	HP14x89
108	108.1	346	370	12.8	12.8	179	288	29	22	320	2.15	1.78	HP14x73
125	124.7	312	312	17.4	17.4	147	245	34	24	277	1.84	1.53	HP12x84
110	110.5	308	310	15.5	15.4	147	244	32	23	277	1.83	1.52	HP12x74
94	93.4	303	308	13.1	13.1	147	243	30	22	277	1.81	1.50	HP12x63
79	78.3	299	306	11.0	11.0	148	244	28	21	277	1.80	1.49	HP12x53
85	85.3	254	260	14.4	14.4	123	192	31	23	225	1.52	1.26	HP10x57
62	62.6	246	256	10.7	10.5	123	191	27	21	225	1.50	1.24	HP10x42
54	53.5	204	207	11.3	11.3	98	157	24	17	181	1.21	1.01	HP8x36

M SHAPES

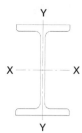

PROPERTIES

Designation[‡]	Dead Load	Area	Axis X-X				Axis Y-Y				Torsional Constant	Warping Constant
			I_x	S_x	r_x	Z_x	I_y	S_y	r_y	Z_y	J	C_w
	kN/m	mm²	10^6 mm⁴	10^3 mm³	mm	10^3 mm³	10^6 mm⁴	10^3 mm³	mm	10^3 mm³	10^3 mm⁴	10^9 mm⁶
M310												
x17.6	0.173	2 240	29.8	197	115	234	0.453	11.6	14.2	18.8	21.5	10.0
x16.1	0.157	2 040	27.0	179	115	213	0.413	10.6	14.2	17.0	16.5	9.03
M250												
x13.4	0.132	1 710	15.8	126	96.1	149	0.274	8.05	12.7	13.0	14.0	4.10
x11.9	0.116	1 510	14.0	112	96.2	132	0.242	7.12	12.7	11.4	9.84	3.61
M200												
x9.7	0.095 4	1 240	7.51	75.5	77.8	88.1	0.153	5.29	11.1	8.45	8.94	1.45
M130												
x28.1	0.276	3 580	10.0	158	52.9	181	3.27	51.5	30.2	81.9	144	11.1
M100												
x8.9	0.087 5	1 140	1.96	40.9	41.6	45.2	0.663	13.4	24.2	20.3	7.00	1.40

‡ Nominal depth in millimetres and mass in kilograms per metre

Note: These shapes have been soft-converted from Imperial to SI dimensions.

M SHAPES

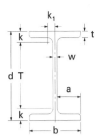

DIMENSIONS AND SURFACE AREAS

Nominal Mass	Theo-retical Mass	Depth d	Flange Width b	Flange Thick-ness t	Web Thick-ness w	Distances				Surface Area (m²) per metre of length		Imperial Designation
						a	T	k	k₁	Total	Minus Top of Top Flange	
kg/m	kg/m	mm	mm	mm	mm	mm	mm	mm	mm			
17.6	17.6	303	78	5.7	4.5	37	276	14	9	0.909	0.831	M12x11.8
16.1	16.1	301	78	5.2	4.1	37	275	13	9	0.906	0.828	M12x10.8
13.4	13.4	250	68	5.2	4.0	32	224	13	9	0.764	0.696	M10x9
11.9	11.8	249	68	4.6	3.5	32	224	13	9	0.763	0.695	M10x8
9.7	9.7	199	58	4.7	3.4	27	172	13	9	0.623	0.565	M8x6.5
28.1	28.1	127	127	10.6	8.0	60	85	21	12	0.746	0.619	M5x18.9
8.9	8.9	96	99	4.1	3.3	48	73	12	8	0.581	0.482	M4x6

SUPER LIGHT BEAMS
SLB100 - SLB75

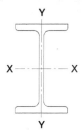

PROPERTIES

Designation[‡]	Dead Load	Area	Axis X-X				Axis Y-Y				Torsional Constant	Warping Constant
			I_x	S_x	r_x	Z_x	I_y	S_y	r_y	Z_y	J	C_w
	kN/m	mm^2	10^6 mm^4	10^3 mm^3	mm	10^3 mm^3	10^6 mm^4	10^3 mm^3	mm	10^3 mm^3	10^3 mm^4	10^9 mm^6
SLB100												
x5.4	0.053 4	694	1.17	23.0	41.1	26.4	0.098 8	3.46	11.9	5.48	2.85	0.239
x4.8	0.047 3	615	1.11	21.9	42.5	24.5	0.098 6	3.45	12.7	5.36	2.12	0.238
SLB75												
x4.5	0.044 4	577	0.613	16.1	32.6	17.9	0.106	3.72	13.6	5.73	2.28	0.140
x4.3	0.042 1	546	0.581	15.2	32.6	16.9	0.099 1	3.47	13.5	5.35	1.92	0.132

‡ Nominal depth in millimetres and mass in kilograms per metre

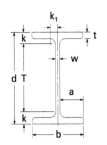

SUPER LIGHT BEAMS
SLB100 - SLB75

DIMENSIONS AND SURFACE AREAS

Nominal Mass	Theo-retical Mass	Depth d	Flange Width b	Flange Thick-ness t	Web Thick-ness w	Distances				Surface Area (m²) per metre of length		Imperial Designation
						a	T	k	k_1	Total	Minus Top of Top Flange	
kg/m	kg/m	mm	mm	mm	mm	mm	mm	mm	mm			
5.4	5.44	101.6	57.15	3.28	3.17	27	81	10	6	0.425	0.368	SLB4x3.64
4.8	4.83	101.6	57.15	3.28	2.34	27	81	10	6	0.427	0.370	SLB4x3.20
4.5	4.53	76.2	57.15	3.53	2.29	27	55	10	5	0.376	0.319	SLB3x3.05
4.3	4.29	76.2	57.15	3.30	2.21	27	56	10	5	0.377	0.319	SLB3x2.90

S SHAPES
S610 - S200

PROPERTIES

Designation‡	Dead Load	Area	Axis X-X				Axis Y-Y				Torsional Constant	Warping Constant
			I_x	S_x	r_x	Z_x	I_y	S_y	r_y	Z_y	J	C_w
	kN/m	mm²	10^6 mm⁴	10^3 mm³	mm	10^3 mm³	10^6 mm⁴	10^3 mm³	mm	10^3 mm³	10^3 mm⁴	10^9 mm⁶
S610												
x180	1.77	22 900	1 310	4 220	239	5 020	34.7	340	38.9	591	5 330	3 070
x158	1.55	20 100	1 220	3 940	247	4 580	32.4	324	40.2	544	4 210	2 860
S610												
x149	1.46	18 900	996	3 270	229	3 930	20.1	218	32.5	392	3 150	1 730
x134	1.32	17 100	939	3 080	234	3 650	18.9	209	33.3	366	2 520	1 640
x119	1.17	15 200	879	2 880	241	3 360	17.9	201	34.3	341	2 030	1 540
S510												
x143	1.40	18 200	700	2 710	196	3 250	21.2	231	34.1	409	3 500	1 280
x128	1.26	16 400	658	2 550	200	3 010	19.6	219	34.6	377	2 770	1 190
S510												
x112	1.09	14 200	532	2 090	194	2 500	12.5	155	29.7	273	1 910	745
x98.2	0.965	12 500	497	1 960	199	2 290	11.7	148	30.6	252	1 490	698
S460												
x104	1.03	13 300	387	1 690	170	2 050	10.3	129	27.7	237	1 740	495
x81.4	0.800	10 400	335	1 470	180	1 720	8.77	115	29.1	198	986	423
S380												
x74	0.732	9 500	203	1 060	146	1 270	6.60	92.3	26.4	163	884	220
x64	0.627	8 150	187	980	151	1 140	6.11	87.3	27.4	149	641	204
S310												
x74	0.730	9 470	127	833	116	1 000	6.60	94.9	26.4	168	1 160	137
x60.7	0.595	7 730	113	744	121	869	5.67	85.3	27.1	144	723	118
S310												
x52	0.512	6 650	95.8	629	120	736	4.16	64.5	25.0	111	450	88.2
x47	0.465	6 040	91.1	597	123	690	3.94	62.1	25.5	105	374	83.6
S250												
x52	0.513	6 660	61.6	485	96.1	584	3.56	56.5	23.1	102	541	51.9
x38	0.371	4 820	51.4	405	103	465	2.84	48.2	24.3	81.1	251	41.5
S200												
x34	0.336	4 370	27.0	266	78.6	316	1.81	34.2	20.4	60.2	229	16.8
x27	0.270	3 500	24.0	237	82.9	272	1.59	31.1	21.3	52.2	140	14.7

‡ Designation consists of nominal depth in millimetres and nominal mass in kilograms per metre.
  For sections formerly available from Canadian mills, the nominal mass has been rounded to the nearest kg/m.

Note: These shapes have been soft-converted from Imperial to SI dimensions.

S SHAPES
S610 - S200

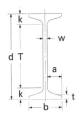

DIMENSIONS AND SURFACE AREAS

Nominal Mass	Theoretical Mass	Depth d	Flange Width b	Flange Mean Thickness t	Web Thickness w	Distances			Surface Area (m²) per metre of length		Imperial Designation
						a	T	k	Total	Minus Top of Top Flange	
kg/m	kg/m	mm	mm	mm	mm	mm	mm	mm			
180	180.0	622	204	27.7	20.3	92	523	50	2.02	1.82	S24x121
158	157.8	622	200	27.7	15.7	92	523	50	2.01	1.81	S24x106
149	148.7	610	184	22.1	18.9	83	524	43	1.92	1.73	S24x100
134	134.4	610	181	22.1	15.9	83	524	43	1.91	1.73	S24x90
119	119.1	610	178	22.1	12.7	83	524	43	1.91	1.73	S24x80
143	143.0	516	183	23.4	20.3	81	427	44	1.72	1.54	S20x96
128	128.6	516	179	23.4	16.8	81	427	44	1.71	1.54	S20x86
112	111.4	508	162	20.2	16.1	73	427	40	1.63	1.47	S20x75
98.2	98.4	508	159	20.2	12.8	73	427	40	1.63	1.47	S20x66
104	104.7	457	159	17.6	18.1	70	383	37	1.51	1.35	S18x70
81.4	81.6	457	152	17.6	11.7	70	383	37	1.50	1.35	S18x54.7
74	74.6	381	143	15.8	14.0	65	314	34	1.31	1.16	S15x50
64	64.0	381	140	15.8	10.4	65	314	34	1.30	1.16	S15x42.9
74	74.4	305	139	16.7	17.4	61	236	35	1.13	0.992	S12x50
60.7	60.6	305	133	16.7	11.7	61	236	35	1.12	0.986	S12x40.8
52	52.2	305	129	13.8	10.9	59	247	29	1.10	0.975	S12x35
47	47.4	305	127	13.8	8.9	59	247	29	1.10	0.973	S12x31.8
52	52.3	254	126	12.5	15.1	55	200	27	0.982	0.856	S10x35
38	37.8	254	118	12.5	7.9	55	200	27	0.964	0.846	S10x25.4
34	34.3	203	106	10.8	11.2	47	156	24	0.808	0.702	S8x23
27	27.5	203	102	10.8	6.9	48	156	24	0.800	0.698	S8x18.4

S SHAPES
S150 - S75

PROPERTIES

Designation‡	Dead Load	Area	Axis X-X				Axis Y-Y				Torsional Constant	Warping Constant
			I_x	S_x	r_x	Z_x	I_y	S_y	r_y	Z_y	J	C_w
	kN/m	mm²	10^6 mm⁴	10^3 mm³	mm	10^3 mm³	10^6 mm⁴	10^3 mm³	mm	10^3 mm³	10^3 mm⁴	10^9 mm⁶
S150												
x26	0.252	3 270	10.9	144	57.8	173	0.981	21.6	17.3	38.7	155	5.01
x19	0.183	2 370	9.19	121	62.2	139	0.776	18.2	18.1	30.5	70.1	3.96
S130												
x15	0.146	1 890	5.12	80.6	52.0	92.8	0.508	13.4	16.4	22.2	47.4	1.79
S100												
x14.1	0.139	1 800	2.85	55.8	39.7	66.5	0.376	10.6	14.4	18.3	49.9	0.842
x11	0.113	1 460	2.56	50.2	41.8	57.9	0.324	9.53	14.9	15.8	30.3	0.725
S75												
x11	0.110	1 430	1.22	32.0	29.2	38.7	0.249	7.77	13.2	13.5	38.2	0.299
x8	0.083	1 070	1.04	27.4	31.2	31.9	0.190	6.43	13.3	10.6	18.3	0.228

‡ Designation consists of nominal depth in millimetres and nominal mass in kilograms per metre.
  For sections formerly available from Canadian mills, the nominal mass has been rounded to the nearest kg/m.

Note: These shapes have been soft-converted from Imperial to SI dimensions.

S SHAPES
S150 - S75

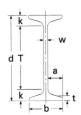

DIMENSIONS AND SURFACE AREAS

Nominal Mass	Theoretical Mass	Depth	Flange Width	Flange Mean Thickness	Web Thickness	Distances			Surface Area (m^2) per metre of length		Imperial Designation
		d	b	t	w	a	T	k	Total	Minus Top of Top Flange	
kg/m	kg/m	mm	mm	mm	mm	mm	mm	mm			
26	25.7	152	91	9.1	11.8	40	111	20	0.644	0.553	S6x17.25
19	18.6	152	85	9.1	5.9	40	111	20	0.632	0.547	S6x12.5
15	14.8	127	76	8.3	5.4	35	89	19	0.547	0.471	S5x10
14.1	14.2	102	71	7.4	8.3	31	68	17	0.471	0.400	S4x9.5
11	11.5	102	68	7.4	4.9	32	68	17	0.466	0.398	S4x7.7
11	11.2	76	64	6.6	8.9	28	45	16	0.390	0.326	S3x7.5
8	8.4	76	59	6.6	4.3	27	45	16	0.379	0.320	S3x5.7

STANDARD CHANNELS (C SHAPES)

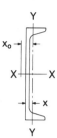

PROPERTIES

Designation†	Dead Load	Total Area	Axis X-X			Axis Y-Y				Shear Centre	Torsional Constant	Warping Constant
			I_x	S_x	r_x	I_y	S_y	r_y	x	x_o	J	C_w
	kN/m	mm²	10⁶mm⁴	10³mm³	mm	10⁶mm⁴	10³mm³	mm	mm	mm	10³mm⁴	10⁹mm⁶
C380												
X74*	0.730	9 480	168	881	133	4.60	62.4	22.0	20.3	34.9	1 100	131
X60*	0.583	7 570	145	760	138	3.84	55.5	22.5	19.7	39.1	603	109
X50*	0.495	6 430	131	687	143	3.39	51.4	23.0	20.0	42.5	421	95.2
C310												
X45*	0.438	5 690	67.3	442	109	2.12	33.6	19.3	17.0	32.4	361	39.9
X37*	0.363	4 720	59.9	393	113	1.85	30.9	19.8	17.1	35.9	223	34.6
X31*	0.302	3 920	53.5	351	117	1.59	28.2	20.1	17.5	39.2	153	29.3
C250												
X45*	0.437	5 670	42.8	337	86.9	1.60	26.8	16.8	16.3	25.3	508	20.5
X37	0.365	4 750	37.9	299	89.4	1.40	24.3	17.1	15.7	28.1	289	18.2
X30	0.291	3 780	32.7	257	93.0	1.16	21.5	17.5	15.3	31.3	153	15.0
X23	0.221	2 880	27.8	219	98.2	0.922	18.8	17.9	15.9	35.6	86.4	11.7
C230												
X30	0.292	3 800	25.5	222	81.9	1.01	19.3	16.3	14.8	27.7	179	10.5
X22	0.219	2 840	21.3	186	86.6	0.806	16.8	16.8	14.9	32.2	86.4	8.33
X20	0.195	2 530	19.8	173	88.6	0.716	15.6	16.8	15.1	33.6	69.3	7.35
C200												
X28	0.274	3 560	18.2	180	71.6	0.825	16.6	15.2	14.4	25.2	182	6.67
X21	0.200	2 600	14.9	147	75.8	0.627	13.9	15.5	14.0	29.1	77.2	5.04
X17	0.167	2 170	13.5	133	78.8	0.544	12.8	15.8	14.5	31.9	54.0	4.34
C180												
X22	0.214	2 780	11.3	127	63.7	0.568	12.8	14.3	13.5	24.5	110	3.47
X18	0.178	2 310	10.0	113	65.9	0.476	11.4	14.3	13.2	26.5	66.8	2.90
X15	0.142	1 850	8.86	99.6	69.3	0.405	10.3	14.8	13.8	30.2	41.4	2.46
C150												
X19	0.189	2 450	7.12	93.7	53.9	0.425	10.3	13.2	12.9	22.2	99.4	1.84
X16	0.152	1 980	6.22	81.9	56.1	0.351	9.13	13.3	12.6	24.6	53.8	1.53
X12	0.118	1 540	5.36	70.6	59.1	0.279	7.93	13.5	12.8	27.6	30.8	1.21
C130												
X13	0.131	1 700	3.66	57.6	46.5	0.252	7.20	12.2	11.9	22.3	45.2	0.746
X10	0.097	1 260	3.09	48.6	49.5	0.195	6.14	12.5	12.2	26.0	22.6	0.580
C100												
X11	0.106	1 370	1.91	37.4	37.3	0.174	5.52	11.3	11.5	20.9	34.1	0.320
X9	0.092	1 190	1.77	34.6	38.5	0.158	5.18	11.5	11.6	22.9	23.1	0.293
X8	0.079	1 020	1.61	31.6	39.7	0.132	4.65	11.4	11.6	24.1	16.6	0.246
C75												
X9	0.087	1 120	0.846	22.3	27.4	0.123	4.31	10.5	11.4	19.4	29.7	0.118
X7	0.072	933	0.749	19.7	28.3	0.0960	3.67	10.1	10.8	20.2	17.4	0.0934
X6	0.059	763	0.670	17.6	29.6	0.0775	3.21	10.1	10.9	22.2	10.9	0.0767

* Not available from Canadian mills.
† For sections available from Canadian mills, the nominal mass has been rounded to the nearest kg/m.
  Designation consists of nominal depth in millimetres by nominal mass in kilograms per metre.
Note: These shapes have been soft converted from Imperial to SI dimensions.

STANDARD CHANNELS (C SHAPES)

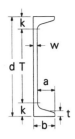

DIMENSIONS AND SURFACE AREAS

Nominal Mass	Theo-retical Mass	Depth	Flange Width	Flange Mean Thickness	Web Thickness	Distances			Surface Area (m²) per metre of length	
		d	b	t	w	a	T	k	Total	Minus Top of Top Flange
kg/m	kg/m	mm	mm	mm	mm	mm	mm	mm		
74	74.4	381	94	16.5	18.2	76	311	35	1.10	1.01
60	59.4	381	89	16.5	13.2	76	311	35	1.09	1.00
50	50.5	381	86	16.5	10.2	76	311	35	1.09	1.00
45	44.7	305	80	12.7	13.0	67	250	27	0.904	0.824
37	37.1	305	77	12.7	9.8	67	250	27	0.898	0.821
31	30.8	305	74	12.7	7.2	67	250	27	0.892	0.818
45	44.5	254	76	11.1	17.1	59	206	24	0.778	0.702
37	37.3	254	73	11.1	13.4	60	206	24	0.773	0.700
30	29.6	254	69	11.1	9.6	59	206	24	0.765	0.696
23	22.6	254	65	11.1	6.1	59	206	24	0.756	0.691
30	29.8	229	67	10.5	11.4	56	183	23	0.703	0.636
22	22.3	229	63	10.5	7.2	56	183	23	0.696	0.633
20	19.8	229	61	10.5	5.9	55	183	23	0.690	0.629
28	27.9	203	64	9.9	12.4	52	159	22	0.637	0.573
21	20.4	203	59	9.9	7.7	51	159	22	0.627	0.568
17	17.0	203	57	9.9	5.6	51	159	22	0.623	0.566
22	21.9	178	58	9.3	10.6	47	136	21	0.567	0.509
18	18.2	178	55	9.3	8.0	47	136	21	0.560	0.505
15	14.5	178	53	9.3	5.3	48	136	21	0.557	0.504
19	19.2	152	54	8.7	11.1	43	113	20	0.498	0.444
16	15.5	152	51	8.7	8.0	43	113	20	0.492	0.441
12	12.1	152	48	8.7	5.1	43	113	20	0.486	0.438
13	13.3	127	47	8.1	8.3	39	90	18	0.425	0.378
10	9.9	127	44	8.1	4.8	39	90	19	0.420	0.376
11	10.8	102	43	7.5	8.2	35	67	17	0.360	0.317
9	9.4	102	42	7.5	6.3	36	67	17	0.359	0.317
8	8.0	102	40	7.5	4.7	35	67	17	0.355	0.315
9	8.8	76	40	6.9	9.0	31	44	16	0.294	0.254
7	7.3	76	37	6.9	6.6	30	44	16	0.287	0.250
6	6.0	76	35	6.9	4.3	31	44	16	0.283	0.248

MISCELLANEOUS CHANNELS
MC460 - MC200

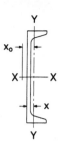

PROPERTIES

Designation†	Dead Load	Total Area	Axis X-X			Axis Y-Y				Shear Centre	Torsional Constant	Warping Constant
			I_x	S_x	r_x	I_y	S_y	r_y	x	x_o	J	C_w
	kN/m	mm²	10^6mm⁴	10^3mm³	mm	10^6mm⁴	10^3mm³	mm	mm	mm	10^3mm⁴	10^9mm⁶
MC460												
X86	0.852	11 100	283	1 240	160	7.45	87.6	26.0	22.0	39.7	1 190	291
X77.2	0.759	9 860	262	1 150	163	6.81	82.8	26.3	21.8	42.0	854	265
X68.2	0.672	8 740	243	1 060	167	6.35	79.5	27.0	22.1	45.4	619	245
X63.5	0.625	8 120	232	1 010	169	5.94	76.3	27.0	22.2	46.7	522	228
MC330												
X74	0.731	9 490	131	793	117	6.92	79.4	27.0	24.8	45.5	1 240	150
X60	0.583	7 570	113	687	122	5.73	70.4	27.5	24.5	50.7	645	123
X52	0.511	6 640	105	635	126	5.14	65.8	27.8	24.9	54.0	472	110
X47.3	0.465	6 040	99.8	605	128	4.86	63.7	28.3	25.7	57.5	392	103
MC310												
X74	0.732	9 510	113	739	109	7.25	92.4	27.6	26.6	45.4	1 360	111
X67	0.660	8 570	105	692	111	6.60	87.3	27.8	26.3	47.8	989	101
X60	0.587	7 630	98.2	644	113	5.97	82.1	28.0	26.3	50.6	718	91.3
X52	0.515	6 680	90.9	596	117	5.32	76.8	28.2	26.8	54.0	530	81.3
X46	0.455	5 900	84.7	555	120	4.69	71.5	28.2	27.4	57.0	424	71.8
MC310												
X15.8	0.153	1 990	23.0	151	107	0.159	5.11	8.94	6.84	14.0	24.5	3.11
MC250												
X61.2	0.603	7 830	66.1	520	91.9	6.73	81.9	29.3	27.8	49.9	949	73.1
X50	0.493	6 400	58.3	459	95.4	5.57	73.0	29.5	27.7	54.6	506	60.5
X42.4	0.418	5 430	53.0	417	98.8	4.79	66.9	29.7	28.4	59.0	333	52.0
MC250												
X37	0.367	4 770	46.1	363	98.3	3.03	48.9	25.2	24.0	49.7	272	33.3
X33	0.323	4 190	43.0	339	101	2.71	46.0	25.4	25.0	53.3	217	30.0
MC250												
X12.5	0.122	1 580	13.3	105	91.7	0.138	4.49	9.35	7.25	15.7	17.1	1.87
MC230												
X37.8	0.374	4 850	37.2	324	87.5	3.20	49.8	25.7	24.6	49.8	296	28.4
X35.6	0.353	4 580	36.0	315	88.6	3.05	48.4	25.8	24.9	51.4	259	27.1
MC200												
X33.9	0.334	4 340	26.7	263	78.4	2.94	46.4	26.0	25.6	52.2	243	20.4
X31.8	0.314	4 080	25.9	255	79.6	2.79	45.0	26.2	26.0	54.0	212	19.4
MC200												
X29.8	0.294	3 830	22.9	225	77.3	1.89	33.9	22.2	21.4	42.8	189	13.1
X27.8	0.276	3 590	22.1	218	78.5	1.79	32.9	22.3	21.7	44.4	163	12.4

† For sections formerly available from Canadian mills, the nominal mass has been rounded to the nearest kg/m.
  Designation consists of nominal depth in millimetres by nominal mass in kilograms per metre.
Note: These shapes have been soft converted from Imperial to SI dimensions.

MISCELLANEOUS CHANNELS
MC460 - MC200

DIMENSIONS AND SURFACE AREAS

Nominal Mass	Theo-retical Mass	Depth d	Flange Width b	Flange Mean Thickness t	Web Thickness w	Distances			Surface Area (m²) per metre of length	
						a	T	k	Total	Minus Top of Top Flange
kg/m	kg/m	mm	mm	mm	mm	mm	mm	mm		
86	86.8	457	107	15.9	17.8	89	386	35	1.31	1.20
77.2	77.4	457	104	15.9	15.2	89	386	35	1.30	1.20
68.2	68.6	457	102	15.9	12.7	89	386	35	1.30	1.19
63.5	63.7	457	100	15.9	11.4	89	386	35	1.29	1.19
74	74.5	330	112	15.5	20.0	92	262	34	1.07	0.956
60	59.5	330	106	15.5	14.2	92	263	34	1.06	0.950
52	52.1	330	103	15.5	11.4	92	263	34	1.05	0.946
47.3	47.4	330	102	15.5	9.5	93	262	34	1.05	0.947
74	74.7	305	105	17.8	21.2	84	239	33	0.988	0.883
67	67.3	305	102	17.8	18.1	84	239	33	0.982	0.880
60	59.9	305	99	17.8	15.0	84	239	33	0.976	0.877
52	52.5	305	96	17.8	11.9	84	239	33	0.970	0.874
46	46.3	305	93	17.8	9.4	84	239	33	0.963	0.870
15.8	15.6	305	38	7.8	4.8	33	271	17	0.752	0.714
61.2	61.5	254	110	14.6	20.2	90	190	32	0.908	0.798
50	50.2	254	104	14.6	14.6	89	190	32	0.895	0.791
42.4	42.6	254	100	14.6	10.8	89	190	32	0.886	0.786
37	37.4	254	86	14.6	9.7	76	190	32	0.833	0.747
33	32.9	254	84	14.6	7.4	77	190	32	0.829	0.745
12.5	12.4	254	38	7.1	4.3	34	222	16	0.651	0.613
37.8	38.1	229	89	14.0	11.4	78	168	31	0.791	0.702
35.6	36.0	229	88	14.0	10.2	78	168	31	0.790	0.702
33.9	34.1	203	89	13.3	10.8	78	144	30	0.740	0.651
31.8	32.0	203	88	13.3	9.5	79	144	30	0.739	0.651
29.8	30.0	203	77	12.7	10.2	67	147	28	0.694	0.617
27.8	28.2	203	76	12.7	9.0	67	147	28	0.692	0.616

MISCELLANEOUS CHANNELS
MC200 - MC150

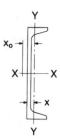

PROPERTIES

Designation†	Dead Load	Total Area	Axis X-X			Axis Y-Y				Shear Centre	Torsional Constant	Warping Constant
			I_x	S_x	r_x	I_y	S_y	r_y	x	x_o	J	C_w
	kN/m	mm²	10⁶mm⁴	10³mm³	mm	10⁶mm⁴	10³mm³	mm	mm	mm	10³mm⁴	10⁹mm⁶
MC200 X12.6	0.124	1 610	9.73	95.9	77.7	0.270	7.31	12.9	11.1	25.1	24.5	2.25
MC180 X33.8	0.335	4 350	20.0	225	67.9	3.06	46.7	26.5	26.5	52.4	270	16.1
X28.4	0.281	3 650	18.2	204	70.6	2.55	42.0	26.4	27.2	56.7	175	13.6
MC150 X26.8	0.265	3 440	12.4	163	60.1	2.51	41.4	27.0	28.5	58.3	161	9.38
MC150 X24.3	0.239	3 110	10.8	143	59.1	1.58	30.1	22.6	23.5	47.0	144	5.95
X22.5	0.223	2 890	10.5	138	60.2	1.48	29.0	22.6	23.9	49.1	123	5.62
MC150 X17.9	0.176	2 280	7.76	102	58.3	0.761	16.8	18.3	17.6	35.8	65.9	2.98

† For sections formerly available from Canadian mills, the nominal mass has been rounded to the nearest kg/m.
  Designation consists of nominal depth in millimetres by nominal mass in kilograms per metre.
Note: These shapes have been soft converted from Imperial to SI dimensions.

MISCELLANEOUS CHANNELS
MC200 - MC150

DIMENSIONS AND SURFACE AREAS

Nominal Mass	Theo-retical Mass	Depth	Flange Width	Flange Mean Thickness	Web Thickness	Distances			Surface Area (m²) per metre of length	
		d	b	t	w	a	T	k	Total	Minus Top of Top Flange
kg/m	kg/m	mm	mm	mm	mm	mm	mm	mm		
12.6	12.7	203	48	7.9	4.5	44	167	18	0.589	0.541
33.8	34.1	178	92	12.7	12.8	79	121	28	0.698	0.606
28.4	28.7	178	88	12.7	8.9	79	121	28	0.690	0.602
26.8	27.0	152	89	12.1	9.6	79	99	27	0.641	0.552
24.3	24.4	152	76	12.1	9.5	67	98	27	0.589	0.513
22.5	22.7	152	75	12.1	8.0	67	98	27	0.588	0.513
17.9	17.9	152	63	9.5	7.9	55	109	21	0.540	0.477

ANGLES - Imperial Series
L203 - L152

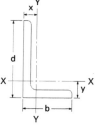

PROPERTIES ABOUT GEOMETRIC AXES

Designation	Dead Load	Area	Axis X-X				Axis Y-Y				Torsional Constant	Warping Constant
			I_x	S_x	r_x	y	I_y	S_y	r_y	x	J	C_w
	kN/m	mm²	10⁶ mm⁴	10³ mm³	mm	mm	10⁶ mm⁴	10³ mm³	mm	mm	10³ mm⁴	10⁹ mm⁶
L203x203												
x29*	0.831	10 800	40.7	287	61.4	61.2	40.7	287	61.4	61.2	2 940	8.73
x25*	0.744	9 670	36.9	258	61.8	60.1	36.9	258	61.8	60.1	2 080	6.27
x22*	0.656	8 520	33.0	229	62.2	58.9	33.0	229	62.2	58.9	1 400	4.30
x19*	0.569	7 390	29.0	200	62.6	57.8	29.0	200	62.6	57.8	899	2.80
x16*	0.478	6 200	24.7	169	63.1	56.6	24.7	169	63.1	56.6	523	1.66
x13*	0.385	4 990	20.2	137	63.6	55.5	20.2	137	63.6	55.5	269	0.865
L203x152												
x25*	0.645	8 370	33.5	247	63.3	67.4	16.0	145	43.7	41.9	1 800	4.37
x19*	0.494	6 420	26.4	191	64.1	65.1	12.7	113	44.5	39.6	780	1.96
x13*	0.335	4 350	18.4	131	65.0	62.8	8.96	78.1	45.4	37.3	234	0.609
x11*	0.294	3 820	16.3	115	65.3	62.2	7.94	68.9	45.6	36.7	157	0.412
L203x102												
x25*	0.547	7 100	29.0	230	63.8	77.2	4.90	65.1	26.3	26.7	1 530	3.46
x19*	0.421	5 460	22.9	178	64.7	74.8	3.95	50.8	26.9	24.3	664	1.55
x13*	0.286	3 710	16.0	123	65.7	72.4	2.84	35.4	27.6	21.9	200	0.482
L178x102												
x19*	0.384	4 980	15.8	139	56.4	63.8	3.82	50.1	27.7	25.8	606	1.08
x16*	0.323	4 200	13.6	118	56.8	62.6	3.31	42.7	28.1	24.6	354	0.642
x13	0.261	3 390	11.1	95.6	57.3	61.4	2.75	35.0	28.5	23.4	183	0.338
x11*	0.230	2 980	9.88	84.3	57.5	60.8	2.45	31.0	28.7	22.8	123	0.229
x9.5	0.198	2 580	8.60	73.0	57.8	60.2	2.15	26.9	28.9	22.2	78.0	0.147
L152x152												
x25*	0.545	7 080	14.6	140	45.5	47.2	14.6	140	45.5	47.2	1 520	2.46
x22*	0.482	6 260	13.2	124	45.9	46.1	13.2	124	45.9	46.1	1 030	1.70
x19*	0.419	5 440	11.6	109	46.3	45.0	11.6	109	46.3	45.0	662	1.12
x16	0.353	4 580	9.99	92.3	46.7	43.9	9.99	92.3	46.7	43.9	386	0.668
x14*	0.319	4 140	9.12	83.8	46.9	43.3	9.12	83.8	46.9	43.3	282	0.494
x13	0.285	3 700	8.22	75.2	47.1	42.7	8.22	75.2	47.1	42.7	199	0.352
x11*	0.250	3 250	7.29	66.4	47.4	42.1	7.29	66.4	47.4	42.1	134	0.239
x9.5	0.216	2 810	6.36	57.5	47.6	41.5	6.36	57.5	47.6	41.5	85.0	0.153
x7.9*	0.181	2 350	5.38	48.4	47.8	41.0	5.38	48.4	47.8	41.0	49.4	0.090 2
x6.4*	0.146	1 890	4.37	39.1	48.1	40.4	4.37	39.1	48.1	40.4	25.4	0.046 9

See page 6-39 for qualification of properties of angles produced by cold-forming.
* Not available from Canadian mills

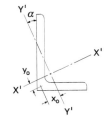

ANGLES - Imperial Series
L203 - L152

DIMENSIONS AND PROPERTIES ABOUT PRINCIPAL AXES

Mass	d	b	t	Axis X'-X'		Axis Y'-Y'		$\bar{r}_o$	β	$\tan \alpha$
				r_x	y_o	r_y	x_o			
kg/m	mm	mm	mm	mm	mm	mm	mm	mm		
84.7	203	203	28.6	77.3	0.00	39.6	66.3	109	0.631	1.00
75.9	203	203	25.4	77.9	0.00	39.7	67.0	110	0.630	1.00
66.9	203	203	22.2	78.5	0.00	39.8	67.6	111	0.629	1.00
58.0	203	203	19.1	79.1	0.00	40.0	68.2	112	0.628	1.00
48.7	203	203	15.9	79.7	0.00	40.1	68.8	113	0.627	1.00
39.2	203	203	12.7	80.3	0.00	40.3	69.5	114	0.626	1.00
65.7	203	152	25.4	69.7	34.2	32.4	51.7	98.8	0.606	0.541
50.4	203	152	19.1	70.9	34.2	32.6	53.1	100	0.604	0.549
34.1	203	152	12.7	72.1	34.3	33.0	54.5	102	0.603	0.556
30.0	203	152	11.1	72.4	34.3	33.1	54.8	103	0.603	0.558
55.7	203	102	25.4	65.6	59.2	21.6	29.2	95.5	0.523	0.249
42.9	203	102	19.1	66.6	59.5	21.7	30.7	96.9	0.523	0.260
29.1	203	102	12.7	67.7	59.8	22.1	32.2	98.4	0.524	0.269
39.1	178	102	19.1	58.9	46.5	21.9	32.2	84.5	0.552	0.325
33.0	178	102	15.9	59.4	46.6	22.1	33.0	85.3	0.552	0.331
26.6	178	102	12.7	60.0	46.7	22.2	33.7	86.1	0.552	0.336
23.4	178	102	11.1	60.3	46.8	22.3	34.1	86.5	0.552	0.339
20.2	178	102	9.53	60.6	46.8	22.4	34.4	86.9	0.553	0.341
55.6	152	152	25.4	57.1	0.00	29.6	48.8	80.8	0.634	1.00
49.1	152	152	22.2	57.7	0.00	29.6	49.5	81.6	0.632	1.00
42.7	152	152	19.1	58.3	0.00	29.7	50.1	82.4	0.630	1.00
36.0	152	152	15.9	58.9	0.00	29.8	50.8	83.3	0.628	1.00
32.5	152	152	14.3	59.2	0.00	29.9	51.1	83.7	0.628	1.00
29.0	152	152	12.7	59.5	0.00	30.0	51.4	84.2	0.627	1.00
25.5	152	152	11.1	59.8	0.00	30.1	51.7	84.6	0.627	1.00
22.0	152	152	9.53	60.1	0.00	30.2	52.0	85.1	0.626	1.00
18.5	152	152	7.94	60.5	0.00	30.3	52.3	85.5	0.626	1.00
14.8	152	152	6.35	60.8	0.00	30.4	52.6	85.9	0.626	1.00

See CAN / CSA S16.1 - 94 Appendix D for definition of x_o, y_o, $\bar{r}_o$ and β.
The axis of symmetry Y-Y of equal-leg (singly-symmetric) angles in Appendix D corresponds to axis X'-X' in the above table.

ANGLES - Imperial Series
L152 - L127

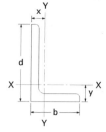

PROPERTIES ABOUT GEOMETRIC AXES

Designation	Dead Load	Area	Axis X-X				Axis Y-Y				Torsional Constant	Warping Constant
			I_x	S_x	r_x	y	I_y	S_y	r_y	x	J	C_w
	kN/m	mm²	10^6 mm⁴	10^3 mm³	mm	mm	10^6 mm⁴	10^3 mm³	mm	mm	10^3 mm⁴	10^9 mm⁶
L152x102												
x22*	0.396	5 150	11.5	117	47.2	53.7	4.10	55.9	28.2	28.7	845	1.08
x19*	0.346	4 490	10.2	102	47.6	52.5	3.66	49.2	28.6	27.5	546	0.712
x16	0.292	3 790	8.73	86.8	48.0	51.4	3.17	41.9	28.9	26.4	319	0.427
x14*	0.264	3 430	7.98	78.8	48.2	50.8	2.91	38.2	29.1	25.8	234	0.316
x13	0.236	3 060	7.20	70.7	48.5	50.2	2.64	34.4	29.3	25.2	165	0.226
x11*	0.208	2 700	6.39	62.4	48.7	49.6	2.35	30.4	29.6	24.6	111	0.153
x9.5	0.179	2 330	5.58	54.2	48.9	49.1	2.06	26.5	29.8	24.1	70.5	0.098 8
x7.9	0.150	1 950	4.72	45.6	49.2	48.5	1.76	22.4	30.0	23.5	41.1	0.058 2
L152x89												
x16	0.275	3 580	8.31	84.7	48.2	53.9	2.12	31.8	24.3	22.4	301	0.393
x13	0.223	2 900	6.86	69.1	48.6	52.7	1.77	26.1	24.7	21.2	156	0.208
x9.5	0.170	2 200	5.32	52.9	49.1	51.6	1.39	20.2	25.1	20.0	66.8	0.091 1
x7.9	0.142	1 850	4.50	44.6	49.3	51.0	1.19	17.1	25.3	19.4	38.9	0.053 6
L127x127												
x22*	0.396	5 150	7.39	84.7	37.9	39.8	7.39	84.7	37.9	39.8	845	0.946
x19*	0.346	4 490	6.57	74.4	38.3	38.7	6.57	74.4	38.3	38.7	546	0.627
x16	0.292	3 790	5.66	63.3	38.7	37.6	5.66	63.3	38.7	37.6	319	0.377
x13	0.236	3 060	4.68	51.7	39.1	36.4	4.68	51.7	39.1	36.4	165	0.200
x11*	0.208	2 700	4.17	45.7	39.3	35.8	4.17	45.7	39.3	35.8	111	0.136
x9.5	0.179	2 330	3.64	39.7	39.5	35.3	3.64	39.7	39.5	35.3	70.5	0.087 8
x7.9	0.150	1 950	3.09	33.5	39.8	34.7	3.09	33.5	39.8	34.7	41.1	0.051 8
x6.4*	0.121	1 570	2.52	27.1	40.0	34.1	2.52	27.1	40.0	34.1	21.1	0.027 0
L127x89												
x19*	0.289	3 760	5.80	70.2	39.3	44.4	2.31	36.4	24.8	25.3	457	0.410
x16*	0.245	3 180	5.01	59.8	39.7	43.2	2.01	31.1	25.2	24.2	268	0.248
x13	0.199	2 580	4.16	48.9	40.1	42.1	1.68	25.6	25.6	23.0	139	0.132
x9.5	0.151	1 970	3.24	37.6	40.6	40.9	1.33	19.8	26.0	21.9	59.5	0.058 2
x7.9	0.127	1 650	2.75	31.7	40.8	40.3	1.13	16.7	26.2	21.3	34.7	0.034 4
x6.4	0.102	1 330	2.24	25.7	41.0	39.7	0.928	13.6	26.4	20.7	17.9	0.018 0
L127x76												
x13	0.186	2 420	3.93	47.7	40.3	44.5	1.07	18.8	21.1	19.1	130	0.119
x11*	0.164	2 130	3.51	42.2	40.6	43.9	0.963	16.7	21.3	18.5	87.6	0.081 5
x9.5	0.142	1 850	3.07	36.7	40.8	43.3	0.849	14.6	21.5	17.9	55.9	0.052 7
x7.9	0.119	1 550	2.61	30.9	41.0	42.7	0.727	12.3	21.7	17.3	32.6	0.031 1
x6.4	0.096 3	1 250	2.13	25.0	41.2	42.1	0.598	10.1	21.9	16.7	16.8	0.016 3

See page 6-39 for qualification of properties of angles produced by cold-forming.

* Not available from Canadian mills

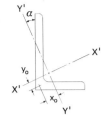

ANGLES - Imperial Series
L152 - L127

DIMENSIONS AND PROPERTIES ABOUT PRINCIPAL AXES

Mass	d	b	t	Axis X'-X'		Axis Y'-Y'		$\bar{r}_o$	β	tan α
				r_x	y_o	r_y	x_o			
kg/m	mm	mm	mm	mm	mm	mm	mm	mm		
40.4	152	102	22.2	50.5	32.2	21.9	32.9	71.7	0.588	0.427
35.2	152	102	19.1	51.0	32.3	21.9	33.6	72.5	0.586	0.434
29.7	152	102	15.9	51.6	32.3	22.0	34.4	73.3	0.585	0.440
26.9	152	102	14.3	51.8	32.4	22.1	34.7	73.7	0.585	0.443
24.1	152	102	12.7	52.1	32.4	22.2	35.1	74.1	0.585	0.446
21.2	152	102	11.1	52.4	32.4	22.3	35.5	74.5	0.584	0.449
18.3	152	102	9.53	52.7	32.4	22.4	35.8	74.9	0.584	0.451
15.3	152	102	7.94	53.0	32.5	22.5	36.1	75.3	0.584	0.454
28.1	152	88.9	15.9	50.5	38.9	19.1	28.4	72.4	0.557	0.339
22.8	152	88.9	12.7	51.0	39.0	19.3	29.2	73.1	0.556	0.345
17.3	152	88.9	9.53	51.6	39.1	19.5	29.9	73.9	0.557	0.351
14.5	152	88.9	7.94	51.9	39.1	19.6	30.2	74.3	0.557	0.354
40.4	127	127	22.2	47.5	0.00	24.7	40.6	67.2	0.635	1.00
35.2	127	127	19.1	48.1	0.00	24.8	41.3	68.0	0.632	1.00
29.7	127	127	15.9	48.7	0.00	24.8	41.9	68.9	0.630	1.00
24.1	127	127	12.7	49.3	0.00	25.0	42.5	69.8	0.628	1.00
21.2	127	127	11.1	49.6	0.00	25.0	42.8	70.2	0.627	1.00
18.3	127	127	9.53	49.9	0.00	25.1	43.2	70.6	0.627	1.00
15.3	127	127	7.94	50.3	0.00	25.2	43.5	71.1	0.626	1.00
12.3	127	127	6.35	50.6	0.00	25.4	43.7	71.5	0.626	1.00
29.5	127	88.9	19.1	42.4	24.9	19.0	29.0	60.2	0.597	0.464
25.0	127	88.9	15.9	43.0	25.0	19.1	29.7	61.0	0.594	0.472
20.3	127	88.9	12.7	43.5	25.0	19.2	30.5	61.8	0.593	0.479
15.4	127	88.9	9.53	44.1	25.0	19.3	31.2	62.6	0.592	0.486
13.0	127	88.9	7.94	44.4	25.1	19.4	31.5	63.0	0.592	0.489
10.4	127	88.9	6.35	44.7	25.1	19.6	31.8	63.4	0.592	0.492
19.0	127	76.2	12.7	42.4	31.6	16.5	24.8	60.7	0.562	0.357
16.7	127	76.2	11.1	42.7	31.7	16.5	25.1	61.1	0.562	0.361
14.5	127	76.2	9.53	43.0	31.7	16.6	25.5	61.5	0.562	0.364
12.2	127	76.2	7.94	43.3	31.7	16.7	25.9	61.9	0.562	0.368
9.81	127	76.2	6.35	43.6	31.8	16.8	26.2	62.3	0.562	0.371

See CAN / CSA S16.1 - 94 Appendix D for definition of x_o, y_o, $\bar{r}_o$ and β.
The axis of symmetry Y-Y of equal-leg (singly-symmetric) angles in Appendix D corresponds to axis X'-X' in the above table.

ANGLES - Imperial Series
L102 - L89

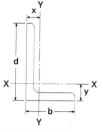

PROPERTIES ABOUT GEOMETRIC AXES

Designation	Dead Load	Area	Axis X-X				Axis Y-Y				Torsional Constant	Warping Constant
			I_x	S_x	r_x	y	I_y	S_y	r_y	x	J	C_w
	kN/m	mm²	10^6 mm⁴	10^3 mm³	mm	mm	10^6 mm⁴	10^3 mm³	mm	mm	10^3 mm⁴	10^9 mm⁶
L102x102												
x19*	0.272	3 530	3.24	46.5	30.3	32.4	3.24	46.5	30.3	32.4	429	0.306
x16*	0.230	2 990	2.81	39.8	30.7	31.3	2.81	39.8	30.7	31.3	252	0.186
x13	0.187	2 430	2.34	32.6	31.1	30.2	2.34	32.6	31.1	30.2	131	0.0996
x11*	0.165	2 140	2.09	28.9	31.3	29.6	2.09	28.9	31.3	29.6	87.9	0.0682
x9.5	0.143	1 850	1.84	25.2	31.5	29.0	1.84	25.2	31.5	29.0	56.1	0.0442
x7.9	0.120	1 560	1.57	21.3	31.7	28.4	1.57	21.3	31.7	28.4	32.7	0.0262
x6.4	0.096 7	1 260	1.28	17.3	31.9	27.9	1.28	17.3	31.9	27.9	16.9	0.0137
L102x89												
x13	0.174	2 260	2.24	32.0	31.5	31.9	1.58	24.9	26.4	25.4	122	0.0818
x9.5	0.133	1 730	1.76	24.7	31.9	30.8	1.24	19.2	26.8	24.2	52.3	0.0364
x7.9	0.112	1 450	1.50	20.9	32.1	30.2	1.06	16.3	27.1	23.6	30.5	0.0216
x6.4	0.090 2	1 170	1.23	16.9	32.3	29.6	0.872	13.2	27.3	23.1	15.8	0.0113
L102x76												
x16*	0.199	2 580	2.54	38.0	31.4	35.0	1.20	22.2	21.6	22.1	217	0.128
x13	0.162	2 100	2.12	31.2	31.8	33.9	1.01	18.3	21.9	21.0	113	0.0692
x11*	0.143	1 850	1.90	27.7	32.0	33.3	0.907	16.3	22.1	20.4	76.2	0.0475
x9.5	0.124	1 610	1.67	24.1	32.2	32.7	0.800	14.2	22.3	19.8	48.7	0.0309
x7.9	0.104	1 350	1.42	20.4	32.4	32.1	0.686	12.0	22.5	19.2	28.4	0.0183
x6.4	0.084 0	1 090	1.17	16.5	32.7	31.6	0.565	9.81	22.7	18.7	14.7	0.009 63
L89x89												
x13	0.161	2 100	1.51	24.4	26.9	26.9	1.51	24.4	26.9	26.9	113	0.0640
x11*	0.142	1 850	1.36	21.7	27.1	26.3	1.36	21.7	27.1	26.3	76.0	0.0440
x9.5	0.123	1 600	1.19	18.9	27.3	25.7	1.19	18.9	27.3	25.7	48.5	0.0286
x7.9	0.104	1 350	1.02	16.0	27.5	25.2	1.02	16.0	27.5	25.2	28.3	0.0170
x6.4	0.083 8	1 090	0.837	13.0	27.7	24.6	0.837	13.0	27.7	24.6	14.6	0.008 96
L89x76												
x13	0.149	1 940	1.44	23.8	27.3	28.6	0.969	18.0	22.4	22.2	104	0.0514
x9.5	0.114	1 480	1.13	18.5	27.7	27.4	0.769	14.0	22.8	21.1	44.9	0.0231
x7.9	0.096 1	1 250	0.970	15.6	27.9	26.9	0.659	11.8	23.0	20.5	26.2	0.0138
x6.4	0.077 6	1 010	0.796	12.7	28.1	26.3	0.543	9.65	23.2	19.9	13.5	0.007 25
L89x64												
x13	0.137	1 770	1.35	23.1	27.6	30.6	0.568	12.5	17.9	17.9	95.4	0.0426
x9.5	0.105	1 360	1.07	17.9	28.0	29.5	0.454	9.71	18.3	16.8	41.2	0.0192
x7.9	0.088 3	1 150	0.912	15.2	28.2	28.9	0.391	8.26	18.5	16.2	24.1	0.0115
x6.4	0.071 4	927	0.749	12.4	28.4	28.3	0.323	6.75	18.7	15.6	12.5	0.006 04

See page 6-39 for qualification of properties of angles produced by cold-forming.
* Not available from Canadian mills

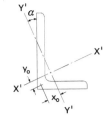

ANGLES - Imperial Series
L102 - L89

DIMENSIONS AND PROPERTIES ABOUT PRINCIPAL AXES

Mass	d	b	t	Axis X'-X'		Axis Y'-Y'		$\bar{r}_o$	β	tan α
				r_x	y_o	r_y	x_o			
kg/m	mm	mm	mm	mm	mm	mm	mm	mm		
27.7	102	102	19.1	37.9	0.00	19.8	32.3	53.7	0.637	1.00
23.5	102	102	15.9	38.5	0.00	19.9	33.0	54.5	0.633	1.00
19.1	102	102	12.7	39.1	0.00	19.9	33.7	55.3	0.630	1.00
16.8	102	102	11.1	39.4	0.00	20.0	34.0	55.8	0.629	1.00
14.5	102	102	9.53	39.7	0.00	20.1	34.3	56.2	0.628	1.00
12.2	102	102	7.94	40.1	0.00	20.2	34.6	56.6	0.627	1.00
9.85	102	102	6.35	40.4	0.00	20.3	34.9	57.1	0.626	1.00
17.8	102	88.9	12.7	36.8	9.16	18.4	30.5	52.0	0.625	0.744
13.6	102	88.9	9.53	37.4	9.15	18.5	31.2	52.8	0.622	0.749
11.4	102	88.9	7.94	37.7	9.15	18.6	31.5	53.3	0.621	0.751
9.20	102	88.9	6.35	38.0	9.15	18.7	31.8	53.7	0.621	0.753
20.3	102	76.2	15.9	34.5	17.3	16.2	25.2	48.8	0.609	0.529
16.5	102	76.2	12.7	35.0	17.3	16.2	25.9	49.6	0.606	0.538
14.6	102	76.2	11.1	35.3	17.3	16.3	26.3	50.0	0.605	0.543
12.6	102	76.2	9.53	35.6	17.3	16.4	26.6	50.4	0.604	0.547
10.6	102	76.2	7.94	35.9	17.3	16.5	27.0	50.9	0.603	0.550
8.57	102	76.2	6.35	36.2	17.3	16.6	27.3	51.3	0.603	0.554
16.5	88.9	88.9	12.7	33.8	0.00	17.3	29.0	47.8	0.632	1.00
14.5	88.9	88.9	11.1	34.1	0.00	17.4	29.3	48.2	0.630	1.00
12.6	88.9	88.9	9.53	34.4	0.00	17.4	29.7	48.7	0.629	1.00
10.6	88.9	88.9	7.94	34.7	0.00	17.5	30.0	49.1	0.627	1.00
8.55	88.9	88.9	6.35	35.0	0.00	17.6	30.3	49.5	0.627	1.00
15.2	88.9	76.2	12.7	31.5	8.86	15.8	25.8	44.6	0.625	0.714
11.6	88.9	76.2	9.53	32.1	8.85	15.9	26.5	45.4	0.622	0.721
9.80	88.9	76.2	7.94	32.4	8.84	15.9	26.8	45.9	0.621	0.724
7.91	88.9	76.2	6.35	32.7	8.84	16.0	27.1	46.3	0.620	0.727
13.9	88.9	63.5	12.7	29.9	16.8	13.6	21.0	42.4	0.600	0.486
10.7	88.9	63.5	9.53	30.5	16.8	13.6	21.7	43.2	0.597	0.496
9.00	88.9	63.5	7.94	30.8	16.8	13.7	22.1	43.7	0.596	0.501
7.28	88.9	63.5	6.35	31.1	16.8	13.8	22.4	44.1	0.596	0.506

See CAN / CSA S16.1 - 94 Appendix D for definition of x_o, y_o, $\bar{r}_o$ and β.

The axis of symmetry Y-Y of equal-leg (singly-symmetric) angles in Appendix D corresponds to axis X'-X' in the above table.

ANGLES - Imperial Series
L76 - L51

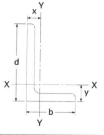

PROPERTIES ABOUT GEOMETRIC AXES

Designation	Dead Load	Area	Axis X-X				Axis Y-Y				Torsional Constant	Warping Constant
			I_x	S_x	r_x	y	I_y	S_y	r_y	x	J	C_w
	kN/m	mm²	10^6 mm⁴	10^3 mm³	mm	mm	10^6 mm⁴	10^3 mm³	mm	mm	10^3 mm⁴	10^9 mm⁶
L76x76												
x13	0.137	1 770	0.923	17.6	22.8	23.7	0.923	17.6	22.8	23.7	95.4	0.038 8
x11	0.121	1 570	0.830	15.6	23.0	23.1	0.830	15.6	23.0	23.1	64.4	0.026 8
x9.5	0.105	1 360	0.733	13.7	23.2	22.5	0.733	13.7	23.2	22.5	41.2	0.017 5
x7.9	0.088 3	1 150	0.629	11.6	23.4	22.0	0.629	11.6	23.4	22.0	24.1	0.010 5
x6.4	0.071 4	927	0.518	9.45	23.6	21.4	0.518	9.45	23.6	21.4	12.5	0.005 54
x4.8	0.054 1	703	0.400	7.22	23.9	20.8	0.400	7.22	23.9	20.8	5.31	0.002 41
L76x64												
x13*	0.124	1 610	0.867	17.1	23.2	25.4	0.542	12.2	18.3	19.1	86.7	0.030 0
x9.5	0.095 5	1 240	0.690	13.3	23.6	24.3	0.434	9.52	18.7	17.9	37.6	0.013 6
x7.9	0.080 6	1 050	0.592	11.3	23.8	23.7	0.374	8.10	18.9	17.4	22.0	0.008 17
x6.4	0.065 2	847	0.488	9.20	24.0	23.1	0.309	6.62	19.1	16.8	11.4	0.004 33
x4.8*	0.049 5	642	0.377	7.04	24.2	22.6	0.240	5.08	19.3	16.2	4.85	0.001 89
L76x51												
x13	0.112	1 450	0.800	16.4	23.5	27.5	0.280	7.77	13.9	14.8	78.0	0.024 4
x9.5	0.086 2	1 120	0.638	12.8	23.9	26.4	0.226	6.09	14.2	13.7	33.9	0.011 1
x7.9	0.072 8	945	0.548	10.9	24.1	25.8	0.196	5.20	14.4	13.1	19.9	0.006 67
x6.4	0.059 0	766	0.453	8.88	24.3	25.2	0.163	4.26	14.6	12.5	10.3	0.003 54
x4.8	0.044 8	582	0.350	6.79	24.5	24.6	0.128	3.28	14.8	11.9	4.39	0.001 55
L64x64												
x13	0.112	1 450	0.511	11.9	18.8	20.5	0.511	11.9	18.8	20.5	78.0	0.021 2
x9.5	0.086 2	1 120	0.410	9.28	19.1	19.4	0.410	9.28	19.1	19.4	33.9	0.009 74
x7.9	0.072 8	945	0.353	7.90	19.3	18.8	0.353	7.90	19.3	18.8	19.9	0.005 87
x6.4	0.059 0	766	0.293	6.46	19.5	18.2	0.293	6.46	19.5	18.2	10.3	0.003 12
x4.8	0.044 8	582	0.227	4.96	19.8	17.6	0.227	4.96	19.8	17.6	4.39	0.001 37
L64x51												
x9.5	0.076 9	998	0.380	8.96	19.5	21.1	0.214	5.94	14.6	14.8	30.2	0.007 22
x7.9	0.065 0	844	0.328	7.64	19.7	20.6	0.186	5.08	14.8	14.2	17.7	0.004 36
x6.4	0.052 8	685	0.272	6.25	19.9	20.0	0.155	4.17	15.0	13.6	9.21	0.002 33
x4.8	0.040 2	521	0.212	4.80	20.1	19.4	0.121	3.21	15.2	13.1	3.94	0.001 02
L51x51												
x9.5	0.067 6	877	0.199	5.76	15.1	16.2	0.199	5.76	15.1	16.2	26.6	0.004 69
x7.9	0.057 3	744	0.173	4.92	15.3	15.6	0.173	4.92	15.3	15.6	15.6	0.002 86
x6.4	0.046 6	605	0.145	4.04	15.5	15.0	0.145	4.04	15.5	15.0	8.13	0.001 54
x4.8	0.035 5	461	0.113	3.12	15.7	14.5	0.113	3.12	15.7	14.5	3.48	0.000 680
x3.2	0.024 1	313	0.079 2	2.14	15.9	13.9	0.079 2	2.14	15.9	13.9	1.05	0.000 213
L51x38												
x6.4+	0.040 4	524	0.131	3.87	15.8	16.9	0.063 0	2.28	11.0	10.5	7.05	0.001 07
x4.8+	0.030 8	401	0.103	2.99	16.0	16.3	0.049 9	1.77	11.2	9.93	3.02	0.000 477

See page 6-39 for qualification of properties of angles produced by cold-forming.

* Not available from Canadian mills
+ Check availability

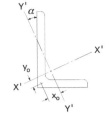

ANGLES - Imperial Series
L76 - L51

DIMENSIONS AND PROPERTIES ABOUT PRINCIPAL AXES

Mass	d	b	t	Axis X'-X'		Axis Y'-Y'		$\bar{r}_o$	β	$\tan \alpha$
				r_x	y_o	r_y	x_o			
kg/m	mm	mm	mm	mm	mm	mm	mm	mm		
13.9	76.2	76.2	12.7	28.6	0.00	14.8	24.5	40.5	0.634	1.00
12.3	76.2	76.2	11.1	28.9	0.00	14.9	24.8	40.9	0.632	1.00
10.7	76.2	76.2	9.53	29.2	0.00	14.9	25.1	41.3	0.630	1.00
9.00	76.2	76.2	7.94	29.5	0.00	15.0	25.5	41.8	0.628	1.00
7.28	76.2	76.2	6.35	29.8	0.00	15.0	25.8	42.2	0.627	1.00
5.52	76.2	76.2	4.76	30.2	0.00	15.1	26.1	42.6	0.626	1.00
12.7	76.2	63.5	12.7	26.4	8.81	13.2	21.1	37.4	0.625	0.667
9.74	76.2	63.5	9.53	27.0	8.79	13.3	21.8	38.2	0.620	0.676
8.21	76.2	63.5	7.94	27.3	8.79	13.3	22.2	38.6	0.619	0.680
6.65	76.2	63.5	6.35	27.6	8.79	13.4	22.5	39.1	0.618	0.684
5.04	76.2	63.5	4.76	27.9	8.79	13.5	22.8	39.5	0.617	0.688
11.4	76.2	50.8	12.7	25.0	16.3	10.9	15.9	35.5	0.589	0.414
8.79	76.2	50.8	9.53	25.5	16.4	10.9	16.7	36.3	0.585	0.428
7.42	76.2	50.8	7.94	25.8	16.4	11.0	17.1	36.7	0.584	0.435
6.01	76.2	50.8	6.35	26.1	16.4	11.0	17.5	37.1	0.583	0.440
4.57	76.2	50.8	4.76	26.4	16.4	11.1	17.8	37.5	0.583	0.446
11.4	63.5	63.5	12.7	23.5	0.00	12.4	20.0	33.2	0.639	1.00
8.79	63.5	63.5	9.53	24.1	0.00	12.4	20.6	34.0	0.632	1.00
7.42	63.5	63.5	7.94	24.4	0.00	12.4	21.0	34.4	0.630	1.00
6.01	63.5	63.5	6.35	24.7	0.00	12.5	21.3	34.9	0.628	1.00
4.57	63.5	63.5	4.76	25.0	0.00	12.6	21.6	35.3	0.627	1.00
7.84	63.5	50.8	9.53	21.9	8.70	10.7	17.1	31.0	0.618	0.614
6.63	63.5	50.8	7.94	22.2	8.70	10.7	17.4	31.4	0.616	0.620
5.38	63.5	50.8	6.35	22.5	8.70	10.8	17.8	31.9	0.614	0.626
4.09	63.5	50.8	4.76	22.8	8.70	10.9	18.1	32.3	0.612	0.631
6.89	50.8	50.8	9.53	18.9	0.00	9.89	16.1	26.7	0.637	1.00
5.84	50.8	50.8	7.94	19.2	0.00	9.90	16.4	27.1	0.633	1.00
4.75	50.8	50.8	6.35	19.5	0.00	9.93	16.8	27.6	0.630	1.00
3.62	50.8	50.8	4.76	19.8	0.00	10.0	17.1	28.0	0.628	1.00
2.46	50.8	50.8	3.18	20.1	0.00	10.1	17.4	28.4	0.626	1.00
4.11	50.8	38.1	6.35	17.5	8.53	8.12	13.0	24.7	0.606	0.543
3.14	50.8	38.1	4.76	17.8	8.53	8.18	13.3	25.1	0.604	0.551

See CAN / CSA S16.1 - 94 Appendix D for definition of x_o, y_o, $\bar{r}_o$ and β.
The axis of symmetry Y-Y of equal-leg (singly-symmetric) angles in Appendix D corresponds to axis X'-X' in the above table.

ANGLES - Imperial Series
L44 - L19

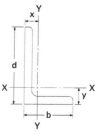

PROPERTIES ABOUT GEOMETRIC AXES

Designation	Dead Load	Area	Axis X-X				Axis Y-Y				Torsional Constant	Warping Constant
			I_x	S_x	r_x	y	I_y	S_y	r_y	x	J	C_w
	kN/m	mm^2	10^6 mm^4	10^3 mm^3	mm	mm	10^6 mm^4	10^3 mm^3	mm	mm	10^3 mm^4	10^9 mm^6
L44x44												
x6.4	0.040 4	525	0.094 9	3.06	13.4	13.4	0.094 9	3.06	13.4	13.4	7.05	0.001 00
x4.8	0.030 9	401	0.074 8	2.36	13.7	12.9	0.074 8	2.36	13.7	12.9	3.03	0.000 448
x3.2	0.021 0	273	0.052 5	1.63	13.9	12.3	0.052 5	1.63	13.9	12.3	0.920	0.000 141
L38x38												
x6.4	0.034 2	444	0.057 7	2.20	11.4	11.8	0.057 7	2.20	11.4	11.8	5.96	0.000 606
x4.8	0.026 2	340	0.045 8	1.71	11.6	11.3	0.045 8	1.71	11.6	11.3	2.57	0.000 273
x3.2	0.017 9	232	0.032 4	1.18	11.8	10.7	0.032 4	1.18	11.8	10.7	0.783	0.000 087
L32x32												
x6.4	0.028 0	364	0.032 1	1.49	9.40	10.2	0.032 1	1.49	9.40	10.2	4.89	0.000 334
x4.8	0.021 6	280	0.025 7	1.16	9.58	9.69	0.025 7	1.16	9.58	9.69	2.12	0.000 153
x3.2	0.014 8	192	0.018 4	0.812	9.79	9.12	0.018 4	0.812	9.79	9.12	0.648	0.000 049
L25x25												
x6.4	0.021 7	282	0.015 3	0.915	7.37	8.62	0.015 3	0.915	7.37	8.62	3.79	0.000 156
x4.8	0.016 9	219	0.012 5	0.719	7.54	8.07	0.012 5	0.719	7.54	8.07	1.66	0.000 073
x3.2	0.011 7	151	0.009 05	0.506	7.73	7.52	0.009 1	0.506	7.73	7.52	0.510	0.000 024
L19x19												
x3.2	0.008 6	111	0.003 64	0.276	5.72	5.93	0.003 6	0.276	5.72	5.93	0.375	0.000 010

See page 6-39 for qualification of properties of angles produced by cold-forming.

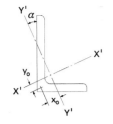

ANGLES - Imperial Series
L44 - L19

DIMENSIONS AND PROPERTIES ABOUT PRINCIPAL AXES

Mass	d	b	t	Axis X'-X'		Axis Y'-Y'		$\bar{r}_o$	β	tan α
				r_x	y_o	r_y	x_o			
kg/m	mm	mm	mm	mm	mm	mm	mm	mm		
4.12	44.5	44.5	6.35	16.9	0.00	8.68	14.5	23.9	0.632	1.00
3.15	44.5	44.5	4.76	17.2	0.00	8.73	14.8	24.4	0.629	1.00
2.14	44.5	44.5	3.18	17.5	0.00	8.82	15.2	24.8	0.627	1.00
3.48	38.1	38.1	6.35	14.3	0.00	7.42	12.2	20.2	0.634	1.00
2.67	38.1	38.1	4.76	14.6	0.00	7.45	12.6	20.7	0.630	1.00
1.82	38.1	38.1	3.18	14.9	0.00	7.52	12.9	21.1	0.627	1.00
2.85	31.8	31.8	6.35	11.8	0.00	6.19	10.0	16.6	0.639	1.00
2.20	31.8	31.8	4.76	12.0	0.00	6.20	10.3	17.0	0.632	1.00
1.51	31.8	31.8	3.18	12.4	0.00	6.25	10.7	17.5	0.628	1.00
2.22	25.4	25.4	6.35	9.17	0.00	4.98	7.70	13.0	0.647	1.00
1.72	25.4	25.4	4.76	9.45	0.00	4.94	8.05	13.4	0.637	1.00
1.19	25.4	25.4	3.18	9.74	0.00	4.97	8.38	13.8	0.630	1.00
0.874	19.1	19.1	3.18	7.18	0.00	3.72	6.14	10.2	0.634	1.00

See CAN / CSA S16.1 - 94 Appendix D for definition of x_o, y_o, $\bar{r}_o$ and β.

The axis of symmetry Y-Y of equal-leg (singly-symmetric) angles in Appendix D corresponds to axis X'-X' in the above table.

STRUCTURAL TEES
Cut from W Shapes
WT460 - WT345

PROPERTIES

Designation	Dead Load	Total Area	Axis X-X				Axis Y-Y			Torsional Constant	Warping Constant
			I_x	S_x	r_x	y	I_y	S_y	r_y	J	C_w
	kN/m	mm²	10^6 mm⁴	10^3 mm³	mm	mm	10^6 mm⁴	10^3 mm³	mm	10^3 mm⁴	10^9 mm⁶
WT460											
x223	2.19	28 500	508	1 410	134	105	270	1 280	97.4	13 300	74.7
x208.5	2.05	26 600	474	1 310	133	103	250	1 190	97.0	10 900	60.9
x193.5	1.90	24 600	438	1 230	133	103	226	1 080	95.9	8 610	48.4
x182.5	1.79	23 200	413	1 160	133	102	211	1 010	95.2	7 180	40.5
x171	1.68	21 800	389	1 100	134	102	195	933	94.6	5 920	33.6
WT460											
x156.5	1.53	19 900	410	1 200	143	124	85.2	551	65.4	5 780	32.0
x144.5	1.42	18 400	376	1 100	143	122	78.2	508	65.2	4 600	24.9
x135.5	1.33	17 300	353	1 040	143	121	72.6	473	64.7	3 830	20.9
x126.5	1.24	16 200	329	969	143	120	66.8	437	64.3	3 120	17.1
x119	1.17	15 200	309	916	143	121	61.4	403	63.6	2 560	14.4
x111.5	1.10	14 300	292	874	143	122	56.1	369	62.7	2 100	12.4
x100.5	0.988	12 800	266	814	144	126	47.2	311	60.7	1 450	10.0
WT420											
x179.5	1.76	22 900	363	1 080	126	97.6	195	965	92.2	7 470	39.3
x164.5	1.62	21 000	333	997	126	97.0	174	870	91.2	5 720	30.4
x149.5	1.47	19 100	303	912	126	96.2	156	780	90.4	4 270	22.9
WT420											
x113	1.11	14 400	247	778	131	108	56.9	387	62.8	2 570	11.5
x105	1.03	13 400	230	734	131	109	51.3	350	61.8	2 020	9.57
x96.5	0.949	12 300	213	688	131	111	45.1	309	60.5	1 530	7.81
x88	0.864	11 200	197	646	132	114	39.1	268	59.0	1 110	6.35
WT380											
x157	1.54	20 000	254	828	113	86.3	158	822	88.8	5 800	26.0
x142	1.39	18 100	229	750	112	85.0	140	733	88.0	4 280	19.1
x128.5	1.26	16 400	206	680	112	83.8	125	656	87.3	3 180	14.2
WT380											
x98	0.965	12 500	175	613	118	99.0	40.9	305	57.1	2 020	7.63
x92.5	0.906	11 800	165	580	118	99.1	37.5	281	56.5	1 660	6.44
x86.5	0.851	11 100	156	554	119	100	34.4	257	55.7	1 340	5.54
x80.5	0.786	10 200	145	523	119	102	30.4	228	54.5	1 030	4.62
x73.5	0.721	9 360	133	490	119	104	26.4	200	53.2	777	3.81
WT345											
x132.5	1.30	16 900	172	624	101	77.4	116	646	82.8	4 050	15.5
x120	1.18	15 300	155	564	101	75.9	103	580	82.2	3 030	11.5
x108.5	1.07	13 900	140	514	101	74.9	92.6	522	81.7	2 280	8.57

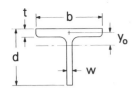

STRUCTURAL TEES
Cut from W Shapes
WT460 - WT345

PROPERTIES AND DIMENSIONS

Nominal Mass	Theoretical Mass	Depth d	Flange Width b	Flange Thickness t	Stem Thickness w	y_o	$\bar{r}_o$	β
kg/m	kg/m	mm	mm	mm	mm	mm	mm	
223	223.5	466	423	42.7	24.0	83.4	185	0.797
208.5	209.0	464	422	39.9	22.5	83.4	185	0.796
193.5	193.4	460	420	36.6	21.3	84.4	185	0.791
182.5	182.3	458	419	34.3	20.3	85.1	185	0.788
171	171.2	456	418	32.0	19.3	85.8	185	0.784
156.5	156.4	466	309	34.5	21.1	107	190	0.686
144.5	144.4	464	308	32.0	19.4	106	190	0.688
135.5	135.9	462	307	30.0	18.4	106	189	0.685
126.5	126.9	460	306	27.9	17.3	107	189	0.683
119	119.2	458	305	25.9	16.5	108	190	0.678
111.5	112.2	456	304	23.9	15.9	110	191	0.669
100.5	100.7	452	304	20.1	15.2	116	194	0.645
179.5	179.7	434	403	35.6	21.1	79.8	175	0.793
164.5	164.7	431	401	32.4	19.7	80.8	175	0.787
149.5	149.8	428	400	29.2	18.2	81.6	175	0.783
113	113.4	426	294	26.8	16.1	94.9	173	0.700
105	105.4	423	293	24.4	15.4	96.9	174	0.691
96.5	96.8	420	292	21.7	14.7	99.9	176	0.677
88	88.1	418	292	18.8	14.0	104	179	0.659
157	157.2	393	384	33.4	19.7	69.6	159	0.809
142	142.0	390	382	30.1	18.0	69.9	159	0.806
128.5	128.7	386	381	27.1	16.6	70.2	158	0.803
98	98.4	385	268	25.4	15.6	86.3	157	0.698
92.5	92.4	383	267	23.6	14.9	87.3	157	0.693
86.5	86.8	381	267	21.6	14.4	89.3	159	0.683
80.5	80.2	379	266	19.3	13.8	92.2	160	0.669
73.5	73.5	376	265	17.0	13.2	95.3	162	0.653
132.5	132.3	353	358	30.2	18.4	62.3	145	0.815
120	119.9	350	356	27.4	16.8	62.2	144	0.813
108.5	109.0	348	355	24.8	15.4	62.5	144	0.811

See CAN/CSA S16.1-94 Appendix D for definition of y_o, $\bar{r}_o$ and β.

STRUCTURAL TEES
Cut from W Shapes
WT345 - WT265

PROPERTIES

Designation	Dead Load	Total Area	Axis X-X				Axis Y-Y			Torsional Constant	Warping Constant
			I_x	S_x	r_x	y	I_y	S_y	r_y	J	C_w
	kN/m	mm^2	10^{6}mm^4	10^{3}mm^3	mm	mm	10^{6}mm^4	10^{3}mm^3	mm	10^{3}mm^4	10^{9}mm^6
WT345											
x85	0.833	10 800	120	463	105	86.7	33.1	259	55.3	1 520	4.69
x76	0.746	9 690	107	415	105	85.8	28.9	227	54.6	1 100	3.38
x70	0.686	8 910	99.3	389	106	86.5	25.9	204	53.9	833	2.72
x62.5	0.616	8 000	90.0	359	106	88.3	22.0	174	52.5	586	2.10
WT305											
x120.5	1.19	15 400	123	491	89.2	68.6	92.1	560	77.3	3 840	11.8
x108.5	1.07	13 900	110	444	88.8	67.4	81.6	497	76.7	2 790	8.58
x97.5	0.959	12 500	99.4	408	89.3	67.4	71.2	435	75.6	1 980	6.23
x87	0.854	11 100	88.3	366	89.2	66.5	61.9	381	74.7	1 400	4.40
x77.5	0.760	9 870	78.9	329	89.4	66.1	53.9	333	73.9	975	3.10
WT305											
x70	0.687	8 920	77.1	332	93.0	75.9	22.6	196	50.3	1 090	2.56
x62.5	0.613	7 970	69.0	299	93.1	75.4	19.7	172	49.7	769	1.84
x56.5	0.556	7 220	63.2	278	93.5	76.3	17.1	150	48.7	559	1.43
x50.5	0.499	6 480	57.4	256	94.1	77.8	14.7	129	47.7	389	1.09
WT305											
x46	0.453	5 880	54.8	256	96.5	87.9	7.20	80.5	35.0	354	1.05
x41	0.402	5 220	48.7	231	96.6	89.1	6.04	67.9	34.0	243	0.785
WT265											
x109.5	1.07	13 900	85.0	388	78.1	60.8	78.4	493	75.0	3 200	8.74
x98	0.964	12 500	75.2	345	77.5	59.0	69.3	438	74.4	2 340	6.28
x91	0.891	11 600	69.3	317	77.3	57.9	63.6	404	74.1	1 860	4.94
x82.5	0.811	10 500	62.2	288	76.9	56.7	56.8	363	73.4	1 410	3.70
x75	0.739	9 600	56.5	261	76.7	55.5	51.4	330	73.2	1 080	2.79
WT265											
x69	0.677	8 800	59.6	291	82.3	69.2	19.3	181	46.9	1 250	2.49
x61.5	0.604	7 850	52.6	258	81.9	67.6	16.9	159	46.4	899	1.75
x54.5	0.535	6 950	46.2	227	81.5	66.1	14.8	140	46.1	630	1.20
x50.5	0.497	6 450	42.6	210	81.2	65.5	13.5	128	45.7	507	0.966
x46	0.453	5 880	38.9	194	81.3	65.7	11.9	114	45.0	380	0.748
x41	0.404	5 250	35.1	178	81.8	66.7	10.1	97.1	44.0	265	0.555
WT265											
x42.5	0.416	5 400	37.8	194	83.7	72.7	6.32	76.1	34.2	367	0.675
x37	0.366	4 760	33.3	176	83.7	74.4	5.21	62.7	33.1	239	0.511
x33	0.322	4 180	29.5	158	84.0	75.7	4.29	52.0	32.0	159	0.376

STRUCTURAL TEES
Cut from W Shapes
WT345 - WT265

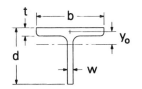

PROPERTIES AND DIMENSIONS

Nominal Mass	Theoretical Mass	Depth d	Flange Width b	Flange Thickness t	Stem Thickness w	y_o	$\bar{r}_o$	β
kg/m	kg/m	mm	mm	mm	mm	mm	mm	
85	84.9	346	256	23.6	14.5	74.9	141	0.716
76	76.1	344	254	21.1	13.1	75.2	140	0.713
70	69.9	342	254	18.9	12.4	77.1	141	0.703
62.5	62.8	339	253	16.3	11.7	80.2	143	0.686
120.5	120.9	318	329	31.0	17.9	53.1	129	0.832
108.5	108.9	314	328	27.7	16.5	53.5	129	0.828
97.5	97.8	311	327	24.4	15.4	55.2	129	0.818
87	87.1	308	325	21.6	14.0	55.7	129	0.813
77.5	77.5	306	324	19.0	12.7	56.6	129	0.808
70	70.0	308	230	22.2	13.1	64.8	124	0.727
62.5	62.5	306	229	19.6	11.9	65.6	124	0.721
56.5	56.7	304	228	17.3	11.2	67.7	125	0.708
50.5	50.9	302	228	14.9	10.5	70.3	127	0.692
46	46.2	302	179	15.0	10.9	80.4	130	0.620
41	41.0	300	178	12.8	10.0	82.7	132	0.606
109.5	109.5	280	318	29.2	18.3	46.2	118	0.846
98	98.3	277	316	26.3	16.5	45.9	117	0.846
91	90.9	276	315	24.4	15.2	45.7	116	0.846
82.5	82.7	273	313	22.2	14.0	45.6	116	0.845
75	75.4	272	312	20.3	12.7	45.4	115	0.845
69	69.1	274	214	23.6	14.7	57.4	111	0.731
61.5	61.6	272	212	21.2	13.1	57.0	110	0.731
54.5	54.6	270	211	18.8	11.6	56.7	109	0.732
50.5	50.7	268	210	17.4	10.9	56.8	109	0.729
46	46.2	266	209	15.6	10.2	57.9	109	0.721
41	41.2	264	209	13.3	9.5	60.1	111	0.705
42.5	42.4	268	166	16.5	10.3	64.4	111	0.663
37	37.3	264	166	13.6	9.7	67.6	113	0.640
33	32.8	262	165	11.4	8.9	70.0	114	0.622

See CAN/CSA S16.1-94 Appendix D for definition of y_o, $\bar{r}_o$ and β.

STRUCTURAL TEES
Cut from W Shapes
WT230 - WT205

PROPERTIES

Designation	Dead Load	Total Area	Axis X-X				Axis Y-Y			Torsional Constant	Warping Constant
			I_x	S_x	r_x	y	I_y	S_y	r_y	J	C_w
	kN/m	mm²	10^6mm⁴	10^3mm³	mm	mm	10^6mm⁴	10^3mm³	mm	10^3mm⁴	10^9mm⁶
WT230											
x88.5	0.869	11 300	49.4	260	66.1	51.4	52.5	367	68.2	2 200	4.66
x79	0.773	10 000	43.5	231	65.8	50.1	45.7	322	67.4	1 550	3.25
x72	0.709	9 210	39.0	208	65.1	48.4	41.8	295	67.4	1 220	2.49
x64	0.630	8 190	34.5	185	64.9	47.2	36.7	260	66.9	856	1.74
x56.5	0.555	7 210	30.2	162	64.7	46.0	31.7	226	66.3	590	1.18
WT230											
x53	0.518	6 730	32.4	183	69.4	57.1	12.6	130	43.2	726	1.07
x48.5	0.474	6 150	29.4	166	69.1	55.8	11.4	118	43.1	563	0.802
x44.5	0.438	5 690	27.0	152	68.8	54.8	10.5	109	42.9	453	0.630
x41	0.402	5 220	24.8	141	68.9	54.8	9.31	97.5	42.2	345	0.493
x37	0.363	4 720	22.1	127	68.4	53.7	8.30	87.4	41.9	258	0.363
x33.5	0.330	4 290	20.5	119	69.1	54.7	7.27	76.6	41.2	188	0.281
x30.5	0.295	3 830	18.6	111	69.7	56.5	6.09	64.4	39.9	129	0.215
WT230											
x34	0.336	4 370	21.8	128	70.6	59.2	4.70	61.1	32.8	254	0.323
x30	0.292	3 800	18.8	111	70.4	58.3	3.98	52.0	32.4	167	0.213
x26	0.255	3 310	16.7	102	71.0	60.8	3.17	41.7	30.9	105	0.160
WT205											
x74.5	0.733	9 520	32.2	188	58.2	44.9	38.8	293	63.9	1 600	2.79
x66	0.647	8 410	27.6	164	57.3	43.0	33.7	256	63.3	1 120	1.91
x57	0.562	7 290	23.8	141	57.2	41.6	28.6	219	62.6	741	1.24
x50	0.489	6 350	20.3	121	56.6	39.8	24.8	191	62.5	496	0.810
WT205											
x42.5	0.416	5 410	20.1	126	60.9	49.0	9.02	99.6	40.8	461	0.531
x37	0.367	4 770	17.6	111	60.7	47.9	7.79	86.6	40.4	318	0.363
x33.5	0.331	4 300	15.8	100	60.7	47.3	6.90	77.0	40.0	234	0.265
x30	0.292	3 800	13.9	87.8	60.4	46.1	6.02	67.7	39.8	164	0.180
x27	0.262	3 410	12.8	83.3	61.4	48.0	5.05	57.0	38.5	113	0.139
WT205											
x23	0.227	2 950	11.5	76.3	62.4	51.5	2.57	36.7	29.5	96.0	0.099 0
x19.5	0.192	2 500	9.96	67.9	63.1	53.5	2.02	28.8	28.4	55.3	0.067 5

STRUCTURAL TEES
Cut from W Shapes
WT230 - WT205

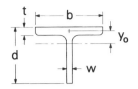

PROPERTIES AND DIMENSIONS

Nominal Mass	Theoretical Mass	Depth	Flange Width	Flange Thickness	Stem Thickness	y_o	$\bar{r}_o$	β
		d	b	t	w			
kg/m	kg/m	mm	mm	mm	mm	mm	mm	
88.5	88.6	241	286	26.9	16.6	38.0	102	0.862
79	78.8	238	284	23.9	15.0	38.1	102	0.859
72	72.3	236	283	22.1	13.6	37.3	101	0.863
64	64.3	234	282	19.6	12.2	37.4	100	0.861
56.5	56.6	232	280	17.3	10.8	37.4	99.9	0.860
53	52.8	234	194	20.6	12.6	46.8	94.2	0.753
48.5	48.3	233	193	19.0	11.4	46.3	93.7	0.756
44.5	44.7	232	192	17.7	10.5	45.9	93.2	0.757
41	41.0	230	191	16.0	9.9	46.8	93.4	0.749
37	37.1	228	190	14.5	9.0	46.5	92.8	0.749
33.5	33.7	227	190	12.7	8.5	48.3	93.8	0.735
30.5	30.1	225	189	10.8	8.1	51.1	95.2	0.712
34	34.3	230	154	15.4	9.1	51.5	93.4	0.696
30	29.8	228	153	13.3	8.0	51.7	93.1	0.692
26	26.0	225	152	10.8	7.6	55.4	95.2	0.662
74.5	74.7	216	265	25.0	14.9	32.4	92.3	0.877
66	66.0	212	263	22.2	13.3	31.9	91.2	0.878
57	57.3	210	261	19.3	11.6	31.9	90.6	0.876
50	49.8	208	260	16.9	10.0	31.4	89.9	0.878
42.5	42.5	208	181	18.2	10.9	39.9	83.5	0.772
37	37.4	206	180	16.0	9.7	39.9	83.1	0.770
33.5	33.8	205	179	14.4	8.8	40.1	83.0	0.767
30	29.8	204	178	12.8	7.7	39.7	82.5	0.769
27	26.7	202	177	10.9	7.5	42.6	84.0	0.743
23	23.1	202	140	11.2	7.0	45.9	82.9	0.694
19.5	19.6	200	140	8.8	6.4	49.1	84.8	0.666

See CAN/CSA S16.1-94 Appendix D for definition of y_o, $\bar{r}_o$ and β.

STRUCTURAL TEES
Cut from W Shapes
WT180

PROPERTIES

Designation	Dead Load	Total Area	Axis X-X				Axis Y-Y			Torsional Constant	Warping Constant
			I_x	S_x	r_x	y	I_y	S_y	r_y	J	C_w
	kN/m	mm²	10⁶mm⁴	10³mm³	mm	mm	10⁶mm⁴	10³mm³	mm	10³mm⁴	10⁹mm⁶
WT180											
x543	5.33	69 300	305	1 560	66.4	88.0	981	4 320	119	299 000	1 410
x495	4.86	63 100	259	1 350	64.1	82.7	867	3 870	117	232 000	1 060
x450	4.43	57 500	219	1 160	61.7	77.5	767	3 470	115	180 000	791
x409	4.02	52 200	184	997	59.4	72.4	678	3 100	114	138 000	585
x372	3.65	47 400	156	864	57.4	67.9	600	2 780	112	106 000	434
x338.5	3.33	43 200	134	754	55.8	63.9	534	2 500	111	81 600	325
WT180											
x317	3.11	40 400	119	677	54.3	61.0	491	2 320	110	68 300	266
x296	2.90	37 700	107	615	53.2	58.4	451	2 140	109	56 400	215
x275.5	2.70	35 100	95.9	557	52.3	55.8	412	1 970	108	46 000	172
x254.5	2.50	32 500	84.8	499	51.1	53.0	377	1 810	108	36 700	135
x231.5	2.27	29 500	73.8	439	50.0	50.1	335	1 630	107	28 100	100
x210.5	2.07	26 800	63.3	384	48.6	47.1	300	1 470	106	21 600	75.4
x191	1.87	24 400	55.4	338	47.7	44.5	268	1 320	105	16 300	56.0
x173.5	1.70	22 100	48.5	300	46.8	42.1	240	1 190	104	12 400	41.6
x157	1.54	20 000	42.6	266	46.2	39.9	213	1 060	103	9 210	30.3
x143.5	1.41	18 300	37.0	234	45.0	37.7	194	972	103	7 220	23.4
x131	1.29	16 700	33.9	215	45.0	36.4	175	880	102	5 510	17.6
x118.5	1.16	15 000	29.1	187	44.0	34.3	155	786	102	4 080	12.8
x108	1.06	13 800	26.2	169	43.6	32.9	141	717	101	3 160	9.79
WT180											
x98	0.964	12 500	24.0	157	43.8	32.7	114	611	95.5	2 560	7.17
x89.5	0.879	11 400	21.6	141	43.5	31.4	103	555	95.2	1 950	5.40
x81	0.794	10 300	18.8	124	42.7	29.8	92.8	500	94.8	1 470	4.00
x73.5	0.723	9 400	17.0	113	42.6	28.9	83.6	452	94.3	1 110	2.98
x67	0.657	8 530	15.2	101	42.2	27.8	75.4	409	94.0	840	2.22
WT180											
x61	0.597	7 760	17.3	118	47.2	35.5	30.7	239	62.9	1 050	1.51
x55	0.540	7 020	15.0	102	46.2	33.5	27.9	218	63.0	801	1.12
x50.5	0.496	6 440	13.5	92.6	45.7	32.5	25.3	199	62.7	627	0.860
x45.5	0.445	5 780	11.9	82.6	45.4	31.5	22.4	177	62.3	457	0.614
WT180											
x39.5	0.389	5 050	11.5	81.2	47.8	35.0	12.1	118	48.9	406	0.394
x36	0.351	4 550	10.3	73.1	47.5	34.2	10.7	105	48.5	301	0.286
x32	0.314	4 080	9.17	65.2	47.4	33.4	9.42	92.8	48.1	219	0.202

STRUCTURAL TEES
Cut from W Shapes
WT180

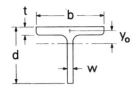

PROPERTIES AND DIMENSIONS

Nominal Mass	Theoretical Mass	Depth d	Flange Width b	Flange Thickness t	Stem Thickness w	y_o	$\bar{r}_o$	β
kg/m	kg/m	mm	mm	mm	mm	mm	mm	
543	543.6	284	454	125	78.0	25.5	139	0.966
495	495.5	275	448	115	71.9	25.2	136	0.966
450	451.3	266	442	106	65.9	24.5	133	0.966
409	409.5	257	437	97.0	60.5	23.9	131	0.966
372	372.1	249	432	88.9	55.6	23.5	128	0.967
338.5	339.1	242	428	81.5	51.2	23.1	127	0.967
317	317.1	237	424	77.1	47.6	22.4	125	0.968
296	296.1	232	421	72.3	45.0	22.2	124	0.968
275.5	275.5	228	418	67.6	42.0	22.0	122	0.968
254.5	254.7	223	416	62.7	39.1	21.6	121	0.968
231.5	231.6	218	412	57.4	35.8	21.4	120	0.968
210.5	210.7	212	409	52.6	32.8	20.8	118	0.969
191	191.2	208	406	48.0	29.8	20.5	117	0.969
173.5	173.6	204	404	43.7	27.2	20.2	116	0.970
157	156.8	200	401	39.6	24.9	20.1	115	0.969
143.5	143.7	196	399	36.6	22.6	19.4	114	0.971
131	131.4	194	398	33.3	21.1	19.8	113	0.970
118.5	118.1	190	395	30.2	18.9	19.2	112	0.971
108	108.2	188	394	27.7	17.3	19.0	112	0.971
98	98.3	186	374	26.2	16.4	19.6	107	0.966
89.5	89.6	184	373	23.9	15.0	19.5	106	0.966
81	81.0	182	371	21.8	13.3	18.9	106	0.968
73.5	73.8	180	370	19.8	12.3	19.0	105	0.967
67	67.0	178	369	18.0	11.2	18.8	105	0.968
61	60.9	182	257	21.7	13.0	24.6	82.4	0.911
55	55.1	180	256	19.9	11.4	23.6	81.6	0.916
50.5	50.6	178	255	18.3	10.5	23.4	81.0	0.917
45.5	45.4	176	254	16.4	9.5	23.3	80.5	0.916
39.5	39.6	177	205	16.8	9.4	26.6	73.4	0.868
36	35.8	175	204	15.1	8.6	26.7	72.9	0.866
32	32.0	174	203	13.5	7.7	26.6	72.6	0.865

See CAN/CSA S16.1-94 Appendix D for definition of y_o, $\bar{r}_o$ and β.

STRUCTURAL TEES
Cut from W Shapes
WT180 - WT155

PROPERTIES

Designation	Dead Load	Total Area	Axis X-X				Axis Y-Y			Torsional Constant	Warping Constant
			I_x	S_x	r_x	y	I_y	S_y	r_y	J	C_w
	kN/m	mm^2	10^{6}mm^4	10^{3}mm^3	mm	mm	10^{6}mm^4	10^{3}mm^3	mm	10^{3}mm^4	10^{9}mm^6
WT180											
x28.5	0.278	3 610	9.70	69.4	51.9	39.2	5.56	64.7	39.3	167	0.150
x25.5	0.248	3 230	8.73	62.8	52.0	39.0	4.84	56.6	38.7	119	0.107
x22.5	0.221	2 870	7.96	58.6	52.7	40.2	4.09	47.8	37.8	79.7	0.0784
WT180											
x19.5	0.192	2 490	7.16	54.1	53.7	43.5	1.88	29.3	27.4	75.2	0.0558
x16.5	0.160	2 080	6.10	47.1	54.1	44.5	1.45	22.9	26.4	42.8	0.0352
WT155											
x250	2.46	31 900	79.7	513	50.0	58.7	247	1 450	88.0	50 100	130
x227	2.23	28 900	68.2	446	48.6	55.2	218	1 300	86.8	38 300	95.7
x207.5	2.04	26 500	59.6	397	47.4	52.2	195	1 170	85.9	29 500	71.9
x187.5	1.84	23 900	50.5	343	46.0	48.9	172	1 040	84.8	22 300	52.5
x171	1.68	21 900	43.8	302	44.8	46.1	155	946	84.2	17 400	40.0
x156.5	1.54	20 000	38.5	269	43.9	43.8	139	852	83.3	13 400	30.1
WT155											
x141.5	1.39	18 000	32.5	231	42.5	40.9	123	764	82.6	10 100	22.1
x126.5	1.24	16 100	28.1	202	41.8	38.6	107	673	81.6	7 350	15.6
x113	1.11	14 400	24.3	176	41.0	36.4	94.6	597	81.0	5 360	11.1
x101	0.993	12 900	21.0	155	40.3	34.4	82.9	527	80.2	3 850	7.81
x89.5	0.876	11 400	17.8	134	39.6	32.3	71.9	459	79.5	2 680	5.29
x79	0.772	10 000	15.2	114	38.9	30.3	62.4	402	78.8	1 880	3.63
x71.5	0.702	9 120	13.5	101	38.4	28.9	56.3	365	78.6	1 430	2.72
x64.5	0.636	8 260	12.0	91.8	38.2	27.9	50.2	326	78.0	1 060	1.98
x59	0.576	7 490	10.7	82.0	37.8	26.8	45.1	294	77.6	799	1.46
x53.5	0.525	6 820	9.71	74.7	37.7	26.0	40.6	265	77.2	607	1.09
x48.5	0.475	6 170	8.59	66.6	37.3	25.0	36.4	239	76.9	455	0.804
WT155											
x43	0.424	5 500	7.93	61.5	38.0	26.1	22.3	175	63.6	438	0.559
x39.5	0.387	5 030	7.38	58.1	38.3	26.1	20.0	157	63.0	328	0.413
WT155											
x37	0.365	4 740	7.81	62.3	40.6	29.7	11.7	114	49.7	372	0.332
x33.5	0.327	4 250	6.88	55.3	40.2	28.7	10.3	101	49.3	272	0.236
x30	0.293	3 800	6.05	48.6	39.9	27.6	9.14	90.1	49.0	198	0.167
WT155											
x26	0.256	3 330	6.54	52.3	44.3	32.8	5.13	61.4	39.2	153	0.117
x22.5	0.219	2 840	5.54	44.6	44.2	31.9	4.27	51.5	38.8	95.4	0.0718
x19.5	0.190	2 470	4.82	39.0	44.2	31.4	3.63	44.0	38.4	62.8	0.0468
WT155											
x16.5	0.160	2 080	4.83	42.2	48.2	41.3	0.959	18.8	21.5	60.6	0.0366
x14	0.139	1 800	4.19	37.3	48.2	41.7	0.790	15.5	20.9	37.7	0.0253
x12	0.117	1 520	3.60	33.4	48.7	44.2	0.578	11.4	19.5	21.1	0.0182
x10.5	0.104	1 350	3.25	30.3	49.1	45.0	0.491	9.73	19.1	14.6	0.0136

STRUCTURAL TEES
Cut from W Shapes
WT180 - WT155

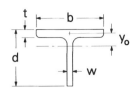

PROPERTIES AND DIMENSIONS

Nominal Mass	Theoretical Mass	Depth d	Flange Width b	Flange Thickness t	Stem Thickness w	y_o	$\bar{r}_o$	β
kg/m	kg/m	mm	mm	mm	mm	mm	mm	
28.5	28.3	179	172	13.1	7.9	32.6	72.8	0.799
25.5	25.3	178	171	11.6	7.2	33.2	72.8	0.793
22.5	22.5	176	171	9.8	6.9	35.3	73.8	0.771
19.5	19.5	176	128	10.7	6.5	38.1	71.3	0.714
16.5	16.4	174	127	8.5	5.8	40.2	72.4	0.691
250	250.4	214	340	75.1	45.1	21.1	103	0.958
227	227.1	208	336	68.7	41.3	20.8	102	0.958
207.5	207.7	202	334	62.7	38.9	20.8	100	0.957
187.5	187.5	196	330	57.2	35.4	20.3	98.6	0.958
171	171.6	191	328	52.6	32.6	19.8	97.4	0.959
156.5	156.7	187	325	48.3	30.0	19.6	96.2	0.958
141.5	141.4	182	322	44.1	26.9	18.9	94.8	0.960
126.5	126.5	178	319	39.6	24.4	18.8	93.6	0.960
113	113.4	174	317	35.6	22.1	18.6	92.6	0.960
101	101.2	170	315	31.8	20.1	18.5	91.7	0.959
89.5	89.3	166	313	28.1	18.0	18.3	90.7	0.959
79	78.8	164	310	25.1	15.5	17.8	89.7	0.961
71.5	71.6	162	309	22.9	14.0	17.5	89.2	0.962
64.5	64.8	159	308	20.6	13.1	17.6	88.6	0.960
59	58.8	157	307	18.7	11.9	17.4	88.1	0.961
53.5	53.5	156	306	17.0	10.9	17.5	87.7	0.960
48.5	48.4	154	305	15.4	9.9	17.3	87.2	0.961
43	43.2	155	254	16.3	9.1	18.0	76.2	0.944
39.5	39.5	153	254	14.6	8.8	18.8	76.1	0.939
37	37.2	155	205	16.3	9.4	21.5	67.7	0.899
33.5	33.4	153	204	14.6	8.5	21.4	67.1	0.898
30	29.8	152	203	13.1	7.5	21.1	66.7	0.900
26	26.1	158	167	13.2	7.6	26.2	64.7	0.837
22.5	22.3	156	166	11.2	6.6	26.3	64.4	0.833
19.5	19.4	155	165	9.7	5.8	26.5	64.3	0.830
16.5	16.4	156	102	10.8	6.6	35.9	63.8	0.683
14	14.2	154	102	8.9	6.0	37.3	64.4	0.666
12	11.9	152	101	6.7	5.6	40.9	66.6	0.622
10.5	10.6	152	101	5.7	5.1	42.2	67.5	0.609

See CAN/CSA S16.1-94 Appendix D for definition of y_o, $\bar{r}_o$ and β.

STRUCTURAL TEES
Cut from W Shapes
WT125 - WT100

PROPERTIES

Designation	Dead Load	Total Area	Axis X-X				Axis Y-Y			Torsional Constant	Warping Constant
			I_x	S_x	r_x	y	I_y	S_y	r_y	J	C_w
	kN/m	mm²	10^6mm⁴	10^3mm³	mm	mm	10^6mm⁴	10^3mm³	mm	10^3mm⁴	10^9mm⁶
WT125											
x83.5	0.820	10 700	11.9	105	33.4	30.6	49.4	373	68.1	3 140	4.57
x74.5	0.730	9 490	10.2	91.3	32.9	28.8	43.1	328	67.4	2 250	3.19
x65.5	0.643	8 360	8.73	78.7	32.3	27.0	37.2	285	66.7	1 560	2.15
x57.5	0.562	7 300	7.17	65.8	31.3	25.0	32.0	247	66.2	1 060	1.43
x50.5	0.496	6 440	6.17	57.0	31.0	23.6	27.7	216	65.6	741	0.973
x44.5	0.439	5 700	5.39	50.2	30.7	22.5	24.2	189	65.1	517	0.664
x40	0.393	5 100	4.61	43.2	30.1	21.2	21.6	169	65.0	377	0.477
x36.5	0.357	4 640	4.08	38.6	29.6	20.3	19.4	153	64.7	287	0.356
WT125											
x33.5	0.329	4 270	4.24	40.4	31.5	23.0	11.1	109	51.0	311	0.262
x29	0.286	3 710	3.68	35.5	31.5	22.2	9.42	92.8	50.4	204	0.167
x24.5	0.241	3 130	3.25	32.0	32.2	22.3	7.56	74.9	49.2	120	0.094 9
WT125											
x22.5	0.220	2 860	3.86	36.7	36.7	27.8	3.52	47.5	35.1	130	0.074 1
x19.5	0.190	2 460	3.26	31.3	36.4	26.7	2.97	40.4	34.7	84.1	0.046 7
x16.5	0.160	2 080	2.85	28.1	37.0	27.3	2.36	32.4	33.7	49.1	0.028 4
WT125											
x14	0.140	1 810	2.79	28.7	39.2	32.6	0.888	17.4	22.1	48.2	0.021 6
x12.5	0.124	1 610	2.50	26.4	39.4	33.3	0.746	14.6	21.5	32.4	0.016 3
x11	0.110	1 430	2.27	24.5	39.9	34.6	0.613	12.0	20.7	21.6	0.012 6
x9	0.088	1 140	1.83	20.0	40.1	34.8	0.457	9.04	20.0	11.2	0.006 8
WT100											
x50	0.486	6 310	4.49	49.7	26.7	23.7	18.3	174	53.9	1 020	0.946
x43	0.424	5 510	3.80	42.8	26.3	22.2	15.7	150	53.4	682	0.617
x35.5	0.349	4 540	2.86	32.5	25.1	19.8	12.7	123	52.9	399	0.349
x29.5	0.290	3 760	2.39	27.7	25.2	18.7	10.2	99.5	52.1	225	0.191
x26	0.255	3 310	2.00	23.4	24.6	17.5	8.92	87.4	51.9	156	0.130
x23	0.224	2 910	1.79	21.1	24.8	17.0	7.67	75.6	51.3	106	0.086 6
WT100											
x21	0.203	2 640	1.73	20.8	25.6	18.5	4.50	54.2	41.3	107	0.061 4
x18	0.174	2 270	1.44	17.4	25.2	17.5	3.82	46.3	41.1	69.4	0.038 8
WT100											
x15.5	0.154	2 000	1.63	19.4	28.5	21.1	2.05	30.6	32.0	59.4	0.025 0
x13.5	0.131	1 700	1.43	17.3	29.1	21.3	1.65	24.8	31.2	35.5	0.015 1
WT100											
x11	0.110	1 430	1.36	17.5	30.9	25.3	0.710	13.9	22.3	28.1	0.010 2
x9.5	0.096	1 240	1.22	16.0	31.3	26.1	0.577	11.3	21.6	18.0	0.007 2
x7.5	0.073	952	0.885	11.7	30.5	24.1	0.434	8.69	21.4	8.75	0.003 0

STRUCTURAL TEES
Cut from W Shapes
WT125 - WT100

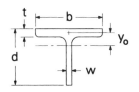

PROPERTIES AND DIMENSIONS

Nominal Mass	Theoretical Mass	Depth d	Flange Width b	Flange Thickness t	Stem Thickness w	y_o	$\bar{r}_o$	β
kg/m	kg/m	mm	mm	mm	mm	mm	mm	
83.5	83.6	144	265	31.8	19.2	14.7	77.2	0.964
74.5	74.5	141	263	28.4	17.3	14.6	76.4	0.963
65.5	65.6	138	261	25.1	15.4	14.5	75.6	0.963
57.5	57.3	134	259	22.1	13.5	14.0	74.6	0.965
50.5	50.6	132	257	19.6	11.9	13.8	73.9	0.965
44.5	44.8	130	256	17.3	10.7	13.9	73.4	0.964
40	40.1	128	255	15.6	9.4	13.4	72.9	0.966
36.5	36.4	126	254	14.2	8.6	13.2	72.4	0.967
33.5	33.5	128	204	15.7	8.9	15.1	61.8	0.940
29	29.1	126	203	13.5	8.0	15.5	61.4	0.937
24.5	24.6	124	202	11.0	7.4	16.8	61.1	0.925
22.5	22.5	133	148	13.0	7.6	21.3	55.1	0.851
19.5	19.3	131	147	11.2	6.6	21.1	54.5	0.850
16.5	16.4	129	146	9.1	6.1	22.7	54.9	0.829
14	14.2	130	102	10.0	6.4	27.6	52.9	0.727
12.5	12.6	128	102	8.4	6.1	29.1	53.5	0.705
11	11.2	127	102	6.9	5.8	31.1	54.7	0.676
9	8.9	126	101	5.3	4.8	32.1	55.1	0.660
50	49.6	114	210	23.7	14.5	11.9	61.3	0.962
43	43.2	111	209	20.6	13.0	11.9	60.7	0.962
35.5	35.6	108	206	17.4	10.2	11.1	59.6	0.966
29.5	29.5	105	205	14.2	9.1	11.6	59.0	0.961
26	26.0	103	204	12.6	7.9	11.2	58.5	0.964
23	22.9	102	203	11.0	7.2	11.5	58.1	0.961
21	20.7	102	166	11.8	7.2	12.6	50.3	0.937
18	17.8	100	165	10.2	6.2	12.4	49.7	0.938
15.5	15.7	105	134	10.2	6.4	16.0	45.8	0.877
13.5	13.3	104	133	8.4	5.8	17.1	45.9	0.862
11	11.2	103	102	8.0	6.2	21.3	43.6	0.761
9.5	9.7	102	102	6.5	5.8	22.8	44.3	0.734
7.5	7.5	100	100	5.2	4.3	21.5	43.0	0.750

See CAN/CSA S16.1-94 Appendix D for definition of y_o, $\bar{r}_o$ and β.

STRUCTURAL TEES
Cut from W Shapes
WT75 - WT50

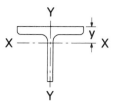

PROPERTIES

Designation	Dead Load	Total Area	Axis X-X				Axis Y-Y			Torsional Constant	Warping Constant
			I_x	S_x	r_x	y	I_y	S_y	r_y	J	C_w
	kN/m	mm²	10^6mm⁴	10^3mm³	mm	mm	10^6mm⁴	10^3mm³	mm	10^3mm⁴	10^9mm⁶
WT75											
x18.5	0.182	2 370	0.947	14.5	20.0	15.5	3.53	45.9	38.6	96.0	0.045 9
x15	0.146	1 900	0.725	11.3	19.6	14.2	2.78	36.3	38.3	50.2	0.023 2
x11	0.110	1 420	0.581	9.38	20.2	14.1	1.93	25.4	36.9	20.9	0.009 1
WT75											
x12	0.118	1 530	0.708	11.3	21.5	17.3	0.913	17.9	24.4	46.4	0.011 4
x9	0.088	1 140	0.544	9.16	21.8	17.1	0.629	12.3	23.5	18.6	0.004 7
x7	0.067	866	0.395	6.68	21.4	15.8	0.459	9.18	23.0	8.50	0.002 0
WT65											
x14	0.138	1 790	0.426	8.02	15.4	12.4	1.91	29.8	32.6	63.9	0.020 8
x12	0.116	1 510	0.350	6.74	15.2	11.6	1.55	24.5	32.1	38.3	0.012 0
WT50											
x9.5	0.095	1 240	0.221	5.28	13.4	11.2	0.803	15.6	25.5	31.5	0.006 3

STRUCTURAL TEES
Cut from W Shapes
WT75 - WT50

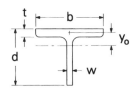

PROPERTIES AND DIMENSIONS

Nominal Mass	Theoretical Mass	Depth d	Flange Width b	Flange Thickness t	Stem Thickness w	y_o	$\bar{r}_o$	β
kg/m	kg/m	mm	mm	mm	mm	mm	mm	
18.5	18.6	81.0	154	11.6	8.1	9.68	44.6	0.953
15	14.9	78.5	153	9.3	6.6	9.51	44.0	0.953
11	11.2	76.0	152	6.6	5.8	10.8	43.4	0.938
12	12.0	80.0	102	10.3	6.6	12.1	34.7	0.878
9	9.0	76.5	102	7.1	5.8	13.5	34.8	0.849
7	6.8	75.0	100	5.5	4.3	13.0	34.0	0.853
14	14.0	65.5	128	10.9	6.9	6.96	36.8	0.964
12	11.8	63.5	127	9.1	6.1	7.07	36.3	0.962
9.5	9.7	53.0	103	8.8	7.1	6.80	29.6	0.947

See CAN/CSA S16.1-94 Appendix D for definition of y_o, $\bar{r}_o$ and β.

HOLLOW STRUCTURAL SECTIONS

General

Hollow structural sections are produced in Canada to requirements of the CSA-G40.20 Standard to either Class C or Class H, from steel meeting the requirements of the CSA-G40.21 material Standard. The normal grade of steel used is G40.21 350W. See pages 6-15 and 6-16 for descriptions of the class of hollow section and for data on manufacturing tolerances. Round sections produced in accordance with common pipe specifications may sometimes be used as structural members, but are not classified as hollow structural sections.

Properties and Dimensions

The tables of properties and dimensions on pages 6-98 to 6-103 include the majority of the standard Canadian rectangular, square and round hollow structural sections covered by the CAN3-G312.3-M92 Standard, plus some additional HSS sizes currently produced in Canada. The designations (e.g. HSS 127 x 76 x 6.4 Class C) used throughout this Handbook should be used by designers and detailers when calling up hollow structural sections, as non-Canadian hollow sections may differ in name, size, designation, tolerances and material grade from those produced in Canada.

Section properties given in the following tables for square and rectangular sections are based on an interior corner radius taken equal to the wall thickness, and on an exterior corner radius taken equal to twice the wall thickness.

Manufacture

Hollow structural sections produced to the CSA-G40.20 Standard may be manufactured using either a seamless or welding process. Seamless products are produced by piercing solid material to form a tube or by an extrusion type process. Welded products are manufactured from flat-rolled steel which is formed and joined by various welding processes into a tubular shape. The tubular shape is then either cold-formed or hot-formed to the final shape and, if cold-formed, may be subsequently stress relieved. Class H sections are either hot-formed to final shape, or are cold-formed to final shape and then stress relieved. Class C sections are generally more readily available than class H sections, though Class H sections have greater resistance in direct compression. Outside dimensions for hollow structural sections are constant for all sizes in the same size range, the inside dimensions changing with material thickness.

HSS manufactured to ASTM A500

HSS manufactured to ASTM Standard A500 grade C are not equivalent to HSS meeting the requirements of CSA G40.21 grade 350W. If HSS produced to A500 are offered as a substitute, it would be prudent to assess the influence of the differences that arise from a possible difference in wall thickness and material strengths. Information on design issues related to this substitution may be found on CISC's web site (www.cisc-icca.ca).

Tables of properties and dimensions for A500 sections are given in a publication entitled: "Hollow Structural Sections to ASTM A500 Grade C", available from CISC. This publication also provides design tables for A500 sections used as beams and columns in accordance with CAN/CSA S16.1-94.

Availability

Some sections are marked with a + sign to indicate that their availability should be checked before selecting them at the design stage. The sections so marked may be either:

– produced in only one geographical market area of Canada;

– subject to tonnage accumulation;

– produced infrequently; or,

– not kept in stock by the producer.

Since the sections listed in this Handbook are those best suited for structural applications, designers may wish to consult the catalogues of HSS producers supplying HSS to their region of the country for sections not listed herein.

Pipe

Properties and dimension for pipe provided in the table on page 6-104 have been soft converted to metric from the ASTM A53 "Standard Specification for Pipe, Steel, Black and Hot-Dipped, Zinc-Coated, Welded and Seamless". Although not a normal structural quality steel, pipe produced in accordance with the ASTM A53 Standard is available in three types and two grades, with specified minimum yield strengths of 205 or 240 MPa. These are considerably lower than 350 MPa generally used for hollow structural sections.

Size and wall thickness of seamless and welded mill pipe up to 323.9 mm outside diameter permit threading the ends for joining lengths with couplings or other connectors. The wall thickness of mill pipe has been expressed in terms of "standard wall" (STD), "extra strong" (XS), "double extra strong" (XXS), and in terms of "schedule numbers" (Sch). STD is the same as Sch 40 for all sizes up to and including 273.1 mm outside diameter, XS is the same as Sch 80 for all sizes up to and including 219.1 mm outside diameter, and XXS is the next heavier pipe to the Sch 160 pipe for all sizes up to and including 168.3 mm outside diameter. For pipe designation, see page vii.

HOLLOW STRUCTURAL SECTIONS
Rectangular

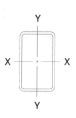

PROPERTIES AND DIMENSIONS

Designation*	Wall Thickness	Mass	Dead Load	Area	Axis X-X				Axis Y-Y				Torsional Constant	Shear Constant
					I_x	S_x	r_x	Z_x	I_y	S_y	r_y	Z_y	J	C_{rt}
mm x mm x mm	mm	kg/m	kN/m	mm²	10^6 mm⁴	10^3 mm³	mm	10^3 mm³	10^6 mm⁴	10^3 mm³	mm	10^3 mm³	10^3 mm⁴	mm²
HSS 305x203														
x13	12.70	93.0	0.912	11 800	147	964	111	1 190	78.2	769	81.2	897	167 000	6 450
x11+	11.13	82.4	0.808	10 500	132	867	112	1 060	70.5	694	82.0	802	149 000	5 790
x9.5	9.53	71.3	0.700	9 090	116	762	113	926	62.1	611	82.7	701	130 000	5 080
x8.0+	7.95	60.1	0.590	7 660	99.4	652	114	787	53.3	525	83.4	596	111 000	4 340
x6.4	6.35	48.6	0.476	6 190	81.5	535	115	640	43.8	431	84.1	486	89 800	3 550
HSS 254x152														
x13	12.70	72.7	0.713	9 260	75.2	592	90.1	747	33.6	442	60.3	522	78 200	5 160
x11+	11.13	64.6	0.634	8 230	68.2	537	91.0	671	30.6	402	61.0	470	70 200	4 660
x9.5	9.53	56.1	0.551	7 150	60.4	475	91.9	589	27.2	357	61.7	413	61 600	4 110
x8.0	7.95	47.5	0.466	6 050	52.0	410	92.7	503	23.6	309	62.4	354	52 600	3 530
x6.4	6.35	38.4	0.377	4 900	42.9	338	93.6	411	19.5	256	63.1	290	43 000	2 900
HSS 203x152														
x13	12.70	62.6	0.614	7 970	43.0	423	73.4	528	27.3	359	58.6	432	56 400	3 870
x11+	11.13	55.7	0.547	7 100	39.2	385	74.3	476	25.0	327	59.3	390	50 800	3 530
x9.5	9.53	48.5	0.476	6 180	34.8	343	75.1	420	22.3	292	60.0	344	44 600	3 150
x8.0	7.95	41.1	0.403	5 240	30.2	297	75.9	360	19.3	254	60.8	295	38 200	2 730
x6.4	6.35	33.4	0.327	4 250	25.0	246	76.7	295	16.1	211	61.5	243	31 200	2 260
x4.8	4.78	25.5	0.250	3 250	19.5	192	77.5	228	12.6	165	62.2	188	24 100	1 760
HSS 203x102														
x13+	12.70	52.4	0.515	6 680	31.3	308	68.4	405	10.2	201	39.1	246	27 000	3 870
x11+	11.13	46.9	0.460	5 970	28.7	283	69.4	368	9.48	187	39.8	224	24 600	3 530
x9.5	9.53	40.9	0.401	5 210	25.8	254	70.3	326	8.57	169	40.5	199	21 900	3 150
x8.0	7.95	34.8	0.341	4 430	22.5	221	71.2	281	7.54	148	41.3	172	18 900	2 730
x6.4	6.35	28.3	0.278	3 610	18.8	185	72.2	232	6.35	125	42.0	143	15 600	2 260
x4.8	4.78	21.7	0.213	2 760	14.7	145	73.1	180	5.03	99.0	42.7	111	12 200	1 760
HSS 178x127														
x13	12.70	52.4	0.515	6 680	26.4	297	62.9	378	15.5	244	48.1	298	33 600	3 230
x11+	11.13	46.9	0.460	5 970	24.2	273	63.7	343	14.3	225	48.9	271	30 400	2 970
x9.5	9.53	40.9	0.401	5 210	21.7	244	64.6	303	12.8	202	49.6	240	26 900	2 660
x8.0	7.95	34.8	0.341	4 430	19.0	213	65.4	261	11.2	177	50.3	207	23 100	2 320
x6.4	6.35	28.3	0.278	3 610	15.8	178	66.2	216	9.40	148	51.1	171	19 000	1 940
x4.8	4.78	21.7	0.213	2 760	12.4	140	67.1	168	7.41	117	51.8	133	14 700	1 520
HSS 152x102														
x13	12.70	42.3	0.415	5 390	14.7	193	52.2	252	7.67	151	37.7	189	17 800	2 580
x11+	11.13	38.0	0.373	4 840	13.6	179	53.1	230	7.15	141	38.4	173	16 300	2 400
x9.5	9.53	33.3	0.327	4 240	12.4	162	54.0	206	6.51	128	39.2	155	14 500	2 180
x8.0	7.95	28.4	0.279	3 620	10.9	143	54.8	179	5.76	113	39.9	135	12 600	1 920
x6.4	6.35	23.2	0.228	2 960	9.19	121	55.7	148	4.88	96.2	40.6	112	10 500	1 610
x4.8	4.78	17.9	0.175	2 280	7.28	95.6	56.5	116	3.89	76.6	41.3	87.8	8 160	1 270

* Depth x Width x Thickness, see page 6-19.
+ Check availability
Note: C_{rt} is the ratio of applied shear force to maximum shear stress for the strong bending axis.

HOLLOW STRUCTURAL SECTIONS
Rectangular

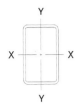

PROPERTIES AND DIMENSIONS

Designation*	Wall Thick-ness	Mass	Dead Load	Area	Axis X-X				Axis Y-Y				Torsional Constant	Shear Constant
					I_x	S_x	r_x	Z_x	I_y	S_y	r_y	Z_y	J	C_{rt}
mm x mm x mm	mm	kg/m	kN/m	mm²	10^6 mm⁴	10^3 mm³	mm	10^3 mm³	10^6 mm⁴	10^3 mm³	mm	10^3 mm³	10^3 mm⁴	mm²
HSS 152x76														
x9.5	9.53	29.5	0.290	3 760	9.89	130	51.3	171	3.24	85.0	29.4	104	8 560	2 180
x8.0	7.95	25.3	0.248	3 220	8.79	115	52.3	149	2.91	76.3	30.1	91.2	7 510	1 920
x6.4	6.35	20.7	0.203	2 640	7.47	98.0	53.2	125	2.50	65.5	30.8	76.6	6 300	1 610
x4.8	4.78	16.0	0.157	2 040	5.96	78.2	54.1	98.1	2.02	52.9	31.5	60.5	4 970	1 270
HSS 127x76														
x9.5	9.53	25.7	0.252	3 280	6.13	96.5	43.3	126	2.70	70.8	28.7	87.8	6 600	1 690
x8.0+	7.95	22.1	0.217	2 820	5.49	86.5	44.2	111	2.44	63.9	29.4	77.4	5 810	1 510
x6.4	6.35	18.2	0.178	2 320	4.70	74.1	45.1	93.4	2.10	55.2	30.1	65.3	4 890	1 290
x4.8	4.78	14.1	0.138	1 790	3.78	59.6	45.9	73.8	1.71	44.8	30.8	51.8	3 860	1 030
x3.8+	3.81	11.4	0.112	1 450	3.14	49.4	46.5	60.6	1.42	37.3	31.3	42.6	3 170	852
HSS 102x76														
x9.5	9.53	21.9	0.215	2 790	3.42	67.4	35.0	87.9	2.16	56.6	27.8	71.6	4 710	1 210
x8.0+	7.95	18.9	0.186	2 410	3.10	61.1	35.9	77.9	1.96	51.5	28.5	63.6	4 170	1 110
x6.4	6.35	15.6	0.153	1 990	2.69	52.9	36.7	66.0	1.71	44.8	29.3	54.0	3 530	968
x4.8	4.78	12.2	0.119	1 550	2.18	43.0	37.5	52.6	1.39	36.6	30.0	43.1	2 800	789
x3.8+	3.81	9.89	0.097	1 260	1.82	35.9	38.0	43.3	1.17	30.7	30.5	35.6	2 300	658
x3.2	3.18	8.35	0.082	1 060	1.57	30.8	38.4	37.0	1.01	26.4	30.7	30.4	1 960	565
HSS 102x51														
x8.0+	7.95	15.8	0.155	2 010	2.21	43.6	33.2	59.0	0.714	28.1	18.9	35.6	1 950	1 110
x6.4	6.35	13.1	0.129	1 670	1.95	38.5	34.2	50.7	0.640	25.2	19.6	30.8	1 690	968
x4.8	4.78	10.3	0.101	1 310	1.61	31.8	35.1	40.8	0.537	21.1	20.3	25.0	1 370	789
x3.8+	3.81	8.37	0.082	1 070	1.36	26.8	35.7	33.9	0.457	18.0	20.7	20.8	1 140	658
x3.2	3.18	7.09	0.070	903	1.17	23.1	36.1	29.0	0.397	15.6	21.0	17.9	979	565
HSS 89x64														
x8.0+	7.95	15.8	0.155	2 010	1.88	42.2	30.6	55.1	1.09	34.4	23.3	43.3	2 450	908
x6.4+	6.35	13.1	0.129	1 670	1.65	37.1	31.4	47.2	0.968	30.5	24.1	37.3	2 100	806
x4.8+	4.78	10.3	0.101	1 310	1.36	30.6	32.3	38.0	0.803	25.3	24.8	30.1	1 690	667
x3.8+	3.81	8.37	0.082	1 070	1.15	25.8	32.8	31.5	0.679	21.4	25.2	25.0	1 400	561
x3.2+	3.18	7.09	0.070	903	0.990	22.3	33.1	27.0	0.588	18.5	25.5	21.4	1 190	485
HSS 76x51														
x8.0+	7.95	12.6	0.124	1 600	1.02	26.7	25.2	36.1	0.527	20.7	18.1	26.9	1 270	706
x6.4	6.35	10.6	0.104	1 350	0.919	24.1	26.1	31.5	0.479	18.9	18.9	23.6	1 110	645
x4.8	4.78	8.35	0.082	1 060	0.775	20.3	27.0	25.8	0.408	16.1	19.6	19.4	911	546
x3.8+	3.81	6.85	0.067	872	0.660	17.3	27.5	21.6	0.350	13.8	20.0	16.3	762	465
x3.2	3.18	5.82	0.057	741	0.575	15.1	27.8	18.6	0.306	12.0	20.3	14.0	655	404
HSS 51x25														
x3.2	3.18	3.28	0.032	418	0.122	4.81	17.1	6.34	0.0400	3.15	9.78	3.85	106	242

* Depth x Width x Thickness, see page 6-19.
+ Check availability
Note: C_{rt} is the ratio of applied shear force to maximum shear stress for the strong bending axis.

HOLLOW STRUCTURAL SECTIONS
Square

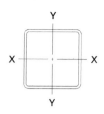

PROPERTIES AND DIMENSIONS

Designation*	Wall Thickness	Mass	Dead Load	Area	I	S	r	Z	Torsional Constant J	Shear Constant C_{rt}	Surface Area
mm x mm x mm	mm	kg/m	kN/m	mm²	10^6 mm⁴	10^3 mm³	mm	10^3 mm³	10^3 mm⁴	mm²	m²/m
HSS 305x305											
x13	12.70	113	1.11	14 400	202	1 330	118	1 560	324 000	6 450	1.18
x11+	11.13	100	0.982	12 800	181	1 190	119	1 390	288 000	5 790	1.18
x9.5	9.53	86.5	0.849	11 000	158	1 040	120	1 210	250 000	5 080	1.19
x8.0+	7.95	72.8	0.714	9 280	135	886	121	1 030	211 000	4 340	1.19
x6.4	6.35	58.7	0.576	7 480	110	723	121	833	171 000	3 550	1.20
HSS 254x254											
x13	12.70	93.0	0.912	11 800	113	889	97.6	1 060	183 000	5 160	0.972
x11+	11.13	82.4	0.808	10 500	102	800	98.4	946	163 000	4 660	0.978
x9.5	9.53	71.3	0.700	9 090	89.3	703	99.1	825	142 000	4 110	0.983
x8.0+	7.95	60.1	0.590	7 660	76.5	602	99.9	702	121 000	3 530	0.989
x6.4	6.35	48.6	0.476	6 190	62.7	494	101	571	97 900	2 900	0.994
HSS 203x203											
x13	12.70	72.7	0.713	9 260	54.7	538	76.9	651	90 700	3 870	0.769
x11+	11.13	64.6	0.634	8 230	49.6	488	77.6	585	81 200	3 530	0.775
x9.5	9.53	56.1	0.551	7 150	43.9	432	78.4	513	71 000	3 150	0.780
x8.0	7.95	47.5	0.466	6 050	37.9	373	79.2	439	60 500	2 730	0.786
x6.4	6.35	38.4	0.377	4 900	31.3	308	79.9	359	49 300	2 260	0.791
HSS 178x178											
x13	12.70	62.6	0.614	7 970	35.2	396	66.5	484	59 200	3 230	0.668
x11+	11.13	55.7	0.547	7 100	32.1	361	67.2	437	53 200	2 970	0.673
x9.5	9.53	48.5	0.476	6 180	28.6	322	68.0	385	46 700	2 660	0.678
x8.0+	7.95	41.1	0.403	5 240	24.8	279	68.8	330	39 900	2 320	0.684
x6.4	6.35	33.4	0.327	4 250	20.6	231	69.6	271	32 700	1 940	0.689
x4.8	4.78	25.5	0.250	3 250	16.1	181	70.3	210	25 200	1 520	0.695
HSS 152x152											
x13	12.70	52.4	0.515	6 680	21.0	276	56.1	342	36 000	2 580	0.566
x11+	11.13	46.9	0.460	5 970	19.3	253	56.9	310	32 500	2 400	0.571
x9.5	9.53	40.9	0.401	5 210	17.3	227	57.6	275	28 700	2 180	0.577
x8.0	7.95	34.8	0.341	4 430	15.1	198	58.4	237	24 600	1 920	0.582
x6.4	6.35	28.3	0.278	3 610	12.6	166	59.2	196	20 300	1 610	0.588
x4.8	4.78	21.7	0.213	2 760	9.93	130	59.9	152	15 700	1 270	0.593
HSS 127x127											
x13	12.70	42.3	0.415	5 390	11.3	177	45.7	225	19 800	1 940	0.464
x11+	11.13	38.0	0.373	4 840	10.5	165	46.5	205	18 000	1 840	0.470
x9.5	9.53	33.3	0.327	4 240	9.48	149	47.3	183	16 000	1 690	0.475
x8.0	7.95	28.4	0.279	3 620	8.36	132	48.0	159	13 900	1 510	0.481
x6.4	6.35	23.2	0.228	2 960	7.05	111	48.8	132	11 500	1 290	0.486
x4.8	4.78	17.9	0.175	2 280	5.60	88.1	49.6	103	8 920	1 030	0.492

* Depth x Width x Thickness, see page 6-19.
+ Check availability
Note: C_{rt} is the ratio of applied shear force to maximum shear stress.

HOLLOW STRUCTURAL SECTIONS
Square

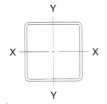

PROPERTIES AND DIMENSIONS

Designation*	Wall Thickness	Mass	Dead Load	Area	I	S	r	Z	Torsional Constant J	Shear Constant C_{rt}	Surface Area
mm x mm x mm	mm	kg/m	kN/m	mm^2	10^6 mm^4	10^3 mm^3	mm	10^3 mm^3	10^3 mm^4	mm^2	m^2/m
HSS 102x102											
x9.5	9.53	25.7	0.252	3 280	4.45	87.6	36.9	110	7 740	1 210	0.374
x8.0	7.95	22.1	0.217	2 820	3.99	78.5	37.6	96.8	6 780	1 110	0.379
x6.4	6.35	18.2	0.178	2 320	3.42	67.3	38.4	81.4	5 670	968	0.385
x4.8	4.78	14.1	0.138	1 790	2.75	54.2	39.2	64.3	4 450	789	0.390
x3.8+	3.81	11.4	0.112	1 450	2.29	45.0	39.7	52.8	3 640	658	0.393
x3.2	3.18	9.62	0.094	1 230	1.96	38.5	40.0	44.9	3 090	565	0.395
HSS 89x89											
x9.5	9.53	21.9	0.215	2 790	2.80	63.0	31.7	80.5	4 970	968	0.323
x8.0+	7.95	18.9	0.186	2 410	2.54	57.1	32.4	71.4	4 390	908	0.328
x6.4	6.35	15.6	0.153	1 990	2.20	49.5	33.2	60.5	3 700	806	0.334
x4.8	4.78	12.2	0.119	1 550	1.79	40.3	34.0	48.2	2 930	667	0.339
x3.8+	3.81	9.89	0.097	1 260	1.50	33.7	34.5	39.8	2 400	561	0.343
x3.2+	3.18	8.35	0.082	1 060	1.29	29.0	34.8	33.9	2 040	485	0.345
HSS 76x76											
x9.5	9.53	18.1	0.178	2 310	1.61	42.4	26.5	55.5	2 940	726	0.272
x8.0+	7.95	15.8	0.155	2 010	1.49	39.1	27.2	49.8	2 630	706	0.278
x6.4	6.35	13.1	0.129	1 670	1.31	34.5	28.0	42.8	2 250	645	0.283
x4.8	4.78	10.3	0.101	1 310	1.08	28.5	28.8	34.4	1 800	546	0.288
x3.8+	3.81	8.37	0.082	1 070	0.914	24.0	29.3	28.6	1 490	465	0.292
x3.2	3.18	7.09	0.070	903	0.790	20.7	29.6	24.5	1 270	404	0.294
HSS 64x64											
x6.4	6.35	10.6	0.104	1 350	0.703	22.2	22.8	28.1	1 230	484	0.232
x4.8	4.78	8.35	0.082	1 060	0.594	18.7	23.6	23.0	1 000	424	0.238
x3.8+	3.81	6.85	0.067	872	0.506	16.0	24.1	19.2	836	368	0.241
x3.2	3.18	5.82	0.057	741	0.441	13.9	24.4	16.6	717	323	0.243
HSS 51x51											
x6.4	6.35	8.05	0.079	1 030	0.319	12.6	17.6	16.4	580	323	0.181
x4.8	4.78	6.45	0.063	821	0.279	11.0	18.4	13.8	485	303	0.187
x3.8+	3.81	5.33	0.052	679	0.243	9.55	18.9	11.7	410	271	0.190
x3.2	3.18	4.55	0.045	580	0.214	8.42	19.2	10.2	355	242	0.192
HSS 38x38											
x4.8	4.78	4.54	0.045	578	0.101	5.30	13.2	6.95	184	181	0.136
x3.8+	3.81	3.81	0.037	485	0.091 2	4.79	13.7	6.06	160	174	0.139
x3.2	3.18	3.28	0.032	418	0.082 2	4.31	14.0	5.35	141	161	0.141

* Depth x Width x Thickness, see page 6-19.
+ Check availability
Note: C_{rt} is the ratio of applied shear force to maximum shear stress.

HOLLOW STRUCTURAL SECTIONS
Round

PROPERTIES AND DIMENSIONS

Designation*	Wall Thickness	Mass	Dead Load	Area	I	S	r	Z	Torsional Constant J	Shear Constant C_{rt}	Surface Area
mm x mm	mm	kg/m	kN/m	mm²	10^6 mm⁴	10^3 mm³	mm	10^3 mm³	10^3 mm⁴	mm²	m²/m
HSS 610											
x13+	12.70	187	1.83	23 800	1 060	3 480	211	4 530	2 120 000	11 900	1.92
x11+	11.13	164	1.61	20 900	937	3 070	212	3 990	1 870 000	10 500	1.92
x9.5+	9.53	141	1.38	18 000	809	2 650	212	3 430	1 620 000	8 980	1.92
x8.0+	7.95	118	1.16	15 000	680	2 230	213	2 880	1 360 000	7 510	1.92
x6.4+	6.35	94.5	0.927	12 000	547	1 800	213	2 310	1 090 000	6 020	1.92
HSS 559											
x13+	12.70	171	1.68	21 800	813	2 910	193	3 790	1 630 000	10 900	1.76
x11+	11.13	150	1.47	19 100	718	2 570	194	3 340	1 440 000	9 580	1.76
x9.5+	9.53	129	1.27	16 400	620	2 220	194	2 880	1 240 000	8 220	1.76
x8.0+	7.95	108	1.06	13 800	522	1 870	195	2 410	1 040 000	6 880	1.76
x6.4+	6.35	86.5	0.849	11 000	421	1 510	195	1 940	841 000	5 510	1.76
HSS 508											
x13+	12.70	155	1.52	19 800	606	2 390	175	3 120	1 210 000	9 890	1.60
x11+	11.13	136	1.34	17 400	536	2 110	176	2 750	1 070 000	8 690	1.60
x9.5+	9.53	117	1.15	14 900	464	1 830	176	2 370	927 000	7 460	1.60
x8.0+	7.95	98.0	0.962	12 500	390	1 540	177	1 990	781 000	6 250	1.60
x6.4+	6.35	78.6	0.771	10 000	315	1 240	177	1 600	630 000	5 000	1.60
HSS 406											
x13+	12.70	123	1.21	15 700	305	1 500	139	1 970	609 000	7 860	1.28
x11+	11.13	108	1.06	13 800	270	1 330	140	1 740	540 000	6 910	1.28
x9.5+	9.53	93.3	0.915	11 900	234	1 150	140	1 500	468 000	5 940	1.28
x8.0+	7.95	78.1	0.766	9 950	198	972	141	1 260	395 000	4 980	1.28
x6.4+	6.35	62.6	0.615	7 980	160	786	141	1 020	319 000	3 990	1.28
HSS 356											
x13+	12.70	107	1.05	13 700	201	1 130	121	1 490	403 000	6 850	1.12
x11+	11.13	94.6	0.928	12 000	179	1 010	122	1 320	358 000	6 030	1.12
x9.5+	9.53	81.3	0.798	10 400	155	873	122	1 140	310 000	5 180	1.12
x8.0+	7.95	68.2	0.669	8 680	131	738	123	961	262 000	4 340	1.12
x6.4+	6.35	54.7	0.537	6 970	106	598	123	775	213 000	3 480	1.12
HSS 324											
x13+	12.70	97.5	0.956	12 400	151	930	110	1 230	301 000	6 220	1.02
x11+	11.13	85.8	0.842	10 900	134	827	111	1 090	268 000	5 470	1.02
x9.5+	9.53	73.9	0.725	9 410	116	719	111	942	233 000	4 710	1.02
x8.0+	7.95	61.9	0.608	7 890	98.5	608	112	794	197 000	3 950	1.02
x6.4+	6.35	49.7	0.488	6 330	79.9	493	112	640	160 000	3 170	1.02
HSS 273											
x13+	12.70	81.6	0.800	10 400	88.3	646	92.2	862	177 000	5 200	0.858
x11+	11.13	71.9	0.705	9 160	78.7	577	92.7	764	157 000	4 590	0.858
x9.5+	9.53	61.9	0.608	7 890	68.6	502	93.2	662	137 000	3 950	0.858
x8.0+	7.95	52.0	0.510	6 620	58.2	427	93.8	559	116 000	3 310	0.858
x6.4+	6.35	41.8	0.410	5 320	47.4	347	94.3	452	94 700	2 660	0.858

* Diameter x Thickness, see page 6-19.
+ Check availability
Note: C_{rt} is the ratio of applied shear force to maximum shear stress.

HOLLOW STRUCTURAL SECTIONS
Round

PROPERTIES AND DIMENSIONS

Designation*	Wall Thickness	Mass	Dead Load	Area	I	S	r	Z	Torsional Constant J	Shear Constant C_{rt}	Surface Area
mm x mm	mm	kg/m	kN/m	mm²	10⁶ mm⁴	10³ mm³	mm	10³ mm³	10³ mm⁴	mm²	m²/m
HSS 219											
x13	12.70	64.6	0.634	8 230	44.0	402	73.1	542	88 000	4 130	0.688
x11+	11.13	57.1	0.560	7 270	39.4	360	73.6	482	78 900	3 640	0.688
x9.5	9.53	49.3	0.483	6 270	34.5	315	74.2	419	69 000	3 140	0.688
x8.0	7.95	41.4	0.406	5 270	29.4	269	74.7	355	58 900	2 640	0.688
x6.4	6.35	33.3	0.327	4 240	24.0	219	75.3	288	48 100	2 120	0.688
x4.8	4.78	25.3	0.248	3 220	18.5	169	75.8	220	37 000	1 610	0.688
HSS 168											
x9.5+	9.53	37.3	0.366	4 750	15.0	179	56.2	241	30 100	2 380	0.529
x8.0+	7.95	31.4	0.308	4 000	12.9	153	56.8	205	25 800	2 010	0.529
x6.4+	6.35	25.4	0.249	3 230	10.6	126	57.3	167	21 200	1 620	0.529
x4.8+	4.78	19.3	0.189	2 460	8.21	97.6	57.8	128	16 400	1 230	0.529
HSS 141											
x9.5+	9.53	31.0	0.304	3 950	8.61	122	46.7	166	17 200	1 980	0.444
x8.0+	7.95	26.1	0.256	3 330	7.43	105	47.2	142	14 900	1 670	0.444
x6.4+	6.35	21.1	0.207	2 690	6.14	86.9	47.8	116	12 300	1 350	0.444
x4.8+	4.78	16.1	0.158	2 050	4.78	67.7	48.3	89.1	9 560	1 030	0.444
HSS 114											
x8.0	7.95	20.9	0.205	2 660	3.78	66.1	37.7	90.1	7 550	1 330	0.359
x6.4	6.35	16.9	0.166	2 150	3.15	55.1	38.2	74.1	6 300	1 080	0.359
x4.8	4.78	12.9	0.127	1 640	2.47	43.2	38.8	57.4	4 940	823	0.359
HSS 102											
x8.0	7.95	18.4	0.180	2 340	2.58	50.8	33.2	69.9	5 170	1 180	0.319
x6.4	6.35	14.9	0.146	1 900	2.16	42.6	33.8	57.7	4 330	953	0.319
x4.8	4.78	11.4	0.112	1 450	1.71	33.6	34.3	44.8	3 420	728	0.319
x3.8	3.81	9.19	0.090	1 170	1.40	27.6	34.6	36.5	2 800	586	0.319
HSS 89											
x8.0	7.95	15.9	0.156	2 020	1.67	37.6	28.8	52.3	3 340	1 020	0.279
x6.4	6.35	12.9	0.127	1 650	1.41	31.7	29.3	43.4	2 820	827	0.279
x4.8	4.78	9.92	0.097	1 260	1.12	25.2	29.8	33.9	2 240	633	0.279
x3.8	3.81	8.00	0.078	1 020	0.924	20.8	30.1	27.6	1 850	510	0.279
HSS 73											
x6.4+	6.35	10.4	0.102	1 330	0.745	20.4	23.7	28.3	1 490	669	0.229
x4.8+	4.78	8.04	0.079	1 020	0.599	16.4	24.2	22.3	1 200	514	0.229
x3.8+	3.81	6.50	0.064	828	0.497	13.6	24.5	18.3	994	415	0.229
x3.2+	3.18	5.48	0.054	698	0.426	11.7	24.7	15.5	852	349	0.229
HSS 60											
x6.4	6.35	8.45	0.083	1 080	0.397	13.2	19.2	18.6	794	543	0.189
x4.8	4.78	6.54	0.064	834	0.324	10.7	19.7	14.8	647	419	0.189
x3.8	3.81	5.31	0.052	676	0.271	8.99	20.0	12.2	542	339	0.189
x3.2	3.18	4.48	0.044	571	0.233	7.74	20.2	10.4	467	286	0.189
HSS 48											
x4.8	4.78	5.13	0.050	654	0.157	6.48	15.5	9.09	313	329	0.152
x3.8	3.81	4.18	0.041	533	0.133	5.50	15.8	7.56	265	268	0.152
x3.2	3.18	3.54	0.035	451	0.115	4.77	16.0	6.48	231	226	0.152

* Diameter x Thickness, see page 6-19.
+ Check availability
Note: C_{rt} is the ratio of applied shear force to maximum shear stress.

PIPE PROPERTIES AND DIMENSIONS

ASTM - A53

Designation[†]	Outside Dia.	Thickness	Weight Class[*]	Mass	Dead Load	Area	I	S	r	Z	Shear Constant	J	Surface Area
mm	mm	mm		kg/m	kN/m	mm^2	10^6 mm^4	10^3 mm^3	mm	10^3 mm^3	mm^2	10^6 mm^4	m^2/m
DN15	21.3	2.77	STD	1.27	0.0124	161	0.00708	0.664	6.62	0.958	81.8	0.0142	0.0669
	21.3	3.73	XS	1.62	0.0158	206	0.00830	0.780	6.35	1.17	106	0.0166	0.0669
	21.3	7.47	XXS	2.55	0.0250	325	0.0100	0.941	5.56	1.57	191	0.0200	0.0669
DN20	26.7	2.87	STD	1.69	0.0165	215	0.0155	1.16	8.49	1.64	108	0.0309	0.0839
	26.7	3.91	XS	2.20	0.0216	280	0.0187	1.40	8.18	2.05	143	0.0374	0.0839
	26.7	7.82	XXS	3.64	0.0357	464	0.0242	1.81	7.23	2.95	257	0.0484	0.0839
DN25	33.4	3.38	STD	2.50	0.0245	319	0.0364	2.18	10.7	3.06	161	0.0727	0.105
	33.4	4.55	XS	3.24	0.0317	412	0.0440	2.63	10.3	3.82	210	0.0879	0.105
	33.4	9.09	XXS	5.45	0.0534	694	0.0585	3.50	9.18	5.62	378	0.117	0.105
DN32	42.2	3.56	STD	3.39	0.0333	432	0.0813	3.85	13.7	5.33	217	0.163	0.133
	42.2	4.85	XS	4.47	0.0438	569	0.101	4.78	13.3	6.80	288	0.202	0.133
	42.2	9.70	XXS	7.77	0.0762	990	0.142	6.75	12.0	10.5	524	0.285	0.133
DN40	48.3	3.68	STD	4.05	0.0397	516	0.129	5.35	15.8	7.34	259	0.259	0.152
	48.3	5.08	XS	5.41	0.0531	690	0.163	6.76	15.4	9.53	348	0.327	0.152
	48.3	10.16	XXS	9.56	0.0937	1220	0.237	9.82	14.0	15.1	637	0.474	0.152
DN50	60.3	3.91	STD	5.44	0.0533	693	0.277	9.18	20.0	12.5	347	0.553	0.189
	60.3	5.54	XS	7.48	0.0734	953	0.361	12.0	19.5	16.7	480	0.722	0.189
	60.3	11.07	XXS	13.4	0.132	1710	0.545	18.1	17.8	27.3	884	1.09	0.189
DN65	73.0	5.16	STD	8.63	0.0847	1100	0.636	17.4	24.1	23.8	552	1.27	0.229
	73.0	7.01	XS	11.4	0.112	1450	0.800	21.9	23.5	30.6	732	1.60	0.229
	73.0	14.02	XXS	20.4	0.200	2600	1.19	32.7	21.4	49.7	1350	2.39	0.229
DN80	88.9	5.49	STD	11.3	0.111	1440	1.26	28.3	29.6	38.3	721	2.51	0.279
	88.9	7.62	XS	15.3	0.150	1950	1.62	36.5	28.9	50.5	979	3.24	0.279
	88.9	15.24	XXS	27.7	0.271	3530	2.49	56.1	26.6	83.9	1810	4.99	0.279
DN90	101.6	5.74	STD	13.6	0.133	1730	1.99	39.2	34.0	52.8	866	3.99	0.319
	101.6	8.08	XS	18.6	0.183	2370	2.61	51.5	33.2	70.8	1190	5.23	0.319
DN100	114.3	6.02	STD	16.1	0.158	2050	3.01	52.7	38.3	70.7	1030	6.02	0.359
	114.3	8.56	XS	22.3	0.219	2840	4.00	70.0	37.5	95.9	1430	8.00	0.359
	114.3	17.12	XXS	41.0	0.402	5230	6.36	111	34.9	163	2670	12.7	0.359
DN125	141.3	6.55	STD	21.8	0.213	2770	6.31	89.3	47.7	119	1390	12.6	0.444
	141.3	9.52	XS	30.9	0.303	3940	8.60	122	46.7	166	1980	17.2	0.444
	141.3	19.05	XXS	57.4	0.563	7320	14.0	198	43.7	287	3720	28.0	0.444
DN150	168.3	7.11	STD	28.3	0.277	3600	11.7	139	57.0	185	1800	23.4	0.529
	168.3	10.97	XS	42.6	0.417	5420	16.9	200	55.8	272	2720	33.7	0.529
	168.3	21.95	XXS	79.2	0.777	10100	27.6	328	52.3	474	5120	55.3	0.529
DN200	219.1	8.18	STD	42.5	0.417	5420	30.2	276	74.6	364	2710	60.4	0.688
	219.1	12.70	XS	64.6	0.634	8230	44.0	402	73.1	542	4130	88.0	0.688
	219.1	22.23	XXS	108	1.06	13700	67.5	616	70.0	865	6930	135	0.688
DN250	273.1	9.27	STD	60.3	0.591	7680	66.9	490	93.3	646	3840	134	0.858
	273.1	12.70	XS	81.6	0.800	10400	88.3	646	92.2	862	5200	177	0.858
	273.1	25.40	XXS	155	1.52	19800	153	1120	88.0	1560	9950	306	0.858
DN300	323.9	9.52	STD	73.8	0.724	9400	116	718	111	941	4700	233	1.02
	323.9	12.70	XS	97.5	0.956	12400	151	930	110	1230	6220	301	1.02
	323.9	25.40	XXS	187	1.83	23800	267	1650	106	2270	12000	534	1.02

[*] Class refers to: Standard Weight - STD, Extra Strong - XS, Double Extra Strong - XXS.
[†] This designation has been suggested by the U.S. National Institute of Building Sciences.

COLD FORMED CHANNELS

General

While various proprietary cold formed channels (CFC) are available from Canadian roll formers, the CFC series listed on pages 6-106 to 6-109 is that included in CSA Standard G312.3-M92. CFC sections listed have elements with width-thickness ratios exceeding the Class 3 limits of CAN/CSA-S16.1-94. Therefore, their factored resistances are determined according to the requirements of CSA Standard CSA-S136-94. Many of the resistances are determined using effective cross-sectional properties computed on the basis of some portion of the element(s) in compression notionally removed. The tables on pages 6-106 to 6-109 list both the gross cross-sectional and the effective cross-sectional properties — all computed assuming the inside bend radius to be twice the thickness. For deflection calculations, the effective moment of inertia, I_{xe}, can be conservatively used.

Material

The values in the table are based on CFC sections roll formed from steel meeting the requirement of ASTM A570M Grade 50 (F_y = 345 MPa). Where appropriate, the enhanced yield stress, F'_y, due to the effects of cold working is listed. Factored resistances of a member at a welded zone should be based on F_y and not on F'_y since welding can anneal the steel eliminating the enhancing effect of cold working.

Tables

The designation used is that approved for CSA Standard G312.3 and consists of the letters CFC, the nominal depth, the flange width and sheet steel thickness, all in millimetres. The following symbols are used for the CFC sections in addition to those used for hot rolled shapes:

F'_y = specified minimum yield strength including effects of cold work and forming, (MPa)

I_{xe} = effective moment of inertia about X-X axis, (10^6 mm^4)

S_{xe} = effective section modulus of the tension flange, (10^3 mm^3)

y_e = distance from the neutral axis to the extreme compressive fibre, (mm)

I_{ye} = effective moment of inertia about Y-Y axis assuming the stiffener lips are in tension, (10^6 mm^4)

S_{ye} = effective section modulus of the tension flange (i.e. stiffener lips), (10^3 mm^3)

x_e = distance from the neutral axis to the extreme compressive fibre, (mm)

M_{rxe} = effective factored moment resistance about the X-X axis based on F'_y, (kN·m)

L_u = maximum unbraced length of compression flange beyond which appropriate values in Table must be reduced for lateral buckling, (mm)

d_i = depth of stiffener lip, (mm)

x_o = distance from shear centre to Y-Y axis, (mm)

r_o = polar radius of gyration, (mm)

J = St. Venant's torsional constant, (10^3 mm^4)

j = torsional-flexural buckling parameter, (mm)

C_w = warping torsional constant, (10^9 mm^6)

These tables have been prepared by Dr. R. Schuster, University of Waterloo.

COLD FORMED CHANNELS
Properties and Dimensions
A570M Grade 50 (F_y = 345 MPa)

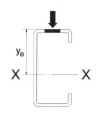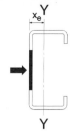

Designation	Mass	Area	F'_y	Effective Section Properties						M_{rxe}	L_u
				X-X Axis			Y-Y Axis				
				I_{xe}	S_{xe}	y_e	I_{ye}	S_{ye}	x_e		
	kg/m	mm²	MPa	10^6mm⁴	10^3mm³	mm	10^6mm⁴	10^3mm³	mm	kNm	mm
CFC 460x127											
x5.69	33.6	4 280	388	127	554	229	7.87	89.5	39.1	194	2 244
x4.93	29.4	3 740	383	110	478	231	6.67	78.7	42.2	165	2 271
x4.17	25.0	3 190	345	93.9	401	234	5.43	67.1	46.0	125	2 404
CFC 380x127											
x5.69	30.2	3 850	388	81.9	430	191	7.69	88.8	40.4	150	2 283
x4.93	26.4	3 370	383	72.2	379	191	6.55	78.2	43.2	131	2 308
x4.17	22.5	2 870	345	61.9	323	192	5.35	66.7	46.8	100	2 442
CFC 380x89											
x5.69	25.7	3 270	405	63.3	332	191	2.77	41.6	22.3	121	1 493
x4.93	22.5	2 870	398	56.0	294	191	2.43	37.3	23.9	105	1 518
x4.17	19.2	2 450	390	47.9	248	193	2.04	32.5	26.2	87.2	1 545
x3.43	16.0	2 040	383	39.5	201	197	1.63	27.3	29.3	69.2	1 571
x2.67	12.6	1 600	345	30.4	150	203	1.18	21.4	33.6	46.7	1 668
CFC 310x89											
x4.93	19.5	2 490	398	32.6	214	152	2.36	37.0	25.2	76.5	1 555
x4.17	16.7	2 130	390	28.2	185	152	1.99	32.2	27.2	65.0	1 579
x3.43	13.9	1 780	383	23.6	154	154	1.59	27.1	30.0	53.0	1 604
x2.67	11.0	1 400	345	18.4	117	158	1.17	21.3	34.0	36.2	1 701
CFC 310x70											
x4.93	17.6	2 240	411	27.4	180	152	1.16	21.9	16.8	66.4	1 161
x4.17	15.1	1 920	401	23.8	156	152	1.00	19.4	18.1	56.5	1 185
x3.43	12.6	1 600	392	20.0	130	154	0.824	16.6	20.1	45.8	1 210
x2.67	9.91	1 260	382	15.6	98.6	158	0.617	13.2	23.3	33.9	1 238
CFC 250x89											
x4.93	17.6	2 240	398	21.0	166	127	2.26	36.5	26.8	59.3	1 586
x4.17	15.1	1 920	390	18.2	144	127	1.94	32.0	28.2	50.5	1 608
x3.43	12.6	1 600	383	15.4	121	127	1.57	26.9	30.8	41.8	1 631
x2.67	9.91	1 260	345	12.1	94.1	129	1.15	21.2	34.5	29.2	1 728
x2.29	8.55	1 090	345	10.1	75.3	134	0.945	18.1	36.8	23.4	1 732
CFC 250x70											
x4.17	13.4	1 710	401	15.3	120	127	0.977	19.2	19.0	43.4	1 210
x3.43	11.2	1 430	392	12.9	102	127	0.808	16.5	20.8	36.0	1 234
x2.67	8.84	1 130	382	10.2	79.5	129	0.608	13.2	23.7	27.4	1 260
x2.29	7.64	973	345	8.82	67.7	130	0.503	11.4	25.7	21.0	1 332
x1.90	6.41	817	345	7.08	52.1	136	0.397	9.48	28.0	16.2	1 338

Note: For an explanation of this table, see page 6-105.

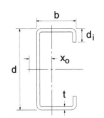

COLD FORMED CHANNELS
Properties and Dimensions

Depth	Flange Width	Stiff'r Depth	Thickness	Gross Section Properties						x_o	r_o	J	j	C_w
				X-X Axis			Y-Y Axis							
d	b	d_i	t	I_x	S_x	r_x	I_y	S_y	r_y					
mm	mm	mm	mm	$10^6 mm^4$	$10^3 mm^3$	mm	$10^6 mm^4$	$10^3 mm^3$	mm	mm	mm	$10^3 mm^4$	mm	$10^9 mm^6$
457	127	44.4	5.69	127	554	172	8.57	91.9	44.7	87.5	198	46.2	262	377
457	127	44.4	4.93	112	488	173	7.65	82.1	45.2	88.0	199	30.3	259	336
457	127	44.4	4.17	95.8	419	173	6.66	71.5	45.7	88.5	200	18.4	256	292
381	127	44.4	5.69	81.9	430	146	8.10	90.4	45.9	94.8	180	41.5	214	255
381	127	44.4	4.93	72.2	379	147	7.23	80.7	46.4	95.2	181	27.2	212	227
381	127	44.4	4.17	62.1	326	147	6.30	70.4	46.8	95.7	182	16.6	210	197
381	89	31.7	5.69	63.3	332	139	2.82	41.8	29.4	55.7	153	35.3	244	86.8
381	89	31.7	4.93	56.0	294	140	2.56	37.9	29.9	56.2	154	23.2	238	78.1
381	89	31.7	4.17	48.4	254	141	2.26	33.5	30.4	56.6	155	14.2	232	68.5
381	89	31.7	3.43	40.7	214	141	1.94	28.8	30.8	57.1	156	7.98	228	58.4
381	89	31.7	2.67	32.3	170	142	1.57	23.3	31.3	57.6	157	3.80	223	47.1
305	89	31.7	4.93	32.6	214	114	2.40	37.2	31.0	62.4	134	20.2	177	47.7
305	89	31.7	4.17	28.2	185	115	2.12	32.9	31.5	62.9	135	12.3	173	42.0
305	89	31.7	3.43	23.8	156	116	1.82	28.2	32.0	63.3	136	6.96	171	35.9
305	89	31.7	2.67	18.9	124	116	1.47	22.9	32.5	63.8	137	3.31	168	29.0
305	70	25.4	4.93	27.4	180	111	1.16	21.9	22.8	43.3	121	18.1	200	23.0
305	70	25.4	4.17	23.8	156	111	1.04	19.6	23.3	43.8	122	11.1	194	20.4
305	70	25.4	3.43	20.1	132	112	0.904	17.1	23.8	44.3	123	6.27	188	17.6
305	70	25.4	2.67	16.1	106	113	0.742	14.0	24.2	44.8	124	2.99	183	14.3
254	89	31.7	4.93	21.0	166	96.9	2.26	36.5	31.8	67.6	122	18.1	146	32.3
254	89	31.7	4.17	18.2	144	97.5	2.00	32.3	32.3	68.0	123	11.1	144	28.4
254	89	31.7	3.43	15.4	121	98.1	1.72	27.8	32.8	68.4	124	6.27	142	24.4
254	89	31.7	2.67	12.3	96.8	98.7	1.39	22.5	33.2	68.9	125	2.99	140	19.7
254	89	31.7	2.29	10.7	84.0	99.0	1.22	19.7	33.5	69.1	125	1.90	139	17.3
254	70	25.4	4.17	15.3	120	94.5	0.988	19.3	24.1	47.8	109	9.88	150	13.6
254	70	25.4	3.43	12.9	102	95.2	0.859	16.8	24.5	48.3	110	5.59	147	11.8
254	70	25.4	2.67	10.4	81.6	95.9	0.705	13.8	25.0	48.8	111	2.67	143	9.62
254	70	25.4	2.29	9.01	70.9	96.2	0.621	12.2	25.3	49.0	111	1.69	142	8.45
254	70	25.4	1.90	7.62	60.0	96.6	0.531	10.4	25.5	49.2	111	0.988	140	7.21

COLD FORMED CHANNELS
Properties and Dimensions
A570M Grade 50 (F_y = 345 MPa)

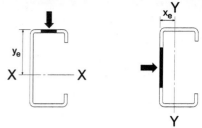

Designation	Mass	Area	F'_y	Effective Section Properties						M_{rxe}	L_u
				X-X Axis			Y-Y Axis				
				I_{xe}	S_{xe}	y_e	I_{ye}	S_{ye}	x_e		
	kg/m	mm^2	MPa	10^{6}mm^4	10^{3}mm^3	mm	10^{6}mm^4	10^{3}mm^3	mm	kNm	mm
CFC 200x89											
x4.17	13.4	1 710	390	10.8	106	102	1.85	31.5	30.0	37.2	1 647
x3.43	11.2	1 430	383	9.11	89.6	102	1.52	26.7	31.9	30.9	1 668
x2.67	8.84	1 130	345	7.24	70.9	102	1.13	21.0	35.2	22.0	1 764
x2.29	7.64	973	345	6.07	57.9	105	0.931	18.0	37.3	18.0	1 767
x1.90	6.41	817	345	4.89	45.0	109	0.734	15.0	39.8	14.0	1 771
CFC 200x70											
x3.43	9.83	1 250	392	7.57	74.5	102	0.783	16.3	21.8	26.3	1 262
x2.67	7.78	991	382	6.09	59.9	102	0.595	13.1	24.4	20.6	1 286
x2.29	6.72	857	345	5.29	51.9	102	0.494	11.3	26.2	16.1	1 359
x1.90	5.65	720	345	4.28	40.4	106	0.392	9.44	28.4	12.5	1 364
CFC 150x70											
x3.43	8.46	1 080	392	3.83	50.2	76.2	0.726	15.9	24.2	17.7	1 305
x2.67	6.72	856	382	3.09	40.6	76.2	0.573	12.9	25.5	14.0	1 325
x2.29	5.81	741	345	2.70	35.4	76.2	0.479	11.2	27.0	11.0	1 398
x1.90	4.89	623	345	2.22	28.5	77.9	0.383	9.36	29.0	8.85	1 401
x1.52	3.95	503	345	1.72	21.2	81.0	0.287	7.45	31.3	6.59	1 404
CFC 150x50											
x2.67	5.71	727	395	2.45	32.1	76.2	0.245	7.04	15.9	11.4	932
x2.29	4.95	630	389	2.14	28.1	76.2	0.209	6.18	17.0	9.84	944
x1.90	4.17	531	382	1.83	24.0	76.2	0.169	5.24	18.5	8.23	956
x1.52	3.37	430	345	1.47	18.9	77.5	0.128	4.22	20.4	5.88	1 011
x1.22	2.72	347	345	1.11	13.6	81.9	0.0956	3.36	22.4	4.22	1 015

Note: For an explanation of this table, see page 6-105.

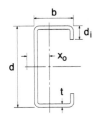

COLD FORMED CHANNELS
Properties and Dimensions

Depth	Flange Width	Stiff'r Depth	Thick-ness	Gross Section Properties										
				X-X Axis			Y-Y Axis			x_o	r_o	J	j	C_w
d	b	d_i	t	I_x	S_x	r_x	I_y	S_y	r_y					
mm	mm	mm	mm	$10^6 mm^4$	$10^3 mm^3$	mm	$10^6 mm^4$	$10^3 mm^3$	mm	mm	mm	$10^3 mm^4$	mm	$10^9 mm^6$
203	89	31.7	4.17	10.8	106	79.4	1.85	31.5	33.0	74.2	114	9.88	121	18.0
203	89	31.7	3.43	9.11	89.6	79.9	1.59	27.1	33.4	74.5	114	5.59	120	15.5
203	89	31.7	2.67	7.29	71.7	80.4	1.29	22.0	33.9	74.9	115	2.67	119	12.6
203	89	31.7	2.29	6.34	62.4	80.7	1.13	19.3	34.1	75.1	115	1.69	118	11.0
203	89	31.7	1.90	5.35	52.7	81.0	0.962	16.4	34.3	75.3	116	0.988	118	9.39
203	70	25.4	3.43	7.57	74.5	77.8	0.802	16.4	25.3	53.2	97.6	4.91	115	7.31
203	70	25.4	2.67	6.09	59.9	78.4	0.658	13.5	25.8	53.7	98.4	2.35	113	5.99
203	70	25.4	2.29	5.30	52.2	78.7	0.580	11.9	26.0	53.9	98.8	1.49	112	5.27
203	70	25.4	1.90	4.49	44.2	79.0	0.496	10.2	26.2	54.1	99.3	0.87	111	4.50
152	70	25.4	3.43	3.83	50.2	59.6	0.726	15.9	25.9	59.5	88.1	4.23	92.7	4.08
152	70	25.4	2.67	3.09	40.6	60.1	0.597	13.1	26.4	59.9	88.9	2.03	91.7	3.36
152	70	25.4	2.29	2.70	35.4	60.4	0.526	11.6	26.6	60.1	89.2	1.29	91.2	2.96
152	70	25.4	1.90	2.29	30.1	60.6	0.450	9.89	26.9	60.2	89.6	0.75	90.8	2.54
152	70	25.4	1.52	1.87	24.5	60.9	0.369	8.13	27.1	60.4	90.0	0.39	90.4	2.09
152	51	20.3	2.67	2.45	32.1	58.0	0.250	7.07	18.5	39.3	72.5	1.72	85.3	1.31
152	51	20.3	2.29	2.14	28.1	58.3	0.222	6.30	18.8	39.5	72.9	1.10	84.2	1.16
152	51	20.3	1.90	1.83	24.0	58.6	0.192	5.45	19.0	39.7	73.3	0.64	83.2	1.00
152	51	20.3	1.52	1.49	19.6	58.9	0.159	4.52	19.2	39.9	73.7	0.33	82.3	0.83
152	51	20.3	1.22	1.21	15.9	59.2	0.131	3.72	19.4	40.1	74.1	0.17	81.7	0.68

NOTES

BUILT-UP SECTIONS

Built-up sections may be fabricated from plate and shapes in various configurations to produce efficient and economical structural sections. Generally the components are joined by welding, though bolting may also be used for some combinations. Frequently used built-up sections include double angles back-to-back, two channels back-to-back or toe-to-toe, and a channel or C shape in combination with a W shape.

Properties and dimensions of double angles are provided for imperial series angles on pages 6-112 to 6-119. Included are equal leg angles and unequal leg angles with long legs back-to-back and with short legs back-to-back. Properites and dimensions of double channels and of C shapes combined with W shapes for some representative combinations are provided in tables on pages 6-120 to 6-123.

Many other combinations of built-up members are possible. Diagrams with expressions and formulae for computing properties of some possible combinations are provided on pages 6-125 to 6-128.

TWO ANGLES EQUAL LEGS - Imperial Series
Back-to-Back

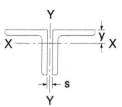

PROPERTIES OF SECTIONS

Designation*	Mass of 2 angles	Dead Load	Area of 2 angles	Axis X-X				Radii of Gyration about Axis Y-Y					
				I	S	r	y	Back-to-back spacing, s, millimetres					
	kg/m	kN/m	mm^2	10^6 mm^4	10^3 mm^3	mm	mm	0	8	10	12	16	20
L203x203													
x29	169	1.66	21 600	81.4	574	61.4	61.2	86.7	89.6	90.3	91.0	92.5	94.0
x25	152	1.49	19 300	73.8	517	61.8	60.1	86.2	89.0	89.7	90.5	91.9	93.4
x22	134	1.31	17 000	66.0	458	62.2	58.9	85.7	88.5	89.2	89.9	91.4	92.8
x19	116	1.14	14 800	58.0	399	62.6	57.8	85.2	88.0	88.7	89.4	90.9	92.3
x16	97.4	0.955	12 400	49.4	337	63.1	56.6	84.8	87.5	88.2	88.9	90.3	91.8
x13	78.4	0.769	9 990	40.4	274	63.6	55.5	84.4	87.0	87.7	88.4	89.8	91.2
L152x152													
x25	111	1.09	14 200	29.3	279	45.5	47.2	65.6	68.5	69.3	70.0	71.5	73.1
x22	98.2	0.964	12 500	26.3	249	45.9	46.1	65.0	67.9	68.7	69.4	70.9	72.5
x19	85.4	0.838	10 900	23.3	218	46.3	45.0	64.5	67.4	68.1	68.9	70.4	71.9
x16	71.9	0.706	9 160	20.0	185	46.7	43.9	64.1	66.9	67.6	68.3	69.8	71.3
x14	65.0	0.638	8 290	18.2	168	46.9	43.3	63.8	66.6	67.3	68.0	69.5	71.0
x13	58.1	0.570	7 400	16.4	150	47.1	42.7	63.6	66.3	67.1	67.8	69.2	70.7
x11	51.0	0.501	6 500	14.6	133	47.4	42.1	63.4	66.1	66.8	67.5	69.0	70.4
x9.5	44.1	0.432	5 610	12.7	115	47.6	41.5	63.2	65.9	66.6	67.3	68.7	70.1
x7.9	36.9	0.362	4 700	10.8	96.8	47.8	41.0	63.0	65.6	66.3	67.0	68.4	69.9
x6.4	29.7	0.291	3 780	8.73	78.2	48.1	40.4	62.8	65.4	66.1	66.8	68.2	69.6
L127x127													
x22	80.8	0.793	10 300	14.8	169	37.9	39.8	55.0	57.9	58.7	59.4	61.0	62.6
x19	70.4	0.691	8 970	13.1	149	38.3	38.7	54.4	57.3	58.1	58.8	60.4	61.9
x16	59.4	0.583	7 570	11.3	127	38.7	37.6	53.9	56.8	57.5	58.3	59.8	61.3
x13	48.1	0.472	6 130	9.37	103	39.1	36.4	53.4	56.2	57.0	57.7	59.2	60.7
x11	42.3	0.415	5 390	8.33	91.4	39.3	35.8	53.2	56.0	56.7	57.4	58.9	60.4
x9.5	36.6	0.359	4 660	7.28	79.4	39.5	35.3	53.0	55.7	56.4	57.2	58.6	60.1
x7.9	30.7	0.301	3 910	6.18	66.9	39.8	34.7	52.8	55.5	56.2	56.9	58.3	59.8
x6.4	24.7	0.242	3 150	5.03	54.2	40.0	34.1	52.6	55.2	55.9	56.6	58.1	59.5
L102x102													
x19	55.4	0.544	7 060	6.48	93.1	30.3	32.4	44.4	47.4	48.1	48.9	50.5	52.1
x16	47.0	0.461	5 980	5.62	79.5	30.7	31.3	43.8	46.7	47.5	48.3	49.8	51.4
x13	38.1	0.374	4 860	4.69	65.3	31.1	30.2	43.3	46.2	46.9	47.7	49.2	50.8
x11	33.6	0.330	4 280	4.19	57.8	31.3	29.6	43.0	45.9	46.6	47.4	48.9	50.4
x9.5	29.1	0.285	3 710	3.68	50.4	31.5	29.0	42.8	45.6	46.4	47.1	48.6	50.1
x7.9	24.4	0.240	3 110	3.13	42.6	31.7	28.4	42.6	45.4	46.1	46.8	48.3	49.8
x6.4	19.7	0.193	2 510	2.56	34.5	31.9	27.9	42.4	45.1	45.8	46.5	48.0	49.5
L89x89													
x13	32.9	0.323	4 190	3.03	48.8	26.9	26.9	38.0	40.9	41.7	42.4	44.0	45.6
x11	29.1	0.285	3 700	2.71	43.3	27.1	26.3	37.7	40.6	41.4	42.1	43.7	45.3
x9.5	25.2	0.247	3 210	2.39	37.8	27.3	25.7	37.5	40.3	41.1	41.8	43.4	45.0
x7.9	21.2	0.208	2 700	2.04	32.0	27.5	25.2	37.3	40.1	40.8	41.6	43.1	44.6
x6.4	17.1	0.168	2 180	1.67	26.0	27.7	24.6	37.0	39.8	40.5	41.3	42.8	44.3

* Designation consists of nominal leg sizes and thickness. For angle size, see Angles - Imperial Series.

Note: The properties of angles currently produced by cold forming are up to 7 percent less than the properties shown in the above tables. Check manufacturer's catalog for the exact properties and dimensions. See also page 6-39.

TWO ANGLES EQUAL LEGS - Imperial Series
Back-to-Back

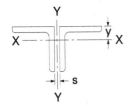

PROPERTIES OF SECTIONS

Designation*	Mass of 2 angles	Dead Load	Area of 2 angles	Axis X-X				Radii of Gyration about Axis Y-Y					
				I	S	r	y	Back-to-back spacing, s, millimetres					
	kg/m	kN/m	mm^2	10^6 mm^4	10^3 mm^3	mm	mm	0	8	10	12	16	20
L76x76													
x13	27.9	0.273	3 550	1.85	35.1	22.8	23.7	32.9	35.9	36.6	37.4	39.0	40.7
x11	24.6	0.242	3 140	1.66	31.2	23.0	23.1	32.6	35.5	36.3	37.1	38.7	40.3
x9.5	21.4	0.210	2 720	1.47	27.3	23.2	22.5	32.3	35.3	36.0	36.8	38.4	40.0
x7.9	18.0	0.177	2 290	1.26	23.2	23.4	22.0	32.1	35.0	35.7	36.5	38.0	39.6
x6.4	14.6	0.143	1 850	1.04	18.9	23.6	21.4	31.9	34.7	35.4	36.2	37.7	39.3
x4.8	11.0	0.108	1 410	0.800	14.4	23.9	20.8	31.7	34.4	35.2	35.9	37.4	39.0
L64x64													
x13	22.8	0.224	2 900	1.02	23.7	18.8	20.5	27.8	30.8	31.6	32.4	34.1	35.8
x9.5	17.6	0.172	2 240	0.819	18.6	19.1	19.4	27.2	30.2	31.0	31.8	33.4	35.0
x7.9	14.8	0.146	1 890	0.707	15.8	19.3	18.8	27.0	29.9	30.7	31.4	33.0	34.7
x6.4	12.0	0.118	1 530	0.585	12.9	19.5	18.2	26.7	29.6	30.3	31.1	32.7	34.3
x4.8	9.14	0.089 6	1 160	0.455	9.92	19.8	17.6	26.5	29.3	30.1	30.8	32.4	34.0
L51x51													
x9.5	13.8	0.135	1 750	0.399	11.5	15.1	16.2	22.1	25.2	26.0	26.8	28.5	30.2
x7.9	11.7	0.115	1 490	0.347	9.84	15.3	15.6	21.8	24.8	25.6	26.4	28.1	29.8
x6.4	9.50	0.093 2	1 210	0.289	8.09	15.5	15.0	21.6	24.5	25.3	26.1	27.7	29.4
x4.8	7.24	0.071 0	922	0.227	6.24	15.7	14.5	21.3	24.2	25.0	25.8	27.4	29.1
x3.2	4.91	0.048 2	626	0.158	4.29	15.9	13.9	21.1	23.9	24.7	25.5	27.1	28.7
L44x44													
x6.4	8.24	0.080 8	1 050	0.190	6.11	13.4	13.4	19.0	22.0	22.8	23.6	25.3	27.0
x4.8	6.30	0.061 8	802	0.150	4.73	13.7	12.9	18.8	21.7	22.5	23.3	24.9	26.6
x3.2	4.28	0.042 0	546	0.105	3.26	13.9	12.3	18.5	21.4	22.2	23.0	24.6	26.3
L38x38													
x6.4	6.96	0.068 3	887	0.115	4.39	11.4	11.8	16.4	19.5	20.3	21.2	22.9	24.6
x4.8	5.34	0.052 4	680	0.091 5	3.41	11.6	11.3	16.2	19.2	20.0	20.8	22.5	24.2
x3.2	3.65	0.035 8	464	0.064 8	2.37	11.8	10.7	15.9	18.9	19.6	20.5	22.1	23.8
L32x32													
x6.4	5.71	0.056 0	727	0.064 2	2.98	9.40	10.2	13.9	17.1	17.9	18.8	20.5	22.3
x4.8	4.40	0.043 1	560	0.051 4	2.33	9.58	9.69	13.6	16.7	17.5	18.4	20.1	21.9
x3.2	3.02	0.029 6	384	0.036 8	1.62	9.79	9.12	13.4	16.4	17.2	18.0	19.7	21.5
L25x25													
x6.4	4.43	0.043 5	565	0.030 7	1.83	7.37	8.62	11.3	14.6	15.5	16.4	18.2	20.0
x4.8	3.44	0.033 8	438	0.024 9	1.44	7.54	8.07	11.0	14.2	15.1	16.0	17.8	19.6
x3.2	2.38	0.023 3	303	0.018 1	1.01	7.73	7.52	10.8	13.9	14.7	15.6	17.3	19.1
L19x19													
x3.2	1.75	0.017 2	223	0.007 3	0.55	5.72	5.93	8.2	11.5	12.3	13.2	15.1	16.9

* Designation consists of nominal leg sizes and thickness. For angle size, see Angles - Imperial Series.

Note: The properties of angles currently produced by cold forming are up to 7 percent less than the properties shown in the above tables. Check manufacturer's catalog for the exact properties and dimensions. See also page 6-39.

TWO ANGLES UNEQUAL LEGS - Imperial Series
Long Legs Back-to-Back

PROPERTIES OF SECTIONS

Designation*	Mass of 2 angles	Dead Load	Area of 2 angles	Axis X-X				Radii of Gyration about Axis Y-Y					
				I	S	r	y	Back-to-back spacing, s, millimetres					
	kg/m	kN/m	mm²	10⁶ mm⁴	10³ mm³	mm	mm	0	8	10	12	16	20
L203x152													
x25	131	1.29	16 700	67.0	494	63.3	67.4	60.6	63.4	64.1	64.9	66.3	67.9
x19	101	0.988	12 800	52.7	382	64.1	65.1	59.6	62.3	63.0	63.7	65.2	66.7
x13	68.3	0.670	8 690	36.8	262	65.0	62.8	58.7	61.4	62.0	62.7	64.1	65.5
x11	59.9	0.588	7 630	32.5	231	65.3	62.2	58.5	61.1	61.8	62.5	63.9	65.3
L203x102													
x25	111	1.09	14 200	57.9	460	63.8	77.2	37.4	40.4	41.2	41.9	43.5	45.1
x19	85.7	0.841	10 900	45.7	357	64.7	74.8	36.3	39.1	39.8	40.5	42.1	43.6
x13	58.3	0.572	7 420	32.0	245	65.7	72.4	35.3	37.9	38.6	39.3	40.7	42.2
L178x102													
x19	78.2	0.767	9 970	31.7	277	56.4	63.8	37.8	40.6	41.4	42.1	43.7	45.2
x16	65.9	0.647	8 400	27.1	235	56.8	62.6	37.3	40.0	40.8	41.5	43.0	44.5
x13	53.3	0.523	6 790	22.3	191	57.3	61.4	36.8	39.5	40.2	40.9	42.4	43.9
x11	46.9	0.460	5 970	19.8	169	57.5	60.8	36.6	39.2	39.9	40.6	42.1	43.6
x9.5	40.5	0.397	5 160	17.2	146	57.8	60.2	36.4	39.0	39.7	40.4	41.8	43.3
L152x102													
x22	80.8	0.793	10 300	22.9	233	47.2	53.7	40.2	43.2	43.9	44.7	46.3	47.9
x19	70.4	0.691	8 970	20.3	204	47.6	52.5	39.7	42.6	43.3	44.1	45.6	47.2
x16	59.4	0.583	7 570	17.5	174	48.0	51.4	39.2	42.0	42.7	43.4	44.9	46.5
x14	53.8	0.528	6 860	16.0	158	48.2	50.8	38.9	41.7	42.4	43.1	44.6	46.2
x13	48.1	0.472	6 130	14.4	141	48.5	50.2	38.7	41.4	42.1	42.8	44.3	45.8
x11	42.3	0.415	5 390	12.8	125	48.7	49.6	38.5	41.1	41.9	42.6	44.0	45.5
x9.5	36.6	0.359	4 660	11.2	108	48.9	49.1	38.3	40.9	41.6	42.3	43.7	45.2
x7.9	30.7	0.301	3 910	9.44	91.2	49.2	48.5	38.1	40.7	41.3	42.0	43.5	44.9
L152x89													
x16	56.2	0.551	7 160	16.6	169	48.2	53.9	33.0	35.9	36.6	37.4	38.9	40.5
x13	45.5	0.446	5 800	13.7	138	48.6	52.7	32.5	35.3	36.0	36.7	38.2	39.8
x9.5	34.6	0.340	4 410	10.6	106	49.1	51.6	32.1	34.7	35.4	36.2	37.6	39.1
x7.9	29.0	0.285	3 700	9.01	89.1	49.3	51.0	31.9	34.5	35.2	35.9	37.3	38.8
L127x89													
x19	59.0	0.579	7 520	11.6	140	39.3	44.4	35.5	38.4	39.2	40.0	41.5	43.2
x16	49.9	0.490	6 360	10.0	120	39.7	43.2	34.9	37.8	38.5	39.3	40.8	42.4
x13	40.5	0.397	5 160	8.31	97.9	40.1	42.1	34.4	37.2	37.9	38.7	40.2	41.8
x9.5	30.9	0.303	3 930	6.48	75.2	40.6	40.9	33.9	36.6	37.4	38.1	39.6	41.1
x7.9	25.9	0.254	3 300	5.50	63.5	40.8	40.3	33.7	36.4	37.1	37.8	39.3	40.8
x6.4	20.9	0.205	2 660	4.48	51.4	41.0	39.7	33.5	36.2	36.8	37.5	39.0	40.5

* Designation consists of nominal leg sizes and thickness. For angle size, see Angles - Imperial Series.

Note: The properties of angles currently produced by cold forming are up to 7 percent less than the properties shown in the above tables. Check manufacturer's catalog for the exact properties and dimensions. See also page 6-39.

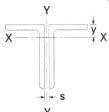

TWO ANGLES UNEQUAL LEGS - Imperial Series
Long Legs Back-to-Back

PROPERTIES OF SECTIONS

Designation*	Mass of 2 angles	Dead Load	Area of 2 angles	Axis X-X				Radii of Gyration about Axis Y-Y					
				I	S	r	y	Back-to-back spacing, s, millimetres					
	kg/m	kN/m	mm²	10⁶ mm⁴	10³ mm³	mm	mm	0	8	10	12	16	20
L127x76													
x13	38.0	0.373	4 840	7.87	95.3	40.3	44.5	28.4	31.2	32.0	32.7	34.3	35.9
x11	33.5	0.328	4 260	7.01	84.4	40.6	43.9	28.2	30.9	31.7	32.4	33.9	35.5
x9.5	29.0	0.284	3 690	6.14	73.3	40.8	43.3	27.9	30.6	31.4	32.1	33.6	35.2
x7.9	24.3	0.239	3 100	5.21	61.9	41.0	42.7	27.7	30.4	31.1	31.8	33.3	34.8
x6.4	19.6	0.193	2 500	4.25	50.1	41.2	42.1	27.5	30.1	30.8	31.5	33.0	34.5
L102x89													
x13	35.5	0.349	4 530	4.48	63.9	31.5	31.9	36.6	39.5	40.2	41.0	42.6	44.1
x9.5	27.1	0.266	3 460	3.52	49.4	31.9	30.8	36.1	38.9	39.7	40.4	41.9	43.5
x7.9	22.8	0.224	2 910	3.00	41.7	32.1	30.2	35.9	38.7	39.4	40.1	41.6	43.2
x6.4	18.4	0.180	2 340	2.45	33.9	32.3	29.6	35.7	38.4	39.1	39.9	41.3	42.9
L102x76													
x16	40.5	0.397	5 160	5.09	75.9	31.4	35.0	30.9	33.9	34.6	35.4	37.0	38.7
x13	33.0	0.324	4 200	4.25	62.4	31.8	33.9	30.3	33.2	34.0	34.8	36.3	37.9
x11	29.1	0.286	3 710	3.80	55.3	32.0	33.3	30.1	32.9	33.7	34.4	36.0	37.6
x9.5	25.2	0.248	3 210	3.34	48.2	32.2	32.7	29.8	32.6	33.4	34.1	35.7	37.2
x7.9	21.2	0.208	2 700	2.85	40.7	32.4	32.1	29.6	32.4	33.1	33.8	35.4	36.9
x6.4	17.1	0.168	2 180	2.33	33.1	32.7	31.6	29.4	32.1	32.8	33.6	35.0	36.6
L89x76													
x13	30.4	0.298	3 870	2.87	47.7	27.3	28.6	31.5	34.5	35.2	36.0	37.6	39.2
x9.5	23.3	0.228	2 970	2.27	36.9	27.7	27.4	31.0	33.9	34.6	35.4	36.9	38.5
x7.9	19.6	0.192	2 500	1.94	31.3	27.9	26.9	30.8	33.6	34.3	35.1	36.6	38.2
x6.4	15.8	0.155	2 020	1.59	25.4	28.1	26.3	30.6	33.3	34.1	34.8	36.3	37.9
L89x64													
x13	27.9	0.273	3 550	2.70	46.2	27.6	30.6	25.3	28.3	29.1	29.8	31.5	33.1
x9.5	21.4	0.210	2 720	2.13	35.9	28.0	29.5	24.8	27.6	28.4	29.2	30.8	32.4
x7.9	18.0	0.177	2 290	1.82	30.4	28.2	28.9	24.5	27.4	28.1	28.9	30.4	32.0
x6.4	14.6	0.143	1 850	1.50	24.7	28.4	28.3	24.3	27.1	27.8	28.6	30.1	31.7
L76x64													
x13	25.3	0.248	3 230	1.73	34.1	23.2	25.4	26.4	29.5	30.2	31.0	32.7	34.4
x9.5	19.5	0.191	2 480	1.38	26.6	23.6	24.3	25.9	28.8	29.6	30.4	32.0	33.6
x7.9	16.4	0.161	2 090	1.18	22.6	23.8	23.7	25.7	28.5	29.3	30.0	31.6	33.3
x6.4	13.3	0.130	1 690	0.977	18.4	24.0	23.1	25.4	28.2	29.0	29.7	31.3	32.9
x4.8	10.1	0.098 9	1 280	0.755	14.1	24.2	22.6	25.2	28.0	28.7	29.4	31.0	32.6
L76x51													
x13	22.8	0.224	2 900	1.60	32.9	23.5	27.5	20.3	23.4	24.2	25.0	26.7	28.4
x9.5	17.6	0.172	2 240	1.28	25.6	23.9	26.4	19.7	22.7	23.5	24.3	25.9	27.6
x7.9	14.8	0.146	1 890	1.10	21.8	24.1	25.8	19.5	22.4	23.1	23.9	25.6	27.2
x6.4	12.0	0.118	1 530	0.905	17.8	24.3	25.2	19.2	22.1	22.8	23.6	25.2	26.8
x4.8	9.14	0.089 6	1 160	0.700	13.6	24.5	24.6	19.0	21.8	22.5	23.3	24.8	26.5

* Designation consists of nominal leg sizes and thickness. For angle size, see Angles - Imperial Series.

Note: The properties of angles currently produced by cold forming are up to 7 percent less than the properties shown in the above tables. Check manufacturer's catalog for the exact properties and dimensions. See also page 6-39.

TWO ANGLES UNEQUAL LEGS - Imperial Series
Long Legs Back-to-Back

PROPERTIES OF SECTIONS

Designation*	Mass of 2 angles	Dead Load	Area of 2 angles	Axis X-X				Radii of Gyration about Axis Y-Y					
				I	S	r	y	Back-to-back spacing, s, millimetres					
	kg/m	kN/m	mm^2	10^6 mm^4	10^3 mm^3	mm	mm	0	8	10	12	16	20
L64x51													
x9.5	15.7	0.154	2 000	0.760	17.9	19.5	21.1	20.8	23.8	24.6	25.4	27.1	28.8
x7.9	13.3	0.130	1 690	0.656	15.3	19.7	20.6	20.5	23.5	24.3	25.1	26.7	28.4
x6.4	10.8	0.106	1 370	0.544	12.5	19.9	20.0	20.3	23.2	23.9	24.7	26.3	28.0
x4.8	8.19	0.080 3	1 040	0.423	9.60	20.1	19.4	20.1	22.9	23.6	24.4	26.0	27.6
L51x38													
x6.4	8.23	0.080 7	1 050	0.263	7.74	15.8	16.9	15.2	18.2	19.0	19.8	21.5	23.3
x4.8	6.29	0.061 7	801	0.206	5.97	16.0	16.3	14.9	17.9	18.6	19.5	21.1	22.8

* Designation consists of nominal leg sizes and thickness. For angle size, see Angles - Imperial Series.

Note: The properties of angles currently produced by cold forming are up to 7 percent less than the properties shown in the above tables. Check manufacturer's catalog for the exact properties and dimensions. See also page 6-39.

TWO ANGLES UNEQUAL LEGS - Imperial Series
Short Legs Back-to-Back

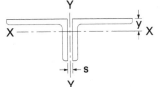

PROPERTIES OF SECTIONS

Designation*	Mass of 2 angles	Dead Load	Area of 2 angles	Axis X-X				Radii of Gyration about Axis Y-Y					
				I	S	r	y	Back-to-back spacing, s, millimetres					
	kg/m	kN/m	mm^2	10^6 mm^4	10^3 mm^3	mm	mm	0	8	10	12	16	20
L203x152													
x25	131	1.29	16 700	32.0	291	43.7	41.9	92.4	95.4	96.1	96.9	98.4	100
x19	101	0.988	12 800	25.4	226	44.5	39.6	91.4	94.3	95.0	95.7	97.2	98.8
x13	68.3	0.670	8 690	17.9	156	45.4	37.3	90.4	93.2	93.9	94.6	96.1	97.6
x11	59.9	0.588	7 630	15.9	138	45.6	36.7	90.1	92.9	93.7	94.4	95.8	97.3
L203x102													
x25	111	1.09	14 200	9.81	130	26.3	26.7	100	103	104	105	106	108
x19	85.7	0.841	10 900	7.90	102	26.9	24.3	98.9	102	103	104	105	107
x13	58.3	0.572	7 420	5.67	70.9	27.6	21.9	97.8	101	102	102	104	105
L178x102													
x19	78.2	0.767	9 970	7.64	100	27.7	25.8	85.1	88.1	88.9	89.7	91.3	92.8
x16	65.9	0.647	8 400	6.61	85.4	28.1	24.6	84.5	87.5	88.3	89.1	90.6	92.2
x13	53.3	0.523	6 790	5.50	69.9	28.5	23.4	84.0	86.9	87.7	88.5	90.0	91.5
x11	46.9	0.460	5 970	4.90	61.9	28.7	22.8	83.7	86.6	87.4	88.2	89.7	91.2
x9.5	40.5	0.397	5 160	4.30	53.9	28.9	22.2	83.4	86.4	87.1	87.9	89.4	90.9
L152x102													
x22	80.8	0.793	10 300	8.20	112	28.2	28.7	71.5	74.5	75.3	76.1	77.6	79.2
x19	70.4	0.691	8 970	7.32	98.3	28.6	27.5	70.9	73.9	74.7	75.5	77.0	78.6
x16	59.4	0.583	7 570	6.34	83.8	28.9	26.4	70.3	73.3	74.1	74.8	76.4	77.9
x14	53.8	0.528	6 860	5.82	76.4	29.1	25.8	70.1	73.0	73.8	74.5	76.1	77.6
x13	48.1	0.472	6 130	5.28	68.7	29.3	25.2	69.8	72.7	73.5	74.2	75.8	77.3
x11	42.3	0.415	5 390	4.71	60.9	29.6	24.6	69.5	72.4	73.2	73.9	75.5	77.0
x9.5	36.6	0.359	4 660	4.13	53.0	29.8	24.1	69.3	72.2	72.9	73.7	75.2	76.7
x7.9	30.7	0.301	3 910	3.51	44.7	30.0	23.5	69.0	71.9	72.6	73.4	74.9	76.4
L152x89													
x16	56.2	0.551	7 160	4.23	63.6	24.3	22.4	72.3	75.3	76.1	76.9	78.5	80.1
x13	45.5	0.446	5 800	3.54	52.2	24.7	21.2	71.7	74.7	75.5	76.3	77.8	79.4
x9.5	34.6	0.340	4 410	2.78	40.4	25.1	20.0	71.2	74.2	74.9	75.7	77.2	78.7
x7.9	29.0	0.285	3 700	2.37	34.1	25.3	19.4	70.9	73.9	74.6	75.4	76.9	78.4
L127x89													
x19	59.0	0.579	7 520	4.63	72.8	24.8	25.3	59.3	62.3	63.1	63.9	65.5	67.1
x16	49.9	0.490	6 360	4.03	62.2	25.2	24.2	58.7	61.7	62.5	63.2	64.8	66.4
x13	40.5	0.397	5 160	3.37	51.2	25.6	23.0	58.1	61.1	61.9	62.6	64.2	65.7
x9.5	30.9	0.303	3 930	2.65	39.6	26.0	21.9	57.6	60.5	61.3	62.0	63.6	65.1
x7.9	25.9	0.254	3 300	2.26	33.5	26.2	21.3	57.4	60.3	61.0	61.7	63.3	64.8
x6.4	20.9	0.205	2 660	1.86	27.2	26.4	20.7	57.1	60.0	60.7	61.5	63.0	64.5

* Designation consists of nominal leg sizes and thickness. For angle size, see Angles - Imperial Series.

Note: The properties of angles currently produced by cold forming are up to 7 percent less than the properties shown in the above tables. Check manufacturer's catalog for the exact properties and dimensions. See also page 6-39.

TWO ANGLES UNEQUAL LEGS - Imperial Series
Short Legs Back-to-Back

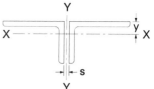

PROPERTIES OF SECTIONS

Designation*	Mass of 2 angles	Dead Load	Area of 2 angles	Axis X-X				Radii of Gyration about Axis Y-Y					
				I	S	r	y	Back-to-back spacing, s, millimetres					
	kg/m	kN/m	mm²	10⁶ mm⁴	10³ mm³	mm	mm	0	8	10	12	16	20
L127x76													
x13	38.0	0.373	4 840	2.15	37.6	21.1	19.1	60.0	63.0	63.8	64.6	66.2	67.8
x11	33.5	0.328	4 260	1.93	33.4	21.3	18.5	59.7	62.7	63.5	64.3	65.8	67.4
x9.5	29.0	0.284	3 690	1.70	29.1	21.5	17.9	59.5	62.4	63.2	64.0	65.5	67.1
x7.9	24.3	0.239	3 100	1.45	24.7	21.7	17.3	59.2	62.1	62.9	63.7	65.2	66.8
x6.4	19.6	0.193	2 500	1.20	20.1	21.9	16.7	58.9	61.9	62.6	63.4	64.9	66.4
L102x89													
x13	35.5	0.349	4 530	3.16	49.7	26.4	25.4	44.8	47.7	48.5	49.3	50.8	52.4
x9.5	27.1	0.266	3 460	2.49	38.5	26.8	24.2	44.3	47.2	47.9	48.7	50.2	51.8
x7.9	22.8	0.224	2 910	2.13	32.6	27.1	23.6	44.1	46.9	47.6	48.4	49.9	51.4
x6.4	18.4	0.180	2 340	1.74	26.5	27.3	23.1	43.8	46.6	47.4	48.1	49.6	51.1
L102x76													
x16	40.5	0.397	5 160	2.40	44.3	21.6	22.1	47.0	50.1	50.9	51.6	53.2	54.9
x13	33.0	0.324	4 200	2.02	36.6	21.9	21.0	46.5	49.4	50.2	51.0	52.6	54.2
x11	29.1	0.286	3 710	1.81	32.5	22.1	20.4	46.2	49.1	49.9	50.7	52.2	53.8
x9.5	25.2	0.248	3 210	1.60	28.4	22.3	19.8	45.9	48.9	49.6	50.4	51.9	53.5
x7.9	21.2	0.208	2 700	1.37	24.1	22.5	19.2	45.7	48.6	49.3	50.1	51.6	53.2
x6.4	17.1	0.168	2 180	1.13	19.6	22.7	18.7	45.4	48.3	49.0	49.8	51.3	52.9
L89x76													
x13	30.4	0.298	3 870	1.94	35.9	22.4	22.2	39.5	42.5	43.2	44.0	45.6	47.2
x9.5	23.3	0.228	2 970	1.54	27.9	22.8	21.1	39.0	41.9	42.6	43.4	45.0	46.6
x7.9	19.6	0.192	2 500	1.32	23.7	23.0	20.5	38.7	41.6	42.3	43.1	44.6	46.2
x6.4	15.8	0.155	2 020	1.09	19.3	23.2	19.9	38.5	41.3	42.1	42.8	44.3	45.9
L89x64													
x13	27.9	0.273	3 550	1.14	24.9	17.9	17.9	41.2	44.2	45.0	45.8	47.4	49.1
x9.5	21.4	0.210	2 720	0.908	19.4	18.3	16.8	40.6	43.6	44.4	45.2	46.8	48.4
x7.9	18.0	0.177	2 290	0.782	16.5	18.5	16.2	40.4	43.3	44.1	44.9	46.4	48.0
x6.4	14.6	0.143	1 850	0.647	13.5	18.7	15.6	40.1	43.0	43.8	44.5	46.1	47.7
L76x64													
x13	25.3	0.248	3 230	1.08	24.4	18.3	19.1	34.4	37.4	38.2	39.0	40.7	42.3
x9.5	19.5	0.191	2 480	0.868	19.0	18.7	17.9	33.8	36.8	37.6	38.4	40.0	41.6
x7.9	16.4	0.161	2 090	0.748	16.2	18.9	17.4	33.6	36.5	37.3	38.1	39.6	41.3
x6.4	13.3	0.130	1 690	0.619	13.2	19.1	16.8	33.3	36.2	37.0	37.8	39.3	40.9
x4.8	10.1	0.098 9	1 280	0.480	10.2	19.3	16.2	33.1	36.0	36.7	37.5	39.0	40.6
L76x51													
x13	22.8	0.224	2 900	0.559	15.5	13.9	14.8	36.2	39.3	40.1	40.9	42.6	44.3
x9.5	17.6	0.172	2 240	0.452	12.2	14.2	13.7	35.6	38.6	39.4	40.2	41.9	43.5
x7.9	14.8	0.146	1 890	0.392	10.4	14.4	13.1	35.3	38.3	39.1	39.9	41.5	43.2
x6.4	12.0	0.118	1 530	0.326	8.52	14.6	12.5	35.0	38.0	38.8	39.6	41.2	42.8
x4.8	9.14	0.089 6	1 160	0.255	6.56	14.8	11.9	34.8	37.7	38.5	39.3	40.8	42.5

* Designation consists of nominal leg sizes and thickness. For angle size, see Angles - Imperial Series.

Note: The properties of angles currently produced by cold forming are up to 7 percent less than the properties shown in the above tables. Check manufacturer's catalog for the exact properties and dimensions. See also page 6-39.

TWO ANGLES UNEQUAL LEGS - Imperial Series
Short Legs Back-to-Back

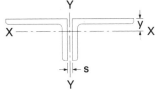

PROPERTIES OF SECTIONS

Designation*	Mass of 2 angles	Dead Load	Area of 2 angles	Axis X-X				Radii of Gyration about Axis Y-Y					
				I	S	r	y	Back-to-back spacing, s, millimetres					
	kg/m	kN/m	mm²	10⁶ mm⁴	10³ mm³	mm	mm	0	8	10	12	16	20
L64x51													
x9.5	15.7	0.154	2 000	0.428	11.9	14.6	14.8	28.7	31.8	32.6	33.4	35.0	36.7
x7.9	13.3	0.130	1 690	0.372	10.2	14.8	14.2	28.5	31.5	32.3	33.1	34.7	36.4
x6.4	10.8	0.106	1 370	0.310	8.34	15.0	13.6	28.2	31.2	32.0	32.7	34.3	36.0
x4.8	8.19	0.080 3	1 040	0.242	6.42	15.2	13.1	28.0	30.9	31.6	32.4	34.0	35.6
L51x38													
x6.4	8.23	0.080 7	1 050	0.126	4.57	11.0	10.5	23.1	26.2	27.0	27.8	29.5	31.2
x4.8	6.29	0.061 7	801	0.100	3.54	11.2	9.93	22.9	25.9	26.6	27.5	29.1	30.8

* Designation consists of nominal leg sizes and thickness. For angle size, see Angles - Imperial Series.

Note: The properties of angles currently produced by cold forming are up to 7 percent less than the properties shown in the above tables. Check manufacturer's catalog for the exact properties and dimensions. See also page 6-39.

TWO CHANNELS
Toe to Toe

PROPERTIES OF SECTIONS

Channel Size	For Two Channels			Axis X-X			Axis Y-Y					
	Mass	Dead Load	Area	I_x	S_x	r_x	Toe to Toe			c = d		
							I_y	S_y	r_y	I_y	S_y	r_y
	kg/m	kN/m	mm²	10⁶mm⁴	10³mm³	mm	10⁶mm⁴	10³mm³	mm	10⁶mm⁴	10³mm³	mm
MC460												
x86*	174	1.70	22 100	567	2 480	160	175	1 630	88.9	959	4 200	208
x77.2*	155	1.52	19 700	524	2 290	163	147	1 410	86.3	857	3 750	208
x68.2*	137	1.34	17 500	486	2 130	167	124	1 220	84.3	757	3 310	208
x63.5*	127	1.25	16 200	463	2 030	169	110	1 100	82.4	703	3 080	208
C380												
x74*	149	1.46	19 000	336	1 760	133	112	1 190	76.9	558	2 930	172
x60*	119	1.17	15 100	290	1 520	138	80.3	902	72.8	449	2 360	172
x50*	101	0.990	12 900	262	1 370	143	62.8	730	69.9	381	2 000	172
C310												
x45*	89.4	0.877	11 400	135	883	109	49.4	617	65.9	213	1 400	137
x37*	74.2	0.728	9 440	120	785	113	37.6	488	63.1	177	1 160	137
x31*	61.6	0.604	7 830	107	702	117	28.2	380	59.9	146	957	136
C250												
x45*	89.0	0.873	11 300	85.6	674	86.9	43.6	573	62.0	142	1 120	112
x37	74.6	0.732	9 490	75.9	598	89.4	34.0	466	59.8	120	948	113
x30	59.2	0.581	7 550	65.4	515	93.0	24.1	349	56.4	96.5	760	113
x23	45.2	0.443	5 750	55.5	437	98.2	15.7	242	52.3	72.9	574	113
C230												
x30	59.6	0.584	7 590	50.9	445	81.9	22.7	339	54.7	77.5	677	101
x22	44.6	0.437	5 680	42.6	372	86.6	14.7	234	50.9	57.9	506	101
x20	39.6	0.388	5 050	39.6	346	88.6	12.1	198	48.9	51.4	448	101
C200												
x28	55.8	0.547	7 110	36.5	360	71.6	19.2	300	51.9	55.7	548	88.5
x21	40.8	0.400	5 190	29.8	294	75.8	11.8	200	47.6	41.0	404	88.9
x17	34.0	0.333	4 340	26.9	265	78.8	8.95	157	45.4	34.0	335	88.5
C180												
x22	43.8	0.430	5 570	22.6	254	63.7	12.2	210	46.8	32.9	370	76.9
x18	36.4	0.357	4 630	20.1	226	65.9	9.05	165	44.2	27.6	310	77.2
x15	29.0	0.284	3 690	17.7	199	69.3	6.49	122	41.9	21.7	244	76.7
C150												
x19	38.4	0.377	4 900	14.2	187	53.9	9.11	169	43.1	20.3	268	64.4
x16	31.0	0.304	3 960	12.4	164	56.1	6.55	128	40.7	16.6	219	64.8
x12	24.2	0.237	3 070	10.7	141	59.1	4.36	90.9	37.7	12.8	169	64.6
C130												
x13	26.6	0.261	3 390	7.32	115	46.5	4.67	99.4	37.1	9.52	150	53.0
x10	19.8	0.194	2 520	6.18	97.3	49.5	2.94	66.7	34.1	7.02	110	52.8
C100												
x11	21.6	0.212	2 740	3.81	74.7	37.3	3.07	71.5	33.5	4.63	90.8	41.1
x9	18.8	0.184	2 380	3.53	69.3	38.5	2.52	60.0	32.5	4.02	78.8	41.1
x8	16.0	0.157	2 040	3.22	63.2	39.7	1.92	47.9	30.6	3.44	67.5	41.0
C75												
x9	17.6	0.173	2 250	1.69	44.5	27.4	2.08	52.0	30.4	†	†	†
x7	14.6	0.143	1 870	1.50	39.4	28.3	1.47	39.7	28.1	1.57	41.3	29.0
x6	12.0	0.118	1 530	1.34	35.2	29.6	1.04	29.8	26.2	1.28	33.6	28.9

† The condition of c = d cannot be met for this section.
* Not available from Canadian mills.

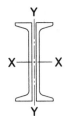

TWO CHANNELS
Back to Back

PROPERTIES OF SECTIONS

Channel Size	For Two Channels			Axis X-X			Radii of Gyration about Axis Y-Y					
	Mass	Dead Load	Area	I_x	S_x	r_x	Back to Back Channels, millimetres					
	kg/m	kN/m	mm²	10^6mm⁴	10^3mm³	mm	0	8	10	12	16	20
MC460												
x86*	174	1.70	22 100	567	2 480	160	34.0	36.7	37.4	38.2	39.6	41.2
x77.2*	155	1.52	19 700	524	2 290	163	34.1	36.8	37.5	38.2	39.7	41.2
x68.2*	137	1.34	17 500	486	2 130	167	34.8	37.5	38.2	38.9	40.4	41.9
x63.5*	127	1.25	16 200	463	2 030	169	35.0	37.6	38.3	39.0	40.5	42.0
C380												
x74*	149	1.46	19 000	336	1 760	133	29.9	32.8	33.5	34.3	35.8	37.4
x60*	119	1.17	15 100	290	1 520	138	30.0	32.7	33.5	34.2	35.7	37.3
x50*	101	0.990	12 900	262	1 370	143	30.5	33.2	33.9	34.7	36.2	37.8
C310												
x45*	89.4	0.877	11 400	135	883	109	25.7	28.5	29.3	30.0	31.6	33.2
x37*	74.2	0.728	9 440	120	785	113	26.1	28.9	29.6	30.4	31.9	33.5
x31*	61.6	0.604	7 830	107	702	117	26.7	29.5	30.2	31.0	32.5	34.1
C250												
x45*	89.0	0.873	11 300	85.6	674	86.9	23.4	26.4	27.2	27.9	29.6	31.2
x37	74.6	0.732	9 490	75.9	598	89.4	23.2	26.1	26.9	27.6	29.2	30.9
x30	59.2	0.581	7 550	65.4	515	93.0	23.3	26.1	26.8	27.6	29.2	30.8
x23	45.2	0.443	5 750	55.5	437	98.2	23.9	26.8	27.5	28.3	29.8	31.5
C230												
x30	59.6	0.584	7 590	50.9	445	81.9	22.0	24.9	25.6	26.4	28.0	29.7
x22	44.6	0.437	5 680	42.6	372	86.6	22.5	25.3	26.1	26.9	28.5	30.1
x20	39.6	0.388	5 050	39.6	346	88.6	22.6	25.5	26.2	27.0	28.6	30.2
C200												
x28	55.8	0.547	7 110	36.5	360	71.6	20.9	23.9	24.6	25.4	27.1	28.7
x21	40.8	0.400	5 190	29.8	294	75.8	20.9	23.8	24.5	25.3	26.9	28.6
x17	34.0	0.333	4 340	26.9	265	78.8	21.4	24.3	25.1	25.9	27.5	29.1
C180												
x22	43.8	0.430	5 570	22.6	254	63.7	19.6	22.6	23.3	24.1	25.8	27.5
x18	36.4	0.357	4 630	20.1	226	65.9	19.5	22.4	23.2	24.0	25.6	27.3
x15	29.0	0.284	3 690	17.7	199	69.3	20.2	23.1	23.9	24.7	26.3	28.0
C150												
x19	38.4	0.377	4 900	14.2	187	53.9	18.5	21.4	22.2	23.1	24.7	26.4
x16	31.0	0.304	3 960	12.4	164	56.1	18.3	21.3	22.0	22.8	24.5	26.2
x12	24.2	0.237	3 070	10.7	141	59.1	18.6	21.5	22.3	23.1	24.8	26.5
C130												
x13	26.6	0.261	3 390	7.32	115	46.5	17.1	20.1	20.9	21.7	23.4	25.1
x10	19.8	0.194	2 520	6.18	97.3	49.5	17.4	20.4	21.2	22.1	23.7	25.5
C100												
x11	21.6	0.212	2 740	3.81	74.7	37.3	16.1	19.1	20.0	20.8	22.5	24.3
x9	18.8	0.184	2 380	3.53	69.3	38.5	16.3	19.4	20.2	21.0	22.7	24.5
x8	16.0	0.157	2 040	3.22	63.2	39.7	16.2	19.3	20.1	20.9	22.6	24.4
C75												
x9	17.6	0.173	2 250	1.69	44.5	27.4	15.5	18.7	19.5	20.3	22.1	23.9
x7	14.6	0.143	1 870	1.50	39.4	28.3	14.8	18.0	18.8	19.7	21.4	23.2
x6	12.0	0.118	1 530	1.34	35.2	29.6	14.8	18.0	18.8	19.7	21.4	23.2

* Not available from Canadian mills.

W SHAPES AND CHANNELS

PROPERTIES OF SECTIONS

Beam	Channel	Dead Load	Total Area	Axis X-X					
				I	$S_1 = I/Y_1$	$S_2 = I/Y_2$	r	Y_1	Y_2
		kN/m	mm²	10^6 mm⁴	10^3 mm³	10^3 mm³	mm	mm	mm
W920x289*	MC460x63.5*	3.46	44 900	6 410	11 800	16 300	378	545	393
	C380x50*	3.33	43 200	6 170	11 600	15 200	378	531	406
x271*	MC460x63.5*	3.29	42 700	6 060	11 100	15 600	377	547	387
	C380x50*	3.16	41 000	5 830	11 000	14 500	377	532	401
x253*	MC460x63.5*	3.11	40 400	5 690	10 400	14 900	375	550	381
	C380x50*	2.98	38 700	5 460	10 200	13 800	376	534	395
x238*	MC460x63.5*	2.96	38 500	5 350	9 700	14 300	373	552	375
	C380x50*	2.83	36 800	5 130	9 580	13 200	373	536	389
x223*	MC460x63.5*	2.82	36 700	5 020	9 060	13 600	370	554	368
	C380x50*	2.69	35 000	4 810	8 950	12 500	371	537	384
W840x226*	MC460x63.5*	2.85	37 000	4 500	8 710	13 000	349	517	346
	C380x50*	2.72	35 300	4 310	8 600	12 000	349	501	360
x210*	MC460x63.5*	2.69	35 000	4 170	8 040	12 300	345	519	339
	C380x50*	2.56	33 300	4 000	7 960	11 300	347	503	353
x193*	MC460x63.5*	2.52	32 800	3 810	7 310	11 500	341	521	330
	C380x50*	2.39	31 100	3 650	7 230	10 600	343	505	345
W760x196*	MC460x63.5*	2.55	33 200	3 270	6 860	10 700	314	477	305
	C380x50*	2.42	31 500	3 120	6 760	9 790	315	462	319
x185*	MC460x63.5*	2.44	31 700	3 070	6 420	10 300	311	478	299
	C380x50*	2.31	30 000	2 930	6 330	9 360	313	463	313
x173*	MC460x63.5*	2.33	30 200	2 880	5 990	9 830	309	480	293
	C380x50*	2.20	28 500	2 750	5 920	8 940	311	465	308
x161*	MC460x63.5*	2.20	28 500	2 650	5 480	9 280	305	484	286
	C380x50*	2.07	26 900	2 530	5 410	8 410	307	467	301
W690x170*	C380x50*	2.16	28 100	2 270	5 360	8 120	284	424	280
	C310x31*	1.97	25 600	2 080	5 230	6 880	285	398	302
x152*	C380x50*	1.99	25 800	2 050	4 800	7 570	282	427	271
	C310x31*	1.79	23 300	1 870	4 670	6 340	283	400	295
x140*	C380x50*	1.87	24 200	1 890	4 390	7 160	280	430	264
	C310x31*	1.67	21 700	1 720	4 280	5 940	282	402	289
W610x125	C380x50*	1.72	22 400	1 390	3 550	6 020	249	391	231
	C310x31*	1.53	19 800	1 260	3 460	4 940	252	364	255
x113	C380x50*	1.61	20 900	1 260	3 190	5 640	246	395	224
	C310x31*	1.41	18 400	1 140	3 110	4 590	249	367	249
W530x101	C380x50*	1.49	19 400	908	2 560	4 710	216	354	193
	C310x31*	1.30	16 800	819	2 490	3 800	221	329	216
x92	C380x50*	1.40	18 200	830	2 320	4 460	214	357	186
	C310x31*	1.21	15 700	747	2 260	3 560	218	330	210
W460x74	C380x50*	1.22	15 900	519	1 640	3 460	181	317	150
	C310x31*	1.03	13 400	466	1 590	2 710	187	292	172
W410x54	C380x50*	1.02	13 200	311	1 060	2 620	154	295	119
	C310x31*	0.826	10 700	279	1 030	2 010	162	271	139
W360x45	C310x31*	0.743	9 650	187	769	1 610	139	243	116
	C250x23	0.663	8 610	176	760	1 390	143	232	127
W310x39	C310x31*	0.682	8 850	132	603	1 340	122	219	98.2
	C250x23	0.601	7 810	124	595	1 150	126	208	108
W250x33	C250x23	0.542	7 050	74.0	417	856	103	178	86.4
	C200x17	0.488	6 340	70.0	411	749	105	170	93.5
W200x27	C200x17	0.428	5 560	38.2	272	530	82.9	140	72.1

* Not available from Canadian mills.

W SHAPES AND CHANNELS

PROPERTIES OF SECTIONS

Mass	Axis Y-Y			Shear Centre	Torsional Constant	Warping Constant	Monosymmetry Constant †
	I	S	r	y_0	J	C_w	β_x
kg/m	10^6 mm^4	10^3 mm^3	mm	mm	10^3 mm^4	10^9 mm^6	mm
352.3	388	1 700	93.0	212	9 750	53 500	524
339.1	287	1 510	81.6	156	9 650	48 200	395
335.4	377	1 650	93.9	216	8 210	50 100	538
322.2	276	1 450	82.1	161	8 110	45 100	411
317.4	365	1 600	95.1	220	6 780	46 600	554
304.2	265	1 390	82.7	167	6 680	42 000	428
302.0	354	1 550	95.9	225	5 670	43 200	570
288.7	254	1 330	83.0	173	5 560	39 100	445
288.0	344	1 500	96.8	230	4 740	39 800	587
274.7	243	1 280	83.4	179	4 640	36 100	464
290.4	345	1 510	96.6	214	5 670	35 000	545
277.1	245	1 280	83.3	167	5 570	31 700	430
274.5	334	1 460	97.7	219	4 580	31 800	562
261.3	234	1 230	83.7	174	4 480	28 900	450
257.3	322	1 410	99.1	224	3 580	28 200	582
244.1	221	1 160	84.3	182	3 480	25 800	474
260.5	313	1 370	97.1	215	4 570	21 600	548
247.3	213	1 120	82.2	178	4 460	19 800	452
248.6	307	1 340	98.4	217	3 850	19 900	558
235.3	206	1 080	82.9	182	3 750	18 300	465
237.4	300	1 310	99.7	219	3 210	18 300	569
224.1	200	1 050	83.7	186	3 110	16 900	479
224.1	292	1 280	101	222	2 590	16 300	583
210.8	192	1 010	84.4	192	2 490	15 200	498
220.4	197	1 030	83.8	172	3 470	13 500	442
200.7	120	785	68.4	115	3 200	11 400	293
202.6	189	991	85.5	176	2 620	12 000	460
182.9	111	730	69.1	122	2 360	10 200	314
190.3	183	959	86.9	179	2 090	10 800	474
170.6	105	690	69.7	128	1 820	9 250	331
175.5	170	894	87.2	173	1 960	6 820	457
155.8	92.9	609	68.5	133	1 690	5 910	338
163.9	165	867	88.9	175	1 540	6 000	469
144.2	87.8	576	69.1	139	1 270	5 240	357
151.9	158	829	90.2	162	1 440	3 790	434
132.2	80.5	528	69.2	136	1 170	3 350	347
142.9	155	812	92.2	162	1 180	3 360	439
123.2	77.3	507	70.2	139	915	2 990	360
124.7	148	775	96.3	143	939	1 800	384
104.9	70.1	460	72.3	131	670	1 630	342
103.9	141	740	103	125	647	897	315
84.2	63.6	417	77.1	124	378	831	331
75.8	61.7	405	80.0	111	313	533	287
67.6	35.9	283	64.6	100	246	493	268
69.5	60.8	399	82.9	98.2	279	377	232
61.3	35.0	276	67.0	89.9	212	349	237
55.3	32.5	256	67.9	83.8	185	168	193
49.8	18.2	179	53.6	71.7	152	153	188
43.6	16.8	165	54.9	65.1	125	74.1	148

† β_x is positive when the larger flange is in compression, and negative otherwise.
See "Buckling properties of monosymmetric I-beams", Kitipornchai, S. and Trahair, N.S., ASCE J. Struct. Div., May 1980.

NOTES

BUILT-UP SECTIONS

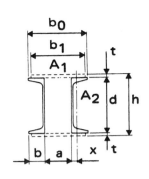

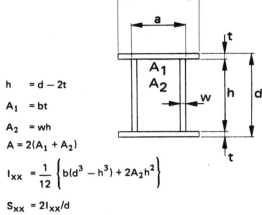

$h = d + 2t$ $b_0 = a + 2b$ $S_{yy} = 2I_{yy}/b_0$ if $b_1 < b_0$

$A_1 = b_1 t$ $S_{yy} = 2I_{yy}/b_1$ if $b_1 \geqslant b_0$

$A = 2(A_1 + A_2)$

$I_{xx} = 2I_{xc} + \dfrac{b_1}{12}\left[h^3 - d^3\right]$ $r_{yy} = \sqrt{I_{yy}/A}$

$S_{xx} = 2I_{xx}/h$

$r_{xx} = \sqrt{I_{xx}/A}$

$I_{yy} = 2I_{yc} + \dfrac{A_1}{6} b_1^2 + 2A_2 (x + a/2)^2$

$h = d - 2t$

$A_1 = bt$

$A_2 = wh$

$A = 2(A_1 + A_2)$

$I_{xx} = \dfrac{1}{12}\left\{b(d^3 - h^3) + 2A_2 h^2\right\}$

$S_{xx} = 2I_{xx}/d$

$Z_{xx} = \dfrac{b}{4}\left\{d^2 - h^2\right\} + \dfrac{A_2 h}{2}$

$r_{xx} = \sqrt{I_{xx}/A}$ $Z_{yy} = \dfrac{h}{4}\left\{a^2 - c^2\right\} + \dfrac{A_1 b}{2}$

$c = a - 2w$ $r_{yy} = \sqrt{I_{yy}/A}$

$I_{yy} = \dfrac{1}{12}\left\{2A_1 b^2 + h(a^3 - c^3)\right\}$

$S_{yy} = 2I_{yy}/b$

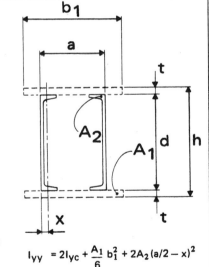

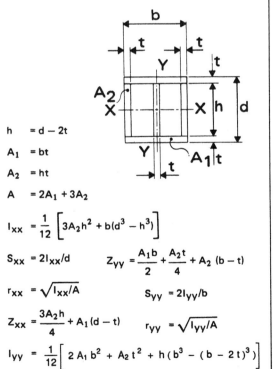

$h = d + 2t$

$A_1 = b_1 t$ $I_{yy} = 2I_{yc} + \dfrac{A_1}{6} b_1^2 + 2A_2 (a/2 - x)^2$

$A = 2(A_1 + A_2)$

$I_{xx} = 2I_{xc} + \dfrac{b_1}{12}(h^3 - d^3)$ $S_{yy} = 2I_{yy}/b_1$ if $a < b_1$

                               $S_{yy} = 2I_{yy}/a$ if $a \geqslant b_1$

$S_{xx} = 2I_{xx}/h$

$r_{xx} = \sqrt{I_{xx}/A}$ $r_{yy} = \sqrt{I_{yy}/A}$

$h = d - 2t$

$A_1 = bt$

$A_2 = ht$

$A = 2A_1 + 3A_2$

$I_{xx} = \dfrac{1}{12}\left[3A_2 h^2 + b(d^3 - h^3)\right]$

$S_{xx} = 2I_{xx}/d$ $Z_{yy} = \dfrac{A_1 b}{2} + \dfrac{A_2 t}{4} + A_2 (b - t)$

$r_{xx} = \sqrt{I_{xx}/A}$ $S_{yy} = 2I_{yy}/b$

$Z_{xx} = \dfrac{3A_2 h}{4} + A_1 (d - t)$ $r_{yy} = \sqrt{I_{yy}/A}$

$I_{yy} = \dfrac{1}{12}\left[2A_1 b^2 + A_2 t^2 + h(b^3 - (b - 2t)^3)\right]$

Elements of the shape which are shown in dotted outline are optional and if omitted the variable defining their size should be set equal to zero.

All elements of the shape are assumed to be continuous along the length of the shape.

BUILT-UP SECTIONS

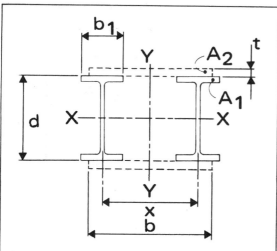

$A_2 = bt$

$A = 2(A_1 + A_2)$

$I_{xx} = 2I_{xw} + \frac{1}{12} b \left[(d + 2t)^3 - d^3\right]$

$S_{xx} = 2I_{xx}/(d + 2t)$

$r_{xx} = \sqrt{I_{xx}/A}$

For $(x + b_1) > b$:
$S_{yy} = 2I_{yy}/(x + b_1)$

For $(x + b_1) \leq b$:
$S_{yy} = 2I_{yy}/b$

$I_{yy} = 2I_{yw} + \frac{1}{6} A_2 b^2 + \frac{1}{2} A_1 x^2$

$r_{yy} = \sqrt{I_{yy}/A}$

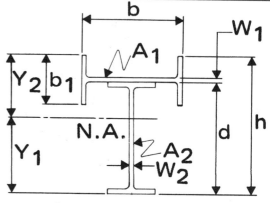

$h = d + \frac{1}{2}(b_1 + w_1)$

$Y_1 = \dfrac{A_1(d + W_1/2) + A_2 d/2}{A_1 + A_2}$

$Y_2 = h - Y_1$ $I_{yy} = I_{x1} + I_{y2}$

$A = A_1 + A_2$ *$I_{yT} = I_{x1} + I_{y2}/2 - (Y_1 - d/2)w_2^3/12$

$I_{xx} = I_{y1} + I_{x2} + A_1(Y_2 - b_1/2)^2 + A_2(Y_1 - d/2)^2$

$S_{x1} = I_{xx}/Y_1$ $S_{yy} = 2I_{yy}/b$

$S_{x2} = I_{xx}/Y_2$

$r_{xx} = \sqrt{I_{xx}/A}$ $r_{yy} = \sqrt{I_{yy}/A}$

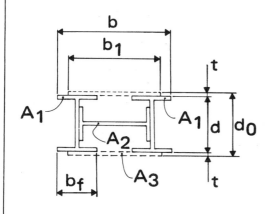

$d_o = d + 2t$ $I_{yy} = I_{x2} + 2I_{y1} + \dfrac{A_3}{6}b_1^2 + A_1(b - b_f)^2/2$

$A_3 = b_1 t$

$A = 2(A_1 + A_3) + A_2$

$I_{xx} = 2I_{x1} + I_{y2} + \dfrac{b_1}{12}(d_o^3 - d^3)$

$S_{yy} = 2I_{yy}/b$ if $b \geq b_1$

$S_{xx} = 2I_{xx}/d_o$ $S_{yy} = 2I_{yy}/b_1$ if $b < b_1$

$r_{xx} = \sqrt{I_{xx}/A}$ $r_{yy} = \sqrt{I_{yy}/A}$

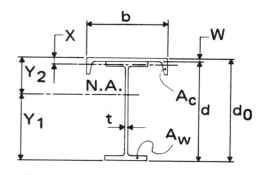

$Y_1 = \dfrac{A_w d/2 + A_c(d_o - x)}{A}$ $A = A_c + A_w$

$d_o = d + w$

$Y_2 = d_o - Y_1$

$I_{xx} = I_{xw} + I_{yc} + A_w(Y_1 - d/2)^2 + A_c(Y_2 - x)^2$

$S_{x1} = I_{xx}/Y_1$ $I_{yy} = I_{yw} + I_{xc}$

$S_{x2} = I_{xx}/Y_2$ *$I_{yT} = I_{xc} + \dfrac{I_{yw}}{2} - (Y_1 - d/2)\dfrac{t^3}{12}$

$r_{xx} = \sqrt{I_{xx}/A}$ $S_{yy} = 2I_{yy}/b$

$r_{yy} = \sqrt{I_{yy}/A}$

*I_{yT} is the moment of inertia of the T section above the neutral axis.

BUILT-UP SECTIONS

Note: Centres of gravity of both channels are on the same vertical line.

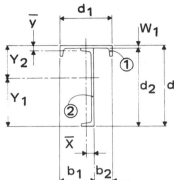

$d = d_2 + w_1$
$A = A_1 + A_2$
$b_1 = (d_1/2) + \bar{x}$
$b_2 = d_1 - b_1$

$y_1 = \dfrac{A_1(d - \bar{y}) + \dfrac{A_2}{2}d_2}{A}$

$I_{xx} = I_{1y} + I_{2x} + A_1(y_2 - \bar{y})^2 + A_2(y_1 - \dfrac{d_2}{2})^2$

$S_{x1} = \dfrac{I_{xx}}{y_1}$

$S_{x2} = \dfrac{I_{xx}}{y_2}$

$y_2 = d - y_1$

$I_{yy} = I_{x1} + I_{y2}$

$S_y = 2I_{yy}/d_1$

$r_{xx} = \sqrt{\dfrac{I_{xx}}{A}}$

$r_{yy} = \sqrt{\dfrac{I_{yy}}{A}}$

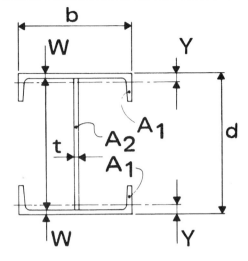

$h = d - 2w$
$A_2 = ht$
$A = 2A_1 + A_2$

$I_{xx} = 2I_{yc} + \dfrac{1}{12}A_2 h^2 + 2A_1(d/2 - y)^2$

$I_{yy} = 2I_{xc} + \dfrac{1}{12}A_2 t^2$

$S_{xx} = 2I_{xx}/d$

$S_{yy} = 2I_{yy}/b$

$r_{yy} = \sqrt{I_{yy}/A}$

$r_{xx} = \sqrt{I_{xx}/A}$

Note: a and b are the angle leg lengths and b_1 is the width of channel flange.

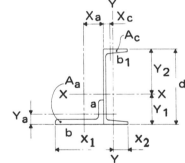

$A = A_a + A_c$

$x_1 = \dfrac{A_a(b - x_a) + A_c(b + x_c)}{A}$; $y_1 = \dfrac{A_a y_a + A_c d/2}{A}$

$x_2 = b_1 + b - x_1$

$y_2 = d - y_1$

$I_{xx} = I_{ya} + I_{xc} + A_a(y_1 - y_a)^2 + A_c(\dfrac{d}{2} - y_1)^2$

$I_{yy} = I_{xa} + I_{yc} + A_a(x_1 - b + x_a)^2 + A_c(b_1 - x_2 - x_c)^2$

$S_{y1} = I_{yy}/x_1$
$S_{y2} = I_{yy}/x_2$
$S_{x1} = I_{xx}/y_1$
$S_{x2} = I_{xx}/y_2$

$r_{yy} = \sqrt{I_{yy}/A}$
$r_{xx} = \sqrt{I_{xx}/A}$

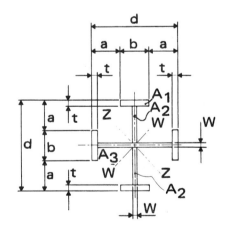

$A_1 = bt$
$A_2 = (d - w - 2t)w/2$
$A_3 = 2A_2 + w^2$
$A = 4A_1 + 2A_2 + A_3$

$I_x = I_y = \dfrac{1}{12}\left\{b(d^3 - E^3) + wE^3 + 2tb^3 + Ew^3 - w^4\right\}$

$S_x = S_y = 2I_x/d$

$r_x = r_y = \sqrt{I_x/A}$

$E = (d - 2t)$

6-127

BUILT-UP SECTIONS

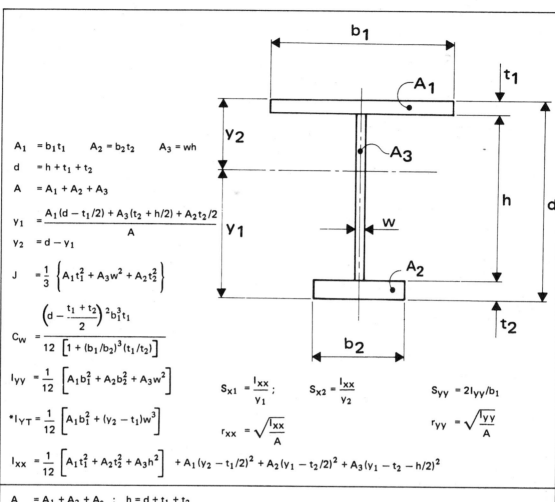

$A_1 = b_1 t_1 \quad A_2 = b_2 t_2 \quad A_3 = wh$

$d = h + t_1 + t_2$

$A = A_1 + A_2 + A_3$

$y_1 = \dfrac{A_1(d - t_1/2) + A_3(t_2 + h/2) + A_2 t_2/2}{A}$

$y_2 = d - y_1$

$J = \dfrac{1}{3}\left\{ A_1 t_1^2 + A_3 w^2 + A_2 t_2^2 \right\}$

$C_w = \dfrac{\left(d - \dfrac{t_1 + t_2}{2}\right)^2 b_1^3 t_1}{12\left[1 + (b_1/b_2)^3 (t_1/t_2)\right]}$

$I_{yy} = \dfrac{1}{12}\left[A_1 b_1^2 + A_2 b_2^2 + A_3 w^2 \right]$

*$I_{YT} = \dfrac{1}{12}\left[A_1 b_1^2 + (y_2 - t_1) w^3 \right]$

$I_{xx} = \dfrac{1}{12}\left[A_1 t_1^2 + A_2 t_2^2 + A_3 h^2 \right] + A_1(y_2 - t_1/2)^2 + A_2(y_1 - t_2/2)^2 + A_3(y_1 - t_2 - h/2)^2$

$S_{x1} = \dfrac{I_{xx}}{y_1} \ ; \quad S_{x2} = \dfrac{I_{xx}}{y_2} \quad\quad S_{yy} = 2 I_{yy}/b_1$

$r_{xx} = \sqrt{\dfrac{I_{xx}}{A}} \quad\quad r_{yy} = \sqrt{\dfrac{I_{yy}}{A}}$

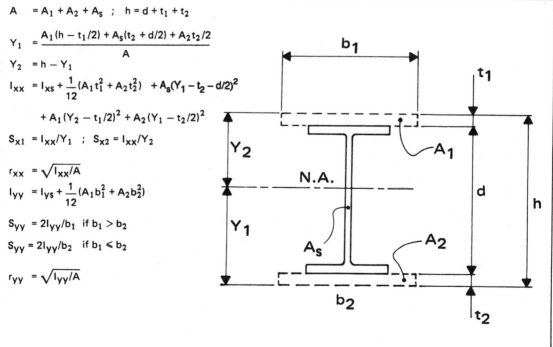

$A = A_1 + A_2 + A_s \ ; \quad h = d + t_1 + t_2$

$Y_1 = \dfrac{A_1(h - t_1/2) + A_s(t_2 + d/2) + A_2 t_2/2}{A}$

$Y_2 = h - Y_1$

$I_{xx} = I_{xs} + \dfrac{1}{12}(A_1 t_1^2 + A_2 t_2^2) + A_s(Y_1 - t_2 - d/2)^2$
$\quad\quad + A_1(Y_2 - t_1/2)^2 + A_2(Y_1 - t_2/2)^2$

$S_{x1} = I_{xx}/Y_1 \ ; \quad S_{x2} = I_{xx}/Y_2$

$r_{xx} = \sqrt{I_{xx}/A}$

$I_{yy} = I_{ys} + \dfrac{1}{12}(A_1 b_1^2 + A_2 b_2^2)$

$S_{yy} = 2 I_{yy}/b_1 \quad \text{if } b_1 > b_2$

$S_{yy} = 2 I_{yy}/b_2 \quad \text{if } b_1 \leq b_2$

$r_{yy} = \sqrt{I_{yy}/A}$

*I_{yt} is the moment of inertia of the T section above the neutral axis.

BARS AND PLATES

Bars

The term "Bars" means:

(a) Rounds, squares and hexagons of all sizes;

(b) Flat-rolled steel up to 200 mm inclusive in width and over 5.0 mm in thickness, except for widths from 150 mm to 200 mm up to 6.0 mm in thickness;

(c) Bar-size shapes under 75 mm in maximum dimension.

Hot rolled bar flat products are available in most widths and thickness according to CAN3-G312.1 and CAN3-G312.2-M Standards.

Plates

The term "Plates" means:

Flat hot-rolled steel, when ordered to thickness;

(a) Over 200 mm in width and over 6.0 mm in thickness, and

(b) Over 1200 mm in width and over 4.5 mm in thickness.

Slabs, sheet, bars, strip and skelp, although frequently falling within these size ranges, are not classified as plate. The following table, Standard Product Classification for Flat Hot-Rolled Steel Products and Bars, summarizes the ranges for plate, bar, strip and sheet products.

Plates may be further defined as "Universal Mill Plates" or "Sheared Plates". Sheared plates are rolled on a mill with horizontal rolls only, producing a product with uneven edges which must be sheared (or, at the option of the producer, flame cut) to ordered dimensions.

Universal mill plates are rolled to the ordered width on a mill having side rollers to control the width. Slab or ingot on a universal mill plate are not cross rolled, but are only elongated during the rolling process. The mill order must specify universal mill plate when it is required.

Extreme plate sizes produced by mills vary greatly with the size of various mills, and individual mills should be consulted for this information. Canadian mills can produce sheared and flame cut plate in widths ranging from 200 mm to 3900 mm and in thicknesses ranging from 4.5 mm to 150 mm. First and second preference thickness for plate according to CAN3-G312.1 is provided in the table on page 6-131. Where possible, designs should generally be based on first preference thickness for economy.

Standard mill practice is to invoice the purchaser for actual scale weight at point of shipment. Allowable overweight above theoretical that may be charged, is limited in accordance with either the CSA-G40.20 or ASTM A6 Standards, depending upon whether the steel is furnished to Canadian or ASTM material specifications.

Various extras for thickness, width, length, cutting, quality, quantity, (or quantity discounts), and for other special requirements are added to the base price of plates. Particulars of these extras should be obtained from the producing mills.

Sketch Plates

Sketch plates of special or unusual shape usually require flame cutting, for which flame cutting extras apply. Some mills can supply sketch plates of certain shapes by shearing to size.

Floor Plates

Floor plates in different styles, patterns, and extreme dimensions are produced by different mills. The nominal, or ordered, thickness is that of the flat plate exclusive of the raised pattern. Individual producers should be consulted for more details.

Bearing Plates

Rolled steel bearing plates are used for column bases, and other bearing plates. Depending on the thickness required by design, bearing plates may require additional thickness for machining to ensure proper bearing. Plates up to and including 50 mm in thickness are rolled flat with surfaces sufficiently smooth to receive, without machining or flattening, the milled or machine cut ends of column shafts. Bearing plates 100 mm and under in thickness can sometimes be flattened by press to within required flatness tolerances. Rolled steel bearing plates greater than 100 mm thick usually require machining of the top surface in contact with the column shaft, but do not require flattening of the bottom surface when bearing on a grouted base. When bearing plates over 100 mm thick bear on a steel surface, both the top and bottom surfaces will require to be machined over the bearing areas.

Tables

The following Tables are included in this section:

    Standard Product Classification of Flat Hot-Rolled Steel Products and Bars

    Flat Metal Products - Sheet & Plate

    Mass (kg/m) for Rectangular Steel Products

    SI Wire Size - Wire Gauges Comparison

    SI Thickness - Imperial Gauge Comparisons

    Mass or Round Bars and Square Bars

STANDARD PRODUCT CLASSIFICATION of Flat Hot-Rolled Steel Products and Bars

Width, w (mm)	Thickness, t (mm)					
	$t > 6$	$6 \geq t > 5$	$5 \geq t > 4.5$	$4.5 \geq t > 1.2$	$1.2 \geq t > 0.9$	$0.9 \geq t > 0.65$
$w \leq 100$	BAR	BAR	STRIP	STRIP	STRIP	STRIP
$100 < w \leq 150$	BAR	BAR	STRIP	STRIP	STRIP	
$150 < w \leq 200$	BAR	STRIP	STRIP	STRIP		
$200 < w \leq 300$	PLATE	STRIP	STRIP	STRIP		
$300 < w \leq 1200$	PLATE	SHEET**	SHEET**	SHEET**		
$1200 < w$	PLATE	PLATE	PLATE	SHEET		

**For alloy steels, sheet begins at widths over 600 mm.

FLAT METAL PRODUCTS* – Sheet & Plate

Nominal Thickness,** mm		Mass†	Dead Load	Nominal Thickness,** mm		Mass†	Dead Load
First preference	Second preference	$\dfrac{kg}{m^2}$	$\dfrac{kN}{m^2}$	First preference	Second preference	$\dfrac{kg}{m^2}$	$\dfrac{kN}{m^2}$
0.050		0.393	0.003 85		14	110	1.08
0.060		0.471	0.004 62	16		126	1.23
0.080		0.628	0.006 16		18	141	1.39
0.10		0.785	0.007 70	20		157	1.54
0.12		0.942	0.009 24		22	173	1.69
	0.14	1.10	0.010 8	25		196	1.92
0.16		1.26	0.012 3		28	220	2.16
	0.18	1.41	0.013 9	30		236	2.31
0.20		1.57	0.015 4		32	251	2.46
	0.22	1.73	0.016 9	35		275	2.69
0.25		1.96	0.019 2		38	298	2.93
	0.28	2.20	0.021 6	40		314	3.08
0.30		2.36	0.023 1		45	353	3.46
	0.35	2.75	0.026 9	50		393	3.85
0.40		3.14	0.030 8		55	432	4.23
	0.45	3.53	0.034 6	60		471	4.62
0.50		3.93	0.038 5		70	550	5.39
	0.55	4.32	0.042 3	80		628	6.16
0.60		4.71	0.046 2		90	707	6.93
	0.65	5.10	0.050 0	100		785	7.70
	0.70	5.50	0.053 9		110	864	8.47
0.80		6.28	0.061 6	120		942	9.24
	0.90	7.07	0.069 3		130	1020	10.0
1.0		7.85	0.077 0	140		1100	10.8
	1.1	8.64	0.084 7		150	1180	11.5
1.2		9.42	0.092 4	160		1260	12.3
	1.4	11.0	0.108	180		1410	13.9
1.6		12.6	0.123	200		1570	15.4
	1.8	14.1	0.139	250		1960	19.2
2.0		15.7	0.154	300		2360	23.1
	2.2	17.3	0.169				
2.5		19.6	0.192				
	2.8	22.0	0.216				
3.0		23.6	0.231				
	3.2	25.1	0.246				
3.5		27.5	0.269				
	3.8	29.8	0.293				
4.0		31.4	0.308				
	4.2	33.0	0.323				
4.5		35.3	0.346				
	4.8	37.7	0.370				
5.0		39.3	0.385				
	5.5	43.2	0.423				
6.0		47.1	0.462				
7.0		55.0	0.539				
8.0		62.8	0.616				
	9.0	70.7	0.693				
10		78.5	0.770				
	11	86.4	0.847				
12		94.2	0.924				

*Sizes are those listed in CAN3-G312.1-75
**For coated structural sheet, the nominal thickness applies to the base metal. For metric thickness dimensions for zinc coated structural quality sheet steel, see Part 7, Structural Sheet Steel Products.
† Computed using steel density of 7 850 kg/m^3.

MASS (kg/m) for RECTANGULAR STEEL PRODUCTS*

Width (mm) Second	Width (mm) First	Thickness (mm) 5	6	8	10	12	14	16	18	20	22	25	28	30	35	40
	25	0.981	1.18	1.57	1.96	2.36	2.75	3.14	3.53	3.93	4.32					
28			1.10	1.32	1.76	2.20	2.64	3.08	3.52	3.96	4.40	4.84	5.50			
	30	1.18	1.41	1.88	2.36	2.83	3.30	3.77	4.24	4.71	5.18	5.89	6.59			
35			1.37	1.65	2.20	2.75	3.30	3.85	4.40	4.95	5.50	6.04	6.87	7.69	8.24	
	40	1.57	1.88	2.51	3.14	3.77	4.40	5.02	5.65	6.28	6.91	7.85	8.79	9.42	11.0	
45			1.77	2.12	2.83	3.53	4.24	4.95	5.65	6.36	7.07	7.77	8.83	9.89	10.6	12.4	14.1
	50	1.96	2.36	3.14	3.93	4.71	5.50	6.28	7.07	7.85	8.64	9.81	11.0	11.8	13.7	15.7
55			2.16	2.59	3.45	4.32	5.18	6.04	6.91	7.77	8.64	9.50	10.8	12.1	13.0	15.1	17.3
	60	2.36	2.83	3.77	4.71	5.65	6.59	7.54	8.48	9.42	10.4	11.8	13.2	14.1	16.5	18.8
70			2.75	3.30	4.40	5.50	6.59	7.69	8.79	9.89	11.0	12.1	13.7	15.4	16.5	19.2	22.0
	80	3.14	3.77	5.02	6.28	7.54	8.79	10.0	11.3	12.6	13.8	15.7	17.6	18.8	22.0	25.1
90			3.53	4.24	5.65	7.07	8.48	9.89	11.3	12.7	14.1	15.5	17.7	19.8	21.2	24.7	28.3
	100	3.93	4.71	6.28	7.85	9.42	11.0	12.6	14.1	15.7	17.3	19.6	22.0	23.6	27.5	31.4
110			4.32	5.18	6.91	8.64	10.4	12.1	13.8	15.5	17.3	19.0	21.6	24.2	25.9	30.2	34.5
	120	4.71	5.65	7.54	9.42	11.3	13.2	15.1	17.0	18.8	20.7	23.6	26.4	28.3	33.0	37.7
140			5.50	6.59	8.79	11.0	13.2	15.4	17.6	19.8	22.0	24.2	27.5	30.8	33.0	38.5	44.0
	160		7.54	10.0	12.6	15.1	17.6	20.1	22.6	25.1	27.6	31.4	35.2	37.7	44.0	50.2
180				8.48	11.3	14.1	17.0	19.8	22.6	25.4	28.3	31.1	35.3	39.6	42.4	49.5	56.5
	200		9.42	12.6	15.7	18.8	22.0	25.1	28.3	31.4	34.5	39.3	44.0	47.1	55.0	62.8
220				10.3	13.8	17.3	20.7	24.2	27.6	31.1	34.5	38.0	43.2	48.4	51.8	60.4	69.1
	250		11.8	15.7	19.6	23.6	27.5	31.4	35.3	39.3	43.2	49.1	55.0	58.9	68.7	78.5
300				14.1	18.8	23.6	28.3	33.0	37.7	42.4	47.1	51.8	58.9	65.9	70.7	82.4	94.2

Note: The mass has been computed using a steel density of 7 850 kg/m^3.
* Sizes are those listed in Table 2 of CAN3-G312.2-M76

SI WIRE SIZE – WIRE GAUGES COMPARISON

SI Wire Size Preferred Diam*. (mm)	The United States Steel Wire Gauge	American or Brown & Sharpe Wire Gauge	British Imperial or English Legal Standard Wire Gauge	Birmingham or Stubs Iron Wire Gauge	SI Wire Size Preferred Diam*. (mm)	The United States Steel Wire Gauge	American or Brown & Sharpe Wire Gauge	British Imperial or English Legal Standard Wire Gauge	Birmingham or Stubs Iron Wire Gauge
25.0					6.0				
24.0						4	3	4	5
23.0					5.6				
22.0								5	
21.0					5.3	5			
20.0							4		6
19.0					5.0				
18.0						6		6	
17.0					4.8				
16.0							5		
15.0					4.6				
		6/0's				7		7	7
14.0					4.4				
		5/0's			4.2				
13.0						8	6	8	8
			7/0's	5/0's	4.0				
12.5					3.8				
	7/0's					9	7	9	9
12.0					3.6				
11.8						10			
	6/0's	4/0's	6/0's	4/0's	3.4				10
11.2							8	10	
11.0					3.2				
	5/0's		5/0's	3/0's		11			11
10.6					3.0				
		3/0's	4/0's				9	11	
10.0	4/0's				2.8				
				2/0's		12		12	12
9.5					2.6				
	3/0's	2/0's	3/0's				10		13
9.0					2.4				
			2/0's	1/0		13	11	13	
8.5					2.3				
	2/0's	1/0	1/0		2.2				
8.0									14
	1/0		1	1	2.1				
7.5						14	12	14	
	1	1	2	2	2.0				
7.0					1.90				
6.7	2					15	13	15	15
		2		3	1.80				
6.5					1.70				
				3			14	16	16
6.3					1.60				
	3			4		16			
6.0					1.50				

*From CAN3–G312.2-M76.

SI THICKNESS – IMPERIAL GAUGE COMPARISONS†

SI Preferred Thickness		United States Standard Gauge*				New Birmington Sheet Gauge		
		Weight	Ga. No.	Approximate thickness		Gauge Number	Thickness	
First mm	Second mm	Oz. per sq. ft.		Inches	mm		Inches	mm
	18							
16						7/0's	0.6666	16.932
						6/0's	0.6250	15.875
	14					5/0's	0.5883	14.943
						4/0's	0.5416	13.757
12						3/0's	0.5000	12.700
						2/0's	0.4452	11.562
	11							
						0	0.3964	10.069
10								
	9.0							
						1	0.3532	8.971
8.0						2	0.3147	7.993
						3	0.2804	7.122
7.0								
		160	3	0.2391	6.073	4	0.2500	6.350
6.0								
		150	4	0.2242	5.695	5	0.2225	5.652
	5.5							
		140	5	0.2092	5.314	6	0.1981	5.032
5.0								
		130	6	0.1943	4.935			
	4.8							
		120	7	0.1793	4.554			
4.5								
						7	0.1764	4.481
	4.2							
		110	8	0.1644	4.176			
4.0								
						8	0.1570	3.988
	3.8	100	9	0.1495	3.797			
						9	0.1398	3.551
3.5								
		90	10	0.1345	3.416			
	3.2							
		80	11	0.1196	3.038	10	0.1250	3.175
3.0								
						11	0.1113	2.827
	2.8							
		70	12	0.1046	2.657	12	0.0991	2.517
2.5								

†Preferred thicknesses are as per CAN3-G312.1-75.
*U.S. Standard Gauge is officially a weight gauge, in oz. per sq. ft. as tabulated. The Approx. thickness shown is the "Manufacturers' Standard" of the AISI based on a steel density of 501.81 lb. per ft.³

SI THICKNESS – IMPERIAL GAUGE COMPARISONS†

SI Preferred Thickness		United States Standard Gauge*				New Birmingham Sheet Gauge		
		Weight	Ga. No.	Approximate thickness		Gauge Number	Thickness	
First mm	Second mm	Oz. per sq. ft.		Inches	mm		Inches	mm
		60	13	0.0897	2.278	13	0.0882	2.240
	2.2							
2.0						14	0.0785	1.994
		50	14	0.0747	1.897			
	1.8							
		45	15	0.0673	1.709	15	0.0699	1.775
1.6								
		40	16	0.0598	1.519	16	0.0625	1.588
						17	0.0556	1.412
	1.4							
		36	17	0.0538	1.367			
		32	18	0.0478	1.214	18	0.0495	1.257
1.2								
						19	0.0440	1.118
	1.1							
		28	19	0.0418	1.062			
1.0						20	0.0392	0.996
		24	20	0.0359	0.912			
	0.90							
		22	21	0.0329	0.836	21	0.0349	0.886
0.80						22	0.0313	0.795
		20	22	0.0299	0.759			
	0.70					23	0.0278	0.706
		18	23	0.0269	0.683			
	0.65							
						24	0.0248	0.630
0.60		16	24	0.0239	0.607			
						25	0.0220	0.559
	0.55							
		14	25	0.0209	0.531			
0.50						26	0.0196	0.498
	0.45	12	26	0.0179	0.455	27	0.0175	0.445
		11	27	0.0164	0.417			
0.40						28	0.0156	0.396
		10	28	0.0149	0.378			
	0.35	9	29	0.0135	0.343	29	0.0139	0.353
						30	0.0123	0.312
0.30		8	30	0.0120	0.305			
	0.28					31	0.0110	0.279
		7	31	0.0105	0.267			
0.25								

†Preferred thicknesses are as per CAN3-G312.1-75.
*U.S. Standard Gauge is officially a weight gauge, in oz. per sq. ft. as tabulated. The Approx. thickness shown is the "Manufacturers' Standard" of the AISI based on a steel density of 501.81 lb. per ft.^{3}

MASS OF BARS
Round Bars

Preferred Nominal Size, mm		Mass*	Area	Preferred Nominal Size, mm		Mass*	Area	Preferred Nominal Size, mm		Mass*	Area
First	Second	kg/m	mm²	First	Second	kg/m	mm²	First	Second	kg/m	mm²
3.0		0.055 5	7.07		21	2.72	346		65	26.0	3 320
	3.5	0.075 5	9.62	22		2.98	380		70	30.2	3 850
4.0		0.098 6	12.6		23	3.26	415		72†	32.0	4 070
	4.5	0.125	15.9		24	3.55	452		75	34.7	4 420
5.0		0.154	19.6	25		3.85	491	80		39.5	5 030
	5.5	0.187	23.7		26	4.17	531		90	49.9	6 360
6.0		0.223	28.3		27†	4.49	573	100		61.7	7 850
	6.5	0.260	33.2		28	4.83	616		110	74.6	9 500
	7.0	0.302	38.5	30		5.55	707	120		88.8	11 300
8.0		0.395	50.3		32	6.31	804		130	104	13 300
	9.0	0.499	63.6	35		7.55	962	140		121	15 400
10		0.617	78.5		36†	7.99	1 020		150	139	17 700
	11	0.746	95.0		38	8.90	1 130	160		158	20 100
12		0.888	113	40		9.86	1 260		170	178	22 700
	13	1.04	133		42	10.9	1 390	180		200	25 400
14		1.21	154	45		12.5	1 590		190	223	28 400
	15	1.39	177		48	14.2	1 810	200		247	31 400
16		1.58	201	50		15.4	1 960		220	298	38 000
	17	1.78	227		55	18.7	2 380	250		385	49 100
18		2.00	254		56†	19.3	2 460		280	483	61 600
	19	2.23	284	60		22.2	2 830	300		555	70 700
20		2.47	314		64†	25.3	3 220		320	631	80 400

† Screw Stock, not listed in CAN3-G312.2-M76.

Square Bars

Preferred Nominal Size, mm		Mass*	Area	Preferred Nominal Size, mm		Mass*	Area	Preferred Nominal Size, mm		Mass*	Area
First	Second	kg/m	mm²	First	Second	kg/m	mm²	First	Second	kg/m	mm²
3.0		0.070 6	9.00	25		4.91	625	100		78.5	10 000
4.0		0.126	16.0		28	6.15	784		110	93.0	12 100
5.0		0.196	25.0	30		7.06	900	120		113	14 400
6.0		0.283	36.0		35	9.62	1 220		140	154	19 600
8.0		0.502	64.0	40		12.6	1 600	160		201	25 600
10		0.785	100		45	15.9	2 020		180	254	32 400
12		1.13	144	50		19.6	2 500	200		314	40 000
	14	1.54	196		55	23.7	3 020		220	380	48 400
16		2.01	256	60		28.3	3 600	250		491	62 500
	18	2.54	324		70	38.5	4 900	300		706	90 000
20		3.14	400	80		50.2	6 400				
	22	3.80	484		90	63.6	8 100				

*Computed using a steel density of 7 850 kg/m³

CRANE RAILS

General

Crane rails are designated by their mass in pounds per yard, with bolt sizes, hole diameters, and washer sizes dimensioned in inches. The SI metric dimensions and properties for crane rails and their accessories given on the following pages are soft converted from manufacturers' catalogues.

ASCE 40, 60 and 85 pound rails require special drilling and punching at time of purchase, while all other rail sizes are predrilled and punched by the manufacturer. The manufacturer should be consulted as to the availability of metric fasteners and metric size holes.

Rails listed in this handbook are the most popular sizes used for crane runways. For dimensions and properties not provided in the tables, consult the manufacturer.

ASCE 40 and 60 pound rails are usully supplied in 9140 mm lengths, ASCE 60 and 85 pound rails are available in 10 100 mm lengths and ASCE 85 pound and heavier rails are available in 11 900 mm lengths. If bolted rail bar splices are to be used, the number of rail lengths required, plus one short length in each run should be specified, to permit staggering of the joints. Orders must clearly specify that "THESE RAILS ARE INTENDED FOR CRANE SERVICE".

The ends of rails are often hardened to better resist rail end batter. Most manufacturers will chamfer the top and sides of the rail head at the ends, unless specified otherwise by the purchaser. Chamfering permits mild deformations to occur and minimizes chipping of the running surfaces.

When selecting a rail for crane service, the characteristics of operation must be considered. Some common variables which affect service life are:

- Frequency of operation
- Crane carriage speed and impact – rate of loading and unloading
- Corrosion – acidic mill conditions
- Abrasion
- Alignment of crane and supporting members
- Crane operating procedures

Crane rails are joined together end-to-end by either mechanical fasteners or welding. When bolting is used, special joint bars are employed as shown on page 6-140 and the rails are usually predrilled to $1/16$-inch larger than the specified bolt size. If welded, manual arc welding is usually used and joint bars are not required. Welding has the advantage of eliminating mechanical joints, thus reducing the problem of aligning the top of rails.

PROPERTIES AND DIMENSIONS

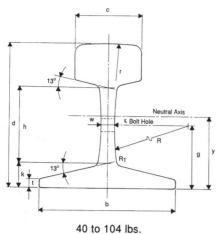

40 to 104 lbs.

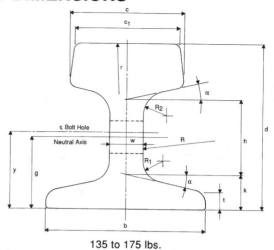

135 to 175 lbs.

Dimensions

Rail Type	Depth d	Head c	Head c_1	Base b	Base t	Web w	Web Gauge g	k	h	r	R	R_1	R_2	α
	mm	mm	mm	mm	mm	mm	mm	mm	mm	mm	mm	mm	mm	deg
ASCE 40	89	48	48	89	5.6	9.9	40	16	47	305	305	6.4	6.4	13
ASCE 60	108	60	60	108	7.1	12	48	19	58	305	305	6.4	6.4	13
ASCE 85	132	65	65	132	7.5	14	57	23	70	305	305	6.4	6.4	13
Beth. 104	127	64	64	127	13	25	62	27	62	305	89	12.7	12.7	13
Beth. 135	146	87	76	132	12	32	63	27	71	356	305	19	19	13
Beth. 171	152	109	102	152	16	32	67	32	70	Flat	Vert.	19	22	12
Beth. 175	152	108	102	152	13	38	68	29	79	457	Vert.	29	51	12

Properties

Rail Type	Mass	Dead Load	Area	I_x	S_x Head	S_x Base	y
	kg/m	kN/m	mm^2	10^6mm^4	10^3mm^3	10^3mm^3	mm
ASCE 40	19.8	0.195	2 540	2.72	58.8	63.7	42.7
ASCE 60	29.8	0.292	3 830	6.06	108	116	52.1
ASCE 85	42.2	0.413	5 370	12.5	182	199	62.7
Beth. 104	51.6	0.506	6 640	12.4	175	221	56.1
Beth. 135	67.0	0.657	8 590	21.1	282	295	71.1
Beth. 171	84.9	0.832	10 800	30.6	402	400	76.5
Beth. 175	86.8	0.851	11 000	29.2	386	381	76.7

Rail fasteners

Hook bolts are primarily used when the flange of the crane beam is too narrow to permit the use of rail clamps. Hook bolts are used in groups of 2, located 75 to 100 mm apart, at 600 mm centres, and may be adjusted plus of minus 12 mm. Rails require special preparation either in the fabricator's shop or by the crane rail supplier.

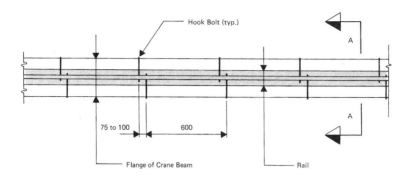

Suggested rail clamp dimensions are shown in Section B-B. For prefabricated rail clamps, reference should be made to Manufacturers' catalogues of track accessories. Two types of clamps are available; the tight clamp and the floating clamp. Floating clamps are used when longitudinal and controlled transverse movement is required for thermal expansion and alignment. Rail clamps are fabricated from pressed or forged steel and usually have single or double bolts.

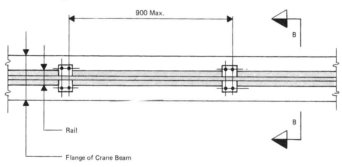

RAIL FASTENERS

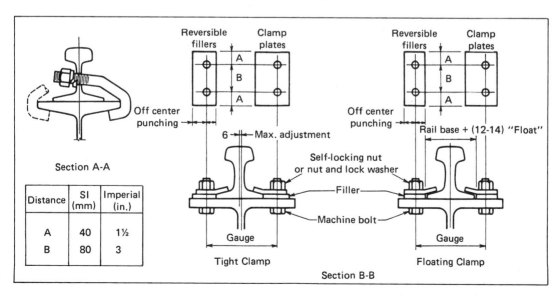

Distance	SI (mm)	Imperial (in.)
A	40	1½
B	80	3

6-139

RAIL SPLICES

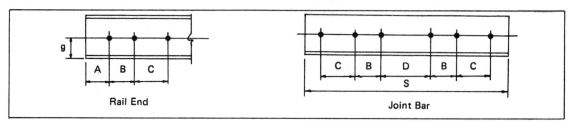

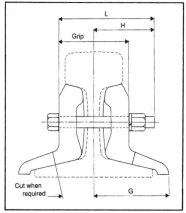

40 to 105 lbs.

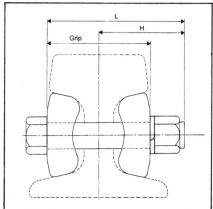

135 to 175 lbs.

Rail Type	Rail					Joint Bar					
	g	Hole dia.	A	B	C	Hole dia.	D	B	C	S	G
	mm	inch.	mm	mm	mm	inch.	mm	mm	mm	mm	mm
40	39.5	*13/16	63.5	127	—	*13/16	125	127	—	508	55.6
60	48.2	*13/16	63.5	127	—	*13/16	125	127	—	610	68.3
85	57.5	*15/16	63.5	127	—	*15/16	125	127	—	610	84.9
104	61.9	1-1/16	102	127	152	1-1/16	202	127	152	864	88.9
135	62.7	1-3/16	102	127	152	1-3/16	202	127	152	864	—
171	66.7	1-3/16	102	127	152	1-3/16	202	127	152	864	—
175	67.5	1-3/16	102	127	152	1-3/16	202	127	152	864	—

*special rail drilling and joint bar punching.

Rail Type	Bolt				Spring Washer		Mass of Ass'y	
	diam.	Grip	L	H	Hole dia.	Thk. & width	With Flg.	Without Flg.
	in.	mm	mm	mm	in.	in. in.	kg.	kg.
40	3/4	49.2	88.9	63.5	13/16	7/16 x 3/8	9.07	7.48
60	3/4	65.9	102	68.3	13/16	7/16 x 3/8	16.56	13.43
85	7/8	80.2	121	81.0	15/16	7/16 x 3/8	25.67	20.55
104	1	88.9	133	88.9	1-1/16	7/16 x 1/2	33.34	25.13
135	1-1/8	92.1	140	93.7	1-3/16	7/16 x 1/2	—	34.16
171	1-1/8	113	159	103	1-3/16	7/16 x 1/2	—	41.19
175	1-1/8	105	152	100	1-3/16	7/16 x 1/2	—	39.78

† For complete description of rail type, see Properties and Dimensions on page 6-138

Splices

Rail drilling and joint bar punching as supplied for track work is not recommended for crane rails, since oversize holes may allow too much movement at the rail ends and result in failure. Tight joints which require special rail and joint bar drilling (see table on page 6-140) and squaring of the rail ends are recommended.

Light rails are not finished at the mill and are usually finished at the fabricator's shop or at the erection site. This may require reaming of holes for proper fit of bolts if dimensional tolerances are cumulative.

Joint bars are provided for crane service to match the rails ordered, and may be ordered blank. Under no circumstances should these joint bars be used as welding straps. Manufacturers' catalogues should be consulted for joint bar material specifications, dimensions and identification necessary to match the crane rail specified.

Joint bar bolts for crane service are readily identified from those used for track work, as they have straight shanks and are manufactured to ASTM A449 specification. Matching nuts are manufactured to ASTM A563 Grade B. The bolted assembly includes an alloy spring washer which is furnished to American Railway Engineers Association (AREA) specification. Bolts and nuts manufactured to ASTM A325 may also be acceptable.

To prolong the life of the runway, bolts should be retightened within 30 days after installation and every 3 months thereafter.

FASTENERS

General

The information on fasteners, provided herein, is based on standards, specifications and publications of the;

> Canadian Standards Association
>
> American National Standards Institute
>
> Industrial Fasteners Institute
>
> Research Council on Structural Connections

Additional fastener information can be obtained from the various manufacturers and from the Canadian Fasteners Institute.

Availability

The more commonly used fasteners for structural purposes in Canada have included the following:

> ⅝-inch ASTM A307 bolts for light steel framing such as girts, purlins, etc.
>
> ¾-inch ASTM A325 bolts for building structures
>
> ⅞-inch ASTM A325 bolts for bridge structures

While other diameters and types of bolts have been used on specific projects in Canada, larger sizes of ASTM A325 bolts, all sizes of ASTM A490 bolts, and all sizes of metric bolts (A325M and A490M) have not been in common use in Canada, and designers contemplating their use should first check for their availability.

At the time of printing, ASTM A325M Type 1 bolts had been produced in the following sizes: M20, M22, and M24.

Clause 23.3.2(e) of CAN/CSA-S16.1-94 permits the use of matching imperial size and metric size bolts in the appropriate metric size holes for the more commonly used high strength bolts. Designers and detailers should therefore base their calculations on the smaller size of matching fastener as appropriate, and allow substitutions as dictated by availability of metric or imperial size fasteners at the time of fabrication.

Definitions

Body Length means the distance from the underside of the head bearing surface to either the last scratch of thread or the top of the extrusion angle, whichever is the closest to the head.

Bolt Length means the length from the underside of the head bearing surface to the extreme point.

Finished Fastener means a fastener made to close tolerances and having surfaces other than the threads and bearing surface finished to provide a general high grade appearance.

Grip means the thickness of material or parts which the fastener is designed to secure when fully assembled.

Height of bolt Head means the overall distance, measured parallel to the fastener axis, from the extreme top (excluding raised identification marks) to the bearing surface and including the thickness of the washer face where provided.

Natural Finish means the as-processed finish, unplated or uncoated of the bolt or nut.

Nominal size means the designation used for the purpose of general identification.

Proof Load means a specified test load which a fastener must withstand without any indication of significant deformation or failure.

Thickness of Nut means the overall distance from the top of the nut to the bearing surface, measured parallel to the axis of the nut.

Thread Length of a Bolt means the distance from the extreme point to the last complete thread.

Transition Thread Length means the distance from the last complete thread to either the last scratch of thread or the top of the extrusion angle, whichever is the closest to the head.

Washer Face means a circular boss on the bearing surface of a bolt or nut.

Tables

The following Tables are included in this section:

- ASTM A325 and ASTM A490 High Strength Bolts and Nuts
- ASTM A325M and ASTM A490M High Strength bolts and Nuts
- Bolts Lengths for Various Grips — ASTM A325 and A490 bolts
- Minimum and Maximum Grips for Metric Heavy Hex. Structural Bolts
- Mass of ASTM A325 Bolts, Nuts and washers
- Mass of ASTM A325M Bolts, Nuts and Washers
- Metric Washer Dimensions
- ASTM A307 Bolts and Nuts
- Fasteners — Miscellaneous Detailing Data
- Usual Gauges — W, M, S, C shapes, Angles
- Erection Clearances — Bolt Impact Wrenches

ASTM A325 AND ASTM A490
HIGH STRENGTH BOLTS AND NUTS

REQUIRED MARKINGS

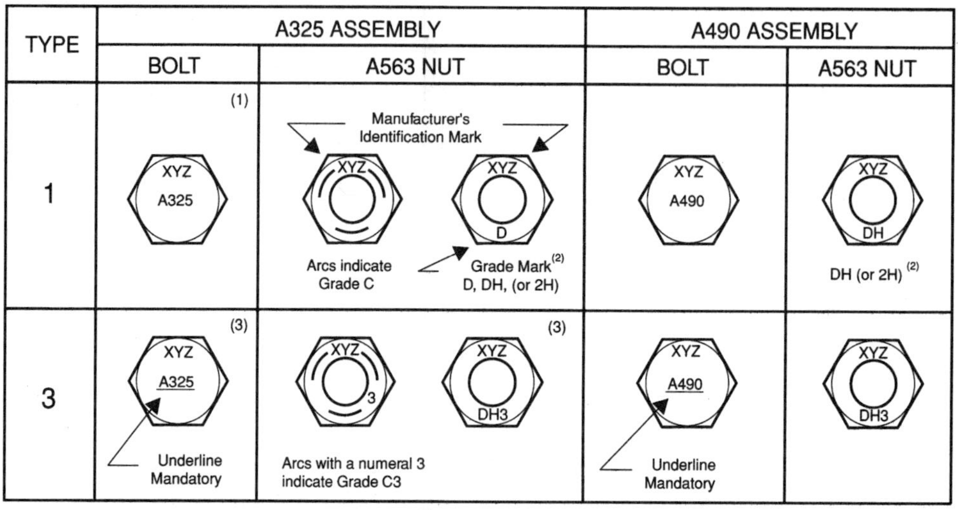

TYPE	A325 ASSEMBLY		A490 ASSEMBLY	
	BOLT	A563 NUT	BOLT	A563 NUT
1	(1) XYZ / A325	XYZ (Arcs indicate Grade C) — Manufacturer's Identification Mark — XYZ / D (Grade Mark D, DH (or 2H))[2]	XYZ / A490	XYZ / DH (DH (or 2H))[2]
3	(3) XYZ / A325 (Underline Mandatory)	XYZ / 3 XYZ / DH3 Arcs with a numeral 3 indicate Grade C3	(3) XYZ / A490 (Underline Mandatory)	XYZ / DH3

(1) Additional Optional 3 Radial Lines at 120° may be added.
(2) Type 3 also acceptable. Nuts according to ASTM A194 Grade 2H, plain finish, may also be used.
(3) Additional Optional Mark indicating Weathering Grade may be added.

DIMENSIONS

	Dimensions						Coarse Thread Series—UNC**		
Nominal Bolt Size, Inches D	Bolt Dimensions*, Inches Heavy Hex Structural Bolts			Nut Dimensions*, Inches Heavy Hex nuts		Threads Per In.	Basic Pitch Dia	Section at Minor Dia	Tensile Stress Area
	Width across flats, F	Height, H	Thread length	Width across flats, W	Height, H		In.	In.²	In.²
½	⅞	⁵⁄₁₆	1	⅞	³¹⁄₆₄	13	0.4500	0.1257	0.1419
⅝	1¹⁄₁₆	²⁵⁄₆₄	1¼	1¹⁄₁₆	³⁹⁄₆₄	11	0.5660	0.202	0.226
¾	1¼	¹⁵⁄₃₂	1⅜	1¼	⁴⁷⁄₆₄	10	0.6850	0.302	0.334
⅞	1⁷⁄₁₆	³⁵⁄₆₄	1½	1⁷⁄₁₆	⁵⁵⁄₆₄	9	0.8028	0.419	0.462
1	1⅝	³⁹⁄₆₄	1¾	1⅝	⁶³⁄₆₄	8	0.9188	0.551	0.606
1⅛	1¹³⁄₁₆	¹¹⁄₁₆	2	1¹³⁄₁₆	1⁷⁄₆₄	7	1.0322	0.693	0.763
1¼	2	²⁵⁄₃₂	2	2	1⁷⁄₃₂	7	1.1572	0.890	0.969
1⅜	2³⁄₁₆	²⁷⁄₃₂	2¼	2³⁄₁₆	1¹¹⁄₃₂	6	1.2667	1.054	1.155
1½	2⅜	¹⁵⁄₁₆	2¼	2⅜	1¹⁵⁄₃₂	6	1.3917	1.294	1.405

*Dimensions according to ANSI B18.2.1
**Thread dimensions according to ANSI B1.1

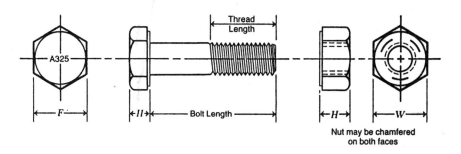

Nut may be chamfered on both faces

ASTM A325M AND ASTM A490M**
HIGH STRENGTH BOLTS AND NUTS†

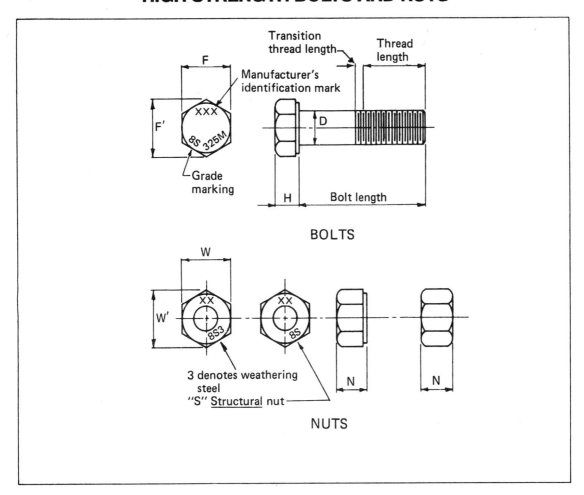

DIMENSIONS

Nominal Bolt Size	Heavy Hex Bolt or Nut Dimension				Heavy Hex Nut Max. Height N	Heavy Hex Structural Bolt			
	Across Flats F or W		Across Corners F' or W'			Max Head Height H	Thread Length*		Max. Transition Thread Length
	Max.	Min.	Max.	Min.			Bolt Lengths ≤100	Bolt Lengths >100	
mm	mm	mm	mm	mm	mm	mm	mm	mm	mm
M16 x 2	27.00	26.16	31.18	29.56	17.1	10.75	31	38	6.0
M20 x 2.5	34.00	33.00	39.26	37.29	20.7	13.40	36	43	7.5
M22 x 2.5	36.00	35.00	41.57	39.55	23.6	14.90	38	45	7.5
M24 x 3	41.00	40.00	47.34	45.20	24.2	15.90	41	48	9.0
M27 x 3	46.00	45.00	53.12	50.85	27.6	17.90	44	51	9.0
M30 x 3.5	50.00	49.00	57.74	55.37	30.7	19.75	49	56	10.5
M36 x 4	60.00	58.80	69.28	66.44	36.6	23.55	56	63	12.0

*Does not include transition thread length.
Bolt dimensions conform to those listed in ANSI B18.2.3.7M-1979 "Metric Heavy Hex Structural Bolts", and the nut dimensions conform to those listed in CSA Standard B18.2.4.6-M1980 "Metric Heavy Hex Nuts".
** Strength requirements are based on ASTM Specifications A325M and A490M. See page 3-5.

BOLT LENGTHS* for VARIOUS GRIPS**
ASTM A325 and A490 Bolts

Grip mm	Grip in.	1/2	5/8	3/4	7/8	1	1 1/8	1 1/4	1 3/8	1 1/2
19	3/4	1 1/2								
21	13/16		1 3/4		2			2 1/2		2 3/4
22	7/8			2		2 1/4	2 1/2		2 3/4	
24	15/16	1 3/4								
25	1		2		2 1/4			2 3/4		3
27	1 1/16									
29	1 1/8			2 1/4		2 1/2	2 3/4		3	
30	1 3/16	2								
32	1 1/4		2 1/4		2 1/2			3		3 1/4
33	1 5/16									
35	1 3/8			2 1/2		2 3/4	3		3 1/4	
37	1 7/16	2 1/4								
38	1 1/2		2 1/2		2 3/4			3 1/4		3 1/2
40	1 9/16									
41	1 5/8			2 3/4		3	3 1/4		3 1/2	
43	1 11/16	2 1/2								
44	1 3/4		2 3/4		3			3 1/2		3 3/4
46	1 13/16									
48	1 7/8			3		3 1/4	3 1/2		3 3/4	
49	1 15/16	2 3/4								
51	2		3		3 1/4			3 3/4		4
52	2 1/16									
54	2 1/8			3 1/4		3 1/2	3 3/4		4	
56	2 3/16	3								
57	2 1/4		3 1/4		3 1/2			4		4 1/4
59	2 5/16									
60	2 3/8			3 1/2		3 3/4	4		4 1/4	
62	2 7/16	3 1/4								
64	2 1/2		3 1/2		3 3/4			4 1/4		4 1/2
65	2 9/16									
67	2 5/8			3 3/4		4	4 1/4		4 1/2	
68	2 11/16	3 1/2								
70	2 3/4		3 3/4		4			4 1/2		4 3/4
71	2 13/16									
73	2 7/8			4		4 1/4	4 1/2		4 3/4	
75	2 15/16	3 3/4								
76	3		4		4 1/4			4 3/4		5
78	3 1/16									
79	3 1/8			4 1/4		4 1/2	4 3/4		5	
81	3 3/16	4								
83	3 1/4		4 1/4		4 1/2			5		5 1/4
84	3 5/16									
86	3 3/8			4 1/2		4 3/4	5		5 1/4	
87	3 7/16	4 1/4								
89	3 1/2		4 1/2		4 3/4			5 1/4		5 1/2
90	3 9/16									
92	3 5/8			4 3/4		5	5 1/4		5 1/2	
94	3 11/16	4 1/2								
95	3 3/4		4 3/4		5			5 1/2		5 3/4
97	3 13/16									
98	3 7/8			5		5 1/4	5 1/2		5 3/4	
100	3 15/16	4 3/4								
102	4		5		5 1/4			5 3/4		6
103	4 1/16									
105	4 1/8			5 1/4		5 1/2	5 3/4		6	
106	4 3/16	5								
108	4 1/4		5 1/4		5 1/2			6		6 1/4
109	4 5/16									
111	4 3/8			5 1/2		5 3/4	6		6 1/4	
113	4 7/16	5 1/4								
114	4 1/2		5 1/2		5 3/4			6 1/4		6 1/2
116	4 9/16									
117	4 5/8			5 3/4		6	6 1/4		6 1/2	
119	4 11/16	5 1/2								
121	4 3/4		5 3/4		6			6 1/2		6 3/4
122	4 13/16									
124	4 7/8			6		6 1/4	6 1/2		6 3/4	
125	4 15/16	5 3/4								
127	5		6		6 1/4			6 3/4		7
129	5 1/16									
130	5 1/8			6 1/4		6 1/2	6 3/4		7	
132	5 3/16	6								
133	5 1/4		6 1/4		6 1/2			7		7 1/4
135	5 5/16									

*Bolt lengths must be specified in inches for ASTM A325 and A490 bolts.
**Grip is thickness of material to be connected exclusive of washers.
For each flat washer, add 5 mm (3/16-inch) to grip.
For each beveled washer, add 8 mm (5/16-inch) to grip.

MINIMUM AND MAXIMUM GRIPS FOR METRIC HEAVY HEX. STRUCTURAL BOLTS, IN MILLIMETRES

Nominal Bolt Size	M16		M20		M22		M24		M27		M30		M36	
L Nominal Length (mm)	Min. Grip	Max. Grip	Min. Grip	Max. Grip	Min. Grip	Max. Grip	Min. Grip	Max. Grip	Min. Grip	Max. Grip	Min. Grip	Max. Grip	Min. Grip	Max. Grip
45	14	26		23		20								
50	19	31	14	28		25		24						
55	24	36	19	32	17	29		29		25				
60	29	41	24	37	22	34	19	34		30				
65	34	46	29	42	27	39	24	39	21	35		32		
70	39	51	34	47	32	44	29	44	26	40	21	37		31
75	44	56	39	52	37	49	34	49	31	45	26	42		36
80	49	61	44	57	42	54	39	54	36	50	31	47	24	41
85	54	66	49	62	47	59	44	59	41	55	36	52	29	46
90	59	71	54	67	52	64	49	64	46	60	41	57	34	51
95	64	76	59	72	57	69	54	69	51	65	46	62	39	56
100	69	81	64	77	62	74	59	74	56	70	51	67	44	61
110	72	91	67	87	65	84	62	84	59	80	54	77	47	71
120	82	101	77	97	75	94	72	94	69	90	64	87	57	81
130	92	110	87	107	85	104	82	103	79	100	74	97	67	91
140	102	120	97	117	95	114	92	113	89	110	84	107	77	101
150	112	130	107	127	105	124	102	123	99	120	94	117	87	111
160	122	138	117	135	115	132	112	131	109	128	104	125	97	119
170	132	148	127	145	125	142	122	141	119	138	114	135	107	129
180	142	158	137	155	135	152	132	151	129	148	124	145	117	139
190	152	168	147	165	145	162	142	161	139	158	134	155	127	149
200	162	178	157	175	155	172	152	171	149	168	144	165	137	159
210	172	188	167	185	165	182	162	181	159	178	154	175	147	169
220	182	198	177	195	175	192	172	191	169	188	164	135	157	179
230	192	208	187	205	185	202	182	201	179	198	174	195	167	189
240	202	218	197	215	195	212	192	211	189	208	184	205	177	199
250	212	228	207	225	205	222	202	221	199	218	194	215	187	209
260	222	238	217	235	215	232	212	231	209	228	204	225	197	219
270	232	248	227	245	225	242	222	241	219	238	214	235	207	229
280	242	258	237	255	235	252	232	251	229	248	224	245	217	239
290	252	268	247	265	245	262	242	261	239	258	234	255	227	249
300	262	278	257	275	255	272	252	271	249	268	244	265	237	259

1. This table is based on ANSI B18.2.3.7M-1979.
2. Bolts with lengths above the heavy solid line are threaded full length.

MASS OF ASTM A325 BOLTS, NUTS AND WASHERS

APPROXIMATE WEIGHTS IN POUNDS PER 100 UNITS

| Length Under Head Inches | HEAVY HEX STRUCTURAL BOLTS WITH HEAVY HEX NUTS (WITHOUT WASHERS) |||||||||
| | Diameter of Bolt, Inches |||||||||
	½	⅝	¾	⅞	1	1⅛	1¼	1⅜	1½
1½	19.2	33.1	52.2	78.0	109	148	197		
1¾	20.5	35.3	55.3	81.9	114	154	205	261	333
2	21.9	37.4	58.4	86.1	119	160	212	270	344
2¼	23.3	39.8	61.6	90.3	124	167	220	279	355
2½	24.7	41.7	64.7	94.6	130	174	229	290	366
2¾	26.1	43.9	67.8	98.8	135	181	237	300	379
3	27.4	46.1	70.9	103	141	188	246	310	391
3¼	28.8	48.2	74.0	107	146	195	255	321	403
3½	30.2	50.4	77.1	111	151	202	263	332	416
3¾	31.6	52.5	80.2	116	157	209	272	342	428
4	33.0	54.7	83.3	120	162	216	280	353	441
4¼	34.3	56.9	86.4	124	168	223	289	363	453
4½	35.7	59.0	89.5	128	173	230	298	374	465
4¾	37.1	61.2	92.7	133	179	237	306	384	478
5	38.5	63.3	95.8	137	184	244	315	395	490
5¼	39.9	65.5	98.9	141	190	251	324	405	503
5½	41.2	67.7	102	146	196	258	332	416	515
5¾	42.6	69.8	105	150	201	265	341	426	527
6	44.0	71.9	108	154	207	272	349	437	540
6¼	45.4	74.1	111	158	212	279	358	447	552
6½	46.8	76.3	114	163	218	286	367	458	565
6¾	48.1	78.5	118	167	223	293	375	468	577
7	49.5	80.6	121	171	229	300	384	479	589
7¼	50.9	82.8	124	175	234	307	392	489	602
7½	52.3	84.9	127	179	240	314	401	500	614
7¾	53.6	87.1	130	183	246	321	410	510	626
8	55.0	89.2	133	187	251	328	418	521	639
8¼	56.4	91.4	136	192	257	335	427	531	651
8½	57.8	93.5	139	196	262	342	435	542	664
8¾	59.1	95.7	142	200	268	349	444	552	676
9	60.5	97.8	145	204	273	356	453	563	689
9¼	62.0	100	149	209	279	363	461	574	701
9½	63.3	102	152	213	284	370	470	584	713
9¾	64.7	104	155	217	290	377	478	595	726
10	66.1	107	158	221	295	384	487	605	738
Per Inch Additional	5.5	8.6	12.4	16.9	22.1	28.0	34.4	42.5	49.7

	½	⅝	¾	⅞	1	1⅛	1¼	1⅜	1½
Plain Round Washers	2.0	3.9	4.5	7.1	9.3	11.3	14.1	16.8	20.0
Bevelled Square Washers	24.9	23.8	22.6	21.0	19.2	34.0	31.6	—	—
Heavy Hex Nuts	6.54	11.9	19.3	29.7	42.5	59.2	78.6	102	131

MASS OF ASTM A325M BOLTS, NUTS AND WASHERS

APPROXIMATE MASS* OF 1000 UNITS (kg)

HEAVY HEX STRUCTURAL BOLTS WITH HEAVY HEX NUTS (WITHOUT WASHERS)

Bolt Length mm	Bolt Size (number following the letter M is the nominal bolt diameter in millimetres)						
	M16	M20	M22	M24	M27	M30	M36
40	170						
45	177	321	396				
50	184	331	409	539	744		
55	191	342	423	554	764		
60	198	353	436	570	784	1000	
65	204	364	449	586	804	1030	
70	211	375	462	601	824	1050	1710
75	218	386	476	617	845	1000	1750
80	225	397	489	632	865	1100	1780
85	232	407	502	648	885	1120	1820
90	239	418	515	664	905	1150	1850
95	246	429	529	679	925	1170	1890
100	253	440	542	695	945	1200	1920
110	267	462	569	726	985	1250	2000
120	281	483	595	757	1020	1300	2070
130	295	505	622	788	1070	1350	2140
140	309	527	648	820	1110	1400	2210
150	322	548	675	851	1150	1440	2280
160	336	570	701	882	1190	1490	2350
170	350	592	728	913	1230	1540	2420
180	364	613	754	945	1270	1590	2490
190	378	635	781	976	1310	1640	2570
Extra per 10 mm	13.9	21.7	26.5	31.2	40.1	49.2	71.3
Mass of Extra Heavy Hex Nuts	61.0	118	145	201	286	371	635
Mass of Plain Washers	18	27	33	45	57	65	100
Mass of Bevel Washers	101	91	85	78	154	142	115

*Computed theoretical mass using a steel density of 7 850 kg/m^3.

ASTM F436M METRIC WASHER DIMENSIONS

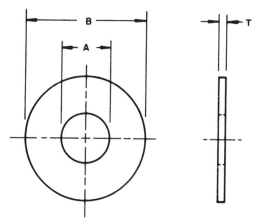

PLAIN CIRCULAR WASHERS

Metric Bolt Size	B Outside Diameter		A Hole Diameter		T Thickness	
	Max	Min	Max	Min	Max	Min
M16 x 2	34.0	32.4	18.4	18.0	4.6	3.1
M20 x 2.5	41.0	39.4	22.5	22.0	4.6	3.1
M22 x 2.5	44.0	42.4	24.5	24.0	4.6	3.4
M24 x 3	50.0	48.4	26.5	26.0	4.6	3.4
M27 x 3	56.0	54.1	30.5	30.0	4.6	3.4
M30 x 3.5	60.0	58.1	33.5	33.0	4.6	3.4
M36 x 4	72.0	70.1	39.5	39.0	4.6	3.4

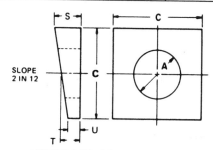

BEVELLED SQUARE WASHERS

Metric Bolt Size	C Width		A Hole Diameter		S Thick Side ±0.5	T Mean Nom.	U Thin Side ±0.5
	Max	Min	Max	Min			
M16 x 2	45.0	43.0	18.4	18.0	11.7	8	4.3
M20 x 2.5	45.0	43.0	22.5	22.0	11.7	8	4.3
M22 x 2.5	45.0	43.0	24.5	24.0	11.7	8	4.3
M24 x 3	45.0	43.0	26.5	26.0	11.7	8	4.3
M27 x3	59.0	57.0	30.5	30.0	12.8	8	3.2
M30 x 3.5	59.0	57.0	33.5	33.0	12.8	8	3.2
M36 x 4	59.0	57.0	39.5	39.0	12.8	8	3.2

ASTM A307 BOLTS AND NUTS
DIMENSIONS IN IMPERIAL UNITS

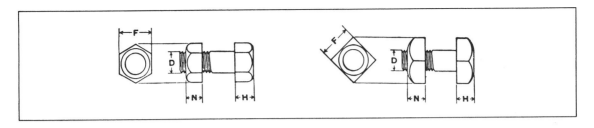

Nominal Diameter of Bolt	Regular Square Bolts		Finished Hex Bolts		Minimum Thread Lengths		Regular Square Nuts		Finished Hex Nuts	
	Nominal Width F	Nominal Height H	Basic Width F	Basic Height H	Length Under Head		Nominal Width F	Nominal Height N	Basic Width F	Basic Height N
					6 In. and Under	Over 6 In.				
Inches	Inches	Inches	Inches	Inches	Inches	Inches	Inches	Inches	Inches	Inches
1/4	3/8	11/64	7/16	5/32	¾	1	7/16	7/32	7/16	7/32
3/8	9/16	1/4	9/16	15/64	1	1¼	5/8	21/64	9/16	21/64
1/2	3/4	21/64	3/4	5/16	1¼	1½	13/16	7/16	3/4	7/16
5/8	15/16	27/64	15/16	25/64	1½	1¾	1	35/64	15/16	35/64
3/4	1 1/8	1/2	1 1/8	15/32	1¾	2	1 1/8	21/32	1 1/8	41/64
7/8	1 5/16	19/32	1 5/16	35/64	2	2¼	1 5/16	49/64	1 5/16	3/4
1	1 1/2	21/32	1 1/2	39/64	2¼	2½	1 1/2	7/8	1 1/2	55/64
1 1/8	1 11/16	3/4	1 11/16	11/16	2½	2¾	1 11/16	1	1 11/16	31/32
1 1/4	1 7/8	27/32	1 7/8	25/32	2¾	3	1 7/8	1 3/32	1 7/8	1 1/16

The dimensions for Regular Square Bolts conform to those for Square Head Bolts, Regular Series listed in CSA Standard B33.1-1961, "Square and Hexagon Bolts and Nuts, Studs and Wrench Openings", and to those for Square Bolts listed in ANSI Standard B18.2.1-1965, "Square and Hex Bolts and Screws".

The dimensions for Finished Hexagon Bolts (Hex Cap Screws) conform to those for Hexagon Head Bolts, Finished Grade, Regular Series listed in CSA Standard B33.1-1961, and to those for Hex Cap Screws (Finished Hex Bolts) listed in ANSI Standard B18.2.1-1965.

The minimum thread lengths are in agreement with the requirements of CSA Standard B33.1-1961 and with ANSI Standard B18.2.1-1965. In general, these requirements are as follows:

Bolts 6 inches or less in length — twice diameter plus ¼-inch.
Bolts longer than 6 inches — twice diameter plus ½-inch.
Bolts too short for the above thread lengths shall be threaded as close to the head as practicable.

The dimensions for Regular Square Nuts conform to those for Square Nuts, Regular Series listed in CSA Standard B33.1-1961, and to those for Square Nuts listed in ANSI Standard B18.2.2-1965, "Square and Hex Nuts".

The dimensions for Finished Hexagon Nuts conform to those for Hexagon Nuts, Finished Grade, Regular Series listed in CSA Standard B33.1-1961 and to those for Hex Nuts listed in ANSI Standard B18.2.2-1965.

Note: Square head bolts are used with either square or hexagon nuts. However, the use of Finished Hexagon Bolts and Finished Hexagon Nuts is gradually replacing the use of square head bolts and nuts in sizes up to one inch in diameter and six inches in length.
A307 bolts and nuts are manufactured in Imperial dimensions only.

FASTENERS — MISCELLANEOUS DETAILING DATA

Metric Fastener Designations

THREAD DATA

Diameter Pitch Combinations

Nominal dia. (mm)	Thread pitch (mm)	Nominal dia. (mm)	Thread pitch (mm)
1.6	0.35	20	2.5
2	0.4	22	2.5
2.5	0.45	24	3
3	0.5	27	3
3.5	0.6	30	3.5
4	0.7	36	4
5	0.8	42	4.5
6.0	1.0	48	5
8	1.25	56	5.5
10	1.5	64	6
12	1.75	72	6
14	2	80	6
16	2	90	6
		100	6

Basic Metric Thread designation: Metric screw threads are designated by the letter "M" followed by the nominal size (basic major diameter) in millimetres and the pitch in millimetres separated by the symbol "X".

M12	X	1.75	–6g
Size (mm)		Thread (pitch in mm)	Standard class of fit

Note: In the metric system, the pitch of the thread is given in mm instead of threads per inch — thus a M12 x 1.75 thread has a nominal diameter of 12 mm and the pitch of the thread is 1.75 mm.

PRODUCT DESIGNATION

Metric Bolt designation: The standard method of designating a metric bolt is by specifying (in sequence) the product name, nominal diameter and thread pitch, nominal length, type, steel property class, and protective coating (if required).

Heavy Hex Structural Bolt, M22 x 2.5 x 160, Type 2, ASTM A325M-79, Zinc Galvanized

Metric Nut designation: The standard method of designating a metric nut is by specifying (in sequence) the product name, nominal diameter and pitch, steel property class or material identification, and protective coating (if required).

Heavy Hex Nut, M30 x 3.5, ASTM A563M class 105, hot dipped galvanized

Note: It is common practice to omit the thread pitch from the product designation.

Slotted Hole Dimensions

See Clause 23.3.2, S16.1-M regarding use provisions.

A, nominal diameter + 2mm

SHORT SLOT DIMENSIONS

Nominal Bolt Diameter	Slot Dimensions	
	Width, A	Length, B
mm	mm	mm
16	18	22
20	22	26
22	24	28
24	26	32
27	29	37
30	32	40
36	38	46

LONG SLOT DIMENSIONS

Nominal Bolt Diameter	Slot Dimensions	
	Width, A	Length, B
mm	mm	mm
16	18	40
20	22	50
22	24	55
24	26	60
27	29	67.5
30	32	75
36	38	90

FASTENERS — MISCELLANEOUS DETAILING DATA
Diagonal Distance for Staggered Fasteners

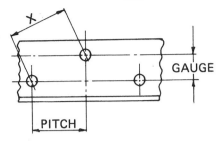

Pitch mm	GAUGE, mm																	
	25	30	35	40	45	50	55	60	65	70	75	80	85	90	95	100	105	110
5	25	30	35	40	45	50	55	60	65	70	75	80	85	90	95	100	105	110
10	27	32	36	41	46	51	56	61	66	71	76	81	86	91	96	100	105	110
15	29	34	38	43	47	52	57	62	67	72	76	81	86	91	96	101	106	111
20	32	36	40	45	49	54	59	63	68	73	78	82	87	92	97	102	107	112
25	35	39	43	47	51	56	60	65	70	74	79	84	89	93	98	103	108	113
30	39	42	46	50	54	58	63	67	72	76	81	85	90	95	100	104	109	114
35	43	46	49	53	57	61	65	69	74	78	83	87	92	97	101	106	111	115
40	47	50	53	57	60	64	68	72	76	81	85	89	94	98	103	108	112	117
45	51	54	57	60	64	67	71	75	79	83	87	92	96	101	105	110	114	119
50	56	58	61	64	67	71	74	78	82	86	90	94	99	103	107	112	116	121
55	60	63	65	68	71	74	78	81	85	89	93	97	101	105	110	114	119	123
60	65	67	69	72	75	78	81	85	88	92	96	100	104	108	112	117	121	125
65	70	72	74	76	79	82	85	88	92	96	99	103	107	111	115	119	123	128
70	74	76	78	81	83	86	89	92	96	99	103	106	110	114	118	122	126	130
75	79	81	83	85	87	90	93	96	99	103	106	110	113	117	121	125	129	133
80	84	85	87	89	92	94	97	100	103	106	110	113	117	120	124	128	132	136
85	89	90	92	94	99	99	101	104	107	110	113	117	120	124	127	131	135	139
90	93	95	97	98	101	103	105	108	111	114	117	120	124	127	131	135	138	142

BOLT LENGTH TOLERANCES

Nominal Length	Nominal Bolt Dia M16 thru M36
to 50 mm	± 1.2
over 50 to 80 mm	± 1.5
over 80 to 120 mm	± 1.8
over 120 to 150 mm	± 2.0
over 150 mm	± 4.0

MINIMUM EDGE DISTANCE FOR BOLT HOLES

Bolt Diameter mm	At sheared Edge mm	At Rolled or Gas Cut Edge † mm
16	28	22
20	34	26
22	38	28
24	42	30
27	48	34
30	52	38
36	64	46
over 36	1¾ x Diameter	1¼ x Diameter

†Gas cut edges shall be smooth and free from notches. Edge distance in this column may be decreased 3 mm when hole is at a point where computed stress under factored loads is not more than 0.3 of the yield stress.

USUAL GAUGES

USUAL GAUGES for W, M, S and C Shapes, Millimetres

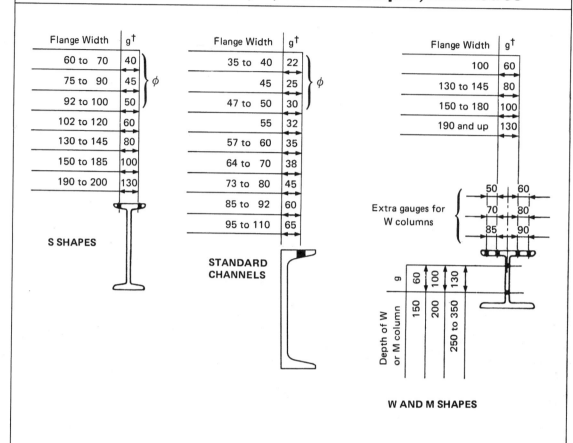

φ Holes usually drilled due to size of punch die block.
† Dependent on edge distance.

USUAL GAUGES for Angles, millimetres

Notes:
Those values shown above the dashed line allow for full socket wrench clearance requirements.

The bolt sizes shown in italics to the left of g and g_1 are the maximum bolt sizes permissible for the dimensions shown.

$g_2 \geqslant$ 2-2/3 bolt diameters.

Gauge Leg	g		g_1		g_2
200	M36	115	M30	80	80
150	M36	90	M24	55	65
125	M30	80	M20	45	54
100	M27	65			
90	M24	60			
80	M24	50			
75	M24	45			
65	M24	35			
60	M24	30			
55	M22	27			
50	M16	28			
45	M16	23			

ERECTION CLEARANCES
Bolt Impact Wrenches

METRIC

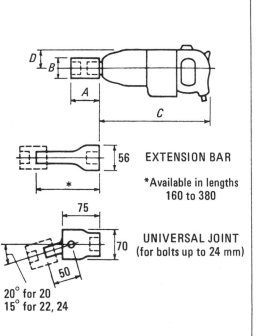

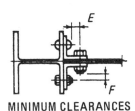

EXTENSION BAR — *Available in lengths 160 to 380

UNIVERSAL JOINT (for bolts up to 24 mm)

20° for 20
15° for 22, 24

MINIMUM CLEARANCES

	Size	C	D
Light Wrenches	16 to 24	337 to 356	54
Heavy Wrenches	24 to 36	375 to 438	64

Sockets			Min. Clear.	
Bolt Size	A	B	E	F
16	80	45	25	28
20	85	54	30	34
22	90	57	32	36
24	95	60	34	38
27	100	70	38	42
30	110	75	41	45
36	130	90	48	52

IMPERIAL

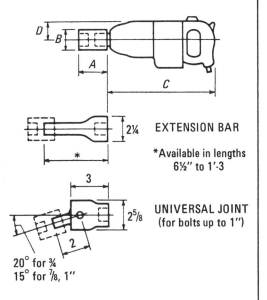

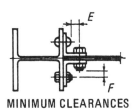

EXTENSION BAR — *Available in lengths 6½" to 1'-3

UNIVERSAL JOINT (for bolts up to 1")

20° for ¾
15° for ⅞, 1"

MINIMUM CLEARANCES

	Size	C	D
Light Wrenches	5/8 to 1	1-1¼ to 1-2	$2^{1}/_{8}$
Heavy Wrenches	1 to 1½	1-2¾ to 1-5¼	2½

Sockets			Min. Clear.	
Bolt Size	A	B	E	F
5/8	2-5/8	1-3/4	1-1/8	1-1/4
3/4	3	2-1/4	1-1/4	1-3/8
7/8	3-1/4	2-1/2	1-5/16	1-7/16
1	3-1/2	2-5/8	1-7/16	1-9/16
1-1/8	3-3/4	2-7/8	1-9/16	1-11/16
1-1/4	4	3-1/8	1-5/8	1-3/4
1-3/8	4-1/4	3-1/4	2-1/8	2-1/4
1-1/2	4-3/8	4-1/4		

WELDING

The welding of steel shapes and plates for structural purposes is governed by CAN/CSA-S16.1, Limit States Design of Steel Structures, and CSA Standard W59, Welded Steel Construction (Metal-Arc Welding).

While both standards provide design information on the resistance of welds, CSA Standard W59 extensively covers workmanship, inspection, and acceptance criteria for welded joints in both statically and dynamically loaded structures.

Welding is a process used to join two or more pieces of material together. Arc welding is a process which produces coalescence of metals by heating them with an arc, with or without the application of pressure, and with or without the use of filler metal.

Welding processes used primarily for structural steel work are:

Shielded Metal Arc Welding	SMAW
Flux Cored Arc Welding	FCAW
Gas Metal Arc Welding	GMAW
Submerged Arc Welding	SAW
Electroslag Welding	ESW
Electrogas Welding	EGW
Stud Welding	STW

Welding Definitions

Arc Cutting: means a group of cutting processes which melts the metal to be cut with the heat of an arc between an electrode and the base metal.

Arc Spot Weld: means a weld made by arc welding between or upon overlapping members in which coalescence may start and occur on the faying surfaces or may proceed from the surface of one member.

Base Metal: means the metal to be welded or cut.

Bevel Angle: means the angle formed between the prepared edge of a member and a plane perpendicular to the surface of the member.

Chain Intermittent Welds: means intermittent welds on both sides of a joint in which the weld increments on one side are approximately opposite those on the other side.

Coalescence: means the growing together or growth into one body of the material being welded.

Complete Joint Penetration: means when the weld metal completely fills the groove and is fused to the base metal throughout its total thickness.

Edge Joint: means a joint between the edge of two or more parallel or nearly parallel members.

Effective Weld Length: means the length of weld throughout which the correctly proportioned cross section exists. In a curved weld, it is measured along the axis of the weld.

Effective Throat: means the minimum distance from the root of a weld to its face less any reinforcement.

End Return (Boxing): means the continuation of a fillet weld around a corner of a member as an extension of the principal weld.

Face of Weld: means the exposed surface of a weld on the side from which the welding was done.

Fillet Weld: means a weld of approximately triangular cross section joining two surfaces approximately at right angles to each other in a lap joint, T-joint, or corner joint.

Groove Weld: means a weld made in a groove between two members to be joined.

Intermittent Weld: means a weld in which the continuity is broken by recurring unwelded spaces.

Joint Design: means the joint geometry together with the required dimensions of the welded joint.

Joint Penetration: means the minimum depth a groove weld extends from its face into a joint, exclusive of reinforcement, but including, if present, root penetration.

Leg of a Fillet Weld: is the distance from the root of the joint to the toe of the fillet weld.

Partial Joint Penetration: means a joint penetration which is less than complete.

Procedure Qualification: means a demonstration that welds made by a specific procedure can meet prescribed standards.

Root of Joint: means that portion of a joint to be welded where the members approach closest to each other. In cross section, the root of the joint may be either a point, a line or an area.

Root of Weld: means the points, as shown in cross section, at which the back of the weld intersects the base metal surfaces.

Root Penetration: means the depth that a weld extends into the root of a joint measured on the centreline of the root cross section.

Size of Weld:

> **Groove Weld:** means the joint penetration (depth of bevel plus the root penetration when specified). The size of a groove weld and its effective throat are one and the same.
>
> **Fillet Weld:**
>
> For equal leg fillet welds, the leg lengths of the largest isosceles right triangle which can be inscribed within the fillet weld cross section.
>
> For unequal leg fillet welds, the leg lengths of the largest right triangle which can be inscribed within the fillet weld cross section.
>
> Note: When one member makes an angle with the other member greater than 105 degrees, the leg length (size) is of less significance than the effective throat which is the controlling factor for the strength of a weld.

Tack Weld: means a weld made to hold parts of a weldment in proper alignment until the final welds are made.

Throat of a Fillet Weld.

> **Theoretical Throat:** means the distance from the beginning of the root of the joint perpendicular to the hypotenuse of the largest right triangle that can be inscribed within the fillet weld cross section. This dimension is based on the assumption that the root opening is equal to zero.
>
> **Actual Throat:** means the shortest distance from the root of weld to its face.
>
> **Effective Throat:** means the minimum distance minus any reinforcement from the root of weld to its face.

WELDING PRACTICE

Fillet Welds

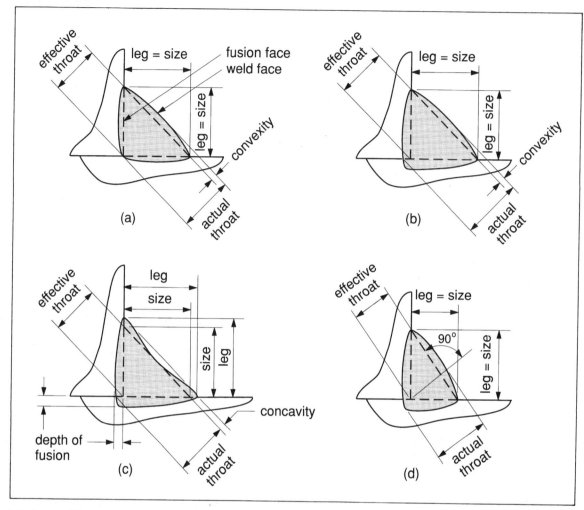

Minimum Size

- The minimum size of a fillet weld as measured should not be less than the values shown on the right, except that it need not exceed the thickness of the thinner part.

  When welding attachments to non-load carrying members the values on the right need not apply.

Material thickness of thicker part joined	Minimum Size of fillet weld, mm
to 12 mm incl.	5
over 12 mm to 20	6
over 20	8

- The minimum effective length of a fillet weld should be 40 mm or 4 times the size of the fillet, whichever is larger.

Maximum Size of Weld

- The maximum fillet weld size, D_{max}, recommended by good practice along a sheared edge is:

$$D_{max} \leqslant t \quad \text{when } t \leqslant 6 \text{ mm}$$
$$D_{max} = t-2 \quad \text{when } t > 6 \text{ mm}$$

- Material with rolled edges:

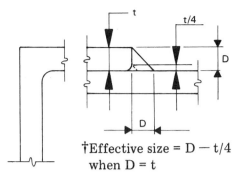
†Effective size = D − t/4 when D = t

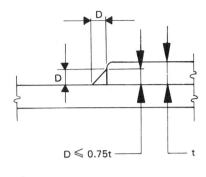
D ≤ 0.75t

- When fillet welds are used in holes or slots, the diameter of the hole or the width of the slot should not be less than the thickness of the member containing it plus 8 mm.

Lap Joints

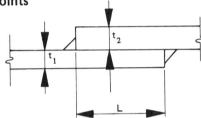

$L_{min} = 5t_1 \geq 25$ mm when $t_1 \leq t_2$
$L_{min} = 5t_2 \geq 25$ mm $\quad t_2 < t_1$

†This detail is not part of W59, but rather is a detail often used by the steel fabricating industry.

Partial Penetration Groove Welds

Minimum Groove Depth for Partial Joint Penetration Groove Welds†

Thickness of Thicker Part Joined (Millimetres)	Minimum Groove Depth*, mm	
	Groove Angle, α, at Root $45° \leq α < 60°$ (V-, Bevel Grooves)	Groove Angle, α, at Root $α \geq 60°$ (V-, Bevel, J-, U- Grooves)
Total 12 incl.	8	5
Over 12 − 20	10	6
Over 20 − 40	12	8
Over 40 − 60	14	10
Over 60	16	12

†Not combined with fillet welds.
*See Notes 2a and 2b of diagram on next page.

Flare Bevel and Flare V-Welds

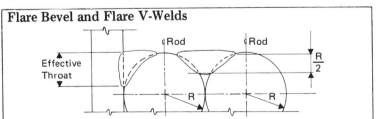

The effective throat thickness for flare groove welds on solid bars, when filled flush to the surface of the solid section of the bar is 5/16 R for Flare Bevel Groove welds and 1/2 R for Flare Vee Groove welds.

WELDED JOINTS
Standard Symbols

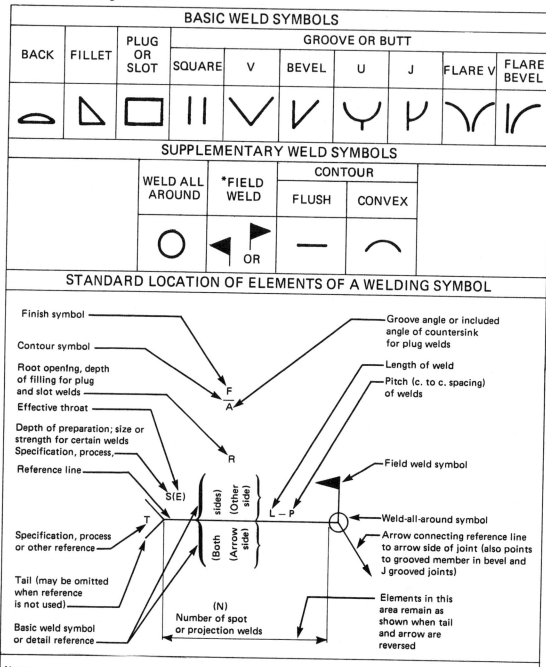

Note:
Size, weld symbol, length of weld and spacing must read in that order from left to right along the reference line. Neither orientation of reference line nor location of the arrow alter this rule.

The perpendicular leg of △, V, Y, ⟨ weld symbols must be at left.

Size and spacing of fillet welds must be shown on both the Arrow Side and the Other Side Symbol.

Symbols apply between abrupt changes in direction of welding unless governed by the "all around" symbol or otherwise dimensioned.

These symbols do not explicitly provide for the case that frequently occurs in structural work, where duplicate material (such as stiffeners) occurs on the far side of a web or gusset plate. The fabricating industry has adopted this convention; that when the billing of the detail material discloses the identity of far side with near side, the welding shown for the near side shall also be duplicated on the far side.

*Pennant points away from arrow.

6-160

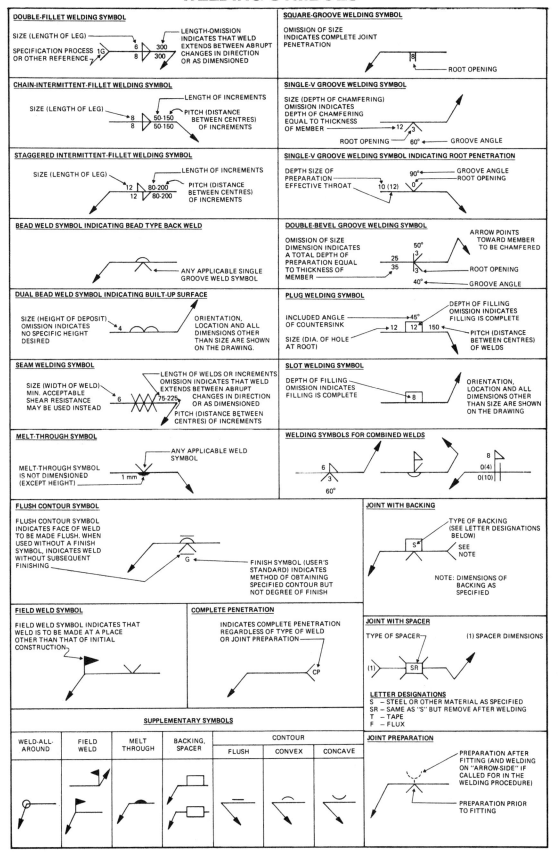

SAMPLE GROOVE WELDS

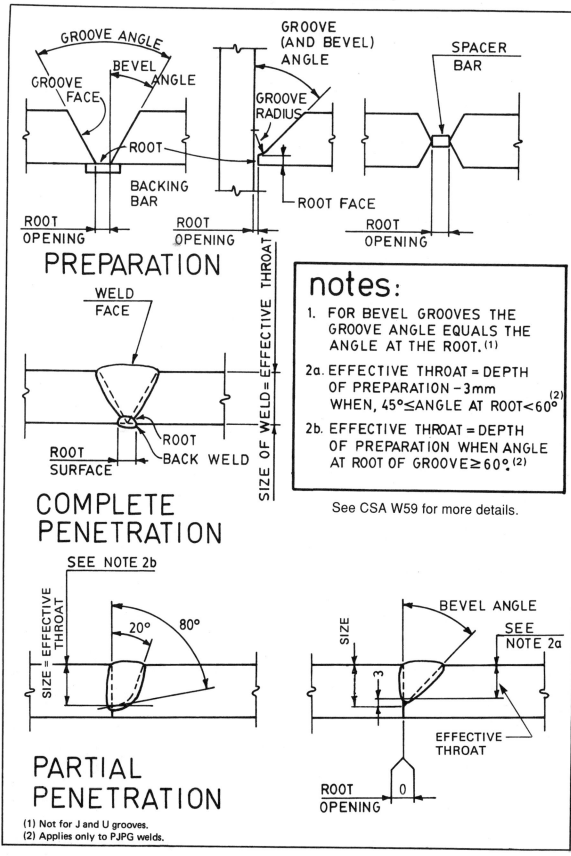

STEEL PRODUCTS — RECORD OF CHANGES

Following is a chronological record of changes to the list of steel sections included in the CISC Handbook of Steel Construction since the first printing of the Third Edition.

1983   No longer produced by Algoma are —

      M100 × 19   S100 × 11
      S150 × 26, and 19   S75 × 11, and 8
      S130 × 22, and 15   All angles except 8" × 8" leg sizes

1985   No longer produced by Algoma are —

      WWF 550 × 217   WWF 350 × 385

     New shapes and sections produced by Algoma —

      WWF 1800 × 632, and 548   WWF 1400 × 491, and 407
      WWF 1600 × 579, and 495   WWF 550 × 280

     Welded Reduced Flange (WRF) shapes with top flanges narrower than the bottom flanges and intended primarily for composite bridge girders —

      WRF 1800 × 543, 480, and 416   WRF 1200 × 373, 309, and 244
      WRF 1600 × 491, 427, and 362   WRF 1000 × 340, 275, and 210
      WRF 1400 × 413, 348, and 284

1986   New shapes and sections produced by Algoma —

      W 610 × 91   W 310 × 31
      W 610 × 84   W 250 × 24
      W 530 × 72   W 200 × 21

1989   Sections produced by Algoma

Sections deleted	Sections added
WWF1800 × 632 and 548	WWF2000 × 732–542
WWF1600 × 579 and 495	WWF1800 × 700–510
WWF1400 × 491 and 407	WWF1600 × 622–431
WWF1200 × 403 and 364	WWF1400 × 597–358
WWF1100 × 335, 291, 255, and 220	WWF1200 × 418, 380, and 333
WWF1000 × 324, 280, and 244	WWF1100 × 351, 304, 273, and 234
WWF900 × 293, 249, and 213	WWF1000 × 340, 293, 262, and 223
WWF800 × 332–154	WWF900 × 309, 262, and 231
WWF700 × 222–141	WWF800 × 339–161
	WWF700 × 245–152
	WWF650 × 864–400
	WWF600 × 793–369

     Sections not available from Canadian mills added

      W1000 – All   W690 × 802–289
      W920 × 1262–488   W610 × 732–262
      W840 × 922–392   W530 × 599–248
      W760 × 865–350   W460 × 464–193
      W760 × 134

1991 Sections no longer available from Canadian mills.

    W310x283 and W310x253

    C380x74–50 and C310x45–31

1993 The following shapes are no longer produced.

    HP330x149*– 89* incl.

    M150x29.8*– 6.5* incl.

    M100x19*

1995 Sections deleted

    W1000x488–286, 976, 790–483; W920x1072, 876, 722; W840x922–577

    W760x865, 783, 644, 531; W690x735, 605, 500, 419; W610x670, 551, 455

    W530x599–331

    HP330x149–89

    M150x29.8, 6.5; M100x19

    S180x30, 22.8; S130x22

    C130x17

    MC250x9.7; MC180x26.2; MC150x22.8

    L152x102x4.8; L127x127x4.8; L127x89x11, 4.8; L127x76x16, 4.8

    L102x102x4.8; L102x89x16, 11, 4.8; L102x76x4.8; L89x89x16, 4.8

    L89x76x16, 11; L89x64x16, 11; L76x76x16; L76x64x16, 11; L76x51x16, 11

    L64x64x3.2; L64x51x3.2; L51x38x9.5, 3.2; L32x32x9.5; L25x25x9.5, 7.9

    L200–L25 (All metric angles)

   Sections added

    W1100x499–342; W1000x749–478, 259, 693–314; W920x381, 345; W840x251

    W760x220; W690x192; W610x153; W360x1202

    M310x16.1, M250x11.9, M100x8.9

    SLB100x5.4, 4.8; SLB75x4.5, 4.3

    L203x102x22, 16, 11; L178x102x11; L19x19x3.2

1997 Sections deleted

    W1000x478, 259, 693; W920x1262; W760x710; W690x667; W610x732, 608

    W460x464–286; W360x1202

    L203x203x14; L203x152x22, 16, 14; L203x102x22, 16, 14, 11; L152x102x6.4

    L152x89x6.4; L89x76x4.8; L89x64x4.8; L64x38x7.9–4.8; L51x38x7.9

L44x44x9.5, 7.9; L38x38x9.5, 7.9, 4.0; L32x32x7.9

HSS51x51x2.5; HSS38x38x2.5; HSS32x32x3.8–2.5; HSS25x25x3.2, 2.5

HSS127x64x9.5–4.8; HSS127x51x9.5–4.8; HSS51x25x2.5

HSS48x2.8; HSS42x3.2, 2.5; HSS33x3.2, 2.5; HSS27x3.2, 2.5

Sections added:

W1000x591, 539, 486, 483; W840x576; W760x531; W690x500, 419

W610x551, 455; W150x13

L152x152x6.4

HSS127x127x13; HSS102x102x3.8, 3.2; HSS89x89x3.8, 3.2

HSS76x76x9.5, 3.8, 3.2

HSS152x102x13; HSS152x76x9.5–4.8; HSS127x76x3.8

HSS102x76x3.8, 3.2; HSS76x51x3.2

HSS610x13–6.4; HSS559x13–6.4; HSS508x13–6.4

NOTES

PART SEVEN
MISCELLANEOUS

CISC Code of Standard Practice	7–3
Structural Sheet Steel Products	7–37
Mass and Forces for Materials	7–40
M/D Ratios	7–42
Coefficients of Thermal Expansion	7–47
1997 Electronic Aids	7–48
Check List for Design Drawings	7–50
Properties of Geometric Sections	7–52
Properties of Geometric Sections and Structural Shapes	7–59
Properties of the Circle	7–63
Properties of Parabola and Ellipse	7–64
Properties of Solids	7–65
Trigonometric Formulae	7–67
Bracing Formulae	7–68
Length of Circular Arcs	7–69
SI Summary	7–70
Millimetre Equivalents	7–75
Miscellaneous Conversion Factors	7–76

CISC
CODE OF STANDARD PRACTICE
for
Structural Steel

Sixth Edition

Published by the
CANADIAN INSTITUTE OF STEEL CONSTRUCTION
300 - 201 Consumers Road ■ Willowdale
Ontario ■ M2J 4G8

TABLE OF CONTENTS

Preface 7-6

1. General Provisions 7-7
   1.1 Scope 7-7
   1.2 Definitions 7-7
   1.3 Governing Technical Standards 7-9
   1.4 Responsibility for Design 7-9
   1.5 Responsibility for Erection Procedure 7-9
   1.6 Patented Devices 7-10
   1.7 Scheduling 7-10

2. Classification of Material 7-10
   2.1 Structural Steel 7-10
   2.2 Field Connection Material 7-11
   2.3 Items Supplied by Others 7-11

3. Quotations and Contracts 7-12
   3.1 Standard Form of Contract 7-12
   3.2 Types of Contracts 7-13
   3.3 Revisions to Contract Documents. 7-13
   3.4 Discrepances 7-13
   3.5 Computation of Units 7-13
   3.6 Contract Price Adjustments. 7-14
   3.7 Scheduling 7-15

4. Contract Documents 7-15
   4.1 Tender Documents – Tender Drawings and Tender Specifications . . . 7-15
   4.2 Architectural, Electrical and Mechanical Drawings 7-16
   4.3 Construction Drawings and Construction Specifications. 7-16

5. Fabrication and Erection Documents 7-16
   5.1 Erection Diagrams 7-16
   5.2 Connection Design Details 7-16
   5.3 Shop Details 7-17
   5.4 Erection Procedures 7-17
   5.5 Field Work Details 7-17
   5.6 Review and Approval 7-17
   5.7 Additions, Deletions or Changes 7-18

6. Material, Fabrication, Inspection, Painting and Delivery. 7-18
   6.1 Material. 7-18
   6.2 Identification 7-18
   6.3 Preparation of Material 7-18
   6.4 Fitting and Fastening 7-18
   6.5 Dimensional Tolerances 7-19

	6.6	Inspection of Steelwork	7-19
	6.7	Surface Preparation	7-19
	6.8	Paint	7-19
	6.9	Marking and Shipping	7-19
	6.10	Delivery of Materials	7-20
7.	Erection	7-20
	7.1	Method of Erection	7-20
	7.2	Erection Safety	7-20
	7.3	Site Conditions	7-20
	7.4	Foundations	7-21
	7.5	Bearing Surfaces	7-21
	7.6	Building Lines and Bench Marks	7-21
	7.7	Installation of Anchor Rods and Embedded Items	7-21
	7.8	Bearing Devices	7-21
	7.9	Examination by Erector	7-22
	7.10	Adjustable Shelf Angles and Sash Angles	7-22
	7.11	Loose Lintels and Wall Bearing Members	7-22
	7.12	Tolerances	7-22
	7.13	Checking Erected Steelwork	7-22
	7.14	Removal of Bracing	7-22
	7.15	Correction of Errors when Material is Not Erected by the Fabricator	7-22
	7.16	Field Assembly	7-22
	7.17	Accommodation of Other Trades	7-22
	7.18	Temporary Floors and Access Stairs	7-23
	7.19	Touch-Up of Shop Paint	7-23
	7.20	Final Painting	7-23
	7.21	Final Clean-Up	7-23

APPENDIX A – Structural Steel in Buildings 7-24

APPENDIX B – Guideline for Unit Price Application for Changes 7-26

APPENDIX C – A Suggested Format for Price Per Unit of Mass or Price Per Item Contracts 7-28

APPENDIX D – Tolerances on Anchor Rod Placement 7-31

APPENDIX E – Conversion of SI Units to Imperial Units 7-32

APPENDIX F – Miscellaneous Steel 7-33

APPENDIX G – A Suggested Format for a Monthly Progress Payment Claim Form 7-35

APPENDIX H – Suggested Definitions for Progress Invoicing and Substantial Performance 7-36

CISC CODE OF STANDARD PRACTICE FOR STRUCTURAL STEEL

PREFACE

The CISC Code of Standard Practice for Structural Steel is a compilation of usual industry practices relating to the design, fabrication and erection of structural steel. These practices evolve over a period of time and are subject to change as improved methods replace those of an earlier period. The Code is revised whenever a sufficient number of changes have occurred to warrant a new edition.

The first edition of the Code was adopted and published in November 1958. A second edition incorporating minor revisions was published in October 1962. The third edition, published in September 1967 and revised in May 1970, incorporated minor changes throughout with principal changes in Section 2 — Definition of Structural Steel and Section 3 — Computation of Weights for Unit Price Bids.

The fourth edition adopted in June 1980, revised December 1980, broadened the scope to include bridges and other structures. It also incorporated the CISC "Guide to Tendering Procedures" into Section 3 and Appendices B & C. The Code was converted to SI (metric) units and provided conversion factors and Imperial units in Appendix E.

The fifth edition (1991) reflected the steel standard's recognition of the preparation of five types of fabrication and erection documents which may be produced in fulfilling a steel construction contract. These documents may be in the form of drawings, diagrams, sketches, computer output, hand calculations and other data which can be supplied by the fabricator/erector. This data is generally referred to in contract documents as "shop drawings". The computation of mass has been changed by deleting the mass of welds and the allowances for paint and other coatings. Appendix B, Guideline for Unit Price Application for Changes, and Appendix C, A Suggested Format for Price Per Unit of Mass or Price Per Item Contracts were substantially revised. To foster uniformity, two new appendices were added; Miscellaneous Steel and A Suggested Format for a Monthly Progress Claim Form.

This sixth edition (1999) continues to clarify the role of the fabricator, the information required, and where that information is expected, as stipulated in the governing technical standards. Added are: definitions of Design Drawings, and Quotations, clauses on Quotations, Discrepancies, shims for bearing surfaces, the allowance for return of documents, the information required when painting is specified, and Appendix H, Suggested Definitions for Progress Invoicing and Substantial Performance. Changes have been made to Appendix C, the terminology for Unit Price contracts, connection types, and anchor rods— the latter two to be consistent with the changes proposed for CSA Standard S16-2001.

By documenting standard practices the Code aims to promote a clear understanding between the Canadian structural steel fabrication and erection industry and its clients.

Canadian Institute of Steel Construction

Adopted November 19, 1999

CISC CODE OF STANDARD PRACTICE FOR STRUCTURAL STEEL

1. General Provisions

1.1 Scope. This Code covers standard industry practice with respect to the furnishing of structural steel. In the absence of provisions to the contrary contained in contracts to which members of the Canadian Institute of Steel Construction are contracting parties, members will abide by the practices described herein.

1.2 Definitions.

Approved for Construction Drawings	Drawings and other documents approved by the client authorizing work to proceed. (May also be called "Issued for Construction Drawings")
Architect	As defined under the appropriate Architect's Act.
Client	A person, corporation, or authority with whom the fabricator has contracted.
Connection Design Details	Documents which provide details of standard and non-standard connections and other data necessary for the preparation of shop details.
Construction Drawings	Drawings used to govern the construction of the works.
Construction Specifications	Specifications used to govern the construction of the works.
Contract	The agreement between the fabricator and/or erector, and the client.
Contract Documents	The documents which define the responsibilities of the parties involved in tendering, purchasing, supplying, fabricating and erecting structural steel, including tender drawings and tender specifications and applicable revisions in effect and agreed to at the time of contract award.
Cost Plus a Fee Contract	An Agreement whereby the fabricator and/or erector agrees to fulfil the contract for a consideration which is calculated on the basis of the fabricator's costs plus a specified fee as defined in the contract.
Design Drawings	Drawings, including computer output, electronic and other data, as prepared by the designer showing member sizes and dimensions and all required forces for connection design i.e. shears, axial forces, moments and torsions. (See governing technical standard).
Designer	The designer of the structure. See Engineer of Record.
Engineer	As defined under the appropriate Professional Engineer's Act.
Engineer of Record	Professional Engineer who designs the structure, as defined under the appropriate Professional Engineer's Act.
Erection Bracing	Bracing materials or members which are used to plumb, align and stabilize structural members or the structure during construction and are removed when the structural members or the structure is secured by bolting or welding of structural members.
Erection Diagrams	Are general arrangement drawings showing the principal dimen-

	sions and elevations of the steel structure, sizes of the steel members, piece marks, size (diameter) and type of bolts, bolt installation requirements, elevations of column bases, all necessary dimensions and details for setting anchor rods, and all other information necessary for the assembly of the structure.
Erection Procedures	Outline the construction methods, erection sequence, erection and temporary bracing requirements, and other engineering details necessary for shipping, handling, erecting, and maintaining the stability of the structural steel frame.
Erector	Means the party responsible for erection of the steelwork.
Fabricator	Means the party responsible for furnishing the structural steel.
Field Work Details	Are details that provide complete information for modifying fabricated members in the field (e.g. prepare existing steel to receive new steel).
General Contractor, Constructor or Construction Manager	The person or corporation who constructs, coordinates, and supervises the construction of the work.
General Terminology e.g. Beams, Joists, Columns, etc.	These terms have the meanings stated or implied in CAN/CSA S16.1 (latest edition), CAN/CSA-S6-M (latest edition) and Appendix A of this Code.
Lump Sum Price Contract	Also called Stipulated Price Contract. An agreement whereby the fabricator and/or erector contracts to fulfil the contract terms for a lump sum (stipulated price) consideration.
Miscellaneous Steel	Steel items described and listed in Appendix F of this code.
Others	Means a party or parties other than the fabricator and/or erector.
Owner	Means the owner of a structure and shall include his authorized agent and any person taking possession of a structure on the owner's behalf. Depending on the circumstances an authorized agent may be the architect, engineer, general contractor, construction manager, public authority or other designated representative of the owner.
Price Per Unit Contract	Also called Unit Price Contract. An agreement whereby the fabricator and/or erector contracts to fulfil the contract terms for a consideration which is based on the units of steel calculated in accordance with the CISC Code of Standard Practice for Structural Steel.
Quotations	Proposals by the fabricator based on structural steel as defined in Clause 2.1 and as included in the tender documents, and in accordance with the documents outlined in Clause 3.3.1.
Revision	A change in the contract documents.
Shop Details	Documents which provide complete information for the fabrication of various members and components of the structure, including the required material and product standards; the location, type, and size of all mechanical fasteners; bolt installation requirements; and welds.

Stipulated Price Contract	See Lump Sum Price Contract.
Structural Drawings	Drawings showing the structural steel required.
Structural Steel	Those items listed under Clause 2.1
Structural Steel Frame	An assemblage of structural steel components (beams, columns, purlins, girts, etc.) for the purpose of resisting loads and forces. See Clause 2.1.
Structural Steel Specifications	The portion of the tender specifications containing the requirements for the fabrication and erection of the structural steel.
Temporary Bracing	Members designed by the Engineer of Record to be removed at a later date at his or her instruction.
Tender Documents	Drawings, specifications, general conditions, addenda, etc., used as the basis for preparing a tender.
Tender Drawings	Drawings used as the basis for preparing a tender.
Tender Specifications	Specifications used as the basis for preparing a tender.
Unit Price Contract	See Price Per Unit Contract.

1.3 Governing Technical Standards. The provisions of the latest edition of CAN/CSA-S16.1 "Limits States Design of Steel Structures", shall govern the design, fabrication and erection of steel structures except bridges. The provisions of the latest edition of CAN/CSA-S6-M "Design of Highway Bridges", "The Ontario Highway Bridge Design Code" (in Ontario) or the American Railway Engineering Association's "Specifications for Steel Railway Bridges" shall govern the design, fabrication and erection of structural steel for bridges. The provisions of the latest edition of CSA Standard W59 "Welded Steel Construction (Metal-Arc Welding)" shall govern arc welding design and practice. The provisions of other standards shall be applicable if called for in the tender drawings and tender specifications.

1.4 Responsibility for Design. When the client provides the structural drawings and specifications, the fabricator and the erector shall not be responsible for determining the adequacy of the design nor liable for the loss or damage resulting from an inadequate design. Should the client desire the fabricator to assume any responsibility for design beyond that of proposing adequate connections and details, and, when required, components, members, or assemblies standardized by the fabricator, the client shall state clearly his requirements in the invitation to tender or in the accompanying tender drawings and tender specifications. Even though proposed connections and design details may be prepared by the fabricator's technical staff, the over-all behaviour of the structure remains the responsibility of the designer of the structure. (See also Clause 5.6).

1.5 Responsibility for Erection Procedure. When the erection of structural steel is part of his contract, the fabricator shall be responsible for determining the erection procedure, for checking the adequacy of the connections for the uncompleted structure and for providing erection bracing or connection details. When the erection of the structural steel is not part of his contract, the fabricator shall not be responsible for determining the erection procedure, for checking the adequacy of the connections for the uncompleted structure, or for providing erection bracing or connection details not included in the contract documents, nor shall the fabricator be liable for loss or damage resulting from faulty erection. However, the steel fabricator shall be informed by the client of the erection sequence to be used which may influence the sequence and process of the manufacturing. (See also Clause 5.1 and 5.4).

1.6 Patented Devices. Except when the contract documents call for the design to be furnished by the fabricator or erector, the fabricator and erector assume that all necessary patent rights have been obtained by the client and that the fabricator and erector will be fully protected by the client in the use of patented designs, devices or parts required by the contract documents.

1.7 Scheduling. The client should provide a construction schedule in the tender documents. In the absence of such a schedule, one should be mutually agreed upon between the contracting parties, prior to the contract award.

2. Classification of Material

2.1 Structural Steel. Unless otherwise specified in the tender documents, a contract to supply, fabricate and deliver structural steel shall include only those items from the following list which are clearly indicated as being required by the structural drawings and tender specifications. (See Appendix A)

2.1.1

> Anchors for structural steel.
> Base plates and bearings for structural steel members.
> Beams, purlins, girts forming part of the structural steel frame.
> Bearing plates and angles for structural steel members and steel deck.
> Bins and hoppers of 6 mm plate or heavier, attached to the structural steel frame.
> Bracing for steel members, trusses or frames.
> Brackets attached to the structural steel.
> Bridge bearings connected to the structural steel members.
> Cables for permanent bracing or suspension systems.
> Canopy framing if attached to the structural steel frame.
> Cold formed channels when used as structural members as listed in the CISC Handbook of Steel Construction
> Columns.
> Conveyor galleries and supporting bents (exclusive of conveyor stringers, deck plate and supporting posts which are normally part of the conveyor assembly).
> Crane rails and stops, excluding unless otherwise noted final alignment of the rails.
> Curb angles and plates attached to the structural steel frame if shown on the structural steel drawings.
> Diaphragms for bridges.
> Deck support angles at columns, walls, if shown on the structural steel drawings.
> Door frame supports attached to the structural steel frame.
> Embedded items connecting structural steel.
> Expansion joints connected to the structural steel frame (excluding expansion joints for bridges).
> Field bolts to connect structural steel components.
> Floor plates, roof plates (raised pattern or plain) and steel grating connected to the structural steel frame.
> Girders.
> Grillage beams of structural steel.
> Hangers supporting structural steel framing.
> Jacking girders.

Lintels if attached to steel frame and shown on the structural drawings.

Mechanical roof support and floor opening framing shown on structural drawings.

Monorail beams of standard structural steel shapes.

Open-web steel joists, including anchors, bridging, headers and trimmers; also, when specified to be included in the structural steel contract documents, light-gauge forms and temperature reinforcement.

Sash angles attached to the structural steel frame.

Separators, angles, tees, clips and other detail fittings essential to the structural steel frame.

Shear connectors/studs, except when installed through sheet steel deck by deck installer.

Shelf angles attached to the structural steel frame if shown on structural drawings.

Shop fasteners or welds, and fasteners required to assemble parts for shipment.

Steel tubes or cores for composite columns.

Steel window sills attached to the structural steel frame.

Struts.

Suspended ceiling supports of structural steel shapes at least 75 mm in depth.

Temporary components to facilitate transportation to the site.

Tie, hanger and sag rods forming part of the structural steel frame.

Trusses.

2.1.2 Only if shown on the structural drawings and specifically noted by the structural engineer to be supplied by the structural fabricator:

Steel stairs, walkways, ladders and handrails forming part of the structural steelwork. [See Appendix A]

2.2 Field Connection Material.

2.2.1 When the fabricator erects the structural steel, he shall supply all material required for temporary and for permanent connection of the component parts of the structural steel.

2.2.2 When the erection of the structural steel is not part of the fabricator's contract, unless otherwise specified, the fabricator shall furnish appropriate bolts and nuts (plus washers, if required) or special fasteners, of suitable size and in sufficient quantity for all field connections of steel to steel which are specified to be thus permanently connected, plus an over-allowance of two per cent of each size to cover waste.

Unless otherwise specified in the tender documents, welding electrodes, back-up bars, temporary shims, levelling plates, fitting-up bolts and drift pins required for the structural steel shall not be furnished by the fabricator when the erection of the structural steel is not part of the fabricator's contract.

2.3 Items Supplied by Others.
Unless otherwise specified in the tender documents, the following steel or other items shall not be supplied by the structural steel fabricator.

Bolts for wood lagging.

Bins and hoppers not covered in Clause 2.1 of this Code.

Bridge bearings not connected to structural steel items.

Canopy framing not attached to structural steel.

Catch basin frames.

Concrete for filling HSS columns. Concrete is to be supplied and poured in columns by others in the shop or field with the co-operation of the fabricator/erector.

Connection material for other trades (e.g. precast concrete).

Conveyor stringers, deck plate and supporting posts.

Drain pipes.

Door and corner guards.

Door frames not covered in Clause 2.1 of this Code.

Drilling of holes into masonry or concrete, including core drilling of anchor rods for bridges and drilling for deck support angles.

Edge forming less than 3.2 mm thick for steel deck and not covered in Clause 2.1 of this code.

Embedded steel parts not required for structural steel or deck.

Embedded steel parts in precast concrete.

Flagpoles and supports.

Floor plates, roof plates and grating not covered in Clause 2.1 of this Code.

Grout.

Hoppers and chutes.

Hose and tire storage brackets.

Installation of embedded parts.

Lag bolts, machine bolts and shields or inserts for attaching shelf angles, trimmer angles and channels to masonry or concrete.

Lintels over wall recesses.

Lintels which are either an integral part of door frames or not attached to the structural steel frame.

Machine bases, rollers and pulleys.

Members made from gauge material except cold-formed channels indicated in Clause 2.1.

Metal-clad doors and frames.

Miscellaneous Steel, See Appendix "F".

Shear connectors through sheet steel deck by deck installer.

Sheet steel cladding.

Sheet steel deck.

Sheet steel flashing.

Shelf angles not covered in Clause 2.1 of this Code.

Shoring under composite floors and stub girders.

Steel doors.

Steel sash.

Steel stacks.

Steel stairs, landings, walkways, ladders and handrails, not covered in Clause 2.1.2 of this Code.

Steel tanks and pressure vessels.

Steel window sills not covered in Clause 2.1 of this Code.

Support for sheet steel deck at column cut outs and for openings not requiring framing connected to structural steel.

Temporary bracing for other trades.

Trench covers.

Trim angles, eave angles or fascia plates not directly attached to the structural steel frame.

3. Quotations and Contracts

3.1 Standard Form of Contract. Unless otherwise agreed upon, a contract to fabricate, deliver and/or erect structural steel shall be the appropriate Standard Construction Documents approved by the Canadian Construction Documents Committee, or the Canadian Construction Association.

3.1.1 Quotations

Quotations from fabricators are based on the following documents:

(1) *A Standard Form Of Contract.* The generally accepted standard form of contract is the CCA L-1 1995 Stipulated Price Subcontract-Long Form. (CCA S-1 1994 Stipulated Price Subcontract-Short Form may also be acceptable), and

(2) Canadian Institute of Steel Construction (CISC) Code of Standard Practice for Structural Steel, latest edition

3.1.2 Progress Payment Claim Form.
A suggested format for a progress payment claim form, is provided in Appendix G

3.1.3 Progress Invoicing and Substantial Completion.
For suggested definitions, see Appendix H.

3.2 Types of Contracts.

3.2.1 For contracts stipulating a "lump sum price", the work required to be performed by the fabricator and/or erector must be completely defined by the contract documents.

3.2.2 For contracts stipulating a "price per unit", the scope of the work, type of materials, character of fabrication, and conditions of erection are based upon the contract documents which must be representative of the work to be performed. For methods of computing mass, area, or quantity, see Clause 3.5. See Appendix C of this Code.

3.2.3 For contracts stipulating "cost plus fee", the work required to be performed by the fabricator and/or erector is indefinite in nature at the time the contract documents are prepared. Consequently the contract documents should define the method of measurement of work performed, and the fee to be paid in addition to the fabricator's costs.

3.3 Revisions to Contract Documents.

3.3.1 Revisions to the contract shall be made by the issue of dated new or revised documents. All revisions shall be clearly indicated. Such revisions should be issued by a Detailed Change Notice.

3.3.2 The fabricator shall advise the client or client representative of any impact such revision or change will have on the existing agreement between the two parties.

3.3.3 Upon agreement between the fabricator and the client or client representative as to the revision's impact, the client or his representative shall issue a change order or extra work order for the revisions.

3.3.4 Unless specifically stated to the contrary, the issue of revision documents or changes indicated on drawing approvals is authorization by the client to release these revisions for construction.

3.4 Discrepancies

In case of discrepancies between the structural drawings and specifications for buildings, the specifications govern. In case of discrepancies between the structural drawings and specifications for bridges, the structural drawings govern. In case of discrepancies between scale dimensions on the structural drawings and figures written on them, the figures govern. In case of discrepancies between the structural drawings and plans for other trades, the structural drawings govern.

3.5 Computation of Units

Unless another method is specified and fully described at the time tenders are requested, the computed mass of steel required for the structure shall be determined by the method of computation described herein. (Although the method of computation described does not result in the actual mass of fabricated structural steel and other items its relative simplicity results in low computational cost and it is based on quantities which can be readily computed and checked by all parties involved to establish the basis of payment). No additional mass for welds, or mass allowance for painting, galvanizing, and metallizing is to be included in the computation of mass.

> (a) *Mass Density.* The mass density of steel is assumed to be 7850 kilograms per cubic metre.
>
> (b) *Shapes, Bars and Hollow Structural Sections.* The mass of shapes, bars and hollow structural sections is computed using the finished dimensions shown on shop details. No deductions shall be made for holes created by cutting, punching or drilling, for material removed by coping or clipping, or for material removed by weld joint preparation. No cutting, milling or planing allowance shall be added to the finished dimensions. The mass per metre of length for shapes and hollow structural sections is the published mass. The mass per metre of length for bars is the published mass, or if no mass is published, the mass computed from the specified cross-sectional area.
>
> (c) *Plates and Slabs.* The mass/area of plates and slabs is computed using the rectangular dimensions of plates or slabs from which the finished plate or slab pieces shown on the shop details can be cut. No burning, cutting, trimming or planing allowance shall be added. When it is practical and economical to do so, several irregularly-shaped pieces may be cut from the same plate or slab. In this case, the mass shall be computed using the rectangular dimensions of the plate or slab from which the pieces can be cut. No cutting or trimming allowance shall be added. In all cases, the specified plate or slab thickness is to be used to compute the mass. The mass of raised-pattern rolled plate is that published by the manufacturer.
>
> (d) *Bolts.* The mass of shop and field bolts, nuts and washers is computed on the basis of the shop details and the nominal published mass of the applicable types and sizes of fastener.
>
> (e) *Studs.* If not included in the contract on a "price per unit basis", the mass of studs is computed on the basis of the shop details and/or erection diagrams and the published mass of the studs.
>
> (f) *Grating.* The mass/area of grating is computed on the basis of the shop details and/or erection diagrams and published mass of the grating. The area to be used is the minimum rectangular area from which the piece of grating can be cut.
>
> (g) Where supplied, such items as shims, levelling plates, temporary connection material, back-up bars and certain field "consumables" shall be considered as part of the structural steel whether or not indicated specifically in the contract documents. Such items then will be added to, and become a part of, computed mass of steel for the structure.

3.6 Contract Price Adjustments

3.6.1 When the responsibility of the fabricator and/or erector is changed from that previously established by the contract documents, an appropriate modification of the contract price shall be made. In computing the contract price adjustment, the fabricator and/or erector shall consider the quantity of work added or deleted, modifications in the character of the work, the timeliness of the change with respect to the status of material ordering, the detailing, fabrication and erection operations and related impact costs.

3.6.2 Requests for contract price adjustments shall be presented by the fabricator and/or erector and shall be accompanied by a description of the change in sufficient detail to permit evaluation and prompt approval by the client.

3.6.3 Price Per Unit Contracts generally provide for minor revisions to the quantity of work prior to the time work is approved for construction. Minor revisions to the quantity of work should be limited to an increase or decrease in the quantity of any category not exceeding ten percent. Should the quantity of steel of any category vary by more than ten percent, then the contract unit price of that category may require adjustment. Changes to the character of the work or the mix of the work, at any time or changes to the quantity of the work after the work is approved for construction, may require a contract price adjustment.

3.6.4 A suggested format for accommodating contract price adjustments is contained in Appendix B.

3.7 Scheduling

3.7.1 The contract documents should specify the schedule for the performance of the work. This schedule should state when the approved for construction drawings will be issued and when the job site, foundations, piers and abutments will be ready, free from obstructions and accessible to the erector, so that erection can start at the designated time and continue without interference or delay caused by the client or other trades.

3.7.2 The fabricator and/or erector has the responsibility to advise the client of the effect any revision may have on the contract schedule.

3.7.3 If the fabrication and erection schedule is significantly delayed due to revisions, or for other reasons which are the client's responsibility, the fabricator and erector shall be compensated for additional costs incurred. Changes to the scope of the work shall provide additional time to the schedule, if required.

4. Contract Documents

4.1 Tender Documents – Tender Drawings and Tender Specifications

4.1.1 At the time tenders are called, the steel fabricator shall receive a complete set of structural drawings and a complete set of tender specifications. In order to ensure adequate and complete tenders for Lump Sum Price Contracts[*], these documents shall include complete structural drawings, conforming to the requirements for design drawings of the governing technical standard. Structural steel specifications should include any special requirements controlling the fabrication and erection of the structural steel, surface preparation and coating, and should indicate the extent of non-destructive examination, if any, to be carried out.

4.1.2 Design drawings shall be drawn to a scale adequate to convey the required information. The drawings shall show a complete design of the structure with members suitably designated and located, including such dimensions and detailed description as necessary to permit the preparation of fabrication and erection documents. Floor levels, column centres, and offsets shall be dimensioned. The term "drawings" may include computer output and other data. Stiffeners and doubler plates required to maintain stability and which are an integral part of the main member shall be shown and dimensioned.

4.1.3 Design drawings shall designate the design standards used, shall show clearly the type or types of construction to be employed, shall show the category of the structural system used for seismic design, and shall designate the material or product standards applicable to the members and details depicted. Drawings shall give the governing combinations of shears, moments, and axial forces to be resisted by the connections.

---

[*] *For other types of contracts, it is desirable for the contract documents to be as complete as possible.*

4.1.4 Where connections are not shown, the connections shall be assumed to be in accordance with the requirements of the governing technical standard (see Clause 1.3).

4.2 Architectural, Electrical and Mechanical Drawings. Architectural, electrical and mechanical drawings may be used as a supplement to the structural drawing to define detail configurations and construction information, *provided all requirements for the structural steel are noted on the structural drawings.*

4.3 Construction Drawings and Construction Specifications

4.3.1 At the time specified in the tender documents or pre-award negotiations (if different), the client shall furnish the fabricator with a plot plan of the construction site, and a set of complete drawings and specifications approved for construction consistent with the tender drawings and tender specifications. These construction drawings and specifications are required by the fabricator for ordering the material and for the preparation and completion of fabrication and erection documents. The approved for construction drawings shall show:

(a) all changes or revisions to the tender drawings, clearly indicated on the construction drawings,

(b) the complete design of the structure with members suitably designated and located, including such dimensions and detailed description as necessary to permit preparation of the fabrication and erection documents. Floor levels, column centres, and offsets shall be dimensioned;

(c) all materials to be furnished by the fabricator, together with sufficient information to prepare fabrication and erection documents, including the design standards used, the type or types of construction to be employed, the category of the system used for seismic design, the applicable material or product standards, and the governing combinations of shears, moments and axial forces to be resisted by connections.

The fabricator shall receive a complete set of the tender drawings and tender specifications.

5. Fabrication and Erection Documents

Note: *The term "shop drawings", frequently used in the construction industry, is replaced in this Code of Practice by the terms "fabrication and erection documents". These terms more correctly describe the following five separate and distinct documents that may be prepared by a fabricator/erector. See also Clause 1.2 for definitions. Not all of these documents will be required for every project.*

5.1 Erection Diagrams. Unless provided by the client, the fabricator will prepare erection diagrams from the approved construction drawings. In this regard, the fabricator may request reproducible copies of the structural drawings which may be altered for use as erection diagrams. When using reproducible copies of the structural drawings, the structural engineer's name and seal shall be removed. Erection diagrams shall be submitted to the designer for review and approval. Erection diagrams are general arrangement drawings showing the principal dimensions of the structure, piece marks, sizes of the members, size (diameter) and type of bolts, bolt installation requirements, elevations of column bases, all necessary dimensions and details for setting anchor rods, and all other information necessary for the assembly of the structure. Only one reproducible copy of each diagram will be submitted for review and approval unless a larger number of copies is required by the client as part of the tender documents.

5.2 Connection Design Details. Connection design details shall be prepared in advance of preparing shop details and submitted to the designer for confirmation that the intent of the design is met. Connection design details shall provide details of standard and non-standard connections, and other data necessary for the preparation of shop details. Connection design details shall be referenced to the design drawings, and/or erection drawings.

5.3.1 Shop Details. Unless provided by the client, shop details shall be prepared in advance of fabrication from the information on the approved construction drawings, the connection design details, and the erection diagrams. Shop details shall provide complete information for the fabrication of various members and components of the structure, including the required material and product standards; the location, type, and size of all attachments, mechanical fasteners, and welds. When shop details are required to be submitted for review and approval, only one reproducible copy of each shop detail will be submitted, unless a larger number of copies is required by the client as part of the tender specifications.

5.3.2 Shop Details Furnished by the Client. When the shop details are furnished by the client he shall deliver them in time to permit fabrication to proceed in an orderly manner according to the time schedule agreed upon. The client shall prepare these shop details, in so far as practicable, in accordance with the detailing standards of the fabricator. The client shall be responsible for the completeness and accuracy of shop details so prepared.

5.3.3 Clipped Double Connections. Where two beams or girders, framing at right angles from opposite sides of a supporting member, share the same bolts, a clipped double connection shall be used unless a seated connection or other detail is used to facilitate safe erection of the beams or girders. A clipped double connection is not applicable to a two-bolt connection or when the beams are equal to or deeper than half the depth of the girder. For a description of a clipped double connection, see Appendix A.

5.4 Erection Procedures. Erection procedures shall outline the construction methods, erection sequence, erection bracing, temporary bracing if required, and other engineering details necessary for shipping, erecting, and maintaining the stability of the steel frame. Erection procedures shall be supplemented by drawings and sketches to identify the location of stabilizing elements. Erection procedures shall be submitted for review when so specified.

5.5 Field Work Details. Field work details shall be submitted to the designer for review and approval. Field work details shall provide complete information for modifying fabricated members on the job site. All operations required to modify the member shall be shown on the field work details. If extra materials are necessary to make modifications, shop details may be required.

5.6 Review and Approval. Erection diagrams, non-standard connection design details, shop details and field work details, are normally submitted for review and approval. The fabricator includes a maximum allowance of fourteen (14) calendar days in his schedule for the return of all documents submitted for approval. Approval, by the designer, of shop details prepared by the fabricator and/or erector, indicates that the fabricator has interpreted correctly the contract requirements. Approval by the designer of shop details prepared by the fabricator does not relieve the fabricator of the responsibility for accuracy of the detail dimensions on shop details, nor of the general fit up of parts to be assembled.

The preparation of fabrication and erection documents is governed by the following sequence of procedures.

> (a) Sufficient information must be indicated in the construction drawings as stipulated in Clause 4.3 of this Code to permit the completion of fabrication and erection documents.
>
> (b) The fabrication and erection documents are prepared by skilled technicians

using industry and company standards, and represent the fabricator's interpretation of intent of the contract documents, particularly as described by the construction drawings. Connection design details are reviewed by the fabricator's technical staff prior to submission to the designer.

(c) The connection design details, shop details, erection diagrams and field work details are submitted to the designer for review and approval. Erection procedures are submitted when so specified.

(d) It is assumed by the fabricator that the fabrication and erection documents, when approved, have been reviewed by the client for accuracy in the interpretation of the contract requirements. Connection design details and shop details are reviewed and approved by the designer for structural adequacy and to ensure conformance with the loads, forces and special instructions contained in the contract documents.

(e) Shop details are prepared from the approved connection design details and erection diagrams.

If the client does not wish to review and approve the fabricator's fabrication and erection documents, the basis for interpreting the contract requirements, as well as the adequacy of connection details, is limited to the information contained in the structural steel specifications and shown on the structural drawings. This information shall be sufficient, as indicated in Clause 4.3, to permit proper execution of the work. However, the Engineer of Record is ultimately responsible for the structural integrity of the structure and the connections.

5.7 Additions, Deletions or Changes. Additions, deletions or changes, when approved, will be considered as contract revisions and constitute the client's authorization to release the additions, deletions or revisions for construction. See also Clauses 3.3 and 3.6.

6. Material, Fabrication, Inspection, Painting and Delivery

6.1 Material. Materials used by the fabricator for structural use shall conform to structural steel material standards of the Canadian Standards Association, or the American Society for Testing and Materials, or to other published material specifications, in accordance with the requirements of the construction drawings and construction specifications.

6.2 Identification

6.2.1 The method of identification stipulated in CAN/CSA-S16.1 shall form the basis for a fabricator's identification of material. Control and identification procedures may differ to some extent from fabricator to fabricator.

6.3 Preparation of Material

6.3.1 Flame cutting of structural steel may be by hand or mechanically guided means.

6.3.2 Surfaces noted as "finished" on the drawings are defined as having a roughness height rating not exceeding 500 (12.5 µm) as defined in CSA Standard B95, Surface Texture (Roughness, Waviness and Lay), unless otherwise specified. Any fabricating technique such as friction sawing, cold sawing, milling, etc., that produces such a finish may be used.

6.4 Fitting and Fastening

6.4.1 Projecting elements of connection attachments need not be straightened in the connecting plane if it can be demonstrated that installation of the connectors or fitting aids will provide adequate contact between faying surfaces.

6.4.2 When runoff tabs are used, the fabricator or erector need not remove them unless specified in the contract documents, required by the governing technical standard or the steel is exposed to view. When their removal is required, they may be hand flame-cut close to the edge of the finished member with no more finishing required, unless other finishing is specifically called for in the contract documents or governing technical standard.

6.5 Dimensional Tolerances. Tolerances on fabricated members shall be those prescribed in the applicable governing technical standard. Tolerances on steel material supplied by the fabricator shall meet those prescribed in Canadian Standards Association Standard G40.20.

6.6 Inspection of Steelwork. Should the client wish to have an independent inspection and non-destructive examination of the steelwork, he shall reserve the right to do so in the tender documents. Arrangements should be made with the fabricator for inspection of steelwork at the fabrication shop by the client's inspectors. The cost of this inspection and testing is the responsibility of the client. Inspectors are to be appointed prior to start of fabrication and client is to advise the fabricator of the arrangment made.

6.7 Surface Preparation. If paint is specified, the fabricator shall clean all steel surfaces to be painted of loose rust, loose mill scale, prominent spatter, slag or flux deposit, oil, dirt and other foreign matter by wire brushing or other suitable means. Unless specified, the fabricator shall not be obliged to blast-clean, pickle or perform any specific surface preparation operation aimed at total or near-total removal of tight mill scale, rust or non-deleterious matter.

6.8 Paint. When structural steel is specified to receive a shop coat of paint, the fabricator shall be responsible only to the extent of performing the surface preparation and painting in the specified manner. The painting requirements specified in the tender documents for the shop coat should include the identification of the members to be painted, surface preparation, paint specification, if applicable, the manufacturer's product identification, and the required minimum and maximum dry film thickness. Unless otherwise agreed upon as part of the contract documents the fabricator shall not be responsible for the deterioration of the paint that may result from exposure to the weather for more than ninety days after completion of the painting.

6.9 Marking and Shipping

6.9.1 Except for weathering steel surfaces exposed to view and for architecturally exposed steel, erection marks shall be painted or otherwise legibly marked on the members. Preferably, members which are heavy enough to require special erection equipment shall be marked to indicate the computed or scale mass and the centre of gravity for lifting.

6.9.2 Bolts of the same length and diameter, and loose nuts and washers of each size shall be packaged separately. Pins, bolts, nuts, washers and other small parts shall be shipped in boxes, crates, kegs or barrels, none of which exceed 135 kg gross mass. A list and description of material contained therein shall be marked plainly on the outside of each container.

6.9.3 When requested by the erector, long girders shall be loaded and marked so that they will arrive at the job site in position for handling without turning. Instructions for such delivery shall be given to the carrying agency when required.

6.9.4 For each shipment, the fabricator shall furnish a shipping bill listing the items in the shipment. Such bill shall show the erection mark, the approximate length, the description (whether beam, column, angle, etc.) of each item. Such bill shall be signed by the receiver and returned to the fabricator within 48 hours of receipt of the shipment with a note regarding shortages or damages, if any, and the bill shall act as a receipt for the shipment. When the shipments are made by truck transport, the bills should accompany the shipment. When shipments are made by rail or water, the bills shall be sent to the receiver to arrive on or before receipt of the shipment.

6.9.5 Unless otherwise specified at time of tender, steel during shipment will not be covered by tarpaulins or otherwise protected. When such protection is specified the shipper is to notify the carrier of the protection requirements.

6.10 Delivery of Materials

6.10.1 Fabricated structural steel shall be delivered in a sequence which will permit the most efficient and economical performance of shop fabrication and erection. If the client contracts separately for delivery and erection he must coordinate planning between the fabricator, erector and general contractor.

6.10.2 Anchor rods, washers and other anchorages, grillages, or materials to be built into masonry or concrete should be shipped so that they will be on hand when needed. The client must give the fabricator sufficient notice to permit fabrication and shipping of materials before they are needed.

6.10.3 The quantities of material shown by the shipping bill are customarily accepted by the client, fabricator and erector as correct. If any shortage or damage is claimed, the client or erector should, within 48 hours, notify the carrier and the fabricator in order that the claim may be investigated.

6.10.4 The size and mass of structural steel assemblies may be limited by the shop capabilities, the permissible mass and clearance dimensions of available transportation or government regulations and the job site conditions. The fabricator determines the number of field splices consistent with economy.

6.10.5 On supply only contracts the unloading of steel is the responsibility of others. Unless stated otherwise the unloading of steel is part of the steel erection.

7. Erection

7.1 Method of Erection. Unless otherwise specified or agreed upon, erection shall proceed according to the most efficient and economical method available to the erector on the basis of continuous operation consistent with the drawings and specifications.

7.1.1 Temporary Bracing. Temporary bracing of the steel frame shall only be removed on instruction from the Engineer of Record.

7.2 Erection Safety. Erection shall be done in a safe manner and in accordance with applicable provincial legislation.

7.3 Site Conditions. The client shall provide and maintain adequate, all-weather access roads cleared of snow and ice and other material that impedes entry into and through the site for the safe delivery of derricks, cranes, other necessary equipment, and the material to be erected. The client shall provide for the erector a firm, properly graded, drained, convenient and adequate space and laydown area for steel of sufficient load carrying capacity at the site for the operation of erection equipment and shall remove at the client's cost all overhead obstructions such as power lines, telephone lines, etc., in order to provide a safe and adequate working area for erection of the steelwork. The erector shall provide and install the safety protection required for his own operations or for his work forces to meet the safety requirements of applicable Acts or Codes. The general contractor shall install protective covers to all protruding rebar, machinery anchor rods, etc., which are a hazard to workers. Any protection for pedestrians, property, other trades, etc., not essential to the steel erection activity is the responsibility of the client. When the structure does not occupy the full available site, the client shall provide adequate storage space to enable the fabricator and erector to operate at maximum practicable speed and efficiency. Cleaning of steelwork required because of site conditions, mud, site worker traffic, etc., shall not be to the fabricator's/erector's account.

7.4 Foundations. The tender specifications preferably shall specify the time that foundations will be ready, free from obstruction and accessible to the erector. Unless otherwise agreed upon, the work of erection shall be tendered on the basis that it will start at the time designated in the tender specifications without interference or delay caused by others. Neither the fabricator nor the erector shall be responsible for the accurate location, strength and suitability of foundations.

7.5 Bearing Surfaces.

Levelling plates shall be set by others true, level and to the correct elevation.

7.6 Building Lines and Bench Marks. The erector shall be provided with a plot plan accurately locating building lines and bench marks at the site of the structure.

7.7 Installation of Anchor Rods and Embedded Items

7.7.1 Anchor rods and foundation rods shall be set by the client in accordance with the erection diagrams. They must not vary from the dimensions shown on the erection diagrams by more than the following (see also Appendix D):

- (a) 3 mm centre to centre of any two rods within an anchor rod group, where an anchor rod group is defined as the set of anchor rods which receives a single fabricated steel shipping piece;

- (b) 6 mm centre-to-centre of adjacent anchor rod groups;

- (c) Maximum accumulation of 6 mm per 30 000 mm along the established column line of multiple anchor rod groups, but not to exceed a total of 25 mm. The established column line is the actual field line most representative of the centres of the as-built anchor rod groups along a line of columns;

- (d) 6 mm from the centre of any anchor rod group to the established column line through that group.

- (e) Shims: the finished tops of all footings shall be at the specified level which will not exceed the maximum specified grouting allowance to predetermine the amount of shimming that will be required.

The tolerances of paragraphs b,c, and d also apply to offset dimensions, shown on the construction drawings, measured parallel and perpendicular to the nearest established column line for individual columns shown on the drawings to be offset from established column lines.

7.7.2 Unless shown otherwise, anchor rods shall be set perpendicular to the theoretical bearing surface, threads shall be protected, free of concrete, and nuts should run freely on the threads. Shear pockets shall be cleaned of debris, formwork, ice and snow by the client prior to steel erection.

7.7.3 Other embedded items or connection materials between the structural steel and the work of others shall be located and set by the client in accordance with approved erection diagrams. Accuracy of these items must satisfy the erection tolerance requirements of Clause 7.12.

7.7.4 All work performed by the client shall be completed so as not to delay or interfere with the erection of the structural steel.

7.8 Bearing Devices. The client shall set to lines and grades all levelling plates and loose bearing plates which can be handled without a derrick or crane. All other bearing devices supporting structural steel shall be set and wedged, shimmed or adjusted with levelling screws by the erector to lines and grades established by the client. The fabricator and/or erector shall provide the wedges, shims or levelling screws that are required, and shall scribe clearly the bearing devices with working lines to facilitate proper alignment. Promptly after the setting of any bearing devices, the client shall check lines and grades, and grout as required. The final location and proper grouting of bearing devices are the responsibility of the client.

When steel columns, girders or beams which will be supported on concrete or masonry have base plates or bearing plates fabricated as an integral part of the member, the bearing area of the support shall be suitably prepared by others so as to be at exact grade and level to receive the steelwork.

7.9 Examination by Erector. Prior to field erection, the erector shall do a random check to examine the work of all others on which his work is in any way dependent and shall report to the client any errors or discrepancies as discovered that may affect erection of structural steel before or during erection. The accurate placement and integrity of all anchor rods/embeddment etc., remains the responsibility of the client.

7.10 Adjustable Shelf Angles and Sash Angles. The erector shall position at time of erection all adjustable shelf angles and sash angles attached to the steel frame true and level within the tolerances permitted by the governing technical standard. Any subsequent adjustment that may be necessary to accommodate the work of others shall be performed by others.

7.11 Loose Lintels and Wall Bearing Members. Unless otherwise specified, loose lintels, shelf angles, wall bearing members and other pieces not attached to the structural steel frame shall be received and set by others.

7.12 Tolerances. Unless otherwise specified, tolerances on erected structural steel shall be those prescribed in the applicable governing technical standard.

7.13 Checking Erected Steelwork. Prior to placing or applying any other materials, the owner is responsible for determining that the location of the structural steel is acceptable for plumbness, level and alignment within tolerances with bolts correctly installed and welds inspected. The erector is given timely notice of acceptance by the owner or a listing of specific items to be corrected in order to obtain acceptance. Such notice is rendered immediately upon completion of any part of the work and prior to the start of work by other trades that may be supported, attached or applied to the structural steelwork.

7.14 Removal of Bracing.

7.14.1 Removal of Erection Bracing. Guys, braces and falsework or cribbing supplied by the erector shall remain the property of the erector. The erector shall remove them when the steel structure is otherwise adequately braced unless other arrangements are made. Guys and braces temporarily left in place under such other arrangements shall be removed by others provided prior permission by the erector for their removal has been given, and returned to the erector in good condition, see Clause 7.14.2

7.14.2 Removal of Temporary Bracing. Temporary bracing required by the designer shall only be removed on instruction from the Engineer of Record.

7.15 Correction of Errors when Material is Not Erected by the Fabricator. Correction of minor misfits and a moderate amount of cutting, welding, and reaming shall be considered a part of the erection; in the same manner as if the Fabricator would be erecting the work. Any major rework required due to incorrect shop work shall be immediately reported to the Fabricator, before rework commences. The Fabricator shall then either correct the error, resupply the item within a reasonable time period, or approve the method of correction including applicable costs, whichever is the most economical.

7.16 Field Assembly. Unless otherwise specified, the fabricator shall provide for suitable field connections that will, in his opinion, afford the greatest overall economy.

7.17 Accommodation of Other Trades. Neither the fabricator nor the erector shall cut, drill or otherwise alter the work of others or his own work to accommodate other trades unless such work is clearly defined in the tender drawings and tender specifications and unless detailed information is provided before the erection diagrams are approved. Any subsequent cutting, drilling or other alteration of the structural steel performed by the fabricator or the erector for the accommodation of other trades, shall be specifically agreed upon and authorized by the client before such work is commenced.

7.18 Temporary Floors and Access Stairs. Unless otherwise required by law, all temporary access stairs shall be provided by others, except for the floor upon which erecting equipment is located. On this floor the erector shall provide such temporary flooring as he requires, moving his planking, etc., as the work progresses.

7.19 Touch-Up of Shop Paint. Unless so specified, the fabricator/erector will not spot-paint field fasteners and field welds nor touch-up abrasions to the shop paint.

7.20 Final Painting. Unless so specified, the fabricator/erector will not be responsible for cleaning the steel after erection in preparation for field painting, nor for any general field painting that may be required.

7.21 Final Clean-Up. Except as provided in Clause 7.14, upon completion of erection and before final acceptance, the erector shall remove all falsework, rubbish and temporary building furnished by him.

APPENDIX A - Structural Steel in Buildings

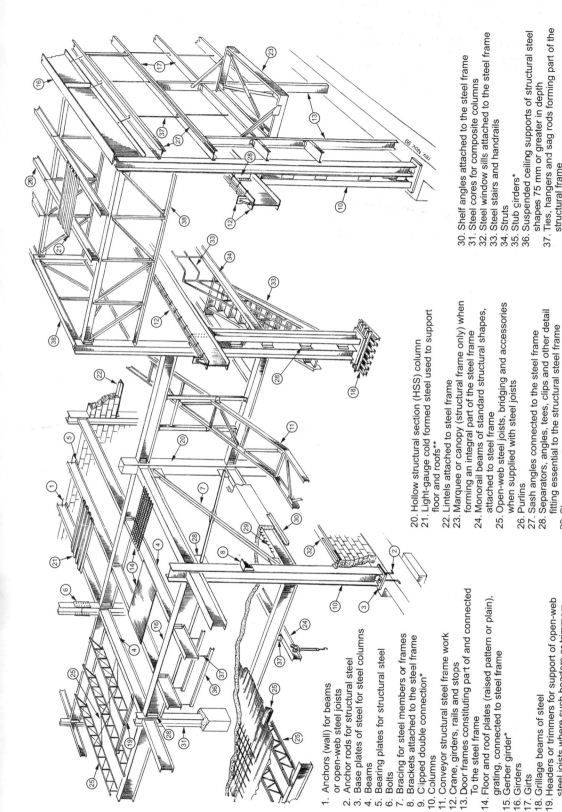

1. Anchors (wall) for beams or open-web steel joists
2. Anchor rods for structural steel
3. Base plates of steel for steel columns
4. Beams
5. Bearing plates for structural steel
6. Bolts
7. Bracing for steel members or frames
8. Brackets attached to the steel frame
9. Clipped double connection*
10. Columns
11. Conveyor structural steel frame work
12. Crane, girders, rails and stops
13. Door frames constituting part of and connected to the steel frame
14. Floor and roof plates (raised pattern or plain), grating, connected to steel frame
15. Gerber girder*
16. Girders
17. Girts
18. Grillage beams of steel
19. Headers or trimmers for support of open-web steel joists where such headers or trimmers frame into structural steel members
20. Hollow structural section (HSS) column
21. Light-gauge cold formed steel used to support floor and roofs**
22. Lintels attached to steel frame
23. Marquee or canopy (structural frame only) when forming an integral part of the steel frame
24. Monorail beams of standard structural shapes, attached to steel frame
25. Open-web steel joists, bridging and accessories when supplied with steel joists
26. Purlins
27. Sash angles connected to the steel frame
28. Separators, angles, tees, clips and other detail fitting essential to the structural steel frame
29. Shear connectors
30. Shelf angles attached to the steel frame
31. Steel cores for composite columns
32. Steel window sills attached to the steel frame
33. Steel stairs and handrails
34. Struts
35. Stub girders*
36. Suspended ceiling supports of structural steel shapes 75 mm or greater in depth
37. Ties, hangers and sag rods forming part of the structural frame
38. Trusses and brace frames

*see separate diagram

**supplied by others

APPENDIX A - Structural Steel in Buildings (continued)

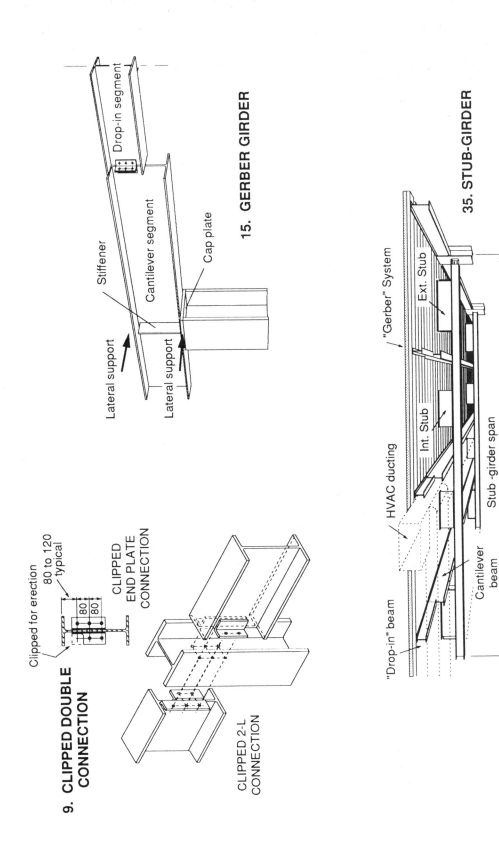

APPENDIX B
Guideline for Unit Price Application for Changes

1. Unit rates shall apply on their own, only up until commencement of material order or shop detail drawings, whichever is the earlier.

2. The following amounts, additional to the unit rate, shall be charged on additions at the various stages of the contract.

    (a) If the addition affects drawings (e.g. of support members) already in progress or complete, then the changes to such drawings or redetailing shall be charged extra at an agreed hourly rate.

    (b) If the addition requires additional work to material manufacture or erection (e.g. supporting members) in progress or complete, then such additional work shall be charged extra at an agreed hourly rate.

    (c) "Detail" or "Connection" materials added to existing or supporting members, whether due to an additional member or not, shall be charged on a cost plus basis.

    (d) If the timing of the addition causes the added material to be shipped as a part load, then transportation shall be charged extra at cost plus an agreed percentage markup.

3. The following amounts, additional to the unit rate, shall be charged for deletions at the various stages of the contract.

    (a) If the deleted material has been ordered or delivered and cannot be used elsewhere then a restocking charge shall be levied.

    (b) If the deleted member has been detailed or drawings are in progress, then the cost of such drawings shall be charged extra at an agreed hourly rate.

    (c) If the deletion affects drawings already completed or in progress, then the changes to such drawings or the redetailing shall be charged extra at an agreed hourly rate.

    (d) If the deleted member has been manufactured or erected or manufacture or erection is in progress, then the cost of such manufacture or erection shall be charged extra at an agreed hourly rate.

    (e) If the deletion affects members already manufactured (e.g. supporting members) then the changes to such members shall be charged extra at an agreed hourly rate.

    (f) If the deleted member has already been shipped, then no credit shall be given.

4. All unit rates shall be applied in accordance with CISC Code of Standard Practice Clause 3.5.

5. Hourly Rates are as follows:

    (a) Engineering Design - $ /labour hour
    (b) Drawing Office Labour - $ /labour hour
    (c) Shop Labour - $ /labour hour
    (d) Field Labour - $ /labour hour

(e) Equipment used for revisions will be charged at negotiated rental rates, according to Canadian Construction Association standard practice.

6. Revisions involving the use of grades of steel, sources of supply, or types of sections, other than specified, will be subject to price adjustments.

7. Units will be computed in accordance with Clause 3.5 of the CISC Code of Standard Practice for Structural Steel.

APPENDIX C
A Suggested Format for Price Per Unit Contracts

The following is a list of suggested categories for which unit prices could be tendered, *such categories being selected or added to,* depending upon the nature of the project.

For payment purposes, the connection material required to connect an individual member to its supporting member is assumed to be part of the member to which it is attached for shipping purposes.

A. STRUCTURAL STEEL

1. COLUMNS
   (a) Rolled Shapes
       1. Up to and including 30 kg/m
       2. Over 30 — up to and including 60 kg/m
       3. Over 60 — up to and including 90 kg/m
       4. Over 90 — up to and including 150 kg/m
       5. Over 150 kg/m
   (b) WWF or Plate Fabricated
       1. Up to and including 90 kg/m
       2. Over 90 — up to and including 150 kg/m
       3. Over 150 kg/m
   (c) Hollow Structural Sections
       1. Up to and including 30 kg/m
       2. Over 30 — up to and including 60 kg/m
       3. Over 60 — up to and including 90 kg/m
       4. Over 90 — up to and including 150 kg/m
       5. Over 150 kg/m

2. BEAMS
   (a) Rolled Shapes
       1. Up to and including 30 kg/m
       2. Over 30 — up to and including 60 kg/m
       3. Over 60 — up to and including 90 kg/m
       4. Over 90 — up to and including 150 kg/m
       5. Over 150 kg/m
   (b) WWF or Plate Fabricated
       1. Up to and including 90 kg/m
       2. Over 90 — up to and including 150 kg/m
       3. Over 150 kg/m
   (c) Stud Shear Connectors
       ___diam. ___mm long

3. CRANE RAILS
   (a) Rail complete with Rail Clips
       1. Up to and including 30 kg/m
       2. Over 30 kg/m
   (b) Stops
   (c) Monorails

1. Straight
2. Curved

4.1 TRUSSES - where principal members are:
   (a) Tees, Angles or W Shapes
   (b) Hollow Structural Sections

4.2 OPEN WEB STEEL JOISTS

5. BRACING
   (a) Rolled Shapes
      1. Up to and including 30 kg/m
      2. Over 30 kg/m
   (b) Hollow Structural Sections
      1. Up to and including 30 kg/m
      2. Over 30 kg/m
   (c) WT Sections
      1. Up to and including 30 kg/m
      2. Over 30 kg/m

6. PURLINS, GIRTS, and SAG RODS
   (a) Rolled Shapes
      1. Up to and including 30 kg/m
      2. Over 30 kg/m
   (b) Cold-Formed Sections
      1. Up to and including 5.75 kg/m
      2. Over 5.75 kg/m
   (c) Hollow Structural Sections
      1. Up to and including 30 kg/m
      2. Over 30 kg/m
   (d) Sag Rods

7. FRAMING
   Wall and Roof Openings

8. STAIR STRINGERS

9. LADDERS (Galvanized or Painted)
   (a) Without safety cage
   (b) With safety cage

10. UTILITY BRIDGES (Plus Interior Framing)
   (a) Tees, Angles or W Shapes
   (b) Hollow Structural Sections

11. CONVEYOR GALLERIES
   (a) Tees, Angles or W Shapes
   (b) Hollow Structural Sections
   (c) Open Conveyor Trusses

12. MISCELLANEOUS PLATFORMS (Tees, Angles or W Shapes)
      1. Up to and including 30 kg/m
      2. Over 30 kg/m

B. MISCELLANEOUS STEEL & GRATING

1. **GRATING**
    (a) Floors
        1. Galvanized
        2. Painted
    (b) Stair Landings
        1. Galvanized
        2. Painted

2. **CHECKERED PLATE (6mm thick)**
    (a) Steel – Galvanized
    (b) Steel – Painted

3. **STAIR TREADS**
    (Maximum 1000 mm long)
    (a) Grating – Galvanized
    (b) Grating – Painted

4. **HANDRAIL (without kickplate)**
    (a) Horizontal
        1. Steel – Galvanized
        2. Steel – Painted
    (b) Sloping
        1. Steel – Galvanized
        2. Steel – Painted

5. **KICKPLATE**
    (a) Plate attached to handrail
        1. Steel – Galvanized
        2. Steel – Painted
    (b) Banding - attached to grating
        1. Steel – Galvanized
        2. Steel – Painted

C. MOBILIZATION AND DEMOBILIZATION

1. Mobilization for erection and demobilization

D. UNIT RATES FOR EXTRA WORK

1. For extra engineering design _____/hr.
2. For extra detailing work _____/hr. all inclusive (composite rate)
3. For extra shop labour _____/hr. all inclusive (composite rate)
4. For extra field erection _____/hr. all inclusive (composite rate)
5. Field labour work week _____/hrs. per week

NOTE: "All inclusive" shall mean, all labour cost including overheads and profits and in the field all small tools (up to $1000 value).

APPENDIX D
Tolerances on Anchor Rod Placement

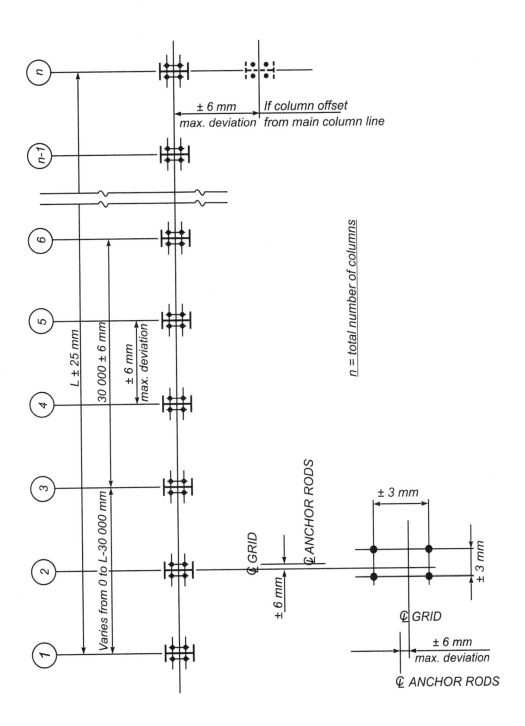

7-31

APPENDIX E
Conversion of SI Units to Imperial Units

When Imperial units are used in contract documents, unless otherwise stipulated, the SI units used in the CISC Code of Standard Practice for Structural Steel shall be replaced by the Imperial units shown, for the clause as noted.

Clause 3.5(a). Unit Weight. The unit weight of steel is assumed to be 0.2833 pounds per cubic inch.

For other clauses, the standard conversion factors (for length, mass, etc.) stipulated in CISC handbooks should be used.

Note: Imperial projects should be entirely in the imperial designation including shape sizes. Metric projects should be entirely in the S.I. designation, including, shape sizes. Units should not be intermixed on the same project.

APPENDIX F
Miscellaneous Steel

Miscellaneous Steel. Unless otherwise specified in the tender documents, the following items are considered miscellaneous steel of ferrous metal only, fabricated from 2.0 mm (14 ga.) and more of metal, including galvanizing, cadmium and chrome plating, but not stainless steel and cast iron items.

This list of items is to be read in conjunction with Clause 2.1 Structural Steel and Clause 2.3 Items Supplied by Others, and shall include all steel items not included in Clauses 2.1 and 2.3 unless specified otherwise.

- Access doors and frames — except trade-name items and those required for servicing mechanical and electrical equipment.
- Angles and channel frames for doors and wall openings — drilling and tapping to be specified as being done by others.
- Benches and brackets.
- Bolts — only includes those bolts and anchors required for anchoring miscellaneous steel supplied under this list.
- Bumper posts and rails.
- Burglar/security bars.
- Clothes line poles, custom fabricated types only.
- Coat rods, custom fabricated types only.
- Corner protection angles.
- Expansion joint angles, plates custom fabricated, etc., including types made from steel, or a combination of steel and non-ferrous metal.
- Fabricated convector frames and enclosures.
- Fabricated items where clearly detailed or specified and made from 2.0 mm (14ga.) and heavier steel, except where included in another division.
- Fabricated steel framing for curtain walls and storefronts where not detailed on structural drawings and not enclosed by architectural metal.
- Fabricated wire mesh and expanded metal partitions and screens.
- Fire escapes.
- Flag poles — steel custom fabricated. (Excluding hardware)
- Custom fabricated footscrapers, mud and foot grilles, including pans, but less drains.
- Frames, grating and plate covers for manholes, catch basins, sumps, trenches, hatches, pits, etc., except cast iron, frames and covers and trade-name floor and roof drains.
- Gates, grilles, grillwork and louvres, excluding baked enamel or when forming part of mechanical system.
- Grating type floors and catwalks — excluding those forming part of mechanical system.
- Handrails, balusters and any metal brackets attached to steel rail including plastic cover, excluding steel handrails forming part of structural steel framing.
- Joist hangers, custom fabricated types only.
- Joist strap anchors.
- Lintels, unless shown on structural drawings.
- Mat recess frames, custom fabricated types only.
- Mobile chalk and tackboard frames, custom fabricated types only.
- Monorail beams of standard shapes, excluding trade name items, unless shown on structural drawings.
- Shop drawings and/or erection diagrams.

Shop preparation and/or priming.

Sleeves if specified, except for mechanical and electrical division.

Stair nosings, custom fabricated types only.

Steel ladders and ladder rungs not forming part of structural steel or mechanical work.

Steel stairs and landings not forming part of structural steel.

Table and counter legs, frames and brackets, custom fabricated types only.

Thresholds and sills, custom fabricated types only.

Vanity and valance brackets, custom fabricated types only.

Weatherbars — steel.

Miscellaneous Steel Items Excluded

Bases and supports for mechanical and electrical equipment where detailed on mechanical or electrical drawings.

Bolts other than for anchoring items of miscellaneous steel.

Cast iron frames and covers for manhole and catch basins.

Chain link and woven wire mesh.

Glulam connections and anchorages.

Joist hangers, trade-name types

Metal cladding and covering, less than 2.0 mm (14 ga.)

Precast concrete connections and anchorages in building structure.

Reinforcing steel or mesh.

Roof and floor hatches when trade-name items.

Sheet metal items, steel decking and siding and their attachments, closures, etc., less than 2.0 mm (14 ga.)

Shoring under composite floors and stub-girders.

Steel stacks.

Steel reinforcement for architectural metal storefronts, curtainwalls and windows.

Stud shear connectors when used with steel deck.

Stone anchors.

Temporary bracing for other trades.

Thimbles and breeching, also mechanical fire dampers.

Window and area wells.

When miscellaneous steel fabricator erects miscellaneous steel, all material required for temporary and/or permanent connections of the component parts of the miscellaneous steel shall be supplied.

APPENDIX G
A Suggested Format for a
Monthly Progress Payment Claim Form

MONTHLY PROGRESS CLAIM FORM

PROJECT: _____
CONTRACT NO: _____
PROGRESS CLAIM NO: _____
DATE: _____

_____ (FIRM NAME)

ITEM		ORIGINAL BASE CONTRACT	APPROVED CHANGES TO-DATE	REVISED BASE CONTRACT	PROGRESS TO-DATE	PREVIOUS AMOUNT CLAIMED	THIS PROGRESS CLAIM	% COMPLETE
1.	ENGINEERING & DRAWING PREPARATION							
2.	RAW MATERIALS IN YARD							
3.	FABRICATION							
4.	FREIGHT TO SITE							
5.	ERECTION							
6.	PLUMB / BOLT / CLEAN UP							
7.	TOTAL GROSS AMOUNT							
8.	HOLDBACK ____ %							
9.	NET AMOUNT							
10.	G S T ____ % OF LINE 9							
11.	TOTAL AMOUNT DUE							

APPROVED CHANGE ORDER(S) TO-DATE: _____

Appendix H
Suggested Definitions for Progress Invoicing and Substantial Performance

1. **Progress Invoicing**

    (a) The submission of erection diagrams and/or shop details will initiate progress invoicing. Payments of invoices will be in accordance with the terms of the contract.

    (b) Any and all materials fabricated will initiate additional progress invoicing with payment as per contract.

    (c) Any and all materials shipped to the site and/or erected in place will initiate additional progress invoicing with payment as per contract.

    (d) Final invoicing will be made after all steel has been delivered to site, erected and all work completed.

    (e) Substantial completion is based on the completion of the WORK of the steel Fabricator or Erector and therefore release of holdback will be 41 days (or as per applicable lien legislation) after issuance of the Certificate of Completion of this work.

2. **Substantial Performance**

    (a) The word "WORK" shall be defined as the product or services provided by the steel Fabricator or Erector.

    (b) Substantial performance and total performance shall be directly related to product or services provided by the steel Fabricator or Erector.

    (c) Certification of completion by the owner's representative applies to the work performed by the steel Fabricator or Erector.

    (d) Payment shall be governed by certified completion of the work.

STRUCTURAL SHEET STEEL PRODUCTS

General

Structural sheet steel products such as roof deck, floor deck and cladding complement the structural steel frame of a building. These large-surface elements often

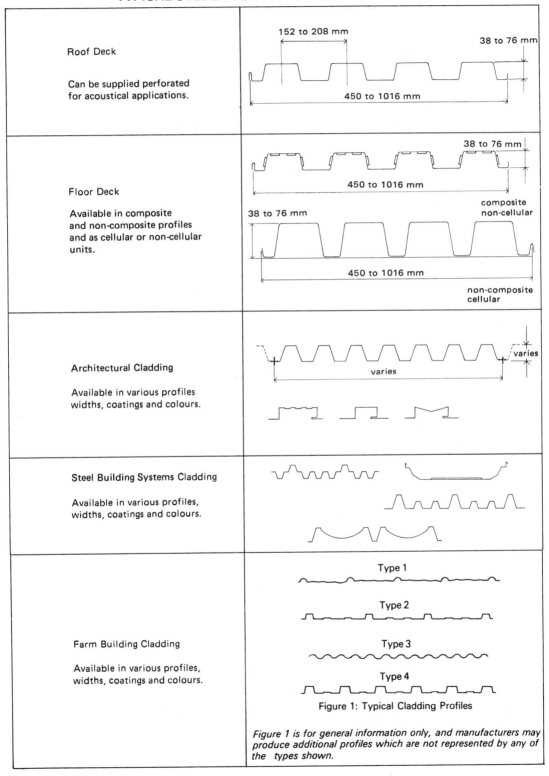

Figure 1: Typical Cladding Profiles

Figure 1 is for general information only, and manufacturers may produce additional profiles which are not represented by any of the types shown.

perform both structural and non-structural functions, thereby enhancing the overall economy of the design.

Many of the sheet steel products used in Canada are supplied by members of the Canadian Sheet Steel Building Institute, a national association of steel producers, zinc producers, coil coaters, fastener manufacturers and fabricators of steel building products, steel building systems and lightweight steel framing components. The Institute promotes the use of sheet steel in building construction by encouraging good design, pleasing form and greater economy.

Sheet steel materials for building construction are metallic coated (zinc or aluminum-zinc alloy) and can be prefinished for extra corrosion protection and aesthetics. Consult fabricators' catalogues for details of available products, profiles, widths, lengths, thicknesses, load capacities and other characteristics. The table below relates base steel thickness increments with the corresponding overall zinc coated thickness for various zinc coating designations applicable to structural quality sheets. For structural design calculations the base steel design thickness is used.

METALLIC COATED STRUCTURAL QUALITY SHEET STEEL THICKNESSES

Base Steel Nominal Thickness[1], mm	Overall Metallic Coated Nominal Thickness, mm					
	Metallic Coating Designation[2]					
	ZF75[3]	Z275 AZ150	Z350[4] AZ180	Z450[4]	Z600[4]	Z700[4]
2.67	2.67	2.71	2.72	2.74	2.76	2.77
1.91	1.91	1.95	1.96	1.98	2.00	2.01
1.52	1.52	1.56	1.57	1.59	1.61	1.62
1.22	1.22	1.26	1.27	1.29	1.31	1.32
0.91	0.91	0.95	0.96	0.98	1.00	1.01
0.76	0.76	0.80	0.81	0.83	0.85	0.86
0.61	0.61	0.65	0.66	0.68	0.70	0.71
0.46	0.46	0.50	0.51	0.53	0.55	0.56

Notes:
1. Base steel thickness is used to establish section properties and for structural design calculations.
2. The listed metallic coating designations apply to metric material from Canadian producers.
3. The small thickness increment for ZF75 (wiped coat) is usually disregarded.
4. Enquire as to delivery date, if time is critical.
■ Enquire as to availability of thickness.

CSSBI PUBLICATIONS

CSSBI publications include industry product standards, informational bulletins and special publications as well as non-technical promotional material. A selection of current publications is listed below.

CSSBI Standards

Steel Roof Deck — covers design, fabrication and erection of steel roof deck with flutes not more than 200 mm on centre and a nominal 77 mm maximum profile depth, intended for use with built-up roofing or other suitable weather-resistant cover on top of the deck. (CSSBI 10M)

Composite Steel Deck — covers design, fabrication and erection of composite steel deck with a nominal 77 mm maximum profile depth, intended for use with a concrete cover slab on top of the deck to create a composite slab. (CSSBI 12M)

Sheet Steel Cladding for Architectural/Industrial Applications — covers design, fabrication and erection of weather-tight wall and roof cladding made from metallic coated, prefinished sheet steel for use on buildings with low internal humidity. (CSSBI 20M)

Steel Building Systems — covers the design, fabrication and erection of steel building systems (SBS). Includes definitions, classification of SBS by type, checklist of items normally furnished, criteria for load combinations, design standards, and certification by a registered engineer. (CSSBI 30M)

Steel Farm Cladding — covers the manufacture, load carrying capacity, handling and installation of sheet steel cladding intended for application to walls and/or roofs of farm buildings. (CSSBI 21M)

Bulletins and Special Publications

Criteria for the Testing of Composite Slabs — provides the criteria for conducting a series of shear-bond tests necessary to determine the structural capacity of a composite slab. (CSSBI S2)

Criteria for the Design of Composite Slabs — contains design criteria, based on limit states design, for composite slabs made of a structural concrete placed permanently over a composite steel deck. (CSSBI S3)

Design of Steel Deck Diaphragms — offers a simple and practical approach to the design of steel deck diaphragms supported by horizontal steel framing. (CSSBI B13)

Criteria for the Design and Installation of Double Skin Insulated Steel Roofs — provides guidelines on good design and installation practices for roof systems consisting of a steel liner and profiled exterior sheet separated by sub-purlins and containing thermal insulation. (CSSBI B11)

Wind, Snow and Earthquake Load Design Criteria for Steel Building Systems — illustrates NBCC roof snow load, wind load and earthquake load provisions for the design of all structural components of steel building systems. (CSSBI B14)

Lightweight Steel Framing Manual — is an introduction to lightweight steel framing (LSF) and describes the applications for LSF in floors, roofs, curtain walls and axial load bearing walls. A suggested Guide Specification is also included. (CSSBI 50M)

Lightweight Steel Framing Design Manual — shows through examples how to design lightweight steel framing structural systems. Detailed calculations are shown for curtain walls, infill walls, and axial load bearing systems as well as all connections. (CSSBI 51M)

Low-Rise Residential Construction Details — provides typical framing and connection details needed for the construction of residential buildings using lightweight steel framing. (CSSBI 53)

Introduction to Residential Steel Framing — describes the application of lightweight steel framing to residential construction. Steel studs, joists, rafters and trusses can be used instead of wood for the construction of traditional framed house construction. This publication introduces the concept and describes some of the applications. (CSSBI 54)

Contact CSSBI at the address below for a complete listing of publications, copies of publications, or other information concerning sheet steel in construction.

Canadian Sheet Steel Building Institute
652 Bishop St. N., Unit 2A
Cambridge, Ontario N3H 4V6
Tel (519) 650-1285
Fax (519) 650-8081
Internet Web Site: www.cssbi.ca

MASS AND FORCES FOR MATERIALS

MATERIAL	Mass (kg/m³)	Force (kN/m³)	MATERIAL	Mass (kg/m³)	Force (kN/m³)
METALS, ALLOYS, ORES			**TIMBER, AIR-DRY**		
Aluminum	2 640	25.9	Birch	689	6.76
Brass	8 550	83.8	Cedar	352	3.45
Bronze, 7.9-14% tin	8 150	79.9	Fir, Douglas, seasoned	545	5.34
Bronze, aluminum	7 700	75.5	Fir, Douglas, unseasoned	641	6.29
Copper	8 910	87.4	Fir, Douglas, wet	801	7.86
Copper ore, pyrites	4 200	41.2	Fir, Douglas, glue laminated	545	5.34
Gold	19 300	189	Hemlock	481	4.72
Iron, cast, pig	7 210	70.7	Larch, tamarack	561	5.50
Iron, wrought	7 770	76.2	Larch, western	609	5.97
Iron, spiegel-eisen	7 500	73.5	Maple	737	7.23
Iron, ferro-silicon	7 000	68.6	Oak, red	689	6.76
Iron ore, hematite	5 210	51.1	Oak, white	753	7.38
Iron ore, hematite in bank	2 560-2 880	25.1-28.2	Pine, jack	481	4.72
Iron ore, hematite, loose	2 080-2 560	20.4-25.1	Pine, ponderosa	513	5.03
Iron ore, limonite	3 800	37.3	Pine, red	449	4.40
Iron ore, magnetite	5 050	49.5	Pine, white	416	4.08
Iron slag	2 760	27.1	Poplar	481	4.72
Lead	11 400	112	Spruce	449	4.40
Lead ore, galena	7 450	73.1	For pressure treated timber		
Magnesium	1 790	17.6	add retention to mass of		
Manganese	7 610	74.6	air-dry material.		
Manganese ore	4 150	40.7			
Mercury	13 600	133	**LIQUIDS**		
Monel	8 910	87.4	Alcohol, pure	785	7.70
Nickel	9 050	88.8	Gasoline	673	6.60
Platinum	21 300	209	Oils	929	9.11
Silver	10 500	103	Water, fresh at 4°C (max.		
Steel, rolled	7 850	77.0	density)	1 000	9.81
Tin	7 350	72.1	Water, fresh at 100°C	961	9.42
Tin ore, cassiterite	6 700	65.7	Water, salt	1 030	10.1
Zinc	7 050	69.1			
Zinc ore, blende	4 050	39.7	**EARTH, ETC. EXCAVATED**		
			Earth, wet	1 600	15.7
MASONRY			Earth, dry	1 200	11.8
Ashlar	2 240-2 560	22.0-25.1	Sand and gravel, wet	1 920	18.8
Brick, soft	1 760	17.3	Sand and gravel, dry	1 680	16.5
Brick, common	2 000	19.6			
Brick, pressed	2 240	22.0	**VARIOUS BUILDING**		
Clay tile, average	961	9.42	**MATERIALS**		
Rubble	2 080-2 480	20.4-24.3	Cement, portland, loose	1 510	14.8
Concrete, cinder, haydite	1 600-1 760	15.7-17.3	Cement, portland, set	2 930	28.7
Concrete, slag	2 080	20.4	Lime, gypsum, loose	849-1 030	8.33-10.1
Concrete, stone	2 310	22.7	Mortar, cement-lime, set	1 650	16.2
Concrete, stone, reinforced	2 400	23.5	Quarry stone, piled	1 440-1 760	14.1-17.3
SOLID FUELS			**MISCELLANEOUS**		
Coal, anthracite, piled	753-929	7.38-9.11	Asphaltum	1 300	12.7
Coal, bituminous, piled	641-865	6.29-8.48	Tar, bituminous	1 200	11.8
Coke, piled	368-513	3.61-5.03	Glass, common	2 500	24.5
Charcoal, piled	160-224	1.57-2.20	Glass, plate or crown	2 580	25.3
Peat, piled	320-416	3.14-4.08	Glass, crystal	2 950	28.9
			Paper	929	9.11
ICE AND SNOW					
Ice	897	8.80			
Snow, dry, fresh fallen	128	1.26			
Snow, dry, packed	192-400	1.88-3.92			
Snow, wet	432-641	4.24-6.29			

DESIGN DEAD LOAD (kPa) OF MATERIALS

STEEL DECKS	
Steel deck* 38 mm deep	
(up to 0.91 mm thick)	0.10
(1.22 to 1.52 mm thick)	0.15
Steel deck* 76 mm deep (Narrow-Rib)	
(up to 0.91 mm thick)	0.15
(1.22 to 1.91 mm thick)	0.30
Steel deck* 76 mm deep (Wide-Rib)	
(up to 0.91 mm thick)	0.10
(1.22 to 1.52 mm thick)	0.15
* for cellular deck, add	0.08
CONCRETE, per 100 mm	
- 2350 kg/m^3 (N.D.)	2.31
- 2000 kg/m^3 (slag aggregate)	1.96
- 1850 kg/m^3 (S.L.D.)	1.82
HOLLOW CORE PRECAST (no topping)	
- 200 mm deep (N.D.)	2.60
- 300 mm deep (N.D.)	3.50
WOOD JOISTS (at 400 mm centres)	
- 38 mm x 184 mm joists	0.09
- 38 mm x 235 mm joists	0.12
- 38 mm x 286 mm joists	0.14
PLYWOOD	
- 11 mm thick	0.06
- 14 mm thick	0.08
- 19 mm thick	0.11
CHIPBOARD	
- 12.7 mm thick	0.07
- 15.9 mm thick	0.09
- 19.0 mm thick	0.11
WALLS AND CLADDING	
- Solid brick wall (concrete)	
- 100 mm thick (S.L.D.)	1.40
- 100 mm thick (N.D.)	1.90
- Hollow block (S.L.D.)	
- 100 mm thick	1.10
- 200 mm thick	1.60
- 300 mm thick	2.30
- Hollow block (N.D.)	
- 100 mm thick	1.40
- 200 mm thick	2.10
- 300 mm thick	2.90
- P.C. wall plus glazing	2.40 - 3.80
- Metal curtain wall	0.74 - 1.50
- Insulated sheet steel wall	
(exclude girts)	0.25 - 0.40
- 38 x 89 wood studs @ 400 mm	0.05
- Gypsum wallboard per 10 mm	0.08
- Stone veneer per 25 mm	0.40

FLOOR FINISHING	
- Vinyl, linoleum or asphalt tile	0.07
- Softwood subfloor per 10 mm	0.06
- Hardwood per 10 mm	0.08
- Carpeting	0.10
- Asphaltic concrete per 10 mm	0.23
- 20 mm Ceramic or quarry tiles on	
12 mm mortar bed	0.80
- Terrazzo per 10 mm	0.24
- Mastic floor (20 mm)	0.45
ROOFING	
- 3 ply asphalt, no gravel	0.15
- 4 ply asphalt, no gravel	0.20
- 3 ply asphalt and gravel	0.27
- 4 ply asphalt and gravel	0.32
- Asphalt strip shingles	0.15
- Gypsum wallboard per 10 mm	0.08
INSULATION (per 100 mm thick)	
- Glass fibre, batts	0.05
- Glass fibre, blown	0.04
- Glass fibre, rigid	0.07
- Urethane, rigid foam	0.03
- Insulating concrete	0.06
CEILINGS	
- Gypsum wallboard per 10 mm	0.08
- Tiled ceiling & suspension system,	
with fixtures, average	0.20
- 20 mm plaster on lath/furring	0.40
- Sprayed fire protect'n, aver'g	0.07
- Ducts/pipes/wiring allowance	0.25
(average condition)	
DECK-SLABS (average condition)	
- 38 mm deck with	
- 65 mm N.D. cover#	1.95
- 90 mm N.D. cover#	2.55
- 65 mm S.L.D. cover##	1.55
- 85 mm S.L.D. cover##	1.90
- 75 mm (or 76 mm) "wide-rib" deck with	
- 65 mm N.D. cover#	2.55
- 90 mm N.D. cover#	3.15
- 65 mm S.L.D. cover##	2.15
- 85 mm S.L.D. cover##	2.50
- 76 mm "narrow-rib" deck with	
- 65 mm N.D. cover#	2.20
- 90 mm N.D. cover#	2.80
- 65 mm S.L.D. cover##	1.90
- 85 mm S.L.D. cover##	2.25
# assume 2350 kg/m^3 concrete	
## assume 1850 kg/m^3 concrete	

M / D RATIOS

Designation	SI (kg/m)/m		Imperial (lb./ft.)/in.		Designation	SI (kg/m)/m		Imperial (lb./ft.)/in.	
	Beam[1]	Column[2]	Beam[1]	Column[2]		Beam[1]	Column[2]	Beam[1]	Column[2]
WWF2000					WWF800				
x732	130		2.23		x339	110		1.88	
x648	116		1.97		x300	97.7		1.67	
x607	108		1.85		x253	91.3		1.56	
x542	99.2		1.69		x223	80.3		1.37	
					x184	74.1		1.27	
WWF1800					x161	64.9		1.11	
x700	134		2.29						
x659	126		2.16		WWF700				
x617	118		2.02		x245	95.0		1.62	
x575	110		1.89		x214	83.1		1.42	
x510	101		1.72		x196	75.8		1.29	
					x175	76.8		1.31	
WWF1600					x152	66.9		1.14	
x622	129		2.20						
x580	120		2.05		WWF650				
x538	112		1.91		x864		228		3.90
x496	103		1.75		x739		192		3.28
x431	92.2		1.57		x598		155		2.65
					x499		129		2.21
WWF1400					x400		104		1.77
x597	135		2.30						
x513	116		1.98		WWF600				
x471	106		1.82		x793		228		3.89
x405	94.9		1.62		x680		192		3.28
x358	90.2		1.54		x551		155		2.64
					x460		129		2.21
WWF1200					x369		104		1.77
x487	121		2.07						
x418	108		1.84		WWF550				
x380	98.2		1.68		x721		227		3.88
x333	93.3		1.59		x620		192		3.27
x302	84.7		1.45		x503		155		2.64
x263	80.5		1.37		x420		129		2.20
					x280		85.5		1.46
WWF1100									
x458	120		2.05		WWF500				
x388	106		1.81		x651		226		3.86
x351	95.6		1.63		x561		191		3.26
x304	90.1		1.54		x456		154		2.63
x273	81.0		1.38		x381		129		2.20
x234	76.1		1.30		x343		116		1.98
					x306		103		1.77
WWF1000					x276		93.1		1.59
x447	124		2.11		x254		85.4		1.46
x377	109		1.86		x223		75.3		1.28
x340	97.9		1.67		x197		66.4		1.13
x293	92.3		1.58						
x262	82.7		1.41		WWF450				
x223	77.6		1.32		x503		191		3.25
x200	70.0		1.19		x409		154		2.63
					x342		129		2.20
WWF900					x308		116		1.98
x417	122		2.08		x274		103		1.76
x347	106		1.81		x248		93.0		1.59
x309	94.4		1.61		x228		85.3		1.46
x262	88.0		1.50		x201		75.2		1.28
x231	77.8		1.33		x177		66.2		1.13
x192	71.8		1.23						
x169	63.3		1.08						

[1] M/D = mass / (surface area - top of top flange)
[2] M/D = mass / surface area

M / D RATIOS

Designation	SI (kg/m)/m		Imperial (lb./ft.)/in.		Designation	SI (kg/m)/m		Imperial (lb./ft.)/in.	
	Beam[1]	Column[2]	Beam[1]	Column[2]		Beam[1]	Column[2]	Beam[1]	Column[2]
WWF400					W920				
x444		190		3.24	x1188	350		5.98	
x362		154		2.62	x967	293		5.00	
x303		128		2.19	x784	243		4.15	
x273		116		1.98	x653	206		3.52	
x243		103		1.76	x585	186		3.18	
x220		92.8		1.58	x534	171		2.92	
x202		85.2		1.45	x488	157		2.69	
x178		75.1		1.28	x446	145		2.47	
x157		66.4		1.13	x417	136		2.32	
					x387	126		2.16	
WWF350					x365	119		2.04	
x315		153		2.61	x342	113		1.92	
x263		128		2.19					
x238		115		1.97	W920				
x212		103		1.76	x381	137		2.34	
x192		92.6		1.58	x345	125		2.13	
x176		85.0		1.45	x313	114		1.94	
x155		75.0		1.28	x289	105		1.80	
x137		66.1		1.13	x271	99.5		1.70	
					x253	93.2		1.59	
					x238	88.0		1.50	
					x223	83.1		1.42	
W1100					x201	74.8		1.28	
x499	146		2.50						
x432	129		2.19		W840				
x390	117		1.99		x576	192		3.28	
x342	102		1.75		x527	177		3.03	
					x473	160		2.74	
W1000					x433	148		2.52	
x883	264		4.50		x392	134		2.30	
x749	227		3.87		x359	124		2.11	
x641	196		3.35		x329	114		1.95	
x591	182		3.11		x299	104		1.78	
x554	172		2.93						
x539	167		2.86		W840				
x483	151		2.57		x251	97.8		1.67	
x443	139		2.37		x226	88.9		1.52	
x412	130		2.21		x210	83.1		1.42	
x371	117		2.00		x193	76.7		1.31	
x321	102		1.74		x176	69.8		1.19	
x296	94.8		1.62						
					W760				
W1000					x582	208		3.54	
x583	196		3.35		x531	191		3.26	
x493	168		2.87		x484	176		3.00	
x486	166		2.82		x434	158		2.70	
x414	143		2.44		x389	143		2.44	
x393	136		2.32		x350	130		2.21	
x350	121		2.07		x314	117		2.00	
x314	110		1.87		x284	107		1.82	
x272	95.7		1.63		x257	97.0		1.65	
x249	88.0		1.50						
x222	79.1		1.35		W760				
					x220	95.0		1.62	
					x196	85.2		1.45	
					x185	80.1		1.37	
					x173	75.6		1.29	
					x161	70.0		1.20	
					x147	64.5		1.10	
					x134	58.8		1.00	

[1] M/D = mass / (surface area - top of top flange)
[2] M/D = mass / surface area

M / D RATIOS

Designation	SI (kg/m)/m		Imperial (lb./ft.)/in.		Designation	SI (kg/m)/m		Imperial (lb./ft.)/in.	
	Beam[1]	Column[2]	Beam[1]	Column[2]		Beam[1]	Column[2]	Beam[1]	Column[2]
W690					W530				
x802	295		5.04		x138	80.8		1.38	
x548	211		3.61		x123	72.6		1.24	
x500	194		3.32		x109	64.6		1.10	
x457	179		3.06		x101	60.2		1.03	
x419	165		2.82		x92	55.4		0.945	
x384	153		2.61		x82	49.5		0.845	
x350	141		2.40		x72	43.5		0.742	
x323	130		2.23						
x289	117		2.00		W530				
x265	108		1.84		x85	54.8		0.935	
x240	98.6		1.68		x74	48.6		0.830	
x217	89.7		1.53		x66	43.0		0.734	
W690					W460				
x192	89.7		1.53		x260	141		2.41	
x170	79.8		1.36		x235	129		2.20	
x152	72.1		1.23		x213	118		2.01	
x140	66.4		1.13		x193	108		1.84	
x125	60.0		1.02		x177	99.2		1.69	
					x158	88.9		1.52	
W610					x144	81.8		1.40	
x551	231		3.94		x128	73.3		1.25	
x498	212		3.61		x113	64.8		1.11	
x455	195		3.32						
x415	180		3.06		W460				
x372	162		2.77		x106	70.9		1.21	
x341	150		2.56		x97	64.9		1.11	
x307	136		2.33		x89	60.4		1.03	
x285	128		2.18		x82	55.4		0.946	
x262	117		2.01		x74	50.6		0.864	
x241	109		1.86		x67	46.0		0.786	
x217	98.9		1.69		x61	41.4		0.707	
x195	89.1		1.52						
x174	80.0		1.37		W460				
x155	71.3		1.22		x68	50.3		0.859	
					x60	44.0		0.752	
W610					x52	38.8		0.663	
x153	80.8		1.38						
x140	74.0		1.26		W410				
x125	66.1		1.13		x149	91.7		1.56	
x113	60.2		1.03		x132	81.8		1.40	
x101	54.6		0.932		x114	71.6		1.22	
x91	49.0		0.837		x100	62.7		1.07	
x84	45.4		0.774						
					W410				
W610					x85	62.6		1.07	
x92	53.8		0.919		x74	55.6		0.949	
x82	47.7		0.814		x67	50.4		0.860	
					x60	44.7		0.762	
W530					x54	40.4		0.689	
x300	144		2.46						
x272	131		2.24		W410				
x248	120		2.05		x46	38.1		0.651	
x219	107		1.83		x39	32.5		0.555	
x196	97.0		1.66						
x182	89.9		1.53						
x165	82.7		1.41						
x150	75.5		1.29						

[1] M/D = mass / (surface area - top of top flange)
[2] M/D = mass / surface area

M / D RATIOS

Designation	SI (kg/m)/m		Imperial (lb./ft.)/in.		Designation	SI (kg/m)/m		Imperial (lb./ft.)/in.	
	Beam[1]	Column[2]	Beam[1]	Column[2]		Beam[1]	Column[2]	Beam[1]	Column[2]
W360					W310				
x1086		390		6.66	x283		144		2.46
x990		360		6.14	x253		130		2.22
x900		335		5.71	x226		118		2.02
x818		307		5.25	x202		106		1.82
x744		285		4.86	x179		95.1		1.62
x677		263		4.49	x158		84.3		1.44
					x143		77.1		1.32
W360					x129	84.4	70.3	1.44	1.20
x634		249		4.25	x118	77.2	64.3	1.32	1.10
x592		235		4.01	x107	70.3	58.5	1.20	0.999
x551		220		3.76	x97	63.9	53.2	1.09	0.907
x509		206		3.51					
x463		189		3.23	W310				
x421		174		2.97	x86	63.3	53.4	1.08	0.911
x382		160		2.72	x79	57.9	48.7	0.988	0.832
x347		146		2.49					
x314		133		2.27	W310				
x287		123		2.10	x74	61.2	52.4	1.04	0.894
x262		113		1.93	x67	55.3	47.3	0.944	0.807
x237		103		1.75	x60	49.7	42.5	0.847	0.725
x216		94.2		1.61					
					W310				
W360					x52	46.7		0.797	
x196		88.9		1.52	x45	40.2		0.686	
x179		81.4		1.39	x39	35.1		0.600	
x162	89.1	74.0	1.52	1.26	x31	28.8		0.492	
x147	81.7	67.8	1.40	1.16					
x134	74.7	62.0	1.28	1.06	W310				
					x33	35.7		0.610	
W360					x28	31.1		0.530	
x122	82.7	70.4	1.41	1.20	x24	26.5		0.452	
x110	75.0	63.9	1.28	1.09	x21	23.5		0.401	
x101	69.5	59.1	1.19	1.01					
x91	62.8	53.5	1.07	0.913	W250				
					x167		105		1.78
W360					x149		94.3		1.61
x79	60.8	52.5	1.04	0.897	x131		83.9		1.43
x72	55.2	47.7	0.942	0.814	x115		74.1		1.26
x64	49.6	42.9	0.847	0.732	x101	79.4	66.1	1.36	1.13
					x89	70.7	58.8	1.21	1.00
W360					x80	63.6	52.9	1.09	0.903
x57	46.5		0.794		x73	58.2	48.4	0.994	0.826
x51	41.8		0.714						
x45	37.4		0.638		W250				
					x67	60.6	51.1	1.03	0.873
W360					x58	53.1	44.8	0.906	0.765
x39	36.3		0.620		x49	45.2	38.1	0.772	0.651
x33	30.7		0.523						
					W250				
W310					x45	46.7		0.798	
x500		235		4.02	x39	40.6		0.693	
x454		217		3.70	x33	34.8		0.593	
x415		201		3.43	x24	26.2		0.448	
x375		184		3.15					
x342		171		2.91	W250				
x313		158		2.69	x28	35.0		0.598	
					x25	31.4		0.536	
					x22	27.9		0.476	
					x18	22.4		0.382	

[1] M/D = mass / (surface area - top of top flange)
[2] M/D = mass / surface area

M/D RATIOS

Designation	SI (kg/m)/m Beam[1]	SI (kg/m)/m Column[2]	Imperial (lb./ft.)/in. Beam[1]	Imperial (lb./ft.)/in. Column[2]	Designation	SI (kg/m)/m Beam[1]	SI (kg/m)/m Column[2]	Imperial (lb./ft.)/in. Beam[1]	Imperial (lb./ft.)/in. Column[2]
W200					S380				
x100	93.4	77.9	1.59	1.33	x74	64.1		1.09	
x86	82.6	68.9	1.41	1.18	x64	55.1		0.940	
x71	69.2	57.6	1.18	0.984					
x59	58.1	48.4	0.992	0.826	S310				
x52	51.5	42.9	0.880	0.732	x74	74.9		1.28	
x46	45.7	38.0	0.779	0.648	x60.7	61.6		1.05	
W200					S310				
x42	46.4	39.1	0.792	0.668	x52	53.5		0.914	
x36	40.3	34.0	0.688	0.580	x47	48.7		0.832	
W200					S250				
x31	38.8	33.3	0.662	0.568	x52	61.1		1.04	
x27	33.2	28.5	0.567	0.486	x38	44.7		0.763	
x21	26.8		0.457						
W200					S200				
x22	31.8		0.543		x34	48.9		0.835	
x19	27.8		0.474		x27	39.4		0.672	
x15	21.6		0.368		S150				
					x26	46.4		0.792	
W150					x19	34.0		0.580	
x37	48.2	40.2	0.823	0.686					
x30	39.2	32.6	0.668	0.556	S130				
x22	29.8	24.8	0.508	0.423	x15	31.5		0.537	
W150					S100				
x24	39.1	33.5	0.667	0.572	x14.1	35.3		0.602	
x18	29.8	25.5	0.509	0.435	x11	28.8		0.491	
x14	23.0	19.6	0.392	0.335	S75				
x13	21.5	18.4	0.367	0.314	x11	34.4		0.587	
W130					x8	26.2		0.447	
x28		37.0		0.631					
x24		31.5		0.538	M310				
					x17.6	21.2		0.361	
W100					x16.1	19.3		0.330	
x19		31.8		0.543	M250				
					x13.4	19.3		0.329	
S610					x11.9	17.1		0.291	
x180	99.0		1.69		M200				
x158	87.0		1.49		x9.7	17.2		0.294	
S610					M130				
x149	85.6		1.46		x28.1		37.7		0.643
x134	77.5		1.32		M100				
x119	69.0		1.18		x8.9		15.4		0.263
S510									
x143	92.7		1.58		SLB100				
x128	83.8		1.43		x5.4	14.8		0.252	
S510					x4.8	13.0		0.223	
x112	75.8		1.29		SLB75				
x98.2	66.9		1.14		x4.5	14.2		0.242	
S460					x4.3	13.4		0.229	
x104	77.1		1.32						
x81.4	60.6		1.03						

[1] M/D = mass / (surface area - top of top flange)
[2] M/D = mass / surface area

COEFFICIENTS OF THERMAL EXPANSION

(Linear, per degree x 10^{-6})

METALS	c per °C	c per °F	NON-METALS	c per °C	c per °F
Aluminum	23	13	Cement, Portland	13	7
Brass	19	10.4	Concrete, Stone	10	5.7
Bronze	18	10.1	Glass	7	4
Copper	16.7	9.3	Granite	8.3	4.6
Iron, Gray Cast	11	5.9	Limestone	7.9	4.4
Iron, Wrought	12.0	6.7	Marble	9	5
Lead	28.7	15.9	Masonry, Ashlar	6.3	3.5
Magnesium	28.8	16	Masonry, Brick	6.1	3.4
Nickel	12.6	7	Masonry, Rubble	6.3	3.5
Steel, Cast	11.3	6.3	Plaster	16	9
Steel, Stainless	17.8	9.9	Sandstone	11	6
Steel, Structural	11.7	6.5	Slate	10	5.8
Zinc, Rolled	31	17.3	Fir (parallel to fibre)	3.8	2.1
			Fir (perpendicular to fibre)	58	32

NOTE: Coefficients of thermal expansion indicated are average values from various sources. Minor variations may be expected in metals. Large variations may be expected in concrete and masonry due to the many combinations of constituents possible.

Coefficients apply in general to a temperature range from 0 to 100 degrees Celsius.

The coefficient of linear thermal expansion (c) is the change in length per unit of length for a change of one degree of temperature. The coefficient for surface expansion is approximately two times, and the coefficient of volume expansion is approximately three times, the linear coefficient.

Change in length = cL x change in temperature, if member is free to elongate or contract.

Change in unit stress = cE x change in temperature, if member is not permitted to elongate or contract.

2000 ELECTRONIC AIDS

Internet

CISC maintains a site on the Internet's World Wide Web at the following URL address http://www.cisc-icca.ca. Since this is a very dynamic area of information transfer, a full description of the site's contents is unwarranted in this publication. In general terms, however, the site provides information on CISC's members, publications, videos, computer programs, hot topics, what's new and frequently asked questions (FAQs).

Computer Programs

Since 1967, the Canadian steel industry has provided the engineering profession with computer programs to assist in the design of steel structures. Much of the early effort was devoted to filling a need for tools to increase the efficiency of designing steel structures and to looking at various alternate solutions to determine the most economical one. Currently, there are numerous commercial software packages available for the analysis and design of steel structures. Thus, there is a diminished need for specialized programs. This section describes the program currently available. All material is distributed from the Canadian Institute of Steel Construction on 3.5" media. Alternatives can be arranged at time of order. Units are SI metric unless noted. Software support and program updates are included in the selling price of the software up until the next release of CAN/CSA-S16.1. An updating fee may be charged at that time. User support is available by facsimile transmission.

Current program users and potential users are encouraged to indicate areas of specific interest as guidance in the future planning of software development.

Gravity Frame Design

GFD aids in the design of gravity load carrying structural steel components. It includes a state-of-the-art graphical user interface and uses a graphical approach to structural modelling.

Version 4 of the Gravity Frame Design program designs gravity loaded steel framing members to CAN/CSA-S16.1-94. Construction types include composite beams and trusses, stub girders, cantilevers, cantilever spans, columns and hangers. Columns may be tiered and all members of a type may be grouped to force repetitiveness. Single component or total building design capability are supported. Load types are superimposed dead, deck-slab and live with live load reduction with cumulative tributary area. Checks member strength, stability and deflection at various construction stages. Correlates shear stud design with steel deck profile selected. Includes quantity take-off by member type, and steel cost estimate.

The graphical user interface, which operates within the Microsoft Windows environment, permits full graphical input and editing of the framing geometry and loads. In creating a frame, a column grid is defined, typical floors are built and then stacked. The creation of framing members parallels the construction sequence and follows practical interconnection rules. Members can be grouped and design parameters applied globally or to specific members. The deck/slab system is defined and its boundaries located graphically. A set of area loads, one of each load type, can be associated with the deck/slab. Loads and tributary areas are determined and distributed to framing members automatically. Floor framing geometry may be output to any Windows supported printer.

Microsoft Windows 3.1/95/98 and a compatible computer, graphics card/monitor combination, pointing device and printer are required. A pentium-based computer with SVGA graphics, 12 MB of memory, 10 MB of hard disk space and mouse is recommended.

Structural Section Tables – SST File

The SST file contains the North American database of structural steel sections listed in this publication. Metric dimensions and section properties for design and detailing are contained in a 160-character string. Sections are grouped as follows:

W, S, M, HP, WWF, C, MC, L, WT, WWT, 2L short legs back to back, 2L long legs back to back, 2L equal legs back to back, WRF, HSS square, HSS rectangular, HSS round and SLB (super light beams).

The database includes two sets of HSS shapes: those produced in accordance with CSA G40.20 and those produced in accordance with ASTM A500.

All dimensions and properties are contained in a single ASCII fixed-width format file that may be imported into spreadsheet applications. Microsoft Windows users may also view the SST file contents using a database browser program.

CHECK LIST FOR DESIGN DRAWINGS

General

A design does not provide a satisfactory structure unless sufficient information is conveyed to the builder to facilitate that the designer's intentions are clearly understood. Furthermore, attempting to prepare an estimate for a structure from plans and specifications which contain insufficient information involves risks which tend to increase the tendered price. Clause 4.1 of CAN/CSA-S16.1 governs the minimum requirements of design drawings. The following items are suggested as a check list of information to be included on design drawings to avoid unnecessary and costly uncertainty at the time of bidding:

1. The type or types of design as defined in CAN/CSA-S16.1. If plastic analysis is employed it should be stated. Show the category of the structural system used for seismic design.
2. The grade(s) of structural steel, grade(s) and diameters of bolts.
3. All structural drawings to be adequately dimensioned, preferably in SI metric units. Do not intermix Metric and Imperial systems of units.
4. Centre-to-centre distances for all columns.
5. Outside dimensions of rigid frames and offset dimensions from grid lines to outside of rigid frames.
6. Out-to-out dimension of trusses and offset dimensions from centre line of chords to outside of chords–include any camber requirements.
7. Offset dimensions from centre of column lines to centre of beams for all beams that are not on the grid lines.
8. Relation of outside of exterior walls to centre lines of columns.
9. Relation of the top surfaces of beams to finished floor elevations.
10. Length of bearing for all beams bearing on exterior walls, including the dimension from the outside of the wall to the end of the steel beam and size of bearing plate.
11. Elevations of underside of column base plates.
12. Dimensions of all clear openings for doorways, ducts, stair wells, roof openings, etc., and their relation to adjacent steel members.
13. Indicate whether loads and forces shown on drawings are factored or unfactored.
14. Axial loads in beams, columns and bracing members and joint pass-through forces.
15. Forces in truss members including moments when members are loaded between panel points.
16. Minimum end reactions required for all connections.
17. Moments for restrained beams and cantilevers. Governing combinations of shears, moments, and axial forces to be resisted by the connections.
18. All information necessary to design and manufacture the open-web steel joists and steel deck to suit the loading conditions.
19. When a particular type of connection is required, the location and type of connection.
20. Type of beam-to-column connection when beams frame over top of columns, including type and location of stiffeners.
21. Any bearing-type connections that are required to be pretensioned.

22. For composite beams, the size and location of shear studs and which beams, if any, must be shored.
23. Size of column base plates and size and location of anchors. (Four anchor bolts should be considered to facilitate erection when practical.)
24. Size and location of stiffeners, web doubler plates, reinforcement, and bracing required for stability of compression elements.
25. Details and location of built-up lintels.
26. Identify roof cladding systems that do not provide lateral restraint to the roof structure.
27. Reinforcement, where necessary for openings through beam webs.
28. Ledger angles complete with method of attachment.
29. Members requiring prime paint or galvanizing.
30. Identify architecturally exposed structural steel elements requiring special tolerances and finishes.
31. Treatment of steel encased in concrete.
32. Fabrication and erection tolerances if other than those specified in CAN/CSA-S16.1. Special tolerances when interfacing with other materials, i.e., steel attached to concrete.
33. A note that all structural welding is to be performed only by companies certified to Division 1 or 2.1 of CSA W47.1.
34. When weld symbols are shown, refer to "WELDED JOINTS Standard Symbols" from Part 6.

Allow as much time as possible (three weeks for an average job) for preparing bids. During the time allotted for preparing tenders, only those changes necessary to clarify bidding instructions should be issued by addendum. If major changes are included in an addendum, an extension of the tender closing should be considered.

PROPERTIES OF GEOMETRIC SECTIONS
Definitions

Neutral Axis

The line, in any given section of a member subject to bending, on which there is neither tension nor compression.

For pure elastic bending of a straight beam, the neutral axis at any cross-section is coincident with the centroidal axis of the cross-section.

In the case of fully plastic bending, the neutral axis divides the sectional area equally. Therefore, the neutral axis for elastic and plastic bending coincide only in the case of sections symmetrical about the neutral axis.

Moment of Inertia I

The sum of the products obtained by multiplying each of the elementary areas, of which the section is composed, by the square of its perpendicular distance from the axis about which the moment of inertia is being calculated.

Elastic Section Modulus S

The moment of inertia divided by the perpendicular distance from the axis about which the moment of inertia has been calculated to the most remote part of the section.

The elastic section modulus is used to determine the bending stress in the extreme fibre of a section by dividing the bending moment by the section modulus, referred to the neutral axis perpendicular to the plane of bending, both values being expressed in like units of measure.

Radius of Gyration r

The perpendicular distance from a neutral axis to the centre of gyration (i.e., the point where the entire area is considered to be concentrated so as to have the same moment of inertia as the actual area). The square of the radius of gyration of a section is equal to the moment of inertia (referred to the appropriate axis) divided by the area.

The radius of gyration of a section is used to ascertain the load this section will sustain when used in compression as a strut or column. The ratio of the effective unsupported length of the section divided by the least radius of gyration applicable to this length is called the slenderness ratio.

Plastic Modulus Z

The modulus of resistance to bending of a completely yielded cross-section, calculated by taking the combined statical moment, about the neutral axis, of the cross-sectional areas above and below that axis.

In general, the plastic modulus is calculated by simple statics and has been included for only a few of the shapes listed.

PROPERTIES OF GEOMETRIC SECTIONS

SQUARE
Axis of moments through center

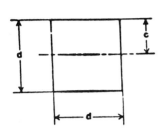

$A = d^2$

$c = \dfrac{d}{2}$

$I = \dfrac{d^4}{12}$

$S = \dfrac{d^3}{6}$

$r = \dfrac{d}{\sqrt{12}}$

$Z = \dfrac{d^3}{4}$

SQUARE
Axis of moments on base

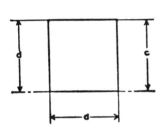

$A = d^2$

$c = d$

$I = \dfrac{d^4}{3}$

$S = \dfrac{d^3}{3}$

$r = \dfrac{d}{\sqrt{3}}$

SQUARE
Axis of moments on diagonal

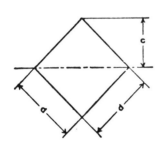

$A = d^2$

$c = \dfrac{d}{\sqrt{2}}$

$I = \dfrac{d^4}{12}$

$S = \dfrac{d^3}{6\sqrt{2}}$

$r = \dfrac{d}{\sqrt{12}}$

$Z = \dfrac{2c^3}{3} = \dfrac{d^3}{3\sqrt{2}}$

RECTANGLE
Axis of moments through center

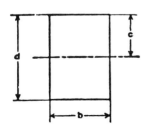

$A = bd$

$c = \dfrac{d}{2}$

$I = \dfrac{bd^3}{12}$

$S = \dfrac{bd^2}{6}$

$r = \dfrac{d}{\sqrt{12}}$

$Z = \dfrac{bd^2}{4}$

PROPERTIES OF GEOMETRIC SECTIONS

RECTANGLE
Axis of moments on base

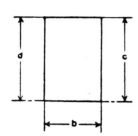

$A = bd$

$c = d$

$I = \dfrac{bd^3}{3}$

$S = \dfrac{bd^2}{3}$

$r = \dfrac{d}{\sqrt{3}}$

RECTANGLE
Axis of moments on diagonal

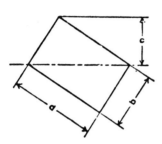

$A = bd$

$c = \dfrac{bd}{\sqrt{b^2 + d^2}}$

$I = \dfrac{b^3 d^3}{6(b^2 + d^2)}$

$S = \dfrac{b^2 d^2}{6\sqrt{b^2 + d^2}}$

$r = \dfrac{bd}{\sqrt{6(b^2 + d^2)}}$

RECTANGLE
Axis of moments any line through center of gravity

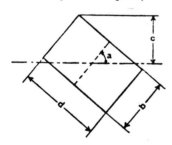

$A = bd$

$c = \dfrac{b \sin a + d \cos a}{2}$

$I = \dfrac{bd(b^2 \sin^2 a + d^2 \cos^2 a)}{12}$

$S = \dfrac{bd(b^2 \sin^2 a + d^2 \cos^2 a)}{6(b \sin a + d \cos a)}$

$r = \sqrt{\dfrac{b^2 \sin^2 a + d^2 \cos^2 a}{12}}$

HOLLOW RECTANGLE
Axis of moments through center

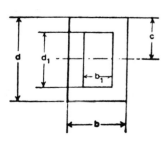

$A = bd - b_1 d_1$

$c = \dfrac{d}{2}$

$I = \dfrac{bd^3 - b_1 d_1^3}{12}$

$S = \dfrac{bd^3 - b_1 d_1^3}{6d}$

$r = \sqrt{\dfrac{bd^3 - b_1 d_1^3}{12A}}$

$Z = \dfrac{1}{4}(bd^2 - b_1 d_1^2)$

PROPERTIES OF GEOMETRIC SECTIONS

EQUAL RECTANGLES

Axis of moments through center of gravity

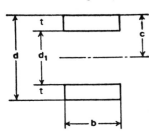

$A = b(d - d_1)$

$c = \dfrac{d}{2}$

$I = \dfrac{b(d^3 - d_1^3)}{12}$

$S = \dfrac{b(d^3 - d_1^3)}{6d}$

$r = \sqrt{\dfrac{d^3 - d_1^3}{12(d - d_1)}}$

$Z = \dfrac{b}{4}(d^2 - d_1^2) = bt(d - t)$

UNEQUAL RECTANGLES

Axis of moments through center of gravity

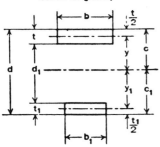

$A = bt + b_1 t_1$

$c = \dfrac{\tfrac{1}{2} bt^2 + b_1 t_1 (d - \tfrac{1}{2} t_1)}{A}$

$I = \dfrac{bt^3}{12} + bty^2 + \dfrac{b_1 t_1^3}{12} + b_1 t_1 y_1^2$

$S = \dfrac{I}{c} \qquad S_1 = \dfrac{I}{c_1}$

$r = \sqrt{\dfrac{I}{A}}$

TRIANGLE

Axis of moments through center of gravity

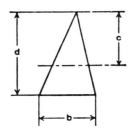

$A = \dfrac{bd}{2}$

$c = \dfrac{2d}{3}$

$I = \dfrac{bd^3}{36}$

$S = \dfrac{bd^2}{24}$

$r = \dfrac{d}{\sqrt{18}}$

TRIANGLE

Axis of moments on base

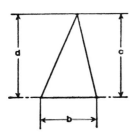

$A = \dfrac{bd}{2}$

$c = d$

$I = \dfrac{bd^3}{12}$

$S = \dfrac{bd^2}{12}$

$r = \dfrac{d}{\sqrt{6}}$

PROPERTIES OF GEOMETRIC SECTIONS

TRAPEZOID

Axis of moments through center of gravity

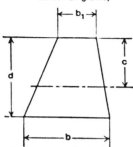

$$A = \frac{d(b + b_1)}{2}$$

$$c = \frac{d(2b + b_1)}{3(b + b_1)}$$

$$I = \frac{d^3(b^2 + 4bb_1 + b_1^2)}{36(b + b_1)}$$

$$S = \frac{d^2(b^2 + 4bb_1 + b_1^2)}{12(2b + b_1)}$$

$$r = \frac{d}{6(b + b_1)}\sqrt{2(b^2 + 4bb_1 + b_1^2)}$$

CIRCLE

Axis of moments through center

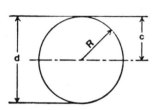

$$A = \frac{\pi d^2}{4} = \pi R^2$$

$$c = \frac{d}{2} = R$$

$$I = \frac{\pi d^4}{64} = \frac{\pi R^4}{4}$$

$$S = \frac{\pi d^3}{32} = \frac{\pi R^3}{4}$$

$$r = \frac{d}{4} = \frac{R}{2}$$

$$Z = \frac{d^3}{6}$$

HOLLOW CIRCLE

Axis of moments through center

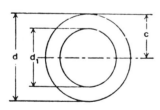

$$A = \frac{\pi(d^2 - d_1^2)}{4}$$

$$c = \frac{d}{2}$$

$$I = \frac{\pi(d^4 - d_1^4)}{64}$$

$$S = \frac{\pi(d^4 - d_1^4)}{32d}$$

$$r = \frac{\sqrt{d^2 + d_1^2}}{4}$$

$$Z = \frac{1}{6}(d^3 - d_1^3)$$

HALF CIRCLE

Axis of moments through center of gravity

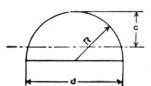

$$A = \frac{\pi R^2}{2}$$

$$c = R\left(1 - \frac{4}{3\pi}\right)$$

$$I = R^4\left(\frac{\pi}{8} - \frac{8}{9\pi}\right)$$

$$S = \frac{R^3(9\pi^2 - 64)}{24(3\pi - 4)}$$

$$r = R\frac{\sqrt{9\pi^2 - 64}}{6\pi}$$

PROPERTIES OF GEOMETRIC SECTIONS

PARABOLA

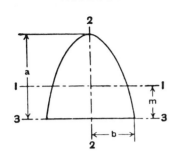

$A = \dfrac{4}{3}ab$

$m = \dfrac{2}{5}a$

$I_1 = \dfrac{16}{175}a^3 L$

$I_2 = \dfrac{4}{15}ab^3$

$I_3 = \dfrac{32}{105}a^3 b$

HALF PARABOLA

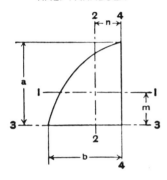

$A = \dfrac{2}{3}ab$

$m = \dfrac{2}{5}a$

$n = \dfrac{3}{8}b$

$I_1 = \dfrac{8}{175}a^3 b$

$I_2 = \dfrac{19}{480}ab^3$

$I_3 = \dfrac{16}{105}a^3 b$

$I_4 = \dfrac{2}{15}ab^3$

COMPLEMENT OF HALF PARABOLA

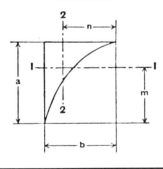

$A = \dfrac{1}{3}ab$

$m = \dfrac{7}{10}a$

$n = \dfrac{3}{4}b$

$I_1 = \dfrac{37}{2100}a^3 b$

$I_2 = \dfrac{1}{80}ab^3$

PARABOLIC FILLET IN RIGHT ANGLE

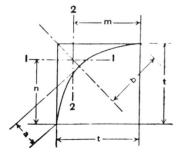

$a = \dfrac{t}{2\sqrt{2}}$

$b = \dfrac{t}{\sqrt{2}}$

$A = \dfrac{1}{6}t^2$

$m = n = \dfrac{4}{5}t$

$I_1 = I_2 = \dfrac{11}{2100}t^4$

PROPERTIES OF GEOMETRIC SECTIONS

*HALF ELLIPSE

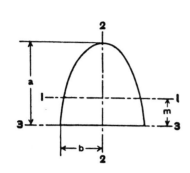

$$A = \frac{1}{2}\pi ab$$

$$m = \frac{4a}{3\pi}$$

$$I_1 = a^3 b\left(\frac{\pi}{8} - \frac{8}{9\pi}\right)$$

$$I_2 = \frac{1}{8}\pi ab^3$$

$$I_3 = \frac{1}{8}\pi a^3 b$$

*QUARTER ELLIPSE

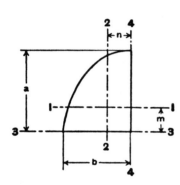

$$A = \frac{1}{4}\pi ab$$

$$m = \frac{4a}{3\pi}$$

$$n = \frac{4b}{3\pi}$$

$$I_1 = a^3 b\left(\frac{\pi}{16} - \frac{4}{9\pi}\right)$$

$$I_2 = ab^3\left(\frac{\pi}{16} - \frac{4}{9\pi}\right)$$

$$I_3 = \frac{1}{16}\pi a^3 b$$

$$I_4 = \frac{1}{16}\pi ab^3$$

*ELLIPTIC COMPLEMENT

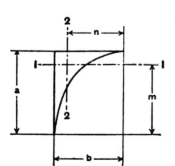

$$A = ab\left(1 - \frac{\pi}{4}\right)$$

$$m = \frac{a}{6\left(1 - \frac{\pi}{4}\right)}$$

$$n = \frac{b}{6\left(1 - \frac{\pi}{4}\right)}$$

$$I_1 = a^3 b\left(\frac{1}{3} - \frac{\pi}{16} - \frac{1}{36\left(1 - \frac{\pi}{4}\right)}\right)$$

$$I_2 = ab^3\left(\frac{1}{3} - \frac{\pi}{16} - \frac{1}{36\left(1 - \frac{\pi}{4}\right)}\right)$$

*To obtain properties of half circle, quarter circle and circular complement substitute $a = b = R$.

PROPERTIES OF GEOMETRIC SECTIONS AND STRUCTURAL SHAPES

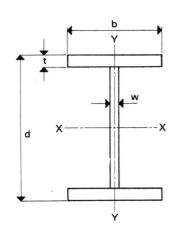

$A = 2bt + (d - 2t)w$

$I = \dfrac{1}{12}\left[bd^3 - (b-w)(d-2t)^3\right]$

$S = \dfrac{1}{6d}\left[bd^3 - (b-w)(d-2t)^3\right]$

$r = \sqrt{\dfrac{I}{A}}$

$Z = \dfrac{1}{4}\left[bd^2 - (b-w)(d-2t)^2\right]$

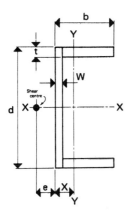

$A = dw + 2(b-w)t$

$I = \dfrac{1}{12}\left[bd^3 - (b-w)(d-2t)^3\right]$

$S = \dfrac{1}{6d}\left[bd^3 - (b-w)(d-2t)^3\right]$

$r = \sqrt{\dfrac{I}{A}}$

$e = \dfrac{b^2 d^2 t}{4I} - \dfrac{w}{2}$

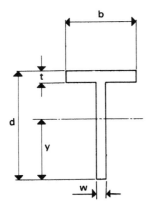

$A = bt + w(d-t)$

$y = \dfrac{1}{2}\left(\dfrac{bdt}{A} + d - t\right)$

$I = \dfrac{1}{12}\left[bt^3 + w(d-t)^3 + \dfrac{3\,bwtd^2(d-t)}{A}\right]$

$S_1 = \dfrac{I}{y}$

$S_2 = \dfrac{I}{d-y}$

$r = \sqrt{\dfrac{I}{A}}$

7-59

PROPERTIES OF GEOMETRIC SECTIONS AND STRUCTURAL SHAPES

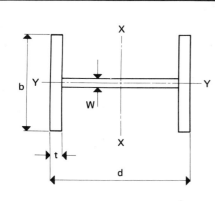

$A = 2bt + w(d - 2t)$

$I = \dfrac{1}{12}\left[2tb^3 + (d - 2t)w^3\right]$

$S = \dfrac{1}{6b}\left[2tb^3 + (d - 2t)w^3\right]$

$r = \sqrt{\dfrac{I}{A}}$

$Z = \dfrac{1}{4}\left[2t(b^2 - w^2) + dw^2\right]$

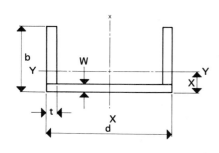

$A = dw + 2(b - w)t$

$x = \dfrac{1}{2A}\left[(d - 2t)w^2 + 2tb^2\right]$

$I = \dfrac{1}{3}\left[dx^3 + 2t(b-x)^3 - (d-2t)(x-w)^3\right]$

$S_1 = \dfrac{I}{b - x}$; $S_2 = \dfrac{I}{x}$

$r = \sqrt{\dfrac{I}{A}}$

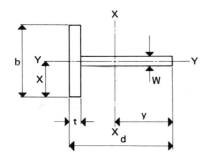

$A = bt + (d - t)w$

$x = b/2$

$I = \dfrac{1}{12}\left[tb^3 + (d - t)w^3\right]$

$S = \dfrac{2I}{b}$

$r = \sqrt{\dfrac{I}{A}}$

PROPERTIES OF GEOMETRIC SECTIONS AND STRUCTURAL SHAPES

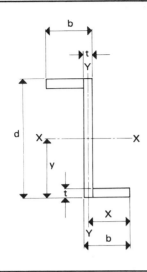

$A = t\left[d + 2(b-t)\right]$

$y = d/2$

$I = \dfrac{bd^3 - (b-t)(d-2t)^3}{12}$

$S = \dfrac{I}{y}$

$r = \sqrt{\dfrac{bd^3 - (b-t)(d-2t)^3}{12t\left[d + 2(b-t)\right]}}$

ANGLE
Axis of moments through center of gravity

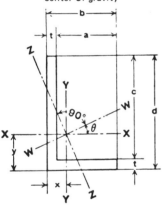

Z-Z is axis of minimum I

$\tan 2\theta = \dfrac{2K}{I_y - I_x}$

$A = t(b+c) \quad x = \dfrac{b^2 + ct}{2(b+c)} \quad y = \dfrac{d^2 + at}{2(b+c)}$

K = Product of Inertia about X-X & Y-Y

$\quad = \mp \dfrac{abcdt}{4(b+c)}$

$I_x = \dfrac{1}{3}\left(t(d-y)^3 + by^3 - a(y-t)^3\right)$

$I_y = \dfrac{1}{3}\left(t(b-x)^3 + dx^3 - c(x-t)^3\right)$

$I_z = I_x \sin^2\theta + I_y \cos^2\theta + K \sin 2\theta$

$I_w = I_x \cos^2\theta + I_y \sin^2\theta - K \sin 2\theta$

K is negative when heel of angle, with respect to c. g., is in 1st or 3rd quadrant, positive when in 2nd or 4th quadrant.

BEAMS AND CHANNELS
Transverse force oblique through center of gravity

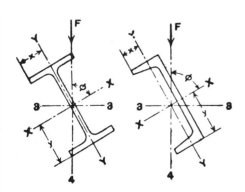

$I_3 = I_x \sin^2\phi + I_y \cos^2\phi$

$I_4 = I_x \cos^2\phi + I_y \sin^2\phi$

$f = M\left(\dfrac{y}{I_x}\sin\phi + \dfrac{x}{I_y}\cos\phi\right)$

where M is bending moment due to force F.

PROPERTIES OF GEOMETRIC SECTIONS AND STRUCTURAL SHAPES

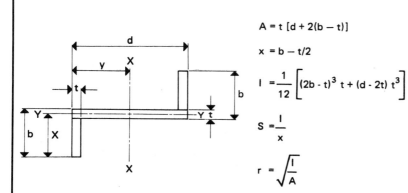

$A = t\,[d + 2(b - t)]$

$x = b - t/2$

$I = \dfrac{1}{12}\left[(2b - t)^3\, t + (d - 2t)\, t^3\right]$

$S = \dfrac{I}{x}$

$r = \sqrt{\dfrac{I}{A}}$

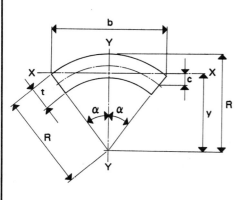

$r = R - t$

$A = \dfrac{\pi\alpha}{180}(R^2 - r^2)$

$b = 2R\sin\alpha$

$c = \dfrac{120\sin\alpha\,(R^3 - r^3)}{\pi\alpha\,(R^2 - r^2)} - \dfrac{(R + r)\cos\alpha}{2}$

$y = \dfrac{120\sin\alpha\,(R^3 - r^3)}{\pi\alpha\,(R^2 - r^2)}$

$I_x = \dfrac{1}{4}(R^4 - r^4)\left(\dfrac{\pi\alpha}{180} + \sin\alpha\cos\alpha\right) - \dfrac{80\sin^2\alpha\,(R^3 - r^3)^2}{\pi\alpha\,(R^2 - r^2)}$

$I_y = \dfrac{1}{4}(R^4 - r^4)\left(\dfrac{\pi\alpha}{180} - \sin\alpha\cos\alpha\right)$

PROPERTIES OF THE CIRCLE

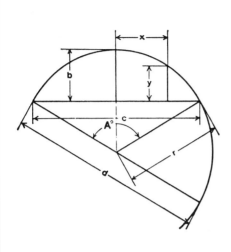

Circumference = 6.28318 r = 3.14159 d
Diameter = 0.31831 circumference
Area = 3.14159 r^2

Arc $\quad a = \dfrac{\pi r A^\circ}{180^\circ} = 0.017453 \, r A^\circ$

Angle $A^\circ = \dfrac{180^\circ \, a}{\pi r} = 57.29578 \dfrac{a}{r}$

Radius $r = \dfrac{4b^2 + c^2}{8b}$

Chord $c = 2\sqrt{2br - b^2} = 2r \sin \dfrac{A}{2}$

Rise $b = r - \tfrac{1}{2}\sqrt{4r^2 - c^2} = \dfrac{c}{2} \tan \dfrac{A}{4}$

$\qquad = 2r \sin^2 \dfrac{A}{4} = r + y - \sqrt{r^2 - x^2}$

$y = b - r + \sqrt{r^2 - x^2}$

$x = \sqrt{r^2 - (r + y - b)^2}$

Diameter of circle of equal periphery as square = 1.27324 side of square
Side of square of equal periphery as circle = 0.78540 diameter of circle
Diameter of circle circumscribed about square = 1.41421 side of square
Side of square inscribed in circle = 0.70711 diameter of circle

CIRCULAR SECTOR

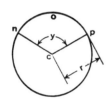

r = radius of circle $\qquad$ y = angle ncp in degrees

Area of Sector ncpo = ½ (length of arc nop × r)

$\qquad = $ Area of Circle $\times \dfrac{y}{360}$

$\qquad = 0.0087266 \times r^2 \times y$

CIRCULAR SEGMENT

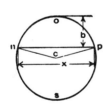

r = radius of circle $\qquad$ x = chord $\qquad$ b = rise

Area of Segment nop = Area of Sector ncpo − Area of triangle ncp

$\qquad = \dfrac{(\text{Length of arc nop} \times r) - x(r-b)}{2}$

Area of Segment nsp = Area of Circle − Area of Segment nop

VALUES FOR FUNCTIONS OF π

$\pi = 3.14159265359, \quad \log = 0.4971499$

$\pi^2 = 9.8696044, \log = 0.9942997 \qquad \dfrac{1}{\pi} = 0.3183099, \log = \overline{1}.5028501 \qquad \sqrt{\dfrac{1}{\pi}} = 0.5641896, \log = \overline{1}.7514251$

$\pi^3 = 31.0062767, \log = 1.4914496 \qquad \dfrac{1}{\pi^2} = 0.1013212, \log = 1.0057003 \qquad \dfrac{\pi}{180} = 0.0174533, \log = 2.2418774$

$\sqrt{\pi} = 1.7724539, \log = 0.2485749 \qquad \dfrac{1}{\pi^3} = 0.0322515, \log = 2.5085500 \qquad \dfrac{180}{\pi} = 57.2957795, \log = 1.7581226$

Note: Logs of fractions such as $\overline{1}.5028501$ and $\overline{2}.5085500$ may also be written 9.5028501 − 10 and 8.5085500 − 10 respectively.

PROPERTIES OF PARABOLA AND ELLIPSE

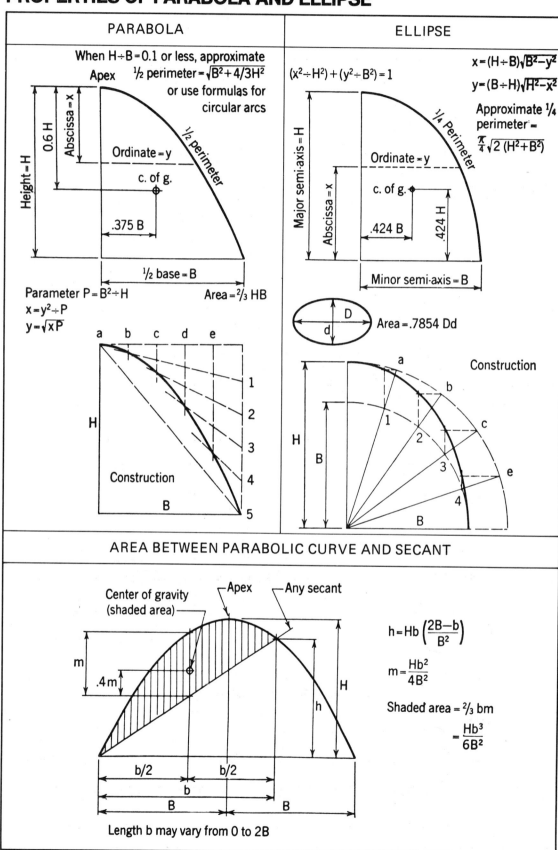

PROPERTIES OF SOLIDS

RECTANGULAR PARALLELEPIPED

Volume = *abc*

Surface area = 2(*ab* + *ac* + *bc*)

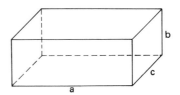

PARALLELEPIPED

Volume = *Ah* = *abc* sin θ

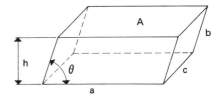

PYRAMID

Volume = $\frac{1}{3}$*Ah*

The centroid of a pyramid is located y-distance from the base on the line joining the centre of gravity of area A and the apex.

$y = \frac{h}{4}$

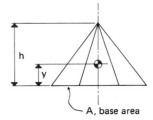

FRUSTUM OF PYRAMID

$V = \frac{h}{3}(A_1 + A_2 + \sqrt{A_1 A_2})$

The centroid is located y-distance up from area A_2 on the line joining the centres of gravity of areas A_1 and A_2

$y = \frac{h(A_1 + 2\sqrt{A_1 A_2} + 3A_2)}{4(A_1 + \sqrt{A_1 A_2} + A_2)}$

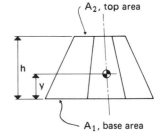

WEDGE

$V = \frac{(2a + c)bh}{6}$

The centroid is located y-distance from the base on the line joining the centre of gravity of the base area and the mid point of edge, c.

$y = \frac{h(a + c)}{2(2a + c)}$

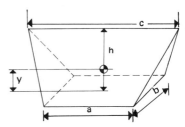

PROPERTIES OF SOLIDS

RIGHT CIRCULAR CYLINDER

Volume = $\pi r^2 h$

Lateral surface area = $2\pi rh$

$y = \dfrac{h}{2}$

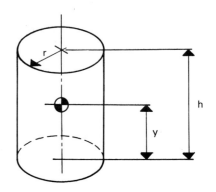

RIGHT CIRCULAR CONE

Volume = $\tfrac{1}{3}\pi r^2 h$

Lateral surface area = $\pi r \sqrt{r^2 + h^2} = \pi r l$

$y = \dfrac{h}{4}$

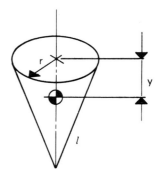

FRUSTRUM OF RIGHT CIRCULAR CONE

Volume = $\tfrac{1}{3}\pi h (a^2 + ab + b^2)$

Lateral surface area = $\pi(a+b)\sqrt{h^2 + (b-a)^2}$

$ = \pi(a+b)l$

$y = \dfrac{h}{4} \dfrac{(b^2 + 2ab + 3a^2)}{(b^2 + ab + a^2)}$

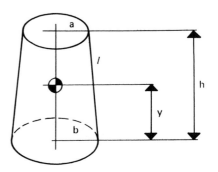

SPHERE

Volume = $\dfrac{4}{3}\pi r^3$

Surface area = $4\pi r^2$

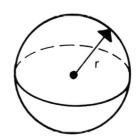

TRIGONOMETRIC FORMULAE

TRIGONOMETRIC FUNCTIONS

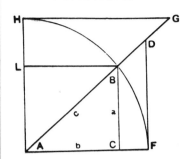

Radius AF $= 1$
$= \sin^2 A + \cos^2 A = \sin A \csc A$
$= \cos A \sec A = \tan A \cot A$

Sine A $= \dfrac{\cos A}{\cot A} = \dfrac{1}{\csc A} = \cos A \tan A = \sqrt{1 - \cos^2 A} = BC$

Cosine A $= \dfrac{\sin A}{\tan A} = \dfrac{1}{\sec A} = \sin A \cot A = \sqrt{1 - \sin^2 A} = AC$

Tangent A $= \dfrac{\sin A}{\cos A} = \dfrac{1}{\cot A} = \sin A \sec A \qquad = FD$

Cotangent A $= \dfrac{\cos A}{\sin A} = \dfrac{1}{\tan A} = \cos A \csc A \qquad = HG$

Secant A $= \dfrac{\tan A}{\sin A} = \dfrac{1}{\cos A} \qquad = AD$

Cosecant A $= \dfrac{\cot A}{\cos A} = \dfrac{1}{\sin A} \qquad = AG$

RIGHT ANGLED TRIANGLES

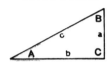

$a^2 = c^2 - b^2$
$b^2 = c^2 - a^2$
$c^2 = a^2 + b^2$

Known	Required					
	A	B	a	b	c	Area
a, b	$\tan A = \dfrac{a}{b}$	$\tan B = \dfrac{b}{a}$			$\sqrt{a^2 + b^2}$	$\dfrac{ab}{2}$
a, c	$\sin A = \dfrac{a}{c}$	$\cos B = \dfrac{a}{c}$		$\sqrt{c^2 - a^2}$		$\dfrac{a\sqrt{c^2 - a^2}}{2}$
A, a		$90° - A$		$a \cot A$	$\dfrac{a}{\sin A}$	$\dfrac{a^2 \cot A}{2}$
A, b		$90° - A$	$b \tan A$		$\dfrac{b}{\cos A}$	$\dfrac{b^2 \tan A}{2}$
A, c		$90° - A$	$c \sin A$	$c \cos A$		$\dfrac{c^2 \sin 2A}{4}$

OBLIQUE ANGLED TRIANGLES

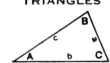

$s = \dfrac{a + b + c}{2}$

$K = \sqrt{\dfrac{(s-a)(s-b)(s-c)}{s}}$

$a^2 = b^2 + c^2 - 2bc \cos A$
$b^2 = a^2 + c^2 - 2ac \cos B$
$c^2 = a^2 + b^2 - 2ab \cos C$

Known	Required					
	A	B	C	b	c	Area
a, b, c	$\tan \tfrac{1}{2} A = \dfrac{K}{s-a}$	$\tan \tfrac{1}{2} B = \dfrac{K}{s-b}$	$\tan \tfrac{1}{2} C = \dfrac{K}{s-c}$			$\sqrt{s(s-a)(s-b)(s-c)}$
a, A, B			$180° - (A+B)$	$\dfrac{a \sin B}{\sin A}$	$\dfrac{a \sin C}{\sin A}$	
a, b, A		$\sin B = \dfrac{b \sin A}{a}$			$\dfrac{b \sin C}{\sin B}$	
a, b, C	$\tan A = \dfrac{a \sin C}{b - a \cos C}$				$\sqrt{a^2 + b^2 - 2ab \cos C}$	$\dfrac{ab \sin C}{2}$

BRACING FORMULAE

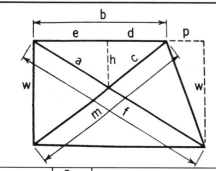 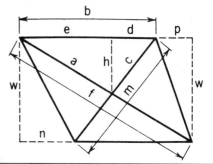

Given	To Find	Formula
bpw	f	$\sqrt{(b+p)^2 + w^2}$
bw	m	$\sqrt{b^2 + w^2}$
bp	d	$b^2 \div (2b + p)$
bp	e	$b(b+p) \div (2b+p)$
bfp	a	$bf \div (2b+p)$
bmp	c	$bm \div (2b+p)$
bpw	h	$bw \div (2b+p)$
afw	h	$aw \div f$
cmw	h	$cw \div m$

Given	To Find	Formula
bpw	f	$\sqrt{(b+p)^2 + w^2}$
bnw	m	$\sqrt{(b-n)^2 + w^2}$
bnp	d	$b(b-n) \div (2b+p-n)$
bnp	e	$b(b+p) \div (2b+p-n)$
bfnp	a	$bf \div (2b+p-n)$
bmnp	c	$bm \div (2b+p-n)$
bnpw	h	$bw \div (2b+p-n)$
afw	h	$aw \div f$
cmw	h	$cw \div m$

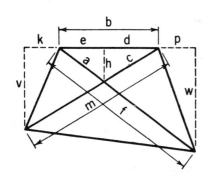

Given	To Find	Formula
bpw	f	$\sqrt{(b+p)^2 + w^2}$
bkv	m	$\sqrt{(b+k)^2 + v^2}$
bkpvw	d	$bw(b+k) \div [v(b+p) + w(b+k)]$
bkpvw	e	$bv(b+p) \div [v(b+p) + w(b+k)]$
bfkpvw	a	$fbv \div [v(b+p) + w(b+k)]$
bkmpvw	c	$bmw \div [v(b+p) + w(b+k)]$
bkpvw	h	$bvw \div [v(b+p) + w(b+k)]$
afw	h	$aw \div f$
cmv	h	$cv \div m$

PARALLEL BRACING

$k = (\log B - \log T) \div$ no. of panels. Constant k plus the logarithm of any line equals the log of the corresponding line in the next panel below.

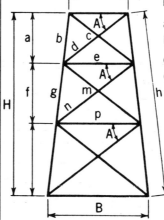

$a = TH \div (T + e + p)$
$b = Th \div (T + e + p)$
$c = \sqrt{(½T + ½e)^2 + a^2}$
$d = ce \div (T + e)$

$\log e = k + \log T$
$\log f = k + \log a$
$\log g = k + \log b$
$\log m = k + \log c$
$\log n = k + \log d$
$\log p = k + \log e$

The above method can be used for any number of panels.
In the formulas for "a" and "b" the sum in parenthesis, which in the case shown is (T + e + p), is always composed of all the horizontal distances except the base.

LENGTH OF CIRCULAR ARCS FOR UNIT RADIUS

By the use of this table, the length of any arc may be found if the length of the radius and the angle of the segment are known.

Example: Required the length of arc of segment 32° 15' 27" with radius of 8 000 mm.
From table: Length of arc (Radius 1) for 32° = .5585054
15' = .0043633
27" = .0001309
.5629996

.5629996 X 8 000 (length of radius) = 4504 mm

For the same arc but with the radius expressed as 24 feet 3 inches, the length of arc would be 0.5629996 X 24.25 = 13.65 feet

DEGREES						MINUTES		SECONDS	
1	.017 4533	61	1.064 6508	121	2.111 8484	1	.000 2909	1	.000 0048
2	.034 9066	62	1.082 1041	122	2.129 3017	2	.000 5818	2	.000 0097
3	.052 3599	63	1.099 5574	123	2.146 7550	3	.000 8727	3	.000 0145
4	.069 8132	64	1.117 0107	124	2.164 2083	4	.001 1636	4	.000 0194
5	.087 2665	65	1.134 4640	125	2.181 6616	5	.001 4544	5	.000 0242
6	.104 7198	66	1.151 9173	126	2.199 1149	6	.001 7453	6	.000 0291
7	.122 1730	67	1.169 3706	127	2.216 5682	7	.002 0362	7	.000 0339
8	.139 6263	68	1.186 8239	128	2.234 0214	8	.002 3271	8	.000 0388
9	.157 0796	69	1.204 2772	129	2.251 4747	9	.002 6180	9	.000 0436
10	.174 5329	70	1.221 7305	130	2.268 9280	10	.002 9089	10	.000 0485
11	.191 9862	71	1.239 1838	131	2.286 3813	11	.003 1998	11	.000 0533
12	.209 4395	72	1.256 6371	132	2.303 8346	12	.003 4907	12	.000 0582
13	.226 8928	73	1.274 0904	133	2.321 2879	13	.003 7815	13	.000 0630
14	.244 3461	74	1.291 5436	134	2.338 7412	14	.004 0724	14	.000 0679
15	.261 7994	75	1.308 9969	135	2.356 1945	15	.004 3633	15	.000 0727
16	.279 2527	76	1.326 4502	136	2.373 6478	16	.004 6542	16	.000 0776
17	.296 7060	77	1.343 9035	137	2.391 1011	17	.004 9451	17	.000 0824
18	.314 1593	78	1.361 3568	138	2.408 5544	18	.005 2360	18	.000 0873
19	.331 6126	79	1.378 8101	139	2.426 0077	19	.005 5269	19	.000 0921
20	.349 0659	80	1.396 2634	140	2.443 4610	20	.005 8178	20	.000 0970
21	.366 5191	81	1.413 7167	141	2.460 9142	21	.006 1087	21	.000 1018
22	.383 9724	82	1.431 1700	142	2.478 3675	22	.006 3995	22	.000 1067
23	.401 4257	83	1.448 6233	143	2.495 8208	23	.006 6904	23	.000 1115
24	.418 8790	84	1.466 0766	144	2.513 2741	24	.006 9813	24	.000 1164
25	.436 3323	85	1.483 5299	145	2.530 7274	25	.007 2722	25	.000 1212
26	.453 7856	86	1.500 9832	146	2.548 1807	26	.007 5631	26	.000 1261
27	.471 2389	87	1.518 4364	147	2.565 6340	27	.007 8540	27	.000 1309
28	.488 6922	88	1.535 8897	148	2.583 0873	28	.008 1449	28	.000 1357
29	.506 1455	89	1.553 3430	149	2.600 5406	29	.008 4358	29	.000 1406
30	.523 5988	90	1.570 7963	150	2.617 9939	30	.008 7266	30	.000 1454
31	.541 0521	91	1.588 2496	151	2.635 4472	31	.009 0175	31	.000 1503
32	.558 5054	92	1.605 7029	152	2.652 9005	32	.009 3084	32	.000 1551
33	.575 9587	93	1.623 1562	153	2.670 3538	33	.009 5993	33	.000 1600
34	.593 4119	94	1.640 6095	154	2.687 8070	34	.009 8902	34	.000 1648
35	.610 8652	95	1.658 0628	155	2.705 2603	35	.010 1811	35	.000 1697
36	.628 3185	96	1.675 5161	156	2.722 7136	36	.010 4720	36	.000 1745
37	.645 7718	97	1.692 9694	157	2.740 1669	37	.010 7629	37	.000 1794
38	.663 2251	98	1.710 4227	158	2.757 6202	38	.011 0538	38	.000 1842
39	.680 6784	99	1.727 8760	159	2.775 0735	39	.011 3446	39	.000 1891
40	.698 1317	100	1.745 3293	160	2.792 5268	40	.011 6355	40	.000 1939
41	.715 5850	101	1.762 7825	161	2.809 9801	41	.011 9264	41	.000 1988
42	.733 0383	102	1.780 2358	162	2.827 4334	42	.012 2173	42	.000 2036
43	.750 4916	103	1.797 6891	163	2.844 8867	43	.012 5082	43	.000 2085
44	.767 9449	104	1.815 1424	164	2.862 3400	44	.012 7991	44	.000 2133
45	.785 3982	105	1.832 5957	165	2.879 7933	45	.013 0900	45	.000 2182
46	.802 8515	106	1.850 0490	166	2.897 2466	46	.013 3809	46	.000 2230
47	.820 3047	107	1.867 5023	167	2.914 6999	47	.013 6717	47	.000 2279
48	.837 7580	108	1.884 9556	168	2.932 1531	48	.013 9626	48	.000 2327
49	.855 2113	109	1.902 4089	169	2.949 6064	49	.014 2535	49	.000 2376
50	.872 6646	110	1.919 8622	170	2.967 0597	50	.014 5444	50	.000 2424
51	.890 1179	111	1.937 3155	171	2.984 5130	51	.014 8353	51	.000 2473
52	.907 5712	112	1.954 7688	172	3.001 9663	52	.015 1262	52	.000 2521
53	.925 0245	113	1.972 2221	173	3.019 4196	53	.015 4171	53	.000 2570
54	.942 4778	114	1.989 6753	174	3.036 8729	54	.015 7080	54	.000 2618
55	.959 9311	115	2.007 1286	175	3.054 3262	55	.015 9989	55	.000 2666
56	.977 3844	116	2.024 5819	176	3.071 7795	56	.016 2897	56	.000 2715
57	.994 8377	117	2.042 0352	177	3.089 2328	57	.016 5806	57	.000 2763
58	1.012 2910	118	2.059 4885	178	3.106 6861	58	.016 8715	58	.000 2812
59	1.029 7443	119	2.076 9418	179	3.124 1394	59	.017 1624	59	.000 2860
60	1.047 1976	120	2.094 3951	180	3.141 5927	60	.017 4533	60	.000 2909

SI SUMMARY

General

The following information on SI units is provided to assist those involved in the planning, design, fabrication and erection of steel structures prepared in SI units. Information related to the metric system in general is to be found in CAN3-Z234.1-79, "Canadian Metric Practice Guide" and for terms related to the steel industry in the "Industry Practice Guide for SI Metric Units in the Canadian Iron and Steel Industry". The latter is available from the Task Force for Metric Conversion in the Canadian Iron and Steel Industry, P.O. Box 4248, Station "D", Hamilton, Ontario, L8V 4L6.

The eleventh General Conference of Weights and Measures, in 1960, adopted the name International System of Units for a coherent system which includes the metre as the base unit of length and the kilogram as the base unit of mass. The international abbreviation of the name of this system, in all languages, is SI.

Canada is a signatory to the General Conference on Weights and Measures, and in 1970, the Canadian government stated that the eventual conversion to the metric system is an objective of Canadian policy. Since that time, metric conversion activity in Canada has developed to the point where material and design standards, building codes and technical literature are available in SI units.

The SI system is based on the seven base units listed in Table 7–1. Decimal multiples and sub-multiples of the SI base units are formed by the addition of the prefixes given in Table 7-2.

SI BASE UNITS
Table 7-1

Quantity	Name	Symbol
length	metre	m
mass	kilogram	kg
time	second	s
electric current	ampere	A
thermodynamic temperature	kelvin	K
amount of substance	mole	mol
luminous intensity	candela	cd

SI PREFIXES
Table 7-2

Multiplying Factor	Prefix	Symbol
$1\,000\,000\,000\,000 = 10^{12}$	tera	T
$1\,000\,000\,000 = 10^{9}$	giga	G
$1\,000\,000 = 10^{6}$	mega	M
$1\,000 = 10^{3}$	kilo	k
$100 = 10^{2}$	hecto	h
$10 = 10^{1}$	deca	da
$0.1 = 10^{-1}$	deci	d
$0.01 = 10^{-2}$	centi	c
$0.001 = 10^{-3}$	milli	m
$0.000\,001 = 10^{-6}$	micro	μ
$0.000\,000\,001 = 10^{-9}$	nano	n
$0.000\,000\,000\,001 = 10^{-12}$	pico	p
$0.000\,000\,000\,000\,001 = 10^{-15}$	femto	f
$0.000\,000\,000\,000\,000\,001 = 10^{-18}$	atto	a

In choosing the appropriate decimal multiple or sub-multiple, the Canadian Metric Practice Guide recommends the use of prefixes representing 10 raised to a power that is a multiple of 3, a ternary power. Thus, common structural steel design units would be:

Force — newton (N), kilonewton (kN)
Stress — pascal (Pa), kilopascal (kPa), megapascal (MPa)
Length — millimetre (mm), metre (m)
Mass — kilogram (kg), megagram (Mg)

The tonne is a special unit, equal to 1 000 kg (or 1 Mg) that will be used in the basic steel industry, but should not be used in structural design calculations.

Designers using SI units must transform loads given in mass (kilograms) to forces, using the relationship force = mass times acceleration. In the design of structures on earth, acceleration is the acceleration due to gravity, designated by "g" and established as 9.806 65 metres per second per second at the third General Conference on Weights and Measures in 1901.

The unit of force to be used in design is the newton (N) (or multiples thereof) where a newton is defined as the force that, when applied to a body having a mass of one kilogram (kg), gives the body an acceleration of one metre (m) per second squared (s^2). The unit of stress is the pascal (Pa), which is one newton per square metre (m^2). Since this is a very small unit, designers of steel structures will generally use megapascals (MPa), where one megapascal is one million pascals and equals one newton per square millimetre (N/mm^2). See also "Structural Loads, Mass and Force".

Properties and dimensions of steel sections are given, in this book, in millimetre units, tabulated to an appropriate ternary power of 10, and millimetres should be used for dimensioning steel structures. Some relationships and values of interest to steel designers are shown below:

SI PREFIXES
Table 7-3

Density of Steel		7 850 kg/m^3
Modulus of Elasticity	E	200 000 MPa
Shear Modulus of Steel	G	77 000 MPa
Coefficient of Thermal Expansion		11.7×10^{-6} /°C
Acceleration due to Earth's Gravity	g	9.806 65 m/s^2

For a more complete description of SI, the Canadian Metric Practice Guide should be consulted; however, Table 7–4 provides a convenient summary listing selected SI units, the quantity represented, the unit name and typical application.

Structural Loads, Mass and Force

Since most civil engineers have been accustomed to designing structures on earth to withstand loads more variable than the acceleration due to gravity, the pound-force and the kilogram-force have been used as standard units of force. These units were assumed to be numerically equal to their mass counter-parts, the pound-mass and the kilogram-mass respectively.

In SI, the units of mass and force, the kilogram and the newton respectively, are distinctly different both in name and in value. The two are related through the famous Newtonian equation, force = mass times acceleration, or

$$F = ma.$$

Thus a newton (N) is defined as the force required to give one kilogram (kg) mass an acceleration of one metre (m) per second (s) squared, or

$$1 \text{ N} = 1 \text{ kg} \cdot \text{m/s}^2.$$

The standard international value of acceleration due to gravity is 9.806 65 m/s^2.

However, for hand calculations in Canada a value of

$$g = 9.81 \text{ m/s}^2$$

may be more acceptable as it retains three significant figures (adequate for most structural design) and produces a numerical value of force distinctly different from the value of mass. Thus, whether or not the mass has been converted to a force will be readily apparent, and errors will tend to be reduced.

SELECTED SI UNITS
Table 7-4

Quantity	Preferred Units	Unit Name	Typical Applications	Remarks
Area	mm^2	square millimetre	Area of cross section for structural sections	Avoid cm^2
	m^2	square metre	Areas in general	
Bending Moment	kN·m	kilonewton metre	Bending moment in structural sections	
Coating mass	g/m^2	gram per square metre	Mass of zinc coating on steel deck	
Coefficient of Thermal Expansion	$1/°C$*	reciprocal (of) degree Celsius	Expansion of materials subject to temperature change (generally expressed as a ratio per degree Celsius)	$11.7 \times 10^{-6}/°C$ for steel
Density, mass	kg/m^3	kilogram per cubic metre	Density of materials in general; mass per unit volume	$7\,850\ kg/m^3$ for steel
Force	N	newton	Unit of force used in structural calculations	$1 N = 1\ kg \cdot m/s^2$
	kN	kilonewton	Force in structural elements such as columns; concentrated forces; axial forces; reactions; shear force; gravitational force	
Force per Unit Length	N/m	newton per metre	Unit for use in calculations	$1\ kg/m \times 9.81\ m/s^2$ $= (9.81\ kg \cdot m/s^2) \times \frac{1}{m}$ $= 9.81\ N/m$
	kN/m	kilonewton per metre	Transverse force per unit length on a beam, column etc.; dead load of a beam for stress calculations	$(1\ kg/m \times 9.81\ m/s^2) \times \frac{1}{1\,000}$ $= (9.81\ kg \cdot m/s^2) \times \frac{1}{m} \times \frac{1\,000}{1\,000}$ $= (9.81\ N/m) \times \frac{1\,000}{1\,000}$ $= 9.81\ kN/m \times 1/1\,000$ $= 0.009\,81\ kN/m$
Force per Unit Area (See Pressure)				
Frequency	Hz	hertz	Frequency of vibration	$1\ Hz = 1/s = s^{-1}$ replaces cycle per second (cps)
Impact energy	J	joule	Charpy V-notch test	$1\ N \cdot m = 1\ J$
Length	mm	millimetre	Dimensions on all drawings; dimensions of sections, spans, deflection, elongations, eccentricity	
	m	metre	Overall dimensions; in calculations; contours; surveys	
	km	kilometre	Distances for transportation purposes	
	μm	micrometre	Thickness of coatings (paint)	
Mass	kg	kilogram	Mass of materials, structural elements and machinery	A metric tonne, t $1t = 10^3\ kg = 1Mg = 1\,000\ kg$
Mass per Unit Length	kg/m	kilogram per metre	Mass per unit length of section, bar, or similar items of uniform cross section.	Also known as "linear density"
Mass per Unit Area	kg/m^2	kilogram per square metre	Mass per unit area of plates, slabs, or similar items of uniform thickness; rating for load-carrying capacities on floors (display on notices only)	DO NOT USE IN STRESS CALCULATION
Mass Density	kg/m^3	kilogram per cubic metre	Density of materials in general; mass per unit volume	$7\,850\ kg/m^3$ for steel
Modulus of Elasticity (Young's)	MPa	megapascal	Modulus of elasticity; Young's modulus	200 000 MPa for carbon, high-strength low alloy and low-alloy wrought steels
Modulus, Shear	MPa	megapascal	Shear Modulus	77 000 MPa assumed for steel
Modulus, Section	mm^3	millimetre to third power	First moment of area of cross section of structural section, such as plastic section modulus, elastic section modulus	

* The preferred unit is 1/K, however $1/°C$ is an acceptable unit for the construction industry.

SELECTED SI UNITS
Table 7-4

Quantity	Preferred Units	Unit Name	Typical Applications	Remarks
Moment of Inertia	mm^4	millimetre to fourth power	Second moment of area; moment of inertia of a section; torsional constant of cross section	
Moment of Force	kN·m	kilonewton metre	Bending moment (in structural sections); overturning moment	
	N·m	newton metre		
Pressure (see also Stress)	Pa	pascal	Unit used in calculation	$1\ Pa = 1\ N/m^2$
	kPa	kilopascal	Uniformly distributed loads on floors; soil pressure; wind loads; snow loads; dead loads; live loads.	$1\ kPa = 1\ kN/m^2$
Section Modulus (see Modulus)				
Stress	MPa	megapascal	Stress (yield, ultimate, permitted, calculated) in structural steel	$1\ MPa = 1\ MN/m^2$ $= 1\ N/mm^2$
Structural Load (see Force)				
Temperature	°C	degree Celsius	Ambient temperature	$0°C \simeq 273.15K$ However, for temperature intervals $1°C = 1K$
Thickness	mm	millimetre	Thickness of web, flange, plate, etc.	
	μm	micrometre	Thickness of paint	
Torque	kN·m	kilonewton metre	Torsional moment on a cross section; inspection torque for high strength bolts.	
Volume	m^3	cubic metre	Volume; volume of earthworks, excavation, concrete, sand, all bulk materials.	$1\ m^3 = 1\ 000\ L$ The cubic metre is the preferred unit of volume for engineering purposes
	L	litre	Volume of fluids and containers for fluids	
Work, Energy	J	joule	Energy absorbed in impact testing of materials; energy in general	$1\ kWh = 3.6\ MJ$ where kWh is a kilowatt hour.

There are two common areas where the designer of a structure must be alert to the distinction between mass and force:

1. dead loads due to the mass of the structural elements, permanent equipment etc.,
2. superimposed, or live loads due to storage of materials.

In these, and other cases where mass is well known since it is the unit of commerce, the designer must convert mass to force by multiplying by g.

COMMON CONVERSION FACTORS
Table 7-5

Item	Imperial — SI	SI — Imperial
Acceleration	1 ft./s^2 = 0.304 8 m/s^2	1 m/s^2 = 3.2808 ft./s^2
Area	1 acre = 0.404 685 6 ha 1 ft.^{2} = 0.092 903 04 m^2 1 in.^{2} = 645.16 mm^2 1 mi.^{2} = 2.589 988 km^2 1 yd.^{2} = 0.836 127 4 m^2	1 ha = 2.471 acres 1 m^2 = 10.764 ft.^{2} 1 mm^2 = 1.55×10^{-3} in.^{2} 1 km^2 = 0.3861 mi.^{2} 1 m^2 = 1.20 yd.^{2}
Capacity (Canadian Legal Units)	1 oz. = 28.413 062 mL 1 gal. = 4.546 090 L 1 pt. = 0.568 261 L 1 qt. = 1.136 522 L	1 mL = 35.2×10^{-3} oz. 1 L = 0.220 gal. 1 L = 1.76 pt. 1 L = 0.880 qt.
Density, Mass	1 lb./ft. = 1.488 16 kg/m 1 lb./yd. = 0.496 055 kg/m 1 oz./ft.^{2} = 305.152 g/m^2 1 lb./ft.^{2} = 4.882 43 kg/m^2 1 lb./in.^{2} = 703.069 6 kg/m^2 1 lb./ft.^{3} = 16.018 46 kg/m^3 1 lb./in.^{3} = 27.679 90 Mg/m^3	1 kg/m = 0.672 lb./ft. 1 kg/m = 2.016 lb./yd. 1 g/m^2 = 3.277×10^{-3} oz./ft.^{2} 1 kg/m^2 = 0.205 lb./ft.^{2} 1 kg/m^2 = 1.42×10^{-3} lb./in.^{2} 1 kg/m^3 = 62.4×10^{-3} lb./ft.^{3} 1 Mg/m^3 = 0.0361 lb./in.^{3}
Force	1 kip = 4.448 222 kN	1 kN = 0.225 kip
Length	1 ft. = 0.304 8 m = 304.8 mm 1 in. = 25.4 mm 1 mile = 1.609 344 km 1 yd. = 0.914 4 m	1 m = 3.28 ft. 1 mm = 0.0394 in. 1 km = 0.622 mi. 1 m = 1.09 yd.
Mass	1 lb. = 0.453 592 37 kg 1 ton (2000 lb.) = 0.907 184 74 Mg	1 kg = 2.20 lb. 1 Mg = 1.10 ton = 2200 lb.
Mass per Unit Area	1 lb./ft.^{2} = 4.882 43 kg/m^2	1 kg/m^2 = 0.205 lb./ft.^{2}
Mass per Unit Length	1 lb./ft. = 1.488 16 kg/m	1 kg/m = 0.672 lb./ft.
Moment of Inertia a) Second Moment of Area b) Section Modulus	1 in.^{4} = 416 231.4 mm^4 1 in.^{3} = 16 387.064 mm^3	1 mm^4 = 2.4×10^{-6} in.^{4} 1 mm^3 = 0.061×10^{-3} in.^{3}
Pressure or Stress	1 ksi = 6.894 757 MPa 1 psf = 47.880 26 Pa 1 psi = 6.894 757 kPa	1 MPa = 0.145 ksi 1 Pa = 0.0209 psf 1 kPa = 0.145 psi
Torque or Moment of Force	1 ft.•kipf = 1.355 818 kN•m	1 kN•m = 0.738 ft.•kipf
Volume	1 in.^{3} = 16 387.064 mm^3 1 ft.^{3} = 28.316 85 dm^3 1 yd.^{3} = 0.764 555 m^3	1 mm^3 = 0.061×10^{-3} in.^{3} 1 dm^3 = 0.0353 ft.^{3} 1 m^3 = 1.308 yd.^{3}
Costs	1 $/ft. = 3.28 $/m 1 $/ft.^{2} = 10.764 $/m^2 1 $/yd.^{2} = 1.20 $/m^2 1 $/ft.^{3} = 35.34 $/m^3 1 $/yd.^{3} = 1.307 $/m^3	1 $/m = 0.305 $/ft. 1 $/m^2 = 0.0929 $/ft.^{2} 1 $/m^2 = 0.836 $/yd.^{2} 1 $/m^3 = 0.0283 $/ft.^{3} 1 $/m^3 = 0.765 $/yd.^{3}

MILLIMETRE EQUIVALENTS
DECIMALS AND EACH 64TH OF AN INCH

FRACTIONS		INCHES—mm
1/64		.015625 — .397
	1/32	.03125 — .794
		.03937 — (1)
3/64		.046875 — 1.191
	1/16	.0625 — 1.588
5/64		.078125 — 1.984
		.07874 — (2)
	3/32	.09375 — 2.381
7/64		.109375 — 2.778
		.11811 — (3)
	1/8	.125 — 3.175
9/64		.140625 — 3.572
	5/32	.15625 — 3.969
		.15748 — (4)
11/64		.171875 — 4.366
	3/16	.1875 — 4.763
		.19685 — (5)
13/64		.203125 — 5.159
	7/32	.21875 — 5.556
15/64		.234375 — 5.953
		.23622 — 6
	(¼)	.25 — 6.350
17/64		.265625 — 6.747
		.27559 — 7
	9/32	.28125 — 7.144
19/64		.296875 — 7.541
	5/16	.3125 — 7.938
		.31496 — 8
21/64		.328125 — 8.334
	11/32	.34375 — 8.731
		.35433 — 9
23/64		.359375 — 9.128
	3/8	.375 — 9.525
25/64		.390625 — 9.922
		.3937 — (10)
	13/32	.40625 — 10.319
27/64		.421875 — 10.716
		.43307 — 11
	7/16	.4375 — 11.113
29/64		.453125 — 11.509
	15/32	.46875 — 11.906
		.47244 — 12
31/64		.484375 — 12.303
	(½)	.5 — 12.700

FRACTIONS		INCHES—mm
		.51181 — 13
33/64		.515625 – 13.097
	17/32	.53125 — 13.494
35/64		.546875 – 13.891
		.55118 — 14
	9/16	.5625 — 14.288
37/64		.578125 – 14.684
		.59055 — (15)
	19/32	.59375 — 15.081
39/64		.609375 – 15.478
	5/8	.625 — 15.875
		.62992 — 16
41/64		.640625 – 16.272
	21/32	.65625 — 16.669
		.66929 — 17
43/64		.671875 – 17.066
	11/16	.6875 — 17.463
45/64		.703125 – 17.859
		.70866 — 18
	23/32	.71875 — 18.256
47/64		.734375 – 18.653
		.74893 — 19
	(¾)	.75 — 19.050
49/64		.765625 – 19.447
	25/32	.781250 – 19.844
		.7874 — (20)
51/64		.796875 – 20.241
	13/16	.8125 — 20.638
		.82677 — 21
53/64		.828125 – 21.034
	27/32	.84375 — 21.431
55/64		.859375 – 21.828
		.86614 — 22
	7/8	.875 — 22.225
57/64		.890625 – 22.622
		.90551 — 23
	29/32	.90625 — 23.019
59/64		.921875 – 23.416
	15/16	.9375 — 23.813
		.94488 — 24
61/64		.953125 – 24.209
	31/32	.96875 — 24.606
		.98425 — (25)
63/64		.984375 – 25.003
(1)		1.000 — 25.4

MISCELLANEOUS CONVERSION FACTORS

Area
1 acre	= 0.404 685 6 ha
1 hectare	= 1 hm^2
1 legal subdivision (40 acres)	= 0.161 874 2 km^2
1 section (1 mile square, 640 acres)	= 2.589 988 km^2
1 square foot	= **929.0304** cm^2
1 square inch	= **645.16** mm^2
1 square mile	= 2.589 988 km^2
1 square yard	= 0.836 127 4 m^2
1 township (36 sections)	= 93.239 57 km^2

Linear Density (Mass per Unit Length)
1 pound per inch	= 17.858 kg/m
1 pound per foot	= 1.488 16 kg/m
1 pound per yard	= 0.496 055 kg/m

Area Density (Mass per Unit Area)
1 ounce per square foot	= 305.152 g/m^2
1 pound per square foot	= 4.882 43 kg/m^2
1 pound per square inch	= 703.0696 kg/m^2

Mass Density (Mass per Unit Volume)
1 pound per cubic foot	= 16.018 46 kg/m^3
1 pound per cubic inch	= 27.679 90 Mg/m^3
1 ton (long) per cubic yard	= 1.328 939 Mg/m^3
1 ton (short) per cubic yard	= 1.186 553 Mg/m^3

Energy
1 British thermal unit (Btu) (International Table)	= **1.055 056** kJ
1 foot pound-force	= 1.355 818 J
1 horsepower hour	= 2.684 52 MJ
1 kilowatt hour	= **3.6** MJ

Force
1 kilogram-force	= **9.806 65** N
1 kip (thousand pounds force)	= 4.448 222 kN
1 pound-force	= 4.448 222 N

Heat
1 Btu *foot per (square foot hour °F)	= 1.730 74 W/(m•K)	k - value
1 Btu per (square foot hour °F)	= 5.678 29 W/(m^2•K)	U - value
1 square foot hour °F per Btu	= 0.176 109 m^2•K/W	R - value

Based on the Btu IT.

Length
1 chain (66 feet)	= **20.1168** m
1 foot	= **0.3048** m
1 inch	= **25.4** mm
1 microinch	= **25.4** nm
1 micron	= **1** μm
1 mil (0.001 inch)	= **25.4** μm
1 mile	= **1.609 344** km
1 mile (International nautical)	= **1.852** km
1 mile (UK nautical)	= **1.853 184** km
1 mile (US nautical)	= **1.852** km
1 yard	= **0.9144** m

MISCELLANEOUS CONVERSION FACTORS

Mass
1 hundredweight (100 lb) = **45.359 237** kg
1 hundredweight (long) (112 lb, UK) = 50.802 345 kg
1 pennyweight = 1.555 174 g
1 pound (avoirdupois) = **0.453 592 37** kg
1 ton (long, 2240 lb, UK) = **1.016 046 908 8** Mg
1 ton (short, 2000 lb) = **0.907 184 74** Mg

Mass Concentration
1 pound per cubic foot = 16.018 46 kg/m^3

Second Moment of Area (Moment of Inertia)
1 inch^4 = 0.416 231 4 × 10^6 mm^4

Section Modulus
1 inch^3 = **16.387 064 × 10^3 mm^3**

Momentum
1 pound foot per second = 0.138 255 kg·m/s

Power. See also Energy.
1 Btu (IT)* per hour = 0.293 072 W
1 foot pound-force per hour = 0.376 616 1 mW
1 foot pound-force per minute = 22.596 97 mW
1 foot pound-force per second = 1.355 818 W
1 horsepower (550 ft·lbf/s) = 745.6999 W

International Tables.

Pressure or Stress (Force per Area)
1 atmosphere, standard = **101.325** kPa
1 inch of mercury (conventional, 32°F) = 3.386 39 kPa
1 inch of water (conventional) = 249.089 Pa
1 ksi (1000 lbf/in^2) = 6.894 757 MPa
1 mm mercury (conventional, 0°C) = 133.322 Pa
1 pound-force per square foot = 47.880 26 Pa
1 pound-force per square inch (psi) = 6.894 757 kPa
1 ton-force per square inch = 13.789 514 MPa
1 ton-force (UK) per square inch = 15.4443 MPa

Temperature

Scales
Celsius * temperature = temperature in kelvins − **273.15**
Fahrenheit temperature = **1.8** (Celsius temperature) + **32**
Fahrenheit temperature = **1.8** (temperature in kelvins) − **459.67**
Rankine temperature = **1.8** (temperature in kelvins)

Intervals
1 degree Celsius* = 1 K
1 degree Fahrenheit = 5/9 K
1 degree Rankine = 5/9 K

"Celsius" replaced "Centigrade" in 1948 to eliminate confusion with the word centigrade, associated with centesimal angular measure.

MISCELLANEOUS CONVERSION FACTORS

Time
1 day (mean solar)	= **86.4** ks
1 hour (mean solar)	= **3.6** ks
1 minute (mean solar)	= **60** s
1 month (mean calendar, 365/12 days)	= **2.628** Ms
1 year (calendar, 365 days)	= **31.536** Ms

Torque (Moment of Force)
1 pound-force foot	= 1.355 818 N•m
1 pound-force inch	= 0.112 985 N•m

Volume
1 acre foot	= 1233.482 m^3
1 barrel (oil, 42 US gallons)	= **0.158 987 3** m^3
1 board foot*	= 2.359 737 dm^3
1 cubic foot	= 28.316 85 dm^3
1 cubic inch	= **16.387 064** cm^3
1 cubic yard	= 0.764 555 m^3
1 gallon‡	= **4.546 09** dm^3
1 gallon (UK)§	= 4.546 092 dm^3
1 gallon (US)	= 3.785 412 dm^3

*The board foot is nominally 1 x 12 x 12 = 144 in^3. However, the actual volume of wood is about 2/3 of the nominal quantity.

§ Also referred to as the "Imperial gallon."

Volume Rate of Flow
1 cubic foot per minute	= 0.471 947 4 dm^3/s
1 cubic foot per second	= 28.316 85 dm^3/s
1 cubic yard per minute	= 12.742 58 dm^3/s
1 gallon per minute	= 75.768 17 cm^3/s
1 gallon (UK) per minute	= 75.7682 cm^3/s
1 gallon (US) per minute	= 63.0902 cm^3/s
1 million gallons per day	= 52.6168 dm^3/s

Notes:

1. The conversion factors give the relationship between SI units and other Canadian legal units as well as commonly encountered units of measure of United Kingdom and USA origin. The yard and the pound are the same throughout the world; by definition they are specified fractions of the metre and the kilogram. The gallons of Canada and Australia, which are identical, differ by a relatively insignificant amount from the gallon of the United Kingdom, whereas that of the USA is a much smaller measure.

2. The conversion factors given in tables apply to Canadian units unless stated otherwise.

3. Conversion factors that are exact are shown in boldface type. Other factors are given to more than sufficient accuracy for most general and scientific work.

4. Conversions are those listed in CAN3-Z234.1-79

PART 8
SELECTED TABLES BASED ON CSA-G40.21 300W

General Information	8–3
Prying Action	8–4
Class of Sections for Beam-Columns	8–6
Factored Axial Compressive Resistances	8–10
WWF Columns	8–10
W Columns	8–17
Composite Beams – Trial Selection Tables	8–23
75 mm Deck with 65 mm Slab, 20 MPa, 2300 kg/m^3 concrete	8–24
75 mm Deck with 85 mm Slab, 25 MPa, 1850 kg/m^3 concrete	8–32
38 mm Deck with 65 mm Slab, 20 MPa, 2300 kg/m^3 concrete	8–40
Beam Selection Table – WWF and W Shapes	8–48
Beam Load Tables	8–52
WWF Shapes	8–52
W Shapes	8–60

GENERAL INFORMATION

General

With the introduction on May 1, 1997 by Algoma Steel Inc. of CSA-G40.21 350W as the basic steel grade for W and HP shapes produced by Algoma, tables for design of new steel structures based on CSA-G40.21 300W for W shapes became redundant. However, as CSA-G40.21 300W had been the basic steel grade in Canada for over three decades, selected design tables for W shapes in CSA-G40.21 300W have been retain in this Part to facilitate evaluation of existing structures to the requirements of CSA Standard S16.1-94 when the steel used was CSA-G40.21 300W.

In addition, the tables for WWF shapes based on CSA-G40.21 300W have been placed in this Part. Because WWF shapes are based on the production of plate, WWF sections in CSA-G40.21 300W steel depends the production of CSA-G40.21 300W plate. Thus, the tables for WWF shapes in this Part may be used for the design of new structures as well as for the evaluation of existing structures when the steel used was CSA-G40.21 300W.

Prying Action

The table on page 8-4 gives the range of flange thickness, t, and the figure on page 8-5 provides amplified bolt forces for various spacings of bolts and flange thicknesses for flanges of CSA-G40.21 300W steel. For more information, see the section Bolts in Tension and Prying Action in Part Three of this handbook.

Class of Sections for Beam-Columns

This table provides the values of C_f/C_y at which the webs of W sections used as beam-columns change class based on CSA-G40.21 300W steel. See the corresponding table in Part Four for a further explanation.

Factored Axial Compressive Resistances

Pages 8-10 to 8-22 provide tables of the Factored Axial Compressive Resistances of WWF and W shapes used as columns based on CSA-G40.21 300W steel. See the corresponding section in Part Four for a further explanation of the contents of like tables.

Composite Beams – Trial Selection Tables

Pages 8-23 to 8-47 provide Composite Beam Tables of WWF and W shapes used as composite beams based on CSA-G40.21 300W steel. See the corresponding section in Part Five for a further explanation of the contents of like tables.

Beam Selection Table

Pages 8-48 to 8-51 provide Beam Selection Tables of WWF and W shapes used as beams based on CSA-G40.21 300W steel. See the corresponding section in Part Five for a further explanation of the contents of like tables.

Beam Load Tables

Pages 8-52 to 8-67 provide Beam Load Tables for WWF and W shapes used as beams based on CSA-G40.21 300W steel. See the corresponding section in Part Five for a further explanation of the contents of like tables.

RANGE OF t

$$t = \left[\frac{KP_f}{(1+\delta\alpha)}\right]^{1/2}$$

t_{min} when $\alpha = 1.0$ and t_{max} when $\alpha = 0.0$

b (mm)	Bolt Size	P_f = 60 kN pitch p (mm)			P_f = 80 kN pitch p (mm)			P_f = 100 kN pitch p (mm)			P_f = 120 kN pitch p (mm)		
		80	90	100	80	90	100	80	90	100	80	90	100
35	3/4	12.8 / 16.8	11.9 / 15.9	11.2 / 15.0	14.7 / 19.4	13.8 / 18.3	13.0 / 17.4	16.5 / 21.7	15.4 / 20.5	14.5 / 19.4			
	M20	12.7 / 16.7	11.9 / 15.7	11.2 / 14.9	14.7 / 19.2	13.7 / 18.1	12.9 / 17.2	16.4 / 21.5	15.3 / 20.3	14.4 / 19.2	17.9 / 23.6	16.8 / 22.2	15.8 / 21.1
40	3/4	14.0 / 18.4	13.1 / 17.3	12.3 / 16.5	16.1 / 21.2	15.1 / 20.0	14.2 / 19.0	18.0 / 23.8	16.9 / 22.4	15.9 / 21.2			
	M20	13.9 / 18.3	13.0 / 17.2	12.2 / 16.3	16.1 / 21.1	15.0 / 19.9	14.1 / 18.9	17.9 / 23.6	16.8 / 22.2	15.8 / 21.1	19.7 / 25.8	18.4 / 24.3	17.3 / 23.1
45	3/4	15.1 / 19.9	14.1 / 18.7	13.3 / 17.8	17.4 / 22.9	16.3 / 21.6	15.3 / 20.5	19.4 / 25.6	18.2 / 24.2	17.1 / 22.9			
	M20	15.0 / 19.7	14.0 / 18.6	13.2 / 17.6	17.3 / 22.8	16.2 / 21.5	15.3 / 20.4	19.4 / 25.5	18.1 / 24.0	17.1 / 22.8	21.2 / 27.9	19.8 / 26.3	18.7 / 24.9
50	3/4	16.1 / 21.2	15.0 / 20.0	14.2 / 19.0	18.6 / 24.5	17.4 / 23.1	16.4 / 21.9	20.8 / 27.4	19.4 / 25.8	18.3 / 24.5			
	M20	16.1 / 21.1	15.0 / 19.9	14.1 / 18.9	18.5 / 24.3	17.3 / 23.0	16.3 / 21.8	20.7 / 27.2	19.4 / 25.7	18.2 / 24.3	22.7 / 29.8	21.2 / 28.1	20.0 / 26.7
55	3/4	17.1 / 22.5	15.9 / 21.2	15.0 / 20.1	19.7 / 26.0	18.4 / 24.5	17.4 / 23.2	22.0 / 29.0	20.6 / 27.4	19.4 / 26.0			
	M20	17.0 / 22.4	15.9 / 21.1	15.0 / 20.0	19.7 / 25.8	18.4 / 24.3	17.3 / 23.1	22.0 / 28.9	20.5 / 27.2	19.4 / 25.8	24.1 / 31.6	22.5 / 29.8	21.2 / 28.3

b (mm)	Bolt Size	P_f = 100 kN pitch p (mm)			P_f = 120 kN pitch p (mm)			P_f = 140 kN pitch p (mm)			P_f = 160 kN pitch p (mm)		
		90	100	110	90	100	110	90	100	110	90	100	110
40	M22	16.6 / 21.8	15.6 / 20.7	14.8 / 19.8	18.2 / 23.9	17.1 / 22.7	16.2 / 21.6	19.6 / 25.9	18.5 / 24.5	17.5 / 23.4			
	7/8	16.6 / 21.8	15.6 / 20.7	14.8 / 19.7	18.2 / 23.9	17.1 / 22.7	16.2 / 21.6	19.6 / 25.8	18.5 / 24.5	17.5 / 23.3	21.0 / 27.6	19.7 / 26.2	18.7 / 24.9
45	M22	18.0 / 23.7	16.9 / 22.4	16.0 / 21.4	19.7 / 25.9	18.5 / 24.6	17.6 / 23.4	21.3 / 28.0	20.0 / 26.6	19.0 / 25.3			
	7/8	18.0 / 23.6	16.9 / 22.4	16.0 / 21.4	19.7 / 25.9	18.5 / 24.5	17.5 / 23.4	21.2 / 27.9	20.0 / 26.5	18.9 / 25.3	22.7 / 29.9	21.4 / 28.3	20.3 / 27.0
50	M22	19.2 / 25.3	18.1 / 24.0	17.2 / 22.9	21.1 / 27.8	19.8 / 26.3	18.8 / 25.1	22.8 / 30.0	21.4 / 28.4	20.3 / 27.1			
	7/8	19.2 / 25.3	18.1 / 24.0	17.2 / 22.9	21.1 / 27.7	19.8 / 26.3	18.8 / 25.1	22.8 / 29.9	21.4 / 28.4	20.3 / 27.1	24.3 / 32.0	22.9 / 30.4	21.7 / 28.9
55	M22	20.4 / 26.9	19.2 / 25.5	18.2 / 24.3	22.4 / 29.5	21.1 / 28.0	20.0 / 26.7	24.2 / 31.8	22.8 / 30.2	21.6 / 28.8			
	7/8	20.4 / 26.9	19.2 / 25.5	18.2 / 24.3	22.4 / 29.4	21.1 / 27.9	20.0 / 26.6	24.2 / 31.8	22.8 / 30.2	21.6 / 28.8	25.8 / 34.0	24.3 / 32.3	23.1 / 30.8

b (mm)	Bolt Size	P_f = 140 kN pitch p (mm)			P_f = 160 kN pitch p (mm)			P_f = 180 kN pitch p (mm)			P_f = 200 kN pitch p (mm)		
		100	110	120	100	110	120	100	110	120	100	110	120
40	M24	18.3 / 24.1	17.3 / 23.0	16.5 / 22.0	19.5 / 25.8	18.5 / 24.6	17.6 / 23.5	20.7 / 27.3	19.6 / 26.1	18.7 / 24.9			
	1	18.1 / 23.8	17.1 / 22.7	16.3 / 21.7	19.4 / 25.4	18.3 / 24.3	17.4 / 23.2	20.5 / 27.0	19.4 / 25.7	18.5 / 24.6	21.6 / 28.4	20.5 / 27.1	19.5 / 26.0
45	M24	19.8 / 26.2	18.8 / 24.9	17.9 / 23.9	21.2 / 28.0	20.1 / 26.7	19.1 / 25.5	22.5 / 29.7	21.3 / 28.3	20.3 / 27.1			
	1	19.7 / 25.9	18.7 / 24.7	17.8 / 23.6	21.1 / 27.7	19.9 / 26.4	19.0 / 25.3	22.3 / 29.3	21.1 / 28.0	20.1 / 26.8	23.5 / 30.9	22.3 / 29.5	21.2 / 28.2
50	M24	21.3 / 28.1	20.2 / 26.8	19.2 / 25.6	22.8 / 30.0	21.5 / 28.6	20.5 / 27.4	24.1 / 31.8	22.9 / 30.4	21.8 / 29.1			
	1	21.2 / 27.8	20.0 / 26.5	19.1 / 25.4	22.6 / 29.7	21.4 / 28.4	20.4 / 27.1	24.0 / 31.5	22.7 / 30.1	21.6 / 28.8	25.3 / 33.2	24.0 / 31.7	22.8 / 30.3
55	M24	22.6 / 29.9	21.4 / 28.5	20.4 / 27.3	24.2 / 31.9	22.9 / 30.4	21.8 / 29.1	25.7 / 33.9	24.3 / 32.3	23.1 / 30.9			
	1	22.5 / 29.6	21.3 / 28.2	20.3 / 27.0	24.1 / 31.7	22.8 / 30.2	21.7 / 28.9	25.6 / 33.6	24.2 / 32.0	23.0 / 30.7	26.9 / 35.4	25.5 / 33.8	24.3 / 32.3

$K = 4b' \, 10^3 / (\phi p \, F_y)$ where $\phi = 0.90$ and $F_y = 300$ MPa

AMPLIFIED BOLT FORCE, T_f (kN)

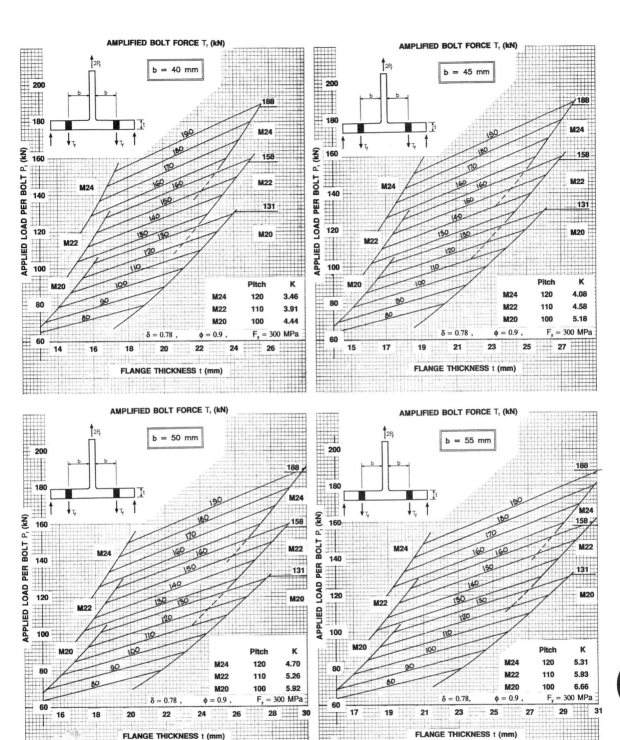

CLASS OF SECTIONS FOR BEAM-COLUMNS
G40.21-M 300W

Designation	Web 1 $C_f/C_y \leq$	Web 2 $C_f/C_y \leq$	Web 3 $C_f/C_y \leq$	Flange	Designation	Web 1 $C_f/C_y \leq$	Web 2 $C_f/C_y \leq$	Web 3 $C_f/C_y \leq$	Flange
WWF2000x732	—	0.053	0.206	1	WWF700x245	0.215	0.668	0.722	1
x648	—	0.036	0.192	1	x214	0.178	0.652	0.710	1
x607	—	0.028	0.185	1	x196	0.156	0.643	0.702	2
x542	—	0.019	0.178	1	x175	0.178	0.652	0.710	1
					x152	0.142	0.637	0.697	1
WWF1800x700	—	0.220	0.346	1					
x659	—	0.211	0.339	1	WWF650x864	1.0	—	—	1
x617	—	0.203	0.332	1	x739	1.0	—	—	1
x575	—	0.195	0.325	1	x598	1.0	—	—	1
x510	—	0.186	0.318	1	x499	1.0	—	—	1
					x400	1.0	—	—	3
WWF1600x622	—	0.073	0.224	1					
x580	—	0.063	0.215	1	WWF600x793	1.0	—	—	1
x538	—	0.053	0.206	1	x680	1.0	—	—	1
x496	—	0.042	0.197	1	x551	1.0	—	—	1
x431	—	0.032	0.189	1	x460	1.0	—	—	1
					x369	1.0	—	—	3
WWF1400x597	—	0.282	0.399	1					
x513	—	0.261	0.381	1	WWF550x721	1.0	—	—	1
x471	—	0.251	0.373	1	x620	1.0	—	—	1
x405	—	0.241	0.364	1	x503	1.0	—	—	1
x358	—	0.241	0.364	1	x420	1.0	—	—	1
					x280	1.0	—	—	3
WWF1200x487	—	0.470	0.557	1					
x418	—	0.460	0.548	1	WWF500x651	1.0	—	—	1
x380	—	0.449	0.539	1	x561	1.0	—	—	1
x333	—	0.449	0.539	1	x456	1.0	—	—	1
x302	—	0.439	0.530	1	x381	1.0	—	—	1
x263	—	0.439	0.530	1	x343	1.0	—	—	1
					x306	1.0	—	—	1
WWF1100x458	—	0.422	0.517	1	x276	1.0	—	—	2
x388	—	0.411	0.507	1	x254	1.0	—	—	3
x351	—	0.399	0.497	1	x223	1.0	—	—	3
x304	—	0.399	0.497	1	x197	0.876	0.941	0.952	4
x273	—	0.387	0.487	1					
x234	—	0.387	0.487	1	WWF450x503	1.0	—	—	1
					x409	1.0	—	—	1
WWF1000x447	—	0.542	0.617	1	x342	1.0	—	—	1
x377	—	0.530	0.607	1	x308	1.0	—	—	1
x340	—	0.518	0.597	1	x274	1.0	—	—	1
x293	—	0.518	0.597	1	x248	1.0	—	—	1
x262	—	0.506	0.587	1	x228	1.0	—	—	2
x223	—	0.506	0.587	1	x201	1.0	—	—	3
x200	—	0.494	0.577	1	x177	1.0	—	—	3
WWF900x417	—	0.394	0.493	1	WWF400x444	1.0	—	—	1
x347	—	0.379	0.480	1	x362	1.0	—	—	1
x309	—	0.364	0.467	1	x303	1.0	—	—	1
x262	—	0.364	0.467	1	x273	1.0	—	—	1
x231	—	0.349	0.455	1	x243	1.0	—	—	1
x192	—	0.349	0.455	1	x220	1.0	—	—	1
x169	—	0.334	0.442	1	x202	1.0	—	—	1
					x178	1.0	—	—	2
WWF800x339	—	0.531	0.608	1	x157	1.0	—	—	3
x300	—	0.516	0.595	1					
x253	—	0.516	0.595	1	WWF350x315	1.0	—	—	1
x223	—	0.501	0.582	1	x263	1.0	—	—	1
x184	—	0.501	0.582	1	x238	1.0	—	—	1
x161	—	0.485	0.569	1	x212	1.0	—	—	1
					x192	1.0	—	—	1
					x176	1.0	—	—	1
					x155	1.0	—	—	1
					x137	1.0	—	—	2

See Table 4-2, page 4-6, for width-thickness criteria.
— Indicates web is never that class.

CLASS OF SECTIONS FOR BEAM-COLUMNS
G40.21-M 300W

Designation	Web 1 $C_f/C_y \leq$	Web 2 $C_f/C_y \leq$	Web 3 $C_f/C_y \leq$	Flange	Designation	Web 1 $C_f/C_y \leq$	Web 2 $C_f/C_y \leq$	Web 3 $C_f/C_y \leq$	Flange
W1100x499	0.962	0.976	0.982	1	W840x251	0.671	0.856	0.881	1
x432	0.686	0.863	0.886	1	x226	0.564	0.812	0.844	1
x390	0.500	0.785	0.821	1	x210	0.474	0.775	0.812	1
x342	0.258	0.685	0.737	1	x193	0.376	0.734	0.778	1
					x176	0.265	0.688	0.740	1
W1000x883	1.0	—	—	1					
x749	1.0	—	—	1	W760x710	1.0	—	—	1
x641	1.0	—	—	1	x582	1.0	—	—	1
x554	1.0	—	—	1	x484	1.0	—	—	1
x478	1.0	—	—	1	x434	1.0	—	—	1
x443	0.976	0.982	0.987	1	x389	1.0	—	—	1
x412	0.788	0.905	0.922	1	x350	1.0	—	—	1
x371	0.593	0.824	0.854	1	x314	1.0	—	—	1
x321	0.293	0.700	0.750	1	x284	0.952	0.972	0.978	1
x296	0.294	0.700	0.750	1	x257	0.816	0.916	0.931	1
x259	0.294	0.700	0.750	2					
					W760x220	0.805	0.912	0.927	1
W1000x693	1.0	—	—	1	x196	0.703	0.869	0.892	1
x583	1.0	—	—	1	x185	0.616	0.834	0.862	1
x493	1.0	—	—	1	x173	0.549	0.806	0.838	1
x414	1.0	—	—	1	x161	0.459	0.769	0.807	1
x393	1.0	—	—	1	x147	0.365	0.730	0.775	1
x350	0.788	0.905	0.922	1	x134	0.125	0.630	0.691	2
x314	0.602	0.828	0.857	1					
x272	0.293	0.700	0.750	1	W690x802	1.0	—	—	1
x249	0.293	0.700	0.750	1	x667	1.0	—	—	1
x222	0.223	0.671	0.725	1	x548	1.0	—	—	1
					x457	1.0	—	—	1
W920x1262	1.0	—	—	1	x384	1.0	—	—	1
x1188	1.0	—	—	1	x350	1.0	—	—	1
x967	1.0	—	—	1	x323	1.0	—	—	1
x784	1.0	—	—	1	x289	1.0	—	—	1
x653	1.0	—	—	1	x265	1.0	—	—	1
x585	1.0	—	—	1	x240	1.0	—	—	1
x534	1.0	—	—	1	x217	0.872	0.939	0.951	1
x488	1.0	—	—	1					
x446	1.0	—	—	1	W690x192	0.881	0.943	0.954	1
x417	1.0	—	—	1	x170	0.766	0.895	0.914	1
x387	0.957	0.975	0.980	1	x152	0.574	0.816	0.847	1
x365	0.879	0.942	0.953	1	x140	0.460	0.769	0.808	1
x342	0.790	0.905	0.922	1	x125	0.337	0.718	0.765	1
W920x381	1.0	—	—	1	W610x732	1.0	—	—	1
x345	0.987	0.987	0.991	1	x608	1.0	—	—	1
x313	0.913	0.956	0.965	1	x498	1.0	—	—	1
x289	0.768	0.896	0.915	1	x415	1.0	—	—	1
x271	0.670	0.856	0.881	1	x372	1.0	—	—	1
x253	0.550	0.806	0.839	1	x341	1.0	—	—	1
x238	0.452	0.766	0.805	1	x307	1.0	—	—	1
x223	0.372	0.733	0.777	1	x285	1.0	—	—	1
x201	0.272	0.691	0.742	1	x262	1.0	—	—	1
					x241	1.0	—	—	1
W840x527	1.0	—	—	1	x217	1.0	—	—	1
x473	1.0	—	—	1	x195	1.0	—	—	1
x433	1.0	—	—	1	x174	0.912	0.956	0.965	1
x392	1.0	—	—	1	x155	0.742	0.886	0.906	2
x359	1.0	—	—	1					
x329	0.930	0.963	0.971	1	W610x153	0.911	0.955	0.964	1
x299	0.797	0.908	0.925	1	x140	0.799	0.909	0.925	1
					x125	0.621	0.835	0.863	1
					x113	0.497	0.784	0.820	1
					x101	0.360	0.728	0.773	1
					x91	0.181	0.653	0.711	2
					x84	—	0.577	0.646	2

See Table 4-2, page 4-6, for width-thickness criteria.
— Indicates web is never that class.

CLASS OF SECTIONS FOR BEAM-COLUMNS
G40.21-M 300W

Designation	Web 1 $C_t/C_y \leq$	Web 2 $C_t/C_y \leq$	Web 3 $C_t/C_y \leq$	Flange	Designation	Web 1 $C_t/C_y \leq$	Web 2 $C_t/C_y \leq$	Web 3 $C_t/C_y \leq$	Flange
W610x92	0.442	0.761	0.801	1	W410x46	0.369	0.731	0.776	1
x82	0.249	0.682	0.734	1	x39	0.158	0.644	0.703	1
W530x300	1.0	—	—	1	W360x1202	1.0	—	—	1
x272	1.0	—	—	1	x1086	1.0	—	—	1
x248	1.0	—	—	1	x990	1.0	—	—	1
x219	1.0	—	—	1	x900	1.0	—	—	1
x196	1.0	—	—	1	x818	1.0	—	—	1
x182	1.0	—	—	1	x744	1.0	—	—	1
x165	1.0	—	—	1	x677	1.0	—	—	1
x150	0.967	0.979	0.984	1					
					W360x634	1.0	—	—	1
W530x138	1.0	—	—	1	x592	1.0	—	—	1
x123	1.0	—	—	1	x551	1.0	—	—	1
x109	0.819	0.917	0.932	1	x509	1.0	—	—	1
x101	0.704	0.870	0.892	1	x463	1.0	—	—	1
x92	0.578	0.818	0.849	1	x421	1.0	—	—	1
x82	0.433	0.758	0.798	1	x382	1.0	—	—	1
x72	0.286	0.697	0.747	2	x347	1.0	—	—	1
					x314	1.0	—	—	1
W530x85	0.596	0.825	0.855	1	x287	1.0	—	—	1
x74	0.475	0.775	0.813	1	x262	1.0	—	—	1
x66	0.286	0.697	0.747	1	x237	1.0	—	—	1
					x216	1.0	—	—	1
W460x464	1.0	—	—	1					
x421	1.0	—	—	1	W360x196	1.0	—	—	1
x384	1.0	—	—	1	x179	1.0	—	—	1
x349	1.0	—	—	1	x162	1.0	—	—	2
x315	1.0	—	—	1	x147	1.0	—	—	2
x286	1.0	—	—	1	x134	1.0	—	—	3
x260	1.0	—	—	1					
x235	1.0	—	—	1	W360x122	1.0	—	—	1
x213	1.0	—	—	1	x110	1.0	—	—	1
x193	1.0	—	—	1	x101	1.0	—	—	1
x177	1.0	—	—	1	x91	1.0	—	—	1
x158	1.0	—	—	1					
x144	1.0	—	—	1	W360x79	1.0	—	—	1
x128	1.0	—	—	1	x72	1.0	—	—	1
x113	0.963	0.977	0.982	1	x64	0.886	0.945	0.956	1
W460x106	1.0	—	—	1	W360x57	0.868	0.938	0.949	1
x97	1.0	—	—	1	x51	0.704	0.870	0.892	1
x89	0.920	0.959	0.967	1	x45	0.619	0.835	0.863	2
x82	0.819	0.917	0.932	1					
x74	0.644	0.845	0.872	1	W360x39	0.504	0.787	0.823	1
x67	0.528	0.797	0.831	1	x33	0.253	0.683	0.736	1
x61	0.429	0.756	0.797	2					
					W310x500	1.0	—	—	1
W460x68	0.664	0.853	0.879	1	x454	1.0	—	—	1
x60	0.402	0.745	0.787	1	x415	1.0	—	—	1
x52	0.288	0.698	0.748	1	x375	1.0	—	—	1
					x342	1.0	—	—	1
W410x149	1.0	—	—	1	x313	1.0	—	—	1
x132	1.0	—	—	1					
x114	1.0	—	—	1					
x100	1.0	—	—	1					
W410x85	1.0	—	—	1					
x74	0.978	0.983	0.988	1					
x67	0.815	0.916	0.931	1					
x60	0.564	0.812	0.844	1					
x54	0.512	0.790	0.826	1					

See Table 4-2, page 4-6, for width-thickness criteria.
— Indicates web is never that class.

CLASS OF SECTIONS FOR BEAM-COLUMNS
G40.21-M 300W

Designation	Web 1 $C_f/C_y \le$	Web 2 $C_f/C_y \le$	Web 3 $C_f/C_y \le$	Flange	Designation	Web 1 $C_f/C_y \le$	Web 2 $C_f/C_y \le$	Web 3 $C_f/C_y \le$	Flange
W310x283	1.0	—	—	1	W200x31	1.0	—	—	1
x253	1.0	—	—	1	x27	1.0	—	—	1
x226	1.0	—	—	1	x21	1.0	—	—	3
x202	1.0	—	—	1					
x179	1.0	—	—	1	W200x22	1.0	—	—	1
x158	1.0	—	—	1	x19	1.0	—	—	1
x143	1.0	—	—	1	x15	0.784	0.903	0.920	2
x129	1.0	—	—	1					
x118	1.0	—	—	1	W150x37	1.0	—	—	1
x107	1.0	—	—	2	x30	1.0	—	—	1
x97	1.0	—	—	3	x22	1.0	—	—	3
W310x86	1.0	—	—	1	W150x24	1.0	—	—	1
x79	1.0	—	—	2	x18	1.0	—	—	1
					x14	1.0	—	—	2
W310x74	1.0	—	—	1					
x67	1.0	—	—	1	W130x28	1.0	—	—	1
x60	1.0	—	—	1	x24	1.0	—	—	1
W310x52	1.0	—	—	1	W100x19	1.0	—	—	1
x45	0.786	0.904	0.921	1					
x39	0.541	0.802	0.836	2					
x31	0.213	0.667	0.722	3					
W310x33	0.782	0.902	0.919	1					
x28	0.605	0.829	0.858	1					
x24	0.462	0.770	0.808	1					
x21	0.256	0.684	0.737	2					
W250x167	1.0	—	—	1					
x149	1.0	—	—	1					
x131	1.0	—	—	1					
x115	1.0	—	—	1					
x101	1.0	—	—	1					
x89	1.0	—	—	1					
x80	1.0	—	—	1					
x73	1.0	—	—	2					
W250x67	1.0	—	—	1					
x58	1.0	—	—	1					
x49	1.0	—	—	2					
W250x45	1.0	—	—	1					
x39	1.0	—	—	1					
x33	0.977	0.983	0.987	1					
x24	0.625	0.837	0.865	3					
W250x28	1.0	—	—	1					
x25	0.974	0.982	0.986	1					
x22	0.892	0.948	0.958	1					
x18	0.542	0.803	0.836	2					
W200x100	1.0	—	—	1					
x86	1.0	—	—	1					
x71	1.0	—	—	1					
x59	1.0	—	—	1					
x52	1.0	—	—	1					
x46	1.0	—	—	2					
W200x42	1.0	—	—	1					
x36	1.0	—	—	1					

See Table 4-2, page 4-6, for width-thickness criteria.
— Indicates web is never that class.

WWF COLUMNS
Factored Axial Compressive Resistances, C_r, in kN

G40.21-M300W

$\phi = 0.90$

Designation	WWF650				
Mass (kg/m)	864 ‡	739 ‡	598	499	400 *
Effective length (KL) in millimetres with respect to least radius of gyration					
0	29 700	25 400	20 600	17 200	13 800
2 000	29 700	25 400	20 600	17 200	13 800
2 250	29 700	25 400	20 600	17 200	13 800
2 500	29 700	25 400	20 600	17 200	13 800
2 750	29 700	25 400	20 600	17 200	13 800
3 000	29 700	25 400	20 600	17 200	13 800
3 250	29 700	25 400	20 600	17 200	13 800
3 500	29 700	25 400	20 600	17 200	13 800
3 750	29 600	25 400	20 500	17 100	13 700
4 000	29 600	25 400	20 500	17 100	13 700
4 250	29 600	25 400	20 500	17 100	13 700
4 500	29 600	25 300	20 500	17 100	13 700
4 750	29 500	25 300	20 500	17 100	13 700
5 000	29 500	25 300	20 500	17 100	13 700
5 250	29 500	25 300	20 500	17 100	13 700
5 500	29 400	25 200	20 400	17 000	13 700
6 000	29 300	25 100	20 400	17 000	13 600
6 500	29 100	25 000	20 300	16 900	13 500
7 000	28 900	24 900	20 200	16 800	13 400
7 500	28 600	24 700	20 000	16 700	13 300
8 000	28 200	24 500	19 900	16 500	13 200
8 500	27 800	24 200	19 700	16 400	13 000
9 000	27 300	23 900	19 400	16 100	12 800
9 500	26 800	23 600	19 200	15 900	12 600
10 000	26 200	23 200	18 800	15 600	12 300
10 500	25 500	22 700	18 500	15 300	12 100
11 000	24 800	22 200	18 100	15 000	11 700
11 500	24 000	21 600	17 600	14 600	11 400
12 000	23 200	21 100	17 200	14 200	11 100

PROPERTIES AND DESIGN DATA					
Area (mm^2)	110 000	94 100	76 200	63 600	51 000
Z_x (10^3 mm^3)	27 300	25 200	21 100	17 500	13 900
S_x (10^3 mm^3)	23 300	22 100	18 900	15 900	12 600
r_x (mm)	262	277	284	285	284
Z_y (10^3 mm^3)	13 200	12 800	10 600	8 510	6 400
S_y (10^3 mm^3)	8 480	8 450	7 040	5 630	4 230
r_y (mm)	158	171	173	170	164
r_x/r_y	1.66	1.62	1.64	1.68	1.73
M_{rx} (kN·m) (L < L_u)	7 370	6 800	5 700	4 730	3 400
M_{ry} (kN·m)	3 560	3 460	2 860	2 300	1 140
J (10^3 mm^4)	132 000	98 400	55 600	29 300	13 300
C_w (10^9 mm^6)	240 000	239 000	206 000	170 000	132 000
L_u (mm)	14 800	14 300	12 800	11 600	11 300
F_y (MPa)	300	300	300	300	300

IMPERIAL SIZE AND MASS					
Mass (lb./ft.)	580	497	402	336	269
Nominal Depth and Width (in.)	26 x 26				

‡ Welding does not fully develop web strength.
* Class 3 flanges.

G40.21-M300W
$\phi = 0.90$

WWF COLUMNS
Factored Axial Compressive Resistances, C_r, in kN

Designation	WWF600				
Mass (kg/m)	793 ‡	680 ‡	551	460	369 *
Effective length (KL) in millimetres with respect to least radius of gyration					
0	27 300	23 400	19 000	15 800	12 700
2 000	27 300	23 400	19 000	15 800	12 700
2 250	27 300	23 400	19 000	15 800	12 700
2 500	27 300	23 400	18 900	15 800	12 700
2 750	27 300	23 400	18 900	15 800	12 700
3 000	27 200	23 400	18 900	15 800	12 700
3 250	27 200	23 400	18 900	15 800	12 700
3 500	27 200	23 400	18 900	15 800	12 700
3 750	27 200	23 300	18 900	15 800	12 700
4 000	27 200	23 300	18 900	15 800	12 700
4 250	27 200	23 300	18 900	15 800	12 600
4 500	27 100	23 300	18 900	15 800	12 600
4 750	27 100	23 300	18 900	15 700	12 600
5 000	27 000	23 200	18 800	15 700	12 600
5 250	27 000	23 200	18 800	15 700	12 600
5 500	26 900	23 100	18 800	15 700	12 500
6 000	26 700	23 000	18 700	15 600	12 500
6 500	26 500	22 900	18 600	15 500	12 400
7 000	26 200	22 700	18 400	15 400	12 300
7 500	25 900	22 500	18 300	15 200	12 100
8 000	25 500	22 200	18 100	15 000	12 000
8 500	25 000	21 900	17 800	14 800	11 700
9 000	24 400	21 500	17 500	14 500	11 500
9 500	23 800	21 100	17 200	14 200	11 200
10 000	23 100	20 600	16 800	13 900	11 000
10 500	22 300	20 100	16 400	13 500	10 600
11 000	21 500	19 500	15 900	13 100	10 300
11 500	20 700	18 900	15 400	12 700	9 930
12 000	19 900	18 200	14 900	12 300	9 560
PROPERTIES AND DESIGN DATA					
Area (mm²)	101 000	86 600	70 200	58 600	47 000
Z_x (10^3 mm³)	22 900	21 200	17 800	14 800	11 800
S_x (10^3 mm³)	19 400	18 500	15 900	13 400	10 700
r_x (mm)	240	253	261	262	261
Z_y (10^3 mm³)	11 200	10 900	9 050	7 250	5 460
S_y (10^3 mm³)	7 230	7 200	6 000	4 800	3 600
r_y (mm)	147	158	160	157	152
r_x/r_y	1.63	1.60	1.63	1.67	1.72
M_{rx} (kN·m) (L < L_u)	6 180	5 720	4 810	4 000	2 890
M_{ry} (kN·m)	3 020	2 940	2 440	1 960	972
J (10^3 mm⁴)	121 000	90 700	51 300	27 000	12 200
C_w (10^9 mm⁶)	158 000	158 000	136 000	113 000	87 800
L_u (mm)	14 600	14 000	12 400	11 000	10 600
F_y (MPa)	300	300	300	300	300
IMPERIAL SIZE AND MASS					
Mass (lb./ft.)	531	456	371	309	248
Nominal Depth and Width (in.)	24 x 24				

‡ Welding does not fully develop web strength.
* Class 3 flanges.

WWF COLUMNS
Factored Axial Compressive Resistances, C_r, in kN

G40.21-M300W
$\phi = 0.90$

Designation	WWF550					WWF500		
Mass (kg/m)	721 ‡	620 ‡	503	420	280 *	651 ‡	561 ‡	456
0	24 800	21 400	17 300	14 500	9 610	22 400	19 300	15 700
2 000	24 800	21 400	17 300	14 500	9 610	22 400	19 300	15 700
2 250	24 800	21 400	17 300	14 500	9 610	22 400	19 300	15 700
2 500	24 800	21 300	17 300	14 500	9 610	22 400	19 300	15 700
2 750	24 800	21 300	17 300	14 500	9 600	22 400	19 300	15 700
3 000	24 800	21 300	17 300	14 500	9 600	22 400	19 300	15 700
3 250	24 800	21 300	17 300	14 500	9 600	22 300	19 300	15 700
3 500	24 800	21 300	17 300	14 400	9 590	22 300	19 300	15 700
3 750	24 700	21 300	17 300	14 400	9 580	22 300	19 300	15 700
4 000	24 700	21 300	17 300	14 400	9 570	22 200	19 200	15 600
4 250	24 700	21 300	17 300	14 400	9 560	22 200	19 200	15 600
4 500	24 600	21 200	17 200	14 400	9 540	22 100	19 200	15 600
4 750	24 600	21 200	17 200	14 400	9 530	22 100	19 100	15 500
5 000	24 500	21 200	17 200	14 300	9 510	22 000	19 100	15 500
5 250	24 400	21 100	17 100	14 300	9 480	21 900	19 000	15 500
5 500	24 300	21 000	17 100	14 300	9 450	21 700	18 900	15 400
6 000	24 100	20 900	17 000	14 200	9 380	21 500	18 700	15 200
6 500	23 800	20 700	16 800	14 000	9 280	21 100	18 500	15 100
7 000	23 500	20 500	16 700	13 900	9 160	20 600	18 200	14 800
7 500	23 000	20 200	16 400	13 700	9 010	20 100	17 800	14 500
8 000	22 500	19 900	16 200	13 400	8 830	19 500	17 400	14 200
8 500	21 900	19 400	15 900	13 100	8 630	18 800	16 900	13 800
9 000	21 300	19 000	15 500	12 800	8 400	18 000	16 300	13 400
9 500	20 600	18 500	15 100	12 500	8 140	17 200	15 700	12 900
10 000	19 800	17 900	14 700	12 100	7 870	16 400	15 100	12 400
10 500	19 000	17 300	14 200	11 700	7 580	15 600	14 500	11 900
11 000	18 100	16 700	13 700	11 200	7 270	14 800	13 800	11 400
11 500	17 300	16 000	13 200	10 800	6 960	14 000	13 100	10 900
12 000	16 500	15 400	12 600	10 300	6 650	13 200	12 500	10 300

Effective length (KL) in millimetres with respect to least radius of gyration

PROPERTIES AND DESIGN DATA

Area (mm²)	92 000	79 100	64 200	53 600	35 600	83 000	71 600	58 200
Z_x (10^3 mm³)	19 000	17 600	14 800	12 400	8 250	15 400	14 300	12 100
S_x (10^3 mm³)	16 000	15 200	13 100	11 100	7 530	12 800	12 300	10 600
r_x (mm)	218	230	237	239	241	196	207	214
Z_y (10^3 mm³)	9 470	9 180	7 610	6 100	3 810	7 850	7 590	6 290
S_y (10^3 mm³)	6 080	6 050	5 040	4 030	2 520	5 030	5 000	4 170
r_y (mm)	135	145	147	144	140	123	132	134
r_x/r_y	1.61	1.59	1.61	1.66	1.72	1.59	1.57	1.60
M_{rx} (kN·m) ($L < L_u$)	5 130	4 750	4 000	3 350	2 030	4 160	3 860	3 270
M_{ry} (kN·m)	2 560	2 480	2 050	1 650	680	2 120	2 050	1 700
J (10^3 mm⁴)	110 000	83 100	47 000	24 700	6 410	99 400	75 400	42 700
C_w (10^9 mm⁶)	100 000	99 900	86 700	72 100	47 800	60 800	60 500	52 700
L_u (mm)	14 400	13 800	12 000	10 500	9 480	14 400	13 700	11 700
F_y (MPa)	300	300	300	300	300	300	300	300

IMPERIAL SIZE AND MASS

Mass (lb./ft.)	484	416	338	282	188	437	377	306
Nominal Depth and Width (in.)	22 x 22					20 x 20		

‡ Welding does not fully develop web strength.
* Class 3 flanges.

G40.21-M300W
$\phi = 0.90$

WWF COLUMNS
Factored Axial Compressive Resistances, C_r, in kN

Designation		WWF500						
Mass (kg/m)		381	343	306	276	254 *	223 *	197 **
Effective length (KL) in millimetres with respect to least radius of gyration	0	13 100	11 800	10 500	9 500	8 720	7 700	6 220
	2 000	13 100	11 800	10 500	9 500	8 720	7 690	6 220
	2 250	13 100	11 800	10 500	9 500	8 720	7 690	6 220
	2 500	13 100	11 800	10 500	9 500	8 710	7 690	6 220
	2 750	13 100	11 800	10 500	9 490	8 710	7 690	6 210
	3 000	13 100	11 800	10 500	9 490	8 710	7 680	6 210
	3 250	13 100	11 800	10 500	9 480	8 700	7 680	6 210
	3 500	13 100	11 800	10 500	9 470	8 690	7 670	6 210
	3 750	13 100	11 800	10 500	9 460	8 680	7 660	6 200
	4 000	13 000	11 800	10 500	9 450	8 670	7 650	6 190
	4 250	13 000	11 700	10 400	9 430	8 650	7 630	6 190
	4 500	13 000	11 700	10 400	9 410	8 630	7 610	6 180
	4 750	13 000	11 700	10 400	9 380	8 600	7 590	6 160
	5 000	12 900	11 600	10 400	9 350	8 570	7 560	6 150
	5 250	12 900	11 600	10 300	9 320	8 540	7 530	6 130
	5 500	12 800	11 500	10 300	9 280	8 500	7 500	6 120
	6 000	12 700	11 400	10 100	9 170	8 400	7 410	6 070
	6 500	12 500	11 200	9 980	9 040	8 270	7 290	6 010
	7 000	12 300	11 000	9 790	8 880	8 110	7 150	5 930
	7 500	12 100	10 800	9 560	8 680	7 920	6 990	5 840
	8 000	11 800	10 500	9 300	8 450	7 700	6 790	5 730
	8 500	11 400	10 200	9 000	8 190	7 450	6 570	5 600
	9 000	11 000	9 840	8 670	7 910	7 180	6 330	5 460
	9 500	10 600	9 460	8 320	7 600	6 890	6 080	5 310
	10 000	10 200	9 060	7 950	7 280	6 590	5 810	5 140
	10 500	9 740	8 650	7 580	6 950	6 280	5 540	4 940
	11 000	9 290	8 230	7 200	6 620	5 960	5 260	4 710
	11 500	8 840	7 820	6 830	6 280	5 660	4 990	4 490
	12 000	8 390	7 410	6 470	5 960	5 360	4 730	4 260

PROPERTIES AND DESIGN DATA							
Area (mm²)	48 600	43 800	39 000	35 200	32 300	28 500	25 200
Z_x (10^3 mm³)	10 100	9 100	8 060	7 420	6 780	6 010	5 410
S_x (10^3 mm³)	9 010	8 140	7 240	6 740	6 160	5 500	4 650
r_x (mm)	215	216	215	218	218	219	223
Z_y (10^3 mm³)	5 040	4 420	3 800	3 530	3 160	2 770	2 510
S_y (10^3 mm³)	3 330	2 920	2 500	2 330	2 080	1 830	1 420
r_y (mm)	131	129	127	129	127	127	129
r_x/r_y	1.64	1.67	1.69	1.69	1.72	1.72	1.73
M_{rx} (kN·m) (L < L_u)	2 730	2 460	2 180	2 000	1 660	1 490	1 260
M_{ry} (kN·m)	1 360	1 190	1 030	953	562	494	383
J (10^3 mm⁴)	22 500	15 400	10 200	7 920	5 820	3 970	2 870
C_w (10^9 mm⁶)	44 100	39 400	34 500	32 500	29 400	26 200	24 000
L_u (mm)	10 000	9 320	8 740	8 630	8 840	8 530	8 750
F_y (MPa)	300	300	300	300	300	300	300

IMPERIAL SIZE AND MASS							
Mass (lb./ft.)	256	230	205	185	170	150	132
Nominal Depth and Width (in.)	20 x 20						

* Class 3 flanges.
** Class 4: C_r calculated according to CAN/CSA-S16.1-94 Clause 13.3.3; S_x, S_y, M_{rx}, M_{ry} according to Clause 13.5(c)(iii).

WWF COLUMNS
Factored Axial Compressive Resistances, C_r, in kN

G40.21-M300W
$\phi = 0.90$

	Designation				WWF450					
	Mass (kg/m)	503 ‡	409	342	308	274	248	228	201 *	177 *
Effective length (KL) in millimetres with respect to least radius of gyration	0	17 300	14 100	11 800	10 600	9 450	8 530	7 830	6 910	6 100
	2 000	17 300	14 100	11 800	10 600	9 450	8 530	7 830	6 910	6 100
	2 250	17 300	14 100	11 800	10 600	9 440	8 530	7 820	6 910	6 100
	2 500	17 300	14 100	11 800	10 600	9 440	8 520	7 820	6 900	6 090
	2 750	17 300	14 100	11 800	10 600	9 430	8 520	7 810	6 900	6 090
	3 000	17 300	14 100	11 700	10 600	9 420	8 510	7 810	6 890	6 090
	3 250	17 200	14 000	11 700	10 600	9 410	8 500	7 800	6 880	6 080
	3 500	17 200	14 000	11 700	10 600	9 400	8 490	7 790	6 870	6 070
	3 750	17 200	14 000	11 700	10 500	9 380	8 470	7 770	6 860	6 060
	4 000	17 200	14 000	11 700	10 500	9 350	8 450	7 750	6 840	6 040
	4 250	17 100	13 900	11 600	10 500	9 320	8 430	7 720	6 820	6 030
	4 500	17 100	13 900	11 600	10 400	9 290	8 400	7 700	6 790	6 000
	4 750	17 000	13 900	11 600	10 400	9 240	8 360	7 660	6 760	5 980
	5 000	16 900	13 800	11 500	10 300	9 190	8 320	7 620	6 720	5 950
	5 250	16 800	13 700	11 400	10 300	9 130	8 270	7 570	6 680	5 910
	5 500	16 700	13 600	11 400	10 200	9 070	8 210	7 510	6 630	5 870
	6 000	16 500	13 500	11 200	10 000	8 900	8 070	7 370	6 510	5 770
	6 500	16 100	13 200	10 900	9 820	8 690	7 890	7 200	6 360	5 640
	7 000	15 700	12 900	10 700	9 550	8 440	7 680	7 000	6 180	5 490
	7 500	15 300	12 500	10 300	9 240	8 150	7 430	6 760	5 960	5 320
	8 000	14 700	12 100	9 980	8 900	7 830	7 150	6 490	5 730	5 120
	8 500	14 200	11 700	9 580	8 520	7 490	6 850	6 200	5 470	4 900
	9 000	13 500	11 200	9 150	8 120	7 120	6 530	5 900	5 210	4 670
	9 500	12 900	10 700	8 700	7 710	6 740	6 200	5 590	4 930	4 430
	10 000	12 200	10 100	8 250	7 300	6 370	5 870	5 280	4 660	4 200
	10 500	11 600	9 620	7 800	6 890	6 000	5 540	4 970	4 390	3 960
	11 000	10 900	9 100	7 360	6 480	5 640	5 210	4 670	4 130	3 730
	11 500	10 300	8 590	6 930	6 100	5 300	4 900	4 390	3 870	3 510
	12 000	9 710	8 100	6 520	5 730	4 970	4 610	4 120	3 640	3 300

PROPERTIES AND DESIGN DATA

Area (mm^2)		64 100	52 200	43 600	39 300	35 000	31 600	29 000	25 600	22 600
Z_x (10^3 mm^3)		11 400	9 640	8 100	7 290	6 470	5 960	5 450	4 840	4 360
S_x (10^3 mm^3)		9 620	8 380	7 150	6 480	5 770	5 380	4 920	4 400	4 000
r_x (mm)		184	190	192	193	193	196	196	197	200
Z_y (10^3 mm^3)		6 150	5 100	4 090	3 580	3 080	2 860	2 560	2 250	2 040
S_y (10^3 mm^3)		4 050	3 380	2 700	2 360	2 030	1 890	1 690	1 490	1 350
r_y (mm)		119	121	118	116	114	116	114	114	116
r_x/r_y		1.55	1.57	1.63	1.66	1.69	1.69	1.72	1.73	1.72
M_{rx} (kN·m) ($L < L_u$)		3 080	2 600	2 190	1 970	1 750	1 610	1 470	1 190	1 080
M_{ry} (kN·m)		1 660	1 380	1 100	967	832	772	691	402	365
J (10^3 mm^4)		67 800	38 400	20 200	13 900	9 140	7 120	5 230	3 570	2 580
C_w (10^9 mm^6)		34 700	30 400	25 500	22 900	20 100	18 900	17 200	15 300	14 000
L_u (mm)		13 700	11 500	9 610	8 850	8 180	8 010	7 710	7 890	7 730
F_y (MPa)		300	300	300	300	300	300	300	300	300

IMPERIAL SIZE AND MASS

Mass (lb./ft.)		337	275	229	207	184	166	152	134	119
Nominal Depth and Width (in.)					18 x 18					

‡ Welding does not fully develop web strength.
* Class 3 flanges.

G40.21-M300W
φ = 0.90

WWF COLUMNS
Factored Axial Compressive Resistances, C_r, in kN

Designation	WWF400								
Mass (kg/m)	444 ‡	362	303	273	243	220	202	178	157 *
Effective length (KL) in millimetres with respect to least radius of gyration									
0	15 300	12 500	10 400	9 400	8 370	7 560	6 940	6 130	5 430
2 000	15 300	12 500	10 400	9 390	8 360	7 550	6 930	6 120	5 420
2 250	15 300	12 500	10 400	9 380	8 360	7 550	6 930	6 120	5 420
2 500	15 300	12 500	10 400	9 380	8 350	7 540	6 920	6 120	5 420
2 750	15 200	12 400	10 400	9 370	8 340	7 540	6 920	6 110	5 410
3 000	15 200	12 400	10 400	9 360	8 330	7 530	6 910	6 100	5 400
3 250	15 200	12 400	10 400	9 340	8 310	7 510	6 890	6 090	5 390
3 500	15 200	12 400	10 300	9 320	8 290	7 490	6 870	6 070	5 380
3 750	15 100	12 400	10 300	9 290	8 260	7 470	6 850	6 050	5 360
4 000	15 100	12 300	10 300	9 250	8 230	7 440	6 820	6 030	5 340
4 250	15 000	12 300	10 200	9 210	8 190	7 400	6 790	6 000	5 310
4 500	14 900	12 200	10 200	9 160	8 140	7 360	6 750	5 960	5 280
4 750	14 800	12 100	10 100	9 090	8 080	7 310	6 700	5 910	5 240
5 000	14 700	12 000	10 000	9 020	8 010	7 240	6 640	5 860	5 200
5 250	14 600	12 000	9 930	8 930	7 920	7 170	6 570	5 800	5 150
5 500	14 400	11 800	9 830	8 840	7 830	7 090	6 490	5 730	5 090
6 000	14 100	11 600	9 580	8 610	7 610	6 900	6 310	5 570	4 950
6 500	13 700	11 200	9 280	8 330	7 350	6 670	6 090	5 380	4 790
7 000	13 200	10 900	8 930	8 000	7 040	6 400	5 840	5 160	4 590
7 500	12 600	10 400	8 530	7 640	6 700	6 100	5 560	4 910	4 380
8 000	12 000	9 930	8 110	7 250	6 340	5 780	5 260	4 650	4 150
8 500	11 300	9 430	7 660	6 840	5 970	5 450	4 950	4 370	3 910
9 000	10 700	8 900	7 210	6 430	5 600	5 120	4 640	4 100	3 670
9 500	10 000	8 380	6 770	6 030	5 240	4 790	4 340	3 840	3 440
10 000	9 400	7 870	6 330	5 640	4 890	4 470	4 050	3 580	3 210
10 500	8 790	7 370	5 920	5 260	4 550	4 170	3 780	3 340	3 000
11 000	8 220	6 900	5 520	4 910	4 240	3 890	3 520	3 110	2 790
11 500	7 670	6 450	5 160	4 580	3 950	3 630	3 280	2 890	2 600
12 000	7 170	6 030	4 810	4 270	3 680	3 380	3 050	2 700	2 430
PROPERTIES AND DESIGN DATA									
Area (mm^2)	56 600	46 200	38 600	34 800	31 000	28 000	25 700	22 700	20 100
Z_x (10^3 mm^3)	8 770	7 480	6 300	5 680	5 050	4 660	4 260	3 790	3 420
S_x (10^3 mm^3)	7 300	6 410	5 500	5 000	4 470	4 170	3 830	3 430	3 120
r_x (mm)	161	166	169	170	170	173	173	174	176
Z_y (10^3 mm^3)	4 870	4 030	3 230	2 840	2 440	2 260	2 020	1 780	1 610
S_y (10^3 mm^3)	3 200	2 670	2 130	1 870	1 600	1 490	1 330	1 170	1 070
r_y (mm)	106	108	105	104	102	103	102	102	103
r_x/r_y	1.52	1.54	1.61	1.63	1.67	1.68	1.70	1.71	1.71
M_{rx} (kN·m) (L < L_u)	2 370	2 020	1 700	1 530	1 360	1 260	1 150	1 020	842
M_{ry} (kN·m)	1 310	1 090	872	767	659	610	545	481	289
J (10^3 mm^4)	60 100	34 100	17 900	12 300	8 110	6 320	4 640	3 170	2 290
C_w (10^9 mm^6)	18 500	16 300	13 800	12 400	11 000	10 300	9 380	8 390	7 700
L_u (mm)	13 800	11 400	9 310	8 450	7 710	7 460	7 120	6 840	7 080
F_y (MPa)	300	300	300	300	300	300	300	300	300
IMPERIAL SIZE AND MASS									
Mass (lb./ft.)	298	243	203	183	163	147	135	119	105
Nominal Depth and Width (in.)	16 x 16								

‡ Welding does not fully develop web strength.
* Class 3 flanges.

WWF COLUMNS
Factored Axial Compressive Resistances, C_r, in kN

G40.21-M300W
$\phi = 0.90$

Designation		WWF350							
Mass (kg/m)		315	263	238	212	192	176	155	137
Effective length (KL) in millimetres with respect to least radius of gyration	0	10 900	9 070	8 180	7 290	6 590	6 050	5 350	4 730
	2 000	10 800	9 060	8 170	7 280	6 580	6 040	5 340	4 720
	2 250	10 800	9 050	8 160	7 270	6 570	6 030	5 330	4 710
	2 500	10 800	9 040	8 150	7 260	6 560	6 030	5 330	4 710
	2 750	10 800	9 030	8 140	7 250	6 550	6 010	5 310	4 700
	3 000	10 800	9 010	8 120	7 230	6 540	6 000	5 300	4 690
	3 250	10 800	8 980	8 090	7 200	6 510	5 980	5 280	4 670
	3 500	10 700	8 940	8 050	7 170	6 480	5 950	5 260	4 650
	3 750	10 700	8 890	8 010	7 120	6 450	5 910	5 220	4 620
	4 000	10 600	8 840	7 950	7 070	6 400	5 870	5 180	4 590
	4 250	10 500	8 770	7 890	7 010	6 350	5 820	5 140	4 550
	4 500	10 400	8 690	7 810	6 930	6 280	5 750	5 080	4 510
	4 750	10 300	8 590	7 720	6 840	6 210	5 680	5 020	4 450
	5 000	10 200	8 480	7 610	6 740	6 120	5 600	4 940	4 390
	5 250	10 100	8 350	7 490	6 630	6 020	5 500	4 860	4 320
	5 500	9 900	8 210	7 350	6 500	5 910	5 400	4 760	4 240
	6 000	9 530	7 880	7 040	6 210	5 660	5 160	4 550	4 060
	6 500	9 100	7 500	6 690	5 880	5 370	4 890	4 300	3 850
	7 000	8 620	7 080	6 300	5 520	5 060	4 590	4 040	3 620
	7 500	8 110	6 650	5 890	5 150	4 730	4 290	3 770	3 390
	8 000	7 590	6 200	5 490	4 780	4 400	3 980	3 500	3 150
	8 500	7 070	5 760	5 090	4 430	4 080	3 680	3 240	2 920
	9 000	6 570	5 340	4 700	4 090	3 770	3 400	2 990	2 700
	9 500	6 090	4 940	4 340	3 770	3 480	3 140	2 750	2 490
	10 000	5 640	4 570	4 010	3 470	3 210	2 890	2 540	2 300
	10 500	5 220	4 220	3 700	3 200	2 970	2 670	2 340	2 120
	11 000	4 840	3 910	3 420	2 960	2 740	2 470	2 160	1 960
	11 500	4 480	3 620	3 170	2 740	2 540	2 280	2 000	1 820
	12 000	4 160	3 360	2 940	2 540	2 350	2 110	1 850	1 680
PROPERTIES AND DESIGN DATA									
Area (mm^2)		40 200	33 600	30 300	27 000	24 400	22 400	19 800	17 500
Z_x (10^3 mm^3)		5 580	4 730	4 280	3 810	3 520	3 220	2 870	2 590
S_x (10^3 mm^3)		4 710	4 070	3 720	3 330	3 120	2 870	2 580	2 350
r_x (mm)		143	146	146	147	150	150	151	153
Z_y (10^3 mm^3)		3 090	2 480	2 170	1 870	1 740	1 550	1 360	1 240
S_y (10^3 mm^3)		2 040	1 630	1 430	1 230	1 140	1 020	899	817
r_y (mm)		94.2	92.3	90.8	89.2	90.5	89.4	89.0	90.4
r_x/r_y		1.52	1.58	1.61	1.65	1.66	1.68	1.70	1.69
M_{rx} (kN·m) (L < L_u)		1 510	1 280	1 160	1 030	950	869	775	699
M_{ry} (kN·m)		834	670	586	505	470	419	367	335
J (10^3 mm^4)		29 800	15 700	10 800	7 070	5 520	4 060	2 760	2 000
C_w (10^9 mm^6)		8 040	6 870	6 210	5 490	5 190	4 720	4 230	3 890
L_u (mm)		11 500	9 120	8 100	7 260	6 990	6 590	6 230	6 070
F_y (MPa)		300	300	300	300	300	300	300	300
IMPERIAL SIZE AND MASS									
Mass (lb./ft.)		211	177	159	142	128	118	104	92
Nominal Depth and Width (in.)		14 x 14							

G40.21-M300W
$\phi = 0.90$

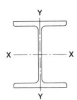

W COLUMNS
Factored Axial Compressive Resistances, C_r, in kN

Designation		W360			W310		
Mass (kg/m)		79	72	64 **	226	202	179
Effective length (KL) in millimetres with respect to least radius of gyration	0	2 730	2 460	2 110	7 800	6 970	6 160
	2 000	2 440	2 200	1 900	7 570	6 750	5 960
	2 250	2 350	2 120	1 840	7 490	6 680	5 900
	2 500	2 250	2 030	1 770	7 390	6 590	5 820
	2 750	2 150	1 930	1 690	7 280	6 490	5 730
	3 000	2 040	1 830	1 610	7 160	6 380	5 620
	3 250	1 930	1 730	1 530	7 020	6 250	5 510
	3 500	1 820	1 630	1 450	6 870	6 120	5 390
	3 750	1 720	1 540	1 360	6 720	5 970	5 260
	4 000	1 610	1 440	1 280	6 550	5 820	5 130
	4 250	1 510	1 350	1 200	6 370	5 660	4 980
	4 500	1 420	1 270	1 120	6 190	5 500	4 840
	4 750	1 330	1 190	1 050	6 010	5 330	4 690
	5 000	1 240	1 110	982	5 820	5 160	4 540
	5 250	1 170	1 040	919	5 630	4 990	4 380
	5 500	1 090	975	861	5 440	4 820	4 230
	6 000	963	858	757	5 070	4 480	3 930
	6 500	852	758	669	4 700	4 160	3 640
	7 000	756	672	593	4 350	3 850	3 370
	7 500	674	599	528	4 030	3 550	3 110
	8 000	603	536	472	3 720	3 280	2 870
	8 500	542	482	424	3 440	3 030	2 650
	9 000	490	435	383	3 180	2 800	2 450
	9 500	444	394	347	2 950	2 590	2 260
	10 000				2 730	2 400	2 090
	10 500				2 530	2 230	1 940
	11 000				2 360	2 070	1 800
	11 500				2 190	1 930	1 680
	12 000				2 040	1 790	1 560
PROPERTIES AND DESIGN DATA							
Area (mm²)		10 100	9 110	8 140	28 900	25 800	22 800
Z_x (10^3 mm³)		1 430	1 280	1 140	3 980	3 510	3 050
S_x (10^3 mm³)		1 280	1 150	1 030	3 420	3 050	2 680
r_x (mm)		150	149	148	144	142	140
Z_y (10^3 mm³)		362	322	284	1 830	1 610	1 400
S_y (10^3 mm³)		236	210	186	1 190	1 050	919
r_y (mm)		48.9	48.5	48.1	81.0	80.2	79.5
r_x/r_y		3.07	3.07	3.08	1.78	1.77	1.76
M_{rx} (kN·m) (L < L_u)		386	346	308	1 070	948	824
M_{ry} (kN·m)		97.7	86.9	76.7	494	435	378
J (10^3 mm⁴)		814	603	438	10 800	7 740	5 380
C_w (10^9 mm⁶)		687	600	524	4 620	3 960	3 340
L_u (mm)		3 270	3 180	3 110	7 540	6 970	6 480
F_y (MPa)		300	300	300	300	300	300
IMPERIAL SIZE AND MASS							
Mass (lb./ft.)		53	48	43	152	136	120
Nominal Depth and Width (in.)			14 x 8			12 x 12	

** Class 4: C_r calculated according to CAN/CSA-S16.1-94 Clause 13.3.3.

W COLUMNS
Factored Axial Compressive Resistances, C_r, in kN

G40.21-M300W
$\phi = 0.90$

Designation		W310						W310	
Mass (kg/m)		158	143	129	118	107	97 ‡	86	79
Effective length (KL) in millimetres with respect to least radius of gyration	0	5 430	4 910	4 460	4 050	3 670	3 320	2 970	2 730
	2 000	5 250	4 760	4 310	3 920	3 550	3 210	2 810	2 570
	2 250	5 190	4 700	4 260	3 870	3 500	3 170	2 750	2 520
	2 500	5 120	4 640	4 200	3 810	3 450	3 120	2 690	2 460
	2 750	5 040	4 560	4 130	3 750	3 400	3 070	2 620	2 390
	3 000	4 950	4 480	4 050	3 680	3 330	3 010	2 540	2 320
	3 250	4 850	4 390	3 970	3 600	3 260	2 950	2 460	2 250
	3 500	4 740	4 290	3 880	3 520	3 180	2 880	2 370	2 160
	3 750	4 630	4 180	3 780	3 430	3 100	2 800	2 280	2 080
	4 000	4 500	4 070	3 680	3 340	3 020	2 720	2 190	2 000
	4 250	4 380	3 960	3 570	3 240	2 930	2 640	2 100	1 910
	4 500	4 250	3 840	3 460	3 140	2 840	2 560	2 000	1 830
	4 750	4 110	3 720	3 350	3 040	2 750	2 480	1 910	1 740
	5 000	3 980	3 590	3 240	2 940	2 650	2 390	1 830	1 660
	5 250	3 840	3 470	3 130	2 830	2 560	2 310	1 740	1 580
	5 500	3 710	3 350	3 020	2 730	2 460	2 220	1 660	1 510
	6 000	3 440	3 110	2 800	2 530	2 280	2 060	1 500	1 360
	6 500	3 180	2 870	2 580	2 340	2 110	1 900	1 360	1 230
	7 000	2 940	2 650	2 380	2 150	1 940	1 750	1 230	1 110
	7 500	2 710	2 450	2 200	1 990	1 790	1 610	1 110	1 010
	8 000	2 500	2 260	2 030	1 830	1 650	1 480	1 010	914
	8 500	2 310	2 080	1 870	1 690	1 520	1 370	919	832
	9 000	2 130	1 920	1 720	1 550	1 400	1 260	839	758
	9 500	1 970	1 780	1 590	1 440	1 290	1 160	767	693
	10 000	1 830	1 640	1 470	1 330	1 190	1 070	703	635
	10 500	1 690	1 520	1 360	1 230	1 110	994	646	584
	11 000	1 570	1 410	1 270	1 140	1 030	921	596	538
	11 500	1 460	1 310	1 180	1 060	953	856	550	497
	12 000	1 360	1 220	1 090	986	887	797	510	460
PROPERTIES AND DESIGN DATA									
Area (mm^2)		20 100	18 200	16 500	15 000	13 600	12 300	11 000	10 100
Z_x (10^3 mm^3)		2 670	2 420	2 160	1 950	1 770	1 590	1 420	1 280
S_x (10^3 mm^3)		2 360	2 150	1 940	1 750	1 590	1 440	1 280	1 160
r_x (mm)		139	138	137	136	135	134	134	133
Z_y (10^3 mm^3)		1 220	1 110	991	893	806	725	533	478
S_y (10^3 mm^3)		805	729	652	588	531	478	351	314
r_y (mm)		78.9	78.6	78.0	77.6	77.2	76.9	63.6	63.0
r_x/r_y		1.76	1.76	1.76	1.75	1.75	1.74	2.11	2.11
M_{rx} (kN·m) (L < L_u)		721	653	583	527	478	389	383	346
M_{ry} (kN·m)		329	300	268	241	218	129	144	129
J (10^3 mm^4)		3 780	2 870	2 130	1 600	1 220	912	877	657
C_w (10^9 mm^6)		2 840	2 540	2 220	1 970	1 760	1 560	961	847
L_u (mm)		6 080	5 820	5 580	5 380	5 220	5 410	4 260	4 140
F_y (MPa)		300	300	300	300	300	300	300	300
IMPERIAL SIZE AND MASS									
Mass (lb./ft.)		106	96	87	79	72	65	58	53
Nominal Depth and Width (in.)		12 x 12						12 x 10	

‡ Class 3 in bending.

G40.21-M300W
$\phi = 0.90$

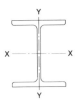

W COLUMNS
Factored Axial Compressive Resistances, C_r, in kN

Designation		W310			W250				
Mass (kg/m)		74	67	60	167	149	131	115	101
Effective length (KL) in millimetres with respect to least radius of gyration	0	2 560	2 300	2 050	5 750	5 130	4 510	3 940	3 480
	2 000	2 300	2 060	1 840	5 480	4 890	4 290	3 750	3 310
	2 250	2 220	1 990	1 770	5 390	4 800	4 210	3 680	3 240
	2 500	2 130	1 910	1 700	5 290	4 710	4 130	3 600	3 180
	2 750	2 040	1 820	1 620	5 170	4 600	4 030	3 510	3 100
	3 000	1 940	1 730	1 540	5 040	4 480	3 920	3 420	3 010
	3 250	1 840	1 640	1 460	4 890	4 350	3 810	3 320	2 920
	3 500	1 740	1 550	1 370	4 740	4 210	3 690	3 210	2 820
	3 750	1 640	1 460	1 290	4 590	4 070	3 560	3 100	2 720
	4 000	1 540	1 370	1 220	4 420	3 920	3 430	2 980	2 620
	4 250	1 450	1 290	1 140	4 260	3 770	3 300	2 860	2 510
	4 500	1 360	1 210	1 070	4 090	3 620	3 160	2 750	2 410
	4 750	1 270	1 130	1 000	3 930	3 470	3 030	2 630	2 310
	5 000	1 190	1 060	940	3 760	3 330	2 900	2 520	2 200
	5 250	1 120	994	882	3 600	3 180	2 770	2 400	2 100
	5 500	1 050	932	827	3 440	3 040	2 650	2 290	2 010
	6 000	927	822	728	3 140	2 770	2 410	2 090	1 820
	6 500	821	727	644	2 860	2 520	2 190	1 900	1 660
	7 000	729	645	572	2 610	2 290	1 990	1 720	1 500
	7 500	650	575	510	2 380	2 090	1 810	1 570	1 360
	8 000	583	515	456	2 170	1 900	1 650	1 420	1 240
	8 500	524	463	410	1 980	1 740	1 510	1 300	1 130
	9 000	474	419	371	1 810	1 590	1 380	1 190	1 030
	9 500	430	380	336	1 660	1 460	1 260	1 090	947
	10 000				1 530	1 340	1 160	999	869
	10 500				1 410	1 230	1 070	919	800
	11 000				1 300	1 140	986	848	738
	11 500				1 200	1 050	912	785	682
	12 000				1 120	977	846	727	632
PROPERTIES AND DESIGN DATA									
Area (mm^2)		9 490	8 510	7 590	21 300	19 000	16 700	14 600	12 900
Z_x (10^3 mm^3)		1 190	1 060	941	2 430	2 130	1 850	1 600	1 400
S_x (10^3 mm^3)		1 060	949	849	2 080	1 840	1 610	1 410	1 240
r_x (mm)		132	131	130	119	117	115	114	113
Z_y (10^3 mm^3)		350	310	275	1 140	1 000	870	753	656
S_y (10^3 mm^3)		229	203	180	746	656	571	495	432
r_y (mm)		49.7	49.3	49.1	68.1	67.4	66.8	66.2	65.6
r_x/r_y		2.66	2.66	2.65	1.75	1.74	1.72	1.72	1.72
M_{rx} (kN·m) (L < L_u)		321	286	254	656	575	500	432	378
M_{ry} (kN·m)		94.5	83.7	74.3	308	270	235	203	177
J (10^3 mm^4)		745	545	397	6 310	4 510	3 120	2 130	1 490
C_w (10^9 mm^6)		505	439	384	1 630	1 390	1 160	976	829
L_u (mm)		3 380	3 280	3 200	6 670	6 160	5 670	5 270	4 950
F_y (MPa)		300	300	300	300	300	300	300	300
IMPERIAL SIZE AND MASS									
Mass (lb./ft.)		50	45	40	112	100	88	77	68
Nominal Depth and Width (in.)		12 x 8			10 x 10				

W COLUMNS
Factored Axial Compressive Resistances, C_r, in kN

G40.21-M300W
$\phi = 0.90$

Designation	W250			W250		
Mass (kg/m)	89	80	73	67	58	49
Effective length (KL) in millimetres with respect to least radius of gyration						
0	3 080	2 750	2 510	2 310	2 000	1 690
2 000	2 920	2 610	2 370	2 090	1 810	1 510
2 250	2 860	2 560	2 330	2 020	1 750	1 460
2 500	2 800	2 510	2 280	1 940	1 680	1 400
2 750	2 730	2 440	2 220	1 860	1 610	1 340
3 000	2 650	2 370	2 150	1 770	1 530	1 270
3 250	2 570	2 300	2 090	1 690	1 450	1 200
3 500	2 480	2 220	2 020	1 600	1 370	1 130
3 750	2 400	2 140	1 940	1 510	1 300	1 070
4 000	2 300	2 060	1 870	1 420	1 220	1 000
4 250	2 210	1 980	1 790	1 340	1 150	942
4 500	2 120	1 890	1 710	1 260	1 080	884
4 750	2 020	1 810	1 640	1 180	1 010	828
5 000	1 930	1 730	1 560	1 110	951	776
5 250	1 850	1 650	1 490	1 040	893	728
5 500	1 760	1 570	1 420	982	838	683
6 000	1 600	1 430	1 290	868	740	602
6 500	1 450	1 290	1 170	770	656	532
7 000	1 310	1 170	1 060	685	583	472
7 500	1 190	1 070	960	612	521	421
8 000	1 080	968	873	549	467	377
8 500	989	882	795	494	420	339
9 000	903	806	726	447	380	306
9 500	826	737	664	406	345	278
10 000	758	677	609	369	314	
10 500	698	622	560			
11 000	643	574	517			
11 500	595	531	478			
12 000	551	492	442			
PROPERTIES AND DESIGN DATA						
Area (mm²)	11 400	10 200	9 280	8 550	7 420	6 250
Z_x (10^3 mm³)	1 230	1 090	985	901	770	633
S_x (10^3 mm³)	1 100	982	891	806	693	572
r_x (mm)	112	111	110	110	108	106
Z_y (10^3 mm³)	574	513	463	332	283	228
S_y (10^3 mm³)	378	338	306	218	186	150
r_y (mm)	65.1	65.0	64.6	51.0	50.4	49.2
r_x/r_y	1.72	1.71	1.70	2.16	2.14	2.15
M_{rx} (kN·m) (L < L_u)	332	294	266	243	208	171
M_{ry} (kN·m)	155	139	125	89.6	76.4	61.6
J (10^3 mm⁴)	1 040	757	575	625	409	241
C_w (10^9 mm⁶)	713	623	553	324	268	211
L_u (mm)	4 690	4 520	4 380	3 580	3 410	3 230
F_y (MPa)	300	300	300	300	300	300
IMPERIAL SIZE AND MASS						
Mass (lb./ft.)	60	54	49	45	39	33
Nominal Depth and Width (in.)	10 x 10			10 x 8		

G40.21-M300W
φ = 0.90

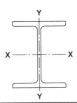

W COLUMNS
Factored Axial Compressive Resistances, C_r, in kN

Designation		W200						W200	
Mass (kg/m)		100	86	71	59	52	46	42	36
Effective length (KL) in millimetres with respect to least radius of gyration	0	3 400	2 970	2 450	2 030	1 790	1 570	1 430	1 230
	2 000	3 120	2 720	2 240	1 850	1 630	1 420	1 210	1 030
	2 250	3 030	2 640	2 170	1 790	1 570	1 380	1 140	979
	2 500	2 920	2 550	2 090	1 730	1 520	1 330	1 080	922
	2 750	2 810	2 450	2 010	1 660	1 450	1 270	1 010	863
	3 000	2 700	2 340	1 920	1 580	1 390	1 210	941	804
	3 250	2 570	2 240	1 830	1 510	1 320	1 150	875	748
	3 500	2 450	2 130	1 740	1 430	1 250	1 090	812	693
	3 750	2 320	2 020	1 650	1 350	1 190	1 030	753	642
	4 000	2 200	1 910	1 560	1 280	1 120	974	697	594
	4 250	2 080	1 800	1 470	1 210	1 060	917	645	550
	4 500	1 970	1 700	1 390	1 140	994	863	598	509
	4 750	1 850	1 610	1 310	1 070	935	811	554	471
	5 000	1 750	1 510	1 230	1 010	880	762	514	437
	5 250	1 650	1 430	1 160	946	827	717	477	406
	5 500	1 550	1 340	1 090	890	778	674	444	377
	6 000	1 380	1 190	971	788	689	596	386	328
	6 500	1 230	1 060	863	700	612	529	337	286
	7 000	1 100	948	770	624	545	471	296	252
	7 500	984	849	689	558	488	421	262	223
	8 000	885	763	619	501	438	377	233	198
	8 500	799	689	559	452	394	340		
	9 000	724	624	506	409	357	307		
	9 500	658	567	459	371	324	279		
	10 000	600	517	419	338	295	254		
	10 500	549	473	383					
	11 000								
	11 500								
	12 000								
PROPERTIES AND DESIGN DATA									
Area (mm²)		12 600	11 000	9 070	7 530	6 620	5 820	5 280	4 540
Z_x (10^3 mm³)		1 150	978	800	650	566	492	442	376
S_x (10^3 mm³)		987	851	707	580	509	445	396	340
r_x (mm)		94.6	92.6	91.7	89.9	89.0	88.1	87.7	86.7
Z_y (10^3 mm³)		533	458	374	302	265	229	165	141
S_y (10^3 mm³)		349	300	246	199	175	151	108	92.6
r_y (mm)		53.8	53.4	52.9	52.1	51.9	51.3	41.3	41.0
r_x/r_y		1.76	1.73	1.73	1.73	1.71	1.72	2.12	2.11
M_{rx} (kN·m) ($L < L_u$)		311	264	216	176	153	133	119	102
M_{ry} (kN·m)		144	124	101	81.5	71.6	61.8	44.6	38.1
J (10^3 mm⁴)		2 060	1 370	801	452	314	213	215	139
C_w (10^9 mm⁶)		386	318	250	196	167	141	84.0	69.5
L_u (mm)		4 990	4 600	4 140	3 770	3 610	3 450	2 850	2 730
F_y (MPa)		300	300	300	300	300	300	300	300
IMPERIAL SIZE AND MASS									
Mass (lb./ft.)		67	58	48	40	35	31	28	24
Nominal Depth and Width (in.)		8 x 8						8 x 6½	

8-21

W COLUMNS
Factored Axial Compressive Resistances, C_r, in kN

G40.21-M300W
$\phi = 0.90$

Designation		W200		W150		
Mass (kg/m)		31	27	37	30	22 ‡
Effective length (KL) in millimetres with respect to least radius of gyration	0	1 080	915	1 280	1 020	770
	2 000	799	666	1 050	838	619
	2 250	733	608	987	787	579
	2 500	668	552	922	735	538
	2 750	606	500	857	682	497
	3 000	549	452	794	631	458
	3 250	497	408	733	582	421
	3 500	450	368	675	536	386
	3 750	408	333	622	494	354
	4 000	371	302	573	454	325
	4 250	338	275	528	418	298
	4 500	308	251	487	386	274
	4 750	282	229	449	356	253
	5 000	258	210	415	329	233
	5 250	238	193	384	304	215
	5 500	219	178	357	282	199
	6 000	188	152	308	244	172
	6 500			269	212	149
	7 000			235	186	131
	7 500			208	164	
	8 000					
	8 500					
	9 000					
	9 500					
	10 000					
	10 500					
	11 000					
	11 500					
	12 000					
PROPERTIES AND DESIGN DATA						
Area (mm^2)		4 000	3 390	4 730	3 790	2 850
Z_x (10^3 mm^3)		335	279	310	244	176
S_x (10^3 mm^3)		299	249	274	219	159
r_x (mm)		88.6	87.3	68.5	67.3	65.1
Z_y (10^3 mm^3)		93.8	76.1	140	111	77.6
S_y (10^3 mm^3)		61.1	49.6	91.8	72.6	50.9
r_y (mm)		32.0	31.2	38.6	38.3	36.9
r_x/r_y		2.77	2.80	1.77	1.76	1.76
M_{rx} (kN·m) (L < L_u)		90.5	75.3	83.7	65.9	42.9
M_{ry} (kN·m)		25.3	20.5	37.8	30.0	13.7
J (10^3 mm^4)		119	71.3	193	101	42.0
C_w (10^9 mm^6)		40.9	32.5	40.0	30.3	20.4
L_u (mm)		2 150	2 050	2 910	2 680	2 590
F_y (MPa)		300	300	300	300	300
IMPERIAL SIZE AND MASS						
Mass (lb./ft.)		21	18	25	20	15
Nominal Depth and Width (in.)		8 x 5¼		6 x 6		

‡ Class 3 in bending.

COMPOSITE BEAMS

The following tables are based on using beams of CSA-G40.21 300W steel. A complete description of tables similar to the following, but based on CSA-G40.21 350W steel, is given in Part Five of this publication, except that the combinations of deck and slab thickness vary, as do concrete strengths and density from those used in these tables.

COMPOSITE MEMBERS
Trial Selection Table

75 mm Deck with 65 mm Slab
$\phi = 0.9$, $\phi_c = 0.6$, $\phi_{sc} = 0.8$

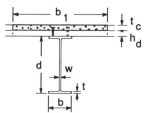

300W
20 MPa

Steel Shape	b_1 mm	Composite Beam*						Steel Shape Data		Non-Composite Shape			
		Factored Resistances								Unbraced Condition			
		M_{rc} (kN·m) for Shear Connections =			Q_r (kN) for	I_t ×10⁶	S_t ×10³			L'	M'r	L'	M'r
		100%	70%	40%	100%	mm⁴	mm³			mm	kN·m	mm	kN·m
WWF900X169 b=300 t=20 d=900	5 000 4 130 3 250 2 380 1 500	2 960 2 890 2 800 2 650 2 450	2 850 2 760 2 650 2 510 2 330	2 600 2 510 2 410 2 310 2 190	3 320 2 740 2 150 1 580 995	7 070 6 760 6 360 5 830 5 110	8 920 8 810 8 660 8 450 8 100	M_r V_r L_u I_x S_x	1 990 1 260 4 130 2 930 6 510	5 000 6 000 7 000 8 000 10 000	1 860 1 680 1 480 1 250 845	12 000 14 000 16 000 18 000 20 000	625 491 402 339 293
WWF800X161 b=300 t=20 d=800	5 000 4 130 3 250 2 380 1 500	2 560 2 480 2 410 2 290 2 120	2 450 2 390 2 290 2 170 2 020	2 250 2 180 2 090 2 000 1 890	3 320 2 740 2 150 1 580 995	5 570 5 330 5 020 4 600 4 020	7 760 7 670 7 540 7 360 7 060	M_r V_r L_u I_x S_x	1 710 1 260 4 220 2 250 5 610	5 000 6 000 7 000 8 000 10 000	1 610 1 470 1 310 1 130 778	12 000 14 000 16 000 18 000 20 000	581 460 379 322 280
WWF700X152 b=300 t=20 d=700	4 500 3 710 2 930 2 140 1 350	2 140 2 070 2 010 1 910 1 770	2 050 1 990 1 910 1 810 1 690	1 880 1 820 1 750 1 670 1 590	2 980 2 460 1 940 1 420 895	4 170 3 980 3 740 3 420 2 980	6 600 6 520 6 410 6 240 5 980	M_r V_r L_u I_x S_x	1 440 1 260 4 330 1 660 4 760	5 000 6 000 7 000 8 000 9 000	1 370 1 260 1 140 1 000 842	10 000 12 000 14 000 16 000 18 000	711 537 429 356 305
W610X174 W24X117 b=325 t=21.6 d=616	4 500 3 710 2 930 2 140 1 350	2 140 2 070 2 000 1 900 1 760	2 050 1 980 1 900 1 800 1 680	1 870 1 810 1 740 1 670 1 590	2 980 2 460 1 940 1 420 895	3 710 3 530 3 300 3 010 2 610	6 750 6 660 6 530 6 340 6 050	M_r V_r L_u I_x S_x	1 450 1 540 4 830 1 470 4 780	5 000 6 000 7 000 8 000 9 000	1 430 1 350 1 250 1 150 1 040	10 000 12 000 14 000 16 000 18 000	924 709 574 482 415
W610X155 W24X104 b=324 t=19 d=612	4 000 3 300 2 600 1 900 1 200	1 890 1 830 1 770 1 680 1 550	1 810 1 750 1 680 1 590 1 480	1 650 1 590 1 530 1 470 1 400	2 650 2 190 1 720 1 260 796	3 260 3 100 2 900 2 640 2 290	5 970 5 880 5 770 5 600 5 340	M_r V_r L_u I_x S_x	1 280 1 380 4 740 1 290 4 220	5 000 6 000 7 000 8 000 9 000	1 260 1 180 1 080 988 886	10 000 11 000 12 000 14 000 16 000	762 659 579 465 388
W610X140 W24X94 b=230 t=22.2 d=618	4 000 3 300 2 600 1 900 1 200	1 760 1 700 1 630 1 530 1 400	1 670 1 610 1 530 1 440 1 330	1 500 1 450 1 380 1 320 1 250	2 650 2 190 1 720 1 260 796	3 010 2 870 2 680 2 440 2 100	5 360 5 290 5 180 5 020 4 780	M_r V_r L_u I_x S_x	1 120 1 440 3 320 1 120 3 630	4 000 5 000 6 000 7 000 8 000	1 050 946 829 695 573	9 000 10 000 12 000 14 000 16 000	486 422 334 277 237
W610X125 W24X84 b=229 t=19.6 d=612	4 000 3 300 2 600 1 900 1 200	1 590 1 530 1 470 1 380 1 260	1 510 1 450 1 380 1 300 1 200	1 360 1 300 1 250 1 180 1 110	2 650 2 190 1 720 1 260 796	2 720 2 590 2 430 2 220 1 920	4 780 4 720 4 630 4 500 4 290	M_r V_r L_u I_x S_x	991 1 300 3 250 985 3 220	4 000 5 000 6 000 7 000 8 000	923 821 708 575 470	9 000 10 000 12 000 14 000 16 000	396 342 269 222 189
W610X113 W24X76 b=228 t=17.3 d=608	4 000 3 300 2 600 1 900 1 200	1 460 1 410 1 350 1 270 1 160	1 380 1 330 1 270 1 190 1 090	1 250 1 190 1 140 1 080 1 010	2 650 2 190 1 720 1 260 796	2 490 2 380 2 230 2 040 1 770	4 330 4 270 4 200 4 080 3 890	M_r V_r L_u I_x S_x	888 1 210 3 180 875 2 880	4 000 5 000 6 000 7 000 8 000	818 719 610 481 391	9 000 10 000 12 000 14 000 16 000	328 282 220 180 153

Note: * 20 MPa, 2 300 kg/m³ Concrete. G40.21-M 300W Steel.

Units: d, b, t, L_u (mm) I_x =10⁶ mm⁴ S_x = 10³ mm³ M_r = kN·m V_r = kN

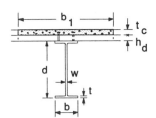

300W
20 MPa

COMPOSITE MEMBERS
Trial Selection Table

75 mm Deck with 65 mm Slab
$\phi = 0.9, \phi_c = 0.6, \phi_{sc} = 0.8$

Steel Shape	b_1 mm	Composite Beam* Factored Resistances M_{rc} (kN·m) for Shear Connections = 100%	70%	40%	Q_r (kN) for 100%	I_t ×10⁶ mm⁴	S_t ×10³ mm³	Steel Shape Data		Non-Composite Shape Unbraced Condition L' mm	M'r kN·m	L' mm	M'r kN·m
W610X101	3 500	1 300	1 240	1 100	2 320	2 190	3 850	M_r	783	4 000	713	9 000	267
W24X68	2 890	1 260	1 190	1 060	1 920	2 090	3 800	V_r	1 130	5 000	619	10 000	228
b=228	2 280	1 200	1 130	1 010	1 510	1 960	3 730	L_u	3 110	6 000	512	11 000	199
t=14.9	1 660	1 130	1 050	949	1 100	1 790	3 620	I_x	764	7 000	396	12 000	176
d=604	1 050	1 020	961	889	696	1 550	3 450	S_x	2 530	8 000	320	14 000	144
W610X91	3 500	1 180	1 120	996	2 320	1 970	3 430	M_r	691	4 000	619	9 000	217
W24X61	2 890	1 140	1 080	951	1 920	1 890	3 390	V_r	1 030	5 000	530	10 000	185
b=227	2 280	1 090	1 020	901	1 510	1 780	3 330	L_u	3 020	6 000	421	11 000	161
t=12.7	1 660	1 020	945	845	1 100	1 620	3 240	I_x	667	7 000	325	12 000	142
d=598	1 050	916	857	785	696	1 410	3 090	S_x	2 230	8 000	261	14 000	115
W610X84	3 500	1 110	1 040	935	2 320	1 840	3 180	M_r	637	3 000	636	8 000	232
W24X56	2 890	1 060	1 010	892	1 920	1 760	3 140	V_r	881	4 000	566	9 000	192
b=226	2 280	1 020	955	844	1 510	1 670	3 090	L_u	2 990	5 000	482	10 000	163
t=11.7	1 660	953	886	789	1 100	1 530	3 010	I_x	613	6 000	377	12 000	125
d=596	1 050	858	801	730	696	1 330	2 870	S_x	2 060	7 000	289	14 000	101
W530X123	3 500	1 390	1 310	1 170	2 320	2 140	4 260	M_r	867	4 000	794	9 000	361
W21X83	2 890	1 340	1 260	1 130	1 920	2 030	4 200	V_r	1 270	5 000	706	10 000	316
b=212	2 280	1 280	1 200	1 080	1 510	1 900	4 110	L_u	3 100	6 000	613	11 000	281
t=21.2	1 660	1 190	1 120	1 020	1 100	1 720	3 980	I_x	761	7 000	505	12 000	253
d=544	1 050	1 090	1 030	964	696	1 480	3 770	S_x	2 800	8 000	421	14 000	211
W530X109	3 500	1 250	1 180	1 060	2 320	1 930	3 790	M_r	764	4 000	692	9 000	291
W21X73	2 890	1 200	1 140	1 020	1 920	1 840	3 740	V_r	1 110	5 000	608	10 000	254
b=211	2 280	1 150	1 080	970	1 510	1 720	3 660	L_u	3 040	6 000	517	11 000	225
t=18.8	1 660	1 080	1 010	918	1 100	1 570	3 550	I_x	667	7 000	413	12 000	202
d=540	1 050	984	928	862	696	1 350	3 380	S_x	2 480	8 000	342	14 000	168
W530X101	3 500	1 180	1 110	999	2 320	1 820	3 530	M_r	707	3 000	707	8 000	301
W21X68	2 890	1 130	1 070	957	1 920	1 730	3 490	V_r	1 040	4 000	635	9 000	255
b=210	2 280	1 080	1 020	911	1 510	1 630	3 420	L_u	2 990	5 000	553	10 000	222
t=17.4	1 660	1 020	951	860	1 100	1 480	3 320	I_x	617	6 000	462	12 000	176
d=538	1 050	924	870	805	696	1 280	3 160	S_x	2 300	7 000	365	14 000	146
W530X92	3 500	1 100	1 030	925	2 320	1 670	3 220	M_r	637	3 000	633	8 000	253
W21X62	2 890	1 050	992	885	1 920	1 600	3 180	V_r	969	4 000	565	9 000	214
b=209	2 280	1 000	943	840	1 510	1 500	3 120	L_u	2 930	5 000	486	10 000	185
t=15.6	1 660	941	878	789	1 100	1 370	3 040	I_x	552	6 000	393	12 000	146
d=534	1 050	852	800	735	696	1 180	2 890	S_x	2 070	7 000	309	14 000	120
W530X82	3 500	997	932	835	2 320	1 500	2 860	M_r	559	3 000	551	8 000	204
W21X55	2 890	951	896	797	1 920	1 440	2 830	V_r	894	4 000	487	9 000	172
b=209	2 280	903	851	754	1 510	1 350	2 780	L_u	2 860	5 000	412	10 000	148
t=13.3	1 660	848	790	705	1 100	1 240	2 710	I_x	479	6 000	321	12 000	115
d=528	1 050	765	714	651	696	1 070	2 580	S_x	1 810	7 000	251	14 000	94.8

Note: * 20 MPa, 2 300 kg/m³ Concrete. G40.21-M 300W Steel.

Units: d, b, t, L_u (mm) I_x =10⁶ mm⁴ S_x = 10³ mm³ M_r = kN·m V_r = kN

COMPOSITE MEMBERS
Trial Selection Table

75 mm Deck with 65 mm Slab
$\phi = 0.9$, $\phi_c = 0.6$, $\phi_{sc} = 0.8$

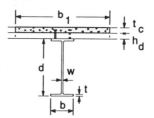

300W
20 MPa

Steel Shape	b_1 mm	Composite Beam* Factored Resistances M_{rc} (kN·m) for Shear Connections =			Q_r (kN) for 100%	I_t ×10⁶ mm⁴	S_t ×10³ mm³	Steel Shape Data		Non-Composite Shape Unbraced Condition			
		100%	70%	40%						L' mm	M'r kN·m	L' mm	M'r kN·m
W530X72 W21X48 b=207 t=10.9 d=524	3 500 2 890 2 280 1 660 1 050	897 853 806 755 678	835 800 759 702 629	745 709 667 619 567	2 320 1 920 1 510 1 100 696	1 320 1 270 1 200 1 100 957	2 490 2 460 2 420 2 360 2 250	M_r V_r L_u I_x S_x	475 831 2 750 402 1 530	3 000 4 000 5 000 6 000 7 000	462 402 331 247 191	8 000 9 000 10 000 12 000 14 000	155 129 111 85.9 70.1
W460X106 W18X71 b=194 t=20.6 d=470	3 500 2 890 2 280 1 660 1 050	1 100 1 050 997 934 847	1 030 987 936 872 797	918 879 836 788 737	2 320 1 920 1 510 1 100 696	1 510 1 440 1 350 1 220 1 040	3 330 3 280 3 210 3 110 2 940	M_r V_r L_u I_x S_x	645 1050 2 910 488 2 080	3 000 4 000 5 000 6 000 7 000	640 579 512 444 366	8 000 9 000 10 000 12 000 14 000	308 266 235 190 160
W460X97 W18X65 b=193 t=19 d=466	3 500 2 890 2 280 1 660 1 050	1 020 970 921 864 784	950 912 867 808 737	851 815 774 728 678	2 320 1 920 1 510 1 100 696	1 400 1 330 1 250 1 140 976	3 050 3 010 2 950 2 860 2 720	M_r V_r L_u I_x S_x	589 947 2 870 445 1 910	3 000 4 000 5 000 6 000 7 000	581 522 457 389 314	8 000 9 000 10 000 12 000 14 000	264 227 200 161 135
W460X89 W18X60 b=192 t=17.7 d=464	3 500 2 890 2 280 1 660 1 050	961 914 865 812 737	894 857 816 760 691	801 767 727 683 634	2 320 1 920 1 510 1 100 696	1 310 1 250 1 180 1 070 923	2 830 2 800 2 750 2 670 2 540	M_r V_r L_u I_x S_x	543 866 2 830 410 1 770	3 000 4 000 5 000 6 000 7 000	534 477 414 343 276	8 000 9 000 10 000 12 000 14 000	231 198 174 140 117
W460X82 W18X55 b=191 t=16 d=460	3 500 2 890 2 280 1 660 1 050	894 848 800 750 680	830 793 754 703 637	742 709 672 629 582	2 320 1 920 1 510 1 100 696	1 210 1 160 1 090 995 859	2 590 2 560 2 510 2 440 2 330	M_r V_r L_u I_x S_x	494 812 2 770 370 1 610	3 000 4 000 5 000 6 000 7 000	482 427 365 292 234	8 000 9 000 10 000 12 000 14 000	195 167 146 117 97.3
W460X74 W18X50 b=190 t=14.5 d=458	3 500 2 890 2 280 1 660 1 050	833 788 742 692 629	771 734 697 651 588	688 657 622 580 534	2 320 1 920 1 510 1 100 696	1 110 1 070 1 010 923 800	2 360 2 330 2 290 2 230 2 130	M_r V_r L_u I_x S_x	446 733 2 720 333 1 460	3 000 4 000 5 000 6 000 7 000	433 380 320 249 198	8 000 9 000 10 000 12 000 14 000	164 140 122 96.9 80.6
W460X67 W18X45 b=190 t=12.7 d=454	3 000 2 480 1 950 1 430 900	745 706 666 624 561	691 660 626 581 523	615 586 553 516 474	1 990 1 640 1 290 948 597	991 950 895 820 707	2 140 2 120 2 080 2 030 1 930	M_r V_r L_u I_x S_x	400 688 2 660 296 1 300	3 000 4 000 5 000 6 000 7 000	390 339 281 214 169	8 000 9 000 10 000 11 000 12 000	140 119 103 91.4 82
W460X61 W18X41 b=189 t=10.8 d=450	3 000 2 480 1 950 1 430 900	685 647 607 567 507	632 602 570 527 470	560 533 500 464 423	1 990 1 640 1 290 948 597	890 855 808 743 642	1 910 1 880 1 850 1 810 1 720	M_r V_r L_u I_x S_x	348 650 2 590 255 1 130	3 000 4 000 5 000 6 000 7 000	336 288 231 172 135	8 000 9 000 10 000 11 000 12 000	111 93.9 81.3 71.7 64.1

Note: * 20 MPa, 2 300 kg/m³ Concrete. G40.21-M 300W Steel.

Units: d, b, t, L_u (mm) I_x =10⁶ mm⁴ S_x = 10³ mm³ M_r = kN·m V_r = kN

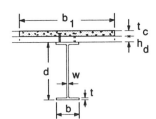

COMPOSITE MEMBERS
Trial Selection Table

300W
20 MPa

75 mm Deck with 65 mm Slab
$\phi = 0.9, \phi_c = 0.6, \phi_{sc} = 0.8$

Steel Shape	b_1 mm	Composite Beam* Factored Resistances M_{rc} (kN·m) for Shear Connections = 100%	70%	40%	Q_r (kN) for 100%	I_t ×10⁶ mm⁴	S_t ×10³ mm³	Steel Shape Data		Non-Composite Shape Unbraced Condition L' mm	M'_r kN·m	L' mm	M'_r kN·m
W410X85 W16X57 b=181 t=18.2 d=418	3 500 2 890 2 280 1 660 1 050	857 810 762 711 644	791 754 715 665 603	704 672 636 595 550	2 320 1 920 1 510 1 100 696	1 070 1 020 962 876 752	2 500 2 470 2 430 2 350 2 240	M_r V_r L_u I_x S_x	467 810 2 730 315 1 510	3 000 4 000 5 000 6 000 7 000	455 406 354 297 243	8 000 9 000 10 000 12 000 14 000	205 178 157 127 107
W410X74 W16X50 b=180 t=16 d=414	3 500 2 890 2 280 1 660 1 050	783 737 690 641 581	720 683 646 601 542	637 608 575 535 492	2 320 1 920 1 510 1 100 696	960 921 868 794 685	2 220 2 190 2 160 2 100 2 000	M_r V_r L_u I_x S_x	408 714 2 670 275 1 330	3 000 4 000 5 000 6 000 7 000	394 348 297 239 194	8 000 9 000 10 000 12 000 14 000	163 140 124 99.8 83.8
W410X67 W16X45 b=179 t=14.4 d=410	3 500 2 890 2 280 1 660 1 050	725 681 635 587 532	664 628 592 551 496	584 558 527 490 447	2 320 1 920 1 510 1 100 696	874 840 794 730 632	2 010 1 980 1 950 1 900 1 820	M_r V_r L_u I_x S_x	367 643 2 610 246 1 200	3 000 4 000 5 000 6 000 7 000	352 307 258 201 161	8 000 9 000 10 000 12 000 14 000	135 116 102 81.7 68.4
W410X60 W16X40 b=178 t=12.8 d=408	3 000 2 480 1 950 1 430 900	631 593 554 513 465	579 548 517 483 434	510 488 461 429 392	1 990 1 640 1 290 948 597	762 732 691 635 549	1 760 1 740 1 710 1 670 1 600	M_r V_r L_u I_x S_x	321 558 2 580 216 1 060	3 000 4 000 5 000 6 000 7 000	306 264 217 165 131	8 000 9 000 10 000 11 000 12 000	109 93.2 81.4 72.2 65
W410X54 W16X36 b=177 t=10.9 d=404	3 000 2 480 1 950 1 430 900	574 548 509 469 423	523 504 473 440 393	459 445 419 388 352	1 840 1 640 1 290 948 597	684 659 624 575 499	1 570 1 560 1 530 1 490 1 430	M_r V_r L_u I_x S_x	284 539 2 480 186 924	3 000 4 000 5 000 6 000 7 000	266 225 176 132 104	8 000 9 000 10 000 11 000 12 000	86.1 73.2 63.6 56.3 50.5
W410X46 W16X31 b=140 t=11.2 d=404	3 000 2 480 1 950 1 430 900	503 494 460 420 376	455 451 424 393 348	396 395 373 343 307	1 590 1 590 1 290 948 597	602 582 553 513 448	1 360 1 350 1 330 1 300 1 240	M_r V_r L_u I_x S_x	239 503 1 930 156 773	2 000 3 000 4 000 5 000 6 000	236 195 142 99.9 76.4	7 000 8 000 9 000 10 000 12 000	61.7 51.8 44.6 39.2 31.6
W410X39 W16X26 b=140 t=8.8 d=400	3 000 2 480 1 950 1 430 900	428 422 408 370 329	386 383 374 344 303	333 332 327 299 265	1 350 1 350 1 290 948 597	514 498 476 444 391	1 150 1 140 1 120 1 100 1 050	M_r V_r L_u I_x S_x	197 448 1 860 127 634	2 000 3 000 4 000 5 000 6 000	193 155 105 73.1 55.2	7 000 8 000 9 000 10 000 12 000	44.1 36.6 31.3 27.4 21.9
W360X79 W14X53 b=205 t=16.8 d=354	3 000 2 480 1 950 1 430 900	694 654 612 570 519	638 606 574 537 489	566 543 515 484 448	1 990 1 640 1 290 948 597	784 747 699 634 540	2 130 2 100 2 060 2 000 1 890	M_r V_r L_u I_x S_x	386 593 3 270 227 1 280	4 000 4 500 5 000 6 000 7 000	364 348 331 298 264	8 000 9 000 10 000 11 000 12 000	225 195 172 154 139

Note : * 20 MPa, 2 300 kg/m³ Concrete. G40.21-M 300W Steel.
Units : d, b, t, L_u (mm) I_x = 10⁶ mm⁴ S_x = 10³ mm³ M_r = kN·m V_r = kN

COMPOSITE MEMBERS
Trial Selection Table

75 mm Deck with 65 mm Slab
$\phi = 0.9$, $\phi_c = 0.6$, $\phi_{sc} = 0.8$

300W
20 MPa

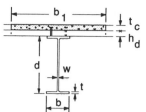

Steel Shape	b_1 mm	Composite Beam* M_{rc} (kN·m) for Shear Connections = 100%	70%	40%	Q_r (kN) for 100%	I_t ×10⁶ mm⁴	S_t ×10³ mm³	Steel Shape Data		Non-Composite Shape Unbraced Condition L' mm	M'_r kN·m	L' mm	M'_r kN·m
W360X72 W14X48 b=204 t=15.1 d=350	3 000 2 480 1 950 1 430 900	643 604 563 522 475	589 558 526 492 446	519 497 471 441 406	1 990 1 640 1 290 948 597	713 681 639 582 498	1 920 1 900 1 860 1 810 1 720	M_r V_r L_u I_x S_x	346 536 3 180 201 1 150	4 000 4 500 5 000 6 000 7 000	322 306 290 257 222	8 000 9 000 10 000 11 000 12 000	186 161 141 126 114
W360X64 W14X43 b=203 t=13.5 d=348	3 000 2 480 1 950 1 430 900	596 558 518 477 433	543 512 481 449 407	474 455 431 402 369	1 990 1 640 1 290 948 597	648 621 584 535 459	1 720 1 700 1 670 1 630 1 550	M_r V_r L_u I_x S_x	308 476 3 110 178 1 030	4 000 4 500 5 000 6 000 7 000	283 268 252 220 183	8 000 9 000 10 000 11 000 12 000	153 131 115 102 92.4
W360X57 W14X38 b=172 t=13.1 d=358	2 500 2 060 1 630 1 190 750	527 494 461 427 385	482 456 430 400 359	424 404 381 355 326	1 660 1 370 1 080 789 497	585 559 526 479 410	1 540 1 510 1 490 1 450 1 370	M_r V_r L_u I_x S_x	273 504 2 550 161 897	3 000 3 500 4 000 4 500 5 000	259 243 225 207 189	6 000 7 000 8 000 9 000 10 000	147 119 99.8 86 75.7
W360X51 W14X34 b=171 t=11.6 d=356	2 500 2 060 1 630 1 190 750	488 456 424 390 351	444 419 393 365 327	388 370 349 324 295	1 660 1 370 1 080 789 497	530 508 479 438 377	1 370 1 360 1 330 1 300 1 240	M_r V_r L_u I_x S_x	241 455 2 500 141 796	3 000 3 500 4 000 4 500 5 000	227 212 195 178 159	6 000 7 000 8 000 9 000 10 000	121 97 81 69.5 60.9
W360X45 W14X30 b=171 t=9.8 d=352	2 500 2 060 1 630 1 190 750	442 419 387 354 317	400 382 358 331 294	347 335 315 290 262	1 550 1 370 1 080 789 497	474 455 431 396 342	1 220 1 200 1 180 1 150 1 100	M_r V_r L_u I_x S_x	210 433 2 430 122 691	3 000 3 500 4 000 4 500 5 000	195 181 165 148 128	6 000 7 000 8 000 9 000 10 000	96.1 76.5 63.4 54.1 47.2
W360X39 W14X26 b=128 t=10.7 d=354	2 500 2 060 1 630 1 190 750	391 383 354 321 285	351 348 324 298 263	302 301 283 259 231	1 340 1 340 1 080 789 497	421 405 385 356 309	1 060 1 050 1 030 1 000 960	M_r V_r L_u I_x S_x	179 409 1 790 102 580	2 000 2 500 3 000 4 000 5 000	173 157 139 97.2 69.8	6 000 7 000 8 000 9 000 10 000	54.2 44.3 37.5 32.5 28.8
W360X33 W14X22 b=127 t=8.5 d=350	2 500 2 060 1 630 1 190 750	330 324 313 281 247	294 292 285 259 228	251 251 247 224 198	1 130 1 130 1 080 789 497	357 345 329 306 269	886 877 864 845 811	M_r V_r L_u I_x S_x	146 361 1 720 82.7 474	2 000 2 500 3 000 4 000 5 000	139 124 108 70.3 49.7	6 000 7 000 8 000 9 000 10 000	38.1 30.8 25.9 22.3 19.6
W310X86 W12X58 b=254 t=16.3 d=310	3 500 2 890 2 280 1 660 1 050	708 662 615 566 516	644 608 571 533 489	563 540 515 484 449	2 320 1 920 1 510 1 100 696	728 694 650 590 502	2 170 2 140 2 100 2 040 1 940	M_r V_r L_u I_x S_x	383 503 4 260 199 1 280	5 000 6 000 7 000 8 000 9 000	367 344 320 297 273	10 000 11 000 12 000 13 000 14 000	248 221 200 182 167

Note: * 20 MPa, 2 300 kg/m³ Concrete. G40.21-M 300W Steel.

Units: d, b, t, L_u (mm) I_x =10⁶ mm⁴ S_x = 10³ mm³ M_r = kN·m V_r = kN

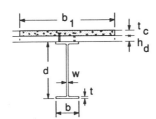

COMPOSITE MEMBERS
Trial Selection Table

300W
20 MPa

75 mm Deck with 65 mm Slab
$\phi = 0.9$, $\phi_c = 0.6$, $\phi_{sc} = 0.8$

Steel Shape	b_1 mm	Composite Beam* Factored Resistances M_{rc} (kN·m) for Shear Connections = 100%	70%	40%	Q_r (kN) for 100%	I_t ×10⁶ mm⁴	S_t ×10³ mm³	Steel Shape Data		Non-Composite Shape Unbraced Condition L' mm	M'$_r$ kN·m	L' mm	M'$_r$ kN·m
W310X79 W12X53 b=254 t=14.6 d=306	3 000 2 480 1 950 1 430 900	629 590 549 508 463	574 543 511 479 437	505 485 461 433 401	1 990 1 640 1 290 948 597	645 614 573 519 440	1 970 1 940 1 900 1 840 1 750	M_r V_r L_u I_x S_x	346 480 4 140 177 1 160	5 000 5 500 6 000 6 500 7 000	327 316 305 293 282	8 000 9 000 10 000 11 000 12 000	258 235 207 184 166
W310X74 W12X50 b=205 t=16.3 d=310	3 000 2 480 1 950 1 430 900	609 570 529 487 441	554 523 491 458 414	484 463 438 410 378	1 990 1 640 1 290 948 597	621 592 553 502 426	1 850 1 830 1 790 1 740 1 650	M_r V_r L_u I_x S_x	321 519 3 380 165 1 060	4 000 4 500 5 000 6 000 7 000	307 295 282 258 233	8 000 9 000 10 000 11 000 12 000	206 179 159 142 129
W310X67 W12X45 b=204 t=14.6 d=306	3 000 2 480 1 950 1 430 900	565 526 486 445 402	511 481 449 417 377	442 423 400 373 341	1 990 1 640 1 290 948 597	561 537 503 458 391	1 660 1 640 1 610 1 570 1 490	M_r V_r L_u I_x S_x	286 463 3 280 145 949	4 000 4 500 5 000 6 000 7 000	270 258 246 222 198	8 000 9 000 10 000 11 000 12 000	169 147 129 116 105
W310X60 W12X40 b=203 t=13.1 d=304	2 500 2 060 1 630 1 190 750	489 456 423 389 352	444 418 393 366 331	387 370 351 327 301	1 660 1 370 1 080 789 497	490 467 438 397 338	1 470 1 450 1 430 1 390 1 310	M_r V_r L_u I_x S_x	254 405 3 200 129 849	4 000 4 500 5 000 5 500 6 000	237 226 214 203 191	6 500 7 000 8 000 9 000 10 000	179 166 139 120 106
W310X52 W12X35 b=167 t=13.2 d=318	2 500 2 060 1 630 1 190 750	464 432 400 365 329	420 394 369 342 307	363 347 327 304 278	1 660 1 370 1 080 789 497	466 445 419 382 327	1 330 1 310 1 290 1 250 1 190	M_r V_r L_u I_x S_x	227 431 2 570 119 750	3 000 3 500 4 000 4 500 5 000	216 203 189 176 162	6 000 7 000 8 000 9 000 10 000	130 106 89.4 77.4 68.4
W310X45 W12X30 b=166 t=11.2 d=314	2 500 2 060 1 630 1 190 750	410 388 356 323 288	368 351 327 301 270	315 305 288 267 242	1 540 1 370 1 080 789 497	405 388 367 337 291	1 140 1 120 1 100 1 080 1 030	M_r V_r L_u I_x S_x	191 368 2 490 99.2 634	3 000 3 500 4 000 4 500 5 000	180 168 155 142 128	6 000 7 000 8 000 9 000 10 000	98.2 79.3 66.5 57.3 50.4
W310X39 W12X26 b=165 t=9.7 d=310	2 500 2 060 1 630 1 190 750	359 351 323 290 256	319 316 294 268 240	272 271 257 237 213	1 330 1 330 1 080 789 497	356 343 326 300 261	991 980 965 942 902	M_r V_r L_u I_x S_x	165 320 2 440 85.1 549	3 000 3 500 4 000 4 500 5 000	153 142 130 117 103	6 000 7 000 8 000 9 000 10 000	77.7 62.2 51.8 44.3 38.8
W250X80 W10X54 b=255 t=15.6 d=256	3 000 2 480 1 950 1 430 900	564 525 484 443 400	509 478 446 414 378	440 420 399 375 347	1 990 1 640 1 290 948 597	512 486 452 407 342	1 790 1 770 1 730 1 670 1 580	M_r V_r L_u I_x S_x	294 429 4 520 126 982	5 000 5 500 6 000 6 500 7 000	287 280 272 265 257	8 000 9 000 10 000 11 000 12 000	242 227 212 197 179

Note: * 20 MPa, 2 300 kg/m³ Concrete. G40.21-M 300W Steel.

Units: d, b, t, L_u (mm) I_x = 10⁶ mm⁴ S_x = 10³ mm³ M_r = kN·m V_r = kN

COMPOSITE MEMBERS
Trial Selection Table

75 mm Deck with 65 mm Slab
$\phi = 0.9$, $\phi_c = 0.6$, $\phi_{sc} = 0.8$

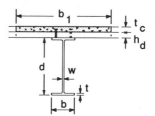

300W
20 MPa

Steel Shape	b_1 mm	Composite Beam* Factored Resistances M_{rc} (kN·m) for Shear Connections =			Q_r (kN) for	I_t ×10⁶ mm⁴	S_t ×10³ mm³	Steel Shape Data		Non-Composite Shape Unbraced Condition			
		100%	70%	40%	100%					L' mm	M'r kN·m	L' mm	M'r kN·m
W250X73 W10X49 b=254 t=14.2 d=254	3 000 2 480 1 950 1 430 900	531 492 452 411 369	477 446 415 384 349	409 389 369 347 319	1 990 1 640 1 290 948 597	473 450 420 379 320	1 640 1 610 1 580 1 530 1 450	M_r V_r L_u I_x S_x	266 388 4 380 113 891	5 000 5 500 6 000 6 500 7 000	257 250 242 235 227	8 000 9 000 10 000 11 000 12 000	212 198 183 165 149
W250X67 W10X45 b=204 t=15.7 d=258	3 000 2 480 1 950 1 430 900	511 473 432 391 349	457 427 395 363 328	388 369 349 325 297	1 990 1 640 1 290 948 597	450 429 401 364 307	1 520 1 500 1 470 1 420 1 350	M_r V_r L_u I_x S_x	243 408 3 580 104 806	4 000 4 500 5 000 6 000 7 000	237 229 221 205 189	8 000 9 000 10 000 11 000 12 000	174 157 139 125 114
W250X58 W10X39 b=203 t=13.5 d=252	2 500 2 060 1 630 1 190 750	430 397 365 331 295	385 360 334 308 277	328 312 295 275 251	1 660 1 370 1 080 789 497	376 358 335 303 255	1 300 1 280 1 260 1 220 1 150	M_r V_r L_u I_x S_x	208 359 3 410 87.3 693	4 000 4 500 5 000 5 500 6 000	199 191 184 176 168	6 500 7 000 8 000 9 000 10 000	160 153 137 119 105
W250X49 W10X33 b=202 t=11 d=248	2 500 2 060 1 630 1 190 750	387 356 324 290 256	344 319 294 268 239	289 273 257 237 214	1 660 1 370 1 080 789 497	322 308 290 264 224	1 100 1 080 1 060 1 030 979	M_r V_r L_u I_x S_x	171 326 3 230 70.6 572	4 000 4 500 5 000 5 500 6 000	160 153 146 138 130	6 500 7 000 8 000 9 000 10 000	123 115 97.2 84 74.1
W250X45 W10X30 b=148 t=13 d=266	2 500 2 060 1 630 1 190 750	375 352 320 287 252	332 315 290 264 234	279 269 252 232 208	1 540 1 370 1 080 789 497	326 312 294 269 230	1 040 1 020 1 010 979 930	M_r V_r L_u I_x S_x	163 360 2 350 71.1 534	3 000 3 500 4 000 4 500 5 000	151 142 132 122 112	6 000 7 000 8 000 9 000 10 000	90.6 75.2 64.4 56.3 50.1
W250X39 W10X26 b=147 t=11.2 d=262	2 500 2 060 1 630 1 190 750	325 318 290 257 223	286 283 261 235 208	239 238 224 206 183	1 330 1 330 1 080 789 497	284 273 259 238 205	897 886 871 849 811	M_r V_r L_u I_x S_x	139 308 2 280 60.1 459	3 000 3 500 4 000 4 500 5 000	126 117 108 98.6 88	6 000 7 000 8 000 9 000 10 000	69.2 57 48.6 42.3 37.6
W250X33 W10X22 b=146 t=9.1 d=258	2 000 1 650 1 300 950 600	272 263 237 211 185	240 234 214 194 171	200 198 185 169 150	1 130 1 090 862 630 398	233 224 211 194 167	750 740 727 708 675	M_r V_r L_u I_x S_x	114 280 2 190 48.9 379	3 000 3 500 4 000 4 500 5 000	102 93.1 84 74.1 63.6	5 500 6 000 6 500 7 000 8 000	55.6 49.4 44.4 40.3 34.1
W200X71 W8X48 b=206 t=17.4 d=216	3 000 2 480 1 950 1 430 900	478 439 399 357 314	424 393 361 329 295	354 335 314 293 267	1 990 1 640 1 290 948 597	378 359 334 300 250	1 470 1 440 1 410 1 360 1 280	M_r V_r L_u I_x S_x	216 393 4 140 76.3 707	5 000 5 500 6 000 6 500 7 000	208 203 198 193 188	8 000 9 000 10 000 11 000 12 000	178 168 158 148 137

Note : * 20 MPa, 2 300 kg/m³ Concrete.   G40.21-M 300W Steel.
Units : d, b, t, L_u (mm) $I_x = 10^6$ mm⁴ $S_x = 10^3$ mm³ M_r = kN·m V_r = kN

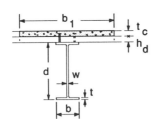

COMPOSITE MEMBERS
Trial Selection Table

300W
20 MPa

75 mm Deck with 65 mm Slab
$\phi = 0.9, \phi_c = 0.6, \phi_{sc} = 0.8$

Steel Shape	b_1 mm	Composite Beam*						Non-Composite Shape				
		Factored Resistances						Unbraced Condition				
		M_{rc} (kN·m) for Shear Connections =			Q_r (kN) for	I_t ×10^6	S_t ×10^3	Steel Shape Data				
		100%	70%	40%	100%	mm^4	mm^3			L' mm	M'$_r$ kN·m	L' mm	M'$_r$ kN·m
W200X59 W8X40 b=205 t=14.2 d=210	3 000 2 480 1 950 1 430 900	428 390 351 311 269	376 346 315 284 251	308 289 270 249 225	1 990 1 640 1 290 948 597	318 304 284 258 217	1 220 1 200 1 180 1 140 1 080	M_r 176 V_r 341 L_u 3 770 I_x 60.9 S_x 580	4 000 4 500 5 000 6 000 7 000	174 169 164 154 144	8 000 9 000 10 000 11 000 12 000	134 124 114 103 93.4
W200X52 W8X35 b=204 t=12.6 d=206	2 500 2 060 1 630 1 190 750	363 331 299 265 231	319 294 269 243 216	264 247 232 214 194	1 660 1 370 1 080 789 497	272 259 242 218 183	1 070 1 050 1 030 997 941	M_r 153 V_r 290 L_u 3 610 I_x 52.5 S_x 509	4 000 4 500 5 000 5 500 6 000	150 145 140 135 131	6 500 7 000 8 000 9 000 10 000	126 121 111 101 89.7
W200X46 W8X31 b=203 t=11 d=204	2 500 2 060 1 630 1 190 750	334 308 276 243 209	291 272 247 221 195	238 226 210 194 174	1 580 1 370 1 080 789 497	244 233 218 198 168	943 929 912 885 839	M_r 133 V_r 260 L_u 3 450 I_x 45.2 S_x 445	4 000 4 500 5 000 5 500 6 000	129 124 119 114 109	6 500 7 000 8 000 9 000 10 000	105 99.8 90.1 78.6 69.6
W200X42 W8X28 b=166 t=11.8 d=206	2 000 1 650 1 300 950 600	290 265 239 212 184	255 235 215 194 172	210 198 185 170 153	1 330 1 090 862 630 398	216 205 192 173 145	846 833 816 790 745	M_r 119 V_r 263 L_u 2 850 I_x 40.6 S_x 396	3 000 3 500 4 000 4 500 5 000	119 114 109 104 98.6	5 500 6 000 6 500 7 000 8 000	93.5 88.4 83.4 77.7 66.7
W200X36 W8X24 b=165 t=10.2 d=202	2 000 1 650 1 300 950 600	261 242 217 191 164	227 213 193 173 152	185 177 164 151 135	1 240 1 090 862 630 398	189 180 169 154 130	733 722 709 688 652	M_r 102 V_r 222 L_u 2 730 I_x 34.1 S_x 340	3 000 3 500 4 000 4 500 5 000	100 95.3 90.4 85.5 80.5	5 500 6 000 6 500 7 000 8 000	75.5 70.6 64.7 59.2 50.6
W200X31 W8X21 b=134 t=10.2 d=210	2 000 1 650 1 300 950 600	236 230 206 180 153	205 202 183 162 141	166 165 153 140 123	1 080 1 080 862 630 398	176 169 159 146 125	654 645 633 616 585	M_r 90.5 V_r 240 L_u 2 150 I_x 31.4 S_x 299	3 000 3 500 4 000 4 500 5 000	81.1 75.3 69.5 63.6 57	5 500 6 000 6 500 7 000 8 000	50.6 45.6 41.5 38 32.7
W200X27 W8X18 b=133 t=8.4 d=208	2 000 1 650 1 300 950 600	203 198 188 162 136	174 172 165 145 125	141 140 137 124 108	915 915 862 630 398	152 147 139 128 111	557 549 540 526 502	M_r 75.3 V_r 214 L_u 2 050 I_x 25.8 S_x 249	3 000 3 500 4 000 4 500 5 000	65.4 59.7 54 47.5 41.2	5 500 6 000 6 500 7 000 8 000	36.4 32.6 29.6 27.1 23.1
W150X30 W6X20 b=153 t=9.3 d=158	2 500 2 060 1 630 1 190 750	204 199 193 165 133	170 168 165 144 119	133 132 131 118 102	1 020 1 020 1 020 789 497	131 126 120 110 95.1	566 558 548 534 510	M_r 65.9 V_r 185 L_u 2 680 I_x 17.2 S_x 219	3 000 3 500 4 000 4 500 5 000	64.1 61.2 58.3 55.4 52.5	6 000 7 000 8 000 9 000 10 000	46.8 40.4 34.8 30.6 27.3

Note : * 20 MPa, 2 300 kg/m^3 Concrete. G40.21-M 300W Steel.

Units : d, b, t, L_u (mm) I_x =10^6 mm^4 S_x = 10^3 mm^3 M_r = kN·m V_r = kN

COMPOSITE MEMBERS
Trial Selection Table

75 mm Deck with 85 mm Slab
$\phi = 0.9$, $\phi_c = 0.6$, $\phi_{sc} = 0.8$

300W
25 MPa

Steel Shape	b_1 mm	Composite Beam* Factored Resistances M_{rc} (kN·m) for Shear Connections = 100%	70%	40%	Q_r (kN) for 100%	I_t ×10⁶ mm⁴	S_t ×10³ mm³	Steel Shape Data		Non-Composite Shape Unbraced Condition L' mm	M'$_r$ kN·m	L' mm	M'$_r$ kN·m
WWF900X169	5 000	3 260	3 110	2 880	5 420	7 320	9 080	M_r	1 990	5 000	1 860	12 000	625
	4 130	3 140	3 010	2 780	4 480	6 990	8 970	V_r	1 260	6 000	1 680	14 000	491
b=300	3 250	3 020	2 910	2 650	3 520	6 580	8 810	L_u	4 130	7 000	1 480	16 000	402
t=20	2 380	2 890	2 760	2 500	2 580	6 030	8 590	I_x	2 930	8 000	1 250	18 000	339
d=900	1 500	2 690	2 530	2 330	1 630	5 270	8 230	S_x	6 510	10 000	845	20 000	293
WWF800X161	5 000	2 850	2 700	2 490	5 420	5 780	7 910	M_r	1 710	5 000	1 610	12 000	581
	4 130	2 740	2 600	2 410	4 480	5 530	7 820	V_r	1 260	6 000	1 470	14 000	460
b=300	3 250	2 620	2 510	2 300	3 520	5 200	7 690	L_u	4 220	7 000	1 310	16 000	379
t=20	2 380	2 490	2 390	2 170	2 580	4 770	7 500	I_x	2 250	8 000	1 130	18 000	322
d=800	1 500	2 320	2 190	2 010	1 630	4 160	7 180	S_x	5 610	10 000	778	20 000	280
WWF700X152	4 500	2 410	2 270	2 080	4 880	4 340	6 750	M_r	1 440	5 000	1 370	10 000	711
	3 710	2 300	2 180	2 010	4 020	4 140	6 660	V_r	1 260	6 000	1 260	12 000	537
b=300	2 930	2 190	2 090	1 920	3 180	3 890	6 540	L_u	4 330	7 000	1 140	14 000	429
t=20	2 140	2 080	1 990	1 820	2 320	3 560	6 370	I_x	1 660	8 000	1 000	16 000	356
d=700	1 350	1 940	1 830	1 690	1 460	3 090	6 090	S_x	4 760	9 000	842	18 000	305
W610X174	4 500	2 420	2 270	2 070	4 880	3 880	6 920	M_r	1 450	5 000	1 430	10 000	924
W24X117	3 710	2 310	2 180	2 000	4 020	3 690	6 820	V_r	1 540	6 000	1 350	12 000	709
b=325	2 930	2 200	2 100	1 910	3 180	3 450	6 680	L_u	4 830	7 000	1 250	14 000	574
t=21.6	2 140	2 080	1 980	1 800	2 320	3 140	6 490	I_x	1 470	8 000	1 150	16 000	482
d=616	1 350	1 920	1 820	1 680	1 460	2 710	6 180	S_x	4 780	9 000	1 040	18 000	415
W610X155	4 000	2 130	2 010	1 830	4 340	3 410	6 110	M_r	1 280	5 000	1 260	10 000	762
W24X104	3 300	2 040	1 930	1 770	3 580	3 240	6 020	V_r	1 380	6 000	1 180	11 000	659
b=324	2 600	1 940	1 850	1 690	2 820	3 030	5 910	L_u	4 740	7 000	1 080	12 000	579
t=19	1 900	1 840	1 750	1 590	2 060	2 750	5 730	I_x	1 290	8 000	988	14 000	465
d=612	1 200	1 700	1 610	1 480	1 300	2 380	5 460	S_x	4 220	9 000	886	16 000	388
W610X140	4 000	2 000	1 880	1 690	4 340	3 150	5 500	M_r	1 120	4 000	1 050	9 000	486
W24X94	3 300	1 910	1 800	1 620	3 580	3 000	5 420	V_r	1 440	5 000	946	10 000	422
b=230	2 600	1 810	1 720	1 540	2 820	2 800	5 310	L_u	3 320	6 000	829	12 000	334
t=22.2	1 900	1 700	1 610	1 440	2 060	2 550	5 150	I_x	1 120	7 000	695	14 000	277
d=618	1 200	1 560	1 460	1 330	1 300	2 190	4 890	S_x	3 630	8 000	573	16 000	237
W610X125	4 000	1 820	1 700	1 530	4 290	2 850	4 910	M_r	991	4 000	923	9 000	396
W24X84	3 300	1 730	1 630	1 470	3 580	2 710	4 840	V_r	1 300	5 000	821	10 000	342
b=229	2 600	1 640	1 550	1 390	2 820	2 550	4 750	L_u	3 250	6 000	708	12 000	269
t=19.6	1 900	1 540	1 460	1 300	2 060	2 320	4 610	I_x	985	7 000	575	14 000	222
d=612	1 200	1 410	1 320	1 200	1 300	2 000	4 390	S_x	3 220	8 000	470	16 000	189
W610X113	4 000	1 660	1 540	1 380	3 890	2 600	4 440	M_r	888	4 000	818	9 000	328
W24X76	3 300	1 600	1 500	1 350	3 580	2 490	4 390	V_r	1 210	5 000	719	10 000	282
b=228	2 600	1 510	1 420	1 280	2 820	2 340	4 300	L_u	3 180	6 000	610	12 000	220
t=17.3	1 900	1 410	1 340	1 190	2 060	2 140	4 190	I_x	875	7 000	481	14 000	180
d=608	1 200	1 290	1 210	1 090	1 300	1 850	3 990	S_x	2 880	8 000	391	16 000	153

Note: * 25 MPa, 1 850 kg/m³ Concrete. G40.21-M 300W Steel.

Units: d, b, t, L_u (mm) I_x =10⁶ mm⁴ S_x = 10³ mm³ M_r = kN·m V_r = kN

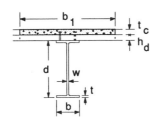

COMPOSITE MEMBERS
Trial Selection Table

300W
25 MPa

75 mm Deck with 85 mm Slab
$\phi = 0.9$, $\phi_c = 0.6$, $\phi_{sc} = 0.8$

Steel Shape	b_1 mm	Composite Beam*						Steel Shape Data		Non-Composite Shape			
		Factored Resistances								Unbraced Condition			
		M_{rc} (kN·m) for Shear Connections =			Q_r (kN) for	I_t ×10⁶	S_t ×10³			L'	M'r	L'	M'r
		100%	70%	40%	100%	mm⁴	mm³			mm	kN·m	mm	kN·m
W610X101 W24X68 b=228 t=14.9 d=604	3 500 2 890 2 280 1 660 1 050	1 480 1 430 1 350 1 260 1 150	1 380 1 340 1 270 1 190 1 070	1 240 1 200 1 140 1 050 962	3 510 3 130 2 470 1 800 1 140	2 290 2 190 2 060 1 870 1 620	3 960 3 900 3 830 3 720 3 530	M_r V_r L_u I_x S_x	783 1 130 3 110 764 2 530	4 000 5 000 6 000 7 000 8 000	713 619 512 396 320	9 000 10 000 11 000 12 000 14 000	267 228 199 176 144
W610X91 W24X61 b=227 t=12.7 d=598	3 500 2 890 2 280 1 660 1 050	1 330 1 300 1 230 1 140 1 040	1 230 1 220 1 150 1 080 963	1 100 1 090 1 030 949 858	3 130 3 130 2 470 1 800 1 140	2 060 1 980 1 860 1 700 1 470	3 530 3 480 3 420 3 330 3 170	M_r V_r L_u I_x S_x	691 1 030 3 020 667 2 230	4 000 5 000 6 000 7 000 8 000	619 530 421 325 261	9 000 10 000 11 000 12 000 14 000	217 185 161 142 115
W610X84 W24X56 b=226 t=11.7 d=596	3 500 2 890 2 280 1 660 1 050	1 230 1 210 1 150 1 070 972	1 140 1 130 1 080 1 010 903	1 010 1 010 965 890 802	2 890 2 890 2 470 1 800 1 140	1 930 1 850 1 750 1 600 1 390	3 270 3 230 3 170 3 090 2 940	M_r V_r L_u I_x S_x	637 881 2 990 613 2 060	3 000 4 000 5 000 6 000 7 000	636 566 482 377 289	8 000 9 000 10 000 12 000 14 000	232 192 163 125 101
W530X123 W21X83 b=212 t=21.2 d=544	3 500 2 890 2 280 1 660 1 050	1 600 1 520 1 430 1 340 1 210	1 490 1 420 1 350 1 260 1 140	1 330 1 270 1 200 1 120 1 030	3 790 3 130 2 470 1 800 1 140	2 250 2 140 1 990 1 800 1 540	4 390 4 320 4 220 4 090 3 870	M_r V_r L_u I_x S_x	867 1 270 3 100 761 2 800	4 000 5 000 6 000 7 000 8 000	794 706 613 505 421	9 000 10 000 11 000 12 000 14 000	361 316 281 253 211
W530X109 W21X73 b=211 t=18.8 d=540	3 500 2 890 2 280 1 660 1 050	1 460 1 380 1 300 1 210 1 100	1 350 1 290 1 220 1 140 1 030	1 210 1 160 1 090 1 020 930	3 750 3 130 2 470 1 800 1 140	2 030 1 940 1 810 1 640 1 410	3 900 3 850 3 770 3 650 3 460	M_r V_r L_u I_x S_x	764 1 110 3 040 667 2 480	4 000 5 000 6 000 7 000 8 000	692 608 517 413 342	9 000 10 000 11 000 12 000 14 000	291 254 225 202 168
W530X101 W21X68 b=210 t=17.4 d=538	3 500 2 890 2 280 1 660 1 050	1 360 1 300 1 220 1 140 1 040	1 260 1 220 1 150 1 070 968	1 120 1 090 1 030 956 872	3 480 3 130 2 470 1 800 1 140	1 910 1 820 1 710 1 560 1 340	3 640 3 590 3 520 3 410 3 240	M_r V_r L_u I_x S_x	707 1 040 2 990 617 2 300	3 000 4 000 5 000 6 000 7 000	707 635 553 462 365	8 000 9 000 10 000 12 000 14 000	301 255 222 176 146
W530X92 W21X62 b=209 t=15.6 d=534	3 500 2 890 2 280 1 660 1 050	1 250 1 220 1 140 1 050 959	1 150 1 130 1 070 996 895	1 020 1 010 954 884 801	3 190 3 130 2 470 1 800 1 140	1 760 1 680 1 580 1 440 1 240	3 320 3 280 3 220 3 120 2 970	M_r V_r L_u I_x S_x	637 969 2 930 552 2 070	3 000 4 000 5 000 6 000 7 000	633 565 486 393 309	8 000 9 000 10 000 12 000 14 000	253 214 185 146 120
W530X82 W21X55 b=209 t=13.3 d=528	3 500 2 890 2 280 1 660 1 050	1 110 1 090 1 040 955 866	1 020 1 010 968 901 807	899 896 864 796 717	2 840 2 840 2 470 1 800 1 140	1 580 1 510 1 420 1 300 1 130	2 950 2 910 2 860 2 790 2 650	M_r V_r L_u I_x S_x	559 894 2 860 479 1 810	3 000 4 000 5 000 6 000 7 000	551 487 412 321 251	8 000 9 000 10 000 12 000 14 000	204 172 148 115 94.8

Note : * 25 MPa, 1 850 kg/m³ Concrete. G40.21-M 300W Steel.

Units : d, b, t, L_u (mm) I_x = 10⁶ mm⁴ S_x = 10³ mm³ M_r = kN·m V_r = kN

COMPOSITE MEMBERS
Trial Selection Table

75 mm Deck with 85 mm Slab
$\phi = 0.9$, $\phi_c = 0.6$, $\phi_{sc} = 0.8$

300W
25 MPa

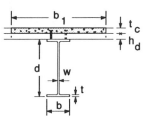

Steel Shape	b_1 mm	Composite Beam* Factored Resistances M_{rc} (kN·m) for Shear Connections =			Q_r (kN) for 100%	I_t ×10⁶ mm⁴	S_t ×10³ mm³	Steel Shape Data		Non-Composite Shape Unbraced Condition			
		100%	70%	40%						L' mm	M'r kN·m	L' mm	M'r kN·m
W530X72 W21X48 b=207 t=10.9 d=524	3 500 2 890 2 280 1 660 1 050	975 961 938 857 772	889 882 871 805 718	778 776 772 709 631	2 470 2 470 2 470 1 800 1 140	1 390 1 330 1 260 1 160 1 010	2 560 2 530 2 490 2 430 2 320	M_r V_r L_u I_x S_x	475 831 2 750 402 1 530	3 000 4 000 5 000 6 000 7 000	462 402 331 247 191	8 000 9 000 10 000 12 000 14 000	155 129 111 85.9 70.1
W460X106 W18X71 b=194 t=20.6 d=470	3 500 2 890 2 280 1 660 1 050	1 290 1 220 1 140 1 050 952	1 190 1 130 1 060 990 889	1 040 1 000 947 878 800	3 650 3 130 2 470 1 800 1 140	1 600 1 520 1 420 1 290 1 100	3 440 3 390 3 310 3 210 3 030	M_r V_r L_u I_x S_x	645 1050 2 910 488 2 080	3 000 4 000 5 000 6 000 7 000	640 579 512 444 366	8 000 9 000 10 000 12 000 14 000	308 266 235 190 160
W460X97 W18X65 b=193 t=19 d=466	3 500 2 890 2 280 1 660 1 050	1 180 1 140 1 060 974 881	1 080 1 050 987 917 824	948 932 880 815 740	3 320 3 130 2 470 1 800 1 140	1 480 1 410 1 320 1 200 1 030	3 150 3 110 3 050 2 950 2 800	M_r V_r L_u I_x S_x	589 947 2 870 445 1 910	3 000 4 000 5 000 6 000 7 000	581 522 457 389 314	8 000 9 000 10 000 12 000 14 000	264 227 200 161 135
W460X89 W18X60 b=192 t=17.7 d=464	3 500 2 890 2 280 1 660 1 050	1 100 1 080 1 000 918 829	1 000 992 931 862 776	878 874 829 767 694	3 080 3 080 2 470 1 800 1 140	1 390 1 330 1 250 1 130 973	2 930 2 890 2 840 2 750 2 610	M_r V_r L_u I_x S_x	543 866 2 830 410 1 770	3 000 4 000 5 000 6 000 7 000	534 477 414 343 276	8 000 9 000 10 000 12 000 14 000	231 198 174 140 117
W460X82 W18X55 b=191 t=16 d=460	3 500 2 890 2 280 1 660 1 050	1 010 988 936 852 766	914 904 865 798 718	798 795 769 711 640	2 810 2 810 2 470 1 800 1 140	1 280 1 220 1 150 1 050 906	2 680 2 640 2 600 2 520 2 400	M_r V_r L_u I_x S_x	494 812 2 770 370 1 610	3 000 4 000 5 000 6 000 7 000	482 427 365 292 234	8 000 9 000 10 000 12 000 14 000	195 167 146 117 97.3
W460X74 W18X50 b=190 t=14.5 d=458	3 500 2 890 2 280 1 660 1 050	920 904 875 793 708	831 824 806 740 666	724 722 713 659 592	2 550 2 550 2 470 1 800 1 140	1 180 1 130 1 070 977 845	2 440 2 410 2 370 2 310 2 200	M_r V_r L_u I_x S_x	446 733 2 720 333 1 460	3 000 4 000 5 000 6 000 7 000	433 380 320 249 198	8 000 9 000 10 000 12 000 14 000	164 140 122 96.9 80.6
W460X67 W18X45 b=190 t=12.7 d=454	3 000 2 480 1 950 1 430 900	835 820 780 711 637	757 750 721 666 594	660 657 638 588 525	2 340 2 340 2 110 1 550 975	1 050 1 010 948 868 747	2 220 2 190 2 150 2 090 1 990	M_r V_r L_u I_x S_x	400 688 2 660 296 1 300	3 000 4 000 5 000 6 000 7 000	390 339 281 214 169	8 000 9 000 10 000 11 000 12 000	140 119 103 91.4 82
W460X61 W18X41 b=189 t=10.8 d=450	3 000 2 480 1 950 1 430 900	749 737 718 652 580	676 670 661 608 540	586 584 581 534 473	2 100 2 100 2 100 1 550 975	944 906 856 787 679	1 980 1 950 1 920 1 870 1 780	M_r V_r L_u I_x S_x	348 650 2 590 255 1 130	3 000 4 000 5 000 6 000 7 000	336 288 231 172 135	8 000 9 000 10 000 11 000 12 000	111 93.9 81.3 71.7 64.1

Note : * 25 MPa, 1 850 kg/m³ Concrete.   G40.21-M 300W Steel.

Units : d, b, t, L_u (mm) I_x =10⁶ mm⁴ S_x = 10³ mm³ M_r = kN·m V_r = kN

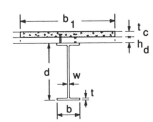

COMPOSITE MEMBERS
Trial Selection Table

300W
25 MPa

75 mm Deck with 85 mm Slab
$\phi = 0.9$, $\phi_c = 0.6$, $\phi_{sc} = 0.8$

Steel Shape	b_1 mm	Composite Beam* Factored Resistances M_{rc} (kN·m) for Shear Connections = 100%	70%	40%	Q_r (kN) for 100%	I_t ×10⁶ mm⁴	S_t ×10³ mm³	Steel Shape Data		Non-Composite Shape Unbraced Condition L' mm	M'$_r$ kN·m	L' mm	M'$_r$ kN·m
W410X85 W16X57 b=181 t=18.2 d=418	3 500 2 890 2 280 1 660 1 050	981 961 899 814 727	885 876 828 759 681	765 762 730 674 606	2 920 2 920 2 470 1 800 1 140	1 140 1 090 1 020 930 796	2 600 2 560 2 510 2 440 2 310	M_r V_r L_u I_x S_x	467 810 2 730 315 1 510	3 000 4 000 5 000 6 000 7 000	455 406 354 297 243	8 000 9 000 10 000 12 000 14 000	205 178 157 127 107
W410X74 W16X50 b=180 t=16 d=414	3 500 2 890 2 280 1 660 1 050	872 856 824 742 657	783 775 755 689 617	675 672 662 611 546	2 580 2 580 2 470 1 800 1 140	1 020 979 923 844 726	2 310 2 280 2 240 2 170 2 070	M_r V_r L_u I_x S_x	408 714 2 670 275 1 330	3 000 4 000 5 000 6 000 7 000	394 348 297 239 194	8 000 9 000 10 000 12 000 14 000	163 140 124 99.8 83.8
W410X67 W16X45 b=179 t=14.4 d=410	3 500 2 890 2 280 1 660 1 050	787 774 755 686 602	704 698 688 634 566	605 603 600 561 501	2 320 2 320 2 320 1 800 1 140	930 893 845 776 671	2 080 2 060 2 020 1 970 1 880	M_r V_r L_u I_x S_x	367 643 2 610 246 1 200	3 000 4 000 5 000 6 000 7 000	352 307 258 201 161	8 000 9 000 10 000 12 000 14 000	135 116 102 81.7 68.4
W410X60 W16X40 b=178 t=12.8 d=408	3 000 2 480 1 950 1 430 900	690 679 661 598 526	618 612 603 554 495	532 530 527 491 438	2 050 2 050 2 050 1 550 975	811 778 735 675 583	1 830 1 810 1 780 1 730 1 650	M_r V_r L_u I_x S_x	321 558 2 580 216 1 060	3 000 4 000 5 000 6 000 7 000	306 264 217 165 131	8 000 9 000 10 000 11 000 12 000	109 93.2 81.4 72.2 65
W410X54 W16X36 b=177 t=10.9 d=404	3 000 2 480 1 950 1 430 900	621 612 598 553 482	554 550 542 510 452	475 474 471 449 397	1 840 1 840 1 840 1 550 975	729 701 664 612 531	1 640 1 620 1 590 1 550 1 480	M_r V_r L_u I_x S_x	284 539 2 480 186 924	3 000 4 000 5 000 6 000 7 000	266 225 176 132 104	8 000 9 000 10 000 11 000 12 000	86.1 73.2 63.6 56.3 50.5
W410X46 W16X31 b=140 t=11.2 d=404	3 000 2 480 1 950 1 430 900	543 536 525 503 433	482 478 473 461 404	410 409 407 402 352	1 590 1 590 1 590 1 550 975	643 620 589 546 476	1 420 1 400 1 380 1 350 1 290	M_r V_r L_u I_x S_x	239 503 1 930 156 773	2 000 3 000 4 000 5 000 6 000	236 195 142 99.9 76.4	7 000 8 000 9 000 10 000 12 000	61.7 51.8 44.6 39.2 31.6
W410X39 W16X26 b=140 t=8.8 d=400	3 000 2 480 1 950 1 430 900	461 456 449 435 383	408 405 401 395 355	345 344 343 341 307	1 350 1 350 1 350 1 350 975	550 532 508 473 416	1 200 1 190 1 170 1 140 1 100	M_r V_r L_u I_x S_x	197 448 1 860 127 634	2 000 3 000 4 000 5 000 6 000	193 155 105 73.1 55.2	7 000 8 000 9 000 10 000 12 000	44.1 36.6 31.3 27.4 21.9
W360X79 W14X53 b=205 t=16.8 d=354	3 000 2 480 1 950 1 430 900	822 798 729 659 583	737 725 668 612 550	630 624 588 546 492	2 730 2 690 2 110 1 550 975	839 799 747 678 575	2 220 2 190 2 140 2 080 1 970	M_r V_r L_u I_x S_x	386 593 3 270 227 1 280	4 000 4 500 5 000 6 000 7 000	364 348 331 298 264	8 000 9 000 10 000 11 000 12 000	225 195 172 154 139

Note: * 25 MPa, 1 850 kg/m³ Concrete. G40.21-M 300W Steel.
Units: d, b, t, L_u (mm) I_x =10⁶ mm⁴ S_x = 10³ mm³ M_r = kN·m V_r = kN

COMPOSITE MEMBERS
Trial Selection Table

75 mm Deck with 85 mm Slab
$\phi = 0.9$, $\phi_c = 0.6$, $\phi_{sc} = 0.8$

300W
25 MPa

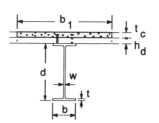

Steel Shape	b_1 mm	Composite Beam*						Non-Composite Shape			
		Factored Resistances				I_t ×10⁶ mm⁴	S_t ×10³ mm³	Steel Shape Data	Unbraced Condition		
		M_{rc} (kN·m) for Shear Connections =			Q_r (kN) for							
		100%	70%	40%	100%				L' mm	M'r kN·m	L' mm	M'r kN·m
W360X72 W14X48 b=204 t=15.1 d=350	3 000 2 480 1 950 1 430 900	745 728 678 609 535	665 657 619 563 504	565 563 539 501 450	2 460 2 460 2 110 1 550 975	764 730 684 623 531	2 000 1 980 1 940 1 880 1 790	M_r 346 V_r 536 L_u 3 180 I_x 201 S_x 1 150	4 000 4 500 5 000 6 000 7 000	322 306 290 257 222	8 000 9 000 10 000 11 000 12 000	186 161 141 126 114
W360X64 W14X43 b=203 t=13.5 d=348	3 000 2 480 1 950 1 430 900	671 658 631 563 490	596 589 573 518 461	505 503 495 459 411	2 200 2 200 2 110 1 550 975	695 665 626 572 490	1 800 1 780 1 740 1 700 1 610	M_r 308 V_r 476 L_u 3 110 I_x 178 S_x 1 030	4 000 4 500 5 000 6 000 7 000	283 268 252 220 183	8 000 9 000 10 000 11 000 12 000	153 131 115 102 92.4
W360X57 W14X38 b=172 t=13.1 d=358	2 500 2 060 1 630 1 190 750	601 589 556 498 437	536 530 507 461 410	457 455 442 406 363	1 950 1 950 1 770 1 290 813	626 598 563 512 437	1 600 1 580 1 550 1 500 1 430	M_r 273 V_r 504 L_u 2 550 I_x 161 S_x 897	3 000 3 500 4 000 4 500 5 000	259 243 225 207 189	6 000 7 000 8 000 9 000 10 000	147 119 99.8 86 75.7
W360X51 W14X34 b=171 t=11.6 d=356	2 500 2 060 1 630 1 190 750	541 531 516 460 401	480 475 468 424 375	408 406 404 373 331	1 740 1 740 1 740 1 290 813	568 544 513 469 402	1 430 1 410 1 390 1 350 1 280	M_r 241 V_r 455 L_u 2 500 I_x 141 S_x 796	3 000 3 500 4 000 4 500 5 000	227 212 195 178 159	6 000 7 000 8 000 9 000 10 000	121 97 81 69.5 60.9
W360X45 W14X30 b=171 t=9.8 d=352	2 500 2 060 1 630 1 190 750	482 474 462 423 365	426 422 416 388 341	361 359 357 338 298	1 550 1 550 1 550 1 290 813	508 488 462 424 366	1 270 1 260 1 240 1 200 1 150	M_r 210 V_r 433 L_u 2 430 I_x 122 S_x 691	3 000 3 500 4 000 4 500 5 000	195 181 165 148 128	6 000 7 000 8 000 9 000 10 000	96.1 76.5 63.4 54.1 47.2
W360X39 W14X26 b=128 t=10.7 d=354	2 500 2 060 1 630 1 190 750	425 419 410 390 331	374 371 366 354 308	314 313 312 306 266	1 340 1 340 1 340 1 290 813	452 435 413 381 331	1 110 1 090 1 080 1 050 1 000	M_r 179 V_r 409 L_u 1 790 I_x 102 S_x 580	2 000 2 500 3 000 4 000 5 000	173 157 139 97.2 69.8	6 000 7 000 8 000 9 000 10 000	54.2 44.3 37.5 32.5 28.8
W360X33 W14X22 b=127 t=8.5 d=350	2 500 2 060 1 630 1 190 750	357 353 347 335 292	313 310 307 302 269	261 261 260 258 232	1 130 1 130 1 130 1 130 813	384 371 354 329 288	930 918 904 883 846	M_r 146 V_r 361 L_u 1 720 I_x 82.7 S_x 474	2 000 2 500 3 000 4 000 5 000	139 124 108 70.3 49.7	6 000 7 000 8 000 9 000 10 000	38.1 30.8 25.9 22.3 19.6
W310X86 W12X58 b=254 t=16.3 d=310	3 500 2 890 2 280 1 660 1 050	837 816 750 667 582	742 731 680 614 547	623 620 587 545 495	2 970 2 970 2 470 1 800 1 140	785 748 700 634 538	2 270 2 240 2 200 2 130 2 020	M_r 383 V_r 503 L_u 4 260 I_x 199 S_x 1 280	5 000 6 000 7 000 8 000 9 000	367 344 320 297 273	10 000 11 000 12 000 13 000 14 000	248 221 200 182 167

Note : * 25 MPa, 1 850 kg/m³ Concrete. G40.21-M 300W Steel.

Units : d, b, t, L_u (mm) $I_x = 10^6$ mm⁴ $S_x = 10^3$ mm³ M_r = kN·m V_r = kN

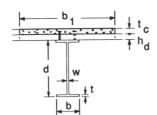

COMPOSITE MEMBERS
Trial Selection Table

300W
25 MPa

75 mm Deck with 85 mm Slab
$\phi = 0.9$, $\phi_c = 0.6$, $\phi_{sc} = 0.8$

Steel Shape	b_1 mm	Composite Beam*						Steel Shape Data		Non-Composite Shape Unbraced Condition			
		M_{rc} (kN·m) for Shear Connections =			Q_r (kN) for 100%	I_t ×10^6 mm^4	S_t ×10^3 mm^3			L' mm	M'r kN·m	L' mm	M'r kN·m
		100%	70%	40%									
W310X79 W12X53 b=254 t=14.6 d=306	3 000 2 480 1 950 1 430 900	756 733 664 594 521	673 660 604 549 491	566 561 525 489 442	2 730 2 690 2 110 1 550 975	695 662 617 558 471	2 060 2 030 1 980 1 920 1 820	M_r V_r L_u I_x S_x	346 480 4 140 177 1 160	5 000 5 500 6 000 6 500 7 000	327 316 305 293 282	8 000 9 000 10 000 11 000 12 000	258 235 207 184 166
W310X74 W12X50 b=205 t=16.3 d=310	3 000 2 480 1 950 1 430 900	721 703 645 575 500	639 631 585 529 470	537 534 505 467 419	2 560 2 560 2 110 1 550 975	669 638 596 540 457	1 940 1 910 1 870 1 810 1 710	M_r V_r L_u I_x S_x	321 519 3 380 165 1 060	4 000 4 500 5 000 6 000 7 000	307 295 282 258 233	8 000 9 000 10 000 11 000 12 000	206 179 159 142 129
W310X67 W12X45 b=204 t=14.6 d=306	3 000 2 480 1 950 1 430 900	650 636 600 531 458	573 566 541 486 429	479 477 463 427 381	2 300 2 300 2 110 1 550 975	606 579 543 494 420	1 740 1 720 1 690 1 640 1 550	M_r V_r L_u I_x S_x	286 463 3 280 145 949	4 000 4 500 5 000 6 000 7 000	270 258 246 222 198	8 000 9 000 10 000 11 000 12 000	169 147 129 116 105
W310X60 W12X40 b=203 t=13.1 d=304	2 500 2 060 1 630 1 190 750	574 559 519 460 400	507 500 470 423 375	425 423 404 373 335	2 050 2 050 1 770 1 290 813	529 504 472 428 363	1 550 1 520 1 490 1 450 1 370	M_r V_r L_u I_x S_x	254 405 3 200 129 849	4 000 4 500 5 000 5 500 6 000	237 226 214 203 191	6 500 7 000 8 000 9 000 10 000	179 166 139 120 106
W310X52 W12X35 b=167 t=13.2 d=318	2 500 2 060 1 630 1 190 750	524 513 494 436 376	461 456 446 399 352	387 385 381 350 311	1 800 1 800 1 770 1 290 813	502 480 451 411 351	1 390 1 370 1 340 1 300 1 240	M_r V_r L_u I_x S_x	227 431 2 570 119 750	3 000 3 500 4 000 4 500 5 000	216 203 189 176 162	6 000 7 000 8 000 9 000 10 000	130 106 89.4 77.4 68.4
W310X45 W12X30 b=166 t=11.2 d=314	2 500 2 060 1 630 1 190 750	450 442 430 392 334	394 390 384 357 310	329 328 326 309 274	1 540 1 540 1 540 1 290 813	437 419 396 363 312	1 190 1 180 1 160 1 130 1 070	M_r V_r L_u I_x S_x	191 368 2 490 99.2 634	3 000 3 500 4 000 4 500 5 000	180 168 155 142 128	6 000 7 000 8 000 9 000 10 000	98.2 79.3 66.5 57.3 50.4
W310X39 W12X26 b=165 t=9.7 d=310	2 500 2 060 1 630 1 190 750	392 386 377 358 301	342 339 334 323 278	284 283 282 277 244	1 330 1 330 1 330 1 290 813	384 370 351 324 281	1 040 1 030 1 010 986 943	M_r V_r L_u I_x S_x	165 320 2 440 85.1 549	3 000 3 500 4 000 4 500 5 000	153 142 130 117 103	6 000 7 000 8 000 9 000 10 000	77.7 62.2 51.8 44.3 38.8
W250X80 W10X54 b=255 t=15.6 d=256	3 000 2 480 1 950 1 430 900	694 668 599 529 456	610 595 539 484 426	503 496 460 425 384	2 750 2 690 2 110 1 550 975	558 530 492 443 370	1 890 1 860 1 820 1 760 1 650	M_r V_r L_u I_x S_x	294 429 4 520 126 982	5 000 5 500 6 000 6 500 7 000	287 280 272 265 257	8 000 9 000 10 000 11 000 12 000	242 227 212 197 179

Note : * 25 MPa, 1 850 kg/m^3 Concrete. G40.21-M 300W Steel.

Units : d, b, t, L_u (mm) $I_x = 10^6$ mm^4 $S_x = 10^3$ mm^3 M_r = kN·m V_r = kN

COMPOSITE MEMBERS
Trial Selection Table

75 mm Deck with 85 mm Slab
$\phi = 0.9$, $\phi_c = 0.6$, $\phi_{sc} = 0.8$

300W
25 MPa

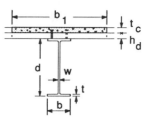

Steel Shape	b_1 mm	Composite Beam*								Non-Composite Shape			
		Factored Resistances				I_t ×10⁶ mm⁴	S_t ×10³ mm³	Steel Shape Data		Unbraced Condition			
		M_{rc} (kN·m) for Shear Connections =			Q_r (kN) for					L' mm	M'r kN·m	L' mm	M'r kN·m
		100%	70%	40%	100%								
W250X73 W10X49 b=254 t=14.2 d=254	3 000 2 480 1 950 1 430 900	637 620 566 497 424	557 548 507 452 395	457 454 429 394 355	2 510 2 510 2 110 1 550 975	515 490 457 412 346	1 730 1 700 1 660 1 610 1 520	M_r V_r L_u I_x S_x	266 388 4 380 113 891	5 000 5 500 6 000 6 500 7 000	257 250 242 235 227	8 000 9 000 10 000 11 000 12 000	212 198 183 165 149
W250X67 W10X45 b=204 t=15.7 d=258	3 000 2 480 1 950 1 430 900	597 583 546 477 404	520 513 487 432 375	426 423 409 374 333	2 310 2 310 2 110 1 550 975	490 467 437 395 333	1 600 1 580 1 540 1 500 1 410	M_r V_r L_u I_x S_x	243 408 3 580 104 806	4 000 4 500 5 000 6 000 7 000	237 229 221 205 189	8 000 9 000 10 000 11 000 12 000	174 157 139 125 114
W250X58 W10X39 b=203 t=13.5 d=252	2 500 2 060 1 630 1 190 750	510 497 460 402 341	444 438 411 364 317	364 362 346 316 282	2 000 2 000 1 770 1 290 813	411 391 365 329 277	1 370 1 350 1 320 1 280 1 210	M_r V_r L_u I_x S_x	208 359 3 410 87.3 693	4 000 4 500 5 000 5 500 6 000	199 191 184 176 168	6 500 7 000 8 000 9 000 10 000	160 153 137 119 105
W250X49 W10X33 b=202 t=11 d=248	2 500 2 060 1 630 1 190 750	435 425 411 360 301	375 371 364 324 278	305 304 302 277 244	1 690 1 690 1 690 1 290 813	353 337 316 287 244	1 160 1 140 1 120 1 090 1 030	M_r V_r L_u I_x S_x	171 326 3 230 70.6 572	4 000 4 500 5 000 5 500 6 000	160 153 146 138 130	6 500 7 000 8 000 9 000 10 000	123 115 97.2 84 74.1
W250X45 W10X30 b=148 t=13 d=266	2 500 2 060 1 630 1 190 750	415 407 395 357 298	359 355 349 320 274	293 292 290 273 239	1 540 1 540 1 540 1 290 813	355 340 320 292 250	1 100 1 080 1 060 1 030 978	M_r V_r L_u I_x S_x	163 360 2 350 71.1 534	3 000 3 500 4 000 4 500 5 000	151 142 132 122 112	6 000 7 000 8 000 9 000 10 000	90.6 75.2 64.4 56.3 50.1
W250X39 W10X26 b=147 t=11.2 d=262	2 500 2 060 1 630 1 190 750	359 353 344 326 268	308 305 301 290 244	251 250 248 244 212	1 330 1 330 1 330 1 290 813	310 298 282 259 223	950 936 920 895 853	M_r V_r L_u I_x S_x	139 308 2 280 60.1 459	3 000 3 500 4 000 4 500 5 000	126 117 108 98.6 88	6 000 7 000 8 000 9 000 10 000	69.2 57 48.6 42.3 37.6
W250X33 W10X22 b=146 t=9.1 d=258	2 000 1 650 1 300 950 600	301 295 287 266 220	258 256 252 238 202	210 210 208 201 174	1 130 1 130 1 130 1 030 650	255 244 230 211 181	794 783 768 747 710	M_r V_r L_u I_x S_x	114 280 2 190 48.9 379	3 000 3 500 4 000 4 500 5 000	102 93.1 84 74.1 63.6	5 500 6 000 6 500 7 000 8 000	55.6 49.4 44.4 40.3 34.1
W200X71 W8X48 b=206 t=17.4 d=216	3 000 2 480 1 950 1 430 900	580 564 513 444 370	500 492 454 398 340	401 398 375 339 301	2 460 2 460 2 110 1 550 975	417 395 367 330 274	1 560 1 530 1 500 1 450 1 360	M_r V_r L_u I_x S_x	216 393 4 140 76.3 707	5 000 5 500 6 000 6 500 7 000	208 203 198 193 188	8 000 9 000 10 000 11 000 12 000	178 168 158 148 137

Note : * 25 MPa, 1 850 kg/m³ Concrete. G40.21-M 300W Steel.

Units : d, b, t, L_u (mm) I_x =10⁶ mm⁴ S_x = 10³ mm³ M_r = kN·m V_r = kN

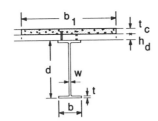

COMPOSITE MEMBERS
Trial Selection Table

75 mm Deck with 85 mm Slab
$\phi = 0.9$, $\phi_c = 0.6$, $\phi_{sc} = 0.8$

300W
25 MPa

Steel Shape	b_1 mm	Composite Beam*								Non-Composite Shape			
		Factored Resistances						Steel Shape Data		Unbraced Condition			
		M_{rc} (kN·m) for Shear Connections =			Q_r (kN) for	I_t ×10⁶ mm⁴	S_t ×10³ mm³			L' mm	M'r kN·m	L' mm	M'r kN·m
		100%	70%	40%	100%								
W200X59 W8X40 b=205 t=14.2 d=210	3 000 2 480 1 950 1 430 900	486 475 457 395 324	415 409 400 351 295	329 328 325 294 257	2 040 2 040 2 040 1 550 975	352 336 314 284 239	1 300 1 280 1 250 1 210 1 150	M_r V_r L_u I_x S_x	176 341 3 770 60.9 580	4 000 4 500 5 000 6 000 7 000	174 169 164 154 144	8 000 9 000 10 000 11 000 12 000	134 124 114 103 93.4
W200X52 W8X35 b=204 t=12.6 d=206	2 500 2 060 1 630 1 190 750	422 411 393 336 276	360 355 345 299 252	287 285 281 252 221	1 800 1 800 1 770 1 290 813	301 286 267 241 202	1 140 1 120 1 090 1 060 997	M_r V_r L_u I_x S_x	153 290 3 610 52.5 509	4 000 4 500 5 000 5 500 6 000	150 145 140 135 131	6 500 7 000 8 000 9 000 10 000	126 121 111 101 89.7
W200X46 W8X31 b=203 t=11 d=204	2 500 2 060 1 630 1 190 750	375 367 354 313 254	318 314 308 277 231	252 251 249 230 200	1 580 1 580 1 580 1 290 813	270 257 242 219 185	1 010 991 971 941 890	M_r V_r L_u I_x S_x	133 260 3 450 45.2 445	4 000 4 500 5 000 5 500 6 000	129 124 119 114 109	6 500 7 000 8 000 9 000 10 000	105 99.8 90.1 78.6 69.6
W200X42 W8X28 b=166 t=11.8 d=206	2 000 1 650 1 300 950 600	337 328 313 268 221	287 283 275 239 202	229 227 224 201 176	1 430 1 430 1 410 1 030 650	239 227 212 191 160	903 888 868 839 790	M_r V_r L_u I_x S_x	119 263 2 850 40.6 396	3 000 3 500 4 000 4 500 5 000	119 114 109 104 98.6	5 500 6 000 6 500 7 000 8 000	93.5 88.4 83.4 77.7 66.7
W200X36 W8X24 b=165 t=10.2 d=202	2 000 1 650 1 300 950 600	293 286 277 246 199	248 245 240 217 181	196 195 194 180 156	1 240 1 240 1 240 1 030 650	209 200 187 170 144	784 771 755 732 692	M_r V_r L_u I_x S_x	102 222 2 730 34.1 340	3 000 3 500 4 000 4 500 5 000	100 95.3 90.4 85.5 80.5	5 500 6 000 6 500 7 000 8 000	75.5 70.6 64.7 59.2 50.6
W200X31 W8X21 b=134 t=10.2 d=210	2 000 1 650 1 300 950 600	263 258 251 234 189	222 220 216 206 170	176 175 174 169 145	1 080 1 080 1 080 1 030 650	195 187 176 161 137	699 688 674 655 621	M_r V_r L_u I_x S_x	90.5 240 2 150 31.4 299	3 000 3 500 4 000 4 500 5 000	81.1 75.3 69.5 63.6 57	5 500 6 000 6 500 7 000 8 000	50.6 45.6 41.5 38 32.7
W200X27 W8X18 b=133 t=8.4 d=208	2 000 1 650 1 300 950 600	225 222 216 207 171	189 187 185 180 153	149 148 148 146 129	915 915 915 915 650	169 163 154 142 122	597 588 576 560 534	M_r V_r L_u I_x S_x	75.3 214 2 050 25.8 249	3 000 3 500 4 000 4 500 5 000	65.4 59.7 54 47.5 41.2	5 500 6 000 6 500 7 000 8 000	36.4 32.6 29.6 27.1 23.1
W150X30 W6X20 b=153 t=9.3 d=158	2 500 2 060 1 630 1 190 750	228 225 219 210 176	187 185 183 178 154	141 141 140 139 124	1 020 1 020 1 020 1 020 813	149 143 135 124 107	618 606 593 576 548	M_r V_r L_u I_x S_x	65.9 185 2 680 17.2 219	3 000 3 500 4 000 4 500 5 000	64.1 61.2 58.3 55.4 52.5	6 000 7 000 8 000 9 000 10 000	46.8 40.4 34.8 30.6 27.3

Note: * 25 MPa, 1 850 kg/m³ Concrete. G40.21-M 300W Steel.

Units: d, b, t, L_u (mm) I_x =10⁶ mm⁴ S_x = 10³ mm³ M_r = kN·m V_r = kN

8-39

COMPOSITE MEMBERS
Trial Selection Table

38 mm Deck with 65 mm Slab
$\phi = 0.9$, $\phi_c = 0.6$, $\phi_{sc} = 0.8$

300W
20 MPa

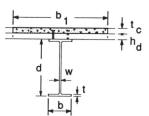

Steel Shape	b_1 mm	Composite Beam* Factored Resistances M_{rc} (kN·m) for Shear Connections = 100%	70%	40%	Q_r (kN) for 100%	I_t ×10⁶ mm⁴	S_t ×10³ mm³	Steel Shape Data		Non-Composite Shape Unbraced Condition L' mm	M'$_r$ kN·m	L' mm	M'$_r$ kN·m
WWF900X169 b=300 t=20 d=900	5 000 4 130 3 250 2 380 1 500	2 840 2 790 2 720 2 600 2 420	2 770 2 690 2 590 2 460 2 310	2 550 2 470 2 380 2 290 2 180	3 320 2 740 2 150 1 580 995	6 540 6 270 5 920 5 460 4 830	8 490 8 400 8 280 8 100 7 810	M$_r$ V$_r$ L$_u$ I$_x$ S$_x$	1 990 1 260 4 130 2 930 6 510	5 000 6 000 7 000 8 000 10 000	1 860 1 680 1 480 1 250 845	12 000 14 000 16 000 18 000 20 000	625 491 402 339 293
WWF800X161 b=300 t=20 d=800	5 000 4 130 3 250 2 380 1 500	2 430 2 380 2 330 2 240 2 080	2 370 2 320 2 230 2 130 1 990	2 200 2 140 2 060 1 970 1 880	3 320 2 740 2 150 1 580 995	5 110 4 900 4 630 4 270 3 780	7 350 7 270 7 170 7 010 6 770	M$_r$ V$_r$ L$_u$ I$_x$ S$_x$	1 710 1 260 4 220 2 250 5 610	5 000 6 000 7 000 8 000 10 000	1 610 1 470 1 310 1 130 778	12 000 14 000 16 000 18 000 20 000	581 460 379 322 280
WWF700X152 b=300 t=20 d=700	4 500 3 710 2 930 2 140 1 350	2 030 1 980 1 940 1 860 1 740	1 970 1 930 1 860 1 780 1 670	1 840 1 780 1 720 1 650 1 570	2 980 2 460 1 940 1 420 895	3 780 3 620 3 420 3 150 2 780	6 210 6 140 6 050 5 920 5 710	M$_r$ V$_r$ L$_u$ I$_x$ S$_x$	1 440 1 260 4 330 1 660 4 760	5 000 6 000 7 000 8 000 9 000	1 370 1 260 1 140 1 000 842	10 000 12 000 14 000 16 000 18 000	711 537 429 356 305
W610X174 W24X117 b=325 t=21.6 d=616	4 500 3 710 2 930 2 140 1 350	2 030 1 980 1 930 1 850 1 730	1 970 1 920 1 850 1 760 1 660	1 820 1 770 1 710 1 640 1 570	2 980 2 460 1 940 1 420 895	3 330 3 180 2 990 2 740 2 410	6 310 6 230 6 130 5 980 5 750	M$_r$ V$_r$ L$_u$ I$_x$ S$_x$	1 450 1 540 4 830 1 470 4 780	5 000 6 000 7 000 8 000 9 000	1 430 1 350 1 250 1 150 1 040	10 000 12 000 14 000 16 000 18 000	924 709 574 482 415
W610X155 W24X104 b=324 t=19 d=612	4 000 3 300 2 600 1 900 1 200	1 790 1 750 1 710 1 630 1 520	1 740 1 690 1 630 1 560 1 460	1 610 1 560 1 510 1 450 1 390	2 650 2 190 1 720 1 260 796	2 930 2 790 2 630 2 410 2 120	5 570 5 500 5 410 5 280 5 080	M$_r$ V$_r$ L$_u$ I$_x$ S$_x$	1 280 1 380 4 740 1 290 4 220	5 000 6 000 7 000 8 000 9 000	1 260 1 180 1 080 988 886	10 000 11 000 12 000 14 000 16 000	762 659 579 465 388
W610X140 W24X94 b=230 t=22.2 d=618	4 000 3 300 2 600 1 900 1 200	1 660 1 620 1 570 1 490 1 370	1 600 1 550 1 490 1 410 1 310	1 460 1 410 1 360 1 300 1 230	2 650 2 190 1 720 1 260 796	2 690 2 570 2 420 2 210 1 940	5 000 4 930 4 840 4 720 4 520	M$_r$ V$_r$ L$_u$ I$_x$ S$_x$	1 120 1 440 3 320 1 120 3 630	4 000 5 000 6 000 7 000 8 000	1 050 946 829 695 573	9 000 10 000 12 000 14 000 16 000	486 422 334 277 237
W610X125 W24X84 b=229 t=19.6 d=612	4 000 3 300 2 600 1 900 1 200	1 490 1 450 1 410 1 340 1 240	1 440 1 400 1 340 1 270 1 170	1 320 1 270 1 220 1 160 1 100	2 650 2 190 1 720 1 260 796	2 420 2 320 2 190 2 010 1 760	4 450 4 400 4 320 4 220 4 050	M$_r$ V$_r$ L$_u$ I$_x$ S$_x$	991 1 300 3 250 985 3 220	4 000 5 000 6 000 7 000 8 000	923 821 708 575 470	9 000 10 000 12 000 14 000 16 000	396 342 269 222 189
W610X113 W24X76 b=228 t=17.3 d=608	4 000 3 300 2 600 1 900 1 200	1 360 1 320 1 280 1 220 1 130	1 310 1 280 1 230 1 160 1 070	1 210 1 160 1 110 1 060 996	2 650 2 190 1 720 1 260 796	2 210 2 120 2 000 1 840 1 610	4 020 3 970 3 910 3 820 3 660	M$_r$ V$_r$ L$_u$ I$_x$ S$_x$	888 1 210 3 180 875 2 880	4 000 5 000 6 000 7 000 8 000	818 719 610 481 391	9 000 10 000 12 000 14 000 16 000	328 282 220 180 153

Note: * 20 MPa, 2 300 kg/m³ Concrete. G40.21-M 300W Steel.
Units: d, b, t, L_u (mm) $I_x = 10^6$ mm⁴ $S_x = 10^3$ mm³ M_r = kN·m V_r = kN

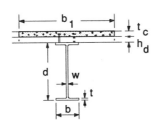

COMPOSITE MEMBERS
Trial Selection Table

38 mm Deck with 65 mm Slab

$\phi = 0.9$, $\phi_c = 0.6$, $\phi_{sc} = 0.8$

300W
20 MPa

Steel Shape	b_1 mm	Composite Beam*								Non-Composite Shape			
		Factored Resistances								Unbraced Condition			
		M_{rc} (kN·m) for Shear Connections =			Q_r (kN) for	I_t ×10⁶	S_t ×10³	Steel Shape Data		L' mm	M'_r kN·m	L' mm	M'_r kN·m
		100%	70%	40%	100%	mm⁴	mm³						
W610X101 W24X68 b=228 t=14.9 d=604	3 500 2 890 2 280 1 660 1 050	1 220 1 180 1 150 1 090 996	1 180 1 140 1 090 1 020 943	1 070 1 030 984 933 878	2 320 1 920 1 510 1 100 696	1 940 1 860 1 760 1 610 1 410	3 570 3 530 3 470 3 380 3 240	M_r V_r L_u I_x S_x	783 1 130 3 110 764 2 530	4 000 5 000 6 000 7 000 8 000	713 619 512 396 320	9 000 10 000 11 000 12 000 14 000	267 228 199 176 144
W610X91 W24X61 b=227 t=12.7 d=598	3 500 2 890 2 280 1 660 1 050	1 100 1 070 1 030 977 891	1 060 1 030 979 916 839	962 923 879 829 775	2 320 1 920 1 510 1 100 696	1 750 1 670 1 580 1 460 1 280	3 180 3 150 3 100 3 020 2 900	M_r V_r L_u I_x S_x	691 1 030 3 020 667 2 230	4 000 5 000 6 000 7 000 8 000	619 530 421 325 261	9 000 10 000 11 000 12 000 14 000	217 185 161 142 115
W610X84 W24X56 b=226 t=11.7 d=596	3 500 2 890 2 280 1 660 1 050	1 020 992 960 913 832	985 957 916 857 783	901 864 821 773 720	2 320 1 920 1 510 1 100 696	1 630 1 570 1 480 1 370 1 200	2 940 2 910 2 870 2 800 2 690	M_r V_r L_u I_x S_x	637 881 2 990 613 2 060	3 000 4 000 5 000 6 000 7 000	636 566 482 377 289	8 000 9 000 10 000 12 000 14 000	232 192 163 125 101
W530X123 W21X83 b=212 t=21.2 d=544	3 500 2 890 2 280 1 660 1 050	1 300 1 260 1 220 1 150 1 060	1 250 1 210 1 160 1 090 1 010	1 140 1 100 1 050 1 000 954	2 320 1 920 1 510 1 100 696	1 890 1 800 1 690 1 540 1 340	3 930 3 880 3 810 3 700 3 540	M_r V_r L_u I_x S_x	867 1 270 3 100 761 2 800	4 000 5 000 6 000 7 000 8 000	794 706 613 505 421	9 000 10 000 11 000 12 000 14 000	361 316 281 253 211
W530X109 W21X73 b=211 t=18.8 d=540	3 500 2 890 2 280 1 660 1 050	1 170 1 130 1 100 1 040 958	1 120 1 090 1 040 983 910	1 030 989 948 901 851	2 320 1 920 1 510 1 100 696	1 700 1 620 1 530 1 400 1 220	3 490 3 450 3 390 3 300 3 160	M_r V_r L_u I_x S_x	764 1 110 3 040 667 2 480	4 000 5 000 6 000 7 000 8 000	692 608 517 413 342	9 000 10 000 11 000 12 000 14 000	291 254 225 202 168
W530X101 W21X68 b=210 t=17.4 d=538	3 500 2 890 2 280 1 660 1 050	1 090 1 060 1 030 976 898	1 050 1 020 979 922 852	964 929 889 844 795	2 320 1 920 1 510 1 100 696	1 590 1 530 1 440 1 320 1 150	3 250 3 210 3 160 3 080 2 950	M_r V_r L_u I_x S_x	707 1 040 2 990 617 2 300	3 000 4 000 5 000 6 000 7 000	707 635 553 462 365	8 000 9 000 10 000 12 000 14 000	301 255 222 176 146
W530X92 W21X62 b=209 t=15.6 d=534	3 500 2 890 2 280 1 660 1 050	1 010 979 945 900 826	970 943 904 850 782	890 856 817 773 725	2 320 1 920 1 510 1 100 696	1 460 1 400 1 330 1 220 1 070	2 960 2 930 2 880 2 810 2 700	M_r V_r L_u I_x S_x	637 969 2 930 552 2 070	3 000 4 000 5 000 6 000 7 000	633 565 486 393 309	8 000 9 000 10 000 12 000 14 000	253 214 185 146 120
W530X82 W21X55 b=209 t=13.3 d=528	3 500 2 890 2 280 1 660 1 050	911 880 847 808 739	872 846 812 762 696	801 768 731 688 641	2 320 1 920 1 510 1 100 696	1 310 1 260 1 190 1 100 961	2 630 2 600 2 560 2 500 2 400	M_r V_r L_u I_x S_x	559 894 2 860 479 1 810	3 000 4 000 5 000 6 000 7 000	551 487 412 321 251	8 000 9 000 10 000 12 000 14 000	204 172 148 115 94.8

Note: * 20 MPa, 2 300 kg/m³ Concrete. G40.21-M 300W Steel.

Units: d, b, t, L_u (mm) $I_x = 10^6$ mm⁴ $S_x = 10^3$ mm³ M_r = kN·m V_r = kN

COMPOSITE MEMBERS
Trial Selection Table

38 mm Deck with 65 mm Slab
$\phi = 0.9$, $\phi_c = 0.6$, $\phi_{sc} = 0.8$

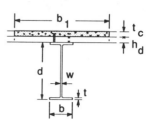

300W
20 MPa

Steel Shape	b_1 mm	Composite Beam*						Non-Composite Shape				
		Factored Resistances				I_t ×10⁶ mm⁴	S_t ×10³ mm³	Steel Shape Data	Unbraced Condition			
		M_{rc} (kN·m) for Shear Connections =			Q_r (kN) for				L' mm	M'r kN·m	L' mm	M'r kN·m
		100%	70%	40%	100%							
W530X72 W21X48 b=207 t=10.9 d=524	3 500 2 890 2 280 1 660 1 050	811 782 750 715 652	775 750 720 673 611	710 680 645 603 557	2 320 1 920 1 510 1 100 696	1 140 1 100 1 050 969 852	2 280 2 250 2 220 2 170 2 080	M_r 475 V_r 831 L_u 2 750 I_x 402 S_x 1 530	3 000 4 000 5 000 6 000 7 000	462 402 331 247 191	8 000 9 000 10 000 12 000 14 000	155 129 111 85.9 70.1
W460X106 W18X71 b=194 t=20.6 d=470	3 500 2 890 2 280 1 660 1 050	1 010 977 941 893 821	967 937 897 844 779	883 850 813 772 727	2 320 1 920 1 510 1 100 696	1 300 1 250 1 170 1 070 930	3 030 2 990 2 930 2 850 2 720	M_r 645 V_r 1050 L_u 2 910 I_x 488 S_x 2 080	3 000 4 000 5 000 6 000 7 000	640 579 512 444 366	8 000 9 000 10 000 12 000 14 000	308 266 235 190 160
W460X97 W18X65 b=193 t=19 d=466	3 500 2 890 2 280 1 660 1 050	933 899 865 823 758	890 863 828 780 719	817 786 751 712 668	2 320 1 920 1 510 1 100 696	1 200 1 150 1 090 995 867	2 770 2 740 2 690 2 620 2 510	M_r 589 V_r 947 L_u 2 870 I_x 445 S_x 1 910	3 000 4 000 5 000 6 000 7 000	581 522 457 389 314	8 000 9 000 10 000 12 000 14 000	264 227 200 161 135
W460X89 W18X60 b=192 t=17.7 d=464	3 500 2 890 2 280 1 660 1 050	875 843 809 771 711	834 807 776 732 673	767 738 705 666 624	2 320 1 920 1 510 1 100 696	1 130 1 080 1 020 937 818	2 580 2 550 2 500 2 440 2 340	M_r 543 V_r 866 L_u 2 830 I_x 410 S_x 1 770	3 000 4 000 5 000 6 000 7 000	534 477 414 343 276	8 000 9 000 10 000 12 000 14 000	231 198 174 140 117
W460X82 W18X55 b=191 t=16 d=460	3 500 2 890 2 280 1 660 1 050	808 777 744 709 654	769 743 715 674 619	708 681 650 613 571	2 320 1 920 1 510 1 100 696	1 030 994 941 866 758	2 350 2 330 2 290 2 240 2 150	M_r 494 V_r 812 L_u 2 770 I_x 370 S_x 1 610	3 000 4 000 5 000 6 000 7 000	482 427 365 292 234	8 000 9 000 10 000 12 000 14 000	195 167 146 117 97.3
W460X74 W18X50 b=190 t=14.5 d=458	3 500 2 890 2 280 1 660 1 050	748 717 686 652 603	710 685 658 622 570	653 629 600 564 524	2 320 1 920 1 510 1 100 696	951 915 868 801 703	2 140 2 120 2 090 2 040 1 960	M_r 446 V_r 733 L_u 2 720 I_x 333 S_x 1 460	3 000 4 000 5 000 6 000 7 000	433 380 320 249 198	8 000 9 000 10 000 12 000 14 000	164 140 122 96.9 80.6
W460X67 W18X45 b=190 t=12.7 d=454	3 000 2 480 1 950 1 430 900	672 646 618 588 539	639 617 593 557 507	586 562 534 502 465	1 990 1 640 1 290 948 597	848 816 772 712 622	1 940 1 920 1 890 1 850 1 770	M_r 400 V_r 688 L_u 2 660 I_x 296 S_x 1 300	3 000 4 000 5 000 6 000 7 000	390 339 281 214 169	8 000 9 000 10 000 11 000 12 000	140 119 103 91.4 82
W460X61 W18X41 b=189 t=10.8 d=450	3 000 2 480 1 950 1 430 900	612 586 559 532 485	581 559 536 503 455	530 508 481 450 414	1 990 1 640 1 290 948 597	759 731 693 642 562	1 730 1 710 1 680 1 640 1 580	M_r 348 V_r 650 L_u 2 590 I_x 255 S_x 1 130	3 000 4 000 5 000 6 000 7 000	336 288 231 172 135	8 000 9 000 10 000 11 000 12 000	111 93.9 81.3 71.7 64.1

Note: * 20 MPa, 2 300 kg/m³ Concrete. G40.21-M 300W Steel.

Units: d, b, t, L_u (mm) I_x = 10⁶ mm⁴ S_x = 10³ mm³ M_r = kN·m V_r = kN

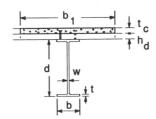

300W
20 MPa

COMPOSITE MEMBERS
Trial Selection Table

38 mm Deck with 65 mm Slab
$\phi = 0.9, \phi_c = 0.6, \phi_{sc} = 0.8$

Steel Shape	b_1 mm	Composite Beam* M_{rc} (kN·m) for Shear Connections = 100%	70%	40%	Q_r (kN) for 100%	I_t ×10⁶ mm⁴	S_t ×10³ mm³	Steel Shape Data		Non-Composite Shape Unbraced Condition L' mm	M'$_r$ kN·m	L' mm	M'$_r$ kN·m
W410X85	3 500	771	731	669	2 320	906	2 250	M_r	467	3 000	455	8 000	205
W16X57	2 890	739	704	644	1 920	869	2 230	V_r	810	4 000	406	9 000	178
b=181	2 280	706	676	614	1 510	821	2 190	L_u	2 730	5 000	354	10 000	157
t=18.2	1 660	670	637	579	1 100	753	2 130	I_x	315	6 000	297	12 000	127
d=418	1 050	618	585	540	696	656	2 040	S_x	1 510	7 000	243	14 000	107
W410X74	3 500	697	659	602	2 320	810	2 000	M_r	408	3 000	394	8 000	163
W16X50	2 890	667	634	580	1 920	779	1 970	V_r	714	4 000	348	9 000	140
b=180	2 280	634	607	552	1 510	738	1 940	L_u	2 670	5 000	297	10 000	124
t=16	1 660	600	573	519	1 100	680	1 900	I_x	275	6 000	239	12 000	99.8
d=414	1 050	555	524	481	696	595	1 820	S_x	1 330	7 000	194	14 000	83.8
W410X67	3 500	640	604	549	2 320	736	1 800	M_r	367	3 000	352	8 000	135
W16X45	2 890	610	579	530	1 920	709	1 780	V_r	643	4 000	307	9 000	116
b=179	2 280	579	553	505	1 510	673	1 760	L_u	2 610	5 000	258	10 000	102
t=14.4	1 660	546	523	473	1 100	623	1 720	I_x	246	6 000	201	12 000	81.7
d=410	1 050	506	478	437	696	547	1 650	S_x	1 200	7 000	161	14 000	68.4
W410X60	3 000	558	527	481	1 990	641	1 580	M_r	321	3 000	306	8 000	109
W16X40	2 480	533	506	464	1 640	618	1 570	V_r	558	4 000	264	9 000	93.2
b=178	1 950	506	483	442	1 290	586	1 540	L_u	2 580	5 000	217	10 000	81.4
t=12.8	1 430	478	458	415	948	542	1 510	I_x	216	6 000	165	11 000	72.2
d=408	900	443	419	384	597	475	1 450	S_x	1 060	7 000	131	12 000	65
W410X54	3 000	506	475	432	1 840	574	1 410	M_r	284	3 000	266	8 000	86.1
W16X36	2 480	487	461	421	1 640	554	1 400	V_r	539	4 000	225	9 000	73.2
b=177	1 950	461	439	400	1 290	526	1 370	L_u	2 480	5 000	176	10 000	63.6
t=10.9	1 430	434	415	374	948	489	1 350	I_x	186	6 000	132	11 000	56.3
d=404	900	401	377	343	597	429	1 290	S_x	924	7 000	104	12 000	50.5
W410X46	3 000	444	414	373	1 590	504	1 220	M_r	239	2 000	236	7 000	61.7
W16X31	2 480	435	410	371	1 590	487	1 200	V_r	503	3 000	195	8 000	51.8
b=140	1 950	412	391	354	1 290	465	1 190	L_u	1 930	4 000	142	9 000	44.6
t=11.2	1 430	385	368	329	948	433	1 160	I_x	156	5 000	99.9	10 000	39.2
d=404	900	354	332	299	597	383	1 120	S_x	773	6 000	76.4	12 000	31.6
W410X39	3 000	379	351	313	1 350	428	1 030	M_r	197	2 000	193	7 000	44.1
W16X26	2 480	372	348	312	1 350	416	1 020	V_r	448	3 000	155	8 000	36.6
b=140	1 950	361	341	308	1 290	398	1 000	L_u	1 860	4 000	105	9 000	31.3
t=8.8	1 430	335	320	285	948	373	982	I_x	127	5 000	73.1	10 000	27.4
d=400	900	307	288	256	597	332	947	S_x	634	6 000	55.2	12 000	21.9
W360X79	3 000	620	586	537	1 990	650	1 890	M_r	386	4 000	364	8 000	225
W14X53	2 480	593	564	518	1 640	622	1 870	V_r	593	4 500	348	9 000	195
b=205	1 950	565	540	496	1 290	585	1 840	L_u	3 270	5 000	331	10 000	172
t=16.8	1 430	535	513	470	948	536	1 790	I_x	227	6 000	298	11 000	154
d=354	900	497	473	439	597	465	1 710	S_x	1 280	7 000	264	12 000	139

Note: * 20 MPa, 2 300 kg/m³ Concrete. G40.21-M 300W Steel.
Units: d, b, t, L_u (mm) $I_x = 10^6$ mm⁴ $S_x = 10^3$ mm³ M_r = kN·m V_r = kN

COMPOSITE MEMBERS
Trial Selection Table

38 mm Deck with 65 mm Slab
$\phi = 0.9$, $\phi_c = 0.6$, $\phi_{sc} = 0.8$

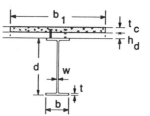

300W
20 MPa

Steel Shape	b_1 mm	Composite Beam*						Non-Composite Shape				
		Factored Resistances						Unbraced Condition				
		M_{rc} (kN·m) for Shear Connections =			Q_r (kN) for	I_t ×10⁶	S_t ×10³	Steel Shape Data				
		100%	70%	40%	100%	mm⁴	mm³			L' mm	M'r kN·m	L' mm	M'r kN·m
W360X72 W14X48 b=204 t=15.1 d=350	3 000 2 480 1 950 1 430 900	570 543 515 487 453	537 515 492 467 430	489 473 452 427 397	1 990 1 640 1 290 948 597	589 565 533 490 426	1 710 1 690 1 660 1 620 1 550	M_r 346 V_r 536 L_u 3 180 I_x 201 S_x 1 150	4 000 4 500 5 000 6 000 7 000	322 306 290 257 222	8 000 9 000 10 000 11 000 12 000	186 161 141 126 114
W360X64 W14X43 b=203 t=13.5 d=348	3 000 2 480 1 950 1 430 900	522 497 470 442 411	491 470 447 425 391	445 430 411 388 360	1 990 1 640 1 290 948 597	534 514 486 448 391	1 530 1 510 1 490 1 450 1 400	M_r 308 V_r 476 L_u 3 110 I_x 178 S_x 1 030	4 000 4 500 5 000 6 000 7 000	283 268 252 220 183	8 000 9 000 10 000 11 000 12 000	153 131 115 102 92.4
W360X57 W14X38 b=172 t=13.1 d=358	2 500 2 060 1 630 1 190 750	465 443 421 397 366	439 420 402 379 347	399 384 365 344 318	1 660 1 370 1 080 789 497	484 464 439 403 350	1 360 1 350 1 330 1 290 1 240	M_r 273 V_r 504 L_u 2 550 I_x 161 S_x 897	3 000 3 500 4 000 4 500 5 000	259 243 225 207 189	6 000 7 000 8 000 9 000 10 000	147 119 99.8 86 75.7
W360X51 W14X34 b=171 t=11.6 d=356	2 500 2 060 1 630 1 190 750	427 406 384 361 333	401 383 365 345 315	363 350 333 312 288	1 660 1 370 1 080 789 497	437 420 398 367 320	1 220 1 200 1 180 1 160 1 110	M_r 241 V_r 455 L_u 2 500 I_x 141 S_x 796	3 000 3 500 4 000 4 500 5 000	227 212 195 178 159	6 000 7 000 8 000 9 000 10 000	121 97 81 69.5 60.9
W360X45 W14X30 b=171 t=9.8 d=352	2 500 2 060 1 630 1 190 750	385 368 347 325 299	360 347 330 310 281	324 315 299 279 255	1 550 1 370 1 080 789 497	389 375 357 330 289	1 080 1 070 1 050 1 030 986	M_r 210 V_r 433 L_u 2 430 I_x 122 S_x 691	3 000 3 500 4 000 4 500 5 000	195 181 165 148 128	6 000 7 000 8 000 9 000 10 000	96.1 76.5 63.4 54.1 47.2
W360X39 W14X26 b=128 t=10.7 d=354	2 500 2 060 1 630 1 190 750	341 333 314 291 266	316 313 296 277 250	282 281 267 247 224	1 340 1 340 1 080 789 497	345 333 317 295 259	934 924 911 890 855	M_r 179 V_r 409 L_u 1 790 I_x 102 S_x 580	2 000 2 500 3 000 4 000 5 000	173 157 139 97.2 69.8	6 000 7 000 8 000 9 000 10 000	54.2 44.3 37.5 32.5 28.8
W360X33 W14X22 b=127 t=8.5 d=350	2 500 2 060 1 630 1 190 750	288 283 273 252 229	265 263 257 239 215	235 234 231 213 191	1 130 1 130 1 080 789 497	291 282 270 252 224	781 773 762 747 719	M_r 146 V_r 361 L_u 1 720 I_x 82.7 S_x 474	2 000 2 500 3 000 4 000 5 000	139 124 108 70.3 49.7	6 000 7 000 8 000 9 000 10 000	38.1 30.8 25.9 22.3 19.6
W310X86 W12X58 b=254 t=16.3 d=310	3 500 2 890 2 280 1 660 1 050	622 591 559 525 490	584 558 532 505 471	529 512 493 468 439	2 320 1 920 1 510 1 100 696	592 566 534 488 424	1 910 1 890 1 850 1 810 1 730	M_r 383 V_r 503 L_u 4 260 I_x 199 S_x 1 280	5 000 6 000 7 000 8 000 9 000	367 344 320 297 273	10 000 11 000 12 000 13 000 14 000	248 221 200 182 167

Note: * 20 MPa, 2 300 kg/m³ Concrete. G40.21-M 300W Steel.

Units: d, b, t, L_u (mm) $I_x = 10^6$ mm⁴ $S_x = 10^3$ mm³ M_r = kN·m V_r = kN

COMPOSITE MEMBERS
Trial Selection Table

38 mm Deck with 65 mm Slab
$\phi = 0.9, \phi_c = 0.6, \phi_{sc} = 0.8$

300W
20 MPa

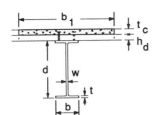

Steel Shape	b_1 mm	Composite Beam* Factored Resistances M_{rc} (kN·m) for Shear Connections = 100%	70%	40%	Q_r (kN) for 100%	I_t ×10⁶ mm⁴	S_t ×10³ mm³	Steel Shape Data		Non-Composite Shape Unbraced Condition L' mm	M'$_r$ kN·m	L' mm	M'$_r$ kN·m
W310X79 W12X53 b=254 t=14.6 d=306	3 000 2 480 1 950 1 430 900	555 529 501 473 441	522 501 478 455 422	475 460 442 419 392	1 990 1 640 1 290 948 597	523 500 470 430 371	1 730 1 700 1 670 1 630 1 560	M_r V_r L_u I_x S_x	346 480 4 140 177 1 160	5 000 5 500 6 000 6 500 7 000	327 316 305 293 282	8 000 9 000 10 000 11 000 12 000	258 235 207 184 166
W310X74 W12X50 b=205 t=16.3 d=310	3 000 2 480 1 950 1 430 900	536 509 481 452 419	503 481 457 433 399	455 439 419 396 369	1 990 1 640 1 290 948 597	503 481 453 414 358	1 630 1 600 1 580 1 540 1 470	M_r V_r L_u I_x S_x	321 519 3 380 165 1 060	4 000 4 500 5 000 6 000 7 000	307 295 282 258 233	8 000 9 000 10 000 11 000 12 000	206 179 159 142 129
W310X67 W12X45 b=204 t=14.6 d=306	3 000 2 480 1 950 1 430 900	491 466 438 410 380	460 438 415 393 361	413 398 381 359 333	1 990 1 640 1 290 948 597	453 435 410 377 326	1 460 1 440 1 410 1 380 1 320	M_r V_r L_u I_x S_x	286 463 3 280 145 949	4 000 4 500 5 000 6 000 7 000	270 258 246 222 198	8 000 9 000 10 000 11 000 12 000	169 147 129 116 105
W310X60 W12X40 b=203 t=13.1 d=304	2 500 2 060 1 630 1 190 750	428 406 383 360 334	401 383 365 345 318	362 350 335 316 293	1 660 1 370 1 080 789 497	396 379 357 327 283	1 290 1 280 1 250 1 220 1 170	M_r V_r L_u I_x S_x	254 405 3 200 129 849	4 000 4 500 5 000 5 500 6 000	237 226 214 203 191	6 500 7 000 8 000 9 000 10 000	179 166 139 120 106
W310X52 W12X35 b=167 t=13.2 d=318	2 500 2 060 1 630 1 190 750	403 382 360 336 310	377 359 341 322 295	339 326 311 292 270	1 660 1 370 1 080 789 497	377 362 342 315 273	1 160 1 150 1 130 1 100 1 050	M_r V_r L_u I_x S_x	227 431 2 570 119 750	3 000 3 500 4 000 4 500 5 000	216 203 189 176 162	6 000 7 000 8 000 9 000 10 000	130 106 89.4 77.4 68.4
W310X45 W12X30 b=166 t=11.2 d=314	2 500 2 060 1 630 1 190 750	353 337 316 294 270	328 316 299 280 257	292 285 272 255 234	1 540 1 370 1 080 789 497	327 314 299 276 241	996 984 970 947 909	M_r V_r L_u I_x S_x	191 368 2 490 99.2 634	3 000 3 500 4 000 4 500 5 000	180 168 155 142 128	6 000 7 000 8 000 9 000 10 000	98.2 79.3 66.5 57.3 50.4
W310X39 W12X26 b=165 t=9.7 d=310	2 500 2 060 1 630 1 190 750	309 302 283 261 238	285 281 266 248 227	253 251 241 226 206	1 330 1 330 1 080 789 497	286 276 264 245 215	867 858 846 828 797	M_r V_r L_u I_x S_x	165 320 2 440 85.1 549	3 000 3 500 4 000 4 500 5 000	153 142 130 117 103	6 000 7 000 8 000 9 000 10 000	77.7 62.2 51.8 44.3 38.8
W250X80 W10X54 b=255 t=15.6 d=256	3 000 2 480 1 950 1 430 900	491 464 436 408 378	458 436 413 390 363	410 396 380 361 338	1 990 1 640 1 290 948 597	402 384 359 327 280	1 540 1 520 1 490 1 450 1 380	M_r V_r L_u I_x S_x	294 429 4 520 126 982	5 000 5 500 6 000 6 500 7 000	287 280 272 265 257	8 000 9 000 10 000 11 000 12 000	242 227 212 197 179

Note : * 20 MPa, 2 300 kg/m³ Concrete. G40.21-M 300W Steel.
Units : d, b, t, L_u (mm) $I_x = 10^6$ mm⁴ $S_x = 10^3$ mm³ M_r = kN·m V_r = kN

8-45

COMPOSITE MEMBERS
Trial Selection Table

38 mm Deck with 65 mm Slab
$\phi = 0.9$, $\phi_c = 0.6$, $\phi_{sc} = 0.8$

300W
20 MPa

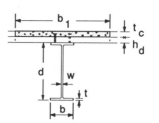

Steel Shape	b_1 mm	Composite Beam*						Non-Composite Shape				
		Factored Resistances						Unbraced Condition				
		M_{rc} (kN·m) for Shear Connections =			Q_r (kN) for	I_t ×10⁶ mm⁴	S_t ×10³ mm³	Steel Shape Data		L' mm	M'r kN·m	L' mm	M'r kN·m
		100%	70%	40%	100%								
W250X73 W10X49 b=254 t=14.2 d=254	3 000 2 480 1 950 1 430 900	457 431 404 376 347	425 404 381 359 334	379 365 350 333 310	1 990 1 640 1 290 948 597	370 354 332 303 260	1 410 1 390 1 360 1 320 1 260	M_r 266 V_r 388 L_u 4 380 I_x 113 S_x 891		5 000 5 500 6 000 6 500 7 000	257 250 242 235 227	8 000 9 000 10 000 11 000 12 000	212 198 183 165 149
W250X67 W10X45 b=204 t=15.7 d=258	3 000 2 480 1 950 1 430 900	438 412 384 356 327	406 384 361 339 313	359 345 329 311 288	1 990 1 640 1 290 948 597	352 337 317 290 249	1 300 1 290 1 260 1 230 1 170	M_r 243 V_r 408 L_u 3 580 I_x 104 S_x 806		4 000 4 500 5 000 6 000 7 000	237 229 221 205 189	8 000 9 000 10 000 11 000 12 000	174 157 139 125 114
W250X58 W10X39 b=203 t=13.5 d=252	2 500 2 060 1 630 1 190 750	369 347 325 301 277	342 324 306 287 264	304 292 279 263 244	1 660 1 370 1 080 789 497	294 280 264 241 207	1 110 1 100 1 080 1 050 1 000	M_r 208 V_r 359 L_u 3 410 I_x 87.3 S_x 693		4 000 4 500 5 000 5 500 6 000	199 191 184 176 168	6 500 7 000 8 000 9 000 10 000	160 153 137 119 105
W250X49 W10X33 b=202 t=11 d=248	2 500 2 060 1 630 1 190 750	326 305 284 261 238	301 284 266 248 227	264 252 241 225 207	1 660 1 370 1 080 789 497	250 240 226 208 179	938 926 910 886 846	M_r 171 V_r 326 L_u 3 230 I_x 70.6 S_x 572		4 000 4 500 5 000 5 500 6 000	160 153 146 138 130	6 500 7 000 8 000 9 000 10 000	123 115 97.2 84 74.1
W250X45 W10X30 b=148 t=13 d=266	2 500 2 060 1 630 1 190 750	318 301 280 257 234	292 280 262 244 222	256 248 236 220 201	1 540 1 370 1 080 789 497	255 245 232 214 185	892 881 866 845 808	M_r 163 V_r 360 L_u 2 350 I_x 71.1 S_x 534		3 000 3 500 4 000 4 500 5 000	151 142 132 122 112	6 000 7 000 8 000 9 000 10 000	90.6 75.2 64.4 56.3 50.1
W250X39 W10X26 b=147 t=11.2 d=262	2 500 2 060 1 630 1 190 750	276 269 250 228 205	252 248 233 215 195	219 218 208 194 176	1 330 1 330 1 080 789 497	222 213 203 188 164	770 761 749 731 701	M_r 139 V_r 308 L_u 2 280 I_x 60.1 S_x 459		3 000 3 500 4 000 4 500 5 000	126 117 108 98.6 88	6 000 7 000 8 000 9 000 10 000	69.2 57 48.6 42.3 37.6
W250X33 W10X22 b=146 t=9.1 d=258	2 000 1 650 1 300 950 600	230 222 206 188 170	210 206 192 178 161	184 182 172 160 145	1 130 1 090 862 630 398	181 174 165 153 133	642 634 623 608 582	M_r 114 V_r 280 L_u 2 190 I_x 48.9 S_x 379		3 000 3 500 4 000 4 500 5 000	102 93.1 84 74.1 63.6	5 500 6 000 6 500 7 000 8 000	55.6 49.4 44.4 40.3 34.1
W200X71 W8X48 b=206 t=17.4 d=216	3 000 2 480 1 950 1 430 900	405 379 351 322 292	372 350 327 304 279	325 310 295 279 258	1 990 1 640 1 290 948 597	285 272 254 231 196	1 230 1 210 1 190 1 150 1 090	M_r 216 V_r 393 L_u 4 140 I_x 76.3 S_x 707		5 000 5 500 6 000 6 500 7 000	208 203 198 193 188	8 000 9 000 10 000 11 000 12 000	178 168 158 148 137

Note: * 20 MPa, 2 300 kg/m³ Concrete. G40.21-M 300W Steel.

Units: d, b, t, L_u (mm) $I_x = 10^6$ mm⁴ $S_x = 10^3$ mm³ M_r = kN·m V_r = kN

300W
20 MPa

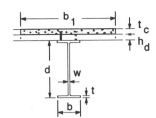

COMPOSITE MEMBERS
Trial Selection Table

38 mm Deck with 65 mm Slab
$\phi = 0.9, \phi_c = 0.6, \phi_{sc} = 0.8$

Steel Shape	b_1 mm	Composite Beam* Factored Resistances M_{rc} (kN·m) for Shear Connections = 100%	70%	40%	Q_r (kN) for 100%	I_t ×10⁶ mm⁴	S_t ×10³ mm³	Steel Shape Data		Non-Composite Shape Unbraced Condition L' mm	M'r kN·m	L' mm	M'r kN·m
W200X59	3 000	355	324	279	1 990	239	1 020	M_r	176	4 000	174	8 000	134
W8X40	2 480	330	303	265	1 640	229	1 010	V_r	341	4 500	169	9 000	124
b=205	1 950	303	281	250	1 290	215	989	L_u	3 770	5 000	164	10 000	114
t=14.2	1 430	276	259	235	948	196	961	I_x	60.9	6 000	154	11 000	103
d=210	900	247	236	216	597	168	914	S_x	580	7 000	144	12 000	93.4
W200X52	2 500	302	276	239	1 660	203	891	M_r	153	4 000	150	6 500	126
W8X35	2 060	281	259	227	1 370	194	878	V_r	290	4 500	145	7 000	121
b=204	1 630	259	241	216	1 080	183	861	L_u	3 610	5 000	140	8 000	111
t=12.6	1 190	236	222	203	789	166	836	I_x	52.5	5 500	135	9 000	101
d=206	750	213	203	186	497	142	795	S_x	509	6 000	131	10 000	89.7
W200X46	2 500	275	250	215	1 580	182	787	M_r	133	4 000	129	6 500	105
W8X31	2 060	257	236	206	1 370	174	776	V_r	260	4 500	124	7 000	99.8
b=203	1 630	236	219	194	1 080	164	762	L_u	3 450	5 000	119	8 000	90.1
t=11	1 190	214	201	182	789	150	741	I_x	45.2	5 500	114	9 000	78.6
d=204	750	191	182	167	497	129	706	S_x	445	6 000	109	10 000	69.6
W200X42	2 000	241	221	191	1 330	161	706	M_r	119	3 000	119	5 500	93.5
W8X28	1 650	224	207	182	1 090	154	695	V_r	263	3 500	114	6 000	88.4
b=166	1 300	207	192	172	862	144	681	L_u	2 850	4 000	109	6 500	83.4
t=11.8	950	188	178	161	630	131	661	I_x	40.6	4 500	104	7 000	77.7
d=206	600	170	161	147	398	112	628	S_x	396	5 000	98.6	8 000	66.7
W200X36	2 000	215	195	167	1 240	140	611	M_r	102	3 000	100	5 500	75.5
W8X24	1 650	202	185	161	1 090	134	602	V_r	222	3 500	95.3	6 000	70.6
b=165	1 300	185	171	151	862	127	591	L_u	2 730	4 000	90.4	6 500	64.7
t=10.2	950	167	157	141	630	116	574	I_x	34.1	4 500	85.5	7 000	59.2
d=202	600	149	142	129	398	99.8	547	S_x	340	5 000	80.5	8 000	50.6
W200X31	2 000	196	177	150	1 080	132	547	M_r	90.5	3 000	81.1	5 500	50.6
W8X21	1 650	190	174	149	1 080	127	539	V_r	240	3 500	75.3	6 000	45.6
b=134	1 300	174	160	141	862	120	529	L_u	2 150	4 000	69.5	6 500	41.5
t=10.2	950	156	146	130	630	110	515	I_x	31.4	4 500	63.6	7 000	38
d=210	600	138	131	117	398	95.5	492	S_x	299	5 000	57	8 000	32.7
W200X27	2 000	169	151	128	915	114	465	M_r	75.3	3 000	65.4	5 500	36.4
W8X18	1 650	165	148	127	915	109	459	V_r	214	3 500	59.7	6 000	32.6
b=133	1 300	156	143	124	862	104	450	L_u	2 050	4 000	54	6 500	29.6
t=8.4	950	139	129	114	630	96.1	439	I_x	25.8	4 500	47.5	7 000	27.1
d=208	600	121	115	102	398	84	420	S_x	249	5 000	41.2	8 000	23.1
W150X30	2 500	166	144	117	1 020	92.7	460	M_r	65.9	3 000	64.1	6 000	46.8
W6X20	2 060	161	142	117	1 020	89.1	452	V_r	185	3 500	61.2	7 000	40.4
b=153	1 630	155	139	116	1 020	84.5	443	L_u	2 680	4 000	58.3	8 000	34.8
t=9.3	1 190	136	124	106	789	78	430	I_x	17.2	4 500	55.4	9 000	30.6
d=158	750	114	106	94.2	497	67.9	411	S_x	219	5 000	52.5	10 000	27.3

Note: * 20 MPa, 2 300 kg/m³ Concrete.   G40.21-M 300W Steel.
Units: d, b, t, L_u (mm) I_x =10⁶ mm⁴ S_x = 10³ mm³ M_r = kN·m V_r = kN

8-47

BEAM SELECTION TABLE
WWF and W Shapes

G40.21-M 300W

$\phi = 0.90$

Designation*	V_r	I_x	b	L_u	M_r	\multicolumn{6}{c}{Factored Moment Resistance (kN·m) M_r' Unbraced length (mm)}					
	kN	10^6mm^4	mm	mm	$\leq L_u$	8 000	10 000	12 000	14 000	16 000	20 000
WWF2000X732	3 640	63 900	550	7 770	**19 400**	19 200	17 600	15 700	13 600	11 100	7 610
WWF1800X700	4 070	50 400	550	7 920	**16 900**	....	15 500	14 000	12 300	10 200	7 120
WWF2000X648	3 600	54 200	550	7 460	**16 700**	16 300	14 800	13 000	10 800	8 510	5 760
WWF1800X659	4 050	46 600	550	7 770	15 700	15 600	14 300	12 800	11 100	8 990	6 190
WWF2000X607	3 590	49 300	550	7 270	**15 300**	14 800	13 300	11 600	9 360	7 330	4 920
WWF1800X617	4 020	42 700	550	7 600	14 500	14 300	13 000	11 500	9 840	7 810	5 330
WWF1600X622	2 360	37 600	550	8 270	14 000	....	13 100	12 000	10 700	9 360	6 580
WWF1800X575	4 000	38 800	550	7 420	**13 300**	12 900	11 700	10 300	8 540	6 700	4 540
WWF2000X542	3 570	41 400	500	6 300	**13 100**	11 900	10 200	8 100	6 070	4 750	3 200
WWF1600X580	2 350	34 600	550	8 130	12 900	....	12 000	10 800	9 610	8 170	5 680
WWF1400X597	2 670	28 100	550	8 480	11 900	....	11 300	10 400	9 400	8 380	6 140
WWF1600X538	2 330	31 600	550	7 970	**11 800**	....	10 800	9 760	8 540	7 050	4 850
WWF1800X510	3 980	32 400	500	6 440	**11 300**	10 400	9 020	7 350	5 530	4 340	2 940
WWF1600X496	2 320	28 400	550	7 800	**10 700**	10 600	9 720	8 660	7 480	6 010	4 090
WWF1400X513	2 670	23 500	550	8 150	9 990	....	9 290	8 430	7 480	6 400	4 460
WWF1400X471	2 660	21 100	550	7 980	**9 020**	9 010	8 300	7 470	6 540	5 410	3 730
WWF1600X431	2 300	23 400	500	6 830	**8 940**	8 460	7 480	6 340	4 920	3 870	2 620
WWF1200X487	2 670	16 700	550	8 370	8 260	....	7 790	7 140	6 440	5 690	4 090
WWF1400X405	2 640	17 300	500	6 990	**7 510**	7 170	6 400	5 500	4 390	3 470	2 380
WWF1100X458	2 050	13 700	550	8 580	7 320	....	6 970	6 440	5 860	5 250	3 900
WWF1200X418	2 670	13 800	500	7 370	6 910	6 740	6 130	5 430	4 680	3 780	2 670
WWF1000X447	2 050	11 100	550	8 750	6 530	....	6 270	5 830	5 350	4 850	3 730
WWF1400X358	2 640	14 500	400	5 410	**6 400**	5 340	4 310	3 130	2 400	1 920	1 360
WWF1200X380	2 670	12 300	500	7 180	6 160	5 940	5 360	4 690	3 890	3 100	2 160
WWF1100X388	2 050	11 200	500	7 570	6 050	5 950	5 450	4 890	4 280	3 550	2 530
WWF900X417	1 260	8 680	550	9 080	5 620	....	5 460	5 120	4 750	4 370	3 550
WWF1000X377	2 050	9 120	500	7 700	5 400	5 340	4 920	4 440	3 940	3 360	2 420
WWF1100X351	2 050	9 940	500	7 380	**5 370**	5 240	4 760	4 210	3 610	2 900	2 030

*Nominal depth in millimetres and mass in kilograms per metre. For imperial designations, see page 6-19.

G40.21-M 300W
$\phi = 0.90$

BEAM SELECTION TABLE
WWF and W Shapes

Designation*	V_r	I_x	b	L_u	Factored Moment Resistance (kN·m)						
					M_r	M_r'					
						Unbraced length (mm)					
	kN	$10^6 mm^4$	mm	mm	$\leq L_u$	5 000	6 000	8 000	10 000	12 000	16 000
WWF1200X333	2 670	10 200	400	5 590	5 210	....	5 100	4 460	3 700	2 780	1 740
WWF1000X340	2 050	8 060	500	7 500	4 780	....	....	4 690	4 280	3 820	2 720
WWF900X347	1 260	7 100	500	7 970	4 620	....	....	4 610	4 280	3 920	3 130
WWF1200X302	2 670	8 970	400	5 390	4 620	....	4 460	3 840	3 100	2 250	1 390
WWF1100X304	2 050	8 220	400	5 760	4 510	....	4 460	3 940	3 330	2 590	1 630
WWF900X309	1 260	6 240	500	7 780	4 050	....	....	4 020	3 700	3 350	2 520
WWF800X339	1 260	5 500	500	8 180	4 020	....	....	....	3 780	3 490	2 880
WWF1000X293	2 050	6 640	400	5 860	4 000	....	3 970	3 540	3 030	2 420	1 550
WWF1100X273	2 050	7 160	400	5 580	3 970	....	3 880	3 380	2 790	2 080	1 280
WWF1200X263	2 670	7 250	300	3 850	3 830	3 460	3 070	2 100	1 430	1 060	684
WWF800X300	1 260	4 840	500	7 930	3 540	....	....	3 530	3 270	2 980	2 360
WWF1000X262	2 050	5 780	400	5 670	3 510	....	3 450	3 030	2 540	1 930	1 210
WWF900X262	1 260	5 110	400	6 120	3 350	....	....	3 040	2 650	2 240	1 450
WWF1100X234	2 050	5 720	300	4 010	3 240	2 990	2 680	1 930	1 320	978	632
WWF800X253	1 260	3 950	400	6 240	2 920	....	....	2 680	2 370	2 040	1 380
WWF900X231	1 260	4 410	400	5 930	2 920	....	2 910	2 590	2 220	1 760	1 110
WWF1000X223	2 050	4 590	300	4 100	2 840	2 640	2 390	1 790	1 230	920	602
WWF800X223	1 260	3 410	400	6 050	2 520	....	....	2 270	1 970	1 630	1 050
WWF700X245	1 250	2 950	400	6 420	2 490	....	....	2 320	2 080	1 840	1 310
WWF1000X200	2 050	3 940	300	3 910	2 480	2 250	2 010	1 390	945	699	449
WWF900X192	1 260	3 460	300	4 310	2 320	2 210	2 030	1 600	1 120	842	555
WWF700X214	1 260	2 540	400	6 190	2 150	....	....	1 960	1 730	1 480	982
W610X241	2 030	2 150	329	5 200	2 070	....	1 990	1 770	1 540	1 300	911
WWF800X184	1 260	2 660	300	4 410	2 000	1 920	1 770	1 440	1 040	791	529
WWF900X169	1 260	2 930	300	4 130	1 990	1 860	1 680	1 240	845	625	402
WWF700X196	1 260	2 300	400	6 060	1 940	....	....	1 750	1 520	1 260	812
W610X217	1 850	1 910	328	5 080	1 850	....	1 760	1 550	1 320	1 070	745
WWF800X161	1 260	2 250	300	4 230	1 710	1 610	1 470	1 140	778	581	379
WWF700X175	1 260	1 970	300	4 520	1 690	1 640	1 530	1 270	965	741	505
W610X195	1 710	1 680	327	4 940	1 640	1 630	1 540	1 340	1 120	870	598
W610X174	1 540	1 470	325	4 830	1 450	1 430	1 350	1 150	924	709	482
WWF700X152	1 260	1 660	300	4 330	1 440	1 370	1 260	1 000	711	537	356
W610X155	1 390	1 290	324	4 740	1 280	1 260	1 180	988	762	579	388

*Nominal depth in millimetres and mass in kilograms per metre. For imperial designations, see page 6-19.

BEAM SELECTION TABLE
WWF and W Shapes

G40.21-M 300W

$\phi = 0.90$

Designation*	V_r	I_x	b	L_u	Factored Moment Resistance (kN·m)						
					M_r	M_r'					
						Unbraced length (mm)					
	kN	$10^6 mm^4$	mm	mm	$\leq L_u$	3 000	4 000	5 000	6 000	8 000	10 000
W610X140	1 440	1 120	230	3 320	**1 120**	....	1 050	946	829	573	422
W610X125	1 300	985	229	3 250	**991**	...	923	821	708	470	342
W530X138	1 440	861	214	3 180	975	...	906	815	720	515	390
W610X113	1 210	875	228	3 180	**888**	...	818	719	610	391	282
W530X123	1 270	761	212	3 100	867	...	794	706	613	421	316
W610X101	1 130	764	228	3 110	**783**	...	713	619	512	320	228
W530X109	1 120	667	211	3 040	764	...	692	608	517	342	254
W530X101	1 050	617	210	2 990	**707**	...	635	553	462	301	222
W610X91	1 030	667	227	3 020	**691**	...	619	530	421	261	185
W460X106	1 060	488	194	2 910	645	640	579	512	444	308	235
W530X92	971	552	209	2 930	637	633	565	486	393	253	185
W610X84	881	613	226	2 990	**637**	636	566	482	377	232	163
W460X97	947	445	193	2 870	589	581	522	457	389	264	200
W530X82	894	479	209	2 860	**559**	551	487	412	321	204	148
W460X89	868	410	192	2 830	543	534	477	414	343	231	174
W460X82	812	370	191	2 770	**494**	482	427	365	292	195	146
W530X72	831	402	207	2 750	**475**	462	402	331	247	155	111
W410X85	812	315	181	2 730	467	455	406	354	297	205	157
W460X74	735	333	190	2 730	446	433	380	320	249	164	122
W410X74	716	275	180	2 670	408	394	348	297	239	163	124
W460X67	688	296	190	2 670	**400**	385	334	276	209	136	99.8
W360X79	593	227	205	3 270	386	...	364	331	298	225	172
W310X86	503	199	254	4 250	383	...	...	367	344	297	248
W410X67	643	246	179	2 610	**367**	352	307	258	201	135	102
W460X61	650	255	189	2 590	**348**	332	284	228	168	108	78.5
W310X79	480	177	254	4 140	346	...	...	327	305	258	207
W360X72	536	201	204	3 190	346	...	322	290	257	186	141
W310X74	519	165	205	3 380	321	...	307	282	258	206	159
W410X60	560	216	178	2 580	**321**	306	264	217	165	109	81.4
W360X64	478	178	203	3 110	308	...	283	252	220	153	115
W310X67	463	145	204	3 280	286	...	270	246	222	169	129

*Nominal depth in millimetres and mass in kilograms per metre. For imperial designations, see page 6-19.

G40.21-M 300W
φ = 0.90

BEAM SELECTION TABLE
WWF and W Shapes

Designation*	V_r	I_x	b	L_u	\multicolumn{7}{c}{Factored Moment Resistance (kN·m)}
					M_r	\multicolumn{6}{c}{M'_r}
						\multicolumn{6}{c}{Unbraced length (mm)}
	kN	10^6mm^4	mm	mm	≤L_u	2 000	3 000	4 000	5 000	6 000	8 000
W410X54	540	186	177	2 480	**284**	...	266	225	176	132	86.1
W360X57	504	161	172	2 550	273	...	259	225	189	147	99.8
W310X60	406	129	203	3 200	254	...	...	237	214	191	139
W250X67	409	104	204	3 570	243	...	...	237	221	205	174
W360X51	457	141	171	2 500	**241**	...	227	195	159	121	81.0
W410X46	504	156	140	1 930	**239**	236	195	142	99.9	76.4	51.8
W310X52	431	119	167	2 570	226	...	216	189	162	130	89.4
W360X45	433	122	171	2 430	**210**	...	195	165	128	96.1	63.4
W250X58	359	87.3	203	3 410	208	...	...	199	184	168	137
W410X39	447	127	140	1 860	**197**	193	155	105	73.1	55.2	36.6
W310X45	369	99.2	166	2 490	191	...	180	155	128	98.2	66.5
W360X39	410	102	128	1 790	**179**	173	139	97.2	69.8	54.2	37.5
W250X49	327	70.6	202	3 240	171	...	...	160	146	130	97.2
W310X39	320	85.1	165	2 440	**165**	...	153	130	103	77.7	51.8
W250X45	360	71.1	148	2 360	163	...	151	132	112	90.6	64.4
W360X33	362	82.7	127	1 720	**146**	139	108	70.3	49.7	38.1	25.9
W250X39	308	60.1	147	2 280	139	...	126	108	88.0	69.2	48.6
W200X42	264	40.6	166	2 850	120	...	119	109	98.6	88.4	66.7
+**W310X31**	264	67.2	164	2 460	**118**	...	110	93.4	72.6	54.3	35.6
W250X33	280	48.9	146	2 180	114	...	102	84.0	63.6	49.4	34.1
W200X36	223	34.1	165	2 730	103	...	100	90.4	80.5	70.6	50.6
W200X31	240	31.4	134	2 150	**90.4**	...	81.1	69.5	57.0	45.6	32.7
W150X37	234	22.2	154	2 910	83.7	...	83.2	77.2	71.3	65.4	53.3
W200X27	215	25.8	133	2 050	**75.3**	...	65.4	54.0	41.2	32.6	23.1
+**W250X24**	226	34.7	145	2 190	**74.2**	...	65.6	53.0	38.2	29.0	19.4
W150X30	186	17.2	153	2 680	65.9	...	64.1	58.3	52.5	46.8	34.8
+**W200X21**	182	19.8	133	2 080	**52.6**	...	45.7	37.1	27.4	21.4	14.8
+W150X22	157	12.1	152	2 590	42.9	...	41.2	36.8	32.2	27.3	19.3

+Class 3 Section.

*Nominal depth in millimetres and mass in kilograms per metre. For imperial designations, see page 6-19.

BEAM LOAD TABLES
WWF Shapes

G40.21-M
300W

Total Uniformly Distributed Factored Loads
for Laterally Supported Beams – kN

$\phi = 0.90$

Designation		WWF2000			Approx. Deflect. (mm)		WWF1800				Approx. Deflect. (mm)
Mass (kg/m)	732	648	607	542		700	659	617	575	510	
11 000					14					7 960	15
11 500					15					7 850	16
12 000					16					7 520	18
12 500					18					7 220	19
13 000					19				8 000	6 940	21
13 500					20				7 860	6 690	23
14 000					22			8 050	7 580	6 450	24
14 500				7 140	24			8 000	7 310	6 230	26
15 000				6 980	25			7 730	7 070	6 020	28
15 500				6 760	27		8 090	7 480	6 840	5 820	30
16 000				6 550	29		7 860	7 250	6 630	5 640	32
16 500				6 350	30	8 140	7 620	7 030	6 430	5 470	34
17 000			7 170	6 160	32	7 970	7 400	6 820	6 240	5 310	36
17 500			6 990	5 990	34	7 740	7 180	6 630	6 060	5 160	38
18 000		7 210	6 790	5 820	36	7 520	6 980	6 440	5 890	5 020	40
18 500		7 200	6 610	5 660	38	7 320	6 800	6 270	5 730	4 880	43
19 000		7 010	6 440	5 510	40	7 130	6 620	6 100	5 580	4 750	45
19 500		6 830	6 270	5 370	43	6 940	6 450	5 950	5 440	4 630	47
20 000		6 660	6 110	5 240	45	6 770	6 290	5 800	5 300	4 510	50
20 500		6 500	5 960	5 110	47	6 610	6 130	5 660	5 170	4 400	52
21 000	7 280	6 350	5 820	4 990	49	6 450	5 990	5 520	5 050	4 300	55
21 500	7 210	6 200	5 690	4 870	52	6 300	5 850	5 400	4 930	4 200	58
22 000	7 050	6 060	5 560	4 760	54	6 160	5 710	5 270	4 820	4 100	60
22 500	6 890	5 920	5 430	4 660	57	6 020	5 590	5 160	4 710	4 010	63
23 000	6 740	5 790	5 320	4 560	59	5 890	5 470	5 040	4 610	3 930	66
23 500	6 600	5 670	5 200	4 460	62	5 760	5 350	4 940	4 510	3 840	69
24 000	6 460	5 550	5 090	4 360	65	5 640	5 240	4 830	4 420	3 760	72
24 500	6 330	5 440	4 990	4 280	67	5 530	5 130	4 730	4 330	3 680	75
25 000	6 200	5 330	4 890	4 190	70	5 420	5 030	4 640	4 240	3 610	78
25 500	6 080	5 230	4 790	4 110	73	5 310	4 930	4 550	4 160	3 540	81
26 000	5 960	5 130	4 700	4 030	76	5 210	4 840	4 460	4 080	3 470	84
26 500	5 850	5 030	4 610	3 950	79	5 110	4 740	4 380	4 000	3 410	87
27 000	5 740	4 940	4 530	3 880	82	5 020	4 660	4 300	3 930	3 340	91
27 500	5 640	4 850	4 450	3 810	85	4 920	4 570	4 220	3 860	3 280	94
28 000	5 540	4 760	4 370	3 740	88	4 840	4 490	4 140	3 790	3 220	98

Span in Millimetres

DESIGN DATA AND PROPERTIES

V_r (kN)	3 640	3 600	3 590	3 570		4 070	4 050	4 020	4 000	3 980	
L_u (mm)	7 770	7 460	7 270	6 300		7 920	7 770	7 600	7 420	6 440	
d (mm)	2 000	2 000	2 000	2 000		1 800	1 800	1 800	1 800	1 800	
b (mm)	550	550	550	500		550	550	550	550	500	
t (mm)	50.0	40.0	35.0	30.0		50.0	45.0	40.0	35.0	30.0	
w (mm)	20.0	20.0	20.0	20.0		20.0	20.0	20.0	20.0	20.0	
k (mm)	61	51	46	41		61	56	51	46	41	

IMPERIAL SIZE AND MASS

Mass (lb./ft.)	490	436	408	364		470	442	415	388	344	
Nominal Depth (in.)		79						71			

G40.21-M 300W
$\phi = 0.90$

BEAM LOAD TABLES
WWF Shapes

Total Uniformly Distributed Factored Loads for Laterally Supported Beams – kN

Designation	WWF1600					Approx. Deflect. (mm)	WWF1400					Approx. Deflect. (mm)
Mass (kg/m)	622	580	538	496	431		597	513	471	405	358	
Span in Millimetres												
9 000						11						13
9 500						13					5 290	14
10 000						14					5 120	16
10 500						15					4 880	18
11 000						17				5 290	4 650	19
11 500						19				5 220	4 450	21
12 000						20				5 000	4 270	23
12 500						22				4 800	4 100	25
13 000						24				4 620	3 940	27
13 500						26			5 330	4 450	3 790	29
14 000						27			5 150	4 290	3 660	31
14 500						29		5 350	4 980	4 140	3 530	34
15 000						32		5 330	4 810	4 000	3 410	36
15 500					4 600	34		5 160	4 650	3 870	3 300	38
16 000					4 470	36		5 000	4 510	3 750	3 200	41
16 500					4 330	38		4 840	4 370	3 640	3 100	44
17 000					4 210	40		4 700	4 240	3 530	3 010	46
17 500					4 080	43	5 350	4 570	4 120	3 430	2 920	49
18 000				4 630	3 970	45	5 280	4 440	4 010	3 340	2 840	52
18 500				4 620	3 860	48	5 140	4 320	3 900	3 250	2 770	55
19 000				4 500	3 760	51	5 000	4 210	3 800	3 160	2 690	58
19 500				4 390	3 670	53	4 870	4 100	3 700	3 080	2 620	61
20 000			4 660	4 280	3 580	56	4 750	4 000	3 610	3 000	2 560	64
20 500			4 600	4 170	3 490	59	4 640	3 900	3 520	2 930	2 500	67
21 000			4 500	4 070	3 400	62	4 530	3 810	3 440	2 860	2 440	71
21 500		4 690	4 390	3 980	3 320	65	4 420	3 720	3 360	2 790	2 380	74
22 000		4 680	4 290	3 890	3 250	68	4 320	3 630	3 280	2 730	2 330	77
22 500		4 580	4 200	3 800	3 180	71	4 220	3 550	3 210	2 670	2 280	81
23 000		4 480	4 100	3 720	3 110	74	4 130	3 480	3 140	2 610	2 230	85
23 500	4 720	4 380	4 020	3 640	3 040	77	4 040	3 400	3 070	2 560	2 180	88
24 000	4 660	4 290	3 930	3 560	2 980	81	3 960	3 330	3 010	2 500	2 130	92
24 500	4 570	4 200	3 850	3 490	2 920	84	3 880	3 260	2 940	2 450	2 090	96
25 000	4 480	4 120	3 780	3 420	2 860	88	3 800	3 200	2 890	2 400	2 050	100
25 500	4 390	4 040	3 700	3 350	2 800	91	3 730	3 130	2 830	2 360	2 010	104
26 000	4 300	3 960	3 630	3 290	2 750	95	3 660	3 070	2 780	2 310	1 970	108

DESIGN DATA AND PROPERTIES

V_r (kN)	2 360	2 350	2 330	2 320	2 300		2 670	2 670	2 660	2 640	2 640	
L_u (mm)	8 270	8 130	7 970	7 800	6 830		8 480	8 150	7 980	6 990	5 410	
d (mm)	1 600	1 600	1 600	1 600	1 600		1 400	1 400	1 400	1 400	1 400	
b (mm)	550	550	550	550	500		550	550	550	500	400	
t (mm)	50.0	45.0	40.0	35.0	30.0		50.0	40.0	35.0	30.0	30.0	
w (mm)	16.0	16.0	16.0	16.0	16.0		16.0	16.0	16.0	16.0	16.0	
k (mm)	61	56	51	44	39		61	51	44	39	39	

IMPERIAL SIZE AND MASS

Mass (lb./ft.)	419	388	361	333	289		402	344	316	272	240	
Nominal Depth (in.)			63						55			

BEAM LOAD TABLES
WWF Shapes

Total Uniformly Distributed Factored Loads
for Laterally Supported Beams – kN

G40.21-M
300W

$\phi = 0.90$

Designation	WWF1200						Approximate Deflection (mm)
Mass (kg/m)	487	418	380	333	302	263	
Span in Millimetres							
5 000							5
5 500						5 350	6
6 000						5 110	7
6 500					5 350	4 720	8
7 000					5 280	4 380	9
7 500				5 350	4 920	4 090	11
8 000				5 210	4 620	3 830	12
8 500				4 900	4 340	3 610	13
9 000			5 350	4 630	4 100	3 410	15
9 500			5 180	4 390	3 890	3 230	17
10 000		5 350	4 920	4 170	3 690	3 070	19
10 500		5 270	4 690	3 970	3 520	2 920	21
11 000		5 030	4 480	3 790	3 360	2 790	23
11 500		4 810	4 280	3 620	3 210	2 670	25
12 000	5 350	4 610	4 100	3 470	3 080	2 560	27
12 500	5 290	4 420	3 940	3 340	2 960	2 450	29
13 000	5 080	4 250	3 790	3 210	2 840	2 360	32
13 500	4 900	4 100	3 650	3 090	2 740	2 270	34
14 000	4 720	3 950	3 520	2 980	2 640	2 190	37
14 500	4 560	3 810	3 400	2 880	2 550	2 120	39
15 000	4 410	3 690	3 280	2 780	2 460	2 040	42
15 500	4 260	3 570	3 180	2 690	2 380	1 980	45
16 000	4 130	3 460	3 080	2 610	2 310	1 920	48
16 500	4 010	3 350	2 980	2 530	2 240	1 860	51
17 000	3 890	3 250	2 900	2 450	2 170	1 800	54
17 500	3 780	3 160	2 810	2 380	2 110	1 750	57
18 000	3 670	3 070	2 740	2 320	2 050	1 700	60
18 500	3 570	2 990	2 660	2 250	2 000	1 660	64
19 000	3 480	2 910	2 590	2 190	1 940	1 610	67
19 500	3 390	2 840	2 530	2 140	1 890	1 570	71
20 000	3 300	2 760	2 460	2 080	1 850	1 530	75
20 500	3 220	2 700	2 400	2 030	1 800	1 500	78
21 000	3 150	2 630	2 340	1 980	1 760	1 460	82
21 500	3 070	2 570	2 290	1 940	1 720	1 430	86
22 000	3 000	2 510	2 240	1 900	1 680	1 390	90
DESIGN DATA AND PROPERTIES							
V_r (kN)	2 670	2 670	2 670	2 670	2 670	2 670	
L_u (mm)	8 370	7 370	7 180	5 590	5 390	3 850	
d (mm)	1 200	1 200	1 200	1 200	1 200	1 200	
b (mm)	550	500	500	400	400	300	
t (mm)	40.0	35.0	30.0	30.0	25.0	25.0	
w (mm)	16.0	16.0	16.0	16.0	16.0	16.0	
k (mm)	51	44	39	39	34	34	
IMPERIAL SIZE AND MASS							
Mass (lb./ft.)	326	281	255	224	203	176	
Nominal Depth (in.)	47						

G40.21-M
300W
$\phi = 0.90$

BEAM LOAD TABLES
WWF Shapes

Total Uniformly Distributed Factored Loads
for Laterally Supported Beams – kN

Designation	WWF1100						Approximate Deflection (mm)
Mass (kg/m)	458	388	351	304	273	234	
6 000						4 090	7
6 500						3 990	9
7 000						3 700	10
7 500					4 090	3 460	11
8 000					3 970	3 240	13
8 500				4 090	3 740	3 050	15
9 000				4 010	3 530	2 880	16
9 500				3 800	3 340	2 730	18
10 000			4 090	3 610	3 180	2 590	20
10 500			4 090	3 440	3 020	2 470	22
11 000			3 910	3 280	2 890	2 360	25
11 500		4 090	3 740	3 140	2 760	2 250	27
12 000		4 030	3 580	3 010	2 650	2 160	29
12 500		3 870	3 440	2 890	2 540	2 070	32
13 000		3 720	3 310	2 780	2 440	1 990	34
13 500		3 580	3 180	2 670	2 350	1 920	37
14 000	4 090	3 460	3 070	2 580	2 270	1 850	40
14 500	4 040	3 340	2 960	2 490	2 190	1 790	43
15 000	3 900	3 230	2 870	2 400	2 120	1 730	46
15 500	3 780	3 120	2 770	2 330	2 050	1 670	49
16 000	3 660	3 020	2 690	2 250	1 980	1 620	52
16 500	3 550	2 930	2 600	2 190	1 920	1 570	55
17 000	3 440	2 850	2 530	2 120	1 870	1 520	59
17 500	3 340	2 760	2 460	2 060	1 810	1 480	62
18 000	3 250	2 690	2 390	2 000	1 760	1 440	66
18 500	3 160	2 620	2 320	1 950	1 720	1 400	70
19 000	3 080	2 550	2 260	1 900	1 670	1 360	74
19 500	3 000	2 480	2 200	1 850	1 630	1 330	77
20 000	2 930	2 420	2 150	1 800	1 590	1 300	81
20 500	2 860	2 360	2 100	1 760	1 550	1 260	86
21 000	2 790	2 300	2 050	1 720	1 510	1 230	90
21 500	2 720	2 250	2 000	1 680	1 480	1 210	94
22 000	2 660	2 200	1 950	1 640	1 440	1 180	99
22 500	2 600	2 150	1 910	1 600	1 410	1 150	103
23 000	2 540	2 100	1 870	1 570	1 380	1 130	108

Span in Millimetres

DESIGN DATA AND PROPERTIES							
V_r (kN)	2 050	2 050	2 050	2 050	2 050	2 050	
L_u (mm)	8 580	7 570	7 380	5 760	5 580	4 010	
d (mm)	1 100	1 100	1 100	1 100	1 100	1 100	
b (mm)	550	500	500	400	400	300	
t (mm)	40.0	35.0	30.0	30.0	25.0	25.0	
w (mm)	14.0	14.0	14.0	14.0	14.0	14.0	
k (mm)	51	44	39	39	34	34	

IMPERIAL SIZE AND MASS							
Mass (lb./ft.)	307	260	236	204	184	157	
Nominal Depth (in.)	43						

BEAM LOAD TABLES
WWF Shapes

Total Uniformly Distributed Factored Loads
for Laterally Supported Beams – kN

G40.21-M
300W

$\phi = 0.90$

Designation			WWF1000					Approx. Deflect. (mm)
Mass (kg/m)	447	377	340	293	262	223	200	
Span in Millimetres								
4 000								4
4 500							4 090	5
5 000							3 960	6
5 500						4 090	3 600	7
6 000						3 780	3 300	8
6 500					4 090	3 490	3 050	9
7 000					4 010	3 240	2 830	11
7 500				4 090	3 740	3 020	2 640	13
8 000				4 000	3 510	2 840	2 480	14
8 500				3 760	3 300	2 670	2 330	16
9 000			4 090	3 550	3 120	2 520	2 200	18
9 500			4 020	3 360	2 960	2 390	2 080	20
10 000			3 820	3 200	2 810	2 270	1 980	22
10 500		4 090	3 640	3 040	2 670	2 160	1 890	25
11 000		3 930	3 480	2 910	2 550	2 060	1 800	27
11 500		3 760	3 320	2 780	2 440	1 970	1 720	30
12 000		3 600	3 190	2 660	2 340	1 890	1 650	32
12 500	4 090	3 460	3 060	2 560	2 250	1 810	1 580	35
13 000	4 020	3 320	2 940	2 460	2 160	1 740	1 520	38
13 500	3 870	3 200	2 830	2 370	2 080	1 680	1 470	41
14 000	3 730	3 090	2 730	2 280	2 010	1 620	1 420	44
14 500	3 600	2 980	2 640	2 200	1 940	1 560	1 370	47
15 000	3 480	2 880	2 550	2 130	1 870	1 510	1 320	50
15 500	3 370	2 790	2 470	2 060	1 810	1 460	1 280	54
16 000	3 270	2 700	2 390	2 000	1 760	1 420	1 240	57
16 500	3 170	2 620	2 320	1 940	1 700	1 380	1 200	61
17 000	3 080	2 540	2 250	1 880	1 650	1 330	1 160	65
17 500	2 990	2 470	2 180	1 830	1 600	1 300	1 130	69
18 000	2 900	2 400	2 120	1 780	1 560	1 260	1 100	73
18 500	2 830	2 340	2 070	1 730	1 520	1 230	1 070	77
19 000	2 750	2 270	2 010	1 680	1 480	1 190	1 040	81
19 500	2 680	2 220	1 960	1 640	1 440	1 160	1 020	85
20 000	2 610	2 160	1 910	1 600	1 400	1 130	990	90
20 500	2 550	2 110	1 860	1 560	1 370	1 110	966	94
21 000	2 490	2 060	1 820	1 520	1 340	1 080	943	99

DESIGN DATA AND PROPERTIES								
V_r (kN)	2 050	2 050	2 050	2 050	2 050	2 050	2 050	
L_u (mm)	8 750	7 700	7 500	5 860	5 670	4 100	3 910	
d (mm)	1 000	1 000	1 000	1 000	1 000	1 000	1 000	
b (mm)	550	500	500	400	400	300	300	
t (mm)	40.0	35.0	30.0	30.0	25.0	25.0	20.0	
w (mm)	14.0	14.0	14.0	14.0	14.0	14.0	14.0	
k (mm)	51	44	39	39	34	34	29	

IMPERIAL SIZE AND MASS								
Mass (lb./ft.)	300	253	228	197	176	150	134	
Nominal Depth (in.)				39				

G40.21-M
300W
$\phi = 0.90$

BEAM LOAD TABLES
WWF Shapes

Total Uniformly Distributed Factored Loads
for Laterally Supported Beams – kN

Designation	WWF900							Approx. Deflect. (mm)
Mass (kg/m)	417	347	309	262	231	192	169	
Span in Millimetres								
6 000							2 530	9
6 500							2 450	11
7 000						2 530	2 270	12
7 500						2 480	2 120	14
8 000						2 320	1 990	16
8 500						2 180	1 870	18
9 000					2 530	2 060	1 770	20
9 500					2 460	1 960	1 680	22
10 000					2 330	1 860	1 590	25
10 500				2 530	2 220	1 770	1 520	27
11 000				2 440	2 120	1 690	1 450	30
11 500				2 330	2 030	1 620	1 380	33
12 000				2 230	1 940	1 550	1 330	36
12 500			2 530	2 140	1 870	1 490	1 270	39
13 000			2 490	2 060	1 790	1 430	1 220	42
13 500			2 400	1 980	1 730	1 380	1 180	45
14 000			2 310	1 910	1 670	1 330	1 140	49
14 500		2 530	2 230	1 850	1 610	1 280	1 100	52
15 000		2 460	2 160	1 790	1 560	1 240	1 060	56
15 500		2 380	2 090	1 730	1 500	1 200	1 030	60
16 000		2 310	2 020	1 670	1 460	1 160	995	64
16 500		2 240	1 960	1 620	1 410	1 130	965	68
17 000		2 170	1 910	1 580	1 370	1 090	936	72
17 500	2 530	2 110	1 850	1 530	1 330	1 060	910	76
18 000	2 500	2 050	1 800	1 490	1 300	1 030	884	81
18 500	2 430	2 000	1 750	1 450	1 260	1 000	860	85
19 000	2 360	1 940	1 700	1 410	1 230	978	838	90
19 500	2 300	1 890	1 660	1 370	1 200	953	816	95
20 000	2 250	1 850	1 620	1 340	1 170	929	796	100
20 500	2 190	1 800	1 580	1 310	1 140	906	777	105
21 000	2 140	1 760	1 540	1 280	1 110	885	758	110
21 500	2 090	1 720	1 510	1 250	1 080	864	740	115
22 000	2 040	1 680	1 470	1 220	1 060	844	724	120
22 500	2 000	1 640	1 440	1 190	1 040	826	708	126
23 000	1 950	1 610	1 410	1 160	1 010	808	692	132

DESIGN DATA AND PROPERTIES								
V_r (kN)	1 260	1 260	1 260	1 260	1 260	1 260	1 260	
L_u (mm)	9 080	7 970	7 780	6 120	5 930	4 310	4 130	
d (mm)	900	900	900	900	900	900	900	
b (mm)	550	500	500	400	400	300	300	
t (mm)	40.0	35.0	30.0	30.0	25.0	25.0	20.0	
w (mm)	11.0	11.0	11.0	11.0	11.0	11.0	11.0	
k (mm)	51	44	39	39	34	34	29	

IMPERIAL SIZE AND MASS								
Mass (lb./ft.)	279	233	208	176	156	128	113	
Nominal Depth (in.)				35				

BEAM LOAD TABLES
WWF Shapes

Total Uniformly Distributed Factored Loads
for Laterally Supported Beams – kN

G40.21-M
300W

$\phi = 0.90$

Designation	\multicolumn{6}{c}{WWF800}	Approximate Deflection (mm)
Mass (kg/m)	339	300	253	223	184	161	
Span in Millimetres							
5 000						2 530	7
5 500						2 480	8
6 000					2 530	2 280	10
6 500					2 460	2 100	12
7 000					2 290	1 950	14
7 500				2 530	2 130	1 820	16
8 000				2 520	2 000	1 710	18
8 500				2 370	1 880	1 610	20
9 000			2 530	2 240	1 780	1 520	23
9 500			2 460	2 120	1 680	1 440	25
10 000			2 330	2 020	1 600	1 360	28
10 500			2 220	1 920	1 520	1 300	31
11 000		2 530	2 120	1 830	1 460	1 240	34
11 500		2 460	2 030	1 750	1 390	1 190	37
12 000		2 360	1 940	1 680	1 330	1 140	40
12 500	2 530	2 260	1 870	1 610	1 280	1 090	44
13 000	2 480	2 180	1 790	1 550	1 230	1 050	47
13 500	2 380	2 100	1 730	1 490	1 190	1 010	51
14 000	2 300	2 020	1 670	1 440	1 140	975	55
14 500	2 220	1 950	1 610	1 390	1 100	941	59
15 000	2 150	1 890	1 560	1 340	1 070	910	63
15 500	2 080	1 830	1 500	1 300	1 030	881	67
16 000	2 010	1 770	1 460	1 260	1 000	853	72
16 500	1 950	1 720	1 410	1 220	970	827	76
17 000	1 890	1 660	1 370	1 190	942	803	81
17 500	1 840	1 620	1 330	1 150	915	780	86
18 000	1 790	1 570	1 300	1 120	889	758	91
18 500	1 740	1 530	1 260	1 090	865	738	96
19 000	1 690	1 490	1 230	1 060	842	718	101
19 500	1 650	1 450	1 200	1 040	821	700	106
20 000	1 610	1 420	1 170	1 010	800	683	112
20 500	1 570	1 380	1 140	984	781	666	118
21 000	1 530	1 350	1 110	961	762	650	123
21 500	1 500	1 320	1 080	938	744	635	129
22 000	1 460	1 290	1 060	917	728	621	136

\multicolumn{7}{c}{DESIGN DATA AND PROPERTIES}
V_r (kN)	1 260	1 260	1 260	1 260	1 260	1 260	
L_u (mm)	8 180	7 930	6 240	6 050	4 410	4 230	
d (mm)	800	800	800	800	800	800	
b (mm)	500	500	400	400	300	300	
t (mm)	35.0	30.0	30.0	25.0	25.0	20.0	
w (mm)	11.0	11.0	11.0	11.0	11.0	11.0	
k (mm)	44	39	39	34	34	29	

\multicolumn{7}{c}{IMPERIAL SIZE AND MASS}
Mass (lb./ft.)	228	202	170	150	123	108	
Nominal Depth (in.)	\multicolumn{6}{c}{31}	

**G40.21-M
300W
φ = 0.90**

**BEAM LOAD TABLES
WWF Shapes**

Total Uniformly Distributed Factored Loads
for Laterally Supported Beams – kN

Designation	WWF700					Approximate Deflection (mm)
Mass (kg/m)	245	214	196	175	152	
Span in Millimetres						
4 000						5
4 500					2 530	6
5 000				2 530	2 300	8
5 500				2 460	2 090	10
6 000			2 530	2 260	1 920	12
6 500		2 530	2 390	2 080	1 770	14
7 000		2 450	2 220	1 940	1 640	16
7 500	2 510	2 290	2 070	1 810	1 530	18
8 000	2 490	2 150	1 940	1 690	1 440	20
8 500	2 340	2 020	1 830	1 590	1 350	23
9 000	2 210	1 910	1 730	1 500	1 280	26
9 500	2 090	1 810	1 640	1 430	1 210	29
10 000	1 990	1 720	1 550	1 350	1 150	32
10 500	1 900	1 640	1 480	1 290	1 090	35
11 000	1 810	1 560	1 410	1 230	1 040	39
11 500	1 730	1 490	1 350	1 180	999	42
12 000	1 660	1 430	1 290	1 130	958	46
12 500	1 590	1 370	1 240	1 080	919	50
13 000	1 530	1 320	1 200	1 040	884	54
13 500	1 470	1 270	1 150	1 000	851	58
14 000	1 420	1 230	1 110	967	821	63
14 500	1 370	1 180	1 070	934	792	67
15 000	1 330	1 140	1 040	903	766	72
15 500	1 280	1 110	1 000	874	741	77
16 000	1 240	1 070	971	846	718	82
16 500	1 210	1 040	941	821	696	87
17 000	1 170	1 010	914	797	676	92
17 500	1 140	981	887	774	657	98
18 000	1 100	954	863	752	638	104
18 500	1 080	928	839	732	621	110
19 000	1 050	904	817	713	605	116
19 500	1 020	881	796	695	589	122
20 000	995	859	777	677	575	128
20 500	970	838	758	661	561	134
21 000	947	818	740	645	547	141

DESIGN DATA AND PROPERTIES						
V_r (kN)	1 250	1 260	1 260	1 260	1 260	
L_u (mm)	6 420	6 190	6 060	4 520	4 330	
d (mm)	700	700	700	700	700	
b (mm)	400	400	400	300	300	
t (mm)	30.0	25.0	22.0	25.0	20.0	
w (mm)	11.0	11.0	11.0	11.0	11.0	
k (mm)	39	34	31	34	29	

IMPERIAL SIZE AND MASS						
Mass (lb./ft.)	164	144	132	117	102	
Nominal Depth (in.)	28					

BEAM LOAD TABLES
W Shapes

Total Uniformly Distributed Factored Loads
for Laterally Supported Beams – kN

G40.21-M
300W
$\phi = 0.90$

Designation					W610							Approx. Deflect. (mm)
Mass (kg/m)	241	217	195	174	155	140	125	113	101	91	84	
Span in Millimetres												
2 000												1
2 500									2 430	2 260	2 050	1 760	2
3 000						2 880	2 600	2 370	2 090	1 840	1 700	3
3 500			3 410	3 070	2 770	2 560	2 260	2 030	1 790	1 580	1 460	4
4 000	4 060	3 690	3 280	2 890	2 550	2 240	1 980	1 780	1 570	1 380	1 270	6
4 500	3 680	3 290	2 910	2 570	2 270	1 990	1 760	1 580	1 390	1 230	1 130	7
5 000	3 310	2 960	2 620	2 320	2 040	1 790	1 580	1 420	1 250	1 110	1 020	9
5 500	3 010	2 690	2 380	2 100	1 860	1 630	1 440	1 290	1 140	1 000	927	11
6 000	2 760	2 470	2 180	1 930	1 700	1 490	1 320	1 180	1 040	922	850	13
6 500	2 550	2 280	2 020	1 780	1 570	1 380	1 220	1 090	964	851	784	16
7 000	2 370	2 110	1 870	1 650	1 460	1 280	1 130	1 020	895	790	728	18
7 500	2 210	1 970	1 750	1 540	1 360	1 200	1 060	948	835	737	680	21
8 000	2 070	1 850	1 640	1 450	1 280	1 120	991	888	783	691	637	24
8 500	1 950	1 740	1 540	1 360	1 200	1 060	933	836	737	651	600	27
9 000	1 840	1 640	1 460	1 290	1 140	996	881	790	696	614	566	30
9 500	1 740	1 560	1 380	1 220	1 080	944	834	748	659	582	537	33
10 000	1 660	1 480	1 310	1 160	1 020	896	793	711	626	553	510	37
10 500	1 580	1 410	1 250	1 100	973	854	755	677	597	527	485	40
11 000	1 510	1 340	1 190	1 050	929	815	721	646	569	503	463	44
11 500	1 440	1 290	1 140	1 010	888	779	689	618	545	481	443	49
12 000	1 380	1 230	1 090	965	851	747	661	592	522	461	425	53
12 500	1 320	1 180	1 050	926	817	717	634	569	501	442	408	57
13 000	1 270	1 140	1 010	891	786	690	610	547	482	425	392	62
13 500	1 230	1 100	971	858	757	664	587	526	464	410	378	67
14 000	1 180	1 060	937	827	730	640	566	508	447	395	364	72
14 500	1 140	1 020	904	798	705	618	547	490	432	381	352	77
15 000	1 100	986	874	772	681	598	528	474	418	369	340	83
15 500	1 070	955	846	747	659	578	511	458	404	357	329	88
16 000	1 040	925	819	724	639	560	495	444	392	346	319	94
16 500	1 000	897	795	702	619	543	480	431	380	335	309	100
17 000	975	870	771	681	601	527	466	418	368	325	300	106
17 500	947	845	749	662	584	512	453	406	358	316	291	112
18 000	920	822	728	643	568	498	440	395	348	307	283	119
18 500	896	800	709	626	552	485	428	384	339	299	276	126
19 000	872	779	690	609	538	472	417	374	330	291	268	133

DESIGN DATA AND PROPERTIES												
V_r (kN)	2 030	1 850	1 710	1 540	1 390	1 440	1 300	1 210	1 130	1 030	881	
V_{r1} (kN)	1 790	1 640	1 530	1 380	1 260	1 300	1 170	1 100	1 030	944	812	
V_{r2} (kN)	1 550	1 430	1 340	1 230	1 130	1 150	1 050	992	939	862	743	
R (kN)	1 190	1 010	867	714	585	637	524	458	397	336	289	
G (kN)	20.4	18.1	16.9	14.5	12.4	11.5	9.85	9.37	9.02	8.43	7.33	
L_u (mm)	5 200	5 080	4 940	4 830	4 740	3 320	3 250	3 180	3 110	3 020	2 990	
d (mm)	635	628	622	616	611	617	612	608	603	598	596	
b (mm)	329	328	327	325	324	230	229	228	228	227	226	
t (mm)	31.0	27.7	24.4	21.6	19.0	22.2	19.6	17.3	14.9	12.7	11.7	
w (mm)	17.9	16.5	15.4	14.0	12.7	13.1	11.9	11.2	10.5	9.7	9.0	
k (mm)	50	47	44	41	38	42	39	37	34	32	31	

IMPERIAL SIZE AND MASS												
Mass (lb./ft.)	162	146	131	117	104	94	84	76	68	62	55	
Nominal Depth (in.)						24						

G40.21-M 300W
$\phi = 0.90$

BEAM LOAD TABLES
W Shapes

Total Uniformly Distributed Factored Loads for Laterally Supported Beams – kN

Designation	W530							Approx. Deflect. (mm)
Mass (kg/m)	138	123	109	101	92	82	72	
Span in Millimetres								
2 000						1 790	1 660	2
2 500	2 880	2 540	2 230	2 090	1 940	1 790	1 520	3
3 000	2 600	2 310	2 040	1 890	1 700	1 490	1 270	4
3 500	2 230	1 980	1 750	1 620	1 460	1 280	1 090	5
4 000	1 950	1 730	1 530	1 420	1 270	1 120	950	7
4 500	1 730	1 540	1 360	1 260	1 130	994	845	9
5 000	1 560	1 390	1 220	1 130	1 020	894	760	11
5 500	1 420	1 260	1 110	1 030	927	813	691	13
6 000	1 300	1 160	1 020	943	850	745	634	15
6 500	1 200	1 070	940	871	784	688	585	18
7 000	1 110	991	873	808	728	639	543	21
7 500	1 040	924	815	755	680	596	507	24
8 000	975	867	764	707	637	559	475	27
8 500	917	816	719	666	600	526	447	31
9 000	866	770	679	629	566	497	422	34
9 500	821	730	643	596	537	471	400	38
10 000	780	693	611	566	510	447	380	42
10 500	743	660	582	539	485	426	362	47
11 000	709	630	556	514	463	406	346	51
11 500	678	603	532	492	443	389	331	56
12 000	650	578	509	472	425	373	317	61
12 500	624	555	489	453	408	358	304	66
13 000	600	533	470	435	392	344	292	71
13 500	578	514	453	419	378	331	282	77
14 000	557	495	437	404	364	319	272	83
14 500	538	478	422	390	352	308	262	89
15 000	520	462	408	377	340	298	253	95
15 500	503	447	394	365	329	288	245	102
16 000	487	433	382	354	319	279	238	108

DESIGN DATA AND PROPERTIES

V_r (kN)	1 440	1 270	1 120	1 050	971	894	831	
V_{r1} (kN)	1 280	1 130	1 000	943	878	820	767	
V_{r2} (kN)	1 110	997	887	841	785	747	703	
R (kN)	812	647	508	448	390	335	292	
G (kN)	17.2	13.7	10.8	9.70	8.94	8.56	8.66	
L_u (mm)	3 180	3 100	3 040	2 990	2 930	2 860	2 750	
d (mm)	549	544	539	537	533	528	524	
b (mm)	214	212	211	210	209	209	207	
t (mm)	23.6	21.2	18.8	17.4	15.6	13.3	10.9	
w (mm)	14.7	13.1	11.6	10.9	10.2	9.5	8.9	
k (mm)	42	39	37	35	34	29	27	

IMPERIAL SIZE AND MASS

Mass (lb./ft.)	93	83	73	68	62	55	48	
Nominal Depth (in.)	21							

BEAM LOAD TABLES
W Shapes

Total Uniformly Distributed Factored Loads
for Laterally Supported Beams – kN

G40.21-M
300W

$\phi = 0.90$

Designation	W460							Approx. Deflect. (mm)
Mass (kg/m)	106	97	89	82	74	67	61	
Span in Millimetres								
2 000	2 110	1 890		1 620	1 470	1 380	1 300	2
2 500	2 060	1 880	1 740	1 580	1 430	1 280	1 110	3
3 000	1 720	1 570	1 450	1 320	1 190	1 070	930	4
3 500	1 480	1 340	1 240	1 130	1 020	913	796	6
4 000	1 290	1 180	1 080	988	891	799	697	8
4 500	1 150	1 050	965	878	792	710	619	10
5 000	1 030	942	868	791	713	639	557	12
5 500	939	856	789	719	648	581	507	15
6 000	860	785	724	659	594	533	464	18
6 500	794	724	668	608	548	492	429	21
7 000	737	673	620	565	509	457	398	24
7 500	688	628	579	527	475	426	372	27
8 000	645	589	543	494	446	400	348	31
8 500	607	554	511	465	419	376	328	35
9 000	574	523	482	439	396	355	310	39
9 500	543	496	457	416	375	337	293	44
10 000	516	471	434	395	356	320	279	49
10 500	492	448	413	376	339	304	265	54
11 000	469	428	395	359	324	291	253	59
11 500	449	409	378	344	310	278	242	64
12 000	430	392	362	329	297	266	232	70
12 500	413	377	347	316	285	256	223	76
13 000	397	362	334	304	274	246	214	82
13 500	382	349	322	293	264	237	206	89
14 000	369	336	310	282	255	228	199	95
DESIGN DATA AND PROPERTIES								
V_r (kN)	1 060	947	868	812	735	688	650	
V_{r1} (kN)	927	837	770	724	658	620	589	
V_{r2} (kN)	799	727	672	637	581	551	528	
R (kN)	620	510	434	384	317	281	254	
G (kN)	14.5	11.7	9.89	9.25	7.70	7.47	7.67	
L_u (mm)	2 910	2 870	2 830	2 770	2 730	2 670	2 590	
d (mm)	469	466	463	460	457	454	450	
b (mm)	194	193	192	191	190	190	189	
t (mm)	20.6	19.0	17.7	16.0	14.5	12.7	10.8	
w (mm)	12.6	11.4	10.5	9.9	9.0	8.5	8.1	
k (mm)	37	36	34	33	31	25	23	
IMPERIAL SIZE AND MASS								
Mass (lb./ft.)	71	65	60	55	50	45	41	
Nominal Depth (in.)	18							

G40.21-M 300W
$\phi = 0.90$

BEAM LOAD TABLES
W Shapes

Total Uniformly Distributed Factored Loads
for Laterally Supported Beams – kN

Designation	W410							Approx. Deflect. (mm)
Mass (kg/m)	85	74	67	60	54	46	39	
Span in Millimetres								
1 000								1
1 500						1 010	895	1
2 000	1 620	1 430	1 290	1 120	1 080	956	788	2
2 500	1 500	1 300	1 180	1 030	907	765	631	3
3 000	1 250	1 090	979	857	756	637	526	5
3 500	1 070	932	839	734	648	546	451	7
4 000	934	815	734	643	567	478	394	9
4 500	830	725	653	571	504	425	350	11
5 000	747	652	588	514	454	382	315	14
5 500	679	593	534	467	412	348	287	17
6 000	623	544	490	428	378	319	263	20
6 500	575	502	452	395	349	294	243	23
7 000	534	466	420	367	324	273	225	27
7 500	498	435	392	343	302	255	210	31
8 000	467	408	367	321	284	239	197	35
8 500	440	384	346	302	267	225	186	39
9 000	415	362	326	286	252	212	175	44
9 500	393	343	309	271	239	201	166	49
10 000	374	326	294	257	227	191	158	55
10 500	356	311	280	245	216	182	150	60
11 000	340	297	267	234	206	174	143	66
11 500	325	284	255	224	197	166	137	72
12 000	311	272	245	214	189	159	131	79
12 500	299	261	235	206	181	153	126	85
13 000	287	251	226	198	174	147	121	92

DESIGN DATA AND PROPERTIES

V_r (kN)	812	716	643	560	540	504	447
V_{r1} (kN)	710	630	568	498	484	452	404
V_{r2} (kN)	608	544	492	436	428	399	360
R (kN)	478	379	313	240	226	198	165
G (kN)	11.9	9.67	8.10	6.13	6.72	5.32	5.22
L_u (mm)	2 730	2 670	2 610	2 580	2 480	1 930	1 860
d (mm)	417	413	410	407	403	403	399
b (mm)	181	180	179	178	177	140	140
t (mm)	18.2	16.0	14.4	12.8	10.9	11.2	8.8
w (mm)	10.9	9.7	8.8	7.7	7.5	7.0	6.4
k (mm)	35	33	31	30	28	28	26

IMPERIAL SIZE AND MASS

Mass (lb./ft.)	57	50	45	40	36	31	26
Nominal Depth (in.)	16						

BEAM LOAD TABLES
W Shapes

Total Uniformly Distributed Factored Loads
for Laterally Supported Beams – kN

G40.21-M
300W
$\phi = 0.90$

Designation	W360								Approx. Deflect. (mm)
Mass (kg/m)	79	72	64	57	51	45	39	33	
1 000									1
1 500						866	820	723	1
2 000				1 010	914	841	715	585	2
2 500	1 190	1 070	955	873	772	673	572	468	4
3 000	1 030	922	821	727	644	561	477	390	6
3 500	883	790	704	623	552	481	409	334	8
4 000	772	691	616	545	483	421	357	293	10
4 500	686	614	547	485	429	374	318	260	13
5 000	618	553	492	436	386	337	286	234	16
5 500	562	503	448	397	351	306	260	213	19
6 000	515	461	410	364	322	280	238	195	22
6 500	475	425	379	336	297	259	220	180	26
7 000	441	395	352	312	276	240	204	167	30
7 500	412	369	328	291	257	224	191	156	35
8 000	386	346	308	273	241	210	179	146	40
8 500	363	325	290	257	227	198	168	138	45
9 000	343	307	274	242	215	187	159	130	50
9 500	325	291	259	230	203	177	151	123	56
10 000	309	276	246	218	193	168	143	117	62
10 500	294	263	235	208	184	160	136	111	69
11 000	281	251	224	198	176	153	130	106	75

Span in Millimetres

DESIGN DATA AND PROPERTIES

V_r (kN)	593	536	478	504	457	433	410	362	
V_{r1} (kN)	503	458	410	441	401	383	361	321	
V_{r2} (kN)	412	380	342	377	345	333	313	281	
R (kN)	374	314	252	262	217	199	178	141	
G (kN)	9.80	8.44	6.82	7.37	6.34	6.68	5.09	4.60	
L_u (mm)	3 270	3 190	3 110	2 550	2 500	2 430	1 790	1 720	
d (mm)	354	350	347	358	355	352	353	349	
b (mm)	205	204	203	172	171	171	128	127	
t (mm)	16.8	15.1	13.5	13.1	11.6	9.8	10.7	8.5	
w (mm)	9.4	8.6	7.7	7.9	7.2	6.9	6.5	5.8	
k (mm)	36	34	33	26	24	22	23	21	

IMPERIAL SIZE AND MASS

Mass (lb./ft.)	53	48	43	38	34	30	26	22	
Nominal Depth (in.)	14								

G40.21-M 300W
$\phi = 0.90$

BEAM LOAD TABLES
W Shapes

Total Uniformly Distributed Factored Loads for Laterally Supported Beams – kN

Designation				W310						Approx. Deflect. (mm)
Mass (kg/m)	86	79	74	67	60	52	45	39	31*	
Span in Millimetres										
1 000										1
1 500									527	2
2 000			1 040	927	813	861	739	641	473	3
2 500		960	1 030	916	813	723	612	527	378	5
3 000	1 000	922	857	763	678	603	510	439	315	7
3 500	876	790	734	654	581	517	437	376	270	9
4 000	767	691	643	572	508	452	382	329	237	12
4 500	682	614	571	509	452	402	340	293	210	15
5 000	613	553	514	458	407	362	306	264	189	18
5 500	558	503	467	416	370	329	278	240	172	22
6 000	511	461	428	382	339	301	255	220	158	26
6 500	472	425	395	352	313	278	235	203	146	31
7 000	438	395	367	327	290	258	218	188	135	35
7 500	409	369	343	305	271	241	204	176	126	41
8 000	383	346	321	286	254	226	191	165	118	46
8 500	361	325	302	269	239	213	180	155	111	52
9 000	341	307	286	254	226	201	170	146	105	59
9 500	323	291	271	241	214	190	161	139	100	65
10 000	307	276	257	229	203	181	153	132	95	72

DESIGN DATA AND PROPERTIES

V_r (kN)	503	480	519	463	406	431	369	320	264
V_{r1} (kN)	420	405	434	391	346	370	320	279	231
V_{r2} (kN)	337	329	348	318	286	309	271	237	199
R (kN)	364	340	387	318	248	251	190	147	105
G (kN)	10.5	10.7	11.5	9.64	7.43	7.34	5.73	4.55	3.64
L_u (mm)	4 250	4 140	3 380	3 280	3 200	2 570	2 490	2 440	2 460
d (mm)	310	306	310	306	303	317	313	310	306
b (mm)	254	254	205	204	203	167	166	165	164
t (mm)	16.3	14.6	16.3	14.6	13.1	13.2	11.2	9.7	7.4
w (mm)	9.1	8.8	9.4	8.5	7.5	7.6	6.6	5.8	4.9
k (mm)	34	32	34	32	31	27	25	24	22

IMPERIAL SIZE AND MASS

Mass (lb./ft.)	58	53	50	45	40	35	30	26	21
Nominal Depth (in.)					12				

* Class 3 section.

BEAM LOAD TABLES
W Shapes

Total Uniformly Distributed Factored Loads
for Laterally Supported Beams – kN

G40.21-M
300W
$\phi = 0.90$

Designation				W250				Approx. Deflect. (mm)
Mass (kg/m)	67	58	49	45	39	33	24*	
Span in Millimetres								
1 000							453	1
1 500				720	616	561	396	2
2 000	818	719	654	650	554	458	297	4
2 500	778	665	547	520	443	366	238	6
3 000	649	554	456	433	369	305	198	8
3 500	556	475	391	372	317	262	170	11
4 000	487	416	342	325	277	229	149	14
4 500	432	370	304	289	246	204	132	18
5 000	389	333	273	260	222	183	119	22
5 500	354	302	249	236	201	167	108	27
6 000	324	277	228	217	185	153	99	32
6 500	299	256	210	200	170	141	91	38
7 000	278	238	195	186	158	131	85	44
7 500	259	222	182	173	148	122	79	50
8 000	243	208	171	163	139	114	74	57
DESIGN DATA AND PROPERTIES								
V_r (kN)	409	359	327	360	308	280	226	
V_{r1} (kN)	331	293	272	314	271	249	205	
V_{r2} (kN)	252	227	216	267	234	218	184	
R (kN)	368	300	260	266	201	174	120	
G (kN)	12.2	10.6	10.4	8.90	6.87	6.78	5.39	
L_u (mm)	3 570	3 410	3 240	2 360	2 280	2 180	2 190	
d (mm)	257	252	247	266	262	258	253	
b (mm)	204	203	202	148	147	146	145	
t (mm)	15.7	13.5	11.0	13.0	11.2	9.1	6.4	
w (mm)	8.9	8.0	7.4	7.6	6.6	6.1	5.0	
k (mm)	33	31	28	23	21	19	16	
IMPERIAL SIZE AND MASS								
Mass (lb./ft.)	45	39	33	30	26	22	16	
Nominal Depth (in.)				10				

* Class 3 section.

G40.21-M
300W
$\phi = 0.90$

BEAM LOAD TABLES
W Shapes

Total Uniformly Distributed Factored Loads
for Laterally Supported Beams – kN

Designation	W200					Approximate Deflection (mm)
Mass (kg/m)	42	36	31	27	21*	

Span in Millimetres	42	36	31	27	21*	Deflection (mm)
1 000				430	364	1
1 500	529	446	479	402	281	3
2 000	482	410	362	301	211	4
2 500	385	328	289	241	168	7
3 000	321	274	241	201	140	10
3 500	275	235	207	172	120	14
4 000	241	205	181	151	105	18
4 500	214	182	161	134	94	23
5 000	193	164	145	121	84	28
5 500	175	149	132	110	77	34
6 000	161	137	121	100	70	40
6 500	148	126	111	93	65	47
7 000	138	117	103	86	60	55

DESIGN DATA AND PROPERTIES

V_r (kN)	264	223	240	215	182	
V_{r1} (kN)	216	185	205	187	160	
V_{r2} (kN)	168	147	171	159	139	
R (kN)	263	196	207	173	133	
G (kN)	10.8	8.11	8.58	7.83	6.72	
L_u (mm)	2 850	2 730	2 150	2 050	2 080	
d (mm)	205	201	210	207	203	
b (mm)	166	165	134	133	133	
t (mm)	11.8	10.2	10.2	8.4	6.4	
w (mm)	7.2	6.2	6.4	5.8	5.0	
k (mm)	24	22	20	18	16	

IMPERIAL SIZE AND MASS

Mass (lb./ft.)	28	24	21	18	14	
Nominal Depth (in.)	8					

* Class 3 section.

PART NINE
GENERAL INDEX

Amplification factors
    U 4-28, 4-97
    U_1 . . . . 1-32, 4-94
    U_2 4-18
Anchor bolts / rods 1-88, 4-135
Angles, double
    beam connections 3-58
    properties and dimensions, imperial series 6-112
    struts, factored compressive resistance, imperial series 4-109
    struts, information 4-3
Angles, single
    availability 6-18
    beam connections 3-68
    designations 6-19, 6-25
    gauge distances 6-154
    permissible variations in sectional dimensions 6-15
    properties and dimensions, imperial series 6-72
    shape size groupings 6-6
Arcs of circles, length for unit radius 7-69
Areas
    cross-section of rolled shapes   (*see tables for specific shapes in* Part Six)
    effective net area 3-99
    effective net area, reduced for shear lag 3-99, 3-105
    gross and net 1-25
    hole diameters for effective net area 3-106
    reduction of area for holes 3-106
    surface of structural shapes   (*see tables for specific shapes in* Part Six)
Availability of structural sections 6-17
Bars
    area 6-136
    designations vii
    general information 6-129
    mass 6-132, 6-136
Base plates, columns 1-87, 4-130
Beams
    beam-columns   (*see* Beam-columns)
    bearing plates 5-151
    bending, laterally supported and laterally unsupported 1-30
    camber 1-15, 6-12
    class of sections in bending 1-22, 5-4
    composite   (*see* Composite beams)
    connections   (*see* Connections)
    continuous 5-145
    copes 1-46, 3-59, 5-74
    deflections 1-15, 5-68
    diagrams and formulae 5-132
    factored load tables 5-73, 5-75
        WWF shapes, 350W 5-96
        W shapes, 350W 5-104
        WWF shapes, 300W 8-52
        W shapes, 300W 8-60
        Rectangular HSS, 350W 5-126
    factored resistance 5-72

  general information 5-3
  selection tables 5-72
   WWF and W shapes, 350W 5-78
   C shapes, 300W 5-92
   S shapes, A572 grade 50 5-94
   WWF and W shapes, 300W 8-48
  shear 1-28, 5-74
  stiffeners (*see* Stiffeners)
  web crippling and yielding 1-49, 5-75
  web openings 1-49, 5-154
Beam-columns 1-31, 4-94
  amplification factors (U), (U_1) 4-28, 4-94
  bending factors (B_x, B_y) 4-23
  bending coefficient (ω_1) 4-26
  biaxial bending 1-31, 4-102
  class of sections 4-7
  factored moment resistance of columns 4-105
  sugested design procedure 4-99
  trial selection 4-97
Bearing 1-33
  factored resistances
   bolted connections 1-33, 3-7, 3-9
    expansion rollers or rockers 1-33
    machined, sawn, or fitted parts 1-33
  joints in compression members 1-77, 1-99, 1-100, 1-103
  open web steel joists 1-55
  piles (*see* HP shapes)
  plates, beams 5-151
  stiffeners 1-47
Bending coefficient (ω_1) 1-33, 4-26
Bending factors (B_x, B_y) 4-23
Biaxial bending 1-31, 4-98
Bolts (*see also* Connections) 6-142
  ASTM A307 . . . . 1-78, 6-151
  ASTM A325 and A490 . . . . 1-80, 6-144, 6-148
  ASTM A325M and A490M 1-80, 6-145, 6-149, 6-152
  data 3-5
  erection clearances 6-155
  grips 6-146
  holes 1-81, 3-106
  inspection 1-84
  installation 1-83
  pretensioned 1-78
  resistances
   factored bearing (for connected material) 1-33, 3-9
   factored shear 1-34, 3-8
   factored tension 1-34, 3-8
   slip 1-34, 3-14
  shear and tension (combined), bearing-type connections 1-34, 3-12
  shear and tension (combined), slip-critical connections 1-35, 3-16
  slots 1-82, 6-152
  snug-tightened 1-78
  tension and prying action 1-80, 3-19
Bracing formulae 7-68
Bracing requirements 1-31, 1-74, 4-128

Building materials	7-40
Built-up members	1-70, 6-111, 6-125
C shapes (*see* Channels)	
Camber	1-15, 6-12
Changes, steel products	6-163
Channels	
cold formed channels	6-105
properties and dimensions	6-106
designations	6-19, 6-24
double channels, properties and dimensions	6-120
miscellaneous (MC), properties and dimensions	6-68
permissible variations in sectional dimensions	6-14
shape size grouping	6-6
standard (C), properties and dimensions	6-66
W shape and channel, properties and dimensions	6-122
Check list for design drawings	7-50
Chemical composition (heat analysis), structural steels	6-8
Circle, properties	7-63
Circular arcs, length for unit radius	7-69
CISC Commentary on standard CAN/CSA-S16.1-94	2-3
Class of sections	1-22
beam-columns	4-7
bending	5-4
Code of Standard Practice for Structural Steel	7-3
Coefficients	
bending (ω_1)	1-33, 4-26
thermal expansion	7-47
Cold formed channels	6-105
properties and dimensions	6-106
Colour code, steel marking	6-9
Columns	
anchor bolts	1-88, 4-135
axially loaded columns	4-29
base plates	1-87, 4-130
beam-columns (*see* Beam-columns)	
biaxial bending	1-31, 4-102
class of sections	4-7
compresssion and bending (*see* Beam-columns)	
effective lengths	1-21
Euler buckling load per unit area	4-27
properties and dimensions	
HSS shapes	6-97
M shapes	6-58
pipe	6-104
W shapes	6-40
WWF	6-32
resistances, factored axial compressive	1-27, 4-29
double angle struts, 300W, imperial series	4-109
HSS, Class C, 350W	4-46
HSS, Class H, 350W	4-70
W shapes, 350W	4-39
WWF shapes, 350W	4-32
W shapes, 300W	8-17
WWF shapes, 300W	8-10
resistances, factored moment, 350W	4-105

WWF and W shapes shapes, 350W	4-106
resistances, unit factored compressive	4-12
resistances, unit factored compressive for WWF and Class H HSS	4-17
stability	1-20, 1-74, 1-127, 4-18
stiffeners	1-76, 3-81
width-thickness limits	1-22, 4-5
Combination factor, load (Ψ)	1-8, 1-19
Composite beams	5-20
CAN/CSA-S16.1-94, Clause 17; CISC commentary	1-61, 2-66
effective width of concrete slab	1-63, 5-21
deflections	5-22
shear connectors	1-65, 5-21
shear studs	5-23
shrinkage strain	1-62, 1-135
trial selection tables, 350W beams and	
75 mm deck + 65 mm slab, 20 MPa, 2300 kg/m^3 concrete	5-28
75 mm deck + 75 mm slab, 25 MPa, 2300 kg/m^3 concrete	5-36
75 mm deck + 85 mm slab, 25 MPa, 2000 kg/m^3 concrete	5-52
75 mm deck + 85 mm slab, 25 MPa, 1850 kg/m^3 concrete	5-44
75 mm deck + 90 mm slab, 20 MPa, 2300 kg/m^3 concrete	5-60
trial selection tables, 300W beams and	
75 mm deck + 65 mm slab, 20 MPa, 2300 kg/m^3 concrete	8-24
75 mm deck + 85 mm slab, 25 MPa, 1850 kg/m^3 concrete	8-32
38 mm deck + 65 mm slab, 20 MPa, 2300 kg/m^3 concrete	8-40
Composite columns	1-68
Compression and bending (*see* Beam-columns)	
Compression members (*see* Columns)	
Computer programs	7-48
Connections (*see also* Bolts, Welds)	
bearing-type	1-33, 3-7
CAN/CSA-S16.1-94, Clause 21; CISC Commentary	1-75, 2-75
eccentric loads	
bolted	3-26
welded	3-42
framed beam shear connections	3-56
double angle	3-58
end plate	3-66
seated	3-74
seated, stiffened	3-78
shear tabs	3-70
single angle	3-68
tee	3-72
hollow structural section connections	3-89
moment	3-80
pin-connected	1-27
resistances	
bolted	3-8, 3-9, 3-13, 3-15, 3-17
welded	3-39
slip-critical	1-34, 3-14
tension and prying action, bolts	1-80, 3-19
Continuous spans, formulae and flexure diagrams	5-145
Conversion factors, imperial — SI — imperial	7-74
Crane rails	6-137
Crippling, web	1-49
CSA G40.20 and G40.21	6-3

CSSBI standards 7-37
Cutting tolerances 6-13
Definitions of section properties 7-52
Deflections 1-15, 5-68
    graphs 5-70
    recommended maximum values 1-125
    table 5-71
Deformations, permanent 1-16
Designations . . . . vii, 6-19
Dimensions (see the specific shapes)
Distances, centre to centre between staggered fasteners 6-153
Double angles
    properties and dimensions, imperial series 6-112
Double angle struts 4-4
    factored axial compressive resistances, imperial series 4-109
Double angles, beam connections 3-58
Double channels, properties and dimensions, back-to-back 6-121
Double channels, properties and dimensions, toe-to-toe 6-120
Drawings, check list for design 7-50
Dynamic effects 1-16
Eccentric loads
    on bolt groups 3-26
    on weld groups 3-42
Effect of factored loads 1-18
Effective lengths of compression members 1-21, 1-106, 1-107
Effective net area 1-25, 3-99
    hole diameters for 3-106
    reduced for shear lag 1-25, 3-99, 3-104
Effective concrete slab width 1-63, 5-21
Electronic aids 7-48
Ellipse, properties 7-64
End distance 1-80
End plate connections 3-66
Equal leg angles (see Angles)
Equivalent uniform bending coefficient 4-26
Erection 1-102, 7-19
    clearances (bolts) 6-155
    tolerances 1-102
Euler buckling load per unit area 4-27
Expansion, coefficients 7-47
Fabrication 1-98
Factored resistances
    bearing
        bolted connections 1-33, 3-7, 3-9
        expansion rollers or rockers 1-33
        machined, sawn, or fitted parts 1-33
    bolts
        shear 1-34, 3-8
        shear and tension (combined) 1-34, 1-35, 3-12, 3-16
        tension 1-34, 3-8
    compression
        columns 1-27, 4-29
        double angle struts 4-109
        unit compressive resistances 4-12
    moment
        beams 1-30, 5-72

9-5

```
        columns ................................................. 4-105
        composite beams ................................... 1-66, 5-28
        girders with thin webs .................................... 1-46
    shear ............................................... 1-28, 5-74
        connecting elements ...................................... 1-29
        girder webs, factored ultimate shear stress ................ 5-9
        shear studs .............................................. 5-23
    tension ............................................ 1-27, 3-99
    welds .............................................. 1-36, 3-39
Factors
    amplification (U) ................................... 4-28, 4-97
    amplification ($U_1$) ...................................... 4-94
    amplification ($U_2$) ...................................... 4-94
    bending ($B_x$, $B_y$) ..................................... 4-12
    effective length (K) ..................... 1-21, 1-106, 1-107
    importance factor ($\gamma$) ......................... 1-2, 1-19
    load ($\alpha$) ...................................... 1-2, 1-19
    load combination ($\Psi$) ............................ 1-2, 1-19
    resistance (definition, $\phi$, $\phi_b$, $\phi_w$, $\phi_{sc}$, $\phi_c$) ...... 1-2, 1-27, 1-34, 1-35, 1-65, 1-66, 2-5
Fasteners  (see Bolts, Welds)
Fatigue ........................................... 1-37, 1-129
Flexural members  (see Beams, Girders)
Floor vibrations ................................... 1-16, 1-116
Formulae
    beam ................................................... 5-132
    bracing .................................................. 7-68
    properties of geometric sections ......................... 7-52
    trigonometric ............................................ 7-63
Framed beam shear connections ............................ 3-56
Friction-type connections  (see Connections, slip-critical)
Gauge distances, usual ................................. 6-154
General nomenclature ...................................... viii
Geometric sections, properties .......................... 7-52
Girders
    CAN/CSA-S16.1-94, Clause 15; CISC Commentary .... 1-46, 2-46
    design ................................................. 5-12
    factored shear stress in web ............................. 5-9
    shear and moment (combined) ............................ 1-29
    stiffeners  (see Stiffeners)
    web crippling and yielding ............................. 1-49
    web openings ..................................... 1-49, 5-154
Grip, bolts ............................................ 6-150
HP shapes
    designations ..................................... 6-19, 6-26
    permissible variations in sectional dimensions ......... 6-13
    properties and dimensions ............................... 6-56
    shape size groupings ..................................... 6-6
High strength bolts  (see Bolts)
Holes ................................................... 1-81
    diameters for effective net area, reduction of area for holes ......... 3-106
    staggered holes .................................. 3-107, 6-153
Hollow structural sections (HSS) ......................... 6-97
    beam load tables ....................................... 5-126
    bending factors ($B_x$, $B_y$) for beam-columns ........ 4-25
    Class C and Class H definitions ......................... 6-15
    connections ............................................ 3-89
```

designations	6-19, 6-26
factored compressive resistances	
Class C	4-46
Class H	4-70
permissible variations in sectional dimensions	6-15
properties and dimensions	6-97
unit factored compressive resistances	
Class C	4-12
Class H	4-17
welding details	3-97
I shapes (see S shapes)	
Imperial designations	6-19
Importance factor (γ)	1-2, 1-19
Index for:	
CAN/CSA-S16.1-94	1-iii
CISC Code of Standard Practice	7-4
compression members	4-1
connections and tension members	3-1
flexural members	5-1
properties and dimensions	6-1
Initial out-of-straightness	2-26, 4-128, 6-12
Interaction equations, beam-columns	1-31, 4-94
ISO steel grades	6-10
Joists, open-web steel	1-51, 2-50
Junior beams (see M shapes)	
Junior channels (see Channels, miscellaneous)	
K factors	1-21, 1-106, 1-107
L shapes (see Angles)	
Lateral bracing	1-31, 1-74, 4-128
Length of circular arcs for unit radius	7-69
Lengths of members, design	1-21
Light beams (see W shapes)	
Limit states	1-2, 1-15, 2-4
Load combination factor (Ψ)	1-2, 1-19
Load factors (α)	1-2, 1-19
Load ratios	3-4
M shapes	
designations	6-19, 6-24
properties and dimensions	6-58
shape size groupings	6-6
MC shapes (see Channels)	
Materials, mass and forces	7-40
Mechanical, properties summary	6-7
Metric conversion (see SI summary)	
Metric designations	6-19
Mill practice, standard	6-11
Miscellaneous channels (see Channels)	
Moment connections	1-76, 3-80
Moment diagrams for beams	5-132, 5-145
Moment resistances, factored	
beams	1-30, 5-78
columns	4-105
composite beams	1-66, 5-28
girders with thin webs	1-46
Moving concentrated loads	5-143
M/D ratios	7-42

Net area of tension members 1-25, 3-99
Nomenclature, general viii

Nuts
    dimensions, A307 6-151
    dimensions, A325 and A490 6-144
    dimensions, A325M and A490M 6-145
    rotation 1-86
Open web steel joists 1-51, 2-50
Parabola, properties 7-64
Parallel bracing formulae 7-68
Permanent deformations 1-16
Piles (see HP shapes)
Pin connected connections 1-27
Pipe 6-97
    properties and dimensions 6-104
Plastic analysis 1-20
Plate girders
    CAN/CSA-S16.1-94, Clause 15; CISC commentary 1-46, 2-46
    design 5-12
    factored shear stress in web 5-9
    stiffeners 5-13
Plates
    beam bearing 5-151
    column base 4-124
    designation vii
    general information 6-129
    mass 6-131, 6-132
Properties and dimensions (see the specific shapes)
Properties of geometric sections 7-52
Properties of sections, definitions 7-52
Prying action and tension, bolts 3-19
PΔ effects 1-20, 4-18
Rectangular hollow structural sections (see Hollow structural sections)
Reduction of area for holes 3-106
Reduction of area for shear lag 3-99
Residual stress 2-26
Resistances (see Factored resistances)
Resistance factors (definition, ϕ, ϕ_b, ϕ_w, ϕ_{sc}, ϕ_c) 1-2, 1-27, 1-34, 1-35, 1-65, 1-66, 2-5
Rolled structural shapes 6-38
Roof deck, steel 7-37
Round bars, dimensions and mass 6-136
Round hollow structural sections (see Hollow structural sections)
S shapes
    designations 6-19, 6-24
    permissible variations in sectional dimensions 6-14
    properties and dimensions 6-62
    shape size groupings 6-6
Safety discussion 2-5
Seated beam shear connections 3-74
Seated beam shear connections, stiffened 3-78
Seismic design requirements
    CSA/CAN-S16.1-94, Clause 27 1-89
    CISC commentary 2-81
Selection tables
    beams 5-78
    composite beams 5-28

Shape size groupings 6-6
Shear 1-28, 5-74
    bolts, bearing-type connections 1-34, 3-8
    connecting elements 1-29
    girder webs, factored ultimate shear stress 5-9
    shear connectors 1-65, 5-21
    shear studs 5-23
    walls 1-138
Shear and moment (combined) in girders 1-29, 2-29, 5-17
Shear and tension (combined) in bolts
    bearing-type connections 1-34, 3-12
    slip-critical connections 1-35, 3-16
Shear connectors for composite beams 1-65
Shear lag 1-25, 2-24, 3-99
Sheet steel products 7-37
SI summary 7-70
Single angle beam connections 3-68
Slenderness ratios 1-22
Slip-resistant connections   (*see* Connections, slip-critical)
Slots, bolt 1-82, 6-152
Source, principal, of sections 6-17
Square bars, dimensions and mass 6-136
Square hollow structural sectional   (*see* Hollow structural sections)
Stability 1-20, 1-74, 1-127, 4-18
Staggered holes in tension members 3-107
Standard beams   (*see* S shapes)
Standard channels   (*see* Channels)
Standard mill practice 6-11
Standard practice, code of 7-3
Starred angles, factored axial compressive resistances 4-121
Steel, grade, types, strength levels 6-6
Steel marking colour code 6-9
Steel plate shear walls 1-138
Steel products — record of changes 6-163
Stiffeners 5-13
    bearing 1-47, 5-18
    intermediate 1-48, 5-16
    moment connection 1-76, 3-80
Structural shapes
    availability 6-17
    general information 6-11
Structural sheet steel products 7-37
Structural steels 6-3
Structural tees   (*see* Tees)
Struts, double angle 4-3
    factored axial compressive resistances, imperial series 4-109
Super light beams 6-60
Sweep 1-3, 6-12
Symbols, welds 6-160
Tees, properties and dimensions 6-82
Tee-type beam connections 3-72
Temperature, coefficients of expansion 7-47
Tensile strengths (mechanical properties) 6-7
Tension and prying action, bolts 3-19
Tension and shear (combined) in bolts   (*see* Shear and tension (combined) in bolts)
Tension members 3-99
    hole diameters for effective net area 1-25, 3-106

    reduction of area for holes 3-106
    resistance, factored axial 1-27
    shear lag 1-25, 3-99, 3-107
    staggered holes 3-107
    tension and bending (combined) 1-33
Thicknesses, comparison of imperial gauges and SI 6-134
Tolerances
    erection 1-102
    fabrication 1-99
Torsion 1-50
Transverse stiffeners 1-47
Truss connections (HSS) 3-94
Two angles (*see* Angles, double)
Two channels, properties and dimensions 6-120
Unequal leg angles (*see* Angles)
Vibrations
    floor 1-116
    wind 1-124
W shapes
    beam load tables, 350W 5-104
    beam load tables, 300W 8-60
    beam selection tables, 350W 5-78
    beam selection tables, 300W 8-48
    designations 6-19, 6-22
    factored axial compressive resistance, 350W 4-39
    factored axial compressive resistance, 300W 8-17
    permissible variations in sectional dimensions 6-13
    properties and dimensions 6-40
    shape size groupings 6-6
    W shape and channel, properties and dimensions 6-122
WWF shapes
    beam load tables, 350W 5-96
    beam load tables, 300W 8-52
    beam selection tables, 350W 5-78
    beam selection tables, 300W 8-48
    designations 6-19, 6-21
    factored axial compressive resistances, 350W 4-32
    factored axial compressive resistances, 300W 8-10
    manufacturing tolerances 6-12
    properties and dimensions, WRF shapes 6-30
    properties and dimensions, WWF shapes 6-32
Walls, steel plate shear 1-138
Washers 6-150
Web crippling and yielding 1-49, 5-75
Web openings 1-50, 5-154
Welded wide flange shapes (*see* WWF shapes)
Welds (*see also* Connections)
    eccentric loads on weld groups [in plane] 3-42
    eccentric loads on weld groups, shear and moment [out of plane] 3-53
    electrode classification and unit factored weld resistance 3-40
    factored resistance of welds 1-36, 3-39
    factored shear resistance of fillet welds 3-41
    general information 6-156
    hollow structural sections 3-97
    symbols 6-160
Width-thickness limits 1-23, 4-5

Wind vibrations 1-124
Wire gauges, comparisons with SI sizes 6-133
Yield strengths, mechanical properties 6-7

NOTES

NOTES

NOTES

NOTES

NOTES